Iron-Sulfur Clusters

Iron-Sulfur Clusters

Biogenesis and Biochemistry

Volume 1

Edited by Silke Leimkühler, Günter Schwarz, Oliver Lenz, and Oliver Einsle

WILEY VCH

Editors

Prof. Silke Leimkühler
Universität Potsdam
Karl-Liebknecht-Str. 24-25
H25
Potsdam
GolM, 14476

Prof. Günter Schwarz
Universität zu Köln
Zülpicher Str. 47
Köln
GM, 50674

Dr. Oliver Lenz
Technische Universität Berlin
Max-Volmer-Laboratorium, PC 14
Strasse des 17. Juni 135
Berlin
GM, 10623

Prof. Oliver Einsle
Albert-Ludwigs-Universität
Albertstraße 21
Freiburg
GM, 79104

Cover Image: Pixabay, The chemical structures featured on the cover were created by Oliver Lenz

Typesetting: Straive, Chennai, India
Printing and Binding: CPI Group (UK) Ltd, Croydon, CR0 4YY

Library of Congress Card No.: applied for

British Library Cataloguing-in-Publication Data A catalogue record for this book is available from the British Library.

Bibliographic information published by the Deutsche Nationalbibliothek
The Deutsche Nationalbibliothek lists this publication in the Deutsche Nationalbibliografie; detailed bibliographic data are available on the Internet at <http://dnb.d-nb.de>.

Print ISBN: 978-3-527-35256-2
ePDF ISBN: 978-3-527-84357-2
ePub ISBN: 978-3-527-84358-9
oBook ISBN: 978-3-527-84359-6

C9783527352562_280825

This book is dedicated to five outstanding people in our careers, who have been our esteemed teachers, mentors, supervisors, and friends: Ralf Mendel, Werner Klipp, K. V. Rajagopalan, Dough Rees, and Bärbel Friedrich. We are grateful for their continuous support, encouragement, patience, motivation, and for kindling in us the enduring flame of a scientist.

Contents

Volume 1

Preface *xv*

1 **ISC-Based Fe–S Protein Biogenesis in Bacteria** *1*
Béatrice Py and Frédéric Barras
Chapter Goals *1*
Main Chapter Points *1*
1.1 Introduction *1*
1.2 Discovery of the ISC Fe–S Protein Biogenesis Machinery *4*
1.3 Assessing the Importance of the Individual ISC Components *5*
1.4 The Fe–S Cluster *de novo* Assembly Step *8*
1.5 The Fe–S Cluster Delivery Step *11*
1.6 Regulation of the ISC Machinery *12*
1.7 Distribution and Evolution of the ISC Machinery *14*
1.8 Conclusion *15*
Acknowledgments *15*
References *16*

2 **Unraveling the Complexity of the Suf-Based Fe–S Biogenesis** *25*
Ingie Elchennawi, Claire E. Fisher, Franklin Wayne Outten, and Ollagnier de Choudens Sandrine
Chapter Goals *25*
Main Chapter Points *26*
2.1 Introduction *26*
2.2 Suf Function: The Sulfur Delivery Step *28*
2.2.1 The SufS Cysteine Desulfurase *28*
2.2.2 SufU and SufE Sulfur Transfer Partners *30*
2.2.3 Is a Cysteine Desulfurase Always Needed? Organisms That Lack SufS *35*
2.3 Suf Function: Fe–S Cluster Assembly on the Scaffold *35*
2.3.1 The SufB Scaffold *35*

2.3.2 SufC and SufD Scaffold Partners *36*
2.3.3 The Minimal Suf Scaffold, SMS *37*
2.3.4 Iron Acquisition and Fe–S Cluster Binding *37*
2.3.5 The Redox Flavin Cofactor *38*
2.4 Suf Function: Fe–S Cluster Trafficking *39*
2.4.1 The A-Type Carrier SufA *39*
2.4.2 SufT and DUF59 Proteins *41*
2.5 Suf Regulation *42*
2.5.1 Stress-Responsive Suf Regulation in Bacteria *43*
2.5.2 SufR and RisR Regulation of Suf *44*
2.5.3 SufB Intein Splicing *45*
2.6 Conclusions and Future Directions *45*
Acknowledgments *46*
References *46*

3 Biogenesis of Mitochondrial Iron–Sulfur Proteins *57*
Ulrich Mühlenhoff, Oliver Stehling, and Roland Lill
Chapter Goals *57*
Main Chapter Points *57*
3.1 Introduction *57*
3.2 Cellular Roles of Eukaryotic Fe–S Proteins *58*
3.3 Components and Mechanism of the Mitochondrial ISC System *62*
3.3.1 Overview *62*
3.3.2 Step 1: *De novo* [2Fe–2S] Cluster Formation on the Core ISC Complex *64*
3.3.3 Step 2: Chaperone-Dependent Transfer of the ISCU2-Bound Cluster to [2Fe–2S] Apoproteins *69*
3.3.4 Step 3: Reductive Fusion of [2Fe–2S] Clusters on the Late ISC Complex Generates [4Fe–4S] Clusters for Insertion into Mitochondrial Apoproteins *71*
3.4 Conclusions and Outlook *74*
Acknowledgments *75*
References *75*

4 Iron–Sulfur Protein Maturation in the Cytosol of Eukaryotes *87*
Joseph J. Braymer and Antonio J. Pierik
Chapter Goals *87*
Main Chapter Points *87*
4.1 Introduction *87*
4.2 Methodology and Challenges of Assigning Cytosolic Fe–S Proteins *95*
4.3 [2Fe–2S] Protein Assembly in the Cytosol and the Early CIA Machinery *99*
4.3.1 Export of X-S from Mitochondria and Essential GSH Requirement in Fe–S Biogenesis *99*

4.3.2 [2Fe–2S] Clusters in Yeast Iron Homeostasis and [2Fe–2S] Chaperones *100*
4.3.3 [2Fe–2S] Target Proteins *102*
4.3.4 The Early CIA and Involvement of [2Fe–2S] Clusters in [4Fe–4S] Cluster Biogenesis *103*
4.4 [2Fe–2S] and [4Fe–4S] Protein Assembly at the Late CIA Machinery *105*
4.4.1 Nar1: Between the Early CIA and Late CIA Machinery *105*
4.4.2 The CTC *106*
4.4.3 Bipartite Recognition of Targets *109*
4.4.4 Potential [2Fe–2S] Cluster Insertion into Proteins at the CTC *113*
4.5 Relevance for Disease and for Biotechnology *115*
4.5.1 Diseases Related to Cytosolic and Nuclear Fe–S Proteins *115*
4.5.2 Diseases Related to Maturation of Cytosolic and Nuclear Fe–S Proteins *116*
4.5.3 Biotechnology *117*
References *118*

5 Mammalian Aldehyde Oxidase *135*
Maria Joao Romao, Guilherme Vilela-Alves, and Cristiano Mota
Chapter Goals *135*
Main Chapter Points *135*
5.1 Introduction *136*
5.2 The Xanthine Oxidase Family of Mononuclear Mo Enzymes *136*
5.2.1 The Fe–S Domain and Structural/Spectroscopic Assignment of the Iron–Sulfur Centers *138*
5.3 Aldehyde Oxidase Crystal Structure *139*
5.3.1 The Mammalian AOX Crystal Structures *139*
5.3.2 The Mo Active Site *141*
5.3.3 The [2Fe–2S] Clusters *145*
5.3.4 The FAD-Binding Site *146*
5.4 Aldehyde Oxidase and Its Role in Drug Metabolism *147*
5.4.1 Reactions Catalyzed by hAOX1 *148*
5.4.1.1 Oxidation of Aldehydes, N-aromatic Heterocycles, and Iminium Ions *148*
5.4.1.2 Reductions of Nitro, N-oxide, S-oxide, and Isothiazol Compounds *149*
5.4.1.3 Amide Hydrolysis Reactions *149*
5.4.2 Inhibition of hAOX1 Activity *149*
5.4.3 Single-Nucleotide Polymorphisms *150*
5.5 Conclusions *154*
List of Abbreviations *154*
References *154*

6 Biological Formation of Sulfide *159*
Marion Jespersen and Tristan Wagner
Chapter Goals *159*
Main Chapter Points *159*
6.1 Sulfur Is a Versatile Element and an Essential Cellular Component *161*
6.1.1 Chemical Properties of Sulfur *161*
6.1.2 Sulfur as an Essential Cellular Component *162*
6.1.3 Sulfur Bioavailability in the Environment *164*
6.2 A Sulfate-Reduction Metabolism to Obtain Cellular Sulfur: The Assimilatory Pathway *165*
6.2.1 Sulfate Is Activated by the ATP-Sulfurylase *165*
6.2.2 APS Is Further Phosphorylated by the APS-Kinase *167*
6.2.3 Sulfite Production and Hydrolysis of the Poisonous By-Product PAP *167*
6.2.4 Sulfite Reduction *168*
6.3 A Sulfate-Reduction Metabolism to Conserve Energy: The Dissimilatory Pathway *168*
6.3.1 Ecological Impact of the Dissimilatory Reduction of Sulfate and Its Bioenergetics *169*
6.3.2 Sulfate Capture and Activation *170*
6.3.3 A Catabolic APS-Reductase *170*
6.3.4 Sulfite Reduction Is Coupled to Energy Conservation *171*
6.3.5 Sulfur Assimilation and Dissimilatory Sulfate Reduction Are Potentially Uncoupled *172*
6.4 The Unconventional Sulfate-Reduction Pathway of *M. thermolithotrophicus* *172*
6.4.1 Methanogens and Sulfur *173*
6.4.2 *M. thermolithotrophicus* Exhibits Remarkable Growth on Sulfate *174*
6.4.3 A Conserved Sulfate Activation Mechanism Produces PAPS *174*
6.4.4 A Dissimilatory-Like PAPS-Reductase and a New PAP-Phosphatase *175*
6.4.5 A F_{420}-Dependent Sulfite Reductase Catalyzes the Last Step of the Pathway *177*
6.5 Evolution and Biotechnological Applications *178*
6.5.1 Fsr as a Structural Template for a Primordial Sulfite Reductase *179*
6.5.2 Sulfate Utilization as a Safer Route for Methanogenic Biotechnologies *180*
6.6 Conclusions *181*
References *181*

7 The Molybdenum Cofactor, Its Biosynthesis, and Relation to Fe–S Clusters in Bacteria *193*
Paolo Olivieri and Silke Leimkühler
Chapter Goals *193*
Main Chapter Points *193*
7.1 Molybdenum in Biology *193*

7.2 The Discovery of Molybdoenzymes: Studies of Mutant Strains from *Neurospora crassa* Show the Universality of the Molybdenum Cofactor *194*
7.3 Mutagenesis of Bacteria Help(s) to Discover Biosynthesis Pathways *194*
7.4 Chemical Investigation on the Structure of the Molybdenum Cofactor *196*
7.4.1 Identification of the First Intermediate, Precursor Z *198*
7.5 Moco Biosynthesis *199*
7.6 Differences in the Molybdenum Cofactor of Prokaryotes and Eukaryotes *202*
7.7 Discovery of the Sulfido Ligand *205*
7.8 The Role of Molybdoenzymes in *E. coli* *207*
7.9 Role of Fe–S Clusters for the Insertion of Moco into Molybdoenzymes *209*
7.10 The Reaction Catalyzed by Molybdoenzymes *211*
7.11 The Involvement of Fe–S Clusters in the Regulation of Bacterial Molybdoenzyme Expression *212*
7.12 Conclusions *213*
Acknowledgments *214*
References *214*

8 Molybdenum Cofactor Biosynthesis in Eukaryotes *227*
Guenter Schwarz, Lukas Flohr, Emanuel Bruckisch, and Katrin Fischer-Schrader
Chapter Goals *227*
Main Chapter Points *227*
8.1 Introduction *228*
8.2 Eukaryotic Molybdenum Enzymes *229*
8.3 Molybdenum Cofactor Biosynthesis – An Overview *231*
8.4 Synthesis of Cyclic Pyranopterin Monophosphate *233*
8.5 Thiolation of cPMP by Molybdopterin Synthase *236*
8.6 Molybdenum Insertion into Molybdopterin *239*
8.7 Deficiency of Moco *242*
8.8 The Intimate Relation Among Moco Synthesis, Moco Enzymes, and Fe–S Clusters *244*
References *245*

9 NifUS Is a Two-component Toolbox Involved in Assembling Fe–S Clusters Associated with Nitrogen Fixation and Beyond *257*
Julia S. Martin del Campo, Shervin Shaybani, Dennis R. Dean, and Patricia C. Dos Santos
Chapter Goals *257*
Main Chapter Points *257*
9.1 Introduction *258*

9.2 Structural and Catalytic Features of NifU 259
9.2.1 NifU N′-Terminal Domain 259
9.2.2 NifU Central Domain 261
9.2.3 NifU C′-Terminal Domain 262
9.3 Structural and Catalytic Features of NifS 263
9.3.1 NifS Is a Cys:NifU Sulfurtransferase (EC 2.8.1.7) 265
9.4 Synthesis of Fe–S Clusters by the NifUS System 267
9.5 Dedicated Role of NifUS in Aerobic Nitrogen Fixation 269
9.5.1 Direct Activation of Nitrogenase Reductase, NifH 271
9.5.2 *Ex Situ* Assembly of the FeMo cofactor 272
9.5.3 *In Situ* Assembly of P-cluster on NifDK 273
9.5.4 Involvement of NifU and NifU-Like Proteins in the Synthesis of Metalloclusters for Vnf and Anf Nitrogenases 273
9.5.5 Cross Talk Between NifUS and the Housekeeping IscUS System 274
9.5.6 Inclusion of NifUS in Biotechnological Efforts to Expand Diazotrophy to Other Species 275
9.5.7 Occurrence of NifUS-Like in Non-Nitrogen-Fixing Species 276
9.6 Concluding Remarks 278
Acknowledgments 279
References 279

10 Alternative Substrates of Nitrogenase *287*
Frederik V. Schmidt and Johannes G. Rebelein
Chapter Goals 287
Main Chapter Points 287
10.1 Nitrogenase Introduction 287
10.1.1 Nitrogenase Isozymes 288
10.1.2 Nitrogenase Mechanism 290
10.1.3 Alternative Substrates of Nitrogenase 292
10.2 Carbon Monoxide and Carbon Dioxide 294
10.2.1 Carbon Monoxide 295
10.2.1.1 CO Reduction by the Mo Nitrogenase 295
10.2.1.2 CO Reduction by the V Nitrogenase 297
10.2.1.3 CO Reduction by the Fe Nitrogenase 299
10.2.1.4 Conclusions on CO Reduction by Nitrogenases 299
10.2.2 Carbon Dioxide 301
10.2.2.1 CO_2 Reduction by the Mo Nitrogenase 301
10.2.2.2 CO_2 Reduction by the V Nitrogenase 302
10.2.2.3 CO_2 Reduction by the Fe Nitrogenase 302
10.2.2.4 Conclusions on CO_2 Reduction by Nitrogenases 303
10.3 Hydrocarbon Substrates 303
10.3.1 Acetylene 305
10.3.2 Other Alkynes and Alkenes 306
10.4 Alternative Nitrogen-Containing Substrates 306
10.4.1 Azide 306
10.4.2 Cyanide, Nitriles, and Isonitriles 307

10.5 Conclusions *308*
References *308*

Volume 2

Preface *xvii*

11 Electron-Bifurcating Hydrogenases *317*
Gregory E. Vansuch, Effie C. Kisgeropoulos, Jonathan R. Humphreys, Carolyn E. Lubner, David W. Mulder, and Paul W. King

12 Inhibition of [FeFe]-Hydrogenases by Small Molecules *381*
Claudia Brocks and Thomas Happe

13 Biosynthesis of [NiFe]-Hydrogenase *407*
Oliver Lenz and Giorgio Caserta

14 [Fe]-Hydrogenase and the FeGP Cofactor Involved in the CO_2-Reducing Hydrogenotrophic Methanogenic Pathway *447*
Seigo Shima, Joao Pedro Fernandes-Queiroz, and Masanori Kaneko

15 Catalysis by Hydrogenase *489*
Seigo Shima, James A. Birrell, Sven T. Stripp, Giorgio Caserta, and Oliver Lenz

16 Macromolecular Crystallography *541*
Konstantin Bikbaev and Ingrid Span

17 Vibrational and Mössbauer Spectroscopic Techniques to Study Iron–Sulfur Clusters *571*
Christian Lorent, Giorgio Caserta, Volker Schünemann, and Ingo Zebger

18 The Multisite Microstate Model: A Theoretical Approach for Analyzing Charge Transfer in Proteins *617*
G. Matthias Ullmann and Rajeev Ranjan Roy

19 Structural and Functional Bioinorganic Model Chemistry of Fe–S Clusters – Synthesis and Analysis *649*
Benedict Josua Elvers and Carola Schulzke

20 Metal-Dependent Formate Dehydrogenases and Their Interplay and Relationship to Iron–Sulfur Clusters *705*
Benjamin R. Duffus

Index *747*

Preface

The study of metalloenzymes began more than a century ago. In 1897, Gabriel Bertrand, working on laccase (a polyphenol oxidase), suggested for the first time that a metal ion was essential for the catalytic activity of an enzyme. In 1926, James Sumner crystallized the first protein, jack bean urease, thereby demonstrating that urease is a protein and that the dissolved crystals catalyze the hydrolysis of urea. Only 49 years after Sumner's pioneering crystallization of an enzyme, urease was eventually found to be a metalloenzyme, containing metal ions at its active site that are essential for enzyme catalysis. Today we know that almost half of the proteins require a metal ion to function, and more than one-third of the known enzymes are metalloenzymes, requiring stoichiometric amounts of metal ions as cofactors, which are typically transition metal ions. Metal ions are essential for life, as metalloproteins are involved in most important biological processes, including respiration; photosynthesis; nitrogen, sulfur, carbon, and hydrogen metabolism; biosynthesis of antibiotics; gene regulation; replication and repair of DNA; antioxidant defense; and neurotransmission. However, even more than 100 years after Bertrand's first discovery of a metalloenzyme, we are only just beginning to understand the role of metals in the catalysis of biological reactions.

The second most abundant metal on earth is iron, which plays an important role in redox reactions in biological systems and is mostly present in either cytochromes or iron–sulfur (Fe–S) clusters. Fe–S clusters occur in more than 200 different types of enzymes or proteins and constitute one of the most ancient, ubiquitous, and structurally diverse classes of biological prosthetic groups. Fe–S clusters were first discovered by Helmut Beinert and others in the early 1960s. The corresponding iron–sulfur enzymes are functionally quite diverse and have both redox and nonredox functions. They are prevalent in central metabolism and notably under anaerobic conditions. Therefore, they have frequently been considered to be of very ancient origin. Since a great variety of Fe–S clusters spontaneously form in solution from Fe^{2+}/Fe^{3+}, $R\text{-}S^-$, S^{2-} or from Fe^{2+}, R-SH, and sulfur, it has been suggested that the first iron–sulfur enzymes arose through the incorporation of preexisting inorganic Fe–S clusters into polypeptides. However, in the 1990s, it was finally discovered that the process of Fe–S cluster biosynthesis is very complex and requires a set of dedicated proteins that are remarkably conserved in prokaryotic and eukaryotic organisms.

Fe–S-containing proteins are also involved in the assembly of metal sites that contain noniron metals. Examples are the assembly of the FeMoco of nitrogenase, the catalytic centers of hydrogenases, the molybdenum cofactor, and heme. Additionally, iron–sulfur clusters are involved in electron transport, gene regulation, DNA replication and repair, and tRNA modification. Thus, it seems that the pathways for the assembly of these metal sites cannot be studied without the understanding of the assembly and regulation of Fe–S clusters in the cell. We mainly describe the assembly and role of metalloproteins that either contain an iron atom in a mononuclear or binuclear active site or for which the assembly and catalytic activity of the enzyme depends on a Fe–S cluster-containing protein. The assembly of metal sites and the role of different metal ions in polynuclear metalloclusters will be described in this book. Furthermore, the availability of metal ions in the cell is tightly regulated, and metal homeostasis and trafficking in a cellular context are also covered.

December 2024
Potsdam

Silke Leimkühler
Günter Schwarz
Oliver Lenz
Oliver Einsle

1

ISC-Based Fe–S Protein Biogenesis in Bacteria

Béatrice Py[1] and Frédéric Barras[2]

[1]*Aix-Marseille Université, Laboratoire de Chimie Bactérienne (UMR7283), Institut de Microbiologie de la Méditerranée, Institut Microbiologie Bioénergies et Biotechnologie, Centre National de la Recherche Scientifique, 31 chemin Joseph Aiguier, 13009 Marseille, France*
[2]*Institut Pasteur, Université Paris Cité, CNRS UMR6047, Department of Microbiology, Unit Stress Adaptation and Metabolism in Enterobacteria, 25-28 rue du Dr Roux, 75015 Paris, France*

Chapter Goals

- Provide a simple overview of the current state of the art on the iron–sulfur cluster (ISC) machinery in bacteria.

Main Chapter Points

- Fe–S (iron–sulfur) proteins are fundamental to life, playing critical roles in processes like electron transport, enzymatic functions, and regulation of gene expression.
- Our article delves into the discovery of one of the most studied multiprotein machineries, iron–sulfur cluster (ISC), which is able to ensure the biogenesis of a group of vital proteins.
- By exploring the latest discoveries, we will also unravel how cells orchestrate the precise construction of Fe–S clusters and integrate them into target proteins.
- Whether you are a student or an experienced researcher new to the field, this article provides a comprehensive overview, highlighting the necessary concepts needed to understand Fe–S protein biogenesis and how this can pave the way for advancements in fundamental research, biotechnology, and medicine.

1.1 Introduction

Fe–S clusters have been "recruited" by living systems very early in evolution [1]. It is likely that, under primitive anaerobic conditions, the abundance and availability of

Iron-Sulfur Clusters: Biogenesis and Biochemistry, Volume 1, First Edition.
Edited by Silke Leimkühler, Günter Schwarz, Oliver Lenz, and Oliver Einsle.

iron, then in ferrous form (Fe^{2+}), and sulfur, then in sulfide form (S^{2-}), contributed to their acquisition by emerging enzymatic systems. However, with the advent of an aerobic atmosphere, maintaining a biology partly based on Fe–S clusters became both more challenging – as iron exists in insoluble oxidized form – and more dangerous – as iron can act as a catalyst for the production of reactive oxygen species (ROS), which are destructive to Fe–S clusters.

There are several types of Fe–S centers. The most common ones are [2Fe–2S] and [4Fe–4S] clusters (Figure 1.1). Fe–S clusters are associated with polypeptides via noncovalent bonds between iron atoms and the side chains of amino acid residues, mostly Cys residues, although other residues, such as Asp, His, and Arg, can also serve as ligands. The different electron valences of iron and sulfur confer properties to Fe–S clusters that allow them to endow host proteins with a wide spectrum of activity. Thus, Fe–S clusters are used frequently as electron relays in electron transfer chains, sensors for modification of the redox potential of the surrounding environment, and also as Lewis acid by directly participating in catalytic reactions [1, 3, 4].

For over 30 years, the formation of Fe–S clusters was considered a spontaneous event. Discovery of the iron–sulfur cluster (ISC) machinery in the 1990s and since then, general concepts for *de novo* Fe–S protein maturation (here referred to as Fe–S protein biogenesis) have emerged. First, *de novo* Fe–S clusters are built onto a scaffold protein whose role is to acquire both elements separately, sulfur and iron, and to provide an adequate platform for assembly of the elements into an Fe–S cluster. Sulfur is produced by a cysteine desulfurase, an enzyme admitting L-cysteine as a substrate. Second, once assembled, the cluster is delivered to one of the many target cellular proteins via one or several Fe–S cluster carriers, which can form a highly plastic network depending on the environmental conditions and/or the target proteins.

Studies in model organisms such as *Escherichia coli* and *Azotobacter vinelandii* for prokaryotes, and *Saccharomyces cerevisiae*, *Arabidopsis thaliana*, and humans for eukaryotes, have identified three machineries necessary for the formation of Fe–S centers and their insertion into "client" cellular proteins: the ISC machinery, present in bacteria and mitochondria, the SUF (Sulfur mobilization) machinery, present in bacteria and chloroplasts, and the NIF (Nitrogen fixation) machinery that is present in nitrogen-fixing bacteria. The construction of Fe–S centers by the ISC, NIF, and SUF machineries follows relatively similar logics, i.e. assembling the Fe and S elements in Fe–S clusters and then delivering Fe–S clusters to the cell's "client" proteins (Figure 1.2). The ISC and SUF systems are referred to as general Fe–S biogenesis machineries, meaning that they are able to furnish Fe–S clusters for all Fe–S proteins of the cell. In contrast, the NIF machinery is specialized in the assembly of Fe–S clusters for only one enzyme, the nitrogenase. Eukaryotic systems responsible for Fe–S cluster formation, ISC within mitochondria and SUF within photosynthetic lineages, have evolutionary origins traceable to endosymbiotic events involving Alphaproteobacteria and Cyanobacteria, respectively [5]. Eukaryotes possess an additional machinery, the cytosolic iron–sulfur clusters assembly (CIA), which could be described as an "extension of the distribution line" of the ISC

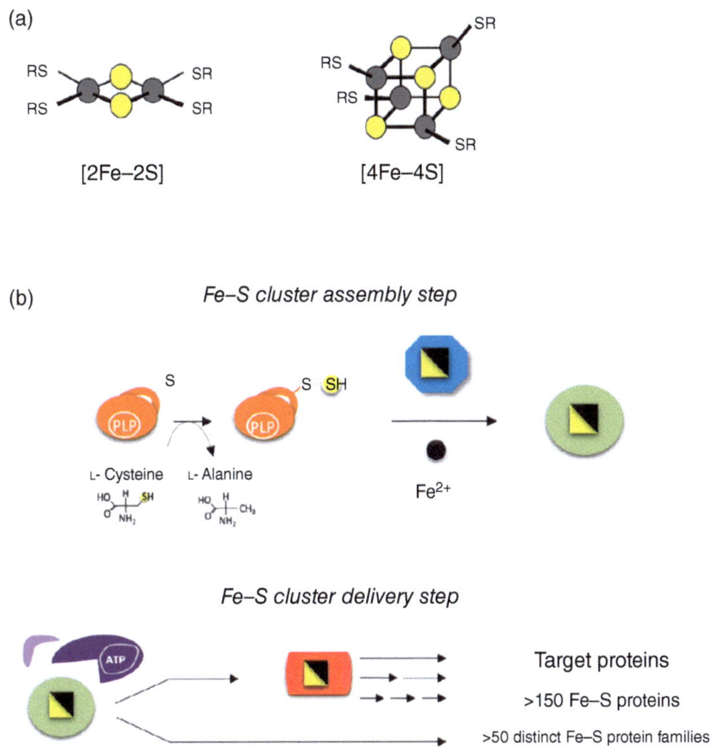

Figure 1.1 Main steps of Fe–S cluster-containing protein biogenesis. (a) Most common Fe–S clusters found in proteins. Schematic representation of the structure of the rhombic [2Fe–2S] cluster (left) and the cubane [4Fe–4S] cluster (right). Yellow and gray circles represent sulfur and iron atoms, respectively. RS/SR indicates the cysteinyl ligation via cysteine residues of the protein. (b) General principles of *de novo* Fe–S protein biogenesis by the ISC machinery. In the first step, the Fe–S cluster (yellow/black square) is assembled on a scaffold protein (green), which receives sulfur (yellow circles) from a cysteine desulfurase (orange) carrying a PLP cofactor, and iron (black circles) from an as-yet nonidentified source. Electrons are provided from ferredoxin (blue). Then, the newly formed Fe–S cluster is released from the scaffold, by a chaperone/co-chaperone (dark and light purple) assisted process that is ATP dependent, and transferred to one or several carrier proteins (red) that deliver the cluster to the final apo-target [2].

machinery, located in the cytoplasm [6]. The CIA is required for the maturation of cytoplasmic and nuclear Fe–S proteins.

Recently, an exhaustive bioinformatic and phylogenomic search of 10 000 archaeal and bacterial genomes, representative of the prokaryotic biodiversity, has unearthed two additional "minimal" Fe–S cluster assembly machineries, minimal iron–sulfur (MIS) and SUF-like minimal system (SMS) [2]. These "minimal" Fe–S cluster assembly machineries have been inferred to be present in the last universal common ancestor (LUCA) [2].

Here, we will focus on the prokaryotic ISC machinery whose functioning is evolutionarily conserved in eukaryotes, notably in humans in which deleterious

(a)

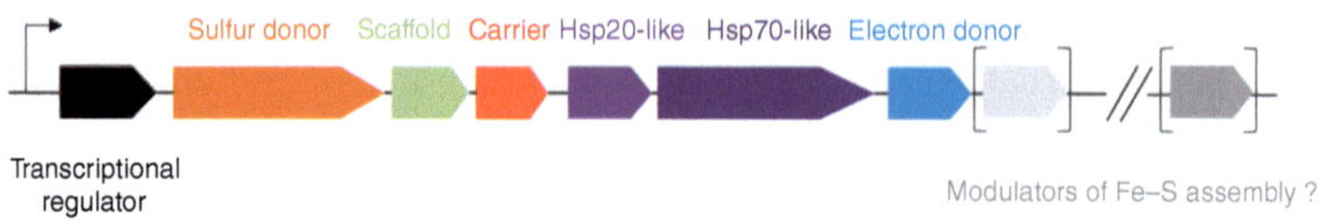

(b)

Characteristics and functionally important residues of the *E. coli* ISC components

IscS	Lys206	PLP binding
	His104 Arg354	L-cysteine binding
	Cys328	Persulfurated Cys
IscU	Tyr3	Homodimer interaction
	Cys106	Persulfide final acceptor
	Asp39, Cys63, Cys106	Essential for sulfur transfer
	Cys63 Asp39	Transient iron ion coordination
	Cys37, Cys63, Cys106	Ligands of a transient 2Fe–2S cluster
	Ile108	Influences CyaY requirement
	His105 Cys63	Positions in the structure modified by binding of CyaY
	L99-PPV-K103	HscA interaction
IscA	CX61-65CXC	Ligands of a « transient » Fe–S cluster
HscB	Leu92, Leu96, Phe153	Interaction with IscU
	His32, Pro33, Asp34	Interaction with HscA
HscA	Gly67-Thr78/Asp260-Lys275	Nucleotide binding cleft
Fdx	Cys42, Cys48, Cys51, Cys87	Ligands of a stable 2Fe–2S cluster
IscX	Highly acidic protein	Iron binding
CyaY	Glu18, Glu19, Asp22	Iron binding

Figure 1.2 The *E. coli* ISC Fe–S protein biogenesis machinery. (a) Genetic organization of the *E. coli isc* operon. Genes are color-coded depending on the function ensured as indicated in Figure 1.1. The *cyaY* gene that is not part of the *isc* operon and *iscX*, the last gene of the *isc* operon, are color coded as a putative modulator of Fe–S cluster assembly by the ISC machinery, in dark and light gray, respectively. The IscR transcriptional regulator (back). The transcription start of the *iscSUA-hscBA-fdx-iscX* operon is indicated by a black arrow. (b) Functionally important residues of the ISC components are indicated.

mutations can lead to severe pathologies [7]. We will detail how this machinery has been discovered, as well as its functioning, regulation, and evolution.

1.2 Discovery of the ISC Fe–S Protein Biogenesis Machinery

The existence of Fe–S protein biogenesis systems was revealed by genetic approaches. This major breakthrough did not emerge from the most widely used model organism, *E. coli*, but from the nitrogen-fixing bacteria *Klebsiella pneumoniae* and *A. vinelandii*. Nitrogenase activity was, and still is, the focus of a tremendous interest as it enables fixation of atmospheric dinitrogen (N_2), under ammonia, a reaction of vital importance for plants and more widely for the terrestrial biogeochemical nitrogen cycle [8, 9]. The nitrogenase is composed of two proteins, which requires a [4Fe–4S] cluster together with the P cluster 8Fe–7S and the Fe–Moco cluster [7FeMo9SC].

Mutants of *K. pneumoniae* incapable of utilizing N_2 as their primary nitrogen source were searched following nitrosoguanidine treatment. Based on variation of the colony aspects (size and pigmentation), when grown on minimal agar medium under N_2 atmosphere conditions, nitrogenase-deficient mutants were isolated. Multiple loci of *K. pneumoniae* were pinpointed and genetically mapped [11]. Parallel studies in *A. vinelandii* also reached the similar conclusion that the *nif* gene clusters identified were much larger than expected given that nitrogenase function relied on a pair of polypeptides [10]. This opened the possibility that among these genes some could encode ancillary factors required for nitrogenase activity. Next, biochemistry came into play with the purification of a protein, NifS, exhibiting cysteine desulfurase activity from *A. vinelandii* [12].

Surprisingly, a low but significant level of nitrogenase activity remained in the *nif* mutants, suggesting that a second system might arise that could assist nitrogenase activity though with reduced efficiency. Hence, the IscS protein that catalyzes the same enzymatic activity as NifS was isolated from extracts prepared from an *A. vinelandii strain* deleted for the *nifS* gene [13]. The oligopeptide sequences obtained by microsequencing of IscS were used to design oligonucleotides employed for PCR screening of an *A. vinelandii* genomic library. Upon cloning and sequencing of a 2.5 kb DNA fragment, the *iscS* gene, encoding cysteine desulfurase, alongside *iscU* and *iscA* genes were identified [13]. Based on their sequence similarity with corresponding *A vinelandii* genes, the *iscS, iscU, iscA*, and *fdx* genes were identified in *E. coli*. Eventually, the ISC system encoded by the *iscRSUA-hscBA-fdx-iscX* operon was described (Figure 1.2). Located outside of the *iscRSUA-hscBA-fdx-iscX* operon, the *cyaY* gene encodes a component that contributes also to the ISC machinery functioning. The initial indication that CyaY might participate in Fe–S protein biogenesis came from studies in eukaryotes. Hence, CyaY is the bacterial ortholog of frataxin, a small protein whose depletion leads to a progressive neurodegenerative disease called Friedreich's ataxia. Among the molecular makers of this disease, a hallmark is the deficiency of the activity of Fe–S proteins [14, 15]. The phylogenetic co-occurrence of *cyaY* with the *hscA* and *hscB* genes of the *isc* operon was further supporting its participation in the Fe–S protein biogenesis process [16]. The *in vivo* contribution and the role of each of these ISC components to the Fe–S protein biogenesis process will be detailed below.

1.3 Assessing the Importance of the Individual ISC Components

Attempts to isolate mutant strains of *A. vinelandii* in which the *isc* genes were inactivated, by in-frame deletion or insertion mutations, were unsuccessful suggesting that the *isc* genes were essential in this organism [13]. To assess the physiological consequences of compromising the activity of individual proteins encoded within the *isc* gene cluster, Dean and colleagues developed a method for the conditional expression of the *isc* genes [17]. To do so, the genomic copy of the *iscRSUA-hscBA-fdx-iscX* operon was duplicated [17]. Expression of one copy was driven by the wild-type *isc* regulatory elements (P_{isc}), while the other was controlled by the glucose-repressed *scrX* promoter (P_{scrX}) (Figure 1.3). Subsequent deletion of individual *isc* genes at the P_{isc}-controlled loci revealed a null growth phenotype in the presence of glucose, indicating the essential role of the IscS, IscU, HscA/HscB, and Fdx proteins [17]. The null growth phenotype of HscBA-depleted cells of *A. vinelandii,* which is exhibited at 20% O_2, could be partially suppressed by lowering the oxygen concentration to 5% [17]. A null growth phenotype of IscA-depleted cells was observed only under conditions of elevated oxygen 40%.

E. coli isc mutants show phenotypes linked to defects in growth rate, nutritional requirements (nicotinic acid, thiamine, valine, and isoleucine), and defects in activities of Fe–S enzymes (respiratory complex I, glutamate synthase GltS, and succinate dehydrogenase SDH) [18–21]. Altogether, these studies repeatedly showed that inactivation of *iscS* always led to more severe phenotype compared to the other genes of the *isc* operon. This is very likely due to the fact that IscS is the sulfur donor for several processes in addition to Fe–S biogenesis, namely tRNA thiolation, molybdenum cofactor, and thiamine synthesis [22–24]. These studies have also revealed several Fe–S proteins whose maturation strongly relies on the ISC machinery. The molecular basis for such dependency remains to be determined. A main limitation however is that these defects are usually dampened down by the presence of SUF, which share the same client proteins as ISC, and that inactivating both ISC and SUF is lethal. However, a genetic approach was designed such as to assess the contribution of *isc* genes to Fe–S cluster homeostasis.

The reason Fe–S biogenesis is essential for *E. coli* growth in rich media in aerobiosis is due to the essentiality of maturation of two key Fe–S proteins, namely IspG and IspH [25]. These proteins are part of the biosynthesis pathway of isopentenyl diphosphate (IPP) from 2-*C*-methyl-D-erythritol 4-phosphate (MEP). In eukaryotes and certain bacteria, it is via an Fe–S-independent mevalonate (MVA)-dependent pathway that IPP is synthesized. Then, an MVA-based genetic bypass, existing in a chromosomic or a plasmidic version, has been designed to circumvent the essentiality of IspGH [25, 26]. In the resulting *E. coli* MVA$^+$ strains, growth is permitted even in the absence of Fe–S biogenesis, when MVA is added in the medium. By utilizing an *E. coli* MVA$^+$ Δ*suf* background, it becomes feasible to systematically delete each gene within the *isc* operon and test its importance for ISC function by evaluating growth in the presence or absence of MVA. This strategy has permitted to show that IscS, IscU, HscA, and HscB are strictly essential in aerobic and anaerobic conditions, while Fdx and IscA appear to be required in aerobiosis but are not fully necessary in anaerobiosis (Figure 1.3) [26]. In fact, previous work had shown that IscA is redundant with SufA under aerobiosis and with ErpA under anaerobiosis [27]. Regarding the *iscX* gene that is present in a minority of bacterial *isc* operons, it appears to be nonessential in *E. coli* strain lacking SUF [2, 26, 28, 29]. Similarly, the *iscR* mutant showed no growth defect when combined with the deletion of the *suf* operon [21].

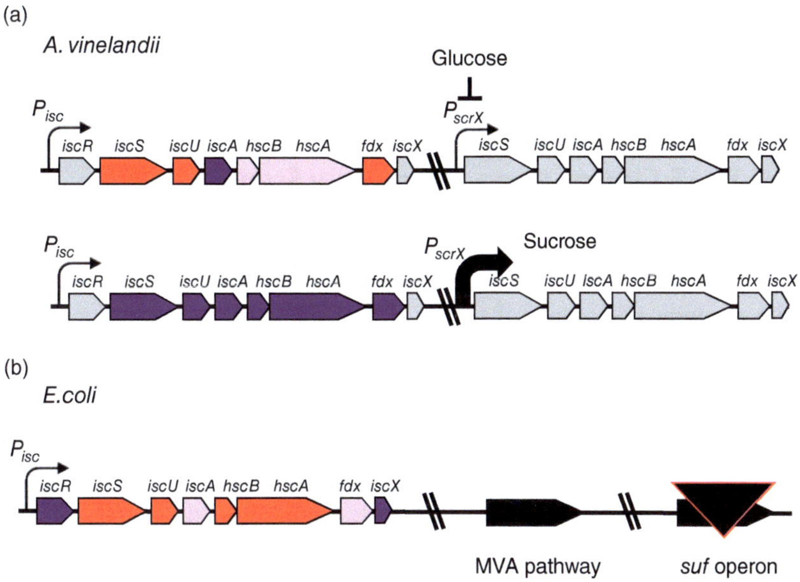

Figure 1.3 Example of genetic strategies to investigate the individual contribution of components to the ISC machinery. (a) In *A. vinelandii*, the chromosomic copy of the *iscSUA-hscBA-fdx-iscX* operon had been duplicated. Expression of one copy was controlled by the WT *isc* regulatory elements (P_{isc}), while expression of the second copy was driven by the glucose-repressed *scrX* promoter (P_{scrX}) [17]. (Top) Deletion of individual *isc* genes was then introduced at the P_{isc}-controlled loci. Under ambient oxygen (~20%), inactivated genes that conducted to a null growth phenotype in the presence of glucose were concluded to be essential for the ISC machinery (represented in red). Inactivated genes that had no or a modest impact on growth in the presence of glucose are represented in pink and purple, respectively. It was concluded that the function exerted by the corresponding proteins might be totally or partly bypassed. (Bottom) In control, when grown in the presence of sucrose, deletion of the individual *isc* genes (indicated purple) did not cause growth defects. Genes in gray were not inactivated. Source: Adapted from [17]. (b) Schematic representation to study the contribution of the *isc* genes in an *E. coli* strain possessing the mevalonate pathway (MVA^+) (here plasmidic but can also be chromosomic) and lacking the SUF system (deletion of the *sufABCDSE* operon). Inactivated genes that conducted to a null growth phenotype in the absence of mevalonate (represented in red) were concluded to be essential for the ISC machinery. Inactivated genes in mutants that were able to grow in the absence of mevalonate during anaerobiosis, but not in aerobiosis are indicated in pink. Genes in purple are those that do not impact growth in anaerobiosis or in aerobiosis.

The importance of the ISC Fe–S cluster biogenesis factors has also been pointed out in screening procedures for antibiotic sensitivity [18, 30, 31], swarming motility [32], or even lambda phage infection [33]. In some cases, the direct implication of Fe–S proteins in these general phenotypes have been identified such as Fe–S proteins involved in tRNA modification for the lambda phage infection [33] and the respiratory complexes (Nuo and Sdh) for the tolerance to aminoglycosides [18], while for others the connection remains to be established.

Altogether, these approaches provide us with a good insight into the fundamental significance of Fe–S clusters in cellular physiology. Studies in *A. vinelandii* and

E. coli have also defined the ISC core components, IscS and IscU, and those for which the contribution can vary depending on the burden imposed on Fe–S protein by the growth conditions and/or the presence of redundant function in the cell.

1.4 The Fe–S Cluster *de novo* Assembly Step

The scaffold protein IscU is the central piece of the ISC machinery, since it is where iron and sulfur assemble into a cluster. The concept of scaffold, meaning that Fe–S clusters are assembled elsewhere than onto the client protein, came from early studies on FeMo cofactor biosynthesis for the nitrogenase [34, 35]. Hence, it has been shown that the MoFe protein subunits of the nitrogenase (*nifD* and *nifK* gene products) are not required for FeMo cofactor biosynthesis [36, 37]. These results implied that the FeMo cofactor is synthesized and delivered into the apo-protein rather than being synthesized stepwise into the MoFe protein.

The sulfur element is provided by the cysteine desulfurase IscS that first abstracts sulfur from L-cysteine desulfurase and second transfers the sulfur atom to IscU. The L-cysteine desulfurase activity of IscS requires the pyridoxal phosphate (PLP) cofactor, commonly known as vitamin B6. Thanks to the PLP cofactor, IscS is able to break the terminal C—S bond of L-cysteine and to capture the released sulfur atom (S^0 redox state) on a functionally essential cysteine residue (Cys328 in *E. coli* IscS), in the form of a cysteine persulfide intermediate (Cys-S-SH). The PLP cofactor is covalently bound to a lysine residue (Lys206 in *E. coli* IscS), leading to a PLP-Lys internal aldimine. The PLP moiety of this aldimine covalently binds the substrate, L-cysteine, forming a PLP-L-Cys external aldimine. Two conserved residues of *E. coli* IscS, His104 and Arg354, interact with the SH and carboxy group of the cysteine part of the PLP-L-Cys, respectively. Based on studies on IscS and other cysteine desulfurases, a complex catalytic mechanism for this large family of proteins has been proposed, during which Cys-S-SH is formed and ends with the release of L-alanine (for review, see the following recent publication [38]). Once the persulfurated cysteine (Cys-S-SH) is formed, the sulfur is activated, mobilizable, and nontoxic. The catalytic Cys328 residue of IscS is located in a loop proposed to be mobile in order to reach IscU to deliver directly the sulfur, as shown by ^{35}S transfer assay and mass spectrometry. As observed in the X-ray crystal structure of the *E. coli* apo-IscU–IscS complex, IscU binds near a loop of IscS that contains the active-site Cys328 [39, 40]. The structure of the apo-IscU–IscS complex shows that IscS is forming a homodimer on which two monomers of IscU bind independently.

While direct sulfur transfer from IscS to IscU has been well characterized, the precise site on IscU where it is received has been debated. Initially, analysis of *A. vinelandii* IscU variants containing individual Cys-to-Ala substitutions by electrospray-ionization mass spectrometry indicated that sulfur acquisition involved none of the three strictly conserved Cys residues of IscU (Cys37, Cys63, and Cys106 in *E. coli* IscU) [41]. However, the study of the mammalian ISC assembly complex by a novel assay that makes use of sulfur-reactive maleimide-peptide compounds, coupled with mass spectrometry analysis, and by the recent cryo-EM analysis of the complex, both concluded that the Cys residue equivalent to Cys106 of *E. coli* IscU was the preferential persulfide final acceptor [42, 43].

Electrons are another ingredient necessary for Fe–S cluster assembly. This has been postulated very early due to the presence of a gene encoding a [2Fe–2S] ferredoxin encoding gene (*fdx*) in the *isc* operon of *A. vinelandii* and *E. coli*. The necessity of Fdx was also evidenced during *in vitro* reconstitution assays [44]. It is important to note that despite artificial reductants, such as dithiothreitol (DTT), provide electrons during *in vitro* Fe–S cluster formation, they promoted it via a reaction that does not reflect the physiological reaction [45, 46]. *E. coli* Fdx was shown to interact with IscS, while the yeast homolog of Fdx interacts strongly with the scaffold and not significantly with the desulfurase [21, 44, 46, 47]. *In vitro* protein persulfidation assay showed that the human homolog of Fdx, FDX2, was required for reduction of the persulfurated form of IscU and concomitant Fe–S cluster formation, but did not affect persulfide transfer from the desulfurase to the scaffold [45]. Interestingly, the influence of FDX2 appeared to hinge on the iron binding to the IscU, presumably facilitating the immediate binding of nascent sulfide to nearby iron sites [45].

CyaY was proposed to regulate persulfide transfer from the desulfurase to the scaffold. First, the structural characterization of prokaryotic and eukaryotic ISC assembly complex has revealed that CyaY is adequately positioned, close to both IscU and the loop of IscS that contains the catalytic Cys328 residue [43, 48, 49]. Second, mammalian frataxin enhances the rate of sulfur transfer from the desulfurase to the scaffold *in vitro* [42]. Third, high-resolution cryo-EM structures of the human ISC assembly complex confirmed the long distance (6.5A) between the sulfur acceptor cysteine of the scaffold (Cys106 in *E. coli* IscU) and the sulfur donor cysteine of the desulfurase, when frataxin is absent [43, 48, 50]. Furthermore, the His137 (His105 in *E. coli* IscU) of the human ISC scaffold seems to prevent a closer approach of the two Cys residues [43]. Upon frataxin binding, significant local conformational changes occurred at the site of Fe–S cluster assembly on the scaffold [43]. Notably, the histidine is shifted substantially away from the assembly site, thereby removing the steric separation between the sulfur donor and the sulfur acceptor cysteine-, which exhibited a distance (3.6 A) more favorable for a direct sulfur transfer [43].

IscX physically interacts with IscS and IscU but how exactly it exert its positive role in *in vivo* Fe–S protein biogenesis is still unknown [28, 51, 52]. IscX exhibits a low-affinity binding for iron and competes with CyaY for binding to IscS [53–55]. Overall, IscS is an important hub for proteins involved in Fe–S cluster biogenesis (IscU, Fdx, CyaY, IscX), but also plays a role in other processes. Hence, *E. coli* IscS is also able to interact and donate sulfur to other partner proteins such as TusA for both tRNA-s2U34 thiolation and molybdenum cofactor (Moco) synthesis and ThiI for both thiamine biosynthesis and tRNA-s4U8 modification [22–24, 40, 56]. IscU is able to compete with TusA bound to IscS and to displace it, suggesting that the sulfur flux from IscS seems to be primarily directed for Fe–S protein biogenesis.

It is thought that iron should be acquired before sulfur, but how IscU accepts iron remains obscure. Fe–S protein biogenesis is an essential process and highly iron demanding. Surprisingly, no dedicated iron donor has been identified as yet. One proposed candidate was CyaY, as it binds iron via an acidic "face." However, in the absence of CyaY, Fe–S protein biogenesis is still functioning in *E. coli* and *Salmonella enterica*, albeit in a decreased manner, showing that, at the best, CyaY is not the sole iron source [28, 57–60]. Polyphosphate, DNA, and nucleotides could well act as iron storage devices, which might feed the ISC system.

In the recent cryo-EM structure of the eukaryotic ISC assembly complex, an iron ion is hosted in the Fe–S assembly site and is coordinated by Asp71 and Cys95 of the scaffold (Asp39 and Cys63 in *E. coli* IscU) and by the active Cys of the desulfurase [43]. The site of Fe–S cluster assembly of IscU can also host a zinc ion [45, 48, 61]. In the presence of metal (Fe or Zn) in the ISC assembly complex, the active Cys residue of IscS is located in the vicinity of the Fe–S assembly site, while in the absence of metal, the active Cys residue is in the vicinity of the PLP binding site [43, 50]. *In vitro* investigations indicated that the presence of zinc inhibits the assembly of Fe–S clusters [45, 48]. These findings align perfectly with *in vivo* studies demonstrating that an excess of intracellular Zn hampers Fe–S cluster biogenesis by impeding the maturation activity of Fe–S proteins such as IscU [64]. Zn mis-metallation likely occurs when IscU is its apo-form, because Zn does not seem to destabilize Fe–S clusters already present in holo-proteins.

IscU is able to assemble [2Fe–2S] cluster, yet the molecular mechanistic details of the Fe–S assembly process remain to be determined. A possibility is that after one round of sulfur and iron acquisition, two [1Fe–1S]-containing IscU interact to form a transient dimer enabling [2Fe–2S] cluster formation [62, 63]. IscU is also able to assemble [4Fe–4S] cluster *in vitro*. Hence, examining the time course of Fe–S cluster assembly on IscU revealed the sequential assembly of a [2Fe–2S] cluster and then a [4Fe–4S] cluster [65]. This was achieved by the reductive coupling of two [2Fe–2S] clusters, one of each subunit of the IscU dimer, into one [4Fe–4S] cluster at the dimer interface [65, 66]. *In vitro*, both IscU [2Fe–2S] and [4Fe–4S] bound forms have the capacity to transfer their cluster to target proteins [67]. The physiological relevance of the [4Fe–4S] bound is questioned by *in vivo* studies in eukaryotes, which showed that IscU is endowed with the properties to assemble [2Fe–2S], while that of [4Fe–4S] clusters is ensured by the Isa1–Isa2–Iba57 complex that would accept two [2Fe–2S] cluster and perform reductive coupling [68, 69].

The exact residues of IscU involved in Fe–S cluster ligation are still not known. They very likely involve the three conserved cysteines, as revealed by mutagenesis studies combined with spectroscopic analyses. These three Cys residues are strictly required for IscU function and could not be replaced by other amino acids even those able to act as a cluster ligand [17, 70]. There is a debate regarding what could be the fourth ligand of the cluster. Three potential candidates were proposed by different approaches: Asp39 and His105 in *E. coli* IscU, and Cys 328 the catalytic cysteine of IscS.

Before concluding on the assembly step, we must underline that NMR methods have revealed another aspect of IscU. Hence, it was repeatedly pointed out that IscU has a highly dynamic structure. Early observations with the *Thermotoga maritima* enzyme suggested that it was largely structured [61, 71]. Studies on *E. coli* apo-IscU further supported the view that it could exist in two slowly interconverting conformational states: one disordered (D) and one structured (S). Importantly, the dynamics of the S/D interconversion could be modified by adding metals such as Zn^{2+} that stabilize the site of Fe–S assembly [61]. Such stabilization has recently been proposed to be a prerequisite for sulfur transfer [72]. IscU is the central piece for the ISC-mediated Fe–S protein biogenesis and shuttles from the assembly step to the delivery step, then the S/D interconversion could be instrumental in orchestrating its sequence of interaction with its multiple ISC partners. The D state of apo-IscU would be the primary substrate for IscS, while the S state of [2Fe–2S]-IscU would be the substrate of cochaperone to initiate the delivery step [73–76].

1.5 The Fe–S Cluster Delivery Step

Release of the cluster from IscU is a task ensured by HscA and HscB that belong to the Hsp70 chaperone and J-protein families, respectively [77]. HscA, as other Hsp70 chaperones, exerts its role by cyclic interaction with short hydrophobic stretches present in their substrates. The only substrate known for HscA is IscU, in which the LPPVK sequence motif has been extensively shown, by pull-down assays and peptide arrays, as being the interaction site with HscA [78–82]. The crystal structure of the substrate binding domain (SBD) of HscA complexed to the IscU recognition peptide revealed that the peptide is located in a hydrophobic cleft under an extended conformation [78].

The activity of Hsp70 chaperones is associated with their ability to bind and hydrolyze ATP to ADP. They work together with two types of cochaperones, the J-domain proteins that stimulate ATP hydrolysis by Hsp70 and the nucleotide exchange factors that promote replacement of Hsp70-bound ADP with ATP. A notable characteristic of bacterial HscA, distinguishing them from other members of the Hsp70 family, is that the IscU-binding cycle operates independently of nucleotide-exchange factors. In a more conventional way, HscB, the J-domain protein that partners with HscA binds and delivers the substrate, IscU, to HscA, and also stimulates HscA ATPase activity [83–85]. In IscU, the binding site for HscB includes residues that are also involved in the interaction with IscS [49, 86]. Also, a direct interaction between HscB and IscS has been shown, suggesting that HscB might recruit IscU while it is bound to IscS [87]. All these findings suggest that IscU interacts first with IscS and subsequently with HscB once the Fe–S cluster has been assembled. Despite extensive studies on the HscA binding cycle on IscU, the mechanism by which HscA facilitates the release of the cluster remains an open question. It is conceivable that the conformation of holo-IscU undergoes "remodeling" in a manner that renders it unable to bind the Fe–S cluster. *In vitro* kinetic studies showed that HscAB of *E. coli* and *A. vinelandii* drastically increased the rate of transfer from [2Fe–2S]-IscU to apo-acceptor proteins [88, 89]. *In vivo*, it is thought that the released Fe–S cluster is captured by IscA, an Fe–S carrier that delivers it to the diverse client proteins.

IscA belongs to the A-type carrier (ATC) protein. This class of proteins is thought to receive preformed Fe–S clusters from the Fe–S core biogenesis machinery and to transfer them to the client cellular apo-proteins. Other ATCs synthesized by *E. coli* are SufA and ErpA. *In vitro*, ATCs can bind Fe–S cluster via three conserved cysteine residues and transfer them to apo-proteins such as ferredoxin, a [2Fe–2S] protein, and BioB, a [4Fe–4S] protein [90–95]. Eukaryote homologs have been proposed to receive [2Fe–2S] cluster and convert them to [4Fe–4S] cluster [68, 69]. Observation consistent with ATCs able to convert [2Fe–2S] cluster into [4Fe–4S] cluster was obtained in the *E. coli* system as well [95–97]. Another possibility is that ATCs would actually act as iron donors [98, 99]. Although entirely possible, this would not rule out the role of carriers, as as-isolated ATCs from different origins have been purified in an Fe–S-bound form.

IscA, SufA, and ErpA share approximately 30% sequence similarity. They exhibit complex functional redundancy that varies with growth conditions. Thus, IscA appears to be redundant with SufA in aerobiosis, as *iscA* and *sufA* deletion mutations are lethal synthetic under aerobic conditions, but redundant with ErpA in anaerobiosis, as *iscA* mutation is synthetic lethal with *erpA* mutation under anaerobic conditions [27]. In the conditions used to investigate the effect of the mutations, namely rich medium, 37 °C, the cause of the lethality is due to the lack of maturation of the essential isoprenoid synthesizing enzymes IspG/H [25, 27]. The conclusion is that under aerobiosis, IspG/H can obtain their cluster from either IscA or SufA, but under anaerobiosis the IspG/H proteins get their cluster from either ErpA or IscA. Interestingly, consistent with the observations carried out under aerobiosis, phylogenetic analysis classified ErpA as of the ATC-I type and IscA and SufA as of the ATC-II type. However, the situation appears to be less clear-cut as under aerobiosis ErpA is essential for IspG/H maturation. Altogether, this led to the proposal that under aerobiosis IscA or SufA receives their cluster from either ISC or SUF machinery and transfer them to ErpA, which transfers them to IspG/H. Under anaerobiosis, either one of the ATCs could transfer their cluster directly to IspG/H. Similar types of studies were carried out using a series of different apo-proteins as targets whose maturation was studied in different genetic backgrounds and/or cultural conditions, such as IscR, NsrR, Fnr, Nuo, and Hyd [19, 97, 100]. The conclusion is that a given cluster can take different routes to reach its target and the choice of a given route is mostly controlled by genetic regulation. Intrinsic features of the apo-target being matured, such as stability of its Fe–S cluster, and overall cellular demand for Fe–S clusters under the conditions considered are also likely to influence the choice of maturation pathway used.

1.6 Regulation of the ISC Machinery

In numerous bacterial species, the pivotal controller governing ISC expression is the transcriptional factor IscR. Our comprehension of the role of IscR primarily derives from investigations encompassing genetics, molecular biology, biochemistry, and structural analyses of *E. coli* IscR [101]. IscR belongs to the Rrf2 family of winged helix-turn-helix transcription factors that are highly versatile in sensing environmental inputs, which are often at the heart of host–pathogen interactions. Therefore, it is not surprising that Rrf2 members have been shown to be important for the virulence of several human pathogens, such as IscR of *Yersinia pseudotuberculosis* and of *S. enterica* [102, 103]. Via its [2Fe–2S] cluster, IscR is sensing aerobiosis, oxidative stress, iron scarcity, and potentially nitrous stress [104–106]. Through mutagenesis and structural studies, residues Cys92, Cys98, Cys104, and His107 have been identified as ligands for the cluster [107]. The presence of a histidine ligand is notably uncommon and may render the cluster susceptible to destabilization, thereby making it responsive to the aforementioned stress signals. Both the apo- and holo-forms of IscR exhibit DNA-binding capabilities and regulatory functions. IscR-regulated promoters contain two distinct binding sites, termed Type 1 and 2 [108, 109]. While the Type 1 site exclusively accommodates the holo-form of IscR, the Type 2 site can be bound by both the holo- and apo-forms [110].

In *E. coli*, the *iscR* gene is the first gene of the *isc* operon and a relatively extensive intergenic region (111 bp) is present between *iscR* and *iscS* (Figure 1.4). This region encompasses a binding site for RyhB, a small regulatory RNA (sRNA) expressed in iron-limiting conditions thanks to the alleviation of Fur repression [111]. Thus, during iron deficiency, RyhB promotes degradation of the *iscSUA* mRNA part while the *iscR* mRNA part is protected from RNase by a stem loop structure, which then causes apo-IscR accumulation (Figure 1.4) [111]. Recently, it was discovered that two other sRNAs, FnrS and OxyS, control *iscR* expression by base pairing to the 5′ UTR of the *iscR* mRNA [112]. These sRNAs act in opposite ways and in opposite conditions: FnrS, expressed in anaerobiosis, represses the expression of *iscR*, while OxyS, expressed during oxidative stress, activates it. Such fine tuning sRNA control represents a sophisticated way to coordinate expression of the two machineries. Hence, as *suf* expression is activated by apo-IscR via a Type 2 site containing promoter, the action of RyhB is to increase *suf* expression while decreasing *isc* expression in iron-limiting conditions [106, 110, 111, 113, 114]. OxyS increases translation of IscR that presumably accumulates under its apo-form since the ISC machinery is less efficient under oxidative stress conditions (Figure 1.4). Therefore, OxyS increases *suf* expression, as RyhB does. Interestingly, through their indirect positive effect on *suf* expression, RyhB and OxyS induce enhanced tolerance to aminoglycosides during low iron and oxidative stress conditions, respectively [112, 113].

The *isc* operon of *Myxococcus xanthus* is regulated by an Rrf2-type regulator, RisR [115]. Phylogenomic analysis indicated that *E. coli* IscR and *M. xanthus* RisR emerged independently through a very complex evolutionary history. Members of the RisR family have been recruited several times independently of the type of Fe–S cluster biosynthesis machinery to be regulated, ISC, SUF, or MIS [115]. *E. coli* IscR and *M. xanthus* RisR belong to phylogenetically distinct groups. RisR binds an Fe–S cluster, likely a [4Fe–4S] cluster, and under its holo-form it represses the *isc* operon by direct binding to a DNA sequence motif that is distinct from the IscR type 1 binding site. A main difference between RisR and IscR lies in the fact that the apo-form of RisR seems to be devoid of regulatory activity. This study illustrated that bacteria make a wide usage of Rrf2 family members as a tunable solution for adapting the Fe–S cluster supply in adequation with their lifestyle. Interestingly, in contrast to *E. coli*, *M. xanthus* is a strictly aerobic bacterium that uses cooperative group movements to predate on prey and to form multicellular fruiting bodies, where the cells differentiate into dormant spores [116].

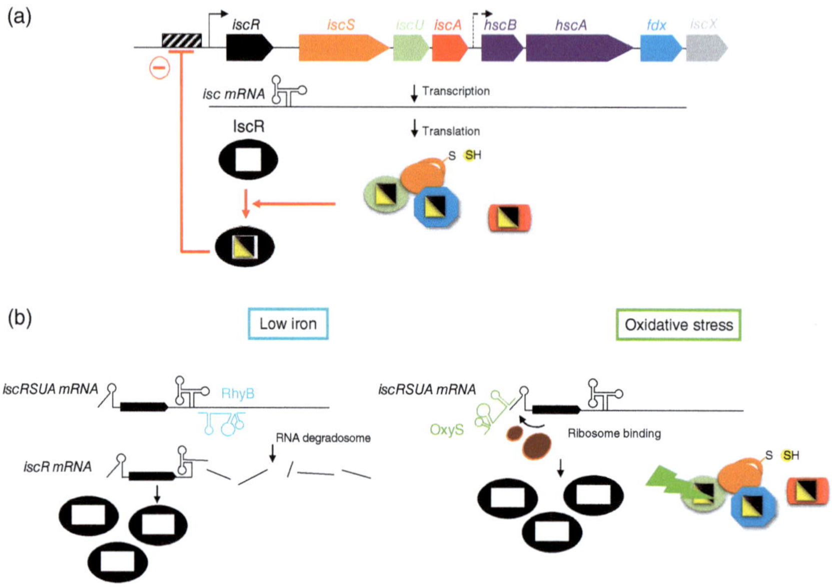

Figure 1.4 Regulation of the *E. coli isc* operon. (a) Regulation at the transcriptional level via IscR. Under optimal growth conditions for Fe–S clusters, IscR is maturated by the ISC machinery. Holo-IscR binds to the P_{isc} promoter and represses *isc* expression. A putative transcriptional start site upstream *hscB* is indicated by a dotted arrow. (b) Regulation at the post-transcriptional level via two sRNAs, RyhB and OxyS. Under iron limitation (left), Fur repression of *ryhB* expression is alleviated. RyhB targets the *iscSUA* mRNA for degradation. Then, IscR accumulates in its apo-form that no longer represses *isc* expression. Under oxidative stress condition (right), the presence of hydrogen peroxide is sensed by OxyR that activates *oxyS* expression. OxyS is proposed to increase accessibility of the ribosome and enhance *iscR* expression. Then, IscR accumulates in its apo-form since the ISC machinery is known to be inactivated by oxidative stress.

In *E. coli*, a distinct transcription start site was identified upstream of *hscA* and *hscB* genes, indicating the possibility of separate transcription units from the upstream *iscRSUA* genes [117, 118]. Interestingly, *hscA* and *hscB* gene expression was observed to increase approximately 11-fold three hours after transitioning from 37 to 10 °C or upon chloramphenicol administration [117]. Utilizing two-dimensional electrophoresis, it was also shown that HscA production increased significantly three hours after a cold shock [117]. This led to speculation regarding the potential role of HscA as a molecular chaperone in response to cold shock [117]. Whether this putative role in cold shock adaptation involves particular Fe–S proteins remains to be established.

1.7 Distribution and Evolution of the ISC Machinery

A recent phylogenomic analysis of over 10 000 prokaryotic genomes revealed that the ISC system arises in a very restrained number of phyla in bacteria and is absent in Archaea [2, 120]. ISC is mainly found in *Proteobacteria* (*Alpha*-, *Beta*-, and *Gamma-proteobacteria*) and closely related phyla (*Nitrospinae*, *Nitrospirae*, *Rokubacteria*, and *Acidobacteria*). In the vast majority, *isc* genes remain clustered with a few exceptions in the *Chlorobi* and the *Bacteroidetes*, wherein the *isc* genes are split into two regions.

The ISC system was proposed to emerge during the diversification of bacteria especially in the ancestor of *Proteobacteria*. ISC would descend from the ancient ancestral two-component MIS system already present in the LUCA [2, 120]. Variants of proteins with general cellular roles, such as DnaK/DnaJ chaperones-like and ferredoxin, would have joined the MIS system and evolved toward specialization for Fe–S biogenesis, giving rise to ISC. The ISC system was transmitted vertically in bacteria and was lost in several lineages.

Enterobacteriaceae such as *E. coli*, *S. enterica* subsp. typhimurium, *Shigella dysenteriae*, and *Dickeya dadantii* have ISC and SUF. The reason for having retained the two systems is thought to be guided by the changing redox conditions met in their environment and the need to develop highly versatile Fe–S protein-rich respiratory systems. One must, however, note that an opportunistic pathogen such as *Pseudomonas aeruginosa*, which can multiply in widely diverse environments including soil and human lungs and can face conditions highly detrimental to Fe–S biology such as oxidative stress and iron limitation, has only one system, ISC [121]. This is also true for *Vibrio cholerae*.

1.8 Conclusion

Over three decades, the study of Fe–S cluster protein biogenesis mechanisms has transformed our vision from one spontaneous phenomenon to one requiring numerous proteins to assist, catalyze, or regulate this process. The basic line followed by ISC and by the other bacterial Fe–S protein biogenesis has been established: acquisition of iron, production of sulfur, assembly of the Fe–S cluster, and delivery to a wide array of different "client" proteins. Furthermore, functioning of the ISC machinery has been shown to be largely conserved from bacteria to humans and plants. Our understanding of *in vivo* Fe–S protein biogenesis has been enriched, and our appreciation of the role of Fe–S cluster proteins has increased considerably over the years. Initially identified by Professor H. Beinert (University of Wisconsin, Madison, USA) in the study of electron transfer chains, proteins with Fe–S clusters have invaded the entire spectrum of cellular processes and of the viral world offering new contexts for their study, oriented towards biomedical and biotechnological issues [121–124].

Acknowledgments

We wish to thank the "Fe-S clusters biogenesis and homeostasis" team (Marseille) and the SAMe unit (Paris) for discussion. This work was supported by grants from Agence Nationale de la Recherche (ANR) (ANR-17-CE12-0033-FRACOL;

ANR-21-CE12-002-KINEBIOTICS; ANR-22-CE44-0043-First-FeS; ANR-10-LABX-62-IBEID), the Centre National de la Recherche Scientifique (CNRS), Aix-Marseille University (AMU), the Institute of Microbiology, Bioenergies and Biotechnology (IM2B), the Fondation de la Recherche Médicale (FDT202304016784), and the Institut Pasteur.

References

1 Beinert, H. (2000). Iron–sulfur proteins: ancient structures, still full of surprises. *J. Biol. Inorg. Chem.* 5: 2–15.

2 Garcia, P.S., D'Angelo, F., Ollagnier de Choudens, S. et al. (2022). An early origin of iron-sulfur cluster biosynthesis machineries before earth oxygenation. *Nat. Ecol. Evol.* 6: 1564–1572.

3 Fontecave, M. (2006). Iron-sulfur clusters: ever-expanding roles. *Nat. Chem. Biol.* 2: 171–174.

4 Mettert, E.L. and Kiley, P.J. (2015). Fe–S proteins that regulate gene expression. *Biochim. Biophys. Acta* 1853: 1284–1293.

5 Garcia, P.S., Barras, F., and Gribaldo, S. (2023). Components of iron-sulfur cluster assembly machineries are robust phylogenetic markers to trace the origin of mitochondria and plastids. *PLoS Biol.* 21: e3002374.

6 Braymer, J.J., Freibert, S.A., Rakwalska-Bange, M., and Lill, R. (2021). Mechanistic concepts of iron-sulfur protein biogenesis in biology. *Biochim. Biophys. Acta Mol. Cell Res.* 1868: 118863.

7 Stehling, O., Wilbrecht, C., and Lill, R. (2014). Mitochondrial iron–sulfur protein biogenesis and human disease. *Biochimie* 100: 61–77.

8 Burén, S., Jiménez-Vicente, E., Echavarri-Erasun, C., and Rubio, L.M. (2020). Biosynthesis of nitrogenase cofactors. *Chem. Rev.* 120: 4921–4968.

9 Seefeldt, L.C., Hoffman, B.M., and Dean, D.R. (2009). Mechanism of Mo-dependent nitrogenase. *Annu. Rev. Biochem.* 78: 701–722.

10 Jacobson, M.R., Cash, V.L., Weiss, M.C. et al. (1989). Biochemical and genetic analysis of the nifUSVWZM cluster from *Azotobacter vinelandii. Mol. Gen. Genet.* 219: 49–57.

11 Streicher, S., Gurney, E., and Valentine, R.C. (1971). Transduction of the nitrogen-fixation genes in *Klebsiella pneumoniae. Proc. Natl. Acad. Sci. USA* 68: 1174–1177.

12 Zheng, L., White, R.H., Cash, V.L. et al. (1993). Cysteine desulfurase activity indicates a role for NIFS in metallocluster biosynthesis. *Proc. Natl. Acad. Sci. USA* 90: 2754–2758.

13 Zheng, L., Cash, V.L., Flint, D.H., and Dean, D.R. (1998). Assembly of iron-sulfur clusters. Identification of an iscSUA-hscBA-fdx gene cluster from *Azotobacter vinelandii. J. Biol. Chem.* 273: 13264–13272.

14 Campuzano, V., Montermini, L., Moltò, M.D. et al. (1996). Friedreich's ataxia: autosomal recessive disease caused by an intronic GAA triplet repeat expansion. *Science* 271: 1423–1427.

15 Rötig, A., de Lonlay, P., Chretien, D. et al. (1997). Aconitase and mitochondrial iron-sulphur protein deficiency in *Friedreich ataxia*. *Nat. Genet.* 17: 215–217.

16 Huynen, M.A., Snel, B., Bork, P., and Gibson, T.J. (2001). The phylogenetic distribution of frataxin indicates a role in iron-sulfur cluster protein assembly. *Hum. Mol. Genet.* 10: 2463–2468.

17 Johnson, D.C., Unciuleac, M.-C., and Dean, D.R. (2006). Controlled expression and functional analysis of iron-sulfur cluster biosynthetic components within *Azotobacter vinelandii*. *J. Bacteriol.* 188: 7551–7561.

18 Ezraty, B., Vergnes, A., Banzhaf, M. et al. (2013). Fe–S cluster biosynthesis controls uptake of aminoglycosides in a ROS-less death pathway. *Science* 340: 1583–1587.

19 Pinske, C., Jaroschinsky, M., and Sawers, R.G. (2013). Levels of control exerted by the Isc iron-sulfur cluster system on biosynthesis of the formate hydrogenlyase complex. *Microbiology* 159: 1179–1189.

20 Schwartz, C.J., Djaman, O., Imlay, J.A., and Kiley, P.J. (2000). The cysteine desulfurase, IscS, has a major role in in vivo Fe–S cluster formation in *Escherichia coli*. *Proc. Natl. Acad. Sci. USA* 97: 9009–9014.

21 Tokumoto, U. and Takahashi, Y. (2001). Genetic analysis of the Isc operon in *Escherichia coli* involved in the biogenesis of cellular iron-sulfur proteins. *J. Biochem.* 130: 63–71.

22 Black, K.A. and Dos Santos, P.C. (2015). Shared-intermediates in the biosynthesis of thio-cofactors: mechanism and functions of cysteine desulfurases and sulfur acceptors. *Biochim. Biophys. Acta* 1853: 1470–1480.

23 Leimkühler, S., Bühning, M., and Beilschmidt, L. (2017). Shared sulfur mobilization routes for tRNA thiolation and molybdenum cofactor biosynthesis in prokaryotes and eukaryotes. *Biomol. Ther.* 7: 5.

24 Zheng, C., Black, K.A., and Dos Santos, P.C. (2017). Diverse mechanisms of sulfur decoration in bacterial tRNA and their cellular functions. *Biomol. Ther.* 7: 33.

25 Loiseau, L., Gerez, C., Bekker, M. et al. (2007). ErpA, an iron-sulfur (Fe–S) protein of the A-type essential for respiratory metabolism in *Escherichia coli*. *Proc. Natl. Acad. Sci. USA* 104: 13626–13631.

26 Tanaka, N., Kanazawa, M., Tonosaki, K. et al. (2016). Novel features of the ISC machinery revealed by characterization of *Escherichia coli* mutants that survive without iron-sulfur clusters. *Mol. Microbiol.* 99: 835–848.

27 Vinella, D., Brochier-Armanet, C., Loiseau, L. et al. (2009). Iron-sulfur (Fe–S) protein biogenesis: phylogenomic and genetic studies of A-type carriers. *PLoS Genet.* 5: e1000497.

28 Roche, B., Huguenot, A., Barras, F., and Py, B. (2015b). The iron-binding CyaY and IscX proteins assist the ISC-catalyzed Fe–S biogenesis in *Escherichia coli*. *Mol. Microbiol.* 95: 605–623.

29 Takahashi, Y. and Nakamura, M. (1999). Functional assignment of the ORF2-iscS-iscU-iscA-hscB-hscA-fdx-ORF3 gene cluster involved in the assembly of Fe–S clusters in *Escherichia coli*. *J. Biochem.* 126: 917–926.

30 Nichols, R.J., Sen, S., Choo, Y.J. et al. (2011). Phenotypic landscape of a bacterial cell. *Cell* 144: 143–156.

31 Tamae, C., Liu, A., Kim, K. et al. (2008). Determination of antibiotic hypersensitivity among 4,000 single-gene-knockout mutants of *Escherichia coli*. *J. Bacteriol.* 190: 5981–5988.

32 Inoue, T., Shingaki, R., Hirose, S. et al. (2007). Genome-wide screening of genes required for swarming motility in *Escherichia coli* K-12. *J. Bacteriol.* 189: 950–957.

33 Maynard, N.D., Birch, E.W., Sanghvi, J.C. et al. (2010). A forward-genetic screen and dynamic analysis of lambda phage host-dependencies reveals an extensive interaction network and a new anti-viral strategy. *PLoS Genet.* 6: e1001017.

34 Allen, R.M., Chatterjee, R., Ludden, P.W., and Shah, V.K. (1995). Incorporation of iron and sulfur from NifB cofactor into the iron-molybdenum cofactor of dinitrogenase. *J. Biol. Chem.* 270: 26890–26896.

35 Brigle, K.E., Weiss, M.C., Newton, W.E., and Dean, D.R. (1987). Products of the iron-molybdenum cofactor-specific biosynthetic genes, nifE and nifN, are structurally homologous to the products of the nitrogenase molybdenum-iron protein genes, nifD and nifK. *J. Bacteriol.* 169: 1547–1553.

36 Robinson, A.C., Dean, D.R., and Burgess, B.K. (1987). Iron-molybdenum cofactor biosynthesis in *Azotobacter vinelandii* requires the iron protein of nitrogenase. *J. Biol. Chem.* 262: 14327–14332.

37 Ugalde, R.A., Imperial, J., Shah, V.K., and Brill, W.J. (1984). Biosynthesis of iron-molybdenum cofactor in the absence of nitrogenase. *J. Bacteriol.* 159: 888–893.

38 Fujishiro, T., Nakamura, R., Kunichika, K., and Takahashi, Y. (2022). Structural diversity of cysteine desulfurases involved in iron-sulfur cluster biosynthesis. *Biophys. Physicobiol.* 19: 1–18.

39 Marinoni, E.N., de Oliveira, J.S., Nicolet, Y. et al. (2012). $(IscS\text{-}IscU)^2$ complex structures provide insights into Fe_2S_2 biogenesis and transfer. *Angew. Chem. Int. Ed. Engl.* 51: 5439–5442.

40 Shi, R., Proteau, A., Villarroya, M. et al. (2010). Structural basis for Fe–S cluster assembly and tRNA thiolation mediated by IscS protein-protein interactions. *PLoS Biol.* 8: e1000354.

41 Smith, A.D., Frazzon, J., Dean, D.R., and Johnson, M.K. (2005). Role of conserved cysteines in mediating sulfur transfer from IscS to IscU. *FEBS Lett.* 579: 5236–5240.

42 Parent, A., Elduque, X., Cornu, D. et al. (2015). Mammalian frataxin directly enhances sulfur transfer of NFS1 persulfide to both ISCU and free thiols. *Nat. Commun.* 6: 5686.

43 Schulz, V., Steinhilper, R., Oltmanns, J. et al. (2024). Mechanism and structural dynamics of sulfur transfer during de novo [2Fe–2S] cluster assembly on ISCU2. *Nat. Commun.* 15: 3269.

44 Yan, R., Adinolfi, S., and Pastore, A. (2015). Ferredoxin, in conjunction with NADPH and ferredoxin-NADP reductase, transfers electrons to the IscS/IscU complex to promote iron-sulfur cluster assembly. *Biochim. Biophys. Acta* 1854: 1113–1117.

45 Gervason, S., Larkem, D., Mansour, A.B. et al. (2019). Physiologically relevant reconstitution of iron-sulfur cluster biosynthesis uncovers persulfide-processing functions of ferredoxin-2 and frataxin. *Nat. Commun.* 10: 3566.

46 Webert, H., Freibert, S.-A., Gallo, A. et al. (2014). Functional reconstitution of mitochondrial Fe–S cluster synthesis on Isu1 reveals the involvement of ferredoxin. *Nat. Commun.* 5: 5013.

47 Yan, R., Konarev, P.V., Iannuzzi, C. et al. (2013). Ferredoxin competes with bacterial frataxin in binding to the desulfurase IscS. *J. Biol. Chem.* 288: 24777–24787.

48 Fox, N.G., Yu, X., Feng, X. et al. (2019). Structure of the human frataxin-bound iron-sulfur cluster assembly complex provides insight into its activation mechanism. *Nat. Commun.* 10: 2210.

49 Prischi, F., Konarev, P.V., Iannuzzi, C. et al. (2010). Structural bases for the interaction of frataxin with the central components of iron-sulphur cluster assembly. *Nat. Commun.* 1: 95.

50 Boniecki, M.T., Freibert, S.A., Mühlenhoff, U. et al. (2017). Structure and functional dynamics of the mitochondrial Fe–S cluster synthesis complex. *Nat. Commun.* 8: 1287. https://www.ncbi.nlm.nih.gov/pmc/articles/PMC5668364/ (accessed 11 Nov 2019).

51 Kim, J.H., Bothe, J.R., Frederick, R.O. et al. (2014). Role of IscX in iron-sulfur cluster biogenesis in *Escherichia coli*. *J. Am. Chem. Soc.* 136: 7933–7942.

52 Tokumoto, U., Nomura, S., Minami, Y. et al. (2002). Network of protein-protein interactions among iron-sulfur cluster assembly proteins in *Escherichia coli*. *J. Biochem.* 131: 713–719.

53 Adinolfi, S., Puglisi, R., Crack, J.C. et al. (2017). The molecular bases of the dual regulation of bacterial iron-sulfur cluster biogenesis by CyaY and IscX. *Front. Mol. Biosci.* 4: 97.

54 Pastore, C., Adinolfi, S., Huynen, M.A. et al. (2006). YfhJ, a molecular adaptor in iron-sulfur cluster formation or a frataxin-like protein? *Structure* 14: 857–867.

55 Shimomura, Y., Takahashi, Y., Kakuta, Y., and Fukuyama, K. (2005). Crystal structure of *Escherichia coli* YfhJ protein, a member of the ISC machinery involved in assembly of iron-sulfur clusters. *Proteins* 60: 566–569.

56 Hasnat, M.A. and Leimkühler, S. (2024). Shared functions of Fe–S cluster assembly and Moco biosynthesis. *Biochim. Biophys. Acta, Mol. Cell Res.* 1871: 119731.

57 Li, D.S., Ohshima, K., Jiralerspong, S. et al. (1999). Knock-out of the cyaY gene in *Escherichia coli* does not affect cellular iron content and sensitivity to oxidants. *FEBS Lett.* 456: 13–16.

58 Roche, B., Agrebi, R., Huguenot, A. et al. (2015). Turning *Escherichia coli* into a frataxin-dependent organism. *PLoS Genet.* 11: e1005134.

59 Velayudhan, J., Karlinsey, J.E., Frawley, E.R. et al. (2014). Distinct roles of the *Salmonella enterica* serovar Typhimurium CyaY and YggX proteins in the biosynthesis and repair of iron-sulfur clusters. *Infect. Immun.* 82: 1390–1401.

60 Vivas, E., Skovran, E., and Downs, D.M. (2006). *Salmonella enterica* strains lacking the frataxin homolog CyaY show defects in Fe–S cluster metabolism in vivo. *J. Bacteriol.* 188: 1175–1179.

61 Ramelot, T.A., Cort, J.R., Goldsmith-Fischman, S. et al. (2004). Solution NMR structure of the iron-sulfur cluster assembly protein U (IscU) with zinc bound at the active site. *J. Mol. Biol.* 344: 567–583.

62 Freibert, S.-A., Boniecki, M.T., Stümpfig, C. et al. (2021). N-terminal tyrosine of ISCU2 triggers [2Fe–2S] cluster synthesis by ISCU2 dimerization. *Nat. Commun.* 12: 6902.

63 Gervason, S., Dutkiewicz, R., Want, K., Benazza, R. et al. (2025). The ISC machinery assembles [2Fe–2S] clusters by formation and fusion of [1Fe–1S] precursors. *Nat Chem Biol.* 21: 767–778 doi: 10.1038/s41589-024-01818-8.

64 Li, J., Ren, X., Fan, B. et al. (2019). Zinc toxicity and iron-sulfur cluster biogenesis in *Escherichia coli*. *Appl. Environ. Microbiol.* 85: e01967-18.

65 Agar, J.N., Krebs, C., Frazzon, J. et al. (2000). IscU as a scaffold for iron-sulfur cluster biosynthesis: sequential assembly of [2Fe–2S] and [4Fe–4S] clusters in IscU. *Biochemistry* 39: 7856–7862.

66 Chandramouli, K., Unciuleac, M.-C., Naik, S. et al. (2007). Formation and properties of [4Fe–4S] clusters on the IscU scaffold protein. *Biochemistry* 46: 6804–6811.

67 Unciuleac, M.-C., Chandramouli, K., Naik, S. et al. (2007). In vitro activation of apo-aconitase using a [4Fe–4S] cluster-loaded form of the IscU [Fe–S] cluster scaffolding protein. *Biochemistry* 46: 6812–6821.

68 Mühlenhoff, U., Richter, N., Pines, O. et al. (2011). Specialized function of yeast Isa1 and Isa2 proteins in the maturation of mitochondrial [4Fe–4S] proteins. *J. Biol. Chem.* 286: 41205–41216.

69 Sheftel, A.D., Wilbrecht, C., Stehling, O. et al. (2012). The human mitochondrial ISCA1, ISCA2, and IBA57 proteins are required for [4Fe–4S] protein maturation. *Mol. Biol. Cell* 23: 1157–1166.

70 Tanaka, N., Yuda, E., Fujishiro, T. et al. (2019). Identification of IscU residues critical for de novo iron–sulfur cluster assembly. *Mol. Microbiol.* 112 (6): 1769–1783.

71 Mansy, S.S., Wu, S., and Cowan, J.A. (2004). Iron-sulfur cluster biosynthesis: biochemical characterization of the conformational dynamics of *Thermotoga maritima* IscU and the relevance for cellular cluster assembly. *J. Biol. Chem.* 279: 10469–10475.

72 Bennett, S.P., Crack, J.C., Puglisi, R. et al. (2022). Native mass spectrometric studies of IscSU reveal a concerted, sulfur-initiated mechanism of iron-sulfur cluster assembly. *Chem. Sci.* 14: 78–95.

73 Kim, J.H., Füzéry, A.K., Tonelli, M. et al. (2009). Structure and dynamics of the iron-sulfur cluster assembly scaffold protein IscU and its interaction with the cochaperone HscB. *Biochemistry* 48: 6062–6071.

74 Kim, J.H., Tonelli, M., Frederick, R.O. et al. (2012a). Specialized Hsp70 chaperone (HscA) binds preferentially to the disordered form, whereas J-protein

(HscB) binds preferentially to the structured form of the iron-sulfur cluster scaffold protein (IscU). *J. Biol. Chem.* 287: 31406–31413.
75 Kim, J.H., Tonelli, M., and Markley, J.L. (2012b). Disordered form of the scaffold protein IscU is the substrate for iron-sulfur cluster assembly on cysteine desulfurase. *Proc. Natl. Acad. Sci. USA* 109: 454–459.
76 Markley, J.L., Kim, J.H., Dai, Z. et al. (2013). Metamorphic protein IscU alternates conformations in the course of its role as the scaffold protein for iron-sulfur cluster biosynthesis and delivery. *FEBS Lett.* 587: 1172–1179.
77 Mayer, M.P. and Gierasch, L.M. (2019). Recent advances in the structural and mechanistic aspects of Hsp70 molecular chaperones. *J. Biol. Chem.* 294: 2085–2097.
78 Cupp-Vickery, J.R., Peterson, J.C., Ta, D.T., and Vickery, L.E. (2004). Crystal structure of the molecular chaperone HscA substrate binding domain complexed with the IscU recognition peptide ELPPVKIHC. *J. Mol. Biol.* 342: 1265–1278.
79 Dutkiewicz, R., Schilke, B., Cheng, S. et al. (2004). Sequence-specific interaction between mitochondrial Fe–S scaffold protein Isu and Hsp70 Ssq1 is essential for their in vivo function. *J. Biol. Chem.* 279: 29167–29174.
80 Hoff, K.G., Cupp-Vickery, J.R., and Vickery, L.E. (2003). Contributions of the LPPVK motif of the iron-sulfur template protein IscU to interactions with the Hsc66-Hsc20 chaperone system. *J. Biol. Chem.* 278: 37582–37589.
81 Hoff, K.G., Ta, D.T., Tapley, T.L. et al. (2002). Hsc66 substrate specificity is directed toward a discrete region of the iron-sulfur cluster template protein IscU. *J. Biol. Chem.* 277: 27353–27359.
82 Tapley, T.L. and Vickery, L.E. (2004). Preferential substrate binding orientation by the molecular chaperone HscA. *J. Biol. Chem.* 279: 28435–28442.
83 Füzéry, A.K., Tonelli, M., Ta, D.T. et al. (2008). Solution structure of the iron-sulfur cluster cochaperone HscB and its binding surface for the iron-sulfur assembly scaffold protein IscU. *Biochemistry* 47: 9394–9404.
84 Hoff, K.G., Silberg, J.J., and Vickery, L.E. (2000). Interaction of the iron-sulfur cluster assembly protein IscU with the Hsc66/Hsc20 molecular chaperone system of *Escherichia coli*. *Proc. Natl. Acad. Sci. USA* 97: 7790–7795.
85 Silberg, J.J., Tapley, T.L., Hoff, K.G., and Vickery, L.E. (2004). Regulation of the HscA ATPase reaction cycle by the co-chaperone HscB and the iron-sulfur cluster assembly protein IscU. *J. Biol. Chem.* 279: 53924–53931.
86 Majewska, J., Ciesielski, S.J., Schilke, B. et al. (2013). Binding of the chaperone Jac1 protein and cysteine desulfurase Nfs1 to the iron-sulfur cluster scaffold Isu protein is mutually exclusive. *J. Biol. Chem.* 288: 29134–29142.
87 Puglisi, R., Yan, R., Adinolfi, S., and Pastore, A. (2016). A new tessera into the interactome of the Isc operon: a novel interaction between HscB and IscS. *Front. Mol. Biosci.* 3: 48.
88 Chandramouli, K. and Johnson, M.K. (2006). HscA and HscB stimulate [2Fe–2S] cluster transfer from IscU to apoferredoxin in an ATP-dependent reaction. *Biochemistry* 45: 11087–11095.

89 Shakamuri, P., Zhang, B., and Johnson, M.K. (2012). Monothiol glutaredoxins function in storing and transporting [Fe_2–S_2] clusters assembled on IscU scaffold proteins. *J. Am. Chem. Soc.* 134: 15213–15216.

90 Krebs, C., Agar, J.N., Smith, A.D. et al. (2001). IscA, an alternate scaffold for Fe–S cluster biosynthesis. *Biochemistry* 40: 14069–14080.

91 Morimoto, K., Yamashita, E., Kondou, Y. et al. (2006). The asymmetric IscA homodimer with an exposed [2Fe–2S] cluster suggests the structural basis of the Fe–S cluster biosynthetic scaffold. *J. Mol. Biol.* 360: 117–132.

92 Ollagnier-de Choudens, S., Nachin, L., Sanakis, Y. et al. (2003). SufA from *Erwinia chrysanthemi* characterization of a scaffold protein required for iron-sulfur cluster assembly. *J. Biol. Chem.* 278: 17993–18001.

93 Ollagnier-de-Choudens, S., Mattioli, T., Takahashi, Y., and Fontecave, M. (2001). Iron-sulfur cluster assembly: characterization of IscA and evidence for a specific and functional complex with ferredoxin. *J. Biol. Chem.* 276: 22604–22607.

94 Ollagnier-de-Choudens, S., Sanakis, Y., and Fontecave, M. (2004). SufA/IscA: reactivity studies of a class of scaffold proteins involved in [Fe–S] cluster assembly. *J. Biol. Inorg. Chem.* 9: 828–838.

95 Tan, G., Lu, J., Bitoun, J.P. et al. (2009). IscA/SufA paralogues are required for the [4Fe–4S] cluster assembly in enzymes of multiple physiological pathways in *Escherichia coli* under aerobic growth conditions. *Biochem. J.* 420: 463–472.

96 Gerstel, A., Zamarreño Beas, J., Duverger, Y. et al. (2020). Oxidative stress antagonizes fluoroquinolone drug sensitivity via the SoxR-SUF Fe–S cluster homeostatic axis. *PLoS Genet.* 16: e1009198.

97 Vinella, D., Loiseau, L., Ollagnier de Choudens, S. et al. (2013). In vivo [Fe–S] cluster acquisition by IscR and NsrR, two stress regulators in *Escherichia coli*. *Mol. Microbiol.* 87: 493–508.

98 Ding, H., Yang, J., Coleman, L.C., and Yeung, S. (2007). Distinct iron binding property of two putative iron donors for the iron-sulfur cluster assembly: IscA and the bacterial frataxin ortholog CyaY under physiological and oxidative stress conditions. *J. Biol. Chem.* 282: 7997–8004.

99 Wang, W., Huang, H., Tan, G. et al. (2010). In vivo evidence for the iron-binding activity of an iron-sulfur cluster assembly protein IscA in *Escherichia coli*. *Biochem. J.* 432: 429–436.

100 Pinske, C. and Sawers, R.G. (2012). Delivery of iron-sulfur clusters to the hydrogen-oxidizing [NiFe]-hydrogenases in *Escherichia coli* requires the A-type carrier proteins ErpA and IscA. *PLoS One* 7: e31755.

101 Mettert, E.L. and Kiley, P.J. (2024). Fe–S cluster homeostasis and beyond: the multifaceted roles of IscR. *Biochim. Biophys. Acta, Mol. Cell Res.* 1871: 119749.

102 Miller, H.K., Kwuan, L., Schwiesow, L. et al. (2014). IscR is essential for Yersinia pseudotuberculosis type III secretion and virulence. *PLoS Pathog.* 10: e1004194.

103 Vergnes, A., Viala, J.P.M., Ouadah-Tsabet, R. et al. (2017). The iron-sulfur cluster sensor IscR is a negative regulator of Spi1 type III secretion system in *Salmonella enterica*. *Cell. Microbiol.* 19.

104 Rajagopalan, S., Teter, S.J., Zwart, P.H. et al. (2013). Studies of IscR reveal a unique mechanism for metal-dependent regulation of DNA binding specificity. *Nat. Struct. Mol. Biol.* 20: 740–747.

105 Schwartz, C.J., Giel, J.L., Patschkowski, T. et al. (2001). IscR, an Fe–S cluster-containing transcription factor, represses expression of *Escherichia coli* genes encoding Fe–S cluster assembly proteins. *Proc. Natl. Acad. Sci. USA* 98: 14895–14900.

106 Yeo, W.-S., Lee, J.-H., Lee, K.-C., and Roe, J.-H. (2006). IscR acts as an activator in response to oxidative stress for the Suf operon encoding Fe–S assembly proteins. *Mol. Microbiol.* 61: 206–218.

107 Fleischhacker, A.S., Stubna, A., Hsueh, K.-L. et al. (2012). Characterization of the [2Fe–2S] cluster of *Escherichia coli* transcription factor IscR. *Biochemistry* 51: 4453–4462.

108 Giel, J.L., Rodionov, D., Liu, M. et al. (2006). IscR-dependent gene expression links iron-sulphur cluster assembly to the control of O2-regulated genes in *Escherichia coli. Mol. Microbiol.* 60: 1058–1075.

109 Nesbit, A.D., Giel, J.L., Rose, J.C., and Kiley, P.J. (2009). Sequence-specific binding to a subset of IscR-regulated promoters does not require IscR Fe–S cluster ligation. *J. Mol. Biol.* 387: 28–41.

110 Mettert, E.L. and Kiley, P.J. (2014). Coordinate regulation of the Suf and Isc Fe–S cluster biogenesis pathways by IscR is essential for viability of *Escherichia coli. J. Bacteriol.* 196: 4315–4323.

111 Desnoyers, G., Morissette, A., Prévost, K., and Massé, E. (2009). Small RNA-induced differential degradation of the polycistronic mRNA iscRSUA. *EMBO J.* 28: 1551–1561.

112 Baussier, C., Oriol, C., Durand, S. et al. (2024). Small RNA OxyS induces resistance to aminoglycosides during oxidative stress by controlling Fe–S cluster biogenesis in *Escherichia coli. Proc. Natl. Acad. Sci. USA* 121: e2317858121.

113 Chareyre, S., Barras, F., and Mandin, P. (2019). A small RNA controls bacterial sensitivity to gentamicin during iron starvation. *PLoS Genet.* 15: e1008078.

114 Giel, J.L., Nesbit, A.D., Mettert, E.L. et al. (2013). Regulation of iron-sulphur cluster homeostasis through transcriptional control of the Isc pathway by [2Fe–2S]-IscR in *Escherichia coli. Mol. Microbiol.* 87: 478–492.

115 Sourice, M., Askenasy, I., Garcia, P.S. et al. (2023). A diverged transcriptional network for usage of two Fe–S cluster biogenesis machineries in the delta-proteobacterium *Myxococcus xanthus. MBio* 14: e0300122.

116 Mercier, R. and Mignot, T. (2016). Regulations governing the multicellular lifestyle of *Myxococcus xanthus. Curr. Opin. Microbiol.* 34: 104–110.

117 Lelivelt, M.J. and Kawula, T.H. (1995). Hsc66, an Hsp70 homolog in *Escherichia coli*, is induced by cold shock but not by heat shock. *J. Bacteriol.* 177: 4900–4907.

118 Seaton, B.L. and Vickery, L.E. (1994). A gene encoding a DnaK/hsp70 homolog in *Escherichia coli. Proc. Natl. Acad. Sci. USA* 91: 2066–2070.

119 Sciuto A.L., D'Angelo, F., Spinnato, M.C. et al. (2024). A molecular comparison of [Fe–S] cluster-based homeostasis in Escherichia coli and Pseudomonas aeruginosa. *Imperi F.mBio* 15(11): e0120624.

120 Garcia, P.S., Gribaldo, S., and Barras, F. (2024). When iron and sulfur met on an anoxic planet and eventually made clusters essential for life. *Biochim. Biophys. Acta, Mol. Cell Res.* 1871: 119730.

121 Maio, N., Lafont, B.A.P., Sil, D. et al. (2021). Fe–S cofactors in the SARS-CoV-2 RNA-dependent RNA polymerase are potential antiviral targets. *Science* 373: 236–241.

122 Maio, N., Raza, M.K., Li, Y. et al. (2023). An iron-sulfur cluster in the zinc-binding domain of the SARS-CoV-2 helicase modulates its RNA-binding and -unwinding activities. *Proc. Natl. Acad. Sci. USA* 120: e2303860120.

123 Ueda, C., Langton, M., Chen, J., and Pandelia, M.-E. (2022). The HBx protein from hepatitis B virus coordinates a redox-active Fe–S cluster. *J. Biol. Chem.* 298: 101698.

124 Villalta, A., Srour, B., Lartigue, A. et al. (2023). Evidence for [2Fe–2S]2+ and linear [3Fe–4S]1+ clusters in a unique family of glycine/cysteine-rich Fe–S proteins from megavirinae giant viruses. *J. Am. Chem. Soc.* 145: 2733–2738.

2

Unraveling the Complexity of the Suf-Based Fe–S Biogenesis

Ingie Elchennawi[1], *Claire E. Fisher*[2], *Franklin Wayne Outten*[2], *and Ollagnier de Choudens Sandrine*[1]

[1]*CNRS, CEA, IRIG, Laboratoire de Chimie et Biologie des Métaux, Univ. Grenoble Alpes, 17 Avenue des Martyrs, Grenoble 38054, France*
[2]*University of South Carolina, Department of Chemistry and Biochemistry, 631 Sumter Street, Columbia, SC 29205, USA*

Chapter Goals

- The main goal of this chapter is to provide an historic overview of our knowledge of the Suf machinery.
- Suf is involved in an essential process in biology, namely the biogenesis of iron–sulfur clusters, a group of fascinating inorganic cofactors with multiple functions.
- The Suf machinery is present throughout the kingdoms of life in archaea, bacteria, and eukaryotic plastids.
- Suf consists of several proteins and complexes and is tightly regulated.
- We intend to explain how sulfide, which is not freely available in most cells, is mobilized and transported as a nontoxic form to combine with iron for the assembly of iron–sulfur clusters in scaffolding proteins.
- We aim to describe the unique properties of scaffold proteins that allow clusters to be assembled and transferred to targets without being destroyed.
- The purpose of this chapter is also to highlight what we still do not fully understand regarding this fascinating biological process and to be a source of inspiration for the younger generation of researchers.
- In this chapter, the reader will learn how genetics, biochemistry, and structural biology approaches have enabled us to define the mechanisms of the Suf machinery dedicated to iron–sulfur cluster assembly.
- The Suf machinery is essential in several pathogens but lacking in humans and constitutes a good therapeutic target.
- Uncovering the molecular mechanism of Suf will help the reader envision ways to inactivate it through the design of novel drug molecules.

Iron-Sulfur Clusters: Biogenesis and Biochemistry, Volume 1, First Edition.
Edited by Silke Leimkühler, Günter Schwarz, Oliver Lenz, and Oliver Einsle.

Main Chapter Points

The phylogenetic distribution of the Suf system for Fe-S cluster biogenesis is distinct from Isc or Nif systems

- There is a detailed biochemical and structural understanding of Suf sulfur mobilization
- Much less is know about the specific biochemical details of iron donation and cluster assembly by the Suf system
- There is complex iron and oxygen-dependent regulation of Suf in different organisms
- Despite more than two decades of study, there are still many unanswered questions about the Suf system

2.1 Introduction

In 1993, Dean and coworkers identified NifS, a pyridoxal phosphate-dependent cysteine desulfurase dedicated to assembling iron–sulfur (Fe–S) clusters in nitrogenase in the Gram-negative diazotroph *Azotobacter vinelandii*, thereby launching the new field of Fe–S cluster biogenesis [1]. Subsequent work by Dean and Johnson demonstrated an Fe–S cluster biogenesis pathway for nitrogenase that also includes a multidomain scaffold protein for de novo cluster assembly, NifU, and a Fe–S cluster carrier protein, NifA [2–4]. Although NifS and NifU are dedicated to nitrogenase maturation, it did not take long to identify a "housekeeping" Fe–S cluster biogenesis system for other Fe–S proteins in both *A. vinelandii* and *Escherichia coli* [5–8]. The Isc (*i*ron–*s*ulfur *c*luster) system is homologous to but distinct from Nif. Isc contains a cysteine desulfurase (IscS), a single domain scaffold protein (IscU), a carrier protein (IscA), as well as molecular chaperones (HscA, HscB), a ferredoxin (Fdx), and small regulatory protein (IscX). Concurrent with discovery of bacterial Isc, evidence was emerging of a homologous mitochondrial pathway for Fe–S cluster biogenesis, especially from work in the yeast, *Saccharomyces cerevisiae* [9–13].

The discovery of a third system for Fe–S cluster biogenesis beyond Nif or Isc also came from studies in Gram-negative bacteria, as work in several different areas converged on the Suf system. Analysis of cysteine desulfurase enzymes in *E. coli* showed that there were two cysteine desulfurase enzymes in addition to IscS in that organism. One of those enzymes, originally referred to as CsdB but later renamed SufS, was shown to be encoded in a polycistronic mRNA with other genes, including a Fe–S carrier protein SufA that was homologous to IscA [14–16]. In parallel, a genetic screen to identify genes required for using ferrioxamine b as an iron source revealed that mutations in *sufS* or another member of the operon, *sufD*, blocked ferrioxamine b utilization in *E. coli* [17]. It was shown in the same study that the block was likely due to failure to mature the [2Fe–2S] FhuF enzyme that is needed for iron release from ferrioxamine b. Further, multiple groups have demonstrated that the *sufABCDSE* (Figure 2.1) operon is regulated by the Fur iron metalloregulatory protein, leading to its induction under iron starvation stress [17–19]. Analysis

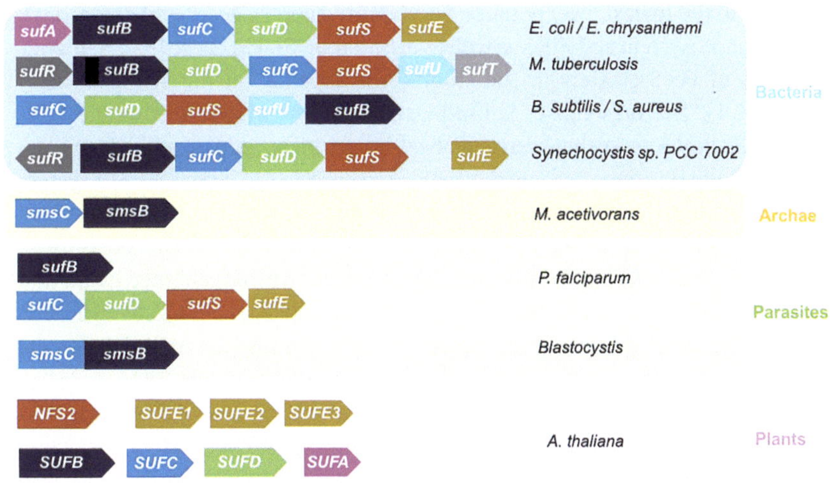

Figure 2.1 Representation of the *suf* and *sms* systems in bacteria, archaea, parasites, and plants. Genes having homologous sequences or similar functions are color coded.

of the oxidative stress response in *E. coli* and *Erwinia chrysanthemi* revealed that *suf* is also transcriptionally induced in response to H_2O_2 stress (by OxyR in *E. coli*) or superoxide stress (by SoxR in *E. chrysanthemi*) [19–21].

Careful genetic analysis of *suf* function in *E. coli* and *E. chrysanthemi* has provided insight into Suf *in vivo* function in Enterobacterales. Combining deletions in the genes of the *suf* operon with gene deletions in the *isc* operon is synthetically lethal, indicating the two systems are partially redundant for essential Fe–S cluster assembly [22, 23]. Furthermore, *suf* mutants show impaired growth under oxidative stress or iron starvation stress in both organisms due to Fe–S cluster enzyme defects [19, 20, 24, 25]. These studies, coupled with the regulation of the *suf* operon, led to the model where Suf is a stress-responsive auxiliary system for Fe–S cluster biogenesis that replaces the housekeeping Isc system under stress.

However, early phylogenetic analysis of Suf distribution provided a more complex picture. In many archaea and bacteria that lack an Isc system, the Suf components appear to provide the only pathway for Fe–S cluster biogenesis [22, 25]. In *E. coli*, the *sufABCDSE* operon is required only under stress conditions because the housekeeping Isc pathway is also present. However, in many Terrabacteria, including important Gram-positive model organisms (*Bacillus subtilis*) and pathogens (*Mycobacterium tuberculosis*, *Staphylococcus aureus*), the Suf system is the only pathway for Fe–S cluster biogenesis and is thus required for viability (Figure 2.1) [26–29]. The distribution of Suf and Isc also diverges in eukaryotes. The Suf machinery is found primarily in plastid-based organelles in Eukaryotes, whereas Isc is localized to the mitochondria [22, 30, 31]. Photosynthetic cyanobacteria, such as *Synechocystis* sp. PCC 6803 and *Synechococcus* sp. PCC 7002, use the Suf system (*sufRBCDSE*) as their major Fe–S cluster assembly pathway, since knockouts cannot be isolated and conditional Suf knockdown strains cannot grow under the nonpermissive condition (Figure 2.1) [32, 33]. In the conditional Suf knockdown strain where Suf expression is minimized to the lowest level possible for viability, there is decreased chlorophyll content and photosystem activities as well as an altered PSI/PSII ratio [33]. In plants, the synthesis of Fe–S clusters in the chloroplast depends on the Suf machinery [34] (Figure 2.1). The presence of Suf in plastid-based organelles likely reflects its presence in *Cyanobacteria* that were involved in a primary endosymbiosis process leading to various lineages of photosynthetic eukaryotes [31]. Since the Suf system was retained in photosynthetic organisms that produce oxygen, including cyanobacteria and the chloroplast of higher plants, this may indicate selective pressure in response to the rise of oxygen and/or increased reactive oxygen species in that organelle. The Suf system is also maintained in the apicoplast, a non-photosynthetic cryptic plastid organelle found in the Apicomplexa, including the malarial parasite *Plasmodium falciparum* and the parasite *Toxoplasma gondii* (Figure 2.1) [35–38].

Recently, a more comprehensive and sophisticated analysis involving many new microbial genomes concluded that a minimal version of the Suf system consisting of separate branches of the SufC and SufB families occurred in the last universal common ancestor (LUCA) [39]. This minimal Suf system, referred to as SmsCB (Figure 2.1), is proposed to be the ancestral core Suf machinery to which other components were recruited throughout evolution [39, 40]. As we shall see, SmsB and SmsC likely represent the minimal Fe–S cluster scaffold complex where clusters are assembled de novo. In that sense, scaffolding to capture iron and sulfide and stabilize the nascent Fe–S cluster may be the most ancient function provided by Fe–S cluster biogenesis systems. Other functionalities, such as sulfide production from L-cysteine and targeted Fe–S cluster trafficking, evolved in response to multiple selective pressures, including global atmospheric oxygenation, organism-specific ecological niches, etc. Nonetheless, for most systems, sulfur donation is the first step of Fe–S cluster biogenesis and historically it has been the most well-studied. Therefore, the sulfur delivery step will be our starting point for in-depth discussion of the Suf pathway.

2.2 Suf Function: The Sulfur Delivery Step

Inorganic sulfide (S^{2-}) within Fe–S clusters is most commonly produced from L-cysteine desulfuration by a two-part cysteine desulfurase–sulfurtransferase system (Figure 2.2). The SufS–SufE (cysteine desulfurase–sulfurtransferase) tandem is encountered in Gram-negative bacteria, cyanobacteria, plastid organelles from plants and parasites while SufS–SufU pair is found in Gram-positive bacteria (including many important pathogens). Most of our knowledge regarding SufSE and SufSU comes from detailed biochemical and genetic studies on the systems from *E. coli* and *B. subtilis*.

2.2.1 The SufS Cysteine Desulfurase

The SufS cysteine desulfurase (CD) is a dimeric enzyme that utilizes pyridoxal-5-phosphate (PLP) as a cofactor to decompose L-cysteine into alanine and an enzyme-bound sulfide [41, 42]. Enzymatic studies of SufS from *B. subtilis*, *Mtb*, and,

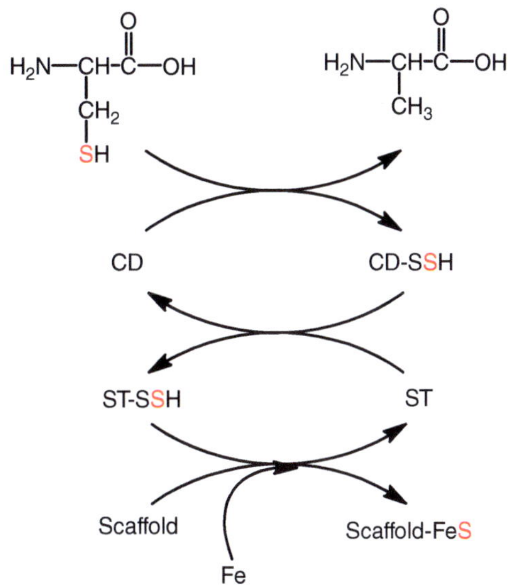

Figure 2.2 Desulfurization of L-cysteine into alanine by cysteine desulfurase (CD)–sulfurtransferase (ST) system. CD-SSH: persulfide intermediate located to the catalytic cysteine of CD; ST-SSH: persulfide intermediate located to the sulfur acceptor cysteine of ST.

more recently, *E. coli* have demonstrated that these enzymes exhibited a half-site reactivity, meaning that the active sites of each monomer alternate their function sequentially. This implies that one active site is accessible at a time [43–46]. The key product of the desulfurase reaction is an enzyme-bound persulfide species, located on a conserved cysteine residue in the SufS polypeptide (*E. coli* Cys364, *B. subtilis* Cys361, *Mtb* Cys373, *S. aureus* Cys389, *E. faecalis* Cys153, and *P. falciparum* Cys497) [43, 45, 47–49] (Figure 2.2). Structural comparison of SufS from *E. coli*, *B. subtilis*, *Mtb*, and *Synechocystis* sp. reveals significant structural homology across these organisms [45, 47, 50, 51]. Additionally, plant SufS (AtNfs2) displays structural features similar to bacterial SufS [52].

In *E. coli* SufS, the cysteine desulfurase reaction requires several crucial amino acid residues proximal to the PLP cofactor. In its resting state, the PLP cofactor is attached to the ε-amino group of Lys226 through a Schiff base, also referred to as an internal aldimine [53]. A short loop contains the active site Cys residue where the persulfide species binds once the desulfurization reaction is done. The short loop is a feature of type II CD and results in limited solvent accessibility and restricted structural flexibility. In contrast, type I CD (IscS and NifS) have a longer, more flexible loop, allowing them to interact with a variety of target proteins for persulfide transfer [15, 50, 54]. The structural rigidity of type II CD accounts for their lower activity in comparison to type I CD. To achieve full activation, type II CD depends on a sulfurtransferase protein (or sulfur shuttle protein), SufE or SufU, depending on the organism (Figure 2.2) [41–43, 45, 55].

Recent studies combining enzymatic, biophysical, and structural approaches have greatly advanced our understanding of the mechanism by which SufS enables cysteine desulfurization. These efforts have successfully identified intermediates, revealing the crucial role of conserved residues within the active site. As a result, a nearly complete reaction mechanism has been outlined (Figure 2.3). The process begins with pyridoxal phosphate (PLP) in its internal aldimine linkage with Lys226,

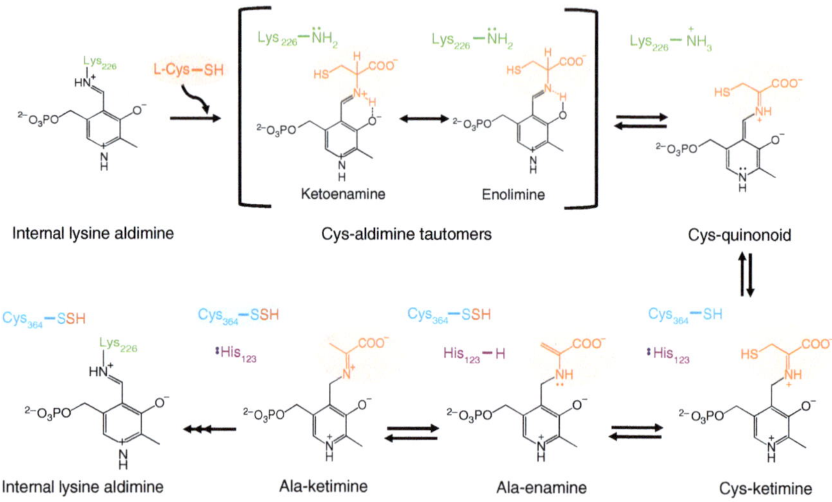

Figure 2.3 SufS cysteine desulfurase mechanism. Pyridoxal-phosphate is depicted in black, cysteine substrate in orange, Cys364 is represented in blue, and His123 in purple.

which absorbs at 422 nm [53, 56]. Once L-cysteine binds, a PLP-cysteine complex forms, namely an external Cys-aldimine [51, 57]. It transits between ketoenamine and enolimine tautomers which are characterized by absorption at 424 nm and 343 nm, respectively [56]. The external aldimine is subsequently converted to a ketimine species that absorbs at 351 nm. The next step involves a nucleophilic attack by a conserved reactive cysteine residue (e.g. Cys364 in *E. coli*) targeting the sulfur atom of the ketamine. This reaction cleaves the C–S bond, producing a persulfide intermediate (Cys-SSH) and an alanine-derived enamine species [51, 56, 57]. Subsequent transformations are hypothesized to involve intermediates such as Ala-ketimine, Ala-quinonoid, and Ala-aldimine that remain to be characterized biophysically. In the end, the enzyme is restored to its initial state, regenerating the internal aldimine of PLP, ready for a new catalytic cycle. Key findings from these investigations emphasize the role of a conserved histidine (e.g. His123), which (i) stabilizes the ketoenamine and enolimine tautomers and (ii) acts as an acid–base catalyst to facilitate conversion from Cys-ketimine to Ala-enamine by forming thiolate group of the catalytic cysteine [56].

2.2.2 SufU and SufE Sulfur Transfer Partners

SufS alone displays a slow attack of the catalytic cysteine on the C–S bond of the substrate L-cysteine [50], leading to an inefficient desulfuration reaction. Optimal activity can be achieved when SufE or SufU proteins bind to the enzyme [41–43, 45, 48, 55, 58, 59]. SufE and SufU are sulfurtransferase proteins required to transfer sulfur from SufS to the SufBCD scaffold complex, through a transient persulfurated form (Figure 2.2). SufE can exist in both monomeric and dimeric forms, depending on its concentration [41]. SufS and SufE can bind in 2 : 2 and 4 : 4 stoichiometries [41] (J. Pérard, unpublished results). Biochemical studies showed that monomeric SufU associates with SufS, generating a complex displaying

Table 2.1 Cysteine desulfurase activity of SufS with their sulfurtransferase (SufE or SufU).

Organism	Sulfurtransferase	Cysteine desulfurase enhancement (fold)	References
E. coli	SufE	40–200	[42, 53, 59]
E. chrysanthemi	SufE	40	[42, 55]
A. thaliana	SUFE1	40	[60]
	SUFE2	40	[61]
	SUFE3	70	[61]
B. subtilis	SufU	35–200	[26, 43, 62]
S. aureus	SufU	1.5	[49]
M. tuberculosis	SufU	180	[45]
P. falciparum	SufE	17	[63]

a 2 : 2 stoichiometry [26, 43, 45]. Enzymatic experiments on both SufSU and SufSE systems from various organisms have demonstrated that SufU and SufE are co-substrates (with L-cysteine) of the desulfurase (SufS), which explains the nonsequential mechanism [43, 45, 53, 59]. Furthermore, the modest activity of SufS alone is boosted by these sulfurtransferase proteins (with variable effects depending on the organism, Table 2.1). The latter accept sulfur from SufS on a conserved cysteine (*Mtb*Cys40, *B. subtilis* Cys41, *S. aureus* Cys43, *E. coli* Cys51, and *P. falciparum* Cys154), allowing SufS to turnover. It seems that SufE and SufU may also allosterically modify residues around the PLP cofactor to enhance activity, but this effect is still under study. Notably, the rate of sulfur transfer from SufS to SufE in *E. coli* is influenced by the sulfuration state of SufE. In its thiolated state, SufE supports the cleavage of the persulfide terminal sulfur on SufS, boosting its catalytic efficiency. On the other hand, when SufE is persulfurated, the turnover slows, showing that excess sulfur accumulation on SufE, through polysulfide chain formation, shutdowns SufS activity and prevents unnecessary sulfur production [53].

Although SufU and SufE serve the same function, they have distinct phylogenetic distribution and are not found in the same organism. They share limited structural homology but are quite divergent at the primary amino acid sequence level. At the level of primary sequence, SufU resembles the scaffold IscU more than it does SufE. In addition to the active site cysteine, which is responsible for persulfide binding in SufU (Cys41 in *B. subtilis* and Cys40 in *Mtb*), IscU and SufU also have two additional cysteines close to the active site. Biochemical studies on wild-type and Cys variants of SufU suggest that these cysteines are Fe–S cluster ligands [64]. The *B. subtilis* SufU binds an Fe–S cluster if chemically reconstituted and can deliver it to target proteins, supporting this view [26]. However, several *in vivo* and *in vitro* observations no longer support this view. First, heterologous complementation experiments involving *B. subtilis* and *E. coli* mutants demonstrated that *B. subtilis* SufSU can replace *E. coli* SufSE, but not IscSU [65]. Second, efforts to define the nature of SufU Fe–S cluster using spectroscopic techniques were not successful, indicating that the Fe–S binding in SufU is likely nonspecific ([26, 44] and I. Elchennawi, unpublished results). Additionally, it is important to note that the LPPVK motif used for binding between U-type scaffolds and the HscA molecular chaperone is missing in SufU proteins. In fact, when purified, SufU associates with a zinc ion (with $K_a = 10^{16-17}$ M^{-1}) [44, 45], which is coordinated by Cys35, Cys41, and Cys128 as well as Asp43 in *B. subtilis* (PDB 6JZV). The zinc ion is essential for the sulfur transfer process between SufS and SufU, as apo SufU, lacking metal, cannot activate SufS [44, 45]. Interestingly, *E. coli* SufE, which is devoid of metal, cannot replace *Mtb* SufU *in vitro* for activating *Mtb* SufS, providing strong evidence for substrate specificity between the desulfurase and sulfurtransferase pair [45].

Structural studies have significantly advanced our understanding of the mechanism by which sulfur is supplied. In the SufSE system, as previously noted, the conserved catalytic cysteine of SufS (Cys364 in *E. coli*) is contained in a short and inflexible loop [47]. Similarly, the sulfur acceptor cysteine of SufE (Cys51 in *E. coli*) is positioned at the tip of a loop, with its side chain protected from solvent exposure [66]. For the transpersulfidation reaction to occur between SufS and SufE, conformational changes in both proteins are essential. Biophysical experiments probing protein conformational changes in the SufSE complex, in combination with X-ray crystallography of SufS, have enabled insights into the changes in the conformation in both proteins. SufE associates near the SufS active site and interacts through two regions (peptide 38–56 that includes Cys51 and amino acids 66–83). Both regions experience conformational alterations once the two proteins associate [58]. A change involves the loop containing Cys51, which becomes more solvent accessible when the desulfurase binds [58]. These changes enhance the ability of SufE to accept sulfur and also make more efficient the binding of SufS to SufE [67]. Additionally, structural changes exist in SufS near the active site (residues 225–236 and 356–366) and at the dimer interface (residues 88–100 and 243–255) [58, 67, 68]. These dynamic shifts of SufS were confirmed by kinetics and structural experiments using variants of the dimer interface [69]. Recent structural investigations have revealed that SufSE activity is also dependent on additional structural elements [70], comprising a β-hairpin (residue 254–272), an α6 helix, a glycine-rich loop preceding the beta-hairpin, and a cis-proline residue located afterward. They form the "β-latch motif", which facilitates the close interaction between SufS and SufE, promoting persulfide transfer [70, 71]. Specifically, the dynamics of the β-hairpin result in an open conformation of SufS active site, allowing sulfur transfer. The importance of thee β-latch motif is emphasized by its high conservation in structures of type II cysteine desulfurases across various species [70].

The mechanism for the SufSU system differs significantly by the presence of zinc ion in SufU [44, 45]. Crystal structures of *Mtb* Zn-SufU-SufS and *B. subtilis* Zn-SufU-SufS determined at high resolution (PDB 8ODQ and PDB 5XT5) [45, 72] revealed the critical role of Zn^{2+} in both the binding of SufU to SufS and sulfur trafficking from the desulfurase to the sulfurtransferase. The key finding was a Zn ligand exchange process during the SufS–SufU complex formation. The conserved cysteine residue of SufU (*B. subtilis* Cys41 or *Mtb* Cys40) exchanges with another conserved residue of SufS (*B. subtilis* His342, *Mtb* His354). Accordingly, mutation of the histidine to a tyrosine led to a dramatic decrease in the SufS–SufU activity [72]. Structural studies also revealed that His354, positioned at the end of a α-helix and extending outward from the SufS surface, interacts with the Zn ion and the Cys40 in SufU [45]. This ligand exchange allows a free thiol from Cys40 to act as a sulfur acceptor. The sulfur transfer mechanism between SufS and SufU in *B. subtilis* was captured in crystallographic studies, showing intermediate catalytic states, including persulfide on SufS followed by its transfer to SufU [72]. The *Mtb* SufS–SufU structure brought further understanding of the changes in SufU conformation upon SufS interaction, particularly the loop comprising the sulfur-accepting cysteine (Cys40) [45]. Comparative structural analysis between *Mtb* SufU from *Mtb* SufS–SufU structure and *B. subtilis* SufU (PDB 6JZV) demonstrated a significant reorientation of SufU's hairpin loop, containing Cys40, upon interaction with SufS. This led to a proposal stepwise mechanism [45]: (i) SufU binds to SufS; (ii) Zn transitions from coordinating with Cys40 to His354; (iii) the hairpin loop of SufU shifts from a closed to an open state; (iv) His354 stabilizes the open conformation of the hairpin via a hydrogen bond with the oxygen of the Cys40 carbonyl group, positioning it inside the SufS cavity; (v) the side chain of Cys40 rotates into proximity with Cys373 of SufS; and (vi) in this position, Cys40 is optimally aligned to take sulfur atoms from the persulfide Cys373 of SufS (Figure 2.4).

The cysteines of SufSU and SufSE displayed different reactivity and sensitivity to oxidation [45]. Specifically, when exposed to H_2O_2, SufE homodimers were formed, regardless of the presence of SufS. On the other side, under the same experimental conditions, SufS–SufU heterodimers were produced. This may account for the higher cysteine desulfurase activity observed for SufSU under oxidative stress. It can be postulated that the Zn ion acts to protect SufU's conserved cysteine residue from oxidation. Therefore, the sulfur uptake mechanism in SufSE and SufSU might

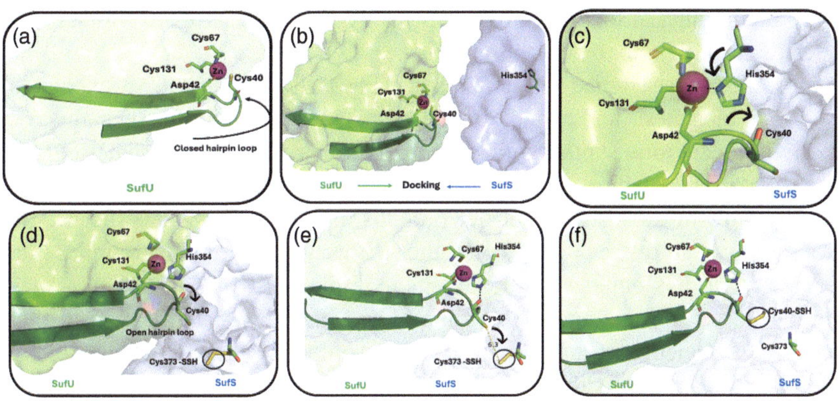

Figure 2.4 Proposed mechanism for the docking of *Mtb* SufS and *Mtb* SufU in sulfur mobilization. (a) In the isolated SufU, zinc is coordinated by four residues (Cys40, Asp42, Cys67 and Cys131), and the hairpin loop adopts a closed conformation (structure adapted from *B. subtilis* SufU). (b) SufS is positioned close to SufU. (c) Zinc ligand exchange: SufS His354 replaces Cys40 of SufU. (d) SufU hairpin loop transitions in an open conformation. (e) His354 forms a new hydrogen bond with the Cys40 of SufU, stabilizing the open conformation of the loop, and the Cys40 side chain flips into a pocket-like region within SufS. (f) Sulfur transfer occurs from the persulfurated catalytic cysteine Cys373 of SufS to the cysteine Cys40 of SufU.

suggest that SufSU have a unique regulatory feature, which ensures that the sulfur shuttle cysteine in SufU is available for accepting and transferring sulfur. This is exclusively possible through interaction with SufS.

While both SufE and SufU can act as sulfurtransferase partners to SufS in bacterial Suf systems, photosynthetic eukaryotes contain only SufE. Cyanobacteria have a single SufE protein, whereas the chloroplast of the flowering plant *A. thaliana* possesses three SufE proteins (SUFE1–3) [60, 73]. Phylogenetic and amino acid sequence analyses show that *A. thaliana's* SUFE1 is related to the ancestral cyanobacterial SUFE protein but contains a BolA domain at the C-terminus. The SUFE2 protein consists of a single SufE domain, while SUFE3 features a SufE domain adjacent to a C-terminal quinolinate syntase domain [61]. The SufE domain in SUFE3 is believed to play a role in forming or repairing the [4Fe–4S] cluster associated with the NadA domain [61]. In *A. thaliana,* NFS2 (formerly CpNifS) is the SufS homologue. It localizes to the chloroplast, and silencing of NFS2 by RNA interference leads to a significant reduction in chloroplast Fe–S proteins, disrupting photosynthesis and chloroplast structure [74–76]. Biochemically, each SUFE protein can remove persulfide from the NFS2 cysteine desulfurase, thereby increasing NFS2 activity and facilitating the sulfur transfer to the scaffold (Table 2.1). The specific functions of the three SUFE proteins in the chloroplast are not fully understood. T-DNA insertion mutants of either SUFE1 or SUFE3 are embryonically lethal in *A. thaliana,* but this has not been determined for SUFE2 [61, 73]. Most plant tissues express SUFE1 and SUFE3; however, SUFE2 is mainly found in flowers and pollen [61, 73]. Translation of SUFE2 and SUFE3 mRNA is light responsive, with peaks level during the light phase of the diurnal cycle [77]. The differential regulation, subcellular localization, and distinct domain structures of these three SUFE proteins suggest they play unique physiological roles in plants, though the full details are still under investigation.

In most organisms studied to date, SufS is completely dedicated to Suf-mediated Fe–S cluster biogenesis and cannot donate sulfur for other biosynthetic pathways, such as thionucleoside synthesis or cyclic pyranopterin monophosphate (cPMP) sulfur insertion [78]. This exclusivity may be due to its dedicated partnership with either the SufU or SufE sulfur transfer proteins in many Suf systems that would limit its ability to transfer persulfide to other pathways. Despite this overall trend, recent evidence has emerged that SufS in the apicoplast of *P. falciparum* may also provide sulfur directly to the enzyme MnmA for tRNA modification in that organelle [79]. Deletion of *sufS* disrupted the apicoplast organelle while deletion of the other *suf* genes, including *sufE,* did not have this effect. The Δ*sufS* apicoplast disruption phenotype could be complemented by supplying the *B. subtilis* MnmA and its partner cysteine desulfurase YrvO. In *B. subtilis,* the MnmA–YrvO pair produce thionucleosides but do not carry out Fe–S cluster biogenesis, which is mediated solely by SufS in the Suf pathway [44, 65, 80]. Expression of the *B. subtilis* MnmA–YrvO heterologous system in the *P. falciparum* Δ*sufS* mutant reversed the disruption of the apicoplast but did not restore Fe–S cluster biogenesis [79]. These *in vivo* experiments suggest that SufS may provide sulfur to MnmA in a pathway independent of SufE.

2.2.3 Is a Cysteine Desulfurase Always Needed? Organisms That Lack SufS

Almost all characterized Fe–S cluster biogenesis pathways involve a cysteine desulfurase enzyme that generates persulfide, which is then reduced and incorporated into the emerging Fe–S cluster as sulfide. However, phylogenetic analysis shows that many anaerobic archaea genomes containing only the Sms system either lack cysteine desulfurase homologues or possess cysteine desulfurase genes that are not located in a conserved genomic context related to Fe–S biogenesis and their function remains unclear [39, 81]. This raises the question of sulfur is provided for Fe–S biogenesis in these organisms without a cysteine desulfurase. One possibility is the use of environmental exogenous sulfide instead of L-cysteine as sulfur source. White and co-workers demonstrated that isotopically labeled sulfide in the medium contributed over 90% of the sulfur incorporated into Fe–S clusters in *Methanococcus maripaludis* [81]. This was consistent even when cysteine biosynthesis was impaired in a Δ*sepS* cysteine auxotroph, proving that cysteine was not a major sulfur contributor, and that a cysteine desulfurase enzyme was unnecessary. The use of sulfide as sulfur source has also been suggested for *Methanocaldococcus. jannaschii* SmsCB, which can complement an *E. coli* mutant lacking ISC and Suf systems, but only when sulfide is provided in the culture medium under anaerobic conditions [82]. This aligns with the environments of *M. maripaludis* and *M. jannaschii*, which are obligate anaerobes living in sulfide-rich habitats. *Archaeoglobus fulgidus*, which can reduce sulfate to sulfide, might also utilize sulfide in a similar way [83]. However, the use of sulfide does not explain the presence of Sms in organisms that do not thrive in sulfur-rich conditions. For organisms that possess both SmsCB and the minimal Isc system, such as *M. acetivorans* or *Methanosarcina barkeri* with MisSU, it could be hypothesized that MisS provides sulfur in the presence of L-cysteine. Yet, recent studies on *Methanococcus voltae* and *M. barkeri* – both harboring the *SmsCB* and *MisSU* genes – show that these methanogens can reduce extracellular Fe_2S (pyrite) and use the products for Fe–S cluster formation [84, 85]. The exact sulfur source in these organisms needs to be determined in the future.

Blastocystis (*Bh*), a single-celled anaerobic parasite, contains a cytoplasmic SmsCB system (Figure 2.1), previously misidentified as SufBC, in addition to the ISC (found in mitochondrion-related organelles) and CIA (located to the cytosol) Fe–S biogenesis systems [86]. Interestingly, SmsC and SmsB are fused into one polypeptide, with both domains interacting. Notably, no SufS protein has been detected in this organism. As such, the sulfur source for Fe–S cluster formation in SmsCB remains unclear.

2.3 Suf Function: Fe–S Cluster Assembly on the Scaffold

2.3.1 The SufB Scaffold

SufB protein is soluble but purifies as a dynamic mixture of monomer, dimer, and trimer forms [87]. As purified under aerobic conditions, SufB does not contain any ligands or cofactors. Early biochemical analysis of the Suf proteins from *E. coli* showed that sulfur mobilized by SufS is transferred through SufE to the SufB protein [41, 87]. Subsequently, mutational analysis of conserved SufB residue Cys254 indicated it is the likely persulfide acceptor site from SufE [87]. SufB can use persulfide provided by SufSE to assemble mixtures of [2Fe–2S] and [4Fe–4S] clusters *in vitro*, although the cluster type and coordination may be altered when SufB is bound to partner proteins SufC and SufD (see below) [87].

2.3.2 SufC and SufD Scaffold Partners

While SufB proteins may be capable of binding Fe–S clusters on their own, in *E. coli* SufB co-purifies with SufC and SufD as a $SufBC_2D$ scaffold complex with a 1 : 2 : 1 stoichiometry [88]. Selective co-expression of only a subset of SufB, SufC, or SufD proteins can also allow the formation of stable sub-complexes, $SufB_2C_2$ and $SufC_2D_2$ [89, 90]. It is possible that these sub-complexes represent actual intermediates in cluster assembly, but they have not been directly observed *in vivo* with the native Suf system. Of the three complexes, both of those containing SufB can bind an Fe–S cluster whereas $SufC_2D_2$ cannot, indicating SufB is the core scaffold protein required for de novo cluster formation [89] (also see SmsCB below).

SufC belongs to the ABC ATPase family and contains conserved Walker A and B motifs as well as D- and Q-loops [20, 24, 91]. Mutations in the ATPase any of the active site residues Lys40 (Walker A motif), Lys152, Glu171 (Walker B motif), Asp173, or His203 (H-motif) abolish SufC ATPase activity *in vitro* and block Suf pathway function *in vivo* [20, 89, 92]. Furthermore, the K40R mutation in SufC reduces $SufBC_2D$ Fe–S cluster binding *in vivo*, apparently by lowering the iron binding without a corresponding reduction in sulfide, suggesting that the mutation alters cluster stability or that SufC ATPase activity is required for *in vivo* iron acquisition [89]. At present, identifying the exact step(s) during Fe–S assembly that are dependent on SufC ATPase activity has remained elusive.

SufD is homologous to SufB and the well-conserved C-terminal helical domain of each interacts with SufC in the various complexes (see below). The core β-helix domain present in both SufB and SufD may be a hallmark of Suf scaffold proteins used for Fe–S cluster biogenesis [92]. SufD can form homodimers that resemble the SufB-SufD heterodimers in $SufBC_2D$ (see below) [93]. The exact role of SufD in Fe–S cluster assembly is unknown although deletion of *sufD* blocks Suf function *in vivo* [19, 24]. Co-expression studies have shown that if SufB and SufC are expressed *in vivo* without SufD, this results in a reduction of the iron content of the subcomplex $SufB_2C_2$, suggestion SufD might also play a role in iron acquisition [89]. Other studies have also suggested a connection between SufD and iron metabolism, although there has been no direct evidence that SufD binds iron [17, 20, 94].

$SufBC_2D$ and a $SufC_2D_2$ sub-complex have been structurally characterized by a number of biophysical techniques [92, 95]. However, thus far neither SufB-associated complexes contain an Fe–S cluster. Due to the extensive homology between SufB and SufD, the $SufBC_2D$ complex and $SufC_2D_2$ sub-complex resemble each other with the two SufC protomers each bound to C-terminal helices in SufB and/or SufD, depending on the complex [95]. An important insight that the structures provided is that the structure of the SufC protomers in the complexes is significantly altered compared to the structure of monomeric SufC crystallized alone. When SufC is not in a complex, the catalytic Glu171 is held by a salt bridge with Lys152 such that it faces away from the ATP binding site [91]. This structural constraint likely explains the very low ATPase activity of SufC alone. When SufC interacts with SufB or SufD, that promotes cleavage of the Glu171–Lys152 salt bridge allowing Glu171 to rotate into the ATP-binding pocket [92, 95]. These SufB/D-dependent changes in SufC reorganize the catalytic site within SufC to promote ATP binding and hydrolysis, similar to other ABC-ATPases, resulting in significant improvement of SufC ATPase activity in the complexes [96]. While one would expect active SufC protomers to dimerize during ATP hydrolysis, structures of the various complexes show that SufC protomers are separated in space by >40 Å. Despite the structures, several solution techniques indicate that SufC in $SufBC_2D$ forms a transient head-to-tail dimer during ATP binding and hydrolysis, possibly causing linked movement of SufB and SufD for a step in cluster assembly [92]. The ATP-dependent conformational changes may also result in putative cluster ligand SufD His360 moving nearer to Cys405 of SufB [92, 95]. Unfortunately, structural analysis has not captured the SufC-dependent conformational changes in SufB–SufD to allow comparison with similar changes seen in other ABC ATPases [92]. SufC ATP hydrolysis likely promotes conformational rearrangement of $SufBC_2D$ to promote Fe–S cluster biogenesis at one or more unknown steps.

2.3.3 The Minimal Suf Scaffold, SMS

As noted above, the Suf-like minimal system (SMS) consists of only two genes that are homologous to *sufB* and *sufC* and that are generally organized as a *smsCB* polycistron in microbial organisms [39]. It was shown that SmsCB can form both homo- and hetero-dimers. The SmsCB complex from the archaeon *Methanosarcina acetivorans* purifies with a Fe–S cluster and can be used to reactivate apo-aconitase *in vitro* [39]. Similar results were obtained using chemically reconstituted Fe–S SmsCB from *M. jannaschii* [39]. Based on these results with SmsCB and previous results with *E. coli* SufB oligomers and the $SufB_2C_2$ sub-complex, it seems that SufB and SmsB can each assemble Fe–S clusters *in vitro* and that, depending on the ecological niche, a minimal complex of SufB and SufC (or SmsCB) can carry out de novo cluster assembly and cluster transfer [39, 87, 89, 90].

2.3.4 Iron Acquisition and Fe–S Cluster Binding

Currently there is no structure of Fe–S cluster bound to SufB, $SufB_2C_2$, or $SufBC_2D$. One characterized $SufBC_2D$ structure contains two Hg^{2+} ions located at the dimer interface between SufB and SufD. SufB Cys405 and SufD Cys358 each coordinate one Hg^{2+} and SufD's His360 is also located close to one of the Hg^{2+} binding sites [92]. It was suggested that the Hg^{2+} binding sites might be the same as iron binding sites in the Fe–S complex, although in this specific structure, the SufB–SufD interface does not allow enough room for a Fe–S cluster to bind. Recent Mössbauer and EXAFS spectroscopic analyses of wild-type and mutant forms of $SufBC_2D$ provide support for a [2Fe–2S] cluster binding to $SufBC_2D$ using SufB Cys405 and His433 as well as SufD Cys358 and His360 as ligands [97]. That same work indicates other cluster types may be present on $SufBC_2D$ as purified, including a tri-iron species different to cubane and linear [3Fe–4S] clusters [97]. This is in contrast to previous studies suggesting multiple cluster types, including a linear [3Fe–4S] cluster, in $SufBC_2D$ co-expressed *in vivo* and purified anaerobically [89]. The observed clusters may represent bona fide intermediates or products of cluster biogenesis or arise from the expression and purification conditions used to isolate the holo-$SufBC_2D$ [89, 97]. *In vivo*, SufB C405A or SufD H360A mutations block Suf Fe–S cluster biogenesis while the SufD C358A mutant had an intermediate phenotype [92, 98].

Despite progress in identifying the cluster binding site, a detailed mechanism for Fe–S cluster assembly on $SufBC_2D$ remains unclear (Figure 2.5). Possibly $SufBC_2D$ cluster synthesis terminates with the formation of a $[2Fe–2S]^{2+}$ cluster located at SufB–SufD interface or in SufB dimer interface within a $SufB_2C_2$ complex. This cluster is then transferred to the SufA A-type carrier protein (see below). This would be in contrast to the IscU scaffold where two $[2Fe–2S]^{2+}$ clusters undergo reductive coupling to form a single $[4Fe–4S]^{2+}$ cluster. Since diferric $[2Fe–2S]^{2+}$ clusters are generally more stable to oxidation, stopping at this cluster type might represent a Suf adaptation to stress. Furthermore, the [2Fe–2S] cluster on SufB has increased stability and resistance to disruption by H_2O_2, O_2, or iron chelators compared to the [2Fe–2S] cluster of IscU [99].

The critical questions of how SufD or the SufC ATPase reaction cycle drives Fe–S cluster assembly on SufB are currently unanswered (Figure 2.5). Possibly SufD and SufC work together for iron acquisition by the $SufBC_2D$ complex. Since we do not yet know the *in vivo* iron donor, it is difficult to test if SufC or SufD directly catalyze iron acquisition. It is also possible that SufC or SufD assist in transferring the Fe–S cluster to SufA or other downstream target proteins.

2.3.5 The Redox Flavin Cofactor

A striking and not yet fully understood characteristic of the $SufBC_2D$ complex is its potential to ligate reduced flavin ($FADH_2$) upon anaerobic isolation of the complex

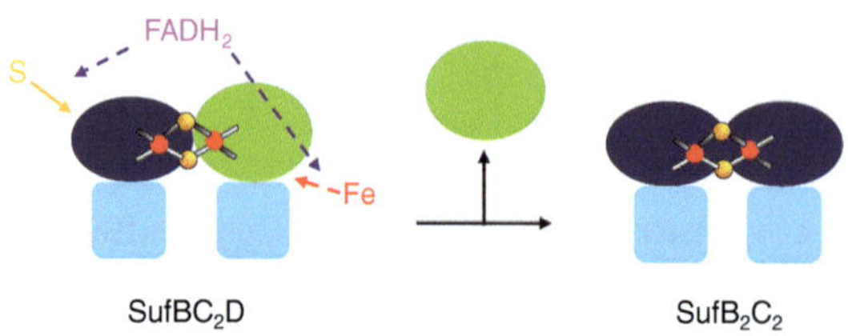

Figure 2.5 Oligomerization states of the scaffold during Fe–S assembly and possible $FADH_2$ functions. Iron (in red) mobilization by the scaffold $SufBC_2D$ involves SufD and/or SufC. Sulfur (in yellow, produces by SufSE or SufSU) enters in the scaffold via SufB. A [2Fe–2S] cluster is formed. Formation of the cluster may induce reorganization of the scaffold ($SufB_2C_2$) with SufD departure. The reduced flavin (violet) may be involved in persulfide reduction or in ferric iron reduction (dashed violet arrows). SufB is navy blue; SufD is green; SufC is light blue.

[88, 89]. Upon exposure to air, the flavin oxidizes (FAD), and no longer binds the complex in this form [88]. Under anaerobic conditions, the apo-SufBC$_2$D can be reloaded with FADH$_2$ at a yield of 0.8–1 FADH$_2$ per complex (k_d = 12 μM) [88]. While SufB alone or in combination with SufC can bind FADH$_2$, the yield is weaker (0.1 mol of FADH$_2$ per mol of SufB; 0.3 mol of FADH$_2$ per mol of SufBC). SufD and SufC alone, however, do not associate with flavin [88]. *In vitro*, the SufBC$_2$D–FADH$_2$ displays a ferric reductase activity on the ferric form of the bacterial frataxin homologue CyaY, indicating that SufBC$_2$D–FADH$_2$ could drive ferric iron reduction into ferrous iron during Fe–S assembly and that one FADH$_2$ can supply the reduction of two ferric iron (Figure 2.5) [88]. Another possibility is that FADH$_2$ may play a role in reducing the persulfide on SufB, a crucial step for the Fe–S biogenesis (Figure 2.5). At the present time, FADH$_2$ function(s) remains to be elucidated. Wollers et al. proposed three key motifs in the SufB sequence associated with potential FADH$_2$ binding the GXXL motif, the P(X)6G(A)XN motif, and the R(X)6EXXYXXXXXG(X)8Y motif. They are typical of those found in the p-cresol-methylhydroxylase family [88]. However, the crystal structure of the SufBC$_2$D complex does not offer conclusive evidence of their involvement in the flavin binding. Indeed, parts of SufB N-terminus (regions 1–33 and 80–156), which includes the P(X)6G(A)XN motif, are not well folded. Interestingly, while these patterns are present in selected SufB proteins (e.g. *Salmonella* and *Yersinia* SufB), they are not ubiquitously conserved. Additionally, mutations in specific residues of the R(X)6EXXYXXXXXG(X)8Y motif (such as R237A or E244A) do not appear to affect the *in vivo* function of SufBC$_2$D [98]. Further research is needed to identify the accurate FADH$_2$ binding site as it may involve contributions from residues in several subunits (Figure 2.5).

2.4 Suf Function: Fe–S Cluster Trafficking

2.4.1 The A-Type Carrier SufA

There are several families of A-type carrier proteins [100]. Type-I ATC (such as ErpA in *E. coli*) is found in a different locus from the Fe–S biogenesis machinery. Type-II such as SufA and IscA are encoded with the Suf and Isc machineries, respectively (Figure 2.1). In some organisms, the Suf operon contains a *sufA* gene (Figure 2.1) [100]. In other organisms that lack a type-II A-type gene in the *suf* operon, it is not clear if the Suf scaffold directly interfaces with a downstream type-I ATC protein or has a novel trafficking network involving SufT (see below) (Figure 2.6). ATC homodimers bind either a $[2Fe–2S]^{2+}$ or a $[4Fe–4S]^{2+}$ cluster and mediate cluster delivery to target proteins *in vitro* [90, 101–105]. ATC proteins may also have the ability to interconvert [2Fe–2S] and [4Fe–4S] clusters in a process regulated by the presence of oxygen and an electron source [106]. Cluster interconversion could enable ATCs to modify the cluster type they are trafficking depending on the requirements of the target protein and/or environmental factors. Fe–S clusters are directed from the SufBC$_2$D and IscU scaffolds to SufA and IscA, respectively (Figure 2.6). The Δ*iscA* Δ*sufA* mutant in *E. coli* shows severe aerobic growth defects due to failure to adequately mature multiple [4Fe–4S] enzymes, while many [2Fe–2S] proteins showed

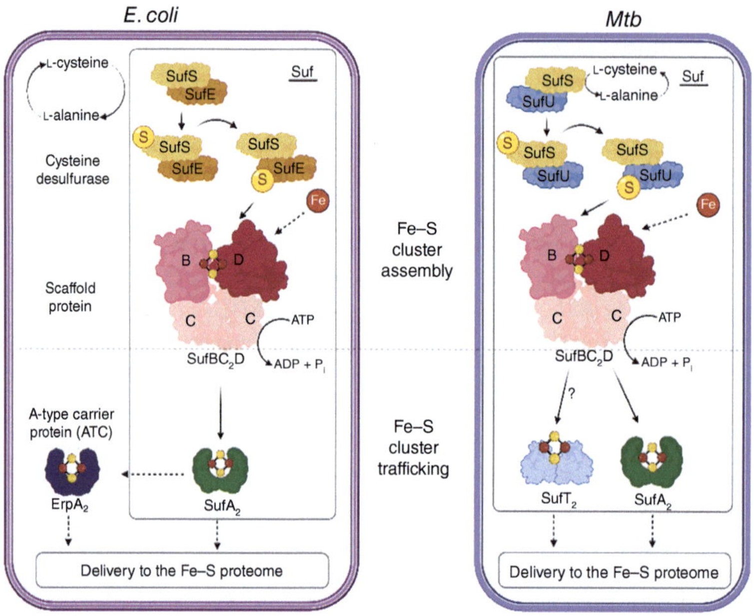

Figure 2.6 De novo Fe–S assembly and trafficking by the Suf pathway. (Left) The *E. coli* Suf system. The separately encoded ATC I ErpA also has genetic and biochemical interactions with SufA. (Right) The Suf system in *Mtb* and other Gram-positive bacteria. The exact role for SufT is currently unknown but, based on homology and phenotypes, it is likely involved in direct cluster Fe–S trafficking or in targeting clusters to specific proteins.

no impairment [100, 107, 108]. These *in vivo* phenotypes indicate that SufA and IscA ATC proteins have a more critical role in [4Fe–4S] cluster trafficking. Similarly, the homologous ATC I and II proteins in eukaryotes are also required for mitochondrial [4Fe–4S] cluster protein maturation, further supporting this role for ATCs [109]. Careful analysis of phenotypes in strains with various combinations of ATC gene deletions in *E. coli* has suggested that IscA and SufA (both type-II ATCs) traffic Fe–S clusters to the ErpA (the only type-I ATC), which directs clusters to essential Fe–S target proteins (Figure 2.6) [100]. Furthermore, *in vitro* experiments show that ErpA can bind IscA and SufA [105]. The current trafficking model suggests IscA and SufA are somewhat redundant. However, subtle phenotype differences as well as differential regulation provide evidence that SufA functions better than IscA under oxidative stress and may be able to circumvent ErpA for cluster delivery under those conditions [108].

The ATC proteins contain three highly conserved cysteine residues in a C-terminal $CX_{63}CX_2C$ motif, corresponding to SufA Cys50, Cys114, and Cys116. It has been clearly shown these three Cys residues are required for function in homologous ATC proteins [109]. Structural characterization of dimeric apo-SufA revealed that these C-terminal regions of the α1 and α2 protomers interact with each other [110]. In the α1 protomer of the apo-structure, Cys50 is distal from Cys114 and Cys116, all three Cys residues are unresolved due to disorder within the α2 protomer. If Cys50 moved closer to the other Cys residues, the total of six cysteines (from the two protomers) would provide adequate ligands to coordinate [2Fe–2S] or [4Fe–4S] clusters [110–112].

ATC proteins are also capable of binding mononuclear Fe^{2+} or Fe^{3+}, with a high affinity for ferric iron [106, 112–115]. The iron coordination site likely consists of Cys residues and additional residues providing an oxygen ligand [112]. *In vitro*, iron bound to *E. coli* IscA can be used for cluster formation on the IscU scaffold [116]. *E. coli* SufA can also bind Fe^{2+} and Fe^{3+} similarly, but a role for SufA as an iron donor for cluster assembly on SufB has not been tested [107]. These intriguing results suggest that ATC proteins may have additional activities beyond Fe–S cluster trafficking, such as iron donation that could be used for biogenesis or repair of Fe–S clusters. Thus far, conclusive evidence for a multifunctional role of ATCs *in vivo* has been lacking.

2.4.2 SufT and DUF59 Proteins

In many bacterial and archaeal species, the *suf* operon also encodes SufT, a DUF59 domain protein [117–119]. The broader DUF59 domain family includes the cytosolic iron–sulfur biogenesis proteins CIAO2A and CIAO2B, and the HCF101 protein required for photosystem I complex maturation in the chloroplast [119]. SufT and other DUF59 proteins contain a reactive Cys residue, within a characteristic "DPE-X_{26-31}-T-$X_{2/3}$-C" motif, that is involved in Fe–S cluster biogenesis [120, 121]. The structures of multiple bacterial, archaeal, and eukaryotic SufT homologues have been solved by X-ray crystallography and NMR [120, 122, 123]. *Thermotoga maritima* SufT has α/β-topology arranged as α1–β1–β2–α2–β3–β4–α3–β5–3_{10}–α4 [122]. In the *T. maritima* structure, the conserved Cys along with Asp, Glu, and Thr are located proximal to each other in space. It was noted that three conserved residues in SufT, including the conserved Cys, were spatially aligned with the Fe–S cluster binding site within the hybrid cluster protein (HCP) from *Desulfovibrio desulfuricans* [122]. While this similarity suggests a possible iron or Fe–S cluster binding site within SufT, the protein does not contain the appropriate Cys residues to form a canonical 4 Cys binding site [120, 122, 124]. One SufT homologue from *Porphyromonas gingivalis* was shown indirectly to bind iron using a gel staining assay but the iron site was not further characterized by spectroscopy or mutagenesis and no metal:protein ratios were reported [125].

SufT proteins can form homo- or hetero-complexes, and such complexes may create a novel composite iron or Fe–S cluster binding site, that possibly includes non-Cys ligands such as oxygen or nitrogen-containing residues. In the eukaryotic cytosol, SufT/DUF59 homologues form heterocomplexes known collectively as CIA-targeting complexes (CTCs) to assist in targeting Fe–S clusters to cytosolic Fe–S proteins as part of the cytosolic iron–sulfur assembly (CIA) machinery [126]. While most non-photosynthetic eukaryotes do not contain the other Suf genes, nonetheless, the structure of a minimal CTC containing the DUF59 homologue CIAO2B provides some support for the hypothesis that multimer formation may create a composite Fe–S cluster binding site in DUF59 proteins including SufT [127]. The CTC that was structurally analyzed consists of the mouse MMS19 and the *Drosophila melanogaster* CIAO1 and CIAO2B. The CTC consisted of a dimer containing two MMS19–CIAO2B–CIAO1 heterotrimers. In each heterotrimer, CIAO2B bridges MMS19 and CIAO1, while the formation of the larger CTC dimer is facilitated by interactions between the two MMS19 protomers as well as between the two CIAO2B protomers. When the two CIAO2B protomers interact, the residues of the conserved motif from each protomer are brought close together, possibly allowing for non-canonical iron or Fe–S cluster binding [127].

SufT within the Suf system is involved in Fe–S cluster biogenesis in a variety of Gram-positive bacteria [120, 124, 128, 129]. A Δ*sufT* mutation in *S. aureus* lowered activity of multiple Fe–S cluster proteins, including aconitase (AcnA), dihydroxy-acid dehydratase (IlvD), lipoyl synthase (LipA), and isopropyl malate isomerase (LeuCD) and the mutation also increased sensitivity to conditions that disrupt Fe–S cluster metabolism [118, 129]. Interestingly, the double mutant Δ*nfu* Δ*sufT* strain showed a more severe synthetic phenotype compared to the single deletion strains and overexpression of *nfu* could partially complement the Δ*sufT* phenotype [118, 129]. A Δ*sufT* mutant of *Sinorhizobium meliloti* showed reduced growth and sensitivity to low pH, high temperature, and iron starvation stress stemming from loss of Fe–S enzymes AcnA, glutamate synthase (GltS), and 6-phosphogluconate dehydratase (6-PGDH) [128]. Similarly, both planktonic and biofilm cultures of the Δ*sufT* strain of *Mycobacterium smegmatis* are sensitive to iron starvation [130]. Even though oxidative stress typically disrupts Fe–S cluster metabolism, the *M. smegmatis* and *S. meliloti* Δ*sufT* mutant strains were not sensitive to it.

Recently, the role of SufT in *Mtb* was thoroughly analyzed [124]. A mycobacterial two-hybrid assay demonstrated that SufT interacts with multiple Fe–S proteins *in vivo*, including Acn, SufS, SufU, and the Fe–S-dependent regulatory SufR. Although a Δ*sufT* mutation is lethal to *Mtb*, a conditional knockout strain was analyzed using CRISPR interference [124]. The depletion of *sufT* caused global metabolic changes linked to loss of Fe–S enzyme activity in pathways such as the tricarboxylic acid cycle, amino acid synthesis, and nucleotide metabolism. The *sufT* knockdown also grew poorly under oxidative stress and nitric oxide stress and could not replicate efficiently inside macrophages. While the exact role of SufT in Fe–S cluster biogenesis by the Suf system is still undefined, based on the published studies and extensive characterization of DUF59 proteins in cytosolic eukaryotic Fe–S biogenesis, SufT may be important for Fe–S cluster trafficking and/or targeting of Fe–S clusters to specific apoproteins rather than in Fe–S assembly (Figure 2.6).

2.5 Suf Regulation

The regulatory pathways that control Suf utilization have been most well-studied in bacterial systems. Therefore, we will focus on those pathways here. However, Suf is likely to be regulated at all steps from transcription to post-translational modification in eukaryotes as well. Further, since some Suf proteins are encoded in the nuclear genome while others are present in organelle genomes, in eukaryotic cells, there is also the possibility of regulating Suf localization to specific compartments (e.g. cytosol vs. chloroplast).

2.5.1 Stress-Responsive Suf Regulation in Bacteria

The *sufABCDSE* operon in *E. coli* and related Gram-negative bacteria is transcriptionally induced by stresses that perturb Fe–S cluster metabolism and diminish Fe–S cluster assembly via the housekeeping Isc system. In *E. coli*, both the Isc and Suf systems are coordinately regulated by the IscR Fe–S cluster transcription factor. [2Fe–2S] IscR binds and represses transcription of *iscRSUA* under Fe–S cluster-replete conditions where cellular demand is met, thereby providing a negative feedback regulation [131]. IscR also regulates Suf but in a more complex manner. When cluster demand increases due to stress, apo-IscR accumulates as Fe–S cluster demand increases, allowing it to bind the *suf* promoter and positively activate *sufABCDSE* transcription [132]. Surprisingly, [2Fe–2S] IscR can also bind the *suf* promoter and activate *suf* transcription [133]. However, the IscR binding site in the *suf* promoter overlaps with a Fur repressor binding site. When there is adequate cellular iron, the *E. coli* Fur protein represses *suf* expression but Fur repression is alleviated under iron starvation conditions due to iron-limited growth media or the addition of a cell-permeable iron chelator such as 2,2′-bipyridyl, or if Fur is inactivated by reactive oxygen species [19, 134]. Under iron-replete conditions, Fur occupancy of the *suf* promoter limits [2Fe–2S] IscR promoter access, thereby limiting activation of *suf* under conditions where Isc is not compromised [133]. Under iron starvation or oxidative stress conditions, Fur no longer binds the *suf* promoter and apo-IscR gains access to activate *suf* transcription (Figure 2.7).

Despite the role of Fur in controlling inappropriate *suf* activation, IscR-dependent activation is absolutely critical for full *suf* expression. Deletion of *iscR* along with the remaining *iscUSA-hscBA-fdx* genes is lethal in *E. coli* because *suf* cannot be sufficiently induced to compensate for Isc loss due to the absence of IscR [133]. Similarly, Δ*iscU* mutations are lethal if combined with *suf* promoter mutations that abolish the apo-IscR binding site [133]. Full induction of *suf* under iron starvation conditions also depends on IscR activation (in addition to loss of Fur repression). Further ancillary regulation of *suf* transcription under hydrogen peroxide stress is also provided by the OxyR activator via a mechanism that also requires the DNA-bending protein IHF in order to bring the distal OxyR binding site proximal to the core *suf*

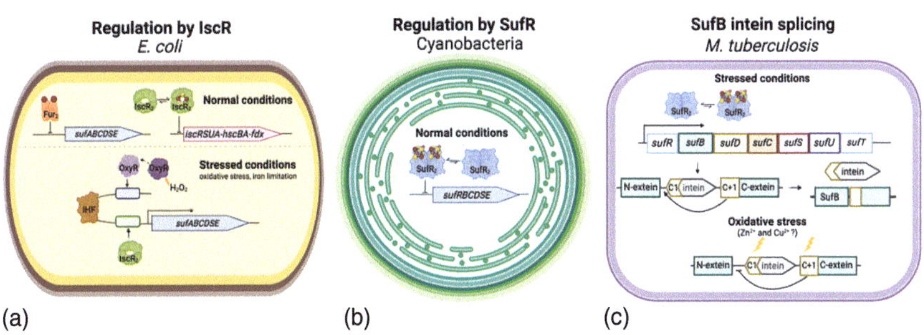

Figure 2.7 Examples of Suf transcriptional and post-translational regulation. (a) Coordinated expression of *E. coli sufABCDSE* by Fur, IscR, OxyR, and IHF in response to stress. (b) Fe–S cluster-dependent regulation of *suf* by SufR in cyanobacteria. (c) Regulation of post-translational splicing of *Mtb* SufB by cysteine reactivity.

promoter [19, 135]. Under H_2O_2 stress, OxyR and apo-IscR appear to work synergistically to fully activate *suf* transcription in order maintain Fe–S cluster biogenesis even though the Isc pathway is inactivated by ROS [132, 136]. The thiophilic metal cobalt can also induce *suf* transcription *in vivo*, through a cascading process that involves disruption of Fe–S and iron homeostasis, leading to downstream oxidative stress (Figure 2.7) [137].

IscR is a member of the Rrf2 protein family, which includes other transcription factors that sense Fe–S clusters in the cell and regulate expression of *suf* and other iron homeostasis genes in response. Among the α-proteobacteria, symbiotic *Rhizobium* and related pathogens like *Agrobacterium* contain the RirA transcriptional repressor that reduces expression of the *sufS2BCDS1XA* operon and many iron homeostasis genes as cellular iron levels increase [138, 139]. RirA requires a labile [4Fe–4S] cluster for high-affinity binding to iron-responsive operator (IRO) sequences present in target promoters [140, 141]. The RirA Fe–S cluster degrades in response to decreasing cellular iron levels *in vivo* and to *in vitro* oxygen exposure, thereby allowing target gene transcription [141, 142]. Deletion of the *sufS2* gene, encoding a cysteine desulfurase, prevented RirA from fully repressing several target genes *in vivo*, suggesting RirA Fe–S cluster acquisition depends on Suf [143]. In the α-proteobacteria discussed here, *suf* expression is also repressed by iron-responsive regulator Irr that responds to changes in cellular heme levels (rather than Fe–S binding) and is homologous to *E. coli* Fur [144]. Recently, RisR, representing a new Rrf2 subfamily, was identified and characterized in Gram-negative δ-proteobacteria Myxobacteria [145]. RisR binds a [4Fe–4S] cluster and represses both Isc and Suf systems in *M. xanthus*.

2.5.2 SufR and RisR Regulation of Suf

Many microbial species, especially those that utilize Suf as the housekeeping system for Fe–S cluster biogenesis, contain a dedicated transcriptional regulator to control Suf expression that does not belong to the Rrf2 family. In *Mtb* and cyanobacteria, this regulator is referred to as SufR [146, 147]. SufR is a member of the DeoR family of regulators that binds two redox-active $[4Fe–4S]^{+1/+2}$ clusters per homodimer (likely with one cluster coordinated by each monomer) [148, 149]. Cluster coordination is required for holo-SufR to bind an inverted repeat target sequence in DNA and repress transcription of the *suf* promoter. Thus, SufR works analogously to IscR as a negative feedback regulator that represses *suf* transcription when cellular Fe–S cluster demand is met (Figure 2.7). Disruption of Fe–S cluster metabolism and increase cluster demand causes apo-SufR to dissociate and allows *suf* transcription to increase to compensate. In the cyanobacterium *Synechocystis* sp. PCC 6803, the redox state of the SufR [4Fe–4S] cluster affects DNA binding, with the reduced $[4Fe–4S]^{+1}$ SufR binding less well than the protein with a partially oxidized $[4Fe–4S]^{+1/+2}$ cluster [148]. In *Mtb*, the SufR repressor was shown to be nitrosylated by NO stress, leading to reduced DNA binding affinity and increased *suf* transcription *in vivo* [149]. In *Mtb* expression of the *sufRBDCSUT* operon increases during iron limitation, under oxidative and nitrosative stress, and during intracellular growth of the pathogen in host macrophages [150, 151]. The *sufR* gene is required for survival in macrophages in cell culture and for persistence during infection in mice [152, 153].

2.5.3 SufB Intein Splicing

In addition to transcriptional regulation, post-translational control of the Suf pathway is utilized in some organisms. In *Mycobacteria*, the *sufB* gene contains an intein insertion, which is a protein embedded in-frame in the coding sequence of *sufB* [154]. Inteins are post-translationally removed from the unspliced polypeptide by autocatalysis, resulting in the fusion of flanking extein peptides to make the mature protein. In *Mtb* if intein removal does not occur, the unspliced SufB cannot associate with SufC or SufD and is nonfunctional (Figure 2.7) [27]. There are two nucleophilic cysteine residues needed for the auto-catalyzed protein splicing mechanism in *Mtb* SufB, intein C1 and C + 1 in the C-terminal extein [155]. The C + 1 cysteine is Cys25, which is one of only two cysteines in mature *Mtb* SufB. It was shown *in vivo* that SufB splicing is inhibited by exposure to oxidative or nitrosative stresses, likely due to the modification of the two cysteine residues required for intein removal [155]. The sensitivity of SufB splicing to cysteine modification may act to downregulate SufB protein function under stress. Interestingly, recent *in vitro* analysis of SufB splicing has shown that Zn^{2+} and Cu^{2+} inhibit SufB splicing, at least in part by coordinating the nucleophilic cysteines and possibly N-terminal histidine residues that also are needed for full splicing [156, 157]. While not yet validated *in vivo*, these *in vitro* results suggest that high levels of thiophilic metals (such as copper) may inhibit SufB splicing and downregulate the Suf pathway (Figure 2.7).

2.6 Conclusions and Future Directions

In closing, it is clear from nearly three decades of research that the Suf system is a critical pathway for Fe–S cluster biogenesis across all kingdoms of life. Despite intense efforts, there are still many gaps in our detailed understanding of Suf function. The specific *in vivo* iron donor for Suf has remained elusive and it is not clear what role the Suf proteins themselves (e.g. SufA, SufC and SufD) may play in iron acquisition. A stepwise structural analysis of Fe–S cluster assembly is also lacking for the Suf system. While there are multiple structures of individual Suf proteins and of Suf protein complexes, there are no structures of relevant Fe–S-bound scaffold complexes. Except for sulfur donation, there are no structures that illuminate intermediate steps during cluster assembly. We also lack information on the possible reorganization of proteins during assembly ($SufBC_2D$ vs. $SufB_2C_2$ for instance). This detailed information is necessary to fully understand how Suf Fe–S cluster assembly proceeds and what adaptations are present in Suf that allow it to be resistant to iron starvation, oxidative stress, and other external disruptors of Fe–S metabolism. Finally, the process by which downstream Fe–S cluster trafficking is coordinated is a key aspect of Fe–S cluster biogenesis that needs further study. We do not fully understand the Fe–S cluster transfer process from scaffold to the SufA carrier and or from SufA carrier to target proteins. Presumably this occurs via a type of ligand exchange mechanism but there are no structures that capture such intermediates. Furthermore, since most organisms contain multiple Fe–S cluster trafficking proteins (ATCs, monothiol glutaredoxins, etc.), it is important to determine how Suf interfaces with this network. Can the SufB scaffold interact with other trafficking proteins, directly with target proteins, or perhaps it is limited to its cognate ATC? Can SufA or SufT access all target proteins or are they dependent on the full trafficking network? As the Suf research field continues to mature, the authors look forward to learning the answer to these and other questions.

Acknowledgments

This work was supported by the French National Agency for Research (Agence Nationale de la Recherche) ANR-22-CE44-0043-First-FeS, ANR-22-CE20-0039-MASTIC, ANR-24-CE44-1662-CoV-FeS, ANR-24-CE44-7921-BAMBI, ANRS0522, the ANR through the LabEx ARCANE program (ANR-11-LABX-0003-01), the LabEx GRAL, the Grenoble Alliance for Integrated Structural and Cell Biology, a program of the Graduate School of Chemistry, Biology, and Health at the University Grenoble Alpes CBH-EUR-GS (ANR-17-EURE-0003). This work is also supported by the European Cooperation in Science and Technology (COST), and authors acknowledge support from COST Action, FeSImmChemNet. This chapter is based on the work from COST Action FeSImmChemNet, CA21115, supported by COST (European Cooperation in Science and Technology). This work was also supported by the US National Institutes of Health (NIH) through Grant GM112919.

References

1 Zheng, L., White, R.H., Cash, V.L. et al. (1993). Cysteine desulfurase activity indicates a role for NIFS in metallocluster biosynthesis. *Proc. Natl. Acad. Sci. USA* 90 (7): 2754–2758.

2 Zheng, L., White, R.H., Cash, V.L., and Dean, D.R. (1994). Mechanism for the desulfurization of L-cysteine catalyzed by the nifS gene product. *Biochemistry* 33 (15): 4714–4720.

3 Fu, W., Jack, R.F., Morgan, T.V. et al. (1994). nifU gene product from *Azotobacter vinelandii* is a homodimer that contains two identical [2Fe-2S] clusters. *Biochemistry* 33 (45): 13455–13463.

4 Yuvaniyama, P., Agar, J.N., Cash, V.L. et al. (2000). NifS-directed assembly of a transient [2Fe-2S] cluster within the NifU protein. *Proc. Natl. Acad. Sci. USA* 97 (2): 599–604.

5 Zheng, L., Cash, V.L., Flint, D.H., and Dean, D.R. (1998). Assembly of iron-sulfur clusters. Identification of an iscSUA-hscBA-fdx gene cluster from *Azotobacter vinelandii*. *J. Biol. Chem.* 273 (21): 13264–13272.

6 Nakamura, M., Saeki, K., and Takahashi, Y. (1999). Hyperproduction of recombinant ferredoxins in *Escherichia coli* by coexpression of the ORF1-ORF2-iscS-iscU-iscA-hscB-hs cA-fdx-ORF3 gene cluster. *J. Biochem. (Tokyo)* 126 (1): 10–18.

7 Takahashi, Y. and Nakamura, M. (1999). Functional assignment of the ORF2-iscS-iscU-iscA-hscB-hscA-fdx-ORF3 gene cluster involved in the assembly of Fe-S clusters in *Escherichia coli*. *J. Biochem. (Tokyo)* 126 (5): 917–926.

8 Agar, J.N., Zheng, L.M., Cash, V.L. et al. (2000). Role of the IscU protein in iron-sulfur cluster biosynthesis: IscS-mediated assembly of a [Fe2S2] cluster in IscU. *J. Am. Chem. Soc.* 122 (9): 2136–2137.

9 Strain, J., Lorenz, C.R., Bode, J. et al. (1998). Suppressors of superoxide dismutase (SOD1) deficiency in *Saccharomyces cerevisiae*. Identification of proteins predicted to mediate iron-sulfur cluster assembly. *J. Biol. Chem.* 273 (47): 31138–31144.

10 Kispal, G., Csere, P., Prohl, C., and Lill, R. (1999). The mitochondrial proteins Atm1p and Nfs1p are essential for biogenesis of cytosolic Fe–S proteins. *EMBO J.* 18 (14): 3981–3989.

11 Schilke, B., Voisine, C., Beinert, H., and Craig, E. (1999). Evidence for a conserved system for iron metabolism in the mitochondria of *Saccharomyces cerevisiae*. *Proc. Natl. Acad. Sci. USA* 96 (18): 10206–10211.

12 Garland, S.A., Hoff, K., Vickery, L.E., and Culotta, V.C. (1999). *Saccharomyces cerevisiae* ISU1 and ISU2: members of a well-conserved gene family for iron-sulfur cluster assembly. *J. Mol. Biol.* 294 (4): 897–907.

13 Li, J., Kogan, M., Knight, S.A. et al. (1999). Yeast mitochondrial protein, Nfs1p, coordinately regulates iron-sulfur cluster proteins, cellular iron uptake, and iron distribution. *J. Biol. Chem.* 274 (46): 33025–33034.

14 Mihara, H., Maeda, M., Fujii, T. et al. (1999). A nifS-like gene, csdB, encodes an *Escherichia coli* counterpart of mammalian selenocysteine lyase. Gene cloning, purification, characterization and preliminary x-ray crystallographic studies. *J. Biol. Chem.* 274 (21): 14768–14772.

15 Fujii, T., Maeda, M., Mihara, H. et al. (2000). Structure of a NifS homologue: X-ray structure analysis of CsdB, an *Escherichia coli* counterpart of mammalian selenocysteine lyase. *Biochemistry* 39 (6): 1263–1273.

16 Lima, C.D. (2002). Analysis of the *E. coli* NifS CsdB protein at 2.0 A reveals the structural basis for perselenide and persulfide intermediate formation. *J. Mol. Biol.* 315 (5): 1199–1208.

17 Patzer, S.I. and Hantke, K. (1999). SufS is a NifS-like protein, and SufD is necessary for stability of the [2Fe-2S] FhuF protein in *Escherichia coli*. *J. Bacteriol.* 181 (10): 3307–3309.

18 Lee, J.-H., Yeo, W.S., and Roe, J.-H. (2003). Regulation of the sufABCDSE operon by fur. *J. Microbiol.* 41: 109–114.

19 Outten, F.W., Djaman, O., and Storz, G. (2004). A suf operon requirement for Fe-S cluster assembly during iron starvation in *Escherichia coli*. *Mol. Microbiol.* 52 (3): 861–872.

20 Nachin, L., Loiseau, L., Expert, D., and Barras, F. (2003). SufC: an unorthodox cytoplasmic ABC/ATPase required for [Fe-S] biogenesis under oxidative stress. *EMBO J.* 22 (3): 427–437.

21 Zheng, M., Wang, X., Templeton, L.J. et al. (2001). DNA microarray-mediated transcriptional profiling of the *Escherichia coli* response to hydrogen peroxide. *J. Bacteriol.* 183 (15): 4562–4570.

22 Takahashi, Y. and Tokumoto, U. (2002). A third bacterial system for the assembly of iron-sulfur clusters with homologs in archaea and plastids. *J. Biol. Chem.* 277 (32): 28380–28383.

23 Tokumoto, U., Kitamura, S., Fukuyama, K., and Takahashi, Y. (2004). Interchangeability and distinct properties of bacterial Fe-S cluster assembly systems: functional replacement of the isc and suf operons in *Escherichia coli* with the nifSU-like operon from *Helicobacter pylori*. *J. Biochem.* 136 (2): 199–209.

24 Nachin, L., El Hassouni, M., Loiseau, L. et al. (2001). SoxR-dependent response to oxidative stress and virulence of *Erwinia chrysanthemi*: the key role of SufC, an orphan ABC ATPase. *Mol. Microbiol.* 39 (4): 960–972.

25 Tokumoto, U., Kitamura, S., Fukuyama, K., and Takahashi, Y. (2004). Interchangeability and distinct properties of bacterial Fe-S cluster assembly systems: functional replacement of the isc and suf operons in *Escherichia coli* with the nifSU-like operon from *Helicobacter pylori*. *J. Biochem. (Tokyo)* 136 (2): 199–209.

26 Albrecht, A.G., Netz, D.J., Miethke, M. et al. (2010). SufU is an essential iron-sulfur cluster scaffold protein in *Bacillus subtilis*. *J. Bacteriol.* 192 (6): 1643–1651.

27 Huet, G., Castaing, J.P., Fournier, D. et al. (2006). Protein splicing of SufB is crucial for the functionality of the *Mycobacterium tuberculosis* SUF machinery. *J. Bacteriol.* 188 (9): 3412–3414.

28 Kobayashi, K., Ehrlich, S.D., Albertini, A. et al. (2003). Essential *Bacillus subtilis* genes. *Proc. Natl. Acad. Sci. USA* 100 (8): 4678–4683.

29 Roberts, C.A., Al-Tameemi, H.M., Mashruwala, A.A. et al. (2017). The Suf iron-sulfur cluster biosynthetic system is essential in *Staphylococcus aureus*, and decreased Suf function results in global metabolic defects and reduced survival in human neutrophils. *Infect. Immun.* 85 (6): e00100-17.

30 Tsaousis, A.D. (2019). On the origin of iron/sulfur cluster biosynthesis in eukaryotes. *Front. Microbiol.* 10: 2478.

31 Garcia, P.S., Barras, F., and Gribaldo, S. (2023). Components of iron-sulfur cluster assembly machineries are robust phylogenetic markers to trace the origin of mitochondria and plastids. *PLoS Biol.* 21 (11): e3002374.

32 Balasubramanian, R., Shen, G., Bryant, D.A., and Golbeck, J.H. (2006). Regulatory roles for IscA and SufA in iron homeostasis and redox stress responses in the cyanobacterium *Synechococcus* sp. strain PCC 7002. *J. Bacteriol.* 188 (9): 3182–3191.

33 Zang, S.S., Jiang, H.B., Song, W.Y. et al. (2017). Characterization of the sulfur-formation (suf) genes in *Synechocystis* sp. PCC 6803 under photoautotrophic and heterotrophic growth conditions. *Planta* 246 (5): 927–938.

34 Przybyla-Toscano, J., Roland, M., Gaymard, F. et al. (2018). Roles and maturation of iron-sulfur proteins in plastids. *J. Biol. Inorg. Chem.* 23 (4): 545–566.

35 Gisselberg, J.E., Dellibovi-Ragheb, T.A., Matthews, K.A. et al. (2013). The suf iron-sulfur cluster synthesis pathway is required for apicoplast maintenance in malaria parasites. *PLoS Pathog.* 9 (9): e1003655.

36 Kumar, B., Chaubey, S., Shah, P. et al. (2011). Interaction between sulphur mobilisation proteins SufB and SufC: evidence for an iron-sulphur cluster biogenesis pathway in the apicoplast of *Plasmodium falciparum*. *Int. J. Parasitol.* 41 (9): 991–999.

37 Haussig, J.M., Matuschewski, K., and Kooij, T.W. (2014). Identification of vital and dispensable sulfur utilization factors in the Plasmodium apicoplast. *PLoS One* 9 (2): e89718.

38 Pamukcu, S., Cerutti, A., Bordat, Y. et al. (2021). Differential contribution of two organelles of endosymbiotic origin to iron-sulfur cluster synthesis and overall fitness in Toxoplasma. *PLoS Pathog.* 17 (11): e1010096.

39 Garcia, P.S., D'Angelo, F., Ollagnier de Choudens, S. et al. (2022). An early origin of iron-sulfur cluster biosynthesis machineries before Earth oxygenation. *Nat. Ecol. Evol.* 6 (10): 1564–1572.

40 Boyd, E.S., Thomas, K.M., Dai, Y. et al. (2014). Interplay between oxygen and Fe-S cluster biogenesis: insights from the Suf pathway. *Biochemistry* 53 (37): 5834–5847.

41 Outten, F.W., Wood, M.J., Munoz, F.M., and Storz, G. (2003). The SufE protein and the SufBCD complex enhance SufS cysteine desulfurase activity as part of a sulfur transfer pathway for Fe-S cluster assembly in *Escherichia coli*. *J. Biol. Chem.* 278 (46): 45713–45719.

42 Loiseau, L., Ollagnier-de-Choudens, S., Nachin, L. et al. (2003). Biogenesis of Fe-S cluster by the bacterial Suf system: SufS and SufE form a new type of cysteine desulfurase. *J. Biol. Chem.* 278 (40): 38352–38359.

43 Selbach, B., Earles, E., and Dos Santos, P.C. (2010). Kinetic analysis of the bisubstrate cysteine desulfurase SufS from *Bacillus subtilis*. *Biochemistry* 49 (40): 8794–8802.

44 Selbach, B.P., Chung, A.H., Scott, A.D. et al. (2014). Fe-S cluster biogenesis in Gram-positive bacteria: SufU is a zinc-dependent sulfur transfer protein. *Biochemistry* 53 (1): 152–160.

45 Elchennawi, I., Carpentier, P., Caux, C. et al. (2023). Structural and biochemical characterization of *Mycobacterium tuberculosis* zinc SufU-SufS complex. *Biomolecules* 13 (5): 732.

46 Gogar, R.K. and Frantom, P.A. (2024). Persulfide transfer to SufE activates the half-sites reactivity of the *E. coli* cysteine desulfurase SufS. *Biochemistry* 63 (12): 1569–1577.

47 Mihara, H., Fujii, T., Kato, S. et al. (2002). Structure of external aldimine of *Escherichia coli* CsdB, an IscS/NifS homolog: implications for its specificity toward selenocysteine. *J. Biochem. (Tokyo)* 131 (5): 679–685.

48 Riboldi, G.P., de Oliveira, J.S., and Frazzon, J. (2011). *Enterococcus faecalis* SufU scaffold protein enhances SufS desulfurase activity by acquiring sulfur from its cysteine-153. *Biochim. Biophys. Acta* 1814 (12): 1910–1918.

49 Hudspeth, J.D., Boncella, A.E., Sabo, E.T. et al. (2022). Structural and biochemical characterization of *Staphylococcus aureus* cysteine desulfurase complex SufSU. *ACS Omega* 7 (48): 44124–44133.

50 Tirupati, B., Vey, J.L., Drennan, C.L., and Bollinger, J.M. Jr., (2004). Kinetic and structural characterization of Slr0077/SufS, the essential cysteine desulfurase from *Synechocystis* sp. PCC 6803. *Biochemistry* 43 (38): 12210–12219.

51 Blauenburg, B., Mielcarek, A., Altegoer, F. et al. (2016). Crystal structure of *Bacillus subtilis* cysteine desulfurase SufS and its dynamic interaction with frataxin and scaffold protein SufU. *PLoS One* 11 (7): e0158749.

52 Roret, T., Pegeot, H., Couturier, J. et al. (2014). X-ray structures of Nfs2, the plastidial cysteine desulfurase from *Arabidopsis thaliana*. *Acta Crystallogr. F Struct. Biol. Commun.* 70 (Pt 9): 1180–1185.

53 Selbach, B.P., Pradhan, P.K., and Dos Santos, P.C. (2013). Protected sulfur transfer reactions by the *Escherichia coli* Suf system. *Biochemistry* 52 (23): 4089–4096.

54 Cupp-Vickery, J.R., Urbina, H., and Vickery, L.E. (2003). Crystal structure of IscS, a cysteine desulfurase from *Escherichia coli*. *J. Mol. Biol.* 330 (5): 1049–1059.

55 Ollagnier-de-Choudens, S., Lascoux, D., Loiseau, L. et al. (2003). Mechanistic studies of the SufS-SufE cysteine desulfurase: evidence for sulfur transfer from SufS to SufE. *FEBS Lett.* 555 (2): 263–267.

56 Blahut, M., Wise, C.E., Bruno, M.R. et al. (2019). Direct observation of intermediates in the SufS cysteine desulfurase reaction reveals functional roles of conserved active-site residues. *J. Biol. Chem.* 294 (33): 12444–12458.

57 Nakamura, R., Hikita, M., Ogawa, S. et al. (2020). Snapshots of PLP-substrate and PLP-product external aldimines as intermediates in two types of cysteine desulfurase enzymes. *FEBS J.* 287 (6): 1138–1154.

58 Singh, H., Dai, Y., Outten, F.W., and Busenlehner, L.S. (2013). *Escherichia coli* SufE sulfur transfer protein modulates the SufS cysteine desulfurase through allosteric conformational dynamics. *J. Biol. Chem.* 288 (51): 36189–36200.

59 Dai, Y. and Outten, F.W. (2012). The *E. coli* SufS-SufE sulfur transfer system is more resistant to oxidative stress than IscS-IscU. *FEBS Lett.* 586 (22): 4016–4022.

60 Ye, H., Abdel-Ghany, S.E., Anderson, T.D. et al. (2006). CpSufE activates the cysteine desulfurase CpNifS for chloroplastic Fe-S cluster formation. *J. Biol. Chem.* 281 (13): 8958–8969.

61 Murthy, U.M.N., Ollagnier-de-Choudens, S., Sanakis, Y. et al. (2007). Characterization of *Arabidopsis thaliana* SuFE2 and SuFE3: Functions in chloroplast iron-sulfur cluster assembly and NAD synthesis. *J. Biol. Chem.* 282 (25): 18254–18264.

62 Albrecht, A.G., Peuckert, F., Landmann, H. et al. (2011). Mechanistic characterization of sulfur transfer from cysteine desulfurase SufS to the iron-sulfur scaffold SufU in *Bacillus subtilis*. *FEBS Lett.* 585 (3): 465–470.

63 Charan, M., Singh, N., Kumar, B. et al. (2014). Sulfur mobilization for Fe-S cluster assembly by the essential SUF pathway in the *Plasmodium falciparum* apicoplast and its inhibition. *Antimicrob. Agents Chemother.* 58 (6): 3389–3398.

64 Riboldi, G.P., Verli, H., and Frazzon, J. (2009). Structural studies of the *Enterococcus faecalis* SufU [Fe-S] cluster protein. *BMC Biochem.* 10: 3.

65 Yokoyama, N., Nonaka, C., Ohashi, Y. et al. (2018). Distinct roles for U-type proteins in iron-sulfur cluster biosynthesis revealed by genetic analysis of the *Bacillus subtilis* sufCDSUB operon. *Mol. Microbiol.* 107 (6): 688–703.

66 Goldsmith-Fischman, S., Kuzin, A., Edstrom, W.C. et al. (2004). The SufE sulfur-acceptor protein contains a conserved core structure that mediates interdomain interactions in a variety of redox protein complexes. *J. Mol. Biol.* 344 (2): 549–565.

67 Dai, Y., Kim, D., Dong, G. et al. (2015). SufE D74R substitution alters active site loop dynamics to further enhance SufE interaction with the SufS cysteine desulfurase. *Biochemistry* 54 (31): 4824–4833.

68 Kim, D., Singh, H., Dai, Y. et al. (2018). Changes in protein dynamics in *Escherichia coli* SufS reveal a possible conserved regulatory mechanism in type II cysteine desulfurase systems. *Biochemistry* 57 (35): 5210–5217.

69 Dunkle, J.A., Bruno, M.R., Outten, F.W., and Frantom, P.A. (2019). Structural evidence for dimer-interface-driven regulation of the type II cysteine desulfurase, SufS. *Biochemistry* 58 (6): 687–696.

70 Dunkle, J.A., Bruno, M.R., and Frantom, P.A. (2020). Structural evidence for a latch mechanism regulating access to the active site of SufS-family cysteine desulfurases. *Acta Crystallogr. D Struct. Biol.* 76 (Pt 3): 291–301.

71 Gogar, R.K., Carroll, F., Conte, J.V. et al. (2023). The beta-latch structural element of the SufS cysteine desulfurase mediates active site accessibility and SufE transpersulfurase positioning. *J. Biol. Chem.* 299 (3): 102966.

72 Fujishiro, T., Terahata, T., Kunichika, K. et al. (2017). Zinc-ligand swapping mediated complex formation and sulfur transfer between SufS and SufU for iron-sulfur cluster biogenesis in *Bacillus subtilis*. *J. Am. Chem. Soc.* 139 (51): 18464–18467.

73 Xu, X.M. and Moller, S.G. (2006). AtSufE is an essential activator of plastidic and mitochondrial desulfurases in *Arabidopsis*. *EMBO J.* 25 (4): 900–909.

74 Farre, E.M., Tiessen, A., Roessner, U. et al. (2001). Analysis of the compartmentation of glycolytic intermediates, nucleotides, sugars, organic acids, amino acids, and sugar alcohols in potato tubers using a nonaqueous fractionation method. *Plant Physiol.* 127 (2): 685–700.

75 Leon, S., Touraine, B., Briat, J.F., and Lobreaux, S. (2002). The AtNFS2 gene from *Arabidopsis thaliana* encodes a NifS-like plastidial cysteine desulphurase. *Biochem. J.* 366 (Pt 2): 557–564.

76 Van Hoewyk, D., Abdel-Ghany, S.E., Cohu, C.M. et al. (2007). Chloroplast iron-sulfur cluster protein maturation requires the essential cysteine desulfurase CpNifS. *Proc. Natl. Acad. Sci. USA* 104 (13): 5686–5691.

77 Zhang, H. and Kramer, U. (2018). Differential diel translation of transcripts with roles in the transfer and utilization of iron-sulfur clusters in *Arabidopsis*. *Front. Plant Sci.* 9: 1641.

78 Buhning, M., Valleriani, A., and Leimkuhler, S. (2017). The role of SufS is restricted to Fe-S cluster biosynthesis in *Escherichia coli*. *Biochemistry* 56 (14): 1987–2000.

79 Swift, R.P., Elahi, R., Rajaram, K. et al. (2023). The *Plasmodium falciparum* apicoplast cysteine desulfurase provides sulfur for both iron-sulfur cluster assembly and tRNA modification. *Elife* 12: e84491.

80 Black, K.A. and Dos Santos, P.C. (2015). Abbreviated pathway for biosynthesis of 2-thiouridine in *Bacillus subtilis*. *J. Bacteriol.* 197 (11): 1952–1962.

81 Liu, Y.C., Sieprawska-Lupa, M., Whitman, W.B., and White, R.H. (2010). Cysteine is not the sulfur source for iron-sulfur cluster and methionine biosynthesis in the methanogenic archaeon *Methanococcus maripaludis*. *J. Biol. Chem.* 285 (42): 31923–31929.

82 Murata, M., Murakami, T., Yuda, E. et al. (2024). The minimal SUF system can substitute for the canonical iron-sulfur cluster biosynthesis systems by using inorganic sulfide as the sulfur source. *bioRxiv*.

83 Habicht, K.S., Salling, L., Thamdrup, B., and Canfield, D.E. (2005). Effect of low sulfate concentrations on lactate oxidation and isotope fractionation during sulfate reduction by *Archaeoglobus fulgidus* strain Z. *Appl. Environ. Microbiol.* 71 (7): 3770–3777.

84 Payne, D., Spietz, R.L., and Boyd, E.S. (2021). Reductive dissolution of pyrite by methanogenic archaea. *ISME J.* 15 (12): 3498–3507.

85 Spietz, R.L., Payne, D., Szilagyi, R., and Boyd, E.S. (2022). Reductive biomining of pyrite by methanogens. *Trends Microbiol.* 30 (11): 1072–1083.

86 Tsaousis, A.D., Ollagnier de Choudens, S., Gentekaki, E. et al. (2012). Evolution of Fe–S cluster biogenesis in the anaerobic parasite *Blastocystis*. *Proc. Natl. Acad. Sci. USA* 109 (26): 10426–10431.

87 Layer, G., Gaddam, S.A., Ayala-Castro, C.N. et al. (2007). SufE transfers sulfur from SufS to SufB for iron-sulfur cluster assembly. *J. Biol. Chem.* 282 (18): 13342–13350.

88 Wollers, S., Layer, G., Garcia-Serres, R. et al. (2010). Iron-sulfur (Fe-S) cluster assembly: the SufBCD complex is a new type of Fe-S scaffold with a flavin redox cofactor. *J. Biol. Chem.* 285 (30): 23331–23341.

89 Saini, A., Mapolelo, D.T., Chahal, H.K. et al. (2010). SufD and SufC ATPase activity are required for iron acquisition during in vivo Fe-S cluster formation on SufB. *Biochemistry* 49 (43): 9402–9412.

90 Chahal, H.K. and Outten, F.W. (2012). Separate Fe–S scaffold and carrier functions for SufB(2)C(2) and SufA during in vitro maturation of [2Fe–2S] Fdx. *J. Inorg. Biochem.* 116: 126–134.

91 Kitaoka, S., Wada, K., Hasegawa, Y. et al. (2006). Crystal structure of *Escherichia coli* SufC, an ABC-type ATPase component of the SUF iron-sulfur cluster assembly machinery. *FEBS Lett.* 580 (1): 137–143.

92 Hirabayashi, K., Yuda, E., Tanaka, N. et al. (2015). Functional dynamics revealed by the structure of the SufBCD complex, a novel ATP-binding cassette (ABC) protein that serves as a scaffold for iron-sulfur cluster biogenesis. *J. Biol. Chem.* 290 (50): 29717–29731.

93 Badger, J., Sauder, J.M., Adams, J.M. et al. (2005). Structural analysis of a set of proteins resulting from a bacterial genomics project. *Proteins* 60 (4): 787–796.

94 Expert, D., Boughammoura, A., and Franza, T. (2008). Siderophore-controlled iron assimilation in the enterobacterium *Erwinia chrysanthemi*: evidence for the involvement of bacterioferritin and the Suf iron-sulfur cluster assembly machinery. *J. Biol. Chem.* 283 (52): 36564–36572.

95 Wada, K., Sumi, N., Nagai, R. et al. (2009). Molecular dynamism of Fe-S cluster biosynthesis implicated by the structure of the SufC(2)-SufD(2) complex. *J. Mol. Biol.* 387 (1): 245–258.

96 Petrovic, A., Davis, C.T., Rangachari, K. et al. (2008). Hydrodynamic characterization of the SufBC and SufCD complexes and their interaction with fluorescent adenosine nucleotides. *Protein Sci.* 17 (7): 1264–1274.

97 Veronesi, G., Perard, J., Clemancey, M. et al. (2024). Multimodal spectroscopic analysis of the Fe-S clusters of the as-isolated *Escherichia coli* SufBC(2)D complex. *Inorg. Chem.* 63 (19): 8730–8738.

98 Yuda, E., Tanaka, N., Fujishiro, T. et al. (2017). Mapping the key residues of SufB and SufD essential for biosynthesis of iron-sulfur clusters. *Sci. Rep.* 7 (1): 9387.

99 Blanc, B., Clemancey, M., Latour, J.M. et al. (2014). Molecular investigation of iron-sulfur cluster assembly scaffolds under stress. *Biochemistry* 53 (50): 7867–7869.

100 Vinella, D., Brochier-Armanet, C., Loiseau, L. et al. (2009). Iron-sulfur (Fe–S) protein biogenesis: phylogenomic and genetic studies of A-type carriers. *PLoS Genet.* 5 (5): e1000497.

101 Chahal, H.K., Dai, Y., Saini, A. et al. (2009). The SufBCD Fe-S scaffold complex interacts with SufA for Fe-S cluster transfer. *Biochemistry* 48 (44): 10644–10653.

102 Gupta, V., Sendra, M., Naik, S.G. et al. (2009). Native *Escherichia coli* SufA, coexpressed with SufBCDSE, purifies as a [2Fe-2S] protein and acts as an Fe-S transporter to Fe-S target enzymes. *J. Am. Chem. Soc.* 131 (17): 6149–6153.

103 Loiseau, L., Gerez, C., Bekker, M. et al. (2007). ErpA, an iron–sulfur (Fe–S) protein of the A-type essential for respiratory metabolism in *Escherichia coli. Proc. Natl. Acad. Sci. USA* 104 (34): 13626–13631.

104 Mapolelo, D.T., Zhang, B., Randeniya, S. et al. (2013). Monothiol glutaredoxins and A-type proteins: partners in Fe-S cluster trafficking. *Dalton Trans.* 42 (9): 3107–3115.

105 Py, B., Gerez, C., Huguenot, A. et al. (2018). The ErpA/NfuA complex builds an oxidation-resistant Fe-S cluster delivery pathway. *J. Biol. Chem.* 293 (20): 7689–7702.

106 Mapolelo, D.T., Zhang, B., Naik, S.G. et al. (2012). Spectroscopic and functional characterization of iron-bound forms of *Azotobacter vinelandii* (Nif)IscA. *Biochemistry* 51 (41): 8056–8070.

107 Lu, J., Yang, J., Tan, G., and Ding, H. (2008). Complementary roles of SufA and IscA in the biogenesis of iron-sulfur clusters in *Escherichia coli. Biochem. J.* 409 (2): 535–543.

108 Vinella, D., Loiseau, L., Ollagnier de Choudens, S. et al. (2013). In vivo [Fe-S] cluster acquisition by IscR and NsrR, two stress regulators in *Escherichia coli. Mol. Microbiol.* 87 (3): 493–508.

109 Jensen, L.T. and Culotta, V.C. (2000). Role of *Saccharomyces cerevisiae* ISA1 and ISA2 in iron homeostasis. *Mol. Cell. Biol.* 20 (11): 3918–3927.

110 Wada, K., Hasegawa, Y., Gong, Z. et al. (2005). Crystal structure of *Escherichia coli* SufA involved in biosynthesis of iron-sulfur clusters: implications for a functional dimer. *FEBS Lett.* 579 (29): 6543–6548.

111 Morimoto, K., Yamashita, E., Kondou, Y. et al. (2006). The asymmetric IscA homodimer with an exposed [2Fe-2S] cluster suggests the structural basis of the Fe-S cluster biosynthetic scaffold. *J. Mol. Biol.* 360 (1): 117–132.

112 Landry, A.P., Cheng, Z., and Ding, H. (2013). Iron binding activity is essential for the function of IscA in iron-sulphur cluster biogenesis. *Dalton Trans.* 42 (9): 3100–3106.

113 Ding, H. and Clark, R.J. (2004). Characterization of iron binding in IscA, an ancient iron-sulphur cluster assembly protein. *Biochem. J.* 379 (Pt 2): 433–440.

114 Ding, H., Clark, R.J., and Ding, B. (2004). IscA mediates iron delivery for assembly of iron-sulfur clusters in IscU under the limited accessible free iron conditions. *J. Biol. Chem.* 279 (36): 37499–37504.

115 Yang, J., Bitoun, J.P., and Ding, H. (2006). Interplay of IscA and IscU in biogenesis of iron-sulfur clusters. *J. Biol. Chem.* 281 (38): 27956–27963.

116 Ding, B., Smith, E.S., and Ding, H. (2005). Mobilization of the iron centre in IscA for the iron-sulphur cluster assembly in IscU. *Biochem. J.* 389 (Pt 3): 797–802.

117 Tsaousis, A.D., Gentekaki, E., Eme, L. et al. (2014). Evolution of the cytosolic iron-sulfur cluster assembly machinery in *Blastocystis* species and other microbial eukaryotes. *Eukaryot. Cell* 13 (1): 143–153.

118 Mashruwala, A.A., Bhatt, S., Poudel, S. et al. (2016). The DUF59 containing protein SufT is involved in the maturation of iron-sulfur (Fe–S) proteins during conditions of high Fe–S cofactor demand in *Staphylococcus aureus*. *PLos Genet.* 12 (8): e1006233.

119 Aubert, C., Mandin, P., and Py, B. (2023). Mrp and SufT, two bacterial homologs of eukaryotic CIA factors involved in Fe-S clusters biogenesis. *Inorganics* 11: 431.

120 Mashruwala, A.A. and Boyd, J.M. (2018). Investigating the role(s) of SufT and the domain of unknown function 59 (DUF59) in the maturation of iron-sulfur proteins. *Curr. Genet.* 64 (1): 9–16.

121 Weerapana, E., Wang, C., Simon, G.M. et al. (2010). Quantitative reactivity profiling predicts functional cysteines in proteomes. *Nature* 468 (7325): 790–795.

122 Almeida, M.S., Herrmann, T., Peti, W. et al. (2005). NMR structure of the conserved hypothetical protein TM0487 from *Thermotoga maritima*: implications for 216 homologous DUF59 proteins. *Protein Sci.* 14 (11): 2880–2886.

123 Ouyang, B., Wang, L., Wan, S. et al. (2013). Solution structure of monomeric human FAM96A. *J. Biomol. NMR* 56 (4): 387–392.

124 Tripathi, A., Anand, K., Das, M. et al. (2022). *Mycobacterium tuberculosis* requires SufT for Fe-S cluster maturation, metabolism, and survival in vivo. *PLoS Pathog.* 18 (4): e1010475.

125 McKenzie, R.M.E., Henry, L.G., Boutrin, M.C. et al. (2016). Role of the *Porphyromonas gingivalis* iron-binding protein PG1777 in oxidative stress resistance. *Microbiology (Reading)* 162 (2): 256–267.

126 Braymer, J.J., Freibert, S.A., Rakwalska-Bange, M., and Lill, R. (2021). Mechanistic concepts of iron-sulfur protein biogenesis in Biology. *Biochim. Biophys. Acta Mol. Cell Res.* 1868 (1): 118863.

127 Kassube, S.A. and Thoma, N.H. (2020). Structural insights into Fe-S protein biogenesis by the CIA targeting complex. *Nat. Struct. Mol. Biol.* 27 (8): 735–742.

128 Sasaki, S., Minamisawa, K., and Mitsui, H. (2016). A *Sinorhizobium meliloti* RpoH-regulated gene is involved in iron-sulfur protein metabolism and effective plant symbiosis under intrinsic iron limitation. *J. Bacteriol.* 198 (17): 2297–2306.

129 Mashruwala, A.A., Roberts, C.A., Bhatt, S. et al. (2016). *Staphylococcus aureus* SufT: an essential iron-sulphur cluster assembly factor in cells experiencing a high-demand for lipoic acid. *Mol. Microbiol.* 102 (6): 1099–1119.

130 Tamuhla, T., Joubert, L., Willemse, D., and Williams, M.J. (2020). SufT is required for growth of *Mycobacterium smegmatis* under iron limiting conditions. *Microbiology (Reading)* 166 (3): 296–305.

131 Mettert, E.L. and Kiley, P.J. (2015). How is Fe-s cluster formation regulated? *Annu. Rev. Microbiol.* 69: 505–526.

132 Yeo, W.S., Lee, J.H., Lee, K.C., and Roe, J.H. (2006). IscR acts as an activator in response to oxidative stress for the suf operon encoding Fe-S assembly proteins. *Mol. Microbiol.* 61 (1): 206–218.

133 Mettert, E.L. and Kiley, P.J. (2014). Coordinate regulation of the Suf and Isc Fe-S cluster biogenesis pathways by IscR is essential for viability of *Escherichia coli*. *J. Bacteriol.* 196 (24): 4315–4323.

134 Lee, K.C., Yeo, W.S., and Roe, J.H. (2008). Oxidant-responsive induction of the suf operon, encoding a Fe-S assembly system, through Fur and IscR in *Escherichia coli*. *J. Bacteriol.* 190 (24): 8244–8247.

135 Lee, J.H., Yeo, W.S., and Roe, J.H. (2004). Induction of the sufA operon encoding Fe-S assembly proteins by superoxide generators and hydrogen peroxide: involvement of OxyR, IHF and an unidentified oxidant-responsive factor. *Mol. Microbiol.* 51 (6): 1745–1755.

136 Jang, S. and Imlay, J.A. (2010). Hydrogen peroxide inactivates the *Escherichia coli* Isc iron-sulphur assembly system, and OxyR induces the Suf system to compensate. *Mol. Microbiol.* 78 (6): 1448–1467.

137 Fantino, J.R., Py, B., Fontecave, M., and Barras, F. (2010). A genetic analysis of the response of *Escherichia coli* to cobalt stress. *Environ. Microbiol.* 12 (10): 2846–2857.

138 Todd, J.D., Wexler, M., Sawers, G. et al. (2002). RirA, an iron-responsive regulator in the symbiotic bacterium *Rhizobium leguminosarum*. *Microbiology (Reading)* 148 (Pt 12): 4059–4071.

139 Chao, T.C., Buhrmester, J., Hansmeier, N. et al. (2005). Role of the regulatory gene rirA in the transcriptional response of *Sinorhizobium meliloti* to iron limitation. *Appl. Environ. Microbiol.* 71 (10): 5969–5982.

140 Yeoman, K.H., Curson, A.R., Todd, J.D. et al. (2004). Evidence that the Rhizobium regulatory protein RirA binds to cis-acting iron-responsive operators (IROs) at promoters of some Fe-regulated genes. *Microbiology (Reading)* 150 (Pt 12): 4065–4074.

141 Pellicer Martinez, M.T., Martinez, A.B., Crack, J.C. et al. (2017). Sensing iron availability via the fragile [4Fe-4S] cluster of the bacterial transcriptional repressor RirA. *Chem. Sci.* 8 (12): 8451–8463.

142 Pellicer Martinez, M.T., Crack, J.C., Stewart, M.Y. et al. (2019). Mechanisms of iron- and O_2-sensing by the [4Fe-4S] cluster of the global iron regulator RirA. *Elife* 8: e47804.

143 Bhubhanil, S., Niamyim, P., Sukchawalit, R., and Mongkolsuk, S. (2014). Cysteine desulphurase-encoding gene sufS2 is required for the repressor function of RirA and oxidative resistance in *Agrobacterium tumefaciens*. *Microbiology (Reading)* 160 (Pt 1): 79–90.

144 Hibbing, M.E. and Fuqua, C. (2011). Antiparallel and interlinked control of cellular iron levels by the Irr and RirA regulators of *Agrobacterium tumefaciens*. *J. Bacteriol.* 193 (14): 3461–3472.

145 Sourice, M., Askenasy, I., Garcia, P.S. et al. (2023). A diverged transcriptional network for usage of two Fe-S cluster biogenesis machineries in the delta-proteobacterium *Myxococcus xanthus*. *Mbio* 14 (1): e0300122.

146 Willemse, D., Weber, B., Masino, L. et al. (2018). Rv1460, a SufR homologue, is a repressor of the suf operon in *Mycobacterium tuberculosis*. *PLoS One* 13 (7): e0200145.

147 Wang, T., Shen, G., Balasubramanian, R. et al. (2004). The sufR gene (sll0088 in *Synechocystis* sp. strain PCC 6803) functions as a repressor of the sufBCDS operon in iron-sulfur cluster biogenesis in cyanobacteria. *J. Bacteriol.* 186 (4): 956–967.

148 Shen, G., Balasubramanian, R., Wang, T. et al. (2007). SufR coordinates two [4Fe-4S]2+, 1+ clusters and functions as a transcriptional repressor of the sufBCDS operon and an autoregulator of sufR in cyanobacteria. *J. Biol. Chem.* 282 (44): 31909–31919.

149 Anand, K., Tripathi, A., Shukla, K. et al. (2021). *Mycobacterium tuberculosis* SufR responds to nitric oxide via its 4Fe-4S cluster and regulates Fe-S cluster biogenesis for persistence in mice. *Redox Biol.* 46: 102062.

150 Rohde, K.H., Abramovitch, R.B., and Russell, D.G. (2007). *Mycobacterium tuberculosis* invasion of macrophages: linking bacterial gene expression to environmental cues. *Cell Host Microbe* 2 (5): 352–364.

151 Rohde, K.H., Veiga, D.F., Caldwell, S. et al. (2012). Linking the transcriptional profiles and the physiological states of *Mycobacterium tuberculosis* during an extended intracellular infection. *PLoS Pathog.* 8 (6): e1002769.

152 Sassetti, C.M. and Rubin, E.J. (2003). Genetic requirements for mycobacterial survival during infection. *Proc. Natl. Acad. Sci. USA* 100 (22): 12989–12994.

153 Pandey, M., Talwar, S., Bose, S., and Pandey, A.K. (2018). Iron homeostasis in *Mycobacterium tuberculosis* is essential for persistence. *Sci. Rep.* 8 (1): 17359.

154 Saves, I., Lewis, L.A., Westrelin, F. et al. (2002). Specificities and functions of the recA and pps1 intein genes of *Mycobacterium tuberculosis* and application for diagnosis of tuberculosis. *J. Clin. Microbiol.* 40 (3): 943–950.

155 Topilina, N.I., Green, C.M., Jayachandran, P. et al. (2015). SufB intein of *Mycobacterium tuberculosis* as a sensor for oxidative and nitrosative stresses. *Proc. Natl. Acad. Sci. USA* 112 (33): 10348–10353.

156 Panda, S., Nanda, A., Nasker, S.S. et al. (2023). Metal regulation of *Mycobacterium tuberculosis* SufB intein splicing at the host-pathogen crossroad. *Clin. Sci. (Lond)* 137 (14): 1027–1048.

157 Panda, S., Nanda, A., Sahu, N. et al. (2022). SufB intein splicing in *Mycobacterium tuberculosis* is influenced by two remote conserved N-extein histidines. *Biosci. Rep.* 42 (3): BSR20212207.

3

Biogenesis of Mitochondrial Iron–Sulfur Proteins

Ulrich Mühlenhoff, Oliver Stehling, and Roland Lill

Philipps-Universität Marburg, Institut für Zytobiologie im Zentrum für Synthetische Mikrobiologie Synmikro, Karl-von-Frisch-Str. 14, 35032 Marburg, Germany

Chapter Goals

- To provide an overview on the variety and physiological functions of iron–sulfur (Fe–S) proteins in a eukaryotic cell
- To define the various proteins needed for the assembly of mitochondrial Fe–S proteins
- To describe both the general principles and detailed mechanisms underlying mitochondrial Fe–S protein biogenesis
- To outline the disease connection for mitochondrial Fe–S protein biogenesis

Main Chapter Points

- Overview on the cellular roles of Fe–S proteins in a eukaryotic cell
- Introduction to the composition of the mitochondrial Fe–S protein assembly (ISC) system
- Detailed description of the three main steps required to assemble mitochondrial Fe–S proteins

3.1 Introduction

Proteins with iron–sulfur (Fe–S) clusters play important roles in a large variety of cellular processes in almost all living organisms. In addition to their well-known role in electron transfer processes, Fe–S clusters are involved in biochemical catalysis, sensing, and regulation or serve to strengthen a protein's fold [1–3]. In contrast

Iron-Sulfur Clusters: Biogenesis and Biochemistry, Volume 1, First Edition.
Edited by Silke Leimkühler, Günter Schwarz, Oliver Lenz, and Oliver Einsle.

to many other protein cofactors that can be taken up from the environment in the form of stable precursors, Fe–S clusters are intrinsically unstable in aqueous solution unless bound to proteins or peptides. Hence, every cell has to provide itself with its own set of Fe–S clusters, and for this purpose cells harbor numerous dedicated biogenesis factors that originally have been evolved in Archaea and Bacteria, and during evolution have been passed on to Eukarya [4–6]. In this review, we concentrate on the biogenesis of nongreen eukaryotes, which typically contain Fe–S proteins in mitochondria, cytosol, and nucleus. Biogenesis is achieved by conserved multiprotein machineries in both mitochondria and cytosol. Plant cells and algae are similar in their mitochondrial and cytosolic Fe–S protein biogenesis systems, yet in their plastids they additionally contain a bacteria-inherited Fe–S protein biogenesis machinery termed sulfur utilization system (SUF; for recent reviews, see [7–9]).

Many of the up to 80 Fe–S proteins present in eukaryotes perform key biochemical functions (Figure 3.1) and hence are essential for the viability of a living cell. As a consequence, Fe–S protein biogenesis is indispensable even in simple unicellular eukaryotes, such as baker's yeast. Not surprisingly, (genetic) defects in the biogenesis factors assisting Fe–S protein maturation in humans are associated with severe metabolic, neurological, or hematological disorders [15–18]. The biosynthesis of Fe–S clusters and their insertion into proteins in eukaryotes is carried out in mitochondria and cytosol by complex proteinaceous machineries. Mitochondria play center stage in this process as they harbor the "iron–sulfur cluster assembly" (ISC) machinery that is required for biogenesis of all cellular Fe–S proteins, whether located in mitochondria, the cytosol, or the nucleus (Figure 3.1). For the biosynthesis of extramitochondrial Fe–S proteins, mitochondria provide a biochemically not fully characterized sulfur- and likely iron- and glutathione-containing metabolite (termed X-S) that is synthesized by the mitochondrial ISC system and exported into the cytosol by the mitochondrial ATP-binding cassette (ABC) transporter ABCB7 (yeast Atm1; Figure 3.1). This metabolite is then utilized by the "cytosolic iron–sulfur protein assembly" (CIA) system for the biosynthesis and insertion of [4Fe–4S] clusters into cytosolic and nuclear target proteins. Further, compound X-S is also essential for the formation of extramitochondrial [2Fe–2S] proteins, yet in almost all cases in a CIA-independent fashion [19]. In this article, we summarize recent developments in the understanding of the molecular mechanisms underlying Fe–S protein biogenesis in mitochondria of eukaryotes. First, we outline the main functions of eukaryotic Fe–S proteins in the different compartments.

3.2 Cellular Roles of Eukaryotic Fe–S Proteins

Iron–sulfur (Fe–S) proteins are ancient protein cofactors. Their structures resemble those of certain Fe–S minerals that have been hypothesized to serve as catalysts for the synthesis of prebiotic molecules at the origin of life [20]. The exact details

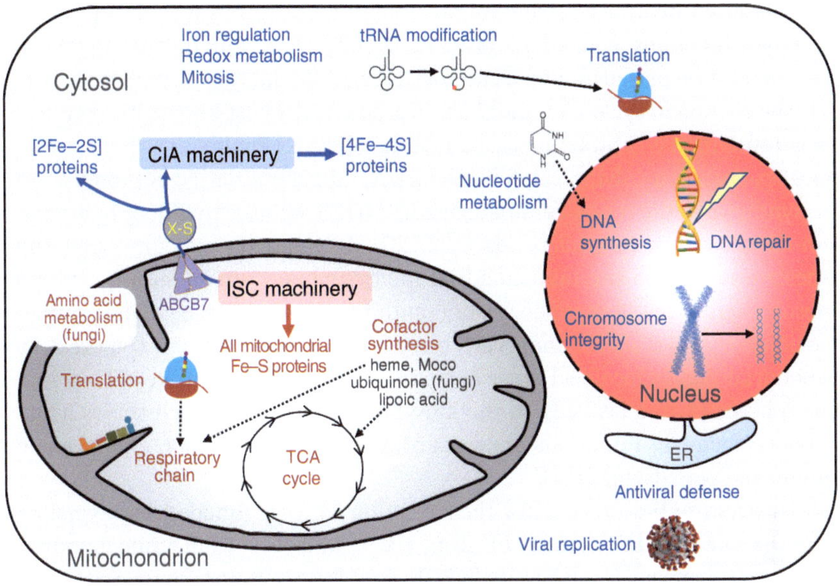

Figure 3.1 Cellular pathways involving Fe–S proteins in (nongreen) eukaryotes. Fe–S proteins can be found in mitochondria, cytosol, nucleus, and at the outer face of the endoplasmic reticulum. Fe–S cluster assembly on these proteins is facilitated by the cooperative action of the mitochondrial ISC and cytosolic CIA machineries. The two biogenesis systems are connected by the mitochondrial ABC transporter ABCB7 exporting the essential sulfur- (and possibly iron-) containing compound (X–S) to the cytosol. Examples of well-known mitochondrial Fe–S proteins include those involved in the respiratory chain (complexes I, II, and III) and the TCA cycle (aconitase). Further mitochondrial Fe–S proteins are involved in the synthesis of other important protein cofactors such as heme (ferrochelatase), heme A (ferredoxin FDX1), and lipoic acid (lipoyl synthase), a cofactor for enzymes of, e.g. the TCA cycle (PDH and α-KGDH). Additionally, Fe–S clusters are essential for the translation of mitochondrial DNA-encoded subunits of the respiratory chain complexes (mitochondrial ribosome, METTL17). In fungi, mitochondrial Fe–S enzymes (ferredoxin) contribute to the synthesis of ubiquinone, an electron carrier in the respiratory chain. Fungi further contain several Fe–S enzymes involved in amino acid metabolism, either in mitochondria (e.g. Aco2, Lys4, and Glt1) or in the cytosol (Leu1). Cytosolic and nuclear Fe–S proteins participate in various cellular processes such as nucleotide metabolism (e.g. GPAT, DPYD, XDH), tRNA modifications (e.g. TYW1, ELP3), cellular redox balance (e.g. GLXR2), mitosis (KIF4A), or iron regulation (mammalian IRP1 and FBXL5, or fungal iron-responsive transcription factors such as HapX, SreA, Aft1, and Yap5 [10, 11]). Fe–S proteins are further crucial for cytosolic translation (ribosome splitting ATPase ABCE1, DPH1/2). Nuclear Fe–S proteins are involved in numerous aspects of genome maintenance such as DNA synthesis and repair (DNA polymerases POLA1, POLD1, etc., primase PRIM2, and helicases XPD, DDX11) and chromosome integrity (FANCJ, RTEL1). The latter proteins and ABCE1 are essential for cell viability and explain why most components of the mitochondrial ISC machinery and the CIA system are indispensable for life. Recently, viral replication-associated proteins with Fe–S clusters have been described, but also the cellular antiviral response depends on a Fe–S protein (viperin aka RSAD2). More factors can be found in an inventory of eukaryotic Fe–S proteins [12]. It seems likely that further Fe–S proteins will be discovered in the future. Identification of such proteins may involve mass spectrometry of proteins bound to targeting factors of the ISC and CIA machineries, or methods utilizing the accessibility of the Fe–S cluster-coordinating Cys ligands under biogenesis-deficient conditions [13, 14]. Source: Alissa Eckert, MSMI, Dan Higgins, MAMS/CDC/Public Domain.

of this hypothesis are under debate, yet the ubiquitous occurrence of Fe–S proteins with catalytic functions in all kingdoms of life is compatible with the idea that Fe–S clusters might have played an important role as versatile catalysts in prebiotic times [5, 21]. The citric acid cycle with aconitase is probably the best-known biochemical pathway involving Fe–S proteins with a role in catalysis (Figure 3.1). Further biochemical pathways involving catalytically active mitochondrial Fe–S proteins are the biosynthesis of the molybdenum cofactor (MOCS1A; note that mainly human protein names are used in this review) [22, 23]. Molybdenum cofactor-containing enzymes are located in the mitochondrial intermembrane space (sulfite oxidase) or cytosol where some of these enzymes (xanthine dehydrogenase [XDH] or aldehyde oxidase [AOX1]) also contain Fe–S clusters as a second cofactor (Figure 3.1). Mitochondrial matrix-localized ferrochelatase catalyzes the insertion of Fe^{2+} into protoporphyrin IX, the last step in the biosynthesis of heme, which serves as an important cofactor in numerous enzymes of mitochondria, cytosol, endoplasmic reticulum, and peroxisomes [24].

Fe–S clusters are best known for their electron transfer function by reversible changes in the oxidative state of iron (Fe^{2+} and Fe^{3+}). In mitochondria, electron-transferring Fe–S proteins include those of the respiratory chain complexes I–III and the flavoprotein-ubiquinone oxidoreductase (ETF dehydrogenase) that couples the oxidation of fatty acids and some amino acids to oxidative phosphorylation (Figure 3.1). In addition, mitochondrial-type [2Fe–2S] ferredoxins serve as versatile one-electron donors for several biosynthetic processes within mitochondria, including the formation of ubiquinone, lipoic acid, and heme *a*, a respiratory chain complex IV cofactor [25–27]. However, what makes mitochondrial ferredoxins essential for the eukaryotic cell is their crucial role as electron donors at the initial step of Fe–S protein biogenesis, i.e. the *de novo* biosynthesis of a [2Fe–2S] cluster by the mitochondrial core ISC complex (see Section 3.3.2). Ferredoxins become reduced by mitochondrial NADPH via the cognate NAD(P)H-dependent ferredoxin reductase ferredoxin reductase (FDXR). Intriguingly, while fungi contain a single all-purpose mitochondrial ferredoxin (Yah1 in yeast), mammals harbor two paralogs (FDX1 and FDX2) with distinct target specificities, despite high structural similarities [26, 27]. While FDX2 is mainly involved in Fe–S cluster formation (and thus is the functional homolog of fungal Yah1), mammalian FDX1 – aka adrenodoxin – is specific for cytochrome P450 reduction during steroid hormone synthesis [28]. Further, FDX1 serves as the specific electron donor for lipoyl synthase (LIAS), a [4Fe–4S] radical S-adenosyl methionine (SAM) protein that catalyzes the last step in the biosynthesis of lipoic acid, an essential cofactor of several α-ketoacid dehydrogenases (pyruvate dehydrogenase (PDH), 2-ketoglutarate dehydrogenase (KGDH), branched-chain ketoacid dehydrogenase (BCKDH), and 2-ketoadipate dehydrogenase (KADH)) and the H-subunit of the glycine cleavage system (GCSH) (Figure 3.1) [29, 30]. As explicated below, impaired lipoylation of these enzymes has served as a versatile tool for the identification and characterization of ISC biogenesis proteins involved in LIAS maturation [12]. LIAS possesses two [4Fe–4S] clusters termed catalytic and auxiliary clusters [31–33]. The synthesis of both [4Fe–4S] clusters of this enzyme depends on the function of FDX2, which cooperates with the other components of the mitochondrial ISC system. The LIAS enzyme sequentially attaches two sulfur atoms at the C6 and C8 positions of a GCSH-bound octanoyl chain to form the mature lipoyl cofactor. The catalytic cluster requires electrons from FDX1 to start the radical chain for reductive cleavage of SAM to methionine and a 5′-deoxyadenosyl radical [26, 27].

Mitochondrial translation is the most recent cellular function known to involve Fe–S clusters (Figure 3.1). The human mitochondrial ribosome contains three [2Fe–2S] clusters, two in the small and one in the large subunit [34]. These clusters are each bound in a bridging fashion between a pair of peripheral mitoribosomal proteins. The three subunit pairs are bS18m–bS6m and mS25–bS16m of the small subunit and mL66–uL10m of the large subunit. The mS25–bS16m pair has been shown to be essential for mitoribosome stability, assembly, and function [35]. Mutations in these [2Fe–2S] cluster-bridging proteins have been linked to human mitochondrial disorders. Yet, the Fe–S clusters are only present in higher eukaryotes (opisthokonts) but absent in, e.g. fungi. The small ribosomal subunit further binds the [4Fe–4S] protein METTL17 that is essential for mitoribosomal function [13]. METTL17 displays some homology to Rsm22 from yeast, a SAM-dependent RNA methyltransferase that may likely play a role as a ribosomal assembly factor [37]. All these mitoribosome-associated Fe–S cluster-binding proteins play an important regulatory role as they mechanistically link mitochondrial translation and function of the mtDNA-encoded core subunits of oxidative phosphorylation (i.e. respiratory chain complexes I and III–V) with the fitness of the mitochondrial ISC system. They also serve as reactive oxygen species (ROS) sensors ablating mitochondrial translation under conditions of oxidative stress [13].

The eukaryotic cytosol harbors many Fe–S proteins, and there are likely still a few more to be discovered. Currently known members perform important roles in nucleotide metabolism (glutaminephosphoribosylpyrophosphate amidotransferase (GPAT), dihydropyrimidine dehydrogenase (DPYD), or XDH), tRNA modifications (TYW1 and ELP3 [38, 39]), iron uptake regulation (IRP1; [40]), redox metabolism (glutaredoxin GLRX2, GLRX3, and AOX1 [41]), mitosis and cytokinesis (kinesin KIF4a; [42]), and translation (ABCE1 [43]) (Figure 3.1). Some cytosolic Fe–S proteins are part of the CIA machinery. CIAPIN1 (aka anamorsin) performs an electron transfer function in the early step of the cytosolic assembly pathway and has an additional role in the assembly of the diferric tyrosyl radical cofactor of ribonucleotide reductase (RNR) [44–47]. Electrons are transferred to CIAPIN1 from NADPH via the FAD- and FMN-containing oxidoreductase NDOR1. The radical SAM Fe–S protein viperin (aka RSAD2) contributes to interferon-stimulated antiviral defense. It catalyzes the formation of didehydro-CTP (ddhCTP), which inhibits viral replication by serving as a specific chain termination reagent of viral RNA-dependent

RNA polymerases [48]. Recently, Fe–S clusters have also been reported for the SARS-CoV-2 RNA polymerase and helicase [49, 50]. In contrast, the proposed regulatory roles of the related [2Fe–2S] proteins CISD1 and CISD2 (aka mitoNEET and Miner1, respectively) in iron and redox homeostasis are still a matter of debate [51]. The proteins are attached to the outer faces of mitochondria and the endoplasmic reticulum.

The nucleus harbors a still-growing number of known Fe–S proteins (e.g. DNA2, *Xeroderma pigmentosum* group D DNA helicase (XPD), NTHL1, FANCJ, RTEL1, and DDX11) with important functions in DNA synthesis and repair, chromosome segregation, and telomere length regulation [52–54] (Figure 3.1). Moreover, the [4Fe–4S] clusters in various DNA polymerases (e.g. POLD1 and POLZ-Rev3) and a primase subunit (PRIM2) are important for DNA replication [55, 56].

Several of the cytosolic and nuclear Fe–S proteins are essential for cell viability because they are involved in fundamental cellular processes such as DNA synthesis (POLD1) or translation (ABCE1). These central functions in turn make the pathway of mitochondrial and cytosolic Fe–S protein biogenesis an indispensable function of the eukaryotic cell and of mitochondria. The importance of properly inserted Fe–S clusters in many cellular proteins explains why most of the ISC and CIA proteins including most of the Fe–S cluster-containing assembly factors (ISCU2, FDX2, CIAPIN1, NUBP1, NUBP2, and CIAO3) are essential for life, even for single-cell organisms such as baker's yeast (Figure 3.1).

3.3 Components and Mechanism of the Mitochondrial ISC System

3.3.1 Overview

The mitochondrial ISC system comprises up to 18 conserved proteins, most of which were inherited from the bacterial endosymbiotic ancestor of mitochondria [4, 12, 18, 57]. The ISC proteins operate in a sequential assembly line that, in a nutshell, can be dissected into three consecutive functional steps (Figure 3.2). In Step 1, a [2Fe–2S] cluster is synthesized *de novo* from ferrous iron (Fe^{2+}) and the sulfur of cysteine (Cys) on the scaffold protein ISCU2 (blue sector in Figure 3.2). Fe^{2+} ions are imported from the cytosol via the inner membrane carriers mitoferrin 1/2 (yeast Mrs3/4) in a proton-motif-force-dependent fashion. The synthesis of the [2Fe–2S] cluster is executed by the core ISC complex that centers around a dimer of the cysteine desulfurase complex NFS1–ISD11–ACP1 that binds to ISCU2 preloaded at its "cluster assembly site" with Fe^{2+}. NFS1 is a pyridoxal-phosphate (PLP)-dependent enzyme and abstracts sulfur from Cys to generate a protein-bound persulfide (-SSH) on a flexible C-terminal loop of NFS1 ("Cys-loop") and alanine as a side product. The Cys loop swinging from the center of the NFS1 dimer into the close vicinity of the

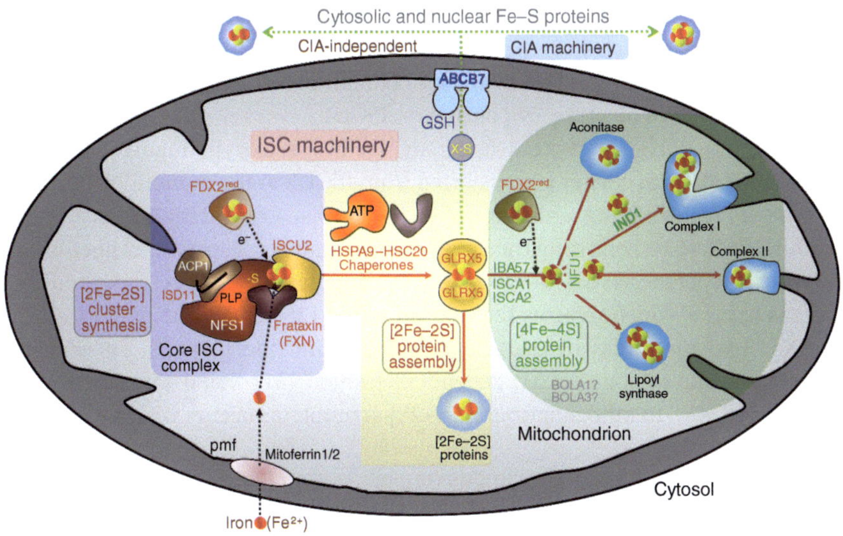

Figure 3.2 General overview of the assembly line of mitochondrial Fe–S proteins. The mitochondrial iron–sulfur cluster assembly (ISC) pathway can be divided into three main steps [12]. In the first step (blue sector), a [2Fe–2S] cluster (red and yellow circles) is synthesized *de novo* from ferrous iron (Fe^{2+}) and a cysteine-derived sulfide on the Fe–S cluster scaffold protein ISCU2. This multistage process is executed by the core ISC complex (see Figure 3.3a,b for structural and mechanistic details) centered around a dimer of the pyridoxal-phosphate (PLP)-dependent cysteine desulfurase complex NFS1–ISD11–ACP1, which binds ISCU2 at its tips (only half of the symmetric complex is shown here for simplicity). Sequential binding of frataxin (FXN) and electron (e^-) supply by reduced (red) mitochondrial ferredoxin FDX2 results in the formation of a [2Fe–2S] cluster on ISCU2. Fe^{2+} is imported into mitochondria by the mitochondrial carriers mitoferrin1/2 (yeast Mrs3/4) in a proton motif force (pmf)-dependent fashion [59]. In the second step (yellow sector), the [2Fe–2S] cluster is released from ISCU2 by assistance of dedicated Hsp40- and Hsp70 chaperones (HSC20, HSPA9) in an ATP-dependent fashion, followed by [2Fe–2S] cluster binding to monothiol glutaredoxin GLRX5. The transiently bound, bridging [2Fe–2S] cluster on dimeric GLRX5 is then inserted into mitochondrial [2Fe–2S] apoproteins without further assistance. All aforementioned ISC components are also required for the synthesis of X-S (see text) exported in an ABC transporter ABCB7- and glutathione (GSH)-dependent fashion to the cytosol for both the CIA-dependent formation of extramitochondrial [4Fe–4S] proteins and CIA-independent assembly of cytosolic [2Fe–2S] proteins. In the third step (green sector), [4Fe–4S] cluster assembly occurs on a complex of the late ISC proteins ISCA1, ISCA2, and IBA57 that receive two [2Fe–2S] clusters from the GLRX5 donor. These clusters are fused to the [4Fe–4S] type by reduction with FDX2, and this cluster is then targeted to mitochondrial [4Fe–4S] apoproteins like aconitase without further assistance. Other [4Fe–4S] proteins such as respiratory complexes I and II or the radical SAM protein lipoyl synthase require late-acting ISC targeting factors (NFU1, IND1), which transiently bind the trafficking [4Fe–4S] clusters. The exact positioning and function of BOLA1–BOLA3 in the late ISC pathway is unknown. Note that the first two assembly steps are essential for the biogenesis of all cellular Fe–S proteins, while step three is specific for mitochondrial [4Fe–4S] proteins.

ISCU2-bound iron, and binding of frataxin (FXN) then facilitates persulfide transfer to a conserved Cys of the cluster assembly site of ISCU2. Subsequent reduction of the ISCU2-bound persulfide to sulfide by the mitochondrial ferredoxin FDX2 results in the *de novo* generation of a [2Fe–2S] cluster on ISCU2.

In Step 2, i.e. [2Fe–2S] protein assembly (yellow sector in Figure 3.2), the ISCU2-bound [2Fe–2S] cluster is transferred from ISCU2 to the monothiol glutaredoxin GLRX5, requiring the assistance of dedicated Hsp40 and Hsp70 chaperones (HSC20, HSPA9) [60]. Following the general mechanism of Hsp40/70 chaperones, the J-type cochaperone HSC20 binds to ISCU2 and guides it to the ATP-dependent HSPA9 chaperone, which induces the transfer of the [2Fe–2S] cluster from ISCU2 to GLRX5. Two molecules of GLRX5 transiently bind the [2Fe–2S] clusters in a bridging and glutathione (GSH)-dependent fashion [61]. Final transfer of the cluster from GLRX5 to mitochondrial [2Fe–2S] apoproteins occurs without further assistance.

In Step 3, i.e. [4Fe–4S] protein assembly (green sector in Figure 3.2), the synthesis of [4Fe–4S] clusters is catalyzed on a complex of the late-acting ISC proteins ISCA1, ISCA2, and IBA57 receiving two [2Fe–2S] clusters from GLRX5 [62–65]. These transiently bound [2Fe–2S] clusters are then fused to a [4Fe–4S] cluster requiring another reductive action of FDX2 [66]. This is followed by cluster release from the late ISC complex and transfer to mitochondrial [4Fe–4S] apoproteins. This partial reaction can occur either without further assistance as in the case of aconitase or requires additional ISC targeting factors (NFU1 or IND1 aka NUBPL) that themselves transiently bind the [4Fe–4S] clusters during target-specific transfer and apoprotein insertion [67–70]. The precise function and location of the BOLA family proteins BOLA1 and BOLA3 within the biosynthetic pathway are not known, but they form a heterodimeric complex with GLRX5 and thereby can hold a bridging [2Fe–2S] cluster [71–73].

Intriguingly, the first two steps of the mitochondrial Fe–S protein biosynthesis pathway are also essential for the assembly of cytosolic and nuclear Fe–S proteins (Figure 3.2). This complex cytosolic assembly pathway depends on the function of the ABC transporter ABCB7 (yeast Atm1) and glutathione, assisting the export of the sulfur- and iron-containing substrate "X-S" whose molecular identity still has to be defined in detail. Maturation of [4Fe–4S] proteins in the cytosol (of both cytosolic and nuclear apoproteins) requires members of the complex CIA pathway [74, 75]. In contrast, most cytosolic [2Fe–2S] proteins are assembled in a CIA-independent fashion [19]. In the following, we will describe the molecular mechanisms underlying the three mitochondrial biogenesis steps in more detail.

3.3.2 Step 1: *De novo* [2Fe–2S] Cluster Formation on the Core ISC Complex

Despite the rather simple structure of [2Fe–2S] clusters, the initiating step of mitochondrial Fe–S cluster biosynthesis requires a complex, coordinated cooperation of seven ISC proteins. *De novo* [2Fe–2S] cluster synthesis takes place on the core ISC complex that centers around a dimer of the cysteine desulfurase NFS1, the largest component of the complex (Figure 3.3a) [76–78]. At the top of its dimer interface, the two NFS1 subunits tightly associate two copies of ISD11, a small triple-helix bundle protein, which in turn interacts with the acyl carrier protein ACP1 via both protein–protein interactions and a phospho-pantothenate-bound fatty acyl chain of 12–16 carbon units [76–78]. Binding of ISD11–ACP1 stabilizes NFS1 and thereby regulates its level, yet the absence of ACP1 has no major influence on the NFS1 catalytic function *per se* [76, 79]. During the various subreactions needed for [2Fe–2S] cluster synthesis, dimeric NFS1–ISD11–ACP1 (termed $(NIA)_2$) dynamically binds two molecules each of ISCU2, FXN, and FDX2 at two opposite sides of NFS1, thereby forming various ISC complexes (Figure 3.3b). The overall reaction has been biochemically and structurally resolved in some detail over the past few years and can be dissected into several subreactions as explained in the following paragraphs.

First, the cysteine desulfurase complex $(NIA)_2$ catalyzes the rapid PLP-dependent desulfuration of free Cys yielding free Ala and a persulfide (-SSH) in which the released sulfur is covalently bound to the catalytic residue $Cys381^{NFS1}$ (Figure 3.3b, step 1) [58, 80]. The strictly conserved and essential $Cys381^{NFS1}$ located on the flexible Cys loop dynamically can move over a distance of more than 20 Å from the persulfide-generating PLP-binding site inside the NFS1 dimer to the cluster assembly site of ISCU2 located at the tip of the NFS1 monomers, and involving the essential residues Cys69, Asp71, Cys95, His137, and Cys138 [58]. Biochemical experiments suggest that Fe^{2+} can apparently either bind to free ISCU2 or enter the dimeric NFS1–ISD11–ACP1–ISCU2 complex to form the $(Fe–N^{S}IAU)_2$ complex (the superscript S indicates a persulfide on NFS1; Figure 3.3b, step 2) [58, 81, 82]. In solution, free ISCU2 binds ferrous Fe^{2+} in a tetrahedral coordination via its invariant residues Cys69, Asp71, Cys95, and His137 [82]. Upon binding to the $(NIAU)_2$ complex (with and without bound SSH), the mode of iron binding by ISCU2 changes in that Fe^{2+} is now coordinated by the catalytic residue $Cys381^{NFS1}$ instead of $Cys69^{ISCU2}$, and $His137^{ISCU2}$ is replaced by water (Figure 3.3b, step 3) [58].

Next, FXN binds to the $(Fe\text{-}N^{S}IAU)_2$ complex and kinetically stimulates the transfer of the NFS1-bound persulfide to ISCU2 (Figure 3.3b, steps 4–5) [58, 78, 80, 83–85]. Cryo-EM structural and Mössbauer spectroscopic analyses of reaction intermediates under anaerobic conditions indicate that the transient binding of FXN to the $(Fe\text{-}N^{S}IAU)_2$ complex induces multiple conformational changes mainly at the cluster assembly site that together position the sulfur-accepting $Cys138^{ISCU2}$ in sufficiently close vicinity to the sulfur-donating $Cys381^{NFS1}$ thereby allowing efficient persulfide transfer. In fact, $Cys138^{ISCU2}$ moves from a 6.5 Å distance to $Cys381^{NFS1}$ in the $(Fe\text{-}NIAU)_2$ cryo-EM structure to a 3.6 Å distance in the $(Fe\text{-}N^{S}IAU^{S}X)_2$ structure [58]. After sulfur transfer from NFS1 to ISCU2, $Cys381^{NFS1}$ is replaced as an Fe^{2+}-coordinating ligand by water and the NFS1 Cys-loop can swing back to the PLP site for another round of rapid persulfidation. The fast rate of NFS1 persulfidation may explain why in a cryo-EM structure two persulfides were present leading to the $(Fe\text{-}N^{S}IAU^{S}X)_2$ complex [58]. A major driving force for the structural changes induced at the ISCU2 assembly site by FXN binding may be the movement of $His137^{ISCU2}$ into a partially hydrophobic pocket of FXN composed of $Trp155^{FXN}$ and $Pro163^{FXN}$ thereby moving the neighboring $Cys138^{ISCU2}$ [58].

In addition, this conformational change by FXN binding, i.e. sliding His137^{ISCU2} away from the ISCU2 assembly site, removes a potential steric hindrance for persulfide transfer from Cys381^{NFS1} to Cys138^{ISCU2}, which may further facilitate sulfur transfer. Overall, these structural insights define the molecular function of FXN in the entire process. No iron-related role of FXN, as previously discussed, is evident from these structures. It should further be mentioned that both sulfur transfer to Cys138^{ISCU2} and [2Fe–2S] cluster formation can occur independently of FXN, yet at an at least 10-fold slower rate [58]. In fact, Mössbauer spectroscopy has defined an equilibrium between the two conformational iron-binding states before and after sulfur transfer. This explains well why yeast Yfh1, as opposed to the other core ISC components, is not essential for viability in yeast [86, 87].

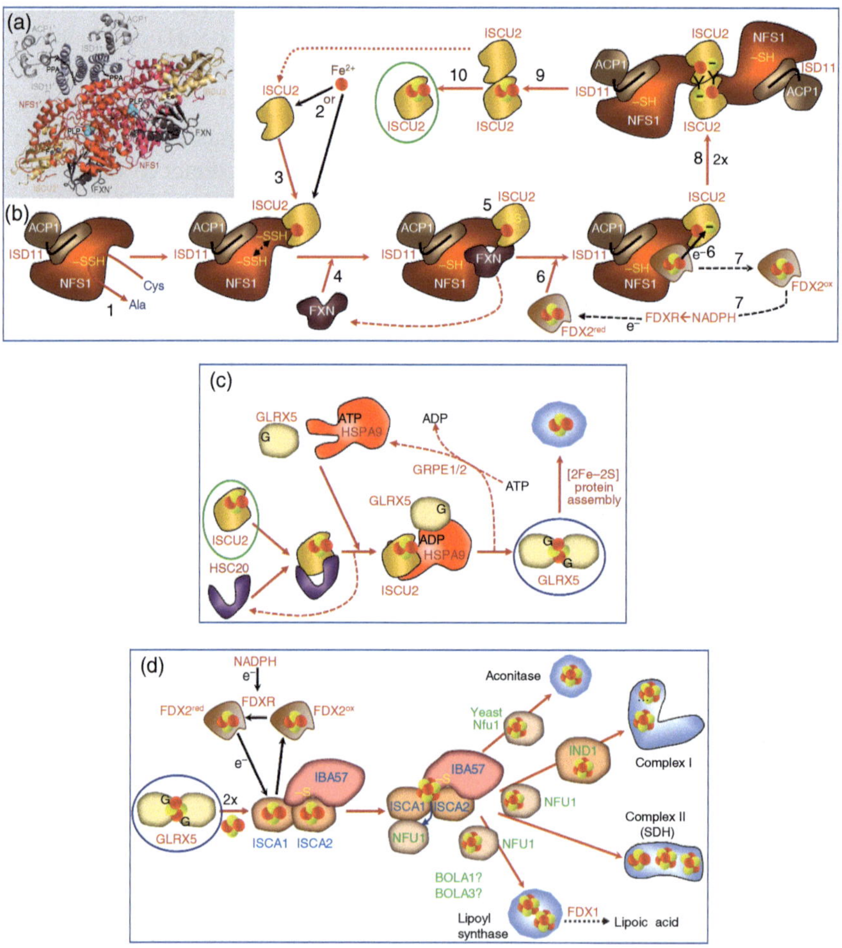

Figure 3.3 Models for the mechanisms of mitochondrial [2Fe–2S] and [4Fe–4S] protein biogenesis. (a) Cryo-EM structure of the dimeric mitochondrial core iron–sulfur cluster assembly (ISC) complex consisting of NFS1–ISD11–ACP1 with bound Fe^{2+}-ISCU2 and FXN (PDB 8PK8; [58]). The binding sites of pyridoxal phosphate (PLP) and the positions of the acyl-phosphopantetheinyl cofactor (PPA) are indicated. The ISCU2 cluster assembly site is identified by the bound iron (Fe). (b) Simplified model for the mechanism of *de novo* [2Fe–2S] cluster synthesis on the ISCU2 scaffold. (1) The cysteine desulfurase complex NFS1–ISD11–ACP1 ($(NIA)_2$; only one half of the dimer shown in part a is depicted here for simplicity) converts free cysteine (Cys) to alanine (Ala) and a $Cys381^{NFS1}$-bound persulfide (-SSH). (2,3) Iron (Fe^{2+}) binds to either free ISCU2 before its binding to $(NIA)_2$ or ISCU2 already bound to $(NIA)_2$. In the $(N^SIAU)_2$ complex, the persulfide can dynamically move between an inward position close to the PLP active site or an outward position close to the Fe^{2+} on ISCU2. (4) Transient FXN binding to $(Fe\text{-}N^SIAU)_2$ induces a conformational change in ISCU2 so that (5) the persulfide can be transferred to $Cys138^{ISCU2}$. (6) After FXN dissociation, reduction of the persulfide to sulfide by electron (e^-) transfer from reduced (red) FDX2 creates a putative [1Fe–1S] intermediate. The binding sites of FXN and FDX2 overlap requiring dissociation of FXN before FDX2 entry. (7) Oxidized (ox) FDX2 is reduced by NADPH and ferredoxin reductase (FDXR). (8) Two $([1Fe\text{–}1S]\text{-}NIAU)_2$ complexes transiently dimerize, facilitated by pi stacking interaction of the two N-terminal $Tyr35^{ISCU2}$ (Y). Note that in the 3D structure the two interacting Tyr35 residues are not in the direct vicinity of the [1Fe–1S] moieties. (9) The interaction of two $([1Fe\text{–}1S]\text{-}NIAU)_2$ complexes may bring their [1Fe–1S] units close enough to facilitate the formation of a [2Fe–2S] cluster by a yet unclear mechanism. (10) The [2Fe–2S] cluster is eventually bound to an ISCU2 monomer to become a target of the Hsp40/Hsp70 chaperones (part c). Phosphopantetheine with its attached acyl chain is represented by black lines on ISD11–ACP1. (c) Model for mitochondrial chaperone- and GLRX5-assisted [2Fe–2S] protein assembly. The Hsp40/DnaJ cochaperone HSC20 (yeast Jac1) transiently binds to holo-ISCU2 and transfers it to the ATP-dependent Hsp70/DnaK chaperone HSPA9 (yeast Ssq1). ATP hydrolysis induces binding of HSPA9 to a hydrophobic motif of ISCU2, an event proposed to eventually induce a conformational change in holo-ISCU2 that releases the [2Fe–2S] cluster. The cofactor can then be transferred to a dimer of GLRX5 (prebound to HSPA9) where it transiently binds via a conserved Cys residue of GLRX5 and a bound glutathione (G). Finally, [2Fe–2S] cluster transfer and insertion into target apoproteins may occur without further assistance. To close the cycle, the ATP-bound form of HSPA9 is regenerated by the nucleotide exchange factors GRPE1/2. (d) Model for the mechanism of mitochondrial [4Fe–4S] protein assembly. Two [2Fe–2S] clusters are transferred from GLRX5 to a complex of ISCA1-ISCA2-IBA57. The precise cluster-binding coordination is unknown. Reduction by FDX2 leads to the fusion of the two [2Fe–2S] clusters and generates the transiently bound [4Fe–4S] cluster, which then is either inserted into apoproteins spontaneously (in the case of human aconitase) or transferred to and transiently bound by the ISC targeting factor NFU1 which directly associates with ISCA1. Holo-NFU1 then facilitates the insertion of the [4Fe–4S] clusters into dedicated recipients. Respiratory complex I additionally and specifically requires the targeting factor IND1 for assembly of at least one of its six [4Fe–4S] clusters. The order of cluster assembly in complex I is unknown, as is that of the late ISC-dependent respiratory complex II (succinate dehydrogenase [SDH]) with [2Fe–2S], [3Fe–4S], and [4Fe–4S] clusters. The enzymatic function of lipoyl synthase further depends on ferredoxin FDX1 as a radical chain starter for lipoyl synthesis. The molecular function and positioning of BOLA1–BOLA3 proteins in the entire [4Fe–4S] protein assembly pathway are mechanistically not clear, as is the physiological role of their complex formation with GLRX5 (not shown).

In the following steps 6–7 (Figure 3.3b), the Cys138^{ISCU2}-bound persulfide is reduced to a sulfide by FDX2 (which, after its own oxidation, itself becomes re-reduced by action of the NADPH-dependent ferredoxin reductase FDXR; [81, 88]) creating a formal [1Fe–1S] unit on ISCU2. The molecular details for this reductive transformation are unknown, as is the coordination structure of the hypothetical [1Fe–1S] moiety. While it is clear that one electron comes from FDX2, another one may be contributed by Fe^{2+} to form Fe^{3+}. However, a repeated second electron donation by FDX2 is not excluded, because electron transfer reactions are typically fast. Recently resolved anaerobic cryo-EM structures of the core ISC complex with bound FDX2 in its reduced state show that FDX2 binds at an overlapping binding site to that of FXN, collectively suggesting that both proteins operate at consecutive steps during [2Fe–2S] cluster biosynthesis, which is consistent with previous biochemical studies [89, 90]. Different cryo-EM structures indicate that FDX2 binds in two conformations, termed "distal" and "proximal" according to the distance of the electron-donating [2Fe–2S] cluster of FDX2 to the ISCU2 cluster assembly site. Only in the proximal conformation, the distance between the FDX2 [2Fe–2S] cluster and the ISCU2 Fe–S cluster assembly site is close enough for rapid electron transfer (14 Å). In contrast, the distal conformer with a cluster-assembly site distance of 23 Å may not support rapid electron transfer. Binding in both conformations occurs via ionic interactions between an Arg-rich patch of NFS1 and negatively charged residues of helix F of FDX2 [89, 90]. Interestingly, the Arg-rich patch of NFS1 has formerly been suggested to function as a nuclear import signal but this motif is also present in bacterial IscS proteins, emphasizing its importance for function rather than cellular localization [91]. Notably, cytosolic versions of yeast and human NFS1 do not support cytosolic Fe–S protein biogenesis in the absence of the mitochondrial copies making a nuclear NFS1 function in Fe–S cluster assembly unlikely [92, 93].

An initial clue of how the final [2Fe–2S] cluster synthesis might be assisted was provided by recent insights into the molecular role of the strictly conserved N-terminal Tyr35^{ISCU2} residue of ISCU-like proteins. This residue is indispensable *in vivo* for function in *E. coli*, yeast, and human cells [94–96] and was found to be essential for enzymatic [2Fe–2S] cluster synthesis [96]. This finding was unexpected, because the residue is not part of the cluster assembly site, and, more importantly, is not required for chemical reconstitution of a [2Fe–2S] cluster [96]. Since the Tyr35^{ISCU2} residues of two $(NIAU)_2$ complexes can be efficiently cross-linked, it was hypothesized that their interaction might facilitate dimerization of two $(NIAU)_2$ complexes with bound [1Fe–1S] moieties and lead to [2Fe–2S] cluster synthesis on ISCU2 (Figure 3.3b, steps 8–10). This scenario is supported by the observation that a mixture of two ISCU2 mutant proteins with opposite charges instead of Tyr (Y35K and Y35D variants) could efficiently support enzymatic [2Fe–2S] cluster synthesis on ISCU2 [96]. Strikingly, the individual Y35K and Y35D variants were completely inactive in *de novo* [2Fe–2S] cluster formation. Experiments with tagged and untagged ISCU2 variants further suggested that two [1Fe–1S]-binding $(NIAU)_2$ complexes rather than a [1Fe–1S]-binding $(NIAU)_2$ complex plus free ISCU2 transiently dimerize to result in [2Fe–2S] cluster formation. Dimer formation might be rather transient, because the ISCU2 monomer can hold the [2Fe–2S] cluster well, and hence represents the end product. Intriguingly, the postulated cytosolic isoform of human ISCU1 lacks this conserved Tyr residue, explaining why it is not functional in cytosolic Fe–S protein assembly [96, 97].

The core ISC complex contains the small protein ISD11 and the acyl carrier protein ACP1, which is bound to ISD11 via direct protein contacts and a fatty acyl chain attached to 4′-phosphopantetheine (Figure 3.3a) [76–78]. No critical catalytic role in the synthesis of the [2Fe–2S] cluster on ISCU2 has been described for ACP1 and likely also not for ISD11. The latter protein is conserved in most eukaryotes, but absent in prokaryotes, and stabilizes the eukaryotic NFS1 structure [98–100]. For instance, yeast Nfs1 aggregates in the absence of Isd11, explaining why it was difficult to purify NFS1 proteins from *E. coli* without this subunit [101]. ISD11 belongs to the family of LYRM proteins that fold into a triple helical bundle [102, 103]. The first two residues of the (often degenerate) LYR (Leu–Tyr–Arg) motif stabilize the triple helical structure of these proteins by hydrophobic interactions, while the second and third residues contribute to the interaction with ACP1 [79, 104, 105]. The 4′-phosphopantetheine cofactor of ACP1 covalently links fatty acyl chains that are inserted into the triple helix of ISD11, and longer acyl chains increase the affinity of ACP1 to ISD11. These propensities intimately link the stability of the NFS1 dimer and thus its function in Fe–S cluster synthesis to the metabolic state of the cell. Short acyl chains produced in mitochondria with low carbon source supply result in ACP1 dissociation and destabilization of the core ISC complex, which dampens the rate of Fe–S protein biogenesis. Conversely, longer acyl chains (typically 12–16 CH_2 groups) lead to optimal ACP1-ISD11 binding, core ISC complex stabilization, and hence higher Fe–S cluster synthesis rates, thereby intimately linking the cellular energy load to the efficiency of *de novo* Fe–S cluster synthesis [79, 104, 106–108].

3.3.3 Step 2: Chaperone-Dependent Transfer of the ISCU2-Bound Cluster to [2Fe–2S] Apoproteins

The journey of [2Fe–2S] clusters from their origin after synthesis on the core ISC complex toward their final destination into [2Fe–2S] apoprotein involves two consecutive partial steps (Figure 3.3c). In the first, the [2Fe–2S] cluster is released from on ISCU2 and transiently inserted into GLRX5 (yeast Grx5), a class II (i.e. monothiol) glutaredoxin that functions as a versatile Fe–S cluster transfer protein. In the second step, the [2Fe–2S] cluster on GLRX5 is further mobilized and inserted into [2Fe–2S] target apoproteins [66]. Since ISCU2 binds its Fe–S cluster relatively tight, its release requires the action of a dedicated chaperone system of the DnaJ/DnaK (Hsp40/Hsp70) family [60, 109, 110]. In most eukaryotes, the J-type cochaperone HSC20 (yeast Jac1) is dedicated to mitochondrial Fe–S protein biogenesis and cooperates with the multipurpose mitochondrial Hsp70 chaperone (human HSPA9), which also functions in mitochondrial protein import and folding. Only a few fungi including *Saccharomyces cerevisiae* employ an ISC-specific Hsp70 (Ssq1 in *S. cerevisiae*) that arouse by gene duplication [60, 111]. In these fungi, the N-terminal J domain of the cochaperone Jac1 is truncated, which likely tunes the cochaperone toward a preferential interaction with the dedicated Hsp70 [109, 111, 112].

In *S. cerevisiae*, Fe–S cluster formation on Grx5 *in vivo* depends on all components of the core ISC complex and the Hsp40-Hsp70 chaperone system Jac1–Ssq1 [61, 113]. Low levels of yeast Grx5 result in impaired maturation of all cellular Fe–S proteins and lead to the accumulation of Fe–S clusters on the scaffold Isu1, a phenotype that is shared with cells depleted for the Hsp70 chaperone Ssq1 or its cochaperone Jac1 [113]. Apparently, Grx5 together with the dedicated chaperone system facilitates Fe–S cluster transfer from Isu1 to all Fe–S recipient apoproteins *in vivo* [114, 115].

The reaction cycle of the ISC chaperone system, which has been elucidated mainly by studies with yeast and bacteria, in many ways is reminiscent to Hsp40/70 chaperone action in protein folding. Since the mechanism has been reviewed expertly, we will only briefly discuss the topic [60, 116]. Initially, the DnaJ cochaperone binds to holo-ISCU2 at a specific hydrophobic motif (Figure 3.3c) [110, 117, 118]. Next, the HSC20 cochaperone–ISCU2 complex recruits the Hsp70 chaperone via specific Hsp40–Hsp70 interactions allowing ISCU2 to associate with the peptide binding domain of the Hsp70 chaperone via a conserved LPPVK motif of ISCU2. ISCU2 binding induces ATP hydrolysis on Hsp70 and DnaJ protein dissociation. This most likely decreases the binding affinity of the ISCU2-bound [2Fe–2S] cluster, thereby favoring its release and transfer to the Fe–S transfer protein GLRX5 (Figure 3.3c) [119]. In *S. cerevisiae,* the cluster recipient Grx5 binds to a second, still unknown part of Ssq1 in an ATP-independent manner [61]. Likely, the simultaneous binding of yeast holo-Isu1 and Grx5 facilitates rapid Fe–S cluster transfer. Finally, the canonical nucleotide exchange factors GRPE1/2 stimulate the exchange of the Hsp70-bound ADP for ATP, which induces opening of the Hsp70 peptide binding pocket and the dissociation of apo-ISCU2.

As evident from Figure 3.2, GLRX5 occupies an important central position in the ISC assembly line, connecting the early part of the ISC system that is dedicated to the *de novo* synthesis of [2Fe–2S] clusters with the later parts involving the generation of [4Fe–4S] proteins. Further, GLRX5 is required for the ABCB7-dependent export of X-S to the cytosol. In that sense, it is surprising that the *GRX5* gene in yeast is not essential for cell viability. Apparently, its Fe–S cluster transfer function can be bypassed, implying that the [2Fe–2S] clusters can be transferred directly from ISCU2 to downstream targets in a chaperone-dependent fashion, even though with lower efficiency.

GLRX5 is a monothiol glutaredoxin with an invariant monothiol CGFS active-site sequence motif and a typical thioredoxin-fold. The Fe–S cluster-coordinating holo-protein forms a symmetrical homo-dimer that accommodates a typical bridging [2Fe–2S] cluster. The cluster is coordinated by four thiol groups, two coming from the two active-site Cys residues and another two from the thiol groups of two GSH molecules that are noncovalently bound in a specific binding pocket [120–122] (Figures 3.2 and 3.3c). The unusual use of nonprotein ligands for cluster coordination most likely contributes to the mobile nature of the GLRX5-bound cluster [122, 123], and, consistent with this view, its rapid transfer to Fe–S apoproteins *in vitro* is well documented [64, 121, 122, 124–130]. Yet, as a note of caution, many *in vitro* studies were carried out in the presence of the artificial thiol-specific reductant dithiothreitol (DTT) that is a cytotoxic chemical and known to support nonphysiological Fe–S cluster-related reactions, including a rapid exchange of Fe ions or artificial Fe–S cluster assembly reactions [81, 83, 131, 132]. Successful *in vitro* Fe–S cluster transfer events therefore require the *in vivo* verification to show the physiological relevance and donor-target specificity of these reactions. Nevertheless, valuable arguments for the specificity of transfer reactions can also be acquired *in vitro*. In the presence of the physiological, thiol-specific GSH/dithiol glutaredoxin (e.g. GLRX1) reducing system, *in vitro* [2Fe–2S] cluster transfer from holo-GLRX5 to apo-FDX1, a physiological [2Fe–2S] target protein, is spontaneous and does not involve any other components [66]. This reaction is finished within less than 15 seconds, and thus is even much faster than a similar reaction in the presence of DTT. Moreover, Fe–S cluster transfer rates from GLRX5 to apo-FDX1 are higher by several orders of magnitude than those from other known mitochondrial ISC assembly factors that can bind Fe–S clusters (i.e. ISCU2, ISCA1, ISCA2, and NFU1). This striking kinetic difference and the donor specificity strongly argue for a physiologically relevant GLRX5 to FDX1 [2Fe–2S] cluster transfer and at the same time exclude a dedicated role of the other ISC factors as physiological donors for mitochondrial [2Fe–2S] proteins. Finally, these conclusions are in full agreement with *in vivo* findings [12, 62, 63], verifying the model of spontaneous cluster transfer from GLRX5 to targets as shown in Figure 3.3c.

3.3.4 Step 3: Reductive Fusion of [2Fe–2S] Clusters on the Late ISC Complex Generates [4Fe–4S] Clusters for Insertion into Mitochondrial Apoproteins

Most mitochondrial Fe–S proteins harbor the [4Fe–4S] cluster type, either alone (such as aconitase), in tandem (lipoyl synthase), or in various combinations with other Fe–S cluster types (respiratory complexes I and II). According to *in vivo* studies in human, pathogen, or yeast cells, the A-type ISC proteins ISCA1–ISCA2 (yeast Isa1-Isa2) and the interacting IBA57 (yeast Iba57) are important for the biogenesis of virtually all mitochondrial [4Fe–4S] proteins [62, 63, 133, 134]. This indicates that these three interacting proteins participate in a crucial biogenesis step that cannot be bypassed. Nevertheless, depletion of these three ISC proteins has no detrimental effect on the biogenesis of mitochondrial [2Fe–2S] proteins by the core ISC system, the ISC chaperones, and GLRX5 (see Sections 3.3.1 and 3.3.2). Conversely, mitochondrial [4Fe–4S] protein biogenesis strictly depends on these early-acting ISC components. Together, these dependencies suggested a model in which the ISCA1–ISCA2–IBA57 proteins act after the core ISC, chaperone, and GLRX5 components (Figure 3.2, green sector). This in turn suggested that the late-acting ISC factors cooperatively facilitate the conversion of the [2Fe–2S] cluster provided by the core ISC system into [4Fe–4S] clusters. The underlying molecular mechanisms have only partially been unraveled to date.

Several *in vivo* studies have identified additional proteins that impact the maturation of a specific subset of [4Fe–4S] proteins, which typically contain more than one Fe–S cluster (see Section 3.2) [12]. The ISC proteins NFU1 and IND1 were characterized to act as Fe–S cluster targeting factors, as they play a role in target-specific [4Fe–4S] cluster insertion into client proteins by operating after

ISCA1–ISCA2–IBA57 (Figures 3.2 and 3.3d) [69–71, 135–137]. Mutations in any of these human late-acting ISC factors cause the "multiple mitochondrial dysfunction syndromes" (MMDS) with severe, often lethal outcomes due to multiple simultaneous defects in mitochondrial [4Fe–4S] proteins, convincingly documenting the importance of these ISC factors for the healthy cell [16, 18, 67, 68, 138–141]. However, in yeasts, the late ISC factors are not essential for life, because they are not involved in the essential mitochondrial export of X-S to the CIA machinery for the assembly of indispensable Fe–S proteins such as ABCE1 involved in translation as well as DNA polymerases and helices required for genome maintenance [56, 142]. A pathogenic mutation was recently found in canine *IBA57* causing hereditary necrotizing myelopathy (HNM) in young Kooiker dogs, a disease characterized by a similar phenotype as in humans, namely progressive ataxia and paralysis [143]. Human IBA57 carrying the canine-derived mutation only partially restored the biochemical deficiency in mitochondrial [4Fe–4S] proteins in IBA57-depleted human cells, demonstrating the functional impairment of this mutant IBA57 protein, and extending the MMDS spectrum to animals.

A physiological *in vitro* reconstitution of the conversion of [2Fe–2S] into [4Fe–4S] clusters and their insertion into the apo-aconitase as a target protein allowed insights into the molecular mechanisms of the [4Fe-4S] cluster biosynthetic step and revealed the minimal set of ISC factors needed (Figure 3.3d) [66]. In these biochemical assays, GLRX5 served as the [2Fe–2S] cluster donor mimicking the *in vivo* situation. Consistent with *in vivo* findings in yeast and human cell culture [62, 63, 133], the reconstitution of aconitase maturation with the holo-GLRX5 cluster donor essentially required the simultaneous presence of ISCA1, ISCA2, and IBA57 [66]. This observation is in contrast to *in vivo* studies suggesting that only ISCA1 might be required in neuronal cells [144, 145], or the claim that these proteins can be bypassed completely [18, 146]. One possible explanation for these observations is that the RNAi knockdown system in primary neurons is not efficient enough and would require the complete knockout of the genes by CRISPR methods.

The *in vitro* reconstitution system for [4Fe–4S] cluster assembly on apo-aconitase further showed a strict requirement of the mitochondrial electron transfer chain NADPH–FDXR–FDX2 (Figure 3.3d) [66]. *In vivo*, the function of these proteins in the late step of mitochondrial iron–sulfur protein biogenesis is occluded by their earlier involvement in the synthesis of [2Fe–2S] clusters (Figure 3.3b). Intriguingly, other *in vitro* studies using only limited sets of the late ISC components also did not reveal this FDX2 requirement [64, 127, 128, 144, 145, 147]. What may be the biochemical explanation for these differences, and what may be the reason for the crucial input of reduction equivalents in this late ISC biosynthetic step? Formally, the fusion of oxidized low potential $[2Fe–2S]^{2+}$ clusters to a low potential $[4Fe–4S]^{2+/3+}$ cluster requires the input of one or two electrons. Reduction may reduce the electronic repulsion between the two $[2Fe–2S]^{2+}$ clusters during fusion [2]. *In vitro*, reductive [2Fe–2S] cluster fusion can artificially be mediated by DTT at reasonable efficiency, explaining why the physiological reductant FDX2 was not required in previous *in vitro* [4Fe–4S] cluster assembly protocols [64, 127, 128, 144, 145, 148–152]. The reductive conversion to a [4Fe–4S] cluster takes place on an ISCA1–ISCA2 heterodimer and further essentially involves

IBA57. Fusion further depends on the physical interaction with reduced FDX2 which cannot be replaced by the second human mitochondrial ferredoxin FDX1 that has been shown to be involved in other reductive functions [27, 66].

The assignment of ISCA1–ISCA2 as the Fe–S cluster fusion site is consistent with many studies that show that A-type ISC proteins from diverse organisms form dimeric or tetrameric complexes that can accommodate [2Fe–2S] and/or [4Fe–4S] clusters upon chemical reconstitution *in vitro* [64, 128, 145, 151–153]. These clusters are ligated by three conserved Cys residues in both ISCA1 and ISCA2 and can be transferred to apoproteins in the presence of DTT. Each of the conserved Cys residues is essential for *in vivo* function, but their precise coordination role during [2Fe–2S] cluster binding and the conversion reaction remains to be resolved, which possibly depends on physiologically relevant 3D structures of these protein complexes. IBA57 is critically involved in these reactions, and belongs to a family of folate-binding proteins, yet despite earlier claims its 3D structure and mutational studies show that this protein is not able to bind or utilize folate because important folate-binding residues have been altered [145, 154–156]. IBA57 forms a Fe–S cluster-binding complex with ISCA2 upon chemical Fe–S cluster reconstitution *in vitro*, which involves the conserved Cys residue within the signature motif of IBA57 that is essential for *in vivo* function (Figure 3.3d). ISCA1 in turn binds the ISC targeting factor NFU1, and both the ISCA1–NFU1 and ISCA2–IBA57 complexes are capable of transferring Fe–S clusters to apoproteins in the presence of DTT [65, 69, 145, 157]. Under conditions better mimicking the physiological situation, ISCA1, ISCA2, and IBA57 may bind two [2Fe–2S] clusters on a heterotrimer, which then facilitates the reductive synthesis of [4Fe–4S] clusters by electron input from NADPH–FDXR–FDX2 (Figure 3.3d). Details need to be worked out.

Final cluster transfer from ISCA1–ISCA2–IBA57 to apo-aconitase may occur independently of NFU1 or, depending on the organism, may be stimulated by NFU1 (Figures 3.2 and 3.3d). *In vivo,* human aconitase does not require any other late ISC factors besides ISCA1–ISCA2–IBA57 for its maturation, yet in yeast a deletion of *NFU1* is associated with a moderate aconitase defect [67, 68, 71, 72]. Consistently, the presence of the late-acting human NFU1 during *in vitro* reconstitution of apo-aconitase does not significantly stimulate its maturation [66]. Notably, the aconitase defect in Nfu1-deficient yeast cells may partially be explained by indirect effects caused by a Nfu1-related impairment of the citric acid cycle and respiration (Figure 3.3d). Deletion of yeast *NFU1* impairs the function of respiratory complex II and LIAS, which supplies pyruvate and ketoglutarate dehydrogenases (PDH and KGDH) with lipoic acid as a cofactor. These mitochondrial enzymes are also compromised in NFU1-depleted human cell culture and in NFU1-defective patient cells, but aconitase activity in these cells remains largely unaffected showing the NFU1 independence of human aconitase maturation [67, 158].

In the radical SAM [4Fe–4S] protein LIAS, the auxiliary [4Fe–4S] cluster functions as a sulfur donor for the lipoyl cofactor synthesis and thus needs constant repair or full replacement [26, 32, 33, 159]. Bacterial and human NFU1 were shown to function in this cluster regeneration step, a finding that would well explain the NFU1 deficiency phenotypes in bacteria and eukaryotes [31, 160]. Consistent with this, NFU1-like proteins readily assemble and transfer [4Fe–4S] clusters to some target apoproteins *in vitro* [152, 161–164].

A similar combined complex II plus LIAS deficiency was observed in human patient cells and yeast cells that lack the BolA family proteins BOLA1 and BOLA3 (yeast Bol1 and Bol3), suggesting that these two proteins may also operate at late steps during the formation of these complex Fe–S proteins [67, 71, 72]. In how far these observations are relevant for BOLA1/3 function as specialized ISC targeting factors remains to be elucidated. Mitochondrial BOLA1/3 proteins tightly interact with the monothiol glutaredoxin GLRX5 [71–73, 122, 165]. Why, in similarity to GLRX5 or ISCA–IBA57 proteins, the BOLA1/3 proteins do not have any impact on cytosolic Fe–S protein assembly or do not show a general [4Fe–4S] protein defect in mitochondria remains unclear. Resolving these mechanistic questions requires the elucidation of the molecular function of the BOLA proteins.

Finally, the function of the ISC targeting factor IND1 (aka NUBPL) appears to be confined to the insertion of one or several of the six [4Fe–4S] clusters of respiratory complex I [70, 135, 136, 166]. Consistently, its gene is missing in *S. cerevisiae*, *S. pombe*, and other yeasts that use type II NADH dehydrogenases lacking Fe–S centers. IND1 is a P-loop ATPase related to the CIA scaffold proteins NUBP1 and NUBP2 (yeast Nbp35 and Cfd1) that transiently bind an [4Fe–4S] cluster on an exposed loop between two monomers that can easily be mobilized [167]. While the role as a complex I-specific targeting factor of IND1 is clear, its precise role of IND1 in the assembly pathway of complex I and the order of events during the insertion of the Fe–S clusters into complex I are unknown.

3.4 Conclusions and Outlook

Since the discovery of eukaryotic Fe–S protein biogenesis 25 years ago, most if not all of the ISC and CIA factors and several associated proteins such as mitochondrial iron transporters have been identified and characterized by genetic and cell biological methods *in vivo*. These investigations suggested a model for the sequential action of these biogenesis proteins, which was corroborated by biochemical approaches including protein interaction studies. Biochemical and biophysical studies with purified proteins defined the mechanisms of some of the partial reactions of the complex overall pathway. These insights were expanded by the recent resolution of the 3D structures of some of the biogenesis components and complexes and unveiled further molecular details of the individual reactions. Nevertheless, there are many riddles of the molecular events that need to be resolved. Concerning mitochondrial Fe–S protein biogenesis, for instance it is unclear how the [2Fe–2S] cluster is precisely generated from a hypothetical [1Fe–1S] intermediate; what decisive chaperone-mediated structural changes underlie the release of the [2Fe–2S] cluster from ISCU2 scaffold; how the ISCA1, ISCA2, and IBA57 proteins exactly cooperate during the fusion of the [2Fe–2S] clusters; how the electron input from FDX2 stimulates the fusion process; how the [4Fe–4S] cluster transfer to the targeting factors occurs; and finally what features these ISC factors recognize in their target apoproteins for specific cluster delivery. Future experiments can use the power of cryo-EM analyses and combine it with modern biochemistry and spectroscopic methods to reveal further molecular details of how the Fe–S clusters are synthesized, trafficked, and inserted into apoproteins. Theoretical methods such as molecular dynamics approaches will provide another important avenue for increasing our mechanistic understanding of the complex biosynthetic events. Since mitochondrial Fe–S protein biogenesis is connected to numerous mitochondriopathies [15, 18, 168], a comprehensive mechanistic understanding of this essential and minimal biochemical process of mitochondria [4, 169, 170] is of general importance.

Acknowledgments

RL acknowledges generous financial support from Deutsche Forschungsgemeinschaft DFG (Koselleck award LI 415/6 and SPP 1927 LI 415/7), EU COST Action FeSImmChemNet, grant CA21115, and Volkswagen Foundation ("Life" program).

References

1 Beinert, H., Holm, R.H., and Münck, E. (1997). Iron–sulfur clusters: nature's modular, multipurpose structures. *Science* 277 (5326): 653–659.

2 Johnson, D.C., Dean, D.R., Smith, A.D., and Johnson, M.K. (2005). Structure, function and formation of biological iron–sulfur clusters. *Annu. Rev. Biochem.* 74: 247–281.

3 Zanello, P. (2019). Structure and electrochemistry of proteins harboring iron–sulfur clusters of different nuclearities. Part IV. Canonical, non-canonical and hybrid iron–sulfur proteins. *J. Struct. Biol.* 205 (2): 103–120.

4 Freibert, S.A., Goldberg, A.V., Hacker, C. et al. (2017). Evolutionary conservation and in vitro reconstitution of microsporidian iron–sulfur cluster biosynthesis. *Nat. Commun.* 8: 13932.

5 Garcia, P.S., D'Angelo, F., Ollagnier de Choudens, S. et al. (2022). An early origin of iron–sulfur cluster biosynthesis machineries before Earth oxygenation. *Nat. Ecol. Evol.* 6 (10): 1564–1572.

6 Braymer, J.J., Freibert, S.A., Rakwalska-Bange, M., and Lill, R. (2021). Mechanistic concepts of iron–sulfur protein biogenesis in biology. *Biochim. Biophys. Acta, Mol. Cell Res.* 1868 (1): 118863.

7 Przybyla-Toscano, J., Christ, L., Keech, O., and Rouhier, N. (2021). Iron–sulfur proteins in plant mitochondria: roles and maturation. *J. Exp. Bot.* 72 (6): 2014–2044.

8 Balk, J. and Schaedler, T.A. (2014). Iron cofactor assembly in plants. *Annu. Rev. Plant Biol.* 65: 125–153.

9 Pedroletti, L., Moseler, A., and Meyer, A.J. (2023). Assembly, transfer, and fate of mitochondrial iron–sulfur clusters. *J. Exp. Bot.* 74 (11): 3328–3344.

10 Gupta, M. and Outten, C.E. (2020). Iron-sulfur cluster signaling: the common thread in fungal iron regulation. *Curr. Opin. Chem. Biol.* 55: 189–201.

11 Misslinger, M., Hortschansky, P., Brakhage, A.A., and Haas, H. (2021). Fungal iron homeostasis with a focus on *Aspergillus fumigatus*. *Biochim. Biophys. Acta, Mol. Cell Res.* 1868 (1): 118885.

12 Lill, R. and Freibert, S.A. (2020). Mechanisms of mitochondrial iron–sulfur protein biogenesis. *Annu. Rev. Biochem.* 89: 471–499.

13 Ast, T., Itoh, Y., Sadre, S. et al. (2024). METTL17 is an Fe–S cluster checkpoint for mitochondrial translation. *Mol. Cell* 84 (2): 359–374.

14 Stehling, O., Mascarenhas, J., Vashisht, A.A. et al. (2013). Human CIA2A-FAM96A and CIA2B-FAM96B integrate iron homeostasis and maturation of different subsets of cytosolic-nuclear iron-sulfur proteins. *Cell Metab.* 18 (2): 187–198.

15 Stehling, O., Wilbrecht, C., and Lill, R. (2014). Mitochondrial iron–sulfur protein biogenesis and human disease. *Biochimie* 100: 61–77.

16 Stehling, O., Paul, V.D., Bergmann, J. et al. (2018). Biochemical analyses of human iron–sulfur protein biogenesis and of related diseases. *Methods Enzymol.* 599: 227–263.

17 Lill, R. and Stehling, O. (2018). Mechanisms of Fe–S protein biogenesis and related diseases. *HemaSphere.* 2 (S2): 76–79.

18 Maio, N. and Rouault, T.A. (2022). Mammalian iron sulfur cluster biogenesis and human diseases. *IUBMB Life* 74 (7): 705–714.

19 Braymer, J.J., Stehling, O., Stümpfig, M. et al. (2024). Requirements for the biogenesis of [2Fe–2S] proteins in the human and yeast cytosol. *Proc. Nat. Acad. Sci.* 121 (21): e2400740121.

20 Camprubi, E., Jordan, S.F., Vasiliadou, R., and Lane, N. (2017). Iron catalysis at the origin of life. *IUBMB Life* 69 (6): 373–381.

21 Huber, C. and Wächtershauser, G. (2006). alpha-Hydroxy and alpha-amino acids under possible Hadean, volcanic origin-of-life conditions. *Science* 314 (5799): 630–632.

22 Schwarz, G., Mendel, R.R., and Ribbe, M.W. (2009). Molybdenum cofactors, enzymes and pathways. *Nature* 460 (7257): 839–847.

23 Mendel, R.R. (2013). The molybdenum cofactor. *J. Biol. Chem.* 288 (19): 13165–13172.

24 Dutt, S., Hamza, I., and Bartnikas, T.B. (2022). Molecular mechanisms of iron and heme metabolism. *Annu. Rev. Nutr.* 42: 311–335.

25 Zanello, P. (2014). The competition between chemistry and biology in assembling iron–sulfur derivatives. Molecular structures and electrochemistry. Part II. {[Fe–2S2](S-Cys[gamma])(4)} proteins. *Coord. Chem. Rev.* 280: 54–83.

26 Schulz, V., Basu, S., Freibert, S.A. et al. (2023). Functional spectrum and specificity of mitochondrial ferredoxins FDX1 and FDX2. *Nat. Chem. Biol.* 19 (2): 206–217.

27 Schulz, V., Freibert, S.A., Boss, L. et al. (2023). Mitochondrial [2Fe–2S] ferredoxins: new functions for old dogs. *FEBS Lett.* 597 (1): 102–121.

28 Ewen, K.M., Kleser, M., and Bernhardt, R. (2011). Adrenodoxin: the archetype of vertebrate-type [2Fe–2S] cluster ferredoxins. *Biochim. Biophys. Acta* 1814 (1): 111–125.

29 Schonauer, M.S., Kastaniotis, A.J., Kursu, V.A. et al. (2009). Lipoic acid synthesis and attachment in yeast mitochondria. *J. Biol. Chem.* 284 (35): 23234–23242.

30 Kastaniotis, A.J., Autio, K.J., Keratar, J.M. et al. (2017). Mitochondrial fatty acid synthesis, fatty acids and mitochondrial physiology. *Biochim. Biophys. Acta Mol. Cell Biol. Lipids* 1862 (1): 39–48.

31 McCarthy, E.L. and Booker, S.J. (2017). Destruction and reformation of an iron–sulfur cluster during catalysis by lipoyl synthase. *Science* 358 (6361): 373–377.

32 McCarthy, E.L. and Booker, S.J. (2018). Biochemical approaches for understanding iron-sulfur cluster regeneration in *Escherichia coli* lipoyl synthase during catalysis. *Methods Enzymol.* 606: 217–239.

33 Landgraf, B.J., McCarthy, E.L., and Booker, S.J. (2016). Radical S-adenosylmethionine enzymes in human health and disease. *Annu. Rev. Biochem.* 85: 485–514.

34 Itoh, Y., Singh, V., Khawaja, A. et al. (2022). Structure of the mitoribosomal small subunit with streptomycin reveals Fe–S clusters and physiological molecules. *elife* 11: e77460.

35 Zhong, H., Janer, A., Khalimonchuk, O. et al. (2023). BOLA3 and NFU1 link mitoribosome iron-sulfur cluster assembly to multiple mitochondrial dysfunctions syndrome. *Nucleic Acids Res.* 51 (21): 11797–11812.

36 Boß, L., Stehling, O., Elsässer, H.-P. et al. (2025). Crucial role and conservation of the three [2Fe–2S] clusters in the human mitochondrial ribosome. *J. Biol. Chem.* 301: 108087. doi: 10.1016/j.jbc.2024.108087

37 Saveanu, C., Fromont-Racine, M., Harington, A. et al. (2001). Identification of 12 new yeast mitochondrial ribosomal proteins including 6 that have no prokaryotic homologues. *J. Biol. Chem.* 276 (19): 15861–15867.

38 Kimura, S. and Suzuki, T. (2015). Iron-sulfur proteins responsible for RNA modifications. *Biochim. Biophys. Acta* 1853 (6): 1272–1283.

39 Suzuki, Y., Noma, A., Suzuki, T. et al. (2007). Crystal structure of the radical SAM enzyme catalyzing tricyclic modified base formation in tRNA. *J. Mol. Biol.* 372 (5): 1204–1214.

40 Galy, B., Conrad, M., and Muckenthaler, M. (2024). Mechanisms controlling cellular and systemic iron homeostasis. *Nat. Rev. Mol. Cell Biol.* 25 (2): 133–155.

41 Berndt, C. and Lillig, C.H. (2017). Glutathione, glutaredoxins, and iron. *Antioxid. Redox Signal.* 27 (15): 1235–1251.

42 Ben-Shimon, L., Paul, V.D., David-Kadoch, G. et al. (2018). Fe–S cluster coordination of the chromokinesin KIF4A alters its subcellular localization during mitosis. *J. Cell Sci.* 131 (12): jcs211433.

43 Mancera-Martinez, E., Brito Querido, J., Valasek, L.S. et al. (2017). ABCE1: a special factor that orchestrates translation at the crossroad between recycling and initiation. *RNA Biol.* 14 (10): 1279–1285.

44 Netz, D.J., Stümpfig, M., Doré, C. et al. (2010). Tah18 transfers electrons to Dre2 in cytosolic iron-sulfur protein biogenesis. *Nat. Chem. Biol.* 6 (10): 758–765.

45 Zhang, Y., Yang, C., Dancis, A., and Nakamaru-Ogiso, E. (2017). EPR studies of wild type and mutant Dre2 identify essential [2Fe–2S] and [4Fe–4S] clusters and their cysteine ligands. *J. Biochem.* 161 (1): 67–78.

46 Li, H., Stumpfig, M., Zhang, C. et al. (2017). The diferric-tyrosyl radical cluster of ribonucleotide reductase and cytosolic iron-sulfur clusters have distinct and similar biogenesis requirements. *J. Biol. Chem.* 292 (27): 11445–11451.

47 Netz, D.J., Genau, H.M., Weiler, B.D. et al. (2016). The conserved protein Dre2 uses essential [2Fe–2S] and [4Fe–4S] clusters for its function in cytosolic iron-sulfur protein assembly. *Biochem. J.* 473 (14): 2073–2085.

48 Gizzi, A.S., Grove, T.L., Arnold, J.J. et al. (2018). A naturally occurring antiviral ribonucleotide encoded by the human genome. *Nature* 558 (7711): 610–614.

49 Maio, N., Lafont, B.A.P., Sil, D. et al. (2021). Fe–S cofactors in the SARS-CoV-2 RNA-dependent RNA polymerase are potential antiviral targets. *Science* 373 (6551): 236–241.

50 Maio, N., Raza, M.K., Li, Y. et al. (2023). An iron-sulfur cluster in the zinc-binding domain of the SARS-CoV-2 helicase modulates its RNA-binding and -unwinding activities. *Proc. Natl. Acad. Sci. USA* 120 (33): e2303860120.

51 Mittler, R., Darash-Yahana, M., Sohn, Y.S. et al. (2019). NEET proteins: a new link between iron metabolism, reactive oxygen species, and cancer. *Antioxid. Redox Signal.* 30 (8): 1083–1095.

52 Shi, R., Hou, W., Wang, Z.Q., and Xu, X. (2021). Biogenesis of iron-sulfur clusters and their role in DNA metabolism. *Front. Cell Dev. Biol.* 9: 735678.

53 Barton, J.K., Silva, R.M.B., and O'Brien, E. (2019). Redox chemistry in the genome: emergence of the [4Fe4S] cofactor in repair and replication. *Annu. Rev. Biochem.* 88: 163–190.

54 Fuss, J.O., Tsai, C.L., Ishida, J.P., and Tainer, J.A. (2015). Emerging critical roles of Fe–S clusters in DNA replication and repair. *Biochim. Biophys. Acta* 1853 (6): 1253–1271.

55 Baranovskiy, A.G., Siebler, H.M., Pavlov, Y.I., and Tahirov, T.H. (2018). Iron-sulfur clusters in DNA polymerases and primases of eukaryotes. *Methods Enzymol.* 599: 1–20.

56 Netz, D.J., Stith, C.M., Stumpfig, M. et al. (2012). Eukaryotic DNA polymerases require an iron-sulfur cluster for the formation of active complexes. *Nat. Chem. Biol.* 8 (1): 125–132.

57 Ciofi-Baffoni, S., Nasta, V., and Banci, L. (2018). Protein networks in the maturation of human iron-sulfur proteins. *Metallomics* 10 (1): 49–72.

58 Schulz, V., Steinhilper, R., Oltmanns, J. et al. (2024). Mechanism and structural dynamics of sulfur transfer during de novo [2Fe–2S] cluster assembly on ISCU2. *Nat. Commun.* 15 (1): 3269.

59 Christenson, E.T., Gallegos, A.S., and Banerjee, A. (2018). In vitro reconstitution, functional dissection, and mutational analysis of metal ion transport by mitoferrin-1. *J. Biol. Chem.* 293 (10): 3819–3828.

60 Dutkiewicz, R. and Nowak, M. (2018). Molecular chaperones involved in mitochondrial iron-sulfur protein biogenesis. *J. Biol. Inorg. Chem.* 23 (4): 569–579.

61 Uzarska, M.A., Dutkiewicz, R., Freibert, S.A. et al. (2013). The mitochondrial Hsp70 chaperone Ssq1 facilitates Fe–S cluster transfer from Isu1 to Grx5 by complex formation. *Mol. Biol. Cell* 24 (12): 1830–1841.

62 Mühlenhoff, U., Richter, N., Pines, O. et al. (2011). Specialized function of yeast Isa1 and Isa2 proteins in the maturation of mitochondrial [4Fe–4S] proteins. *J. Biol. Chem.* 286 (48): 41205–41216.

63 Sheftel, A.D., Wilbrecht, C., Stehling, O. et al. (2012). The human mitochondrial ISCA1, ISCA2, and IBA57 proteins are required for [4Fe–4S] protein maturation. *Mol. Biol. Cell* 23 (7): 1157–1166.

64 Nasta, V., Suraci, D., Gourdoupis, S. et al. (2020). A pathway for assembling [4Fe–4S](2+) clusters in mitochondrial iron-sulfur protein biogenesis. *FEBS J.* 287 (11): 2312–2327.

65 Suraci, D., Saudino, G., Nasta, V. et al. (2021). ISCA1 orchestrates ISCA2 and NFU1 in the maturation of human mitochondrial [4Fe–4S] proteins. *J. Mol. Biol.* 433 (10): 166924.

66 Weiler, B.D., Bruck, M.C., Kothe, I. et al. (2020). Mitochondrial [4Fe–4S] protein assembly involves reductive [2Fe–2S] cluster fusion on ISCA1-ISCA2 by electron flow from ferredoxin FDX2. *Proc. Natl. Acad. Sci. USA* 117 (34): 20555–20565.

67 Cameron, J.M., Janer, A., Levandovskiy, V. et al. (2011). Mutations in iron-sulfur cluster scaffold genes NFU1 and BOLA3 cause a fatal deficiency of multiple respiratory chain and 2-oxoacid dehydrogenase enzymes. *Am. J. Hum. Genet.* 89 (4): 486–495.

68 Navarro-Sastre, A., Tort, F., Stehling, O. et al. (2011). A fatal mitochondrial disease is associated with defective NFU1 function in the maturation of a subset of mitochondrial Fe–S proteins. *Am. J. Hum. Genet.* 89 (5): 656–667.

69 Jain, A., Singh, A., Maio, N., and Rouault, T.A. (2020). Assembly of the [4Fe–4S] cluster of NFU1 requires the coordinated donation of two [2Fe–2S] clusters from the scaffold proteins, ISCU2 and ISCA1. *Hum. Mol. Genet.* 29 (19): 3165–3182.

70 Sheftel, A.D., Stehling, O., Pierik, A.J. et al. (2009). Human ind1, an iron-sulfur cluster assembly factor for respiratory complex I. *Mol. Cell. Biol.* 29 (22): 6059–6073.

71 Melber, A., Na, U., Vashisht, A. et al. (2016). Role of Nfu1 and Bol3 in iron-sulfur cluster transfer to mitochondrial clients. *eLife* 5: e15991.

72 Uzarska, M.A., Nasta, V., Weiler, B.D. et al. (2016). Mitochondrial Bol1 and Bol3 function as assembly factors for specific iron-sulfur proteins. *eLife* 5: e16673.

73 Nasta, V., Giachetti, A., Ciofi-Baffoni, S., and Banci, L. (2017). Structural insights into the molecular function of human [2Fe–2S] BOLA1-GRX5 and [2Fe–2S] BOLA3-GRX5 complexes. *Biochim. Biophys. Acta Gen. Subj.* 1861 (8): 2119–2131.

74 Netz, D.J., Mascarenhas, J., Stehling, O. et al. (2014). Maturation of cytosolic and nuclear iron-sulfur proteins. *Trends Cell Biol.* 24 (5): 303–312.

75 Dancis, A., Pandey, A.K., and Pain, D. (2024). Mitochondria function in cytoplasmic Fe–S protein biogenesis. *Biochim Biophys Acta Mol. Cell Res.* 1871 (5): 119733.

76 Boniecki, M.T., Freibert, S.A., Muhlenhoff, U. et al. (2017). Structure and functional dynamics of the mitochondrial Fe–S cluster synthesis complex. *Nat. Commun.* 8 (1): 1287.

77 Cai, K., Frederick, R.O., Dashti, H., and Markley, J.L. (2018). Architectural features of human mitochondrial cysteine desulfurase complexes from crosslinking mass spectrometry and small-angle X-ray scattering. *Structure* 26 (8): 1127–1136.

78 Fox, N.G., Yu, X., Feng, X. et al. (2019). Structure of the human frataxin-bound iron-sulfur cluster assembly complex provides insight into its activation mechanism. *Nat. Commun.* 10 (1): 2210.

79 Van Vranken, J.G., Jeong, M.Y., Wei, P. et al. (2016). The mitochondrial acyl carrier protein (ACP) coordinates mitochondrial fatty acid synthesis with iron sulfur cluster biogenesis. *eLife* 5: e17828.

80 Patra, S. and Barondeau, D.P. (2019). Mechanism of activation of the human cysteine desulfurase complex by frataxin. *Proc. Natl. Acad. Sci. USA* 116 (39): 19421–19430.

81 Webert, H., Freibert, S.A., Gallo, A. et al. (2014). Functional reconstitution of mitochondrial Fe–S cluster synthesis on Isu1 reveals the involvement of ferredoxin. *Nat. Commun.* 5: 5013.

82 Srour, B., Gervason, S., Hoock, M.H. et al. (2022). Iron insertion at the assembly site of the ISCU Scaffold protein is a conserved process initiating Fe–S cluster biosynthesis. *J. Am. Chem. Soc.* 144 (38): 17496–17515.

83 Gervason, S., Larkem, D., Mansour, A.B. et al. (2019). Physiologically relevant reconstitution of iron-sulfur cluster biosynthesis uncovers persulfide-processing functions of ferredoxin-2 and frataxin. *Nat. Commun.* 10 (1): 3566.

84 Fox, N.G., Das, D., Chakrabarti, M. et al. (2015). Frataxin accelerates [2Fe–2S] cluster formation on the human Fe–S assembly complex. *Biochemistry* 54 (25): 3880–3889.

85 Parent, A., Elduque, X., Cornu, D. et al. (2015). Mammalian frataxin directly enhances sulfur transfer of NFS1 persulfide to both ISCU and free thiols. *Nat. Commun.* 6: 5686.

86 Babcock, M., de Silva, D., Oaks, R. et al. (1997). Regulation of mitochondrial iron accumulation by Yfh1p, a putative homolog of frataxin. *Science* 276 (5319): 1709–1712.

87 Mühlenhoff, U., Richhardt, N., Ristow, M. et al. (2002). The yeast frataxin homolog Yfh1p plays a specific role in the maturation of cellular Fe–S proteins. *Hum. Mol. Genet.* 11 (17): 2025–2036.

88 Shi, Y., Ghosh, M., Kovtunovych, G. et al. (2012). Both human ferredoxins 1 and 2 and ferredoxin reductase are important for iron-sulfur cluster biogenesis. *Biochim. Biophys. Acta* 1823 (2): 484–492.

89 Steinhilper, R., Boß, L., Freibert, S.-A. et al. (2024). Structural evidence for two-stage binding of mitochondrial ferredoxin 2 to the core iron-sulfur cluster assembly complex. *Nat. Commun.* 15: 10559.

90 Uzarska, M.A., Grochowina, I., Soldek, J. et al. (2022). During Fe–S cluster biogenesis, ferredoxin and frataxin use overlapping binding sites on yeast cysteine desulfurase Nfs1. *J. Biol. Chem.* 298 (2): 101570.

91 Nakai, Y., Nakai, M., Hayashi, H., and Kagamiyama, H. (2001). Nuclear localization of Yeast Nfs1p is required for cell survival. *J. Biol. Chem.* 276: 8314–8320.

92 Kispal, G., Csere, P., Prohl, C., and Lill, R. (1999). The mitochondrial proteins Atm1p and Nfs1p are essential for biogenesis of cytosolic Fe–S proteins. *EMBO J.* 18 (14): 3981–3989.

93 Biederbick, A., Stehling, O., Rösser, R. et al. (2006). Role of human mitochondrial Nfs1 in cytosolic iron-sulfur protein biogenesis and iron regulation. *Mol. Cell. Biol.* 26: 5675–5687.

94 Tanaka, N., Yuda, E., Fujishiro, T. et al. (2019). Identification of IscU residues critical for de novo iron-sulfur cluster assembly. *Mol. Microbiol.* 112 (6): 1769–1783.

95 Gerber, J., Neumann, K., Prohl, C. et al. (2004). The yeast scaffold proteins Isu1p and Isu2p are required inside mitochondria for maturation of cytosolic Fe–S proteins. *Mol. Cell. Biol.* 24: 4848–4857.

96 Freibert, S.A., Boniecki, M.T., Stumpfig, C. et al. (2021). N-terminal tyrosine of ISCU2 triggers [2Fe–2S] cluster synthesis by ISCU2 dimerization. *Nat. Commun.* 12 (1): 6902.

97 Crooks, D.R., Jeong, S.Y., Tong, W.H. et al. (2012). Tissue specificity of a human mitochondrial disease: differentiation-enhanced mis-splicing of the Fe–S scaffold gene ISCU renders patient cells more sensitive to oxidative stress in ISCU myopathy. *J. Biol. Chem.* 287 (48): 40119–40130.

98 Adam, A.C., Bornhövd, C., Prokisch, H. et al. (2006). The Nfs1 interacting protein Isd11 has an essential role in Fe–S cluster biogenesis in mitochondria. *EMBO J.* 25: 174–183.

99 Wiedemann, N., Urzica, E., Guiard, B. et al. (2006). Essential role of Isd11 in mitochondrial iron-sulfur cluster synthesis on Isu scaffold proteins. *EMBO J.* 25 (1): 184–195.

100 Lim, S.C., Friemel, M., Marum, J.E. et al. (2013). Mutations in LYRM4, encoding iron-sulfur cluster biogenesis factor ISD11, cause deficiency of multiple respiratory chain complexes. *Hum. Mol. Genet.* 22 (22): 4460–4473.

101 Mühlenhoff, U., Balk, J., Richhardt, N. et al. (2004). Functional characterization of the eukaryotic cysteine desulfurase Nfs1p from *Saccharomyces cerevisiae*. *J. Biol. Chem.* 279: 36906–36915.

102 Angerer, H. (2013). The superfamily of mitochondrial complex1_LYR motif-containing (LYRM) proteins. *Biochem. Soc. Trans.* 41 (5): 1335–1341.

103 Angerer, H., Schonborn, S., Gorka, J. et al. (2017). Acyl modification and binding of mitochondrial ACP to multiprotein complexes. *Biochim. Biophys. Acta, Mol. Cell Res.* 1864 (10): 1913–1920.

104 Van Vranken, J.G., Nowinski, S.M., Clowers, K.J. et al. (2018). ACP acylation is an acetyl-CoA-dependent modification required for electron transport chain assembly. *Mol. Cell* 71 (4): 567–580.

105 Lill, R. (2020). Do Fe–S clusters rule bacterial iron regulation? *J. Biol. Chem.* 295 (46): 15464–15465.

106 Masud, A.J., Kastaniotis, A.J., Rahman, M.T. et al. (2019). Mitochondrial acyl carrier protein (ACP) at the interface of metabolic state sensing and mitochondrial function. *Biochim. Biophys. Acta, Mol. Cell Res.* 1866 (12): 118540.

107 Nowinski, S.M., Van Vranken, J.G., Dove, K.K., and Rutter, J. (2018). Impact of mitochondrial fatty acid synthesis on mitochondrial biogenesis. *Curr. Biol.* 28 (20): R1212–R1219.

108 Majmudar, J.D., Feng, X., Fox, N.G. et al. (2019). 4′-Phosphopantetheine and long acyl chain-dependent interactions are integral to human mitochondrial acyl carrier protein function. *MedChemComm.* 10 (2): 209–220.

109 Craig, E.A. and Marszalek, J. (2017). How do J-proteins get Hsp70 to do so many different things? *Trends Biochem. Sci.* 42 (5): 355–368.

110 Majewska, J., Ciesielski, S.J., Schilke, B. et al. (2013). Binding of the chaperone Jac1 protein and cysteine desulfurase Nfs1 to the iron-sulfur cluster scaffold Isu protein is mutually exclusive. *J. Biol. Chem.* 288 (40): 29134–29142.

111 Delewski, W., Paterkiewicz, B., Manicki, M. et al. (2016). Iron-sulfur cluster biogenesis chaperones: evidence for emergence of mutational robustness of a highly specific protein–protein interaction. *Mol. Biol. Evol.* 33 (3): 643–656.

112 Pukszta, S., Schilke, B., Dutkiewicz, R. et al. (2010). Co-evolution-driven switch of J-protein specificity towards an Hsp70 partner. *EMBO Rep.* 11 (5): 360–365.

113 Mühlenhoff, U., Gerber, J., Richhardt, N., and Lill, R. (2003). Components involved in assembly and dislocation of iron-sulfur clusters on the scaffold protein Isu1p. *EMBO J.* 22: 4815–4825.

114 Rodriguez-Manzaneque, M.T., Tamarit, J., Belli, G. et al. (2002). Grx5 is a mitochondrial glutaredoxin required for the activity of iron/sulfur enzymes. *Mol. Biol. Cell* 13 (4): 1109–1121.

115 Ye, H., Jeong, S.Y., Ghosh, M.C. et al. (2010). Glutaredoxin 5 deficiency causes sideroblastic anemia by specifically impairing heme biosynthesis and depleting cytosolic iron in human erythroblasts. *J. Clin. Invest.* 120 (5): 1749–1761.

116 Kampinga, H.H. and Craig, E.A. (2010). The HSP70 chaperone machinery: J proteins as drivers of functional specificity. *Nat. Rev. Mol. Cell Biol.* 11 (8): 579–592.

117 Ciesielski, S.J., Schilke, B.A., Osipiuk, J. et al. (2012). Interaction of J-protein co-chaperone Jac1 with Fe–S scaffold Isu is indispensable in vivo and conserved in evolution. *J. Mol. Biol.* 417 (1–2): 1–12.

118 Dutkiewicz, R., Schilke, B., Cheng, S. et al. (2004). Sequence-specific interaction between mitochondrial Fe–S scaffold protein Isu and Hsp70 Ssq1 is essential for their *in vivo* function. *J. Biol. Chem.* 279 (28): 29167–29174.

119 Bonomi, F., Iametti, S., Morleo, A. et al. (2008). Studies on the mechanism of catalysis of iron-sulfur cluster transfer from IscU[2Fe2S] by HscA/HscB chaperones. *Biochemistry* 47 (48): 12795–12801.

120 Johansson, C., Roos, A.K., Montano, S.J. et al. (2011). The crystal structure of human GLRX5: iron-sulfur cluster co-ordination, tetrameric assembly and monomer activity. *Biochem. J.* 433 (2): 303–311.

121 Iwema, T., Picciocchi, A., Traore, D.A. et al. (2009). Structural basis for delivery of the intact [Fe2S2] cluster by monothiol glutaredoxin. *Biochemistry* 48 (26): 6041–6043.

122 Berndt, C., Christ, L., Rouhier, N., and Mühlenhoff, U. (2021). Glutaredoxins with iron-sulphur clusters in eukaryotes – structure, function and impact on disease. *BBA-Bioenergetics* 1862 (1): 148317.

123 Lillig, C.H., Berndt, C., and Holmgren, A. (2008). Glutaredoxin systems. *Biochim. Biophys. Acta* 1780 (11): 1304–1317.

124 Jia, M., Sen, S., Wachnowsky, C. et al. (2020). Characterization of [2Fe–2S]-cluster-bridged protein complexes and reaction intermediates by use of native mass spectrometric methods. *Angew. Chem. Int. Ed. Engl.* 59 (17): 6724–6728.

125 Moseler, A., Aller, I., Wagner, S. et al. (2015). The mitochondrial monothiol glutaredoxin S15 is essential for iron-sulfur protein maturation in *Arabidopsis thaliana*. *Proc. Natl. Acad. Sci. USA* 112 (44): 13735–13740.

126 Mapolelo, D.T., Zhang, B., Randeniya, S. et al. (2013). Monothiol glutaredoxins and A-type proteins: partners in Fe–S cluster trafficking. *Dalton Trans.* 42 (9): 3107–3115.

127 Brancaccio, D., Gallo, A., Mikolajczyk, M. et al. (2014). Formation of [4Fe–4S] clusters in the mitochondrial iron-sulfur cluster assembly machinery. *J. Am. Chem. Soc.* 136 (46): 16240–16250.

128 Banci, L., Brancaccio, D., Ciofi-Baffoni, S. et al. (2014). [2Fe–2S] cluster transfer in iron-sulfur protein biogenesis. *Proc. Natl. Acad. Sci. USA* 111 (17): 6203–6208.

129 Bandyopadhyay, S., Gama, F., Molina-Navarro, M.M. et al. (2008). Chloroplast monothiol glutaredoxins as scaffold proteins for the assembly and delivery of [2Fe–2S] clusters. *EMBO J.* 27 (7): 1122–1133.

130 Rouhier, N., Couturier, J., Johnson, M.K., and Jacquot, J.P. (2010). Glutaredoxins: roles in iron homeostasis. *Trends Biochem. Sci.* 35 (1): 43–52.

131 Gokul, G. and Singh, J. (2022). Dithiothreitol causes toxicity in *C. elegans* by modulating the methionine-homocysteine cycle. *eLife* 11: e76021.

132 Bridwell-Rabb, J., Fox, N.G., Tsai, C.L. et al. (2014). Human frataxin activates Fe–S cluster biosynthesis by facilitating sulfur transfer chemistry. *Biochemistry* 53 (30): 4904–4913.

133 Gelling, C., Dawes, I.W., Richhardt, N. et al. (2008). Mitochondrial Iba57p is required for Fe–S cluster formation on aconitase and activation of radical SAM enzymes. *Mol. Cell. Biol.* 28: 1851–1861.

134 Long, S., Changmai, P., Tsaousis, A.D. et al. (2011). Stage-specific requirement for Isa1 and Isa2 proteins in the mitochondrion of *Trypanosoma brucei* and heterologous rescue by human and Blastocystis orthologues. *Mol. Microbiol.* 81 (6): 1403–1418.

135 Bych, K., Kerscher, S., Netz, D.J. et al. (2008). The iron-sulphur protein Ind1 is required for effective complex I assembly. *EMBO J.* 27 (12): 1736–1746.

136 Calvo, S.E., Tucker, E.J., Compton, A.G. et al. (2010). High-throughput, pooled sequencing identifies mutations in NUBPL and FOXRED1 in human complex I deficiency. *Nat. Genet.* 42 (10): 851–858.

137 Cai, K., Liu, G., Frederick, R.O. et al. (2016). Structural/functional properties of human NFU1, an intermediate [4Fe–4S] carrier in human mitochondrial iron-sulfur cluster biogenesis. *Structure* 24 (12): 2080–2091.

138 Ajit Bolar, N., Vanlander, A.V., Wilbrecht, C. et al. (2013). Mutation of the iron-sulfur cluster assembly gene IBA57 causes severe myopathy and encephalopathy. *Hum. Mol. Genet.* 22 (13): 2590–2602.

139 Al-Hassnan, Z.N., Al-Dosary, M., Alfadhel, M. et al. (2015). ISCA2 mutation causes infantile neurodegenerative mitochondrial disorder. *J. Med. Genet.* 52 (3): 186–194.

140 Torraco, A., Stehling, O., Stumpfig, C. et al. (2018). ISCA1 mutation in a patient with infantile-onset leukodystrophy causes defects in mitochondrial [4Fe–4S] proteins. *Hum. Mol. Genet.* 27 (15): 2739–2754.

141 Lebigot, E., Schiff, M., and Golinelli-Cohen, M.P. (2021). A review of multiple mitochondrial dysfunction syndromes, syndromes associated with defective Fe–S protein maturation. *Biomedicines.* 9 (8): 989.

142 Kispal, G., Sipos, K., Lange, H. et al. (2005). Biogenesis of cytosolic ribosomes requires the essential iron-sulphur protein Rli1p and mitochondria. *EMBO J.* 24: 589–598.

143 Mandigers, P.J.J., Stehling, O., Vos-Loohuis, M. et al. (2023). A novel IBA57 variant is associated with mitochondrial iron-sulfur protein deficiency and necrotizing myelopathy in dogs. *Front. Genet.* 14: 1190222.

144 Beilschmidt, L.K., Ollagnier de Choudens, S., Fournier, M. et al. (2017). ISCA1 is essential for mitochondrial Fe4S4 biogenesis in vivo. *Nat. Commun.* 8: 15124.

145 Gourdoupis, S., Nasta, V., Calderone, V. et al. (2018). IBA57 recruits ISCA2 to form a [2Fe–2S] cluster-mediated complex. *J. Am. Chem. Soc.* 140 (43): 14401–14412.

146 Maio, N., Singh, A., Uhrigshardt, H. et al. (2014). Cochaperone binding to LYR motifs confers specificity of iron sulfur cluster delivery. *Cell Metab.* 19 (3): 445–457.

147 Kim, K.D., Chung, W.H., Kim, H.J. et al. (2010). Monothiol glutaredoxin Grx5 interacts with Fe–S scaffold proteins Isa1 and Isa2 and supports Fe–S assembly and DNA integrity in mitochondria of fission yeast. *Biochem. Biophys. Res. Commun.* 392 (3): 467–472.

148 Gao, H., Subramanian, S., Couturier, J. et al. (2013). Arabidopsis thaliana Nfu2 accommodates [2Fe–2S] or [4Fe–4S] clusters and is competent for in vitro maturation of chloroplast [2Fe–2S] and [4Fe–4S] cluster-containing proteins. *Biochemistry* 52 (38): 6633–6645.

149 Zhang, B., Bandyopadhyay, S., Shakamuri, P. et al. (2013). Monothiol glutaredoxins can bind linear [Fe3S4]+ and [Fe4S4]2+ clusters in addition to

[Fe2S2]2+ clusters: spectroscopic characterization and functional implications. *J. Am. Chem. Soc.* 135 (40): 15153–15164.

150 Unciuleac, M.C., Chandramouli, K., Naik, S. et al. (2007). In vitro activation of apo-aconitase using a [4Fe–4S] cluster-loaded form of the IscU [Fe–S] cluster scaffolding protein. *Biochemistry* 46 (23): 6812–6821.

151 Roche, B., Aussel, L., Ezraty, B. et al. (2013). Iron/sulfur proteins biogenesis in prokaryotes: formation, regulation and diversity. *Biochim. Biophys. Acta* 1827 (3): 455–469.

152 Wachnowsky, C., Fidai, I., and Cowan, J.A. (2018). Iron-sulfur cluster biosynthesis and trafficking – impact on human disease conditions. *Metallomics* 10 (1): 9–29.

153 Jensen, L.T. and Culotta, V.C. (2000). Role of *Saccharomyces cerevisiae ISA1* and *ISA2* in iron homeostasis. *Mol. Cell. Biol.* 20: 3918–3927.

154 Waller, J.C., Alvarez, S., Naponelli, V. et al. (2010). A role for tetrahydrofolates in the metabolism of iron-sulfur clusters in all domains of life. *Proc. Natl. Acad. Sci. USA* 107 (23): 10412–10417.

155 Hasnain, G., Waller, J.C., Alvarez, S. et al. (2012). Mutational analysis of YgfZ, a folate-dependent protein implicated in iron/sulphur cluster metabolism. *FEMS Microbiol. Lett.* 326 (2): 168–172.

156 Muhlenhoff, U., Weiler, B.D., Nadler, F. et al. (2022). The iron-sulfur cluster assembly (ISC) protein Iba57 executes a tetrahydrofolate-independent function in mitochondrial [4Fe–4S] protein maturation. *J. Biol. Chem.* 298 (10): 102465.

157 Da Vela, S., Saudino, G., Lucarelli, F. et al. (2023). Structural plasticity of NFU1 upon interaction with binding partners: insights into the mitochondrial [4Fe–4S] cluster pathway. *J. Mol. Biol.* 435 (15): 168154.

158 Baker, P.R. 2nd, Friederich, M.W., Swanson, M.A. et al. (2014). Variant non ketotic hyperglycinemia is caused by mutations in LIAS, BOLA3 and the novel gene GLRX5. *Brain* 137 (Pt 2): 366–379.

159 McCarthy, E.L., Rankin, A.N., Dill, Z.R., and Booker, S.J. (2019). The A-type domain in *Escherichia coli* NfuA is required for regenerating the auxiliary [4Fe–4S] cluster in *Escherichia coli* lipoyl synthase. *J. Biol. Chem.* 294 (5): 1609–1617.

160 Warui, D.M., Sil, D., Lee, K.-H. et al. In vitro demonstration of human lipoyl synthase catalytic activity in the presence of NFU1. *ACS Bio. Med. Chem.* 2 (5): 456–468.

161 Mapolelo, D.T., Zhang, B., Naik, S.G. et al. (2012). Spectroscopic and functional characterization of iron-bound forms of *Azotobacter vinelandii* (Nif) IscA. *Biochemistry* 51 (41): 8056–8070.

162 Bandyopadhyay, S., Naik, S.G., O'Carroll, I.P. et al. (2008). A proposed role for the *Azotobacter vinelandii* NfuA protein as an intermediate iron-sulfur cluster carrier. *J. Biol. Chem.* 283 (20): 14092–14099.

163 Touraine, B., Boutin, J.P., Marion-Poll, A. et al. (2004). Nfu2: a scaffold protein required for [4Fe–4S] and ferredoxin iron-sulphur cluster assembly in *Arabidopsis chloroplasts. Plant J.* 40 (1): 101–111.

164 Tong, W.H., Jameson, G.N., Huynh, B.H., and Rouault, T.A. (2003). Subcellular compartmentalization of human Nfu, an iron-sulfur cluster scaffold protein, and its ability to assemble a [4Fe–4S] cluster. *Proc. Natl. Acad. Sci. USA* 100 (17): 9762–9767.

165 Li, H., Mapolelo, D.T., Dingra, N.N. et al. (2009). The yeast iron regulatory proteins Grx3/4 and Fra2 form heterodimeric complexes containing a [2Fe–2S] cluster with cysteinyl and histidyl ligation. *Biochemistry* 48 (40): 9569–9581.

166 Maclean, A.E., Kimonis, V.E., and Balk, J. (2018). Pathogenic mutations in NUBPL affect complex I activity and cold tolerance in the yeast model *Yarrowia lipolytica*. *Hum. Mol. Genet.* 27 (21): 3697–3709.

167 Stehling, O., Jeoung, J.H., Freibert, S.A. et al. (2018). Function and crystal structure of the dimeric P-loop ATPase CFD1 coordinating an exposed [4Fe–4S] cluster for transfer to apoproteins. *Proc. Natl. Acad. Sci. USA* 115 (39): E9085–E9094.

168 Beilschmidt, L.K. and Puccio, H.M. (2014). Mammalian Fe–S cluster biogenesis and its implication in disease. *Biochimie* 100: 48–60.

169 Goldberg, A.V., Molik, S., Tsaousis, A.D. et al. (2008). Localization and functionality of microsporidian iron-sulphur cluster assembly proteins. *Nature* 452: 624–628.

170 Molik, S. and Lill, R. (2012). Role of mitosomes in cellular iron-sulfur protein biogenesis. *J. Endocyt. Cell Res.* 23: 77–85.

4

Iron–Sulfur Protein Maturation in the Cytosol of Eukaryotes

Joseph J. Braymer[1,2,3] *and Antonio J. Pierik*[4]

[1]*Philipps-Universität Marburg, Institut für Zytobiologie und Zytopathologie, Fachbereich Medizin, Karl-von-Frisch-Straße 14, 35032 Marburg, Germany*
[2]*Philipps-Universität Marburg, Zentrum für Synthetische Mikrobiologie Synmikro, Karl-von-Frisch-Straße 14, 35032 Marburg, Germany*
[3]*Bundesanstalt für Materialforschung und -prüfung (BAM), 12205 Berlin, Germany*
[4]*Rheinland-Pfälzische Technische Universität Kaiserslautern-Landau, Department of Chemistry, Erwin-Schrödinger-Straße 54, 67663 Kaiserslautern, Germany*

Chapter Goals

- Provide an overview of how Fe–S proteins are identified and how they are matured
- Differentiate the pathways for [2Fe–2S] and [4Fe–4S] clusters
- Provide a general perspective on the evolution of the CIA system and convince the young reader that there is yet much to discover from the fundamental to the applied

Main Chapter Points

- Discovering and characterizing cytosolic and nuclear Fe–S proteins in eukaryotes
- Understanding how eukaryotic Fe–S protein biogenesis machineries function
- Implications of cytosolic Fe–S protein biogenesis in disease and biotechnology

4.1 Introduction

The first three chapters have described how iron–sulfur (Fe–S) clusters are biosynthesized, trafficked, and inserted into target proteins by sulfur mobilization (SUF) and Fe–S cluster pathway (ISC) machineries in prokaryotes and within mitochondria. A clear takeaway message can be that Fe–S clusters do not spontaneously form in a cellular context but instead require complex protein machineries

Iron-Sulfur Clusters: Biogenesis and Biochemistry, Volume 1, First Edition.
Edited by Silke Leimkühler, Günter Schwarz, Oliver Lenz, and Oliver Einsle.

for the biogenesis, trafficking, and insertion of Fe–S clusters. Fe–S protein biogenesis in the eukaryotic cytosol is not different and depends on two machineries, the mitochondrial ISC system including the mitochondrial adenosine triphosphate (ATP)-binding cassette (ABC) transporter Atm1 and the cytosolic Fe–S protein assembly (CIA) machinery (Figure 4.1) [4, 5]. Fe–S-binding proteins existing in the eukaryotic cytosol and nucleus had been first described in the 1990s and the first indication of their dependency on an Fe–S cluster biogenesis system was the discovery that the mitochondrial transporter Atm1 in the yeast *Saccharomyces cerevisiae* influenced the maturation of the cytosolic aconitase family member isopropylmalate isomerase (Leu1) [6]. Several years later, the first CIA component Cfd1 was discovered by looking for yeast mutants affecting the [4Fe–4S] cluster assembly of ectopically expressed mammalian cytosolic aconitase [7]. Over the past two decades of research, cytosolic Fe–S protein biogenesis is now known to involve at least 13 cytosolic proteins, as documented for humans (*Homo sapiens*, Table 4.1). Intriguingly, some eukaryotes like *Monocercomonoides exilis* require only a core set of cytosolic CIA proteins [8] (Table 4.1). Importantly, the machineries for Fe–S protein biogenesis in eukaryotes are essential as they produce inorganic cofactors that bind to cytosolic and nuclear enzymes that maintain the blueprints of life, e.g. DNA and RNA [9, 10] among many others.

With the evolution of eukaryotes from an archaeal host and bacterial symbiont, two prokaryotic lineages merged leading to the last eukaryotic common ancestor (LECA) [11]. An ancestral ISC pathway stemming from bacteria and located in mitochondria is clearly recognizable in modern eukaryotes, while the identity and fate of the host archaeal Fe–S machinery remains enigmatic [12, 13]. Phylogenetic analysis has suggested that LECA already contained CIA proteins [12, 13]. Although some CIA proteins, e.g. Cfd1/Nbp35 and Cia2, have homologs in archaea, the prokaryotic lineage of the CIA pathway remains uncertain [14, 15]. The prokaryotic cell merger yet eventually led to the unique situation found in many eukaryotes in which two cellular compartments were linked in their requirement for Fe–S clusters [4, 12, 16]. This was made possible by the presence of the mitochondrial transporter Atm1, which exports an unknown iron- and sulfur-substrate termed X-S, which facilitates the maturation of extra-mitochondrial Fe–S proteins [1, 5, 6, 17, 18]. An alternative model of cytosolic Fe–S protein biogenesis in humans has been proposed that depends on a cytosolic ISC machinery [19]; yet, this system is not functional in Fe–S protein biogenesis [20] and would not account for the essentiality of several CIA components such as Nbp35 and Cfd1 in Fe–S protein biogenesis [7, 21–23]. While the CIA pathway is known to be essential in all eukaryotes, the mitochondrial ISC system is not [24, 25]. Alternatively, various protists and parasites employ SUF or the nitrogen fixation (NIF) Fe–S biogenesis systems, which presumably connect to the CIA system in place of ISC [4, 13, 25–27]. In *Giardia intestinalis*, an Atm1-like transporter has not been found and it remains unclear how the CIA system initiates Fe–S protein biogenesis [28]. Overall, cytosolic Fe–S protein biogenesis throughout the eukaryotic domain is a fascinating crossroad of the various Fe–S protein maturation machineries. Much remains to be explored including the complexity of eukaryotes with more than one cytosol [29].

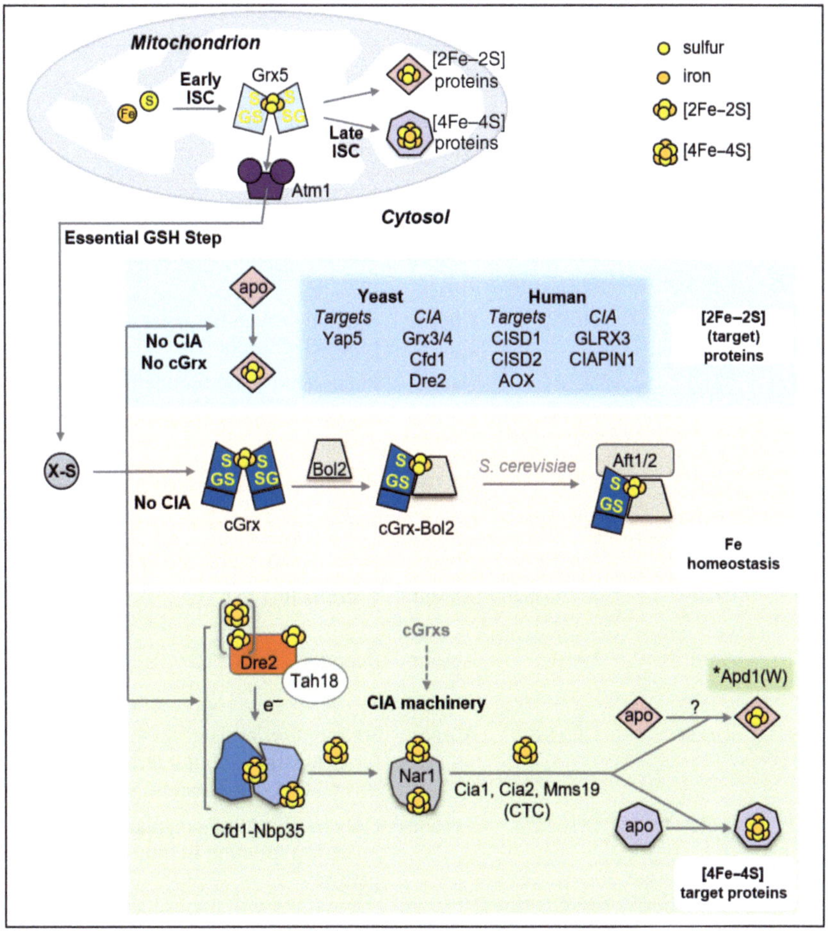

Figure 4.1 Model of eukaryotic Fe–S protein biogenesis with a focus on the maturation of cytosolic [2Fe–2S] and [4Fe–4S] proteins. The early mitochondrial ISC system, the ABC transporter Atm1, and GSH all are required for the export of the unknown sulfur- and iron-containing substrate X-S, which is proposed to be essential for the biogenesis of all cytosolic and nuclear Fe–S proteins. The most parsimonious assignment of X-S would be the export of [2Fe–2S] clusters from mitochondria, potentially in the form of [2Fe–2S] $(GSH)_4$ [1, 2], for the trafficking to cytosolic [2Fe–2S] proteins [3]. In yeast and humans, X-S has been proposed to feed into three pathways involving [2Fe–2S] clusters: (i) CIA and cGrx independent maturation of [2Fe–2S] target proteins (blue box), (ii) maturation of holo cGrx and cGrx–Bol complexes, which specifically in yeast act as sensors for cellular iron status (orange box), and (iii) maturation of the early CIA components Dre2 and Cfd1/Nbp35, which are responsible for the biogenesis of cytosolic [4Fe–4S] clusters (green box). Nar1 traffics [4Fe–4S] clusters from the early CIA to the late CIA proteins Cia1, Cia2, and Mms19 for the final insertion into target proteins. The three latter proteins generally form the core of CIA targeting complexes (CTCs). cGrxs (and cGrx–Bol complexes in humans) influence the CIA machinery yet are not an essential component for the generation of [4Fe–4S] clusters and are therefore considered CIA-related proteins (dotted gray arrow). A subset of [2Fe–2S] proteins are also CIA dependent, via targeting of a C-terminal tryptophan residue to the CTC, *Apd1(W). Source: Braymer et al. [3]/National Academy of Sciences.

Table 4.1 Components of the CIA system and other cytosolic Fe–S components.

Animal (*H. sapiens*)	**Yeast (*S. cerevisiae*)**	**Plant (*A. thaliana*)**	**Protist (*M. exilis*)**	**Function**
NDOR1	Tah18	TAH18	Absent	Electron transfer to CIAPIN1
CIAPIN1	Dre2	DRE2	Absent	Electron acceptor from NDOR1 and donor to scaffold complex
CFD1 (NUBP2)	Cfd1	Absent	Absent	Scaffold protein for generation of [4Fe–4S] clusters
NBP35 (NUBP1)	Nbp35	NBP35	Nbp35	Scaffold protein for generation of [4Fe–4S] clusters
CIAO3	Nar1	NAR1	Nar1	Putative [4Fe–4S] trafficking protein
CIAO1	Cia1	CIA1	Cia1	Docking platform for Fe–S target proteins
CIAO2B	Cia2	AE7	Cia2B	Assists Fe–S transfer to target proteins
CIAO2A	Absent	AEL1/2	Cia2A	Assists Fe–S transfer to target proteins
MMS19	Mms19	MMS19	Absent	Recruits target proteins to CTC
LTO1, ORAOV1	Lto1	AT2G16260	Absent	Recruits Rli1 to CTC
YAE1	Yae1	YAE1	Absent	Recruits Rli1 to CTC
Other				
GLRX3	Grx3/4	GRXS17	Absent	Sensor of Fe levels in yeast: nonessential for CIA pathway, but interacts with several CIA components
BOLA2	Bol2	BOLA2	Absent	Only involved in Fe regulation in yeast: unknown function in human CIA pathway

The inventory of yeast and human cytosolic, nuclear, and endoplasmic reticulum (ER)-associated Fe–S proteins involves the binding of [2Fe–2S], [3Fe–4S], or [4Fe–4S] clusters (Tables 4.2 and 4.3). The origin and trafficking of cytosolic [2Fe–2S] clusters for apoproteins in several cytosolic pathways remain to be described by *in vivo* methods; yet, several lines of evidence currently suggest that mitochondria export binuclear clusters to the cytosol (Figure 4.1) [1, 3, 18, 105]. The putative assignment of X-S involving [2Fe–2S] clusters would be the most parsimonious mechanism considering also the fact that an active biogenesis machinery for [2Fe–2S] clusters in the cytosol has not been identified. For the tetranuclear Fe–S clusters, the CIA pathway is responsible for maturing all cytosolic, nuclear, and ER-associated [4Fe–4S] proteins, which also stem from X-S and most likely [2Fe–2S] clusters.

The CIA model in Figure 4.1 has been developed predominately from genetic studies in the model organisms yeast and human. *In vivo* staging experiments employing the radiolabel ^{55}Fe as a marker for Fe–S clusters and mutants of Fe–S protein biogenesis components initially provided the concept of an assembly

Table 4.2 Characterization of cytosolic proteins involved in Fe–S protein biogenesis and associated proteins/complexes.

CIA component (yeast)	Fe–S clusters	Other cofactors/ functional groups	Representative PDBs	Structure references	Spectroscopy	Spectroscopy references
Dre2	[2Fe–2S] or [4Fe–4S]	—	2KM1, 2LD4, 4M7R, 2YUI	[30–32]	UV–Vis, EPR, Mössbauer	[33–37]
Tah18	—	FMN, FAD, NADPH	4H2D	[38]	UV–Vis	[36, 39, 40]
Cfd1	[2Fe–2S] or [4Fe–4S]	Mg-ATP	6G2G	[22]	UV–Vis, EPR, Mössbauer	[41, 42]
Nbp35	[2Fe–2S] or [4Fe–4S]	Mg-ATP	—	None	UV–Vis, EPR, Mössbauer	[41, 42]
Nar1	2 × [4Fe–4S]	C-terminal W	—	None	UV–Vis, EPR	[43–47]
Cia1	—	—	2HES, 6TBN, 6TC0, 6TBL, 3FM0	[48–51]	None	None
Cia2	?	Reactive Cys	6TBN, 6TC0, 6TBL, 3UX2, 3UX3, 2M5H	[48, 52, 53]	None	None
Mms19	—	—	6TBN, 6TC0, 6TBL, 8USQ, 8USP	[48, 49]	None	None
Lto1	? [4Fe–4S][a)]	C-terminal W	—	None	UV–Vis, EPR	[54]
Yae1	—	—	—	None	None	None
Additional cytosolic Fe–S components						
Grx3/4	[2Fe–2S]	GSH	5Y4T, 5Y4U, 3ZYW	[55]	UV–Vis, EPR, Mössbauer	[56–58]
Grx3/4-Bol2	[2Fe–2S]	GSH	5Y4B	[55]	UV–Vis, EPR, Mössbauer	[56–58]

a) Human Lto1 only contained an Fe–S cluster after reconstitution, no *in vivo* evidence reported. Yeast Lto1 lacks C-terminal cysteine residues.

Table 4.3 Cytosolic and nuclear Fe–S target proteins of *Homo sapiens* and *Saccharomyces cerevisiae*.

Class/proteins (*H. sapiens*/*S. cerevisiae*)	Fe–S cluster	C-terminus	CIA dependent	References
Replication				
POLA1/Pol1	[4Fe–4S][a)]	S/N	Yes	[59]
POLE/Pol2	[4Fe–4S]	H/I	Yes	[59, 60]
POLD1/Pol3	[4Fe–4S]	**W/W**	Yes	[59, 61]
PRIM2A/Pri2	[4Fe–4S]	S[b)]/H[b)]	Yes	[48, 62]
DNA2/Dna2	[4Fe–4S]	E/S	TBD	[48, 63]
KIF4A/absent	[4Fe–4S]	H/–	TBD[c)]	[64]
DNA repair				
XPD1/Rad3	[4Fe–4S]	L/Q	Yes	[65–67]
DDX11/Chl1	[4Fe–4S]	E/R	TBD	[68]
RTEL1/absent	[4Fe–4S]	L/–	Yes	[69, 70]
FANCJ/absent	[4Fe–4S]	K/–	Yes	[67, 70]
REV3L/Rev3	[4Fe–4S]	**F/W**	Yes	[59, 71]
MUTYH/Ntg2	[4Fe–4S]	Q/K	Yes	[72, 73]
NTHL1/non-Fe–S	[4Fe–4S]	L/–	Yes	[69]
Transcription				
XPD1/Rad3	[4Fe–4S]	L/Q	Yes	[65–67]
CPSF4/absent	[2Fe–2S]	Q/–	TBD	[74]
POLR3F/non-Fe–S	[4Fe–4S]	**F**/–	TBD	[75]
Translation				
CTU1/Ncs6	[4Fe–4S]	**F/F**	TBD	[76, 77]
ELP3/Elp3	2 × [4Fe–4S]	K[d)]/I[d)]	TBD	[78]
DPH1/Dph1	[4Fe–4S]	P/E	TBD	[79, 80]
DPH2/Dph2	[4Fe–4S]	G/K	TBD	[79, 80]
TYW1B/Tyw1	2 × [4Fe–4S]	C/A	TBD	[81, 82]
ABCE1/Rli1	2 × [4Fe–4S]	D[e)]/I[e)]	Yes	[83, 84]
CDKAL1/absent	2 × [4Fe–4S]	N/–	TBD	[85, 86]
Regulation				
RSAD1/absent	[4Fe–4S]	**W**/–	Yes	[87]
NARF/absent	2 × [4Fe–4S]	**W**/–	TBD	[88]
Absent/Yap5	[2Fe–2S]	–/H	No	[3, 89]
FBXL5/absent	[2Fe–2S]	E/–	TBD	[90]
NCOA4/absent	[3Fe–4S]	M/–	TBD	[91]

Table 4.3 (Continued)

Class/proteins (*H. sapiens*/*S. cerevisiae*)	Fe–S cluster	C-terminus	CIA dependent	References
Metabolism				
XDH/absent	2×[2Fe–2S]	V/–	No[f)]	[3]
AOXA/absent	2×[2Fe–2S]	I/–	No	[3]
PUR1/non-Fe-S	[4Fe–4S]	**W**/–	Yes	[69]
ACOC/absent	[4Fe–4S]	K/–	Yes	[69]
DPYD/absent	4×[4Fe–4S]	C/–	Yes	[69]
RFESD/absent	[2Fe–2S]	S/–	TBD	none
CMAHP/absent	[2Fe–2S]	A/–	TBD	[92]
CISD1/absent	[2Fe–2S]	T/–	No	[3, 93]
CISD2/absent	[2Fe–2S]	V/–	No	[3, 94]
Absent/Leu1	[4Fe–4S]	–/**W**	Yes	[95, 96]
Absent/Met5	[4Fe–4S]	–/Y	Yes	[65, 95]
Absent/Glt1	[4Fe–4S]	–/V	TBD	[97]
ATE1/Ate1	[2Fe–2S] + [4Fe–4S]	N/M	TBD	[98]
SPRY2/absent	[2Fe–2S]	T/–	TBD	[99]
Absent/Apd1	[2Fe–2S]	–/**W**	Yes	[3, 100]
SNAP25/non-Fe–S	[2Fe–2S]	G/–	TBD	[101]
Absent/Aft1/2	[2Fe–2S]	–/D,Y	No	[102]

TBD: To be determined.
C-termini in bold correspond to the CIA targeting complex recognition signal.
a) Has been detected in yeast [59], but the presence of Fe–S is debated [103].
b) The non-Fe–S subunit (PRIM1/Pri1) acts as an adaptor [96] and has F/F as C-terminus.
c) Upon expression in yeast the human protein localized to mitochondria [64].
d) A non-Fe–S subunit (ELP4/Elp4) acts as an adaptor [96] and has F/Y as C-terminus.
e) Yae1D/Yae1 (L/Q) and LTO1/Lto1 (F/W) act as an adaptor [104].
f) Based on protein homology with AOXA.

line and trafficking system for [4Fe–4S] clusters [4, 95, 106]. In the past 10 years, structural and biophysical studies have started to define the molecular structures of the model (Figures 4.1 and 4.2). Furthermore, with the development of the powerful protein structure prediction tool, Alphafold2 (AF2) [107], components and even protein complexes can be visualized (Figure 4.2). The maturation of cytosolic [4Fe–4S] target proteins by early and late CIA proteins is proposed to follow the three general steps described for the other Fe–S protein biogenesis machineries, e.g. Fe–S cluster biogenesis, trafficking, and insertion (Figure 4.1) [4]. The model of the CIA pathway is summarized as follows: (i) binuclear Fe–S clusters are the source for the biogenesis of [4Fe–4S] clusters on the Cfd1/Nbp35 scaffolding complex, (ii) electrons for the reductive coupling of two [2Fe–2S] clusters are provided by the

NADPH/Tah18/Dre2 electron chain, and (iii) [4Fe–4S] clusters generated by the early CIA components are first trafficked to the late CIA machinery by Nar1, from which they are inserted into target apoproteins by the CIA targeting complex (CTC) involving Cia1, Cia2, and Mms19 [4, 5, 109].

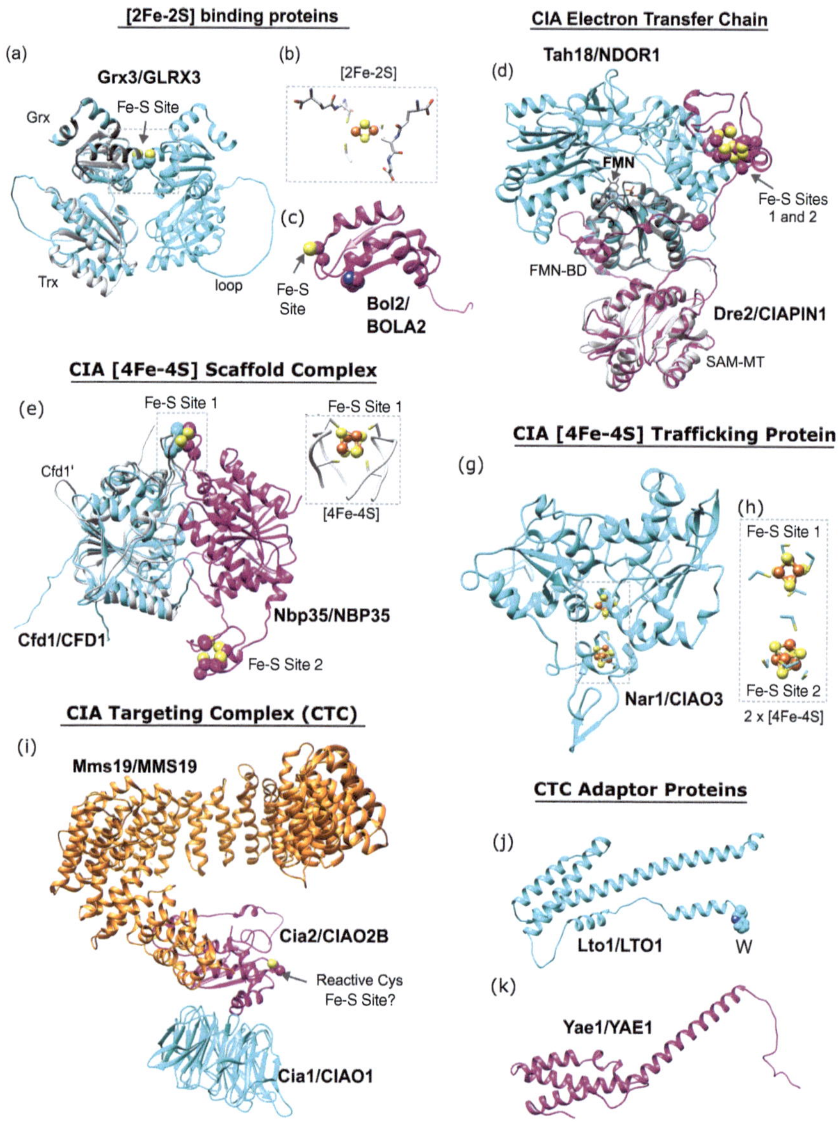

Figure 4.2 Structural overview of CIA components and CIA-related proteins. Structural images were made from either Protein Databank (PDB) entries or the use of modeling programs Alphafold2 (AF2, [107]) or Swiss-Model (SM, [108]). Protein names are given for yeast and human homologs (yeast/human). Residues for Fe–S cluster binding (Cys or His ligands, spheres) are indicated. Fe–S sites that have been structurally characterized or predicted are shown in dashed boxes. (a) Grx3 homodimer (cyan, yeast, AF2) is composed of an N-terminal thioredoxin (Trx) domain (yeast, dark gray, PDB 5Y4T) and glutaredoxin (Grx) domain (yeast, light gray, PDB 5Y4T) connected by a loop. (b) Example of [2Fe–2S] cluster coordinated by two Cys residues and 2 GSH molecules found in monothiol glutaredoxins (human GLRX5, PDB 2WUL). (c) Bol2 contains one Cys and one His ligand for coordinating a [2Fe–2S] cluster in cGrx–Bol complexes (yeast, PDB 5Y4B). (d) Tah18 (cyan) and Dre2 (magenta) form a complex (human homologs, AF2) that provides electrons for [4Fe–4S] cluster biogenesis. Tah18 binds multiple organic cofactors, one of which is structurally characterized (flavin mononucleotide [FMN] in an FMN-binding domain [FMN-BD], light gray, PDB 4H2D). Dre2 homologs contain an N-terminal SAM methyltransferase-like domain (SAM-MT, dark gray, PDB 4M7R) and a C-terminal domain that contains two Fe–S cluster-binding sites consisting of eight cysteine residues. (e) The CIA scaffolding complex involves the heterodimer of Cfd1 (cyan) and Nbp35 (magenta) (yeast homologs, AF2), on which [4Fe–4S] clusters are proposed to bind and/or be assembled at either Fe–S site 1 or 2. (f) A bridging [4Fe–4S] cluster has been observed in a Cfd1 homodimer at Fe–S site 1 (*Chaetomium thermophilum*, PDB 6G2G). (g) Nar1 (*S. cerevisiae* homolog threaded through PDB 1FEH, SM) has two conserved Fe–S-binding sites and presumably traffics [4Fe–4S] clusters from the early CIA to the late CIA. (h) Two [4Fe–4S] clusters found in FeFe-hydrogenase proteins are modeled into the Nar1 structure prediction. (i) The CIA targeting complex (CTC) composed of Mms19 in orange, Cia2 in magenta, and Cia1 in cyan (mouse and *Drosophila* homologs, PDB 6TCO) provides a platform for target proteins to bind and receive Fe–S clusters. (j, k) CTC adaptor proteins Yae1 (yeast, AF2) and Lto1 (yeast, AF2) assist in the maturation of the [4Fe–4S] target protein Rli1. Some target proteins and adaptors contain a C-terminal tryptophan (W) for targeting to the CTC (Lto1).

In this chapter, our main goal is to provide an overview of how Fe–S proteins are identified and which components are required for their maturation in the eukaryotic cytosol. Further in-depth details of the CIA system can be found in the literature [4, 5, 19, 109–111]. For the eukaryotic cytosol, we will consider three types of clusters, binuclear [2Fe–2S] clusters, [3Fe–4S] clusters, and tetranuclear [4Fe–4S] clusters. Each cytosolic Fe–S protein has its own Fe–S cluster biogenesis machinery dependencies (Figure 4.1). The discussion will focus on the yeast (*S. cerevisiae*) and human CIA models, which are evolutionarily very well conserved. Protein nomenclature will refer mostly to the yeast names for the sake of consistency (Table 4.1). At the end, we will discuss the relevance of cytosolic Fe–S protein biogenesis for disease and for biotechnology.

4.2 Methodology and Challenges of Assigning Cytosolic Fe–S Proteins

Yeast and human as model organisms have been fundamental to identifying cytosolic, nuclear, and ER-associated Fe–S proteins and defining the Fe–S protein biogenesis machineries required to mature them [4]. Using these standard models has allowed for the confirmation of cytosolic Fe–S protein biogenesis across various eukaryotic supergroups [4, 8, 12, 16, 29, 112–116]. Candidate Fe–S proteins may be identified from various scenarios including by known protein–protein interactions with an existing Fe–S protein or Fe–S protein biogenesis component (i.e. interaction with the CTC [65, 104, 113]), deletion or mutant strain phenotypes [7, 41, 64, 117], bioinformatic prediction of Fe–S-binding motifs [21, 83] or CTC-interacting residues [100], or the discovery of Fe–S clusters binding to recombinantly produced protein [98, 118]. One of the general approaches of confirming *in vivo* Fe–S cluster binding in yeast and human cells has been the use of ^{55}Fe as a radiolabel for Fe–S clusters bound to proteins suspected of binding Fe–S clusters [95, 106]. A further powerful approach is the determination of protein functionality in activity assays such as phenotypic growth analysis or enzymatic activity of cell lysates [95, 106]. Importantly, in order to be confirmed as Fe–S binding as compared to nonheme, heme-, or nonspecific iron binding, cytosolic protein candidates should show a strong dependency on ^{55}Fe binding and activity on the mitochondrial ISC machinery, the mitochondrial transporter Atm1, and glutathione (GSH) [3, 5]. The identity of cytosolic Fe–S cluster biogenesis and trafficking proteins can also be confirmed by either deleting or depleting the gene of interest and analyzing the activity of a cytosolic Fe–S target protein, e.g. Leu1 or sulfite reductase [95]. The readily genetically tractable and relatively fast-growing yeast *S. cerevisiae* has provided a robust means of testing these dependencies along with the development of various methods in human cells, e.g. CRISPR/Cas9 gene editing and regulation [3, 95, 106].

Other radiolabeling strategies involving the isolation and mixing of cellular compartments of yeast have also emerged as tools in studying the export of iron- and sulfur-containing species that are used for the biogenesis of cytosolic Fe–S proteins [18, 119]. *In vivo* radiolabeling is a powerful method for detecting iron bound to proteins yet has the drawback that methods do not yet exist for resolving the Fe–S cluster nuclearities on proteins *in vivo*. For some proteins, it has not been a trivial task of assigning whether [2Fe–2S] or [4Fe–4S] clusters are the physiological cofactor, for example Dre2 and Ate1 [33, 34, 98, 120]. In order to define the type of Fe–S cluster binding under close-to native environments, Mössbauer and electron paramagnetic resonance (EPR) spectroscopies have been utilized *in cellulo* in *Escherichia coli* cells and with yeast extracts containing overexpressed proteins [100, 120]. *In cellulo* nuclear magnetic resonance (NMR) has also been successfully employed to characterize a [2Fe–2S] protein in a host environment [121]. While these are powerful spectroscopic methods, proteins of interest must be stably overexpressed, in addition to already having had their spectroscopic fingerprints analyzed *in vitro*.

Going hand-in-hand with *in vivo* characterization is the further validation of cytosolic Fe–S proteins by *in vitro* biochemical and biophysical characterization of recombinantly isolated proteins (Table 4.2 and Figure 4.2). A common approach has been the heterologous overexpression of Fe–S proteins in *E. coli* followed by either aerobic or anaerobic purification. Fe–S clusters are often either chemically or semienzymatically reconstituted anaerobically in recombinant proteins due to the lability and/or lack of adequate Fe–S cluster incorporation in *E. coli* [122, 123]. While this strategy is very effective at building Fe–S clusters, interpretations of data produced with chemically reconstituted proteins should be made cautiously due to established pitfalls [123]. Whereas EPR spectroscopy does not suffer so strongly from adventitiously bound excess iron, UV–Vis and Mössbauer spectroscopic are particularly prone to misinterpretation due to background signals. The optimization of expression conditions in *E. coli* [123], the use of eukaryotic expression systems [48], or the expression in the host organism [124] are approaches that should also

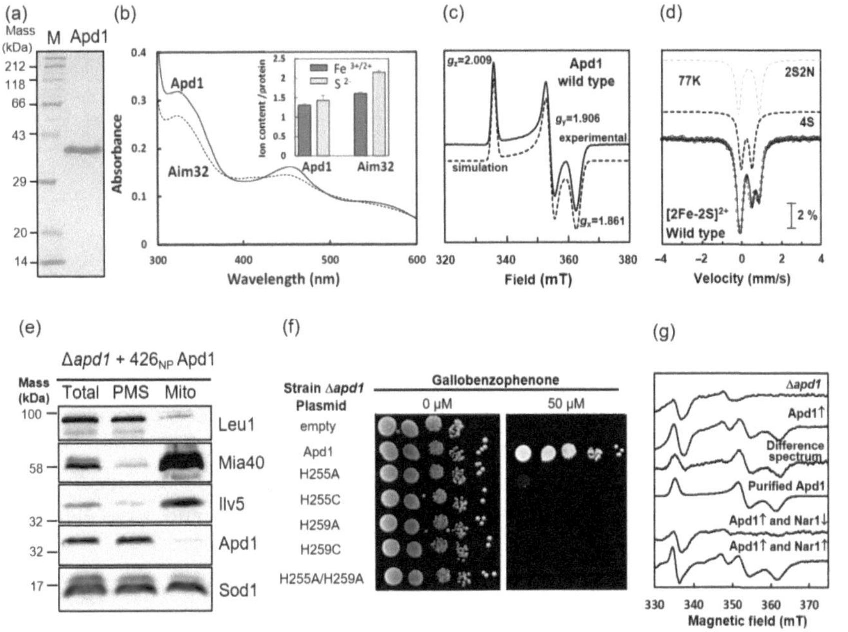

Figure 4.3 The validation of Apd1 as a cytosolic [2Fe–2S] protein. (a–d) *In vitro* validation of [2Fe–2S] cluster binding to recombinantly produced Apd1. (a) Apd1 recombinantly expressed in *E. coli* can be isolated in a highly pure form. (b–d) Spectroscopic analyses reveal [2Fe–2S] cluster binding. (b) UV–Vis absorption features at 325, 455, and 550 nm are typical for $[2Fe–2S]^{2+}$ clusters. (c) EPR detects the presence of a reduced non-Rieske $[2Fe–2S]^{+}$ cluster. (d) The Apd1 Mössbauer parameters, e.g. isomer shift and quadruple splitting, are characteristic for $[2Fe–2S]^{2+}$ clusters. (e–j) *In vivo* characterization of Apd1. (e) A cellular fractionation experiment demonstrates the localization of Apd1 in the cytosol (presence in the postmitochondrial supernatant [PMS] fraction) and not in the mitochondrial (mito) fraction. Apd1 was detected with an anti-Apd1 antibody. Leu1 is a marker for the cytosol, Mia40 for the mitochondrial intermembrane space (IMS), Ilv5 for mitochondria, Sod1 mainly occurs in the cytosol, and to a lesser extent in the IMS. (f) A cell growth activity assay demonstrates the lethality of 50 μM gallobenzophenone in yeast cells lacking Apd1 (*Δapd1*) and the complementation of cell growth with Apd1 expressed on a plasmid. The expression of mutant proteins allows for the determination of important metal binding residues. (g) EPR analysis of yeast extracts containing overexpressed Apd1 matches the *in vitro* data for [2Fe–2S] cluster binding (compare (g) and (c)). Depletion of the CIA component Nar1 (Nar1↓) leads to a disappearance of the Apd1-dependent EPR signal. (h–j) N-terminally HA-tagged Apd1 binds ^{55}Fe *in vivo* in *Saccharomyces cerevisiae*. Depletion (indicated by ↓) of ISC pathway components, Atm1, cGrxs, and CIA components (h) along with the deletion of the biosynthetic gene for GSH (*gsh1Δ*) leads to decreased ^{55}Fe binding in Apd1 (i). (j) ^{55}Fe binding to Apd1 is independent of Bol2. In (h)–(j), Western blots confirm the expression of Apd1 using an anti-HA antibody. Source: (a)–(g) Stegmaier et al., (2019)/ with permission of American Chemical Society (h)–(j) [3]/National Academy of Sciences/CC BY 4.0.

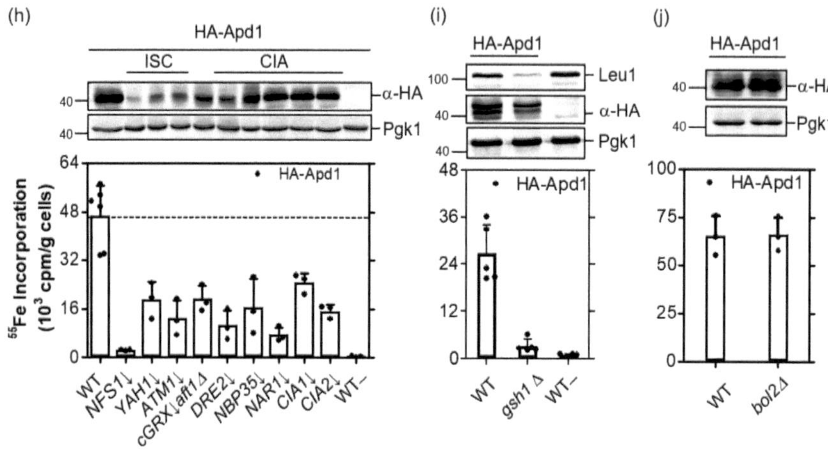

Figure 4.3 (*Continued*)

be considered. Isolation of highly pure cytosolic Fe–S proteins and Fe–S biogenesis components then ensures for successful characterization of Fe–S cofactors by a multitude of methods (succinctly summarized in [125]), including X-ray crystallography (see Chapter 18 and [126]), vibrational spectroscopies (see Chapter 19), EPR [127], native mass spectrometry [128], NMR [129], and cryo-electron microscopy [48, 49].

Figure 4.3 provides an interesting case study of how *in vitro* and *in vivo* methodologies can be employed to identify and study potential cytosolic Fe–S proteins. Bioinformatic analysis identified the yeast protein Apd1 as a potential Fe–S protein. The facile isolation of Apd1 expressed in *E. coli* (Figure 4.3a–d) and the possibility to generate variants with spectroscopically and electrochemically distinct properties allowed an *in vitro* identification of a non-Rieske $[2Fe–2S]^{1+/2+}$ cluster bound in a 2Cys/2His coordination site [100]. These findings and the colocalization in yeast with Leu1 (Figure 4.3e) defined Apd1 as the first nonregulatory, noncomplex, single [2Fe–2S] cluster-containing eukaryotic protein. In the absence of a known biochemical activity assay, the lethality of the nonessential Apd1 protein upon growth on solid media containing gallobenzophenone provided a convenient *in vivo* phenotypic assay for [2Fe–2S] functionality (Figure 4.3f), as well as for maturation. EPR spectroscopic experiments of yeast cell extracts where Apd1 was overexpressed in wild-type cells and cells depleted of the CIA component Nar1 provided not only the confirmation of the [2Fe–2S] cluster *in vivo* but also the first evidence for a CIA-machinery dependence (Figure 4.3g). These initial studies were followed up in a full analysis by *in vivo* ^{55}Fe labeling experiments demonstrating a CIA-machinery dependence (Figure 4.3h), which is unlike all other eukaryotic [2Fe–2S] containing proteins studied thus far [3]. Matching the CIA dependence was also the requirement of the ISC system, the mitochondrial transporter Atm1, and glutathione (Figure 4.3h,i). *In vivo* Fe–S cluster binding though was not dependent on Bol2 (Figure 4.3j). Reasons for the CIA dependence of Apd1 are discussed in Section 4.4.4.

4.3 [2Fe–2S] Protein Assembly in the Cytosol and the Early CIA Machinery

Although not as prominent as cytosolic [4Fe–4S] clusters, cytosolic [2Fe–2S] proteins play important roles in metabolism, transcription, and cytosolic [4Fe–4S] protein biogenesis (Table 4.3) [130]. In the following sections (Sections 4.3.1–4.3.4), we will focus on the iron and sulfur source for cytosolic Fe–S clusters, iron homeostasis in yeast, target proteins binding [2Fe–2S] clusters in a CIA-independent fashion, and the early CIA components that are proposed to be involved in generating [4Fe–4S] clusters (Figure 4.1). Like proteins binding [4Fe–4S] clusters, all cytosolic [2Fe–2S] proteins are dependent on mitochondria, Atm1-facilitated export, and GSH, yet most [2Fe–2S] proteins most likely do not require the CIA system.

4.3.1 Export of X-S from Mitochondria and Essential GSH Requirement in Fe–S Biogenesis

All yeast and mammalian Fe–S proteins residing outside of mitochondria, e.g. cytosolic, nuclear, and ER-associated, require the core mitochondrial ISC pathway (Figure 4.1) [3, 6]; this excludes proteins localized in chloroplasts and plastids that rely on the plastidal SUF pathway [131]. The inner membrane mitochondrial ABC transporter Atm1 acts as a first conduit for connecting the matrix-localized early ISC system and the cytosol via X-S [1, 17, 132]. One leading candidate for the transported substrate, a tetra-glutathione-coordinated [2Fe–2S] cluster, [2Fe–2S] $(GSH)_4$ [1, 2], would provide a preassembled Fe–S cluster to the mitochondrial intermembrane space (IMS) for further trafficking to the cytosol. Precedents for the transport of metal-glutathione complexes via ABC transporters are found in the yeast vacuole and in bacteria [133, 134]. IMS components of the mitochondrial import disulfide relay system, Erv1 and Mia40, have been suggested to influence cytosolic Fe–S protein biogenesis [135, 136] and Mia40 can bind *in vitro* a [2Fe–2S] cluster [137]. The involvement of these proteins remains, however, unconfirmed as other *in vivo* work in yeast has found no influence of the Erv1/Mia40 proteins on Fe–S protein biogenesis or iron homeostasis [138]. Another study has suggested that the mitochondrial inner NEET protein (MiNT or CISD3) passes [2Fe–2S] clusters through porins of the mitochondrial outer membrane to an outer mitochondrial membrane NEET (mNT or CISD1) [139]. CISD3, however, is targeted to the mitochondria matrix and has not yet been demonstrated to be localized within the IMS. The *in vivo* identification of X-S and the molecular mechanisms describing how the ISC and CIA systems are connected remain the most profound questions in eukaryotic Fe–S protein biogenesis to address.

Following the discovery of the essentiality of mitochondria for cytosolic Fe–S protein biogenesis was the observation that cytosolic [4Fe–4S] proteins in yeast also strictly require the cysteine-containing tripeptide GSH [140, 141]. GSH plays at least two key roles in Fe–S protein biogenesis. The tripeptide assists monothiol glutaredoxins in the coordination of [2Fe–2S] clusters in both mitochondria and the cytosol, which may be trafficked to other Fe–S proteins (discussed in Section 4.3.2), and binds to Atm1 [17, 134, 142]. When cellular GSH levels are depleted to critical levels, all cytosolic [2Fe–2S] and [4Fe–4S] proteins are affected while mitochondrial proteins retain their Fe–S clusters [3, 141]. General cytosolic Fe–S protein defects are not due to the lack of GSH binding to cytosolic monothiol glutaredoxins (cGrxs) Grx3/4 as depletion of these proteins did not impair the Fe–S maturation on several cytosolic proteins. Furthermore, the essentiality of cGrxs in *S. cerevisiae* can be alleviated by various mutations that rebalance cellular iron concentrations [3, 143]. Therefore, the essential GSH-dependent step at Atm1 supersedes the cGrxs. The *in vivo* requirement of GSH for the export of X-S bolsters the evidence for the transport of GSH-coordinated [2Fe–2S] complexes from mitochondria [1], as mentioned earlier.

Once X-S reaches the cytosol, it feeds into multiple pathways involving [2Fe–2S] clusters (Figure 4.1). Proteinaceous components involved in the trafficking of [2Fe–2S] clusters to these pathways in the cytosol, as typically found in mitochondria and prokaryotes, are very poorly understood. Alternatively, it has yet to be ruled out that [2Fe–2S] clusters are also trafficked in the form of small molecule complexes, i.e. involving thiol-containing compounds like cysteine and GSH [144]. In yeast and human cells, shutting off one maturation path employing appropriate mutants often leads to increased [2Fe–2S] cluster binding to proteins in the other pathways [3]. The concept of a dynamic flow of [2Fe–2S] clusters also applies physiologically in yeast and humans. Fluctuations in the concentration of [2Fe–2S] clusters in the yeast cytosol determine whether the cell stores excess iron in the vacuole or upregulates high-affinity iron uptake [142, 145]. Human cells also have the ability to rapidly respond to iron availability and produce an expandable pool of cytosolic [2Fe–2S] clusters [146].

4.3.2 [2Fe–2S] Clusters in Yeast Iron Homeostasis and [2Fe–2S] Chaperones

The phenotypes in yeast of mutations of early ISC, Atm1, or biosynthetic genes for GSH all converge on disturbed mitochondrial iron homeostasis, which is a result of X-S not being exported to the cytosol [142, 147]. Mitochondrial iron levels increase under these cases of disturbed Fe–S protein biogenesis because X-S is an essential component of the [2Fe–2S] cluster-dependent regulation of iron homeostasis in yeast (Figure 4.1, orange box) [3, 141, 142, 145]. In other words, when X-S is not exported to the cytosol, yeast perceives this as not having enough iron and try to correct this by high-affinity iron uptake. Iron homeostasis in fungi has been reviewed extensively [142, 145] and we will provide here an overview with a focus on how [2Fe–2S] clusters are involved in iron regulation. The response to varying iron levels in yeast is completely independent of the CIA system (Figure 4.1) [3, 148], which is fundamentally different from iron homeostasis in humans (see Sections 4.4.2 and 4.4.4).

cGrxs play a central role in iron regulation [56, 149, 150]. In contrast to mitochondrial-related Fe–S protein biogenesis disruptions, their depletion in *S. cerevisiae* leads to increased cellular iron content yet lower iron levels in the mitochondria [150]. Therefore, this phenotype, along with the ISC mutants, demonstrates the importance of mitochondria in the general processing of iron and how cGrxs play multiple roles including directing iron to mitochondria [147]. Monothiol cGrxs in yeast entail an N-terminal thioredoxin (Trx) domain connected by a loop to a C-terminal monothiol glutaredoxin domain (Grx) (Figure 4.2a). Homodimers of these proteins bind a bridging [2Fe–2S] cluster via a cysteine of the Grx domain and a cysteine of GSH that is noncovalently bound to the protein (Figure 4.2a,b) (see [151] for [2Fe–2S] cluster coordination to a monothiol glutaredoxin and [56] for *in vitro* characterization of cGrxs). cGrxs also form heterodimeric complexes with the small Bol-like proteins, which also bind [2Fe–2S] clusters via cysteine and histidine residues (Figure 4.2c) [56, 57, 152].

When yeast cells having adequate amounts of iron and mitochondria are producing X-S, cGrx homodimers and cGrx–Bol2 heterodimers bind [2Fe–2S] clusters stemming from X-S (Figure 4.1) [3, 146, 150]. The latter holo-complex enters the nucleus and interacts with the transcriptional regulators Aft1/2 [57, 149]. The cGrx–Bol2–Aft complex represses the high-affinity iron uptake regulon. Repression may occur via the transfer of the [2Fe–2S] cluster to Aft1/2, as Aft2 has a Cys-Asp-Cys motif that binds [2Fe–2S] clusters *in vitro* [102, 153], yet confirmation of *in vivo* [2Fe–2S] cluster binding has not been able to be determined [3]. Upon a drop in cellular iron concentrations, transcription of the high-affinity iron uptake regulon is activated by Aft1/2 due to a lack of [2Fe–2S] cluster-bound cGrx–Bol2.

Elevated levels of iron are also detrimental to cells and yeast also use the status of cytosolic [2Fe–2S] clusters as a sensor of high iron levels. Upon an excess of cytosolic [2Fe–2S] clusters, the transcription factor Yap5 binds two [2Fe–2S] clusters in a C-terminal domain leading to a conformational change and activation of gene transcription [89]. Among the transcribed genes are the vacuolar iron importer Ccc1, Grx4, and the [4Fe–4S] protein Tyw1 [154–156]. Strikingly, Yap5 [2Fe–2S] cluster binding is not dependent on cGrxs nor the CIA pathway but is dependent on GSH levels [3, 89]. This dependency suggests that either [2Fe–2S] clusters are trafficked to the nucleus to bind constitutively DNA-bound Yap5 or [2Fe–2S] clusters bind Yap5 immediately following translation in the cytosol. Although protein families and mechanisms vary among fungi, fungal iron homeostasis generally relies on a combination of cGrxs and [2Fe–2S] clusters [145]. Altogether, cGrx and cGrx–Bol2 complexes act as sensors of iron levels in yeast via the Fe–S cofactor.

Along with their essential roles in iron homeostasis in yeast, homodimeric cGrxs and heterodimeric cGrx–Bol complexes may also function as [2Fe–2S] cluster chaperones, delivering [2Fe–2S] clusters to cytosolic proteins in both yeast and humans (see also discussion below in Section 4.3.4) [145, 146, 150, 157–159]. While such a function matches the role of the monothiol glutaredoxin Grx5 in mitochondria (discussed in Chapter 3 and [160]), *in vivo* experiments in yeast and human have determined that cGrxs are not essential for the maturation of Dre2, Cfd1, nor Nbp35 [3]. Matching these observations is the fact that cGrxs and cytosolic Bol proteins are not essential in various eukaryotic cell types [3]. Therefore, a [2Fe–2S] cluster trafficking function of cGrx and cGrx–Bol complexes is not essential for the CIA pathway, which itself is essential [9]. Although not required for cytosolic Fe–S protein biogenesis, cGrxs clearly impact the CIA pathway as evidenced by their physical interaction *in vivo* with multiple CIA components and strong effects on cytosolic

[4Fe–4S] target proteins [3, 69, 116, 146, 150, 161]. Proteins binding GSH may also promiscuously bind [2Fe–2S] clusters and influence [2Fe–2S] trafficking [162], but such an influence on cytosolic Fe–S protein maturation has not been tested. Interestingly, no other candidate cytosolic [2Fe–2S] chaperones or trafficking proteins have been identified to-date, which would function in the absence of cGrxs complexes.

4.3.3 [2Fe–2S] Target Proteins

A second pathway for the trafficking of cytosolic [2Fe–2S] clusters is the CIA and cGrx independent maturation of [2Fe–2S] target proteins (Figure 4.1 and Table 4.3). This pathway was recently borne out by analyzing the Fe–S cluster biogenesis dependencies of several target proteins in yeast and human cells in a similar manner to dissecting the CIA components themselves [3]. In addition to Yap5 discussed earlier, cytosolic proteins of the NEET and aldehyde oxidase (AOX) families fall into this category. The NEET proteins CISD1 and CISD2 bind a [2Fe–2S] cluster in an initially described Zn-finger site composed of 3Cys/1His ligands [163]. CISD1 is found on the mitochondrial outer membrane and CISD2 is predominately anchored to the endoplasmic reticulum facing the cytosol. The AOX family of proteins, also including xanthine oxidase, binds two [2Fe–2S] clusters in addition to the molybdenum cofactor (Moco) [164]. The CIA components Cfd1 and Dre2 also fit into this category and are discussed in Section 4.3.4. Alternative to the CIA independent maturation of [2Fe–2S] proteins is also a CIA-dependent pathway (as described in the case study of Apd1), which will be further discussed in further detail in Section 4.4.4.

The biogenesis requirements and *in vivo* confirmation of Fe–S binding for some cytosolic target proteins that may bind [2Fe–2S] clusters have yet to be detailed. These include yeast transcription factors involved in iron homeostasis and proteins of various functions. As discussed earlier, cGrx–Bol2 complexes in yeast interact with transcription factors that themselves may bind [2Fe–2S] clusters via cysteine residues. Aft2 has been shown *in vitro* to bind [2Fe–2S] clusters [102, 153] and it has been postulated that other Zn-finger sites in Fep1 [165], HapX [116], SreA, Cir1, and Sfu1 bind [2Fe–2S] clusters *in vivo* (Table 4.4) [145]. Iron homeostasis in humans may also be influenced by [2Fe–2S] clusters (refer to Section 4.4.4).

Various other eukaryotic proteins have been reported binding [2Fe–2S] clusters from *in vitro* studies, which have yet to be fully validated *in vivo* for their Fe–S protein biogenesis dependencies (Table 4.3). The cleavage and polyadenylation specificity factor (CPSF4) important for the processing of pre-mRNA has been demonstrated *in vitro* to bind [2Fe–2S] clusters in a 3Cys/1His metal binding site [170]. Posttranslational modifications of proteins in the cytosol may depend on [2Fe–2S] cluster binding to an arginyl-tRNA transferase (Ate1) [98]. Levels of inositol pyrophosphate (IPP) may be influenced by the presence of a [2Fe–2S] cluster in the dual-functional kinase/phosphatase Asp1 [118, 124]. The physiological relevance of [2Fe–2S] cluster binding to a cysteine normally involved in palmitate binding in SNAP-25 has thus far not been assessed [101]. Other eukaryotic proteins can be found binding [2Fe–2S] clusters with unknown function [99]. Cytidine monophosphate (CMP)-*N*-acetylneuraminic acid hydroxylase binds a Rieske-type [2Fe–2S] cluster [92].

Table 4.4 Cytosolic and nuclear Fe–S target proteins with relevance to the CIA machinery, but not occurring in *Homo sapiens* and *Saccharomyces cerevisiae*.

Class/proteins	Organism (virus)	Fe–S cluster	C-terminus	CIA dependent	References
Regulatory					
Fep1	*Schizosaccharomyces pombe*	[2Fe–2S]?	T	TBD	[165]
Asp1	*Schizosaccharomyces pombe*	[2Fe–2S]?	N	TBD	[124]
HapX	*Cryptococcus neoformans*	[2Fe–2S]?	P	TBD	[116]
Metabolic					
Nitrite reductase	*Aspergillus nidulans* *Fusarium oxysporum*	[4Fe–4S] and [2Fe–2S]	**W**	TBD	[166]
Neverland	*Drosophila melanogaster*	[2Fe–2S]	**W**	TBD	[167]
Choline monooxygenase (Cmo1)	*Scheffersomyces stipitis*	[2Fe–2S]	**W**	TBD	[168]
Viral replication					
Nsp12	SARS-CoV-2	[4Fe–4S]	L	Assumed	[169]

TBD: To be determined.
C-termini in bold correspond to the CIA targeting complex recognition signal.

4.3.4 The Early CIA and Involvement of [2Fe–2S] Clusters in [4Fe–4S] Cluster Biogenesis

A third cytosolic [2Fe–2S] cluster pathway is the maturation of early CIA components Dre2, Cfd1, and Nbp35 (Figure 4.1, green box). Dre2 contains an N-terminal S-adenosyl methionine methyltransferase-like fold (SAM-MT) and a C-terminal domain that is predicted to be largely unfolded in the apo state, which contains two well-conserved Fe–S-binding sites (Figure 4.2d) [30, 32]. *In vivo* in yeast, Dre2 does not require any of the CIA proteins for maturation and function [3, 36, 150]. In-cell EPR and Mössbauer of the human homolog (CIAPIN1 also known as Anamorsin) support the presence of two [2Fe–2S] clusters bound to protein expressed in *E. coli* cells [120]. Contrarily, for the recombinantly purified yeast protein expressed in *E. coli*, the ability to bind both [2Fe–2S] and [4Fe–4S] clusters has been demonstrated [33, 34]. The physiological binding of a [4Fe–4S] cluster would make Dre2 an exception to the rule that all cytosolic [4Fe–4S] clusters depend on CIA components or might indicate that the [4Fe–4S] cluster of Dre2 is generated *in situ* from [2Fe–2S] clusters by Dre2/Tah18/NADPH.

Dre2 forms a complex with the diflavin reductase Tah18 (Figure 4.2d), which reduces the Fe–S clusters of Dre2 using electrons derived from NADPH [36, 38]. The

AF2 model of Tah18 strongly resembles the crystal structure of NADPH-cytochrome P450 reductase (PDB 2BN4) [36, 171]. Only the flavin mononucleotide-binding domain (FMN-BD) has been structurally characterized with bound FMN (Figure 4.2d) [38]. Reduced Dre2 is proposed to provide the reducing equivalents for the formation of [4Fe–4S] clusters on the CIA scaffolding complex composed of Cfd1 and Nbp35 [172]. In plants, Nbp35 and Dre2 have been demonstrated to interact [173]. Dre2 and Tah18, as well as cGrxs, also play important roles in the maturation of di-iron ribonucleotide reductase (RNR), though this does not require the transfer of an Fe–S cluster and is independent of the rest of the CIA components [174].

The CIA [4Fe–4S] scaffolding complex is composed of the two homologous P-loop ATPase proteins Cfd1 and Nbp35 in yeast and humans (Figures 4.1 and 4.2e) [42]. Plants and some protists only contain Nbp35 (Table 4.1) although in yeast and humans, both proteins are essential for the CIA pathway [7, 21, 22, 42]. Whereas ^{55}Fe binding in yeast to Cfd1 is not CIA nor cGrx dependent, binding to Nbp35 and Cfd1/Nbp35 complexes is strongly dependent on Dre2 and Tah18 [3, 36]. In yeast and humans, both proteins form heterodimeric- or tetrameric complexes and contain a bridging C-terminal Fe–S-binding site (Figure 4.2e,f, Fe–S site 1) [41, 42, 175]. Site 1 in the Cfd1 homodimer has been the only structurally characterized Fe–S site for any CIA component showing a [4Fe–4S] cluster bridged between two Cfd1 subunits (PDB 6G2G, Figure 4.2f) [22], which is coordinated by the cysteine residues of the CXXC motif as defined by *in vivo* studies [41]. Nbp35 is distinguished from Cfd1 by the presence of an N-terminal ferredoxin-like Fe–S cluster-binding domain (Figure 4.2e, Fe–S site 2). Biochemical and biophysical analyses indicate that both Fe–S sites can accommodate [4Fe–4S] clusters in proteins that have been either chemically reconstituted or formed from the reductive fusion of [2Fe–2S] clusters [22, 41, 42, 157, 172]. Using various conditions and mutant proteins, [2Fe–2S] clusters have also been spectroscopically observed in both Fe–S sites in protein complexes of the human Nbp35 homolog [157, 172]. Notably, [2Fe–2S] clusters bound at Fe–S site 1 do not readily convert to [4Fe–4S] clusters in the presence of GSH as compared to the conversion observed at Fe–S site 2 [157]. This reductive coupling at the N-terminal site can also be promoted by the reduced [2Fe–2S] cluster at Fe–S site 1 in the human Dre2 homolog [172]. The biogenesis of [4Fe–4S] clusters at Fe–S site 2 in Nbp35 is reminiscent of the reductive coupling of [2Fe–2S] clusters at the late ISC machinery in the mitochondrion [176], yet on an entirely different protein scaffold.

Like the early ISC and SUF systems, cytosolic [4Fe–4S] cluster biogenesis on Cfd1/Nbp35 scaffolding complexes involves the hydrolysis of ATP, yet via distinctly different protein complexes (Table 4.1). Mutation of ATP-binding residues is lethal in yeast cells indicating that ATP binding and hydrolysis are allosterically coupled to either [4Fe–4S] cluster biogenesis or transfer steps [41, 177]. *In vitro* work has confirmed Cfd1–Nbp35 and homodimeric Nbp35 to have ATPase activity, whereas homodimeric Cfd1 alone does not have activity but may be involved in dictating the binding of nucleotides [175, 178]. Since some organisms like *Arabidopsis thaliana* lack Cfd1 (Table 4.1) and thus only rely on Nbp35 as a CIA scaffold complex

[179], it has also been proposed that Cfd1 evolved to help enhance the lability of [4Fe–4S] clusters and transfer to downstream targets [180]. Further studies will have to elucidate the complex choreography of Fe–S binding, Fe–S reduction, and nucleotide binding on Cfd1/Nbp35 complexes and how Fe–S clusters are further passed on down the assembly line to the CIA trafficking component Nar1.

4.4 [2Fe–2S] and [4Fe–4S] Protein Assembly at the Late CIA Machinery

4.4.1 Nar1: Between the Early CIA and Late CIA Machinery

The pursuit of identifying new cytosolic and nuclear proteins in the early 2000s led to the discovery of Nar1 being an Fe–S protein. From staging experiments, it was shown that the acquisition of Fe–S clusters on Nar1 requires the early CIA components Dre2, Tah18, Nbp35, and Cfd1 [36, 42], but not the function of late CIA protein Cia1 [72]. As protein–protein interaction studies firmly established both association of Nar1 with Cfd1–Nbp35 [181] and with the CTC [65, 67, 104, 181–183], the current consensus is that Nar1 links the early and late CIA machineries as a [4Fe–4S] cluster trafficking protein. The existence of higher order complexes containing Cfd1–Nbp35, Nar1, the CTC, and CIA substrates has also been reported [181], pointing to the hypothesis that a CIA supercomplex could (temporarily) exist. The occurrence of a gene encoding a fusion of Cfd1 and Cia1 proteins in *Schizosaccharomyces pombe* (which led to the recognition of Cia1 [72]) would corroborate the close proximity of the early and late CIA-machinery proteins in a supercomplex.

Nar1 has a considerable amino acid sequence identity to [FeFe]-hydrogenases [43, 46]. Primary sequence comparison and EPR spectroscopic studies on yeast Nar1 purified from *E. coli* (containing 3.5 Fe/Nar1) defined two different [4Fe–4S] cluster-binding sites: one cluster with a rhombic spectrum with sharp features and a second cluster that could only be detected at elevated microwave power [43] and upon full photochemical reduction [44]. Four highly conserved cysteine residues in Fe–S site 1 of Nar1 and its homologs align with the residues binding the [4Fe–4S] cluster of the unique H cluster, $[4Fe\text{–}4S]_H$, of [FeFe]-hydrogenases. Since yeast and humans lack the machinery to synthesize the other H-cluster component, ($[Fe_2(S\text{-}CH_2\text{-}NH\text{-}CH_2\text{-}S)(CO)_3(CN)_2]$ or $[2Fe]_H$) [46], this [4Fe–4S] cluster of Nar1 is anticipated to have a void or tunnel adjacent to this cluster. A second Fe–S-binding site contains four cysteine residues at the N-terminus of Nar1, which are arranged in a ferredoxin-like C-X_n-C-XX-C-XX-C motif (Figure 4.2g,h), and corresponds to the median cluster in the *Desulfovibrio desulfuricans* [FeFe]-hydrogenase [184]. In Nar1, the two predicted [4Fe–4S] clusters are expected to be separated by 8.1 Å, as estimated by protein modeling (Figure 4.2g,h). Such distance is in full agreement with the spin–spin coupling between the clusters in EPR spectroscopy. Since the rhombic EPR spectrum was abolished upon mutagenesis of cysteine residues at Fe–S site 1, it was assigned to the $[4Fe\text{–}4S]_H$ site [46]. The broad spin–spin coupled features in fully reduced wild-type Nar1 were assigned to the [4Fe–4S] cluster at

Fe–S site 2. EPR spectroscopy of the *A. thaliana* Nar1 protein has also indicated the presence of [4Fe–4S] clusters in the two Fe–S sites [45]. Thus far, no crystallographic or Cryo-EM structure for Nar1 has been obtained, which presumably correlates with the instability and the low cluster occupancy.

In humans, two Nar1 homologs occur, which are called Iron-Only hydrogenase-like Proteins 1 and 2 (IOP1 and IOP2). Studies on Fe–S protein maturation in human cells and *in vivo* experiments in knock-out mice pointed out that IOP1 (now commonly referred to as CIAO3, Table 4.1) is the homolog that is active in the CIA machinery [65, 88, 185]. CIAO3 has also been shown *in vitro* to form complexes with CIAO1 and CIAO2A of the CTC [186]. In addition to its central role in the CIA pathway, studies have also suggested a potential role of Nar1 and homologs being important in O_2 sensing. In human cells under normoxic conditions, CIAO3 (but not IOP2) was shown to be involved in both the CIA machinery and to participate in the repression of hypoxia-induced genes [88]. Using a high-throughput CRISPR/Cas9 screen to identify human genes with differential fitness between 1% and 21% O_2 levels, five genes encoding proteins for Fe–S cluster biogenesis were found selectively nonessential at low O_2 levels [187], among which *CIAO3* was the only gene of the CIA machinery. Further physiological studies reported the beneficial effect of hypoxia on growth of nematodes, yeast [188], and plants [45, 189, 190] in which Nar1 was depleted. This even led to the name *Gollum* for *A. thaliana* Nar1 (*G*rowth in different *O*xygen *L*eve*L*s infl*U*ences *M*orphogenesis). In addition to the possible connection with oxygen levels, Nar1 is the most susceptible CIA component to thiol redox stress in yeast [191]. It remains to be determined if Nar1 is directly involved in the oxidative stress response or if the observations are a secondary phenotype, similar to the pleiotropic phenotypes observed in *MMS19* mutants [65]. Overall, studies suggest Nar1 to be a very sensitive protein of the CIA pathway playing a central but not yet clearly defined role.

4.4.2 The CTC

The identification of the first CIA components Nbp35, Cfd1, and Nar1 eventually led to uncovering proteins responsible for the final insertion of [4Fe–4S] into target proteins, i.e. the late CIA machinery. This was first demonstrated by the minimal impact of Cia1 depletion in yeast on the incorporation of ^{55}Fe into early CIA-machinery components, such as Dre2, Nbp35, and Nar1, but at the same time significantly affecting [4Fe–4S] target proteins [36, 72]. The further identification of Cia2 as a CIA factor [192], along with studies revealing *CIA1* and *MMS19* as negatively interacting genes [193] or Cia1 and Mms19 as interacting partner proteins [194], led Stehling and colleagues [65] to define the human and yeast CTC (Figures 4.2i and 4.4) as the final component of the CIA machinery. Particularly pleasing was that the puzzling pleiotropic phenotypes of *MMS19* (formerly named *MET18*) mutants could all be interpreted on the sound scientific basis of defective Fe–S protein biogenesis. A methionine auxotrophy derives from defective biosynthesis of the [4Fe–4S] cluster containing enzyme sulfite reductase (hence the name Met18 [195]). Defective Fe–S cluster incorporation into [4Fe–4S] helicases responsible for transcription and DNA repair explains: (i) methyl methanesulfonate sensitivity, (ii) defective nucleotide excision repair (NER) and transcription [196], (iii) UV sensitivity [197], and (iv) genetic instability [198].

The structures of mouse MMS19 [48] and yeast Mms19 [49] revealed that both contain 46 α-helices forming 23 tandem repeats of ~40 amino acids each constituting two α-helices (Figures 4.2i and 4.4b). Such HEAT repeats, the acronym derives from four protein members (*H*untingtin, *e*longation factor 3, protein phosphatase

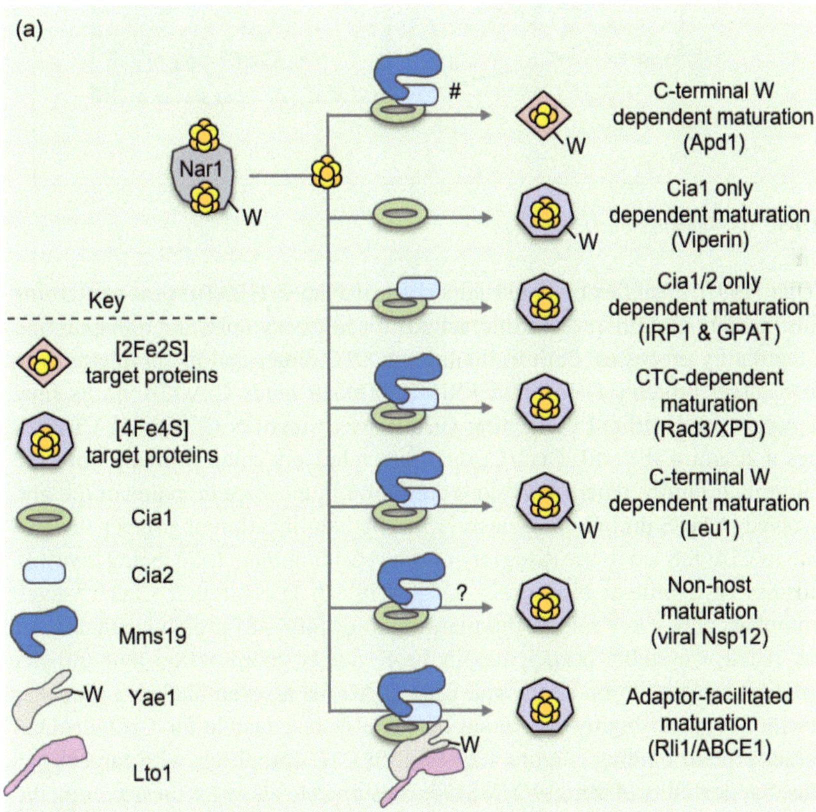

Figure 4.4 Fe–S target protein recognition by the CTC. (a) Overview of the varying dependencies of Fe–S proteins on the late CIA components. Yeast names are given for CIA components for clarity but variations in branching may be applicable in any eukaryote depending on the CIA components present. Proteins for each type of precedent are given in parentheses. Cartoons with –W represent the C-terminal tryptophan important for CTC recognition. #, the exact mechanism of how [2Fe–2S] clusters are inserted into Apd1 at the CTC is unclear. (b) Molecular model of a complex depicting adaptor (Pri1/PRIM1)-facilitated interaction of an Fe–S target protein (Pri2/PRIM2) at the CTC. The inset depicts how the C-terminal phenylalanine of DNA primase subunit 1 (PRIM1) allows recognition by late CIA components (CIAO1 and CIAO2B). A second recognition interface is seen between the Fe–S domain of DNA primase subunit 2 (PRIM2) and N-terminus of MMS19. Source: (b) Marquez et al. [96]/National Academy of Sciences.

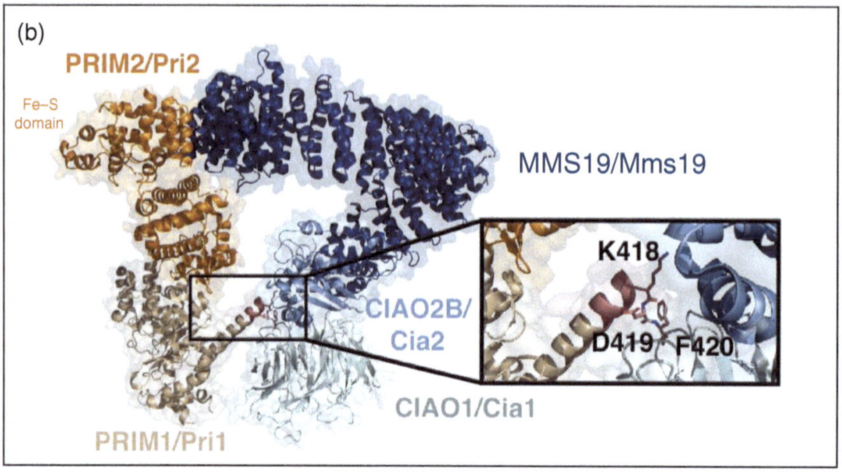

Figure 4.4 (*Continued*)

2*A*, *T*OR1 [199]), form flexible S-like superhelical shapes. HEAT repeat-containing proteins mediate protein–protein interactions for nucleocytoplasmic transport and many regulatory processes. Both in the human CTC dimer and in complexes with the Fe–S target proteins DNA2 and PRIM2, MMS19 binds CIAO2B via its final HEAT repeat and, without contacting CIAO1 itself, recruits CIAO1 via CIAO2B (Figures 4.2i and 4.4b) [48]. Cia2/CIAO2B are relatively small proteins from the DUF59 protein family, where homologs are found in all three domains of life and are involved in Fe–S protein biogenesis [15]. The identification of a hyper-reactive cysteine in CIAO2B led to the discovery of the yeast homolog, Cia2, being involved in cytosolic Fe–S protein biogenesis [192]. While it is clear that Cia2/CIAO2B and homologs play a key role in the maturation of [4Fe–4S] proteins, whether or not the reactive cysteine participates in Fe–S cluster coordination and transfer (Figure 4.2i) remains to be addressed. Cia1/CIAO1 is a seven-bladed ß-propeller protein [50, 51] with highly conserved blades 2–4, responsible for Cia2/CIAO2B and target protein binding (Figure 4.2i) [48]. In CTC complexes with target Fe–S proteins, the flexibility of Mms19/MMS19 is presumed to allow for the accommodation of various Fe–S target proteins of different protein families and to orchestrate the transfer of Fe–S clusters (Figure 4.4a and see discussion in Section 4.4.3). Strikingly, the CTC may even be high-jacked by nonhost proteins for the insertion of [4Fe–4S] clusters, as seen for the viral protein Nsp12 encoded by SARS-CoV-2 (Figure 4.4a and Table 4.4) [169]. Conserved lysine residues of the final HEAT repeat of MMS19, also responsible for CIAO2B binding, can be ubiquitinated by the NSE1 E3 ubiquitin ligase upon melanoma-associated antigen F1 (MAGE-F1) binding, leading to the degradation of MMS19 [70], and consequently impairment of CTC function. Thus, Mms19/MMS19 mediate both protein–protein interactions (with Cia2/CIAO2B and target Fe–S proteins) and regulatory roles, like many other HEAT repeat proteins. Structural data demonstrate a range of higher order oligomeric states of yeast Mms19 (tetramers and hexamers), which disassemble upon Cia2 addition [49]. It is not clear whether the oligomeric states correspond to "resting states," or even function as self-protection against ubiquitination. With the dimeric mouse MMS19 structure, the dimeric mouse/fly hybrid CTC, and the monomeric human CTC–target protein complexes [48], a highly flexible landscape of species is revealed.

In human cells, CTC function appears to be more complex than in yeast. Two Cia2 homologs, CIAO2A (FAM96A) and CIAO2B (FAM96B), are present in humans and form mutually exclusive complexes with CIAO1 (Table 4.1) [69]. Whereas CIAO2B along with CIAO1-MMS19 assembles most cytosolic and nuclear Fe–S proteins, CIAO2A appears to have a divergent function associated with Fe regulation via iron-regulatory proteins 1 and 2 (IRP1 and IRP2). First, maturation of IRP1 (with [4Fe–4S] acting as cytosolic aconitase) is dependent on CIAO2A. Second, IRP2 (lacking the [4Fe–4S] binding site) is stabilized by CIAO1 and CIAO2A [69]. Along these lines, *in vivo* maturation and co-immunoprecipitation data with human cell lines indicated that differential target specificity can occur at the late CIA machinery (Figure 4.4a). The variations in the essential requirements of CTC components are exemplified by the following examples: (i) only CIAO1 in the maturation of Viperin [87], (ii) only CIAO2A-CIAO1 for the maturation of IRP1 [65, 69], (iii) only CIAO2B–CIAO1 for the maturation of GPAT [65, 69, 70], (iv) full CTC complex for the maturation of XPD [65], and (v) full CTC complex with additional adaptor proteins for the maturation of ABCE1, ELP3, and PRIM2 [96, 104]. These findings describe the bare minimum of what is required for the maturation of some target Fe–S proteins, yet in some cases, the most efficient maturation likely occurs with a full CTC as evidenced by Viperin binding additionally to MMS19-CIAO2B and exceptionally also CIAO2A. Furthermore, POLD1 (the catalytic subunit of DNA polymerase δ) maturation is strictly dependent on MMS19 and CIAO1, yet only mildly sensitive to CIAO2B depletion and not at all to CIAO2A depletion [65, 69]. Along these lines of essentiality, many unicellular eukaryotes have CIAO1 and CIAO2B, but lack Mms19 [28, 115]. Moreover, yeast strains lacking Mms19 are viable, yet they are sensitive to UV and DNA damaging agents [200], are methionine auxotrophic, and show retarded growth already at 30 °C [201]. Rad3 overexpression could correct the UV sensitivity of Δ*mms19* yeast, but not the temperature sensitivity and methionine auxotrophy, which can be understood in the light of Mms19 function in cytosolic and nuclear Fe–S proteins maturation. It may be expected that further variations on the theme of differential maturation efficiency are possible. For example, *A. thaliana* contains three Cia2/CIAO2B homologs and *M. exilis* contains two, which remain to be explored functionally (see Table 4.1) [183].

4.4.3 Bipartite Recognition of Targets

Similar to the discovery of bacterial, mitochondrial, and cytosolic Fe–S protein biogenesis machineries nullifying the outdated assumption that Fe–S clusters spontaneously assembly from $Fe^{2+/3+}$ and S^{2-} ions in the cell, the naive assumption that Fe–S clusters find their target proteins by binding to the most thermodynamically favored sites has of late also been challenged. In addition to the selectivity provided by the CTC, the maturation of Fe–S target proteins appears to be carefully orchestrated by protein–protein contacts involving additional specialized adaptor proteins and at least one known recognition motif. Notably, heat shock protein (HSP) chaperones in yeast (e.g. Hsp70 and Hsp90) do not seem to play a general role in the process of cytosolic Fe–S protein biogenesis despite evidence of genetic and physical interactions [202].

Although a C-terminal tryptophan residue was identified as a unique feature of Nar1 [44], was required in Viperin for correct targeting to the CTC and Fe–S cluster insertion [203], and was required in the adaptor protein Lto1 (Figure 4.2j) for the correct maturation of Rli1 [104], the broader role of this primary structural determinant in CTC recruitment remained underappreciated. Further correlation of an aromatic C-terminal residue and Fe–S cluster binding could be found in Apd1, Pol3, Rev3, and Leu1 [6, 59, 100]. Extensive bioinformatic analysis of eukaryotic Fe–S cytosolic and nuclear proteins revealed that 25% of these proteins contain either tryptophan (W) or phenylalanine (F) at their C-terminus (Figure 4.5a) [96]. Included in the analysis were multiple [2Fe–2S] cluster-binding proteins (see Section 4.4.4). Notably, prokaryotic, archaeal, and eukaryotic mitochondrial and plastidial Fe–S proteins, as well as non-Fe–S proteins, showed no C-terminal enrichment of these amino acids. Enrichment was observed not only for the C-terminal W or F but also for the amino acids at positions −1 and −2, leading to the identification of a [LIM]-[DES]-[WF]-COO^- consensus motif (Figure 4.5b,c). Because the motif is restricted to the C-terminus, it exhibits much higher specificity compared to internal motifs, which can occur at any position within the polypeptide. For instance, the LYR consensus motif ([ILV]-[FY]-[KR]), found in mitochondrial Fe–S recipient and/or accessory proteins that are recruited by IscU-Hsc20 ISC machinery components [205], is statistically present in nearly every second eukaryotic protein (51% in yeast and 46% in humans, https://www.genome.jp/tools/motif/MOTIF2.html).

The function of the C-terminal recognition motif was demonstrated *in vitro* through affinity purification of CTC–Leu1 and CTC–Apd1 complexes [96], which critically depended on the presence of the C-terminal tryptophan residue. For at least six structurally and functionally unrelated cytosolic and nuclear Fe–S proteins, including Pol3, Nar1, Apd1, and Leu1 [96], as well as early observations on Viperin [203] and Rli1 (via Lto1/Yae1) [104], the CTC recognition motif has been shown to be operative *in vivo* (Figure 4.4a). The earlier observation of high-copy suppression of *LTO1* mutations by *RLI1* [206] is in full agreement with the adaptor function. Interestingly, Lto1 and Yae1 are not essential for growth of yeast under anaerobic conditions [207]. Possibly, these adaptors increase the efficiency of Rli1 maturation under aerobic conditions at which turnover due to cluster loss of Rli1 is higher [104]. Rli1 levels are known to be critical under oxidative stress [208], a phenomenon that might equally apply to Nar1 levels under aerobic conditions (see Section 4.4.1).

The case of Lto1, acting as an adaptor protein alongside Yae1 (Figure 4.2j,k), is particularly intriguing, as it suggests the possible existence of other, previously unrecognized adaptor proteins (Figure 4.4a). One such candidate that awaits to be studied is the human MLF2 protein, which terminates with a YDW sequence but lacks the deca-GX_3 sequence motif found in Lto1 and Yae1. Bioinformatic analysis

(a)

Function	Protein name		C-terminal residue							
	Human	Yeast	Hs	Sc	An	Nc	Dr	Xl	Dm	At
DNA Pol δ	POLD1	Pol3	W	W	W	W	W	W	W	F
DNA Pol ζ	REV3L	Rev3	F	W	W	W	L	F	F	F
Purine synthesis	PUR1	Ade4	W	&	&	&	W	W	W	%
RNA Pol III	POLR3F	Rpc6	F	&	F	I	F	F	D	F
AA synthesis	*none*	Leu1		W	W	W				%
Antiviral protein	RSAD2	*none*	W				W	W		
Oxidoreductase	*none*	Apd1		W	W	W				%
tRNA thiolation	CTU1	Ncs6	F	F	G	I	F	H	F	F
Neverland	*none*	*none*					F	W	W	
Nitrite reductase	*none*	*none*			W	W				
CIA factor	CIAO3	Nar1	W	W	W	W	W	W	W	W
Rli1 adaptor[a]	LTO1	Lto1	F	W	H	V	F	F	F	F
Primase[a]	PRIM1	Pri1	F	F	F	F	F	F	F	W
Elongator[a]	ELP4	Elp4	F	Y	F	F	F	F	F	F

[a] Non Fe–S or adaptor protein

Hs *H. sapiens*
Sc *S. cerevisiae*
An *A. nidulans*
Nc *N. crassa*
Dr *D. rerio*
Xl *X. laevis*
Dm *D. melanogaster*
At *A. thaliana*
C-terminal amino acid
% Plastidial
& non Fe–S
absent

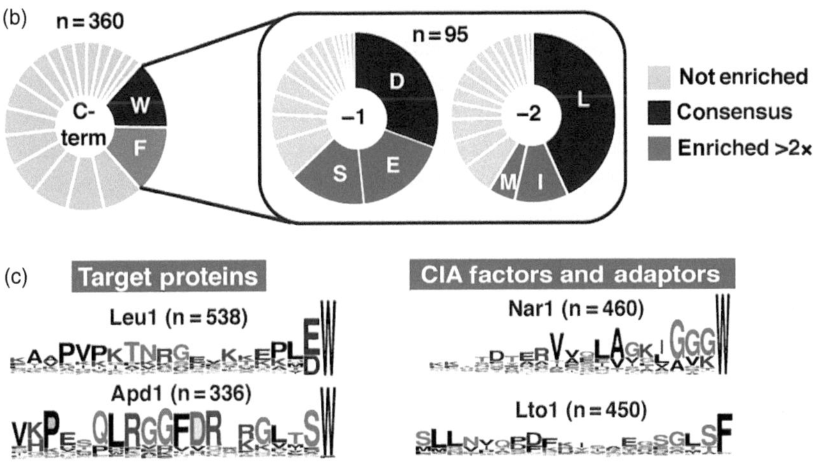

Figure 4.5 A C-terminal tryptophan or phenylalanine residue is conserved in 26% of cytosolic Fe–S proteins in eukaryotes. (a) Inventory of Fe–S proteins with tryptophan (W) or phenylalanine (F) at their C-terminus. Non-Fe–S subunits (PRIM1/Pri1 or ELP4/Elp4) or an adaptor protein (LTO1/Lto1) dedicated to ABCE1/Rli1 maturation also employ this recognition. (b) The recognition motif is a C-terminal tripeptide. Consensus and enriched amino acids in comparison to the amino acid composition of human proteins are shown for 95 C-termini ending with W/F. (c) Weblogos [204], depicting the conservation of the C-terminal sequence by the height of single-letter amino acid codes. Examples for isopropylmalate isomerases (Leu1, [4Fe–4S]) and Apd1 ([2Fe–2S]) target proteins, as well as the CIA protein Nar1 and the adaptor protein Lto1 are shown. The number of primary sequences (n) analyzed is depicted above the Weblogos. Source: Marquez et al. [96]/National Academy of Sciences.

(Figure 4.5a [96]) and the cryoEM structure of the human CTC–PRIM1–PRIM2 complex [48] have already identified two further adaptor proteins: ELP4 and PRIM1. These subunits accompany the [4Fe–4S] cluster containing ELP3 and PRIM2 subunits in the elongator and primase complexes, respectively (Table 4.3). Notably, after remodeling the C-terminal α-helix of PRIM1 (which ends in a

KDF sequence, Figure 4.5a [96]) into the known structure (Electron Microscopy Data Bank, EMDB-11017), it became evident that the C-terminal recognition motif contacts the CIAO2B–CIAO1 heterodimeric interface (Figure 4.4b). Further structural and functional studies are awaited to determine if other C-termini from the 25% of Fe–S proteins ending in W or F interact in a similar position as that of the PRIM1 adaptor protein.

The CryoEM structural studies [48] of the human CTC with the associated human [4Fe–4S] target proteins DNA2 or the primase complex (Figure 4.4b) revealed the first structural insights into the recruitment of apoproteins for Fe–S cluster insertion. In cases involving C-terminal recognition motifs, it is predicted that these motifs interact with CIAO1/Cia1 (most likely via the highly conserved blade 3), not only localizing but also positioning Fe–S target proteins or their relevant adaptor partners at the CTC. Furthermore, additional contacts at the N-terminus of MMS19/Mms19 can facilitate the optimal orientation, although the primary or secondary structural requirements for binding remain unknown. These structures have yet, however, to define how Fe–S clusters are delivered to the cofactor binding sites of Fe–S target proteins. It remains unclear if CIAO3/Nar1 binds to CTC-target protein complexes for direct Fe–S cluster delivery as suggested by the evidence of supercomplexes [181] or if the former transfers Fe–S clusters first to the CTC complex, possibly via the reactive cysteine of Cia2/CIAO2B (Figure 4.2i). For the ~75% of proteins lacking an F/W C-terminus, little is understood about how they are positioned at the CTC. Further high-resolution structures of the CTC with targets and the description of interaction sites are necessary to reveal how positioning of apo targets and acceptor cysteine residues takes place. Such structures along with protein complex predictions by Alphafold will also be important for the development of biochemical studies and reconstitution assays, which will reveal the dynamics of Fe–S cluster transfer at the CTC.

Table 4.3 summarizes the current knowledge regarding cytosolic and nuclear Fe–S proteins in the reference organisms, yeast and humans, including their cluster types. With the Fe–S-containing CIA factors (Table 4.2), the total number of known Fe–S proteins reaches 52. The data indicate that [4Fe–4S] clusters are the most abundant, comprising a total of 39 (including multiple clusters within single proteins), followed by [2Fe–2S] clusters, which account for 17. Four proteins are known to potentially bind both [2Fe–2S] and [4Fe–4S] clusters. For the NCOA4 protein, UV–Vis spectra clearly indicate the presence of a $[2Fe–2S]^{2+}$ cluster; however, mass spectrometry and Mössbauer spectroscopy of *E. coli* cell extracts suggest the existence of a $[3Fe–4S]^{1+}$ cluster [91]. This example further highlights the challenges of accurately identifying cluster types in physiological contexts, as it is currently impossible to rapidly isolate sufficient quantities of protein along with its native, potentially labile cluster for spectroscopic analysis (see Section 4.2). While progress has been made for 19 Fe–S proteins in establishing dependence on the CIA machinery, recent findings have shown that six [2Fe–2S] proteins are independent of this machinery (refer to Section 4.3.3) [3]. The dependence of the CIA machinery for 20 proteins identified as authentic Fe–S proteins, based on robust spectroscopic, bioinformatic, and structural data, remains to be determined (TBD in Table 4.3).

4.4.4 Potential [2Fe–2S] Cluster Insertion into Proteins at the CTC

Analysis of Fe–S proteins not occurring in humans and yeast (Table 4.4) identifies three classes of eukaryotic proteins predicted to contain a Rieske-type [2Fe–2S] center. All harbor a C-terminal tryptophan residue unlike the CMP-*N*-acetylneuraminic acid hydroxylase Rieske protein found in humans (Figure 4.5a and Tables 4.3 and 4.4) and, since they are lacking a mitochondrial import sequence, are predicted to be cytosolic or nuclear proteins. Neverland/DAF36 constitutes a family of 7-cholesterol desaturases, which are responsible for steroid hormone biosynthesis in ciliates, nematodes, and insects [209]. Anchored to membranes via N-terminal transmembrane helices, this class of enzymes has a central Rieske and a C-terminal mononuclear-iron binding catalytic domain [167]. Fungal choline monooxygenases [168] have a similar domain structure, but seem to lack the N-terminal membrane anchoring domain. The third class of cytosolic Rieske-containing proteins are fungal multiple domain nitrite reductases [166]. Their N-terminal flavin-containing domain is joined via a siroheme-[4Fe–4S] catalytic domain to the C-terminal Rieske domain, equipped with LEW as a C-terminal CTC recognition sequence in *Neurospora crassa*. These three Rieske protein classes share the commonality of 2Cys/2His coordination found in Apd1, yet Apd1 belongs to a distinctly different protein family.

Extending the case study of Apd1 in Figure 4.2, Figure 4.6 demonstrates the importance of the C-terminal W in CTC interaction and [2Fe–2S] cluster binding. Apd1 function in the gallobenzophenone growth assay (Figure 4.6a, compared with Figure 4.3f), in an *in vitro* CTC binding assay (Figure 4.6b), and in the *in vivo* ^{55}Fe incorporation assay (Figure 4.6c) all depend on the C-terminal recognition sequence and the CIA machinery [3, 96]. Based on these observations, it is hypothesized that the three cytosolic classes of Rieske proteins will also be CIA dependent. It would be remarkable, if structurally different [2Fe–2S] proteins (i.e. Rieske and Apd1) with bis-histidinyl coordination, contrary to all other cytosolic and nuclear [2Fe–2S] proteins, require the CIA machinery for maturation. Studies on the function of C-terminal target sequences should be approached with caution, as nonspecific Fe–S cluster delivery, in a CIA and C-terminal tryptophan-independent manner has been uncovered for Apd1 that has the C-terminal W capped by a polypeptide extension (Apd1-Cap, Figure 4.6d) [3].

In addition to Apd1, a further cytosolic [2Fe–2S] cluster-binding protein, FBXL5, binds to the CTC independent of a C-terminal W (Table 4.3) [90, 210]. Along with IRPs, FBXL5 is an integral player of Fe homeostasis in humans, which is also oxygen dependent. For more details on the complexity of Fe–S clusters involved in Fe homeostasis, the reader is referred to specialized literature [69, 90, 210, 211]. The dependencies of ISC, Atm1, and CIA components on the maturation of the [2Fe–2S] cluster in FBXL5 though have yet to be investigated. Taken all together, the CTC is a central hub coordinating the biogenesis of cytosolic [2Fe–2S]/[4Fe–4S] cluster-binding proteins and homeostasis of iron and oxygen in humans, in a manner that is regulatable by posttranslational modifications. In addition to the

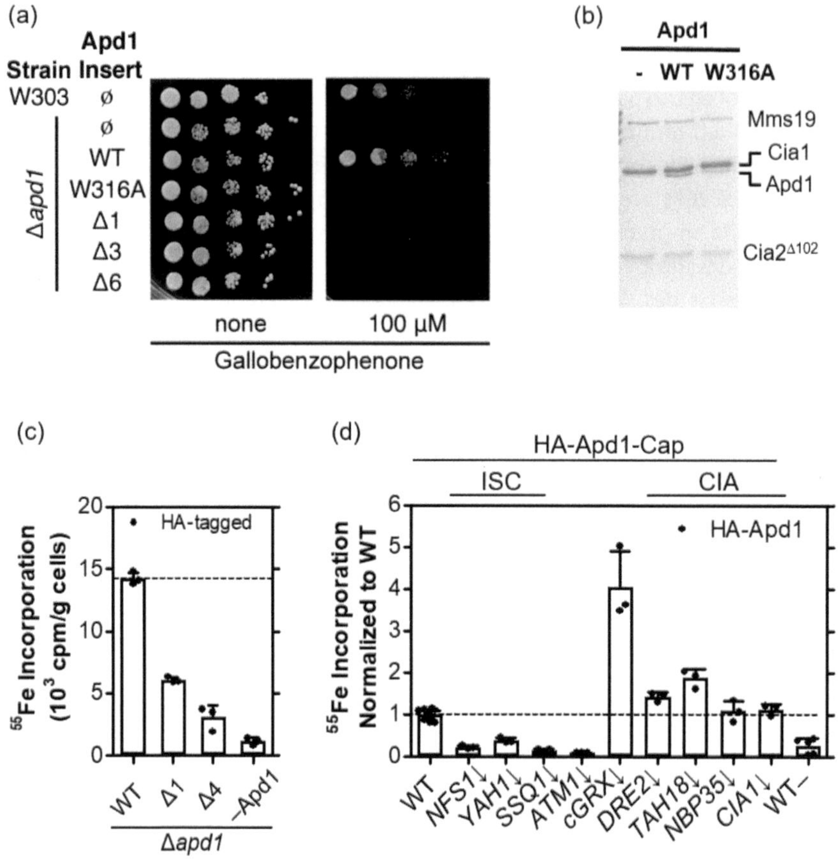

Figure 4.6 The validation of the C-terminal tryptophan being crucial for the binding of a [2Fe–2S] cluster to Apd1. (a) Similar to Apd1 variants lacking the [2Fe–2S] cluster due to removal of ligands (Figure 4.3f), growth of *Δapd1* yeast cells in the presence of gallobenzophenone is abolished when the C-terminus of Apd1 is truncated or the W316A variant is used. (b) The Streptactin affinity copurification of yeast CTC subunits (Mms19, Strep-tagged), Cia1, and Cia2 with Apd1 (wild-type WT) is less efficient for the W316A variant of Apd1. Full-length Cia2 has a similar electrophoretic mobility, so an N-terminally truncated variant of Cia2 (Δ102) was used, which equally bound to Cia1/Mms19 in the absence of Apd1 (–). Complexes were revealed by SDS-PAGE and Coomassie staining. (c) ^{55}Fe incorporation into Apd1 in yeast critically depends on its C-terminus. (d) If the native C-terminus of Apd1 is capped by a 25-residue extension (HA-Apd1-Cap), a 4–10-fold lower ^{55}Fe incorporation in yeast is seen, which is independent of the CIA machinery. Source: (a) and (b) [96]/National Academy of Sciences / CC BY 4.0. (c) and (d) Braymer et al. [3]/National Academy of Sciences.

questions concerning [4Fe–4S] protein maturation, several intriguing questions on the maturation of C-terminal tryptophan-containing [2Fe–2S] proteins also remain. Could oxidative decoupling of [4Fe–4S] clusters at the CTC provide [2Fe–2S] clusters for Fe–S target proteins or could delivery of both [2Fe–2S] and [4Fe–4S] via the CIA components occur at the CTC?

4.5 Relevance for Disease and for Biotechnology

4.5.1 Diseases Related to Cytosolic and Nuclear Fe–S Proteins

NER is a very important pathway to recognize and repair DNA damage, such as UV- or chemical-induced DNA adducts. NER employs numerous proteins, including the [4Fe–4S] cluster containing helicase XPD (xeroderma pigmentosum group D) protein, which is part of the transcription factor IIH complex not only essential for NER but even more so for transcription. Inherited mutations in XPD are associated with diseases such as XPD, Cockayne syndrome, and trichothiodystrophy [212]. Depending on the mutation, patients are sensitive to sunlight and have a skin cancer predisposition, skin, nail, or neurodevelopmental defects. The three further XPD-like [4Fe–4S] helicases (FANCJ, DDX11, and RTEL1) have roles in DNA repair of interstrand crosslinks and function to bypass lesions at stalled replication forks. Mutated alleles of FANCJ are associated with Fanconi anemia and are prominent in breast and ovarian cancers [213]. Biallelic mutations of DDX11 lead to Warsaw Breakage Syndrome, a rare disorder related to Fanconi anemia [214]. Mutations of RTEL1 are thought to cause telomere-related pulmonary fibrosis without bone marrow failure and some forms of dyskeratosis congenita (Hoyeraal–Hreidarsson syndrome) [215]. Some but not all pathogenic variants of these proteins relate to decreased Fe–S cluster incorporation or stability (Arg112His of XPD [65]). Mutations of two [4Fe–4S] cluster containing human DNA glycosylases have been associated with polyposis and colorectal cancer. This applies to NTHL1 [216, 217], which removes oxidized pyrimidine bases, and MUTYH (related to yeast Ntg2) [218], responsible for base excision repair of 8-oxoguanine. Mutations in MUTYH may also contribute to ovarian cancer development [219]. Despite lacking a direct relation to replication and DNA repair, certain variants of the tRNA modifying CDKAL1 enzyme increase the cancer risk [220]. Finally, breast and endometrial cancers and probably brain and ovarian tumors correlate with mutations of POLD1 and POLE [221], the genes encoding the catalytic subunit of the [4Fe–4S] cluster containing replicative polymerases δ and ε [59]. In summary, mutations in at least 9 Fe–S enzymes are known to be associated with formation or development of cancer.

Several other mutations relating to cytosolic Fe–S proteins, not necessarily leading to cancer, have been identified. Mutations in the gene encoding xanthine dehydrogenase can lead to hereditary Xanthinuria Type I [222], which is a rare disorder in comparison to the more prevalent disorders resulting from mutations in molybdopterin biosynthetic genes (see Chapter 9). A severe pharmacogenetic syndrome from the combination of a deficiency of the catabolic enzyme dihydropyrimidine dehydrogenase (DPYD) and treatment with 5-fluorouracil has been recognized for decades, which if not diagnosed can lead to severe toxicity and even death of treated patients [223]. Not only an increase in the cancer risk but also the involvement of CDKAL1 single-nucleotide polymorphisms as risk factor for type 2 diabetes is very well documented (reviewed in [220]). Mutations of CISD2 cause

Wolfram syndrome 2, which is an autosomal recessive neurodegenerative disorder characterized by optic atrophy, neuropathy, diabetes mellitus, and ulcer bleeding. In patients, a G-to-C transversion leads to a splicing error, eliminating the part of the mRNA encoding the protein sequence responsible for [2Fe–2S] binding [94]. With increased sequencing efforts and better methodologies for correlation of disease with polymorphisms, it is anticipated that the consequences of mutations of further cytosolic and nuclear proteins can be better understood and hopefully improve the diagnosis and treatment of patients.

4.5.2 Diseases Related to Maturation of Cytosolic and Nuclear Fe–S Proteins

If dysfunction of Fe–S cluster-containing proteins can cause disease, it is obvious that defective insertion of their essential cofactors by the CIA could equally cause a plethora of diseases. Before the discovery of the function of Lto1 as an adaptor to guide Rli1 to the CTC (see Section 4.4), it was already known that the human homolog ORAOV1 (originally called TAOS1, tumor amplified and overexpressed sequence 1) was overexpressed in solid tumors (hence its current name *ora*l cancer *ov*erexpressed 1) [224]. Amplification of the 11q13 locus (which includes ORAOV1, RRM2, and MIR548K) is frequent in squamous cell carcinomas [225]. Following MYC and PIK3CA, amplification of this locus (6%) is the third most common in cancers. How the oncogenicity relates to the function of Lto1/ORAOV1 in Rli1 maturation is thus an important task to resolve in the future.

For at least three genes of the core of the CIA-machinery, disease associations have been found. A mutation leading to Ser161Ile in CIAO3 (NARFL, Nar1) is associated with diffuse pulmonary arteriovenous malformations in two daughters of a consanguineous family [226]. Follow-up studies reported that Ser161 is well conserved and that this mutant fails to assemble with the CTC [181]. However, no phenotypic analysis of patient fibroblasts or rescue by nonmutated alleles was reported; therefore, the direct evidence for the disease association remains to be established.

Recently, two very extensive studies defined that biallelic loss-of-function variants of CIAO1 led to neuromuscular and neurogenerative disorders [227, 228]. Summed over these two studies in total six patients, the following variants of CIAO1 were found (the number of patients, if >1, allele frequency derived from c. ~800 000 diploid human genomes [https://gnomad.broadinstitute.org]) [229]: Arg65Trp (four patients, 1.64E−3), His302Pro (three patients, 3.28E−5), His193Tyr (5.58E−6), Asp171Gly, Trp184Cys, His251Leu, Phe250–Leu339 deletion (all one patient, <1E−6). The four patients in one study (with distinct neuromuscular deficits and other findings) suffered from an onset of weakness in early childhood to adolescence and posed a wide range of disabilities and clinical manifestations at their current ages of 17–59 years [228]. In the second study, two patients (with low DPYD activity from a diagnostic screening) were identified, who tragically passed away at 18 months and 19 years of age [227]. The authors of the latter study speculated that the severe infections and respiratory failure occurring in the patients might have been related to defective antiviral host defense caused by the defective

CIA machinery. Their study also incorporated a third patient with very low DPYD activity, who had a biallelic deletion of the three nucleotides corresponding to Glu213 of MMS19 (allele frequency < 1E–6 [229]) and sadly died at 13 years of age. Important was that in both studies the causative nature of CIAO1 (and the MMS19) variants in patient fibroblasts could be pinpointed by a range of cell biological and biochemical experiments, including rescue of phenotypes by transfection of patient fibroblasts with wild-type CIAO1.

These two benchmark publications have convincingly described the first CIA-related disorders and it is anticipated that future studies will uncover additional cases involving more CIA components. Together with further understanding of the mechanisms involved in cytosolic Fe–S protein biogenesis, uncovering further CIA-associated diseases will hopefully lead to a more efficient disease recognition, improved diagnoses, and, ultimately, development of beneficial treatments.

4.5.3 Biotechnology

Microorganisms are the efficient workhorses of many biotechnological applications and are expected to play an ever-increasing role in industrial biotechnology [230]. Fe–S enzymes may be an ideal choice for various chemical transformations based on their robust enzymatic reactivity. At the same time, their application must be carefully considered in order to ensure their compatibility with cellular pathways in recombinant hosts (i.e. Fe–S cluster biogenesis and electron transfer chains). Recombinant expression of non-native Fe–S enzymes even in bacterial systems faces the challenge of inactive enzymes. Systematic studies have correlated the probability of functional expression with the phylogenetic position amongst bacteria [231]. If that is correct, expression of bacterial Fe–S enzymes in unicellular eukaryotes will face many challenges. Various other aspects such as Fe–S cluster instability, absence of compatible reductases, or electron transfer components essential for activity can impose serious hurdles to overcome [232]. Considerable progress in synthetic biology has been made on overcoming these hurdles for the expression and efficient function of nitrogenases in eukaryotes, which may in the future help relieve our dependence on the energy-intensive Haber–Bosch process (reviewed in [232]).

When pursuing biotransformations with Fe–S enzymes in the cytosol of eukaryotes, take for example yeast, it will be critical to assess how enzymes will depend on the ISC and CIA machineries. Especially if a noneukaryotic Fe–S enzyme is required, for which recognition by the CTC is inherently missing, application can be seriously hampered. As evidenced even for a native enzyme, nonspecific Fe–S cluster loading in Apd1 results only in a partially loaded enzyme that does not support cellular function (Figure 4.6d). Table 4.5 summarizes a series of trials aimed at establishing biotransformations employing bacterial Fe–S enzymes in yeast (see references in Table 4.5). Due to inactive Fe–S enzymes creating bottlenecks, these studies were not successful. The maturation of bacterial Fe–S enzymes might have failed due to the low efficiency of Fe–S cluster loading into these proteins by the CIA machinery. Yet, the observation that the human CTC can be used for maturing a viral protein [169] provides optimism that optimal protein–protein interactions can be achieved

Table 4.5 Biotechnology: unsuccessful expression of Fe–S enzymes in heterologous hosts.

Pathway	Fe–S enzyme	Product	References
Entner–Doudoroff	6-Phosphogluconate dehydratase	Ethanol	[233]
Entner–Doudoroff	6-Phosphogluconate dehydratase	Ethanol/isobutanol	[234]
Mosaic	Dihydroxyacid dehydratase	Isobutanol	[235]
Ehrlich	Dihydroxyacid dehydratase	Isobutanol	[236]
Dahms (D-Xylose oxidation)	D-Xylonate dehydratase	Ethylene glycol	[237]
Weimberg	D-Xylonate dehydratase	2-Ketoglutarate	[238]
2-C-Methyl-D-erythritol-4-phosphate (MEP)	IspG[a] and IspH[b]	Isoprenoids	[239]
2-C-Methyl-D-erythritol-4-phosphate (MEP)	IspG[a] and IspH[b]	Isoprenoids	[240]
2-C-Methyl-D-erythritol-4-phosphate (MEP)	IspG[a), c] and IspH[b), c]	Isoprenoids	[241]

a) (*E*)-4-Hydroxy-3-methylbut-2-enyl diphosphate synthase.
b) (*E*)-4-hydroxy-3-methyl-but-2-enyl diphosphate reductase.
c) Enough activity to replace the yeast mevalonate pathway, only at low aeration.
Source: Adapted from [96].

(as a result of evolution or synthetic biology). In a first proof-of-principle experiment, Marquez et al. demonstrated that the attachment of the C-terminal sequence of yeast Leu1 (isopropylmalate isomerase) successfully led to the maturation and efficient function of the *E. coli* homolog [96]. Whether this methodology can overcome inefficient Fe–S protein maturation of other bacterial Fe–S enzymes in yeast requires further studies and more structural information on the recognition of eukaryotic Fe–S proteins by the CTC. Though there is still a long way to efficient maturation of Fe–S proteins in different hosts for synthetic biology applications, we expect that further studies on the CIA machinery will reveal the determinants critical for Fe–S cluster insertion, paving the way for green chemistry employing fungal systems.

References

1 Li, P., Hendricks, A.L., Wang, Y. et al. (2022). Structures of Atm1 provide insight into [2Fe-2S] cluster export from mitochondria. *Nat. Commun.* 13 (1): 4339.

2 Pearson, S.A., Wachnowsky, C., and Cowan, J.A. (2020). Defining the mechanism of the mitochondrial Atm1p [2Fe-2S] cluster exporter. *Metallomics* 12 (6): 902–915.

3 Braymer, J.J., Stehling, O., Stümpfig, M. et al. (2024). Requirements for the biogenesis of [2Fe-2S] proteins in the human and yeast cytosol. *Proc. Natl. Acad. Sci. USA* 121 (21): e2400740121.
4 Braymer, J.J., Freibert, S.A., Rakwalska-Bange, M., and Lill, R. (2021). Mechanistic concepts of iron-sulfur protein biogenesis in biology. *Biochim. Biophys. Acta Mol. Cell Res.* 1868 (1): 118863.
5 Dancis, A., Pandey, A.K., and Pain, D. (2024). Mitochondria function in cytoplasmic Fe–S protein biogenesis. *Biochim. Biophys. Acta Mol. Cell Res.* 1871 (5): 119733.
6 Kispal, G., Csere, P., Prohl, C., and Lill, R. (1999). The mitochondrial proteins Atm1p and Nfs1p are essential for biogenesis of cytosolic Fe–S proteins. *EMBO J.* 18 (14): 3981–3989.
7 Roy, A., Solodovnikova, N., Nicholson, T. et al. (2003). A novel eukaryotic factor for cytosolic Fe-S cluster assembly. *EMBO J.* 22 (18): 4826–4835.
8 Vacek, V., Novak, L.V.F., Treitli, S.C. et al. (2018). Fe-S cluster assembly in oxymonads and related protists. *Mol. Biol. Evol.* 35 (11): 2712–2718.
9 Paul, V.D. and Lill, R. (2015). Biogenesis of cytosolic and nuclear iron-sulfur proteins and their role in genome stability. *Biochim. Biophys. Acta* 1853 (6): 1528–1539.
10 Fuss, J.O., Tsai, C.L., Ishida, J.P., and Tainer, J.A. (2015). Emerging critical roles of Fe-S clusters in DNA replication and repair. *Biochim. Biophys. Acta* 1853 (6): 1253–1271.
11 Embley, T.M. and Martin, W. (2006). Eukaryotic evolution, changes and challenges. *Nature* 440 (7084): 623–630.
12 Freibert, S.A., Goldberg, A.V., Hacker, C. et al. (2017). Evolutionary conservation and in vitro reconstitution of microsporidian iron-sulfur cluster biosynthesis. *Nat. Commun.* 8: 13932.
13 Tsaousis, A.D. (2019). On the origin of iron/sulfur cluster biosynthesis in eukaryotes. *Front. Microbiol.* 10: 2478.
14 Boyd, J.M., Drevland, R.M., Downs, D.M., and Graham, D.E. (2009). Archaeal ApbC/Nbp35 homologs function as iron-sulfur cluster carrier proteins. *J. Bacteriol.* 191 (5): 1490–1497.
15 Mashruwala, A.A. and Boyd, J.M. (2018). Investigating the role(s) of SufT and the domain of unknown function 59 (DUF59) in the maturation of iron-sulfur proteins. *Curr. Genet.* 64 (1): 9–16.
16 Maclean, A.E., Sloan, M.A., Renaud, E.A. et al. (2024). The *Toxoplasma gondii* mitochondrial transporter ABCB7L is essential for the biogenesis of cytosolic and nuclear iron-sulfur cluster proteins and cytosolic translation. *mBio* 15: e0087224.
17 Srinivasan, V., Pierik, A.J., and Lill, R. (2014). Crystal structures of nucleotide-free and glutathione-bound mitochondrial ABC transporter Atm1. *Science* 343 (6175): 1137–1140.
18 Shepherd, R.E., Kreinbrink, A.C., Njimoh, C.L. et al. (2023). Yeast mitochondria import aqueous Fe(II) and, when activated for iron-sulfur cluster assembly,

export or release low-molecular-mass iron and also export iron that incorporates into cytosolic proteins. *J. Am. Chem. Soc.* 145 (25): 13556–13569.

19 Maio, N. and Rouault, T.A. (2020). Outlining the complex pathway of mammalian Fe-S cluster biogenesis. *Trends Biochem. Sci.* 45 (5): 411–426.

20 Freibert, S.A., Boniecki, M.T., Stümpfig, C. et al. (2021). N-terminal tyrosine of ISCU2 triggers [2Fe-2S] cluster synthesis by ISCU2 dimerization. *Nat. Commun.* 12 (1): 6902.

21 Hausmann, A., Aguilar Netz, D.J., Balk, J. et al. (2005). The eukaryotic P-loop NTPase Nbp35: an essential component of the cytosolic and nuclear iron-sulfur protein assembly machinery. *Proc. Natl. Acad. Sci. USA* 102: 3266–3271.

22 Stehling, O., Jeoung, J.H., Freibert, S.A. et al. (2018). Function and crystal structure of the dimeric P-loop ATPase CFD1 coordinating an exposed [4Fe-4S] cluster for transfer to apoproteins. *Proc. Natl. Acad. Sci. USA* 115 (39): E9085–E9094.

23 Stehling, O., Netz, D.J., Niggemeyer, B. et al. (2008). Human Nbp35 is essential for both cytosolic iron-sulfur protein assembly and iron homeostasis. *Mol. Cell. Biol.* 28 (17): 5517–5528.

24 Karnkowska, A., Vacek, V., Zubacova, Z. et al. (2016). A eukaryote without a mitochondrial organelle. *Curr. Biol.* 26 (10): 1274–1284.

25 Williams, S.K., Jerlstrom Hultqvist, J., Eglit, Y. et al. (2024). Extreme mitochondrial reduction in a novel group of free-living metamonads. *Nat. Commun.* 15 (1): 6805.

26 Tsaousis, A.D., Ollagnier de Choudens, S., Gentekaki, E. et al. (2012). Evolution of Fe–S cluster biogenesis in the anaerobic parasite *Blastocystis*. *Proc. Natl. Acad. Sci. USA* 109 (26): 10426–10431.

27 Peña-Diaz, P., Braymer, J.J., Vacek, V. et al. (2024). Characterization of the SUF Fe–S cluster synthesis machinery in the amitochondriate eukaryote *Monocercomonoides exilis*. *Curr. Biol.* 34: 3855–3665.

28 Pyrih, J., Pyrihova, E., Kolisko, M. et al. (2016). Minimal cytosolic iron-sulfur cluster assembly machinery of *Giardia intestinalis* is partially associated with mitosomes. *Mol. Microbiol.* 102 (4): 701–714.

29 Grosche, C., Diehl, A., Rensing, S.A., and Maier, U.G. (2018). Iron-sulfur cluster biosynthesis in algae with complex plastids. *Genome Biol. Evol.* 10 (8): 2061–2071.

30 Banci, L., Bertini, I., Ciofi-Baffoni, S. et al. (2011). Anamorsin is a [2Fe-2S] cluster-containing substrate of the Mia40-dependent mitochondrial protein trapping machinery. *Chem. Biol.* 18 (6): 794–804.

31 Soler, N., Craescu, C.T., Gallay, J. et al. (2012). A S-adenosylmethionine methyltransferase-like domain within the essential, Fe-S-containing yeast protein Dre2. *FEBS J.* 279 (12): 2108–2119.

32 Song, G., Cheng, C., Li, Y. et al. (2014). Crystal structure of the N-terminal methyltransferase-like domain of anamorsin. *Proteins* 82 (6): 1066–1071.

33 Netz, D.J., Genau, H.M., Weiler, B.D. et al. (2016). The conserved protein Dre2 uses essential [2Fe-2S] and [4Fe-4S] clusters for its function in cytosolic iron-sulfur protein assembly. *Biochem. J.* 473 (14): 2073–2085.

34 Zhang, Y., Yang, C., Dancis, A., and Nakamaru-Ogiso, E. (2017). EPR studies of wild type and mutant Dre2 identify essential [2Fe–2S] and [4Fe–4S] clusters and their cysteine ligands. *J. Biochem.* 161 (1): 67–78.

35 Zhang, Y., Lyver, E.R., Nakamaru-Ogiso, E. et al. (2008). Dre2, a conserved eukaryotic Fe–S cluster protein, functions in cytosolic Fe–S protein biogenesis. *Mol. Cell. Biol.* 28 (18): 5569–5582.

36 Netz, D.J., Stümpfig, M., Doré, C. et al. (2010). Tah18 transfers electrons to Dre2 in cytosolic iron-sulfur protein biogenesis. *Nat. Chem. Biol.* 6 (10): 758–765.

37 Banci, L., Ciofi-Baffoni, S., Mikolajczyk, M. et al. (2013). Human anamorsin binds [2Fe-2S] clusters with unique electronic properties. *J. Biol. Inorg. Chem.* 18 (8): 883–893.

38 Banci, L., Bertini, I., Calderone, V. et al. (2013). Molecular view of an electron transfer process essential for iron-sulfur protein biogenesis. *Proc. Natl. Acad. Sci. USA* 110 (18): 7136–7141.

39 Olteanu, H. and Banerjee, R. (2003). Redundancy in the pathway for redox regulation of mammalian methionine synthase: reductive activation by the dual flavoprotein, novel reductase 1. *J. Biol. Chem.* 278 (40): 38310–38314.

40 Finn, R.D., Basran, J., Roitel, O. et al. (2003). Determination of the redox potentials and electron transfer properties of the FAD- and FMN-binding domains of the human oxidoreductase NR1. *Eur. J. Biochem.* 270 (6): 1164–1175.

41 Netz, D.J., Pierik, A.J., Stümpfig, M. et al. (2012). A bridging [4Fe-4S] cluster and nucleotide binding are essential for function of the Cfd1-Nbp35 complex as a scaffold in iron-sulfur protein maturation. *J. Biol. Chem.* 287 (15): 12365–12378.

42 Netz, D.J., Pierik, A.J., Stümpfig, M. et al. (2007). The Cfd1-Nbp35 complex acts as a scaffold for iron-sulfur protein assembly in the yeast cytosol. *Nat. Chem. Biol.* 3 (5): 278–286.

43 Balk, J., Pierik, A.J., Netz, D.J. et al. (2004). The hydrogenase-like Nar1p is essential for maturation of cytosolic and nuclear iron-sulphur proteins. *EMBO J.* 23 (10): 2105–2115.

44 Balk, J., Pierik, A.J., Aguilar Netz, D.J. et al. (2005). Nar1p, a conserved eukaryotic protein with similarity to Fe-only hydrogenases, functions in cytosolic iron-sulphur protein biogenesis. *Biochem. Soc. Trans.* 33 (Pt 1): 86–89.

45 Cavazza, C., Martin, L., Mondy, S. et al. (2008). The possible role of an [FeFe]-hydrogenase-like protein in the plant responses to changing atmospheric oxygen levels. *J. Inorg. Biochem.* 102 (5–6): 1359–1365.

46 Urzica, E., Pierik, A.J., Mühlenhoff, U., and Lill, R. (2009). Crucial role of conserved cysteine residues in the assembly of two iron-sulfur clusters on the CIA protein Nar1. *Biochemistry* 48 (22): 4946–4958.

47 Hagedoorn, P.L., van der Weel, L., and Hagen, W.R. (2014). EPR monitored redox titration of the cofactors of *Saccharomyces cerevisiae* Nar1. *J. Vis. Exp.* 93: e51611.

48 Kassube, S.A. and Thomä, N.H. (2020). Structural insights into Fe-S protein biogenesis by the CIA targeting complex. *Nat. Struct. Mol. Biol.* 27 (8): 735–742.

49 Vasquez, S., Marquez, M.D., Brignole, E.J. et al. (2023). Structural and biochemical investigations of a HEAT-repeat protein involved in the cytosolic iron-sulfur cluster assembly pathway. *Commun. Biol.* 6 (1): 1276.

50 Srinivasan, V., Netz, D.J., Webert, H. et al. (2007). Structure of the yeast WD40 domain protein Cia1, a component acting late in iron-sulfur protein biogenesis. *Structure* 15 (10): 1246–1257.

51 Xu, C. and Min, J. (2011). Structure and function of WD40 domain proteins. *Protein Cell* 2 (3): 202–214.

52 Mas, C., Chen, K.E., Brereton, I.M. et al. (2013). Backbone resonance assignments of the monomeric DUF59 domain of human Fam96a. *Biomol. NMR Assign.* 7 (2): 117–120.

53 Chen, K.E., Richards, A.A., Ariffin, J.K. et al. (2012). The mammalian DUF59 protein Fam96a forms two distinct types of domain-swapped dimer. *Acta Crystallogr. D Biol. Crystallogr.* 68 (Pt 6): 637–648.

54 Prusty, N.R., Camponeschi, F., Ciofi-Baffoni, S., and Banci, L. (2021). The human YAE1-ORAOV1 complex of the cytosolic iron-sulfur protein assembly machinery binds a [4Fe-4S] cluster. *Inorg. Chim. Acta* 518: 120252.

55 Chi, C.B., Tang, Y., Zhang, J. et al. (2018). Structural and biochemical insights into the multiple functions of yeast Grx3. *J. Mol. Biol.* 430 (8): 1235–1248.

56 Li, H., Mapolelo, D.T., Dingra, N.N. et al. (2009). The yeast iron regulatory proteins Grx3/4 and Fra2 form heterodimeric complexes containing a [2Fe-2S] cluster with cysteinyl and histidyl ligation. *Biochemistry* 48 (40): 9569–9581.

57 Li, H., Mapolelo, D.T., Dingra, N.N. et al. (2011). Histidine 103 in Fra2 is an iron-sulfur cluster ligand in the [2Fe-2S] Fra2-Grx3 complex and is required for in vivo iron signaling in yeast. *J. Biol. Chem.* 286 (1): 867–876.

58 Li, H., Mapolelo, D.T., Randeniya, S. et al. (2012). Human glutaredoxin 3 forms [2Fe-2S]-bridged complexes with human BolA2. *Biochemistry* 51 (8): 1687–1696.

59 Netz, D.J., Stith, C.M., Stümpfig, M. et al. (2012). Eukaryotic DNA polymerases require an iron-sulfur cluster for the formation of active complexes. *Nat. Chem. Biol.* 8 (1): 125–132.

60 Ter Beek, J., Parkash, V., Bylund, G.O. et al. (2019). Structural evidence for an essential Fe-S cluster in the catalytic core domain of DNA polymerase ε. *Nucleic Acids Res.* 47 (11): 5712–5722.

61 Jain, R., Rice, W.J., Malik, R. et al. (2019). Cryo-EM structure and dynamics of eukaryotic DNA polymerase delta holoenzyme. *Nat. Struct. Mol. Biol.* 26 (10): 955–962.

62 Baranovskiy, A.G., Zhang, Y., Suwa, Y. et al. (2015). Crystal structure of the human primase. *J. Biol. Chem.* 290 (9): 5635–5646.

63 Zhou, C., Pourmal, S., and Pavletich, N.P. (2015). Dna2 nuclease-helicase structure, mechanism and regulation by Rpa. *elife* 4: e09832.

64 Ben-Shimon, L., Paul, V.D., David-Kadoch, G. et al. (2018). Fe-S cluster coordination of the chromokinesin KIF4A alters its subcellular localization during mitosis. *J. Cell Sci.* 131 (12): jcs211433.

65 Stehling, O., Vashisht, A.A., Mascarenhas, J. et al. (2012). MMS19 assembles iron-sulfur proteins required for DNA metabolism and genomic integrity. *Science* 337 (6091): 195–199.

66 van Eeuwen, T., Shim, Y., Kim, H.J. et al. (2021). Cryo-EM structure of TFIIH/Rad4-Rad23-Rad33 in damaged DNA opening in nucleotide excision repair. *Nat. Commun.* 12 (1): 3338.

67 Gari, K., Leon Ortiz, A.M., Borel, V. et al. (2012). MMS19 links cytoplasmic iron-sulfur cluster assembly to DNA metabolism. *Science* 337 (6091): 243–245.

68 Simon, A.K., Kummer, S., Wild, S. et al. (2020). The iron-sulfur helicase DDX11 promotes the generation of single-stranded DNA for CHK1 activation. *Life Sci. Alliance* 3 (3): e201900547.

69 Stehling, O., Mascarenhas, J., Vashisht, A.A. et al. (2013). Human CIA2A-FAM96A and CIA2B-FAM96B integrate iron homeostasis and maturation of different subsets of cytosolic-nuclear iron-sulfur proteins. *Cell Metab.* 18 (2): 187–198.

70 Weon, J.L., Yang, S.W., and Potts, P.R. (2018). Cytosolic iron-sulfur assembly is evolutionarily tuned by a cancer-amplified ubiquitin ligase. *Mol. Cell* 69 (1): 113–125.e6.

71 Malik, R., Kopylov, M., Gomez-Llorente, Y. et al. (2020). Structure and mechanism of B-family DNA polymerase zeta specialized for translesion DNA synthesis. *Nat. Struct. Mol. Biol.* 27 (10): 913–924.

72 Balk, J., Aguilar Netz, D.J., Tepper, K. et al. (2005). The essential WD40 protein Cia1 is involved in a late step of cytosolic and nuclear iron-sulfur protein assembly. *Mol. Cell. Biol.* 25 (24): 10833–10841.

73 Nakamura, T., Okabe, K., Hirayama, S. et al. (2021). Structure of the mammalian adenine DNA glycosylase MUTYH: insights into the base excision repair pathway and cancer. *Nucleic Acids Res.* 49 (12): 7154–7163.

74 Pritts, J.D., Hursey, M.S., Michalek, J.L. et al. (2020). Unraveling the RNA binding properties of the iron-sulfur zinc finger protein CPSF30. *Biochemistry* 59 (8): 970–982.

75 Li, L., Yu, Z., Zhao, D. et al. (2021). Structure of human RNA polymerase III elongation complex. *Cell Res.* 31 (7): 791–800.

76 Arragain, S., Bimai, O., Legrand, P. et al. (2017). Nonredox thiolation in tRNA occurring via sulfur activation by a [4Fe-4S] cluster. *Proc. Natl. Acad. Sci. USA* 114 (28): 7355–7360.

77 Liu, Y., Vinyard, D.J., Reesbeck, M.E. et al. (2016). A [3Fe-4S] cluster is required for tRNA thiolation in archaea and eukaryotes. *Proc. Natl. Acad. Sci. USA* 113 (45): 12703–12708.

78 Jaciuk, M., Scherf, D., Kaszuba, K. et al. (2023). Cryo-EM structure of the fully assembled Elongator complex. *Nucleic Acids Res.* 51 (5): 2011–2032.

79 Zhang, Y., Su, D., Dzikovski, B. et al. (2021). Dph3 enables aerobic diphthamide biosynthesis by donating one iron atom to transform a [3Fe-4S] to a [4Fe-4S] cluster in Dph1-Dph2. *J. Am. Chem. Soc.* 143 (25): 9314–9319.

80 Dong, M., Kathiresan, V., Fenwick, M.K. et al. (2018). Organometallic and radical intermediates reveal mechanism of diphthamide biosynthesis. *Science* 359 (6381): 1247–1250.

81 Grell, T.A.J., Young, A.P., Drennan, C.L., and Bandarian, V. (2018). Biochemical and structural characterization of a schiff base in the radical-mediated biosynthesis of 4-demethylwyosine by TYW1. *J. Am. Chem. Soc.* 140 (22): 6842–6852.

82 Young, A.P. and Bandarian, V. (2021). Eukaryotic TYW1 is a radical SAM flavoenzyme. *Biochemistry* 60 (27): 2179–2185.

83 Kispal, G., Sipos, K., Lange, H. et al. (2005). Biogenesis of cytosolic ribosomes requires the essential iron-sulphur protein Rli1p and mitochondria. *EMBO J.* 24: 589–598.

84 Kratzat, H., Mackens-Kiani, T., Ameismeier, M. et al. (2021). A structural inventory of native ribosomal ABCE1-43S pre-initiation complexes. *EMBO J.* 40 (1): e105179.

85 Landgraf, B.J., McCarthy, E.L., and Booker, S.J. (2016). Radical S-adenosylmethionine enzymes in human health and disease. *Annu. Rev. Biochem.* 85: 485–514.

86 Arragain, S., Handelman, S.K., Forouhar, F. et al. (2010). Identification of eukaryotic and prokaryotic methylthiotransferase for biosynthesis of 2-methylthio-N6-threonylcarbamoyladenosine in tRNA. *J. Biol. Chem.* 285 (37): 28425–28433.

87 Upadhyay, A.S., Stehling, O., Panayiotou, C. et al. (2017). Cellular requirements for iron-sulfur cluster insertion into the antiviral radical SAM protein viperin. *J. Biol. Chem.* 292 (33): 13879–13889.

88 Song, D. and Lee, F.S. (2008). A role for IOP1 in mammalian cytosolic iron-sulfur protein biogenesis. *J. Biol. Chem.* 283 (14): 9231–9238.

89 Rietzschel, N., Pierik, A.J., Bill, E. et al. (2015). The basic leucine zipper stress response regulator Yap5 senses high-iron conditions by coordination of [2Fe-2S] clusters. *Mol. Cell. Biol.* 35 (2): 370–378.

90 Wang, H., Shi, H., Rajan, M. et al. (2020). FBXL5 regulates IRP2 stability in iron homeostasis via an oxygen-responsive [2Fe2S] cluster. *Mol. Cell.* 78 (1): 31–41.e5.

91 Zhao, H., Lu, Y., Zhang, J. et al. (2024). NCOA4 requires a [3Fe-4S] to sense and maintain the iron homeostasis. *J. Biol. Chem.* 300 (2): 105612.

92 Schlenzka, W., Shaw, L., Kelm, S. et al. (1996). CMP-N-acetylneuraminic acid hydroxylase: the first cytosolic Rieske iron-sulphur protein to be described in Eukarya. *FEBS Lett.* 385 (3): 197–200.

93 Paddock, M.L., Wiley, S.E., Axelrod, H.L. et al. (2007). MitoNEET is a uniquely folded 2Fe 2S outer mitochondrial membrane protein stabilized by pioglitazone. *Proc. Natl. Acad. Sci. USA* 104 (36): 14342–14347.

94 Conlan, A.R., Axelrod, H.L., Cohen, A.E. et al. (2009). Crystal structure of Miner1: the redox-active 2Fe-2S protein causative in Wolfram Syndrome 2. *J. Mol. Biol.* 392 (1): 143–153.

95 Pierik, A.J., Netz, D.J.A., and Lill, R. (2009). Analysis of iron-sulfur protein maturation in eukaryotes. *Nat. Protoc.* 4: 753–766.

96 Marquez, M.D., Greth, C., Buzuk, A. et al. (2023). Cytosolic iron-sulfur protein assembly system identifies clients by a C-terminal tripeptide. *Proc. Natl. Acad. Sci. USA* 120 (44): e2311057120.

97 Filetici, P., Martegani, M.P., Valenzuela, L. et al. (1996). Sequence of the *GLT1* gene from *Saccharomyces cerevisiae* reveals the domain structure of yeast glutamate synthase. *Yeast* 12 (13): 1359–1366.

98 Van, V., Brown, J.B., O'Shea, C.R. et al. (2023). Iron-sulfur clusters are involved in post-translational arginylation. *Nat. Commun.* 14 (1): 458.

99 Wu, X., Alexander, P.B., He, Y. et al. (2005). Mammalian sprouty proteins assemble into large monodisperse particles having the properties of intracellular nanobatteries. *Proc. Natl. Acad. Sci. USA* 102 (39): 14058–14062.

100 Stegmaier, K., Blinn, C.M., Bechtel, D.F. et al. (2019). Apd1 and Aim32 are prototypes of bishistidinyl-coordinated non-rieske [2Fe-2S] proteins. *J. Am. Chem. Soc.* 141 (14): 5753–5765.

101 Huang, Q., Hong, X., and Hao, Q. (2008). SNAP-25 is also an iron-sulfur protein. *FEBS Lett.* 582 (10): 1431–1436.

102 Poor, C.B., Wegner, S.V., Li, H. et al. (2014). Molecular mechanism and structure of the *Saccharomyces cerevisiae* iron regulator Aft2. *Proc. Natl. Acad. Sci. USA* 111 (11): 4043–4048.

103 Shaw, A.E., Mihelich, M.N., Whitted, J.E. et al. (2024). Revised mechanism of hydroxyurea-induced cell cycle arrest and an improved alternative. *Proc. Natl. Acad. Sci. USA* 121 (42): e2404470121.

104 Paul, V.D., Mühlenhoff, U., Stümpfig, M. et al. (2015). The deca-GX_3 proteins Yae1-Lto1 function as adaptors recruiting the ABC protein Rli1 for iron-sulfur cluster insertion. *elife* 4: e08231.

105 Pandey, A.K., Pain, J., Dancis, A., and Pain, D. (2019). Mitochondria export iron-sulfur and sulfur intermediates to the cytoplasm for iron-sulfur cluster assembly and tRNA thiolation in yeast. *J. Biol. Chem.* 294 (24): 9489–9502.

106 Stehling, O., Paul, V.D., Bergmann, J. et al. (2018). Biochemical analyses of human iron–sulfur protein biogenesis and of related diseases. *Methods Enzymol.* 599: 227–263.

107 Varadi, M., Anyango, S., Deshpande, M. et al. (2022). AlphaFold protein structure database: massively expanding the structural coverage of protein-sequence space with high-accuracy models. *Nucleic Acids Res.* 50 (D1): D439–D444.

108 Waterhouse, A., Bertoni, M., Bienert, S. et al. (2018). SWISS-MODEL: homology modelling of protein structures and complexes. *Nucleic Acids Res.* 46 (W1): W296–W303.

109 Ciofi-Baffoni, S., Nasta, V., and Banci, L. (2018). Protein networks in the maturation of human iron-sulfur proteins. *Metallomics* 10 (1): 49–72.

110 Peña-Diaz, P. and Lukes, J. (2018). Fe-S cluster assembly in the supergroup Excavata. *J. Biol. Inorg. Chem.* 23 (4): 521–541.

111 Wachnowsky, C., Fidai, I., and Cowan, J.A. (2018). Iron-sulfur cluster biosynthesis and trafficking – impact on human disease conditions. *Metallomics* 10 (1): 9–29.

112 Basu, S., Netz, D.J., Haindrich, A.C. et al. (2014). Cytosolic iron-sulphur protein assembly is functionally conserved and essential in procyclic and bloodstream *Trypanosoma brucei*. *Mol. Microbiol.* 93 (5): 897–910.

113 Tonini, M.L., Pena-Diaz, P., Haindrich, A.C. et al. (2018). Branched late-steps of the cytosolic iron-sulphur cluster assembly machinery of *Trypanosoma brucei*. *PLoS Pathog.* 14 (10): e1007326.

114 Bernard, D.G., Netz, D.J., Lagny, T.J. et al. (2013). Requirements of the cytosolic iron-sulfur cluster assembly pathway in *Arabidopsis*. *Philos. Trans. R. Soc. Lond. Ser. B Biol. Sci.* 368 (1622): 20120259.

115 Tsaousis, A.D., Gentekaki, E., Eme, L. et al. (2014). Evolution of the cytosolic iron-sulfur cluster assembly machinery in *Blastocystis* species and other microbial eukaryotes. *Eukaryot. Cell* 13 (1): 143–153.

116 Misslinger, M., Scheven, M.T., Hortschansky, P. et al. (2019). The monothiol glutaredoxin GrxD is essential for sensing iron starvation in *Aspergillus fumigatus*. *PLoS Genet.* 15 (9): e1008379.

117 Chanet, R. and Heude, M. (2003). Characterization of mutations that are synthetic lethal with *pol3-13*, a mutated allele of DNA polymerase delta in *Saccharomyces cerevisiae*. *Curr. Genet.* 43 (5): 337–350.

118 Wang, H., Nair, V.S., Holland, A.A. et al. (2015). Asp1 from *Schizosaccharomyces pombe* binds a [2Fe-2S](2+) cluster which inhibits inositol pyrophosphate 1-phosphatase activity. *Biochemistry* 54 (42): 6462–6474.

119 Pandey, A.K., Pain, J., and J B, Dancis A, Pain D. (2023). Essential mitochondrial role in iron-sulfur cluster assembly of the cytoplasmic isopropylmalate isomerase Leu1 in *Saccharomyces cerevisiae*. *Mitochondrion* 69: 104–115.

120 Matteucci, S., Camponeschi, F., Clemancey, M. et al. (2021). In-cellulo Mössbauer and EPR studies bring new evidences to the long-standing debate on the iron-sulfur cluster binding in human anamorsin. *Angew. Chem. Int. Ed. Engl.* 60 (27): 14841–14845.

121 Grifagni, D., Silva, J.M., Querci, L. et al. (2024). Biochemical and cellular characterization of the CISD3 protein: molecular bases of cluster release and destabilizing effects of nitric oxide. *J. Biol. Chem.* 300 (3): 105745.

122 Freibert, S.A., Weiler, B.D., Bill, E. et al. (2018). Biochemical reconstitution and spectroscopic analysis of iron-sulfur proteins. *Methods Enzymol.* 599: 197–226.

123 Jansing, M., Mielenbrink, S., Rosenbach, H. et al. (2023). Maturation strategy influences expression levels and cofactor occupancy in Fe-S proteins. *J. Biol. Inorg. Chem.* 28 (2): 187–204.

124 Rosenbach, H., Walla, E., Cutsail, G.E. 3rd, et al. (2021). The Asp1 pyrophosphatase from *S. pombe* hosts a $[2Fe\text{-}2S]^{2+}$ cluster *in vivo*. *J. Biol. Inorg. Chem.* 26 (1): 93–108.

125 Valer, L., Rossetto, D., Scintilla, S. et al. (2022). Methods to identify and characterize iron–sulfur oligopeptides in water. *Can. J. Chem.* 100: 475–483.

126 Tsai, C.L. and Tainer, J.A. (2018). Robust production, crystallization, structure determination, and analysis of [Fe-S] proteins: uncovering control of electron shuttling and gating in the respiratory metabolism of molybdopterin guanine dinucleotide enzymes. *Methods Enzymol.* 599: 157–196.

127 Stich, T. (2021). Characterization of paramagnetic iron-sulfur clusters using electron paramagnetic resonance spectroscopy. In: *Fe-S Proteins: Methods and Protocols* (ed. P.C. Dos Santos), 259–280. New York: Humana Press.

128 Crack, J.C. and Le Brun, N.E. (2021). Native mass spectrometry of iron-sulfur proteins. *Methods Mol. Biol.* 2353: 231–258.

129 Querci, L., Piccioli, M., Ciofi-Baffoni, S., and Banci, L. (2024). Structural aspects of iron-sulfur protein biogenesis: an NMR view. *Biochim. Biophys. Acta Mol. Cell Res.* 1871 (7): 119786.

130 Andreini, C., Banci, L., and Rosato, A. (2016). Exploiting bacterial operons to illuminate human iron-sulfur proteins. *J. Proteome Res.* 15 (4): 1308–1322.

131 Kairis, A., Neves, B.D., Couturier, J. et al. (2024). Iron-sulfur cluster synthesis in plastids by the SUF system: a mechanistic and structural perspective. *Biochim. Biophys. Acta Mol. Cell Res.* 1871: 119797.

132 Miao, R., Kim, H., Koppolu, U.M. et al. (2009). Biophysical characterization of the iron in mitochondria from Atm1p-depleted *Saccharomyces cerevisiae*. *Biochemistry* 48 (40): 9556–9568.

133 Preveral, S., Gayet, L., Moldes, C. et al. (2009). A common highly conserved cadmium detoxification mechanism from bacteria to humans: heavy metal tolerance conferred by the ATP-binding cassette (ABC) transporter SpHMT1 requires glutathione but not metal-chelating phytochelatin peptides. *J. Biol. Chem.* 284 (8): 4936–4943.

134 Lee, J.Y., Yang, J.G., Zhitnitsky, D. et al. (2014). Structural basis for heavy metal detoxification by an Atm1-type ABC exporter. *Science* 343 (6175): 1133–1136.

135 Lange, H., Lisowsky, T., Gerber, J. et al. (2001). An essential function of the mitochondrial sulfhydryl oxidase Erv1p/ALR in the maturation of cytosolic Fe–S proteins. *EMBO Rep.* 2: 715–720.

136 Murari, A., Thiriveedi, V.R., Mohammad, F. et al. (2015). Human mitochondrial MIA40 (CHCHD4) is a component of the Fe-S cluster export machinery. *Biochem. J.* 471 (2): 231–241.

137 Spiller, M.P., Ang, S.K., Ceh-Pavia, E. et al. (2013). Identification and characterization of mitochondrial Mia40 as an iron-sulfur protein. *Biochem. J.* 455 (1): 27–35.

138 Ozer, H.K., Dlouhy, A.C., Thornton, J.D. et al. (2015). Cytosolic Fe-S cluster protein maturation and iron regulation are independent of the mitochondrial Erv1/Mia40 import system. *J. Biol. Chem.* 290 (46): 27829–27840.

139 Karmi, O., Marjault, H.B., Bai, F. et al. (2022). A VDAC1-mediated NEET protein chain transfers [2Fe-2S] clusters between the mitochondria and the cytosol and impacts mitochondrial dynamics. *Proc. Natl. Acad. Sci. USA* 119 (7): e2121491119.

140 Kumar, C., Igbaria, A., D'Autreaux, B. et al. (2011). Glutathione revisited: a vital function in iron metabolism and ancillary role in thiol-redox control. *EMBO J.* 30 (10): 2044–2056.

141 Sipos, K., Lange, H., Fekete, Z. et al. (2002). Maturation of cytosolic iron-sulfur proteins requires glutathione. *J. Biol. Chem.* 277: 26944–26949.

142 Mühlenhoff, U., Braymer, J.J., Christ, S. et al. (2020). Glutaredoxins and iron-sulfur protein biogenesis at the interface of redox biology and iron metabolism. *Biol. Chem.* 401 (12): 1407–1428.

143 Li, G., Nanjaraj Urs, A.N., Dancis, A., and Zhang, Y. (2022). Genetic suppressors of Δ*grx3* Δ*grx4*, lacking redundant multidomain monothiol yeast glutaredoxins, rescue growth and iron homeostasis. *Biosci. Rep.* 42 (6): BSR20212665.

144 Qi, W., Li, J., Chain, C.Y. et al. (2012). Glutathione complexed Fe-S centers. *J. Am. Chem. Soc.* 134 (26): 10745–10748.

145 Gupta, M. and Outten, C.E. (2020). Iron-sulfur cluster signaling: the common thread in fungal iron regulation. *Curr. Opin. Chem. Biol.* 55: 189–201.

146 Frey, A.G., Palenchar, D.J., Wildemann, J.D., and Philpott, C.C. (2016). A glutaredoxinis·BolA complex serves as an iron-sulfur cluster chaperone for the cytosolic cluster assembly machinery. *J. Biol. Chem.* 291 (43): 22344–22356.

147 Lill, R., Hoffmann, B., Molik, S. et al. (2012). The role of mitochondria in cellular iron-sulfur protein biogenesis and iron metabolism. *Biochim. Biophys. Acta* 1823: 1491–1508.

148 Rutherford, J.C., Ojeda, L., Balk, J. et al. (2005). Activation of the iron regulon by the yeast Aft1/Aft2 transcription factors depends on mitochondrial but not cytosolic iron-sulfur protein biogenesis. *J. Biol. Chem.* 280 (11): 10135–10140.

149 Ojeda, L., Keller, G., Mühlenhoff, U. et al. (2006). Role of glutaredoxin-3 and glutaredoxin-4 in the iron regulation of the Aft1 transcriptional activator in *Saccharomyces cerevisiae. J. Biol. Chem.* 281 (26): 17661–17669.

150 Mühlenhoff, U., Molik, S., Godoy, J.R. et al. (2010). Cytosolic monothiol glutaredoxins function in intracellular iron sensing and trafficking via their bound iron-sulfur cluster. *Cell Metab.* 12 (4): 373–385.

151 Johansson, C., Roos, A.K., Montano, S.J. et al. (2011). The crystal structure of human GLRX5: iron-sulfur cluster co-ordination, tetrameric assembly and monomer activity. *Biochem. J.* 433 (2): 303–311.

152 Roret, T., Tsan, P., Couturier, J. et al. (2014). Structural and spectroscopic insights into BolA-glutaredoxin complexes. *J. Biol. Chem.* 289: 24588–24598.

153 Li, H. and Outten, C.E. (2019). The conserved CDC motif in the yeast iron regulator Aft2 mediates iron-sulfur cluster exchange and protein-protein interactions with Grx3 and Bol2. *J. Biol. Inorg. Chem.* 24 (6): 809–815.

154 Pimentel, C., Vicente, C., Menezes, R.A. et al. (2012). The role of the Yap5 transcription factor in remodeling gene expression in response to Fe bioavailability. *PLoS One* 7 (5): e37434.

155 Li, L., Jia, X., Ward, D.M., and Kaplan, J. (2011). Yap5 protein-regulated transcription of the TYW1 gene protects yeast from high iron toxicity. *J. Biol. Chem.* 286 (44): 38488–38497.

156 Li, L., Bagley, D., Ward, D.M., and Kaplan, J. (2008). Yap5 is an iron-responsive transcriptional activator that regulates vacuolar iron storage in yeast. *Mol. Cell. Biol.* 28 (4): 1326–1337.

157 Camponeschi, F., Prusty, N.R., Heider, S.A.E. et al. (2020). GLRX3 acts as a [2Fe-2S] cluster chaperone in the cytosolic iron-sulfur assembly machinery transferring [2Fe-2S] clusters to NUBP1. *J. Am. Chem. Soc.* 142 (24): 10794–10805.

158 Banci, L., Ciofi-Baffoni, S., Gajda, K. et al. (2015). N-terminal domains mediate [2Fe-2S] cluster transfer from glutaredoxin-3 to anamorsin. *Nat. Chem. Biol.* 11 (10): 772–778.

159 Banci, L., Camponeschi, F., Ciofi-Baffoni, S., and Muzzioli, R. (2015). Elucidating the molecular function of human BOLA2 in GRX3-dependent anamorsin maturation pathway. *J. Am. Chem. Soc.* 137 (51): 16133–16143.

160 Lill, R. and Freibert, S.A. (2020). Mechanisms of mitochondrial iron-sulfur protein biogenesis. *Annu. Rev. Biochem.* 89: 471–499.

161 Haunhorst, P., Hanschmann, E.M., Brautigam, L. et al. (2013). Crucial function of vertebrate glutaredoxin 3 (PICOT) in iron homeostasis and hemoglobin maturation. *Mol. Biol. Cell* 24 (12): 1895–1903.

162 Hoff, K.G., Goodlitt, R., Li, R. et al. (2009). Fluorescence detection of a protein-bound 2Fe2S cluster. *ChemBioChem* 10 (4): 667–670.

163 Karmi, O., Marjault, H.B., Pesce, L. et al. (2018). The unique fold and lability of the [2Fe-2S] clusters of NEET proteins mediate their key functions in health and disease. *J. Biol. Inorg. Chem.* 23 (4): 599–612.

164 Terao, M., Garattini, E., Romao, M.J., and Leimkühler, S. (2020). Evolution, expression, and substrate specificities of aldehyde oxidase enzymes in eukaryotes. *J. Biol. Chem.* 295 (16): 5377–5389.

165 Kim, H.J., Lee, K.L., Kim, K.D., and Roe, J.H. (2016). The iron uptake repressor Fep1 in the fission yeast binds Fe-S cluster through conserved cysteines. *Biochem. Biophys. Res. Commun.* 478 (1): 187–192.

166 Iida, Y., Kurata, T., Harimoto, Y., and Tsuge, T. (2008). Nitrite reductase gene upregulated during conidiation is involved in macroconidium formation in *Fusarium oxysporum*. *Phytopathology* 98 (10): 1099–1106.

167 Yoshiyama-Yanagawa, T., Enya, S., Shimada-Niwa, Y. et al. (2011). The conserved Rieske oxygenase DAF-36/Neverland is a novel cholesterol-metabolizing enzyme. *J. Biol. Chem.* 286 (29): 25756–25762.

168 Linder, T. (2014). *CMO1* encodes a putative choline monooxygenase and is required for the utilization of choline as the sole nitrogen source in the yeast *Scheffersomyces stipitis* (syn. *Pichia stipitis*). *Microbiology (Reading)* 160 (Pt 5): 929–940.

169 Maio, N., Lafont, B.A.P., Sil, D. et al. (2021). Fe-S cofactors in the SARS-CoV-2 RNA-dependent RNA polymerase are potential antiviral targets. *Science* 373 (6551): 236–241.

170 Shimberg, G.D., Michalek, J.L., Oluyadi, A.A. et al. (2016). Cleavage and polyadenylation specificity factor 30: an RNA-binding zinc-finger protein with an unexpected 2Fe-2S cluster. *Proc. Natl. Acad. Sci. USA* 113 (17): 4700–4705.

171 Lamb, D.C., Kim, Y., Yermalitskaya, L.V. et al. (2006). A second FMN binding site in yeast NADPH-cytochrome P450 reductase suggests a mechanism of electron transfer by diflavin reductases. *Structure* 14 (1): 51–61.

172 Bargagna, B., Matteucci, S., Ciofi-Baffoni, S. et al. (2023). Unraveling the mechanism of [4Fe-4S] cluster assembly on the N-terminal cluster binding site of NUBP1. *Protein Sci.* 32 (5): e4625.

173 Bastow, E.L., Bych, K., Crack, J.C. et al. (2017). NBP35 interacts with DRE2 in the maturation of cytosolic iron-sulphur proteins in *Arabidopsis thaliana*. *Plant J.* 89 (3): 590–600.

174 Li, H., Stümpfig, M., Zhang, C. et al. (2017). The diferric-tyrosyl radical cluster of ribonucleotide reductase and cytosolic iron-sulfur clusters have distinct and similar biogenesis requirements. *J. Biol. Chem.* 292 (27): 11445–11451.

175 Camire, E.J., Grossman, J.D., Thole, G.J. et al. (2015). The yeast Nbp35-Cfd1 cytosolic iron-sulfur cluster scaffold is an ATPase. *J. Biol. Chem.* 290 (39): 23793–23802.

176 Weiler, B.D., Bruck, M.C., Kothe, I. et al. (2020). Mitochondrial [4Fe-4S] protein assembly involves reductive [2Fe-2S] cluster fusion on ISCA1-ISCA2 by electron flow from ferredoxin FDX2. *Proc. Natl. Acad. Sci. USA* 117 (34): 20555–20565.

177 Grossman, J.D., Gay, K.A., Camire, E.J. et al. (2019). Coupling nucleotide binding and hydrolysis to iron-sulfur cluster acquisition and transfer revealed through genetic dissection of the Nbp35 ATPase site. *Biochemistry* 58 (15): 2017–2027.

178 Grossman, J.D., Camire, E.J., Glynn, C.A. et al. (2019). The Cfd1 subunit of the Nbp35-Cfd1 iron sulfur cluster scaffolding complex controls nucleotide binding. *Biochemistry* 58 (12): 1587–1595.

179 Bych, K., Netz, D.J., Vigani, G. et al. (2008). The essential cytosolic iron-sulfur protein Nbp35 acts without Cfd1 partner in the green lineage. *J. Biol. Chem.* 283 (51): 35797–35804.

180 Pallesen, L.J., Solodovnikova, N., Sharma, A.K., and Walden, W.E. (2013). Interaction with Cfd1 increases the kinetic lability of Fe–S on the Nbp35 scaffold. *J. Biol. Chem.* 288: 23358–23367.

181 Fan, X., Barshop, W.D., Vashisht, A.A. et al. (2022). Iron-regulated assembly of the cytosolic iron-sulfur cluster biogenesis machinery. *J. Biol. Chem.* 298 (7): 102094.

182 Seki, M., Takeda, Y., Iwai, K., and Tanaka, K. (2013). IOP1 protein is an external component of the human cytosolic iron-sulfur cluster assembly (CIA) machinery and functions in the MMS19 protein-dependent CIA pathway. *J. Biol. Chem.* 288 (23): 16680–16689.

183 Luo, D., Bernard, D.G., Balk, J. et al. (2012). The DUF59 family gene AE7 acts in the cytosolic iron-sulfur cluster assembly pathway to maintain nuclear genome integrity in *Arabidopsis*. *Plant Cell* 24 (10): 4135–4148.

184 Nicolet, Y., Piras, C., Legrand, P. et al. (1999). *Desulfovibrio desulfuricans* iron hydrogenase: the structure shows unusual coordination to an active site Fe binuclear center. *Structure* 7 (1): 13–23.

185 Song, D. and Lee, F.S. (2011). Mouse knock-out of IOP1 protein reveals its essential role in mammalian cytosolic iron-sulfur protein biogenesis. *J. Biol. Chem.* 286 (18): 15797–15805.

186 Maione, V., Grifagni, D., Torricella, F. et al. (2020). CIAO3 protein forms a stable ternary complex with two key players of the human cytosolic iron-sulfur cluster assembly machinery. *J. Biol. Inorg. Chem.* 25 (3): 501–508.

187 Jain, I.H., Calvo, S.E., Markhard, A.L. et al. (2020). Genetic screen for cell fitness in high or low oxygen highlights mitochondrial and lipid metabolism. *Cell* 181 (3): 716–27 e11.

188 Fujii, M., Adachi, N., Shikatani, K., and Ayusawa, D. (2009). [FeFe]-hydrogenase-like gene is involved in the regulation of sensitivity to oxygen in yeast and nematode. *Genes Cells* 14 (4): 457–468.

189 Nakamura, M., Buzas, D.M., Kato, A. et al. (2013). The role of *Arabidopsis thaliana* NAR1, a cytosolic iron-sulfur cluster assembly component, in gametophytic gene expression and oxidative stress responses in vegetative tissue. *New Phytol.* 199 (4): 925–935.

190 Mondy, S., Lenglet, A., Cosson, V. et al. (2014). GOLLUM [FeFe]-hydrogenase-like proteins are essential for plant development in normoxic conditions and modulate energy metabolism. *Plant Cell Environ.* 37 (1): 54–69.

191 Braymer, J.J., Stümpfig, M., Thelen, S. et al. (2019). Depletion of thiol reducing capacity impairs cytosolic but not mitochondrial iron-sulfur protein assembly machineries. *Biochim. Biophys. Acta Mol. Cell Res.* 1866 (2): 240–251.

192 Weerapana, E., Wang, C., Simon, G.M. et al. (2010). Quantitative reactivity profiling predicts functional cysteines in proteomes. *Nature* 468 (7325): 790–795.

193 Costanzo, M., VanderSluis, B., Koch, E.N. et al. (2016). A global genetic interaction network maps a wiring diagram of cellular function. *Science* 353 (6306): aaf1420.

194 Ho, Y., Gruhler, A., Heilbut, A. et al. (2002). Systematic identification of protein complexes in *Saccharomyces cerevisiae* by mass spectrometry. *Nature* 415 (6868): 180–183.

195 Masselot, M. and De Robichon-Szulmajster, H. (1975). Methionine biosynthesis in *Saccharomyces cerevisiae*. I. Genetical analysis of auxotrophic mutants. *Mol. Gen. Genet.* 139 (2): 121–132.

196 Lombaerts, M., Tijsterman, M., Verhage, R.A., and Brouwer, J. (1997). *Saccharomyces cerevisiae mms19* mutants are deficient in transcription-coupled and global nucleotide excision repair. *Nucleic Acids Res.* 25 (20): 3974–3979.

197 Prakash, L. and Prakash, S. (1979). Three additional genes involved in pyrimidine dimer removal in *Saccharomyces cerevisiae*: *RAD7*, *RAD14* and *MMS19*. *Mol. Gen. Genet.* 176 (3): 351–359.

198 Lobachev, K.S., Stenger, J.E., Kozyreva, O.G. et al. (2000). Inverted Alu repeats unstable in yeast are excluded from the human genome. *EMBO J.* 19 (14): 3822–3830.

199 Andrade, M.A. and Bork, P. (1995). HEAT repeats in the Huntington's disease protein. *Nat. Genet.* 11 (2): 115–116.

200 Kou, H., Zhou, Y., Gorospe, R.M., and Wang, Z. (2008). Mms19 protein functions in nucleotide excision repair by sustaining an adequate cellular concentration of the TFIIH component Rad3. *Proc. Natl. Acad. Sci. USA* 105 (41): 15714–15719.

201 Lauder, S., Bankmann, M., Guzder, S.N. et al. (1996). Dual requirement for the yeast MMS19 gene in DNA repair and RNA polymerase II transcription. *Mol. Cell. Biol.* 16 (12): 6783–6793.

202 Carvalho, F.A., Mühlenhoff, U., Braymer, J.J. et al. (2023). Hsp90 and metal-binding J-protein family chaperones are not critically involved in cellular iron-sulfur protein assembly and iron regulation in yeast. *FEBS Lett.* 597 (13): 1718–1732.

203 Upadhyay, A.S., Vonderstein, K., Pichlmair, A. et al. (2014). Viperin is an iron-sulfur protein that inhibits genome synthesis of tick-borne encephalitis virus via radical SAM domain activity. *Cell. Microbiol.* 16 (6): 834–848.

204 Crooks, G.E., Hon, G., Chandonia, J.M., and Brenner, S.E. (2004). WebLogo: a sequence logo generator. *Genome Res.* 14 (6): 1188–1190.

205 Maio, N., Singh, A., Uhrigshardt, H. et al. (2014). Cochaperone binding to LYR motifs confers specificity of iron sulfur cluster delivery. *Cell Metab.* 19 (3): 445–457.

206 Zhai, C., Li, Y., Mascarenhas, C. et al. (2014). The function of ORAOV1/LTO1, a gene that is overexpressed frequently in cancer: essential roles in the function and biogenesis of the ribosome. *Oncogene* 33 (4): 484–494.

207 Snoek, I.S. and Steensma, H.Y. (2006). Why does *Kluyveromyces lactis* not grow under anaerobic conditions? Comparison of essential anaerobic genes of *Saccharomyces cerevisiae* with the *Kluyveromyces lactis* genome. *FEMS Yeast Res.* 6 (3): 393–403.

208 Alhebshi, A., Sideri, T.C., Holland, S.L., and Avery, S.V. (2012). The essential iron-sulfur protein Rli1 is an important target accounting for inhibition of cell growth by reactive oxygen species. *Mol. Biol. Cell* 23 (18): 3582–3590.

209 Poklepovich, T.J., Urtasun, N., Miranda, M.V. et al. (2015). Expression and functional characterization of a C-7 cholesterol desaturase from *Tetrahymena thermophila* in an insect cell line. *Steroids* 96: 132–139.

210 Mayank, A.K., Pandey, V., Vashisht, A.A. et al. (2019). An oxygen-dependent interaction between FBXL5 and the CIA-targeting complex regulates iron homeostasis. *Mol. Cell* 75 (2): 382–393.e5.

211 Galy, B., Conrad, M., and Muckenthaler, M. (2024). Mechanisms controlling cellular and systemic iron homeostasis. *Nat. Rev. Mol. Cell Biol.* 25 (2): 133–155.

212 Theil, A.F., Hackes, D., and Lans, H. (2023). TFIIH central activity in nucleotide excision repair to prevent disease. *DNA Repair (Amst)* 132: 103568.

213 Calvo, J.A., Fritchman, B., Hernandez, D. et al. (2021). Comprehensive mutational analysis of the BRCA1-associated DNA helicase and tumor-suppressor FANCJ/BACH1/BRIP1. *Mol. Cancer Res.* 19 (6): 1015–1025.

214 van Schie, J.J.M., Faramarz, A., Balk, J.A. et al. (2020). Warsaw breakage syndrome associated DDX11 helicase resolves G-quadruplex structures to support sister chromatid cohesion. *Nat. Commun.* 11 (1): 4287.

215 Tanshee, R.R., Mahmud, Z., Nabi, A., and Sayem, M. (2024). A comprehensive in silico investigation into the pathogenic SNPs in the RTEL1 gene and their biological consequences. *PLoS One* 19 (9): e0309713.

216 Weren, R.D., Ligtenberg, M.J., Kets, C.M. et al. (2015). A germline homozygous mutation in the base-excision repair gene NTHL1 causes adenomatous polyposis and colorectal cancer. *Nat. Genet.* 47 (6): 668–671.

217 Das, L., Quintana, V.G., and Sweasy, J.B. (2020). NTHL1 in genomic integrity, aging and cancer. *DNA Repair (Amst)* 93: 102920.

218 Peterlongo, P., Mitra, N., Sanchez de Abajo, A. et al. (2006). Increased frequency of disease-causing MYH mutations in colon cancer families. *Carcinogenesis* 27 (11): 2243–2249.

219 Hutchcraft, M.L., Gallion, H.H., and Kolesar, J.M. (2021). MUTYH as an emerging predictive biomarker in ovarian cancer. *Diagnostics (Basel)* 11 (1): 84.

220 Adami, R. and Bottai, D. (2020). S-adenosylmethionine tRNA modification: unexpected/unsuspected implications of former/new players. *Int. J. Biol. Sci.* 16 (15): 3018–3027.

221 Magrin, L., Fanale, D., Brando, C. et al. (2021). POLE, POLD1, and NTHL1: the last but not the least hereditary cancer-predisposing genes. *Oncogene* 40 (40): 5893–5901.

222 Massimo, G., Khambata, R.S., Chapman, T. et al. (2023). Natural mutations of human XDH promote the nitrite (NO_2^--reductase capacity of xanthine oxidoreductase: a novel mechanism to promote redox health? *Redox Biol.* 67: 102864.

223 Diasio, R.B. and Offer, S.M. (2022). Testing for dihydropyrimidine dehydrogenase deficiency to individualize 5-fluorouracil therapy. *Cancers (Basel)* 14 (13): 3207.

224 Huang, X., Gollin, S.M., Raja, S., and Godfrey, T.E. (2002). High-resolution mapping of the 11q13 amplicon and identification of a gene, TAOS1, that is amplified and overexpressed in oral cancer cells. *Proc. Natl. Acad. Sci. USA* 99 (17): 11369–11374.

225 Mahieu, C.I., Mancini, A.G., Vikram, E.P. et al. (2024). ORAOV1, CCND1, and MIR548K are the driver oncogenes of the 11q13 amplicon in squamous cell carcinoma. *Mol. Cancer Res.* 22 (2): 152–168.

226 Liu, H.Z., Du, C.X., Luo, J. et al. (2017). A novel mutation in nuclear prelamin a recognition factor-like causes diffuse pulmonary arteriovenous malformations. *Oncotarget* 8 (2): 2708–2718.

227 van Karnebeek, C.D.M., Tarailo-Graovac, M., Leen, R. et al. (2024). CIAO1 and MMS19 deficiency: a lethal neurodegenerative phenotype caused by cytosolic Fe-S cluster protein assembly disorders. *Genet. Med.* 26 (6): 101104.

228 Maio, N., Orbach, R., Zaharieva, I.T. et al. (2024). CIAO1 loss of function causes a neuromuscular disorder with compromise of nucleocytoplasmic Fe-S enzymes. *J. Clin. Invest.* 134 (12): e179559.

229 Chen, S., Francioli, L.C., Goodrich, J.K. et al. (2024). A genomic mutational constraint map using variation in 76,156 human genomes. *Nature* 625 (7993): 92–100.

230 Nielsen, J., Tillegreen, C.B., and Petranovic, D. (2022). Innovation trends in industrial biotechnology. *Trends Biotechnol.* 40 (10): 1160–1172.

231 D'Angelo, F., Fernandez-Fueyo, E., Garcia, P.S. et al. (2022). Cellular assays identify barriers impeding iron-sulfur enzyme activity in a non-native prokaryotic host. *elife* 11: e70936.

232 Shomar, H. and Bokinsky, G. (2024). Harnessing iron-sulfur enzymes for synthetic biology. *Biochim. Biophys. Acta Mol. Cell Res.* 1871 (5): 119718.

233 Benisch, F. and Boles, E. (2014). The bacterial Entner–Doudoroff pathway does not replace glycolysis in *Saccharomyces cerevisiae* due to the lack of activity of iron-sulfur cluster enzyme 6-phosphogluconate dehydratase. *J. Biotechnol.* 171: 45–55.

234 Morita, K., Nomura, Y., Ishii, J. et al. (2017). Heterologous expression of bacterial phosphoenol pyruvate carboxylase and Entner–Doudoroff pathway in *Saccharomyces cerevisiae* for improvement of isobutanol production. *J. Biosci. Bioeng.* 124 (3): 263–270.

235 Milne, N., Wahl, S.A., van Maris, A.J.A. et al. (2016). Excessive by-product formation: a key contributor to low isobutanol yields of engineered *Saccharomyces cerevisiae* strains. *Metab. Eng .Commun.* 3: 39–51.

236 Gambacorta, F.V., Wagner, E.R., Jacobson, T.B. et al. (2022). Comparative functional genomics identifies an iron-limited bottleneck in a *Saccharomyces cerevisiae* strain with a cytosolic-localized isobutanol pathway. *Synth. Syst. Biotechnol.* 7 (2): 738–749.

237 Salusjarvi, L., Toivari, M., Vehkomaki, M.L. et al. (2017). Production of ethylene glycol or glycolic acid from D-xylose in *Saccharomyces cerevisiae. Appl. Microbiol. Biotechnol.* 101 (22): 8151–8163.

238 Wasserstrom, L., Portugal-Nunes, D., Almqvist, H. et al. (2018). Exploring D-xylose oxidation in *Saccharomyces cerevisiae* through the Weimberg pathway. *AMB Express* 8 (1): 33.

239 Partow, S., Siewers, V., Daviet, L. et al. (2012). Reconstruction and evaluation of the synthetic bacterial MEP pathway in *Saccharomyces cerevisiae. PLoS One* 7 (12): e52498.

240 Carlsen, S., Ajikumar, P.K., Formenti, L.R. et al. (2013). Heterologous expression and characterization of bacterial 2-C-methyl-D-erythritol-4-phosphate pathway in *Saccharomyces cerevisiae. Appl. Microbiol. Biotechnol.* 97 (13): 5753–5769.

241 Kirby, J., Dietzel, K.L., Wichmann, G. et al. (2016). Engineering a functional 1-deoxy-D-xylulose 5-phosphate (DXP) pathway in *Saccharomyces cerevisiae. Metab. Eng.* 38: 494–503.

5

Mammalian Aldehyde Oxidase

Maria Joao Romao[1,2], *Guilherme Vilela-Alves*[1,2], *and Cristiano Mota*[1,2]

[1] *Universidade NOVA de Lisboa, NOVA School of Science and Technology, Associate Laboratory i4HB – Institute for Health and Bioeconomy, 2829-516 Caparica, Portugal*
[2] *Universidade NOVA de Lisboa, NOVA School of Science and Technology, UCIBIO, Applied Molecular Biosciences Unit, Department of Chemistry, 2829-516 Caparica, Portugal*

Chapter Goals

- The goal of this chapter is to describe the structural features of aldehyde oxidase (AOX) a complex metalloenzyme dependent on molybdenum, iron–sulfur clusters and one flavin adenine dinucleotide (FAD) and involved in drug and xenobiotic metabolism in humans and animal models.
- The action of AOX and differences among species have had a profound impact in the pharmaceutical industry, often compromising the discovery of novel drugs at later stages of the development pipeline. AOX is a puzzling and remarkably promiscuous enzyme that accepts a large variety of substrates while performing oxidation and reduction reactions.
- In this chapter, we provide an overview on existing structural data of the XO-related enzyme AOX highlighting its complex structure: a 2×150 kDa homodimer with a molybdenum-active site (Moco), two distinct [2Fe–2S] centers, and one FAD all essential for catalysis.
- The described structures include complexes with substrates and inhibitors that have provided insights into the catalytic mechanism, mechanisms of inhibition, and allosteric inhibitors.

Main Chapter Points

- AOX belongs to the xanthine oxidase family of enzymes, whose prototype, xanthine oxidase/xanthine dehydrogenase (XO/XDH) is a well-known enzyme involved in the final steps of purine catabolism and implicated in diseases such as gout or cardiovascular diseases.
- The identification of reaction mechanisms and molecular determinants for substrate specificity of AOX are key to predict the metabolism of new drug candidates

Iron-Sulfur Clusters: Biogenesis and Biochemistry, Volume 1, First Edition.
Edited by Silke Leimkühler, Günter Schwarz, Oliver Lenz, and Oliver Einsle.

during drug discovery and development and the wide range of reactions catalyzed are briefly overviewed in the context of the prominent role of human AOX in drug metabolism.

- The chapter also includes some insight into the genetic variability of the hAOX (single-nucleotide polymorphisms [SNPs]), their structural implications and the corresponding consequences on the drug-metabolizing activity of these individuals.

5.1 Introduction

Aldehyde oxidase (AOX, EC1.2.3.1) is an enzyme involved in the metabolism of drugs and xenobiotics in humans and animals. It belongs to the XO family of mononuclear molybdenum enzymes [1, 2] and harbors, in separate domains, two [2Fe–2S] clusters, one FAD, and the molybdopterin (Moco) cofactor as its active components. XO and AOXs share structural and functional similarities, using the Moco catalytic site for oxidizing a wide range of substrates while the other cofactors, FAD and the two [2Fe–2S] clusters, play essential roles in electron transfer (ET) processes during substrate oxidation or reduction.

Sections 5.2 and 5.3 detail the structural aspects of XO and AOX, focusing on the mammalian enzymes and emphasizing the differences that account for the high specificity of XO in contrast to the wide catalytic diversity and substrate promiscuity of AOX.

5.2 The Xanthine Oxidase Family of Mononuclear Mo Enzymes

The XO family is a very broad one and comprises enzymes involved in purine metabolism and in oxidation and reduction reactions of different substrates. Members of this family are broadly distributed within eukaryotes, prokaryotes, and archaea and include, among others, XO, AOX, and carbon monoxide dehydrogenase (CODH) [1, 2]. These enzymes share structural and functional similarities and are hydroxylases or oxotransferases, promoting various two-electron redox reactions, whereby an oxygen atom transfer occurs.

At the center of this family is XO, also known as xanthine dehydrogenase (XDH) in its reduced form. While XO uses dioxygen as the physiological electron acceptor, XDH uses NAD^+. XO/XDH catalyzes the conversion of hypoxanthine to xanthine and xanthine to uric acid, a final product of purine catabolism in humans, playing vital roles in various physiological processes, including the regulation of uric acid levels and antioxidant defense. The enzyme has been implicated in pathological conditions such as gout, cardiovascular diseases, and inflammation, making it a target for therapeutic interventions. Understanding the structure, function, and regulation of enzymes of the XO family has been crucial for unraveling their physiological significance with potential therapeutic implications [3].

The first crystal structure for an XO-related enzyme was that of the aldehyde oxidoreductase from the sulfate reducer *Desulfovibrio gigas (Dg)* (Mop) reported in 1995 (PDB_ID: 1VLB) [4, 5]. Crystal structures of XO/XDH from two different sources followed, namely: bovine milk XDH in 2000 (PDB_ID: 1FO4) [6] and *Rhodobacter capsulatus (Rc)* XDH in 2002 (PDB_ID: 1JRO) [7]. The ensemble of these structures provided additional information at an atomic level for this enzyme family, which was crucial to reveal the geometries and ligands of the Mo-active sites, the cofactors (two spectroscopically distinguishable [2Fe–2S] centers and the FAD), and the pathways for ET.

AOX is an enzyme from the XO family of mononuclear Mo enzymes that plays a crucial role in the metabolism of various endogenous and exogenous compounds in many organisms. AOX exhibits a very broad substrate specificity, not only oxidizing aromatic and aliphatic aldehydes into the corresponding carboxylic acids but also hydroxylating a wide range of heteroaromatic rings. In some cases, reduction reactions are also reported (see Section 5.4). Several of these reactions are vital for the detoxification and elimination of xenobiotics and endogenous aldehydes and detailed knowledge is therefore of paramount importance for the pharmaceutical industry.

AOX is highly conserved across species, indicating its evolutionary significance [8] and its widespread presence in mammals highlights its importance in numerous physiological processes, underscoring its potential as a target for pharmacological investigation in drug metabolism and toxicology research. Mammals are characterized by a maximum of four AOX genes coding for a corresponding number of AOX isoenzymes. While humans express a single AOX1 isoform, rodents synthesize four AOX1, AOX2 (also known as AOX3/1), AOX3, and AOX4 isoenzymes [8]. Understanding the structural details of mammalian AOXs at an atomic level and correlating them with biochemical and functional data have provided insights into catalytic and inhibition mechanisms and their significance in drug metabolism. The first crystal structure of a mammalian AOX protein was that of mouse liver AOX3 [9, 10] (mAOX3). Subsequent efforts have led to the first crystal structure of the human AOX1 (hAOX1) that was obtained in its free form as well as in a ternary complex with a substrate molecule (phthalazine) at the Mo-active site and an inhibitor molecule (thioridazine) bound to an allosteric site [11] (see Section 5.3). This constituted an important milestone in the field of drug discovery and development.

Members of the XO family show about 25% sequence identity and 60–70% sequence similarity, with higher conservation in segments involved in the binding of the metal centers and redox-active sites. Enzymes belonging to this class are either homodimers, $\alpha 2$ or dimers of heterodimers $\alpha 2\beta 2$ or of heterotrimers, $\alpha 2\beta 2\gamma 2$. *Dg*Mop, AOXs, and eukaryotic XO are organized as $\alpha 2$ homodimers, with all redox-active cofactors confined within a single polypeptide chain (Figure 5.1). The common folding pattern for this group of enzymes starts with the two [2Fe–2S]

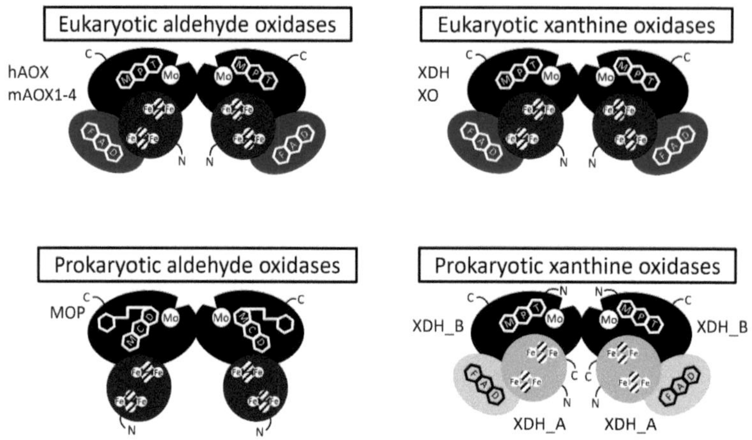

Figure 5.1 Comparison of domain arrangement and quaternary structure of aldehyde oxidases (AOXs) and xanthine oxidases/dehydrogenases (XOs/XDHs) from eukaryotic and prokaryotic organisms. The redox cofactors present in each domain/subunit are schematically represented.

domains, followed by a flavin domain (which is absent in *Dg*Mop and replaced by an extended connecting segment) and finishes with one large catalytic domain that binds the molybdopterin cofactor [1, 4].

The overall fold and architecture of enzymes of the XO family are remarkably similar as well as the arrangement of the prosthetic groups involved in electron transfer whether in a single polypeptide chain with different domains (XO/XDH, Mop, AOX (α2)) or in independent subunits (e.g. *Rc* XDH (dimer of heterodimers ((αβ)2)) that harbor the different cofactors (Figure 5.1)).

In all structures, the Moco-active site lies at the bottom of a hydrophobic funnel at the interface of two subdomains of the catalytic domain. All redox centers are aligned within the protein matrix, defining a suitable intramolecular electron transfer pathway (Figure 5.2), with distances between the prosthetic groups similar in all structures from the family.

5.2.1 The Fe–S Domain and Structural/Spectroscopic Assignment of the Iron–Sulfur Centers

In all enzymes from the XO family, the Fe–S domain is divided into two subdomains, each binding one distinct [2Fe–2S] cluster (Figure 5.2). Type I cluster is approximately 12 Å below the protein surface and its corresponding subdomain exhibits a four-helix bundle topology, a new type of ferredoxin fold first reported for *Dg*Mop [4] and characteristic of all XO-related enzymes. Type II [2Fe–2S] cluster is solvent exposed, and its domain fold and binding motif are typical of plant and cyanobacterial ferredoxins. Both type I and type II Fe–S centers and the respective domains are characteristic of the XO family of enzymes.

The two [2Fe–2S] centers are Electron Paramagnetic Resonance (EPR) active in the reduced state (S = 1/2) and the designation as type I and type II was based on

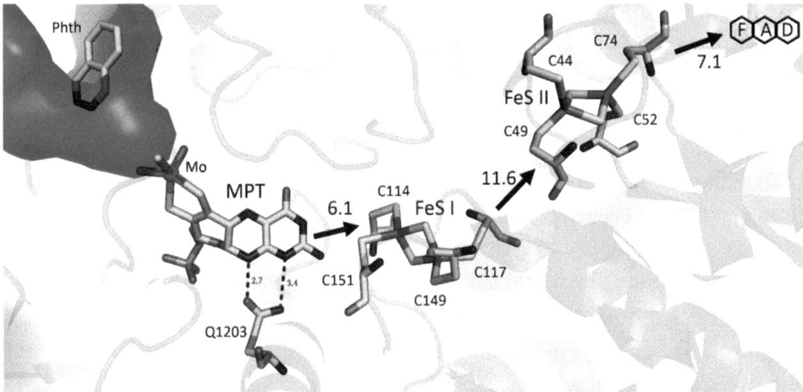

Figure 5.2 Representation of the MPT cofactor and the two adjacent [2Fe–2S] clusters I and II that define an electron transfer pathway (black arrows, showing inter-cofactor distances in Å) in the structure of hAOX1 in complex with phthalazine (Phth) and thioridazine (not shown) (PDB ID: 4UHX). Residues stabilizing the cofactors are shown as sticks. The substrate entry pocket with the substrate molecule phthalazine is shown as a dark gray surface. This figure was created using PyMOL v2.5.4.

their different EPR signals studied on XO samples in the 1970's by Bray and others [12–14]. Fe–S I EPR signal is similar to plant-type ferredoxin, is observable up to higher temperatures, and presents magnetic coupling with Mo(V) signal, while cluster II shows more broadened lines at lower temperatures. In the first reported structure of an XO-family member [4], the two Fe–S clusters were named Fe–S_B and Fe–S_A. In subsequent publications, the assignment of the Fe–S clusters in the structure was based on the protein folding and on EPR data resemblance and was unanimously reported by independent authors working on other members of the XO family that the proximal Fe–S center with a new ferredoxin-type fold corresponds to Fe–S_I and the distal Fe–S center to Fe–S_II (Figure 5.2) [6, 15–17].

5.3 Aldehyde Oxidase Crystal Structure

5.3.1 The Mammalian AOX Crystal Structures

The report of the first crystal structure of a mammalian AOX, the mouse liver mAOX3 (PDB_ID: 3ZYV), included site-directed mutagenesis and kinetic data, and those studies have provided valuable insights into the singularities of the protein's active site and of the putative catalytic mechanism [9]. However, this structure was an insufficient model for studying human metabolism due to the relatively low sequence identity (65%) between the mouse and human homologs (mAOX3 and hAOX1) as well as the presence of different isoforms in mice and other animal species. The existence of isoforms in animal species has posed significant challenges in selecting an appropriate experimental model to mimic human metabolism and this has led to numerous failures in the drug discovery process [18]. The elucidation of the first crystal structure of human AOX1 was therefore a breakthrough due to the profound impact in the field of drug discovery and development [11, 19].

The 3D structures of hAOX1 in its free form (PDB_ID: 4UHW) and in a ternary complex with a substrate and an allosteric inhibitor (PDB_ID: 4UHX) revealed insights into its molecular architecture and its catalytic and inhibition mechanisms. It exhibits a homodimeric (2 × 150 kDa) α2 structure, of approximate dimensions 150 Å × 90 Å × 65 Å, with each monomer comprising three domains: the N-terminal domain (Fe–S domain (I), 25 kDa) harboring two spectroscopically distinguishable iron–sulfur clusters ([2Fe–2S] I and [2Fe–2S] II) in two subdomains, the central domain (FAD domain (II), 40 kDa) binding one flavin adenine dinucleotide (FAD) cofactor, and the C-terminal catalytic domain (Moco domain (III), 85 kDa) enclosing the buried molybdenum cofactor (MPT) and the substrate-binding cavity (Figure 5.3). The Fe–S domain I is connected to domain II by linker 1 (residues 166–231). This is a surface-exposed region and, as in the mAOX3 crystal structure, a total of 32 residues had no interpretable electron density which suggests a mobile region. Interestingly, the same was found in all reported structures of XO and XDHs [11], suggesting that the flexibility of this region is intrinsic to this family of enzymes. Domain II is linked to the Mo domain III by a shorter and ordered linker (residues 516–555).

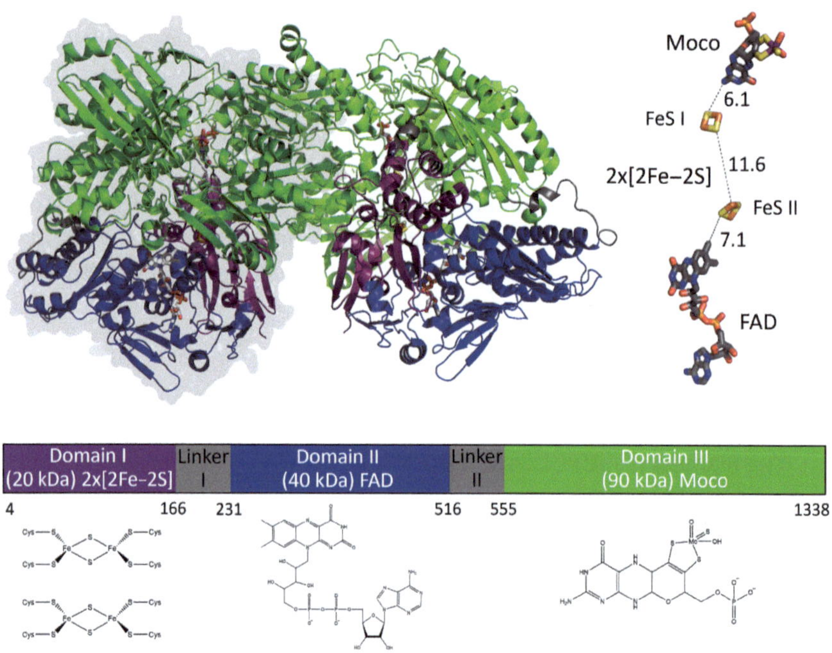

Figure 5.3 Overall structure of the human AOX1 homodimer in ribbon and surface representation (PDB ID: 4UHW). The three domains are colored: violet (domain I), blue (domain II), and green (domain III). Domain I harbors the two [2Fe–2S] centers, domain II contains the FAD cofactor, and the MPT cofactor is buried in domain III. On the right, the protein cofactors are represented and the shortest distances between them are indicated in Å. Below, the chemical 2D structures of the cofactors are presented. This figure was created using PyMOL v2.5.4 and ChemDraw 20.1.

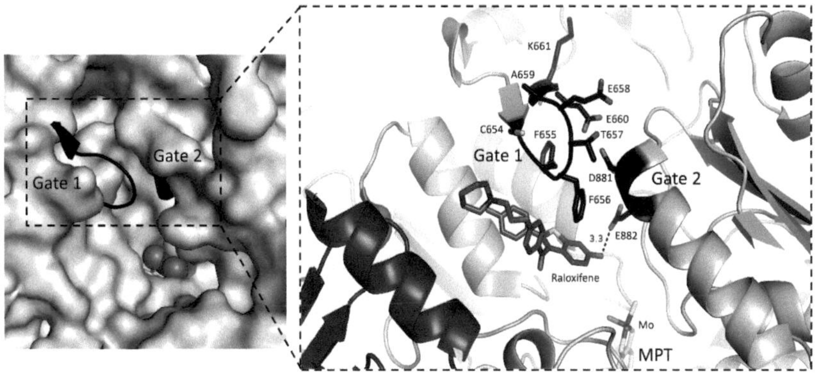

Figure 5.4 Close-up view of the active site of hAOX1 in complex with raloxifene (PDB ID: 7ORC), highlighting the substrate channel Gates 1 and 2 (black cartoon) (left). Gates 1 and 2, shown as black sticks on the right, are stabilized by the interaction with the raloxifene molecule. This figure was created using PyMOL v2.5.4.

The crystal structures define the arrangement of redox centers within the protein matrix, defining a favorable intramolecular electron transfer pathway essential for catalysis (Figure 5.3). Additionally, details of the substrate-binding cavity supported by mutagenesis and kinetic data have led to proposals of the molecular determinants for catalysis, a first step toward the design of potential therapeutics and to the understanding of the hAOX1 prevalent role in drug and xenobiotic metabolism (see Section 5.4).

The crystal structure of substrate-free hAOX1 reveals several residues of the Moco domain, lining up the entrance of the substrate channel that is mobile, with no clear electron density. These residues correspond to two flexible loops that define two gates that were proposed to control the entrance of substrates and the release of products to the Moco-active site (Figure 5.4).

As might be expected, the crystal structures of hAOX1 and mAOX3, despite a high overall similarity, exhibit differences in the FAD-binding region and in the substrate funnel that led into the buried Mo-active site. Since mAOX3 is the only mouse isoform with its crystal structure known, a computational study (homology modeling and molecular dynamics) was used to provide models of the three other mouse isoforms (mAOX1, mAOX2, and mAOX4) [20]. Most of the structural differences among the four isoforms were found at the protein surface and in the substrate access tunnel and it was proposed that access of substrates to the active site is determined by the structural features of the wide deep funnel. At the bottom of the funnel, the active site cavity is conserved in all four isoforms, while a region located above is considered isoform specific. The differences in the substrate-binding site were suggested to influence the substrate specificity of each enzyme.

5.3.2 The Mo Active Site

In all known 3D structures of the XO family, the Mo-active site is located at the bottom of a 15 Å deep hydrophobic channel. The buried Mo cofactor, Moco, includes

(a) (b)

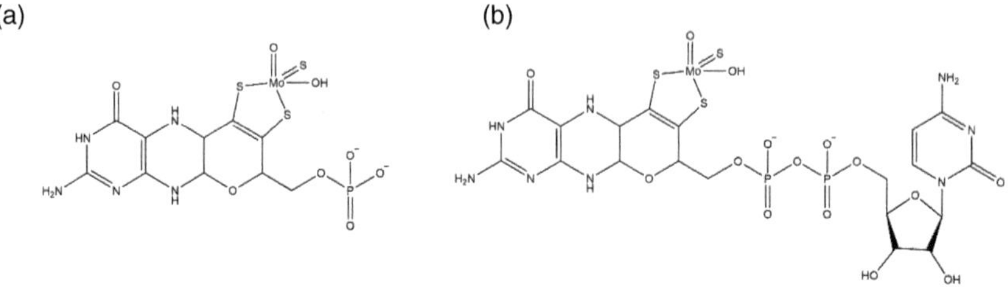

Molybdopterin monophosphate (MPT) Molybdopterin cytosine dinucleotide (MCD)

Figure 5.5 2D chemical structures of the molybdenum cofactor present in enzymes of the xanthine oxidase family, with the family characteristic sulfido ligand at the equatorial position of the Mo coordination sphere. In eukaryotic enzymes of this family, the basic form of Moco is a pyranopterin monophosphate (a), named MPT, which coordinates the Mo atom by one dithiolene group. In bacteria, the MPT core is modified by a cytosine monophosphate nucleotide forming molybdopterin cytosine dinucleotide (MCD) (b). This figure was created using ChemDraw 20.1.

one organic cofactor (pyranopterin or molybdopterin) that coordinates to the metal via its dithiolene moiety and may be present either in a dinucleotide (e.g. molybdopterin cytosine dinucleotide, MCD) (in prokaryotes) or in the monophosphate form (molybdopterin monophosphate, MPT) (usually in eukaryotes) (Figure 5.5).

The primary role of Moco is to position the Mo-active site in the protein interior as well as to modulate the redox potential at the Mo center, being involved, via the tricyclic ring system, in the transfer of electrons to the nearest [2Fe–2S] center I. The geometry at the Mo catalytic site is a distorted square pyramid with no protein ligand bound to the metal. The equatorial plane is constituted by two sulfurs provided by the dithiolene moiety, one hydroxyl and one sulfido ligand. At the apical position an oxo group completes the Mo coordination sphere [Mo,=O,=S,-OH, MPT] (Figure 5.6). The apical Mo = O forms an H-bond with the conserved Gln776. The catalytic labile oxygen is the equatorial Mo-OH ligand that points toward the entrance of the substrate-binding site (Figure 5.6) and forms a H-bond with the strictly conserved glutamate residue (Glu1270). It has been demonstrated by site-directed mutagenesis that Glu1270 is catalytically essential in AOX and XO enzymes [2]. The equatorial Mo = S ligand is also essential for catalysis as extensively reported for XO and AOX, and it has been proven that the rate-limiting step of catalysis is the substrate-hydride transfer to this sulfido ligand [21, 22].

When comparing the Mo-active sites of bovine XO (bXO), hAOX1, and mouse AOX3 (mAOX3), there are several conserved amino acid residues (Figure 5.6) – Gln776, Phe923, Ser1089, and Glu1270 (hAOX1 numbering) – that might suggest similar reaction mechanisms. However, in AOXs, clear differences in the active sites emerge and, for instance, residues in XO involved in substrate recognition and orientation – Glu802, Arg880, and Phe1009 (bXO numbering) – are replaced in hAOX1 by neutral and smaller residues, creating a larger catalytic pocket less specific and able to accommodate a wider diversity of substrates (Figure 5.6).

Interestingly, His884 from bXO is replaced by a conserved lysine in all AOX isoenzymes [20]. In hAOX1, the Lys893 side chain is found in a similar position as

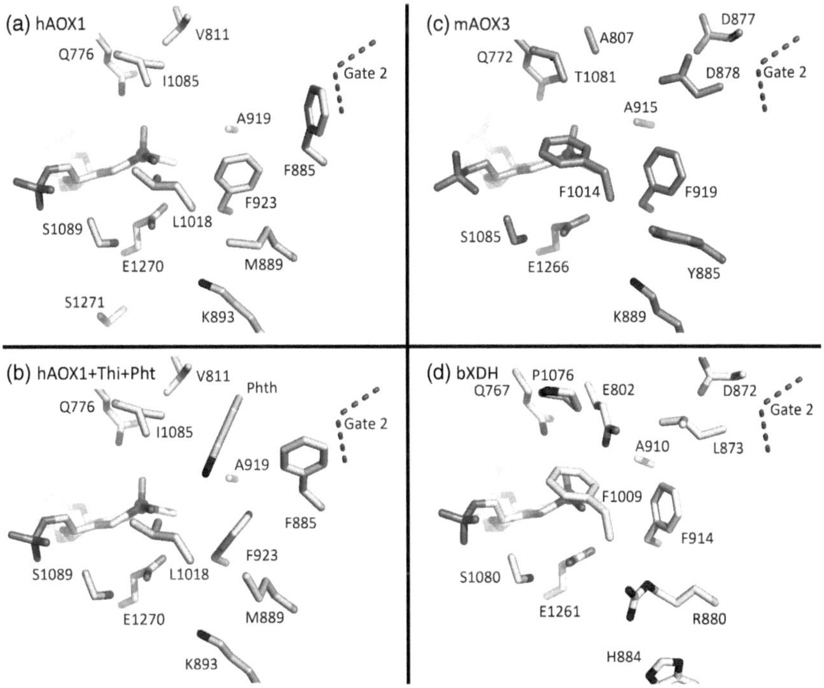

Figure 5.6 Close-up views of AOXs and XOR active sites from (a) hAOX1 (PDB ID: 4UHW); (b) hAOX1 in complex with phthalazine (Pht) and thioridazine (Thi) (PDB ID: 4UHX); (c) mouse mAOX3 (PDB ID: 3ZYV); and (d) bovine bXDH (PDB ID: 3UNC). The Mo ion is represented as a sphere and its S and O ligands as sticks. This figure was created using PyMOL v2.5.4.

Arg880 in bXO structures, a critical residue for substrate orientation, together with Glu802 [6] (Figure 5.7). Lys893 lies ~10 Å away from the Mo center and is 6 Å away from the catalytically essential Glu1270. The oxidation mechanism of hAOX1 was studied with atomistic detail by QM-MM methods that confirmed that Lys893 plays a relevant role in the reaction, not only for anchoring the substrate close to the Mo site but also in the catalytic reaction (Figure 5.7) [23]. The reaction mechanism starts with (i) the positioning of the substrate in the proximity of the metal site and the protonation of N_2 of phthalazine by Lys893, (ii) the deprotonation of Mo-OH by Glu1270, followed by the nucleophilic attack on the carbon atom of the substrate, leading to the formation of the tetrahedral intermediate Mo-O-C. Subsequently, in a concerted manner, (iii) a hydride from the substrate is transferred to the sulfido ligand, thereby reducing the metal to Mo (IV). This computational analysis led to the conclusion that the rate-limiting step involves the transfer of the hydride from the substrate to the essential sulfido ligand (Mo = S). In a subsequent step, and preceding product displacement, (iv) one electron is released, leading to the oxidation of the metal to Mo(V). Finally, (v) the oxidized product is released along with one proton and one electron (Mo(VI)) and the Mo site undergoes hydroxylation by a water molecule, with a proton transfer to Lys893 (Figure 5.7).

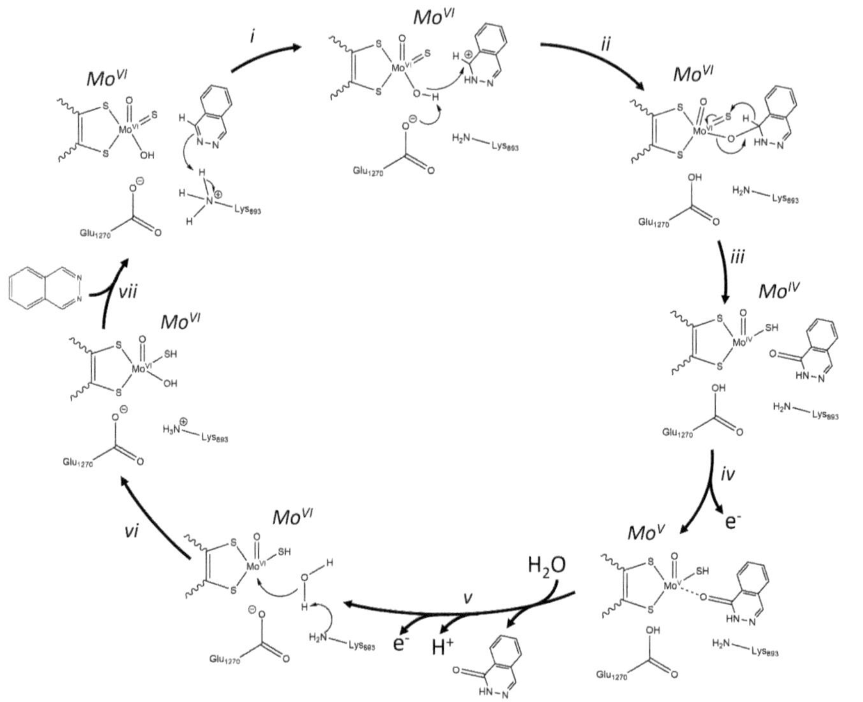

Figure 5.7 Complete mechanism for the oxidation of phthalazine to phthalazin-1(2H)-one by hAOX1 supported by the associated free energy profile for the corresponding reaction (not shown) (Ferreira 2020). This figure was created using ChemDraw 20.1. Source: Adapted from [23].

The entrance of the substrate access funnel presents two flexible loops where some residues could not be modeled in the electron density maps in all structures of hAOX1 without raloxifene (gate 1, residues 654–661 and gate 2, residues 881–882) (Figure 5.4). Those two gates were proposed to play a role in providing access to substrate and product molecules [11]. Gate 1 is composed of hydrophobic residues and, except for Glu660, other residues from this gate are not conserved in other members of the XO family. On gate 2, Asp881 is conserved and followed by a Glu or Asp residue in the AOXs subfamily. The side chains of these amino acids could be modeled in the mAOX3 crystal structure and point toward the substrate channel. Those amino acids may probably orient polar substrates toward the Mo-active site. Phe885, close to gate 2, has a role in substrate orientation and its side chain changes position upon substrate binding in the structure of the ternary complex with the substrate phthalazine and the inhibitor thioridazine (hAOX1_thi_pht) (PDB_ID: 4UHX) [11] (Figure 5.8). Interestingly, the crystal structure of the complex of hAOX1 with the inhibitor raloxifene (the most potent hAOX1 inhibitor known) revealed that the molecule is blocking the entrance of the tunnel to the Mo-active site. The protein structure is more stable and in this inhibited form, the access gates 1 and 2 become ordered and visible in the structure (Figure 5.4) [24].

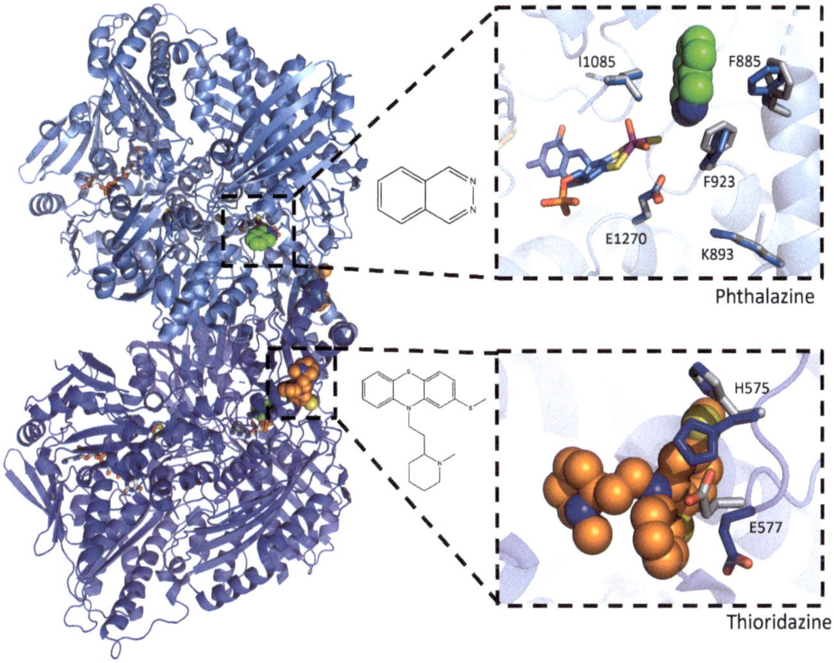

Figure 5.8 Overall structure of the hAOX1 homodimer (blue) in complex with phthalazine (green) and thioridazine (orange) (PDB ID: 4UHX). Top: close-up view of phthalazine-binding site with relevant residues shown as sticks, in a superposition with free hAOX1 (gray) (PDB ID: 4UHW). Bottom: close-up view of thioridazine-binding pocket with relevant residues shown as sticks, in a superposition with free hAOX1 (gray) (PDB ID: 4UHW). This figure was created using PyMOL v2.5.4 and ChemDraw 20.1.

5.3.3 The [2Fe–2S] Clusters

Iron–sulfur clusters are ubiquitous in nature and are present in most Moco-dependent enzymes, with a clear role in electron transfer. The redox centers in the XO family, the Mo center, two [2Fe–2S], and one FAD are aligned along the protein and define a facile intramolecular electron transfer pathway with distances between the prosthetic groups similar in all structures from the family (Figure 5.2). In mammalian AOX and XOs, both [2Fe–2S] centers are located in the N-terminal Fe–S domain. The two Fe–S centers exhibit distinct EPR signals, and studies with XO samples [12–14] led to the classification of Fe–S type I and type II. This nomenclature was assigned in the structure to the proximal Fe–S (type I) and the distal Fe–S cluster (type II) that was kept for all enzymes of the family [6, 15–17] (see Section 2.1). The two reducing equivalents generated during oxidation reactions are transferred to an external electron acceptor and the electron flow is mediated by the intervening cofactors: Moco → [2Fe–2S] I → [2Fe–2S] II → FAD. Both Fe–S clusters are planar with the iron and cysteine sulfur atoms defining a plane orthogonal to the plane of the [2Fe–2S] moiety. All iron atoms are tetrahedrally coordinated by the sulfur atoms of the cysteines, with Fe—S bond lengths of ~2.2 Å and ~2.3 Å for the iron-sulfide and the iron-cysteine Sγ, respectively. The N-terminal plant-type ferredoxin iron cluster (Fe–S_II) has one of the iron atoms coordinated to SCys44 and SCys49, while the other iron atom is coordinated to SCys52 and SCys74. The second iron cluster (Fe–S_I) has one iron atom bound to SCys117 and SCys149 and the other bound to SCys114 and SCys151. Center [2Fe–2S] I (Fe–S_I) is buried and in contact with the molybdopterin cofactor (6.1 Å away from the exocyclic NH_2 of the pterin ring) while [2Fe–2S] II (Fe–S_II) is close to the protein surface and in contact to the FAD site (7.1 Å from the nearest carbon atom of the flavin ring (Figure 5.3)).

The arrangement of the two [2Fe–2S] centers in hAOX1 provides a favorable electron transfer pathway from the Moco site to the FAD cofactor. The two [2Fe–2S] clusters are crucial for enzymatic activity since they facilitate electron transfer reactions, enabling the enzyme to transfer electrons from its substrates to molecular oxygen. During catalysis, the aldehyde substrate is oxidized, and released electrons shuttled through the two [2Fe–2S] clusters mediated by several covalent and hydrogen bonding interactions (Figure 5.2). This electron transfer ultimately leads to the reduction of molecular oxygen to water, a reaction integral to AOX's role in detoxification and metabolism. Additionally, the [2Fe–2S] clusters help stabilize the AOX protein structure, ensuring proper folding and enzymatic function.

The first Fe–S_II subdomain (residues 1–89) has a topology very similar to the plant-type ferredoxin class of iron–sulfur proteins. The chain fold is similar to that of plant (*Spirulina platensis*) [25], alga (*Aphanothece sacrum*) [26], and cyanobacterium *Anabaena* [27] [2Fe–2S] ferredoxins. It shows the topology of a five-stranded beta-half barrel with an alpha-helix running almost perpendicular to the strand's direction. Apart from the absence of a 19-residue loop toward the C-terminus of this domain and a beta-turn, shorter by six-residues, between the first two strands, the overall fold can be superimposed to *Spirulina platensis* ferredoxin [25] with a rmsd of 0.86 Å for 57 Cα atoms. The second Fe–S_I subdomain (residues 98–166) reveals a new ferredoxin type fold first found in the XO family (*Dg*Mop): a twofold symmetric four-helical bundle with two longer central helices flanked by two oblique shorter alpha-helices. The iron–sulfur cluster lies at the N-termini of the two central helices and type I [2Fe–2S] cluster is buried ~12 Å below the protein surface.

The two [2Fe–2S] clusters in hAOX1, although indistinguishable in UV–Vis spectra, exhibit different midpoint redox potentials. While in XO/XDH enzymes, Fe–S_II has the highest redox potential (−235 mV) in comparison to Fe–S I (−310 mV) [28], the order of reduction is reversed in AOXs being −207 and − 310 mV for Fe–S_I and Fe–S_II, respectively, in rabbit liver AOX [29] and − 206 and − 286 mV for Fe–S_I and Fe–S_II in mAOX3, respectively [30].

In terms of EPR signals, Fe–S_I and Fe–S_II show distinct properties in AOX1 as well as in XO enzymes (see Section 2.1) [1, 2].

5.3.4 The FAD-Binding Site

The hAOX1 FAD domain (domain II) is responsible for binding and positioning the FAD cofactor with its isoalloxazine ring stacked between the side chains of two

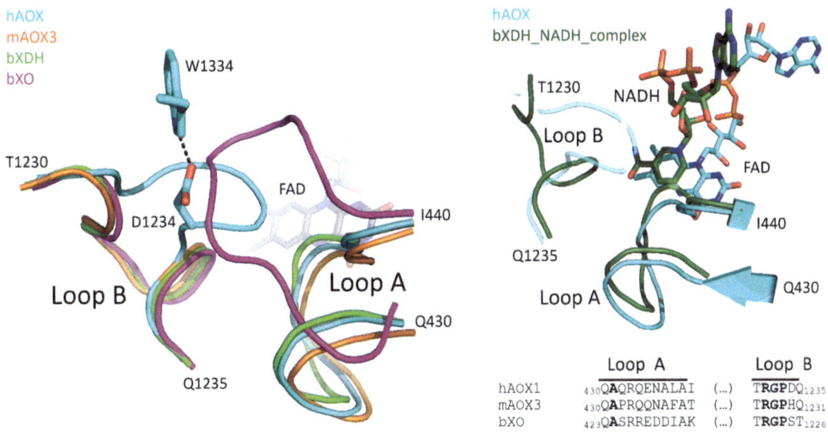

Figure 5.9 Superposition of the AOXs and XOR FAD-binding sites (left) from human AOX1 (cyan) (PDB ID: 4UHW), mouse AOX3 (orange) (PDB ID: 3ZYV), bovine XDH (green) (PDB ID: 3UNC), and bovine XO (magenta) (PDB_ID: 1FIQ), highlighting the variable loops A and B. Superposition of the FAD-binding site from hAOX1 (cyan) (PDB ID: 4UHW) and bXDH in complex with NADH (dark green) (PDB ID: 3UNI) (right). The sequence alignment of FAD variable loops A and B is presented at the bottom, for hAOX1, mAOX3, and bXO. This figure was created using PyMOL v2.5.4.

leucine residues (Leu344 and Leu438). At the entrance of the FAD-binding pocket, the loop T_{1230}-RGPD-Q_{1235} is flipped almost 180° when compared to the corresponding loop in mAOX3 and bXO (Figure 5.9). This loop adopts a conformation so far only found in hAOX1 and lies at a similar position as the nicotinamide ring in the bXDH–NADH complex (PDB_ID: 3UNI) blocking access of large redox cofactors to the FAD pocket in agreement with the fact that O_2 is the unique electron acceptor of human AOX.

The FAD cofactor is crucial for hAOX1's catalytic activity acting as a coenzyme in redox reactions by accepting and donating electrons during substrate oxidation/reduction processes. In addition to its role in the electron transfer chain, the FAD cofactor has been proposed to be involved in catalysis with reports of reduction reactions carried out at this site, instead of the Mo-active site, namely, reduction of organic nitrates by XO [31] and reduction of the nitro group from 5-nitroquinoline by hAOX1 [32] (see Section 5.4).

5.4 Aldehyde Oxidase and Its Role in Drug Metabolism

Human AOX is mainly expressed in the adrenal gland, adipose tissue, liver, and kidney and it is characterized by a broad substrate specificity, playing a significant role in drug metabolism [18, 33–38]. Thus, identification of reaction mechanisms and molecular determinants that define substrate specificity is key to predict the metabolism of new drug candidates during drug discovery and development pipelines. hAOX1 catalyzes a wide range of reactions, which include mostly oxidations of different types of substrates, in addition to reductions and hydrolysis [18, 39]. This enzyme, known for its pronounced promiscuity, metabolizes a broad spectrum of chemical structures, including aldehydes, diverse azaheterocycles, iminium ions, N-oxides, S-oxides, nitro-derivatives, and amides. Considering the reactions diversity, it is reasonable to anticipate different catalytic mechanisms, potentially occurring in different active sites. Although the primary role of AOX is [1] oxidation of aldehydes and of N-aromatic heterocycles and iminium ions, it is also known to catalyze [2] reduction of nitro, N-oxide, S-oxide, or isothiazol compounds and [3] hydrolysis of amide bonds. Such reactions with examples of different substrates and inhibitors of hAOX1 are described in Sections 5.4.1 and 5.4.2. Our aim is to provide a comprehensive understanding of hAOX1 reaction mechanisms, the relationship between the structure and metabolism of substrates, and its impact on drug discovery and development.

5.4.1 Reactions Catalyzed by hAOX1

5.4.1.1 Oxidation of Aldehydes, N-aromatic Heterocycles, and Iminium Ions

Human AOX1 catalyzes the oxidation of aldehydes into the respective carboxylic acids. Several physiological aldehydes are substrates of hAOX1 (e.g. 5-hydroxyindoleacetaldehyde, retinaldehyde, or pyridoxal [40]). While drugs typically do not contain aldehyde groups, they may arise due to secondary phase I metabolism, in combination with other enzymes (e.g. CYPs and MAO). For example, the antidepressant citalopram [41, 42] and the antiestrogen, tamoxifen [41, 43] are drugs metabolized into aldehyde intermediates by CYPs and monoamine oxidase, respectively, which are subsequently oxidized by hAOX1 into the corresponding carboxylic acids.

N-heterocycles frequently appear in chemical libraries of drug candidates because they are typically not metabolized by CYP450 enzymes [35]. However, hAOX1 catalyzes the oxidation of N-aromatic heterocycles to the corresponding oxo N-heterocycles. Most of these compounds are oxidized in the ortho C–H position but metabolism can be modulated by adding substituents to the most electropositive carbon atom [44, 45]. There are several examples of N-heterocycle oxidation reactions some of which led to the suspension of clinical trials due to hAOX1 metabolism [18]. On the other hand, hAOX1 metabolism can also be used as an efficient way to activate prodrugs, such as famciclovir, which is activated by hAOX1 to penciclovir, a viral polymerase inhibitor [46, 47].

Iminium ions arise as intermediates in the metabolic pathways of cyclic amines such as pyrrolidines or piperidines, catalyzed by enzymes like CYP450 or MAO. Subsequently, these iminium ions undergo further oxidation to form cyclic lactams, a reaction catalyzed by hAOX1. For example, the metabolism of nicotine proceeds initially through its conversion to N1-methylnicotinamide by CYP450, followed by its transformation to cotinine by hAOX1 [48].

5.4.1.2 Reductions of Nitro, N-oxide, S-oxide, and Isothiazol Compounds

The reduction reaction catalyzed by hAOX1 is not yet fully understood as well as the determination of which cofactor is directly involved in the reduction of the substrate (Moco or FAD) [32]. In addition to molecular oxygen, hAOX1 can reduce nitro compounds [32, 49], N-oxides [50], sulfoxides [51], isoxazoles [52], nitrite [53], N—S and N—O bonds [54], and hydroxamic acids [55]. Because molecular oxygen is the natural electron acceptor of hAOX1, most of the reduction reactions described were studied *in vitro* in anaerobiosis, with the drug substrate being the only electron acceptor. Nonetheless, the reduction of 5-nitroquinoline [32] and the reduction of the hypnotic agent nitrazepam [49] were confirmed under normal oxygen conditions. Consequently, toxic metabolites such as 7-hydroxyl-nitrazepam may be the cause of nitrazepam-induced liver injury [49].

5.4.1.3 Amide Hydrolysis Reactions

In contrast to oxidation and reduction reactions, amide hydrolysis reactions catalyzed by hAOX1 have recently been documented. This was initially observed in the drug candidate GDC-0834, which is a Bruton's tyrosine kinase inhibitor under investigation for its potential treatment of rheumatoid arthritis [56]. This compound was hydrolyzed by hAOX1, resulting in undetectable levels of the drug in human circulation following oral administration. Docking studies suggest that this reaction likely occurs at the Moco site [56], and an extensive study into the hydrolysis of a large library of amides by hAOX1 was conducted to assess potential correlations of reactivity based on the molecular structure [44]. While hAOX1 exhibits broad substrate specificity, particularly concerning amides, its hydrolysis activity is notably influenced by the amine moiety. The stability in terms of metabolism by hAOX1 is determined by the order of the substituents in the position of the anilide (ortho < meta < para) [44].

5.4.2 Inhibition of hAOX1 Activity

Several families of drugs have been described for their capacity to inhibit hAOX1 activity [57], raising concerns regarding the potential impact on drug–drug interactions. From the family of estrogens, raloxifene was identified as the most potent inhibitor of human AOX, with an inhibition constant (Ki) of 7 nM [24]. The crystallographic structure of hAOX1 in complex with raloxifene shows the molecule bound to the protein at the entrance of the catalytic pocket, stabilized by two entrance gates (Figure 5.4) [24], confirming the relevance of the terminal phenol groups, as previously suggested in biochemical assays [58]. Raloxifene presents an unusual substrate-dependent mechanism of inhibition, since it acts as an uncompetitive inhibitor toward small substrates but competes with bulky substrates [24]. Other estrogenic compounds such as tamoxifen or estradiol, among others, also showed hAOX1 inhibition [57].

The family of phenothiazines, related to antipsychotic drugs, is also a strong inhibitor of hAOX1 [57]. The inhibitor thioridazine could be observed in the crystal structure of the ternary complex hAOX1–phthalazine–thioridazine (Figure 5.8), in which the substrate was found at the substrate-binding cavity, close to the Mo-active site, while the inhibitor was found ca. 34 Å apart, located in a groove at the surface of the protein. Thioridazine shows an allosteric noncompetitive inhibition mode with a Ki of 1.3 μM [11]. Considering the topology of the thioridazine inhibition pocket, we could suggest that phenothiazine drugs will generally bind to this substrate-free site (Figure 5.8) to accomplish hAOX1 inhibition via a noncompetitive manner.

Dibenzazepines and flavonoids are also groups of potential inhibitors of AOX. The tricyclic antidepressant agents derived from dibenzazepine such as amitriptyline or maprotiline present substantial inhibition ranging from 0.26 to 21 μM [57]. Flavonoids are natural compounds obtained from diet or intrinsic metabolism and their ability to inhibit AOX have been extensively studied [59]. While these natural compounds influence the catalysis of numerous enzymes, they can also modulate AOX activity and affect drug metabolism.

5.4.3 Single-Nucleotide Polymorphisms

One of the reasons for the variability in human AOX1 levels and its activity is SNPs. The human AOX gene is a target of numerous SNPs, which were identified in different individuals and reported in NCBI database (http://www.ncbi.nlm.nih.gov/snp). SNPs are classified as synonymous SNPs, which do not change the amino acid sequence due to the degeneracy of the genetic code, or nonsynonymous SNPs (nsSNPs), a subset of SNPs that occur within the coding region of a gene and result in the alteration of the encoded amino acid sequence. These alterations can affect the structure, function, stability, and protein–protein interactions, potentially influencing phenotypic traits or disease susceptibility. While only a small number of these variants have undergone biochemical and structural characterization, a computational study analyzed 526 nsSNPs and their predicted effects on protein stability with putative implications on hAOX1 function [60].

In this section and based on the current knowledge, we aim to provide insights into the impact of some characterized nsSNPs in hAOX1. nsSNPs have been reported to influence the ability of the enzyme to metabolize different substrates or the catalytic efficiency, allowing the classification of the individuals bearing these variants as fast metabolizers, poor metabolizers, and individuals with no relevant effect on the human AOX catalysis [21]. Therefore, such analysis is of particular interest in drug development, personalized medicine, and therapy as well as in pharmacogenetic studies.

Taking into consideration the human AOX1 structure, SNPs were classified according to their location: domain I that holds the Fe–S clusters, domain II containing the FAD, and domain III holding the Moco-binding site. Amino acids that establish contacts with the other monomer were also labeled as amino acids from the dimer interface (Table 5.1).

Table 5.1 The most relevant SNPs in hAOX1 are listed and their respective location within the enzyme structure is indicated.

SNP	Location in hAOX1	Dimer interface	Computational prediction [60]	Experimental data
R32K	Domain I (2x[2Fe–2S])	Yes	Destabilizing	—
C44W[a)]		—	Destabilizing	Yes [61]
G48V		—	Neutral	—
G50D		—	Destabilizing	Yes [62]
C52G[a)]		—	Destabilizing	—
T53I		—	Stabilizing	—
N72K		—	Destabilizing	—
R150H		—	Destabilizing	—
R150G		—	Destabilizing	—
G177E	Linker 1	—	Destabilizing	—
G346R	Domain II (FAD)	—	Destabilizing	Yes [62]
H358P		—	Destabilizing	—
F396C		—	Destabilizing	—
R429Q		—	Destabilizing	—
R433Q		—	Destabilizing	Yes [62]
A439E		—	Destabilizing	—
G595E	Domain III (Moco)	Yes	Destabilizing	—
P762L		Yes	Destabilizing	—
Q776V		—	Stabilizing	—
R802C		Yes	Destabilizing	Yes [21]
R802H		—	Destabilizing	—
A806V		—	Destabilizing	—
R921H		—	Destabilizing	Yes [21]
G924A		—	Destabilizing	—
A1026T		Yes	Destabilizing	—
A1028P		Yes	Destabilizing	—
A1028V		Yes	Destabilizing	—
Y1033C		Yes	Destabilizing	—
S1060R		—	Destabilizing	—
S1089P		—	Destabilizing	—
A1102D		Yes	Destabilizing	—
R1109C		Yes	Destabilizing	—
R1109H		Yes	Destabilizing	—

(Continued)

Table 5.1 (Continued)

SNP	Location in hAOX1	Dimer interface	Computational prediction [60]	Experimental data
R1231H		—	Destabilizing	Yes [62]
K1237N		—	Destabilizing	Yes [62]
G1269R		—	Destabilizing	Yes [61, 63]
S1271L		—	Stabilizing	Yes [61]

a) C44 and C52 are ligands from the [2Fe–2S] center II.
The attribution of an SNP to the dimer interface reflects that it may be relevant to dimerization due to its location; nonetheless, those residues are still located in one of the three domains.

In silico analysis of the database predicted a total of 526 nsSNPs, but from these, only 119 were rated with confidence to be deleterious and analyzed in terms of structural stability (104 destabilizing effects, 5 stabilizing, 4 neutral, and 6 nonconclusive) [60].

In domain I, the two [2Fe–2S] have important roles not only in the pathway of electron flow between Moco and FAD but also as an electron sink. The tuning of the redox potential may control catalysis, and therefore it is expected that SNPs in these regions may also impact catalytic efficiency. The mutation of the conserved cysteines of the [2Fe–2S]-binding motif may have a deleterious effect on protein folding since these amino acids establish covalent bonds with the irons and these interactions may be essential for protein folding. In fact, the SNP C44W was subjected to production and purification attempts, but this variant demonstrated instability and only vestigial amounts of protein were obtained [61]. This outcome was expected due to the involvement of Cys44 as a ligand of [2Fe–2S] center II. The SNP G50D, also in the vicinity of [2Fe–2S] center II, is predicted to have a negative effect of destabilization (Table 5.1) [60]. While biochemical data on this variant display a similar K_m to the wild type, G50D presents 20% reduced activity toward oxidation reactions [62]. The substitution of G50 by a residue with a long and charged side (Asp) is anticipated to induce an effect on adjacent residues, with a direct impact on Cys49 residue, which is coordinating [2Fe–2S] II. Analysis of G46E variant also shows a similar K_m but a twofold increase in k_{cat}, suggesting that this variant does not have a negative effect on the protein [62]. The loop containing G46 is situated within the electron pathway associated with [2Fe–2S] II, and it is hypothesized that the presence of the Glu side chain may accelerate the flow of electrons.

Domain II harbors the FAD cofactor, responsible for transferring electrons to the terminal acceptor, molecular oxygen. Several SNP are predicted in this region, and some were already biochemically characterized. Four are in variable loop A (R433P, A437V, L438V, and A439E) and one in loop B (R1231H) [61]. The variant L438V was the only one that exhibited a slightly increased V_{max} in the phenantridine: O_2 reactivity compared to the wild type. While A437V did not display significant differences,

R1231H displayed a 50% decrease in V_{max} and variants A433P and A439E exhibited an even greater reduction of activity (up to 80%) [62]. Despite the catalytic differences on R1231H, the overall fold of the structure of the complex with raloxifene did not reveal significant differences to the wild type and meaningful differences are found at the FAD site [24]. Contrary to Arg1231, His is now stabilized by the hydrogen bond to Thr1230. In addition, the variant presented similar structural stability to the wild type [63]. The rate of superoxide production on the FAD variants was also probed. While most of these variants showed a comparable superoxide production rate to the wild type, L438V displayed an increased rate by 72%. The role of the hydrophobic side chain of leucine residue in triggering the redox potential of the FAD cofactor may be the reason for this difference in superoxide production. The residue L438 is 3.4 Å away from the N5 of the isoalloxazine ring, which is stacked between two leucine residues (L344 and L438). Further investigation is still required to elucidate the potential impact on drug metabolism and cellular consequences of individuals with enhanced superoxide production caused by hAOX1 SNPs.

In domain III, two variants with mutations on the amino acid residues at the Moco site were successfully purified, characterized, and crystallized (S1271L, G1269R) [61, 63]. The S1271L variant was active, facilitating the characterization of its impact on the hAOX1 activity. The kinetic analysis revealed slight differences compared to the wild type resulting in comparable catalytic efficiencies for different substrates. Hence, the amino acid exchange of a serine by leucine adjacent to the active site glutamate residue did not significantly affect the substrate binding or the turnover of the tested substrates at the active site. In addition, the structure showed meaningful differences when compared to the wild type [61]. On the other hand, the G1269R SNP was shown to be inactive. This outcome was expected since a simple inspection of the active site pocket indicated that it would be very difficult to accommodate an arginine residue at position 1269 without disturbing the insertion or orientation of the Moco in the pocket. Indeed, the crystal structure revealed that this variant was depleted of the whole Moco [63].

Several SNPs are found at the dimerization interface, and these variants may have an impact on the dimer assembly (Table 5.1) [60]. Experimental data on SNP R802C show a shift in the dimer monomer-equilibrium in solution, as the protein was mainly isolated in its monomeric form [21]. Arg802 establishes a salt bridge with Glu765 residue from the second monomer, highlighting the relevance of this interaction for dimerization. This mutation seems not to affect the catalytic efficiency of this variant, suggesting that hAOX1 in monomeric form is also active.

These findings suggest that homozygous individuals with SNPs resulting in amino acid substitutions, which affect the hAOX1 enzymatic activity or protein–protein interactions, may have consequences on the drug-metabolizing activity of these individuals. More research will be required to evaluate the potential influence of SNPs in drug development, personalized medicine, therapy, and pharmacogenetics.

5.5 Conclusions

Structural and functional studies on the drug-metabolizing enzyme AOX are crucial to understand the singularities of AOX and its function and variability among species and individuals. This knowledge will contribute to better predict drug behavior, optimize drug design, minimize drug interaction, and produce better and safer drugs. AOX is a functional dimer and in each monomer, one Moco-active site, two distinct [2Fe–2S] clusters, and one FAD cofactor are critical for the correct functioning of the enzyme. In addition to the catalytic activity (oxidation and reduction reactions), all cofactors have a clear structural role to maintain the structural integrity and stability of the AOX functional dimer. During catalysis, the Fe–S clusters and FAD are involved in the transfer of electrons in a process regulated by the redox potential of the redox centers. Any disruption in one of the redox cofactors can therefore influence AOX activity, potentially leading to changes in drug metabolism and pharmacokinetics.

List of Abbreviations

AOX	aldehyde oxidase
bXO	bovine xanthine oxidase
CYP450	cytochrome P450
DMSOR	dimethylsulfoxide reductase
EPR	electron paramagnetic resonance
FAD	flavin adenine dinucleotide
MAO	monoamine oxidase
mAOX	mouse aldehyde oxidase
Mop	aldehyde oxidoreductase from *Desulfovibrio gigas*
MPT	mononucleotide form of molybdopterin
Moco	molybdopterin cofactor
MCD	molybdopterin cytosine dinucleotide
MGD	molybdopterin guanine dinucleotide
Thi	thioridazine
Pht	phthalazine
XO	xanthine oxidase
XDH	xanthine dehydrogenase

References

1 Romão, M.J. (2009). Molybdenum and tungsten enzymes: a crystallographic and mechanistic overview. *Dalton Trans.* 21: 4053–4068.

2 Hille, R., Hall, J., and Basu, P. (2014). The mononuclear molybdenum enzymes. *Chem. Rev.* 23 (1): 1–7.

3 Nishino, T., Okamoto, K., Eger, B.T. et al. (2008). Mammalian xanthine oxidoreductase – mechanism of transition from xanthine dehydrogenase to xanthine oxidase. *FEBS J.* 275 (13): 3278–3289.

4 Romão, M.J., Archer, M., Moura, I. et al. (1995). Crystal structure of the xanthine oxidase-related aldehyde oxido-reductase from *D. gigas*. *Science* 270 (5239): 1170–1176.

5 Huber, R., Hof, P., Duarte, R.O. et al. (1996). A structure-based catalytic mechanism for the xanthine oxidase family of molybdenum enzymes. *Proc. Natl. Acad. Sci.* 93 (17): 8846–8851.

6 Enroth, C., Eger, B.T., Okamoto, K. et al. (2000). Crystal structures of bovine milk xanthine dehydrogenase and xanthine oxidase: structure-based mechanism of conversion. *Proc. Natl. Acad. Sci.* 97 (20): 10723–10728.

7 Truglio, J.J., Theis, K., Leimkühler, S. et al. (2002). Crystal structures of the active and alloxanthine-inhibited forms of xanthine dehydrogenase from *Rhodobacter capsulatus*. *Structure* 10 (1): 115–125.

8 Kurosaki, M., Bolis, M., Fratelli, M. et al. (2013). Structure and evolution of vertebrate aldehyde oxidases: from gene duplication to gene suppression. *Cell. Mol. Life Sci.* 70 (10): 1807–1830.

9 Coelho, C., Mahro, M., Trincão, J. et al. (2012). The first mammalian aldehyde oxidase crystal structure: insights into substrate specificity. *J. Biol. Chem.* 287 (48): 40690–40702.

10 Mahro, M., Coelho, C., Trincão, J. et al. (2011). Characterization and crystallization of mouse aldehyde oxidase 3: from mouse liver to *Escherichia coli* heterologous protein expression. *Drug Metab. Dispos.* 39 (10): 1939–1945.

11 Coelho, C., Foti, A., Hartmann, T. et al. (2015). Structural insights into xenobiotic and inhibitor binding to human aldehyde oxidase. *Nat. Chem. Biol.* 11 (10): 779–783.

12 Lowe, D.J. and Bray, R.C. (1978). Magnetic coupling of the molybdenum and iron–sulphur centres in xanthine oxidase and xanthine dehydrogenases. *Biochem. J.* 169 (3): 471–479.

13 Bray, R.C. (1975). EPR of molybdenum-containing enzymes. In: *Biological Magnetic Resonance* (ed. L.J. Berliner), 45–84. Boston, MA: Springer US.

14 Lowe, D.J., Lynden-Bell, R.M., and Bray, R.C. (1972). Spin–spin interaction between molybdenum and one of the iron–sulphur systems of xanthine oxidase and its relevance to the enzymic mechanism. *Biochem. J.* 130 (1): 239–249.

15 Rebelo, J., Dias, J., Huber, R. et al. (2001). Structure refinement of the aldehyde oxidoreductase from Desulfovibrio gigas (MOP) at 1.28 Å. *J. Biol. Inorg. Chem.* 6 (8): 791–800.

16 Romão, M. and Moura, J. (2001). *Handbook of Metalloproteins*, 1e, vol. 1 (ed. A. Messerschmidt, R. Huber, T. Poulos, and K. Wieghardt). Chichester: John Wiley & Sons.

17 Gremer, L., Kellner, S., Dobbek, H. et al. (2000). Binding of flavin adenine dinucleotide to molybdenum-containing carbon monoxide dehydrogenase from *Oligotropha carboxidovorans*. *J. Biol. Chem.* 275 (3): 1864–1872.

18 Mota, C., Coelho, C., Leimkühler, S. et al. (2018). Critical overview on the structure and metabolism of human aldehyde oxidase and its role in pharmacokinetics. *Coord. Chem. Rev.* 368: 35–59.

19 Romão, M.J., Coelho, C., Santos-Silva, T. et al. (2017). Structural basis for the role of mammalian aldehyde oxidases in the metabolism of drugs and xenobiotics. *Curr. Opin. Chem. Biol.* 37: 39–47.

20 Cerqueira, N.M.F.S.A., Coelho, C., Brás, N.F. et al. (2015). Insights into the structural determinants of substrate specificity and activity in mouse aldehyde oxidases. *J. Biol. Inorg. Chem.* 20 (2): 209–217.

21 Hartmann, T., Terao, M., Garattini, E. et al. (2012). The impact of single nucleotide polymorphisms on human aldehyde oxidase. *Drug Metab. Dispos.* 40 (5): 856–864.

22 Montefiori, M., Jørgensen, F.S., and Olsen, L. (2017). Aldehyde oxidase: reaction mechanism and prediction of site of metabolism. *ACS Omega.* 2 (8): 4237–4244.

23 Ferreira, P., Cerqueira, N.M.F.S.A., Fernandes, P.A. et al. (2020). Catalytic mechanism of human aldehyde oxidase. *ACS Catal.* 10 (16): 9276–9286.

24 Mota, C., Diniz, A., Coelho, C. et al. (2021). Interrogating the inhibition mechanisms of human aldehyde oxidase by X-ray crystallography and NMR spectroscopy: the raloxifene case. *J. Med. Chem.* 64 (17): 13025–13037.

25 Fukuyama, K., Hase, T., Matsumoto, S. et al. (1980). Structure of *S. platensis* [2Fe–2S] ferredoxin and evolution of chloroplast-type ferredoxins. *Nature* 286 (5772): 522–524.

26 Tsukihara, T., Fukuyama, K., Mizushima, M. et al. (1990). Structure of the [2Fe–2S]ferredoxin I from the blue-green Alga Aphanothece sacrum at 2·2 Å resolution. *J. Mol. Biol.* 216 (2): 399–410.

27 Rypniewski, W.R., Breiter, D.R., Benning, M.M. et al. (1991). Crystallization and structure determination of 2.5-.ANG. resolution of the oxidized iron–sulfur [2Fe–2S] ferredoxin isolated from Anabaena 7120. *Biochemistry* 30 (17): 4126–4131.

28 Nishino, T. and Okamoto, K. (2000). The role of the [2Fe–2S] cluster centers in xanthine oxidoreductase. *J. Inorg. Biochem.* 82 (1–4): 43–49.

29 Barber, M.J., Coughlan, M.P., Rajagopalan, K.V., and Siegel, L.M. (1982). Properties of the prosthetic groups of rabbit liver aldehyde oxidase: a comparison of molybdenum hydroxylase enzymes. *Biochemistry* 21 (15): 3561–3568.

30 Mahro, M., Brás, N.F., Cerqueira, N.M.F.S.A. et al. (2013). Identification of crucial amino acids in mouse aldehyde oxidase 3 that determine substrate specificity. *PLoS ONE* 8 (12): e82285.

31 Doel, J.J., Godber, B.L.J., Eisenthal, R., and Harrison, R. (2001). Reduction of organic nitrates catalysed by xanthine oxidoreductase under anaerobic conditions. *Biochim. Biophys. Acta Gen. Subj.* 1527 (1–2): 81–87.

32 Paragas, E.M., Humphreys, S.C., Min, J. et al. (2017). The two faces of aldehyde oxidase: oxidative and reductive transformations of 5-nitroquinoline. *Biochem. Pharmacol.* 145: 210–217.

33 Garattini, E. and Terao, M. (2013). Aldehyde oxidase and its importance in novel drug discovery: present and future challenges. *Expert Opin. Drug Discov.* 8 (6): 641–654.

34 Hutzler, J.M., Obach, R.S., Dalvie, D., and Zientek, M.A. (2013). Strategies for a comprehensive understanding of metabolism by aldehyde oxidase. *Expert Opin. Drug Metab. Toxicol.* 9 (2): 153–168.

35 Pryde, D.C., Dalvie, D., Hu, Q. et al. (2010). Aldehyde oxidase: an enzyme of emerging importance in drug discovery. *J. Med. Chem.* 53 (24): 8441–8460.

36 Rashidi, M.R. and Soltani, S. (2017). An overview of aldehyde oxidase: an enzyme of emerging importance in novel drug discovery. *Expert Opin. Drug Discov.* 12 (3): 305–316.

37 Sanoh, S., Tayama, Y., Sugihara, K. et al. (2015). Significance of aldehyde oxidase during drug development: effects on drug metabolism, pharmacokinetics, toxicity, and efficacy. *Drug Metab. Pharmacokinet.* 30 (1): 52–63.

38 Garattini, E. and Terao, M. (2011). Increasing recognition of the importance of aldehyde oxidase in drug development and discovery. *Drug Metab. Rev.* 43 (3): 374–386.

39 Dalvie, D. and Di, L. (2019). Aldehyde oxidase and its role as a drug metabolizing enzyme. *Pharmacol. Ther.* 201: 137–180.

40 Garattini, E., Fratelli, M., and Terao, M. (2008). Mammalian aldehyde oxidases: genetics, evolution and biochemistry. *Cell. Mol. Life Sci.* 65 (7–8): 1019–1048.

41 Terao, M., Romão, M.J., Leimkühler, S. et al. (2016). Structure and function of mammalian aldehyde oxidases. *Arch. Toxicol.* 90 (4): 753–780.

42 Rochat, B., Kosel, M., Boss, G. et al. (1998). Stereoselective biotransformation of the selective serotonin reuptake inhibitor citalopram and its demethylated metabolites by monoamine oxidases in human liver. *Biochem. Pharmacol.* 56 (1): 15–23.

43 Ruenitz, P.C. and Bai, X. (1995). Acidic metabolites of tamoxifen: aspects of formation and fate in the female rat. *Drug Metab. Dispos.* 23 (9): 993–998.

44 Lepri, S., Ceccarelli, M., Milani, N. et al. (2017). Structure–metabolism relationships in *human-* AOX: chemical insights from a large database of aza-aromatic and amide compounds. *Proc. Natl. Acad. Sci.* 114 (16): E3178–E3187.

45 Torres, R.A., Korzekwa, K.R., McMasters, D.R. et al. (2007). Use of density functional calculations to predict the regioselectivity of drugs and molecules metabolized by aldehyde oxidase. *J. Med. Chem.* 50 (19): 4642–4647.

46 Clarke, S.E., Harrel, A.W., and Chenery, R.J. (1995). Role of aldehyde oxidase in the in vitro conversion of famciclovir to penciclovir in human liver. *Drug Metab. Dispos.* 23 (2): 251–254.

47 Rashisi, M.R., Smith, J.A., Clarke, S.E., and Beedham, C. (1997). In vitro oxidation of famciclovir and 6-deoxypenciclovir by aldehyde oxidase from human, guinea pig, rabbit, and rat liver. *Drug Metab. Dispos.* 25 (7): 805–813.

48 Brandänge, S. and Lindblom, L. (1979). The enzyme "aldehyde oxidase" is an iminium oxidase. Reaction with nicotine $\Delta 1'(5')$ iminium ion. *Biochem. Biophys. Res. Commun.* 91 (3): 991–996.

49 Konishi, K., Fukami, T., Gotoh, S., and Nakajima, M. (2017). Identification of enzymes responsible for nitrazepam metabolism and toxicity in human. *Biochem. Pharmacol.* 140: 150–160.

50 Kitamura, S. and Tatsumi, K. (1984). Reduction of tertiary amine N-oxides by liver preparations: function of aldehyde oxidase as a major N-oxide reductase. *Biochem. Biophys. Res. Commun.* 121 (3): 749–754.

51 Tatsumi, K., Kitamura, S., and Yamada, H. (1983). Sulfoxide reductase activity of liver aldehyde oxidase. *Biochim. Biophys. Acta* 747 (1–2): 86–92.

52 Stoddart, A.M. and Levine, W.G. (1992). Azoreductase activity by purified rabbit liver aldehyde oxidase. *Biochem. Pharmacol.* 43 (10): 2227–2235.

53 Li, H., Kundu, T.K., and Zweier, J.L. (2009). Characterization of the magnitude and mechanism of aldehyde oxidase-mediated nitric oxide production from nitrite. *J. Biol. Chem.* 284 (49): 33850–33858.

54 Beedham, C., Miceli, J., and Obach, R. (2003). Ziprasidone metabolism, aldehyde oxidase, and clinical implications. *J. Clin. Psychopharmacol.* 23 (3): 229–232.

55 Sugihara, K. and Tatsumi, K. (1986). Participation of liver aldehyde oxidase in reductive metabolism of hydroxamic acids to amides. *Arch. Biochem. Biophys.* 247 (2): 289–293.

56 Sodhi, J.K., Wong, S., Kirkpatrick, D.S. et al. (2015). A novel reaction mediated by human aldehyde oxidase: amide hydrolysis of GDC-0834. *Drug Metab. Dispos.* 43 (6): 908–915.

57 Obach, R.S., Huynh, P., Allen, M.C., and Beedham, C. (2004). Human liver aldehyde oxidase: inhibition by 239 drugs. *J. Clin. Pharmacol.* 44 (1): 7–19.

58 Obach, R.S. (2004). Potent inhibition of human liver aldehyde oxidase by raloxifene. *Drug Metab. Dispos.* 32 (1): 89–97.

59 Barr, J.T., Jones, J.P., Oberlies, N.H., and Paine, M.F. (2015). Inhibition of human aldehyde oxidase activity by diet-derived constituents: structural influence, enzyme-ligand interactions, and clinical relevance. *Drug Metab. Dispos.* 43 (1): 34–41.

60 Coelho, C., Muthukumaran, J., Santos-Silva, T., and João, R.M. (2019). Systematic exploration of predicted destabilizing nonsynonymous single nucleotide polymorphisms (nsSNPs) of human aldehyde oxidase: a bio-informatics study. *Pharmacol. Res. Perspect.* 7 (6): e00538.

61 Foti, A., Hartmann, T., Coelho, C. et al. (2016). Optimization of the expression of human aldehyde oxidase for investigations of single-nucleotide polymorphisms. *Drug Metab. Dispos.* 44 (8): 1277–1285.

62 Foti, A., Dorendorf, F., and Leimkühler, S. (2017). A single nucleotide polymorphism causes enhanced radical oxygen species production by human aldehyde oxidase. *PLoS ONE* 12 (7): e0182061.

63 Mota, C., Esmaeeli, M., Coelho, C. et al. (2019). Human aldehyde oxidase (hAOX1): structure determination of the Moco-free form of the natural variant G1269R and biophysical studies of single nucleotide polymorphisms. *FEBS Open Bio.* 9 (5): 925–934.

6

Biological Formation of Sulfide

Marion Jespersen[1,2] *and Tristan Wagner*[1,3]

[1]*Microbial Metabolism Research Group, Max Planck Institute for Marine Microbiology, Bremen, Germany*
[2]*Greening Lab, Department of Microbiology, Monash Biomedicine Discovery Institute, Melbourne, Australia*
[3]*Microbial Metabolism Laboratory, Université Grenoble Alpes, CEA, CNRS, Institut de Biologie Structurale, Grenoble, France*

Chapter Goals

- The aim of this chapter is to highlight the different acquisition strategies for the assimilation of the essential element sulfur.
- To achieve this, the chapter covers the geochemical sulfur cycle and the conventional strategy of most organisms to acquire sulfur by reducing sulfate to sulfide.
- The enzymes that catalyze these reactions are documented and compared to an unconventional sulfate assimilation pathway from an anaerobic methane-producing archaeon.

Main Chapter Points

1) Introduction to sulfur and its importance
   - *Chemical properties, bioavailability, and sulfur's role in life*:
     - Sulfur is ubiquitous on Earth and significantly impacts the planet's redox state and climate through abiotic processes and microbial metabolic activities.
     - Sulfur exists in multiple redox states and forms, each with distinct reactivity and availability in the environment, such as its most oxidized and reduced forms, sulfate and sulfide, respectively.
     - Sulfur is an essential biological component, including vitamins and cofactors, with key roles in cellular processes and enzyme function.
     - Sulfur needs to be in its most reduced state to be acquired through pathways like cysteine biosynthesis.
     - In the biosphere, sulfur is predominantly present as sulfate, and most organisms must reduce it to sulfide before incorporating sulfur into biomass.

Iron-Sulfur Clusters: Biogenesis and Biochemistry, Volume 1, First Edition.
Edited by Silke Leimkühler, Günter Schwarz, Oliver Lenz, and Oliver Einsle.

2) The assimilatory sulfate-reduction pathway
   - *Organisms reduce sulfate to sulfide with a set of enzymes that perform assimilatory sulfate reduction*:
     - The ATP-sulfurylase (ATPS): Activates sulfate to adenosine 5′-phosphosulfate (APS)
     - APS-reductase (APSR): Reduces APS to sulfite
     - Or alternatively, the APS-kinase (APSK): Phosphorylates APS to 3′-phosphoadenosine-5′-phosphosulfate (PAPS)
     - PAPS-reductase (PAPSR): Reduces PAPS to generate sulfite and 3′-phosphoadenosine-5′-phosphate (PAP)
     - PAP-phosphatase (PAPP): Hydrolyses toxic PAP
     - Assimilatory sulfite-reductase (aSir): Reduces sulfite to sulfide
3) Sulfate-reduction metabolism to conserve energy: The dissimilatory pathway
   - Microorganisms can use sulfate as an electron acceptor to conserve energy via the dissimilatory sulfate-reduction pathway, producing hydrogen sulfide (H_2S).
   - This pathway plays a vital role in the sulfur cycle, particularly in anaerobic marine sediments, contributing significantly to organic matter mineralization.
   - The enzymes of the dissimilatory sulfate-reduction pathway, with the exception of the ATPS, are different from those of the assimilatory pathway.
     - *The dissimilatory APS- and sulfite-reductase*: The dissimilatory APSR and sulfite-reductase are distinct from the assimilatory versions and are indirectly coupled to energy conservation via the generation of a proton motive force.
   - Sulfate reducers may separate sulfur assimilation and energy acquisition pathways to avoid cross-reactions and allow independent regulation.
4) One methanogen harbors an unconventional sulfate-reduction pathway
   - Methanogens are anaerobic archaea that can directly assimilate the reduced sulfur species from their environment.
   - *Methanothermococcus thermolithotrophicus* is the only known methanogen that has been shown to grow robustly on sulfate, using a unique sulfate assimilation pathway:
     - Sulfate activation with an assimilatory ATPS and APSK.
     - A dissimilatory-like APSR reduces PAPS to sulfite and PAP.
     - PAP detoxification with a novel PAPP, unique to archaea.
     - Sulfite, particularly toxic for methanogens, is immediately reduced to sulfide by an $F_{420}H_2$-oxidizing sulfite reductase (Fsr).
5) Evolutionary insights and biotechnological applications of a sulfate-reducing methanogen
   - Methanogens, some of the oldest microbial life forms on Earth, are models for understanding early metabolism. They also provide biotechnological potential by producing the energy carrier molecule methane from CO_2.
   - *Evolutionary insights*:
     - The structure of Fsr from methanogens provides new evolutionary trajectories of sulfite reductases. It supports recent findings from comparative

genomics and phylogenetic analyses that assimilatory and dissimilatory sulfite reductases originated from a common archaeal ancestral enzyme.

- *Biotechnological applications*:
  - Growing methanogens on sulfate instead of toxic and explosive H_2S for biotechnological applications could improve the safety, cost, and efficiency of renewable methane production.

Preamble: This chapter summarizes our scientific contribution to the iron–sulfur for life program, in which we have studied sulfur acquisition by methanogenic archaea. This scientific project is intertwined with the iron–sulfur for Life theme, as sulfur metabolism is a prerequisite for providing the elemental block sulfide for the assembly of Fe- and Mo-containing metalloclusters of enzymes. The project also fits in with the diversity of methods used, from microbiology to structural biology passing by spectroscopy. The presented work initially focused on understanding how methanogens protect themselves from toxic sulfite (SO_3^{2-}) and gradually evolved to unravel the transformation of sulfate into sulfide by one particular methanogen. This story will be the focus of this chapter as it illustrates how a microbe has acquired features of the two main metabolic pathways involved in sulfate reduction.

The chapter is structured as follows: first, a brief overview of the geochemical sulfur cycle on planet Earth and the importance of sulfur for life is provided. Next, the two canonical pathways involved in the reduction of sulfate to sulfide and their respective enzymes are presented in detail. Finally, the archaeon of interest and its intriguing sulfate-reduction metabolism are depicted, followed by a conclusion discussing the evolutionary and biotechnological aspects.

6.1 Sulfur Is a Versatile Element and an Essential Cellular Component

Sulfur is one of the most abundant elements on Earth and is essential for all forms of life. With its 16 electrons, sulfur has similar electron configurations to oxygen, which allows it to form analog oxygen compounds (e.g. OCN^- cyanate and SCN^- thiocyanate). Sulfur can form bonds with carbon, oxygen, and hydrogen or form chains with S—S bonds. Its chemical versatility and the reactivity of reduced sulfur species make it a perfect component of the chemical reactions of life [1, 2].

6.1.1 Chemical Properties of Sulfur

Due to its chemistry, sulfur can exist in eight redox states, ranging from its most oxidized state, sulfate (SO_4^{2-}), to its most reduced state, sulfide (S^{2-}, or hydrogen sulfide H_2S, Figure 6.1) [3]. Sulfur compounds with oxidation states lower than +6, such as sulfite (SO_3^{2-}), trithionate ($S_3O_6^{2-}$), or thiosulfate ($S_2O_3^{2-}$), tend to be quite reactive and might be harmful to living organisms. For example, sulfide in the form of H_2S (easily identified by its distinctive rotten-egg odor) is reactive,

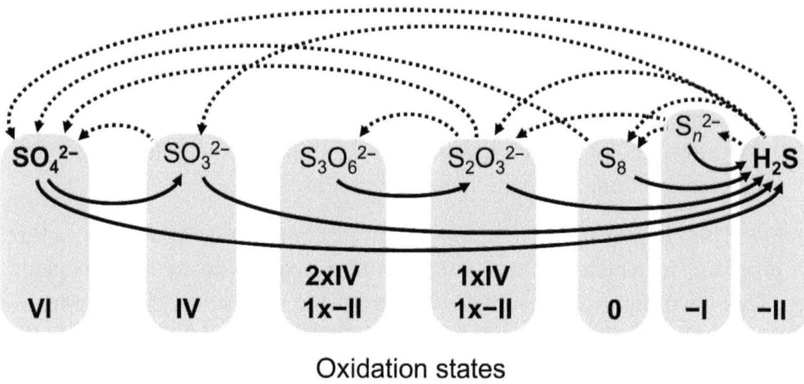

Figure 6.1 The redox species among the biogeochemical sulfur cycle. The reductive pathways are shown in black arrows and dashed lines represent the oxidative pathways. Microbial disproportionation reactions are omitted for simplicity. Source: Adapted from Morrison and Mojzsis [2].

explosive, and toxic to humans [4, 5]. H_2S can diffuse across cell membranes and bind to metal cations, causing them to precipitate and thus interfere with metabolic processes.

Because of its omnipresence on Earth, particularly in volcanic areas, sulfur has had and continues to have a significant planetary impact by regulating the redox state of the planet and influencing the Earth's climate [6]. While geochemical activities contribute to the chemical transformation of sulfur, microorganisms are constantly shaping the planetary sulfur cycle and the biomineralization of organic matter through their metabolic activities, particularly in marine sediments, thereby linking the sulfur to the carbon cycle [2, 6–12]. Conversely, it has been suggested that sulfur played a critical role in the sulfide-reduced world of primeval times, where it supported the first reaction and protometabolic pathways together with iron and nickel [13]. During the earliest stages of life, H_2S served not only as a sulfur source but also as a reactant, catalyst, and sustainable energy source [14], and among the proposed scenarios, life may have emerged from Fe–S bubbles at an alkaline hot spring in an acidic ocean. Here, early microbial life harnessed the chemical reactivity and properties of sulfur to support various physiological functions. In our modern world, these functions have been extended from enzymatic catalysis [15] to transcriptional/translational regulation [16], redox homeostasis, defense/predation (e.g. in antibiotics), or even cell signaling [17–21].

6.1.2 Sulfur as an Essential Cellular Component

Sulfur is an essential nutrient [22], as it is a component of the proteinogenic amino acids cysteine and methionine as well as posttranslationally modified amino acids (e.g. thioglycine [23]). It is also a component of vitamins (e.g. vitamins B1, and B7) and (metallo)cofactors (e.g. FeMo-cofactor/P-cluster of the nitrogenase, coenzymes M and B used in methanogenesis). It can be found in modified nucleic

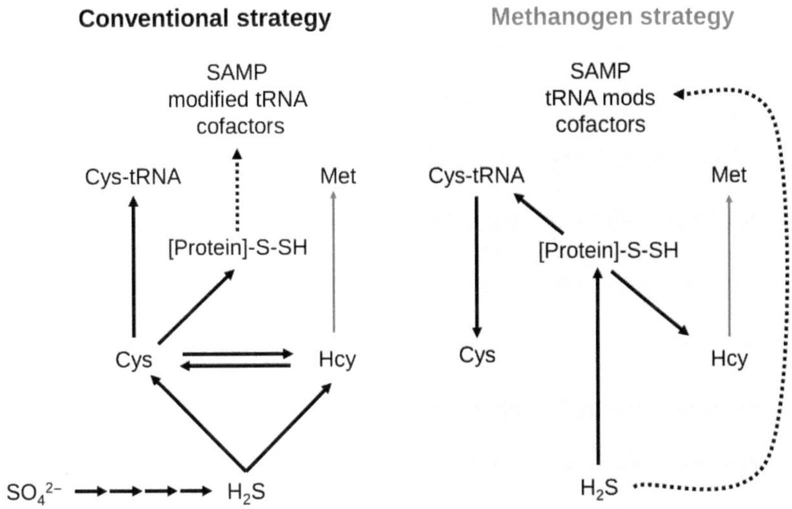

Figure 6.2 Sulfur acquisition, entry point, and example of molecules carrying the sulfur in life. Discrepancy between the canonical pathway and methanogens. Hcy stands for homocysteine and SAMP stands for ubiquitin-like small archaeal modifier protein. Source: This figure was adapted from Perona et al. [29].

acids (DNA/RNA/tRNA) [24–27] and in many antibiotics, such as ampicillin [28]. Therefore, living organisms must rely on an efficient system to acquire sulfur and chemically activate it to synthesize the necessary biomolecules (Figure 6.2) [29–31]. In most cases, the enzyme CysK (EC number 2.5.1.47) [32, 33] binds *O*-acetyl-L-serine (obtained from serine) and the active reduced state of sulfur, S^{2-} (HS^-/H_2S) to produce the sulfide carrier cysteine, which is used for most of the biosynthetic pathways in many prokaryotes (bacteria and archaea) and plants [34]. Two alternative pathways for cysteine biosynthesis have been described [35]: (i) an *O*-phospho-L-serine reacts with H_2S, generating cysteine and P_i (EC number 2.5.1.65, *O*-phosphoserine sulfhydrylase) [36]; (ii) *O*-acetyl-L-serine reacts with thiosulfate ($S_2O_3^{2-}$) to form *S*-sulfo-L-cysteine (EC number 2.5.1.144, CysM, *S*-sulfo-L-cysteine synthase [37]), which allows the production of cysteine. The sulfur-transferase then transfers the sulfide group to a wide range of biosynthetic precursors.

Despite its previously mentioned toxicity, some microorganisms can directly use reactive sulfide to generate sulfur-containing biomolecules, rather than bypassing a cysteine intermediate (Figure 6.2) [29]. This is especially the case for microbes that originated prior to the Great Oxygenation Event (about 2.3 billion years ago), such as methanogenic archaea, which were already present in the sulfidic Archaean world [14, 38]. An example of sulfur incorporation without a required cysteine intermediate is the biosynthesis of vitamin B1, also called thiamine. Thiamine consists of a thiazole ring that contains a sulfur atom [39] and carries out catalytic functions such as (de)carboxylation or carbonization, which are essential for cellular processes (e.g. the tricarboxylic acid cycle) [40, 41]. The thiazole synthase (EC number 2.4.2.60) is the enzyme responsible for sulfur insertion and ring closure, a multistep reaction that has been thoroughly studied [39]. A sacrificial cysteine from the active site of the thiazole synthase provides the sulfur atom. Once donated, the one-turnover enzyme must be eliminated as it cannot provide more sulfur, and another thiazole synthase must be produced, costing cellular energy [42]. In contrast, methanogens have a modified active site in which the catalytic cysteine is replaced by a histidine [43] (EC number 2.4.2.59). Here, the methanogen directly inserts an external sulfide ion to produce the thiazole, saving the enzyme and cellular energy [44].

6.1.3 Sulfur Bioavailability in the Environment

Heterotrophs feed on other life forms to obtain their sulfur from organic molecules through digestion and decomposition. Autotrophs, on the other hand, must acquire sulfur from simple inorganic molecules available in the environment.

Sulfide is readily bioavailable in anaerobic niches, in which the redox potential and absence of oxygen prevent its spontaneous conversion to more oxidized sulfur states. Sulfide originates from geological sources (e.g. volcanic activity [45]) or, more commonly, from biological activities (e.g. sulfate-reducing microbes [1, 14]). It can also be acquired in a mineral form (e.g. pyrite [Fe–S]) by reduction [46–48]. However, the most abundant form of sulfur in the biosphere is sulfate (e.g. a major ion in seawater with a concentration of about 28 mM), which is why many organisms must generate their sulfide through dedicated chemical sulfate-reduction reactions [12, 49, 50]. Most aerobic microorganisms, such as photosynthetic organisms or fungi, depend on sulfate to assimilate sulfur via the assimilatory sulfate-reduction pathway (Figure 6.3).

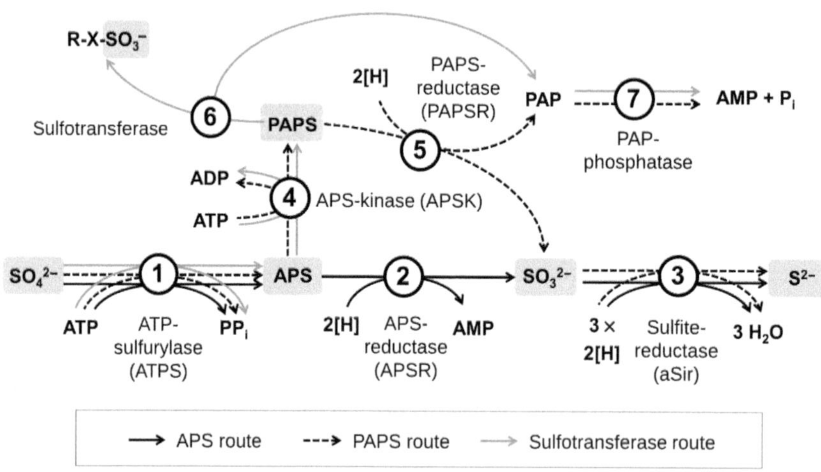

Figure 6.3 The assimilatory sulfate-reduction pathway (reactions and enzymes). "X" represents O or NH for the sulfotransferase reaction.

6.2 A Sulfate-Reduction Metabolism to Obtain Cellular Sulfur: The Assimilatory Pathway

All the enzymatic steps to reduce sulfate to sulfide take place in the cytoplasm or in contact with the inner side of the cytoplasmic membrane. Unlike H_2S, sulfate cannot diffuse freely across hydrophobic membranes. The sulfate anion must be transported inside the cell through specific transporters. Different types of transporters have been reported [51, 52] that rely on passive diffusion (e.g. CysZ permease) [53], chemical energy in the form of an ion gradient (e.g. SulP in prokaryotes or SUL, SULTR, SLC13, and SLC26 in eukaryotes), or adenosine triphosphate (ATP) hydrolysis (e.g. ABC-type complex CysTWA) [54].

Once in the cytoplasm, the sulfate ion must first be reduced to sulfite. However, the tetrahedral arrangement of its oxygen atoms makes sulfate quite an inert ion, and the redox potential of the reduction reaction from sulfate to sulfite ($E_0' = -516$ mV, [55, 56]) is too low to be carried out by biological systems. Therefore, sulfate must first be chemically activated to increase its redox potential sufficiently enough to allow its reduction to sulfite, followed by its ultimate reduction to sulfide [56].

6.2.1 Sulfate Is Activated by the ATP-Sulfurylase

The ATP-sulfurylase (abbreviated as ATPS, EC number 2.7.7.4) binds the sulfate and an ATP. The sulfate is covalently bound to the α-phosphate of ATP, producing adenosine 5′-phosphosulfate (APS) and pyrophosphate (PP_i, reaction 1 in Figures 6.3 and 6.4) [57]. The transfer reaction requires a magnesium or manganese ion in its active site.

To date, two non-homologous ATPS have been discovered: CysDN and Sat. Only one structure has been reported for CysDN, in which the regulatory GTPase subunit CysN is intertwined with the catalytic subunit CysD [58]. CysN hydrolyses a guanosine-5’-triphosphate (GTP) to guanosine diphosphate (GDP) and inorganic phosphate (P_i) to counterbalance the unfavorable reaction catalyzed by CysD to produce APS and PP_i from sulfate and one ATP. While CysDN invests one ATP and GTP, the homo-oligomeric Sat enzyme requires only one ATP. Sat is used for both assimilatory and dissimilatory sulfate reduction and shows overall a high sequence and reaction conservation. From here on, ATPS will refer to Sat enzyme only for simplicity.

ATPS structures have been solved from several sulfate-reducing and sulfide-oxidizing microorganisms (a strategy to conserve energy by oxidizing sulfide to sulfate). All these structures show the same conserved core consisting of three domains [57, 59–66], with domain II carrying the active site and ATP-binding site. Many subtleties have evolved from this conserved core: different dimeric interfaces, extensions, and even the fusion of additional domains. For example, the thermophilic bacterium *Thermus thermophilus* has a Zn-binding motif inserted between the domains II and III, which is involved in the formation of the dimeric interface [64]. Or, as in some eukaryotes, the ATPS can be fused to the APS-kinase

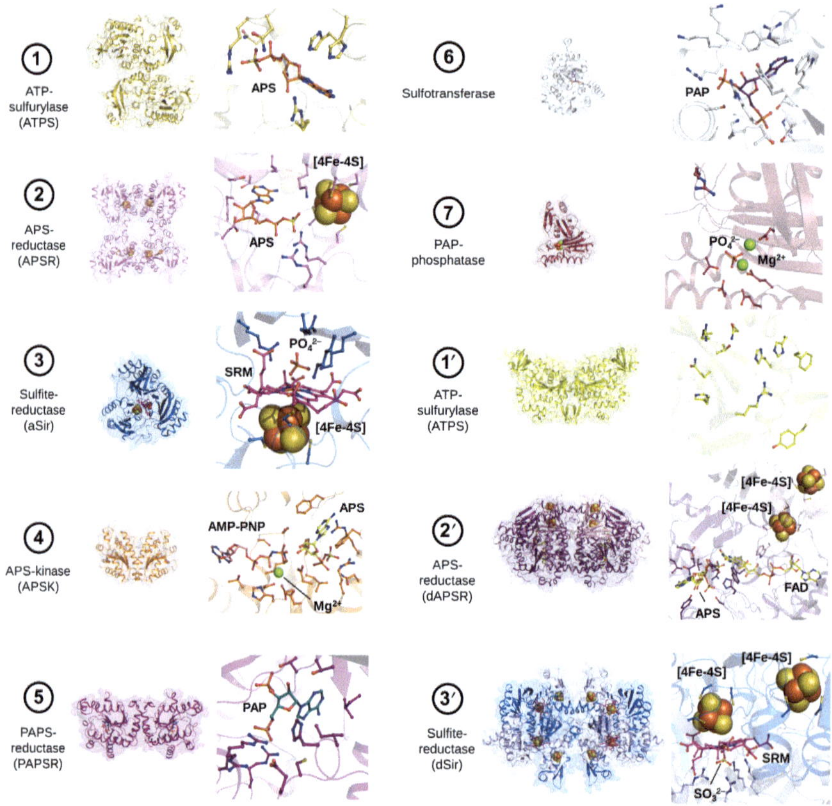

Figure 6.4 Enzymes and their active sites from sulfate assimilation and dissimilation pathways. The overall structure (surface transparency and secondary structure in cartoon, left panel) and closeup of active sites (secondary structure in cartoon, right panel) are displayed for every enzyme mentioned in Figures 6.3 and 6.5. Substrates, products, analogs, (metallo)cofactors, and the protein residues interacting with them are represented as sticks and balls (for Mg and [4Fe–4S] clusters). The protein data bank identifiers for each structure and the organism of origin are the following: (1) 4MAF, *Glycine max*; (2) 2GOY, *Pseudomonas aeruginosa*; (3) 1AOP, *Escherichia coli* (a P_i occupies the sulfite position); (4) 3UIE, *Arabidopsis thaliana*; (5) 2OQ2, *Saccharomyces cerevisiae*; (6) 5MEK, *A. thaliana*; (7) 5DJJ, *Mycobacterium tuberculosis*; (1′) Alphafold model of ATPS from *A. fulgidus*; (2′) 2FJA, *A. fulgidus*; (3′) 3MM5, *A. fulgidus*. In the dAPSR structure, the modeled APS is shown in two alternative conformations. As there is no structure derived from experimental data, the dissimilatory ATPS from *A. fulgidus* structure is predicted by alphafold2 [178]. Adenylyl imidodiphosphate (AMP-PNP) is a nonhydrolyzable analog of ATP, and SRM stands for siroheme.

(abbreviated as APSK), the following enzyme of the assimilatory sulfate-reduction pathway, allowing tunneling of the reaction [61].

ATPS has been shown to be sensitive to molybdate as the latter mimics sulfate and can bind to the active site of the enzyme. Once bound, molybdate forces the hydrolysis of ATP to AMP and PP_i in a process called molybdolysis, which depletes cellular energy without forming APS [67–69]. Because of this property, molybdate can be used as an inhibitor for sulfate-reducing microorganisms [70, 71].

The equilibrium of the ATPS reaction is toward ATP and sulfate formation. In other words, the enzyme prefers to catalyze the reverse reaction. The reaction is pulled toward APS formation by a pyrophosphatase (EC number 3.6.1.1) that hydrolyzes pyrophosphate. However, even with the effective removal of PP_i, the product APS can inhibit the ATPS and must be efficiently scavenged and further transformed by the following enzyme [1].

6.2.2 APS Is Further Phosphorylated by the APS-Kinase

APS creates a branching point in the assimilatory sulfate-reduction pathway. Some organisms directly reduce APS to sulfite and then sulfide (reactions 2 and 3 in Figures 6.3 and 6.4), whereas in various species, APS is further phosphorylated to PAPS by the APSK (reaction 4 in Figures 6.3 and 6.4, EC number 2.7.1.25). The APSK binds APS and ATP to catalyze the transfer of the γ-phosphate from ATP to the 3′-hydroxyl group of the APS. The reaction products are PAPS and ADP. Structural investigations have shown that the APS-kinase is organized as a functional homodimer [72–75] and, like ATPS, APSK contains a magnesium or manganese ion in its active site to catalyze the transfer reaction. The catalytic metal is localized in a cleft, allowing the spatial proximity of the ATP γ-phosphate and the 3′-OH of the APS. Compared to ATPS, APSK has fewer structural variations, such as differences in oligomerization and fusion. Nevertheless, some exotic forms exist, such as the enzyme from *Thiobacillus denitrificans*, which contains an APSK fused to an inactive ATPS-like domain of unknown function [66] or the APSK from *Mycobacterium tuberculosis*, which is fused to the CysN subunit of the CysDN complex [73].

In the sulfotransferase pathway shown in Figure 6.3, the generated PAPS is further transformed by sulfotransferases. These enzymes transfer the sulfo group carried on the PAPS to branch it to an acceptor group of other substrates [76] (reaction 6 in Figures 6.3 and 6.4).

6.2.3 Sulfite Production and Hydrolysis of the Poisonous By-Product PAP

The next step is the reduction of APS or PAPS to sulfite (reaction 2 or 5 in Figure 6.3). APS or PAPS reductases will reduce their respective substrate to sulfite and to the reaction by-product adenosine monophosphate (AMP) or PAP. Assimilatory APSR (APSR, EC number 1.8.4.10) and PAPSR (PAPSR, EC number 1.8.4.8) are structurally similar, but the homo-oligomeric enzymes can differ in their oligomerization [77]. In contrast to PAPSRs, most of the APSRs contain a [4Fe–4S] cluster close to their active site that is essential for catalysis [78, 79], with the exception of an isoform from *Physcomitrella patens* [80].

The substrate is localized at the same position between APSR and PAPSR. However, substitutions precisely accommodate the additional 3′-phosphate group for coordinating PAPS. APSR and PAPSR share a conserved reaction mechanism in which the thiol of a catalytic cysteine located on the C-terminal loop of the protein binds to the sulfo group. The thioredoxin acts as an electron donor to reduce the $S{-}SO_3^-$ bond, forming oxidized thioredoxin, sulfite, and AMP/PAP [79].

PAP is a nucleotide analog with the dangerous properties of inhibiting sulfotransferases, PAPS reductases, exoribonucleases, and disrupting RNA catabolism [81, 82] while exhibiting regulatory roles in plants [83]. Therefore, it must be immediately hydrolyzed to AMP and P_i by the PAPP (reaction 7 in Figures 6.3 and 6.4, EC number 3.1.3.7). This enzyme has been well studied in eukaryotes (human, Protein Data Bank ID: 2WEF) and bacteria [84–86]. The phosphoester bond is hydrolyzed by magnesium cations coordinated in the active site of the PAPP. It has also been shown that some ribonucleases, such as nanoRNAse [87], have the catalytic potential to hydrolyze PAP, albeit as a nonphysiological reaction.

6.2.4 Sulfite Reduction

Sulfite, a reductant that is highly reactive and therefore potentially toxic, must be efficiently removed from the cytoplasm by a sulfite reductase (abbreviated here as aSir for assimilatory sulfite reductase) [88]. The nature of the electron donor makes this reaction favorable and efficient, allowing rapid turnover (reaction 3 in Figure 6.3). In plants, cyanobacteria, and algae, the electron donor for the sulfite reductase is a [2Fe–2S] cluster-containing ferredoxin (EC number 1.8.7.1) [89], whereas most prokaryotic microorganisms studied use an indirect nicotinamide adenine dinucleotide/nicotinamide adenine dinucleotide phosphate (NADH/NADPH) oxidase coupling (EC number 1.8.1.2) [90, 91].

The aSir carries a siroheme covalently bound to a [4Fe–4S] cluster via a cysteine thiolate (reaction 3 in Figure 6.4) [92]. The conserved motif carrying the cluster is $CX_5CX_{37}CX_3C$, a family hallmark. In this reaction mechanism, sulfite is covalently bound to the iron atom of the siroheme. Reduced electrons are transferred through the cluster, allowing the successive sulfite reduction to sulfide in three cycles. Each reduction cycle generates a water molecule by transferring protons, leading to sulfide synthesis.

Few examples in the literature have reported a sulfite-reduction activity for nitrite reductase [93], which shares the same fold and cofactors as the aSir. On the other hand, a close homolog of the nitrite reductase from *M. tuberculosis* exhibits a sulfite reductase activity and a covalent cysteine–tyrosine bond in the active site, which has been shown to be necessary for the activity [94].

6.3 A Sulfate-Reduction Metabolism to Conserve Energy: The Dissimilatory Pathway

Microorganisms have found many ways to use energy conservation to extract chemical energy from the environment for their cellular processes. Some metabolize coal and lignin instead of sugars, and others respire sulfate by the dissimilatory sulfate-reduction metabolism when more energetically favorable electron acceptors like oxygen or nitrate/nitrite are depleted. Microbes that rely on the dissimilatory sulfate-reduction pathway produce high concentrations of hydrogen sulfide. This is distinguishable from assimilatory sulfate reduction, which does not result in a significant sulfide release.

6.3.1 Ecological Impact of the Dissimilatory Reduction of Sulfate and Its Bioenergetics

Dissimilatory microbial sulfate reduction is an essential ecological process at the heart of the sulfur cycle that has shaped and continues to shape the global biochemical sulfur and carbon cycles on our planet [8, 49, 95, 96]. It predominates on the seafloor and in marine sediments at the anaerobic interface [12], where it is considered an essential process for the biomineralization of organic matter (it may even account for up to 50% of organic matter mineralization) [97, 98]. As H_2S is highly reactive, its excretion has a major impact on the chemistry of the environment. It can either be abiotically oxidized or act as an electron donor to a wide variety of aerobic chemotrophic and anoxygenic phototrophic microorganisms [1]. However, in biotechnological applications such as in anaerobic digesters and wastewater treatment plants, H_2S produced by sulfate reducers can become problematic [99, 100]. It contributes to air and water pollution; has undesirable impacts on ecosystems, human health, and our society, including the souring of oil reservoirs; and can cause corrosion of iron, steel, and concrete [56, 96, 99, 101–103]. On the other hand, sulfate-reducing microbes are highly valuable in various biotechnological applications, such as for bioremediation of metals and hydrocarbons, bioelectrochemistry, off-gas treatment, and in wastewater, where they are used to remove carbon, nitrogen, and phosphorus while also reducing sludge production [104].

The standard redox potential of the sulfate/sulfide couple of −220 mV allows prokaryotes to rely on this reaction for energy conservation. In other words, due to its reasonably positive redox potential, anaerobic microbes can transfer electrons to sulfate for cellular respiration [56]. Nevertheless, the energy gain from dissimilatory sulfate reduction is relatively low compared to aerobic respiration, which is reflected in the comparably low growth yields of sulfate reducers [1]. While dissimilatory sulfate-reducing microorganisms have been shown to be quite versatile in terms of electron donors and acceptors [99], they must obtain electrons from a donor in which the oxidation reaction would have a sufficiently low potential to run this metabolism efficiently, such as H_2 ($E_0' = -414$ mV) or formate ($E_0' = -432$ mV) [55]. The oxidation of organic matter, such as sugars, fatty acids, alcohols, and organic acids, can also serve as substrates [99]. Instead of organic molecules, sulfate-reducing microbes can receive reducing equivalents from another organism. In such a syntrophic relationship, the oxidation of a substrate by one organism, such as methane (CH_4) by a methanotroph, results in the liberation of reducing equivalents that are transferred to the sulfate-reducing bacterial partner, where they are used to respire sulfate [105, 106].

While the substrate, product, and some intermediates are the same as in the previously described assimilatory pathway, the main difference between the metabolisms

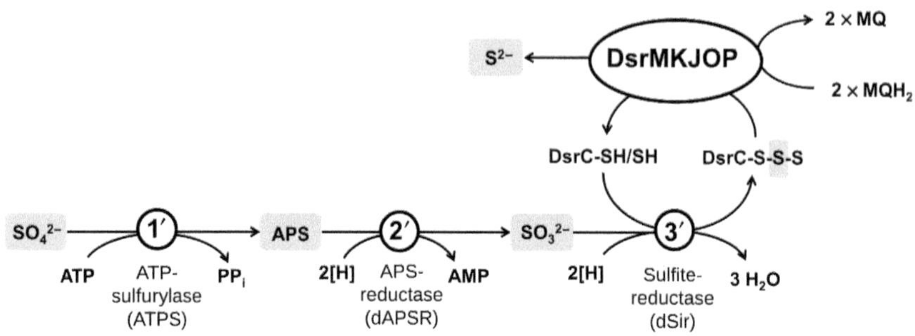

Figure 6.5 The dissimilatory sulfate-reduction pathway (reactions and enzymes).

is that in the dissimilatory one, the reducing reactions are coupled to energy conservation by membrane pumps [56].

6.3.2 Sulfate Capture and Activation

Like in the assimilatory pathway, sulfate needs to be transported inside the cytoplasm [52] and activated by an ATPS (reaction 1′ in Figures 6.4 and 6.5). As the pathway is catabolic, molybdolysis significantly impairs growth. Compared to assimilation, where a sulfate/molybdate 1 : 1 ratio has been shown to be inhibitory, a ratio of only 1 : 0.004 prevents sulfate-reducing bacteria from growing [107–109], while a ratio of 1 : 0.001 inhibits the growth of a sulfate-reducing archaeon [15].

To date, no structural data are available for the ATPS used by dissimilatory sulfate-reducing microorganisms. However, based on sequence analyses of their ATPS, the reaction mechanism is expected to be conserved between the assimilatory and dissimilatory pathways [110].

6.3.3 A Catabolic APS-Reductase

Compared to the assimilatory pathway, APS is the only known sulfonucleotide intermediate of the dissimilatory pathway. To avoid retro-inhibition of the ATPS, the dissimilatory APS-reductase (abbreviated as dAPSR, EC 1.8.99.2, reaction 2′ in Figures 6.4 and 6.5) captures the APS and reduces it to sulfite and AMP. dAPSR and its reaction differ entirely from the assimilatory APSRs [110–113]. dAPSRs are heterotetramers formed by an $(\alpha\beta)_2$ configuration, which dimerize through the α-subunit. The α-subunit has a fold similar to a fumarate reductase and harbors a flavin adenine nucleotide (FAD) cofactor. The β-subunit is a ferredoxin-like domain that harbors two [4Fe–4S] clusters. Its C-terminal domain spans the surface of the α-subunit to anchor the β-subunit to the α-subunit. The reaction mechanism operates through a nucleophilic attack of the N5 atom of reduced FAD on the sulfur atom of APS [112]. The FAD-APS intermediate decays to AMP and FAD-SO_3^-, which can be characterized by a structural snapshot from the dAPSR of the archaeon *Archaeoglobus fulgidus* [112]. The physiological electron donor for the APS reduction is still elusive, but the transfer should occur from the distal [4Fe–4S] cluster (located close to the protein surface) to the FAD via the proximal [4Fe–4S] cluster and a conserved tryptophane. Because of the redox potential of APS/sulfite ($E_0{}' = -65\,mV$), this exergonic reaction can be coupled to the establishment of an ion gradient across the membrane. Therefore, it has been postulated that the dAPSR reaction consumes the menaquinone pool ($E_0{}' = -74\,mV$) [114] by directly interacting with the membrane complex QmoABC [115]. An alternative hypothesis proposed that an electron confurcation system, instead of menaquinone, would fuel QmoABC to allow proton translocation [96, 115–117].

6.3.4 Sulfite Reduction Is Coupled to Energy Conservation

Whereas in the assimilatory pathway, the sulfite reductase produces sulfide that can then be further incorporated into biomass, the purpose of the dissimilatory sulfite reductase is to conserve energy. The sulfite-reduction mechanism of the dissimilatory system differs from that of the aSir, although some structural traits are conserved between the enzymes (reaction 3′ in Figures 6.4 and 6.5).

Compared to standalone aSirs, the dissimilatory sulfite reductase (dSir, EC number 1.8.99.5) is composed of two subunits, DsrA and DsrB, organized in an $(\alpha\beta)_2$ arrangement [118–121]. Both subunits share high structural and phylogenetic similarity to each other and, to some extent, with aSir [119]. aSirs are composed of N- and C-terminal domains that are functionally equivalent to DsrA and DsrB, respectively. DsrA contains a nonfunctional siroheme (also called structural siroheme, with the [4Fe–4S] cluster coordinated by the conserved motif $CX_5CX_{37}CX_3C$), which can be depleted of Fe (called sirohydrochlorin) depending on the organism. DsrB harbors a functional siroheme with a [4Fe–4S] cluster coordinated by the conserved motif $CX_{36}CCX_3C$. In this configuration, the active site is formed at the interface of the DsrA part, which contains the sulfite-binding site, and DsrB, which presents the siroheme linked to the proximal [4Fe–4S] cluster. DsrA and DsrB contain an inserted ferredoxin domain carrying one [4Fe–4S] cluster.

While the aSir performs the unidirectional reduction of sulfite to sulfide in the assimilatory system, dSir produces or can accept intermediates such as trithionate and thiosulfate [118, 122] *in vitro*. This is the so-called trithionate pathway, in which sulfite can be progressively branched to become trithionate, and the trithionate can be progressively reduced to sulfide. The reaction is due to the physiological reaction in which the overall 6-electron reduction process is uncoupled in two subprocesses orchestrated by different enzymatic actors.

Under physiological conditions, dSir binds the sulfite on its reduced siroheme and converts it into an S^{II} intermediate [123]. Next, a small sulfur carrier protein called DsrC binds to the S^{II}-DsrAB complex. DsrC carries two conserved cysteines on its C-terminal arm that will enter the active site of DsrAB and interact with the bound S^{II} intermediate [123]. The sulfur species is transferred to DsrC by establishing disulfide bridges between DsrC cysteines (i.e. R-S-**S**-S-R, where the bold S is the substrate and R stands for the cysteine side chain). During the formation of the trisulfide, protons are transferred to form three water molecules. In the next step, the DsrC-bound S^0 trisulfide is released from DsrAB, and the oxidized state of the siroheme coupled to its [4Fe–4S] cluster is reduced by two electrons by an unknown electron donor. In the following reaction, the DsrC-bound S^0 trisulfide is reduced by a membrane complex (such as DsrMKJOP, present in most organisms that have DsrAB and DsrC) [124, 125] by a total of four electrons from the menaquinone pool. At the end of the reduction process, H_2S and reduced DsrC are liberated. Electron transfer is suspected to be coupled to the proton motive force used for energy conservation. In addition, a proton motive force is indirectly generated by decreasing the intracellular concentration of protons through reduction reactions, leading to additional cellular energy acquisition. The product H_2S can diffuse out of the cell due to its hydrophobicity.

6.3.5 Sulfur Assimilation and Dissimilatory Sulfate Reduction Are Potentially Uncoupled

While it would seem logical that dissimilatory sulfate-reducing microorganisms would use the H_2S produced to generate biological sulfur compounds and not expend energy by expressing an additional assimilatory sulfate-reduction system, this may not always be the case. Intriguingly, the genomes of some sulfate reducers contain part of the assimilatory sulfate machinery (i.e. APSK and PAPSR) [75]. The presence of an APSK might reflect the need for PAPS production for the biosynthesis of particular biological compounds that play critical roles in detoxification, molecular recognition, and cell signaling [75]. The PAPS reductase would provide a control point for the overall sulfur flux in the cell. The separation of intracellular sulfur processing would be metabolically advantageous, as it would allow the pathways to be regulated individually and avoid cross-reactions between the two metabolisms. Sulfur assimilation could then only occur when needed, while allowing for uninterrupted cellular energy acquisition.

Uncoupling of the assimilatory and dissimilatory reduction of sulfate could be achieved in several ways. By spatial separation, both metabolic pathways occur at a specific cellular localization, for example, on the intracellular side of the membrane vs. the cytoplasm, in metabolons or dedicated microcompartments. By metabolic channeling, the enzymes from the same pathway are physically connected, offering a direct production line to transfer intermediates efficiently and avoid their diffusion in the cytoplasm [126]. Advances in cryo-electron tomography, which is particularly suited to studying these huge cellular complexes, will hopefully provide further insights in the coming years.

6.4 The Unconventional Sulfate-Reduction Pathway of *M. thermolithotrophicus*

Methanogens are the main actors in the carbon cycle, producing half of the atmospheric CH_4 annually and thus regulating climate on a global scale [127]. All known methanogens are obligate anaerobes, as they rely on various O_2-sensitive metallo-containing enzymes that are essential for cellular energy conservation (i.e. catabolism) or nutrient assimilation. They are of major ecological importance because of their role in the final degradation of organic matter [128–130]. Three different catabolic methanogenesis pathways have been described, namely hydrogenotrophic (e.g. H_2/CO_2 or formate), methylotrophic (e.g. from methylated compounds such as methanol and methylamines), and aceticlastic (i.e. acetate) [131–133]. The three methanogenic pathways lead to the release of CH_4, and methylotrophic and aceticlastic reactions generate additional CO_2. Because these reactions are close to the thermodynamic equilibrium [134], methanogens are considered energy-limited microorganisms, forcing them to develop adaptive strategies to save cellular energy [135–139].

6.4.1 Methanogens and Sulfur

Assimilation pathways tend to be well conserved between organisms that have inherited enzymes through vertical and horizontal transfer. However, the regulatory networks can sometimes differ dramatically when adapting to microbial physiology [137]. In some microbes, parts of the assimilation pathway itself or its enzymes are modified (e.g. to avoid poisoning of the catabolism by intermediates or to take advantage of alternative reducing power, such as ferredoxin instead of NADPH in plants). The sulfate-reducing pathway used by the methanogen *M. thermolithotrophicus* is an example of a reconfiguration of the sulfate assimilation pathway by different types of enzymes to avoid cross-reaction with its central catabolism: methanogenesis.

Since methanogens evolved in anaerobic ecological niches, their primary sulfur source is sulfide. Some can grow on elemental sulfur or Fe–S minerals, and others can feed on organic sulfur compounds (e.g. cysteine, methionine, short-chain alkylthiols, and dimethylsulfoxide) [140, 141]. Some methanogens also tolerate oxidized sulfur species such as thiosulfate or even sulfite and utilize them as a sulfur source [140, 142]. The latter is contradictory as methanogens are highly sensitive to sulfite, which can damage the methanogenic enzyme machinery [143–145].

Among the known microbial world, only one methanogen has shown robust growth on sulfate as a sole source of sulfur: *M. thermolithotrophicus* [146]. The growth of this organism under laboratory conditions on sulfate is paradoxical to what would be expected in nature [147]. First, methanogens occupy ecological niches in which all electron acceptors other than CO_2 (e.g. oxygen, nitrate, and iron) should be depleted, including sulfate. Second, in the sulfate–methane transition zone, where sulfate ions and methanogens are co-occurring, it has been shown that methanogens are out-competed by sulfate reducers that have a higher affinity for the required electron donors (e.g. H_2 or formate).

The marine thermophile *M. thermolithotrophicus* was isolated from geothermally heated sediments from the shores of Naples [148] at a depth of half a meter. Here, the methanogen may be sporadically exposed to sulfate and may benefit from the ability to assimilate it in order to expand its environmental niche or to adapt to the fluctuating availability of sulfur sources.

M. thermolithotrophicus has the advantage of growing to high biomass under laboratory conditions [135]. Therefore, it represented a suitable model to unravel the unconventional molecular adaptation of sulfate assimilation in methanogens.

6.4.2 *M. thermolithotrophicus* Exhibits Remarkable Growth on Sulfate

M. thermolithotrophicus is a hydrogenotrophic methanogen [149, 150] that thrives in anaerobic flasks containing a gas phase of H_2/CO_2 and minimal mineral medium supplemented with sulfide (Na_2S) as a sulfur source [135]. Alternatively, the microbe can be cultivated in a fermenter where H_2/CO_2 is flushed into the medium and CH_4 is exhausted. However, in such an open system, at a pH of 6.2, most of the sulfide will be present in the gaseous form of H_2S that can escape into the gas phase, slowly depleting the medium of its available sulfur source. This explains why fermenters are usually operated with a constant injection of the explosive and toxic H_2S, as it counteracts sulfide depletion and alkalinization of the medium.

M. thermolithotrophicus was gradually transferred to a medium containing only sulfate as a sulfur source to eliminate residual sulfide and to allow for a smooth adaptation. A minimal concentration of 100 μM sulfate was determined to sustain growth. By measuring the sulfate concentration of the medium, which was monitored over time, it was confirmed that the sulfate concentration decreased during the growth of *M. thermolithotrophicus* cells. Molybdate addition was tested to verify if sulfate assimilation depends on a conventional ATPS system (see Sections 6.2.1 and 6.3.2). A high molybdate/sulfate ratio did not disturb the sulfide-grown cells. However, a ratio of 6.25 : 1 was sufficient to inhibit the growth of *M. thermolithotrophicus* on sulfate, confirming that the first step is likely to depend on an ATPS sensitive to molybdate.

The ATPS homolog encoded in the genome of *M. thermolithotrophicus* was found to be part of a genomic region (Figure 6.6) that contained other candidate genes associated with sulfate-reduction pathways, such as an APSK, a sulfite transporter, and a dissimilatory APSR. In comparison, the genome of *Methanocaldococcus infernus* also contains part of these genes but is not able to assimilate sulfate. The potential enzymes were recombinantly expressed and characterized to validate their physiological role and to map the metabolic road.

6.4.3 A Conserved Sulfate Activation Mechanism Produces PAPS

In the assimilatory sulfate-reduction pathway of *M. thermolithotrophicus*, sulfate is proposed to enter the cell via a DASS/SulP transporter (respectively, divalent anion: Na^+ symporter/solute: cation and solute: solute antiporter). Next, sulfate is activated to APS by an ATPS and further phosphorylated to PAPS by an APSK. The experimental structures of the ATPS and APSK confirmed bioinformatic studies on the protein sequences: the three-dimensional organization and the residues of the active sites are conserved with bacterial homologs. *M. thermolithotrophicus* ATPS harbors the Zn-binding motif inserted between the second and third domain, as previously observed in the ATPS from the thermophilic bacterium *T. thermophilus* [64]. The APSK structure exhibits a more pronounced conservation with bacterial

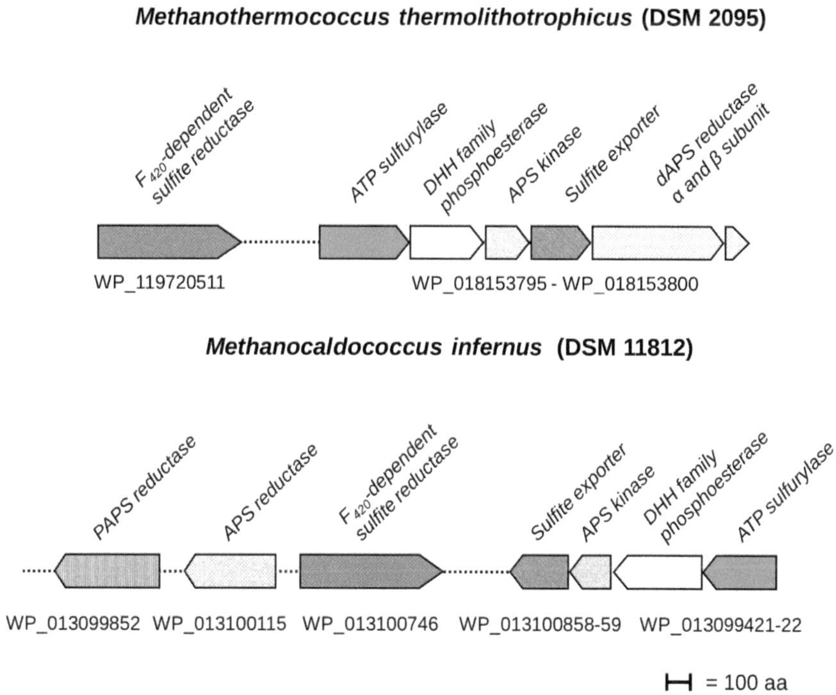

Figure 6.6 Genomic organization of genes involved in the sulfur metabolism from two *Methanococcales*. The annotation of the genes from *M. thermolithotrophicus* is the one before pathway elucidation. The DHH family phosphoesterase was characterized as a PAP-phosphatase, and the putative dAPSR as a PAPSR. The gene annotation for the PAPSR and APSR in *M. infernus* has been derived from the following works: Source: Cho [151] and Lee et al. [152], and improved genome annotation tools allowed a sharpened picture of the organization of genes of interests. Here, the operon continuity of the ATPS/PAPP/APSK/sulfite transporter is hypothetic.

homologs. A coupled enzyme assay of the ATPS and APSK enzymes, monitoring the production of adenosine diphosphate (ADP) from APSK, confirmed that the enzymes were active and required ATP and sulfate to produce PAPS (Figure 6.7). The *in vitro* production of PAPS was puzzling and suggested that PAPS is an intermediate of the sulfate assimilation pathway. Since no known sulfotransferase PAPS-dependent homolog was found in the genome (sulfotransferase route in Figure 6.3), we hypothesized that *M. thermolithotrophicus* uses a PAPSR to reduce the sulfonucleotide (Section 6.2.3) further.

6.4.4 A Dissimilatory-Like PAPS-Reductase and a New PAP-Phosphatase

Since no conventional gene encoding an assimilatory PAPSR was found in *M. thermolithotrophicus* genome, the best candidate for PAPS reduction was the dissimilatory APS reductase-like enzyme. However, this would imply that the enzyme switches substrate specificity from APS to PAPS.

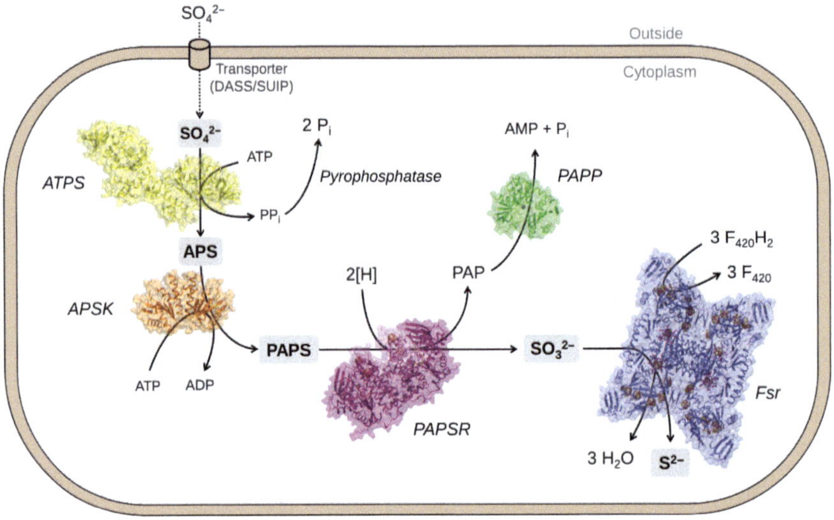

Figure 6.7 The sulfate assimilation pathway in the methanogen *M. thermolithotrophicus*. The different enzymes composing the pathway are shown as transparent surfaces and cartoons. (Metallo)cofactors are displayed as balls and sticks. The protein data bank identifiers for the ATPS, APSK, PAPSR, PAPP, and Fsr are 8A8G, 8A8H, 8A8O, 8A8K, and 7NPA, respectively.

The structure of the APSR-like enzyme from *M. thermolithotrophicus* confirmed a high structural resemblance with the dissimilatory APSR from the hyperthermophilic archaeon *A. fulgidus* or the one from the thermophilic bacterium *Megalodesulfovibrio gigas* [111, 113]. Accordingly, phylogenetic analyses suggest a horizontal gene transfer of the gene from bacteria to *M. thermolithotrophicus* [15]. While the $(\alpha\beta)_2$ assembly and the (metallo)cofactors are the same, some notable differences exist in the active site. Indeed, the loops and the environment surrounding the APS in dAPSR show several substitutions in *M. thermolithotrophicus* enzyme. These substitutions create a positively charged pocket that would allow the binding of an additional 3′-phosphate from the sulfur-carrier nucleotide, making it possible to distinguish between APS and PAPS. A coupled enzyme assay combining all the enzymes required for sulfite formation confirmed that *M. thermolithotrophicus* uses a PAPSR (Figure 6.7). The electron donor for the PAPSR is still unknown, but it has been suggested to be ferredoxin, a central electron carrier in methanogens that is also used for anabolic reactions. In hydrogenotrophic methanogens, the membrane-bound [NiFe]-hydrogenases Eha and Ehb [149, 153, 154] can reduce the ferredoxin pool dedicated to anabolic reactions.

While it was shown that the gene encoding a dAPSR indeed encodes a new type of PAPSR, the PAP produced still needs to be efficiently eliminated to avoid poisoning the cell. This seems to be a dead end since the genome of *M. thermolithotrophicus* does not encode any known PAPP. However, in the genomic environment of the genes encoding for the ATPS and APSK, there is a gene encoding for a putative phosphoesterase belonging to the DHH family. The structural and biochemical investigations of this gene product confirmed that *M. thermolithotrophicus* relies on this phosphoesterase to hydrolyze the 3′-phosphate of PAP at very high rates and high specificity (Figure 6.7). The reaction proceeds through manganese-dependent hydrolysis as described for an esterase homolog (e.g. NanoRNase and RecJ exonuclease) [155–157]. The gene encoding for the PAPP described in *M. thermolithotrophicus* is conserved in the genome of numerous archaea and would represent an alternative system in the sulfur metabolism of these organisms.

6.4.5 A F_{420}-Dependent Sulfite Reductase Catalyzes the Last Step of the Pathway

The sulfite released by the PAPSR must be eliminated to prevent damage to the enzymes involved in the methanogenesis pathway. In 2005, the Mukhopadhyay team found that some methanogens belonging to the order of *Methanococcales* are naturally exposed to sulfite and are able to detoxify it intracellularly by an $F_{420}H_2$-oxidizing sulfite reductase (Fsr) [142]. The electron donor of Fsr is the coenzyme F_{420}, a two-electron carrier with structural similarity to FAD and flavin mononucleotide (FMN), but chemically it resembles NAD(P) [158–160]. However, compared to the latter, its redox potential is more negative (E_0' NADP/NADPH = −324 mV, vs. E_0' $F_{420}/F_{420}H_2$ = −350 mV), allowing it to drive the cellular reaction more favorably. The cofactor, highly abundant in methanogens, serves several physiological purposes. Besides its involvement in sulfite reduction, it is a central electron carrier for hydrogenotrophic and methylotrophic methanogenesis and has been proposed for use in anabolic reactions. The intrinsic fluorescence of F_{420} has been used for decades as a natural marker to identify methanogens by microscopy techniques. In addition to archaea, F_{420} has also been detected in some bacteria, such as mycobacteria, and has been implicated as a respiratory electron donor or to play a role in the persistence response upon oxygen deprivation [161, 162].

In hydrogenotrophic methanogens, the reduced F_{420} pool is mostly reduced to H_2 by the F_{420}-reducing [NiFe]-hydrogenase [163] and alternative systems such as F_{420}-reducing formate oxidase [164]. A high concentration of reduced F_{420} and its low redox potential explain the high turnover rate of the Fsr [165].

A gene encoding for Fsr was found in the genome of *M. thermolithotrophicus* but separate from the sulfate-reducing region (Figure 6.6). When *M. thermolithotrophicus* is exposed to sulfite, a burst of Fsr expression is seen in the cell extract either by native polyacrylamide gel electrophoresis (native PAGE) or by monitoring the sulfite reductase activity. Fsr expression and activity were stimulated under sulfate-grown conditions compared to sulfide-grown conditions, implying that Fsr is the last enzyme in the sulfate-reduction pathway.

Fsr has been structurally characterized from two *Methanococcales*: from *M. thermolithotrophicus,* which was grown on sulfate as the sole sulfur source, and from *Methanocaldococcus jannaschii*, which was grown on sulfite [166]. Both Fsr structures are highly similar and are organized into three functional units: an N-terminal $F_{420}H_2$-oxidase, an intraelectronic wire, and a C-terminal sulfite

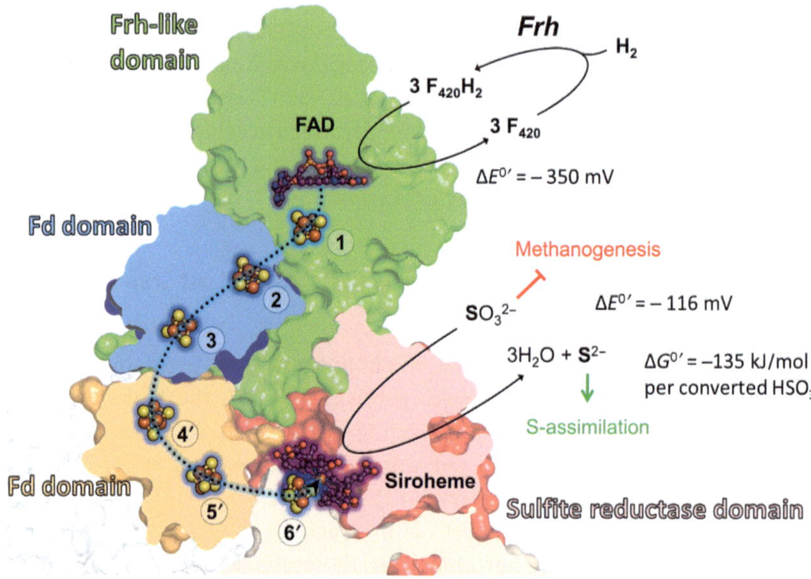

Figure 6.8 Fsr mechanism. One functional unit of Fsr is represented as a surface cut-through with (metallo)cofactors shown as balls and sticks. The [4Fe–4S] clusters are numbered, with the prime referring to the Fsr dimer. Frh and Fd stand for the F_{420}-reducing hydrogenase and ferredoxin, respectively.

reductase. Structurally, Fsr is a dimer of homodimers in which the core sulfite reductase is flanked by the $F_{420}H_2$-oxidase. Both catalytic sites are linked by the intraelectronic wire, which consists of two inserted ferredoxin domains, each containing two [4Fe–4S] clusters (one at the N-terminal of the $F_{420}H_2$-oxidase and the second inserted in the sulfite reductase domain at the same position as observed in DsrA and B). The $F_{420}H_2$-oxidase domain, analogous to the β-subunit of the F_{420}-reducing hydrogenase, harbors an FAD and one [4Fe–4S] cluster. The FAD receives two electrons from the reduced F_{420} and catalyzes the electron transfer to the electron wire (Figure 6.8). The electrons are transferred through the five [4Fe–4S] clusters to the final cluster, which is coupled to the siroheme. Interestingly, the last cluster shares the same coordination motif $CX_5CX_{33}CX_3C$ with that of aSirs, which might explain why Fsr performs three reduction events leading to the unidirectional generation of sulfide as observed for aSirs (Figures 6.7 and 6.8).

As it has been previously reported for other sulfite reductases, Fsr also exhibits an efficient nitrite-reduction activity. This might expand the functionality of the enzyme to the nitrogen metabolism under certain physiological conditions [139].

6.5 Evolution and Biotechnological Applications

Methanogens are one of the oldest microbial life forms found in microfossils (e.g. 3.5 billion years ago) [167–169] and are suitable models for understanding early metabolism and its evolution [170, 171]. Furthermore, their metabolic abilities make methanogens amazing gas transformers, producing CH_4 from CO_2 [172]. Therefore, the elucidation of the sulfate-reduction pathway offers both evolutionary and biotechnological perspectives.

6.5.1 Fsr as a Structural Template for a Primordial Sulfite Reductase

Methanogenesis and sulfate reduction have been suggested to be two of the oldest metabolic pathways on our planet but were so far not assumed to co-occur in the same organism due to previously listed barriers. The sulfate acquisition strategy of *M. thermolithotrophicus* is intriguing for two points. It shows that methanogenesis and sulfate reduction, albeit an assimilatory one, can be used in the same organism, and also because it seems that the microbe picked up the best of two worlds: assimilatory and dissimilatory sulfate reduction. An exciting question is: where and when did this unconventional pathway originate? According to phylogenetic analysis, most of the genes encoding the enzymes of the assimilatory sulfate-reduction pathway of *M. thermolithotrophicus* appeared to be of bacterial origin and were acquired by the methanogen by horizontal gene transfer [15]. However, it is interesting to note that all the genes (except for *fsr*) are located in the same genomic region, which may allow for a tight regulation of their expression. If the genes were transferred progressively through successive generations, why are they all clustered together (except for *fsr*)? If the pathway originated from a methanogenic ancestor, why is *M. thermolithotrophicus* the only one known to carry all the genes?

While the origin of the whole pathway is still a mystery, the structural elucidation of Fsr has provided more pieces to the evolutionary puzzle of sulfite reductases [169].

Modern sulfite reductases are key enzymes in sulfur metabolism with different evolutionary trajectories between aSir and dSir. However, based on sequence and phylogenetic analyses, it has been suggested that sulfite reductases diverged from a common ancestor [93, 119, 173, 174]. While it was long assumed that sulfite reductases likely evolved in bacteria, recent comparative genomics, phylogenetic, and synteny analysis by Neukirchen et al. point to an archaeal origin of the minimal dissimilatory sulfite reductase enzymes [124]. In the accepted model, the primordial sulfite reductase would have been a homodimer carrying one [4Fe–4S] cluster connected to the siroheme (Figure 6.9). In the case of aSir, a duplication followed by a fusion would have led to a single polypeptide containing an N-terminal domain that had lost the metallocofactor but retained the sulfite-binding site and a C-terminal domain that has lost the sulfite-binding site but maintained the metallocofactor. For dSirs, insertion of a ferredoxin carrying a [4Fe–4S] cluster followed by its duplication would have resulted in DsrA and DsrB. DsrA would have retained the metallocofactor to conserve structural integrity, and DsrB would have modified the [4Fe–4S] cluster coordination while degenerating its sulfite-binding site to serve only a structural purpose. Regarding Fsr, a ferredoxin containing two [4Fe–4S] clusters would have been first inserted in the sulfite reductase domain, followed by a fusion with the $F_{420}H_2$-oxidase domain that contained a ferredoxin with two

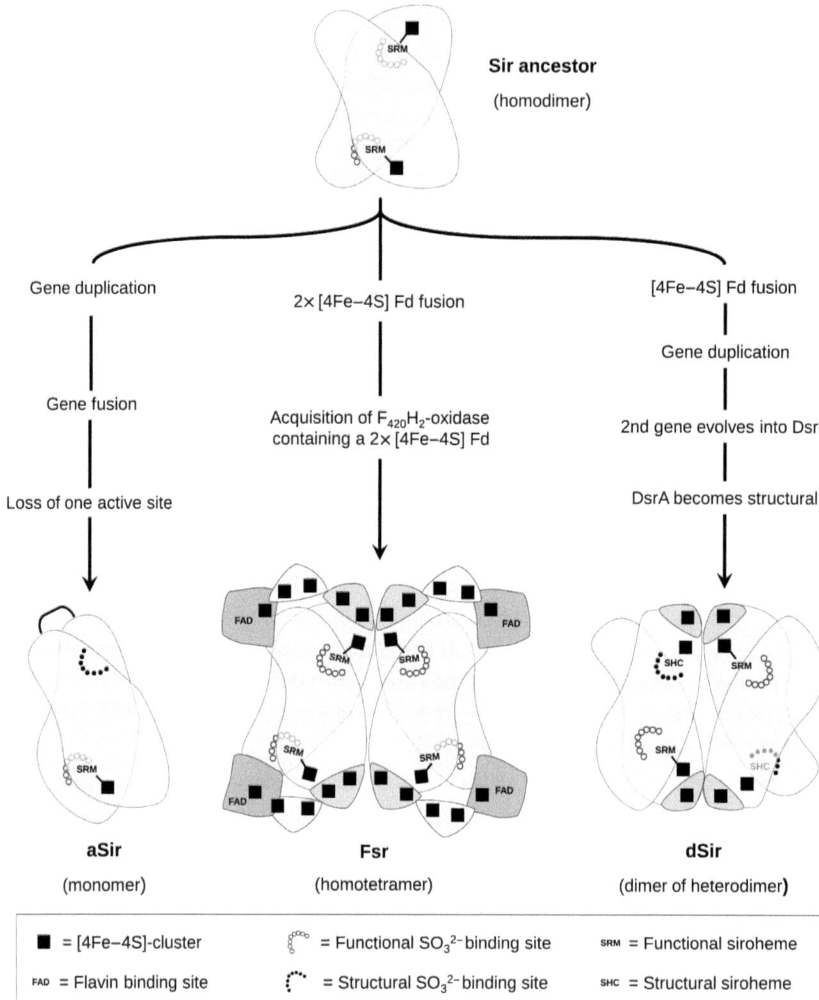

Figure 6.9 A proposed evolutionary trajectory of the sulfite reductase. Fd stands for ferredoxin.

[4Fe–4S] clusters in its N-terminal domain. Due to the simplicity of its sulfite reductase domain, Fsr likely presents a snapshot of the primordial homodimeric sulfite reductase before the duplication event, in which the active sites of the two units constituting the homodimers are functionally independent (Figure 6.9).

6.5.2 Sulfate Utilization as a Safer Route for Methanogenic Biotechnologies

Biotechnologies have recently harnessed the amazing gas conversion capabilities of methanogens. Here, H_2 generated from green energy is concomitantly bubbled with CO_2 in bioreactors to generate CH_4 that can be injected into the gas grid (e.g. Electrochaea GmbH) [172, 175, 176]. One of the challenges in developing this technology is the dependence on sulfide required for biomass production and maintaining the redox balance of the medium. Hydrogen sulfide is toxic, corrosive, flammable, and explosive and generates artificial by-products detrimental to anaerobic biotechnologies [177]. Sulfate, on the other hand, is cheap and safe to handle. Switching to a methanogen capable of growing in bioreactors with sulfate instead of sulfide would be advantageous regarding costs, waste, maintenance, and safety.

As a proof of concept, *M. thermolithotrophicus* was successfully grown in a fermenter with only sulfate as a sulfur source, showing robust growth and reaching high cell yields (about 6–7 of optical density at 600 nm). This means that despite the cellular energy costs (see Section 6.4), *M. thermolithotrophicus* is still capable of maintaining enough sulfide bioproduction for growth while stabilizing the redox balance of the medium.

Now that the entire sulfate assimilation pathway has been elucidated, it would be possible to transplant all the required set of genes (genes coding for: sulfate transporter, ATPS, APSK, PAPSR, PAPP, and Fsr) into genetically engineered methanogens already adapted to CH_4 bioproduction. This would stimulate the development of a renewable CH_4 technology by harnessing the chemical ingenuity of ancient microbes.

6.6 Conclusions

Sulfate reduction is an ancient metabolic pathway that leads to the production of sulfide required for the synthesis of biological material. Until now, the only two pathways known to reduce sulfate to sulfide have been considered to serve two fundamentally different biological roles, namely the formation of biological matter for the assimilatory sulfate-reduction pathway and energy conservation for the dissimilatory sulfate reduction. However, the methanogenic archaeon *M. thermolithotrophicus* combines enzymes from both pathways and has adapted them to allow sulfur assimilation from sulfate and sulfite as well as detoxification of the latter. This suggests that sulfur acquisition strategies of microorganisms may be more modular than previously thought and that future investigations might discover alternative routes.

References

1 Rabus, R., Hansen, T.A., and Widdel, F. (2006). *Dissimilatory Sulfate- and Sulfur-Reducing Prokaryotes. Prokaryotes: A Handbook on the Biology of Bacteria*, 3e, vol. 2, 659–768. Springer.

2 Morrison, P.R. and Mojzsis, S.J. (2021). Tracing the early emergence of microbial sulfur metabolisms. *Geomicrobiol. J.* 38 (1): 66–86.

3 Zopfi, J., Ferdelman, T.G., Fossing, H. et al. (2004). Distribution and fate of sulfur intermediates—sulfite, tetrathionate, thiosulfate, and elemental sulfur—in marine sediments. In: *Sulfur Biogeochemistry – Past and Present* (ed. J. Amend, K. Edwards, and A.T. Lyons), 97–116. Boulder: Geological Society of America.

4 Santana Maldonado, C., Weir, A., and Rumbeiha, W.K. (2023). A comprehensive review of treatments for hydrogen sulfide poisoning: past, present, and future. *Toxicol. Mech. Methods* 33 (3): 183–196.

5 Iciek, M., Bilska-Wilkosz, A., Kozdrowicki, M., and Gorny, M. (2022). Reactive sulfur species and their significance in health and disease. *Biosci. Rep.* 42 (9): BSR20221006.

6 Fike, D.A., Bradley, A.S., and Rose, C.V. (2015). Rethinking the ancient sulfur cycle. *Annu. Rev. Earth Planet. Sci.* 43: 593–622.

7 Wu, B., Liu, F.F., Fang, W.W. et al. (2021). Microbial sulfur metabolism and environmental implications. *Sci. Total Environ.* 778: 146085.

8 Wasmund, K., Mussmann, M., and Loy, A. (2017). The life sulfuric: microbial ecology of sulfur cycling in marine sediments. *Environ. Microbiol. Rep.* 9 (4): 323–344.

9 Park, Y. and Faivre, D. (2022). Diversity of microbial metal sulfide biomineralization. *ChemPlusChem* 87 (1): e202100457.

10 Findlay, A.J. (2016). Microbial impact on polysulfide dynamics in the environment. *FEMS Microbiol. Lett.* 363 (11): fnw103.

11 Jørgensen, B.B. (2019). Unravelling the sulphur cycle of marine sediments. *Environ. Microbiol.* 21 (10): 3533–3538.

12 Jørgensen, B.B., Findlay, A.J., and Pellerin, A. (2019). The biogeochemical sulfur cycle of marine sediments. *Front. Microbiol.* 10: 849.

13 Russell, M.J., Nitschke, W., and Branscomb, E. (2013). The inevitable journey to being. *Philos. Trans. R. Soc. Lond. Ser. B Biol. Sci.* 368 (1622): 20120254.

14 Olson, K.R. and Straub, K.D. (2016). The role of hydrogen sulfide in evolution and the evolution of hydrogen sulfide in metabolism and signaling. *Physiology (Bethesda)* 31 (1): 60–72.

15 Jespersen, M. and Wagner, T. (2023). Assimilatory sulfate reduction in the marine methanogen *Methanothermococcus thermolithotrophicus*. *Nat. Microbiol.* 8 (7): 1227–1239.

16 Tuorto, F. and Lyko, F. (2016). Genome recoding by tRNA modifications. *Open Biol.* 6 (12): 160287.

17 Han, S., Li, Y., and Gao, H. (2022). Generation and physiology of hydrogen sulfide and reactive sulfur species in bacteria. *Antioxidants (Basel)* 11 (12): 2487.

18 Liu, H., Wang, J., Liu, J. et al. (2021). Hydrogen sulfide (H_2S) signaling in plant development and stress responses. *aBIOTECH* 2 (1): 32–63.

19 Zou, J., Yuan, Z., Chen, X. et al. (2024). Hydrogen sulfide responsive nanoplatforms: novel gas responsive drug delivery carriers for biomedical applications. *Asian J. Pharm. Sci.* 19 (1): 100858.

20 Munteanu, C., Turnea, M.A., and Rotariu, M. (2023). Hydrogen sulfide: an emerging regulator of oxidative stress and cellular homeostasis-a comprehensive one-year review. *Antioxidants (Basel)* 12 (9): 1737.

21 Kimura, H., Shibuya, N., and Kimura, Y. (2012). Hydrogen sulfide is a signaling molecule and a cytoprotectant. *Antioxid. Redox Signal.* 17 (1): 45–57.

22 Chaudhary, S., Sindhu, S.S., Dhanker, R., and Kumari, A. (2023). Microbes-mediated sulphur cycling in soil: impact on soil fertility, crop production and environmental sustainability. *Microbiol. Res.* 271: 127340.

23 Wagner, T., Kahnt, J., Ermler, U., and Shima, S. (2016). Didehydroaspartate modification in methyl-coenzyme M reductase catalyzing methane formation. *Angew. Chem. Int. Ed. Engl.* 55 (36): 10630–10633.

24 Chen, S., Wang, L.R., and Deng, Z.X. (2010). Twenty years hunting for sulfur in DNA. *Protein Cell* 1 (1): 14–21.

25 Yang, Y., Xu, G., Liang, J. et al. (2017). DNA backbone sulfur-modification expands microbial growth range under multiple stresses by its anti-oxidation function. *Sci. Rep.* 7 (1): 3516.

26 Zheng, C., Black, K.A., and Dos Santos, P.C. (2017). Diverse mechanisms of sulfur decoration in bacterial tRNA and their cellular functions. *Biomol. Ther.* 7 (1): 33.

27 Zheng, Y.Y., Wu, Y., Begley, T.J., and Sheng, J. (2021). Sulfur modification in natural RNA and therapeutic oligonucleotides. *RSC Chem. Biol.* 2 (4): 990–1003.

28 Burley, J.C., van de Streek, J., and Stephens, P.W. (2006). Ampicillin trihydrate from synchrotron powder diffraction data. *Acta Crystallogr. E.* 62: O797–O799.

29 Perona, J.J., Rauch, B.J., and Driggers, C.M. (2018). Sulfur assimilation and trafficking in methanogens. *Mol. Mech. Microb. Evol.* 371–408.

30 Chatterjee, S. and Hausinger, R.P. (2022). Sulfur incorporation into biomolecules: recent advances. *Crit. Rev. Biochem. Mol.* 57 (5–6): 461–476.

31 Fujishima, K., Wang, K.M., Palmer, J.A. et al. (2018). Reconstruction of cysteine biosynthesis using engineered cysteine-free enzymes. *Sci. Rep.* 8 (1): 1776.

32 Dharavath, S., Vijayan, R., Kumari, K. et al. (2020). Crystal structure of O-acetylserine sulfhydralase (OASS) isoform 3 from *Entamoeba histolytica*: pharmacophore-based virtual screening and validation of novel inhibitors. *Eur. J. Med. Chem.* 192: 112157.

33 Francois, J.A., Kumaran, S., and Jez, J.M. (2006). Structural basis for interaction of *O*-acetylserine sulfhydrylase and serine acetyltransferase in the *Arabidopsis* cysteine synthase complex. *Plant Cell* 18 (12): 3647–3655.

34 Kredich, N.M. (2008). Biosynthesis of cysteine. *EcoSal Plus* 3 (1): 1–30.

35 Kawano, Y., Suzuki, K., and Ohtsu, I. (2018). Current understanding of sulfur assimilation metabolism to biosynthesize L-cysteine and recent progress of its fermentative overproduction in microorganisms. *Appl. Microbiol. Biotechnol.* 102 (19): 8203–8211.

36 Mino, K. and Ishikawa, K. (2003). A novel *O*-phospho-L-serine sulfhydrylation reaction catalyzed by *O*-acetylserine sulfhydrylase from *Aeropyrum pernix* K1. *FEBS Lett.* 551 (1–3): 133–138.

37 Hensel, G. and Trüper, H.G. (1983). *O*-Acetylserine sulfhydrylase and S-sulfocysteine synthase activities of *Rhodospirillum tenue*. *Arch. Microbiol.* 134 (3): 227–232.

38 Wolfe, J.M. and Fournier, G.P. (2018). Horizontal gene transfer constrains the timing of methanogen evolution. *Nat. Ecol. Evol.* 2 (5): 897–903.

39 Begley, T.P., Ealick, S.E., and McLafferty, F.W. (2012). Thiamin biosynthesis: still yielding fascinating biological chemistry. *Biochem. Soc. T.* 40: 555–560.

40 Farghaly, T.A., Alfaifi, G.H., and Gomha, S.M. (2024). Recent literature on the synthesis of thiazole derivatives and their biological activities. *Mini-Rev. Med. Chem.* 24 (2): 196–251.

41 Prajapati, S., Pappenheim, F.R.V., and Tittmann, K. (2022). Frontiers in the enzymology of thiamin diphosphate-dependent enzymes. *Curr. Opin. Struct. Biol.* 76: 102441.

42 Chatterjee, A., Abeydeera, N.D., Bale, S. et al. (2011). THI4p is a suicide thiamine thiazole synthase. *Nature* 478 (7370): 542–546.

43 Zhang, X., Eser, B.E., Chanani, P.K. et al. (2016). Structural basis for iron-mediated sulfur transfer in archael and yeast thiazole synthases. *Biochemistry* 55 (12): 1826–1838.

44 Eser, B.E., Zhang, X., Chanani, P.K. et al. (2016). From suicide enzyme to catalyst: the iron-dependent sulfide transfer in *Methanococcus jannaschii* thiamin thiazole biosynthesis. *J. Am. Chem. Soc.* 138 (11): 3639–3642.

45 Ma, X.X., Zheng, G.D., Liang, M.L. et al. (2019). Occurrence and origin of H_2S from volcanic reservoirs in Niudong area of the Santanghu Basin, NW China. *Geofluids* 2019 (1): 1279658.

46 Spietz, R.L., Payne, D., Kulkarni, G. et al. (2022). Investigating abiotic and biotic mechanisms of pyrite reduction. *Front. Microbiol.* 13: 878387.

47 Spietz, R.L., Payne, D., Szilagyi, R., and Boyd, E.S. (2022). Reductive biomining of pyrite by methanogens. *Trends Microbiol.* 30 (11): 1072–1083.

48 Payne, D., Spietz, R.L., and Boyd, E.S. (2021). Reductive dissolution of pyrite by methanogenic archaea. *ISME J.* 15 (12): 3498–3507.

49 Jørgensen, B.B. (1982). Mineralization of organic matter in the sea bed – the role of sulfate reduction. *Nature* 296 (5858): 643–645.

50 Canfield, D.E. and Farquhar, J. (2009). Animal evolution, bioturbation, and the sulfate concentration of the oceans. *Proc. Natl. Acad. Sci. USA* 106 (20): 8123–8127.

51 Aguilar-Barajas, E., Díaz-Pérez, C., Ramírez-Díaz, M.I. et al. (2011). Bacterial transport of sulfate, molybdate, and related oxyanions. *Biometals* 24 (4): 687–707.

52 Marietou, A., Roy, H., Jørgensen, B.B., and Kjeldsen, K.U. (2018). Sulfate transporters in dissimilatory sulfate reducing microorganisms: a comparative genomics analysis. *Front. Microbiol.* 9: 309.

53 Sanghai, Z.A., Liu, Q., Clarke, O.B. et al. (2018). Structure-based analysis of CysZ-mediated cellular uptake of sulfate. *elife* 7: e27829.

54 Takahashi, H., Buchner, P., Yoshimoto, N. et al. (2012). Evolutionary relationships and functional diversity of plant sulfate transporters. *Front. Plant Sci.* 2: 119.

55 Thauer, R.K., Jungermann, K. and Decker, K. (1977). Energy conservation in chemotrophic anaerobic bacteria. *Bacteriol. Rev.* 41 (1): 100–180.

56 Zhang, Z., Zhang, C.H., Yang, Y. et al. (2022). A review of sulfate-reducing bacteria: metabolism, influencing factors and application in wastewater treatment. *J. Clean. Prod.* 376: 134109.

57 Ullrich, T.C., Blaesse, M., and Huber, R. (2001). Crystal structure of ATP sulfurylase from *Saccharomyces cerevisiae*, a key enzyme in sulfate activation. *EMBO J.* 20 (3): 316–329.

58 Mougous, J.D., Lee, D.H., Hubbard, S.C. et al. (2006). Molecular basis for G protein control of the prokaryotic ATP sulfurylase. *Mol. Cell* 21 (1): 109–122.

59 Parey, K., Demmer, U., Warkentin, E. et al. (2013). Structural, biochemical and genetic characterization of dissimilatory ATP sulfurylase from *Allochromatium vinosum*. *PLoS One* 8 (9): e74707.

60 Beynon, J.D., MacRae, I.J., Huston, S.L. et al. (2001). Crystal structure of ATP sulfurylase from the bacterial symbiont of the hydrothermal vent tubeworm *Riftia pachyptila*. *Biochemistry* 40 (48): 14509–14517.

61 Harjes, S., Bayer, P., and Scheidig, A.J. (2005). The crystal structure of human PAPS synthetase 1 reveals asymmetry in substrate binding. *J. Mol. Biol.* 347 (3): 623–635.

62 Yu, Z., Lansdon, E.B., Segel, I.H., and Fisher, A.J. (2007). Crystal structure of the bifunctional ATP sulfurylase-APS kinase from the chemolithotrophic thermophile *Aquifex aeolicus*. *J. Mol. Biol.* 365 (3): 732–743.

63 Herrmann, J., Ravilious, G.E., McKinney, S.E. et al. (2014). Structure and mechanism of soybean ATP sulfurylase and the committed step in plant sulfur assimilation. *J. Biol. Chem.* 289 (15): 10919–10929.

64 Taguchi, Y., Sugishima, M., and Fukuyama, K. (2004). Crystal structure of a novel zinc-binding ATP sulfurylase from *Thermus thermophilus* HB8. *Biochemistry* 43 (14): 4111–4118.

65 MacRae, I.J., Segel, I.H., and Fisher, A.J. (2001). Crystal structure of ATP sulfurylase from *Penicillium chrysogenum*: insights into the allosteric regulation of sulfate assimilation. *Biochemistry* 40 (23): 6795–6804.

66 Gay, S.C., Segel, I.H., and Fisher, A.J. (2009). Structure of the two-domain hexameric APS kinase from *Thiobacillus denitrificans*: structural basis for the absence of ATP sulfurylase activity. *Acta Crystallogr. D.* 65: 1021–1031.

67 Yu, M., Martin, R.L., Jain, S. et al. (1989). Rat liver ATP-sulfurylase: purification, kinetic characterization, and interaction with arsenate, selenate, phosphate, and other inorganic oxyanions. *Arch. Biochem. Biophys.* 269 (1): 156–174.

68 Seubert, P.A., Hoang, L., Renosto, F., and Segel, I.H. (1983). ATP sulfurylase from *Penicillium chrysogenum*: measurements of the true specific activity of an enzyme subject to potent product inhibition and a reassessment of the kinetic mechanism. *Arch. Biochem. Biophys.* 225 (2): 679–691.

69 Osslund, T., Chandler, C., and Segel, I.H. (1982). ATP sulfurylase from higher plants: purification and preliminary kinetics studies on the cabbage leaf enzyme. *Plant Physiol.* 70 (1): 39–45.

70 Smith, R.L. and Klug, M.J. (1981). Electron donors utilized by sulfate-reducing bacteria in eutrophic lake sediments. *Appl. Environ. Microbiol.* 42 (1): 116–121.

71 Reuveny, Z. (1977). Derepression of ATP sulfurylase by the sulfate analogs molybdate and selenate in cultured tobacco cells. *Proc. Natl. Acad. Sci. USA* 74 (2): 619–622.

72 Ravilious, G.E., Nguyen, A., Francois, J.A., and Jez, J.M. (2012). Structural basis and evolution of redox regulation in plant adenosine-5′-phosphosulfate kinase. *Proc. Natl. Acad. Sci. USA* 109 (1): 309–314.

73 Poyraz, Ö., Brunner, K., Lohkamp, B. et al. (2015). Crystal structures of the kinase domain of the sulfate-activating complex in *Mycobacterium tuberculosis*. *PLoS One* 10 (4): e0121494.

74 Herrmann, J., Nathin, D., Lee, S.G. et al. (2015). Recapitulating the structural evolution of redox regulation in adenosine 5′-phosphosulfate kinase from cyanobacteria to plants. *J. Biol. Chem.* 290 (41): 24705–24714.

75 Kawakami, T., Teramoto, T., and Kakuta, Y. (2023). Crystal structure of adenosine 5′-phosphosulfate kinase isolated from *Archaeoglobus fulgidus*. *Biochem. Biophys. Res. Co.* 643: 105–110.

76 Negishi, M., Pedersen, L.G., Petrotchenko, E. et al. (2001). Structure and function of sulfotransferases. *Arch. Biochem. Biophys.* 390 (2): 149–157.

77 Yu, Z.H., Lemongello, D., Segel, I.H., and Fisher, A.J. (2008). Crystal structure of *Saccharomyces cerevisiae* 3′-phosphoadenosine-5′-phosphosulfate reductase complexed with adenosine 3′,5′-bisphosphate. *Biochemistry* 47 (48): 12777–12786.

78 Chartron, J., Carroll, K.S., Shiau, C. et al. (2006). Substrate recognition, protein dynamics, and iron-sulfur cluster in *Pseudomonas aeruginosa* adenosine 5′-phosphosulfate reductase. *J. Mol. Biol.* 364 (2): 152–169.

79 Feliciano, P.R., Carroll, K.S., and Drennan, C.L. (2021). Crystal structure of the [4Fe-4S] cluster-containing adenosine-5′-phosphosulfate reductase from *Mycobacterium tuberculosis*. *ACS Omega* 6 (21): 13756–13765.

80 Stevenson, C.E.M., Hughes, R.K., McManus, M.T. et al. (2013). The X-ray crystal structure of APR-B, an atypical adenosine 5′-phosphosulfate reductase from *Physcomitrella patens*. *FEBS Lett.* 587 (22): 3626–3632.

81 Dichtl, B., Stevens, A., and Tollervey, D. (1997). Lithium toxicity in yeast is due to the inhibition of RNA processing enzymes. *EMBO J.* 16 (23): 7184–7195.

82 Wakamatsu, T., Kim, K., Uemura, Y. et al. (2011). Role of RecJ-like protein with 5′-3′ exonuclease activity in oligo(deoxy)nucleotide degradation. *J. Biol. Chem.* 286 (4): 2807–2816.

83 Mueller, J.W. and Shafqat, N. (2013). Adenosine-5′-phosphosulfate – a multifaceted modulator of bifunctional 3′-phospho-adenosine-5′-phosphosulfate synthases and related enzymes. *FEBS J.* 280 (13): 3050–3057.

84 Patel, S., Martinez-Ripoll, M., Blundell, T.L., and Albert, A. (2002). Structural enzymology of Li^{+}-sensitive/Mg^{2+}-dependent phosphatases. *J. Mol. Biol.* 320 (5): 1087–1094.

85 Tarique, K.F., Rehman, S.A.A., and Gourinath, S. (2014). Structural elucidation of a dual-activity PAP phosphatase-1 from *Entamoeba histolytica* capable of hydrolysing both 3′-phosphoadenosine 5′-phosphate and inositol 1,4-bisphosphate. *Acta Crystallogr. D Biol. Crystallogr.* 70: 2019–2031.

86 Erickson, A.I., Sarsam, R.D., and Fisher, A.J. (2015). Crystal structures of *Mycobacterium tuberculosis* CysQ, with substrate and products bound. *Biochemistry* 54 (45): 6830–6841.

87 Schmier, B.J., Nelersa, C.M., and Malhotra, A. (2017). Structural basis for the bidirectional activity of nanoRNase NrnA. *Sci. Rep.* 7: 11085.

88 Kappler, U. and Dahl, C. (2001). Enzymology and molecular biology of prokaryotic sulfite oxidation. *FEMS Microbiol. Lett.* 203 (1): 1–9.

89 Kim, J.Y., Nakayama, M., Toyota, H. et al. (2016). Structural and mutational studies of an electron transfer complex of maize sulfite reductase and ferredoxin. *J. Biochem.* 160 (2): 101–109.

90 Sugio, T., Suzuki, H., Tanaka, T., and Tano, T. (1993). NADH-dependent sulfite reductase activity in the periplasmic space of *Thiobacillus ferrooxidans*. *Biosci. Biotech. Bioch.* 57 (8): 1357–1359.

91 Murray, D.T., Weiss, K.L., Stanley, C.B. et al. (2021). Small-angle neutron scattering solution structures of NADPH-dependent sulfite reductase. *J. Struct. Biol.* 213 (2): 107724.

92 Branzanic, A.M.V., Ryde, U., and Silaghi-Dumitrescu, R. (2019). Why does sulfite reductase employ siroheme? *Chem. Commun. (Camb.)* 55 (93): 14047–14049.

93 Crane, B.R. and Getzoff, E.D. (1996). The relationship between structure and function for the sulfite reductases. *Curr. Opin. Struct. Biol.* 6 (6): 744–756.

94 Schnell, R., Sandalova, T., Hellman, U. et al. (2005). Siroheme- and [Fe_4-S_4]-dependent NirA from *Mycobacterium tuberculosis* is a sulfite reductase with a covalent Cys-Tyr bond in the active site. *J. Biol. Chem.* 280 (29): 27319–27328.

95 Zhou, J.Z., He, Q., Hemme, C.L. et al. (2011). How sulphate-reducing microorganisms cope with stress: lessons from systems biology. *Nat. Rev. Microbiol.* 9 (6): 452–466.

96 Rabus, R., Venceslau, S.S., Wohlbrand, L. et al. (2015). A post-genomic view of the ecophysiology, catabolism and biotechnological relevance of sulphate-reducing prokaryotes. *Adv. Microb. Physiol.* 66: 55–321.

97 Zhang, Y., Wang, X., Zhen, Y. et al. (2017). Microbial diversity and community structure of sulfate-reducing and sulfur-oxidizing bacteria in sediment cores from the East China Sea. *Front. Microbiol.* 8: 2133.

98 Plugge, C.M., Zhang, W., Scholten, J.C., and Stams, A.J. (2011). Metabolic flexibility of sulfate-reducing bacteria. *Front. Microbiol.* 2: 81.

99 Muyzer, G. and Stams, A.J.M. (2008). The ecology and biotechnology of sulphate-reducing bacteria. *Nat. Rev. Microbiol.* 6 (6): 441–454.

100 Zwain, H.M., Nile, B.K., Faris, A.M. et al. (2020). Modelling of hydrogen sulfide fate and emissions in extended aeration sewage treatment plant using TOXCHEM simulations. *Sci. Rep.* 10 (1): 22209.

101 Stoeva, M.K. and Coates, J.D. (2019). Specific inhibitors of respiratory sulfate reduction: towards a mechanistic understanding. *Microbiology (Reading)* 165 (3): 254–269.

102 Gao, P.K. and Fan, K.Y. (2023). Sulfur-oxidizing bacteria (SOB) and sulfate-reducing bacteria (SRB) in oil reservoir and biological control of SRB: a review. *Arch. Microbiol.* 205 (5): 162.

103 Kushkevych, I., Cejnar, J., Treml, J. et al. (2020). Recent advances in metabolic pathways of sulfate reduction in intestinal bacteria. *Cells* 9 (3): 698.

104 Ruckert, C. (2016). Sulfate reduction in microorganisms-recent advances and biotechnological applications. *Curr. Opin. Microbiol.* 33: 140–146.

105 Wegener, G., Laso-Pérez, R., Orphan, V.J., and Boetius, A. (2022). Anaerobic degradation of alkanes by marine archaea. *Ann. Rev. Microbiol.* 76: 553–577.

106 Chadwick, G.L., Skennerton, C.T., Laso-Pérez, R. et al. (2022). Comparative genomics reveals electron transfer and syntrophic mechanisms differentiating methanotrophic and methanogenic archaea. *PLoS Biol.* 20 (1): e3001508.

107 de Jesus, E.B. and de Andrade Lima, L.R. (2016). Simulation of the inhibition of microbial sulfate reduction in a two-compartment upflow bioreactor subjected to molybdate injection. *Bioprocess Biosyst. Eng.* 39 (8): 1201–1211.

108 Biswas, K.C., Woodards, N.A., Xu, H., and Barton, L.L. (2009). Reduction of molybdate by sulfate-reducing bacteria. *Biometals* 22 (1): 131–139.

109 Zane, G.M., Wall, J.D., and De Leon, K.B. (2020). Novel mode of molybdate inhibition of *Desulfovibrio vulgaris* hildenborough. *Front. Microbiol.* 11: 610455.

110 Parey, K., Fritz, G., Ermler, U., and Kroneck, P.M.H. (2013). Conserving energy with sulfate around 100 °C – structure and mechanism of key metal enzymes in hyperthermophilic *Archaeoglobus fulgidus*. *Metallomics* 5 (4): 302–317.

111 Fritz, G., Roth, A., Schiffer, A. et al. (2002). Structure of adenylylsulfate reductase from the hyperthermophilic *Archaeoglobus fulgidus* at 1.6-Å resolution. *Proc. Natl. Acad. Sci. USA* 99 (4): 1836–1841.

112 Schiffer, A., Fritz, G., Kroneck, P.M.H., and Ermler, U. (2006). Reaction mechanism of the iron-sulfur flavoenzyme adenosine-5′-phosphosulfate reductase based on the structural characterization of different enzymatic states. *Biochemistry* 45 (9): 2960–2967.

113 Chiang, Y.L., Hsieh, Y.C., Fang, J.Y. et al. (2009). Crystal structure of Adenylylsulfate reductase from *Desulfovibrio gigas* suggests a potential self-regulation mechanism involving the C terminus of the β-subunit. *J. Bacteriol.* 191 (24): 7597–7608.

114 Wenk, C.B., Wing, B.A., and Halevy, I. (2018). Electron carriers in microbial sulfate reduction inferred from experimental and environmental sulfur isotope fractionations. *ISME J.* 12 (2): 495–507.

115 Ramos, A.R., Keller, K.L., Wall, J.D., and Pereira, I.A. (2012). The membrane QmoABC complex interacts directly with the dissimilatory adenosine 5′-phosphosulfate reductase in sulfate reducing bacteria. *Front. Microbiol.* 3: 137.

116 Keller, K.L., Rapp-Giles, B.J., Semkiw, E.S. et al. (2014). New model for electron flow for sulfate reduction in *Desulfovibrio alaskensis* G20. *Appl. Environ. Microbiol.* 80 (3): 855–868.

117 Price, M.N., Ray, J., Wetmore, K.M. et al. (2014). The genetic basis of energy conservation in the sulfate-reducing bacterium *Desulfovibrio alaskensis* G20. *Front. Microbiol.* 5: 577.

118 Parey, K., Warkentin, E., Kroneck, P.M., and Ermler, U. (2010). Reaction cycle of the dissimilatory sulfite reductase from *Archaeoglobus fulgidus*. *Biochemistry* 49 (41): 8912–8921.

119 Schiffer, A., Parey, K., Warkentin, E. et al. (2008). Structure of the dissimilatory sulfite reductase from the hyperthermophilic archaeon *Archaeoglobus fulgidus*. *J. Mol. Biol.* 379 (5): 1063–1074.

120 Oliveira, T.F., Vonrhein, C., Matias, P.M. et al. (2008). The crystal structure of *Desulfovibrio vulgaris* dissimilatory sulfite reductase bound to DsrC provides novel insights into the mechanism of sulfate respiration. *J. Biol. Chem.* 283 (49): 34141–34149.

121 Oliveira, T.F., Franklin, E., Afonso, J.P. et al. (2011). Structural insights into dissimilatory sulfite reductases: structure of desulforubidin from *Desulfomicrobium norvegicum*. *Front. Microbiol.* 2: 71.

122 Hatchikian, E.C. and Zeikus, J.G. (1983). Characterization of a new type of dissimilatory sulfite reductase present in *Thermodesulfobacterium commune*. *J. Bacteriol.* 153 (3): 1211–1220.

123 Santos, A.A., Venceslau, S.S., Grein, F. et al. (2015). A protein trisulfide couples dissimilatory sulfate reduction to energy conservation. *Science* 350 (6267): 1541–1545.

124 Neukirchen, S., Pereira, I.A.C., and Sousa, F.L. (2023). Stepwise pathway for early evolutionary assembly of dissimilatory sulfite and sulfate reduction. *ISME J.* 17 (10): 1680–1692.

125 Barbosa, A.C.C., Venceslau, S.S., and Pereira, I.A.C. (2024). DsrMKJOP is the terminal reductase complex in anaerobic sulfate respiration. *Proc. Natl. Acad. Sci. USA* 121 (6): e2313650121.

126 Sweetlove, L.J. and Fernie, A.R. (2018). The role of dynamic enzyme assemblies and substrate channelling in metabolic regulation. *Nat. Commun.* 9 (1): 2136.

127 Thauer, R.K., Kaster, A.K., Seedorf, H. et al. (2008). Methanogenic archaea: ecologically relevant differences in energy conservation. *Nat. Rev. Microbiol.* 6 (8): 579–591.

128 Smith, K.S. and Ingram-Smith, C. (2007). *Methanosaeta*, the forgotten methanogen? *Trends Microbiol.* 15 (4): 150–155.

129 Sharma, P., Bano, A., Singh, S.P. et al. (2023). Different stages of microbial community during the anaerobic digestion of food waste. *J. Food Sci. Technol.* 60 (8): 2079–2091.

130 Liu, Y. and Whitman, W.B. (2008). Metabolic, phylogenetic, and ecological diversity of the methanogenic archaea. *Ann. N. Y. Acad. Sci.* 1125: 171–189.

131 Costa, K.C. and Whitman, W.B. (2023). Model organisms to study methanogenesis, a uniquely archaeal metabolism. *J. Bacteriol.* 205 (8): e0011523.

132 Kurth, J.M., Op den Camp, H.J.M., and Welte, C.U. (2020). Several ways one goal-methanogenesis from unconventional substrates. *Appl. Microbiol. Biotechnol.* 104 (16): 6839–6854.

133 Costa, K.C. and Leigh, J.A. (2014). Metabolic versatility in methanogens. *Curr. Opin. Biotechnol.* 29: 70–75.

134 Buan, N.R. (2018). Methanogens: pushing the boundaries of biology. *Emerg. Top. Life Sci.* 2 (4): 629–646.

135 Wagner, T., Koch, J., Ermler, U., and Shima, S. (2017). Methanogenic heterodisulfide reductase (HdrABC-MvhAGD) uses two noncubane [4Fe-4S] clusters for reduction. *Science* 357 (6352): 699–702.

136 Wagner, T., Ermler, U., and Shima, S. (2016). The methanogenic CO_2 reducing-and-fixing enzyme is bifunctional and contains 46 [4Fe-4S] clusters. *Science* 354 (6308): 114–117.

137 Müller, M.C., Lemaire, O.N., Kurth, J.M. et al. (2024). Differences in regulation mechanisms of glutamine synthetases from methanogenic archaea unveiled by structural investigations. *Commun. Biol.* 7 (1): 111.

138 Müller, A.L., Gu, W.Y., Patsalo, V. et al. (2021). An alternative resource allocation strategy in the chemolithoautotrophic archaeon *Methanococcus maripaludis*. *Proc. Natl. Acad. Sci. USA* 118 (16): e2025854118.

139 Maslać, N., Sidhu, C., Teeling, H., and Wagner, T. (2022). Comparative transcriptomics sheds light on remodeling of gene expression during diazotrophy in the thermophilic methanogen *Methanothermococcus thermolithotrophicus*. *MBio* 13 (6): e0244322.

140 Lefaou, A., Rajagopal, B.S., Daniels, L., and Fauque, G. (1990). Thiosulfate, polythionates and elemental sulfur assimilation and reduction in the bacterial world. *FEMS Microbiol. Lett.* 75 (4): 351–382.

141 Rajagopal, B.S. and Daniels, L. (1986). Investigation of mercaptans, organic sulfides, and inorganic sulfur-compounds as sulfur sources for the growth of methanogenic bacteria. *Curr. Microbiol.* 14 (3): 137–144.

142 Johnson, E.F. and Mukhopadhyay, B. (2005). A new type of sulfite reductase, a novel coenzyme F_{420}-dependent enzyme, from the methanarchaeon *Methanocaldococcus jannaschii*. *J. Biol. Chem.* 280 (46): 38776–38786.

143 Balderston, W.L. and Payne, W.J. (1976). Inhibition of methanogenesis in salt marsh sediments and whole-cell suspensions of methanogenic bacteria by nitrogen oxides. *Appl. Environ. Microbiol.* 32 (2): 264–269.

144 Becker, D.F. and Ragsdale, S.W. (1998). Activation of methyl-SCoM reductase to high specific activity after treatment of whole cells with sodium sulfide. *Biochemistry* 37 (8): 2639–2647.

145 Mahlert, F., Bauer, C., Jaun, B. et al. (2002). The nickel enzyme methyl-coenzyme M reductase from methanogenic archaea: in vitro induction of the nickel-based MCR-ox EPR signals from MCR-red2. *J. Biol. Inorg. Chem.* 7 (4–5): 500–513.

146 Daniels, L., Belay, N., and Rajagopal, B.S. (1986). Assimilatory reduction of sulfate and sulfite by methanogenic bacteria. *Appl. Environ. Microbiol.* 51 (4): 703–709.

147 Liu, Y.C., Beer, L.L., and Whitman, W.B. (2012). Methanogens: a window into ancient sulfur metabolism. *Trends Microbiol.* 20 (5): 251–258.

148 Huber, H., Thomm, M., König, H. et al. (1982). *Methanococcus thermolithotrophicus*, a novel thermophilic lithotrophic methanogen. *Arch. Microbiol.* 132 (1): 47–50.

149 Huang, G.F., Wagner, T., Ermler, U., and Shima, S. (2020). Methanogenesis involves direct hydride transfer from H_2 to an organic substrate. *Nat. Rev. Chem.* 4 (4): 213–221.

150 Shima, S., Huang, G.F., Wagner, T., and Ermler, U. (2020). Structural basis of hydrogenotrophic methanogenesis. *Annu. Rev. Microbiol.* 74 (74): 713–733.

151 Cho, M.K. (2013). Discovery of novel 3′-phosphoadenosine-5′-phosphosulfate (PAPS) reductase from methanarcheon *Methanocaldococcus jannaschii.* MSc thesis. Baylor University.

152 Lee, J.S., White, E., Kim, S.G. et al. (2011). Discovery of a novel adenosine 5′-phosphosulfate (APS) reductase from the methanarcheon *Methanocaldococcus jannaschii. Process Biochem.* 46: 154–161.

153 Lie, T.J., Costa, K.C., Lupa, B. et al. (2012). Essential anaplerotic role for the energy-converting hydrogenase Eha in hydrogenotrophic methanogenesis. *Proc. Natl. Acad. Sci. USA* 109 (38): 15473–15478.

154 Porat, I., Kim, W., Hendrickson, E.L. et al. (2006). Disruption of the operon encoding Ehb hydrogenase limits anabolic CO_2 assimilation in the archaeon *Methanococcus maripaludis. J. Bacteriol.* 188 (4): 1373–1380.

155 Mechold, U., Fang, G., Ngo, S. et al. (2007). YtqI from *Bacillus subtilis* has both oligoribonuclease and pAp-phosphatase activity. *Nucleic Acids Res.* 35 (13): 4552–4561.

156 Schmier, B.J., Nelersa, C.M., and Malhotra, A. (2017). Structural basis for the bidirectional activity of bacillus nanoRNase NrnA. *Sci. Rep.* 7 (1): 11085.

157 Yamagata, A., Kakuta, Y., Masui, R., and Fukuyama, K. (2002). The crystal structure of exonuclease RecJ bound to Mn^{2+} ion suggests how its characteristic motifs are involved in exonuclease activity. *Proc. Natl. Acad. Sci. USA* 99 (9): 5908–5912.

158 Grinter, R. and Greening, C. (2021). Cofactor F_{420}: an expanded view of its distribution, biosynthesis and roles in bacteria and archaea. *FEMS Microbiol. Rev.* 45 (5): fuab021.

159 Bashiri, G., Antoney, J., Jirgis, E.N.M. et al. (2019). A revised biosynthetic pathway for the cofactor F_{420} in prokaryotes. *Nat. Commun.* 10 (1): 1558.

160 Greening, C., Ahmed, F.H., Mohamed, A.E. et al. (2016). Physiology, biochemistry, and applications of F_{420}- and F_o-dependent redox reactions. *Microbiol. Mol. Biol. Rev.* 80 (2): 451–493.

161 Jirapanjawat, T., Ney, B., Taylor, M.C. et al. (2016). The redox cofactor F_{420} protects mycobacteria from diverse antimicrobial compounds and mediates a reductive detoxification system. *Appl. Environ. Microbiol.* 82 (23): 6810–6818.

162 Ney, B., Ahmed, F.H., Carere, C.R. et al. (2017). The methanogenic redox cofactor F_{420} is widely synthesized by aerobic soil bacteria. *ISME J.* 11 (1): 125–137.

163 Vitt, S., Ma, K., Warkentin, E. et al. (2014). The F_{420}-reducing [NiFe]-hydrogenase complex from *Methanothermobacter marburgensis*, the first X-ray structure of a group 3 family member. *J. Mol. Biol.* 426 (15): 2813–2826.

164 Watanabe, T., Pfeil-Gardiner, O., Kahnt, J. et al. (2021). Three-megadalton complex of methanogenic electron-bifurcating and CO_2-fixing enzymes. *Science* 373 (6559): 1151–1156.

165 Johnson, E.F. and Mukhopadhyay, B. (2008). Coenzyme F_{420}-dependent sulfite reductase-enabled sulfite detoxification and use of sulfite as a sole sulfur source by *Methanococcus maripaludis. Appl. Environ. Microbiol.* 74 (11): 3591–3595.

166 Jespersen, M., Pierik, A.J., and Wagner, T. (2023). Structures of the sulfite detoxifying F_{420}-dependent enzyme from *Methanococcales. Nat. Chem. Biol.* 19 (6): 695–702.

167 Sun, F.N., Hu, W.X., Wang, X.L. et al. (2021). Methanogen microfossils and methanogenesis in Permian lake deposits. *Geology* 49 (1): 13–18.

168 Shen, Y., Buick, R., and Canfield, D.E. (2001). Isotopic evidence for microbial sulphate reduction in the early Archaean era. *Nature* 410 (6824): 77–81.

169 Susanti, D. and Mukhopadhyay, B. (2012). An intertwined evolutionary history of methanogenic archaea and sulfate reduction. *PLoS One* 7 (9): e45313.

170 Lyu, Z., Shao, N., Akinyemi, T., and Whitman, W.B. (2018). Methanogenesis. *Curr. Biol.* 28 (13): R727–R732.

171 Mei, R., Kaneko, M., Imachi, H., and Nobu, M.K. (2023). The origin and evolution of methanogenesis and Archaea are intertwined. *PNAS Nexus* 2 (2): pgad023.

172 Schaaf, T., Grünig, J., Schuster, M.R. et al. (2014). Methanation of CO_2 – storage of renewable energy in a gas distribution system. *Energy Sustain. Soc.* 4: 2.

173 Dhillon, A., Goswami, S., Riley, M. et al. (2005). Domain evolution and functional diversification of sulfite reductases. *Astrobiology* 5 (1): 18–29.

174 Crane, B.R., Siegel, L.M., and Getzoff, E.D. (1995). Sulfite reductase structure at 1.6 Å: evolution and catalysis for reduction of inorganic anions. *Science* 270 (5233): 59–67.

175 Hoogendoorn, A. and van Kasteren, H. (2020). *Biological Conversion of Syngas into Methane. Transportation Biofuels: Pathways for Production*. The Royal Society of Chemistry.

176 Martin, M.R., Fornero, J.J., Stark, R. et al. (2013). A single-culture bioprocess of *Methanothermobacter thermautotrophicus* to upgrade digester biogas by CO_2-to-CH_4 conversion with H_2. *Archaea* 2013: 157529.

177 Shakeri Yekta, S., Svensson, B.H., Skyllberg, U., and Schnurer, A. (2023). Sulfide in engineered methanogenic systems – friend or foe? *Biotechnol. Adv.* 69: 108249.

178 Jumper, J., Evans, R., Pritzel, A. et al. (2021). Highly accurate protein structure prediction with AlphaFold *Nature* 596: 583–589. https://doi.org/10.1038/s41586-021-03819-2.

7

The Molybdenum Cofactor, Its Biosynthesis, and Relation to Fe–S Clusters in Bacteria

Paolo Olivieri and Silke Leimkühler

University of Potsdam, Institute of Biochemistry and Biology, Department of Molecular Enzymology, Karl-Liebknecht Str. 24-25, Potsdam 14476, Germany

Chapter Goals

- to provide an overview of the discovery of the molybdenum cofactor (Moco) and how the structure was solved by chemical methods before three-dimensional data of Moco enzymes were available.
- to describe the connection of the Moco and molybdoenzymes with Fe–S cluster assembly.
- to describe the insertion of the cofactors into enzymes and the hierarchy of their insertion.

Main Chapter Points

- Overview of Moco biosynthesis and the involved proteins
- Structural characterization of the Moco, its intermediates, and degradation products by chemical methods
- Role of Fe–S clusters for Moco insertion, catalysis, and regulation

7.1 Molybdenum in Biology

Molybdenum (Mo) has the atomic number 42 and is a trace element that is present relatively in small amounts in the environment, with 2.4 mg/kg in the lithosphere and 0.2–36 mg/kg in soils [1, 2]. Despite its low abundance in the environment, it is an important and essential micronutrient for almost all living organisms except for certain strains of thermophilic anaerobes (archaea and bacteria that grow optimally at temperatures of 50 °C or higher), which require tungsten instead of Mo for survival [3, 4]. It is bioavailable in the form of the oxyanion molybdate MoO_4^{2-} [5]. Molybdate itself is not biologically active, since molybdate needs to be coordinated by either

Iron-Sulfur Clusters: Biogenesis and Biochemistry, Volume 1, First Edition.
Edited by Silke Leimkühler, Günter Schwarz, Oliver Lenz, and Oliver Einsle.

a pyranopterin scaffold called molybdopterin or metal-binding pterin (MPT) or, in the case of nitrogenase, iron–sulfur clusters [6]. The so-formed prosthetic groups are called molybdenum cofactor (Moco) and iron–molybdenum cofactor (FeMoco), respectively [7–9]. Molybdate is incorporated into these metal cofactors by a complex biosynthetic machinery, serving as the active site of molybdenum-requiring enzymes (molybdoenzymes) [2]. Moco is the ubiquitous form of cofactor found in most molybdoenzymes [10], while FeMoco is unique to the bacterial nitrogenase [6]. Molybdoenzymes exist in nearly all living organisms with the exception such as *Saccharomyces cerevisiae* and thermophilic anaerobes [11]. This chapter will focus entirely on the discovery of Moco and Moco-containing enzymes and Moco biosynthesis in bacteria and their relation to Fe–S clusters.

7.2 The Discovery of Molybdoenzymes: Studies of Mutant Strains from *Neurospora crassa* Show the Universality of the Molybdenum Cofactor

It was first reported in 1930 that the Moco is important for living organisms by studies describing that molybdenum is essential for plant growth and further plays an important role for multiple bacterial and animal enzymes [5].

Later, in 1967, Mudd identified sulfite oxidase deficiency in humans [12]. Cohen and Fridovich, in 1971, characterized the enzyme sulfite oxidase as a protein that binds the metal molybdenum, further proving that molybdenum is necessary for human health [13]. There is no treatment known to date for sulfite oxidase deficiency that leads to death in early childhood. These discoveries were underlined by studies showing that the Moco is universal and essential for most living organisms [14, 15]. In 1974, Nason et al. developed an assay that used the extract of *Neurospora crassa* mutant *nit-1* for the reconstitution of its nitrate reductase activity by using crude extracts from other sources [16]. This assay takes advantage of the Moco present in various sources by using the low molecular weight fraction of denatured Mo-enzymes of plant, mammalian, or bacterial origin [8]. After separation, the protein-free Moco fraction was obtained and inserted into the cofactor-free apo-nitrate reductase from the *nit-1* extract of the fungus *N. crassa*, which then reconstituted the enzyme activity of nitrate reductase. Finally, in 1974, these studies led to the understanding that, except for nitrogenase, Moco is the common cofactor that is mainly available in all molybdoenzymes from the majority of bacterial or eukaryotic sources.

7.3 Mutagenesis of Bacteria Help(s) to Discover Biosynthesis Pathways

The genes coding for proteins involved in the biosynthesis of the Moco were identified by studies using mutant strains, first from bacteria and later from other model organisms, that were unable to synthesize the Moco [17]. The mutant strains were identified mainly by using mutants of the bacterial model organism *Escherichia coli* that were resistant to growth on chlorate [18–20]. While chlorate itself is relatively nontoxic, it is enzymatically reduced to the toxic oxidizing agent, chlorite, by nitrate reductase. Mutant strains that are unable to produce an active nitrate reductase are conclusively resistant to chlorate exposure and can grow on a medium containing chlorate, while strains with an active nitrate reductase are unable to grow on a medium containing chlorate because the chlorite they produce will kill the cells. Using this selection system, several *E. coli* mutant strains that were resistant to chlorate were identified; these strains were initially named and abbreviated as *chl* for chlorate resistance (Table 7.1) [17]. Subsequently, it was identified by genetic mapping that only mutations in genes for nitrate reductase, Moco biosynthesis, or molybdate uptake can yield single-step chlorate-resistant mutants [21]. Assignment of mutations to different *chl* loci was based on the genetic map location and phenotypic characteristics (Table 7.1). Nitrate reductase structural genes at the previously designated *chlC* locus have now been renamed as *nar* genes [22].

Some of the mutants with these mutations, formerly *chlC*, now *narG*, *H*, *I*, *K*, and *L*, were only defective in the nitrate reductase activity or its regulatory genes [23]. Several other mutant classes, *chlA*, *B*, *D*, *E*, and *G*, lacked all four known Mo enzymes [19, 20]. The *chlD* and *chlG* mutants are restored to a wild-type phenotype by growth on very high Mo, suggesting that their defect is in Mo transport or processing rather than in MPT synthesis [17, 24–26]. Strains with mutations in *chlB* were postulated to lack a factor (named factor FA) necessary for insertion of the Mo cofactor into the protein [27, 28], and those with mutations in *chlA* and *chlE* were lacking the Moco [20, 29–31]. The *nit-1* reconstitution assay then established a definite role for *chlA*, *E*, *M*, and *N* in MPT biosynthesis. Later, after all *chl* mutants had been well characterized and most of them were assigned to Moco biosynthesis, a new nomenclature was proposed that replaces the *chl* nomenclature and introduces

Table 7.1 Designation of proteins for Moco biosynthesis in bacteria (+ old nomenclature), plants, and humans and their catalyzed reactions.

Bacteria	Humans	Plants	Old name	Reaction catalyzed
MoaA	MOCS1A	Cnx2	chlA	Formation of (3′,8-cH_2GTP) from 5′GTP
MoaC	MOCS1B	Cnx3	chlA	Formation of cPMP from (3′,8-cH_2GTP)
MoaD	MOCS2A	Cnx6	chlM	Sulfur insertion into cPMP, formation of MPT
MoaE	MOCS2B	Cnx7	chlA	Binding of cPMP, formation of MPT
MoeB	MOCS3	Cnx5	chlN	Adenylylation of MoaD
MogA	GEPHYRIN-G	Cnx1-G	chlG	Formation of MPT-AMP
MoeA	GEPHYRIN-E	Cnx1-E	chlG	Molybdate insertion, formation of Mo-MPT
MobA	–	–	chlB	Formation of bis-Mo-MPT, GMP transfer to bis-Mo-MPT, formation of bis-MGD
MocA	–	–	–	CMP transfer to Mo-MPT, formation of MCD

a new nomenclature based on the atomic symbol for molybdenum [32]. The proposed nomenclature was to name all genes with mo- as the first two letters, and the third letter will be determined using the existing fourth letter of the *chl* nomenclature (Table 7.1). In this way, it was intended to avoid confusion and to allow a simplified carry-over from the *chl*- to the *mo*-nomenclature [32].

7.4 Chemical Investigation on the Structure of the Molybdenum Cofactor

In addition to demonstrating that Moco is universal by using the *nit-1* extract assays, it has been discovered that Moco is very labile with a lifetime of only a few minutes after its release from molybdoenzymes. This makes it challenging to perform chemical characterization that requires higher amounts of active Moco. Therefore, structural characterization of Moco was achieved through the analysis of stable degradation products [33]. The oxidation of Moco under aerobic and acidic conditions generated two fluorescent pterin derivatives that were later named Form A and Form B (Figure 7.1) [33].

Both oxidation products of Moco were stable and thus were used for the investigation of the chemical structure of Moco because they were also obtained in sufficient amounts by using sulfite oxidase purified from livers of rats [34, 35]. The first intermediate, named Form A, was obtained by the oxidation of Moco with acidic iodine [36, 37], while the second intermediate, named Form B, was obtained by oxidation of Moco under air (Figure 7.1) [38]. The dephospho analogs of the two Moco derivatives can be distinguished from each other by differences in their absorption and fluorescence spectra [33]. For the purification and characterization of the two derivatives, HPLC in conjunction with sensitive detection methods provided a yield and purity for chemical studies of the samples. Additionally, NMR studies were used for their characterization. The NMR spectrum obtained by using Form A revealed the presence of the oxidized form of the pterin ring. With these data, the first structure of Form A, which is shown in Figure 7.1, has been proposed. However, as Form A is lacking the molybdenum atom, not much insights were gained on the functionality of the Moco. Therefore, it became obvious that the detailed characterization of a second cofactor derivative is mandatory before a detailed structure for Moco could be suggested. The compound selected for further studies was then Form B, the O_2 oxidation product of Moco, which was also obtained in high amounts in a pure form.

Characterization of Form B revealed that it had properties similar to Form A [33]. It was revealed that Form B also harbors an ester linkage that is sensitive to alkaline phosphatase treatment but becomes sensitive to oxidation by periodate only after its dephosphorylation. However, the oxidation with alkaline permanganate did not give rise to pterin-6-carboxylic acid. Later, by using thin-layer cellulose plate chromatography, the permanganate-treated Form B was separated into two fluorescent products, and one of the products corresponding to a derivative of isoxanthopterin (7-oxopterin) containing an anionic substituent [37]. It had been concluded that

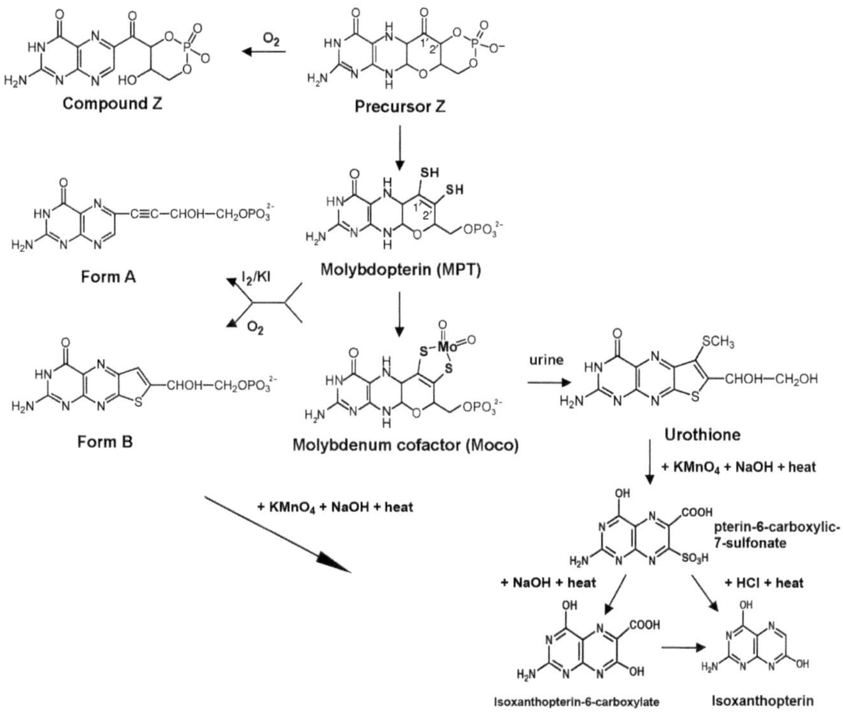

Figure 7.1 Structures of cPMP, MPT, Moco, and their derivatives. Shown are the structures of cyclic pyranopterin monophosphate (cPMP), molybdopterin (MPT), the molybdenum cofactor (Moco), and their degradation products, Form A and Form B. A unique dithiolene group coordinates the molybdenum atom (in Moco), which is shown in its dioxo coordination. Form B is the air-oxidized degradation product of MPT and Moco, while Form A is the iodine-oxidized derivative. Form A and Form B are shown in their phosphorylated forms. Urothione is the degradation product of Moco in humans, which is excreted in the urine. Permanganate oxidation of urothione produces pterin-6-carboxylic-7-sulfonic acid. Heating of urothione in 4 M HCl results in isothanthopterin, or heating in NaOH results in isothanthopterin-6-carboxylic acid that is also obtained by alkaline permanganate treatment of Form B.

this substituent could be a sulfonate group that originated from a sulfur atom present in Form B (Figure 7.1).

An analysis of human urine was reported by Koschara in 1940, which discovered a non-fluorescent sulfur-containing pterin that was named urothione [39]. Later, Goto et al. [40] published the structure of urothione (Figure 7.1). The structural similarities between Form B and urothione were identified by oxidation of urothione with alkaline permanganate that produced an acidic, green-fluorescing compound identified to be pterin-6-carboxylic-7-sulfonate [41, 42]. After heat treatment with 4 M HCl, the 6-carboxylic group was lost and further replacement of the 7-sulfonate group with an –OH finally produced isoxanthopterin (Figure 7.1). On the contrary, heating the sulfonate compound in NaOH gave rise to the formation of isoxanthopterin-6-carboxylate (Figure 7.1). Surprisingly, the treatment of Form B with alkaline permanganate yielded the same sulfonate compound as that obtained from the treatment of urothione, which in Form B had been partially degraded to isoxanthopterin-6-carboxylate (Figure 7.1). Later, mass spectral analyses confirmed a molecular mass difference of 46 Da between dephospho Form B and urothione that corresponded to the presence of a proton instead of the $-SCH_3$ group [43]. The proposed structure of Form B is shown in Figure 7.1. In conclusion, by these studies, it was revealed that Form B and urothione were structurally similar, with the only difference being the $- SCH_3$ group. The identified structural identities between urothione and Form B revealed that a metabolic link must exist in two molecules: an observation that was supported by the fact that in the urine of patients with a defect in Moco biosynthesis, no urothione is present. Therefore, urothione was identified as the degradation product of Moco in the urine. Again, a few decades later, an enzyme required for the biosynthesis of urothione was discovered, *S*-thiopterin methyltransferase [44], a well-known enzyme in drug metabolism of purine analogs used to treat various types of cancer and inflammatory diseases.

Later, based on the chemical studies of the three Moco compounds, Form A, Form B, and urothione, the structure of molybdopterin (MPT) shown in Figure 7.1 was proposed in 1982 [42]. This first proposed structure of MPT consisted of a reduced pterin with a 6-alkyl side chain consisting of four (and not three) carbons, a terminal phosphate ester, and a unique dithiolene group that provides the ligands to bind the molybdenum ion [38]. The structure was basically confirmed by the first crystal structures of several Mo-enzymes from bacterial origin, but the difference revealed the existence of a third pyrano ring between the OH group at C3′ of the side chain and the pterin C7 atom [45, 46]. With the closed pyrano ring, a fully reduced hydrogenated pterin is present. Based on the unique nature of the pterin in Moco, the metal-free form of the cofactor is called molybdopterin or MPT, also referred to as pyranopterin (PPT). These names also demonstrate that pyranopterin scaffold can coordinate not only Mo but also W.

7.4.1 Identification of the First Intermediate, Precursor Z

Studies in the 1980s showed that the oxidation of *E. coli* extracts with acidic iodine, in particular the ones containing deletions in either the *moa*D (*chlM*) or *moeB* (*chlN*) (Table 7.1) genes, did not give rise to Form A. This suggested the absence of MPT in these extracts [47]. More detailed analyses of these *moaD*⁻ and *moeB*⁻ mutant strains revealed the generation of a novel Moco intermediate that was initially named Precursor Z and that could be converted to MPT after its incubation with the cellular fraction of extracts obtained from *moaA*⁻ (*chlA*) mutant cells [48]. First characterization steps of Precursor Z were not possible since the compound was also very labile in the presence of O_2 and was rapidly oxidized. However, the oxidation product was stable and fluorescent, and it was named Compound Z [49]. Using the same method as for the structural determination of Form A and Form B, Compound Z was first structurally characterized in order to relate later its structure to the one of Precursor Z [50]. The combination of chemical and spectral analyses identified Compound Z as being a 6-alkyl pterin containing a 4-carbon side chain. The side chain contained a phosphate moiety as a cyclic phosphate group at C2′ and C4′ of the pterin ring (Figure 7.1). In 2004, Santamaria-Araujo et al. used ^{1}H-NMR analysis to identify that Precursor Z exists in the cyclo pyrano form (Figure 7.1). In this publication, the name cyclic pyranopterin monophosphate (cPMP) was first proposed for Precursor Z [51]. Both names have been used in some publications; however, since cPMP is the more detailed description of the molecule, the term cPMP is the better description of the molecule (Figure 7.1).

7.5 Moco Biosynthesis

Later studies of Compound Z using *moeB*⁻ cells that were grown in a medium containing labeled $^{[14\,C]}$guanosine at the C8 group showed that the label was transferred to the C1′-position of Compound Z [52]. Furthermore, the label present at the C8 and 5 groups in Compound Z purified from cultures grown on $^{[3\,H]}$guanosine was lost after removal of the three terminal side chain carbons [52]. These results identified that the guanosine derivative 5′GTP is the initial substrate for the biosynthesis of Precursor Z. However, in contrast to the synthesis of all other previously studied pterin derivatives (like flavin or tetrahydrofolate) in MPT biosynthesis, the C8 carbon of the initial guanosine precursor is not released but instead incorporated as the first carbon of the MPT side chain. Later, Rieder et al. [53] showed that all five of the guanosine ribose carbons are retained in cPMP and two of them are inserted, which then become the pterin ring with the remaining three building the C2′, C3′, and C4′ groups of the cPMP side chain. In *E. coli*, the *moaA* and *moaC* gene products are responsible for the complicated chemical reactions required to generate cPMP [52, 54] (Figure 7.2). MoaA is a 2× [4Fe–4S] cluster-containing protein [55]. MoaA was identified to belong to the superfamily of radical *S*-adenosyl-L-methionine (SAM) of enzymes and is one of the founding members of this enzyme superfamily that was first defined by Sofia et al. in 2001 [56]. The MoaA product was subsequently characterized as 3′,8-cH_2 GTP (Figure 7.2) [57]. MoaA produces a 5′-dA· radical from SAM that abstracts the H-3′ of GTP to generate a C3′ GTP radical that is then added to C8 of guanine base [57]. The resulting aminyl radical is then reduced by a transfer of a proton and an electron delivered from the second 4Fe–4S cluster which is called the AUX cluster.

The 3′,8-cyclization of purine nucleoside/nucleotide was unprecedented, and the aminyl radical reduction would require a strong reductant. It was suggested that the MoaA active site provides an inert environment for the chemically labile 3′,8-cH_2 GTP intermediate that is transferred to MoaC in the next step and finally converted to the end product cPMP. The structural characterization of MoaC has been used to study its catalytic mechanism. MoaC forms a hexamer composed of a trimer of dimers [58]. The active site is located at the interface of two monomers in each dimer [59]. It has been proposed that the open conformation of MoaC binds the 3′,8-cH_2 GTP intermediate and catalyzes its transformation to another yet unknown intermediate X [56]. Subsequently, MoaC catalyzes the transformation of intermediate X into cPMP. However, some uncertainties remain about the mechanism. Primarily, the structure of intermediate X is not known. More detailed

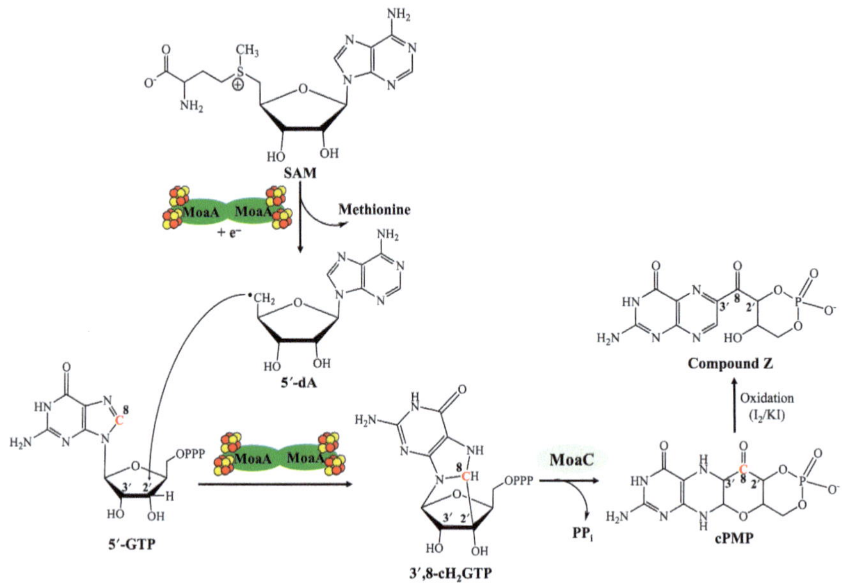

Figure 7.2 The biosynthesis of cPMP from 5′GTP. The conversion of 5′-GTP to cPMP is catalyzed by MoaA and MoaC in bacteria. All carbon atoms of the 5′-GTP are found within cPMP. The C8 atom (colored in red) from the 5′-GTP is inserted between the C2′ and C3′ atoms of the ribose, forming the 3′,8-cH_2GTP intermediate. This reaction is catalyzed by MoaA, which belongs to the superfamily of SAM-dependent enzymes, binding 2× [4Fe–4S] clusters. In the reaction catalyzed by MoaC, the pyrophosphate group is cleaved, forming cPMP. The 2× [4Fe–4S] clusters present on MoaA are shown schematically to highlight them. The air-oxidized product of cPMP is Compound Z.

studies need to resolve in detail the mechanism of conversion of GTP to cPMP (Figure 7.2).

While cPMP is structurally quite similar to MPT, it lacks the dithiolene function essential for molybdenum ligation in Moco. Therefore, for the conversion of cPMP to MPT, two sulfur atoms must be incorporated at the C1′ and C2′ positions of cPMP [23] (Figure 7.3).

In *E. coli*, conversion of cPMP is catalyzed by the protein MPT synthase that forms a heterotetramer. MPT synthase activity was first identified in 1989 by Pitterle et al., who described that the *N. crassa nit-1* strain accumulated cPMP that was then converted to MPT after the incubation with the protein fraction from an *E. coli moaA*$^-$ strain [54]. Later, the purification of the MPT synthase protein revealed a heterotetramer consisting of two small MoaD (~8750 Da) and two larger MoaE (~16,850 Da) protein subunits [53, 60]. Later, Pitterle et al. established an *in vitro* system for MPT synthesis that contained only the enzyme MPT synthase and isolated cPMP [23]. These studies also showed that protein subunit MoaD of MPT synthase contains the sulfur utilized for the generation of MPT [48], which was later confirmed by Gutzke et al. to be a thiocarboxylate group [61]. Electrospray mass spectrometry confirmed a mass difference of 16 Da when active MoaD was compared to inactive MoaD. This difference in mass of 16 Da of the two proteins would

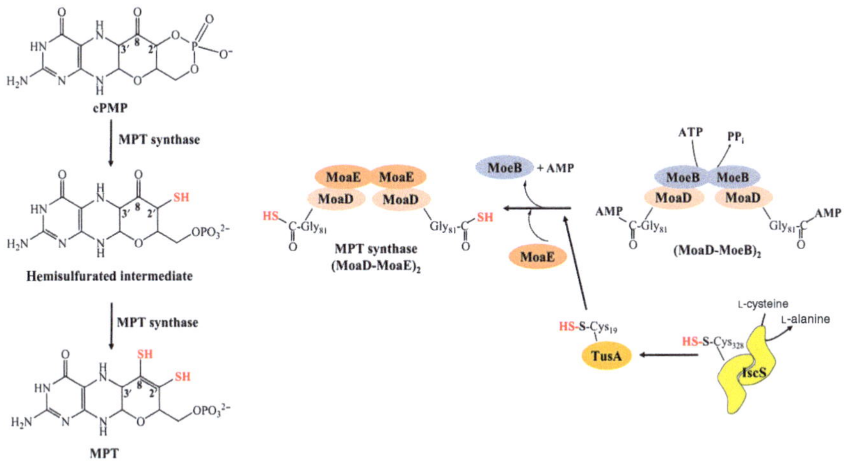

Figure 7.3 The biosynthesis of molybdopterin from cPMP. The MPT synthase tetramer is composed of two MoaE and two MoaD subunits. In the MPT synthase mechanism, cPMP is bound to the MoaE subunit, and sulfur is transferred from the C-terminal thiocarboxylate group of MoaD to cPMP. The initial attack and transfer of the first MoaD-SH sulfur atom occurs at the C2′ position, coupled with the hydrolysis of the Precursor Z cyclic phosphate. A hemisulfurated intermediate is formed. Opening of the cyclic phosphate shifts the location of the intermediate within the complex to a position where C1′ becomes more accessible. A new MoaD thiocarboxylate attacks the C1′ resulting in an intermediate that is converted to MPT. MoaD is regenerated, and a new MoaD-SH thiocarboxylate is formed on MoeB, where MoaD is converted to an MoaD-acyl adenylate under ATP consumption. MoaD-AMP is then sulfurated by a protein-bound persulfide, which is transferred from IscS via TusA to MoaD. After the formation of the thiocarboxylate group, MoaD-SH dissociates from the MoeB dimer and reassociates with MoaE, forming the active MPT synthase.

correspond to the exchange of an oxygen atom by a sulfur atom in MoaD [49, 62]. These results were further confirmed by the first crystal structure of MPT synthase, revealing that the sulfur is present as a thiocarboxylate group at the C-terminal glycine of MoaD [61, 63]. After the reaction, MPT synthase is inactive and to act catalytically, the thiocarboxylate group needs to be regenerated (Figure 7.3) [64].

Since an *E. coli moeB*$^-$ strain is accumulating cPMP, this suggested that the MoeB protein is involved in MPT synthesis and might be involved in the activation of MPT synthase [65]. Therefore, it was proposed that MoeB serves as the protein responsible for regenerating the thiocarboxylate group at the C-terminal glycine of MoaD [64]. This MoeB-dependent activation is ATP dependent [65].

After the formation of the activated MoaD acyl-adenylate in the complex with MoeB, sulfur is inserted to form the terminal MoaD thiocarboxylate group on MoaD-Gly81. It was determined that in *E. coli*, L-cysteine serves as the origin of the MPT dithiolene sulfurs [66] and that the cysteine sulfur is transferred to the activated MoaD acyl-adenylate by the action of a persulfide-containing protein that was identified to be the TusA protein (Figure 7.3) [67]. Initial results showed that this transfer additionally requires the L-cysteine desulfurase IscS, since this protein initially activated sulfur from L-cysteine as a persulfide that can be further transferred via other proteins [68] like TusA, to the final acceptor, cPMP [67].

Summarizing all results on MPT synthesis, a sulfur transfer mechanism for the formation of the MoaD thiocarboxylate group was proposed, which is also similar to the ATP-dependent activation of ubiquitin in protein degradation [69, 70]. In the $(MoaD\text{-}MoeB)_2$ MPT synthase complex, MoeB first activates MoaD and adds an acyladenylate group to the C-terminal glycine, and IscS in its persulfide-bound form with L-cysteine as a sulfur-donor molecule is able to interact with this complex in a reaction that might also involve the TusA-SSH protein as an intermediate sulfur carrier [71]. In this reaction, MoaD is transferred to either the IscS-Cys328 or the TusA-Cys19 persulfide group, forming a disulfide bond. Release of the MoaD thiocarboxylate requires a second thiol group that might be the MoeB-Cys187, but cysteine residues or reducing factors are also possible [64]. After the formation of MPT, molybdate is ligated to the dithiolene group in MPT.

In *E. coli*, insertion of molybdate into the MPT is accomplished by the *moeA* and *mogA* gene products of the *mog* and *moe* loci [72, 73]. The first structure of *E. coli* MogA was published by Liu et al. in 2000 [74]. The crystal structure showed that MogA is a trimer in solution. MogA binds MPT with high affinity [73]. Shortly later, two other crystal structures of *E. coli* MoeA were published in 2001 by Xiang et al. [75] and Schrag et al. [76], revealing that MoeA contains a domain that is structurally related to MogA. Nichols and Rajagopalan [72] reported that the two proteins have distinct functions in the molybdate chelation reaction. While MoeA mediates molybdenum ligation at low concentrations of MoO_4^{2-}. This reaction was strongly inhibited by MogA in the absence of ATP, but in the presence of ATP, MogA doubled the rate of molybdenum ligation [73]. Therefore, MogA is required in an ATP-dependent activation step in the molybdate-ligation reaction. Llamas et al. [77, 78] were the first to verify the catalytic formation of an MPT-AMP intermediate during the reaction (Figure 7.4), which was previously identified in the crystal structure of a Cnx1 variant of the *Arabidopsis thaliana* Cnx1-S583A protein [79].

The accumulation of a comparable MPT-AMP intermediate in *E. coli* $moeA^-$ extracts has also been verified (Leimkühler, unpublished data). Conclusively, MPT is activated by a MogA-mediated adenylylation to MPT-AMP prior to molybdenum insertion by MoeA under physiological molybdenum concentrations, a reaction that can be bypassed in the presence of high molybdenum concentrations *in vitro* [26].

7.6 Differences in the Molybdenum Cofactor of Prokaryotes and Eukaryotes

The initial studies of Moco from various sources using the *N. crassa nit-1* nitrate reductase reconstitution assay led to the hypothesis that Moco is universal and has only one common structure [16]. However, as more purified molybdoenzymes became available and were studied, the possible existence of a different structure of Moco in prokaryotes became more evident. It was reported that the cofactor from formate dehydrogenase of *Methanobacterium formicium* was inactive in the *nit-1* reconstitution assay but was capable of producing Form A after oxidation

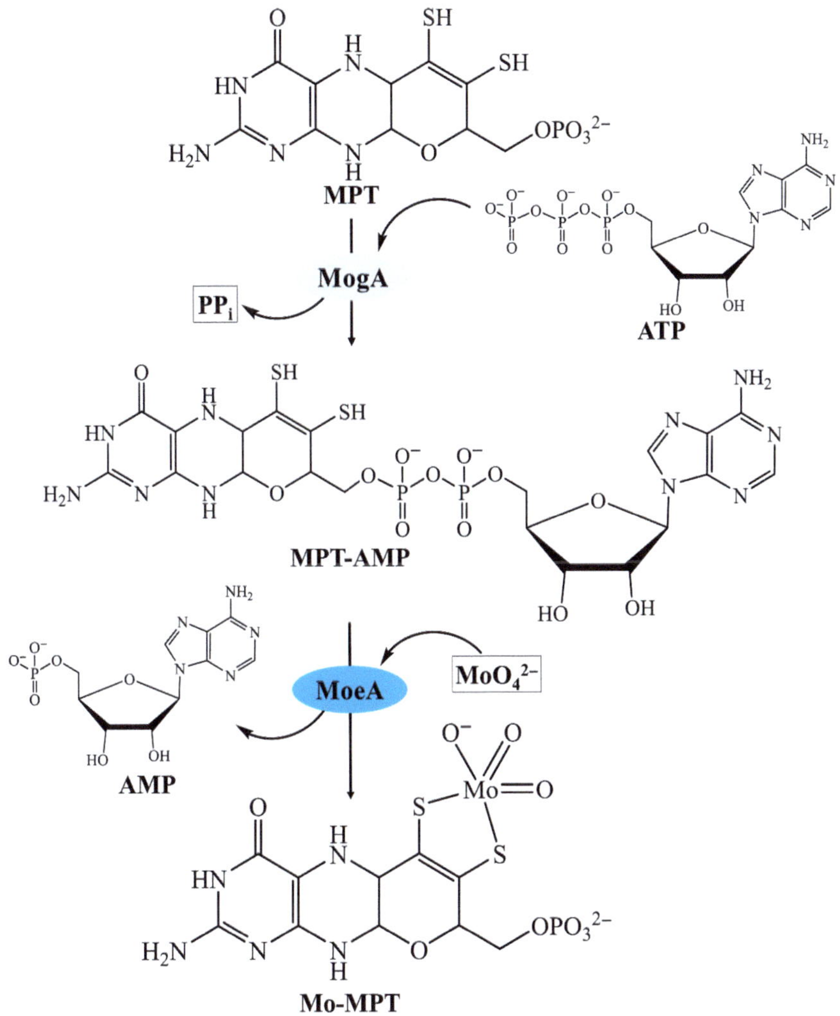

Figure 7.4 Insertion of molybdate into MPT. The bacterial MogA and MoeA proteins catalyze the specific incorporation of molybdate into MPT in a multistep reaction with an adenylated MPT intermediate (MPT-AMP). MogA forms the MPT-adenylate intermediate from Mo-MPT and Mg-ATP and transfers it to MoeA, which mediates molybdate ligation to MPT and cleaves the AMP group.

with acidic iodine and heat treatment [80]. Later, studies by Meyer and coworkers [81, 82] showed that the *Hydrogenophaga pseudoflava* CO dehydrogenase contained a more complex form of the pterin molecule that was named initially bactopterin. The study of alkylated pterin derivatives of molybdoenzymes from various sources, particularly of Moco from dimethylsulfoxide (DMSO) reductase isolated from *Rhodobacter sphaeroides* [83], led to the identification of a new form of MPT referred to as the molybdopterin guanine dinucleotide (MGD) cofactor [83] (Figure 7.5).

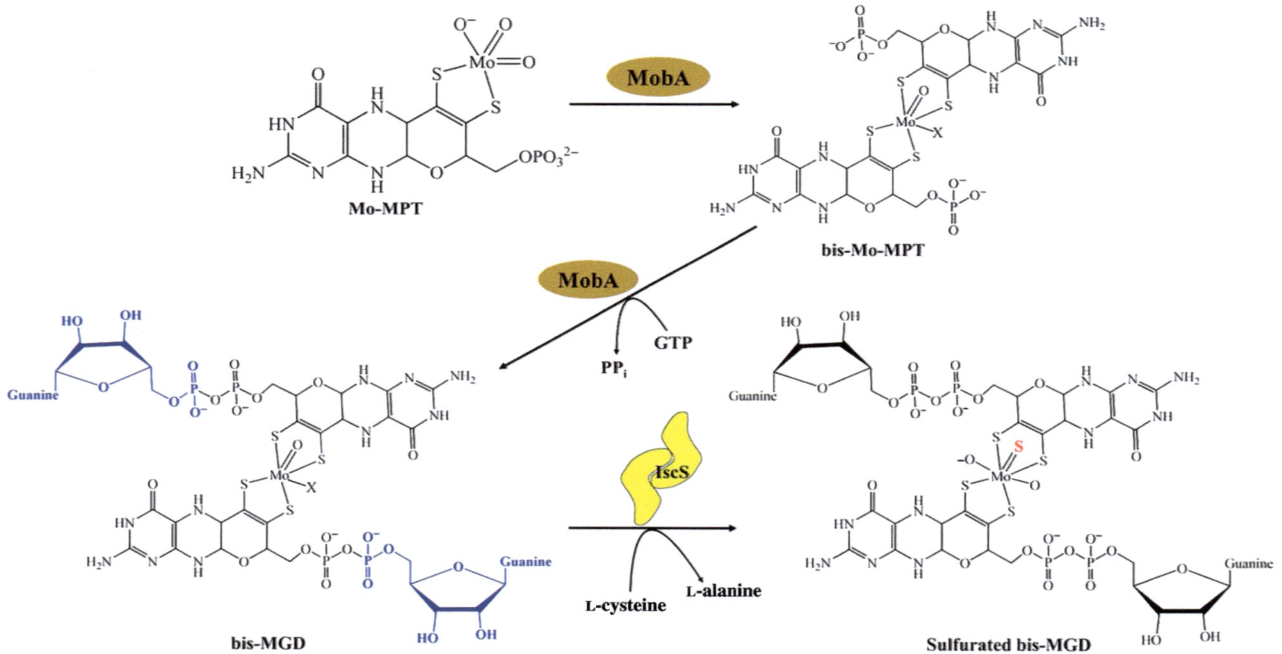

Figure 7.5 Molybdopterin dinucleotide biosynthesis. After its synthesis, either Mo-MPT is inserted into enzymes of the sulfite oxidase family or, alternatively, Mo-MPT can be further modified in prokaryotes to form bis-MGD. In this reaction, first bis-Mo-MPT is formed, a reaction catalyzed by MobA that is converted to Bis-MGD and also by MobA by the addition of two GMP nucleotides. The bis-MGD cofactor is inserted into enzymes of the DMSO reductase family. For enzymes of the xanthine oxidase (XO) family, MCD is formed by the MocA protein, which acts specifically on CTP. Then, MCD is also further modified in *Escherichia coli* by the exchange of the equatorial oxygen to a sulfido ligand, forming sulfurated MCD. Bis-MGD can also be further modified by sulfuration. In the reaction of bis-MGD, sulfuration of IscS is also involved.

Before the identification of YedY in 2004, which was shown to bind either bis-Mo-MPT or bis-W-MPT without any nucleotides [84], it was commonly believed that all *E. coli* molybdoenzymes bind with the bis-MGD form of Moco with two guanine nucleotides [83]. In *E. coli*, MGD is formed by the addition of GMP to the terminal phosphate group of MPT [85]. This reaction is catalyzed by MobA, one of the two proteins encoded by the *mob* locus [86]. While MobA is essential for this reaction [87], the role of MobB of the *mob* locus remains unclear [88].

The structure of MGD was verified when the crystal structure of DMSO reductase was solved from *R. sphaeroides*, but revealed a structure with two MGD molecules, later named bis-MGD [46, 51]. In prokaryotes, including *E. coli*, variants of the cofactor containing other CMP, AMP, or IMP dinucleotide forms of MPT have been identified [80, 89]. In *E. coli*, three enzymes of the xanthine oxidase family were identified, named XdhABC, XdhD, and PaoABC, that contain the MCD form of the cofactor [90, 91]. The existence of MCD in *E. coli* molybdoenzymes was clearly demonstrated by the characterization and structure of the periplasmic aldehyde oxidoreductase PaoABC [92]. Later, the specific protein involved in *E. coli* MCD biosynthesis was identified which was named MocA (for molybdopterin cytosine dinucleotide synthesis) [93] (Figure 7.5). The catalytic reaction of MocA is similar to the reaction of MobA and acts as a CTP:MPT transferase that covalently links MPT and CMP to MCD with the concomitant release of the β- and γ-phosphates of CTP as pyrophosphate. A bis-form of MCD has not been identified to date.

One of the last remaining questions in Moco biosynthesis is why some enzymes bind the dinucleotide variant of the cofactor while others do not. This question has come more into focus with the identification of the bis-Mo-MPT cofactor present in YdhV in *E. coli* [84]. One hypothesis in the past was that the attachment of nucleotides provides a level of regulation by the availability of the *mobAB* gene products for nucleotide insertion [94]. However, this operon was revealed not to be regulated and is rather constitutively expressed [95]. Another suggestion was a stabilizing role of the nucleotides for the bis-MGD-free apo-enzyme [96]. However, even with the guanine moieties present, the apo-enzyme was less stable than the holoenzyme. Therefore, we proposed a model that the nucleotides are inserted first into the apo-molybdoenzyme, which then induces a conformational change that facilitates the binding of the bis-MPT moiety; the charges of the phosphates are also likely crucial for the tight interaction of the bis-MGD in the binding pocket of the molybdoenzyme. Thus, the nucleotides serve as an anchor for bis-MGD insertion [97].

7.7 Discovery of the Sulfido Ligand

In addition to the attachment of GMP or CMP nucleotides to Mo-MPT in prokaryotes, the molybdenum coordination environment can be further modified. The equatorial molybdenum oxygen ligand can be replaced by a sulfur atom present as a sulfido ligand. The result is the sulfurated or mono-oxo form of Moco, as shown in Figures 7.5 and 7.6. Sulfurated Moco is the form of the cofactor and the characteristic of all members of the XO family in prokaryotes and eukaryotes [98–101], and

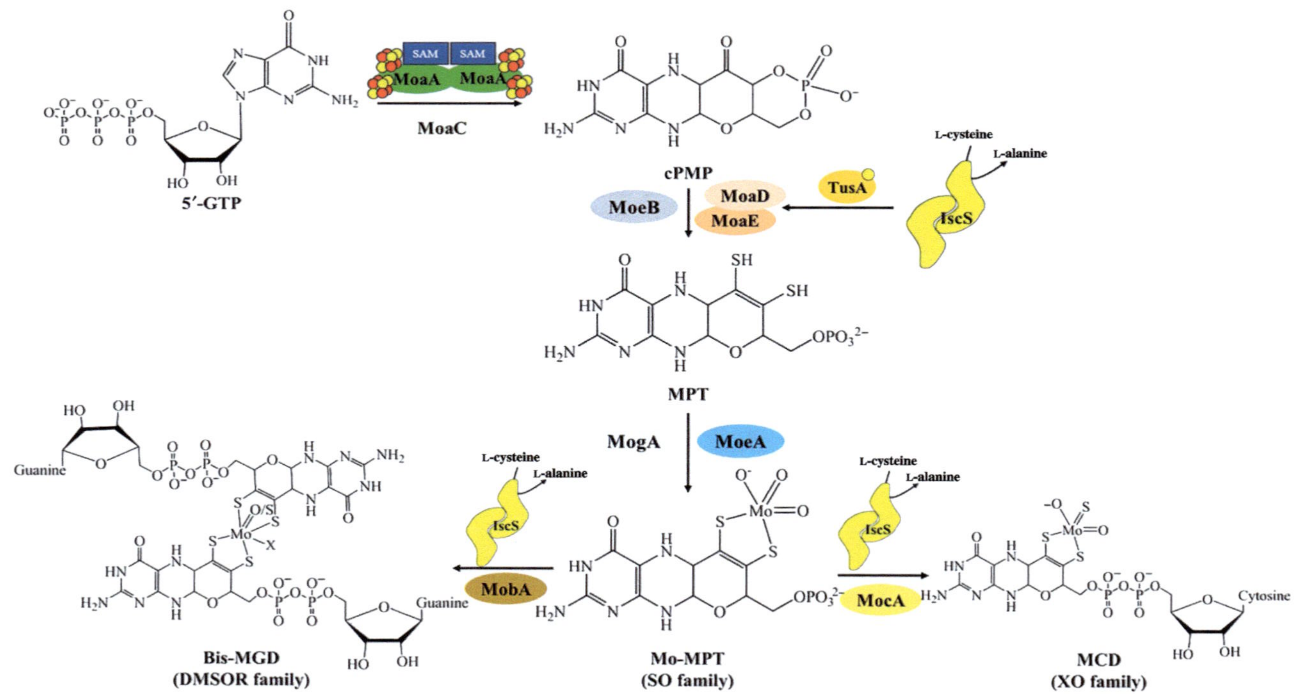

Figure 7.6 Schematic presentation of the biosynthesis for Moco in *E. coli* showing the proteins involved in this pathway. Moco is formed from 5′GTP with cPMP and MPT as stable intermediates. The molybdoenzymes are classified into three different enzyme families, the DMSO reductase family, the sulfite oxidase family, and the xanthine oxidase family, a categorization based on the first coordination sphere of the molybdenum center. The first step of the Moco biosynthesis is the conversion of 5′-GTP to cPMP, catalyzed by MoaA and MoaC in an Fe–S cluster-dependent reaction. In the second step, the MPT-synthase (MoaD-MoaE complex) catalyzes the insertion of two sulfur atoms into the cPMP molecule to form MPT. IscS represents the main sulfur donor, and the sulfur transfer to the MPT-synthase is facilitated by TusA. The third step involves the insertion of molybdate into MPT, catalyzed by MogA and MoeA. The resulting Mo-MPT can be directly inserted into sulfite oxidase (SO) family of molybdoenzymes. In the fourth step, the Mo-MPT can be further modified into MCD by MocA or into bis-MGD by MobA. MCD is used as a cofactor by the xanthine oxidase (XO) family, while bis-MGD is used as a cofactor by the DMSO reductase (DMSOR) family, such as nitrate reductase (NR).

the sulfur is present as sulfido in the equatorial position of the molybdenum atom [100, 102] (Figure 7.6).

Unusual CN inhibition was identified for XO [103], and cyanide is then released as SCN^-, resulting in an inactive form (desulfo form) that is present also naturally in significant amount at least in eukaryotes [98]. Subsequently, it was shown that the sulfur atom is essential for the catalytic activity of bovine milk XO [98] and chicken liver XDH [104] and was definitively shown to be one of the ligands of Mo (Mo = S) by the EPR study [105]. Later, it was revealed that a chemical reconstitution of the sulfido ligand is possible by the incubation of the enzyme with sulfide and/or dithionite under reducing conditions, regaining almost full activity. Fully active bovine milk XOR can be reproducibly isolated by the folate affinity chromatography coupled with allopurinol inhibition [106, 107] and was purified from natural sources of other organisms such as from chicken under particular conditions [104, 108] or bacteria such as *Rhodobacter capsulatus* [109]. The other ligands at the molybdenum atom are the two dithiolene sulfurs from the MPT moiety, an oxo ligand and a hydroxo ligand [110].

More recently, it was identified that the sulfido ligand is also present in enzymes that do not belong to the XO family. The sulfido ligand has also been identified to be present in enzymes from the bacterial DMSO reductase family, like formate dehydrogenase or nitrate reductase [111–113]. Here, this ligand contributes to the oxygen sensitivity of the enzyme, since it is lost upon exposure of the enzyme to oxygen, resulting in an inactive enzyme with oxygen present instead of a sulfido ligand [114]. In FDH, azide stabilizes the sulfido ligand, suggesting that N_3^- binds in the vicinity to the sulfido ligand and hinders thereby the access of O_2 to the active site sulfido group [114]. In *E. coli*, FdhD was reported to be involved in the formation of the essential terminal sulfido ligand at the Mo ion, facilitating the reaction between the L-cysteine desulfurase IscS [115] and FdhF-bound bis-MGD, a mechanism that is essential to yield active FdhF [116, 117]. Initially, it was proposed that the two cysteine residues C121 and C124 in FdhD are involved in the sulfur transfer reaction from IscS further onto the bis-MGD in FdhF [117]. For the sulfuration of bis-MGD, FdhD specifically interacts with IscS in *E. coli*. It has been suggested that IscS transfers the sulfur from L-cysteine to FdhD in the form of a persulfide. Located in a conserved CXXC motif of FdhD, cysteines C121 and C124 were proposed to be involved in the sulfur transfer process from IscS to bis-MGD by Magalon and co-workers, as well as Walburger, Arnoux, and co-workers [117]. FdhD was cocrystallized in complex with GDP, suggesting direct binding of bis-MGD to FdhD [117]; however, it was not possible to prove that the bis-MGD cofactor is bound in an active form [118].

7.8 The Role of Molybdoenzymes in *E. coli*

Respiration is central to the metabolism of many microorganisms. While the properties of molecular oxygen, such as its rapid diffusion across biological membranes and the relatively high redox potential of the O_2/H_2O couple, make it an ideal oxidant for microbial respiration [119]; there are many ecological niches in which oxygen tension is low, or oxygen is absent entirely. In oxygen (O_2) limiting environments, numerous aerobic bacteria have the ability to shift from aerobic to anaerobic respiration to release energy. Anaerobic respiration often uses two such alternative electron acceptors, DMSO and trimethylamine-*N*-oxide (TMAO) [120]. However, the best alternative electron acceptor to oxygen is nitrate (NO_3^-) [119], which has the next best reduction potential after O_2. The main role of molybdoenzymes is to catalyze alternative respiration under anaerobic conditions, using alternative electron acceptors to O_2, such as TMAO, DMSO, or nitrate, by catalyzing nonessential redox reactions [121].

In *E. coli*, the reduction of nitrate can be carried out by at least three enzyme systems that are expressed under different conditions [122]: at high nitrate availability, only the membranous NarGHI system is synthesized [21], and at lower nitrate concentrations, the periplasmic NapAB nitrate reductase is produced (Figure 7.7) [123].

The operon encoding a third system, narZYWV, is expressed during the early stationary phase independent of the presence of nitrate [124]. Furthermore, the formate dehydrogenase FdhGHI is a component of the nitrate respiratory pathway, where, under anaerobic conditions, formate oxidation is coupled to nitrate reduction (NarG) via lipid-soluble quinones [125]. FdhGHI and NarZYV are the corresponding isoenzymes that are additionally present under aerobic conditions to ensure rapid adaptation during a shift from aerobiosis to anaerobiosis [116]. A third formate dehydrogenase is present in *E. coli*, namely FdhF, that is facing the cytoplasm and forms the formate hydrogen lyase (FHL) complex together with the hydrogenase-3 (encoded by hycBCDEFG) (not shown) [126]. The FHL complex is membrane-bound and produces H_2 and CO_2 under fermentative conditions [127]. The three *E. coli* formate dehydrogenases have in common that their catalytic

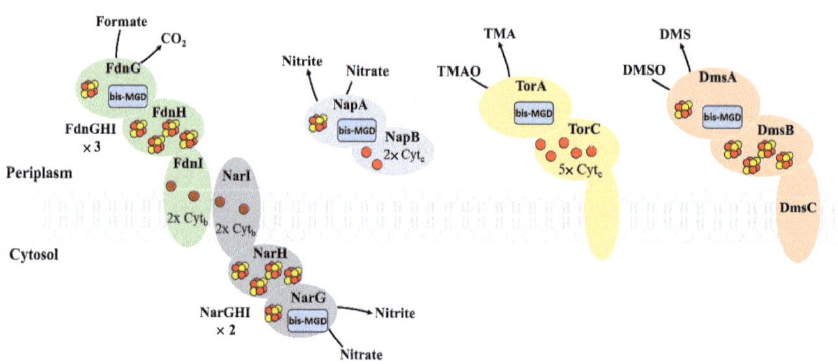

Figure 7.7 Localization of some periplasmic molybdoenzymes in *E. coli*. FdnGHI/NarGHI are components of the nitrate respiratory pathway, in which formate oxidation is coupled to nitrate reduction (NarGHI), with formate being the electron donor of the reaction. The periplasmic nitrate reductase NapAB is produced mainly under nitrate-limiting conditions. In the presence of TMAO, TorCA is produced. TMAO is reduced to TMA (trimethylamine) by TorA that is translocated to the periplasm. The DMSO reductase DmsABC reduces DMSO to DMS and is also present in the periplasm. All enzymes contribute to the anaerobic respiration of *E. coli*. The cofactors present in these proteins are schematically shown: bis-MGD, Fe–S clusters (red and yellow dots), and cytochromes (red dots).

subunits coordinate the bis-MGD cofactor with a selenocysteine ligand and a sulfido ligand, and as an additional cofactor, a [4Fe–4S] cluster is present in the vicinity [128]. FdnGHI expression is induced in the presence of nitrate under anaerobic conditions. In contrast, FdoGHI is also present at low levels under aerobiosis, during fermentative conditions and nitrate respiration.

A further anaerobic molybdoenzyme is DMSO reductase, encoded by *dmsABC* in *E. coli* [129] (Figure 7.7). DMSO reductase reduces DMSO to DMS, and the enzyme from *E. coli* and some other species is an αβγ-heterotrimer that is located in the periplasm and attached to the membrane [130]. *E. coli* DMSO reductase has a broad substrate specificity, being able to reduce S- and N-oxides. Another alternative electron acceptor that is used under aerobic and anaerobic conditions by *E. coli* is TMAO [120]. TMAO reductase is encoded by the *torCAD* operon in *E. coli* [131]. The TorA faces the periplasm and has a high substrate specificity toward TMAO as a substrate, while DMSO is not used (not shown).

7.9 Role of Fe–S Clusters for the Insertion of Moco into Molybdoenzymes

Molybdoenzymes are often composed of different subunits harboring several metal cofactors in addition to Moco, such as cytochromes, Fe–S cluster, or FAD/FMN, that are involved in intramolecular electron transfer reactions [132]. The molybdenum atom exists in various oxidation states under physiological conditions, couples electron and proton transfer, and acts as a transducer between two-electron transfer and one-electron transfer processes to the additional cofactors present in the enzyme.

A well-studied enzyme from the xanthine oxidase family is *R. capsulatus* xanthine dehydrog*enase* [133, 134]. The active form of the enzyme is an $(\alpha\beta)_2$ heterodimer encoded by the *xdhAB* genes, with each of the (aβ) dimers containing two nonidentical [2Fe–2S] clusters, FAD, and the sulfurated form of Mo-MPT as catalytically acting units [135]. The oxidation of xanthine takes place at the molybdenum center, and electrons introduced are rapidly distributed to the other centers according to their relative redox potentials [109]. The reoxidation of the reduced enzyme by the oxidant substrate NAD^+ occurs through FAD. The two [2Fe–2S] clusters (Fe–SI and Fe–SII) are indistinguishable in terms of their absorption spectra, but the midpoint redox potential of Fe–SII is generally more positive than that of the Fe–SI center [136]. The insertion and assembly of *R. capsulatus* XDH has been studied, and it was suggested that insertion of the three different redox centers is a complex process that occurs in an ordered manner [137]. The assembly involves the synthesis of the XdhA and XdhB subunits, the dimerization of both subunits, the insertion of Fe–SI, Fe–SII, and FAD into the XdhA subunit, dimerization of two (αβ) dimers via the XdhB subunit, and, finally, insertion of sulfurated Moco by XdhC into XdhB. The dimerization of the (αβ) subunits is required to stabilize a structure of the protein, which makes the protein suitable for Moco insertion [138–140]. The step of Moco insertion is strictly regulated, since *in vivo* di-oxo Moco without the sulfido ligand is not inserted into *R. capsulatus* XDH. Thus, XdhC has to perform two "quality control" steps: (i) to ensure that Moco is sulfurated before insertion into XDH and (ii) to control that XDH is correctly assembled as an $(\alpha\beta)_2$ heterotetramer that contains the other cofactors before sulfurated Moco is inserted [137]. Since Moco is deeply buried in the protein, it is also believed that XdhC is involved in the proper folding of XDH after Moco insertion [140]. So far, the proteins that specifically insert the Fe–S clusters or FAD into XDH have not been identified in *R. capsulatus*.

Further, several members of the Mo/W-bisPGD superfamily were likely present in the last common universal ancestor (LUCA), with the common presence of an [4Fe–4S] cluster and the Moco in the catalytic subunit, as suggested from both phylogenetics and geochemistry data [141]. This Fe–S cluster is generally called FS0 and is located in the vicinity of the bis-MGD molecule in mostly all enzymes of the DMSO reductase family in bacteria [142, 143]. One exception presents the *E. coli* TorA protein that solely binds bis-MGD without the additional FS0 cluster [144]. The FS0 cluster also contributes to the redox potential of the bis-MGD cofactor and influences the insertion of Bis-MGD into this subunit [145]. A well-studied example presents the cytoplasmically oriented membrane-bound NarGHI nitrate reductase that has been considered as a general model for the assembly of several Mo/W-bisMGD enzymes. This enzyme is widespread in bacteria and archaea and is proposed to have existed in LUCA [146]. The genomic organization typically includes the *narJ* gene within an operon [147], which encodes the three other structural components *narGHI* in the case of this nitrate reductase. The catalytic dimer is composed of a complex between the NarG subunit, which holds the bis-MGD Moco and the His-coordinated [4Fe–4S] cluster (FS0), and the electron transfer subunit NarH, which coordinates three [4Fe–4S] and one [3Fe–4S] clusters [148]. This dimer is anchored to the membrane via the NarI subunit that coordinates two b-type hemes. In the absence of NarJ, a global defect in metal incorporation into NarGHI is associated with the synthesis of an inactive but stable complex mostly attached to the membrane [149]. In the absence of NarJ, not only are the Moco and the Fe–S cluster of the catalytic subunit NarG absent, but also the heme that is positioned at the interface between NarGH and NarI [149–151]. In total, three different types of metal cofactors were not incorporated when the chaperone NarJ was absent. Prior to Moco insertion, NarJ maintains the apoenzyme in a soluble state. However, once attached to the membrane, the apoNarGH complex can no longer incorporate Moco. Here, a second NarJ binding site within the NarG catalytic subunit is responsible for the sequential insertion of the Fe–S cluster (FS0) followed by Mo-bisMGD [149]. Indeed, while the lack of Moco does not preclude FS0 insertion, the absence of NarJ or substitution of cysteine ligands of the FS0 cluster prevents the insertion of both FS0 and Moco into NarG.

In *E. coli* DmsABC, it is the DmsA catalytic subunit that contains the Mo-bisMGD cofactor and an FS0 [4Fe–4S] cluster [129]. The FS0 cluster is the last stepping stone of the electron transfer relay and transfers electrons from FS1 in DmsB to the cofactor in the active site of DmsA [152]. The N-terminal sequence of DmsA contains four highly conserved Cys residues, which form a ferredoxin-like Cys motif that coordinates the FS0 cluster. Maturation of the catalytically competent DmsABC requires efficient cofactor insertion into DmsA [129], association of DmsA and DmsB to form the DmsAB "catalytic dimer" [153], translocation of DmsAB across the cytoplasmic membrane via the tat translocon [154], and anchoring of DmsAB to the membrane through association with DmsC. In the case of DmsABC, the system-specific chaperone DmsD plays a role in targeting DmsAB to the tat translocon and may have additional roles in maturation. DmsD ensures the folding and assembly of both the DmsABC and the two putative selenate reductases. FS0 insertion in the catalytic subunit DmsA is a prerequisite for Mo-bisPGD insertion [155].

The mechanism of FS0 and Mo-bisMGD assembly likely involves [4Fe–4S] cluster insertion, generating a cofactor binding-competent conformation of DmsA. Correctly assembled FS0 dictates a correct conformation that allows subsequent Mo-bisPGD insertion. In the case of DmsABC, DmsD has been shown to bind to the "tat" leader, directing the fully folded cofactor containing DmsAB dimer to the periplasm via the tat translocon without an obligatory role in cofactor insertion. It is therefore notable that NarJ appears to coordinate insertion of both FS0 and Mo-bisMGD [156], whereas DmsD has a more specific role in directing DmsAB to the periplasm via the tat system. It has to be noted that NarGHI is assembled in a tat-independent manner. The maturation of DmsABC is an example in which the apoenzyme can be translocated by the tat translocon across the cytoplasmic membrane and assembled into its correct location in the absence of its Mo-bisPGD cofactor and/or its FS0 [4Fe–4S] cluster. It is therefore notable that NarJ appears to coordinate insertion of both FS0 and Mo-bisMGD and therefore has a distinct role compared to DmsD.

7.10 The Reaction Catalyzed by Molybdoenzymes

In 1990, Holm postulated that all molybdoenzymes catalyze an oxygen atom transfer (OAT) mechanism, in which the oxygen atom from water is transferred to the substrate which is oxidized, or in the opposite direction, from the substrate to yield water; these reactions are coupled to the reversible transfer of two electrons and two protons in the course of the transformation cycles [157]. The electrons are directly transferred to the Mo/W metal ion of the cofactor and the metals cycle between the Mo/W^{IV} and Mo/W^{VI} oxidation states, with Mo/W^{V} being the intermediate state. One exception from this rule has been considered to be formate dehydrogenase, based on a report by Khangulov [158] who investigated product formation in *E. coli* FdhF using GC–MS with ^{13}C-labeled formate in ^{18}O-enriched water, which resulted in the detection of only $^{13}CO_2$ gas without any ^{18}O atoms. Therefore, a "hydride transfer mechanism" was postulated by Hille and coworkers, which considered the binding of formate in the second coordination sphere of the active site of the enzyme without contacting the Mo ion directly [159]. In this proposal, the sulfido ligand abstracts a hydride ion from formate, resulting in a two-electron-reduced intermediate Mo^{IV}-SH, and CO_2 is released. In this mechanism, however, the (seleno-) cysteine would not be involved and the role of the metal ion is only to provide a hydride acceptor by coordinating the sulfido ligand that is then reduced in the following electron transfer reaction. However, in a report by Kumar et al. [160], the OAT mechanism was indeed confirmed for *R. capsulatus* FDH, showing that indeed ^{18}O is detected in CO_2 and that the immediate product is bicarbonate and CO_2 is only produced in the secondary reaction. Once this mechanism is generally accepted, this would mean that indeed all molybdoenzymes catalyze the OAT mechanism.

7.11 The Involvement of Fe–S Clusters in the Regulation of Bacterial Molybdoenzyme Expression

In *E. coli*, O_2 represses all anaerobic respiratory pathways and fermentation, whereas nitrate is the preferred acceptor under anaerobic conditions and represses other anaerobic pathways [119]. The transcriptional regulation is mainly affected by regulators responding to O_2 and nitrate. Regulation by O_2 is mainly affected by the O_2 sensor-regulators FNR and ArcA [119, 161]. FNR and ArcA control different sets of target genes [162]. FNR acts as the major regulator of anaerobic metabolism, and ArcA as the regulator of aerobic metabolism. However, this classification is not strict, and some of the FNR-regulated genes code for enzymes of aerobically required enzymes and *vice versa*. A large variety of FNR-regulated genes are known, which are transcriptionally activated or repressed under anaerobic conditions.

FNR is a dual transcriptional regulator that requires [4Fe–4S] clusters to bind to DNA. It senses oxygen by a functional [4Fe–4S] cluster and activates the transcription of genes involved in anaerobic respiration while the genes involved in aerobic respiration are repressed [163, 164]. The mechanism of its function is well understood and involves transition of [4Fe–4S] clusters into [2Fe–2S] clusters in two steps. FNR is produced under aerobic and anaerobic conditions; however, it is able to bind DNA only under anoxia, in its [4Fe–4S] cluster-bound form. In the presence of oxygen, the [4Fe–4S] cluster is degraded to a [3Fe–4S] cluster and is further transformed into a [2Fe–2S] cluster in the second step. FNR regulates *moaABCDE* and *moeAB* operons [165–167]. It has been shown that FNR regulates the expression of many molybdoenzymes [168]. The only exception is the *torCAD* operon encoding TMAO reductase in *E. coli*, that is not directly regulated by FNR [168]. ArcA was shown to be regulated by the transcriptional factor FNR [168]. Therefore, changes in the iron concentrations during growth influence the availability of ArcA, shown to be directly involved in the regulation of *moeA* and *moeB* gene expression.

A recent study showed that the A-type carrier proteins ErpA and IscA are involved in Fe–S cluster insertion into FNR [169]. Since the genes for Moco biosynthesis are regulated by FNR, the production of Moco is FNR-dependent, and thus, all active molybdoenzymes depend on the transcriptional regulator FNR in an Fe–S cluster-bound form, since without Moco, no active molybdoenzyme will be produced. Since FNR expression depends on ArcA and Fur regulation, Moco biosynthesis is also indirectly regulated by ArcA and Fur [169].

The expression of the genes of the *moaABCDE* operon is regulated at least at two different levels. At the transcriptional level, the regulators ModE and FNR are important to ensure that transcription of the Moco biosynthesis genes only occurs under anaerobic growth conditions, controlled by FNR and under sufficient molybdate availability conditions, controlled by ModE [166]. Additionally, the transcription of the genes of the *moa*-operon underlies a feedback regulation by an accumulation of MPT in the cell via the Moco riboswitch [170]. When MPT accumulates, the riboswitch ensures that transcription of the *moa*-operon is turned off.

Further, the regulation of the *moeAB* operon is surprisingly opposite to that described for the *moaA* operon, since the *moeAB* operon is repressed by FNR and activated by NarL and ArcA [171]. Since both the proteins encoded by the *moaA* operon and the *moeA* operon are required to produce an active Moco, this counteracting regulation by FNR seems unusual but ensures that Moco is produced only under anaerobic conditions. Recently, we were further able to show that ArcA and IscR are major regulators for Moco biosynthesis, since when either of them was overexpressed in *E. coli*, altered Moco levels were obtained [172]. This regulation seems to reveal the bottlenecks of Moco biosynthesis: the first step, which depends on the Fe–S protein MoaA and depends on Fe–S cluster production that is regulated by IscR, and the second step, which depends on MoeB that is activated by ArcA. The fact that an increasing concentration of IscR produces lower Moco levels can be explained by the fact that fewer Fe–S clusters will be produced under these conditions, since IscR is the repressor of the *isc*-operon for Fe–S cluster assembly [172]. Higher ArcA levels, in contrast, will increase MPT production by increasing the levels of MoeA and MoeB [172].

Since FNR is a protein that requires [4Fe–4S] clusters for binding to the DNA, the Fe–S cluster insertion into FNR is consequently crucial for the expression of numerous proteins, including the *moa*-operon for Moco biosynthesis. ErpA and IscA were revealed to be the proteins that insert the [4Fe–4S] cluster into FNR. Here, ErpA had a major role and was able to partially replace the role of IscA in transferring Fe–S clusters to FNR [169].

The important role of ErpA also became obvious while studying the expression of the *narGHIJ* operon encoding nitrate reductase A in *E. coli*. Here, in the absence of ErpA, the *narGHJI* operon is not expressed, which is likely the result of an inactive FNR protein under these conditions in the absence of ErpA [169].

7.12 Conclusions

The biosynthesis of Moco is conserved and important to all organisms, with the exception of some yeast strains. Thus, it seemed that studies of the Moco biosynthetic pathway would have been a relatively easy and determined achievement. However, almost 50 years after the first report of a crystal structure of a Moco-containing enzyme and nearly 45 years after the publication of the first proposed structure for Moco, a number of questions remain about the pathway of Moco biosynthesis and molybdoenzyme catalysis. Now, cryo-EM structures are also available. The catalytic reaction of some molybdoenzymes is still under debate whether they represent an OAT reaction or whether the substrate is only bound in the second coordination sphere. Further, in higher eukaryotes, many individual Moco biosynthesis proteins appear to be involved in several different biosynthetic pathways and perform several roles as a consequence of gene sharing. Gephyrin is a so-called moonlighting protein, but also others like MOCS2B have additional roles in other pathways outside Moco biosynthesis. Although a wide range of variations on the basic Moco structure exists in bacteria, no dinucleotide forms of Moco seem to be present in eukaryotes. The dinucleotide form important for anchoring Moco to the protein was lost during evolution. Additionally, bacteria contain a large variety of molybdoenzymes that catalyze specific, usually nonessential, redox reactions. In humans, only four molybdoenzymes have been identified, and a defect in Moco biosynthesis is lethal due to the essential role of sulfite oxidase. However, the physiological role of mARC and AOX1 and their substrates is still unclear in humans. Further, the dependence of Moco biosynthesis on Fe–S clusters has only been recently discovered and a dependence on several levels of Moco synthesis, regulation, cofactor insertion, and enzymatic activity has been elucidated, which awaits further discovery.

Acknowledgments

This work was funded by the DFG (German Research Foundation) priority program SPP1927 "Fe-S for life."

References

1 Abumrad, N.N. (1984). Molybdenum—is it an essential trace metal? *Bull. N. Y. Acad. Med.* 60: 163–171.

2 Aguilar-Barajas, E., Diaz-Perez, C., Ramirez-Diaz, M.I. et al. (2011). Bacterial transport of sulfate, molybdate, and related oxyanions. *Biometals* 24 (4): 687–707.

3 Hille, R. (2002). Molybdenum and tungsten in biology. *Trends Biochem. Sci.* 27: 360–367.

4 Bevers, L.E. and Hagen, W.R. (2009). One- and two-electron reduction of molybdate reversibly bound to the archaeal tungstate/molybdate transporter WtpA. *Dalton Trans.* 39: 8168–8170.

5 Bortels, H. (1930). Molybdenum as a catalyst in the biological fixation of nitrogen. *Arch Microbial.* 1: 333–342.

6 Schwarz, G., Mendel, R.R., and Ribbe, M.W. (2009). Molybdenum cofactors, enzymes and pathways. *Nature* 460 (7257): 839–847.

7 Mendel, R.R. (2013). Metabolism of molybdenum. *Met. Ions Life Sci.* 12: 503–528.

8 Mendel, R.R. (2013). The molybdenum cofactor. *J. Biol. Chem.* 288 (19): 13165–13172.

9 Hoffman, B.M., Lukoyanov, D., Dean, D.R., and Seefeldt, L.C. (2013). Nitrogenase: a draft mechanism. *Acc. Chem. Res.* 46 (2): 587–595.

10 Hille, R. (1996). The mononuclear Mo enzymes. *Chem. Rev.* 96: 2757–2816.

11 Zhang, Y. and Gladyshev, V.N. (2008). Molybdoproteomes and evolution of molybdenum utilization. *J. Mol. Biol.* 379 (4): 881–899.

12 Mudd, S.H., Irreverre, F., and Laster, L. (1967). Sulfite oxidase deficiency in man: demonstration of the enzymatic defect. *Science* 156 (782): 1599–1602.

13 Cohen, H.J., Fridovich, I., and Rajagopalan, K.V. (1971). Hepatic sulfite oxidase. A functional role for molybdenum. *J. Biol. Chem.* 246: 374–382.

14 Johnson, J.L., Hainline, B.E., Jones, H.P., and Rajagopalan, K.V. (1980). Molybdenum cofactor. In: *The Biology of Nitrogen Fixation. I* (ed. W.E. Newton and W.H. Orme-Johnson), 249–260. Baltimore: University Park Press.

15 Johnson, J.L., Hainline, B.E., and Rajagopalan, K.V. (1980). Characterization of the molybdenum cofactor of sulfite oxidase, xanthine oxidase and nitrate reductase. Identification of a pteridine as a structural component. *J. Biol. Chem.* 255: 1783–1786.

16 Nason, A., Lee, K.-Y., Pan, S.-S., and Erickson, R.H. (1974). Evidence for a molybdenum cofactor common to all molybdenum enzymes based on the *in vitro* assembly of assimilatory NADPH-nitrate reductase using the *Neurospora* mutant *nit-1*. *J. Less Common Metals* 36: 449–459.

17 Stewart, V. and MacGregor, C.H. (1982). Nitrate reductase in *Escherichia coli* K-12: involvement of *chlC*, *chlE*, and *chlG* loci. *J. Bacteriol.* 151: 788–799.

18 Amy, N.K. (1981). Identification of the molybdenum cofactor in chlorate-resistant mutants of *Escherichia coli*. *J. Bacteriol.* 148: 274–282.

19 Boxer, D.H. (1989). Involvement of the chlorate resistance loci and the molybdenum cofactor in the biosynthesis of the *Escherichia coli* respiratory nitrate reductase. In: *Molecular and Genetic Aspects of Nitrate Assimilation* (ed. J.L. Wray and J.R. Kinghorn), 27–39. Oxford: Oxford Science.

20 Giordano, G., Saracino, L., and Grillet, L. (1985). Identification in various chlorate-resistant mutants of a protein involved in the activation of nitrate reductase in the soluble fraction of a *chlA* mutant of *Escherichia coli* K12. *Biochim. Biophys. Acta* 839: 181–190.

21 Stewart, V. (1988). Nitrate respiration in relation to facultative metabolism in enterobacteria. *Microbiol. Rev.* 52: 190–232.

22 Bachmann, B.J. (1990). Linkage map of *Escherichia coli* K-12, edition 8. *Microbiol. Rev.* 54 (2): 130–197.

23 Pitterle, D.M., Johnson, J.L., and Rajagopalan, K.V. (1993). *In vitro* synthesis of molybdopterin from precursor Z using purified converting factor. Role of protein-bound sulfur in formation of the dithiolene. *J. Biol. Chem.* 268: 13506–13509.

24 Glaser, J.H. and DeMoss, J.A. (1971). Phenotypic restoration by molybdate of nitrate reductase activity in *chlD* mutants of *Escherichia coli*. *J. Bacteriol.* 108: 854–860.

25 Sperl, G.T. and DeMoss, J.A. (1975). *chlD* gene function in molybdate activation of nitrate reductase. *J. Bacteriol.* 122: 1230–1238.

26 Joshi, M.S., Johnson, J.L., and Rajagopalan, K.V. (1996). Molybdenum cofactor biosynthesis in *Escherichia coli mod* and *mog* mutants. *J. Bacteriol.* 178: 4310–4312.

27 Boxer, D.H., Low, D.C., Pommier, J., and Giordano, G. (1987). Involvement of a low-molecular weight substance in *in vitro* activation of the molybdoenzyme respiratory nitrate reductase from a *chlB* mutant of *Escherichia coli*. *J. Bacteriol.* 169: 4678–4685.

28 Low, D.C., Pommier, J., Giordano, G., and Boxer, D.H. (1988). Biosynthesis of molybdoenzymes in *E. coli: chlB* is the only chlorate resistance locus required for protein FA activity. *FEMS Microbiol. Lett.* 49: 331–336.

29 Grillet, L. and Giordano, G. (1983). Identification and purification of a protein involved in the activation of nitrate reductase in the soluble fraction of a *chlA* mutant of *Escherichia coli* K12. *Biochim. Biophys. Acta* 749: 115–124.

30 Johnson, M.E. and Rajagopalan, K.V. (1987). Involvement of *chlA*, *E*, *M*, and *N* loci in *Escherichia coli* molybdopterin biosynthesis. *J. Bacteriol.* 169: 117–125.

31 Nohno, T., Kasai, Y., and Saito, T. (1988). Cloning and sequencing of the *Escherichia coli chlEN* operon involved in molybdopterin biosynthesis. *J. Bacteriol.* 170: 4097–4102.

32 Shanmugam, K.T., Stewart, V., Gunsalus, R.P. et al. (1992). Proposed nomenclature for the genes involved in molybdenum metabolism in *Escherichia coli* and *Salmonella typhimurium*. *Mol. Microbiol.* 6: 3452–3454.

33 Johnson, J.L., Hainline, B.E., Rajagopalan, K.V., and Arison, B.H. (1984). The pterin component of the molybdenum cofactor. Structural characterization of two fluorescent derivatives. *J. Biol. Chem.* 259: 5414–5422.

34 Cohen, H.J., Johnson, J.L., and Rajagopalan, K.V. (1974). Molecular basis of the biological function of molybdenum. Developmental patterns of sulfite oxidase and xanthine oxidase in the rat. *Arch. Biochem. Biophys.* 164: 440–446.

35 Johnson, J.L. (1980). The molybdenum cofactor common to nitrate reductase, xanthine dehydrogenase and sulfite oxidase. In: *Molybdenum and Molybdenum-Containing Enzymes* (ed. M.P. Coughlan), 345–383. Oxford: Pergamon Press.

36 Taylor, E.C., Ray, P.S., Darwish, I.S. et al. (1989). Studies on the molybdenum cofactor. Determination of the structure and absolute configuration of Form A. *J. Am. Chem. Soc.* 111: 7664–7665.

37 Taylor, E.C. and Sabb, A.L. (1988). Studies on the molybdenum cofactor: model synthetic routes directed at Form B. *J. Org. Chem.* 53: 5839–5847.

38 Rajagopalan, K.V., Johnson, J.L., and Hainline, B.E. (1982). The pterin of the molybdenum cofactor. *Fed. Proc.* 41: 2608–2612.

39 Koschara, W. (1940). Urothion, ein gelber, schwefelreicher Farbstoff aus Menschenharn. *Hoppe Seylers Z. Physiol. Chem.* 263: 78–79.

40 Goto, M., Sakurai, A., Ohta, K., and Yamakami, H. (1969). Die struktur des urothions. *J. Biochem.* 65: 611–620.

41 Johnson, J.L., Hainline, B.E., and Rajagopalan, K.V. (1983). The pterin component of the molybdenum cofactor (molybdopterin): relationship to urothione. In: *Chemistry and Biology of Pteridines and Folic Acid Derivatives* (ed. J.A. Blair), 1043–1047. Berlin: Walter de Gruyter.

42 Johnson, J.L. and Rajagopalan, K.V. (1982). Structural and metabolic relationship between the molybdenum cofactor and urothione. *Proc. Natl. Acad. Sci. USA* 79: 6856–6860.

43 Taylor, E.C. and Reiter, L.A. (1989). Studies on the molybdenum cofactor. An unequivocal total synthesis of (±)-urothione. *J. Am. Chem. Soc.* 111: 285–291.
44 Pristup, J., Schaeffeler, E., Arjune, S. et al. (2022). Molybdenum cofactor catabolism unravels the physiological role of the drug metabolizing enzyme thiopurine S-methyltransferase. *Clin. Pharmacol. Ther.* 112 (4): 808–816.
45 Kisker, C., Schindelin, H., and Rees, D.C. (1997). Molybdenum-cofactor-containing enzymes: structure and mechanism. *Annu. Rev. Biochem.* 66: 233–267.
46 Schindelin, H., Kisker, C., Hilton, J. et al. (1996). Crystal structure of DMSO reductase: redox-linked changes in molybdopterin coordination. *Science* 272: 1615–1621.
47 Zurick, T.R., Pitterle, D.M., and Rajagopalan, K.V. (1991). Molybdopterin biosynthesis: role of the *chlN* gene product. *FASEB J.* 5: A470.
48 Pitterle, D.M., Johnson, J.L., and Rajagopalan, K.V. (1990). Molybdopterin formation by converting factor of *E. coli chlA1. FASEB J.* 4: A1957.
49 Pitterle, D.M. and Rajagopalan, K.V. (1993). The biosynthesis of molybdopterin in *Escherichia coli*. Purification and characterization of the converting factor. *J. Biol. Chem.* 268: 13499–13505.
50 Johnson, J.L., Wuebbens, M.M., and Rajagopalan, K.V. (1989). The structure of a molybdopterin precursor. Characterization of a stable, oxidized derivative. *J. Biol. Chem.* 264: 13440–13447.
51 Santamaria-Araujo, J.A., Fischer, B., Otte, T. et al. (2004). The tetrahydropyranopterin structure of the sulfur- and metal-free molybdenum cofactor precursor. *J. Biol. Chem.* 279: 15994–15999.
52 Wuebbens, M.M. and Rajagopalan, K.V. (1995). Investigation of the early steps of molybdopterin biosynthesis in *Escherichia coli* through the use of *in vivo* labeling studies. *J. Biol. Chem.* 270: 1082–1087.
53 Rieder, C., Eisenreich, W., O'Brien, J. et al. (1998). Rearrangement reactions in the biosynthesis of molybdopterin. An NMR study with multiply $^{13}C/^{15}N$ labelled precursors. *Eur. J. Biochem.* 255: 24–36.
54 Wuebbens, M.M. and Rajagopalan, K.V. (1993). Structural characterization of a molybdopterin precursor. *J. Biol. Chem.* 268: 13493–13498.
55 Hänzelmann, P. and Schindelin, H. (2004). Crystal structure of the S-adenosylmethionine-dependent enzyme MoaA and its implications for molybdenum cofactor deficiency in humans. *Proc. Natl. Acad. Sci. USA* 101 (35): 12870–12875.
56 Yokoyama, K., Li, D., and Pang, H. (2022). Resolving the multidecade-long mystery in MoaA radical SAM enzyme reveals new opportunities to tackle human health problems. *ACS Bio. Med. Chem. Au.* 2 (2): 94–108.
57 Hover, B.M., Loksztejn, A., Ribeiro, A.A., and Yokoyama, K. (2013). Identification of a cyclic nucleotide as a cryptic intermediate in molybdenum cofactor biosynthesis. *J. Am. Chem. Soc.* 135 (18): 7019–7032.
58 Wuebbens, M.M., Liu, M.T., Rajagopalan, K., and Schindelin, H. (2000). Insights into molybdenum cofactor deficiency provided by the crystal structure

of the molybdenum cofactor biosynthesis protein MoaC. *Structure Fold Des.* 8 (7): 709–718.

59 Hover, B.M., Tonthat, N.K., Schumacher, M.A., and Yokoyama, K. (2015). Mechanism of pyranopterin ring formation in molybdenum cofactor biosynthesis. *Proc. Natl. Acad. Sci. USA* 112 (20): 6347–6352.

60 Pitterle, D.M. and Rajagopalan, K.V. (1991). Purification and characterization of the converting factor from *E. coli chlA1*. *FASEB J.* 5: A468.

61 Gutzke, G., Fischer, B., Mendel, R.R., and Schwarz, G. (2001). Thiocarboxylation of molybdopterin synthase provides evidence for the mechanism of dithiolene formation in metal-binding pterins. *J. Biol. Chem.* 276: 36268–36274.

62 Pitterle, D.M. and Rajagopalan, K.V. (1989). Two proteins encoded at the *chlA* locus constitute the converting factor of *Escherichia coli chlA1*. *J. Bacteriol.* 171: 3373–3378.

63 Rudolph, M.J., Wuebbens, M.M., Turque, O. et al. (2003). Structural studies of molybdopterin synthase provide insights into its catalytic mechanism. *J. Biol. Chem.* 278: 14514–14522.

64 Leimkühler, S., Wuebbens, M.M., and Rajagopalan, K.V. (2001). Characterization of *Escherichia coli* MoeB and its involvement in the activation of MPT synthase for the biosynthesis of the molybdenum cofactor. *J. Biol. Chem.* 276: 34695–34701.

65 Lake, M.W., Wuebbens, M.M., Rajagopalan, K.V., and Schindelin, H. (2001). Mechanism of ubiquitin activation revealed by the structure of a bacterial MoeB-MoaD complex. *Nature* 414: 325–329.

66 Leimkühler, S. and Rajagopalan, K.V. (2001). A sulfurtransferase is required in the transfer of cysteine sulfur in the in vitro synthesis of molybdopterin from precursor Z in *Escherichia coli*. *J. Biol. Chem.* 276 (25): 22024–22031.

67 Dahl, J.U., Radon, C., Buhning, M. et al. (2013). The sulfur carrier protein TusA has a pleiotropic role in *Escherichia coli* that also affects molybdenum cofactor biosynthesis. *J. Biol. Chem.* 288 (8): 5426–5442.

68 Leimkuhler, S., Buhning, M., and Beilschmidt, L. (2017). Shared sulfur mobilization routes for tRNA thiolation and molybdenum cofactor biosynthesis in prokaryotes and eukaryotes. *Biomolecules* 7 (1): 5.

69 Matthies, A., Nimtz, M., and Leimkühler, S. (2005). Molybdenum cofactor biosynthesis in humans: identification of a persulfide group in the rhodanese-like domain of MOCS3 by mass spectrometry. *Biochemistry* 44 (21): 7912–7920.

70 Matthies, A., Rajagopalan, K.V., Mendel, R.R., and Leimkühler, S. (2004). Evidence for the physiological role of a rhodanese-like protein for the biosynthesis of the molybdenum cofactor in humans. *Proc. Natl. Acad. Sci. USA* 101 (16): 5946–5951.

71 Leimkühler, S. and Rajagopalan, K.V. (2001). An *Escherichia coli* NifS-like sulfurtransferase is required for the transfer of cysteine sulfur in the *in vitro* synthesis of molybdopterin from precursor Z. *J. Biol. Chem.* 276: 22024–22031.

72 Nichols, J. and Rajagopalan, K.V. (2002). *Escherichia coli* MoeA and MogA. Function in metal incorporation step of molybdenum cofactor biosynthesis. *J. Biol. Chem.* 277: 24995–25000.

73 Nichols, J.D. and Rajagopalan, K.V. (2005). In vitro molybdenum ligation to molybdopterin using purified components. *J. Biol. Chem.* 280 (9): 7817–7822.

74 Liu, M.T., Wuebbens, M.M., Rajagopalan, K.V., and Schindelin, H. (2000). Crystal structure of the gephyrin-related molybdenum cofactor biosynthesis protein MogA from *Escherichia coli*. *J. Biol. Chem.* 275: 1814–1822.

75 Xiang, S., Nichols, J., Rajagopalan, K.V., and Schindelin, H. (2001). The crystal structure of *Escherichia coli* MoeA and its relationship to the multifunctional protein gephyrin. *Structure* 9: 299–310.

76 Schrag, J.D., Huang, W., Sivaraman, J. et al. (2001). The crystal structure of *Escherichia coli* MoeA, a protein from the molybdopterin synthesis pathway. *J. Mol. Biol.* 310: 419–431.

77 Llamas, A., Mendel, R.R., and Schwarz, G. (2004). Synthesis of adenylated molybdopterin: an essential step for molybdenum insertion. *J. Biol. Chem.* 279 (53): 55241–55246.

78 Llamas, A., Otte, T., Multhaup, G. et al. (2006). The mechanism of nucleotide-assisted molybdenum insertion into molybdopterin. A novel route toward metal cofactor assembly. *J. Biol. Chem.* 281 (27): 18343–18350.

79 Kuper, J., Llamas, A., Hecht, H.J. et al. (2004). Structure of the molybdopterin-bound Cnx1G domain links molybdenum and copper metabolism. *Nature* 430 (7001): 803–806.

80 Börner, G., Karrasch, M., and Thauer, R.K. (1991). Molybdopterin adenine dinucleotide and molybdopterin hypoxanthine dinucleotide in formylmethanofuran dehydrogenase from *Methanobacterium thermoautotrophicum* (Marburg). *FEBS Lett.* 290: 31–34.

81 Krüger, B. and Meyer, O. (1987). Structural elements of bactopterin from *Pseudomonas carboxydoflava* carbon monoxide dehydrogenase. *Biochim. Biophys. Acta* 912: 357–364.

82 Krüger, B. and Meyer, O. (1986). The pterin (bactopterin) of carbon monoxide dehydrogenase from *Pseudomonas carboxydoflava*. *Eur. J. Biochem.* 157: 121–128.

83 Hilton, J.C. and Rajagopalan, K.V. (1996). Identification of the molybdenum cofactor of dimethyl sulfoxide reductase from *Rhodobacter sphaeroides* f. sp. *denitrificans* as bis(molybdopterin guanine dinucleotide)molybdenum. *Arch. Biochem. Biophys.* 325: 139–143.

84 Reschke, S., Duffus, B.R., Schrapers, P. et al. (2019). Identification of YdhV as the first molybdoenzyme binding a bis-Mo-MPT cofactor in *Escherichia coli*. *Biochemistry* 58 (17): 2228–2242.

85 Neumann, M., Seduk, F., Iobbi-Nivol, C., and Leimkuhler, S. (2011). Molybdopterin dinucleotide biosynthesis in *Escherichia coli*: identification of amino acid residues of molybdopterin dinucleotide transferases that determine specificity for binding of guanine or cytosine nucleotides. *J. Biol. Chem.* 286 (2): 1400–1408.

86 Palmer, T., Goodfellow, I.P., Sockett, R.E. et al. (1998). Characterisation of the *mob* locus from *Rhodobacter sphaeroides* required for molybdenum cofactor biosynthesis. *Biochem. Biophys. Acta* 1395: 135–140.

87 Palmer, T., Vasishta, A., Whitty, P.W., and Boxer, D.H. (1994). Isolation of protein FA, a product of the *mob* locus required for molybdenum cofactor biosynthesis in *Escherichia coli*. *Eur. J. Biochem.* 222: 687–692.

88 McLuskey, K., Harrison, J.A., Schuttelkopf, A.W. et al. (2003). Insight into the role of *Escherichia coli* MobB in molybdenum cofactor biosynthesis based on the high resolution crystal structure. *J. Biol. Chem.* 278 (26): 23706–23713.

89 Meyer, O., Frunzke, K., Tachil, J., and Volk, M. (1993). The bacterial molybdenum cofactor. In: *Molybdenum Enzymes, Cofactors, and Model Systems* (ed. E.I. Stiefel, D. Coucouvanis, and W.E. Newton), 50–68. Washington, DC: American Chemical Society.

90 Correia, M.A., Otrelo-Cardoso, A.R., Schwuchow, V. et al. (2016). The *Escherichia coli* periplasmic aldehyde oxidoreductase is an exceptional member of the xanthine oxidase family of molybdoenzymes. *ACS Chem. Biol.* 11 (10): 2923–2935.

91 Leimkuhler, S. (2020). The biosynthesis of the molybdenum cofactors in *Escherichia coli*. *Environ. Microbiol.* Jun; 22 (6): 2007–2026. doi: 10.1111/1462-2920.15003. Epub 2020 Apr 6. PMID: 32239579.

92 Neumann, M., Mittelstadt, G., Iobbi-Nivol, C. et al. (2009). A periplasmic aldehyde oxidoreductase represents the first molybdopterin cytosine dinucleotide cofactor containing molybdo-flavoenzyme from *Escherichia coli*. *FEBS J.* 276 (10): 2762–2774.

93 Neumann, M., Mittelstadt, G., Seduk, F. et al. (2009). MocA is a specific cytidylyltransferase involved in molybdopterin cytosine dinucleotide biosynthesis in *Escherichia coli*. *J. Biol. Chem.* 284 (33): 21891–21898.

94 Leimkühler, S. and Klipp, W. (1999). The molybdenum cofactor biosynthesis protein MobA from *Rhodobacter capsulatus* is required for the activity of molybdenum enzymes containing MGD, but not for xanthine dehydrogenase harboring the MPT cofactor. *FEMS Microbiol. Lett.* 174 (2): 239–246.

95 Solomon, P.S., Shaw, A.L., Lane, I. et al. (1999). Characterization of a molybdenum cofactor biosynthetic gene cluster in *Rhodobacter capsulatus* which is specific for the biogenesis of dimethylsulfoxide reductase. *Microbiology* 145 (Pt 6): 1421–1429.

96 Temple, C.A. and Rajagopalan, K.V. (2000). Mechanism of assembly of the bis(molybdopterin guanine dinucleotide)molybdenum cofactor in *Rhodobacter sphaeroides* dimethyl sulfoxide reductase. *J. Biol. Chem.* 275 (51): 40202–40210.

97 Tiedemann, K., Iobbi-Nivol, C., and Leimkuhler, S. (2022). The role of the nucleotides in the insertion of the bis-molybdopterin guanine dinucleotide cofactor into apo-molybdoenzymes. *Molecules* (9): 27, 2993.

98 Massey, V. and Edmondson, D. (1970). On the mechanism of inactivation of xanthine oxidase by cyanide. *J. Biol. Chem.* 245: 6595–6598.

99 Wahl, R.C., Hageman, R.V., and Rajagopalan, K.V. (1984). The relationship of Mo, molybdopterin, and the cyanolyzable sulfur in the molybdenum cofactor. *Arch. Biochem. Biophys.* 230: 264–273.

100 Wahl, R.C. and Rajagopalan, K.V. (1982). Evidence for the inorganic nature of the cyanolyzable sulfur of molybdenum hydroxylases. *J. Biol. Chem.* 257: 1354–1359.

101 Wahl, R.C., Warner, C.K., Finnerty, V., and Rajagopalan, K.V. (1982). *Drosophila melanogaster ma-l* mutants are defective in the sulfuration of desulfo Mo hydroxylases. *J. Biol. Chem.* 257: 3958–3962.

102 Okamoto, K., Matsumoto, K., Hille, R. et al. (2004). The crystal structure of xanthine oxidoreductase during catalysis: implications for reaction mechanism and enzyme inhibition. *Proc. Natl. Acad. Sci. USA* 101 (21): 7931–7936.

103 Dixon, M. and Thurlow, S. (1924). Studies on xanthine oxidase. *Biochem. J.* 18: 976–988.

104 Nishino, T. (1974). Purification of hepatic xanthine dehydrogenase from chicken fed a high- protein diet. *Biochim. Biophys. Acta* 341 (1): 93–98.

105 Gutteridge, S., Tanner, S.J., and Bray, R.C. (1978). Comparison of the molybdenum centres of native and desulpho xanthine oxidase. The nature of the cyanide-labile sulphur atom and the nature of the proton-accepting group. *Biochem. J.* 175 (3): 887–897.

106 Nishino, T., Nishino, T., and Tsushima, K. (1981). Purification of highly active milk xanthine oxidase by affinity chromatography on sepharose 4B/folate gel. *FEBS Lett.* 131: 369–372.

107 Nishino, T. (1986). Reversible interconversion between sulfo and desulfo xanthine dehydrogenase. *Adv. Exp. Med. Biol.* 195 (Pt B): 259–262.

108 Itoh, R., Nishino, T., Usami, C., and Tsushima, K. (1978). An immunochemical study of the changes in chicken liver xanthine dehydrogenase activity during dietary adaptation. *J. Biochem. (Tokyo)* 84 (1): 19–26.

109 Leimkühler, S., Stockert, A.L., Igarashi, K. et al. (2004). The role of active site glutamate residues in catalysis of *Rhodobacter capsulatus* xanthine dehydrogenase. *J. Biol. Chem.* 279: 40437–40444.

110 Enroth, C., Eger, B.T., Okamoto, K. et al. (2000). Crystal structures of bovine milk xanthine dehydrogenase and xanthine oxidase: structure-based mechanism of conversion. *Proc. Natl. Acad. Sci. USA* 97 (20): 10723–10728.

111 Hartmann, T. and Leimkühler, S. (2013). The oxygen-tolerant and NAD-dependent formate dehydrogenase from *Rhodobacter capsulatus* is able to catalyze the reduction of CO_2 to formate. *FEBS J.* 280 (23): 6083–6096.

112 Coelho, C., Gonzalez, P.J., Moura, J.G. et al. (2011). The crystal structure of *Cupriavidus necator* nitrate reductase in oxidized and partially reduced states. *J. Mol. Biol.* 408 (5): 932–948.

113 Duffus, B.R., Schrapers, P., Schuth, N. et al. (2020). Anion binding and oxidative modification at the molybdenum cofactor of formate dehydrogenase from *Rhodobacter capsulatus* studied by X-ray absorption spectroscopy. *Inorg. Chem.* 59 (1): 214–225.

114 Laun, K., Duffus, B.R., Wahlefeld, S. et al. (2022). Infrared spectroscopy elucidates the inhibitor binding sites in a metal-dependent formate dehydrogenase. *Chemistry* 28 (54): e202201091.

115 Thome, R., Gust, A., Toci, R. et al. (2012). A sulfurtransferase is essential for activity of formate dehydrogenases in *Escherichia coli*. *J. Biol. Chem.* 287 (7): 4671–4678.

116 Abaibou, H., Pommier, J., Benoit, S. et al. (1995). Expression and characterization of the *Escherichia coli* fdo locus and a possible physiological role for aerobic formate dehydrogenase. *J. Bacteriol.* 177 (24): 7141–7149.

117 Arnoux, P., Ruppelt, C., Oudouhou, F. et al. (2015). Sulphur shuttling across a chaperone during molybdenum cofactor maturation. *Nat. Commun.* 6: 6148.

118 Schwanhold, N., Iobbi-Nivol, C., Lehmann, A., and Leimkuhler, S. (2018). Same but different: comparison of two system-specific molecular chaperones for the maturation of formate dehydrogenases. *PLoS One* 13 (11): e0201935.

119 Unden, G., Becker, S., Bongaerts, J. et al. (1995). O_2-sensing and O_2-dependent gene regulation in facultatively anaerobic bacteria. *Arch. Microbiol.* 164 (2): 81–90.

120 McCrindle, S.L., Kappler, U., and McEwan, A.G. (2005). Microbial dimethylsulfoxide and trimethylamine-N-oxide respiration. *Adv. Microb. Physiol.* 50: 147–198.

121 Leimkühler, S. and Iobbi-Nivol, C. (2016). Bacterial molybdoenzymes: old enzymes for new purposes. *FEMS Microbiol. Rev.* 40 (1): 1–18.

122 Cole, J. (1996). Nitrate reduction to ammonia by enteric bacteria: redundancy, or a strategy for survival during oxygen starvation? *FEMS Microbiol. Lett.* 136 (1): 1–11.

123 Brondijk, T.H., Nilavongse, A., Filenko, N. et al. (2004). NapGH components of the periplasmic nitrate reductase of *Escherichia coli* K-12: location, topology and physiological roles in quinol oxidation and redox balancing. *Biochem. J.* 379 (Pt 1): 47–55.

124 Chang, L., Wei, L.I., Audia, J.P. et al. (1999). Expression of the *Escherichia coli* NRZ nitrate reductase is highly growth phase dependent and is controlled by RpoS, the alternative vegetative sigma factor. *Mol. Microbiol.* 34 (4): 756–766.

125 Jormakka, M., Richardson, D., Byrne, B., and Iwata, S. (2004). Architecture of NarGH reveals a structural classification of Mo-bisMGD enzymes. *Structure* 12 (1): 95–104.

126 Sawers, G. and Suppmann, B. (1992). Anaerobic induction of pyruvate formate-lyase gene expression is mediated by the ArcA and FNR proteins. *J. Bacteriol.* 174 (11): 3474–3478.

127 Rossmann, R., Sawers, G., and Bock, A. (1991). Mechanism of regulation of the formate-hydrogenlyase pathway by oxygen, nitrate, and pH: definition of the formate regulon. *Mol. Microbiol.* 5 (11): 2807–2814.

128 Hartmann, T., Schwanhold, N., and Leimkuhler, S. (2015). Assembly and catalysis of molybdenum or tungsten-containing formate dehydrogenases from bacteria. *Biochim. Biophys. Acta* 1854 (9): 1090–1100.

129 Rothery, R.A. and Weiner, J.H. (1993). Topological characterization of *Escherichia coli* DMSO reductase by electron paramagnetic resonance spectroscopy of an engineered [3Fe-4S] cluster. *Biochemistry* 32 (22): 5855–5861.

130 Bilous, P.T. and Weiner, J.H. (1988). Molecular cloning and expression of the *Escherichia coli* dimethyl sulfoxide reductase operon. *J. Bacteriol.* 170 (4): 1511–1518.

131 Mejean, V., Iobbi-Nivol, C., Lepelletier, M. et al. (1994). TMAO anaerobic respiration in *Escherichia coli*: involvement of the tor operon. *Mol. Microbiol.* 11 (6): 1169–1179.

132 Hille, R. (2013). The molybdenum oxotransferases and related enzymes. *Dalton Trans.* 42 (9): 3029–3042.

133 Leimkühler, S., Kern, M., Solomon, P.S. et al. (1998). Xanthine dehydrogenase from the phototrophic purple bacterium *Rhodobacter capsulatus* is more similar to its eukaryotic counterparts than to prokaryotic molybdenum enzymes. *Mol. Microbiol.* 27 (4): 853–869.

134 Yokoyama, K. and Leimkuhler, S. (2014). The role of Fe–S clusters for molybdenum cofactor biosynthesis and molybdoenzymes in bacteria. *Biochim. Biophys. Acta* 1853 (6): 1335–1349.

135 Truglio, J.J., Theis, K., Leimkuhler, S. et al. (2002). Crystal structures of the active and alloxanthine-inhibited forms of xanthine dehydrogenase from *Rhodobacter capsulatus*. *Structure* 10 (1): 115–125.

136 Nishino, T. and Okamoto, K. (2000). The role of the [2Fe-2s] cluster centers in xanthine oxidoreductase. *J. Inorg. Biochem.* 82 (1–4): 43–49.

137 Schumann, S., Saggu, M., Möller, N. et al. (2008). The mechanism of assembly and cofactor insertion into *Rhodobacter capsulatus* xanthine dehydrogenase. *J. Biol. Chem.* 283 (24): 16602–16611.

138 Leimkühler, S. and Klipp, W. (1999). Role of XDHC in Molybdenum cofactor insertion into xanthine dehydrogenase of *Rhodobacter capsulatus*. *J. Bacteriol.* 181 (9): 2745–2751.

139 Neumann, M. and Leimkühler, S. (2011). The role of system-specific molecular chaperones in the maturation of molybdoenzymes in bacteria. *Biochem. Res. Int.* 2011: 850924.

140 Neumann, M., Schulte, M., Jünemann, N. et al. (2006). *Rhodobacter capsulatus* XdhC is involved in molybdenum cofactor binding and insertion into xanthine dehydrogenase. *J. Biol. Chem.* 281 (23): 15701–15708.

141 Schoepp-Cothenet, B., van Lis, R., Philippot, P. et al. (2012). The ineluctable requirement for the trans-iron elements molybdenum and/or tungsten in the origin of life. *Sci. Rep.* 2: 263.

142 Rothery, R.A., Bertero, M.G., Spreter, T. et al. (2010). Protein crystallography reveals a role for the FS0 cluster of *Escherichia coli* nitrate reductase A (NarGHI) in enzyme maturation. *J. Biol. Chem.* 285 (12): 8801–8807.

143 Rothery, R.A., Workun, G.J., and Weiner, J.H. (2008). The prokaryotic complex iron-sulfur molybdoenzyme family. *Biochim. Biophys. Acta* 1778 (9): 1897–1929.

144 Czjzek, M., Dos Santos, J.P., Pommier, J. et al. (1998). Crystal structure of oxidized trimethylamine N-oxide reductase from *Shewanella massilia* at 2.5 A resolution. *J. Mol. Biol.* 284 (2): 435–447.

145 Fourmond, V., Burlat, B., Dementin, S. et al. (2008). Major Mo(V) EPR signature of *Rhodobacter sphaeroides* periplasmic nitrate reductase arising from a dead-end species that activates upon reduction. Relation to other molybdoenzymes from the DMSO reductase family. *J. Phys. Chem. B* 112 (48): 15478–15486.

146 Blasco, F., Iobbi, C., Ratouchniak, J. et al. (1990). Nitrate reductases of *Escherichia coli*: sequence of the second nitrate reductase and comparison with that encoded by the narGHJI operon. *Mol. Gen. Genet.* 222 (1): 104–111.

147 Blasco, F., Dos Santos, J.P., Magalon, A. et al. (1998). NarJ is a specific chaperone required for molybdenum cofactor assembly in nitrate reductase A of *Escherichia coli. Mol. Microbiol.* 28: 435–447.

148 Bertero, M.G., Rothery, R.A., Boroumand, N. et al. (2005). Structural and biochemical characterization of a quinol binding site of *Escherichia coli* nitrate reductase A. *J. Biol. Chem.* 280 (15): 14836–14843.

149 Blasco, F., Guigliarelli, B., Magalon, A. et al. (2001). The coordination and function of the redox centres of the membrane-bound nitrate reductases. *Cell. Mol. Life Sci.* 58 (2): 179–193.

150 Arias-Cartin, R., Lyubenova, S., Ceccaldi, P. et al. (2010). HYSCORE evidence that endogenous mena- and ubisemiquinone bind at the same Q site (Q[D]) of *Escherichia coli* nitrate reductase A. *J. Am. Chem. Soc.* 132 (17): 5942–5943.

151 Guigliarelli, B., Magalon, A., Asso, M. et al. (1996). Complete coordination of the four Fe-S centers of the beta subunit from *Escherichia coli* nitrate reductase. Physiological, biochemical, and EPR characterization of site-directed mutants lacking the highest or lowest potential [4Fe-4S] clusters. *Biochemistry* 35 (15): 4828–4836.

152 Trieber, C.A., Rothery, R.A., and Weiner, J.H. (1996). Consequences of removal of a molybdenum ligand (DmsA-Ser-176) of *Escherichia coli* dimethyl sulfoxide reductase. *J. Biol. Chem.* 271: 27339–27345.

153 Ray, N., Oates, J., Turner, R.J., and Robinson, C. (2003). DmsD is required for the biogenesis of DMSO reductase in *Escherichia coli* but not for the interaction of the DmsA signal peptide with the Tat apparatus. *FEBS Lett.* 534 (1–3): 156–160.

154 Sarfo, K.J., Winstone, T.L., Papish, A.L. et al. (2004). Folding forms of *Escherichia coli* DmsD, a twin-arginine leader binding protein. *Biochem. Biophys. Res. Commun.* 315 (2): 397–403.

155 Cherak, S.J. and Turner, R.J. (2017). Assembly pathway of a bacterial complex iron sulfur molybdoenzyme. *Biomol. Concepts* 8 (3–4): 155–167.

156 Turner, R.J., Papish, A.L., and Sargent, F. (2004). Sequence analysis of bacterial redox enzyme maturation proteins (REMPs). *Can. J. Microbiol.* 50 (4): 225–238.

157 Holm, R.H. (1990). The biologically relevant oxygen atom transfer chemistry of molybdenum: from synthetic analogue systems to enzymes. *Coord. Chem. Rev.* 100: 183–221.

158 Khangulov, S.V., Gladyshev, V.N., Dismukes, G.C., and Stadtman, T.C. (1998). Selenium-containing formate dehydrogenase H from *Escherichia coli:* a molybdopterin enzyme that catalyzes formate oxidation without oxygen transfer. *Biochemistry* 37: 3518–3528.

159 Niks, D. and Hille, R. (2019). Molybdenum- and tungsten-containing formate dehydrogenases and formylmethanofuran dehydrogenases: structure, mechanism, and cofactor insertion. *Protein Sci.* 28 (1): 111–122.

160 Kumar, H., Khosraneh, M., Bandaru, S.S.M. et al. (2023). The mechanism of metal-containing formate dehydrogenases revisited: the formation of bicarbonate as product intermediate provides evidence for an oxygen atom transfer mechanism. *Molecules* 28 (4): 1537.

161 Park, D.M., Akhtar, M.S., Ansari, A.Z. et al. (2013). The bacterial response regulator ArcA uses a diverse binding site architecture to regulate carbon oxidation globally. *PLoS Genet.* 9 (10): e1003839.

162 Crack, J.C., Thomson, A.J., and Le Brun, N.E. (2017). Mass spectrometric identification of intermediates in the O_2-driven [4Fe-4S] to [2Fe-2S] cluster conversion in FNR. *Proc. Natl. Acad. Sci. USA* 114 (16): E3215–E3223.

163 Green, J. and Guest, J.R. (1993). A role for iron in transcriptional activation by FNR. *FEBS Lett.* 329 (1–2): 55–58.

164 Spiro, S. and Guest, J.R. (1990). FNR and its role in oxygen-regulated gene expression in *Escherichia coli. FEMS Microbiol. Rev.* 6 (4): 399–428.

165 Myers, K.S., Yan, H., Ong, I.M. et al. (2013). Genome-scale analysis of *Escherichia coli* FNR reveals complex features of transcription factor binding. *PLoS Genet.* 9 (6): e1003565.

166 Anderson, L.A., McNairn, E., Lubke, T. et al. (2000). ModE-dependent molybdate regulation of the molybdenum cofactor operon moa in *Escherichia coli. J. Bacteriol.* 182 (24): 7035–7043.

167 Partridge, J.D., Sanguinetti, G., Dibden, D.P. et al. (2007). Transition of *Escherichia coli* from aerobic to micro-aerobic conditions involves fast and slow reacting regulatory components. *J. Biol. Chem.* 282 (15): 11230–11237.

168 Zupok, A., Iobbi-Nivol, C., Mejean, V., and Leimkuhler, S. (2019). The regulation of Moco biosynthesis and molybdoenzyme gene expression by molybdenum and iron in bacteria. *Metallomics* 11 (10): 1602–1624.

169 Hasnat, M.A., Zupok, A., Olas, J.J. et al. (2021). A-type carrier proteins are involved in [4Fe-4S] cluster insertion into the radical S-adenosylmethionine protein MoaA for the synthesis of active molybdoenzymes. *J. Bacteriol.* 203 (12): e0008621.

170 Regulski, E.E., Moy, R.H., Weinberg, Z. et al. (2008). A widespread riboswitch candidate that controls bacterial genes involved in molybdenum cofactor and tungsten cofactor metabolism. *Mol. Microbiol.* 68 (4): 918–932.

171 Hasona, A., Self, W.T., and Shanmugam, K.T. (2001). Transcriptional regulation of the moe (molybdate metabolism) operon of *Escherichia coli. Arch. Microbiol.* 175 (3): 178–188.

172 Hasnat, M.A., Zupok, A., Gorka, M. et al. (2024). Iron limitation indirectly reduces the *Escherichia coli* torCAD operon expression by a reduction of molybdenum cofactor availability. *Microbiol. Spectr.* 12 (2): e0348023.

8

Molybdenum Cofactor Biosynthesis in Eukaryotes

Guenter Schwarz, Lukas Flohr, Emanuel Bruckisch, and Katrin Fischer-Schrader

University of Cologne, Institute of Biochemistry, Department of Chemistry and Biochemistry, and Center for Molecular Medicine Cologne, Zülpicher Str. 47, 50674 Köln, Germany

Chapter Goals

- provide an overview of the structure and function of the molybdenum cofactor in eukaryotes
- describe the three-dimensional structures of eukaryotic molybdenum cofactor-synthesizing proteins
- explain the evolutionary origin of novel functions derived from proteins involved in molybdenum cofactor biosynthesis.
- explain the impact of loss of function mutations in molybdenum cofactor-synthesizing proteins and enzymes

Main Chapter Points

- introduction to molybdenum cofactor chemical structures
- protein structures and functions of eukaryotic molybdenum cofactor-dependent enzymes
- biosynthesis of cyclic pyranopterin: chemical structure of the intermediate, reaction mechanism, and subcellular localization of MOCS1 proteins
- biosynthesis of molybdopterin: reaction mechanism, assembly and disassembly of molybdopterin synthase (MOCS2), and thiolation of MOCS2A by MOCS3.
- molybdenum insertion into molybdopterin: reaction mechanism, product-substrate-channeling, and novel functions in vertebrates
- deficiency and therapy of molybdenum cofactor biosynthesis
- links between molybdenum cofactor and iron–sulfur cluster biology

Iron-Sulfur Clusters: Biogenesis and Biochemistry, Volume 1, First Edition.
Edited by Silke Leimkühler, Günter Schwarz, Oliver Lenz, and Oliver Einsle.

8.1 Introduction

Metal-based cofactors play a crucial role in many biological redox reactions. Among these metal cofactors, molybdenum and iron are intimately linked via different evolutionary trajectories, most notably in the complex "iron–molybdenum cofactor" found in nitrogenase [1]. Nitrogen-fixing bacteria, with their high nitrogenase expression, served as a prominent example for the study of iron–sulfur clusters (Fe–S). Basic principles of Fe–S cluster synthesis were discovered in these organisms and later diversified by studies in bacteria and eukaryotes. Eukaryotes possess both mitochondrial and cytosolic Fe–S cluster biogenesis pathways, the latter being strictly dependent on the biosynthesis of mitochondrial [2Fe–2S] clusters [2]. Today, Fe–S clusters are known to be essential for molybdenum cofactor (Moco) biosynthesis [3], and Moco-dependent enzymes utilize either Fe–S clusters or heme cofactors for their internal electron transfer pathways.

In eukaryotes, Moco consists of a reduced (tetrahydro) pterin-based heterocyclic structure with a dithiolene function in the C6-substituted four-carbon side chain that forms a pyran ring to the C5 carbon of the pterin and carries a terminal phosphate at the C4′ position [4]. This structure has been called molybdopterin or metal-binding pterin (MPT), as it can bind both molybdenum and tungsten [5]. The pterin dithiolene binds molybdenum, which can have various ligand configurations depending on the enzyme it is incorporated into [5], for example, three oxo-ligands before insertion into any enzyme, two oxo-ligands, and one cysteine-derived thiol as found in sulfite oxidase, nitrate reductase or amidoxime-reducing component, or two oxo-ligands and a sulfido-ligand as present in xanthine oxidoreductase and aldehyde oxidase (Figure 8.1). In prokaryotes, Moco exhibits greater structural diversity including the eukaryotic form of the cofactor, nucleotide-modified species (with guanosine monophosphate [GMP] or cytidine monophosphate [CMP] attached) as well as the so-called bis-forms where the molybdenum (or tungsten) atom is chelated by two dithiolenes derived from two pterin moieties. The bacterial Moco biosynthetic pathway is well characterized and has been extensively reviewed elsewhere [6] and is covered in Chapter 7 by Olivieri and Leimkühler [editorial office: Add Reference].

This chapter focuses on eukaryotic Moco biosynthesis emphasizing the structural aspects of involved proteins, their evolution, and roles beyond Moco synthesis.

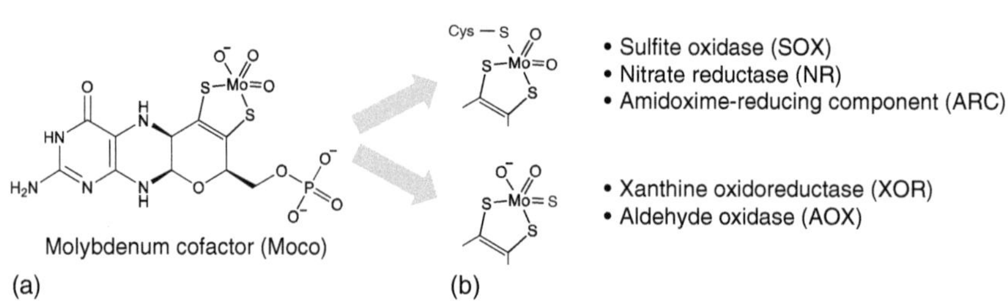

Figure 8.1 Structural variations of the molybdenum cofactor in eukaryotes. Coordination of the molybdenum cofactor (Moco) outside enzymes (a) and when incorporated into the different Moco-dependent enzymes (b).

Moco biosynthesis defects result in the loss of activity of all Moco enzymes with detrimental consequences on the respective organism. In humans, the associated disorder is accompanied by neurodegeneration and early childhood death [7]. Regarding the underlying pathomechanisms, links to Fe–S cluster biology and treatment opportunities will be discussed in this chapter.

8.2 Eukaryotic Molybdenum Enzymes

Moco enzymes play crucial roles in the global biochemical cycles of carbon, nitrogen, and sulfur, both on a global scale and within individual organisms [8]. Eukaryotes possess five distinct types of Moco enzymes, of which four are found in animals including humans. These enzymes were initially classified based on how Moco is coordinated within their active sites and later the respective structures of those enzymes lead to a further categorization into three classes. Generally, either Moco is covalently bound to a cysteine side chain for which sulfite oxidase (SOX) is the prototype or Moco harbors a terminal sulfido-ligand as first reported in xanthine oxidoreductase (XO).

In humans and vertebrates, SOX is considered the most important, as it serves a key enzyme in the last and critical step of cysteine catabolism [9], catalyzing the conversion of sulfite to sulfate. Structurally, SOX exists as a homodimer with each subunit comprising three distinct domains: an N-terminal cytochrome b_5-type heme domain, a central catalytic Moco domain, and a dimerization domain at the C-terminus [4] (Figure 8.2). In vertebrates, SOX is found within the mitochondrial intermembrane space, facilitating the electron transfer from sulfite oxidation to the physiological electron acceptor cytochrome *c* [12]. Interestingly, plants also possess a SOX enzyme, although they can assimilate sulfate directly to sulfide. However, plant SOX differs from vertebrate SOX in several aspects: it is localized to peroxisomes where it utilizes molecular oxygen as an electron acceptor and the enzymes lack a heme domain thus facilitating a direct electron transfer from Moco to oxygen [13].

In plants, there is another Moco enzyme with high homology to SOX, which is nitrate reductase (NR). Plant NR can be classified to the Mo enzyme family of SOX based on its Moco-binding mechanism, which involves a highly conserved cysteine residue in the active site. Each subunit of a homodimeric NR comprises four distinct domains: an N-terminal Moco domain followed by a dimerization domain, which is connected via a linker to a heme domain of the b_5-type, and a flavin adenine dinucleotide (FAD) domain for NADH binding at the C-terminus (Figure 8.2). Localized in the cytosol, NR catalyzes the reduction of nitrate to nitrite by the transfer of two electrons. Nitrite is subsequently transported to the chloroplast, where nitrite reductase further reduces it to ammonia. Similar to SOX, NR relies on its heme domain for intramolecular electron transfer. The activities of both enzymes require domain movements, which are impacted by solution viscosity [14]. NR also undergoes post-translational modifications and binding of a regulatory 14–3–3 protein leading to its inhibition [15, 16].

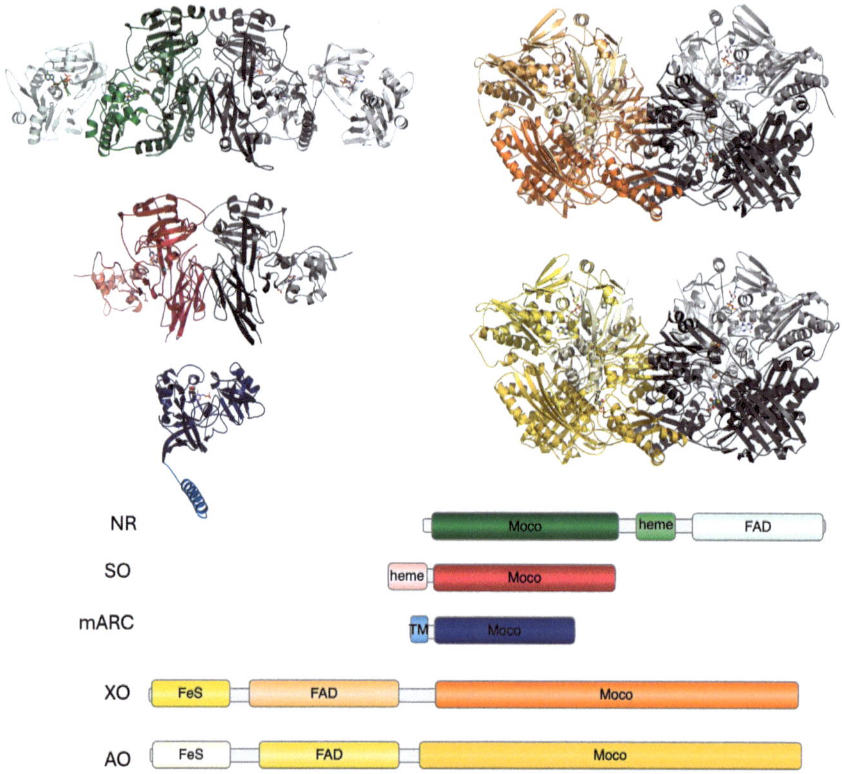

Figure 8.2 Three-dimensional structures and domain structures of eukaryotic Moco enzymes. Except mARC enzymes, all other four enzymes form stable dimers composed of multiple domains. Shown are the domain structures as well as crystal structures of chicken SOX (1SOX), bovine XO (3UNC), and human AO (4UHW) rendered at a comparable size. The three-dimensional structure of NR was generated using AlphaFold 3 [10] derived from *Aspergillus caelatus,* which was presented with the highest pLDDT score (90.06) among 614 structures shown in AlphaFold. The mARC2 structure was generated with AlphaFold 2 [11] despite the fact that the core structure of human mARC1 has been determined using X-ray crystallography (PDB 6FW2). The here presented structure includes the N-terminal transmembrane domain (TM) needed for anchoring the protein in the outer mitochondrial membrane.

Humans, plants, and fungi possess two additional cytosolic Mo-enzymes, xanthine oxidoreductase (XO) and aldehyde oxidase (AO). These closely related Molybdenum–iron–flavin enzymes catalyze hydroxylation reactions of purines and various heterocyclic substrates, respectively [17]. XO is a key enzyme in purine catabolism, catalyzing the final two steps: the conversion of hypoxanthine to xanthine and further oxidation to uric acid. Structurally, XO is a homodimer of approximately 300 kDa size, with each subunit comprising three distinct domains: the N-terminal domain contains two [2Fe–2S] clusters, a central FAD-binding domain, and a C-terminal Moco-containing catalytic domain that also mediates dimerization and is much larger than the respective domain in SOX and NR (Figure 8.2).

XO-bound Moco is characterized by a sulfido-ligand without any covalent binding to the enzyme. The XO form can generate reactive oxygen species, specifically superoxide anions and hydrogen peroxide. This capability suggests that XO has an additional role in cellular stress responses and as a potential risk factor in ischemia–reperfusion injury [18].

Aldehyde oxidase (AO) shares significant structural similarity with XO [19] (Figure 8.2). AO oxidizes various heterocyclic aldehydes to carboxylic acids, generating superoxide and hydrogen peroxide as byproducts, much similar to XO [20]. While the primary physiological substrate is currently unknown for vertebrate AO, the enzyme has an important role in vertebrate drug metabolism [21] and in plant biosynthesis of abscisic acid, which is a critical plant hormone [22].

The mitochondrial amidoxime-reducing component (mARC) is the last discovered Moco-enzyme in vertebrates [23]. It was identified as an enzymatic activity catalyzing the reduction of N-hydroxylated prodrugs as well as endogenous compounds such as N-hydroxy-L-arginine [24, 25]. The latter is an important intermediate in nitric oxide (NO) synthesis. Interestingly, various Moco enzymes exhibit a moonlighting function, namely the reduction of nitrite to NO [26–31]. This process has significant physiological implications, including blood pressure regulation, hypoxic vasodilation [32], cellular cyto-protection, as well as modulation of mitochondrial respiration, and adaptive responses to hypoxic or exercise stress [33].

8.3 Molybdenum Cofactor Biosynthesis – An Overview

The biosynthesis of Moco is an ancient biological process conserved across all domains of life. The chemical transformations involved are highly conserved, with identical and interchangeable intermediates across organisms. The synthesis occurs in three defined steps, each producing stable and isolatable products of cyclic pyranopterin monophosphate (cPMP), molybdopterin (MPT), and Moco. Each individual reaction step is catalyzed either by proteins that assemble into complexes or by multi-domain proteins, which are highly conserved in both structure and function [34].

In humans, the initial step of cPMP synthesis is catalyzed by MOCS1A and MOCS1B proteins, which are expressed from a single *MOCS1* gene following complex alternative splicing [35]. The conversion of cPMP to MPT is carried out by a tetrameric protein complex in the cytosol, consisting of two MOCS2A and two MOCS2B subunits. Similar to MOCS1, these subunits are expressed by a bicistronic transcript from a single *MOCS2* gene, however, here through a leaky scanning mechanism [36, 37]. After each catalytic cycle, the MPT synthase complex undergoes regeneration. This process involves the dissociation of the complex, followed by an ATP-dependent activation of the MOCS2A subunit. Specifically, the carboxyterminal glycine residue of MOCS2A is acylated, resulting in the formation of MOCS2A-AMP. This reaction and the subsequent thiolation of the C-terminus are catalyzed by the protein MOCS3, consisting of two domains [38, 39]. The final step of Moco synthesis is catalyzed by the multidomain protein gephyrin. Initially,

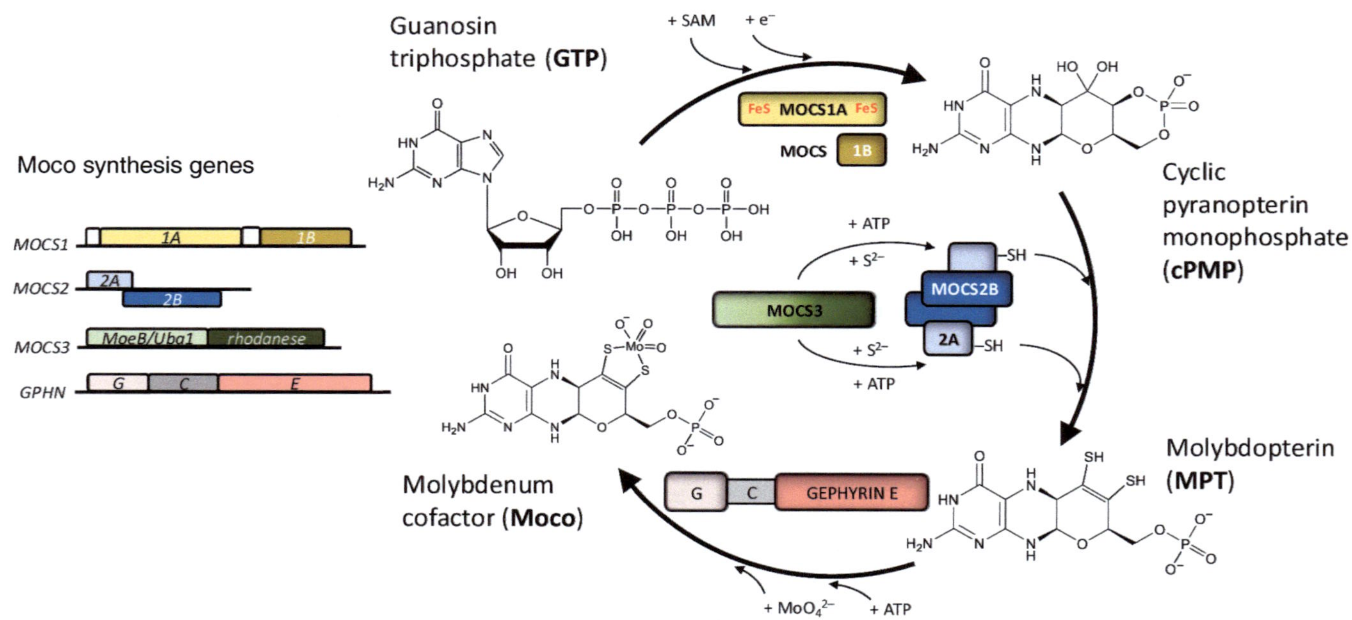

Figure 8.3 Overview of Moco biosynthesis in humans. Intermediates (cPMP and MPT) defining the three biosynthetic steps are shown. Shown are the four genes involved in Moco synthesis (left) and the respective proteins expressed from those open reading frames and their function in the individual steps of Moco synthesis. SAM: S-adenosylmethionine, Fe–S: Fe–S cluster.

the G-domain of gephyrin adenylylates the MPT phosphate, and then the E-domain inserts molybdate thus releasing AMP and Moco (Figure 8.3). Mature Moco is then either directly incorporated into enzymes of the SOX family, such as human SOX or mARC, or undergoes sulfuration by the Moco sulfurase to form the sulfido variant of the cofactor, as found in XO and AO enzymes [40, 41].

While the intermediates in Moco biosynthesis are conserved, the genomic structure of the genes encoding Moco biosynthetic proteins varies significantly across species. Prokaryotes typically require eight genes and plants six, while vertebrates require only four genes encoding six Moco biosynthesis proteins [5]. In certain yeasts and fungi, the ability to utilize molybdenum was lost over the course of evolution, leading to the disappearance of most genes (except MoeB/Mocs3) essential for Moco biosynthesis [42]. Different mechanisms such as bicistronic expression (*MOCS1*, *MOCS2*), alternative splicing (*MOCS1*, *GPHN*) as well as domain fusions (*MOCS3*, *GPHN*) have been manifested to further improve spatio-temporal efficacy in eukaryotic Moco synthesis [43] (Figure 8.3).

8.4 Synthesis of Cyclic Pyranopterin Monophosphate

The initial and most intricate step in Moco biosynthesis is catalyzed by MOCS1A and MOCS1B proteins in humans. These proteins convert GTP to cyclic pyranopterin monophosphate (cPMP) through a complex process involving the formation of a unique cyclic nucleotide, 3′,8-cyclodihydro-GTP, which has been first discovered in bacteria by studies on the homologous MoaA protein [44]. In plants, proteins and protein domains homologous to MOCS1 are named Cnx2 and Cnx3 [45], while in fungi, they are referred to as NIT-7A and NIT-7B (*Neurospora crassa*) [46, 47], respectively (Figure 8.4a). While crystal structures are only available for the bacterial MoaA [48] and MoaC proteins [49], sequences of orthologous proteins are highly conserved (Figure 8.4b) and thereby the use of AlphaFold [10, 11] allows for a reliable prediction of three-dimensional structures of MoaA and MoaC-homologous proteins from humans, fungi, and plants resulting in models with high confidence as evidence by their pLDDT scores (Figure 8.4).

MOCS1A and its orthologs are members of the radical S-adenosylmethionine (SAM) enzyme superfamily [50]. They are classified by a [4Fe–4S] cluster, the radical SAM cluster, which cleaves SAM in a reductive manner, producing an adenosyl radical following oxidation of the [4Fe–4S] cluster from +1 to +2 [51]. These proteins are structurally unique [52] because, along with the N-terminal radical SAM [4Fe–4S] cluster, they also contain an additional auxiliary [4Fe–4S] cluster located at the C-terminus. This structural feature is shared by four out of the eight known human radical SAM enzymes [53–55]. A CX_3CX_2C motif coordinates the radical SAM cluster of MOCS1A, while a $CX_2CX_{13}C$ motif binds the auxiliary C-terminal cluster. Given that three cysteines coordinate each of the [4Fe–4S] clusters, which results in an increased sensitivity to oxygen. As part of the reaction mechanism, the fourth coordination of the N-terminal [4Fe–4S] cluster is provided by SAM while GTP serves this function in the C-terminal cluster [43, 53].

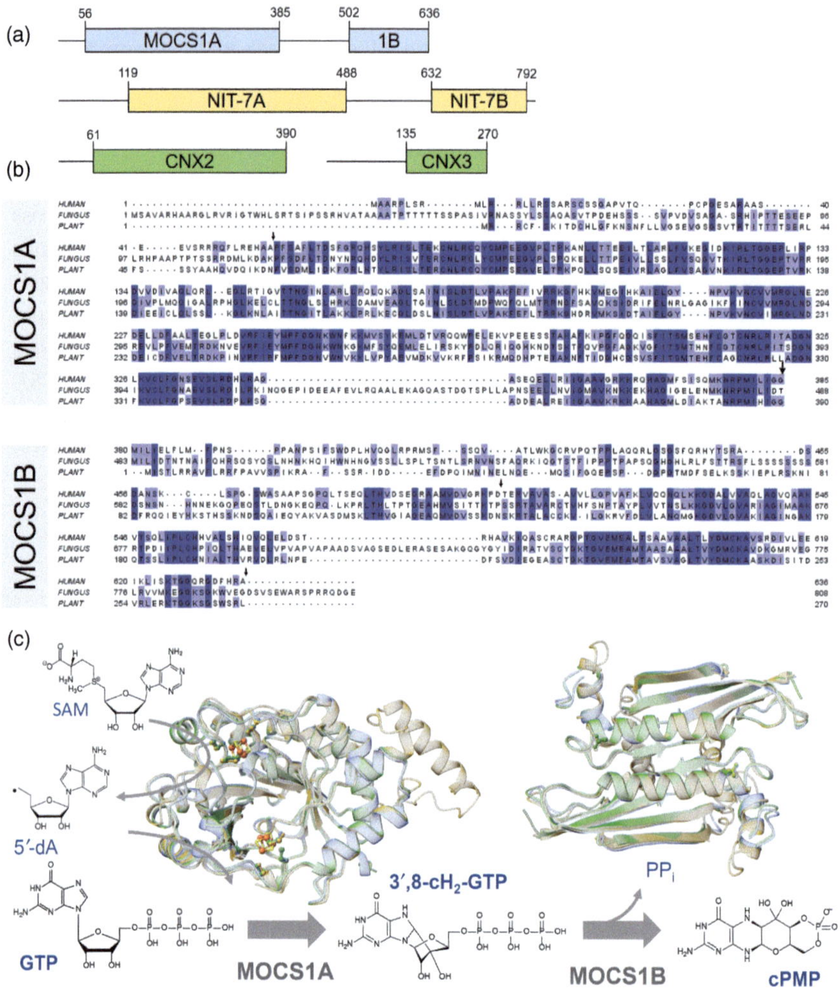

Figure 8.4 Domains structure, sequence alignment, structure, and catalyzed reactions by the human, fungal, and plant proteins involved in cPMP synthesis. (a) Domain structure and (b) sequence alignment of human MOCS1AB, *N. crassa* (fungus) NIT-7AB, *A. thaliana* (plant) Cnx2 and Cnx3. First and last residues of the respective three-dimensional structures are indicated in (a) and highlighted with an arrow in (b). (c) Reactions catalyzed by MOCS1A and MOCS2B are depicted including the cosubstrate S-adenosyl methionine (SAM), the formation of the 5′-desoxyadenosyl radical (5′-dA), and pyrophosphate release (PPi). AlphaFold-derived structures are shown for MOCS2A/MOCS2B (light blue), Cnx2/Cnx3 (light green), and Nit-7A/Nit-7B (light orange). Both [4Fe–4S] clusters are depicted in ball-and-stick representation.

Most of the mechanistic aspects in the cPMP synthesis are derived from studies with the MOCS1A-homologous bacterial MoaA proteins by the Yokoyama Laboratory [3] and they are discussed in detail in Chapter 7. In brief, following structural rearrangements, MoaA-bound SAM is shielded by the protein's C-terminal tail [44], which is required to trigger electron transfer from the [4Fe–4S] cluster to SAM, leading to its reductive cleavage and the formation of the 5′-deoxyadenosyl radical [56]. This is followed by the abstraction of the hydrogen atom from the C3 of bound GTP, followed by the attack of the C3 radical on the C8 atom of the GTP, resulting in the cyclization initiation of the molecule [3]. This reaction places the C8 atom between the C2 and C3 positions of the GTP ribose ring, forming a C3–C8 linkage and leading to the formation of 3′,8-cyclodihydro-GTP (3′,8-cH2-GTP) [57, 58].

The second half of the reaction was studied in detail by MOCS1B-homologous bacterial protein MoaC, which forms two active sites on a functional dimer (Figure 8.4c), which in turn assembles into a hexameric protein complex [49]. After identifying 3′,8-cH_2-GTP, crystal structures of MoaC bound to both substrate and product showed that two loops within MoaC undergo major rearrangements during the reaction [44], positioning residues near the active site to promote an acid/base catalysis [49]. The ultimate formation of the cyclic phosphate is proposed to supply the thermodynamic energy needed for closing the pterin ring [44].

The exceptional chemistry of the cPMP molecule was first described in 2004 indicating a tetrahydropyranopterin instead of the previously reported dihydropterin [59], demonstrating the existence of a geminal diol instead of not a keto function at the C1′ position. Later studies established the in vitro chemical synthesis of cPMP, which also yielded the geminal diol form in the pyran ring underlying the chemical importance of this unique structure for cPMP reactivity and downstream processing [60].

The genomic structure and expression of proteins participating in cPMP synthesis in eukaryotes is quite different to the bacterial situation. The human *MOCS1* gene is composed of 10 exons, with exons 1–9 encoding MOCS1A and exon 10 encoding MOCS1B [61]. The *MOCS1* gene undergoes a complex alternative splice [62] in two regions: (i) either alternative spicing in exon 9 will lead to a bicistronic transcript that allows only for the expression of the MOCS1A protein (full-length exon 9), while following the expression of a transcript with a truncated, or deleted exon 9 will lead to a fusion protein that harbors both MOCS1A and MOCS1B. (ii) The N-terminal sequence of exon 10, preceding the conserved MOCS1 sequence, encodes for an internal mitochondrial targeting sequence that drives the translocation of the MOCS1AB fusion protein to the mitochondria. Subsequent proteolytic cleavage and release of the MOCS1B protein to the mitochondrial matrix completes its maturation, while the MOCS1A part was found to be degraded [35]. In addition to alternative splicing in exon 9, a splice event in exon 1 provides MOCS1A variants with an N-terminal mitochondrial targeting sequence driving the translocation of MOCS1A into the mitochondrial matrix. Collectively, these findings provide strong evidence that the formation of cPMP and thus the first step of Moco synthesis is taking place in mitochondria [35].

In light of the two [4Fe–4S] clusters needed for MOCS1A function, enzymes of the late Fe–S biosynthetic machinery [2, 63] are needed and therefore it is not surprising that cPMP synthesis takes place in mitochondria. Further evidence supporting the mitochondrial localization of step 1 in Moco synthesis comes from the plant orthologues of MOCS1A and MOCS1B. In contrast, plant Cnx2 and Cnx3 are encoded by single genes, however, each of them harbor N-terminal extensions that were proposed to function as mitochondrial targeting sequence [45]. Finally, in the fungus *Neurospora crassa*, again a *MOCS1*-homologous *Nit*-7AB bicistronic gene structure has been reported that results in different transcripts following alternative splice [47]. Again, the bicistronic transcript produces only Nit-7A with a mitochondrial targeting sequence, while the monocistronic mRNA leads to the expression of a Nit-7AB fusion protein, which is targeted by the same N-terminal sequence to mitochondria and undergoes proteolytic cleavage to release Nit-7B [47]. In aggregate, there is strong evidence derived from three distinct model organisms collectively suggesting that cPMP synthesis takes place in mitochondria.

A synthesis of cPMP within the mitochondrial matrix raises the question of how cPMP is transported outside of mitochondria. Based on early co-culturing studies using fibroblasts derived from patients suffering from Moco deficiency, a transfer of cPMP (at that time called precursor Z) through the plasma membrane has been proposed [64]. Cells accumulating cPMP (defective in *MOCS2*) were able to release it and cells unable to synthesize cPMP (defective in *MOCS1*) were able to take it up and convert it to active Moco [64]. Therefore, transporters and/or carriers must exist to facilitate cPMP translocation through membranes.

A study in plants proposed that the export of cPMP out of plant mitochondria depends on the mitochondrial ATP binding cassette (ABC) transporter ATM3 [65] given that *ATM3* mutant plants exhibited reduced activity of enzymes NR and XO and an accumulation of cPMP in the mitochondria. However, both cPMP and Moco enzyme levels were only partially altered raising the question whether ATM3 is indeed involved in cPMP transport given its established role in the translocation of a Fe–S cluster precursor to the cytosol. Various studies have proposed a glutathione (GSH)-complexed [2Fe–2S] cluster as a possible substrate “X” [63], which is clearly different from cPMP [66]. In humans, MOCS1A variants were reported that lack the mitochondrial targeting sequence and might function in the cytosol and would then require the cytosolic Fe–S cluster biosynthetic machinery [67]. In a patient, harboring a mutation affecting the expression of MOCS1B, we found that trace amounts of cytosolic MOCS1B were causative for a mild disease phenotype suggesting either the export of a cPMP precursor from mitochondria or a function of MOCS1B outside mitochondria [68]. Therefore, residual Moco synthesis in ATM3-deficient cells might be also indicative for a dual localized cPMP synthesis in two compartments, mitochondria and cytosol, that may have evolved as functional redundancy due to the critical role of Moco in cellular physiology.

8.5 Thiolation of cPMP by Molybdopterin Synthase

Similar to the first step of Moco biosynthesis, the synthesis of MPT is catalyzed by two proteins which are encoded by a single gene, named *MOCS2*. In contrast to *MOCS1*, *MOCS2* is strictly bicistronic and both encoded proteins (MOCS2A and MOCS2B) are expressed from the same mRNA. The translation of both proteins is controlled via a leaky scanning mechanism leading to the expression of a small MOCS2A and large MOCS2B subunit, which together form a tetrameric

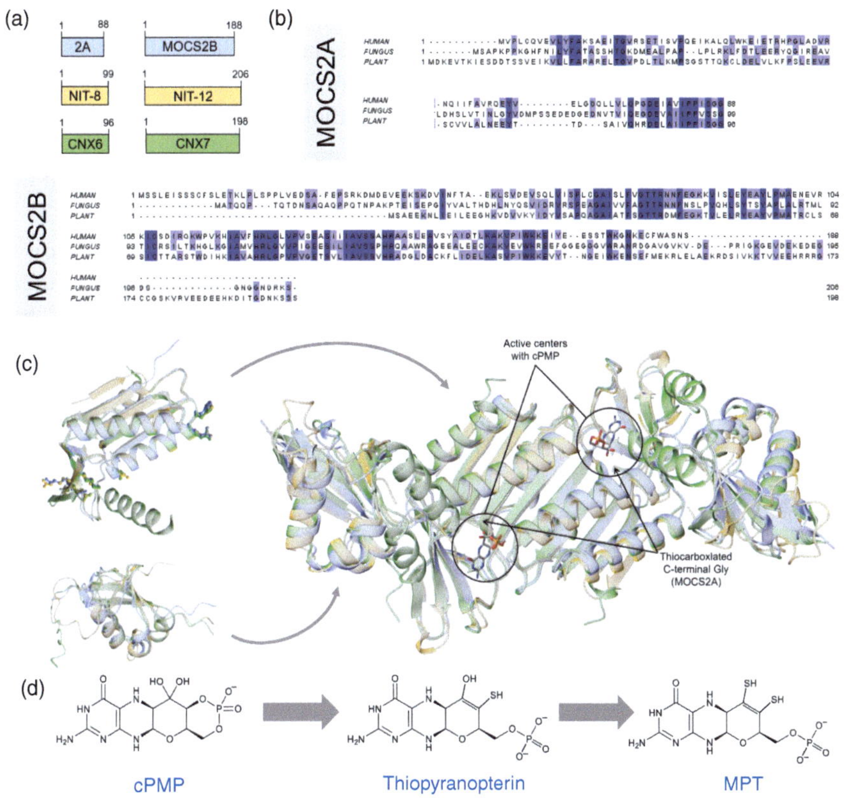

Figure 8.5 Molybdopterin synthesis by MPT synthase. (a) Primary structure and (b,c) sequence alignment of MPT synthase small (b) and large (c) subunits from human (MOCS2A/B), plants (*Arabidopsis thaliana* Cnx6/7), and fungal (*N. crassa* Nit-8/Nit-12). (c) The crystal structures of MOCS2A and MOCS2B (light blue) overlayed with Cnx6 and Cnx7 (light green) and Nit-8 and Nit-12 (light orange) before assembly (left side) into the tetrameric complex are shown. Functional important residues are highlighted in MOCS2B in ball-and-stick. The two active sites are indicated by the co-crystallized cPMP (ball-and-stick). (d) The two-step thiolation reaction is catalyzed by MPT synthase.

MPT-synthase complex (Figure 8.5). Similar to prokaryotic protein expression, both open reading frames overlap on the *MOCS2* mRNA by 77 nucleotides and both proteins are translated in different reading frames [36, 37].

MOCS2A harbors a C-terminal double glycine motif, essential for catalysis (Figure 8.5b), which undergoes a post-translational modification leading to thiocarboxylation [69]. This PTM is crucial for MPT synthesis and provides the sulfur group for the dithiolene formation. The first structure of the MPT synthase complex, revealed by crystallization, was obtained from *E. coli* proteins MoaD and MoaE, representing the small and large subunit, respectively [70]. Analysis of the structure reveals a dimerization of the large subunits within the heterotetrameric complex (Figure 8.5d), while the two small subunits bind via their C-terminal tails to each of the two binding pockets, created by the two large subunits, resulting in the formation of two clearly separated active sites [70]. Whether they function entirely independently remains to be determined. The crystal structure of human MPT synthase MOCS2A/B complex (PDB 5MPO) reveals a remarkable level of conservation to the bacterial MoaD/E complex. Consequently, AlphaFold-based predictions for the plant (Cnx6/Cnx7) and fungal (Nit-8/Nit-12) orthologues represent models with high accuracy (Figure 8.5c). Variabilities are seen in some surface exposed loops and the terminal ends of the small and large subunits given that are elongated by 10 residues in Nit-8 and more than 50 residues in MOCS2B (Figure 8.5a–c).

In MPT synthesis, the formation of the dithiolene group requires two consecutive sulfur transfer steps [69] (Figure 8.5d). The substrate cPMP first attaches to one of the large subunits, close to the double glycine motif found on the small subunit. The geminal diol at the C1′ atom on the cPMP pyran ring [59] allows the first sulfur to attach to the C2′ site, causing the ring structure of the cyclic phosphate to open. Afterward, the second thiolation at the C1′ position is performed by another small subunit MOCS2A, resulting in the completion of the dithiolene function [71].

The MOCS3 protein supplies the sulfur atoms needed for MPT synthase activity by facilitating the thiocarboxylation of the C-terminus of MOCS2A. MOCS3 is a two-domain protein that possesses an N-terminal ATPase activity enabling the adenylation of the C-terminus of MOCS2A [39]. The rhodanese-like domain at the C-terminus of MOCS3 transfers a sulfur from a labile persulfide – derived from a sulfur transferase – to adenylated MOCS2A. Both sulfur atoms are provided by cysteine via the NifS-like sulfurtransferase NFS1 in humans [72, 73] or the mercapto-pyruvate sulfur transferase (MPST), also called TUM1 [74, 75].

In addition to their crucial roles in Moco synthesis, secondary functions have been identified for both MOCS2 and MOCS3 proteins. Recent studies demonstrated that MOCS2B inhibits the antiviral response protein kinase R (PKR) [76]. The control of PKR activity is mediated via binding of the MAP3K12-binding inhibitory protein 1 (MBIP) to PKR, which inhibits the autophosphorylation of PKR. Sequence analysis identified a motif within the MBIP protein that exhibits homology to MOCS2B. Together with the finding that Drosophila MOCS2B proteins represent a fusion with MBIP raised the question whether MOCS2B might have another function in gene expression regulation in the nucleus [77]. Accordingly, it has been reported that MOCS2B can bind to the N-terminus of PKR recruiting MBIP, which subsequently inhibits PKR autophosphorylation [78]. This in turn inhibits the phosphorylation of eIF2α allowing translation initiation even under high PKR expression [79].

It was previously described that MOCS3, important for the regeneration of MOCS2A, and its plant homologue Cnx5 are participating in the thiolation of the carboxylic group of another protein's C-terminus: the ubiquitin-related modifier 1 (URM1). URM1 may either facilitate the transfer of a sulfur atom to the uridine 34 position on tRNA [80, 81] or might promote the URMylation of target proteins in response to oxidative stress [82]. This positions URM1 within the growing family of modifiers related to ubiquitin [83] with diverse cellular targets [84]. Notably, the evolutionary ties between various proteins are suggested by the striking similarities in structural characteristics and mechanisms of activation when comparing the small subunits of MPT synthase and proteins such as ubiquitin and URM1. Evidence for this evolutionary link is seen in how well the adenyltransferase domain of

MOCS3 aligns with the N-terminal domain of UbA4, a ubiquitin-activating enzyme [82]. UBA4, which is the closest yeast equivalent of MOCS3, represents the only "Moco biosynthesis" gene in *Saccharomyces cerevisiae*. Therefore, both the gene fusion and the development of UbA4's new function probably occurred in the early stages of eukaryotic evolution. Furthermore, mutations leading to a complete loss of MOCS3 function have not been found in patients suffering of Moco deficiency [85] suggesting embryonal lethality in case of loss of MOCS3 function (see chapter on Moco deficiency).

Finally, thc thiolation of MOCS2A and related proteins requires an S-mobilization mechanism, which is also used for Fe–S Cluster biogenesis. For example, NFS1 has been primarily studied as part of the mitochondrial Fe–S cluster synthesis machinery [63]. However, NFS1 was found to localize also to the cytosol, where it generates a persulfide on MOCS3 [72, 86], thus connecting again Fe–S cluster and Moco synthesis.

8.6 Molybdenum Insertion into Molybdopterin

The multi-domain protein gephyrin, also referred to as molybdenum insertase, facilitates the last step of Moco biosynthesis in humans where it combines enzymatic functions that, in bacteria, reside on separate proteins [87]. Within the protein gephyrin, the G- and E-domains, which are homologous to the *E. coli* proteins MogA and MoeA, respectively, catalyze two subsequent transformations in Moco synthesis. The G-domain mediates the adenylylation of MPT [88, 89], while the E-domain incorporates molybdate into adenylated MPT thus leading to AMP and Moco release [90] (Figure 8.6). Also in the fungus *N. crassa*, the Nit-9 protein represents the same type of domain fusion [46, 47], while in the plant *A. thaliana*, also a fusion protein exists, however, the order of the G- and E-domain is inversed with a much shorter linker connecting both domains [91]. Therefore, the gene fusions that resulted in *GPHN*/*Nit-9* and *Cnx1* must have happened in two distinct evolutionary occurrences (Figure 8.6). These multiple and autonomous instances of gene consolidation emphasize an adaptive advantage of an unified multifunctional protein expediting both successive steps of Moco biogenesis in eukaryotic organisms.

Crystal structures of the G- and E-domains of gephyrin and Cnx1 have been determined [92–94] demonstrating a high degree of structural similarity for both domains across all domains of life [95–97]. Similar to other Moco-synthesizing proteins, AlphaFold-based predictions of structures of the Nit-9 protein domains reveal a high degree of confidence and structural similarities with additional elements located outside the boundaries of the core domain folding (Figure 8.6b). Not only the domain structure but also the oligomeric state is conserved among all kingdoms of life. The G domains as well as bacterial MogA [96] proteins present as trimers, while orthologous MoaB proteins from thermophilic archaebacteria such as *Pyrococcus furiosus* form dimers of trimers (hexamers) thus increasing thermal stability [95, 98]. The E-domains and bacterial MoeA proteins [97] are dimers harboring two active sites that are formed at the interface of two intertwined dimers

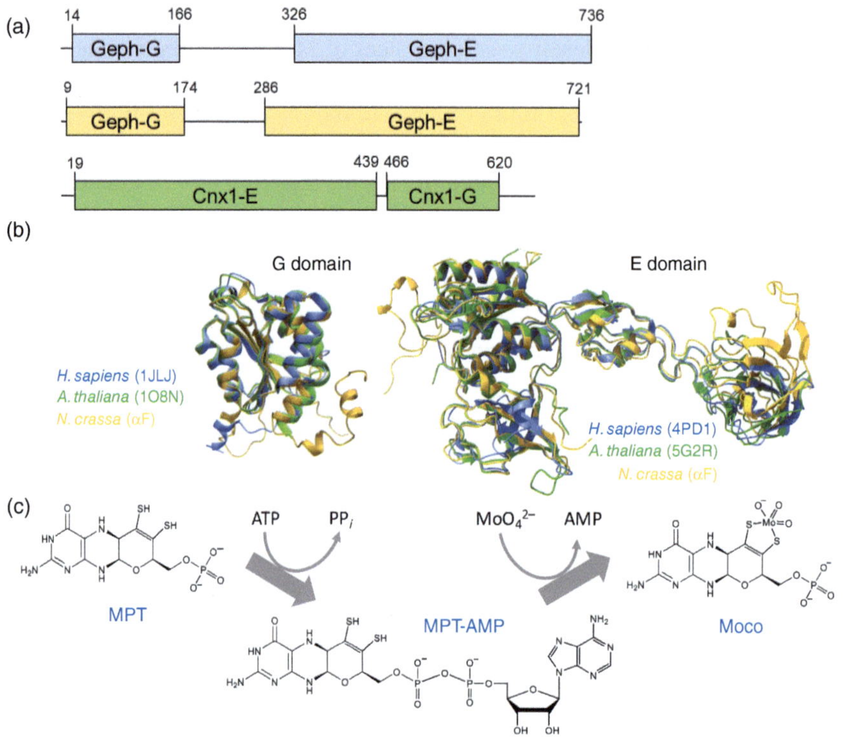

Figure 8.6 ATP-dependent molybdate insertion into molybdopterin. (a) Primary structure and domain boundaries of gephyrin, Cnx1 and Nit-9. (b) Crystal structures of gephyrin and Cnx1 G and E domains as well as AlphaFold-derived structures of Nit-9A and Nit-9B. (c) Reactions catalyzed by the molybdenum insertases.

where one monomer's subdomain III interacts with a subdomain II of another monomer [92]. The latter represents the most mobile part of the E domain, probably indicating different conformations in the catalytic process [93, 99].

The modification of MPT through adenylylation is catalyzed by the G domain of gephyrin/Cnx1 in its trimeric form following MPT binding near an invariant Gly-Gly-Thr-Gly motif [94]. The reaction itself depends on Mg^{2+} and ATP resulting in the formation of MPT-AMP. It is remarkable that the adenylylation of MPT is highly conserved in all kingdoms of life and can be reconstituted by enzymes of very diverse evolutionary origin [100]. While MPT-AMP was first discovered within the crystal structure of the G-domain in Cnx1, there was also a bound copper identified at the MPT dithiolene [88]. However, until today, it remains enigmatic, which function copper might play in Moco biosynthesis despite other indirect links between copper and molybdenum chemistry and biochemistry [101].

The G-domain reaction product (MPT-AMP) is subsequently handed over to the neighboring E-domain where it binds in a cooperative manner with molybdate [89]. It has been postulated that the adenylylation of MPT primarily serves as an anchor on MPT allowing a binding inside the E-domain's active center where the final transfer of molybdate to the MPT dithiolene completes Moco synthesis [88, 90].

Work on plant Cnx1 E-domain disclosed a substrate-bound crystal structure thus providing the structural basis of metal insertion [102]. MPT-AMP binds to the E-domain via a specific adenine-binding pocket, while molybdate is coordinated by four residues at the molybdate binding site [103]. Sensing the two phosphate groups in bound MPT-AMP triggers a conformational change in a loop, facilitating molybdenum insertion and hydrolysis of MPT-AMP leading to AMP and Moco release [103]. An octahedral Zn^{2+}–water complex catalyzes the hydrolysis, coordinated by an aspartate residue (Asp274 in Cnx1 and Asp580 in gephyrin), which aids in the hydrolytic cleavage of the bond between the two phosphate groups [99].

In vitro studies have shown that full-length gephyrin exhibits over 300-fold higher Moco biosynthetic activity compared to its separated G and E domains at comparable concentrations, demonstrating the beneficial effect of domain fusion [104]. Interestingly, a truncation of the C domain markedly reduced the catalytic activity of gephyrin, suggesting the importance of an ideal orientation of both domains to each other to facilitate product–substrate channeling [104]. It has been suggested that the fusion of the ancestral *mogA* and *moeA* genes in *GPHN*, *Cnx1*, and *Nit-9* represents the result of product–substrate channeling in domain-fused proteins [104].

While gephyrin-derived Moco is able to reconstitute apo-SOX *in vitro*, XOR family enzymes require a final sulfuration step of Moco [105]; the human molybdenum cofactor sulfurase (HMCS) (referred to as ABA3 in plants) catalyzes this reaction. HMCS is a homodimeric protein composed of two domains [40], the N-terminal cysteine desulfurase domain that mobilizes sulfur in a pyridoxal phosphate-dependent manner [106] and transfers it to the C-terminal domain by means of an intramolecular persulfide relay [107, 108]. The C-terminal domain of HMCS binds Moco und is proposed to undergo sulfuration prior to its insertion into XOR and AOX.

The domain fusion in gephyrin and homologous eukaryotic proteins, driven by the advantage of channeling the intermediates in Moco biosynthesis, provides the origin for novel functions in higher eukaryotes. For this reason, gephyrin does not adhere to the *MOCS* nomenclature of human Moco-synthesizing genes, as it was initially identified as a glycine receptor-associated protein [109]. Thus, the name gephyrin descends from the Greek word gephyra (*Γέφυρα*) for "bridge", as gephyrin was initially found to bridge components of the cytoskeleton with post-synaptic receptors. Today, it is well established that gephyrin represents a major constituent of the inhibitory postsynaptic density where it plays a crucial structural and signaling role in synaptic biology within the central nervous system [110, 111]. Through its E-domain, it binds to the large intracellular domains of various subunits of gamma-aminobutyric acid (GABA) type A and glycine receptors [92]. It is remarkable that the active site for Moco synthesis is distinct of the receptor binding site while both functions seem to require E-domain dimerization.

Structural information for the full-length protein of any of the eukaryotic molybdenum insertases is lacking due to their high conformational flexibility [112]. The extended C-domain in gephyrin is central to its conformational diversity and has been classified as intrinsically structural disordered sequence with multiple binding motifs for proteins that impact gephyrin's neuronal function. In addition, gephyrin's C domain is the area subjected primarily to different types of post-translational modifications such as nitrosylation [113], phosphorylation [114], and palmitoylation [115] that collectively define gephyrin's function as key regulator of inhibitory synapses [116]. Additionally, alternative splicing that affects the C-domain further amplifies the functional diversity of gephyrin [117, 118]. Overall, gephyrin can be seen as a prototype for an expanding family of moonlighting proteins that have acquired novel functions derived from ancient evolutionary pathways [119].

8.7 Deficiency of Moco

Mutations in *MOCS1*, *MOCS2*, and *GEPH* genes result in Moco deficiency (MoCD), a rare metabolic disorder that was first reported in 1978 [120]. Since that time, more than 200 patients have been documented with MoCD [121] due to mutations affecting primarily the first two biosynthetic steps in Moco biosynthesis while only very few cases were found in step 3 [85]. As a result, patients were categorized into three distinct types (A, B, and C) of MoCD [34]. Moco-deficiency type A is attributed to mutations in *MOCS1* [61] and is the most common form of the disease, followed by type B deficiency, which are caused by mutations in *MOCS2* [85, 121]. Since only a single patient has been documented with a mutation in *MOCS3*, the underlying pathology remains unclear given the residual activity associated with the reported missense variant and the heterozygous nature of the carrier [122]. Hence, mutations in *MOCS3* have not been classified as a subset Type of MoCD. But mutations in *GPHN* have been found to cause a very severe form of MoCD type C. With only two cases reported so far, type C is the rarest class of MoCD [123, 124].

Due to the high prevalence of MoCD type A, an animal model of Mocs1 deficiency was established in mice, which exhibited a severe phenotype with neonatal mortality in homozygous individuals. The absence of cPMP in these mice was remedied by administering *E. coli*-derived cPMP through intrahepatic injections, which led to restored Moco enzyme activity and an increased lifespan [125]. A successful clinical translation into the first patient [126] was followed by a prospective study involving a cohort of patients, where some received cPMP shortly after birth, resulting in a normalization of disease-related biomarkers and clinical improvement allowed some patients to reach slightly delayed age developmental milestones [127].

All three categories of MoCD cause disruption in the functionality of all four Moco-dependent enzymes in humans. Individuals affected by MoCD frequently struggle with food intake, produce piercing cries, and manifest various severe neurological disorders. They often suffer from seizures characterized by muscular rigidity and rhythmic contractions. Some patients may also experience dislocation of the eye lens and abnormal facial characteristics [120]. These clinical signs are predominantly caused by the disruption of SOX function, as evidenced by a comparable disease progression observed in isolated SOX deficiency (ISOD) [128].

The lack of SOX leads to a buildup of harmful sulfite in patients, which is a transient intermediate of cysteine breakdown. Accumulated sulfite disrupts disulfide bonds in proteins and sulfur-containing metabolites, resulting in the formation of S-sulfonated compounds. Examples of these include S-sulfonated transthyretin and

S-sulfocysteine (SSC) [129, 130]. SSC bears a molecular resemblance to glutamate. Early research has proposed that elevated concentrations of SSC can have detrimental effects on the nervous system, potentially leading to cerebral impairment and cognitive deficits in individuals with ISOD and MoCD [131]. Later studies have revealed that S-sulfocysteine (SSC) interacts with N-methyl D-aspartate receptors, promoting Ca^{2+} ion entry into neurons. This influx initiates a cascade of intracellular signals, notably among them the activation of the enzyme calpain. As a result, calpain cleaves multiple neuronal proteins, including gephyrin, leading to a reduction in GABAergic synaptic inhibition. Therefore, the neurotoxic effects associated with SSC can be traced back to hyper-excitation and are considered a significant factor in the neurological decline seen in individuals suffering from MoCD and ISOD [132].

SSC formation is accompanied by cystine depletion in plasma [133, 134]. Cystine levels are critical to maintain sufficient cellular cysteine levels being particularly important for biosynthesis of glutathione, the main low-molecular thiol and antioxidant in the cell. When cysteine becomes scarce, it can lead to ferroptosis, a specific type of non-apoptotic cellular death. This condition arises from the impaired function of glutathione-dependent mechanisms responsible for clearing lipid peroxidation products [135].

Besides reduced or depleted levels of cystine, plasma total homocysteine was also found to be reduced in MoCD patients [136, 137]. The underlying molecular mechanisms for the latter symptom remain unexplored, but potential explanations include the release of protein-bound homocysteine due to sulfite cleavage and/or an enhanced conversion of homocysteine to cysteine through the transsulfuration pathway, aimed at replenishing diminished cysteine levels.

Sulfite is generated through two distinct metabolic routes: the first involves the oxidative pathway, which utilizes cysteine dioxygenase (CDO) and glutamate oxaloacetate transaminase (GOT) [138]. The second pathway allows for the transformation of cysteine into hydrogen sulfide (H_2S) via various enzymatic processes, including cystathionine γ-lyase (CSE), cystathionine β-synthase (CBS), and the combined action of GOT and 3-mercaptopyruvate sulfurtransferase (MPST) [139]. Alternatively, cysteine can be directly converted into cysteine persulfide [140]. While H_2S is maintained at low concentrations at a steady state, the flux is considered high due to constant synthesis and degradation. In mitochondria, H_2S is oxidized by sulfur quinone oxidoreductase (SQR) leading again to persulfides [141], which have been shown to serve protective functions against reactive oxygen species (ROS) [142, 143]. Persulfide-bearing substances, such as glutathione persulfide, are subject to enzymatic modification via two distinct routes. One pathway utilizes persulfide dioxygenase (PDO) to generate sulfite, while the other employs thiosulfate sulfur transferase (TST) to yield thiosulfate as the end product [144]. MoCD and ISOD are characterized by elevated thiosulfate excretion, suggesting that an enhanced sulfur flux through the hydrogen sulfide (H_2S) pathway may play a role in the disease process. Another investigation, noting the shared accumulation of thiosulfate in MoCD, ISOD, and also ethylmalonic encephalopathy deficiency, uncovered comparable patterns of mitochondrial impairment across these conditions [145].

The clinical manifestations of MoCD and ISOD are virtually indistinguishable, necessitating the use of alternative diagnostic markers. A key differentiating factor is the lack of XO function in MoCD, which causes increased concentrations of xanthine and hypoxanthine, concurrent with uric acid deficiency. This specific metabolic pattern provides a dependable means of discerning MoCD from ISOD. Given that patients with xanthinuria type 1, caused by XOR deficiency, show only mild pathologies, that is, kidney failure due to xanthine stones, the overall contribution of XOR to MoCD pathology is rather low. To which extend defects of mARC enzymes and AOX may also cause clinical symptoms remains to be studied in the future.

8.8 The Intimate Relation Among Moco Synthesis, Moco Enzymes, and Fe–S Clusters

Moco biosynthesis is a highly conserved multistep pathway required to ensure catalytic activity of five different enzymes in eukaryotes. Both the biosynthesis of Moco and Moco-dependent enzymes are connected to the biology of Fe–S clusters in multiple ways. The radical SAM enzyme MOCS1A (Cnx2 in plants and Nit-7 in fungi) harbors two highly reactive and oxygen-sensitive [4Fe–4S] clusters [48]. Strong experimental evidence suggests that MOCS1A and homologous proteins are translocated to the mitochondrial matrix [35]. In addition, data from patients suggest that MOCS1A may also function in the cytosol [68], which in turn may require that MOCS1A would be able to interact with both the mitochondrial and the cytosolic Fe–S cluster machinery. Indirect evidence from plants supports this idea because mutant cells defective in ATM3, the transporter of an Fe–S cluster precursor that is required for cytosolic Fe–S cluster biosynthesis and assembly [146], show partially impaired Moco synthesis and Moco enzyme activity [65]. Given the critical importance of Moco enzymes in metabolism, one may consider that duplication of the first, most complex, and rate-limiting Moco synthetic step could provide an evolutionary advantage.

The second step in Moco synthesis involves ATP-dependent thiolation of the small subunit of MPT synthase. This process, like the initial steps of Fe–S cluster biogenesis, requires cysteine desulfuration and transient persulfidation, a reaction catalyzed by PLP-dependent enzymes [147]. Consequently, the availability of cysteine or PLP likely affects both Moco and Fe–S cluster biosynthesis. Defects in Fe–S cluster synthesis can therefore impact downstream proteins such MOCS1A, potentially causing secondary Moco deficiency or XOR deficiency, as XOR activity depends on cytosolic Fe–S cluster synthesis. Conversely, loss of Moco or SOX activity leads to sulfite accumulation, which can reduce cysteine levels [148]. Sulfite, being a strong nucleophile, may target oxidized cysteines, potentially depleting persulfides and forming S-sulfoxides [148]. This could affect NFS1-dependent persulfide formation in Fe–S cluster synthesis, ultimately causing Fe–S cluster deficiency. In essence, these lines may illustrate the complexity of how Moco and Fe–S synthesis and their functions may be interconnected. Understanding these relations could help to explain the severe and complex pathologies observed in MoCD and SOX deficiency on the one hand and aspects in alteration of Fe–S biology on the other hand.

References

1 Spatzal, T., Aksoyoglu, M., Zhang, L. et al. (2011). Evidence for interstitial carbon in nitrogenase FeMo cofactor. *Science* 334: 940. https://doi.org/10.1126/science.1214025.

2 Lill, R. and Muhlenhoff, U. (2005). Iron-sulfur-protein biogenesis in eukaryotes. *Trends Biochem. Sci.* 30: 133–141.

3 Yokoyama, K. (2018). Radical breakthroughs in natural product and cofactor biosynthesis. *Biochemistry* 57: 390–402. https://doi.org/10.1021/acs.biochem.7b00878.

4 Kisker, C., Schindelin, H., Pacheco, A. et al. (1997). Molecular basis of sulfite oxidase deficiency from the structure of sulfite oxidase. *Cell* 91: 973–983.

5 Schwarz, G. (2005). Molybdenum cofactor biosynthesis and deficiency. *Cell. Mol. Life Sci.* 62: 2792–2810.

6 Leimkuhler, S. (2020). The biosynthesis of the molybdenum cofactors in *Escherichia coli*. *Environ. Microbiol.* 22: 2007–2026. https://doi.org/10.1111/1462-2920.15003.

7 Spiegel, R., Schwahn, B.C., Squires, L., and Confer, N. (2022). Molybdenum cofactor deficiency: A natural history. *J. Inherit. Metab. Dis.* 45: 456–469. https://doi.org/10.1002/jimd.12488.

8 Hille, R. (2002). Molybdenum enzymes containing the pyranopterin cofactor: An overview. *Met. Ions Biol. Syst.* 39: 187–226.

9 Hille, R. (1996). The mononuclear molybdenum enzymes. *Chem. Rev.* 96: 2757–2816.

10 Abramson, J., Adler, J., Dunger, J. et al. (2024). Accurate structure prediction of biomolecular interactions with AlphaFold 3. *Nature* 630: 493–500. https://doi.org/10.1038/s41586-024-07487-w.

11 Jumper, J., Evans, R., Pritzel, A. et al. (2021). Highly accurate protein structure prediction with AlphaFold. *Nature* 596: 583–589. https://doi.org/10.1038/s41586-021-03819-2.

12 Klein, J.M. and Schwarz, G. (2012). Cofactor-dependent maturation of mammalian sulfite oxidase links two mitochondrial import pathways. *J. Cell Sci.* 125: 4876–4885. https://doi.org/10.1242/jcs.110114.

13 Schrader, N., Fischer, K., Theis, K. et al. (2003). The crystal structure of plant sulfite oxidase provides insights into sulfite oxidation in plants and animals. *Structure* 11: 1251–1263.

14 Feng, C., Kedia, R.V., Hazzard, J.T. et al. (2002). Effect of solution viscosity on intramolecular electron transfer in sulfite oxidase. *Biochemistry* 41: 5816–5821.

15 Lambeck, I., Chi, J.C., Krizowski, S. et al. (2010). Kinetic analysis of 14-3-3-inhibited Arabidopsis thaliana nitrate reductase. *Biochemistry* 49: 8177–8186. https://doi.org/10.1021/bi1003487.

16 Lambeck, I.C., Fischer-Schrader, K., Niks, D. et al. (2012). Molecular mechanism of 14-3-3 protein-mediated inhibition of plant nitrate reductase. *J. Biol. Chem.* 287: 4562–4571. https://doi.org/10.1074/jbc.M111.323113.

17 Schwarz, G., Mendel, R.R., and Ribbe, M.W. (2009). Molybdenum cofactors, enzymes and pathways. *Nature* 460: 839–847. https://doi.org/10.1038/nature08302.

18 Kelley, E.E., Khoo, N.K., Hundley, N.J. et al. (2010). Hydrogen peroxide is the major oxidant product of xanthine oxidase. *Free Radic. Biol. Med.* 48: 493–498. https://doi.org/10.1016/j.freeradbiomed.2009.11.012.

19 Coelho, C., Mahro, M., Trincao, J. et al. (2012). The first mammalian aldehyde oxidase crystal structure: Insights into substrate specificity. *J. Biol. Chem.* 287: 40690–40702. https://doi.org/10.1074/jbc.M112.390419.

20 Terao, M., Garattini, E., Romao, M.J., and Leimkuhler, S. (2020). Evolution, expression, and substrate specificities of aldehyde oxidase enzymes in eukaryotes. *J. Biol. Chem.* 295: 5377–5389. https://doi.org/10.1074/jbc.REV119.007741.

21 Garattini, E. and Terao, M. (2013). Aldehyde oxidase and its importance in novel drug discovery: Present and future challenges. *Expert Opin. Drug Discov.* 8: 641–654. https://doi.org/10.1517/17460441.2013.788497.

22 Seo, M., Peeters, A.J., Koiwai, H. et al. (2000). The arabidopsis aldehyde oxidase 3 (AAO3) gene product catalyzes the final step in abscisic acid biosynthesis in leaves. *Proc. Nat. Acad. Sci. USA.* 97: 12908–12913.

23 Havemeyer, A., Bittner, F., Wollers, S. et al. (2006). Identification of the missing component in the mitochondrial benzamidoxime prodrug-converting system as a novel molybdenum enzyme. *J. Biol. Chem.* 281: 34796–34802. https://doi.org/10.1074/jbc.M607697200.

24 Klopp, C., Zhang, X., Campbell, M.K. et al. (2024). mARC1 is the main contributor to metabolic reduction of N-hydroxyurea. *J. Med. Chem.* 64 (15): 10967–10980. https://doi.org/10.1021/acs.jmedchem.4c01148.

25 Kotthaus, J., Wahl, B., Havemeyer, A. et al. (2011). Reduction of N(omega)-hydroxy-L-arginine by the mitochondrial amidoxime reducing component (mARC). *Biochem. J.* 433: 383–391. https://doi.org/10.1042/BJ20100960.

26 Bender, D. and Schwarz, G. (2018). Nitrite-dependent nitric oxide synthesis by molybdenum enzymes. *FEBS Lett.* 592: 2126–2139. https://doi.org/10.1002/1873-3468.13089.

27 Bender, D., Tobias Kaczmarek, A., Niks, D. et al. (2019). Mechanism of nitrite-dependent NO synthesis by human sulfite oxidase. *Biochem. J.* 476: 1805–1815. https://doi.org/10.1042/BCJ20190143.

28 Maia, L.B. and Moura, J.J. (2015). Nitrite reduction by molybdoenzymes: A new class of nitric oxide-forming nitrite reductases. *J. Biol. Inorg. Chem.* 20: 403–433. https://doi.org/10.1007/s00775-014-1234-2.

29 Mohn, M.A., Thaqi, B., and Fischer-Schrader, K. (2019). Isoform-specific NO synthesis by arabidopsis thaliana nitrate reductase. *Plants.* 8: 67. https://doi.org/10.3390/plants8030067.

30 Sparacino-Watkins, C.E., Tejero, J., Sun, B. et al. (2014). Nitrite reductase and nitric-oxide synthase activity of the mitochondrial molybdopterin enzymes mARC1 and mARC2. *J. Biol. Chem.* 289: 10345–10358. https://doi.org/10.1074/jbc.M114.555177.

31 Wang, J., Krizowski, S., Fischer-Schrader, K. et al. (2015). Sulfite oxidase catalyzes single-electron transfer at molybdenum domain to reduce nitrite to nitric oxide. *Antioxid. Redox Signal.* 23: 283–294. https://doi.org/10.1089/ars.2013.5397.

32 Christie, I.N., Theparambil, S.M., Braga, A. et al. (2023). Astrocytes produce nitric oxide via nitrite reduction in mitochondria to regulate cerebral blood flow during brain hypoxia. *Cell Rep.* 42: 113514. https://doi.org/10.1016/j.celrep.2023.113514.

33 Gladwin, M.T., Schechter, A.N., Kim-Shapiro, D.B. et al. (2005). The emerging biology of the nitrite anion. *Nat. Chem. Biol.* 1: 308–314. https://doi.org/10.1038/nchembio1105-308.

34 Schwarz, G. (2016). Molybdenum cofactor and human disease. *Curr. Opin. Chem. Biol.* 31: 179–187. https://doi.org/10.1016/j.cbpa.2016.03.016.

35 Mayr, S.J., Roper, J., and Schwarz, G. (2020). Alternative splicing of the bicistronic gene molybdenum cofactor synthesis 1 (MOCS1) uncovers a novel mitochondrial protein maturation mechanism. *J. Biol. Chem.* 295: 3029–3039. https://doi.org/10.1074/jbc.RA119.010720.

36 Hahnewald, R., Leimkuhler, S., Vilaseca, A. et al. (2006). A novel MOCS2 mutation reveals coordinated expression of the small and large subunit of molybdopterin synthase. *Mol. Genet. Metab.* 89: 210–213. https://doi.org/10.1016/j.ymgme.2006.04.008.

37 Stallmeyer, B., Drugeon, G., Reiss, J. et al. (1999a). Human molybdopterin synthase gene: Identification of a bicistronic transcript with overlapping reading frames. *Am. J. Hum. Genet.* 64: 698–705.

38 Matthies, A., Nimtz, M., and Leimkuhler, S. (2005). Molybdenum cofactor biosynthesis in humans: Identification of a persulfide group in the rhodanese-like domain of MOCS3 by mass spectrometry. *Biochemistry* 44: 7912–7920. https://doi.org/10.1021/bi0503448.

39 Matthies, A., Rajagopalan, K.V., Mendel, R.R., and Leimkuhler, S. (2004a). Evidence for the physiological role of a rhodanese-like protein for the biosynthesis of the molybdenum cofactor in humans. *Proc. Natl. Acad. Sci. U S A* 101: 5946–5951. https://doi.org/10.1073/pnas.0308191101.

40 Bittner, F., Oreb, M., and Mendel, R.R. (2001). ABA3 is a molybdenum cofactor sulfurase required for activation of aldehyde oxidase and xanthine dehydrogenase in *Arabidopsis thaliana*. *J. Biol. Chem.* 276: 40381–40384.

41 Ichida, K., Matsumura, T., Sakuma, R. et al. (2001). Mutation of human molybdenum cofactor sulfurase gene is responsible for classical xanthinuria type II. *Biochem. Biophys. Res. Commun.* 282: 1194–1200.

42 Zhang, Y. and Gladyshev, V.N. (2008). Molybdoproteomes and evolution of molybdenum utilization. *J. Mol. Biol.* 379: 881–899. https://doi.org/10.1016/j.jmb.2008.03.051.

43 Mayr, S.J., Mendel, R.R., and Schwarz, G. (2021). Molybdenum cofactor biology, evolution and deficiency. *Biochim. Biophys. Acta, Mol. Cell Res.* 1868: 118883. https://doi.org/10.1016/j.bbamcr.2020.118883.

44 Hover, B.M. and Yokoyama, K. (2015). C-Terminal glycine-gated radical initiation by GTP 3′,8-cyclase in the molybdenum cofactor biosynthesis. *J. Am. Chem. Soc.* 137: 3352–3359. https://doi.org/10.1021/ja512997j.

45 Hoff, T., Schnorr, K.M., Meyer, C., and Caboche, M. (1995). Isolation of two Arabidopsis cDNAs involved in early steps of molybdenum cofactor biosynthesis by functional complementation of *Escherichia coli* mutants. *J. Biol. Chem.* 270: 6100–6107.

46 Heck, I.S. and Ninnemann, H. (1995). Molybdenum cofactor biosynthesis in *Neurospora crassa*: Biochemical characterization of pleiotropic molybdoenzyme mutants nit-7, nit-8, nit- 9A, B and C. *Photochem. Photobiol.* 61: 54–60.

47 Wajmann, S., Hercher, T.W., Buchmeier, S. et al. (2020). The first step of neurospora crassa molybdenum cofactor biosynthesis: Regulatory aspects under N-derepressing and nitrate-inducing conditions. *Microorganisms.* 8: 534. https://doi.org/10.3390/microorganisms8040534.

48 Hänzelmann, P. and Schindelin, H. (2004). Crystal structure of the S-adenosylmethionine-dependent enzyme MoaA and its implications for molybdenum cofactor deficiency in humans. *Proc. Natl. Acad. Sci. U S A* 101: 12870–12875.

49 Wuebbens, M.M., Liu, M.T.W., Rajagopalan, K.V., and Schindelin, H. (2000). Insights into molybdenum cofactor deficiency provided by the crystal structure of the molybdenum cofactor biosynthesis protein MoaC. *Structure* 8: 709–718.

50 Sofia, H.J., Chen, G., Hetzler, B.G. et al. (2001). Radical SAM, a novel protein superfamily linking unresolved steps in familiar biosynthetic pathways with radical mechanisms: Functional characterization using new analysis and information visualization methods. *Nucleic Acids Res.* 29: 1097–1106.

51 Walsby, C.J., Hong, W., Broderick, W.E. et al. (2002). Electron-nuclear double resonance spectroscopic evidence that S-adenosylmethionine binds in contact with the catalytically active [4Fe–4S](+) cluster of pyruvate formate-lyase activating enzyme. *J. Am. Chem. Soc.* 124: 3143–3151. https://doi.org/10.1021/ja012034s.

52 Landgraf, B.J., McCarthy, E.L., and Booker, S.J. (2016). Radical S-adenosylmethionine enzymes in human health and disease. *Annu. Rev. Biochem.* 85: 485–514. https://doi.org/10.1146/annurev-biochem-060713-035504.

53 Haenzelmann, P., Hernandez, H.L., Menzel, C. et al. (2004). Characterization of MOCS1A, an oxygen-sensitive iron-sulfur protein involved in human molybdenum cofactor biosynthesis. *J. Biol. Chem.* 279 (33): 34721–34732.

54 Morikawa, T., Yasuno, R., and Wada, H. (2001). Do mammalian cells synthesize lipoic acid? Identification of a mouse cDNA encoding a lipoic acid synthase located in mitochondria. *FEBS Lett.* 498: 16–21. https://doi.org/10.1016/s0014-5793(01)02469-3.

55 Reiter, V., Matschkal, D.M., Wagner, M. et al. (2012). The CDK5 repressor CDK5RAP1 is a methylthiotransferase acting on nuclear and mitochondrial RNA. *Nucleic Acids Res.* 40: 6235–6240. https://doi.org/10.1093/nar/gks240.

56 Hover, B.M., Loksztejn, A., Ribeiro, A.A., and Yokoyama, K. (2013). Identification of a cyclic nucleotide as a cryptic intermediate in molybdenum cofactor biosynthesis. *J. Am. Chem. Soc.* 135: 7019–7032. https://doi.org/10.1021/ja401781t.

57 Pang, H., Lilla, E.A., Zhang, P. et al. (2020). Mechanism of rate acceleration of radical C–C bond formation reaction by a radical SAM GTP 3′,8-cyclase. *J. Am. Chem. Soc.* 142: 9314–9326. https://doi.org/10.1021/jacs.0c01200.

58 Pang, H., Walker, L.M., Silakov, A. et al. (2021). Mechanism of reduction of an aminyl radical intermediate in the radical SAM GTP 3′,8-cyclase MoaA. *J. Am. Chem. Soc.* 143: 13835–13844. https://doi.org/10.1021/jacs.1c06268.

59 Santamaria-Araujo, J.A., Fischer, B., Otte, T. et al. (2004). The tetrahydropyranopterin structure of the sulfur-free and metal-free molybdenum cofactor precursor. *J. Biol. Chem.* 279: 15994–15999.

60 Clinch, K., Watt, D.K., Dixon, R.A. et al. (2013). Synthesis of cyclic pyranopterin monophosphate, a biosynthetic intermediate in the molybdenum cofactor pathway. *J. Med. Chem.* 56: 1730–1738. https://doi.org/10.1021/jm301855r.

61 Reiss, J., Cohen, N., Dorche, C. et al. (1998). Mutations in a polycistronic nuclear gene associated with molybdenum cofactor deficiency. *Nat. Genet.* 20: 51–53.

62 Gross-Hardt, S. and Reiss, J. (2002). The bicistronic MOCS1 gene has alternative start codons on two mutually exclusive exons. *Mol. Genet. Metab.* 76: 340–343.

63 Braymer, J.J., Freibert, S.A., Rakwalska-Bange, M., and Lill, R. (2021). Mechanistic concepts of iron-sulfur protein biogenesis in biology. *Biochim. Biophys. Acta, Mol. Cell Res.* 1868: 118863. https://doi.org/10.1016/j.bbamcr.2020.118863.

64 Johnson, J.L., Wuebbens, M.M., Mandell, R., and Shih, V.E. (1989). Molybdenum cofactor biosynthesis in humans. Identification of two complementation groups of cofactor-deficient patients and preliminary characterization of a diffusible molybdopterin precursor. *J. Clin. Invest.* 83: 897–903.

65 Teschner, J., Lachmann, N., Schulze, J. et al. (2010). A novel role for Arabidopsis mitochondrial ABC transporter ATM3 in molybdenum cofactor biosynthesis. *Plant Cell* 22: 468–480. https://doi.org/10.1105/tpc.109.068478.

66 Kruse, I., Maclean, A.E., Hill, L., and Balk, J. (2018). Genetic dissection of cyclic pyranopterin monophosphate biosynthesis in plant mitochondria. *Biochem. J.* 475: 495–509. https://doi.org/10.1042/BCJ20170559.

67 Marquez, M.D., Greth, C., Buzuk, A. et al. (2023). Cytosolic iron-sulfur protein assembly system identifies clients by a C-terminal tripeptide. *Proc. Natl. Acad. Sci. U S A* 120: e2311057120. https://doi.org/10.1073/pnas.2311057120.

68 Mayr, S.J., Sass, J.O., Vry, J. et al. (2018). A mild case of molybdenum cofactor deficiency defines an alternative route of MOCS1 protein maturation. *J. Inherit. Metab. Dis.* 41: 187–196. https://doi.org/10.1007/s10545-018-0138-7.

69 Gutzke, G., Fischer, B., Mendel, R.R., and Schwarz, G. (2001). Thiocarboxylation of molybdopterin synthase provides evidence for the mechanism of dithiolene formation in metal-binding pterins. *J. Biol. Chem.* 276: 36268–36274.

70 Rudolph, M.J., Wuebbens, M.M., Rajagopalan, K.V., and Schindelin, H. (2001). Crystal structure of molybdopterin synthase and its evolutionary relationship to ubiquitin activation. *Nat. Struct. Biol.* 8: 42–46.

71 Daniels, J.N., Wuebbens, M.M., Rajagopalan, K.V., and Schindelin, H. (2008). Crystal structure of a molybdopterin synthase-precursor Z complex: Insight into its sulfur transfer mechanism and its role in molybdenum cofactor deficiency. *Biochemistry* 47: 615–626. https://doi.org/10.1021/bi701734g.

72 Marelja, Z., Mullick Chowdhury, M., Dosche, C. et al. (2013). The L-cysteine desulfurase NFS1 is localized in the cytosol where it provides the sulfur for molybdenum cofactor biosynthesis in humans. *PLoS One* 8: e60869. https://doi.org/10.1371/journal.pone.0060869.

73 Marelja, Z., Stocklein, W., Nimtz, M., and Leimkuhler, S. (2008). A novel role for human Nfs1 in the cytoplasm: Nfs1 acts as a sulfur donor for MOCS3, a protein involved in molybdenum cofactor biosynthesis. *J. Biol. Chem.* 283: 25178–25185. https://doi.org/10.1074/jbc.M804064200.

74 Frasdorf, B., Radon, C., and Leimkuhler, S. (2014). Characterization and interaction studies of two isoforms of the dual localized 3-mercaptopyruvate sulfurtransferase TUM1 from humans. *J. Biol. Chem.* 289: 34543–34556. https://doi.org/10.1074/jbc.M114.605733.

75 Ogunkola, M.O., Guiraudie-Capraz, G., Feron, F., and Leimkuhler, S. (2023). The human mercaptopyruvate sulfurtransferase TUM1 is involved in Moco biosynthesis, cytosolic tRNA thiolation and cellular bioenergetics in human embryonic kidney cells. *Biomol. Ther.* 13 (1): 144. https://doi.org/10.3390/biom13010144.

76 Suganuma, T., Swanson, S.K., Florens, L. et al. (2016). Moco biosynthesis and the ATAC acetyltransferase engage translation initiation by inhibiting latent PKR activity. *J. Mol. Cell Biol.* 8: 44–50. https://doi.org/10.1093/jmcb/mjv070.

77 Suganuma, T., Mushegian, A., Swanson, S.K. et al. (2012). A metazoan ATAC acetyltransferase subunit that regulates mitogen-activated protein kinase signaling is related to an ancient molybdopterin synthase component. *Mol. Cell. Proteomics* 11: 90–99. https://doi.org/10.1074/mcp.M111.015818.

78 Suganuma, T., Swanson, S.K., Gogol, M. et al. (2018). MPTAC determines APP fragmentation via sensing sulfur amino acid catabolism. *Cell Rep.* 24: 1585–1596. https://doi.org/10.1016/j.celrep.2018.07.013.

79 Dey, M., Trieselmann, B., Locke, E.G. et al. (2005). PKR and GCN2 kinases and guanine nucleotide exchange factor eukaryotic translation initiation factor 2B (eIF2B) recognize overlapping surfaces on eIF2alpha. *Mol. Cell. Biol.* 25: 3063–3075. https://doi.org/10.1128/MCB.25.8.3063-3075.2005.

80 Chowdhury, M.M., Dosche, C., Lohmannsroben, H.G., and Leimkuhler, S. (2012). Dual role of the molybdenum cofactor biosynthesis protein MOCS3 in tRNA thiolation and molybdenum cofactor biosynthesis in humans. *J. Biol. Chem.* 287: 17297–17307. https://doi.org/10.1074/jbc.M112.351429.

81 Nakai, Y., Harada, A., Hashiguchi, Y. et al. (2012). Arabidopsis molybdopterin biosynthesis protein Cnx5 collaborates with the ubiquitin-like protein Urm11 in the thio-modification of tRNA. *J. Biol. Chem.* 287: 30874–30884. https://doi.org/10.1074/jbc.M112.350090.

82 Schmitz, J., Chowdhury, M.M., Hanzelmann, P. et al. (2008). The sulfurtransferase activity of Uba4 presents a link between ubiquitin-like protein conjugation and activation of sulfur carrier proteins. *Biochemistry* 47: 6479–6489. https://doi.org/10.1021/bi800477u.

83 Cappadocia, L. and Lima, C.D. (2018). Ubiquitin-like protein conjugation: Structures, chemistry, and mechanism. *Chem. Rev.* 118: 889–918. https://doi.org/10.1021/acs.chemrev.6b00737.

84 Khoshnood, B., Dacklin, I., and Grabbe, C. (2017). A proteomics approach to identify targets of the ubiquitin-like molecule Urm1 in Drosophila melanogaster. *PLoS One* 12: e0185611. https://doi.org/10.1371/journal.pone.0185611.

85 Reiss, J. and Hahnewald, R. (2011). Molybdenum cofactor deficiency: Mutations in GPHN, MOCS1, and MOCS2. *Hum. Mutat.* 32: 10–18. https://doi.org/10.1002/humu.21390.

86 Marelja, Z., Leimkuhler, S., and Missirlis, F. (2018). Iron sulfur and molybdenum cofactor enzymes regulate the drosophila life cycle by controlling cell metabolism. *Front. Physiol.* 9: 50. https://doi.org/10.3389/fphys.2018.00050.

87 Stallmeyer, B., Schwarz, G., Schulze, J. et al. (1999b). The neurotransmitter receptor-anchoring protein gephyrin reconstitutes molybdenum cofactor biosynthesis in bacteria, plants, and mammalian cells. *Proc. Natl. Acad. Sci. U S A* 96: 1333–1338.

88 Kuper, J., Llamas, A., Hecht, H.J. et al. (2004). Structure of molybdopterin-bound Cnx1G domain links molybdenum and copper metabolism. *Nature* 430: 806–806.

89 Llamas, A., Mendel, R.R., and Schwarz, G. (2004). Synthesis of adenylated molybdopterin: An essential step for molybdenum insertion. *J. Biol. Chem.* 279: 55241–55246.

90 Llamas, A., Otte, T., Multhaup, G. et al. (2006). The mechanism of nucleotide-assisted molybdenum insertion into molybdopterin: A novel route towards metal cofactor assembly. *J. Biol. Chem.* 281 (27): 18343–18350.

91 Schwarz, G., Schulze, J., Bittner, F. et al. (2000). The molybdenum cofactor biosynthetic protein Cnx1 complements molybdate- repairable mutants, transfers molybdenum to the metal binding pterin, and is associated with the cytoskeleton. *Plant Cell* 12: 2455–2472.

92 Kim, E.Y., Schrader, N., Smolinsky, B. et al. (2006). Deciphering the structural framework of glycine receptor anchoring by gephyrin. *EMBO J.* 25: 1385–1395.

93 Krausze, J., Probst, C., Curth, U. et al. (2017). Dimerization of the plant molybdenum insertase Cnx1E is required for synthesis of the molybdenum cofactor. *Biochem. J.* 474: 163–178. https://doi.org/10.1042/BCJ20160846.

94 Schwarz, G., Schrader, N., Mendel, R.R. et al. (2001). Crystal structures of human gephyrin and plant Cnx1 G domains: Comparative analysis and functional implications. *J. Mol. Biol.* 312: 405–418.

95 Havarushka, N., Fischer-Schrader, K., Lamkemeyer, T., and Schwarz, G. (2014). Structural basis of thermal stability of the tungsten cofactor synthesis protein MoaB from Pyrococcus furiosus. *PLoS One* 9: e86030. https://doi.org/10.1371/journal.pone.0086030.

96 Liu, M.T., Wuebbens, M.M., Rajagopalan, K.V., and Schindelin, H. (2000). Crystal structure of the gephyrin-related molybdenum cofactor biosynthesis protein MogA from *Escherichia coli*. *J. Biol. Chem.* 275: 1814–1822.

97 Xiang, S., Nichols, J., Rajagopalan, K.V., and Schindelin, H. (2001). The crystal structure of *Escherichia coli* MoeA and its relationship to the multifunctional protein gephyrin. *Structure* 9: 299–310.

98 Bevers, L.E., Hagedoorn, P.L., Santamaria-Araujo, J.A. et al. (2008). Function of MoaB proteins in the biosynthesis of the molybdenum and tungsten cofactors. *Biochemistry* 47: 949–956. https://doi.org/10.1021/bi7020487.

99 Hercher, T.W., Krausze, J., Hoffmeister, S. et al. (2020). Insights into the Cnx1E catalyzed MPT-AMP hydrolysis. *Biosci. Rep.* 40: https://doi.org/10.1042/BSR20191806.

100 Kuper, J., Palmer, T., Mendel, R.R., and Schwarz, G. (2000). Mutations in the molybdenum cofactor biosynthetic protein Cnx1G from Arabidopsis thaliana define functions for molybdopterin bind, Mo-insertion and molybdenum cofactor stabilization. *Proc. Natl. Acad. Sci. U S A* 97: 6475–6480.

101 Mercer, J.F. (2001). The molecular basis of copper-transport diseases. *Trends Mol. Med.* 7: 64–69.

102 Krausze, J., Hercher, T.W., Zwerschke, D. et al. (2018). The functional principle of eukaryotic molybdenum insertases. *Biochem. J.* 475: 1739–1753. https://doi.org/10.1042/BCJ20170935.

103 Kasaragod, V.B. and Schindelin, H. (2016). Structural framework for metal incorporation during molybdenum cofactor biosynthesis. *Structure* 24: 782–788. https://doi.org/10.1016/j.str.2016.02.023.

104 Belaidi, A.A. and Schwarz, G. (2013). Metal insertion into the molybdenum cofactor: Product-substrate channelling demonstrates the functional origin of domain fusion in gephyrin. *Biochem. J.* 450: 149–157. https://doi.org/10.1042/BJ20121078.

105 Wahl, R.C., Hageman, R.V., and Rajagopalan, K.V. (1984). The relationship of Mo, molybdopterin, and the cyanolyzable sulfur in the Mo cofactor. *Arch. Biochem. Biophys.* 230: 264–273.

106 Heidenreich, T., Wollers, S., Mendel, R.R., and Bittner, F. (2005). Characterization of the NifS-like domain of ABA3 from *Arabidopsis thaliana* provides insight into the mechanism of molybdenum cofactor sulfuration. *J. Biol. Chem.* 280: 4213–4218.

107 Lehrke, M., Rump, S., Heidenreich, T. et al. (2012). Identification of persulfide-binding and disulfide-forming cysteine residues in the NifS-like domain of the molybdenum cofactor sulfurase ABA3 by cysteine-scanning mutagenesis. *Biochem. J.* 441: 823–832. https://doi.org/10.1042/BJ20111170.

108 Wollers, S., Heidenreich, T., Zarepour, M. et al. (2008). Binding of sulfurated molybdenum cofactor to the C-terminal domain of ABA3 from *Arabidopsis*

thaliana provides insight into the mechanism of molybdenum cofactor sulfuration. *J. Biol. Chem.* 283: 9642–9650. https://doi.org/10.1074/jbc.M708549200.
109 Prior, P., Schmitt, B., Grenningloh, G. et al. (1992). Primary structure and alternative splice variants of gephyrin, a putative glycine receptor-tubulin linker protein. *Neuron* 8: 1161–1170.
110 Fritschy, J.M., Harvey, R.J., and Schwarz, G. (2008). Gephyrin: Where do we stand, where do we go? *Trends Neurosci.* 31: 257–264. https://doi.org/10.1016/j.tins.2008.02.006.
111 Groeneweg, F.L., Trattnig, C., Kuhse, J. et al. (2018). Gephyrin: A key regulatory protein of inhibitory synapses and beyond. *Histochem. Cell Biol.* 150: 489–508. https://doi.org/10.1007/s00418-018-1725-2.
112 Sander, B., Tria, G., Shkumatov, A.V. et al. (2013). Structural characterization of gephyrin by AFM and SAXS reveals a mixture of compact and extended states. *Acta Crystallogr. D Biol. Crystallogr.* 69: 2050–2060. https://doi.org/10.1107/S0907444913018714.
113 Dejanovic, B. and Schwarz, G. (2014). Neuronal nitric oxide synthase-dependent S-nitrosylation of gephyrin regulates gephyrin clustering at GABAergic synapses. *J. Neurosci.* 34: 7763–7768. https://doi.org/10.1523/JNEUROSCI.0531-14.2014.
114 Tyagarajan, S.K., Ghosh, H., Yevenes, G.E. et al. (2011). Regulation of GABAergic synapse formation and plasticity by GSK3beta-dependent phosphorylation of gephyrin. *Proc. Natl. Acad. Sci. U S A* 108: 379–384. https://doi.org/10.1073/pnas.1011824108.
115 Dejanovic, B., Semtner, M., Ebert, S. et al. (2014). Palmitoylation of gephyrin controls receptor clustering and plasticity of GABAergic synapses. *PLoS Biol.* 12: e1001908. https://doi.org/10.1371/journal.pbio.1001908.
116 Tyagarajan, S.K. and Fritschy, J.M. (2014). Gephyrin: A master regulator of neuronal function? *Nat. Rev. Neurosci.* 15: 141–156. https://doi.org/10.1038/nrn3670.
117 Liebsch, F., Eggersmann, F.R., Merkler, Y. et al. (2023). Automated image analysis reveals different localization of synaptic gephyrin C4 splice variants. *eNeuro.* 10: 1–17. https://doi.org/10.1523/ENEURO.0102-22.2022.
118 Smolinsky, B., Eichler, S.A., Buchmeier, S. et al. (2008). Splice-specific functions of gephyrin in molybdenum cofactor biosynthesis. *J. Biol. Chem.* 283: 17370–17379. https://doi.org/10.1074/jbc.M800985200.
119 Copley, S.D. (2003). Enzymes with extra talents: Moonlighting functions and catalytic promiscuity. *Curr. Opin. Chem. Biol.* 7: 265–272.
120 Duran, M., Beemer, F.A., van de Heiden, C. et al. (1978). Combined deficiency of xanthine oxidase and sulphite oxidase: A defect of molybdenum metabolism or transport? *J. Inherit. Metab. Dis.* 1: 175–178.
121 Johannes, L., Fu, C.Y., and Schwarz, G. (2022). Molybdenum cofactor deficiency in humans. *Molecules* 27 (20): 6896. https://doi.org/10.3390/molecules27206896.
122 Huijmans, J.G.M., Schot, R., de Klerk, J.B.C. et al. (2017). Molybdenum cofactor deficiency: Identification of a patient with homozygote mutation in the MOCS3 gene. *Am. J. Med. Genet. A* 173: 1601–1606. https://doi.org/10.1002/ajmg.a.38240.

123 Dejanovic, B., Djemie, T., Grunewald, N. et al. (2015). Simultaneous impairment of neuronal and metabolic function of mutated gephyrin in a patient with epileptic encephalopathy. *EMBO Mol. Med.* 7: 1580–1594. https://doi.org/10.15252/emmm.201505323.

124 Reiss, J., Gross-Hardt, S., Christensen, E. et al. (2001). A mutation in the gene for the neurotransmitter receptor-clustering protein gephyrin causes a novel form of molybdenum cofactor deficiency. *Am. J. Hum. Genet.* 68: 208–213.

125 Schwarz, G., Santamaria-Araujo, J.A., Wolf, S. et al. (2004). Rescue of lethal molybdenum cofactor deficiency by a biosynthetic precursor from Escherichia coli. *Hum. Mol. Genet.* 13: 1249–1255.

126 Veldman, A., Santamaria-Araujo, J.A., Sollazzo, S. et al. (2010). Successful treatment of molybdenum cofactor deficiency type A with cPMP. *Pediatrics* 125: e1249–e1254. https://doi.org/10.1542/peds.2009-2192.

127 Schwahn, B.C., Van Spronsen, F.J., Belaidi, A.A. et al. (2015). Efficacy and safety of cyclic pyranopterin monophosphate substitution in severe molybdenum cofactor deficiency type A: A prospective cohort study. *Lancet* 386: 1955–1963. https://doi.org/10.1016/S0140-6736(15)00124-5.

128 Misko, A.L., Liang, Y., Kohl, J.B., and Eichler, F. (2020). Delineating the phenotypic spectrum of sulfite oxidase and molybdenum cofactor deficiency. *Neurol. Genet.* 6: e486. https://doi.org/10.1212/NXG.0000000000000486.

129 Kishikawa, M., Nakanishi, T., Shimizu, A., and Yoshino, M. (2000). Detection by mass spectrometry of highly increased amount of S-sulfonated transthyretin in serum from a patient with molybdenum cofactor deficiency. *Pediatr. Res.* 47: 492–494. https://doi.org/10.1203/00006450-200004000-00013.

130 Kishikawa, M., Sass, J.O., Sakura, N. et al. (2002). The peak height ratio of S-sulfonated transthyretin and other oxidized isoforms as a marker for molybdenum cofactor deficiency, measured by electrospray ionization mass spectrometry. *Biochim. Biophys. Acta* 1588: 135–138. https://doi.org/10.1016/s0925-4439(02)00156-4.

131 Olney, J.W., Misra, C.H., and de Gubareff, T. (1975). Cysteine-S-sulfate: Brain damaging metabolite in sulfite oxidase deficiency. *J. Neuropathol. Exp. Neurol.* 34: 167–177. https://doi.org/10.1097/00005072-197503000-00005.

132 Kumar, A., Dejanovic, B., Hetsch, F. et al. (2017). S-sulfocysteine/NMDA receptor-dependent signaling underlies neurodegeneration in molybdenum cofactor deficiency. *J. Clin. Invest.* 127: 4365–4378. https://doi.org/10.1172/JCI89885.

133 Barbot, C., Martins, E., Vilarinho, L. et al. (1995). A mild form of infantile isolated sulphite oxidase deficiency. *Neuropediatrics* 26: 322–324. https://doi.org/10.1055/s-2007-979783.

134 Johnson, J.L. and Duran, M. (2001). Molybdenum cofactor deficiency and isolated sulfite oxidase deficiency. In: *The Metabolic and Molecular Bases of Inherited Disease* (ed. C. Scriver, A. Beaudet, W. Sly, and D. Valle), 3163–3177. New York: McGraw-Hill.

135 Stockwell, B.R., Friedmann Angeli, J.P., Bayir, H. et al. (2017). Ferroptosis: A regulated cell death nexus linking metabolism, redox biology, and disease. *Cell* 171: 273–285. https://doi.org/10.1016/j.cell.2017.09.021.

136 Graf, W.D., Oleinik, O.E., Jack, R.M. et al. (1998). Ahomocysteinemia in molybdenum cofactor deficiency. *Neurology* 51: 860–862.

137 Sass, J.O., Kishikawa, M., Puttinger, R. et al. (2003). Hypohomocysteinaemia and highly increased proportion of S-sulfonated plasma transthyretin in molybdenum cofactor deficiency. *J. Inherit. Metab. Dis.* 26: 80–82. https://doi.org/10.1023/a:1024091900547.

138 Bagley, P.J. and Stipanuk, M.H. (1994). The activities of rat hepatic cysteine dioxygenase and cysteinesulfinate decarboxylase are regulated in a reciprocal manner in response to dietary casein level. *J. Nutr.* 124: 2410–2421. https://doi.org/10.1093/jn/124.12.410.

139 Abe, K. and Kimura, H. (1996). The possible role of hydrogen sulfide as an endogenous neuromodulator. *J. Neurosci.* 16: 1066–1071.

140 Akaike, T., Ida, T., Wei, F.Y. et al. (2017). Cysteinyl-tRNA synthetase governs cysteine polysulfidation and mitochondrial bioenergetics. *Nat. Commun.* 8: 1177. https://doi.org/10.1038/s41467-017-01311-y.

141 Kabil, O. and Banerjee, R. (2010). Redox biochemistry of hydrogen sulfide. *J. Biol. Chem.* 285: 21903–21907. https://doi.org/10.1074/jbc.R110.128363.

142 Kabil, H., Kabil, O., Banerjee, R. et al. (2011). Increased transsulfuration mediates longevity and dietary restriction in Drosophila. *Proc. Natl. Acad. Sci. U S A* 108: 16831–16836. https://doi.org/10.1073/pnas.1102008108.

143 Zivanovic, J., Kouroussis, E., Kohl, J.B. et al. (2019). Selective persulfide detection reveals evolutionarily conserved antiaging effects of S-sulfhydration. *Cell Metab.* 30 (1152–1170): e1113. https://doi.org/10.1016/j.cmet.2019.10.007.

144 Hildebrandt, T.M. and Grieshaber, M.K. (2008). Three enzymatic activities catalyze the oxidation of sulfide to thiosulfate in mammalian and invertebrate mitochondria. *FEBS J.* 275: 3352–3361. https://doi.org/10.1111/j.1742-4658.2008.06482.x.

145 Grings, M., Seminotti, B., Karunanidhi, A. et al. (2019). ETHE1 and MOCS1 deficiencies: Disruption of mitochondrial bioenergetics, dynamics, redox homeostasis and endoplasmic reticulum-mitochondria crosstalk in patient fibroblasts. *Sci. Rep.* 9: 12651. https://doi.org/10.1038/s41598-019-49014-2.

146 Zuo, J., Wu, Z., Li, Y. et al. (2017). Mitochondrial ABC transporter ATM3 is essential for cytosolic iron-sulfur cluster assembly. *Plant Physiol.* 173: 2096–2109. https://doi.org/10.1104/pp. 16.01760.

147 Black, K.A. and Dos Santos, P.C. (2015). Shared-intermediates in the biosynthesis of thio-cofactors: Mechanism and functions of cysteine desulfurases and sulfur acceptors. *Biochim. Biophys. Acta* 1853: 1470–1480. https://doi.org/10.1016/j.bbamcr.2014.10.018.

148 Kohl, J.B., Mellis, A.T., and Schwarz, G. (2019). Homeostatic impact of sulfite and hydrogen sulfide on cysteine catabolism. *Br. J. Pharmacol.* 176: 554–570. https://doi.org/10.1111/bph.14464.

9

NifUS Is a Two-component Toolbox Involved in Assembling Fe–S Clusters Associated with Nitrogen Fixation and Beyond

Julia S. Martin del Campo[1], *Shervin Shaybani*[2], *Dennis R. Dean*[1], *and Patricia C. Dos Santos*[2]

[1] *Virginia Polytechnic Institute and State University, Department of Biochemistry, 110 Fralin Hall, Blacksburg, VA 24061, USA*
[2] *Wake Forest University, Department of Chemistry, 455 Vine Street, Wake Downtown, Winston-Salem, NC 27101, USA*

Chapter Goals

- Complexes of iron and sulfur ([Fe–S] clusters) are ubiquitous protein cofactors participating in a wide range of biochemical reactions.
- The synthesis of the [Fe–S] cluster requires dedicated pathways for the recruitment of sulfur and iron, assembly of clusters, and transfer of clusters into Fe–S proteins.
- NifUS is a two-component system initially identified in *Azotobacter vinelandii* due to its role in providing [Fe–S] clusters for the synthesis of nitrogenase cofactors.
- NifS is a cysteine desulfurase involved in sulfur mobilization and NifU is a modular protein that carries out the assembly and transfer of [Fe–S] clusters.
- The distribution of NifU- and NifS-like sequences is not restricted to nitrogen-fixing organisms. Thus, the role of the NifUS system in other species is expanded to supply Fe–S clusters to a general clientele of Fe–S proteins that are not limited to nitrogen fixation.

Main Chapter Points

- Introduction to iron and sulfur [Fe–S] clusters and the NifUS [Fe–S] cluster biosynthetic systems
- Structural and catalytic features of NifU and NifS
- Synthesis of Fe–S clusters by the NifUS system
- The roles of *Azotobacter vinelandii* NifUS in nitrogen fixation
- Occurrence of NifUS-like in non-nitrogen-fixing species

Iron-Sulfur Clusters: Biogenesis and Biochemistry, Volume 1, First Edition.
Edited by Silke Leimkühler, Günter Schwarz, Oliver Lenz, and Oliver Einsle.

9.1 Introduction

The biological synthesis of [Fe–S] clusters involves dedicated enzymes to promote sulfur mobilization, the *ex situ* assembly of [Fe–S] clusters, and the transfer of these clusters to client proteins. Several systems have been described with varying numbers and structural folds of their biosynthetic components [1–3]. These pathways are also subjected to regulatory factors to adjust the production of Fe–S clusters to meet metabolic demands while avoiding the formation of toxic byproducts [4]. The existence of pathways dedicated to the assembly of [Fe–S] clusters was initially disregarded because simple units of [Fe–S] clusters can be assembled in a test tube in reactions with free iron and sulfide in the presence of suitable ligands. However, functional analysis of genes involved in nitrogen fixation (NIF) led to the discovery of the NIF [Fe–S] cluster biosynthetic system composed of NifU and NifS and the subsequent discovery of additional pathways dedicated to the general assembly of Fe–S clusters with roles beyond NIF [5].

The genes coding for NifU and NifS tend to co-occur in close genomic proximity. The *nifUS* genes were first identified in the diazotroph *Klebsiella pneumoniae* [6] and the initial nomenclature of nitrogen-fixing (*nif*) genes was the result of the alphabetic assignment of genes within the nitrogen-fixation gene cluster in that organism. Later genetic studies on the strict aerobe *Azotobacter vinelandii* showed that the deletion of *nifS* or *nifU* led to a drastic reduction of diazotrophy associated with impaired activity of both nitrogenase components, the Fe and MoFe proteins [7]. This initial observation suggested that NifU and NifS could participate in iron and sulfur acquisition since a shared feature between both nitrogenase components is the presence of metallocofactors known to be required for their catalytic functions. While the NifUS pair is required for NIF under aerobic growth [7], it is also known that not all diazotrophs carry genes for the expression of NifU or NifS and that housekeeping systems can replace their functions [8]. Furthermore, sequences similar to NifU and NifS are also present in non-diazotrophs, as in the case of *Helicobacter pylori* and *Entamoeba histolytica* [9, 10], where such system plays a housekeeping role in sustaining the synthesis of [Fe–S] clusters for general purposes.

Biochemical characterization of *A. vinelandii* NifU and NifS established that these proteins constitute the condensed toolbox for the synthesis of Fe–S clusters [11]. NifS promotes the formation and transfer of the persulfide species to be incorporated into nascent [Fe–S] clusters formed on NifU. Serving as a scaffold in this capacity, NifU is able to assemble and coordinate transient species [2Fe–2S] and [4Fe–4S] clusters, providing the initial building blocks for the synthesis of nitrogenase cofactors [12]. Functional units within the NifUS system show sequence similarities with components of the iron–sulfur cluster (ISC) biosynthetic system. The NifS and IscS paralogs are homodimeric pyridoxal-5′-phosphate (PLP) cysteine desulfurases [13]. NifU displays a modular composition containing an N-terminal IscU-like scaffold domain, a central domain coordinating a permanent [2Fe–2S] ferredoxin-like cluster, and a C-terminal NfU-like cluster carrier domain. Despite sequence and functional similarities between components of the ISC and NifUS systems, only very limited functional cross talk exists [14, 15]. Nevertheless, the genetic and biochemical principles of the NifUS system are also observed in other more complex Fe–S cluster biosynthetic systems, making the NifUS system an attractive simplified model for uncovering chemical details involving the biological synthesis of [Fe–S] clusters.

Much of the understanding of the NifUS systems is from genetic and biochemical studies performed in *A. vinelandii*. It is ironic that the first identified biosynthetic system dedicated to the synthesis of simple [Fe–S] clusters was through genetic efforts to understand the assembly of one of Nature's most complex metallocofactors, the nitrogenase FeMo cofactor. The specialized function of the NIF system allowed for clean genetic and functional assignment, as inactivation of *nifUS* in *A. vinelandii* drastically affects the ability of the bacterium to grow under nitrogen-fixing conditions but has no phenotype under conditions that nitrogenase is not required. This feature made possible straightforward genetic manipulation that provided mechanistic insight into biochemical principles of synthesis of [Fe–S] clusters and metallocofactors that are conserved in NIF-independent pathways in most species.

9.2 Structural and Catalytic Features of NifU

NifU is defined by the modular composition of its primary sequence encompassing three distinct functional domains [16]. The *A. vinelandii* protein contains 312 residues, including nine conserved cysteine residues shown to be able to coordinate Fe–S clusters (Figure 9.1a). While these functional domains and Cys ligands are also present in NifU from other species, for simplicity, in this chapter, we refer to residue numbers from *A. vinelandii*, unless otherwise stated. Proteins displaying equivalent functions for each individual domain have been assigned to support the biogenesis of [Fe–S] clusters by other systems. Yet, NifU apparently retains three functions within a single protein, an aspect of which is considered a strategy to optimize the synthesis and transport of labile [Fe–S] clusters.

9.2.1 NifU N′-Terminal Domain

The initial assembly of transient clusters is promoted by the N′-terminal domain scaffold site. This domain shows sequence similarity to IscU, the scaffold protein of the ISC system. In *A. vinelandii*, the N′-terminal domain of NifU consists of the residues M1-T125 (Figure 9.1, shown in yellow), containing the reactive cysteines, C35, C62, and C106 that can coordinate transient [2Fe–2S] and [4Fe–4S] clusters like IscU [5, 22, 23]. The labile nature of these clusters has hindered additional spectroscopic investigation of clusters associated with IscU and the N′-terminal domain of NifU. For instance, attempts to reduce transient clusters with the artificial reductant dithionite have led to cluster destruction. Nevertheless, mutagenesis studies showed that when any of the cysteine residues is substituted by alanine, the resulting phenotype is a decreased growth rate under diazotrophic conditions, and NifU is unable to bind the transient clusters at this site [24]. In addition to the conservation of Cys residues, active site residues include conserved Asp (D37) and His (H105)

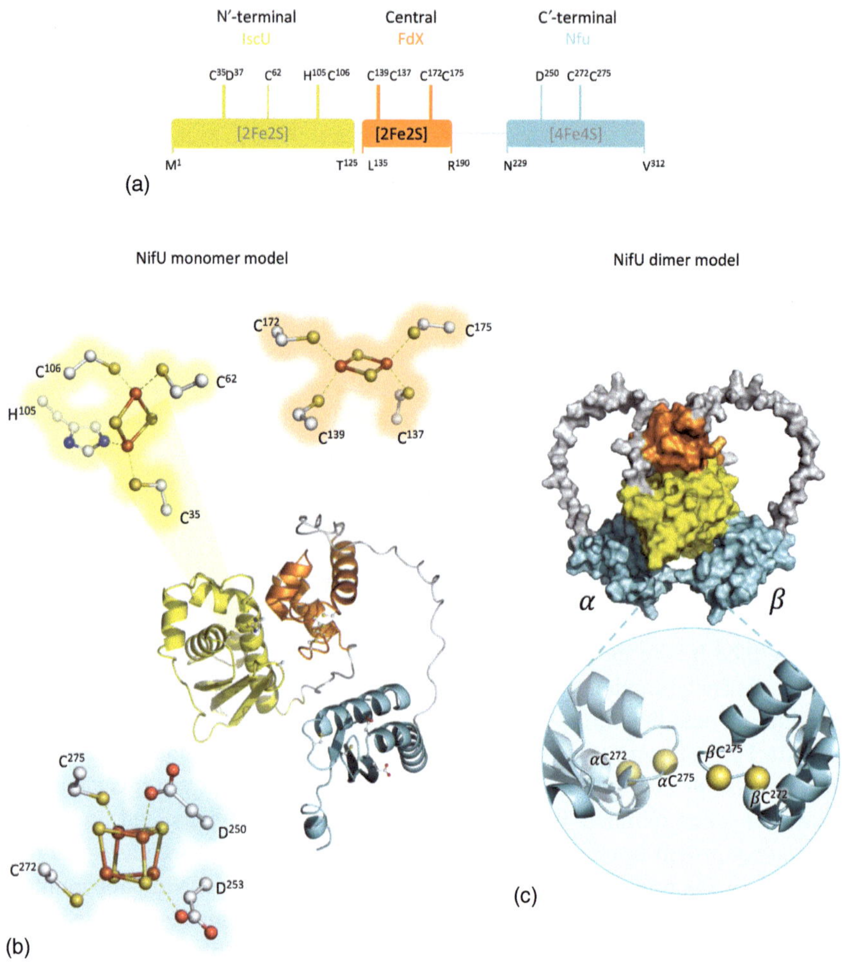

Figure 9.1 Modular composition of NifU. (a) Functional domains of *A. vinelandii* NifU: N′-terminal or IscU (yellow), central domain or FdX (orange), and C′-terminal domain or Nfu (cyan). Transient clusters are labeled in gray. Fe–S cluster coordinating residues conserved in NifU sequences are indicated in each domain. (b) Structural model of NifU generated with AlphaFold [17, 18]; the color code corresponds with code in panel A. The reactive residues are shown in a ball-and-stick representation, where blue is used for nitrogen, red for oxygen, yellow for sulfur, and gray for carbon. Reactive residues that harbor the [2Fe–2S] cluster in the N′-terminal domain of NifU from *A. vinelandii*, the cluster was placed following the geometry in PDB 2Z7E [19]. Residues involved in the [2Fe–2S] cluster in the central domain of NifU using PDB 1I7H as a reference [20]. Hypothesized [4Fe–4S] cluster anchorage to the C′-terminal domain of NifU, assuming one cluster per NifU subunit using a [4Fe–4S] cluster bound to three cysteines and one aspartate residues as a reference (PDB 5N07 [21]). (c) Dimeric structural model of NifU generated with AlphaFold [17, 18], following the color code in (a). The zoom-in view shows the interface of the C′-terminal domain of the NifU dimer, which displays the four reactive cysteines in proximity and in orientation for cluster coordination.

residues. Much of the proposed functionality of the NifU N′-terminal domain is based on the structural and functional analysis of IscU [25–27], which provided insights into the reaction steps involving the N′-terminal domain.

Analysis of IscU structures shows the positioning of the three C35, C62, and C106 in close proximity to D37 and H105. D37 was initially suspected to be the fourth ligand for the [2Fe–2S] cluster, like in the case of IscU from *Methanothrix thermoacetophila* (MtIscU) [28]. However, NifU D37A substitution led to the stabilization of the [2Fe–2S] cluster [24], impacting the function of this scaffold site in the transfer of nascent Fe–S clusters. In the cluster-bound structure of MtIscU, the Asp residue serves as a fourth ligand of the [2Fe–2S] cluster when a dimer of IscU positions the two clusters in route for reductive coupling to form a [4Fe–4S] cluster. Interestingly, in MtIscU holo-structure, the H106 (equivalent to H105 in AvNifU) side chain is positioned away from the cluster. However, substitution of this residue by Ala (MtIscU H106A) hinders the ability of the scaffold to assemble [4Fe–4S] clusters, implicating the role of this residue in the conversion of two [2Fe–2S] clusters into one [4Fe–4S] cluster.

The observed exchange of cluster ligands in different studies could provide a mechanistic clue about steps in cluster formation and subsequent cluster transfer. The structure of *Archaeoglobus fulgidus* IscU D35A in complex with IscS shows the presence of a [2Fe–2S] cluster coordinated by its three conserved Cys residues and the IscS active site residue C321 providing the fourth binding ligand [29]. This structure potentially represents the last step in the construction of [2Fe–2S] clusters prior to complex dissociation. Notably, the structure also shows the interaction of the conserved His residues with one of the cluster-bridging sulfide ions. Furthermore, in the *Aquifex aeolicus* IscU (AaIscU) D38A holo-structure (PDB 2Z7E [19]), the residue H106 is the fourth bridging ligand for the [2Fe–2S] cluster coordinated by residues from two different subunits in the trimer structure. Although no structures have been reported yet for NifU, a structural model of NifU (Figure 9.1b) generated with AlphaFold [17, 18] allows for structural alignment with the structures described above. This computational model suggests that, in *A. vinelandii* NifU (AvNifU), the residue H105 might serve as the fourth ligand for the [2Fe–2S] cluster at some point in the assembly process, similar to the cluster coordination described for AaIscU (Figure 9.1b). Ongoing mechanistic and structural investigation of NifU and IscU should clarify the involvement of the N′-terminal domain NifU residues promoting interactions with the desulfurase, participating in sulfur transfer, formation, and coordination of [2Fe–2S] and [4Fe–4S] clusters, and potentially leading to intramolecular and intermolecular interaction across domains. Despite the lack of detailed information, experimental biochemical and genetic data support the nonessential yet dominant involvement of the NifU N′-terminal domain in promoting the initial steps of cluster assembly through direct engagement with the sulfur donor NifS.

9.2.2 NifU Central Domain

The central domain NifU is characterized by the presence of a stable [2Fe–2S] cluster proposed to serve in a redox role. The internal sequence spanning residues

L135–R190 includes four conserved cluster ligand residues C137, C139, C172, and C175 (Figure 9.1, shown in orange). Substitution of these Cys residues by Ala also resulted in slow diazotrophic growth, as in the case of the substitution of the cysteine residues of the N′-terminal domain of NifU [24]. The central domain is the diagnostic feature for the identity of NifU sequences. Heterologous expression of NifU in *Escherichia coli* leads to the isolation of a dimeric NifU harboring 2 × [2Fe–2S] clusters coordinated by the four Cys residues within each monomer of the dimer structure. The "permanent cluster" designation reflects the stability of the central domain cluster even after treatment with a 40-fold excess of the strong iron chelator 2,2′-bipyridine [16]. The UV–Vis absorption spectrum shows features at 330, 420, 460, and a shoulder at 550 nm, characteristic of [2Fe–2S] cluster coordination with extinction coefficient values of 12, 6.6, 7, and 4 $mM^{-1}\,cm^{-1}$, respectively (values indicated for each monomer). The identity of the physiological redox partner of NifU has yet to be identified, but artificial reductants allowed spectroscopic characterization of the central redox-active $[2Fe–2S]^{2+/+}$ ($E_m = -254 \pm 20\,mV$) [16]. These characteristics, along with EPR (electron paramagnetic resonance) and MCD (magnetic circular dichroism) spectra, indicate shared features to the bacterioferritin-associated ferredoxin; therefore, it is described as the FdX domain of NifU [24]. While the role of this domain has not been fully established, it has been proposed to serve in an equivalent capacity as the ferredoxins found in both prokaryotic and eukaryotic ISC systems. These proposed roles include serving as a site for electron transfer during the reduction of persulfides to iron-bound NifU, to the reduction of Fe^{3+}, and/or reductively coupling of [2Fe–2S] into [4Fe–4S] clusters.

9.2.3 NifU C′-Terminal Domain

A secondary site for the coordination of transient Fe–S clusters is located at the C′-terminal domain of NifU. Connected through a linker sequence, the AvNifU C′-terminal domain extends from S228 to V312 and includes conserved cysteine residues C272 and C275 (Figure 9.1, shown in cyan). In contrast with the N′-terminal and central domains of NifU, the substitution of the cysteine residues in the C′-terminal domain has no effect on cell growth under diazotrophic conditions [24]. However, phenotypes resulting from the inactivation of this domain are revealed in strains that carry Cys to Ala substitutions within both N′- and C′-terminal domain residues, suggesting that the C′-terminal domain operates in a supporting capacity in cluster assembly and transfer [30]. *In vitro* [Fe–S] cluster reconstitution experiments showed that residues of the C′-terminal domain can coordinate a transient [4Fe–4S] cluster that is likely bridged between subunits of the dimer [12, 30, 31].

Proteins displaying primary sequence similarity to the C′-terminal domain within NifU are also found to participate in [Fe–S] cluster metabolism in other systems. Named after shared features and conservation of Cys residues within C′-terminal domain of NifU, the Nfu protein domains are diversely distributed across prokaryotic and eukaryotic domains of life. In *A. vinelandii* and *E. coli*, NfuA

(AvNfuA) is a modular protein having two functional domains: an N′-terminus A-type domain and a C′-terminus NfU domain [32, 33]. The AvNfuA C′-terminal domain Cys residues (equivalent to C272 and C275 of AvNifU) can coordinate one transient [4Fe–4S] cluster in its homodimer interface that can be readily transferred to apo-[Fe–S] target proteins such as aconitase [33]. The ability of NfU domains to receive and transfer clusters has supported a model in which the C′-terminal domain of NifU (NfU domain) operates as an intermediate [4Fe–4S] cluster carrier, facilitating the delivery of clusters from the IscU domain to acceptor proteins. Extensive Mössbauer spectroscopy analyses confirmed the *in vitro* assembly of a [4Fe–4S] in the Nfu domain of NifU [12]. Despite the absence of structural information on NifU or NfuA showing their corresponding cluster environment, AlphaFold structural models of NifU potentially suggest that a [4Fe–4S] cluster might be coordinated by C272, C275, D250, and D253 (Figure 9.1b) [12]. Nonetheless, the possibility of having one [4Fe–4S] cluster per C′-terminal dimer cannot be ruled out either. A dimeric AlphaFold model of NifU (Figure 9.1c) positions C272 and C275 of each monomer facing each other and exposed to the surface, providing an ideal environment for the binding of one [4Fe–4S] per NifU dimer (Figure 9.1c, zoom-in view) as proposed for NfuA [33].

9.3 Structural and Catalytic Features of NifS

NifS is the cysteine desulfurase that provides the sulfur source for the biosynthesis of [Fe–S] clusters. In *A. vinelandii*, this role is proposed to be restricted to pathways involving the synthesis of metalloclusters involved in NIF (Table 9.1). Coexpression of NifU and NifS leads to the isolation of a complex [22] with greater stability than its individual components, denoting the specificity of sulfur donor and acceptor. Like NifU, NifS is required for complete activation of both nitrogenase catalytic components [31] and, therefore, for optimum growth under NIF conditions [7]. The *A. vinelandii* NifS (AvNifS) was the first cysteine desulfurase to be functionally assigned in pioneering studies reported by Dean and coworkers in 1993 [13]. This discovery led to a subsequent search and identification of a paralog enzyme, IscS, performing an equivalent role in sulfur donation for the housekeeping synthesis of Fe–S clusters in metalloenzymes [5, 24]. Orthologs of NifS are found in all three domains of life [39]. While their shared functions involve sulfur mobilization reactions, their physiological roles are not restricted to NIF. In fact, they have broad participation as catalysts in the synthesis of various sulfur-containing cofactors, including thiamine, biotin, lipoic acid, molybdenum cofactor, and thionucleosides [40].

Cysteine desulfurases share mechanistic and structural features, and their specific functions are proposed to be restricted by interactions with sulfur acceptors [3]. Overall, members of this class of enzymes are presented as homodimers with nearly identical topology and the presence of PLP cofactor coordinated to the enzyme via a Schiff base with a conserved Lys at each active site monomer (Figure 9.2) [41]. The presence of a PLP cofactor endows NifS with its characteristic bright

Table 9.1 Enzymes proposed to or known to be involved in nitrogen fixation, sulfur metabolism, and [Fe–S] cluster synthesis and trafficking during diazotrophy in *A. vinelandii*.

Nitrogenase isozyme	Protein	Type of [Fe–S] cluster	Function
Molybdenum	NifH*	[4Fe–4S]	Nitrogenase reductase
	NifDK*	2[8Fe–7S] and 2[7Fe–Mo–9S–C-homocitrate]	Nitrogenase
	NafB	[4Fe–4S]	Unknown
	NifEN*	2[4Fe–4S]	FeMo-cofactor synthase
	NifX	[8Fe–9S–C]	NifB-cofactor carrier
	NafN	[4Fe–4S]	Ferredoxin
	NafO	[2Fe–2S]	Ferredoxin
	NafF	[4Fe–4S]	IscA like protein
	NafG	—	Sulfur metabolism
	NifM*	—	cis-trans peptidyl prolyl isomerase
	NifF	—	Flavodoxin, NifH reductant
	NafY	[7Fe–Mo–9S–C-homocitrate]	FeMo-cofactor carrier
	NafV	[2Fe–2S]	Ferredoxin
	NifO	—	activation of nitrogenase and nitrate reductase
	NifQ	[Mo–3Fe–4S]	Mo carrier for the FeMo-cofactor synthesis
	NafW	—	Sulfurtransferase
	NafZ	[2Fe–2S]	Glutaredoxin
Vanadium	VnfY	[7Fe–V–9S–C-homocitrate]	FeV-cofactor synthesis protein
	VnfDGK*	2[8Fe–7S] and 2[7Fe–V–9S–C-homocitrate]	Nitrogenase
	VnfF	2[4Fe–4S]	Ferredoxin
	VnfH*	[4Fe–4S]	Nitrogenase reductase
	VnfX	[8Fe–9S–C] or [7Fe–V–9S–C-homocitrate]	VFe-cofactor carrier
	VnfEN*	2[4Fe–4S]	FeMo-cofactor synthase
	VnfU	[4Fe–4S]	C′-terminal NifU-like protein
Fe-only	AnfO	[8Fe–9S–C-homocitrate] and [7Fe–Mo–9S–C-homocitrate]	Nitrogenase assembly protein
	AnfDGK*	2[8Fe–7S] and 2[8Fe–9S–C–homocitrate]	Nitrogenase
	AnfH*	[4Fe–4S]	Nitrogenase reductase
	AnfU	[4Fe–4S]	C′-terminal NifU-like protein

Table 9.1 (Continued)

Nitrogenase isozyme	Protein	Type of [Fe–S] cluster	Function
Molybdenum, vanadium, and Fe-only	NifU*	2[2Fe–2S]	Fe–S cluster assembly scaffold protein
	NifS*	—	Cysteine desulfurase
	NifV*	—	Homocitrate synthase
	NifB*	[4Fe–4S]	NifB-cofactor synthase

Only enzymes directly related to nitrogen reduction or enzymes that harbor [Fe–S] clusters located in the nitrogenase gene regions were included. Proteins marked with an asterisk (*) are essential for diazotrophic growth. For a comprehensive list of proteins involved in nitrogen fixation, their function, interactions, and regulation in *A. vinelandii*, see references [34–38].

yellow color. AvNifS is a 402-residue protein containing characteristic structural motifs for (i) a cofactor binding site including the strictly conserved K202 involved in PLP ligation and H201, H100 proposed to stabilize the cofactor and substrate, respectively (Figure 9.2a) and (ii) a catalytic loop extending from V315 to T359 that includes a reactive surface-exposed C325 residue and an internal R351 residue proposed to stabilize substrate binding (Figure 9.2 shown in pink) [42]. While there is no structure available for AvNifS, available structures for *E. coli* IscS [43], *Hydrogenimonas thermophila*, and *H. pylori* NifS [41, 42] captured different reaction stages and provide insight into the NifS catalytic cycle. The *H. pylori* NifS (HpNifS), for example, shows the residue H104 (equivalent to AvNifS H100) located at 3.8 Å from the Cys substrate thiol and the residue R354 (equivalent to R351) forming a salt bridge with the carboxyl moiety of the Cys substrate [42]. These structures also show differences in the catalytic loop containing the active site Cys residue. Structural models fail to resolve these structural elements due to either a disordered conformation or position of the active site Cys residue 23 Å away from the active site cofactor, suggesting the occurrence of large topological movement of the active site loop during catalysis. Furthermore, no structures have been reported for NifS in complex with NifU. Although it is anticipated that the N′-terminal domain provides the primary point of contact to NifS, displaying equivalent interactions to the IscS–IscU complex, the residues promoting this binding interface or those involved in the catalytic mechanism of NifS have not been explored either.

9.3.1 NifS Is a Cys:NifU Sulfurtransferase (EC 2.8.1.7)

Initial studies describing the function of NifS and other orthologous enzymes have been performed *in vitro* and in the presence of artificial reducing agents. Under these conditions, NifS catalyzes a cysteine desulfurase reaction, which is defined as the desulfuration of the free amino acid Cys to form Ala and sulfide (S^{2-}). Characterization of this reaction led to the initial enzyme designation and still commonly used general assignment as Cys desulfurases [44]. However, *in vivo*, NifS and NifS-like

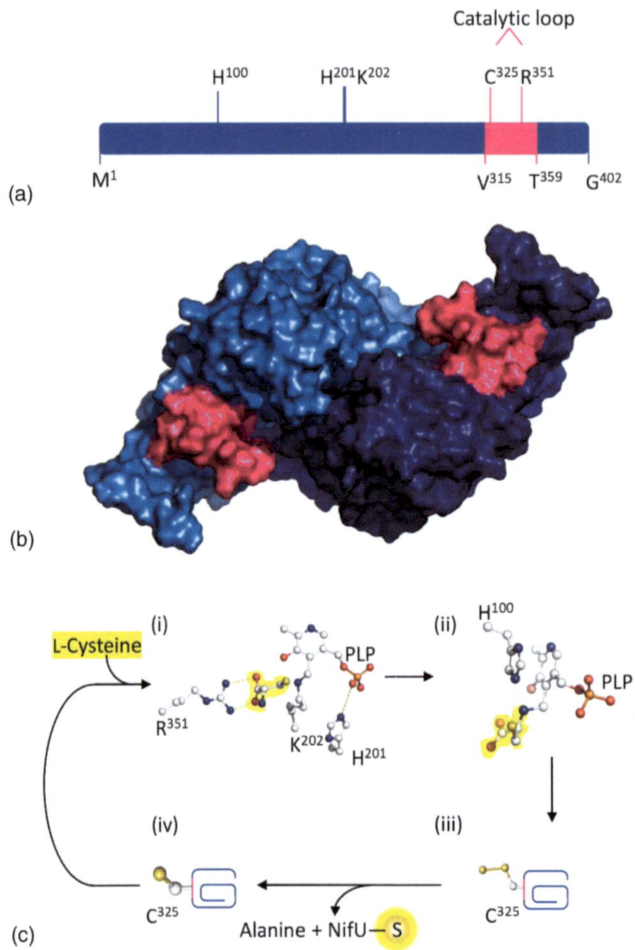

Figure 9.2 NifS functions as PLP-dependent Cys:NifU sulfurtransferase. (a) Functional motifs and conserved residues of *A. vinelandii* NifS. (b) Structural dimeric model of NifS shows each surface monomer colored in a different shade of blue and the catalytic loop is colored pink. (c) Catalytic cycle of NifS generated with extracts from the crystal structures (i) PDB 5WT5, with substitution of L-homocysteine for L-cysteine, (ii) PDB 5ZSS, and (iii) PDB 5ZST.

enzymes are, in fact, sulfurtransferases where the enzyme utilizes a double displacement (ping-pong) mechanism for the direct transfer of sulfur from the substrate Cys to a sulfur acceptor protein or small molecule via the formation of covalent enzyme intermediate. Kinetic and genetic data show the essentiality of the active site Cys residue (AvNifS C325) and dependency on the sulfur acceptor (AvNifU) [30].

The NifS catalytic cycle is marked by the formation of the enzyme persulfide intermediate and the transfer of the activated sulfur to NifU (Figure 9.2c). In the first half of the cycle, the reaction is initiated by displacement of K202-PLP internal aldimine upon the binding of the substrate Cys to form an external Cys-PLP aldimine Schiff base [13, 40]. Changes in the visible absorption spectrum are observed upon substrate binding and convey the reactivity and specificity of the enzyme for L-Cys but not for the other standard 19 amino acids or D-Cys. The reaction mechanism proceeds through the abstraction of the alpha proton of the Cys substrate, leading to the formation of a quinonoid intermediate. Protonation of this intermediate generates a Cys-PLP ketimine intermediate, which poises the thiol group of L-Cys substrate in optimal conformation for the nucleophilic attack by the deprotonated thiol of NifS C325 to form a persulfide covalent enzyme intermediate (R-SSH, perthiol, sulfane sulfur) and an Ala-PLP adduct. Although it is well known that the reactive loop experiences a series of conformational changes [40], the trajectory followed during catalysis is not fully mapped [42]. The first half of the reaction results in the formation and release of the first product, Ala. The enzyme in its persulfurated form (NifS C325-SSH conformation) engages with NifU for Fe–S cluster assembly in the second half of the reaction.

9.4 Synthesis of Fe–S Clusters by the NifUS System

The site for the initial assembly of the Fe–S cluster by NifUS is proposed to be the N′-terminal domain of NifU [12]. The construction of Fe–S clusters requires (i) a source sulfur provided by NifS in its persulfurated form (NifS C325-SSH), (ii) a source Fe that is yet to be identified, and (iii) a source of electrons. The order of addition of these components onto the NifU scaffold has yet to be established. Two models have been proposed to describe individual chemical steps during the formation of initial Fe–S units [44–46]. The "sulfur-first" model involves the persulfide sulfur transfer reaction from NifS to NifU followed by the binding of Fe. Alternatively, the "iron-first" model involves the transfer of the persulfide sulfur directly to the Fe-bound NifU. In both cases, the sulfane sulfur (S^o) must be reduced to S^{2-} during cluster assembly. Thus, the synthesis of Fe–S clusters requires the input of electrons for the reductions of persulfides and for the reductive coupling of [2Fe–2S] to form [4Fe–4S] clusters to result in the overall NifU reaction described in the following equations:

$$2Fe^{2+} + 2x[\text{NifS C325} - \text{SSH}] + 2e^- \rightarrow [2Fe - 2S]^{2+} \tag{9.1}$$

$$2x[2Fe - 2S]^{2+} + 2e^- \rightarrow [4Fe - 4S]^{2+} \tag{9.2}$$

The ultimate electron source(s) required for the transient Fe–S cluster formation has yet to be identified. The permanent [2Fe–2S] ferredoxin cluster within the central domain of NifU is proposed to be involved in one or more of the reduction steps (Eqs. (9.1) and (9.2)). Yet, the identity of the reductase partnering with NifU has not been identified. The genomic context *nifUS* in *A. vinelandii* does not offer a clear candidate, but the genome does encode for two nicotinamide adenine dinucleotide phosphate (NADPH)-dependent ferredoxin reductases (Avin_38470 and Avin_07740), which could serve in this capacity. Nevertheless, the involvement of a one-electron redox [2Fe–2S] cluster intermediate in the two-electron reduction of persulfide presents a conundrum in identifying the physiologically relevant sulfur transfer and reduction mechanisms. It is possible that persulfide transfer is coupled to a shared one-electron transfer from the central cluster and the oxidation of Fe^{+2} to Fe^{+3}. This model is supported by iron binding studies demonstrating the high-affinity metal binding of NifU [24]. Alternatively, it is possible that the ferredoxin-like cluster is not involved at this step and that Cys, a substrate in the overall reaction, or glutathione (GSH), a physiological thiol redox buffer, can serve as a two-electron reductant of persulfides locally liberating free sulfide at the site of cluster synthesis. This second proposal is supported by *in vitro* studies demonstrating that artificial reductants such as dithiothreitol are competent in accelerating the rates of cluster assembly [12].

Protein–protein interactions between NifS and the N′-terminal domain of NifU are expected to drive the initial steps of cluster assembly as suggested by available *in vitro* experiments [12, 22, 28], crystal structures [12, 28, 47], and structural models [17, 18] of NifUS and the homologous IscUS pair. The crystal structure of *E. coli* IscU–IscS complex in its apo form (pDB 3LVL [47]) shows a heterotetramer with a central IscS dimer, in which reactive loops are facing the [2Fe–2S] cluster binding region of IscU. The Alpha Fold model of the dimer NifUS preserves this structural arrangement, with the N′-terminal domain of NifU facing the reactive loop of NifS (Figure 9.3). Furthermore, structural predictions also position the ferredoxin cluster in close proximity to the N′-terminal domain cluster scaffold site, supporting the involvement of this domain in electron transfer in initial cluster formation. Following a similar reaction scheme proposed for IscSU, NifUS complex dissociation is expected to allow for a NifU dimer to form, enabling the formation of a [4Fe–4S] cluster bridged between two the N′-terminal domain monomers. Although there are no available crystal structures of NifU or the cluster-loaded IscUS active complex, the [2Fe–2S] cluster formation in IscU has been visualized in several crystal structures of IscU multimers [19, 28]. The resulting [4Fe–4S] cluster formed at the N′-terminal domain is proposed to be transferred to the C′-terminal domain (Figure 9.3) and subsequently transferred to a target protein (Figures 9.3–9.5) [28, 30].

Heterologous high-level expression of *A. vinelandii* NifU [16] in *E. coli* enabled the study and development of efficient *in vitro* [Fe–S] cluster assembly protocols [22]. To begin, the overexpressed proteins were purified under anoxic conditions, preserving the integrity of its ferredoxin [2Fe–2S] cluster [12, 30]. *In vitro* assembly of [Fe–S] clusters with NifUS achieved the synthesis of [2Fe–2S] cluster in the N′-terminal domain of full-length NifU or a NifU truncated construct that consisted uniquely of its N′-terminal domain [22]. Hence, this experimental evidence suggests the involvement of the N′-terminal domain of NifU as a scaffold for the initial synthesis of [Fe–S] cluster and that NifS drives the assembly of the clusters in NifU. The proposed mechanism for the assembly of [2Fe–2S] and [4Fe–4S] clusters in NifU was established in a series of time-dependent *in vitro* [Fe–S] cluster synthesis monitored by UV–Vis absorption and Mössbauer spectroscopies using full-length and N′-terminal domain truncated NifU constructs. It is worth noting that *in vitro* experiments, while advancing much of our understanding of the reactivity of the

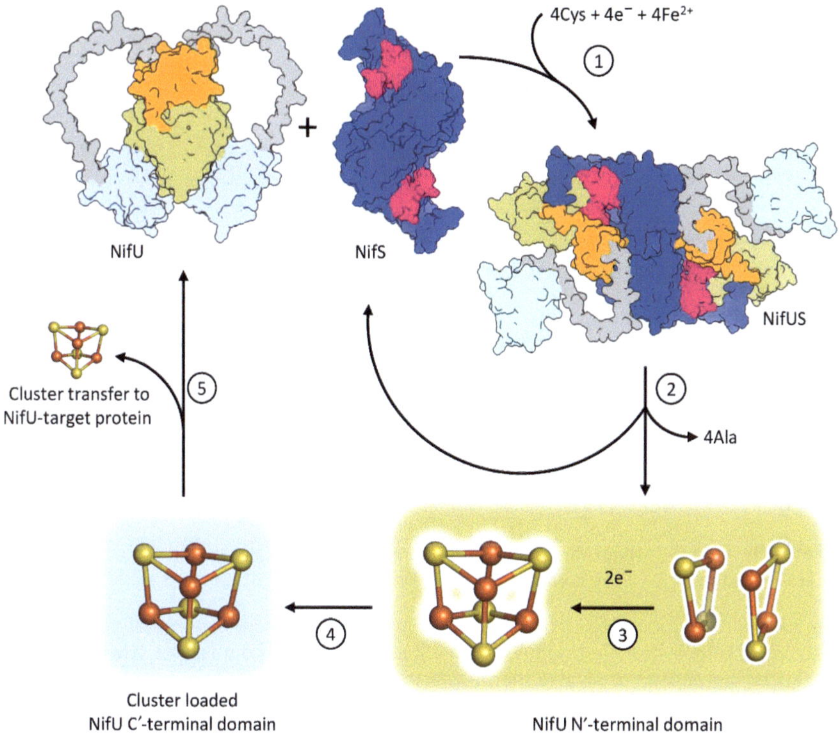

Figure 9.3 Proposed catalytic cycle of [4Fe–4S] cluster synthesis by NifUS [12, 28]. The color code is equivalent to Figures 9.1 and 9.2. (1) NifS generates the persulfide species from L-cysteine and transfers to NifU for the reductive condensation of 2Fe with 2S in the N′-terminal domain of NifU, yielding one [2Fe–2S] cluster per NifU subunit. (2) The [2Fe–2S] cluster-loaded NifU dissociates from NifS to form a homodimer with two [2Fe–2S] clusters in the interface of the N′-terminal domain dimer. (3) Two [2Fe–2S] clusters are fused into one [4Fe–4S] cluster by reductive coupling. (4) The [4Fe–4S] is transferred to the C′-terminal domain of NifU (5). The reaction cycle is completed upon transfer of a [4Fe–4S] cluster to a client protein.

NifUS system, were carried out with catalytic amounts of NifS, excess Fe, and Cys, and in the presence of artificial reducing agents. Under these conditions, NifU dimer is able to accumulate as many as three transient [4Fe–4S] clusters. However, the number and type of clusters that NifU accumulates during the [Fe–S] cluster synthesis under physiological conditions remains to be determined.

9.5 Dedicated Role of NifUS in Aerobic Nitrogen Fixation

The NifUS system supplies Fe–S clusters to a clientele of distinct acceptors involved in nitrogenase fixation (Figure 9.4). Inactivation of *nifU* or *nifS* affects the activity of all three known nitrogenase isozymes Nif, Vnf, and Anf isoenzymes that differ

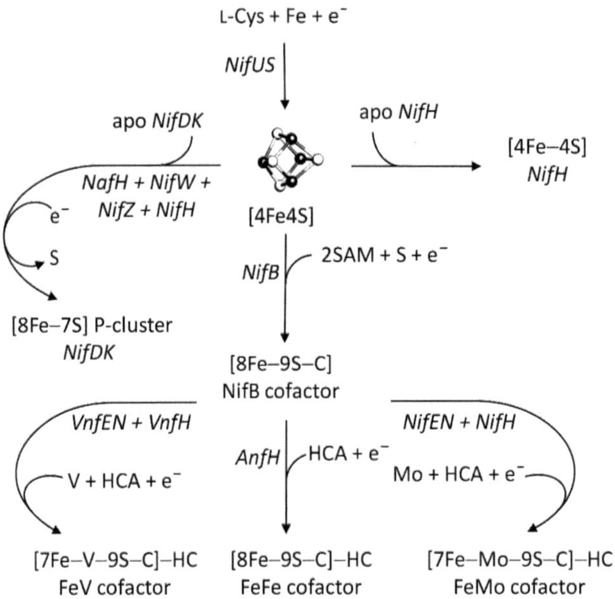

Figure 9.4 NifUS supplies Fe–S cluster for the synthesis of nitrogenase metalloclusters. NifU activates NifH by transferring one [4Fe–4S] cluster per NifH dimer. The synthesis of the P-cluster is proposed to involve the transfer of two [4Fe–4S] clusters from NifU to NifDK in a reaction that requires activated NifH and a source of electrons. In addition, the synthesis of the P-cluster is assisted by NafH, NifW, and NifZ. The synthesis of active site cofactors in all three nitrogenases has a common precursor, the NifB cofactor whose synthesis is carried out by NifB that catalyzes a SAM-driven chemistry reaction that requires the transfer of two [4Fe–4S] clusters, presumably donated from NifU.

mainly in the heterometal (Mo, V, or Fe) composition of their active sites. The Nif nitrogenase consists of molybdenum-iron nitrogenase (also known as MoFe protein, component I, NifDK) and its reductase counterpart Fe–protein (component II, NifH), which is the primordial nitrogenase present in all diazotrophs. The other two systems, also known as the alternative nitrogenases, are the vanadium nitrogenase (VnfDGK) and the iron-only nitrogenase (AnfDGK) with their nitrogenase iron proteins termed VnfH and AnfH, respectively [34]. The NifDK dimer and the alternative DGK trimer each carry an [8Fe–7S] cluster in their exposed DK interface, and an additional active site metallocofactor designated FeMo cofactor for Nif, VFe cofactor for Vnf, and FeFe cofactor for Anf. The nitrogenase reductase component of each of these three isoenzymes contains a [4Fe–4S] cluster bridge across subunits.

The occurrence of *nifUS* genes is dominant in aerobic diazotrophs. The distribution of this dedicated Fe–S cluster biosynthetic system is attributed to a higher demand for Fe–S clusters and associated proteins during NIF (Table 9.1). The nitrogenase isozymes are all relatively slow enzymes (turnover rates of 5–10 s^{-1}) [48]. Consequently, in order to sustain diazotrophic growth, copious amounts of both enzyme components are necessary. The nitrogenase catalytic components MoFe protein and Fe protein can constitute up to 10% of the total intracellular protein pool in *A. vinelandii* [49, 50]. Both enzymatic components contain metallocofactors, and based on the intracellular concentration of each enzyme, it is estimated that 630 μM of Fe is routed to the synthesis of Fe–S clusters on nitrogenase [51]. The accumulation of [Fe–S] clusters is also evidenced by the dark brown coloration of *A. vinelandii* cells expressing nitrogenase.

NifUS sustains NIF in *A. vinelandii* by boosting the capacity for synthesizing and transporting [Fe–S] clusters under conditions of aerobic diazotrophy. An enhanced capacity for [Fe–S] cluster synthesis under nitrogen-fixing conditions is also supported by an additional copy of serine acetyltransferase, termed *nafG*, within the same transcriptional unit of *nifUS* [34]. This enzyme catalyzes the rate-limiting step in cysteine biosynthesis and is proposed to lead to increased levels of cysteine, which is the substrate of NifS. Furthermore, *A. vinelandii* has an elaborate respiratory system that provides protection against oxidative damage to metalloclusters through the presence of terminal oxidases and systems that generate low-potential reducing equivalents that upregulate oxygen consumption while also providing a source of electrons to support nitrogenase catalysis [34].

To produce an active form of the Mo-dependent nitrogenase, NifU is required to supply [4Fe–4S] clusters for at least three distinct roles: (i) the direct activation of the nitrogenase reductase Fe protein (NifH), (ii) the *ex situ* assembly of the [7Fe–Mo–9S–C]-homocitrate cluster called FeMo cofactor with NifD, and (iii) the *in situ* assembly of the [8Fe–7S] cluster known as the P-cluster located within NifDK. The two Nif gene regions also include 16 additional coding sequences, whose protein products contain [Fe–S] cluster binding motifs [52]. The involvement of NifUS in supplying [Fe–S] clusters for the maturation of these proteins is yet to be determined (Table 9.1). NifUS is also known to be required for NIF catalyzed by the V-dependent (Vnf) and Fe-only (Anf) nitrogenases. Given the similarity of their metallocofactors, it is likely that NifUS has equivalent roles as those mentioned above for the maturation of Vnf and Anf nitrogenase isozymes.

9.5.1 Direct Activation of Nitrogenase Reductase, NifH

The nitrogenase iron protein mediates electron transfer for dinitrogen reduction as well as for the assembly of the metalloclusters of the nitrogenase enzyme. The *A. vinelandii* strain deficient of NifH is unable to fix nitrogen [53], resulting in a MoFe protein with an immature P-cluster [54, 55] but also unable to synthesize the FeMo cofactor [34, 53, 56], indicating the involvement of the reductase in the synthesis of both of these metallocofactors. The crystal structure of NifH reveals an exposed [4Fe–4S] cluster on the surface of a homodimer interface [57]. Each subunit is able to bind one MgATP (magnesium adenosine triphosphate) molecule [58], and upon MgATP hydrolysis, NifH follows a conformational change that increases the exposed area surrounding its [4Fe–4S] cluster, allowing for an effective interaction with the P-cluster of the MoFe protein [34, 59, 60]. For each electron transfer from NifH to the MoFe protein, two MgATP molecules are hydrolyzed. To bind MgATP, the [4Fe–4S] cluster of NifH needs to be reduced by flavodoxin [61, 62], the physiological reductant, while *in vitro* dithionite is routinely used as the reducing agent.

The synthesis of a functional NifH requires proper folding and insertion of the [4Fe–4S] cluster. In *A. vinelandii*, the peptidyl-prolyl cis/trans isomerase NifM is required for NifH maturation, presumably assisting protein folding [63], while NifU is in charge of the cluster transfer [30] (Figure 9.4). The role of NifU in the [4Fe–4S] cluster maturation of NifH from *A. vinelandii* was unveiled in a series of *in vitro* cluster transfer assays involving incubation of NifH without its [4Fe–4S] cluster (apo NifH) and cluster-loaded NifU in the presence of MgATP and dithionite under anoxic conditions [12, 30]. The generation of apo NifH was achieved *in vitro* by chelation of its [4Fe–4S] cluster using 2,2′dipyridyl in the presence of MgATP and dithionite. Under reducing conditions while in its MgATP-bound state, the [4Fe–4S] cluster is solvent exposed and therefore susceptible to removal by chelation. *In vitro* Fe–S cluster reconstitution of heterologously expressed NifU followed by purification allows trapping of transient cluster species within NifU. The quantification of cluster transfer upon direct mixing of apo NifH and cluster-loaded NifU was evaluated by recovering NifH activity [12, 30].

9.5.2 *Ex Situ* Assembly of the FeMo cofactor

The assembly of the [7Fe–Mo–9S–C]-homocitrate active site FeMo cofactor requires a dedicated pathway initiated by transferring Fe–S clusters from NifU to the NifB. The maturase NifB harbors a radical SAM [4Fe–4S] cluster to promote the S-adenosyl methionine (SAM)-dependent fusion of two independent [4Fe–4S] clusters donated by NifU and insertion of carbide and sulfur to form a core [8Fe–9S–C] cluster, termed NifB cofactor [64, 65]. Although specific details concerning the NifB-cofactor synthesis are still to be defined, crystallographic evidence points to the fusion of the two [4Fe–4S] clusters into a P-cluster like [8Fe–8S] intermediate, also defined as K-clusters [66–68] (Figure 9.4). The reaction proceeds with the activation of a SAM molecule bound to the NifB permanent [4Fe–4S] cluster that leads to methylation of the proximal belt sulfide of the P-cluster-like intermediate, generating a [8Fe–8S-CH_3] cluster. A second SAM molecule is then incorporated and drives the electronic rearrangement to exchange the central sulfur for carbon, yielding a [8Fe–8S–C] cluster. Finally, a ninth sulfur [69] is incorporated, yielding [8Fe–9S–C] cluster, the NifB cofactor [34, 66–68].

The NifB cofactor is a shared intermediate in the synthesis of active site cofactors for Nif, Vnf, and Anf nitrogenases [34] (Figure 9.4). The synthesis of the FeMo and the FeV cofactor is completed in their respective cofactor biosynthesis protein, termed NifEN and VnfEN, that share high identities with their corresponding nitrogenases NifDK and VnfDGK. NifEN and VnfEN catalyze the exchange of one apical Fe atom, from the NifB cofactor, for either Mo or V and attach a homocitrate molecule to it in a process that requires the participation of NifH or VnfH, respectively (Figure 9.4). In the case of the Fe-only nitrogenase, no nitrogenase cofactor synthase is required for the assembly of the FeFe cofactor and it is likely completed at the Fe-only nitrogenase [70]. Therefore, NifB is an essential enzyme for the activity of all nitrogenase isozymes [34]. Strains carrying a *nifB* deletion generate nitrogenases with fully assembled P-cluster but are unable to reduce N_2 because they lack their active site cofactors [71].

9.5.3 *In Situ* Assembly of P-cluster on NifDK

The formation of the [8Fe–7S] P-cluster is also proposed to involve the transfer to two [4Fe–4S] clusters from NifU and the synthesis of the [8Fe–7S] is completed within the interface of NifDK (Figure 9.4). Although individual steps in P-cluster formation have not been determined, the reaction involves the NifH serving as a reductase to promote the fusion of the two cluster units. Genetic and copurification experiments have identified NafH, NifW, and NifZ to be involved in P-cluster maturation [55]. Although their roles have been defined as assembly factors, the inactivation of these genes does not eliminate the capacity for NIF.

9.5.4 Involvement of NifU and NifU-Like Proteins in the Synthesis of Metalloclusters for Vnf and Anf Nitrogenases

The biosynthesis of metallocofactors associated with alternative nitrogenases is still not fully defined. The involvement of NifU in the synthesis of active site cofactors, VFe cofactor and FeFe cofactor, is anticipated due to its role as a cluster donor to NifB, a shared maturase required for all three nitrogenases [34]. However, the synthesis of the P-cluster within VnfDGK and AnfDGK and the [4Fe–4S] cluster of VnfH and AnfH have not been explored to date and the involvement of additional components cannot be discarded. The fold change in the *nifUS* gene transcripts, induced by diazotrophic growth under Mo, V, and Fe-only conditions, is comparable for the three nitrogenase systems [35], indicating a possible equivalent role of NifUS in the biogenesis of all [Fe–S] clusters for the three nitrogenase isozymes.

In *A. vinelandii*, the *nifUS* genes are located in the *nif* gene cluster [7, 34]. Their transcription is regulated by NifA, the master regulator of the molybdenum nitrogenase, whose activity is independent of the presence of Mo or V. NifA is negatively regulated by association with NifL. The complex formation of NifAL, in turn, is driven by nitrogen, carbon, and oxygen conditions [36, 72]. The transcription of the Vnf and Anf nitrogenases is a complex process regulated by metal availability [36, 37]. Similar to the Nif nitrogenase, both alternative nitrogenase systems have their dedicated transcriptional regulators designated as VnfA and AnfA. Interestingly, in *A. vinelandii*, *vnf* and *anf* gene regions also include *vnfU* and *anfU*, encoding short protein sequences similar to the C′-terminal domain of NifU (Figure 9.5). These genes are cotranscribed [35] with VnfA and AnfA, respectively [34, 36]. Although the involvement of VnfU and AnfU in the maturation or trafficking of [Fe–S] clusters for the alternative nitrogenases has not been determined, the presence of a protein having a cluster carrier sequence motif, including cluster coordinating Cys residues, suggests their potential involvement [73]. One of the many possibilities is that VnfU and AnfU are involved in the activation of their respective nitrogenase reductase VnfH and AnfH. The recruitment of additional cluster carriers guaranteeing the delivery of transient Fe–S units could be justified by the enhanced oxygen sensitivity of alternative enzymes and their lower catalytic efficiency [74, 75]. In agreement with the Fe-only nitrogenase being less efficient than the vanadium nitrogenase [74, 75], transcriptional studies of *A. vinelandii*

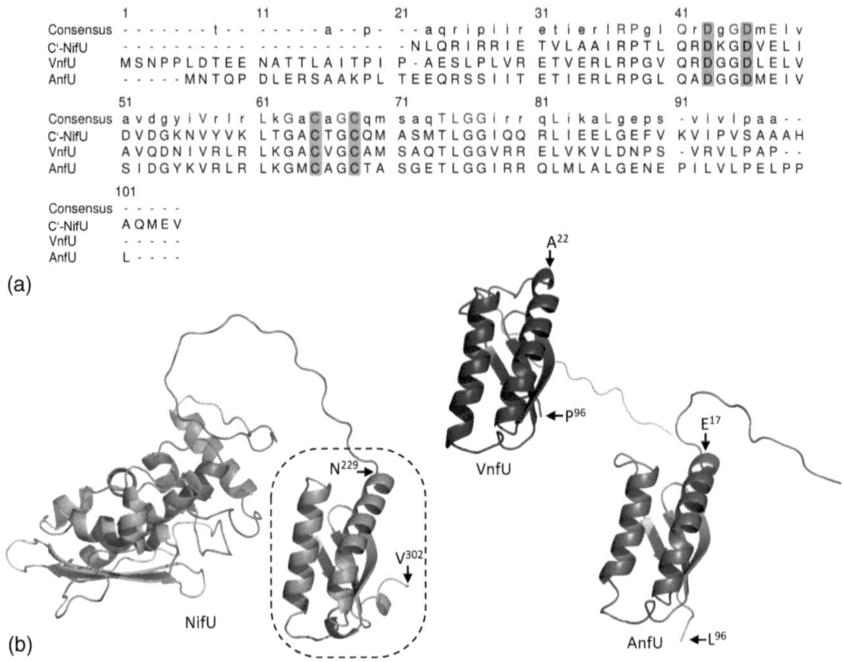

Figure 9.5 VnfU and AnfU show sequence and structural similarity to NifU C′-terminal domain. (a) Sequence alignment of the C′-terminal domain of NifU (N229–V312) with VnfU and AnfU. Gray boxes highlight the cysteine residues involved in the cluster binding as well as the aspartate residues (D250 and D253 in NifU) proposed as candidates for the binding of one [4Fe–4S] cluster per NifU monomer [12]. (b) AlphaFold models [17, 18] of NifU with its C′-terminal domain inside a dotted box, VnfU and AnfU. The arrows indicate the residues that conform to the core structure of the NifU like C′-terminal domain.

indicate that the fold change of AnfU transcripts is about twice the fold change of VnfU, under diazotrophic growth with Fe-only and vanadium, respectively [35].

9.5.5 Cross Talk Between NifUS and the Housekeeping IscUS System

Deletion of the *nifU* and/or *nifS* genes in *A. vinelandii* severely impairs NIF, but yet diazotrophic growth is not entirely eliminated [7, 76]. This observation led to the subsequent discovery of the housekeeping ISC system, which is able to provide a limited functional replacement for clusters destined for NIF proteins. As described above, components of the ISC system, IscS, and IscU show sequence and functional similarity to NifS and to the N′-terminal domain of NifU in sulfur mobilization and cluster assembly, respectively. The ferredoxin (Fdx) involved in electron transfer is functionally equivalent to the central domain NifU. The housekeeping ISC encoded by the *iscRSUA-hscBA-Fdx* [5] also includes the transcriptional regulator IscR [77] and the molecular chaperones *hscBA* that facilitate the cluster transfer to cognate proteins [78, 79]. In addition, NfuA, similar to the C′-terminal domain of NifU, is not part of the *isc* gene cluster, but it plays an important role as an intermediate cluster carrier *in vivo*, especially under oxidative stress conditions [32, 33]. Deletion of any of *isc* housekeeping genes *iscS, iscU, hscBA*, or *fdx* is lethal in *A. vinelandii* [14, 80], making it almost impossible to determine the effect of their contributions to the nitrogenase system. Nevertheless, the development of a system for the controlled expression of target genes in *A. vinelandii* [15] allowed for the investigation of IscUS and NifUS clients of [Fe–S] cluster acceptor proteins. Although each system serves a specific group of [Fe–S] cluster proteins, the ISC and NIF machineries display some degree of cross talk under oxygen-reduced conditions or elevated levels of expression of their components, indicating that either lower demand for [Fe–S] clusters or upregulation of cluster synthesis can overcome their specificities toward target acceptors. Namely, overexpression of IscU can rescue defects in NIF associated with the *nifU* deletion phenotype [14]. Likewise, hyperexpression of NifU bypasses the requirement of IscU to sustain robust diazotrophic growth. Nevertheless, reciprocal cross talk is not observed for the cysteine desulfurases NifS and IscS. Although elevated expression of IscS suppresses the requirement for NifS to sustain diazotrophic growth, NifS cannot replace the requirement of IscS to sustain intermediary metabolism. This feature is likely due to the fact that IscS interacts with additional sulfur acceptors besides IscU in the biosynthesis of other essential sulfur-containing cofactors [40].

9.5.6 Inclusion of NifUS in Biotechnological Efforts to Expand Diazotrophy to Other Species

The involvement of NifUS in activating nitrogenase has also been considered in efforts to develop genetic constructs enabling NIF in non-diazotrophic species. The ultimate goal has been to generate crops capable of fixing nitrogen and avoiding environmental harm caused by fertilizers [81–83]. Initial efforts to enable NIF in non-natural occurring diazotrophs were initially dated from 1972 in studies describing chromosomal integration of *K. pneumoniae* (renamed *Klebsiella oxytoca*) *nif* gene cluster in *E. coli* [84]. These experiments predated genetic mapping of *nif* gene cluster in *K. pneumoniae* but showed that *E. coli* strain containing *K. pneumoniae* genomic DNA showed acetylene reduction and $^{15}N_2$ incorporation when cultured anaerobically and in the absence of ammonia. More recently, these efforts have been revisited with the expression of a minimum *nif* gene region from *Paenibacillus* sp. WLY78 in *E. coli*. In contrast to the *nif*-associated gene clusters from *A. vinelandii,* consisting of more than fifty genes [34], the nif-associated gene cluster from *Paenibacillus* consists of only nine genes (*nifBHDKENXV* and *nafF*) [85]. *E. coli* constructs expressing this gene region were also capable of acetylene reduction and $^{15}N_2$ incorporation under anaerobic conditions, indicating the synthesis of active nitrogenase [85]. However, the activity level was low (only 10% of the *Paenibacillus*) and consequently led to insufficient N_2 reduction to sustain diazotrophic growth. The *Paenibacillus* sp. WLY78 *nif* gene region does not contain the *nifUS* genes, and, in that case, it was presumed that in the heterologous construct, the *E. coli* endogenous [Fe–S] cluster biosynthetic systems ISC and SUF supply [Fe–S] clusters under anoxic conditions [2], demonstrating that elevated expression of [Fe–S] acceptors may overcome the need for a specialized system.

In subsequent studies, the recruitment of NifUS system was investigated for the synthesis and activation of metalloclusters for heterologous expression of *A. vinelandii* nitrogenases. Detailed deletion analysis of Fe-only nitrogenase genes found a minimum gene set for functional nitrogenase expression. It consisted of *anfHDGK*, *nifBUSV*, *nifF*, and *nifJ* that, when inserted in *E. coli*, enabled $^{15}N_2$ incorporation under oxygen-free conditions [86]. In this case, the requirement of *nifUS* highlights the level of specialization of the [Fe–S] cluster system in *A. vinelandii* that could not be replaced by the host's system, like in the case of *Paenibacillus* [85]. Furthermore, attempts to dissect biosynthetic components for synthesizing the nitrogenase active site cofactor using *E. coli* for heterologous expression showed species-specific requirements to match NifUS with an NifB acceptor. For example, the synthesis and accumulation of *Methanosarcina acetivorans* NifB cluster intermediate was higher when co-expressing one of three NifUS distinct copies (designated copy 3) from the same species [87], reinforcing the requirement of specific structural elements restricting interactions between the cluster donor NifU and the NifB acceptor.

Efforts to develop a heterologous expression system for NIF in eukaryotes have also considered the role of the NifUS system. Given the oxygen liability of the nitrogenase components, the current approach consists of the heterologous expression of nitrogenase proteins in the mitochondria of the eukaryote [88–95]. The use of *Saccharomyces cerevisiae* as a eukaryote model for the heterologous production of nitrogenase components demonstrated a varying degree of requirement for the NifUS system. While expression of Fe-protein (NifH) was supported through the mitochondrial expression of the genes *A. vinelandii nifHM* only [90], the NifUS system was required for the accumulation of NifB cofactor on NifB [93]. The expression of NifB from different archaea species in the mitochondria and chloroplast of *Nicotiana benthamiana* also required NifUS [94]. In this case, the *A. vinelandii* NifU could serve as a cluster donor to NifB. Successful production of an *in vitro* active NifB in plant organelles marks the beginning of the heterologous expression of nitrogenase in plants and constitutes a significant step forward for the generation of self-fertilizing crops.

9.5.7 Occurrence of NifUS-Like in Non-Nitrogen-Fixing Species

The presence and function of NifUS-like systems are not limited to diazotrophs. The non-nitrogen-fixing pathogen *H. pylori* lacks genes encoding components of the iron-sulfur cluster (ISC) and sulfur utilization factor (SUF) and relies on the NifUS system (also designated HpMisUS [39]) for the general assembly of [Fe–S] clusters [10]. As anticipated, *H. pylori nifU* and *nifS* are required for cellular viability under a range of O_2 concentrations [10]. HpNifU displays absorption, MCD, and EPR spectral features similar to that of AvNifU, and HpNifS has cysteine desulfurase activity, albeit at 10% of the levels of AvNifS. Other [Fe–S] cluster biosynthetic components have also been identified in *H. pylori*, including NfU, Fdx2, and ApbC. These proteins interact with HpNifUS system and target proteins *in vivo* and potentially assist in the delivery of [Fe–S] clusters from NifUS to final acceptors.

Among them, ApbC is essential and proposed to serve as either an alternate scaffold or an [Fe–S] cluster carrier [96]. Trans-complementation studies showed that HpNifUS is capable of functional replacement of *E. coli* ISC and SUF systems under low oxygen concentrations, demonstrating the functionality of the NifUS system in supplying [Fe–S] clusters for general purposes *in vivo* [97]. This genetic approach also showed that the NifU cannot functionally partner with IscS. Therefore, the function of HpNifUS likely requires specific interactions between sulfur donor and acceptor in the first stage of cluster assembly that are not provided by components of the Isc system.

NifUS-like systems have also been identified in the eukaryotes *E. histolytica* and *Mastigamoeba malamute*. Both organisms lack a typical mitochondrial organelle containing the ISC system characteristic of eukaryotes. Instead, *E. histolytica* contains mitosomes having roles limited to sulfate assimilation and do not carry components associated with respiration and ATP formation [98]. *E. histolytica* NifUS (EhNifUS) system operates in the cytosol and appears to be the sole system for [Fe–S] cluster assembly because its genome lacks ISC and SUF components observed in other eukaryotic species. Similar to HpNifUS, the expression of EhNifUS complements *E. coli isc* and *suf* knockout strains under anaerobic conditions [9]. Another evolutionary-related entamoeba, M*astigamoeba balamuthi* shows the presence of NifUS in the mitochondria and in the cytosol [99]. Interestingly, *M. balamuthi* also contains unusual mitochondria, but it represents an intermediate to *Hathewaya histolytica* mitosomes as it converted its organelle into a hydrogenosome to support its anaerobic metabolism [100]. Proteomic analysis of cytosolic and mitochondrial proteins shows the presence of Fe–S proteins at both cellular locations, justifying the dual-localization of NifUS components.

Bioinformatic analysis of sequenced genomes shows a larger distribution of NifU sequences beyond those few species that have been experimentally studied so far. Recently, phylogenetic analyses showed that differentiation among these sequences into distinct groups termed minimal iron–sulfur (MIS) and NIF, in which members of the MIS system include three-domain NifU-like and single-domain IscU-like sequences that are distinct from NIF and ISC systems [39]. BLAST (Basic Local Alignment Search Tool) search using the NCBI RefSeq Select database (release 224, including 150 742 species) resulted in the identification of 826 unique NifU and NifU-like sequences from both diazotrophic and non-diazotrophic bacterial species. These unique sequences are distributed across 325 distinct genera, representing a larger distribution of NifU across bacterial species than previously reported, and include NifU sequences containing all three NifU domains but discarded IscU single-domain sequences [39, 97]. Analysis of a subset of these sequences showed that while the presence of NifUS in nitrogen-fixing bacteria is dominant in aerobes, the occurrence of NifU in non-diazotrophs does not correlate with a respiratory requirement and also includes anaerobes and facultative species. No NifU sequences were identified in archaeal and eukaryotic genomes of the RefSeq database, possibly due to the low representation of these life domains in this database. The occurrence of NifUS-like systems in *M. balamuthi* indicates that its occurrence is not limited to bacterial species.

Amino acid analysis of this diverse group of NifU and NifU-like sequences showed conservation of Cys residues involved in Fe–S cluster coordination and assembly. The majority of sequences (801 sequences) contained all 9 conserved cysteine residues at the N′-terminal, central, and C′-terminal domains. A small subset of sequences (25 unique sequences) lacked the C′-terminal domain and were represented primarily in either anaerobic or facultative anaerobes. Genomic analysis of representative *nifU* sequences considering one species per genus shows the co-occurrence of an adjacent *nifS* gene [39], including species lacking the C′-terminal domain. Taken together, the strong genomic synteny of *nifUS* gene sequences demonstrates the coevolution of these components and likely the mechanistic dependency of sulfur donor NifS and the initial sulfur acceptor site of NifU, the N′-terminal domain.

Interestingly, despite conservation within individual NifU domains, the inter-domain linker regions lack conservation based on primary structure and length. For example, although the variable linker sequence within *A. vinelandii* NifU connecting the central and C′-terminal domains spans 44 residues (E186 to L230); some species contain a much shorter or longer linker sequence. In the case of *Beggiatoa leptomitoformis*, only 24 amino acid residues separate the NifU central and C′-terminal domain. In comparison, the corresponding linker sequence in NifU produced by *Halochromatium glycolicum* spans 96 residues. Alpha fold predictions of NifU structures containing either short or long linkers display conservation in the folding of individual domains and the orientation of N′-terminal and central domains nearby, positioning the C′-terminal domain away from the other two core domains. Thus, the proposed flexibility of the C′-terminal domain agrees with its proposed role of assisting the delivery of clusters to acceptor proteins, which appears to be a conserved feature of full-length NifU sequences.

9.6 Concluding Remarks

Fe–S clusters are ancient and prevalent prosthetic groups assisting a range of biochemical reactions [101]. Their synthesis in cellular environments requires the participation of dedicated biosynthetic components involved in the recruitment of sulfur and iron, the synthesis of metallocomplexes, and the delivery of nascent [Fe–S] clusters to acceptor proteins. Until now, five [Fe–S] cluster biogenesis systems have been identified: NIF, ISC, SUF, MIS, and SUF-like minimal (SMS) [10]. Most recently, classified MIS and SMS systems are considered more ancient functional fragments of already described systems that display a phylogenetic distinct distribution operating in archaea and bacteria species [39]. NifUS is a two-component system that shares similar structural and functional features when compared to the more complex ISC system and, therefore, serves as an attractive model to study mechanistic steps leading to forming [Fe–S] clusters. NifUS was first identified in *A. vinelandii* and characterized for its dedicated role in supplying [Fe–S] clusters to nitrogen-fixing proteins. While much progress has been made to uncover mechanistic and chemical details on the synthesis of

[Fe–S] clusters, selected aspects of this process are still a topic of investigation and debate. These include the physiological source of iron and electrons, the sequence of individual steps, and the kinetic and regulatory parameters associated with each event. The yet-to-be-determined structure of a NifUS complex could offer mechanistic insight into these unresolved aspects. The apparent flexibility of the C′-terminal domain of NifU has challenged efforts to resolve the architecture of this system through crystallographic approaches. Analysis of an expanded dataset of NifU sequences offers an opportunity to identify candidate sequences suitable for structural investigation.

Acknowledgments

The authors thank the collaboration with Mike K. Johnson for his contributions to understanding the function and chemistry of the NifUS system described in this chapter. The National Science Foundation (MCB2335999 to PDS) and the U.S. Department of Energy (DE-SC0010834 to DRD) support the research in our laboratories.

References

1 Braymer, J.J., Freibert, S.A., Rakwalska-Bange, M., and Lill, R. (2021). Mechanistic concepts of iron-sulfur protein biogenesis in Biology. *Biochim. Biophys. Acta Mol. Cell Res.* 1868 (1): 118863.

2 Esquilin-Lebron, K., Dubrac, S., Barras, F., and Boyd, J.M. (2021). Bacterial approaches for assembling iron-sulfur proteins. *MBio* 12 (6): e02425-21.

3 Zheng, C. and Dos Santos, P.C. (2018). Metallocluster transactions: dynamic protein interactions guide the biosynthesis of Fe–S clusters in bacteria. *Biochem. Soc. Trans.* 46 (6): 1593–1603.

4 Mettert, E.L. and Kiley, P.J. (2024). Fe-S cluster homeostasis and beyond: the multifaceted roles of IscR. *Biochim. Biophys. Acta Mol. Cell Res.* 1871 (6): 119749.

5 Zheng, L., Cash, V.L., Flint, D.H., and Dean, D.R. (1998). Assembly of iron-sulfur clusters. Identification of an iscSUA-hscBA-fdx gene cluster from *Azotobacter vinelandii*. *J. Biol. Chem.* 273 (21): 13264–13272.

6 Houmard, J., Bogusz, D., Bigault, R., and Elmerich, C. (1980). Characterization and kinetics of the biosynthesis of some nitrogen fixation (nif) gene products in *Klebsiella pneumoniae*. *Biochimie* 62 (4): 267–275.

7 Jacobson, M.R., Cash, V.L., Weiss, M.C. et al. (1989). Biochemical and genetic analysis of the nifUSVWZM cluster from *Azotobacter vinelandii*. *Mol. Gen. Genet. MGG* 219 (1): 49–57.

8 Addo, M.A. and Dos Santos, P.C. (2020). Distribution of nitrogen fixation genes in prokaryotes containing alternative nitrogenases. *ChemBioChem* 21 (12): 1749–1759.

9 Ali, V., Shigeta, Y., Tokumoto, U. et al. (2004). An intestinal parasitic protist, *Entamoeba histolytica*, possesses a non-redundant nitrogen fixation-like system for iron-sulfur cluster assembly under anaerobic conditions. *J. Biol. Chem.* 279 (16): 16863–16874.

10 Olson, J.W., Agar, J.N., Johnson, M.K., and Maier, R.J. (2000). Characterization of the NifU and NifS Fe–S cluster formation proteins essential for viability in *Helicobacter pylori. Biochemistry* 39 (51): 16213–16219.

11 Dos Santos, P.C. and Dean, D.R. (2017). Nif system for simple [Fe-S] cluster assembly in nitrogen-fixing bacteria. In: *Metalloprotein Active Site Assembly* (ed. M.K. Johnson and R.A. Scott), 3–16. Wiley.

12 Smith, A.D., Jameson, G.N.L., Dos Santos, P.C. et al. (2005). NifS-mediated assembly of [4Fe-4S] clusters in the N- and C-terminal domains of the NifU scaffold protein. *Biochemistry* 44 (39): 12955–12969.

13 Zheng, L., White, R.H., Cash, V.L. et al. (1993). Cysteine desulfurase activity indicates a role for NifS in metallocluster biosynthesis. *Proc. Natl. Acad. Sci. USA* 90 (7): 2754–2758.

14 Dos Santos, P.C., Johnson, D.C., Ragle, B.E. et al. (2007). Controlled expression of nif and isc iron-sulfur protein maturation components reveals target specificity and limited functional replacement between the two systems. *J. Bacteriol.* 189 (7): 2854–2862.

15 Johnson, D.C., Dos Santos, P.C., and Dean, D.R. (2005). NifU and NifS are required for the maturation of nitrogenase and cannot replace the function of isc-gene products in *Azotobacter vinelandii. Biochem. Soc. Trans.* 33 (Pt 1): 90–93.

16 Fu, W., Jack, R.F., Morgan, T.V. et al. (1994). nifU gene product from *Azotobacter vinelandii* is a homodimer that contains two identical [2Fe-2S] clusters. *Biochemistry* 33 (45): 13455–13463.

17 Jumper, J., Evans, R., Pritzel, A. et al. (2021). Highly accurate protein structure prediction with AlphaFold. *Nature* 596 (7873): 583–589.

18 Varadi, M., Anyango, S., Deshpande, M. et al. (2022). AlphaFold Protein Structure Database: massively expanding the structural coverage of protein-sequence space with high-accuracy models. *Nucleic Acids Res.* 50 (D1): D439–D444.

19 Shimomura, Y., Wada, K., Fukuyama, K., and Takahashi, Y. (2008). The asymmetric trimeric architecture of [2Fe–2S] IscU: implications for its scaffolding during iron–sulfur cluster biosynthesis. *J. Mol. Biol.* 383 (1): 133–143.

20 Kakuta, Y., Horio, T., Takahashi, Y., and Fukuyama, K. (2001). Crystal structure of *Escherichia coli* Fdx, an adrenodoxin-type ferredoxin involved in the assembly of iron–sulfur clusters. *Biochemistry* 40 (37): 11007–11012.

21 Volbeda, A., Dodd, E.L., Darnault, C. et al. (2017). Crystal structures of the NO sensor NsrR reveal how its iron-sulfur cluster modulates DNA binding. *Nat. Commun.* 8 (1): 15052.

22 Yuvaniyama, P., Agar, J.N., Cash, V.L. et al. (2000). NifS-directed assembly of a transient [2Fe-2S] cluster within the NifU protein. *Proc. Natl. Acad. Sci. USA* 97 (2): 599–604.

23 Tanaka, N., Yuda, E., Fujishiro, T. et al. (2019). Identification of IscU residues critical for de novo iron–sulfur cluster assembly. *Mol. Microbiol.* 112 (6): 1769–1783.

24 Agar, J.N., Yuvaniyama, P., Jack, R.F. et al. (2000). Modular organization and identification of a mononuclear iron-binding site within the NifU protein. *J. Biol. Inorg. Chem.* 5 (2): 167–177.

25 Berman, H., Henrick, K., and Nakamura, H. (2003). Announcing the worldwide Protein Data Bank. *Nat. Struct. Mol. Biol.* 10 (12): 980.

26 ww PDBc. (2019). Protein Data Bank: the single global archive for 3D macromolecular structure data. *Nucleic Acids Res.* 47 (D1): D520–D528.

27 Berman, H., Henrick, K., Nakamura, H., and Markley, J.L. (2007). The worldwide Protein Data Bank (wwPDB): ensuring a single, uniform archive of PDB data. *Nucleic Acids Res.* 35 (suppl_1): D301–D303.

28 Kunichika, K., Nakamura, R., Fujishiro, T., and Takahashi, Y. (2021). The structure of the dimeric state of IscU harboring two adjacent [2Fe–2S] clusters provides mechanistic insights into cluster conversion to [4Fe–4S]. *Biochemistry* 60 (20): 1569–1572.

29 Marinoni, E.N., de Oliveira, J.S., Nicolet, Y. et al. (2012). (IscS-IscU)2 complex structures provide insights into Fe_2S_2 biogenesis and transfer. *Angew. Chem. Int. Ed.* 51 (22): 5439–5442.

30 Dos Santos, P.C., Smith, A.D., Frazzon, J. et al. (2004). Iron-sulfur cluster assembly: NifU-directed activation of the nitrogenase Fe protein. *J. Biol. Chem.* 279 (19): 19705–19711.

31 Zhao, D., Curatti, L., and Rubio, L.M. (2007). Evidence for nifU and nifS participation in the biosynthesis of the iron-molybdenum cofactor of nitrogenase*. *J. Biol. Chem.* 282 (51): 37016–37025.

32 Angelini, S., Gerez, C., Choudens, S.O. et al. (2008). NfuA, a new factor required for maturing Fe–S proteins in *Escherichia coli* under oxidative stress and iron starvation conditions. *J. Biol. Chem.* 283 (20): 14084–14091.

33 Bandyopadhyay, S., Naik, S.G., O'Carroll, I.P. et al. (2008). A proposed role for the *Azotobacter vinelandii* NfuA protein as an intermediate iron-sulfur cluster carrier. *J. Biol. Chem.* 283 (20): 14092–14099.

34 Martin del Campo, J.S., Rigsbee, J., Bueno Batista, M. et al. (2023). Overview of physiological, biochemical, and regulatory aspects of nitrogen fixation in *Azotobacter vinelandii. Crit. Rev. Biochem. Mol. Biol.* 57: 492–538.

35 Hamilton, T.L., Ludwig, M., Dixon, R. et al. (2011). Transcriptional profiling of nitrogen fixation in *Azotobacter vinelandii. J. Bacteriol.* 193 (17): 4477–4486.

36 Appia-Ayme, C., Little, R., Chandra, G. et al. (2022). Interactions between paralogous bacterial enhancer-binding proteins enable metal-dependent regulation of alternative nitrogenases in *Azotobacter vinelandii. Mol. Microbiol.* 118 (1–2): 105–124.

37 Luque, F. and Pau, R.N. (1991). Transcriptional regulation by metals of structural genes for *Azotobacter vinelandii* nitrogenases. *Mol. Gen. Genet. MGG* 227 (3): 481–487.

38 Barney, B.M., Plunkett, M.H., Natarajan, V. et al. (2017). Transcriptional analysis of an ammonium-excreting strain of *Azotobacter vinelandii* deregulated for nitrogen fixation. *Appl. Environ. Microbiol.* 83 (20): e01534-17.

39 Garcia, P.S., D'Angelo, F., Ollagnier de Choudens, S. et al. (2022). An early origin of iron–sulfur cluster biosynthesis machineries before Earth oxygenation. *Nat. Ecol. Evol.* 6 (10): 1564–1572.

40 Black, K.A. and Dos Santos, P.C. (2015). Shared-intermediates in the biosynthesis of thio-cofactors: mechanism and functions of cysteine desulfurases and sulfur acceptors. *Biochim. Biophys. Acta Mol. Cell Res.* 1853 (6): 1470–1480.

41 Fujishiro, T., Nakamura, R., Kunichika, K., and Takahashi, Y. (2022). Structural diversity of cysteine desulfurases involved in iron-sulfur cluster biosynthesis. *Biophys. Physicobiol.* 19: 1–18.

42 Nakamura, R., Hikita, M., Ogawa, S. et al. (2020). Snapshots of PLP-substrate and PLP-product external aldimines as intermediates in two types of cysteine desulfurase enzymes. *FEBS J.* 287 (6): 1138–1154.

43 Cupp-Vickery, J.R., Urbina, H., and Vickery, L.E. (2003). Crystal structure of IscS, a cysteine desulfurase from *Escherichia coli*. *J. Mol. Biol.* 330 (5): 1049–1059.

44 Dos Santos, P.C. (2017). *B. subtilis* as a model for studying the assembly of Fe–S clusters in gram-positive bacteria. In: *Methods in Enzymology*, vol. 595 (ed. S.S. David), 185–212. Academic Press.

45 Srour, B., Gervason, S., Monfort, B., and D'Autréaux, B. (2020). Mechanism of iron–sulfur cluster assembly: in the intimacy of iron and sulfur encounter. *Inorganics* 8 (10): 55.

46 Lin, C.-W., McCabe, J.W., Russell, D.H., and Barondeau, D.P. (2020). Molecular mechanism of ISC iron–sulfur cluster biogenesis revealed by high-resolution native mass spectrometry. *J. Am. Chem. Soc.* 142 (13): 6018–6029.

47 Shi, R., Proteau, A., Villarroya, M. et al. (2010). Structural basis for Fe–S cluster assembly and tRNA thiolation mediated by IscS protein–protein interactions. *PLoS Biol.* 8 (4): e1000354.

48 Yang, Z.Y., Ledbetter, R., Shaw, S. et al. (2016). Evidence that the Pi release event is the rate-limiting step in the nitrogenase catalytic cycle. *Biochemistry* 55 (26): 3625–3635.

49 Burgess, B.K., Jacobs, D.B., and Stiefel, E.I. (1980). Large-scale purification of high activity *Azotobacter vinelandii* nitrogenase. *Biochim. Biophys. Acta* 614 (1): 196–209.

50 Dingler, C., Kuhla, J., Wassink, H., and Oelze, J. (1988). Levels and activities of nitrogenase proteins in *Azotobacter vinelandii* grown at different dissolved oxygen concentrations. *J. Bacteriol.* 170 (5): 2148–2152.

51 Jacobs, D., Mitchell, D., and Watt, G.D. (1995). The concentration of cellular nitrogenase proteins in *Azotobacter vinelandii* whole cells as determined by activity measurements and electron paramagnetic resonance spectroscopy. *Arch. Biochem. Biophys.* 324 (2): 317–324.

52 Dos Santos, P.C. and Dean, D.R. (2017). Fe–S cluster assembly: NIF system in nitrogen-fixing bacteria. In: *Encyclopedia of Inorganic and Bioinorganic Chemistry* (ed. M.K. Johnson and R.A. Scott), 1–13. Wiley.

53 Robinson, A.C., Dean, D.R., and Burgess, B.K. (1987). Iron-molybdenum cofactor biosynthesis in *Azotobacter vinelandii* requires the iron protein of nitrogenase. *J. Biol. Chem.* 262 (29): 14327–14332.

54 Ribbe, M.W., Hu, Y., Guo, M. et al. (2002). The FeMoco-deficient MoFe protein produced by a nifH deletion strain of *Azotobacter vinelandii* shows unusual P-cluster features. *J. Biol. Chem.* 277 (26): 23469–23476.

55 Jimenez-Vicente, E., Yang, Z.Y., Ray, W.K. et al. (2018). Sequential and differential interaction of assembly factors during nitrogenase MoFe protein maturation. *J. Biol. Chem.* 293 (25): 9812–9823.

56 Rubio, L.M. and Ludden, P.W. (2008). Biosynthesis of the iron-molybdenum cofactor of nitrogenase. *Ann. Rev. Microbiol.* 62 (1): 93–111.

57 Einsle, O. and Rees, D.C. (2020). Structural enzymology of nitrogenase enzymes. *Chem. Rev.* 120 (12): 4969–5004.

58 Georgiadis, M.M., Komiya, H., Chakrabarti, P. et al. (1992). Crystallographic structure of the nitrogenase iron protein from *Azotobacter vinelandii*. *Science* 257 (5077): 1653–1659.

59 Schmid, B., Einsle, O., Chiu, H.-J. et al. (2002). Biochemical and structural characterization of the cross-linked complex of nitrogenase: comparison to the ADP-AlF_4^--stabilized structure. *Biochemistry* 41 (52): 15557–15565.

60 Spatzal, T., Aksoyoglu, M., Zhang, L. et al. (2011). Evidence for interstitial carbon in nitrogenase FeMo cofactor. *Science* 334 (6058): 940.

61 Pence, N., Tokmina-Lukaszewska, M., Yang, Z.-Y. et al. (2017). Unraveling the interactions of the physiological reductant flavodoxin with the different conformations of the Fe protein in the nitrogenase cycle. *J. Biol. Chem.* 292 (38): 15661–15669.

62 Lowery, T.J., Wilson, P.E., Zhang, B. et al. (2006). Flavodoxin hydroquinone reduces *Azotobacter vinelandii* Fe protein to the all-ferrous redox state with a S = 0 spin state. *Proc. Natl. Acad. Sci. USA* 103 (46): 17131–17136.

63 Gavini, N., Tungtur, S., and Pulakat, L. (2006). Peptidyl-prolyl *cis/trans* isomerase-independent functional NifH mutant of *Azotobacter vinelandii*. *J. Bacteriol.* 188 (16): 6020–6025.

64 Wilcoxen, J., Arragain, S., Scandurra, A.A. et al. (2016). Electron paramagnetic resonance characterization of three iron–sulfur clusters present in the nitrogenase cofactor maturase NifB from *Methanocaldococcus infernus*. *J. Am. Chem. Soc.* 138 (24): 7468–7471.

65 Rettberg, L.A., Wilcoxen, J., Lee, C.C. et al. (2018). Probing the coordination and function of Fe4S4 modules in nitrogenase assembly protein NifB. *Nat. Commun.* 9 (1): 2824.

66 Fajardo, A.S., Legrand, P., Payá-Tormo, L. et al. (2020). Structural insights into the mechanism of the radical SAM carbide synthase NifB, a key nitrogenase cofactor maturating enzyme. *J. Am. Chem. Soc.* 142 (25): 11006–11012.

67 Jenner, L.P., Cherrier, M.V., Amara, P. et al. (2021). An unexpected P-cluster like intermediate en route to the nitrogenase FeMo-co. *Chem. Sci.* 12 (14): 5269–5274.

68 Kang, W., Rettberg, L., Stiebritz, M. et al. (2021). X-Ray crystallographic analysis of NifB with a full complement of clusters: structural insights into the radical SAM-dependent carbide insertion during nitrogenase cofactor assembly. *Angew. Chem. Int. Ed.* 60 (5): 2364–2370.

69 Tanifuji, K., Lee, C.C., Sickerman, N.S. et al. (2018). Tracing the 'ninth sulfur' of the nitrogenase cofactor via a semi-synthetic approach. *Nat. Chem.* 10 (5): 568–572.

70 Pérez-González, A., Jimenez-Vicente, E., Gies-Elterlein, J. et al. (2021). Specificity of NifEN and VnfEN for the assembly of nitrogenase active site cofactors in *Azotobacter vinelandii*. *MBio* 12 (4): e01568-21.

71 Paustian, T.D., Shah, V.K., and Roberts, G.P. (1990). Apodinitrogenase: purification, association with a 20-kilodalton protein, and activation by the iron-molybdenum cofactor in the absence of dinitrogenase reductase. *Biochemistry* 29 (14): 3515–3522.

72 Dixon, R. and Kahn, D. (2004). Genetic regulation of biological nitrogen fixation. *Nat. Rev. Microbiol.* 2 (8): 621–631.

73 Setubal, J.C., dos Santos, P., Goldman, B.S. et al. (2009). Genome sequence of *Azotobacter vinelandii*, an obligate aerobe specialized to support diverse anaerobic metabolic processes. *J. Bacteriol.* 191 (14): 4534–4545.

74 Miller, R.W. and Eady, R.R. (1988). Molybdenum and vanadium nitrogenases of *Azotobacter chroococcum*. Low temperature favours N_2 reduction by vanadium nitrogenase. *Biochem. J.* 256 (2): 429–432.

75 Harris, D.F., Lukoyanov, D.A., Kallas, H. et al. (2019). Mo-, V-, and Fe-nitrogenases use a universal eight-electron reductive-elimination mechanism to achieve N_2 reduction. *Biochemistry* 58 (30): 3293–3301.

76 Kennedy, C. and Dean, D. (1992). The nifU, nifS and nifV gene products are required for activity of all three nitrogenases of *Azotobacter vinelandii*. *Mol. Gen. Genet. MGG* 231 (3): 494–498.

77 Schwartz, C.J., Giel, J.L., Patschkowski, T. et al. (2001). IscR, an Fe-S cluster-containing transcription factor, represses expression of *Escherichia coli* genes encoding Fe-S cluster assembly proteins. *Proc. Natl. Acad. Sci. USA* 98 (26): 14895–14900.

78 Chandramouli, K. and Johnson, M.K. (2006). HscA and HscB stimulate [2Fe-2S] cluster transfer from IscU to apoferredoxin in an ATP-dependent reaction. *Biochemistry* 45 (37): 11087–11095.

79 Vickery, L.E. and Cupp-Vickery, J.R. (2007). Molecular chaperones HscA/Ssq1 and HscB/Jac1 and their roles in iron-sulfur protein maturation. *Crit. Rev. Biochem. Mol. Biol.* 42 (2): 95–111.

80 Johnson, D.C., Unciuleac, M.-C., and Dean, D.R. (2006). Controlled expression and functional analysis of iron-sulfur cluster biosynthetic components within *Azotobacter vinelandii*. *J. Bacteriol.* 188 (21): 7551–7561.

81 Bijay, S. and Craswell, E. (2021). Fertilizers and nitrate pollution of surface and ground water: an increasingly pervasive global problem. *SN Appl. Sci.* 3 (4): 518.

82 Jimenez-Vicente, E. and Dean, D.R. (2017). Keeping the nitrogen-fixation dream alive. *Proc. Natl. Acad. Sci. USA* 114 (12): 3009–3011.

83 Burén, S. and Rubio, L.M. (2018). State of the art in eukaryotic nitrogenase engineering. *FEMS Microbiol. Lett.* 365 (2).

84 Dixon, R.A. and Postgate, J.R. (1972). Genetic transfer of nitrogen fixation from *Klebsiella pneumoniae* to *Escherichia coli*. *Nature* 237 (5350): 102–103.

85 Wang, L., Zhang, L., Liu, Z. et al. (2013). A minimal nitrogen fixation gene cluster from *Paenibacillus* sp. WLY78 enables expression of active nitrogenase in *Escherichia coli*. *PLOS Genet.* 9 (10): e1003865.

86 Yang, J., Xie, X., Wang, X. et al. (2014). Reconstruction and minimal gene requirements for the alternative iron-only nitrogenase in *Escherichia coli*. *Proc. Natl. Acad. Sci. USA* 111 (35): E3718–E3725.

87 Quechol, R., Solomon, J.B., Liu, Y.A. et al. (2023). Heterologous synthesis of the complex homometallic cores of nitrogenase P- and M-clusters in *Escherichia coli*. *Proc. Natl. Acad. Sci. USA* 120 (44): e2314788120.

88 Lill, R. and Mühlenhoff, U. (2005). Iron–sulfur-protein biogenesis in eukaryotes. *Trends Biochem. Sci.* 30 (3): 133–141.

89 Cammack, R. (2013). Iron–sulfur proteins. In: *Encyclopedia of Biological Chemistry*, Seconde (ed. W.J. Lennarz and M.D. Lane), 657–664. Waltham: Academic Press.

90 Lopez-Torrejon, G., Jimenez-Vicente, E., Buesa, J.M. et al. (2016). Expression of a functional oxygen-labile nitrogenase component in the mitochondrial matrix of aerobically grown yeast. *Nat. Commun.* 7: 11426.

91 Braymer, J.J. and Lill, R. (2017). Iron–sulfur cluster biogenesis and trafficking in mitochondria. *J. Biol. Chem.* 292 (31): 12754–12763.

92 Burén, S., Young, E.M., Sweeny, E.A. et al. (2017). Formation of nitrogenase NifDK tetramers in the mitochondria of *Saccharomyces cerevisiae*. *ACS Synth. Biol.* 6 (6): 1043–1055.

93 Burén, S., Pratt, K., Jiang, X. et al. (2019). Biosynthesis of the nitrogenase active-site cofactor precursor NifB-co in *Saccharomyces cerevisiae*. *Proc. Natl. Acad. Sci. USA* 116 (50): 25078–25086.

94 Jiang, X., Coroian, D., Barahona, E. et al. (2022). Functional nitrogenase cofactor maturase NifB in mitochondria and chloroplasts of *Nicotiana benthamiana*. *MBio* 13 (3): e0026822.

95 Yang, J., Xie, X., Xiang, N. et al. (2018). Polyprotein strategy for stoichiometric assembly of nitrogen fixation components for synthetic biology. *Proc. Natl. Acad. Sci. USA* 115 (36): E8509–E8517.

96 Benoit, S.L., Agudelo, S., and Maier, R.J. (2021). A two-hybrid system reveals previously uncharacterized protein–protein interactions within the *Helicobacter pylori* NIF iron–sulfur maturation system. *Sci. Rep.* 11 (1): 10794.

97 Tokumoto, U., Kitamura, S., Fukuyama, K., and Takahashi, Y. (2004). Interchangeability and distinct properties of bacterial Fe-S cluster assembly systems: functional replacement of the isc and suf operons in *Escherichia coli* with the nifSU-like operon from *Helicobacter pylori*. *J. Biochem.* 136 (2): 199–209.

98 Mi-ichi, F., Yousuf, M.A., Nakada-Tsukui, K., and Nozaki, T. (2009). Mitosomes in *Entamoeba histolytica* contain a sulfate activation pathway. *Proc. Natl. Acad. Sci. USA* 106 (51): 21731–21736.

99 Nývltová, E., Šuták, R., Harant, K. et al. (2013). NIF-type iron-sulfur cluster assembly system is duplicated and distributed in the mitochondria and cytosol of *Mastigamoeba balamuthi*. *Proc. Natl. Acad. Sci. USA* 110 (18): 7371–7376.

100 Žárský, V., Klimeš, V., Pačes, J. et al. (2021). The *Mastigamoeba balamuthi* genome and the nature of the free-living ancestor of entamoeba. *Mol. Biol. Evol.* 38 (6): 2240–2259.

101 Hoppert, M. (2011). Metalloenzymes. In: *Encyclopedia of Geobiology* (ed. J. Reitner and V. Thiel), 558–563. Dordrecht: Springer Netherlands.

10

Alternative Substrates of Nitrogenase

Frederik V. Schmidt[1] *and Johannes G. Rebelein*[1,2]

[1] *Max Planck Institute for Terrestrial Microbiology, Karl-von-Frisch Straße 10, 35043 Marburg, Germany*
[2] *Philipps-University Marburg, Center for Synthetic Microbiology (SYNMIKRO), 35043 Marburg, Germany*

Chapter Goals

- To summarize and explain the nitrogenase mechanism
- To provide an overview of the promiscuous nitrogenase activities. Nitrogenases reduce molecular nitrogen as well as hydrocarbons, nitrogen-containing substrates and industrial waste gases including carbon monoxide and carbon dioxide, which are converted into hydrocarbons.
- To highlight nitrogenases as engineering targets for biotechnological applications to recycle carbon monoxide and carbon dioxide into value-added compounds.

Main Chapter Points

- Nitrogenases, a class of metalloenzymes best known for the reduction of molecular nitrogen are able to reduce a wide range of alternative substrates including acetylene, carbon monoxide, and carbon dioxide.
- The promiscuity of nitrogenases is based on reactive hydrides, which are generated on the active site metallocluster and drive the reduction of substrates.
- The outstanding potential of nitrogenases to convert industrial waste gases into hydrocarbons renders them promising candidates for biotechnological applications and a sustainable, circular economy.

10.1 Nitrogenase Introduction

Biological nitrogen fixation is a critical ecological process contributing to the global nitrogen cycle, an essential reaction to all life on Earth. At the heart of this process is nitrogenase, a remarkable enzyme complex unique to bacteria and archaea that fix nitrogen, the diazotrophs. Nitrogenase is the only enzyme capable of reducing

Iron-Sulfur Clusters: Biogenesis and Biochemistry, Volume 1, First Edition.
Edited by Silke Leimkühler, Günter Schwarz, Oliver Lenz, and Oliver Einsle.

atmospheric nitrogen (N_2) to ammonia (NH_3), a form of nitrogen that plants can assimilate and use to synthesize vital organic compounds such as amino acids and nucleic acids [1, 2].

The enzymatic reduction of N_2 by nitrogenases is complemented by the industrial Haber–Bosch process. While the Haber–Bosch process also synthesizes NH_3 from N_2, it requires high temperatures (400–500 °C) and pressures (150–350 bar), making it energy-intensive and environmentally taxing [3]. In contrast to the industrial process, nitrogenase operates under ambient conditions, offering a more sustainable and eco-friendly pathway for nitrogen fixation [4, 5]. Instead of high pressures and temperatures, nitrogenases reduce N_2 by coupling the electron transfer to the hydrolysis of adenosine triphosphate (ATP), as demonstrated by the optimal chemical formula of biological N_2 reduction (Eq. (10.1)) [6].

$$N_2 + 8H^+ + 16MgATP + 8e^- \rightarrow 2NH_3 + H_2 + 16MgADP + 16P_i \tag{10.1}$$

Nitrogenase converts one mole of molecular nitrogen into two moles of ammonia and forms molecular hydrogen (H_2) as an obligate side product. This process requires at least eight protons (H^+) and eight electrons (e^-). Each electron transferred comes at a cost of two ATPs. The exact stoichiometry of the reaction varies between the three naturally occurring isoforms of nitrogenase.

10.1.1 Nitrogenase Isozymes

Nitrogenases are metal-containing enzymes. Based on the metal content of their active site, they are distinguished into three isoforms. These include the molybdenum (Mo) nitrogenase, the most abundant and extensively studied nitrogenase, alongside two alternative isoforms, the vanadium (V) and iron (Fe) nitrogenase (Figure 10.1a) [7, 8]. Though all nitrogenase isoforms can fix N_2, they require varying amounts of ATP to produce ammonia (Eqs. (10.2)–(10.4)) [9].

Mo nitrogenase:

$$N_2 + 10H^+ + 20MgATP + 10e^- \rightarrow 2NH_3 + 2H_2 + 20MgADP + 20P_i \tag{10.2}$$

V nitrogenase:

$$N_2 + 18H^+ + 36MgATP + 18e^- \rightarrow 2NH_3 + 6H_2 + 36MgADP + 36P_i \tag{10.3}$$

Fe nitrogenase:

$$N_2 + 20H^+ + 40MgATP + 20e^- \rightarrow 2NH_3 + 7H_2 + 40MgADP + 40P_i \tag{10.4}$$

Due to its energy efficiency, Mo nitrogenase is present in all nitrogen-fixing species. In contrast, only 6.4% and 8.9% of sequenced diazotrophs harbor the V or Fe isoform, respectively [10]. Thus, alternative nitrogenases are believed to serve as backup systems in environments where Mo is scarce, indicating their evolutionary significance in ensuring nitrogen fixation under various environmental conditions.

All nitrogenases share a similar two-component architecture: a reductase component known as Fe protein and a catalytic component known as the MFe protein (MoFe, VFe, or FeFe protein, depending on the isoform). The reductase component is a homodimer of two H subunits encoded by *nifH*, *vnfH*, or *anfH* for the Mo, V, and Fe nitrogenase, respectively. Regardless of the specific isoform, every homodimer contains two ATP-binding sites and a subunit bridging

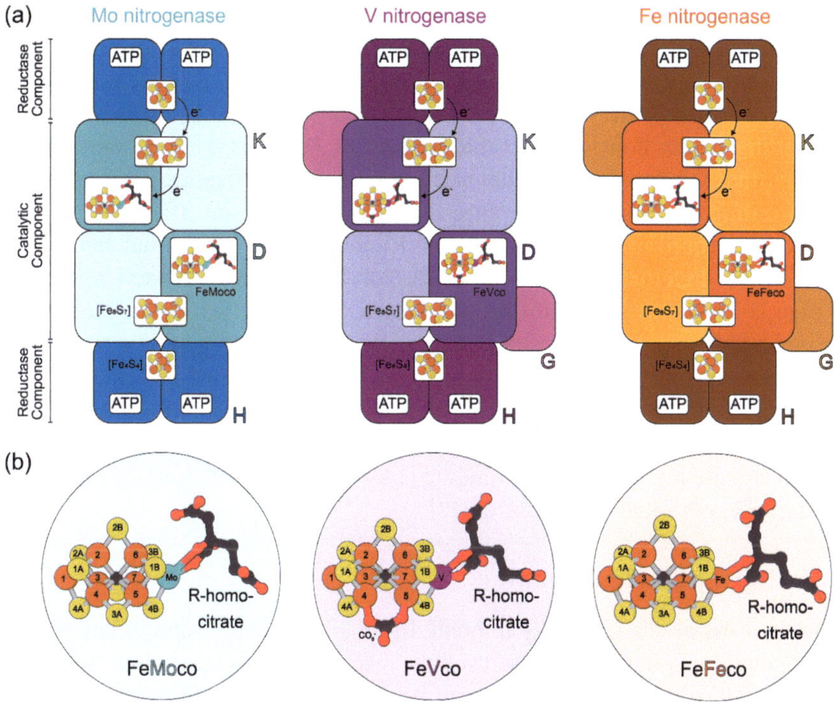

Figure 10.1 Nitrogenase isoforms. (a) Schematic representations of the Mo (left), V (middle), and Fe (right) nitrogenase complexes, each featuring their respective catalytic component (Nif[DK]$_2$, Vnf[DGK]$_2$, Anf[DGK]$_2$) with two bound reductase components (NifH$_2$, VnfH$_2$, AnfH$_2$). Each H-subunit harbors one ATP-binding site. Electrons flow from the [Fe_4–S_4] cluster to the [Fe_8–S_7] cluster (P-cluster) to the respective FeMco (M = Mo, V, or Fe). (b) Close-up view of the FeMoco (left) from PDB 7UTA, FeVco (middle) from PDB 5N6Y, and FeFeco (right) from PDB 8OIE, the atoms of the metallocluster are numbered. Sulfur atoms of the clusters are represented as yellow spheres, iron atoms as orange spheres, molybdenum atoms as cyan spheres, vanadium atoms as purple spheres, and the central carbide ion as a black sphere. The homocitrate coordinated to the FeMco is depicted as a ball-and-stick model with carbon in black and oxygen in red. Source: Frederik Schmidt (Coauthor).

[Fe_4–S_4] cluster. The differences among the catalytic components of the three isozymes are more significant. The MoFe protein is encoded by the *nifDK* genes and forms a (DK)$_2$-heterotetramer structure. Each (DK)$_2$ tetramer contains four metallic cofactors: two [Fe_8–S_7] clusters (P-clusters) at the DK interfaces and two Mo-containing active site cofactors (FeMoco) buried inside the D- subunits. The V and Fe nitrogenases, encoded by *vnfDGK* and *anfDGK*, respectively, form a (DGK)$_2$ heterohexamer. Analogous to the Mo isoform, each (DGK)$_2$ hexamer features two P-clusters and two active site cofactors. The active site cofactors are termed FeVco or FeFeco for the V and Fe nitrogenase, respectively.

The active site cofactors of nitrogenases are among the most complex metallocofactors in nature, highlighting the intricate chemistry involved in nitrogen fixation (Figure 10.1b). First, the FeMoco, found in Mo nitrogenase, has a composition of [$MoFe_7S_9C$-(*R*)-homocitrate], featuring a unique arrangement of Fe atoms around an interstitial carbon atom [11]. Second, the FeVco, present in V nitrogenase, is composed of [$VFe_7S_8C(CO_3)$-(*R*)-homocitrate], with a carbonate ion replacing one of the sulfide ligands found in the FeMoco [12]. Third, the FeFeco of the Fe nitrogenase is a [Fe_8S_9C-(*R*)-homocitrate] cluster where another iron occupies the heterometal position found in the FeMoco or FeVco [13]. This comparison highlights that the three active site cofactors share several features. All of them are iron–sulfur clusters that form cage-like structures around a central carbide ion. They are coordinated by a bidentate (*R*)-homocitrate ligand that binds the clusters at the heterometal position via its 2-hydroxyl and 2-carboxyl moieties. The cage is covalently linked to the protein backbone via two highly conserved cysteine and histidine residues that bind to iron in position 1 and the heterometal position, respectively. The active site of nitrogenases is constituted of further highly conserved residues that are involved in catalysis. In the *Azotobacter vinelandii* Mo nitrogenase, these include, for example R96, Q191, and H195, which are involved in proton channeling and substrate binding [14–16] as well as V70, which regulates substrate access through steric control [17].

10.1.2 Nitrogenase Mechanism

Due to their extraordinary ability to break the stable N_2 triple bond under physiological conditions, the nitrogenase catalytic mechanism has been under scrutiny for decades [18, 19]. The mechanism can be divided into two subprocesses: Electron delivery to the active site and the accumulation of reducing equivalents at the FeMco (M = Mo, V, or Fe).

The electron delivery process is facilitated by the reductase component (Figure 10.2a). Each of the two H-subunits harbors a nucleotide-binding pocket that contains either adenosine diphosphate (ADP) or ATP. The reductase component interacts with specific ferredoxins or flavodoxins, which can reduce its subunit-bridging [Fe_4–S_4] cluster from the [Fe_4–S_4]$^{2+}$ to the [Fe_4–S_4]$^{1+}$ state. Subsequently, ATP binds in both nucleotide-binding pockets, triggering a structural rearrangement that allows the interaction of the reductase component with its partner catalytic component to form the nitrogenase complex [20]. Although cross-reactivity between the individual nitrogenase systems can be observed *in vitro* [25], this interaction is thought to be specific *in vivo*, meaning that $NifH_2$ associates with Nif$(DK)_2$, $VnfH_2$ with Vnf$(DGK)_2$, and $AnfH_2$ with Anf$(DGK)_2$. This specificity is ensured at the transcriptional and protein levels. First, diazotrophs encoding for multiple nitrogenase isoforms produce only one isoform at a time by utilizing nitrogenase building blocks as transcriptional regulators. For example, in *Rhodobacter capsulatus*, which harbors the Mo and Fe nitrogenase, Mo is a potent transcriptional activator for *nif* genes while repressing *anf* genes [26]. Second, the interaction interfaces of Mo, V, or Fe nitrogenase catalytic and reductase components feature distinct surface charge patterns, which guide the electrostatically driven interaction [27]. In contrast to the Mo isoform, alternative nitrogenases feature an additional small subunit (VnfG and AnfG) involved in the reductase component docking process [28]. Thus, the G subunit might help to prevent complex formation between $NifH_2$ and alternative nitrogenase catalytic components, thereby increasing the specificity of the interaction.

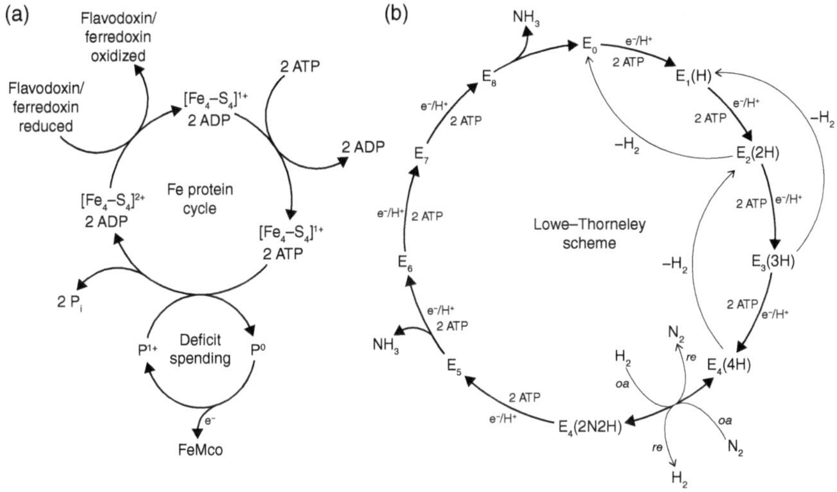

Figure 10.2 Fe protein cycle and Lowe–Thorneley scheme. (a) Electron delivery to the nitrogenase catalytic component by the Fe protein. Flavodoxins or ferredoxins reduce the Fe protein $[Fe_4-S_4]$ cluster. Subsequently, ATP replaces ADP to trigger a conformational change in the reductase component, allowing it to bind to the catalytic component [20]. The $[Fe_4-S_4]^{1+}$ cluster of the bound Fe protein reduces the oxidized P-cluster (P^{1+}), which previously donated one electron to the active site cofactor via the "deficit spending" mechanism [21]. Finally, both molecules of ATP are hydrolyzed and the nitrogenase complex disassembles [22]. (b) A simplified version of the Lowe–Thorneley scheme describing the mechanism of N_2 reduction at the nitrogenase active site cofactor [23]. The cluster cycles through nine redox states (E_0-E_8). The subscript integer represents the amounts of electrons (e^-) and protons (H^+) added to the cluster, which reach the cluster through the "deficit spending" mechanism and H^+ channeling, respectively. Each electron transferred comes at the cost of two ATPs, which are required for electron delivery by the reductase component. E_4 is regarded as the critical step of N_2 reduction, also called the Janus intermediate [24]. Up to this point, reducing equivalents are stored in the reduced state of the cluster (E_1 and E_3) or pairwise in bound hydrides (E_2 and E_4). The H and N stoichiometry is given in parenthesis. E_2-E_4 are kinetically unstable and can lose reducing equivalents through hydrogen (H_2) release, causing the cluster to fall back to the E_{n-2} state ($2 \leq n \leq 4$). At E_4, oxidative addition (*oa*) of N_2 accompanied by reductive elimination (*re*) of two bound hydrides and H_2 release is observed. Through this reversible reaction, the cluster transitions into a hydrazine-bound intermediate E_4 (2N2H) [9]. From this point forward, the continuous addition of e^- and H^+ irreversibly leads to ammonia (NH_3) production and regeneration of the E_0 state. Source: Frederik Schmidt (Co-author).

Upon formation of the nitrogenase complex, structural rearrangements within the catalytic component cause the reduction potential of the P-cluster to decrease, thereby allowing one electron to be transferred to the active site cofactor [21]. As a result, the P-cluster is oxidized to its P^{1+} state and subsequently reduced again by the associated reductase component. Hence, electrons are not flowing directly from the reductase component to the active site but are donated sequentially by the P-cluster. Thus, the P-cluster acts as an electron relay that the $[Fe_4-S_4]$ cluster refills in a process known as the "deficit spending model." Following each electron transfer, the reductase component hydrolyses both molecules of ATP to ADP, causing the nitrogenase complex to dissociate. Eventually, repeated association and dissociation of the two components results in a continuous electron flux toward the nitrogenase active site [22].

The constant delivery of electrons to the catalytic component causes the accumulation of reducing equivalents at the active site cofactor. The Lowe–Thorneley scheme is a model that describes how this process leads to the formation of NH_3 (Figure 10.2b) [23]. The model classifies the active site cofactor in nine redox states (E_0–E_8), corresponding to the number of electrons and protons added to the active site cluster. Electrons reach the cluster through the "deficit spending model" described above; protons enter through specific proton channels. Starting from the most oxidized E_0 state, protons and electrons accumulate in a stepwise fashion to form the E_1–E_4 states, corresponding to one to four associated electrons. In this phase of catalysis, each electron is either stored in the reduced state of the cluster (E_1 and E_3) or pairwise in hydrides bound to the cluster (E_2 and E_4). Once the E_4 state is reached, the two bound hydrides are exchanged for one molecule of N_2, leading to the reductive elimination of two hydrides through H_2 release and oxidative addition of N_2 [29]. Thereby, the cluster transitions into a hydrazine-bound intermediate referred to as E_4 (2N2H) without changing its redox state [9]. From this point on, any further electrons and protons added to the cluster are used to produce two molecules of NH_3, one in the E_5 to E_6 transition and another in the regeneration of the E_0 state. Thus, the E_4 state is regarded as the critical step of nitrogen reduction and is called the Janus intermediate [24]. Furthermore, the Lowe–Thorneley scheme explains the role of H_2 formation in the nitrogen fixation process. First, H_2 is an obligate side product of the reaction because the oxidative addition of N_2 requires the reductive elimination of two hydrides, which causes the release of H_2. Second, the model can explain the varying H_2 to N_2 stoichiometry. Up to the E_4 state, spontaneous release of H_2 can occur, causing a transition from E_n to E_{n-2} ($2 \leq n \leq 4$). In contrast to the obligate H_2 release in the E_4 to E_4 (2N2H) transition, this spontaneous H_2 release is considered a wasteful side reaction because it increases the amount of ATP required to reach the Janus intermediate and thus achieve N_2 reduction.

Even though the Lowe–Thorneley scheme was initially designed exclusively for the Mo nitrogenase, it may be applied to explain the elevated H_2 formation rates and, thus, ATP requirements of alternative nitrogenases [7, 30]. According to this theory, the differences in cofactor composition between the three isoforms contribute to the kinetics at which the FeMoco, FeVco, and FeFeco pass through the individual redox states of the Lowe–Thorneley scheme. For example, the Mo, V, or Fe atom in the characteristic heterometal position has been shown to contribute to spin state fine-tuning of the respective cluster, thereby affecting its overall reactivity [31–33]. Thus, the FeMoco performs best at running continuously through the individual redox states without wasteful H_2 release, followed by the FeVco and the FeFeco, which appears most prone to losing bound hydrides.

10.1.3 Alternative Substrates of Nitrogenase

Nitrogenases possess a very low reduction potential that allows them to break the stable triple bond of the N_2 molecule under physiological conditions. The energy for this reaction is stored in hydrides, which are generated at the active site cofactor [34, 35]. As demonstrated by the spontaneous release of H_2 in the

Lowe–Thorneley kinetic scheme, such hydrides are very reactive and difficult to control. Consequently, these hydrides can reduce any molecule that reaches the active site of nitrogenase. To prevent wasteful side reactivities, the enzyme uses substrate channels to direct the physiological substrates H^+ and N_2 to the active site. On the one hand, protons reach the FeMco through H^+ channels, which are typically highly selective for their substrate because they can utilize proton hopping between specific amino acid residues for transport [36]. On the other hand, other substrates, such as N_2, are typically discriminated based on size and charge [37]. Therefore, molecules of similar size and charge as the N_2 molecules have good chances of reaching the cluster-bound hydrides. For this reason, nitrogenases are very promiscuous enzymes. They can reduce a wide range of alternative substrates, including nonphysiological nitrogen compounds (e.g. azide [38] and nitrite [39]), carbon–nitrogen compounds (e.g. cyanide [40] and nitriles [41, 42]), unsaturated hydrocarbons (e.g. acetylene [43–46] and cyclopropene [47]), and carbon–chalcogen compounds (e.g. carbon monoxide [CO] [48, 49] and carbon dioxide [50, 51]).

While it might seem like these reactivities are solely undesired side reactions, an increasing amount of evidence suggests that these reactions might also be of physiological relevance. For instance, CO is a potent inhibitor of the Mo nitrogenase. In response to this problem, some diazotrophs have developed a protection mechanism and produced CowN, a protein that binds the Mo nitrogenase, thereby shielding it from CO inhibition [52, 53]. Intriguingly, the deletion of CowN did not impair diazotrophic growth of *R. capsulatus* strains that relied on the Fe nitrogenase. In fact, AnfA, a transcriptional regulator of the Fe nitrogenase genes, represses the expression of *cowN*, indicating that CO does not adversely affect Fe nitrogenase activity. Indeed, the Fe nitrogenase and the V isoform were shown to readily reduce CO, converting it to short-chain hydrocarbons such as methane, ethane, and ethylene. [48, 54–56]. Hence, alternative nitrogenases might be involved in detoxifying CO inside cells.

Aside from actively removing toxic substances, these reactions might also be essential to maintain a viable redox balance within the cell. As nitrogenase is an oxygen-sensitive enzyme, diazotrophic organisms are typically found in anaerobic environments [57]. Under these conditions, accumulating excessive amounts of reducing equivalents is a common problem, as oxygen is not available as the terminal electron acceptor of the canonical electron transport chain. Thus, anaerobic bacteria must remove excessive electrons through alternative pathways to maintain a pool of oxidized electron acceptors. Depending on the species, this can typically happen through the reduction of carbon dioxide (CO_2) via the Calvin–Benson–Bassham (CBB) cycle, the reduction of dimethyl sulfoxide (DMSO) via DMSO reductases and nitrogenase-mediated N_2-reduction [58, 59]. In accordance with nitrogenases acting as redox mediators, H_2 formation rates of the Mo nitrogenase were found to increase *in vivo* upon interruption of the CBB cycle through deletion of the CO_2-reducing enzyme RubisCo (ribulose-1,5-bisphosphate carboxylase/oxygenase) [60]. Analogously, reducing alternative nitrogenase substrates, such as the common bacterial metabolite CO_2 [59], could add to this mechanism of maintaining a viable redox homeostasis.

The reduction of CO and CO_2 releases short-chain hydrocarbons and formate into the cellular environment, thereby providing carbon sources for other organisms [61]. While symbiosis of diazotrophic bacteria with non-nitrogen-fixing organisms for NH_3 exchange is a well-known phenomenon [62], the influence of alternative nitrogenase products on local microbiota remains elusive. Recently, methane (CH_4) produced *in vivo* from CO_2 by the Fe nitrogenase of a *Rhodopseudomonas palustris* strain was shown to support the growth of a co-cultivated methanotroph *Methylomonas* sp. LW13 [63, 64]. Thus, the reduction of alternative substrates by nitrogenases can provide additional nutrients to symbiotically living microorganisms, suggesting that nitrogenase side reactivities are also involved in shaping the local microbiota.

In addition to the biological significance, the reduction of alternative substrates by nitrogenases is of great interest to biotechnology. Here, the unique reactivities of nitrogenases might offer novel solutions for the sustainable production of feedstock chemicals. For example, nitrogenases can convert the industrial waste gases CO and CO_2 into short-chain hydrocarbons and formate [61], products that could be used directly as biofuels or further processed into value-added compounds. However, to realize nitrogenase-based biotechnological processes, the enzymes need to be further optimized. Enzyme engineering approaches are a promising route to enhance specific side reactivities of nitrogenases and turn them into the main activities. For example, while wild-type Mo nitrogenase cannot produce CH_4 from CO_2, just two mutations in the *A. vinelandii* NifD subunit (V70A and H195Q) enabled CH_4 production from CO_2 while abolishing the mutant's ability to reduce N_2 [65]. Introducing mutations at homologous positions into the V nitrogenase of *R. palustris* could boost the low CH_4 formation rate from CO_2 by more than 20-fold [63]. These examples demonstrate that nitrogenase has the potential to act as a platform for reductive biocatalysis beyond N_2 reduction.

Below, we will discuss the scope of alternative nitrogenase substrates by elaborating on a selection of nitrogenase-catalyzed reactions. We will start with an in-depth review of the reduction of CO and CO_2 (Section 10.2), reactions that are of great interest for biotechnological applications and thus constitute the focus of this chapter. Next, we will analyze unsaturated hydrocarbon substrates (Section 10.3) that were essential for elucidating the nitrogenase mechanism, most importantly acetylene. Finally, we finish this chapter by discussing a selection of interesting nitrogenase activities on alternative nitrogen-containing substrates (Section 10.4), which consolidate the idea of nitrogenases as reductive biocatalytic platforms.

10.2 Carbon Monoxide and Carbon Dioxide

CO and CO_2 are typical waste products of industrial processes that arise mainly from burning fossil fuels. Most prominently, the ever-rising levels of CO_2 in the Earth's atmosphere are causing climate change. Hence, novel carbon-capture techniques are a promising approach to counteract global warming. Nitrogenases' ability to convert both CO and CO_2 into hydrocarbons and formate might offer new solutions for the sustainable synthesis of feedstock chemicals that could contribute to the transition into a circular economy.

10.2.1 Carbon Monoxide

CO is a colorless, odorless, tasteless, and poisonous gas. It consists of one carbon atom and one oxygen atom, connected by a triple bond that includes two covalent bonds and one dative covalent bond. Due to its low energy π^*-orbital, CO is a strong π-acid. Hence, CO can readily interact with transition metals via their d-orbitals, resulting in strong coordinative bonds. This characteristic causes the toxicity of CO, as it binds strongly to the Fe atom of the heme prosthetic group in hemoglobin, the oxygen carrier protein of red blood cells [66]. Analogously, CO also binds the metal cofactors of nitrogenase. Here, CO inhibits N_2 reduction but can be converted into short-chain hydrocarbons. From a historical perspective, CO played an essential role in deciphering the molecular mechanism of nitrogenase catalysis. Starting from the 1970s, researchers could exploit the strong affinity of CO to the FeMoco for comprehensive spectroscopic and structural studies of substrate binding to the Mo nitrogenase active site cofactor.

10.2.1.1 CO Reduction by the Mo Nitrogenase

CO is a noncompetitive inhibitor for the reduction of N_2 and several other alternative substrates such as N_3^-, C_2H_2, and CN^- by the Mo nitrogenase [67, 68]. The binding of CO to the FeMoco can be monitored under turnover conditions by electron paramagnetic resonance (EPR) spectroscopy. Two signals are characteristic of this interaction: One forms already at low CO partial pressure and is known as the lo-CO signal. The second signal appears only at high CO partial pressure and is consequently denoted as the hi-CO signal [69]. A comprehensive study on the two CO signals through ^{13}C electron nuclear double resonance (ENDOR) spectroscopy could link the lo-CO and hi-CO signals to one and two CO molecules bound to a cluster in the MoFe protein, respectively [70]. Shortly after, ^{57}Fe labeling of the FeMoco identified the active site cofactor as the binding site of CO and, thus, the origin of the two EPR signals [71]. The two states were interchangeable simply by adjusting the CO partial pressure in the experiment without any changes to the FeMoco redox state, suggesting that they originate from the same E_n state within the Lowe–Thorneley scheme. An orientation-selective ^{13}C ENDOR spectroscopy study could finally locate the binding site of CO for both states to the belt region of the FeMoco [72]. The position of CO binding was further refined via nuclear resonance vibrational spectroscopy (NRVS) and extended X-ray absorption fine structure (EXAFS) spectroscopy [73]. Both methods revealed a reduction in symmetry around the Fe_6S_9C core of the FeMoco upon CO binding. These results, together with density functional theory (DFT) calculations, indicated that CO binds to the FeMoco interface composed of Fe(2), Fe(3), Fe(6), and Fe(7), leading the researchers to speculate that CO substitutes one bridging sulfur atom between Fe(2) and Fe(6). An electrochemical study combined with infrared (IR) spectroscopy could further pinpoint the redox state of the FeMoco at which CO binding occurs [74]. Using europium(III/II) polyaminocarboxylate complexes instead of the natural reductase component as an electron mediator, the researchers compared the potentials required for H_2 formation and CO binding to the MoFe protein. In conclusion, the results suggest CO binding during the early H^+ reduction steps of the catalytic cycle, which was hypothesized to correlate to the E_2 state of the Lowe–Thorneley scheme. Direct evidence for the binding of CO to the FeMoco was finally provided by a crystal structure of the CO-inhibited MoFe protein [75]. In the structure, a CO ligand replaces the S2B belt sulfur between Fe(2) and Fe(6) (Figure 10.3a). The inhibition and replacement of the belt sulfur atom was shown to be reversible. Located around 22 Å from the resting state position of sulfur S2B, a potential sulfur-binding site was identified that might act as an intermediate relay of the sulfur atom during turnover. Interestingly, CO could only bind to the FeMoco under catalytic conditions but not if the active site cofactor was in the resting state E_0. Thus, an initial proton and electron transfer must precede the replacement of the S2B sulfur by CO and substantiates the theory that CO binding occurs in the E_2 state. Later, another crystal structure with two CO ligands bound to the

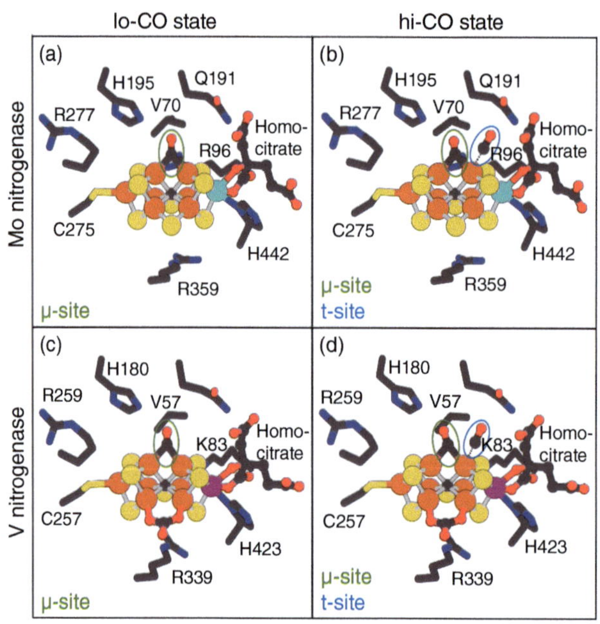

Figure 10.3 Structures of lo-CO and hi-CO states of the FeMoco and FeVco. CO-bound structures of the FeMoco (top) and FeVco (bottom) in the lo-CO state (left) and hi-CO state (right) from PDB 4TKV (a), PDB 7JRF (b), PDB 7ADR (c), and PDB 7AIZ (d). Fe(2) and Fe(6) of each cluster are labeled. CO bound in μ- and t-sites are circled in green and blue, respectively. Sulfur atoms of the clusters are represented as yellow spheres, iron atoms as orange spheres, molybdenum atoms as cyan spheres, vanadium atoms as purple spheres, and the central carbide ion as a black sphere. The surrounding amino acids of the FeMcos are depicted as stick models with carbon in black, nitrogen in blue, oxygen in red, and sulfur in yellow. The homocitrate coordinated to the FeMcos is depicted as a ball-and-stick model with the same color coding as the amino acid residues. Source: Frederik Schmidt (Coauthor).

FeMoco was published [76]. Analogously to the previous study, the new structure contained one CO bound by Fe(2) and Fe(6) on the so-called μ-site of the cluster and an additional CO bound to Fe(6) on the t-site (Figure 10.3b). Structures with one or two CO bound were obtained by applying low and high CO partial pressures to MoFe protein crystals, as described for the EPR studies. Thus, the two different CO-bound structures were assigned to the lo-CO and hi-CO states, which could be verified by EPR spectroscopy.

Interestingly, while CO inhibits N_2 reduction by the Mo nitrogenase, it does not abolish H_2 formation and ATP consumption [67]. This paradox remained a mystery until Ribbe and co-workers could track the reduction of traces of CO to short-chain hydrocarbons, including C_2H_4, C_2H_6, C_3H_6, and C_3H_8 [56]. Conducting the activity assays in deuterated water (D_2O) could stimulate the formation of longer hydrocarbon chains such as α-C_4D_8 and n-C_4D_{10}. Another study could demonstrate that replacing valine 70, located above the CO-binding site in the *A. vinelandii* Mo nitrogenase, with a sterically less demanding alanine or glycine residue improves hydrocarbon formation rates from CO [49]. Importantly, these coupling reactions do not directly involve the interstitial carbide ion of the active site cofactor, as ^{13}C labeling of the FeMoco did not result in the release of ^{13}C-labeled hydrocarbon products [77]. However, the fact that the Mo nitrogenase does reduce CO gives good reason to assume that the structurally and spectroscopically described binding sites of CO are of general relevance for reducing other substrates at the FeMoco. Especially, the μ-site is regarded as a critical position for substrate reduction, which was substantiated for N_2 reduction by a crystal structure containing N_2 molecules at several belt sulfur positions [16].

10.2.1.2 CO Reduction by the V Nitrogenase

Compared to the Mo Nitrogenase, the V nitrogenase is 800 times more active in CO reduction compared to its Mo counterpart and releases CH_4, C_2H_4, C_2H_6, C_3H_6, C_3H_8, α-C_4H_8, and n-C_4H_{10} [48]. The main product of the reaction is C_2H_4, which is formed at a rate of 7.5 nmol (C_2H_4) × nmol (VFe protein)$^{-1}$ × min^{-1} and accounts for 94% of the reaction's electron flux [56]. In contrast to the Mo isoform, the use of D_2O did not significantly increase hydrocarbon formation by the V nitrogenase. However, using a deuterated buffer, the source of hydrogen for the released hydrocarbons could be tracked to protons originating from the buffer [78]. Intriguingly, the ability of V nitrogenase to catalyze C–C coupling reactions from CO also translates *in vivo*. Cultivating V nitrogenase expressing *A. vinelandii* under 15% CO resulted in the release of C_2H_4, C_2H_6, and C_3H_8 to the culture headspace. Again, C_2H_4 was the main product of the reaction formed at a maximum rate of 1 mmol C_2H_4 × g (dry cell)$^{-1}$ × h^{-1} [55]. Furthermore, a recent study successfully tested the production of C_2H_4 from CO via the V nitrogenase of *A. vinelandii* in small bioreactors [79]. Overall, these results demonstrate that biotechnological bio-ethylene production is feasible, but further optimization is needed to increase product yields.

Due to its higher CO reduction rates, the analysis of CO-bound intermediates of the VFe protein has proven to be more difficult than for the Mo isoform. Replacing the V nitrogenase reductase component by europium(II)-diethylenetriaminepenta-acetic acid (Eu(II)-DTPA) as an electron source allowed substrate binding to the resting state VFe protein and thus uncoupling of CO binding from its turnover [80]. This way, an intermediate of the VFe protein was obtained that exhibited a similar EPR spectrum to the lo-CO state of the Mo nitrogenase. Constitution of the intermediate with ^{13}C-labeled CO and subsequent exposure to ^{12}CO under turnover conditions led to the release of partially labeled ethylene ($^{13}CH_2{}^{12}CH_2$), which proved the catalytic relevance of the lo-CO VFe intermediate. A hi-CO VFe species can be obtained by decreasing the electron flux under CO turnover or increasing the CO partial pressure when using Eu(II)-DTPA as a reductant [81, 82]. Notably, the second CO bound to the FeVco does not engage in the C–C coupling reactions, raising doubts about the catalytic relevance of the hi-CO state. Eventually, X-ray crystallography structures reported by Einsle and co-workers revealed the molecular details of the lo- and hi-CO states of the V nitrogenase from *A. vinelandii* [83, 84]. The structures confirm a high similarity of CO binding between FeMoco and FeVco. As for the Mo isoform, the CO ligand corresponding to the lo-CO EPR signal binds to Fe(2) and Fe(6) to replace belt sulfur S2B in the V nitrogenase (Figure 10.3c). In the hi-CO state, an additional CO ligand coordinates the FeVco on the t-site (Figure 10.3d), which is fully reversible, as demonstrated through *in situ* attenuated total reflectance IR difference spectroscopy. Inversely, crystals in the lo-CO state could partially be transformed to the hi-CO state by applying CO overpressure, which agrees with the behavior of the Mo nitrogenase.

Apart from the apparent differences in the FeMoco and FeVco composition, the lo- and hi-CO states of the two isozymes appear highly similar. In the direct cofactor environment, most amino acid residues are conserved between the V and Mo nitrogenases of *A. vinelandii*, except for NifD R96, which corresponds to K83 in the V isoform. The striking differences in the Mo and V nitrogenase CO reduction activity raise the question: Does their cofactor composition or the protein environment determine their reactivity? Intriguingly, both isolated FeMoco and FeVco can reduce CO to hydrocarbons at similar rates in the presence of strong reductants [85–87]. The fact that the CO reduction rate of isolated FeVco corresponds to only 0.05% of the V nitrogenase holoenzyme activity suggests an important role of the nitrogenase protein in the reaction. Testing the CO reduction activity of nitrogenase hybrid enzymes further substantiated this hypothesis. A Mo nitrogenase hybrid containing the FeVco was 640 times less active in hydrocarbon formation from CO relative to the V nitrogenase [88]. Correspondingly, a V nitrogenase hybrid containing the FeMoco showed 100-fold increased CO reduction activity compared to the Mo nitrogenase [89]. However, the protein environment does not solely determine the CO reduction activity, as wild-type V nitrogenase is six times more active in hydrocarbon formation than the FeMoco substituted V nitrogenase hybrid. Overall, this suggests that while the cofactor is undoubtedly the location of the reaction, the molecular mechanism of CO reduction requires a finely tuned interplay of cofactor and protein. Since nitrogenase is a symmetrical enzyme, this could involve both catalytic halves, as an increasing amount of evidence suggests a ping-pong mechanism for substrate reduction [90].

10.2.1.3 CO Reduction by the Fe Nitrogenase

Finally, the Fe nitrogenase also accepts CO as a substrate but surprisingly cannot perform C–C coupling reactions. Hence, the Fe nitrogenase converts CO exclusively to CH_4 at a rate of 4.5 nmol (CH_4) × nmol (FeFe protein)$^{-1}$ × min^{-1} [54]. In contrast to the Mo and V nitrogenases, no spectroscopic or structural elucidations on the CO reduction mechanism of the Fe nitrogenase are available. Thus, the underlying mechanisms for the observed selectivity remain speculative and are currently attributed to the unique composition of the FeFeco. This hypothesis is substantiated by the observation that the isolated L-cluster, the precursor of the FeMoco, also exclusively reduces CO to CH_4 with Eu(II)-DTPA as an electron source [87]. The L-cluster is a [Fe_9S_9C] cluster and thus structurally identical to the FeFeco, except for the missing (*R*)-homocitrate ligand [13, 28]. Therefore, the inability of the Fe nitrogenase to produce hydrocarbon chains from CO might be a direct cause of the missing heterometal in FeFeco. However, further studies are needed to elucidate the exact underlying mechanisms.

10.2.1.4 Conclusions on CO Reduction by Nitrogenases

Despite their differences in reactivity, all three nitrogenase isoforms can reduce CO, indicating that CO binds to the FeMoco, FeVco, and FeFeco. For the Mo and V nitrogenase, this was shown to happen in two interchangeable modes, the lo- and hi-CO states. The overall high similarity between the three isozymes gives good reason to assume that the molecular mechanism of CO binding to the FeFeco happens in a similar fashion. Thus, based on the structures of the lo- and hi-CO states of the Mo and V nitrogenase, a mechanism for CO reduction at the FeMco was proposed (Figure 10.4) [83]. CO binding requires activation of the cluster by adding one hydride to the μ-site, which corresponds to the E_2 state of the Lowe–Thorneley scheme (**2**). At this point, CO can enter and bind to the t-site (**3**), allowing CO to exchange positions with the μ-site bound hydride (**4**). From this equilibrium state, two directions are possible. Protonation of the t-site bound hydride releases H_2 and leaves the cluster in a formal E_0 state with a bridging carbonyl ligand (**5**). This stage corresponds to the lo-CO state, which can be converted to the hi-CO state through further CO exposure (**6**). Alternatively, productive CO reduction initiates through an insertion reaction, yielding a formyl adduct (**7**) that is progressively reduced to hydrocarbons by adding hydrides and further CO molecules (**8–11**). Both hydrides and CO enter the reaction at the t-site and subsequently insert into the μ-site bound carbon moiety. Eventually, the E_2 state is recovered through a β-H elimination reaction, which releases the hydrocarbon product. In contrast, methane formation occurs through protonation of the methyl-bound intermediate (**8**), which appears to be the preferred pathway for the Fe nitrogenase. Notably, this scheme explains why the hi-CO state cannot couple both CO molecules to a hydrocarbon product, as this requires a t-site-bound hydride to form the formyl adduct.

In summary, all nitrogenases reduce CO to hydrocarbons, but the turnover rates and product profiles differ significantly between the three isoforms. The V and Mo nitrogenases catalyze C–C coupling reactions from CO in a Fischer–Tropsch-like reaction. The Fischer–Tropsch reaction is an industrial process in which syngas,

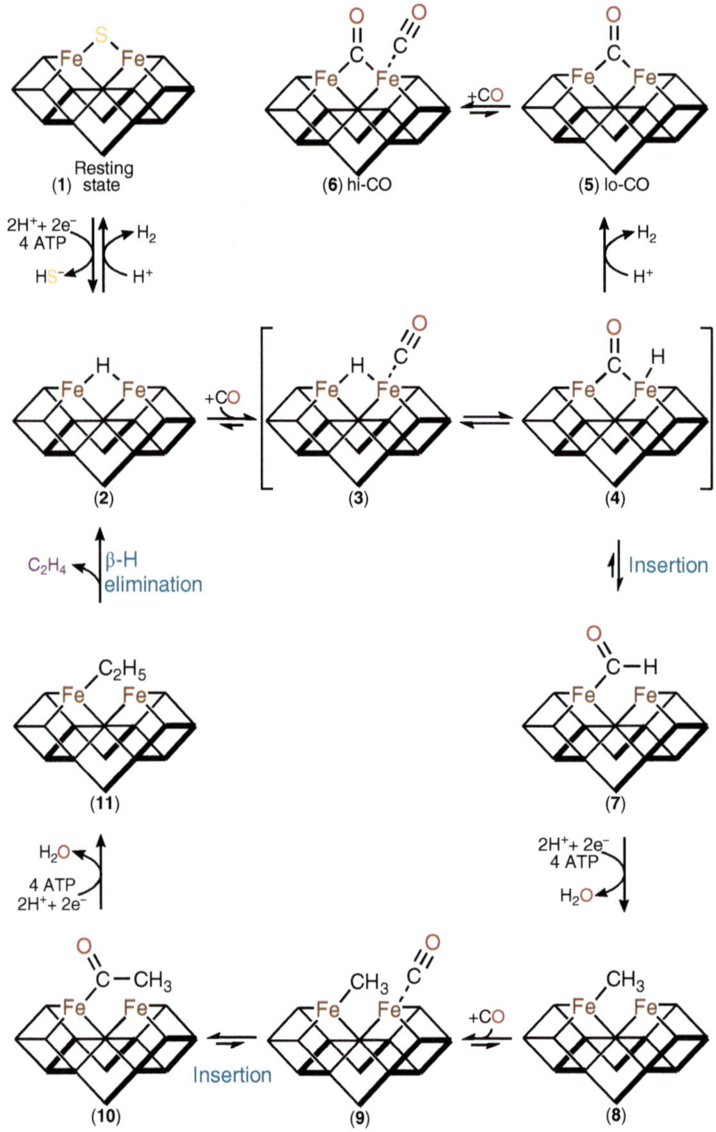

Figure 10.4 Mechanistic model of CO reduction at the nitrogenase active site cofactor. The FeMco acquires two electrons (e^-) and protons (H^+), thereby transitioning into the hydride-activated E_2 state (**2**). At this point, CO binds initially at the t-site (**3**) but can rearrange to the μ-site (**4**). Protonation of the t-site bound hydride releases, leaving CO at the μ-site (**5**). This is the lo-CO state, which can be converted to the hi-CO state through further CO exposure (**6**). Productive CO reduction can be initiated from (**4**) by insertion of hydride into the μ-site bound CO ligand, yielding a formyl adduct (**7**). Reduction by another hydride, entering through the t-site, yields a methyl intermediate (**8**). While protonation of (**8**) would release CH_4, the addition of another CO (**9**) enables the C−C coupling reaction and forms an acetyl intermediate (**10**). Eventually, further reduction to (**11**) allows the release of ethene through a β-H elimination reaction and restores the E_2 state. Source: [91]/John Wiley & Sons/CC BY 4.0.

a mixture of CO and H_2, is reduced to hydrocarbons via heterogeneous metal catalysis at high temperatures (150–300 °C) and pressures (10–20 atm) [92]. In contrast, nitrogenases reduce CO and protons to hydrocarbons at physiological conditions. Thus, nitrogenases might offer green alternatives not only to industrial fertilizer production (Haber–Bosch process) but also to the production of fuels (Fischer–Tropsch process). Interestingly, the Fe nitrogenase cannot catalyze C–C coupling reactions but selectively produces significant amounts of CH_4. The ability of the alternative nitrogenases to produce hydrocarbons from CO is of great interest for biotechnology to develop sustainable solutions for green hydrocarbon formation pathways.

10.2.2 Carbon Dioxide

CO_2 is a colorless, odorless gas composed of one carbon atom covalently double-bonded to two oxygen atoms. CO_2 is a minor component of Earth's atmosphere but plays a significant role in the greenhouse effect, a process essential for maintaining the Earth's temperature in a habitable range. However, human activities, such as burning fossil fuels and deforestation, have significantly increased the concentration of CO_2 in the atmosphere. This increase enhances the greenhouse effect, leading to global warming and climate change, resulting in higher global temperatures, melting ice caps, rising sea levels, and extreme weather patterns. Thus, reducing atmospheric CO_2 levels is crucial for mitigating the impacts of global warming and ensuring a sustainable future of the planet. The recent discoveries on nitrogenases' ability to reduce CO_2 to formate and hydrocarbons have raised interest in the biotechnological potential of this reaction. Analogous to the reduction of CO, nitrogenase isoforms differ significantly in their ability to reduce CO_2, as discussed below.

10.2.2.1 CO_2 Reduction by the Mo Nitrogenase

The reduction of CO_2 by the Mo nitrogenase was first reported by Seefeldt and co-workers, who detected the formation of CO *in vitro* [50]. In this study, hemoglobin was used as a CO probe, which prevented Mo nitrogenase inhibition from CO but failed to detect other reaction products. Later, a ^1H and ^{13}C NMR spectroscopy study confirmed the Mo nitrogenase-dependent formation of CO from CO_2 but identified formate ($HCOO^-$) as the main product of the reaction [93]. Under a pure CO_2 atmosphere, the Mo nitrogenase reduces CO_2 to CO and formate ($HCOO^-$) at maximal rates of 0.4 nmol (CO) × nmol (MoFe protein)$^{-1}$ × min^{-1} and 53 nmol ($HCOO^-$) × nmol (MoFe protein)$^{-1}$ × min^{-1}, respectively. However, under these conditions, most electrons do not end up in the reduction of CO_2 but are released as H_2. This side reactivity accounts for 76% of the enzyme's total electron flux, which renders the Mo nitrogenase a very inefficient CO_2 reductase [61]. The product spectrum of the Mo nitrogenase-catalyzed CO_2 reduction was extended to CH_4 by introducing two mutations in the *A. vinelandii* NifD subunit (V70A and H195Q). The reaction likely proceeds via an intermediate CO, as the addition of deoxyhemoglobin, a CO-binding molecule, led to a 25% decrease in the CH_4 formation rate [65]. An analogous *R. palustris* Mo nitrogenase mutant produced CH_4 from CO_2 *in vivo* [94]. Notably, the phototroph *R. palustris* generates ATP for the nitrogenase reaction photosynthetically, making this a sustainable CH_4 formation pathway. The yield of CH_4 could be increased by cutting off other electron-consuming pathways, such as the CBB cycle, to increase electron flux toward the nitrogenase system.

10.2.2.2 CO_2 Reduction by the V Nitrogenase

The V nitrogenase reduces CO_2 to CO at similar rates as its Mo counterpart [51]. When placed in D_2O, the V nitrogenase converts small amounts of CO_2 to CD_4 and performs C–C coupling reactions to release C_2D_4 and C_2D_6. Notably, while CD_4 partially originates from a CO intermediate, the mechanism for C–C coupling does not involve CO. Replacing the V nitrogenase reductase component by Eu(II)-DTPA enables the formation of hydrocarbons in H_2O, including CH_4, C_2H_4, C_2H_6, C_3H_6, C_3H_8, C_4H_8, and C_4H_{10} [95]. Intriguingly, placing the same reaction in a deuterated buffer boosts the formation of C_2–C_4 products while abolishing CD_4 formation. Overall, this indicates that D_2O positively affects the C–C coupling efficiency of nitrogenase, probably by lowering the number of electrons ending up in H_2 and redirecting them toward CO_2 reduction. The formation of C_2H_4 and C_3H_6 from CO_2 was also achieved electrochemically, with 1,1′-dicarboxycobaltocenium as an electron carrier to drive the V nitrogenase reaction in an ATP- and reductase component-independent manner [96]. The reduction of CO_2 to CH_4 was also described *in vivo*, where V nitrogenase expressing *R. palustris* strains accumulated up to 25 nmol $(CH_4) \times$ mg (total protein)$^{-1}$. This amount increased more than 16-fold in a V nitrogenase VnfD V57A and H180Q double mutant, corresponding to the CH_4-producing Mo nitrogenase mutant of *A. vinelandii* [63].

10.2.2.3 CO_2 Reduction by the Fe Nitrogenase

Finally, the Fe nitrogenase reduces CO_2 to $HCOO^-$, CO, and CH_4 but does not perform C–C coupling reactions. CO and CH_4 are trace products of the reaction that are formed at maximum rates of 0.19 nmol (CO) $\times$ nmol (FeFe protein)$^{-1}$ $\times$ min^{-1} and 0.03 nmol $(CH_4) \times$ nmol (FeFe protein)$^{-1}$ $\times$ min^{-1}, respectively. In contrast, the Fe nitrogenase produces substantial amounts of $HCOO^-$ from CO_2, which is formed at 60 nmol $(HCOO^-) \times$ nmol (FeFe protein)$^{-1}$ $\times$ min^{-1}. Thus, the Fe nitrogenase reaches the same rates for $HCOO^-$ formation as the Mo nitrogenase, an astonishing result given the overall twofold higher electron flux of the Mo isoform. Moreover, the Fe nitrogenase appears to be more efficient in channeling its available electrons into the reduction of CO_2. While the Mo nitrogenase can only direct 24% of its total electron flux into $HCOO^-$ and CO formation, the Fe nitrogenase directs two-thirds of its electrons into CO_2 reduction products. Notably, this electron flux closely matches the amount of electrons flowing into the formation of NH_3 under a N_2 atmosphere, emphasizing the ability of the Fe nitrogenase to function as a N_2 and CO_2 reductase [61]. The CO_2-reducing ability of the Fe nitrogenase was also confirmed electrochemically by immobilizing FeFe protein on an electrode, which achieves substrate reduction without requiring a reductase component or

ATP [97]. Furthermore, a study in a Fe nitrogenase expressing *R. palustris* strain could demonstrate the formation of up to 150 nmol (CH_4) × mg (total protein)$^{-1}$ *in vivo*, which was sufficient to support the growth of a co-cultivated methanotroph *Methylomonas* sp. LW13. Surprisingly, introducing the mutations found to increase CH_4 formation from CO_2 for the Mo and V nitrogenase abolished CH_4 formation by the Fe nitrogenase [63]. Overall, this suggests that CO_2 competes with N_2 reduction by the Fe nitrogenase. When placed under mixed atmospheres of N_2 and CO_2, the Fe nitrogenase starts reducing CO_2 already in a 50 : 50 mixture of the two substrates. At 75% CO_2 mixed with 25% N_2, $HCOO^-$ becomes the main product of the reaction.

In contrast, the Mo nitrogenase has a much higher selectivity for its natural substrate and reduces significant amounts of CO_2 only in the absence of N_2. The lower N_2 selectivity of the Fe nitrogenase observed *in vitro* also translates *in vivo*, where the addition of 20% CO_2 to the headspace of diazotrophically growing *R. capsulatus* cells drastically increases their doubling times if they rely on the Fe nitrogenase. Conversely, the same treatment has no effect if the cells depend on the Mo nitrogenase for N_2 fixation. The CO_2-dependent growth inhibition of Fe nitrogenase-dependent *R. capsulatus* strains is accompanied by the accumulation of CO_2 reduction products in the culture supernatant. For example, Fe nitrogenase expressing *R. capsulatus* cells cultivated under 8% CO_2 accumulated up to 13 mM $HCOO^-$ in the culture medium while releasing significant amounts of CH_4. In line with previous results, CH_4 formation was also detected without adding CO_2 to the *R. capsulatus* culture headspace, indicating that metabolically derived CO_2, thus physiological conditions, are sufficient to drive the CO_2 reduction by the Fe nitrogenase [61].

10.2.2.4 Conclusions on CO_2 Reduction by Nitrogenases

In summary, out of the three nitrogenase enzymes, the Fe isoform is most prone and efficient for CO_2 reduction, as it directs the largest proportion of electrons into the reduction of CO_2. The CO_2 reduction activities of the Mo and V nitrogenase could be increased by introducing two mutations into their respective D-subunits. Overall, nitrogenases can convert CO_2 into $HCOO^-$, CO, and CH_4. The V isoform can even perform C–C coupling reactions to short-chain hydrocarbons. These reactivities, summarized in Table 10.1, provide new pathways for CO_2 capture and highlight the potential of nitrogenases for the biotechnological conversion of CO_2 into value-added compounds.

10.3 Hydrocarbon Substrates

Nitrogenases are best known for their ability to break the stable triple bond of the N_2 molecule. Similarly, nitrogenases can reduce other small molecules that contain triple or double bonds. Among the earliest studies on nitrogenases was the observation that nitrogenase can reduce acetylene. Later, many further unsaturated carbon compounds were discovered as alternative nitrogenase substrates.

Table 10.1 Selected reports on nitrogenase CO_2 reduction *in vivo* and *in vitro*.

Isoform	Organism	Products	Comments	References
Mo	*Azotobacter vinelandii*	CO, HCOO–	• *In vitro* formation of CO [0.03 nmol × nmol (MoFe protein)$^{-1}$ × min^{-1}] and $HCOO^-$ [9.80 nmol × nmol (MoFe protein)$^{-1}$ × min^{-1}] • Electron mediator = $NifH_2$ • Low electron flux conditions	[93]
Mo	*Rhodobacter capsulatus*	CO, HCOO–	• *In vitro* formation of CO [0.40 nmol × nmol (MoFe protein)$^{-1}$ × min^{-1}] and $HCOO^-$ [53.0 nmol × nmol (MoFe protein)$^{-1}$ × min^{-1}] • Electron mediator = $NifH_2$ • High electron flux conditions	[61]
Mo (NifD V70A, H195Q)	*Azotobacter vinelandii*	CH_4	• *In vitro* formation rate = 0.95 nmol × nmol (MoFe protein)$^{-1}$ × min^{-1} • Electron mediator = $NifH_2$ • High electron flux conditions	[65]
Mo (NifD V75A, H201Q)	*Rhodopseudomonas palustris*	CH_4	• *In vivo* formation of max. 30.1 nmol × mg (total protein)$^{-1}$ • Electron mediator = $AnfH_2$	[94]
V	*Azotobacter vinelandii*	CO, CD_4, C_2D_4, C_2D_6	• *In vitro* formation of max. 100 nmol × μmol (protein)$^{-1}$ in H_2O and max 250 nmol × μmol (protein)$^{-1}$ in D_2O • *In vitro* formation of slight amounts of CD_4, C_2D_4, and C_2D_6 in D_2O • Electron mediator = $VnfH_2$ • High electron flux conditions	[51]
V	*Azotobacter vinelandii*	CO, CH_4, C_2H_4, C_2H_6, C_3H_6, C_3H_8, C_4H_8, C_4H_{10}	• *In vitro* formation of CO and CH_4 as well as slight amounts of short-chain hydrocarbons • *In vitro* CO_2 reduction in D_2O yields deuterated hydrocarbons but no CD_4 • Electron mediator = Eu(II)-DTPA	[95]
V	*Rhodopseudomonas palustris*	CH_4	• *In vivo* formation of max. 25 nmol (CH_4) × mg (total protein)$^{-1}$ • VnfD V57A and H180Q mutations increase *in vivo* CH_4 production up to 420 nmol (CH_4) × mg (total protein)$^{-1}$ • Electron mediator = $AnfH_2$	[63]
Fe	*Azotobacter vinelandii*	$HCOO^-$	• *In vitro* formation of $HCOO^-$ both in solution with $AnfH_2$ as the electron mediator and electrochemically with the FeFe protein immobilized on an electrode	[97]

Table 10.1 (Continued)

Isoform	Organism	Products	Comments	References
Fe	*Rhodobacter capsulatus*	$HCOO^-$, CO, CH_4	• *In vitro* formation of $HCOO^-$ [60 nmol × nmol (FeFe protein)$^{-1}$ × min^{-1}], CO [0.19 nmol × nmol (FeFe protein)$^{-1}$ × min^{-1}] and CH4 [0.03 nmol × nmol (FeFe protein)$^{-1}$ × min^{-1}] • Electron mediator = $AnfH_2$ • High electron flux conditions • Accumulation of the same products was also observed *in vivo*	[61]
Fe	*Rhodopseudomonas palustris*	CH_4	• *In vivo* formation of max. 150 nmol (CH_4) × mg (total protein)$^{-1}$ • *In vitro* formation of 1 nmol (CH_4) × mg (FeFe protein)$^{-1}$ using *A. vinelandii* • Electron mediator = $AnfH_2$	[63]

10.3.1 Acetylene

C_2H_2 is the simplest alkyne, a class of hydrocarbons classified by at least one carbon–carbon triple bond. Known for its high reactivity, it is also a substrate for all three nitrogenases. However, turnover rate and product profiles differ between each isoform. The Mo nitrogenase readily accepts C_2H_2 as a substrate, with K_m values reported between 0.3 and 2 kPa. The reaction exclusively yields ethylene (C_2H_4), with specific activities of up to 2000 nmol (C_2H_4) × mg (MoFe protein)$^{-1}$ × min^{-1} [98–101]. The V and Fe nitrogenases also convert C_2H_2 but show lower affinities toward the substrate with apparent K_m values of around 6 kPa for the VFe protein [102, 103] and 12.5–14 kPa for the FeFe protein [30, 104]. Correspondingly, C_2H_2 reduction activities of the V and Fe nitrogenase are generally lower, reaching only ~30% and ~20% of the Mo nitrogenase C_2H_4 formation rate, respectively [105]. Intriguingly, the alternative nitrogenases convert C_2H_2 also to low amounts of ethane (C_2H_6). The reaction accounts for less than 5% of the electron flux observed for C_2H_4 formation [103, 106] but can be exploited to fingerprint V or Fe nitrogenase activity from whole cell assays [107, 108]. Notably, the formation of C_2H_6 does not involve a C_2H_4 intermediate, which suggests distinct reaction mechanisms for the formation of C_2H_4 and C_2H_6 [102]. While C_2H_6 formation is not observed for the wild-type Mo nitrogenase, several mutants with amino acid substitutions around the FeMoco have been described to form low amounts of C_2H_6 from C_2H_2 [109].

Early investigations on the C_2H_2 reduction mechanism have provided valuable insights into the substrate binding and turnover at the nitrogenase active site. Analysis of the stereochemistry with deuterated acetylene (C_2D_2) in H_2O or C_2H_2 in the deuterated buffer revealed *cis*-dideuteroethylene as the primary product of the reaction [44, 46, 102]. Based on this result, a mechanism via a η^2-vinyl intermediate was proposed to add two protons to the cofactor-facing side of the C_2H_2 molecule [110]. In contrast to N_2, C_2H_2 strongly inhibits the formation of H_2 during turnover, so more than 90% of electrons flow into C_2H_4 formation [23, 111]. Further increase in the C_2H_2 partial pressure can completely abolish H_2 formation, whereas H_2 is an obligate side product of N_2 reduction. Hence, these results imply distinct mechanisms for C_2H_2 and N_2 reduction by nitrogenases [68, 112].

10.3.2 Other Alkynes and Alkenes

In addition to C_2H_2, several other alkynes have been tested as nitrogenase substrates. For example, the Mo nitrogenase converts propyne, 1-butyne, and 2-butyne to their according alkenes, though turnover rates are meager [113–115]. Furthermore, the Mo nitrogenase reduces the triple bonds of propargyl alcohol and propargyl amine to double bonds, which was used to probe substrate binding to the FeMoco [116–118].

Nitrogenases can also reduce double bonds, i.e. alkenes. One interesting example is the reduction of cyclopropene. Here, the reduction of the double bond results in the formation of cyclopropane, but the reduction of a single bond of the triangular molecule can also yield propene [47]. Furthermore, the Mo nitrogenase was shown to catalyze the reduction of 3,3-difluorocyclopropene, the first reported halogen-containing substrate of nitrogenase. In contrast to cyclopropene, this molecule can be reduced by four electrons to generate 2-fluoropropene or by six electrons to yield propene [119]. Propene is also the product of propadiene reduction, catalyzed at low rates by the Mo nitrogenase [120].

10.4 Alternative Nitrogen-Containing Substrates

Nitrogenases can catalyze the formation of NH_3 not only from N_2 but also from alternative nitrogen-containing molecules. For example, the Mo nitrogenase-catalyzed reductions of nitrite (NO_2^-) and hydroxylamine (NH_2OH) have both been reported to release NH_3 [39, 121], which yielded valuable insights into the N_2 reduction mechanism as further reviewed in [18]. In the following section, we will briefly discuss two other important species of nitrogen-containing substrates, azide (N_3^-) and cyanide (CN^-), together with the closely related class of nitriles and isonitriles.

10.4.1 Azide

The N_3^- molecule is a linear, inorganic anion that consists of three nitrogen atoms, thereby constituting the logical extension of the N_2 molecule. N_3^- was first reported as a nitrogenase substrate in 1967 by Schöllhorn and co-workers, who observed equimolar production of N_2 and NH_3 when adding N_3^- to nitrogenase-containing cell extracts of *A. vinelandii* and *Clostridium pasteurianum* [122]. Later investigations on the reaction with purified Mo nitrogenase could demonstrate that NH_3 was not formed in equal amounts to N_2 but in excess. This result was accompanied by the observation that hydrazine (N_2H_4) is also a product of the reaction [123, 124]. N_2H_4 had previously been identified as another alternative substrate of Mo nitrogenase that can be reduced to NH_3, leading to the hypothesis that excessive NH_3 formed in the N_3^- assays could originate from N_2H_4 reduction [125]. However, this theory was disproven by adding isotopically labeled $^{15}N_2H_4$ to N_3^- reduction assays, as no ^{15}N labeled NH_3 was detected [124]. Instead, Burgess and co-workers could track the excessive amounts of NH_3 formed in the N_3^--reduction assays to the concurrent reduction of N_2. Conducting the assays in the presence of D_2 gas resulted in an increased production of N_2 while the corresponding amount of NH_3 decreased. This result indicated inhibition of N_2 reduction by D_2, which stands in accordance with previous observations and thus supports the idea of N_2 being the source of excessive NH_3 formation in N_3^- reduction assays [123]. The reaction was also investigated with V nitrogenase from *A. vinelandii*, which catalyzes the same reaction but at lower turnover rates [126]. In contrast, N_3^- reduction by the Fe nitrogenase has not been investigated to date, underscoring the general lack of data for this particular isoform of nitrogenase.

10.4.2 Cyanide, Nitriles, and Isonitriles

CN^- is a highly toxic chemical compound that consists of a carbon atom triple-bonded to a nitrogen atom, making it isoelectronic to the N_2 molecule. While early studies demonstrated that CN^- acts as a substrate and inhibitor of nitrogenase, it was later shown that CN^- is the inhibitor and hydrogen cyanide (HCN) the substrate of Mo nitrogenase [40]. As such, HCN is reduced by two, four, or six electrons to yield methylene imine, methylamine or methane, and ammonia, respectively. The same reactivity was observed for the V nitrogenase, though it was less active and experienced less inhibition from the CN^- anion [126, 127].

Nitriles are a class of organic compounds characterized by a cyano (CN) functional group. Studies on the Mo and V nitrogenase have demonstrated that nitrogenases generally reduce alkyl nitriles to their corresponding alkanes accompanied by NH_3 release [113, 114]. The reaction is most efficient when using acrylonitrile, which can be reduced by six or eight electrons, yielding propene or propane together with NH_3, respectively [41]. In general, nitriles composed of up to four nonbranched carbon atoms were reduced, demonstrating that the nitrogenase active site can fit substrates that are substantially larger than N_2. Similar to acrylonitrile reduction, the Mo and V nitrogenases also catalyze the six- or eight-electron reduction of cyanamide (CN_2H_2), thereby releasing NH_3 together with methylamine or methane, respectively [128].

Lastly, isonitriles (also referred to as isocyanides) are constitution isomers of their corresponding nitriles, with the functional group linked to the alkyl or aryl residue via the nitrogen atom. They were shown to be reduced by the Mo nitrogenase, which generally results in the formation of methane and the corresponding amine [42, 129]. A special case is the reduction of methyl isocyanide (CH_3NC), which is converted to dimethylamine without the release of methane [130]. Alternative nitrogenases have not been tested for their reactivity with isonitriles so far, but based on the results of other alternative substrates they can be expected to follow similar reactivity patterns.

10.5 Conclusions

Nitrogenases are best known for their ability to break the stable triple bond of N_2, a reaction that is vital to all life on Earth. However, nitrogenases are not limited to N_2 but can reduce a wide range of alternative substrates, including CO, CO_2, cyanides, and acetylene. Intriguingly, substrate selectivity and reactivity vary significantly between the three isoforms of nitrogenase and are thus likely to be dependent on their characteristic metalloclusters FeMoco, FeVco, or FeFeco. Nevertheless, the principles for the reduction of diverse substrates remain the same among all isoforms. Nitrogenases generate low potential electrons through an ATP-dependent process, which are stored as FeMco-associated hydrides in the active site. Through a sophisticated and yet not fully understood mechanism, the enzyme directs these electrons preferably to N_2. However, alternative substrates can also enter the active site where they are likely to be reduced by the highly reactive hydrides. In the past, the side reactivities of nitrogenases have been exploited extensively to study their molecular mechanism. More recently, ideas to harness the unique reactivities of nitrogenase for biotechnological applications have emerged, offering exciting perspectives for the transition into a circular economy. Three reactions stand out:

1) The reduction of acetylene has proven to be a valuable tool to fingerprint nitrogenase activity. While the wild-type Mo nitrogenase reduces acetylene exclusively to ethylene, the reaction with alternative nitrogenases additionally yields ethane, allowing for an easy identification of V or Fe nitrogenase activity.
2) Spectroscopic and structural studies with CO-inhibited nitrogenase have provided groundbreaking insights into substrate binding at the nitrogenase active site. Moreover, all three nitrogenases can reduce CO to methane and the V and Mo isoform also forms C—C bonds in a Fischer–Tropsch-like chemistry to release short-chain hydrocarbons.
3) The reduction of CO_2 to formate and short-chain hydrocarbons offers exciting new pathways for the recycling of carbon waste and highlights the potential of nitrogenases to act as biocatalytic platforms for the green production of biofuels and bulk chemicals.

References

1 Fowler, D., Coyle, M., Skiba, U. et al. (2013). The global nitrogen cycle in the twenty-first century. *Philos. Trans. R. Soc. B Biol. Sci.* 368 (1621): 20130164.
2 Zhang, X., Ward, B.B., and Sigman, D.M. (2020). Global nitrogen cycle: critical enzymes, organisms, and processes for nitrogen budgets and dynamics. *Chem. Rev.* 120 (12): 5308–5351.

3 Cherkasov, N., Ibhadon, A., and Fitzpatrick, P. (2015). A review of the existing and alternative methods for greener nitrogen fixation. *Chem. Eng. Process. Process Intensif.* 90: 24–33.

4 Haber, F. (1922). Über die darstellung des ammoniaks aus stickstoff und wasserstoff. *Naturwissenschaften* 10 (49): 1041–1049.

5 Leigh, G. (2004). *Catalysts for Nitrogen Fixation.* Springer.

6 Eady, R.R., Lowe, D.J., and Thorneley, R.N. (1978). Nitrogenase of *Klebsiella pneumoniae*: a pre-steady state burst of ATP hydrolysis is coupled to electron transfer between the component proteins. *FEBS Lett.* 95 (2): 211–213.

7 Eady, R.R. (1996). Structure–function relationships of alternative nitrogenases. *Chem. Rev.* 96 (7): 3013–3030.

8 Hu, Y., Lee, C.C., and Ribbe, M.W. (2012). Vanadium nitrogenase: a two-hit wonder? *Dalton Trans.* 41 (4): 1118–1127.

9 Harris, D.F., Lukoyanov, D.A., Kallas, H. et al. (2019). Mo-, V-, and Fe-nitrogenases use a universal eight-electron reductive-elimination mechanism to achieve N_2 reduction. *Biochemistry* 58 (30): 3293–3301.

10 Usman, N.I. and Wali, M.M. (2024). *Nitrogen Fixation by Rhizobacterial Nif Mechanism: An Advanced Genetic Perspective.* IntechOpen Limited.

11 Kim, J. and Rees, D.C. (1992). Structural models for the metal centers in the nitrogenase molybdenum-iron protein. *Science* 257 (5077): 1677–1682.

12 Sippel, D. and Einsle, O. (2017). The structure of vanadium nitrogenase reveals an unusual bridging ligand. *Nat. Chem. Biol.* 13 (9): 956–960.

13 Trncik, C., Detemple, F., and Einsle, O. (2023). Iron-only Fe-nitrogenase underscores common catalytic principles in biological nitrogen fixation. *Nat. Catal.* 6 (5): 415–424.

14 Benton, P.M., Mayer, S.M., Shao, J. et al. (2001). Interaction of acetylene and cyanide with the resting state of nitrogenase α-96-substituted MoFe proteins. *Biochemistry* 40 (46): 13816–13825.

15 Fisher, K., Dilworth, M.J., and Newton, W.E. (2000). Differential effects on N_2 binding and reduction, HD formation, and azide reduction with α-195His-and α-191Gln-substituted MoFe proteins of *Azotobacter vinelandii* nitrogenase. *Biochemistry* 39 (50): 15570–15577.

16 Kang, W., Lee, C.C., Jasniewski, A.J. et al. (2020). Structural evidence for a dynamic metallocofactor during N_2 reduction by Mo-nitrogenase. *Science* 368 (6497): 1381–1385.

17 Yang, Z.-Y., Seefeldt, L.C., Dean, D.R. et al. (2011). Steric control of the hi-CO MoFe nitrogenase complex revealed by stopped-flow infra-red spectroscopy. *Angew. Chem. Int. Ed.* 50 (1): 272.

18 Seefeldt, L.C., Yang, Z.-Y., Lukoyanov, D.A. et al. (2020). Reduction of substrates by nitrogenases. *Chem. Rev.* 120 (12): 5082–5106.

19 Winter, H. and Burris, R. (1976). Nitrogenase. *Annu. Rev. Biochem.* 45 (1): 409–426.

20 Howard, J.B. and Rees, D.C. (1994). Nitrogenase: a nucleotide-dependent molecular switch. *Annu. Rev. Biochem.* 63 (1): 235–264.

21 Lanzilotta, W.N. and Seefeldt, L.C. (1997). Changes in the midpoint potentials of the nitrogenase metal centers as a result of iron protein–molybdenum-iron protein complex formation. *Biochemistry* 36 (42): 12976–12983.

22 Hageman, R.V. and Burris, R. (1978). Nitrogenase and nitrogenase reductase associate and dissociate with each catalytic cycle. *Proc. Natl. Acad. Sci. USA* 75 (6): 2699–2702.

23 Burgess, B.K. and Lowe, D.J. (1996). Mechanism of molybdenum nitrogenase. *Chem. Rev.* 96 (7): 2983–3012.

24 Rohde, M., Sippel, D., Trncik, C. et al. (2018). The critical E(4) state of nitrogenase Catalysis. *Biochemistry* 57 (38): 5497–5504.

25 Trncik, C., Müller, T., Franke, P., and Einsle, O. (2022). Structural analysis of the reductase component AnfH of iron-only nitrogenase from *Azotobacter vinelandii*. *J. Inorg. Biochem.* 227: 111690.

26 Masepohl, B. and Klipp, W. (1996). Organization and regulation of genes encoding the molybdenum nitrogenase and the alternative nitrogenase in *Rhodobacter capsulatus*. *Arch. Microbiol.* 165 (2): 80–90.

27 Pence, N., Lewis, N., Alleman, A.B. et al. (2021). Revealing a role for the G subunit in mediating interactions between the nitrogenase component proteins. *J. Inorg. Biochem.* 214: 111273.

28 Schmidt, F.V., Schulz, L., Zarzycki, J. et al. (2024). Structural insights into the iron nitrogenase complex. *Nat. Struct. Mol. Biol.* 31 (1): 150–158.

29 Lukoyanov, D., Khadka, N., Yang, Z.Y. et al. (2016). Reversible photoinduced reductive elimination of H_2 from the nitrogenase dihydride state, the E(4)(4H) Janus intermediate. *J. Am. Chem. Soc.* 138 (4): 1320–1327.

30 Harris, D.F., Lukoyanov, D.A., Shaw, S. et al. (2018). Mechanism of N_2 reduction catalyzed by Fe-nitrogenase involves reductive elimination of H_2. *Biochemistry* 57 (5): 701–710.

31 Bjornsson, R., Lima, F.A., Spatzal, T. et al. (2014). Identification of a spin-coupled Mo (III) in the nitrogenase iron–molybdenum cofactor. *Chem. Sci.* 5 (8): 3096–3103.

32 Rees, J.A., Bjornsson, R., Kowalska, J.K. et al. (2017). Comparative electronic structures of nitrogenase FeMoco and FeVco. *Dalton Trans.* 46 (8): 2445–2455.

33 Yang, Z.Y., Jimenez-Vicente, E., Kallas, H. et al. (2021). The electronic structure of FeV-cofactor in vanadium-dependent nitrogenase. *Chem. Sci.* 12 (20): 6913–6922.

34 Hoffman, B.M., Lukoyanov, D., Yang, Z.-Y. et al. (2014). Mechanism of nitrogen fixation by nitrogenase: the next stage. *Chem. Rev.* 114 (8): 4041–4062.

35 Lukoyanov, D., Barney, B.M., Dean, D.R. et al. (2007). Connecting nitrogenase intermediates with the kinetic scheme for N_2 reduction by a relaxation protocol and identification of the N_2 binding state. *Proc. Natl. Acad. Sci. USA* 104 (5): 1451–1455.

36 Wraight, C.A. (2006). Chance and design—proton transfer in water, channels and bioenergetic proteins. *Biochim. Biophys. Acta Bioenerg.* 1757 (8): 886–912.

37 Morrison, C.N., Hoy, J.A., Zhang, L. et al. (2015). Substrate pathways in the nitrogenase MoFe protein by experimental identification of small molecule binding sites. *Biochemistry* 54 (11): 2052–2060.

38 Hardy, R. and Knight, E. Jr., (1967). ATP-dependent reduction of azide and HCN by N_2-fixing enzymes of *Azotobacter vinelandii* and *Clostridium pasteurianum. Biochim. Biophys. Acta Enzymol.*. 139 (1): 69–90.

39 Vaughn, S.A. and Burgess, B.K. (1989). Nitrite, a new substrate for nitrogenase. *Biochemistry* 28 (2): 419–412.

40 Li, J., Burgess, B.K., and Corbin, J.L. (1982). Nitrogenase reactivity: cyanide as substrate and inhibitor. *Biochemistry* 21 (18): 4393–4402.

41 Fuchsman, W. and Hardy, R. (1972). Nitrogenase-catalyzed acrylonitrile reductions. *Bioinorg. Chem.* 1 (3): 195–213.

42 Kelly, M., Postgate, J., and Richards, R. (1967). Reduction of cyanide and isocyanide by nitrogenase of *Azotobacter chroococcum. Biochem. J.* 102 (1): 1.

43 Dilworth, M.J., Eady, R.R., Robson, R.L., and Miller, R.W. (1987). Ethane formation from acetylene as a potential test for vanadium nitrogenase *in vivo*. *Nature* 327 (6118): 167–168.

44 Hardy, R.W., Holsten, R.D., Jackson, E., and Burns, R. (1968). The acetylene-ethylene assay for N_2 fixation: laboratory and field evaluation. *Plant Physiol.* 43 (8): 1185–1207.

45 Schöllhorn, R. and Burris, R. (1967). Acetylene as a competitive inhibitor of N-2 fixation. *Proc. Natl. Acad. Sci. USA* 58 (1): 213–216.

46 Dilworth, M. (1966). Acetylene reduction by nitrogen-fixing preparations from *Clostridium pasteurianum. Biochim. Biophys. Acta Gen. Subj.* 127 (2): 285–294.

47 McKenna, C.E., McKenna, M.C., and Higa, M.T. (1976). Chemical probes of nitrogenase. 1. Cyclopropene. Nitrogenase-catalyzed reduction to propene and cyclopropane. *J. Am. Chem. Soc.* 98 (15): 4657–4659.

48 Lee, C.C., Hu, Y., and Ribbe, M.W. (2010). Vanadium nitrogenase reduces CO. *Science* 329 (5992): 642.

49 Yang, Z.-Y., Dean, D.R., and Seefeldt, L.C. (2011). Molybdenum nitrogenase catalyzes the reduction and coupling of CO to form hydrocarbons. *J. Biol. Chem.* 286 (22): 19417–19421.

50 Seefeldt, L.C., Rasche, M.E., and Ensign, S.A. (1995). Carbonyl sulfide and carbon dioxide as new substrates, and carbon disulfide as a new inhibitor, of nitrogenase. *Biochemistry* 34 (16): 5382–5389.

51 Rebelein, J.G., Hu, Y., and Ribbe, M.W. (2014). Differential reduction of CO_2 by molybdenum and vanadium nitrogenases. *Angew. Chem. Int. Ed.* 126 (43): 11727–11730.

52 Hoffmann, M.-C., Pfänder, Y., Fehringer, M. et al. (2014). NifA-and CooA-coordinated cowN expression sustains nitrogen fixation by *Rhodobacter capsulatus* in the presence of carbon monoxide. *J. Bacteriol.* 196 (19): 3494–3502.

53 Medina, M.S., Bretzing, K.O., Aviles, R.A. et al. (2021). CowN sustains nitrogenase turnover in the presence of the inhibitor carbon monoxide. *J. Biol. Chem.* 296: 100501.

54 Harris, D.F., Jimenez-Vicente, E., Yang, Z.-Y. et al. (2020). CO as a substrate and inhibitor of H^+ reduction for the Mo-, V-, and Fe-nitrogenase isozymes. *J. Inorg. Biochem.* 213: 111278.

55 Rebelein, J.G., Lee, C.C., Hu, Y., and Ribbe, M.W. (2016). The *in vivo* hydrocarbon formation by vanadium nitrogenase follows a secondary metabolic pathway. *Nat. Commun.* 7 (1): 13641.

56 Hu, Y., Lee, C.C., and Ribbe, M.W. (2011). Extending the carbon chain: hydrocarbon formation catalyzed by vanadium/molybdenum nitrogenases. *Science* 333 (6043): 753–755.

57 Gallon, J. (1981). The oxygen sensitivity of nitrogenase: a problem for biochemists and micro-organisms. *Trends Biochem. Sci.* 6: 19–23.

58 Tichi, M.A. and Tabita, F.R. (2000). Maintenance and control of redox poise in *Rhodobacter capsulatus* strains deficient in the Calvin-Benson-Bassham pathway. *Arch. Microbiol.* 174: 322–333.

59 McKinlay, J.B. and Harwood, C.S. (2010). Carbon dioxide fixation as a central redox cofactor recycling mechanism in bacteria. *Proc. Natl. Acad. Sci. USA* 107 (26): 11669–11675.

60 Wang, D., Zhang, Y., Welch, E. et al. (2010). Elimination of Rubisco alters the regulation of nitrogenase activity and increases hydrogen production in *Rhodospirillum rubrum. Int. J. Hydrog. Energy* 35 (14): 7377–7385.

61 Oehlmann, N.N., Schmidt, F.V., Herzog, M. et al. (2023). CO_2 Reduction by the iron nitrogenase competes with N_2 fixation Under physiological conditions. *BioRxiv* https://doi.org/10.1126/sciadv.ado7729.

62 James, E.K. (2000). Nitrogen fixation in endophytic and associative symbiosis. *Field Crop Res.* 65 (2): 197–209.

63 Zheng, Y., Harris, D.F., Yu, Z. et al. (2018). A pathway for biological methane production using bacterial iron-only nitrogenase. *Nat. Microbiol.* 3 (3): 281–286.

64 Harwood, C.S. (2020). Iron-only and vanadium nitrogenases: fail-safe enzymes or something more? *Ann. Rev. Microbiol.* 74: 247–266.

65 Yang, Z.-Y., Moure, V.R., Dean, D.R., and Seefeldt, L.C. (2012). Carbon dioxide reduction to methane and coupling with acetylene to form propylene catalyzed by remodeled nitrogenase. *Proc. Natl. Acad. Sci. USA* 109 (48): 19644–19648.

66 Case, D., Huynh, B.H., and Karplus, M. (1979). Binding of oxygen and carbon monoxide to hemoglobin. An analysis of the ground and excited states. *J. Am. Chem. Soc.* 101 (16): 4433–4453.

67 Hwang, J., Chen, C., and Burris, R. (1973). Inhibition of nitrogenase-catalyzed reductions. *Biochim. Biophys. Acta Bioenerg.* 292 (1): 256–270.

68 Rivera-Ortiz, J.M. and Burris, R.H. (1975). Interactions among substrates and inhibitors of nitrogenase. *J. Bacteriol.* 123 (2): 537–545.

69 Yates, M.G. and Lowe, D.J. (1976). Nitrogenase of *Azotobacter chroococcum*: a new electron-paramagnetic-resonance signal associated with a transient species of the Mo-Fe protein during catalysis. *FEBS Lett.* 72 (1): 121–125.

70 Pollock, R.C., Lee, H.-I., Cameron, L.M. et al. (1995). Investigation of CO bound to inhibited forms of nitrogenase MoFe protein by ^{13}C ENDOR. *J. Am. Chem. Soc.* 117 (33): 8686–8687.

71 Christie, P.D., Lee, H.-I., Cameron, L.M. et al. (1996). Identification of the CO-binding cluster in nitrogenase MoFe protein by ENDOR of ^{57}Fe isotopomers. *J. Am. Chem. Soc.* 118 (36): 8707–8709.

72 Lee, H.-I., Cameron, L.M., Hales, B.J., and Hoffman, B.M. (1997). CO binding to the FeMo cofactor of CO-inhibited nitrogenase: ^{13}CO and ^{1}H Q-band ENDOR investigation. *J. Am. Chem. Soc.* 119 (42): 10121–10126.

73 Scott, A.D., Pelmenschikov, V., Guo, Y. et al. (2014). Structural characterization of CO-inhibited Mo-nitrogenase by combined application of nuclear resonance vibrational spectroscopy, extended X-ray absorption fine structure, and density functional theory: new insights into the effects of CO binding and the role of the interstitial atom. *J. Am. Chem. Soc.* 136 (45): 15942–15954.

74 Paengnakorn, P., Ash, P.A., Shaw, S. et al. (2017). Infrared spectroscopy of the nitrogenase MoFe protein under electrochemical control: potential-triggered CO binding. *Chem. Sci.* 8 (2): 1500–1505.

75 Spatzal, T., Perez, K.A., Einsle, O. et al. (2014). Ligand binding to the FeMo-cofactor: structures of CO-bound and reactivated nitrogenase. *Science* 345 (6204): 1620–1623.

76 Buscagan, T.M., Perez, K.A., Maggiolo, A.O. et al. (2021). Structural characterization of two CO molecules bound to the nitrogenase active site. *Angew. Chem. Int. Ed.* 133 (11): 5768–5771.

77 Wiig, J.A., Lee, C.C., Hu, Y., and Ribbe, M.W. (2013). Tracing the interstitial carbide of the nitrogenase cofactor during substrate turnover. *J. Am. Chem. Soc.* 135 (13): 4982–4983.

78 Lee, C.C., Hu, Y., and Ribbe, M.W. (2011). Tracing the hydrogen source of hydrocarbons formed by vanadium nitrogenase. *Angew. Chem. Int. Ed.* 50 (24): 5545–5547.

79 Natzke, J. and Bruno-Bárcena, J.M. (2020). Two-stage continuous conversion of carbon monoxide to ethylene by whole cells of *Azotobacter vinelandii*. *Appl. Environ. Microbiol.* 86 (11): e00446–e00420.

80 Lee, C.C., Fay, A.W., Weng, T.-C. et al. (2015). Uncoupling binding of substrate CO from turnover by vanadium nitrogenase. *Proc. Natl. Acad. Sci. USA* 112 (45): 13845–13849.

81 Hiller, C.J., Stiebritz, M.T., Lee, C.C. et al. (2017). Tuning electron flux through nitrogenase with methanogen iron protein homologues. *Chemistry* 23 (64): 16152–16156.

82 Lee, C.C., Wilcoxen, J., Hiller, C.J. et al. (2018). Evaluation of the catalytic relevance of the CO-bound states of V-nitrogenase. *Angew. Chem. Int. Ed.* 130 (13): 3469–3472.

83 Rohde, M., Laun, K., Zebger, I. et al. Two ligand-binding sites in CO-reducing V nitrogenase reveal a general mechanistic principle. *Sci. Adv.* 7 (22): eabg4474.

84 Rohde, M., Grunau, K., and Einsle, O. (2020). CO binding to the FeV cofactor of CO-reducing vanadium nitrogenase at atomic resolution. *Angew. Chem. Int. Ed.* 59 (52): 23626–23630.

85 Lee, C.C., Hu, Y., and Ribbe, M.W. (2012). ATP-independent hydrocarbon formation catalyzed by isolated nitrogenase cofactors. *Angew. Chem. Int. Ed.* 51 (8): 1947.

86 Lee, C.C., Hu, Y., and Ribbe, M.W. (2015). Catalytic reduction of CN^-, CO, and CO_2 by nitrogenase cofactors in lanthanide-driven reactions. *Angew. Chem.* 127 (4): 1235–1238.

87 Lee, C.C., Hu, Y., and Ribbe, M.W. (2015). Insights into hydrocarbon formation by nitrogenase cofactor homologs. *MBio* 6 (2): e00307-15.

88 Lee, C.C., Tanifuji, K., Newcomb, M. et al. (2018). A comparative analysis of the CO-reducing activities of MoFe proteins containing Mo-and V-nitrogenase cofactors. *ChemBioChem* 19 (7): 649–653.

89 Rebelein, J., Lee, C., Newcomb, M. et al. (2018). Characterization of an M-cluster-substituted nitrogenase VFe protein. *MBio* 9: e00310-18.

90 Rutledge, H.L., Cook, B.D., Nguyen, H.P.M. et al. (2022). Structures of the nitrogenase complex prepared under catalytic turnover conditions. *Science* 377 (6608): 865–869.

91 Oehlmann, N.N. and Rebelein, J.G. (2022). The conversion of carbon monoxide and carbon dioxide by nitrogenases. *ChemBioChem* 23 (8): e202100453.

92 Dry, M.E. (2002). The Fischer–Tropsch process: 1950–2000. *Catal. Today* 71 (3): 227–241.

93 Khadka, N., Dean, D.R., Smith, D. et al. (2016). CO_2 reduction catalyzed by nitrogenase: pathways to formate, carbon monoxide, and methane. *Inorg. Chem.* 55 (17): 8321–8330.

94 Fixen, K.R., Zheng, Y., Harris, D.F. et al. (2016). Light-driven carbon dioxide reduction to methane by nitrogenase in a photosynthetic bacterium. *Proc. Natl. Acad. Sci. USA* 113 (36): 10163–10167.

95 Rebelein, J.G., Hu, Y., and Ribbe, M.W. (2015). Widening the product profile of carbon dioxide reduction by vanadium nitrogenase. *ChemBioChem* 16 (14): 1993–1996.

96 Cai, R., Milton, R.D., Abdellaoui, S. et al. (2018). Electroenzymatic C–C bond formation from CO_2. *J. Am. Chem. Soc.* 140 (15): 5041–5044.

97 Hu, B., Harris, D.F., Dean, D.R. et al. (2018). Electrocatalytic CO_2 reduction catalyzed by nitrogenase MoFe and FeFe proteins. *Bioelectrochemistry* 120: 104–109.

98 Yates, M.G. and Planqué, K. (1975). Nitrogenase from *Azotobacter chroococcum*. Purification and properties of the component proteins. *Eur. J. Biochem.* 60 (2): 467–476.

99 Davis, L.C., Henzl, M.T., Burris, R.H., and Orme-Johnson, W.H. (1979). Iron-sulfur clusters in the molybdenum-iron protein component of nitrogenase. Electron paramagnetic resonance of the carbon monoxide inhibited state. *Biochemistry* 18 (22): 4860–4869.

100 Hales, B.J., Case, E.E., Morningstar, J.E. et al. (1986). Isolation of a new vanadium-containing nitrogenase from *Azotobacter vinelandii*. *Biochemistry* 25 (23): 7251–7255.

101 Hu, Y., Lee, C.C., Grosch, M. et al. (2023). Enzymatic Fischer–Tropsch-type reactions. *Chem. Rev.* 123 (9): 5755–5797.

102 Dilworth, M.J., Eady, R.R., and Eldridge, M.E. (1988). The vanadium nitrogenase of *Azotobacter chroococcum*. Reduction of acetylene and ethylene to ethane. *Biochem. J.* 249 (3): 745–751.

103 Hales, B.J. (2004). Vanadium nitrogenase. In: *Catalysts for Nitrogen Fixation: Nitrogenases, Relevant Chemical Models and Commercial Processes* (ed. B.E. Smith, R.L. Richards, and W.E. Newton), 255–279. Dordrecht: Springer Netherlands.

104 Schneider, K., Gollan, U., Dröttboom, M. et al. (1997). Comparative biochemical characterization of the iron-only nitrogenase and the molybdenum nitrogenase from *Rhodobacter capsulatus. Eur. J. Biochem.* 244 (3): 789–800.

105 Jasniewski, A.J., Lee, C.C., Ribbe, M.W., and Hu, Y. (2020). Reactivity, mechanism, and assembly of the alternative nitrogenases. *Chem. Rev.* 120 (12): 5107–5157.

106 Schneider, K. and Müller, A. (2004). Iron-only nitrogenase: exceptional catalytic, structural and spectroscopic features. In: *Catalysts for Nitrogen Fixation: Nitrogenases, Relevant Chemical Models and Commercial Processes* (ed. B.E. Smith, R.L. Richards, and W.E. Newton), 281–307. Dordrecht: Springer Netherlands.

107 Schneider, K., Müller, A., Schramm, U., and Klipp, W. (1991). Demonstration of a molybdenum- and vanadium-independent nitrogenase in a nifHDK-deletion mutant of *Rhodobacter capsulatus. Eur. J. Biochem.* 195 (3): 653–661.

108 Oda, Y., Samanta, S.K., Rey, F.E. et al. (2005). Functional genomic analysis of three nitrogenase isozymes in the photosynthetic bacterium *Rhodopseudomonas palustris. J. Bacteriol.* 187 (22): 7784–7794.

109 Fisher, K., Dilworth, M.J., Kim, C.-H., and Newton, W.E. (2000). *Azotobacter vinelandii* nitrogenases containing altered MoFe proteins with substitutions in the FeMo-cofactor environment: effects on the catalyzed reduction of acetylene and ethylene. *Biochemistry* 39 (11): 2970–2979.

110 Benton, P.M., Christiansen, J., Dean, D.R., and Seefeldt, L.C. (2001). Stereospecificity of acetylene reduction catalyzed by nitrogenase. *J. Am. Chem. Soc.* 123 (9): 1822–1827.

111 Lowe, D.J., Fisher, K., and Thorneley, R.N. (1990). *Klebsiella pneumoniae* nitrogenase. Mechanism of acetylene reduction and its inhibition by carbon monoxide. *Biochem. J.* 272 (3): 621–625.

112 Simpson, F.B. and Burris, R.H. (1984). A nitrogen pressure of 50 atmospheres does not prevent evolution of hydrogen by nitrogenase. *Science* 224 (4653): 1095–1097.

113 Hardy, R. and Jackson, E. ed. (1967). Reduction of model substrates-nitriles and acetylenes-by nitrogenase (N_2ase). *Federation Proceedings*. Rockville Pike, Bethesda, MD: Federation of American Societies for Experimental Biology.

114 Hardy, R., Burns, R., and Parshall, G. (1971). *The Biochemistry of N_2 Fixation*. ACS Publications.

115 Dos Santos, P.C., Mayer, S.M., Barney, B.M. et al. (2007). Alkyne substrate interaction within the nitrogenase MoFe protein. *J. Inorg. Biochem.* 101 (11–12): 1642–1648.

116 Mayer, S.M., Niehaus, W.G., and Dean, D.R. (2002). Reduction of short chain alkynes by a nitrogenase α-70 Ala-substituted MoFe protein. *J. Chem. Soc. Dalton Trans.* 5: 802–807.

117 Igarashi, R.Y., Dos Santos, P.C., Niehaus, W.G. et al. (2004). Localization of a catalytic intermediate bound to the FeMo-cofactor of nitrogenase. *J. Biol. Chem.* 279 (33): 34770–34775.

118 Lee, H.-I., Igarashi, R.Y., Laryukhin, M. et al. (2004). An organometallic intermediate during alkyne reduction by nitrogenase. *J. Am. Chem. Soc.* 126 (31): 9563–9569.

119 McKenna, C., Nakajima, T., Jones, J. et al. (1980). Active site probes of nitrogenese. In: *Molybdenum Chemistry of Biological Significance* (ed. W.E. Newton and S. Otsuka), 39–57. Springer.

120 Burns, R., Hardy, R., and Phillips, W. (1975). *Azotobacter* nitrogenase: mechanism and kinetics of allene reduction. *Int. Biol. Programme* 6: 447–452.

121 Shaw, S., Lukoyanov, D., Danyal, K. et al. (2014). Nitrite and hydroxylamine as nitrogenase substrates: mechanistic implications for the pathway of N_2 reduction. *J. Am. Chem. Soc.* 136 (36): 12776–12783.

122 Schöllhorn, R. and Burris, R.H. (1967). Reduction of azide by the N_2-fixing enzyme system. *Proc. Natl. Acad. Sci. USA* 57 (5): 1317–1323.

123 Rubinson, J.F., Burgess, B.K., Corbin, J.L., and Dilworth, M.J. (1985). Nitrogenase reactivity: azide reduction. *Biochemistry* 24 (2): 273–283.

124 Dilworth, M.J. and Thorneley, R.N. (1981). Nitrogenase of *Klebsiella pneumoniae*. Hydrazine is a product of azide reduction. *Biochem. J.* 193 (3): 971–983.

125 Davis, L.C. (1980). Hydrazine as a substrate and inhibitor of *Azotobacter vinelandii* nitrogenase. *Arch. Biochem. Biophys.* 204 (1): 270–276.

126 Fisher, K., Dilworth, M.J., and Newton, W.E. (2006). *Azotobacter vinelandii* vanadium nitrogenase: formaldehyde is a product of catalyzed HCN reduction, and excess ammonia arises directly from catalyzed azide reduction. *Biochemistry* 45 (13): 4190–4198.

127 Fisher, K., Dilworth, M.J., Kim, C.-H., and Newton, W.E. (2000). *Azotobacter vinelandii* nitrogenases with substitutions in the FeMo-cofactor environment of the MoFe protein: effects of acetylene or ethylene on interactions with H^+, HCN, and CN. *Biochemistry* 39 (35): 10855–10865.

128 Miller, R.W. and Eady, R.R. (1988). Cyanamide: a new substrate for nitrogenase. *Biochim. Biophys. Acta* 952 (3): 290–296.

129 Kelly, M. (1968). The kinetics of the reduction of isocyanides, acetylenes and the cyanide ion by nitrogenase preparation from *Azotobacter chroococcum* and the effects of inhibitors. *Biochem. J.* 107 (1): 1–6.

130 Rubinson, J.F., Corbin, J.L., and Burgess, B.K. (1983). Nitrogenase reactivity: methyl isocyanide as substrate and inhibitor. *Biochemistry* 22 (26): 6260–6268.

Iron-Sulfur Clusters

Iron-Sulfur Clusters

Biogenesis and Biochemistry

Volume 2

Edited by Silke Leimkühler, Günter Schwarz, Oliver Lenz, and Oliver Einsle

WILEY-VCH

Editors

Prof. Silke Leimkühler
Universität Potsdam
Karl-Liebknecht-Str. 24-25
H25
Potsdam
GolM, 14476

Prof. Günter Schwarz
Universität zu Köln
Zülpicher Str. 47
Köln
GM, 50674

Dr. Oliver Lenz
Technische Universität Berlin
Max-Volmer-Laboratorium, PC 14
Strasse des 17. Juni 135
Berlin
GM, 10623

Prof. Oliver Einsle
Albert-Ludwigs-Universität
Albertstraße 21
Freiburg
GM, 79104

Cover Image: Pixabay, The chemical structures featured on the cover were created by Oliver Lenz

Typesetting: Straive, Chennai, India
Printing and Binding: CPI Group (UK) Ltd, Croydon, CR0 4YY

Library of Congress Card No.: applied for

British Library Cataloguing-in-Publication Data A catalogue record for this book is available from the British Library.

Bibliographic information published by the Deutsche Nationalbibliothek
The Deutsche Nationalbibliothek lists this publication in the Deutsche Nationalbibliografie; detailed bibliographic data are available on the Internet at <http://dnb.d-nb.de>.

Print ISBN: 978-3-527-35257-9
ePDF ISBN: 978-3-527-84357-2
ePub ISBN: 978-3-527-84358-9
oBook ISBN: 978-3-527-84359-6

C9783527352579_280825

The manufacturer's authorized representative according to the EU General Product Safety Regulation is WILEY-VCH GmbH, Boschstr. 12, 69469 Weinheim, Germany,
e-mail: Product_Safety@wiley.com.

This book is dedicated to five outstanding people in our careers, who have been our esteemed teachers, mentors, supervisors, and friends: Ralf Mendel, Werner Klipp, K. V. Rajagopalan, Dough Rees, and Bärbel Friedrich. We are grateful for their continuous support, encouragement, patience, motivation, and for kindling in us the enduring flame of a scientist.

Contents

Volume 1

Preface *xv*

1 **ISC-Based Fe–S Protein Biogenesis in Bacteria** *1*
Béatrice Py and Frédéric Barras

2 **Unraveling the Complexity of the Suf-Based Fe–S Biogenesis** *25*
Ingie Elchennawi, Claire E. Fisher, Franklin Wayne Outten, and Ollagnier de Choudens Sandrine

3 **Biogenesis of Mitochondrial Iron–Sulfur Proteins** *57*
Ulrich Mühlenhoff, Oliver Stehling, and Roland Lill

4 **Iron–Sulfur Protein Maturation in the Cytosol of Eukaryotes** *87*
Joseph J. Braymer and Antonio J. Pierik

5 **Mammalian Aldehyde Oxidase** *135*
Maria Joao Romao, Guilherme Vilela-Alves, and Cristiano Mota

6 **Biological Formation of Sulfide** *159*
Marion Jespersen and Tristan Wagner

7 **The Molybdenum Cofactor, Its Biosynthesis, and Relation to Fe–S Clusters in Bacteria** *193*
Paolo Olivieri and Silke Leimkühler

8 **Molybdenum Cofactor Biosynthesis in Eukaryotes** *227*
Guenter Schwarz, Lukas Flohr, Emanuel Bruckisch, and Katrin Fischer-Schrader

9 **NifUS Is a Two-component Toolbox Involved in Assembling Fe–S Clusters Associated with Nitrogen Fixation and Beyond** *257*
Julia S. Martin del Campo, Shervin Shaybani, Dennis R. Dean, and Patricia C. Dos Santos

10 **Alternative Substrates of Nitrogenase** *287*
Frederik V. Schmidt and Johannes G. Rebelein

Volume 2

Preface *xvii*

11 **Electron-Bifurcating Hydrogenases** *317*
Gregory E. Vansuch, Effie C. Kisgeropoulos, Jonathan R. Humphreys, Carolyn E. Lubner, David W. Mulder, and Paul W. King
Chapter Goals *317*
Main Chapter Points *317*
11.1 Introduction *317*
11.1.1 Hydrogenases: From "Simple" Prototypical Architectures to Not So Simple Electron-Bifurcating Architectures *317*
11.1.2 Electron Bifurcation: A Brief Overview *320*
11.2 Physiology, Subunit Compositions, and Reactivity *321*
11.2.1 Overview of Physiological Roles: Three Thematic Examples *321*
11.2.2 Subunit Compositions and Reactivities *324*
11.3 Structural and Biophysical Characterizations *337*
11.3.1 Structures and Cofactors of Hyd-type Electron-Bifurcating Hydrogenase *338*
11.3.1.1 HydA and HydSL *339*
11.3.1.2 HydB *340*
11.3.1.3 HydC *341*
11.3.2 Spectroscopy of HydABC and HndABCD *342*
11.3.2.1 EPR Spectroscopy of the Fe–S Clusters in *Tm* HydABC *342*
11.3.2.2 EPR Spectroscopy of HndABCD from *S. fructosivorans* *344*
11.3.2.3 EPR and IR Spectroscopy of the H-Cluster in [FeFe]–HydABC *344*
11.3.3 Structures and Cofactors of MvhAGD-HdrABC *346*
11.3.3.1 Selenocysteine Considerations *348*
11.3.4 Spectroscopy of MvhAGD *349*
11.4 Mechanistic Proposals *349*
11.4.1 Electron Bifurcation: General Mechanistic Considerations *350*
11.4.2 [FeFe]- and [NiFe]–HydABC: Electron Transfer Pathways *352*
11.4.3 [NiFe]–MvhAGD–HdrABC *356*
11.4.3.1 Electron Transfer Pathways *356*
11.4.3.2 CoM-S-S-CoB Binding and Reduction *359*
11.5 Forefronts *360*

11.6 Summary and Outlook *362*
Acknowledgments *363*
Author Contributions *363*
References *364*

12 Inhibition of [FeFe]-Hydrogenases by Small Molecules *381*
Claudia Brocks and Thomas Happe
Chapter Goals *381*
12.1 Historical Perspective of [FeFe]-Hydrogenase Research *381*
12.2 Characteristics of [FeFe]-Hydrogenases: How They Do What They Do *384*
12.3 Small Molecules Inhibit [FeFe]-Hydrogenases *389*
12.3.1 Diffusion of Small Molecules to the H-cluster *390*
12.3.2 Irreversible Attack of O_2 Molecules *392*
12.3.3 Inhibitor Molecules that Protect [FeFe]-Hydrogenases Against O_2 *394*
12.3.3.1 CO Protects Against the O_2-Initiated H-Cluster Degradation *394*
12.3.3.2 Formaldehyde Attacks Catalytically Important Sites of [FeFe]-Hydrogenases *394*
12.3.3.3 The Role of Sulfide in [FeFe]-Hydrogenases *396*
12.3.3.4 Cba5H – A [FeFe]-Hydrogenase that Protects Itself by Its Own Sulfide *396*
12.3.4 Influences of Small Molecules on the Proton Transfer Pathway *398*
12.4 Summary and Outlook *399*
References *399*

13 Biosynthesis of [NiFe]-Hydrogenase *407*
Oliver Lenz and Giorgio Caserta
Chapter Goals *407*
Main Chapter Points *407*
13.1 [NiFe]-Hydrogenases and Their Function *407*
13.2 The Inorganic Catalytic Center in the [NiFe]-Hydrogenase Basic Module *409*
13.3 Genetic Basis of [NiFe]-Hydrogenase Maturation *409*
13.4 Cyanide Synthesis by HypE and HypF *412*
13.5 Assembly of the $Fe(CN)_2CO$ Synthon on the HypCD Complex *414*
13.6 Origin and Synthesis of the Carbon Monoxide Ligand *417*
13.7 Transfer of the $Fe(CN)_2CO$ Synthon from the HypCD Complex to Apo-Hydrogenase *419*
13.8 Mobilization and Insertion of Nickel by HypAB *421*
13.9 Role of the C-Terminal Extension of the Premature Large Subunit *425*
13.10 Hydrogenase Maturation Intermediates Isolated from Living Cells *427*
13.11 Iron–Sulfur Cluster Insertion into the Small Subunit *429*
13.12 Subunit Oligomerization and Transport Across the Cytoplasmic Membrane *431*

13.13 Special Case: Maturation of O_2-Tolerant Membrane-Bound [NiFe]-Hydrogenases *433*
13.14 Conclusions *435*
Acknowledgments *436*
References *436*

14 [Fe]-Hydrogenase and the FeGP Cofactor Involved in the CO_2-Reducing Hydrogenotrophic Methanogenic Pathway *447*
Seigo Shima, Joao Pedro Fernandes-Queiroz, and Masanori Kaneko
Chapter Goals *447*
Main Chapter Points *447*
14.1 Introduction *447*
14.2 Unique Coenzymes Found in Methanogens *448*
14.3 Enzymes Involved in the Hydrogenotrophic Methanogenic Pathway *449*
14.4 [NiFe]-Hydrogenases Involved in the Hydrogenotrophic Methanogenesis *452*
14.5 Function of [Fe]-Hydrogenase (Hmd) *453*
14.6 Reconstitution of the Hmd Holoenzyme with the Extracted Cofactor *454*
14.7 Light Sensitivity of Hmd and Finding of the Functional Iron in the Cofactor *456*
14.8 Structure and Properties of the FeGP Cofactor *457*
14.9 Photolysis Mechanism *459*
14.10 Crystal Structures of Hmd Apoenzymes *459*
14.10.1 Crystal Structure of Reconstituted Hmd Holoenzymes *460*
14.10.2 Crystal Structure of Reconstituted Hmd with the Substrate *460*
14.10.3 The Other Crystal Structures of Hmd and its Homologs *461*
14.11 ESI-MS Analysis for Detection of the CO/Acyl Ligands *462*
14.12 Biosynthesis of the FeGP Cofactor *463*
14.12.1 Stable Isotope Labeling Experiments *463*
14.12.2 The *hcg* Gene Cluster, Sequence Similarities, and Mutational Analysis *465*
14.12.3 Structure to Function Analysis of Hcg Proteins *466*
14.12.3.1 HcgA *467*
14.12.3.2 HcgB *467*
14.12.3.3 HcgC *468*
14.12.3.4 HcgD *470*
14.12.3.5 HcgE *471*
14.12.3.6 HcgF *472*
14.12.3.7 HcgG *473*
14.12.4 *In Vitro* Biosynthesis *473*
14.12.4.1 Design of the *In Vitro* Biosynthesis Assay *474*
14.12.4.2 Confirmation of the Precursors by *In Vitro* Biosynthesis *475*
14.12.4.3 Basis of the *In Vitro* Complementation Assay *476*
14.12.4.4 *In Vitro* Complementation of HcgA *476*

14.12.4.5 *In Vitro* Complementation of HcgG 477
14.12.5 Proposed Biosynthesis Sequence of the FeGP Cofactor 479
14.13 Conclusion 479
Acknowledgment 480
References 480

15 Catalysis by Hydrogenase 489
Seigo Shima, James A. Birrell, Sven T. Stripp, Giorgio Caserta, and Oliver Lenz
Chapter Goals 489
Main Chapter Points 489
15.1 Catalytic Cycle of [Fe]-Hydrogenase 490
15.1.1 Electronic Properties of the Iron Site of the FeGP Cofactor 490
15.1.2 Enzyme Reaction Kinetics 491
15.1.3 Hmd Inhibitors and Their Contribution to the Catalytic Mechanism 492
15.1.3.1 CO and CN^- 492
15.1.3.2 Isocyanides 492
15.1.3.3 Cu 493
15.1.3.4 Fe 494
15.1.3.5 H_2O_2 494
15.1.3.6 O_2 is Reduced to H_2O_2 by the Hmd Reaction 495
15.1.4 Semisynthetic Hmd 496
15.1.4.1 Importance of the 2-OH Group of the Pyridinol Ring 498
15.1.4.2 Semisynthetic [Mn]-Hydrogenase 498
15.1.5 Circular Dichroism Spectroscopy 499
15.1.6 Crystal Structure of Hmd with the Substrate 499
15.1.7 Proposed Catalytic Mechanism of Hmd 500
15.1.8 Conclusion 502
15.2 Toward the Catalytic Mechanism of [FeFe]-Hydrogenase 502
15.2.1 The Active Site Cofactor Is an Iron–Sulfur Cluster 503
15.2.2 The Electronic Structure of the H-cluster 504
15.2.3 Current State of the Catalytic Mechanism of [FeFe]-hydrogenase 507
15.2.4 Toward a Consensus Catalytic Mechanism 511
15.2.5 Conclusion 512
15.3 Catalytic Cycle of [NiFe]-Hydrogenase 512
15.3.1 Classification of [NiFe]-Hydrogenases: The Complex Trait of O_2 Tolerance 514
15.3.2 [NiFe]-Hydrogenase Active Site States 516
15.3.2.1 CO-Bound States 517
15.3.2.2 Hydroxy-Bridged States 517
15.3.2.3 Unusual [NiFe] Site Arrangements 519
15.3.3 Catalytic Intermediates 520
15.3.3.1 Ni_a-S 520
15.3.3.2 Ni_a-SR 522
15.3.3.3 Ni_a-C 523

15.3.3.4 Ni_a-L *524*
15.3.4 Conclusions *525*
Acknowledgment *525*
References *526*

16 Macromolecular Crystallography *541*
Konstantin Bikbaev and Ingrid Span
Chapter Goals *541*
Main Chapter Points *541*
16.1 Introduction *542*
16.2 Crystallization of Macromolecules *544*
16.3 Symmetry and the Unit Cell *548*
16.4 Diffraction and Interpretation of Diffraction Patterns *550*
16.5 Data Collection and Processing *552*
16.6 X-ray Structure Analysis *554*
16.6.1 Isomorphous Replacement *555*
16.6.2 Anomalous Diffraction *556*
16.6.3 Molecular Replacement *557*
16.6.4 Structure Refinement *557*
16.7 Time-Resolved Crystallography *562*
16.8 Nuances of Fe–S Protein Crystallography *565*
16.9 Conclusions *566*
References *566*

17 Vibrational and Mössbauer Spectroscopic Techniques to Study Iron–Sulfur Clusters *571*
Christian Lorent, Giorgio Caserta, Volker Schünemann, and Ingo Zebger
Chapter Goals *571*
Main Chapter Points *571*
17.1 Vibrational Spectroscopy *572*
17.1.1 Normal Modes *573*
17.1.2 Vibrational Spectroscopic Techniques *573*
17.2 Infrared Spectroscopy *574*
17.2.1 MIR Spectroscopy *575*
17.2.2 FIR Spectroscopy *578*
17.2.3 Conclusion *579*
17.3 Raman Spectroscopy *579*
17.3.1 Resonance Raman Spectroscopy *581*
17.3.2 Studying [Fe–S] Clusters by Resonance Raman Spectroscopy *582*
17.3.3 Rubredoxin *582*
17.3.4 [2Fe–2S] Clusters *583*
17.3.5 [3Fe–4S] Clusters *585*
17.3.6 [4Fe–4S] Clusters *586*
17.3.7 Studying Hydrogenases by Resonance Raman Spectroscopy *587*
17.3.8 Conclusion *588*

17.4 Mössbauer-Based Spectroscopic Techniques *588*
17.4.1 Mössbauer Spectroscopy *588*
17.4.2 Mössbauer Spectroscopy from Single Fe(S–Cys)$_4$ Sites up to [4Fe–4S] Clusters *592*
17.4.3 Complex Iron–Sulfur Clusters: Hydrogenases and Nitrogenases *595*
17.4.4 Conclusion *596*
17.5 Nuclear Resonance Vibrational Spectroscopy *596*
17.5.1 Theoretical Background *597*
17.5.2 [Fe(S–Cys)$_4$] Metal Site: The Case Study of Rubredoxin *599*
17.5.3 [2Fe–2S] Clusters: 4xCys, 2xCys–2xHis, and 3xCys–1xHis Ligations *599*
17.5.4 [3Fe–4S] Clusters *601*
17.5.5 [4Fe–4S] Clusters *602*
17.5.6 Binding of NO to [Fe–S] Clusters *602*
17.5.7 The Nitrogenase FeMe Cofactor *603*
17.5.8 [NiFe]- and [FeFe]-Hydrogenases *605*
17.5.9 Conclusion *606*
References *606*

18 The Multisite Microstate Model: A Theoretical Approach for Analyzing Charge Transfer in Proteins *617*
G. Matthias Ullmann and Rajeev Ranjan Roy
Main Chapter Points *617*
18.1 Introduction *617*
18.2 Binding of Ligands to a Receptor *620*
18.3 Analyzing Biomolecular Systems *622*
18.3.1 General Considerations *622*
18.3.2 Thermodynamic Analysis *624*
18.3.3 Kinetic Analysis *625*
18.4 Modeling Protein Electrostatics Using the Poisson–Boltzmann Equation *626*
18.4.1 Conceptual Model *626*
18.4.2 The Mathematical Model *627*
18.4.3 Electrostatic Potentials and Electrostatic Energies *629*
18.4.4 Defining the Low-Dielectric Cavity of the Protein *631*
18.4.5 Fitting Quantum Chemical Electrostatic Potentials to Point Charges *633*
18.4.6 Problems and Shortcomings of the Poisson–Boltzmann Model *635*
18.5 Electrostatic Calculations of the Energy Parameter *636*
18.6 Practical Hints to Construct a Multisite System *639*
18.6.1 Molecular Structures as a Basis of Detailed Calculations *639*
18.6.2 How to Choose Sites, Model Compounds, and Model Compound Energies *640*
18.7 Conclusions *642*
Acknowledgement *643*
References *643*

19 Structural and Functional Bioinorganic Model Chemistry of Fe–S Clusters – Synthesis and Analysis *649*
Benedict Josua Elvers and Carola Schulzke
Chapter Goals *649*
Main Chapter Points *649*
19.1 Introduction *649*
19.2 Fe–S Cluster Motifs: Natural Occurrence, Reactivity, and Artificial Synthesis *652*
19.3 $[FeS_4]$ *653*
19.3.1 Biological Relevance *654*
19.3.2 Synthetic Model Complexes *654*
19.4 $[Fe_2S_2]$ Clusters *657*
19.4.1 Biological Relevance of $[Fe_2S_2]$ Clusters *657*
19.4.2 Synthetic Model Complexes of $[Fe_2S_2]$ Clusters *659*
19.5 $[Fe_4S_4]$ and $[Fe_3S_4]$ Clusters *665*
19.5.1 Biological Relevance of $[Fe_4S_4]$ and $[Fe_3S_4]$ Clusters *665*
19.5.2 Synthetic Model Complexes of $[Fe_4S_4]$ *667*
19.5.3 Synthetic Model Complexes of $[Fe_3S_4]$ *676*
19.6 Clusters of Higher Nuclearity *679*
19.6.1 Hydrogenase and Nitrogenase *680*
19.6.2 Model Chemistry for Clusters of Higher Nuclearity *683*
19.7 Conclusion *691*
References *692*

20 Metal-Dependent Formate Dehydrogenases and Their Interplay and Relationship to Iron–Sulfur Clusters *705*
Benjamin R. Duffus
Chapter Goals *705*
Main Chapter Points *705*
20.1 Formate: General Considerations *706*
20.2 Metal-Independent FDHs *707*
20.3 Metal-Dependent FDHs *708*
20.4 Metal-Dependent Formate Dehydrogenases: Common Modular Catalytic Unit *711*
20.5 FDH and Link of bis-MGD to a [4Fe–4S] Cluster *712*
20.6 Link of the FDH bis-MGD Cofactor with Extended Fe-S Cluster Electron Chains *714*
20.7 FDH that Employ Ferredoxin as Electron Acceptor *715*
20.8 *E. coli* Metal-Dependent Formate Dehydrogenases *715*
20.9 Formate Hydrogenlyase Complex *716*
20.10 NAD^+-Reducing, Metal-Dependent FDHs *717*
20.11 W-Containing FDHs from Sulfate-Reducing Bacteria *719*
20.12 FDH: *D. vulgaris* Hildenborough *719*
20.13 Tungsten vs. Molybdenum FDH *720*

20.14 Metal-Dependent Formate Dehydrogenase from *M. thermoacetica*: Variance and Codependence of Metal Ions *721*
20.15 Formylmethanofuran Dehydrogenases *721*
20.16 FDHs in Methanogens *722*
20.17 FDH O_2 Sensitivity: Inhibition and Enzymatic Activation *724*
20.18 FDH and Electron Bifurcation vs. O_2 Tolerance *725*
20.19 FDH in H_2-Dependent CO_2 Reduction (HDCR) *727*
20.20 Conclusion *728*
Acknowledgments *729*
References *729*

Index *747*

Preface

The study of metalloenzymes began more than a century ago. In 1897, Gabriel Bertrand, working on laccase (a polyphenol oxidase), suggested for the first time that a metal ion was essential for the catalytic activity of an enzyme. In 1926, James Sumner crystallized the first protein, jack bean urease, thereby demonstrating that urease is a protein and that the dissolved crystals catalyze the hydrolysis of urea. Only 49 years after Sumner's pioneering crystallization of an enzyme, urease was eventually found to be a metalloenzyme, containing metal ions at its active site that are essential for enzyme catalysis. Today we know that almost half of the proteins require a metal ion to function, and more than one-third of the known enzymes are metalloenzymes, requiring stoichiometric amounts of metal ions as cofactors, which are typically transition metal ions. Metal ions are essential for life, as metalloproteins are involved in most important biological processes, including respiration; photosynthesis; nitrogen, sulfur, carbon, and hydrogen metabolism; biosynthesis of antibiotics; gene regulation; replication and repair of DNA; antioxidant defense; and neurotransmission. However, even more than 100 years after Bertrand's first discovery of a metalloenzyme, we are only just beginning to understand the role of metals in the catalysis of biological reactions.

The second most abundant metal on earth is iron, which plays an important role in redox reactions in biological systems and is mostly present in either cytochromes or iron–sulfur (Fe–S) clusters. Fe–S clusters occur in more than 200 different types of enzymes or proteins and constitute one of the most ancient, ubiquitous, and structurally diverse classes of biological prosthetic groups. Fe–S clusters were first discovered by Helmut Beinert and others in the early 1960s. The corresponding iron–sulfur enzymes are functionally quite diverse and have both redox and nonredox functions. They are prevalent in central metabolism and notably under anaerobic conditions. Therefore, they have frequently been considered to be of very ancient origin. Since a great variety of Fe–S clusters spontaneously form in solution from Fe^{2+}/Fe^{3+}, R-S^-, S^{2-} or from Fe^{2+}, R-SH, and sulfur, it has been suggested that the first iron–sulfur enzymes arose through the incorporation of preexisting inorganic Fe–S clusters into polypeptides. However, in the 1990s, it was finally discovered that the process of Fe–S cluster biosynthesis is very complex and requires a set of dedicated proteins that are remarkably conserved in prokaryotic and eukaryotic organisms.

Fe–S-containing proteins are also involved in the assembly of metal sites that contain noniron metals. Examples are the assembly of the FeMoco of nitrogenase, the catalytic centers of hydrogenases, the molybdenum cofactor, and heme. Additionally, iron–sulfur clusters are involved in electron transport, gene regulation, DNA replication and repair, and tRNA modification. Thus, it seems that the pathways for the assembly of these metal sites cannot be studied without the understanding of the assembly and regulation of Fe–S clusters in the cell. We mainly describe the assembly and role of metalloproteins that either contain an iron atom in a mononuclear or binuclear active site or for which the assembly and catalytic activity of the enzyme depends on a Fe–S cluster-containing protein. The assembly of metal sites and the role of different metal ions in polynuclear metalloclusters will be described in this book. Furthermore, the availability of metal ions in the cell is tightly regulated, and metal homeostasis and trafficking in a cellular context are also covered.

December 2024
Potsdam

Silke Leimkühler
Günter Schwarz
Oliver Lenz
Oliver Einsle

11

Electron-Bifurcating Hydrogenases

Gregory E. Vansuch, Effie C. Kisgeropoulos, Jonathan R. Humphreys, Carolyn E. Lubner, David W. Mulder, and Paul W. King

National Renewable Energy Laboratory, Biosciences Center, Golden, CO 80401, USA

Chapter Goals

- Overview the biochemistry and general reactivity of electron-bifurcating hydrogenase
- Summarize structural knowledge of electron-bifurcating hydrogenases and spectroscopy of their hydrogenase modules
- Discuss mechanistic proposals of electron-bifurcating hydrogenases within the context of electron transfer and protein dynamics
- Provide motivations and rationalizations for future areas of study

Main Chapter Points

- Electron-bifurcating hydrogenases can be essential for the growth and survival of some microorganisms
- These hydrogenases have very complex architectures, as evidenced from their subunit compositions and manifested by their structures
- The structural studies indicate a need for coordination of electron transfer dynamics to protein conformational changes
- The interested researcher will find opportunities at the forefronts of microbial metabolism and enzyme dynamics tied to Fe–S cluster and flavin-mediated electron transfer

11.1 Introduction

11.1.1 Hydrogenases: From "Simple" Prototypical Architectures to Not So Simple Electron-Bifurcating Architectures

Molecular hydrogen (H_2) is a common biological energy vector. Its activation (typically to generate cellular reducing equivalents) and its production (to recycle

Iron-Sulfur Clusters: Biogenesis and Biochemistry, Volume 2, First Edition.
Edited by Silke Leimkühler, Günter Schwarz, Oliver Lenz, and Oliver Einsle.

cellular reducing equivalents) are primarily catalyzed by two evolutionarily distinct classes of hydrogenases, [FeFe]- and [NiFe]-hydrogenases, which are named after their active site metal content. They have various architectures that facilitate interactions with numerous electron carriers, which link hydrogenase function to specific cellular demands [1–4].

The simplest hydrogenase architectures are called prototypical [3, 5]. Those of the [FeFe] type are monomers that contain a cofactor called the H-cluster: a diiron, H_2 processing active linked to an iron–sulfur (Fe–S) cluster electron transfer (ET) component. Additional Fe–S clusters, often called accessory clusters, can form an extended ET relay (Figure 11.1a–c, top) [1–6]. The prototypical [NiFe] types are heterodimers: the [NiFe] site is located in what is called the large subunit and three Fe–S clusters form an ET relay in the small subunit (Figure 11.1a, bottom) [1–5, 17, 18]. Fe–S clusters near the protein surface oxidize or reduce external electron carriers, namely *b*- and *c*- type cytochromes, ferredoxins, or flavodoxins [1–4].

The scope of electron carriers can be expanded with more elaborate architectures that, in some cases, link hydrogenase function to the balancing of cellular pyridine nucleotide pools. Examples are the [FeFe]-hydrogenase from *Syntrophomonas wolfei* (Figure 11.1d, top) [7], the tetrameric [NiFe]-hydrogenase from *Pyrococcus furiosus* (Figure 11.1b, bottom) [19], and the pentameric Hox-type [NiFe]-hydrogenases (Figure 11.1c, bottom) [20], all of which contain a catalytic flavin cofactor (FMN or FAD) that reacts with NAD(H) or NAD(P)(H).

In 2009, Schut and Adams discovered unexpected reactivity for a trimeric FMN containing [FeFe]-hydrogenase from *Thermotoga maritima* (*Tm* HydABC) [30]. Unlike seemingly similar hydrogenases, it displayed little to no activity for (i) the thermodynamically driven reduction of $NAD(P)^+$ by H_2 [8] and (ii) the thermodynamically driven production of H_2 by reduced ferredoxin [8, 30]. But, incredibly accelerated H_2 production rates were observed when both NADH and reduced ferredoxin were simultaneously used as electron donors. This was unexpected not only because two electron donors were required, but also because the production of H_2 by NADH is not thermodynamically driven. This resulted in *Tm* HydABC being classified as an electron-bifurcating hydrogenase [30], which was in accord with dissimilar enzymes that exhibited analogous, unexpected reactivity. That is, within a single enzyme complex, using a mid-potential, 2 e^- donor (acceptor) to reduce (oxidize) a thermodynamically favorable electron acceptor (donor) and a thermodynamically unfavorable acceptor (donor), with the net process being near thermoneutral [31–33]. Within the following two years, Kaster, Thauer, and coworkers supported the hypothesis that some methanogens use electron-bifurcating hydrogenases [27, 34, 35] by demonstrating the FAD containing [NiFe]-hydrogenase/heterodisulfide reductase MvhAGD-HdrABC from *Methanothermobacter marburgensis* couples H_2 oxidation to the reduction of heterodisulfide (favorable) and ferredoxin (unfavorable) [23].

In the decade and a half since these two seminal reports [23, 30], many other electron-bifurcating hydrogenases have been discovered or predicted. They mostly fall into two types based on their subunit compositions and substrates/products: (i) the FMN containing [FeFe]- and [NiFe]-HydABC type (Figure 11.1e, top and

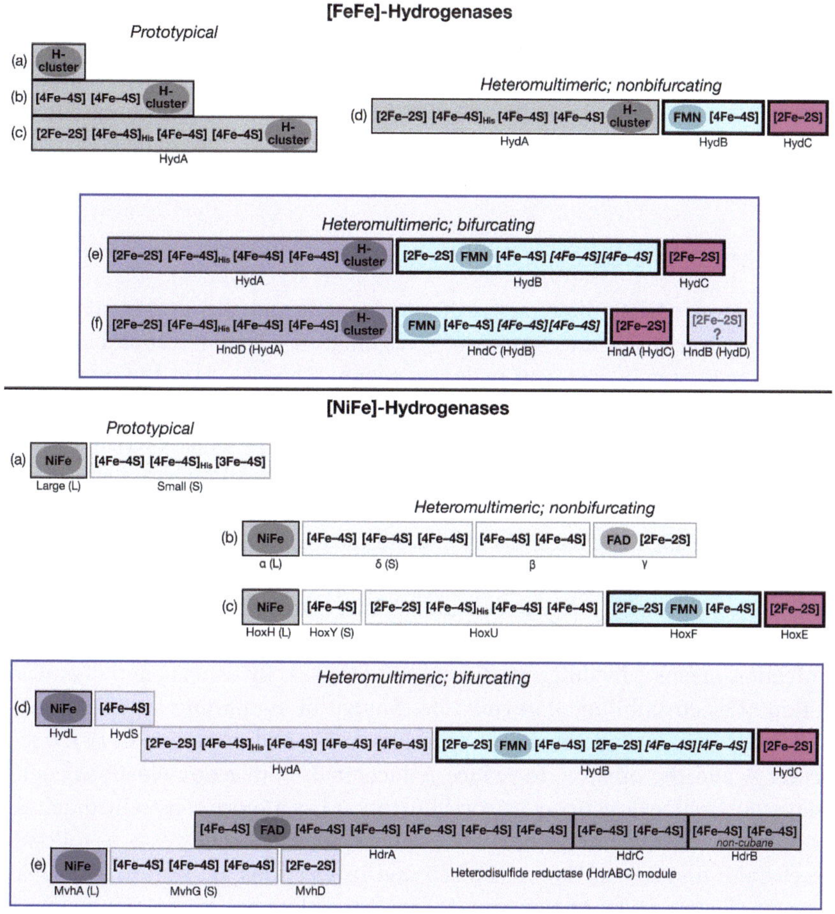

Figure 11.1 Representative subunit compositions of [FeFe]-hydrogenases (top) and [NiFe]-hydrogenases (bottom). Top: (a–c) Prototypical [FeFe]-hydrogenases [1–6]. (d) The trimeric HydABC hydrogenase from *Syntrophomonas wolfei* [7]. (e) Electron-bifurcating HydABC type [3, 5, 8–13]. *Tm* HydABC harbors an additional [2Fe–2S] cluster in the C-terminal domain of HydA [8, 14]. (f) Bifurcating HndABCD type [4, 15, 16]. Bottom: (a) Prototypical [NiFe]-hydrogenases [1–5, 17, 18]. (b) The tetrameric hydrogenase from *Pyrococcus furiosus* [2, 3, 19]. (c) Generalized pentameric Hox hydrogenase. Some Hox hydrogenases are tetrameric [1–4, 12, 20, 21]. (d) Electron-bifurcating HydABCSL type [9, 22]. (e) Generalized electron-bifurcating hydrogenase MvhAGD–heterodisulfide (Hdr) complex [23–26]. There are analogous selenocysteine containing Vhu–Hdr bifurcating hydrogenases [24, 25, 27–29]. Cofactor orderings of each subunit are based on genetic organizations, e.g. References [2–5, 8, 9, 11, 19, 24] and Uniprot code: A0A2D0TCB9. Subunit orderings are not based on genetic organizations. The italicized [4Fe–4S] [4Fe–4S] represent a seemingly important 2[4Fe–4S] ferredoxin-like domain found within the C-terminal of the HydB/HndC subunit of the bifurcating HydABC and HndABCD hydrogenases [9, 12, 15].

Figure 11.1d, bottom) that react with NAD(P)(H) and ferredoxin [3, 5, 9–13] and (ii) the FAD containing [NiFe]-MvhAGD-HdrABC type (Figure 11.1e, bottom), which react with heterodisulfide and ferredoxin [3, 5, 24, 25, 36]. There is a third type, the FMN-containing [FeFe]-HndABCD identified in *Solidesulfovibrio fructosivorans* (formerly *Desulfovibrio fructosovorans*; Figure 11.1f, top), which is rather similar to HydABC [4, 15, 16].

11.1.2 Electron Bifurcation: A Brief Overview

Enzymes often use ET to interconvert thermodynamic and chemical bond energy [37–39]. In standard ET, an electron (e^-) residing on a donor complex reduces an acceptor complex of identical or lower energy. That is, via a thermoneutral or exergonic (downhill) process. A dilemma, particularly when maximal energy conservation is essential to survival, is that excess energy from an exergonic reaction can be lost to the surroundings as heat. Additionally, an endergonic (uphill) ET process requires energy input, e.g. thermal energy [40, 41]. A well-known way to overcome such challenges is by chemiosmosis, which provides a means to conserve energy and can also provide the energy input to drive otherwise endergonic ET [42–44]. Another way to accomplish this is through electron bifurcation [43–45].

Bifurcation means "dividing into two parts" [46, 47]. By extension, a chemical definition of electron bifurcation could be thought of as splitting an e^- pair, e.g. those derived from a single chemical bond like in H_2, and using one e^- to reduce reductant A and the other e^- to reduce reductant B. Both redox reactions could be exergonic. In bioenergetics, electron bifurcation has a more constrained definition because it must be energy efficient. A conceptualization of this is as follows: a two-electron donor (e.g. H_2) is oxidized and its electrons are sent to a bifurcation center (Figure 11.2). At this center, one of the electrons is sent down an ET pathway to reduce a lower-energy acceptor (exergonic reaction). The reduction of a higher-energy acceptor by the second electron (endergonic reaction) is driven by

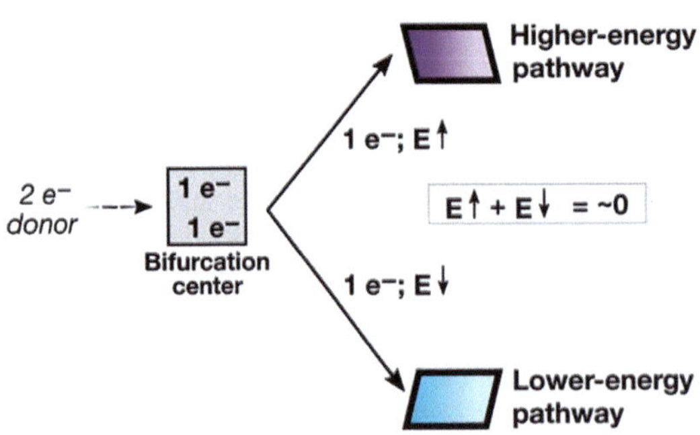

Figure 11.2 General schematic of electron bifurcation where a 2 e^- donor molecule (such as H_2) is oxidized, with both electrons ending up at a bifurcation center. At the center, one electron is sent along an ET pathway to an acceptor of lower-energy relative to the 2 e^- donor and the other electron is sent along an ET pathway to an acceptor of higher-energy relative to the 2 e^- donor. The net reaction is essentially thermoneutral.

energy leftover from the exergonic reaction. In other words, electron bifurcation thermodynamically couples an exergonic ET reaction to an endergonic ET reaction. The energy loss is minimal (Figure 11.2) and no external energy input is required. The reverse process is called electron confurcation [47–52].

There are five known families of electron-bifurcating enzymes [9, 31, 52]. The general name/description of each family, the first enzyme reported for each, the year of discovery, and the corresponding bifurcation reaction are listed in Table 11.1. Hydrogenases are found in the Bfu and the HdrABC-containing families (Table 11.1). The reduction potentials of the thermodynamically favored and thermodynamically unfavored substrates and the H_2/H^+ couple are provided in Table 11.2 at standard and physiological conditions. As further discussed in Section 11.2.2, the physiological conditions are incredibly important to keep in mind because the actual reduction potentials of the substrates depend on the ratio of oxidized vs. reduced species in the cell and on other cellular and environmental conditions like temperature, pH, and hydrogen partial pressure [48, 50, 66].

11.2 Physiology, Subunit Compositions, and Reactivity

11.2.1 Overview of Physiological Roles: Three Thematic Examples

Electron-bifurcating hydrogenases originated from evolutionary events that ensured the survival of some microorganisms in harsh, energy-limiting environments [45, 49, 67–70]. Here, we provide three examples highlighting their importance in metabolic pathways: (1) fermentation of glucose, (2) acetogenesis, and (3) methanogenesis:

(1) During glucose oxidation (part of glycolytic fermentation) NAD^+ and Fd serve as electron acceptors for glyceraldehyde-3-phosphate dehydrogenase and pyruvate:ferredoxin oxidoreductase [42, 71]. When NADH and Fd_{red} are generated during the oxidation of one glucose molecule, they must be re-oxidized to NAD^+ and Fd_{ox} for the oxidation of a second glucose molecule to occur, and so on. Some microorganisms accomplish this by generating H_2 from NADH and Fd_{red} via electron confurcation [30, 49, 50, 72], making electron confurcation an integral component of their growth and survival. The resulting H_2 yields can be essentially that of the Thauer limit [73]: 4 mol H_2 per mol glucose [74], which, on a side note, is of interest for engineering organisms with H_2 yields that meet or surpass this limit [75].

(2) Acetogenic bacteria utilize the Wood–Ljungdahl pathway for growth on single-carbon substrates (autotrophic growth). This pathway is split into two branches: the methyl branch and the carbonyl branch. The methyl branch reduces a CO_2 molecule to methyl-tetrahydrofolate (methyl-THF) via several enzymatic steps and the carbonyl branch reduces a CO_2 molecule to CO via a carbon monoxide dehydrogenase. Afterward, methyl-THF and CO are combined to form acetyl-CoA, which is then transformed into products such as acetate [70, 76].

Table 11.1 The five known classes of electron bifurcating enzymes, the first enzyme reported for each class, the corresponding year of discovery, and the reaction catalyzed. Hydrogenase relevant rows are bolded. Q = quinone.

Family/General description	Enzyme	Year	Reaction
Cytochrome bc_1 and b_6f type	Cytochrome bc_1 [31, 53]	ca. 1975	$2Q_aH_2 + Q_b + 2cyt\ c_{ox} \leftrightarrow 2Q_a + Q_bH_2 + 2cyt\ c_{red} + 2H^+$
Electron Transfer Flavoprotein (Etf/Fix)	Butyryl-CoA Dehydrogenase/Electron Transfer Flavoprotein (Bcd-EtfAB) from *Clostridium kluyveri* [32, 33] in other enzymes of this family, other high-potential acceptors substitute for crotonyl-CoA [52]	2008	$2NADH + Crotonyl\text{-}CoA + Fd_{ox} \leftrightarrow 2NAD^+ + Butyryl\text{-}CoA + Fd_{red}^{2-}$
Bfu	**[FeFe]-Hydrogenase HydABC** **from *Thermotoga maritima*** [30] in other enzymes from this family, other mid-potential donors substitute for H_2 [9]	**2009**	$\mathbf{NAD^+ + 2H_2 + 2Fd_{ox} + H^+ \leftrightarrow NADH + 4\ H^+ + 2Fd_{red}^-}$
NADH-Dependent Ferredoxin:NADP$^+$ Reductase (Nfn)	NADH-dependent ferredoxin-NADP$^+$ oxidoreductase from *Clostridium kluyveri* [54]	2010	$NAD^+ + 2\ NADPH + Fd_{ox} \leftrightarrow NADH + 2\ NADP^+ + H^+ + Fd_{red}^{2-}$
Heterodisulfide Reductase ABC (HdrABC)-containing	**[NiFe]-methyl viologen reducing hydrogenase AGD/heterodisulfide reductase ABC (MvhADG-HdrABC)** **from *Methanothermobacter marburgensis*** [23] in other enzymes of this family, other mid-potential donors substitute for H_2 [24, 25]	**2011**	$\mathbf{CoM\text{-}S\text{-}S\text{-}CoB + 2H_2 + Fd_{ox} \leftrightarrow CoM\text{-}SH + CoB\text{-}SH + 2H^+ + Fd_{red}^{2-}}$

Table 11.2 Generalized electron-bifurcating hydrogenase substrate reduction potentials.

Substrate		E'_0 (mV) generalized	E′ (approx. physiological; mV)[a)]	References
High-energy (lower reduction potential vs. mid-level)	**Ferredoxin (Fd_{ox}/Fd_{red}^-)**	−400	≤−450	[50, 55–57]
Mid-level	**H_2/H^+**	−414 (pH = 7.0)	≤−296 (for H_2 ≥10 Pa)	[58]
Low-energy (higher reduction potential vs. mid-level)	**NAD(H)**	−320	−280	[50, 59–62]
	NAD(P)H	−320	−380	[50, 60, 63]
	Heterodisulfide (CoM-S-S-CoB)	−140 to −281	n.a.[b)]	[64, 65]

a) The physiological values tabulated here are only rough guidelines because they are dependent on precise cellular conditions.
b) We have not been able to find a physiological midpoint potential estimate of heterodisulfide in methanogens, although in a recent report it was hypothesized to be near −272 mV [65].

Fd_{red} is the sole source of reducing power in the carbonyl branch. Elsewhere in the cell, ATP generation is dependent on the oxidation of Fd_{red} by a ferredoxin-NAD^+ oxidoreductase (Rnf), with the net activity of the Rnf resulting in the translocation of Na^+ or H^+ out the cell for use by membrane-bound ATPases. Fd_{red} therefore drives a significant portion of energy transfer in acetogenic cells [70, 76].

Acetogenic bacteria can grow autotrophically on $H_2 + CO_2$ [70], but for many years it was unclear how H_2 could serve as the sole energy source for driving acetogenesis because the production of Fd_{red} from H_2 (to ultimately power CO_2 reduction) is endergonic. The mystery was solved during 2012–2013 by the laboratories of Rudolph Thauer, Gerhard Gottschalk, and Volker Müller, which reported the energetic barrier could be overcome by electron bifurcation in *Moorella thermoacetica* (Thauer [72, 77]) and *Acetobacterium woodii* (Thauer, Gottschalk, Müller [78] and Müller [79]), with H_2 being used to reduce NAD^+ (favorable) and Fd_{ox} (unfavorable).

(3) Methanogens play an integral role in the global methane cycle by reducing specific carbon compounds, such as CO_2, to methane. These organisms are divided into two groups: those with and those without cytochromes [27, 34, 80]. Generally, those without cytochromes grow slower, have a broader substrate range, have a higher pH_2 threshold, and can produce more methane from $CO_2 + H_2$ than strains without cytochromes [34]. It is important to note the only species with cytochromes that can grow only on $CO_2 + H_2$ are from *Methanosarcina* [34, 81], while growth on $CO_2 + H_2$ is common for species without cytochromes [27, 34, 81].

For the most part, both groups have the same set of cytoplasmic and membrane-bound enzymes for reducing CO_2 to CH_4. Differences between the two groups include (i) how they generate Fd_{red} for driving CO_2 reduction to formyl-methanofuran in the first step of methanogenesis and (ii) how they reduce CoM-S-S-CoB to generate coenzymes CoM-SH and CoB-SH, which are required for the final steps of CH_4 generation [34, 80, 81]. Methanogens with cytochromes reduce Fd by establishing a proton gradient with a cytochrome *b*-dependent, membrane-bound hydrogenase, with the gradient ultimately driving Fd_{red} generation by an energy-converting hydrogenase. CoM-S-S-CoB gets reduced by a heterodisulfide reductase [27, 34, 50, 80, 82]. The proton gradient cannot be established in methanogens without cytochromes because the cytochrome *b*-dependent, membrane-bound hydrogenase is absent. Instead, the generation of Fd_{red} is sometimes dependent on an enzyme unique to methanogens without cytochromes: the electron-bifurcating hydrogenase–heterodisulfide reductase complex, Mvh–Hdr or Vhu–Hdr, which reduces both CoM-S-S-CoB and Fd_{ox} [23, 28]. This links the first and final steps of methanogenesis: Fd_{red} generation (first step) and coenzyme M and coenzyme B generation (final steps) [23, 27, 28, 34, 35, 49, 50, 81, 83].

11.2.2 Subunit Compositions and Reactivities

Although each type of electron-bifurcating hydrogenase reacts with NADP(H) and ferredoxin or heterodisulfide and ferredoxin, there are variations of subunit compositions (and other features) within each type. These are highlighted in Table 11.3, column 2.

Summary points, highlights, and additional points about the HydABC type are as follows:

(1) The minimal subunit composition is the HydABC trimer. However, the trimer itself is not indicative of bifurcation. For example, HydABC from *Syntrophomonas wolfei* (Figure 11.1d, top) does not bifurcate [7]. Recent biochemical and phylogenetic studies demonstrated there are at least two differences in the HydB subunit that differentiate bifurcating from nonbifurcating HydABC hydrogenases. The first is the presence of two [4Fe–4S] clusters in a ferredoxin-like C-terminal domain of bifurcating HydABC enzymes not found in the nonbifurcating ones (Figure 11.1) [9, 11, 12]. The second is the occurrence of amino acid differences near the NADH-binding site [11]. In summary, the trimer appears necessary for bifurcation, but other aspects, especially within the HydB subunit, are at play [9, 11, 12, 99].

(2) Sometimes, a HydABCD complex is observed, the HydD protein being encoded by the *hydD* gene that is usually part of the operon containing *hydC*, *hydB*, and *hyd*A [4, 5, 9, 13, 22, 79]. Functional roles of HydD in catalysis are unknown. Müller and coworkers hypothesized that it does not function in ET because it probably lacks redox cofactors [49, 79]. In fact, it may not be part of the functional enzyme and instead be involved in the maturation or regulation process [9, 13, 22]. The homologous protein in HndABCD, HndB, also has unclear roles [4, 15, 100]. Investigations honing in on the functions of HydD and HndB are needed.

Table 11.3 Electron-bifurcating hydrogenases that have been isolated and characterized biochemically and/or structurally.

Type	Enzyme source/Subunit composition	Reactivity (rates)[a),b)]
Hyd [9, 10]	***Thermotoga maritima*** MSB8 [30, 66, 84] $[(HydABC)_2]_2$ [8, 14] *Accession codes*: HydA, AAD36496.1; HydB, AAD36495.1; HydC, AAD36494.1. *PDB*: 7P92, 7P91, 7P8N, 7P5H	***Bifurcation reactions*** $\mathbf{2Fd_{red}^{-} + 3H^+ + NADH \rightarrow 2Fd_{ox} + 2\mathit{H_2} + NAD^+}$ *Report 1* [30]: (< 7 μmol H_2/min/mg)[c)], a value of 1.6 ± 0.6 μmol H_2/min/mg being reported in the main text $\mathbf{\mathit{Fd_{ox}} + 2H_2 + NAD \rightarrow 2Fd_{red}^{-} + 3H^+ + \mathit{NADH}}$ *Report 2* [84]: (16.6 ± 0.2 μmol NADH/min/mg) *Report 2*: (33.5 ± 2 μmol Fd_{red}/min/mg) ***Half-reactions*** $\mathbf{2Fd_{red}^{-} + 2H^+ \rightarrow 2Fd_{ox} + \mathit{H_2}}$ *Report 1*: (~ 0.25 μmol H_2/min/mg)[c)] $\mathbf{H^+ + NADH \rightarrow \mathit{H_2} + NAD^+}$ *Report 1*: (< 0.1 μmol H_2/min/mg) $\mathbf{2\mathit{Fd_{ox}} + H_2 \rightarrow 2Fd_{red}^{-} + 2H^+}$ *Report 2*: (< 0.1 μmol H_2/min/mg) $\mathbf{H_2 + NAD^+ \rightarrow H^+ + \mathit{NADH}}$ *Report 2*: (0.5 ± 0.1 μmol NADH/min/mg) *Conditions*: H_2 formation [30]: pH = 7.5 (KP_i) and 80 °C H_2 oxidation [84]: pH = 8.0 (KP_i) and 70 °C

(Continued)

Table 11.3 (Continued)

Type	Enzyme source/Subunit composition	Reactivity (rates)[a),b)]
	Acetobacterium woodii [78, 79] $(HydABC)_2$ [79, 85] *Accession codes*: HydA, AFA49450.1; HydB, AFA49451.1; HydC, AFA49454.1; HydD, AFA49452.1. *PDB*: 7Q4W, 7Q4V (oxidized), 8A5E (reduced) a fourth subunit of unknown function, HydD, has been reported [79] but was not seen in the cryo-EM structure [85]	***Bifurcation reaction*** $\mathbf{Fd_{ox} + 2H_2 + NAD^+ \rightarrow Fd_{red}^{2-} + 3H^+ + NADH}$ *Report 1* [79]: (3.0 Umg^{-1})[d)] *Report 2* [85]: (4.1 ± 0.7 Umg^{-1}) *U = 2 µmol electrons/min *Conditions*: pH = 8.0 (EPPS) and 30 °C (report 1) or 66 °C (report 2)
	Moorella thermoacetica [72, 77] $(HydABC)_2$ [72] *Accession codes*: HydA, WXJ90217.1; HydB, WXJ90218.1; HydC, WXJ90219.1	***Bifurcation reactions*** $\mathbf{Fd_{red}^{2-} + 3H^+ + NADH \rightarrow Fd_{ox} + 2H_2 + NAD^+}$ 7.8 Umg^{-1} (pH = 7.0)[c)] 4.6 Umg^{-1} (pH = 7.5)[c)] $\mathbf{Fd_{ox} + 2H_2 + NAD^+ \rightarrow Fd_{red}^{2-} + 3H^+ + NADH}$ 49.7 Umg^{-1} (pH = 7.0) 57.5 Umg^{-1} (pH = 7.5) ***Half-reactions*** $\mathbf{Fd_{red}^{2-} + 2H^+ \rightarrow Fd_{ox} + H_2}$ 0.2 Umg^{-1} (pH = 7.0)[c)] $\mathbf{H^+ + NADH \rightarrow H_2 + NAD^+}$ <0.01 Umg^{-1} (pH = 7.0)

	$\boldsymbol{Fd_{ox}} + \mathbf{H_2} \rightarrow \mathbf{Fd_{red}}^{2-} + \mathbf{2H^+}$ <0.01 Umg^{-1} (pH = 7.5) $\mathbf{H_2} + \mathbf{NAD^+} \rightarrow \mathbf{H^+} + \boldsymbol{NADH}$ <0.01 Umg^{-1} (pH = 7.5) * U = 2 μmol electrons/min *Conditions* [72]: pH = 7.5 or 7.0 (KP_i) and 45 °C Rates measured at 55 °C (optimal growth temperature) were noted to double, as would commonly be expected for a 10 °C increase
Clostridium autoethanogenum [86, 87] ($HytABCDE_1E_2$-FdhA)$_2$ [86] Fdh = formate dehydrogenase *Accession codes*: HytA, WP_023162857.1; HytB, WP_013237374.1; HytC, WP_013237373.1	***Bifurcation reactions*** $\mathbf{Fd_{red}}^{2-} + \mathbf{3H^+} + \mathbf{NADPH} \rightarrow \mathbf{Fd_{ox}} + \boldsymbol{2H_2} + \mathbf{NADP^+}$ 26.5 Umg^{-1} (pH = 6.0)[c]; 8.7 Umg^{-1} (pH = 7.5)[c] $\boldsymbol{Fd_{ox}} + \mathbf{2H_2} + \mathbf{NADP^+} \rightarrow \mathbf{Fd_{red}}^{2-} + \mathbf{3H^+} + \boldsymbol{NADPH}$ 32 Umg^{-1} (pH = 6.5); 29.2 Umg^{-1} (pH = 7.5) $\mathbf{CO_2} + \mathbf{Fd_{ox}} + \mathbf{H_2} + \mathbf{NADP^+} \rightarrow$ $\boldsymbol{HCOO^-} + \mathbf{Fd_{red}}^{2-} + \mathbf{NADPH}$ 40 Umg^{-1} (pH = 7.0), see reference for more details ***Half-reactions*** $\mathbf{Fd_{red}}^{2-} + \mathbf{2H^+} \rightarrow \mathbf{Fd_{ox}} + \boldsymbol{H_2}$ 0.9 Umg^{-1} (pH = 6.0**)[c] $\mathbf{2H^+} + \boldsymbol{NADPH} \rightarrow \boldsymbol{H_2} + \mathbf{NADP^+}$ < 0.1 Umg^{-1} (pH = 6.0**)

(Continued)

Table 11.3 (Continued)

Type	Enzyme source/Subunit composition	Reactivity (rates)[a),b)]
		$\mathbf{H_2 + NADP^+ \rightarrow 2H^+} + \boldsymbol{NADPH}$ 1.6 Umg^{-1} (pH = 6.5**) $\boldsymbol{Fd_{ox}} + \mathbf{H_2 \rightarrow Fd_{red}{}^{2-} + 2H^+}$ < 0.2 Umg^{-1} (pH = 6.5**) $\mathbf{HCOO^- + H^+ \rightarrow} \boldsymbol{H_2} + \mathbf{CO_2}$ 40 Umg^{-1} (pH = 6.0); 23 Umg^{-1} (pH = 7.5) $\mathbf{H_2 + CO_2 \rightarrow} \boldsymbol{HCOO^-} + \mathbf{H^+}$ 41 Umg^{-1} (pH = 7.0); 35 Umg^{-1} (pH = 7.5) *U = 2 μmol electrons/min ** we did not find clear specification of these pH values. Another option based on the reference is that these assays were conducted at pH 7.5 *Conditions* [86]: pH = 6.0 – 7.5 (KP_i) and 37 °C
	Ruminococcus albus [88] HydABC *Accession codes*: HydA, ADU23430.1; HydB, ADU23431.1; HydC, ADU23432.1	***Bifurcation reaction*** $\boldsymbol{Fd_{ox}} + \mathbf{2H_2 + NAD^+ \rightarrow Fd_{red}{}^{2-} + 3H^+} + \boldsymbol{NADH}$ 60 Umg^{-1} *U = 1 μmol/min. The stoichiometry implies U is with respect to Fd_{ox} or NAD(H), although this was not specified *Conditions* [88]: pH = 7.5 (Tris) and 37 °C

Thermoanaerobacter kivui [89] $(HydABC)_2$ [85, 89] *Accession codes*: HydA, AIS53102.1; HydB, AIS53103.1; HydC, AIS53104.1 *PDB*: 8BEW (oxidized) and 8A6T (reduced)	***Bifurcation reactions*** $\mathbf{Fd_{red}}^{2-} + \mathbf{3H^+} + \boldsymbol{NADPH} \rightarrow \boldsymbol{Fd_{ox}} + 2H_2 + \mathbf{NADP^+}$ *Report 1* [89]: (9.3 ± 2.0 Umg^{-1}); monitored H_2 production[c)] *Report 2* [57]: (25 ± 1 Umg^{-1})[c)] *Report 2***: (57.4 ± 4.9 Umg^{-1})[c)] $\boldsymbol{Fd_{ox}} + \mathbf{2H_2} + \mathbf{NADP^+} \rightarrow \mathbf{Fd_{red}}^{2-} + \mathbf{3H^+} + \boldsymbol{NADPH}$ *Report 1*: (26.8 ± 2.9 Umg^{-1})[e)] *Report 3* [85]: (20.5 ± 2.8 Umg^{-1}) ***Half-reactions*** $\boldsymbol{Fd_{ox}} + \mathbf{H_2} \rightarrow \mathbf{Fd_{red}}^{2-} + \mathbf{2H^+}$ *Report 1*: < 0.003 μmol Umg^{-1} $\mathbf{H_2} + \mathbf{NADP^+} \rightarrow \mathbf{H^+} + \boldsymbol{NADPH}$ *Report 1*: < 0.005 μmol Umg^{-1} *U = 2 μmol electrons/min ** used a ferredoxin native to *T. kivui* as opposed to one from *Clostridium pasteurianum* *Conditions*: pH = 8.0 (EPPS, report 1 and 3) or pH = 7.5 (Tris, report 2) at 66 °C (reports 1–3)

(Continued)

Table 11.3 (Continued)

Type	Enzyme source/Subunit composition	Reactivity (rates)[a),b)]
	Acetomicrobium mobile [22] $(HydABCSL)_2$ [22] *Accession codes*: HydA, 7T2R_F; HydB, 7T2R_G; HydC, 7T2R_H; HydS, 7T2R_J; HydL, 7T2R_I *PDB*: 7T2R (no FMN) and 7T30 (w/ FMN + NADH) L = large [NiFe] hydrogenase subunit S = small, Hox like [NiFe]-hydrogenase subunit [20–22]	***Bifurcation reaction*** $\mathbf{2Fd_{red}^{-} + 3H^{+} + NADH \rightarrow 2Fd_{ox} + 2\mathit{H_2} + NAD^{+}}$ (1.93 nmol H_2/min/mg)[c)] ***Half-reactions*** $\mathbf{2Fd_{red}^{-} + 2H^{+} \rightarrow 2Fd_{ox} + \mathit{H_2}}$ (0.35 nmol H_2/min/mg)[c)] $\mathbf{H^{+} + NADH \rightarrow \mathit{H_2} + NAD^{+}}$ (0.063 nmol H_2/min/mg) *Conditions* [22]: pH = 7.5 (HEPES) and 50 °C
Hnd[f)] [90]	***Solidesulfovibrio fructosivorans*** [15, 91] $(HndABCD)_2$ [15, 16] *formerly Desulfovibrio fructosivorans* *Accession codes*: HndA, EFL52768.1; HndB, EFL52769.1; HndC, EFL52770.1; HndD, EFL52771.1	***Bifurcation reaction***[g)] $\mathbf{?\mathit{Fd_{ox}} + ?H_2 + ?NAD \rightarrow ?Fd_{red}^{-} + ?H^{+} + ?\mathit{NADH}}$ *Report 1* [15]: (10 ± 1 μmol NADH/min/mg); aerobic purification *Report 2* [92]: (0.72 ± 0.3 μmol H_2/min/mg); anaerobic purification *Report 2*: (0.21 ± 0.12 μmol H_2/min/mg); aerobic purification ***Half-reaction*** $\mathbf{?H_2 + ?NAD^{+} \rightarrow ?H^{+} + ?\mathit{NADH}}$ *Report 1*: 10–20 times less than activity vs. bifurcation *Conditions*: pH = 8.0 (Tris) and 30 °C (reports 1 and 2)

Mvh-Hdr[h)] and Vhu-Hdr[h)] [24, 27]	***Methanothermobacter marburgensis*** [23, 93] (MvhAGD-HdrABC) and/or $(MvhAGD\text{-}HdrABC)_2$ [23, 94, 95] $(MvhAGDB\text{-}HdrABC\text{-}FwdABCDFG)_2$ [93] *formerly Methanobacterium thermoautotrophicum Strain Marburg* *Accession codes*: MvhA, P60227.2; MvhD, ADL59098.1; MvhG, P60239.2; HdrA, Q50756.3 MvhB = polyferredoxin Fwd = formylmethanofuran dehydrogenase	***Bifurcation Reactions*** MvhAGD-HdrABC $\mathbf{Fd_{ox} + 2H_2 + CoM\text{-}S\text{-}S\text{-}CoB \rightarrow}$ $\mathbf{Fd_{red}^{2-} + 2H^+ + CoM\text{-}SH + CoB\text{-}SH}$ (130 μmol Fd/min/mg) [23] $(MvhAGDB\text{-}HdrABC\text{-}FwdABCDFG)_2$ *Reaction Steps*: **a)** $\mathbf{polyFd_{ox} + 2H_2 + CoM\text{-}S\text{-}S\text{-}CoB \rightarrow}$ $\mathbf{polyFd_{red} + 2H^+ + CoM\text{-}SH + CoB\text{-}SH}$ **b)** $\mathbf{polyFd_{red} + 2H^+ + {}^{13}CO_2 + MFR \rightarrow polyFd_{ox}} + \boldsymbol{MFR\text{-}^{13}CHO} + \mathbf{H_2O}$ (complete ^{13}C labeling in 2 minutes) [93] MFR = methanofuran ***Half-reaction*** $\mathbf{2H_2 + CoM\text{-}S\text{-}S\text{-}CoB} \rightarrow \boldsymbol{CoM\text{-}SH + CoB\text{-}SH}$ (1.3 μmol Co(M)(B)-SH/min/mg) [94] *Conditions*: Bifurcation reaction: pH = 7.0 (KP_i) and 70 °C [23] 65 °C [93] Half-reaction: pH = 7.0 (KP_i) and 65 °C [94]

(Continued)

Table 11.3 (Continued)

Type	Enzyme source/Subunit composition	Reactivity (rates)[a),b)]
	Methanococcus maripaludis [28] $(VhuAUGD\text{-}HdrABC)_2$ or (VhuAUGD-HdrABC)(FdhAB-VhuD-HdrABC) [35] *Accession codes*: VhuA, MBG0769290.1; VhuU, CAF31249.1; VhcD, WP_011170765.1; VhuD, WP_258559266.1; VhuG, CAF31251.1; HdrA, CAF31253.1 VhuD = selenocysteine containing subunit analogous to MvhD [24, 35] VhuU = selenocysteine containing. Analogous to part of the C-terminal portion of a typical [NiFe]-hydrogenase large subunit [26, 96–98] Fdh = formate dehydrogenase	***Bifurcation reactions*** $(VhuAUGD\text{-}HdrABC)_2$ $\mathbf{\mathit{Fd}_{ox} + 2H_2 + CoM\text{-}S\text{-}S\text{-}CoB \rightarrow}$ $\mathbf{Fd_{red}{}^{2-} + 2H^+ + CoM\text{-}SH + CoB\text{-}SH}$ (41 ± 7 μmol Fd/min/mg) $\mathbf{\mathit{Fd}_{ox} + H_2 + HCOO^- + CoM\text{-}S\text{-}S\text{-}CoB \rightarrow}$ $\mathbf{Fd_{red}{}^{2-} + H^+ + CO_2 + CoM\text{-}SH + CoB\text{-}SH}$ (36 ± 9 μmol Fd/min/mg) $\mathbf{\mathit{Fd}_{ox} + 2HCOO^- + CoM\text{-}S\text{-}S\text{-}CoB \rightarrow}$ $\mathbf{Fd_{red}{}^{2-} + 2CO_2 + CoM\text{-}SH + CoB\text{-}SH}$ (10 ± 5 μmol Fd/min/mg) (VhuAUGD-HdrABC)(FdhAB-VhuD-HdrABC) $\mathbf{\mathit{Fd}_{ox} + 2H_2 + CoM\text{-}S\text{-}S\text{-}CoB \rightarrow}$ $\mathbf{Fd_{red}{}^{2-} + 2H^+ + CoM\text{-}SH + CoB\text{-}SH}$ (15 ± 5 μmol Fd/min/mg) $\mathbf{\mathit{Fd}_{ox} + H_2 + 2HCOO^- + CoM\text{-}S\text{-}S\text{-}CoB \rightarrow}$ $\mathbf{Fd_{red}{}^{2-} + H^+ + CO_2 + CoM\text{-}SH + CoB\text{-}SH}$ (35.5 ± 9 μmol Fd/min/mg)

	Fd_{ox} + 2HCOO$^-$ + CoM-S-S-CoB → **Fd_{red}^{2-} + 2CO_2 + CoM-SH + CoB-SH** (17 ± 0.5 μmol Fd/min/mg) *Conditions* [28]: pH = 7.0 (MOPS) and 30 °C
Methanothermococcus thermolithotrophicus[i)] $(MvhAGD\text{-}HdrABC)_2$ [26] *Accession codes*: MvhA, 5ODC_L; MvhD, 5ODC_J; MvhG, 5ODC_K; HdrA, 5ODC_G; HdrB, 5ODC_B; HdrC, 5ODC_C *PDB*: 5ODC (w/ FAD); 5ODI (w/ FAD and CoM-SH); 5ODR (w/ FAD and a two minute soaking time with CoM-S-S-CoB); 5ODH (w/ FAD and a three and a half minute reaction time with CoM-S-S-CoB); 5ODQ (w/ FAD and bromoethanesulfonate)	Not reported
Methanothermobacter wolfeii[i)] $(MvhAGD\text{-}HdrABC)_2$ [26] *Accession codes*: MvhA, A0A9E7UMY7; MvhD, A0A9E7UML9; MvhG, A0A9E7UMV4; HdrA, A0A9E7UMU8; HdrB, A0A9E7RTW6; HdrC, A0A9E7RVG9	Not reported

General references are provided in the first column about a given type of bifurcating hydrogenase. References in the second column next to the organism are those that originally reported bifurcation reactivity (or plausible bifurcation function) for a given enzyme, and, if available, a reference regarding physiological relevance is provided (if not already in the report about bifurcation reactivity or plausible function). References in the second column next to the subunit compositions are original reports for these compositions from biochemical analysis and/or structural information. Accession codes for select subunits and PDB codes (for those structurally characterized) are also provided.

Table 11.3 (Continued)

a) If reaction rates were not directly provided in the reference(s), then we estimated rate(s) by visual inspection of graphs in the reference(s). Errors are tabulated if they are directly provided or if they could be estimated by visual inspection of error bars from plots in the reference(s).
Units are those from the references and are listed or defined based on what was deemed most convenient for visual purposes. The reaction stoichiometries are what were determined/written in the reference(s).
The substrate(s)/product(s) measured to determine activity are indicated by *italicized* text based on the references (this is to the best of our knowledge, as sometimes it was not apparent what was being measured).

b) Turnover rates of different enzymes (and even those of the same enzyme) should only be readily compared if the conditions are identical. Extrapolation between different conditions should be made with caution, especially for hyperthermophilic organisms because they do not always display predictable trends of rates and redox cofactor reduction potentials upon going from ambient to higher temperatures [56].

c) A pyruvate ferredoxin oxidoreductase was included in these assays to continually regenerate Fd_{red}.

d) Michaelis–Menten kinetics were reported with respect to [Fd_{ox}], [NAD^+], and [FMN]. The rate we tabulated is the maximal rate from the NAD^+-based plot, which was nearly identical to that for Fd and FMN (see Reference [79] for more details).

e) The report provided a temperature and pH profile for H_2 oxidation activity. Listed here is the maximum rate observed at the pH optimum (pH = 8) and temperature optimum ($T = 66\,°C$) [89].

f) We list this separately because it does not necessarily meet the requirements to be defined as the bifurcating Hyd(Bfu)ABC-type hydrogenase [9] based on current information, e.g. cofactor composition (lack of the N-terminal [2Fe–2S] cluster of HydB) and overall truncated sequence length of HndC (the Hnd version of HydB). See also Figure 11.1e,f, top) [15, 16].

g) Reaction stoichiometries were unusual [15] and activity varied significantly because of factors such as storage/growth conditions [92] and storage time [15, 92]. Tabulated from Report 1 is the maximal rate observed (shortly after enzyme purification) [15].
Prior to investigating its electron-bifurcating capabilities, HndABCD was reported to have $NADP^+$-reducing activity [16, 258] (and not NAD^+-reducing activity [16]). In contrast, the recent reports that investigated bifurcation activity indicated it reduces NAD^+ [15, 92].

h) Formyl-methanofuran dehydrogenase (Fmd) and polyFd (MvhB; referred to in other bifurcating complexes as FmdF or FwdF [101, 112]) might bind to these enzymes even if not reported because is possible Fmd and MvhB/FmdF/FwdF are lost during purification [35, 101, 112]. This suggests high buffer concentrations used for assays (e.g. 1.6 M [23, 28]) that was recently ascribed as being necessary to facilitate the binding of external Fd to Vhu–Hdr from *M. maripaludis* could partly be a consequence of needing to use external Fd when MvhB is not present [28]. Interestingly, the bifurcation assay used for the recently reported Mvh–Hdr–MvhB–formyl-MFR dehydrogenase complex from *M. marburgensis* only used 0.8 M buffer (although it was not clear if this was the optimal concentration). It has also been reported that high buffer concentrations are needed for optimal rates of CoM-S-S-CoB reduction in the absence of Fd in *M. marburgensis* [94]. More work is needed to determine the effect of buffer concentration (and perhaps buffer type) for Mvh–Hdr and Vhu–Hdr with and without the presence of polyFd.

i) It is assumed Mvh–Hdr from *Methanothermococcus thermolithotrophicus* and *Methanothermobacter wolfeii* bifurcate electrons based on their similarity to *Methanothermobacter marburgensis*. While this is not unreasonable, reports of their ability to bifurcate electrons are needed.

(3) The enzyme from *Clostridium autoethanogenum* (encoded by *hyt* instead of *hyd* genes) is incredibly complex. HytA harbors the H-cluster and two additional [4Fe–4S] clusters. HytB and HytC are analogous to HydB and HydC. HytD has homology to HydA and contains one [2Fe–S] and two [4Fe–4S] clusters. In addition to HytABCD, the enzyme has two additional subunits with four [4Fe–4S] clusters each (HytE1 and HytE2) and a formate dehydrogenase module that can also bifurcate electrons [9, 10, 86]. Assays using either H_2 or CO_2 as the mid-potential substrate demonstrated the bifurcating hydrogenase and formate dehydrogenase activities can be independent of one another [86]. The hydrogenase's bifurcation function is balancing cellular redox levels by producing H_2 from NADPH and Fd_{red}, which would otherwise accumulate and be toxic during CO-based growth [86], and vice versa under different growth conditions [87]. It also appears to have a nonbifurcating function: oxidizing H_2 for providing electrons to the formate dehydrogenase module during CO_2 reduction to formate [86, 87].

Regarding the Mvh–Hdr class, key summary points, highlights, and additional points are as follows:

(1) For Vhu–Hdr, with the hydrogenase component being encoded by *vhu* instead of *mvh* genes, Vh(*u*) indicates the presence of selenocysteine (Sec). The VhuD subunit, which is analogous to MvhD (Figure 11.1e, bottom) contains a Cys → Sec substitution at residues 16 and 67 (MvhD numbering from *Methanothermococcus thermolithotrophicus* [26, 101], see also Section 11.3.3.1). Cys67 is a ligand to the [2Fe–2S] cluster in MvhD [26], indicating that the corresponding Sec residue in VhuD could also serve as a ligand. Residue 16 is adjacent to 67 in reported Mvh–Hdr structures (see also Section 11.4.3) [26]. The presence of Sec naturally prompts speculation about special functions of Sec(Cys)67 and 16 [26, 101], e.g. redox activity in the Sec case [102], although any special role(s) remain unclear. Also, unlike most [NiFe]-hydrogenases, which have a single subunit that contains all of the amino acid ligands of the [NiFe] site [1–4], two subunits harbor these ligands in Vhu: VhuA and VhuU [2, 27, 96–98, 103]. More so, VhuU has a Sec → Cys substitution for one of the terminal ligands to the Ni atom, meaning the hydrogenase is best described as a [NiFeSe]-hydrogenase [1, 98]. These are known to have larger catalytic rates than their [NiFe] counterparts, which can broadly be rationalized on the basis of Sec properties like a lowered pK_a and higher nucleophilicity vs. Cys [98, 102, 104, 105]. For the "simpler" dimeric [NiFeSe]-hydrogenases [1], there are evident differences vs. [NiFe]-hydrogenases regarding the mechanism of inactivation/reactivation [98, 104, 106]. Regarding turnover, a computational study identified differences in the gas channels and proton transfer pathways [107]. Nevertheless, it is still largely unclear what distinguishes [NiFeSe] vs. typical [NiFe] hydrogenases regarding H_2 oxidation/production [98, 104, 105, 108] and to the best of our knowledge there are no studies highlighting particular role(s) of the [NiFeSe] site within the context of bifurcation by Vhu–Hdr hydrogenases.
(2) In general, the diversity of HdrABC-containing complexes is thought to correlate with energy sources available to the cell [24, 35, 36]. An interesting

example of this comes from HdrABC-containing complexes isolated from *M. maripaludis* cells cultured on formate: a VhuAUGD-HdrABC homodimer, a VhuAUGD-HdrABC and FdhAB-VhuD-HdrABC heterodimer, and FdhAB-VhuD-HdrABC homodimer were isolated [29, 35]. Assays indicated the bifurcating ability of the hydrogenase and formate dehydrogenase can be independent [28]. It thus appears that the mid-potential donor modules (hydrogenase and/or formate dehydrogenase subunits) that feed electrons to the HdrABC module during bifurcation are adjusted based on precise cellular/environmental conditions, specifically, H_2 and formate concentrations.

(3) A polyferredoxin (polyFd) called MvhB is encoded in the *mvh* operon [109–111] and analogous polyFd's termed FmdF or FwdF are encoded in other HdrABC-containing bifurcating enzymes including Vhu–Hdr [35, 101, 112]. Historically, polyFd has only been observed in small amounts, at best, in purified Vhu/Hdr-HdrABC hydrogenases [23, 29, 35, 112]. Nomura, Shima, and coworkers recently isolated MvhAGD-HdrABC in a complex with MvhB and a formyl-methanofuran dehydrogenase (formyl-MFR dehydrogenase) from *M. marburgensis* [93]. The entire complex catalyzed the conversion of CO_2 and methanofuran to formyl-MFR in the presence of H_2 and CoM-S-S-CoB. No external Fd was added. However, the reduction of CO_2 to formyl-methanofuran by formyl-MFR dehydrogenase in the assay (see Section 11.2.1) needs to be powered by Fd_{red} [23, 34, 93]. This suggests that MvhB played the role of Fd in the assay and supports the notion that polyFd can play the role of Fd *in vivo* [35, 101, 112].

Table 11.3, column 3 lists *in vitro* reaction rates (if published) and, if available, half-reaction rates. The half-reactions lack either the low- or high-energy electron acceptor. In considering the reactivities, there is something we believe is important to keep in mind: electron bifurcation is defined as the thermodynamic coupling of an exergonic redox process to an endergonic redox process, but bifurcation assays do not necessarily demonstrate thermodynamic coupling. As a case point, let us consider HydABC from *T. maritima*, in which reported assays were conducted at pH 7.5 (80 °C) or pH 8.0 (70 °C) with an Fd native to *T. maritima* [30, 84]. Based on this information, reduction potentials were approximately:

- $E^{0\prime}(Tm\ Fd_{ox}/Tm\ Fd_{red})$ = −453 mV (80 °C) or −440 mV (70 °C) [56].
- $E^{0\prime}(H^+/H_2)$ = −544 mV (pH 8.0, 70 °C) for H_2 oxidation, as calculated with the Nernst equation for a saturated H_2 solution [113].

The value of $E^{0\prime}$ is not well defined during H_2 production (pH 7.5, 80 °C) because ρH_2 will start at zero and constantly change during the time course of the assay. Using the amount of H_2 produced during the experiment and the Nernst equation, $E^{0\prime} = 0\,V - [2.3(RT/F) \times pH] - [2.3(RT/2F) * \log(\rho H_2/\rho_0)]$, $E^{0\prime}$ was −413 mV at the end of the experiment [30, 113].

- $E^{0\prime}(NAD^+/NADH)$ = −415 mV, as estimated with published pH- and temperature-dependent data [60–62].

Under such laboratory conditions, H^+/H_2 are not necessarily mid-potential substrates. As emphasized by Furlan [113], this means an assay showing rate enhancements in the presence of all three substrates vs. corresponding half-reactions may not be representative of thermodynamic coupling. A review from Thauer and Buckel [50] and a report from Hunt, Carlson, and coworkers [66] are excellent references that point out how consideration of *in vivo* reduction potentials is often necessary to rationalize thermodynamic coupling. Reduction potentials under such conditions can be (see also Table 11.2)

- $E'(Fd_{ox}/Fd_{red}) \sim -500$ mV, since the ratio of Fd_{ox}/Fd_{red} may be quite low in the cell [50].
- $E'(H^+/H_2) > -420$ mV, given environmental ρH_2 levels can be rather low [50, 58].
- $E'(NAD^+/NADH) > -300$ mV, because the $NAD^+/NADH$ ratio may be quite high in the cell [50].

Regardless, laboratory assays are very informative. Indeed, Losey, McInerney, and coworkers noted that for a HydABC hydrogenase from *Syntrophomonas wolfei* and a similar hydrogenase from *Syntrophus aciditrophicus*, the ability to bifurcate electrons would not be advantageous to metabolism, and *in vitro* assays correspondingly showed these hydrogenases do not have enhanced reactivity in the presence of a third substrate [7, 11]. Thus, an observation of enhanced reactivity in the presence of a third substrate may support the capability and the importance of a hydrogenase to thermodynamically couple reactions *in vivo*.

On another note, there are numerous enzymes with the HydABC architecture that may have the ability to bifurcation electrons [5, 9, 11, 13], two examples being the [FeFe]-hydrogenases from *Caldanaerobacter subterraneus* [5, 11, 30, 114, 115], of which the ability to bifurcate electrons was largely implied from *in vivo* studies using a genetically engineered strain of *Escherichia coli* [116], and *Caldicellulosiruptor bescii* [5, 11, 13, 115, 117]. There are also putative bifurcating hydrogenases in methanogens [5, 25, 118]. Additionally, assays are still needed to verify the bifurcation capability of the hydrogenases used for the Mvh–Hdr structural report (*M. thermolithotrophicus and M. wolfeii*, Table 11.3) [25, 26].

11.3 Structural and Biophysical Characterizations

This section summarizes the progress that has been made toward understanding the structural and biophysical properties of electron-bifurcating hydrogenases. We also utilized this structural information to derive comparisons to hydrogenases that remain structurally uncharacterized. This approach enables the interpretation of the spectral properties of enzymes that lack structures and helps identify sequence determinants that may control biochemical reactivity and substrate preferences.

11.3.1 Structures and Cofactors of Hyd-type Electron-Bifurcating Hydrogenase

Single-particle cryo-EM methods [119] were used to solve atomic-resolution structures of HydABC from *A. woodii* [85], *T. kivui* [85], and HydABCSL from *A. mobile* [22]. The oligomeric structures are comprised of protomer units that interact to form dimers or tetramers (see Table 11.3). The structures for HydABC and HydABCSL protomers are shown in Figure 11.3 to illustrate the organization of the hydrogenase catalytic sites (the H-cluster and [NiFe] cofactor, respectively), Fe–S clusters, and the FMN cofactor.

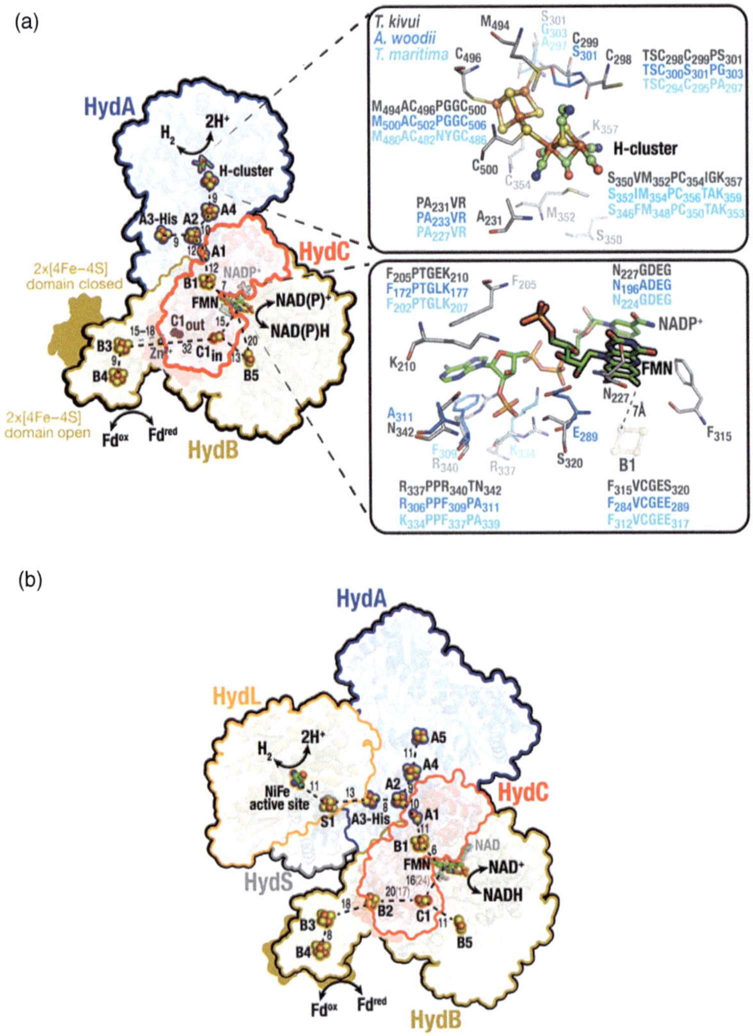

11.3.1.1 HydA and HydSL

Generally speaking, cofactors housed in HydA for bifurcating [FeFe]-hydrogenases include the H-cluster (hydrogenase active site) [1, 6, 104], a [2Fe–2S] cluster (A1), two [4Fe–4S] clusters (A2 and A4), and a noncanonical 3Cys1His ligated [4Fe–4S] cluster (A3) (numbers based on the HydABCSL assignments [9, 22, 122], see Table 11.4). Collectively, these form a "Y" junction cofactor arrangement (Figure 11.3a) [6, 12, 123]. The pathway from the H-cluster to the flavin site in HydB is proposed to be H-cluster → A4→ A2 → A1 (Figure 11.3a, see also Section 11.4.2). A3 appears to be off pathway and its mechanistic function(s), if any, are unclear [14, 85]. We note that Furlan, Chongdar, Blaza, Birrell, and coworkers proposed it has an interprotomer ET function [14].

An exception to the composition of the subunit housing the H-cluster is observed for HytABCDE$_1$E$_2$-FdhA from *C. autoethanogenum* (Table 11.4) HytA contains the H-cluster-binding motifs but lacks the A1, A2, A3, and A4 cluster-binding motifs. Instead, it contains two overlapping cysteine-rich motifs, suggestive of binding two [4Fe–4S] clusters, that are located upstream of the H-cluster-binding motif (Table 11.4) [86]. This illustrates how HytABCDE$_1$E$_2$-FdhA has cofactor compositions/arrangements in HytA that differ from other Hyd-type bifurcating hydrogenases [10], which could be related to its function(s) [86, 87] (see also Section 11.2.2).

In the [NiFe]-hydrogenase HydABCSL (Figure 11.3b), there are differences in the compositions of clusters and the ET pathway to HydB. Specifically, the H-cluster is

Figure 11.3 (a) Subunit organization and cluster/cofactor arrangement of HydABC from *T. kivui* (PDB 8A6T). Alternate conformations of HydC (C1$_{in}$ from *T. kivui* under reducing conditions (PDB 8A6T), C1$_{out}$ from *A. woodii* under oxidizing conditions (PDB 7Q4W) and the 2x[4Fe–4S] ferredoxin C-terminal domain of HydB (closed/open from *T. maritima*) are depicted in solid color shades. Fe–S cluster numbering based on *A. mobile* HydABCSL nomenclature (see Tables 11.4 and 11.5). Edge-to-edge distances (dotted lines) are indicated in Angstroms (Source: Adapted from Refs. [14, 85]). Right: H-cluster protein environment (top) and NADP(H)-binding pocket (bottom) of HydABC from *T. kivui* depicting the variation in surrounding residues and corresponding primary sequence motifs of HydABC from *A. woodii* (PDB 8A5E) and *T. maritima* (PDB 7P5H). Diiron site of the H-cluster cofactor depicted from PDB 6N59. Coloring scheme: Fe atoms, rust; S atoms, yellow-orange; C atoms of H-cluster, FMN, and NADP, green; C atoms of amino acid residues, gray (*T. kivui*), light-blue (*A. woodii*), or cyan (*T. maritima*); N atoms, blue; O atoms, red. (b) Subunit arrangement and cofactor content of [NiFe]-hydrogenase HydABCSL from *A. mobile*. Ribbon structure, cofactors, and edge-to-edge distances are indicated in Angstroms for the flavin-free state (PDB 7T2R). For HydB, the C-terminal and N-terminal domains were not available in the PDB file, so the B3, B4, and B5 clusters are depicted by superimposition of the homologous HydB subunit of HydABC from *T. kivui*. FMN and NAD are depicted from the FMN and NADH bound structure of *A. mobile* (PDB 7T30). Darker color shades of HydC and HydB subunits indicate conformational flexibility in the respective ferredoxin-like C-terminal domains. The distances in gray indicate the FMN + NAD(H) structure state based on Reference [22]. Coloring scheme: Fe atoms, rust; Ni atom, dark cyan; S atoms, yellow-orange; C atoms, green; N atoms, blue; O atoms, red. Fe–S cofactor numbering is according to HydABCSL nomenclature (see Tables 11.4 and 11.5).

Table 11.4 Structurally derived binding motifs of HydA cofactors.

Cofactor	[1]Source and cofactor name			Motif
	A. mobile	*T. maritima*	*A. woodii*	
[2][2Fe–2S]	A1	A1	A4	LCYL X_{6-11} GACRhCIVE X_{7-9} SCs
[2][4Fe–4S]	A2	A2	A2	KCILCtRCV X_{45} CP
[2][4Fe–4S]	A4	A3	A3	sC X_{32} CltCGQCl
[2][4Fe–4S]$_{His}$	A3	A4	A5	HptcChsCs+stsC
[2]H-cluster	N.A.	[4Fe–4S]$_H$	A1	PAVR X_{62-68} TSCsPu X_{48} SIMPCTuK $X_{126-140}$ MuCPGGC
[2][2Fe–2S]	N.A.	A5	N.A.	CLGTSC X_{31} CVENC
[3][4Fe–4S]	A5	N.A.	N.A.	VCPLCAVGCK X_{26} CH
[4]N-terminal [4Fe–4S], [4Fe–4S]			[1]*C. autoeth*	CTGCRRCADVCPV X_{15} EKCVMCGQCIQVC

C or C, coordinating cysteine; X_n, any amino acid; s, small; t, turnlike; l, aliphatic; p, polar; c, charged; h, hydrophobic; +, positive; u, tiny.

[1]*A. mobile* HydABCSL [22]; *T. maritima* (*Tm*) HydABC [85]; *A. woodii* HydABC; *C. autoethanolegenum* (abbreviated as *C. autoeth*) HytABCDE$_1$E$_2$-Fdh [86]. [2–4]Motif, residue numbers: [2]A1, 33–61; A2, 142–197; A3, 152–193; A4, 92–105; H-cluster, 226–486; A5, 575–616 (accession AAD36496.1) [8, 120]. [3]A5, 228-265 (accession 7T2R_F) [22]. [4]N-terminal [4Fe–4S], [4Fe–4S], 15–55, (accession WP_023162857.1) [8]. Alignments prepared in Clustal Omega and visualized in MView using Clustal coloring [121].

replaced by an additional [4Fe–4S] cluster called A5. The hydrogenase active site is found in HydL and is linked to the HydA module via HydS. Specifically, the HydS module exchanges electrons with the A3 cluster, with the overall ET pathway to HydB being HydSL → A3 → A2 → A1 (Figure 11.3b, see also Section 11.4.2) [22].

11.3.1.2 HydB

The HydB subunits of HydABC and HydABCSL contain one FMN, one [2Fe–2S] cluster, three [4Fe–4S] clusters, and an additional cofactor that is either a Zn^{2+} site or an additional [2Fe–2S] cluster [14, 22, 85] (Figure 11.3, Table 11.5). Although the additional [2Fe–2S] cluster could function as a redox cofactor [9, 22], Zn^{2+} sites (with zinc being redox inactive) are not known to function as ET cofactors [124, 125]. Collectively, the cofactors enable HydB to link electrons from H_2 activation at the H-cluster or the [NiFe] cofactor to (i) $NAD(P)^+$ reduction at the FMN site and (ii) Fd reduction. This places HydB as having a central role in the capability of Hyds to bifurcate electrons.

One observation we can make from amino acid motifs and the structures regards the substrate preference for NAD(H) vs. NADP(H) at the FMN site. As noted in the motif comparisons in Table 11.5, there are three positions in the FMN:NAD(P)(H)-binding region that change the occupancy and polarity of the NAD(P)(H)-binding pocket (Figure 11.3a). For example, the HydB residues Glu289 and Phe309 of *A. woodii* and Glu317 and Phe337 of *T. maritima*, with both enzymes having a substrate preference for NAD(H), are substituted for Ser320 and Arg340 in HydB of *T. kivui* and Ser288 and His308 in *C. autoethanogenum*, which are enzymes

Table 11.5 Structurally derived binding motifs of HydB, HydC, and HydS cofactors.

HydB Cofactor	**[1]Source and cofactor name**			**Motif**
	A. mobile	***T. maritima***	***A. woodii***	
[2][2Fe–2S]	B5	B1	B2	CsGTGC X_{31-45} ChGhC
[2]Flavin	FMN	FMN	FMN	FPTGhKWtF/T,G X_{13-15} NuDEG X_{79} GAGAFVCGEE/STAL X_{13} +PPF/R,HPA X_{87-88} GGPSG
[2][4Fe–4S]	B1	B2	B1	CGKCsPCR.G X_{33} suLCuLG
[2]Zn	N.A.	Zn	Zn	CMV- X_{34} TpG X_{45} AHI+-K+CPAtlC
[3][2Fe–2S]	B2	N.A.	N.A.	CMVD X_{34} CNG X_{45} AHIKDKRCPAKIC
[2][4Fe–4S]	B3	B4	B3	CpGCGhCs+ X_{31} C.F
[2][4Fe–4S]	B4	B3	B4	CPs X_{15} pKCIKCGsC
HydC				
[4][2Fe–2S]	C1	C1	C1	ChGTAC X_{35} ClGuC
HydS				
[4][4Fe–4S]	S1	N.A.	N.A.	ACAGCH X_{67} DCA X_{60} PGCPP

C, coordinating cysteine. In FMN motifs; at the F/T,G position, F is conserved in Hyd's with a preference for NAD(H); T or G is conserved in Hyd's with a preference for NADP(H); at the E/S position, E is conserved in Hyd's with a preference for NAD(H); S is conserved in Hyd's with a preference for NADP(H); at the F/R,H position, F is conserved in Hyd's with a preference for NAD(H); R or H is conserved in Hyd's with a preference for NADP(H). X_n, any amino acid; h, hydrophobic; t, turnlike; a, aromatic; s, small; –, acidic; l, aliphatic; p, polar; u, tiny; +, positive. N.A., not assigned.

[1]*A. mobile* HydABCSL [22]; *T. maritima* HydABC [14]; *A. woodii* HydABC [85]. [2–4]Motif, residue numbers. [2]B1, 31-82; FMN, 202-431; B2, 485-534; Zn, 468-565; B4, 578-620; B3, 588-614 (accession AAD36495.1) [8, 120]. [3]B2, 438-536 (accession 7T2R_G) [22]. [4]HydC [2Fe–2S], 81-126; HydS, 14-149 (accession AAD36494.1) [22]. Alignments prepared in Clustal Omega and visualized in MView using Clustal coloring [121].

that prefer for NAD(P)(H). The substitution of the Glu/Phe pair with Ser/Arg creates favorable sterics and charge contacts for the phosphate group of NAD(P)(H) (see Figure 11.3a). Altering the substrate preference in NAD-dependent *Rhodobacter capsulatus* formate dehydrogenase has been demonstrated by structurally guided engineering of substitutions to the NAD/FMN-binding site [126], which included the same residue pair identified in the Hyds. In *R. capsulatus*, the Glu-to-Leu and Gly-to-Arg substitutions that corresponded to HydB residues (i.e. Glu289 and Phe309 of *A. woodii* HydB) resulted in variants that gained reactivity with NADP(H), consistent with a role of these residues for controlling the NAD(P)(H) preferences in bifurcating Hyds.

11.3.1.3 HydC

HydC subunits encode a highly conserved motif with cysteines that coordinate a [2Fe–2S] cluster (C1, Table 11.5). HydC has been implicated as having a key role in regulating electron flow. Notably, structural data [22, 85] and calculations [85] indicated the distances from C1 to B2, C1 to B3, and C1 to the FMN site could fluctuate during turnover, which may be important for ET (see Section 11.4.2).

11.3.2 Spectroscopy of HydABC and HndABCD

The most extensive spectroscopic characterization of the HydABC bifurcating hydrogenases has been on *Tm* HyABC. The HndABCD from *S. fructosivorans* has also undergo quite a bit of spectroscopic characterization, and we have included Hnd here for clarity because it harbors many of the same clusters as the Hyds (see Figure 11.1e,f, top) [4, 15, 16].

11.3.2.1 EPR Spectroscopy of the Fe–S Clusters in *Tm* HydABC

A powerful technique for understanding the electronic properties of Fe–S clusters is electron paramagnetic resonance (EPR) spectroscopy [127–129]. The presence of numerous clusters in *Tm* HydABC complicates the assignment of EPR signals to individual clusters [120, 130], but studies with isolated subunits have provided some resolution in the assignments. Likewise, temperature studies take advantage of the differential relaxation behavior of $[2Fe–2S]^+$ and $[4Fe–4S]^+$ clusters to further distinguish signals from one another, with lifetime broadening of the faster-relaxing $[4Fe–4S]^+$ signals resulting in relatively clean spectra of $[2Fe–2S]^+$ cluster signals at higher temperatures [128, 129, 131]. While this has resulted in the identification of signals that can be assigned to the $[2Fe–2S]^+$ clusters for a given subunit, some uncertainty remains in deciphering one of the $[2Fe–2S]^+$ signals in *Tm* HydA and rigorous descriptions of the $[4Fe–4S]^+$ signals for both HydA and HydB are lacking. Future work incorporating more data simulation should improve the assessment of these signals. The addition of multifrequency EPR analyses, not yet tested, would greatly aid in deconvoluting overlapping cluster signals and assessing the origin of features arising from spin–spin interactions [132–134].

At temperatures >35 K, the spectra of sodium dithionite (NaDT) reduced *Tm* HydA display a $[2Fe–2S]^+$ cluster signal at $g = [2.004, 1.955, 1.923]$ (Table 11.6). A nearly identical signal was observed in samples of apo-*Tm* HydA, which lack the H-cluster's diiron site [84]. This rhombic signal was also observed in the 35 K EPR spectrum of the isolated *Tm* HydA subunit reduced under H_2 at 50 °C (either with or without the addition of NaDT) [8].

Along with the prevalent signal described above, a shoulder feature at $g \sim 2.02$ was observed in the NaDT reduced spectra of *Tm* HydA and apo-*Tm* HydA that was further accompanied by additional intensity around $g \sim 1.94$ in the apo spectrum (Table 11.6) [84]. These features were also evident in the EPR spectrum of H_2 reduced *Tm* HydABC, but were poorly resolved in the corresponding spectrum of the H_2 reduced HydA subunit [8]. This axial-type signal was hypothesized to correspond to the nonconserved, C-terminal [2Fe–2S] cluster of *Tm* HydA (assigned as A5, Table 11.4) [8, 14].

At lower temperatures (9–10 K) a more complex EPR spectrum was observed for *Tm* HydA, which can be attributed to signals from the three [4Fe–4S] clusters, A2–A4

Table 11.6 EPR signals of the HydABC [FeFe]-hydrogenases.

System	Cluster type	*g*-values	Collection parameters	Reference
Tm HydA[a)]	[2Fe–2S]	2.00, (1.95–1.96), 1.92 (2.02–2.03), 1.94	T = 35–40 K; P = 1 mW	[8, 84]
	[4Fe–4S][b)]	2.00, 1.9x, 1.92 >2.04; ~1.88	T = 9–10 K; P = 0.01 mW	[8, 84]
Tm HydB[c)]	[2Fe–2S]	2.00, 1.95, 1.92	T = 40 K; P = 1 mW	[8, 14]
	[4Fe–4S]	2.04, 1.91, 1.89	T = 9–10 K; P = 0.01 mW	
Tm HydC	[2Fe–2S]	2.005, 1.950, 1.919[d)]	T = 40 K; P = 0.01 mW	[135]
		2.000, 1.949, 1.914[e)]	T = 55 K; P = 1 mW	[136]
Sf HndA (HydC)[f)]	[2Fe–2S]	2.000, 1.950, 1.915	T = 80 K; P = 4 mW	[100, 137]
Sf HndD (HydA)	[2Fe–2S]	2.054, 1.961, 1.901	T = 20 K; P = 4 mW	[100]

T, temperature; P, microwave power.

a) Reduced with either (i) 10 mM NaDT or (ii) H_2 in the presence of 2 mM NaDT at pH 8.
b) Approximate *g*-values for a rhombic signal associated with [4Fe–4S] clusters and wing features associated with spin–spin interactions.
c) Reduced with either (i) 10 mM NaDT or (ii) 2 mM NaDT at pH 8.
d) Reduced with 10 mM NaDT at pH 8, *g*-values from data simulations.
e) Reduced with 4 mM NaDT and at pH 6. In the sample prepared at pH 10, an additional feature at g = 1.977 was observed.
f) Reduced with NaDT at pH 8.

(Table 11.4, Figure 11.3). These can be identified as a rhombic signal along with other signals that include broad spectral wings associated with spin–spin interactions between paramagnetic centers (Table 11.6) [133, 138].

The site-differentiated [4Fe–4S] cluster of HydA (A3, Table 11.4) is also found in [FeFe]-hydrogenases from *Clostridium pasteurianum* (*Cp*I) and *Clostridium acetobutylicum* (*Ca*I), which are homologous to the HydA module in bifurcating hydrogenases [1, 2, 6]. Although any functional relevance of the A3 cluster for ET in *Tm* HydABC and other bifurcating [FeFe]-hydrogenases is unclear [14, 85], it does have a functional relevance in [NiFe]-HydABCSL [22] and other Bfu family members [9, 139]. Low-field EPR data collected on reduced *Cp*I identified features at $g = 5.80$ and $g = 4.76$ that were assigned to an $S = 3/2$ spin system of unknown origin [140]. Later work on a truncated form of *Ca*I, which contained only the site-differentiated [4Fe–4S] cluster and a canonical [2Fe–2S] cluster, identified similar EPR signals at $g = 5.6$ and $g = 5.1$ that were assigned to the $\pm 3/2$ doublets of $S = 3/2$ and $S = 7/2$ spin systems, respectively [127, 141]. Valence trapping [127, 128, 142] of canonical $[2Fe–2S]^+$ clusters makes their access to higher-spin states very unlikely, and changes to the low-field signals were observed when the noncanonical histidine ligand was mutated to a cysteine [141]. Thus, the low-field signals in truncated *Ca*I were assigned to the [4Fe–4S] cluster with noncanonical 3Cys1His ligation [141].

Low-field EPR data was reported in the original work on *Tm* HydABC from Juszczak, Aono, and Adams. In this study, several resonances were observed that indicated the presence of cluster(s) with half-integer, $S > 1/2$ spin states. These included features at g ~5.7 and 10.3 that the authors attributed to $S = 3/2$ and $S = 7/2$ forms of a $[4Fe{-}4S]^+$ cluster [120]. Therefore, based on similarity to the low-field EPR signals of *Cp*I/*Ca*I, it appears that the low-field resonances observed in *Tm* HydABC can partially be assigned to HydA's site-differentiated cluster (A3, Table 11.4). However, the presence of additional, unassigned low-field resonances in the *Tm* HydABC spectra also suggests the presence of other cluster(s) with higher spin states. Notably, canonically ligated $[4Fe{-}4S]^+$ clusters can give rise to $S > 1/2$ spin states [128, 143–145]. Low-field EPR studies on isolated subunits would be helpful in addressing origin(s) of these additional low-field resonances.

When isolated HydB is reduced with NaDT, the EPR spectrum collected at $T = 35$ K displays a single rhombic signal, in agreement with the presence of a [2Fe–2S] cluster in this subunit (Table 11.6) [8, 14]. At lower temperatures ($T = 9$–10 K), contributions from the three $[4Fe{-}4S]^+$ clusters of HydB become apparent from additional spectral complexity, with features identified at g ~ [2.04, 1.97, 1.89] (Table 11.6) [8, 14]. The HydC subunit includes the binding motif for a [2Fe–2S] cluster (Table 11.5), which has been characterized using EPR (or electrochemical) methods over a wide range of pH values [135, 136]. Electrochemical measurements have also been conducted over a broad temperature range [136]. The $[2Fe{-}2S]^+$ cluster's rhombic EPR signal was characterized as g = [2.005 1.950 1.919].

11.3.2.2 EPR Spectroscopy of HndABCD from *S. fructosivorans*

The only other EPR data available for electron-bifurcating [FeFe]-hydrogenases comes from earlier work on the [FeFe]-HndABCD hydrogenase from *S. fructosivorans* (Figure 11.1f, top) [100, 137]. EPR spectra of isolated HndA (homologous to HydC) and of HndD (homologous to HydA) showed signals with temperature-dependent behavior and g-values consistent with [2Fe–2S] clusters (Table 11.6) [100, 128]. Interestingly, the [2Fe–2S] signal observed for HndA ($g = 2.000$ 1.950 1.915) was observable up to temperatures >150 K. Detailed temperature-dependent EPR analysis by the authors determined the antiferromagnetic exchange interaction between the two irons was defined by a particularly strong exchange coupling constant [127, 128, 150, 151] of $|J| = 275 \pm 15\,\text{cm}^{-1}$. In contrast to the cluster of HndA, the [2Fe–2S] cluster in HndD was undetectable by 100 K and the two irons were measured to have an antiferromagnetic exchange interaction of $|J| = 90 \pm 10\,\text{cm}^{-1}$ [100]. The different values of $|J|$ begs the question: what are the functional implications? It would be interesting to know how the analogous cluster in HydA (with respect to HndD) and in HydC (with respect to HndA) behave.

11.3.2.3 EPR and IR Spectroscopy of the H-Cluster in [FeFe]–HydABC

EPR and infrared (IR) spectroscopy are powerful techniques for studying the electronic properties of the H-cluster. EPR reports on formal metal oxidation states, whereas IR vibrations of the carbon monoxide (CO) and cyanide ligands

Table 11.7 The stretching frequencies of H-cluster signals from room temperature FTIR.

Signals	H_{ox}		H_{red}		$H_{red}H^+$		H_{Hyd}	
Frequency (cm^{-1}) **(Δcm^{-1} from H_{ox})**[a)]	***t*-CO**	***μ*-CO**	***t*-CO**	***μ*-CO**	***t*-CO**	***μ*-CO**	***t*-CO**	***μ*-CO**
T. maritima[b)]	1964 1939	1802	n.d.[c)]	n.d.	1956 1919 (8) (20)	1887 (-85)	n.d.	n.d.
A. woodii	1967 1942	1802	1965 1917 (-2) (-25)	1896 (-94)	n.d.	n.d.	n.d.	n.d.
T. kivui[d)]	1963 1938	1804	1961 1917 (-2) (-21)	1888 (-84)	n.d.	n.d.	1985 1961 (22) (23)	1858 (54)
C. reinhardtii[e)]	1964 1940	1800	1935 1891 (-29) (-49)	1793 (-7)	1968 1914 (-4) (26)	1891 (-91)	1978 1960 (14) (20)	1860 (60)

μ-CO corresponds to a bridging CO ligand of the diiron site and *t*-CO corresponds to a terminal CO ligand of the diiron site.

a) The stretching frequencies are reported for *t*-CO and *μ*-CO bands. The frequency difference for each state (Δ) vs. H_{ox} is shown in parenthesis.
b) We have used the interpretation that the H_{sred} state assigned in a pre-steady state report [146] is the $H_{red}H^+$ state [84].
c) n.d., not determined.
d) Signal for H_{hyd} obtained by treatment with H_2 at pH 4 [85].
e) The H_{ox} and H_{red} signals from Sialkov and Lubitz [147]. The $H_{red}H^+$ signal is from Sommer and Lubitz [148]. The H_{hyd} signal is from Winkler and Happe [149].

provide information about electron delocalization throughout the H-cluster and indirect information about protonation states (we only discuss the CO vibrations here) [104, 152]. The oxidized (resting state) of the H-cluster, called H_{ox} [104], has been examined in *Tm* HydABC and isolated *Tm* HydA with Fourier transform infrared (FTIR) and EPR spectroscopy [84, 146] and in *Tm* HydABC with transient absorption IR spectroscopy [146] (Table 11.7). Essentially, no difference was present in the H_{ox} EPR signal from the full complex or isolated HydA, with the latter simulated at $g = [2.103, 2.045, 1.999]$ [84]. This signal is almost identical to the H_{ox} species of a putative bifurcating [FeFe]-hydrogenase from *C. subterraneus* [114] and the H_{ox} signal from other nonbifurcating [FeFe]-hydrogenases such as HydA1 from *Chlamydomonas reinhardtii* (*Cr*) [104, 153]. Likewise, the vibrational bands associated with the H_{ox} state are almost identical between *Tm* HydA(BC) and *Cr* HydA1 (Table 11.7).

The IR bands of *Tm* HydA(BC) included features assigned to the protonated, one-electron reduced state of the H-cluster ($H_{red}H^+$, Table 11.7) [84, 146] that are again similar to those of *Cr* HydA1. HydABC from *T. kivui* and *A. woodii* have also been studied by IR and displayed bands assigned to H_{ox} and H_{red} (one-electron reduced state that lacks a proton) (Table 11.7) [85, 148]. As with *T. maritima*, the H_{ox} vibrational bands of *T. kivui* and *A. woodii* aligned with studies of other [FeFe]-hydrogenases. While the H_{red} vibrational bands of *T. kivui* and *A. woodii* are very similar, they are blueshifted from typical H_{red} bands, e.g. that of *Cr*HydA1, especially regarding the bridging CO band (Table 11.7). A two-electron reduced, hydride-bound state (H_{hyd}) of the H-cluster has so far only been identified in *T. kivui* (Table 11.7) [85, 149]. In general, no features have been definitively reported for the other two-electron reduced state of the H-cluster (H_{sred}) which lacks the hydride [104, 148, 149, 154], highlighting the need for more studies of complete active site turnover [146] that also expand to additional bifurcating [FeFe]-hydrogenases. There are currently no reported IR or EPR studies of the [NiFe] cofactor, or even EPR studies of the Fe-S clusters, in [NiFe]–HydABCSL.

11.3.3 Structures and Cofactors of MvhAGD-HdrABC

The crystal structures of Mvh–Hdr from *M. thermolithotrophicus* (Figure 11.4) showed a complex arrangement of redox cofactors that link H_2 activation to Fd and CoM-S-S-CoB reduction. The [NiFe] cofactor and the MG1–MG3 clusters form an ET relay typical of many [NiFe]-hydrogenases [1, 2, 17, 18, 26, 104]. MG3 is located at the interface of MvhG–HdrA and is in close proximity to HA1. From a distance perspective, this results in a straightforward ET route to MvhD's [2Fe–2S] cluster, which is called MD (Figure 11.4). Large distances to other redox cofactors make the ET route(s) beyond MD unclear, and it has been proposed that conformational changes may be involved in tuning ET pathways [26], similar to the Hyds (Section 11.4).

The binding motifs of the [NiFe] cofactor, MG1-MG3, and MD enables structurally informed alignments to the biochemically characterized complexes from *M. marburgensis* and *M. maripaludis* (Tables 11.8 and 11.9). Except for MvhD/VhuD,

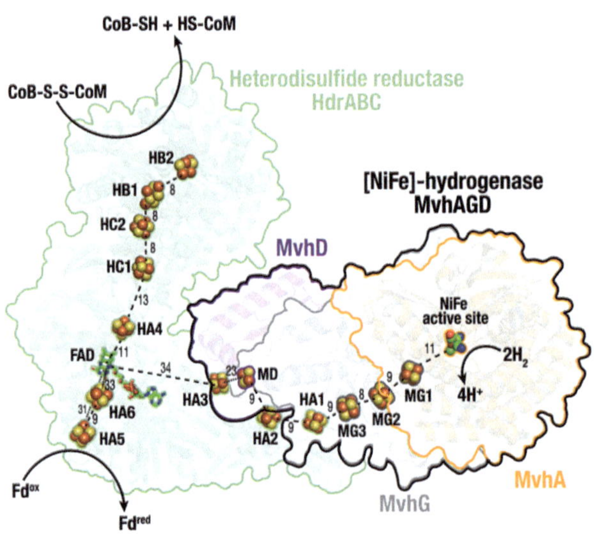

Figure 11.4 Structural depiction of the subunit arrangement and cofactor content of the heterodisulfide reductase [NiFe]-hydrogenase MvhAGD-HdrABC protomer from *M. thermolithotrophicus* (PDB 5ODC) [26]. Edge-to-edge distances indicated in Angstroms (Source: Adapted from Wagner et al. [26]). Coloring scheme: Fe atoms, rust; Ni atom, dark cyan; S atoms, yellow-orange; C atoms, green; N atoms, blue; O atoms, red.

Table 11.8 Structurally derived binding motifs of MvhAGD and VhuAUGD cofactors.

Subunit	Cofactor	Name	Motif
[1]MvhG	[4Fe–4S]	MG1	sGCSGCHlS X_{64-66} oCAs X_{59} PGCPP
	[4Fe–4S]	MG2	CGA.CPS.shPCpGC
	[4Fe–4S]	MG3	C-.CsR X_{21-24} hCLh.QGhhChG
C, coordinating cysteine; s, small; l, aliphatic; o, alcohol; h, hydrophobic; p, polar; –, acidic; X_n, any amino acid.			
[2]MvhD	[2Fe–2S]	MD1	CCNWC X_{21} IRVMCSG X_{21} GCIhGDCHY
[2]VhuD	[2Fe–2S]	MD	VCYQU X_{21} IRVPCTG X_{21} GUKPNECAF
C, coordinating cysteine; U, selenocysteine; h, hydrophobic; X_n, any amino acid.			
[3]MvhA	[NiFe]	[NiFe]	lEGH X_{23} RGFEpF X_{13-14} PRICGICQssH X_{34-37}
[3]HydL			SHuLHFYhLAAPD $X_{210-212}$ paARLIE X_{36-37} APRGTLhH
			X_{59} DPChSCAsH
[4]VhuA	[NiFeSe]	[NiFeSe]	lEGH X_{23} RGFEpF X_{13-14} PRICGICQssH X_{34-37}
			SHuLHFYhLAAPD $X_{210-212}$ paARLIE X_{36-37} APRGTLhH
[4]VhuU SeCys	[NiFeSe]	[NiFeSe]	DPUhSCAsH

C, coordinating cysteine; U, coordinating selenocysteine; s, small; l, aliphatic; o, alcohol; h, hydrophobic; p, polar; –, acidic; X_n, any amino acid; u, tiny; a, aromatic.

[1–4]Motif residue numbers. [1]MG1, 9–151; MG2, 224–243; MG3, 176–217 (accession 5ODC_K) [26]. [2]MD1, 12–74 (accession 5ODC_J) [26] and MD, 10–72 (ref accession WP_258559266.1) [95, 155]. [2]VhuD, 10–199 (accession WP_258559266.1) [95, 155]. [3]MvhA, 12–448 (accession 5ODC_L) [26]; HydL, 15–450 (accession 7T2R_I) [22]. [4]VhuA, 12-379; VhuU, 18–26 (accessions CAF31249.1 and MBG0769290.1, respectively) [95, 155]. Alignments prepared in Clustal Omega and visualized in MView using Clustal coloring [121].

the Fe–S clusters have canonical cysteine ligands. The [NiFe] cofactor-binding motifs share homology with many of the conserved amino acids found in sequences of other [NiFe]-hydrogenases [1, 3–5], including HydL of the HydABCSL complex (Table 11.8).

The Fe–S-binding motifs for HdrA are shown in Table 11.9. The motifs were again derived from the *M. thermolithotrophicus* structures. HA1/HA2 and HA5/HA6 are [4Fe–4S] cluster pairs that are coordinated by overlapping cysteine-rich motifs, which commonly occurs in 2[4Fe–4S] Fd-like binding motifs. This motif is found to be ubiquitous in the subunit structures of many bifurcating complexes, as noted here for Hyds and Mvh–Hdr complexes in Tables 11.4, 11.5, 11.8, and 11.9. It creates a close proximity of the cluster pair, which is probably related to their ET role(s).

A unique cofactor in Mvh–Hdr is HA3, which is located at the protomer–protomer interface formed by a pair of HdrA subunits. In the FAD-bound and reduced form of Mvh–Hdr, an iron atom of HA3 is coordinated by five cysteine residues (Table 11.9), one of them residing in the adjacent HdrA subunit, HdrA′ [26]. In the HdrA sequence of *M. maripaludis*, this residue is substituted by selenocysteine (Table 11.8). When Mvh–Hdr was prepared in an oxidized state, the coordination

Table 11.9 Structurally derived binding motifs of HdrA cofactors.

HdrA subunit		
cofactor	**Name**	**Motif**
[1]**C-terminal [4Fe–4S], [4Fe–4S]**	HA1, HA2	CGGCtVCh..CPa X_{15-19} hCKGCGsCsAuCPS
C, HA1 coordinating cysteine; C, HA2 coordinating cysteine; t, turnlike; h, hydrophobic; a, aromatic; X_n, any amino acid; h, hydrophobic; s, small; u, tiny.		
[2]**Supernumery [4Fe–4S]**	HA3	VCaCG X_{30} hCuD X_{20} AACos X_{121} uhC/UIL
[2]**Disulfide bridge**	HA3	DCuhC/UIL
C, C, coordinating cysteine; HdrA′ subunit; C/U is cysteine, C, in *M. thermoaceticum* and *M. marburgensis* or U (selenosysteine) in *M. maripaludis*; C, bridging cysteine, HdrA′ subunit; C/U is a bridging cysteine, C, in *M. thermoaceticum* and *M. marburgensis*; or bridging selenocysteine, U, in *M. maripaludis*; a, aromatic; o, alcohol; s, small; u, tiny; h, hydrophobic; X_n, any amino acid.		
[3]**FAD proximal [4Fe–4S]**	HA4	CVGSR X_7 YCSpVCC
[4]**[4Fe–4S], [4Fe–4S]**	HA5, HA6	CTGCGtCstsCPl X_{30} DK-aCI-ChLCscVC
C, coordinating cysteine; C, HA5 coordinating cysteine; C, HA6 coordinating cysteine; X_n, any amino acid; p, polar; t, turnlike; s, small; ., nonconserved; –, acidic; a, aromatic; h, hydrophobic; u, tiny; l, aliphatic; X_n, any amino acid; p, polar; s, small; a, aromatic; h, hydrophobic.		
[5]**FAD**		GuGl X_{20} Kp.SluGRMuQLsKTF X_{139} GYD X_{21} LElER X_{43} pVCCMahhK

[1–6]Motif residue numbers. [1]HA1, HA2, 584–629. [2]HA3 supernumerary, 10–199; HA3 disulfide bridge, 192–199. [3]HA4, 386–404. [4]HA5, 246–303. [5]FMN, 150–409 (accession 5ODC_G) [26]. Alignments prepared in Clustal Omega and visualized in MView using Clustal coloring [121].

of HA3 by the HdrA′ residue was absent. The residue instead formed a disulfide bond with an adjacent cysteine at position 194 in HdrA′ (*M. thermolithotrophicus* numbering). Whether the change in coordination has functional role(s) has not been investigated, and it could be the result of oxygen damage [26]. It is possible that the unique coordination of HA3 functions in tuning its oxidation state range in a manner similar to atypical clusters found in some nonbifurcating [NiFe]-hydrogenases [104, 156, 157]. It is also noteworthy that the HA3 (and HA4) motifs are nonstandard and was not predicted based on sequence information alone [23, 26, 27, 50], underscoring the impact that structural information has on resolving cofactor compositions and organizations and their roles in ET and the overall reactivity of Mvh/Vhu–Hdr complexes.

11.3.3.1 Selenocysteine Considerations

The discovery of the first selenocysteine-containing hydrogenase from *Methanococcus vannielii* [158] was around the start of research efforts to understand the structural and biochemical role of selenocysteine in [NiFeSe]-hydrogenases [1, 98, 104] and other enzymes/proteins like formate dehydrogenase [102, 159, 160]. For example, selenocysteine role(s) have been studied in the Vhu–Hdr containing

Table 11.10 EPR signals of the MvhAGD subunits from *M. marburgensis*.

Subunit	Cluster type	*g*-values	Collection parameters	References
MvhD[a)]	[2Fe–2S]	2.006 1.936 1.912	$T = 70$ K; $P = 0.2$ mW	[95]
MvhG[a)]	[4Fe–4S]	2.073 2.036 1.890	$T < 20$ K	
MvhA[b)]	NiFe site – Ni^{3+}-C	2.21 2.13 2.01	$T = 90$ K; $P = 15$ mW	[163, 164]

T, temperature; *P*, microwave power.
a) MvhAGD from *M. marburgensis*, isolated from the HdrABC component and reduced with 5 mM NaDT. Spectra examined under two different temperature regimes provided g-values for the [2Fe–2S] cluster of MvhD and features associated with [4Fe–4S] clusters of MvhG.
b) From redox titrations performed at pH 6 and 45 °C. The signal was originally assigned as a Ni^{1+} state [163, 164]. We assigned the reported g-values to the Ni^{3+}-C state of the NiFe site based on the (i) a review from Albracht designating it as the Ni-C state [164] and (ii) the extensive literature precedent for these g-values representing the formal 3+ redox state of the Ni atom [104].

systems *M. maripaludis* [29, 161] and *Methanococcus voltae* [96, 97, 103, 162]. In Vhu–Hdr complexes, selenocysteine replaces a cysteine of the [NiFe] cofactor, creating a [NiFeSe] cofactor (see Section 11.2.2). The selenocysteine codon is encoded by *vhuU*. VhuU matches the C-terminal end of the [NiFe] cofactor-binding motif in MvhA subunits and other large hydrogenase subunits. Together, VhuA and VhuU contain the complete [NiFeSe] cofactor-binding motif (Table 11.8) [96–98, 103, 162]. Properties of [NiFeSe]-hydrogenases differ from many standard [NiFe]-hydrogenases (see Section 11.2.2), which could have functional implications with regard to the bifurcation reaction. Additionally, selenocysteine is involved in coordination of the MD cluster in VhuD (Table 11.8) [162], which is located at the interface of the VhuD and HdrA subunits [26, 101].

11.3.4 Spectroscopy of MvhAGD

Spectroscopic characterization of the MvhAGD module is limited to EPR studies of the [NiFe] cofactor in MvhA and cluster(s) in MvhD and MvhG (Table 11.10). The [2Fe–2S] cluster signals from MvhD were found using temperature to isolate its EPR features from [4Fe–4S] clusters signals (Table 11.10). Similar to the analysis on *Tm* HydA, features associated with the [4Fe–4S] clusters of MvhG were identified in EPR spectra collected at lower temperatures, although they were not assigned to individual clusters (Table 11.10).

11.4 Mechanistic Proposals

Proposed mechanisms of electron-bifurcating hydrogenases are largely (but not exclusively) based on structural considerations. The structures give us an understanding of cofactor placements and (mostly static) distances between them, which allows for predicting ET pathways and plausible substrate-binding sites. Substrate binding and/or transformations can be demonstrated in favorable circumstances. Additionally, structural analyses can identify flexible regions of a protein that may coordinate short- and long-range motions, which can be essential for function [165–168]. Spectroscopic signatures are also important to have in mind when drawing up mechanistic hypotheses, especially for cases where signature(s) can be assigned to a specific cofactor or a set of cofactor(s) (see Section 11.3), which form a basis for understanding more detailed reactivity patterns of cofactor(s) during ET and/or substrate transformations. In this section, we bring up such signatures as appropriate.

11.4.1 Electron Bifurcation: General Mechanistic Considerations

Electron bifurcation mechanisms find their origins in the concept of inverted (or crossed) reduction potentials, which is a general electrochemical phenomenon [169] that is highly applicable to flavins and quinones (we will mostly refer to flavins for simplicity) [47, 48, 52, 170–172]. What inverted potentials means is that the redox transitions regarding flavin in its oxidized (OX), singly reduced (SQ), and doubly reduced (HQ) forms [38, 173] are ordered as: SQ/HQ < OX/HQ < OX/SQ. Another way to think about this is that the singly reduced flavin (SQ) is higher in energy than the doubly reduced flavin (HQ) [38, 47, 48, 51, 170–172, 174]. Figure 11.5a depicts a normal vs. inverted reduction potential ordering.

The power of inverted potentials in electron bifurcation is manifest by NfnI from *Pyrococcus furiosus* [175, 176]. This enzyme is a Nfn family member, which use a catalytic FAD bifurcation center (Figure 11.2 and Table 11.1). Unlike enzymes in other bifurcating families, it does not undergo large-scale conformational motions [122, 172, 177, 178]. The FAD oxidizes the mid-potential, two-electron donor (NADPH in this case, see also Table 11.1), thus undergoing the OX/HQ transition. In the mechanistic model, one electron residing on this doubly reduced FAD is sent toward the low-energy acceptor (SQ/HQ transition). The resulting singly reduced FAD (SQ) is incredibly unstable and of very high energy (the OX/SQ transition was estimated to be below −900 mV) [176]. This lets the second electron be transferred toward the high-energy acceptor. A general schematic of this is shown in Figure 11.5b with OX/HQ, SQ/HQ, and OX/SQ reduction potentials that correspond to NfnI [175, 176] (the exact potentials are system dependent) [48, 69, 179]. The very unstable, high energy SQ is not the only factor responsible for electron bifurcation in NfnI [180]. Indeed, finely tuned distances between ET cofactors, Fe–S cluster reduction potentials, and communication of the bifurcating FAD with adjacent Fe–S clusters are also important for determining the pathway each electron transverses and for mitigating short-circuit reactions (see references for more details) [171, 175, 176, 181]. Short-circuit reactions being the "leaking" of the electron meant to reduce the high-energy acceptor into the downhill pathway where it instead reduces the low-energy acceptor [48, 182].

Bifurcating enzymes from other families undergo rather large-scale conformational changes [31, 122, 179]. This can lead to mechanisms like in Figure 11.5c,

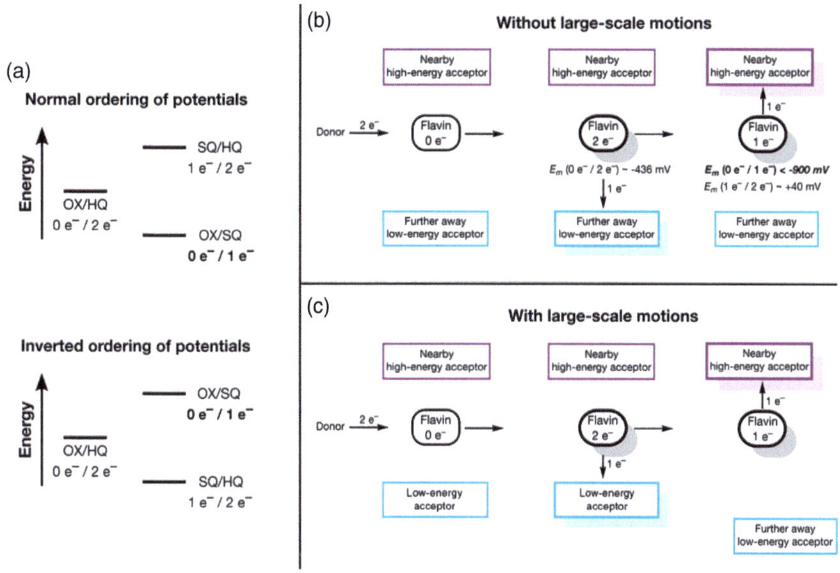

Figure 11.5 (a) Normal vs. inverted ordering of reduction potentials. Source: Adapted in part from Vanoni [38]. (b) Electron bifurcation schematic using the *Pyrococcus furiosus* NfnI energetic landscape for the flavin cofactor. See text for more details. Keep in mind that the low- and high-energy acceptors near the flavin site are not the low- and high-energy substrates but instead are ET cofactors within enzyme. (c) Electron bifurcation schematic similar to (b), but highlighting the role of protein motions which can be advantageous, e.g. to avoid short-circuit reactions.

where sending an electron down a specific pathway is aided by large-scale protein motions. Short-circuit reactions are mitigated because, at the appropriate step of the reaction, the ET distance between the low-energy acceptor and the bifurcation center is very large. Inverted potentials (like in Figure 11.5b) can still be operative [48, 171]. Important to note is that lack of potential inversion has not been demonstrated for a bifurcating enzyme. But, theoretical work suggests that electron bifurcation can occur without potential inversion, provided the right design principles [183, 184].

From a structural perspective (see Figure 11.4), the Hdr-containing class of bifurcating hydrogenases can fit into this general framework because the centralized FAD can accept two electrons to send down different pathways [26, 174]. A lysine residue near the FAD site could transiently stabilize a highly reactive, reduced flavin state [26, 122, 174] (the analogous residue for the Nfn family is an arginine [174, 176]). Interestingly, the structures of bifurcating Hyd enzymes [14, 22, 85] and the structure of another Bfu family member [139] showed that the FMN site lacks a nearby residue to stabilize a highly reactive/unstable flavin state [122]. Because of this, and also because the FMN site (i) is a catalytic cofactor that reduces the low-energy substrate and (ii) also appears to be part of the group of redox cofactors that sends electrons toward the high-energy substrate (Figure 11.3 and Section 11.4.2), it has proven difficult to rationalize this dual FMN functionality into mechanistic models used for other bifurcating enzymes (e.g. Nfn) [9, 14, 22, 85, 122]. Outside-the-box mechanisms have been proposed to account for this conundrum.

11.4.2 [FeFe]- and [NiFe]–HydABC: Electron Transfer Pathways

Reports regarding *T. maritima* [14], *A. woodii* and *T. kivui* [85], and *A. mobile* [9, 22] included mechanistic proposals with many similarities but also notable differences. It is important to keep in mind that it is generally assumed each Fe–S cluster only accepts/donates one electron at a time, an assumption supported by (i) fairly typical cluster architectures and (ii) EPR [8, 120, 130, 136] and resonance Raman studies [185] that indicated the clusters cycle between the formal +2 and +1 redox states in *Tm* HydABC. Additionally, it has been postulated (specifically for *Tm* HydABC [84]) that the H-cluster uses a mechanism for H_2 oxidation/production that is quite similar to other [FeFe]-hydrogenases [104, 149, 154], which is somewhat supported by the fairly typical/expected behavior of the H-cluster from rather limited equilibrium experiments [84, 85] and the sole pre-turnover report [146]. By extension, it could be supposed the [NiFe] site in [NiFe]–HydABC-bifurcating hydrogenases use a mechanism that is similar to other [NiFe]-hydrogenases [104, 186, 187]. We note that it would not be surprising if unusual behaviors are one day discovered at the H-cluster or [NiFe] sites that could be traced back to variations in the secondary/outer coordination spheres (vs. more "standard" active sites), long-range influences on the hydrogenase active site exerted by the rest of the protein and its cofactors, and/or allosteric effects from NAD(P)H and Fd. Such findings would be consistent with typical observations of "nonstandard" behavior at active sites of nonbifurcating hydrogenases [21, 104, 187–190] and long-range communication in enzymes [191–193] and would suggest the active sites are finely tuned for their integration into the electron bifurcation process involving the more conserved cofactors/subunits in the Bfu family. Keeping these considerations in mind, we now discuss ET in the direction of H_2 oxidation.

Electrons derived from H_2 are sent along the Fe–S cluster relay in HydA. In the [FeFe] case, they flow from the H-cluster to A4, A2, and then A1, whereas in the [NiFe] case they flow from HydSL to A3, A2, and then A1 (Figures 11.3 and 11.6). From there, they transfer to B1 and then the FMN site. These pathways are supported by distances between the redox sites being below the 14–20 Å limit for electron tunneling in proteins (Figure 11.3) [37] and the presence of similar cofactor arrangements in nonbifurcating, NAD(P)H-dependent enzymes [12, 21, 123, 194].

NAD(P)(H) binding at the FMN cofactor has been observed [22, 85], supporting FMN as the $NAD(P)^+$ reduction site. The site of Fd binding and reduction is unknown, but likely occurs downstream from the FMN site in HydB or HydC. Indeed, the A3 cluster that could otherwise serve as a surface-exposed Fd docking site in HydA [1, 6] is blocked by dimerization [14, 85] or HydSL [22], supporting the notion that Fd reduction occurs in the HydBC module. The B3 and B4 clusters (Figures 11.3 and 11.6) are currently the favored Fd reduction spots for two main reasons. The first is that bringing electrons to B3/B4 would ensure the electrons for Fd reduction are as far away as possible from the FMN site in the HydBC

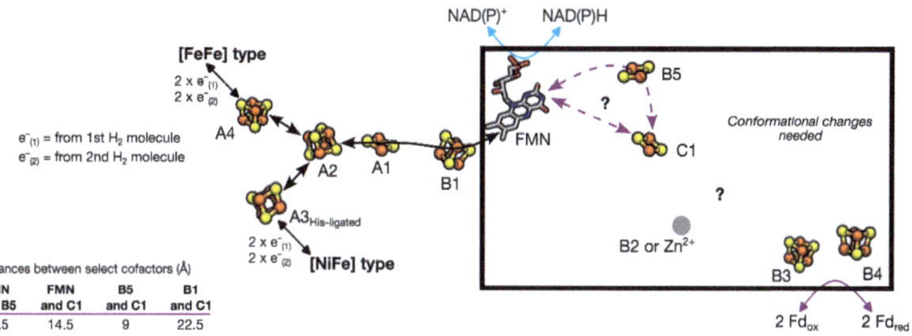

Approximate distances between select cofactors (Å)

	FMN and B5	FMN and C1	B5 and C1	B1 and C1
T. maritima	18.5	14.5	9	22.5
T. kivui	19.5	14	9	21.5

Figure 11.6 General cofactor compositions, proposed ET routes, and substrate-binding sites for [FeFe]–HydABC and [NiFe]–HydABCSL. Also shown are select distances between redox cofactors in the HydBC module. The shaded box highlights the regions where $NAD(P)^+$ and Fd get reduced. Note that the Fd-binding site is putative because Fd has not been resolved in the structures. See the main text for more details [9, 14, 22, 85, 122]. Iron = brown and sulfur = yellow. Source: Partly adapted from Schut et al. [9].

module, which would help avoid short circuits. The second is that the domain harboring B3/B4 has similarities to part of the Fd-binding region in a complex I-like photosynthetic NAD(P)H-dehydrogenase [22, 85, 195, 196].

Getting electrons to B3/B4 requires the use of the B5 and/or C1 cluster (Figure 11.6) [9, 14, 22, 85, 122]. Based on structures viewed in PyMOL [197] and using the distances between cysteine sulfur atoms on the Fe–S clusters for donor-acceptor distance measurements [198–200], the FMN to B5 distance is about 19 Å, the FMN to C1 distance is roughly 14 Å, and the B5 to C1 distance is roughly 9 Å (Figure 11.6). C1 is housed in a flexible C-terminal region of HydC and B5 is in a somewhat flexible N-terminal region of HydB [22, 85]. Such flexibility could mediate donor–acceptor compressions, or more general motions, that fine tune tunneling pathways/distances, which may be advantageous regarding intersubunit ET (e.g. transfer from B5 to C1) [201–203] and electrons traversing fairly large distances (e.g. transfer from FMN to B5) [85, 201, 202, 204–207].

Once electrons are at B5 or C1, the distance to B3/B4 is still >26 Å away. This distance limitation must be alleviated by long-range (domain) motions of the protein [167]. There are two probable locations for these motions. One is the ferredoxin-like, C-terminal domain of HydC that contains C1 [22, 85] and the other is the ferredoxin-like, C-terminal domain of HydB that contains B3/B4 (Figure 11.3) [14, 85]. Their ability to undergo domain motions is postulated from structural information (e.g. implied flexibility) [14, 22, 85, 139], molecular dynamics simulations that indicate FMN-C1 and C1-B3 distance fluctuations [85], and the ability to model different conformations of these domains [14, 22]. Regarding the C-terminal domain of HydB, Furlan, Chongdar, Blaza, Birrell, and coworkers noted the B2/Zn^{2+} site may facilitate large-scale motions that position B3/B4 close to or far away from C1/B5 [14].

Up to now, we have discussed mostly agreed upon aspects of the electron flow. To summarize: electrons travel from the hydrogenase active site to the FMN site in HydB. From there, they are either used to reduce $NAD(P)^+$ or are transferred downstream to B3/B4, the latter requiring conformational dynamics in HydB and/or HydC (Figure 11.3 and Figure 11.6) [9, 14, 22, 85]. We now turn to three more detailed mechanisms of ET that have been proposed, which mostly regard the HydBC module. Each begins with one of two scenarios: electrons from the oxidation of the first H_2 molecule (referred to here as $e^-_{(1)}$) transfer to the FMN, forming a doubly reduced flavin (HQ, see Section 11.4.1) or land at A1 and B1 (Figure 11.7a). Electrons from oxidation of the second H_2 molecule ($e^-_{(2)}$) later move toward HydB and the following scenarios could happen:

- Figure 11.7b [9, 22]: An $e^-_{(1)}$ stored in the HQ and an $e^-_{(2)}$ transfer to C1 and B5, while the remaining $e^-_{(2)}$ lands at B1. The FMN is probably now in a neutral semiquinone (NSQ) form [208]. C-terminal domain motions of HydC, and perhaps HydB, coordinate the transfer of $e^-_{(1)}$ and $e^-_{(2)}$ from C1 and B5 to B3/B4, thus committing them to Fd reduction (reminiscent of conformational gating during bifurcation by cytochrome bc_1 [209–211]). The motions could be initiated by $NAD(P)^+$ binding. $NAD(P)^+$ is reduced at the FMN site by the remaining $e^-_{(1)}$ and $e^-_{(2)}$ via hydride transfer.
- Figure 11.7c [14]: $NAD(P)^+$ binding stabilizes a semiquinone (e.g. NSQ [208]) at the FMN site. This lets FMN function as a one-electron transfer agent by altering between the fully oxidized and SQ forms, allowing it to reduce B5 and C1 with the $e^-_{(1)}$ electrons. At this stage of the turnover, formation of an HQ is not permitted (presumably because of a fairly negative $E_{SQ/HQ}$ value), which avoids the short-circuit reaction of reducing NAD^+ with $e^-_{(1)}$ electrons.
  Concomitantly, domain motions of HydB's C-terminal domain bring B3/B4 within an ET distance of B5 and C1. The authors of the proposal hypothesized that B1 reduction triggers the motions by initiating a hinge movement [167] in the B2/Zn^{2+} region, the communication of B1 and the hinge region being facilitated by a four-helix bundle they are both part of. It was also hypothesized that the presence of $NAD(P)^+$ stabilizes the position of HydB's C-terminal domain near B5 and C1 at this stage of the mechanism. We note that protein motions could also be triggered by just $NADP^+$ binding itself, which would be reminiscent of NADP(H) induced conformational changes in bifurcating Etf/Fix enzymes [179, 212, 213]. Overall: B1 reduction and/or $NADP^+$ binding play important roles in coordinating the motions and/or the position of HydB's C-terminal domain, allowing for C1 and B5 to transfer the $e^-_{(1)}$ electrons to B3 and B4.
  Afterward, Fd binding promotes the movement of HydB's C-terminal domain away from C1/B5, which commits the $e^-_{(1)}$ electrons to Fd reduction (similar to Figure 11.7b). This results in altered FMN reduction potentials that allow access to the HQ form, permitting $NADP^+$ reduction by the $e^-_{(2)}$ electrons via hydride transfer.
- Figure 11.7d [85]: C1 and B5 get reduced by $e^-_{(1)}$ electrons, with FMN probably serving as a one electron carrier during the process (like in Figure 11.7c). B5 reduction is thought to promote $NAD(P)^+$ binding as supported by a stronger $NADP^+$-binding affinity under 5% H_2 in *T. kivui* vs. its binding affinity when B5 was removed via an N-terminal truncation of HydB. This suggests that short-circuit suppression is aided by the notion that B5 must be reduced for

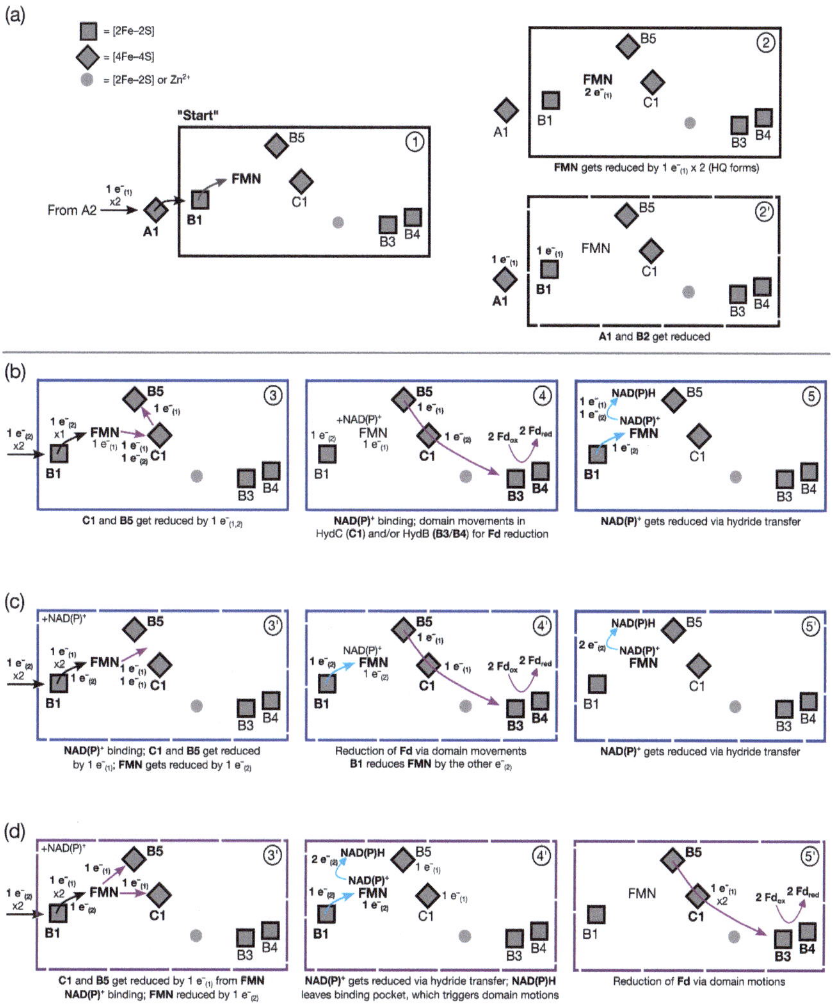

Figure 11.7 Mechanistic proposals of ET routes and substrate reduction in the HydABC-type hydrogenases told in a storyboard format. (a) Starting point(s) in the mechanisms [14, 22, 85]. (b) Model based on reports from Feng, Schut, Adams, Li, and coworkers [9, 22]. (c) Model based on a report from Furlan, Chongdar, Blaza, Birrell, and coworkers [14]. (d) Model based on a reports from Katsyv, Kumar, Saura, Kaila, Müller, Schuller, and coworkers [85]. See also Reference [139]. Some steps in the published models have been combined for simplicity and compactness. The exact content of Zn/B2 site is intentionally ambiguous.

NAD(P)$^+$ to stay bound, making it unlikely for $e^-_{(1)}$ electrons to reduce the pyridine nucleotide.

Once it binds, NAD(P)$^+$ gets reduced by the $e^-_{(2)}$ electrons via hydride transfer. NAD(P)H subsequently diffuses out of the FMN-binding pocket, triggering domain motions of HydC's and/or HydB's C-terminal domain such that $e^-_{(1)}$ electrons transfer from B5/C1 to B3/B4 to reduce Fd (this latter part being somewhat an opposite order of events vs. Figure 11.7c).

11.4.3 [NiFe]-MvhAGD-HdrABC

11.4.3.1 Electron Transfer Pathways

Proposed routes of e^- flow account for protein regions that could undergo conformational changes and for nontrivial ET segments identified in the structures [26]. Here, we cautiously assume each Fe–S cluster transfers/accepts one electron at a time, which can partly be rationalized in a manner similar to HydABC: most of the clusters have typical architectures (HA3 and the heterodisulfide reduction sites in HdrB being exceptions) [26], suggesting they should undergo typical Fe–S cluster redox transitions (HdrB clusters being exceptions [214, 215]). During catalysis, the [NiFe] active site probably uses an H_2 oxidation/production mechanism that is rather similar to other [NiFe]-hydrogenases [104, 164, 216]. However, as we also noted with HydABC, there may be unique active site behaviors yet to be uncovered. It is also important to note that a structural report from Wagner, Shima, Murphy, and coworkers about a somewhat analogous formate dehydrogenase–Hdr–formyl methanofuran (Fdh–Hdr–Fmd) complex, which notably has MvhD and HdrABC modules that are similar to those found in Mvh–Hdr, outlined hypotheses about ET routes that might be applicable to Mvh–Hdr [101]. We have accordingly incorporated these hypotheses into our discussion.

Let us consider ET in the direction of H_2 oxidation. Electrons from H_2 are sent from the hydrogenase active site in MvhA to the Fe–S relay in MvhG, which is analogous to the ET relay in other [NiFe]-hydrogenases (Figure 11.1, bottom and Figure 11.4). Afterward, they are transferred to HA1 and HA2, which are part of HdrA's C-terminal, ferredoxin-like domain [26]. The next electron acceptor is the [2Fe–2S] cluster in MvhD, which is called the MD cluster (Figure 11.8a,b). The C67 ligand of MD and the adjacent C16 residue (Figure 11.8a,b) are selenocysteines in VhuD and other enzymes with MvhD analogs (see also Sections 11.2.2 and 11.3.3.1) [24, 162, 217], suggesting nontrivial properties of these cysteines and/or the cluster [26, 101, 102]. Interestingly, MD is at a peculiar location because ET beyond MD is not feasible without conformational changes and/or the use of Cys residues [26, 101].

Proposed pathways beyond MD implicate ET between Mvh–Hdr protomers, that is, between the Mvh–Hdr components of the $(\text{Mvh–Hdr})_2$ dimer (Table 11.3 and Figure 11.8a,b, the prime symbol (′) in Figure 11.8 is used to differentiate between protomers). These can be classified as either involving FAD (housed in HdrA) or FAD′ (housed in HdrA′):

- Pathways involving FAD (Figure 11.8a): Transfer beyond MD begins with two possible routes, both of which have the same endpoint: the central FAD that is ~32 Å away (Figure 11.8a, both routes could presumably be used at different stages of the catalytic cycle). One route involves the ferredoxin-like domain harboring HA5/HA6, which can be in a conformation amenable to electron exchange with MD or in a conformation amenable to electron exchange with FAD. The other route involves electron transfer from MD to FAD, which would require a conformational change of MvhD and/or the use of Cys residues near the FAD site (Figure 11.8a) [26].

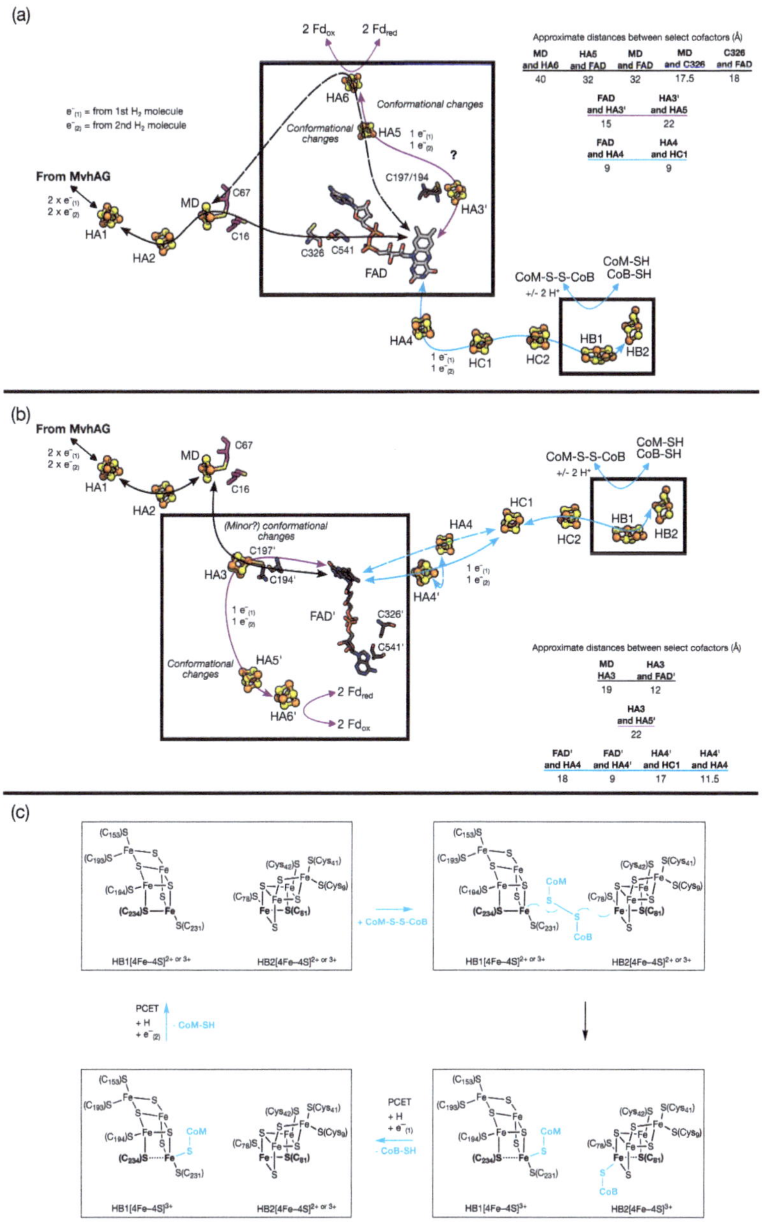

Figure 11.8 (a) ET routes in MvhAGD–HdrABC (starting from HA1) involving the centralized FAD of HdrA [26] and (b) ET routes in MvhAGD–HdrABC (starting from HA1) involving the centralized FAD′ of HdrA′ [26, 101]. General coloring: iron = brown and sulfur = yellow, grey/black = carbon, red = oxygen, orange = phosphorous, blue = oxygen. Exceptions to coloring: (i) C16/67 and C16′/C67′ = side chains are purple to account for these being Sec residues in the Vhu–Hdr case and (ii) C197/C194 and C197′/C194′ sulfurs are dark yellow to account for their ligation to and/or association with a cluster housed in the other Hdr protomer [26]. (c) Proposed mechanism(s) of heterodisulfide reduction by the HB1 and HB2 clusters. See text for more details. Source: Adapted in part from Wagner et al. [26].

Once at the FAD site, electrons transfer to the CoM-S-S-CoB reduction sites in HdrB (Figure 11.8a). ET toward the Fd/poly-Fd reduction site (presumably HA6 based on the Fdh–Hdr–Fmd structure with polyFd [101]) uses HA3′. Large-scale movements of the domain housing HA3′ and/or the domain containing HA5/HA6 facilitate the reduction of the HA5/HA6 clusters. Motions of the domain housing HA3′ could necessitate breaking the bond between HA3′ and its ligand that is harbored in HdrA: C197, meaning an iron atom of HA3′ would thus go from being five- to four-coordinate [26, 101].

We note that transfer from FAD to HA3′ and then HA5/HA6 could be avoided in two ways: (i) motion of the ferredoxin domain harboring HA5/HA6 to within ET distance of FAD, thus bypassing the need for HA3′ and (ii) using the MD to HA5/HA6 pathway without transferring electrons that will be used for Fd reduction to the FAD site in the first place. Both scenarios are very speculative. The latter scenario would require the FAD function solely as an ET agent en route to HdrB.

- Pathways involving FAD′ (Figure 11.8b): Electron flow from MD to the central FAD′ is proposed to involve conformational changes of HdrA's N-terminal domain harboring HA3 [26]. We note that this portion of the pathway (MD → HA3 → FAD′) does not necessarily have major distance concerns since MD and HA3 are only slightly greater than 20 Å apart and the distance between HA3 and FAD′ is ~12 Å (using cysteine sulfur atoms on the clusters for measurement purposes [197–200]). Thus, large-scale motions may not be necessary because more subtle motions could bring MD and HA3 within ET distance. This means HA3 could retain its pentacoordinate Cys ligation at this stage of catalysis, with such coordination perhaps imparting yet undiscovered properties onto HA3 that are relevant to ET.

Subsequent transfer to HdrC involves either the HA4′ and/or the HA4 cluster. From distance arguments, neither route seems better over the other. Because of the shorter distance between FAD′ and HA4′ (~9 Å) vs. HA4 (~18 Å), the FAD′ → HA4′ route seems initially favorable. But, the HA4′ and C1 distance is ~17 Å whereas the HA4 and HC1 distance is about half as far at ~9 Å. The precise pathway may depend on how ET rate(s) are modulated vs. the maximal achievable rate at particular step(s) of catalysis, consistent with the idea that rapid enzymatic ET rates, relative to slower ones, are not always beneficial [218].

Transfer from FAD′ to the Fd/polyFd reduction site requires HA3. Evidence for this is found in Fdh–Hdr–Fmd structures that show the N-terminal domain of HdrA, which houses HA3, adopts two positions: one that is near FAD′, and another that places HA3 within ET distance of HA5′ [101]. For the latter, the otherwise ~22 Å distance between HA3 and HA5′ (as determined structurally) would probably be shortened significantly [26, 101]. Again, a large-scale movement involving HA3 would probably require it to switch from being five- to four-coordinate by breaking the bond with the C197′ ligand [26, 101]. The need to transfer electrons from FAD′ to HA3 for Fd reduction could be avoided by just using a MD → HA3 → HA5′ pathway, but, similar to Figure 11.8a, this would depend on the precise mechanistic requirements of the FAD′ site. The HA3 domain (or the domain harboring HA5′/HA6′) would still need to undergo some motions to shorten the HA3 to HA5′ distance.

A similar mechanism involving HdrC′ and HdrB′ was outlined in a review from Appel, Boll, and coworkers [24].

Two additional points regarding both Figure 11.8a,b are as follows:

(1) The possible role(s) of cysteines in ET indicates thiyl radicals [219] and/or disulfide bonds [220] could be used in the mechanism (the ability to form disulfides has been observed in Mvh–Hdr, albeit under probable trace O_2 exposure [26], and Fdh–Hdr–Fmd [101]). In either case, the reduction potential for such species [221] would probably need tuning by the protein environment to be near that of the Fe–S clusters and/or FAD site. How (and if) certain cysteine residues are used in the ET relay is an open question.
(2) It is possible the reduced FAD forms an anionic semiquinone (ASQ) during ET steps since tentative observation of an ASQ was reported in the isolated Hdr module from *M. marburgensis* [222]. However, this needs to be verified.

11.4.3.2 CoM-S-S-CoB Binding and Reduction

Homolysis of the CoM-S-S-CoB disulfide bond and formation of coenzymes CoM-SH and CoB-SH occurs at the noncubane HB1 and HB2 clusters. Evidence for the binding of heterodisulfide (or its reduced products) to Fe–S cluster(s) came from early studies with HdrABC from *M. marburgensis*, which included results such as: distinct magnetic circular dichroism spectra upon the addition of CoM-SH or CoB-SH [223]; broadening of EPR features in the presence of CoM-^{33}SH [224]; ^{57}Fe ENDOR spectroscopy indicating CoM-SH substrate/product interaction with a cluster [225]; and X-ray absorption spectroscopy detecting a Se–Fe bond in the presence of the product/substrate analog CoM-SeH [226]. Hamann, Hedderich, and coworkers demonstrated HdrB likely contained the heterodisulfide reductase active site via spectroscopic studies of isolated HdrB. Notably, EPR signatures of heterologously expressed HdrB were broadened upon addition of CoM-^{33}SH [214].

The binding and partial transformation of heterodisulfide in HdrB was demonstrated in the structural report. When MvhAGD-HdrABC crystals were soaked with H_2 (mid-potential donor) and CoM-S-S-CoB (low-energy acceptor) for two minutes, density attributed to a CoM-S-Fe species at HC1 and a CoB-S-Fe species at HB2 were identified. A longer soaking time of three and a half minutes led to density attributed to CoM-S-Fe at HB1 and a CoB-SH product. Because Fd (the high-energy acceptor) was not present, it was unsurprising that only one of the two coenzyme products was observed: formation of CoB-SH and CoM-SH should require two rounds of bifurcation. The exclusion of Fd should only permit the transfer of one electron to HdrB, which can only result in the formation of one of the coenzymes [26]. As an aside: the result aligns with the standard definition of electron bifurcation in enzymes, in which the electrons from the mid-potential donor from the first round of bifurcation, the $e^-_{(1)}$ electrons, reduce two different substrates as we have drawn in Figure 11.8a,b (which contrasts with Figure 11.7c,d).

The structural report proposed a mechanistic scheme for CoM-S-S-CoB reduction to coenzymes CoM-SH and CoB-SH [26]. This is shown in Figure 11.7c, where we also emphasize that the formation of both CoM-SH and CoB-SH are proton-coupled electron transfer (PCET) processes that are not well understood, although these have been tackled in a recent theoretical study [227]. Also, we kept rather vague the redox states of HB1 and HB2 and whether or not there is a five-coordinate iron when CoM-S and/or CoB-S are bound for the following reasons:

(1) Regarding their redox states, current evidence suggests that HB1 and HB2 can be thought of as $[4Fe–4S]^{3+}$ clusters after binding CoM-S and/or CoB-S (Figure 11.8c) [223, 228]. There is also evidence the oxidized (resting state) of the clusters are $[4Fe–4S]^{3+}$ instead of $[4Fe–4S]^{2+}$), because the oxidized forms might have electronic configurations that are similar to the coenzyme-bound form as reported for isolated HdrB from *M. marburgensis* (albeit with notable caveats regarding the possible binding of cluster reconstitution reagents to the oxidized state [214]) and from the nonmethanogen *Desulfovibrio vulgaris* [215]. We have thus elected to assign the redox states of oxidized HB1/B2 as either +2 or +3 (Figure 11.8c), the former being the original assignment [222, 228].
(2) Regarding the presence of a five-coordinate iron when CoM-S and/or CoB-S are bound: the structural report suggests that, following homolysis of the disulfide bond, the iron atoms that bind CoM-S and CoB-S go from being four-coordinate to five-coordinate [26]. Vibrational studies coupled to density-functional theory calculations on HdrB from *Desulfovibrio vulgaris* indicated one of the Fe–S bonds breaks following substrate binding, such that the iron atoms remain four-coordinate [215]. To acknowledge both interpretations, we have drawn these bonds with dashed lines when CoM-S and CoB-S are bound to a cluster (Figure 11.8c).

11.5 Forefronts

Within the context of Sections 11.2–11.4, we now provide a list of research goals we believe are important for a more comprehensive understanding of electron-bifurcating hydrogenases:

- Physiological (cellular) conditions determine a bifurcating enzyme's environment and thus set the reaction conditions. A better understanding of what these conditions are for a given enzyme, especially the ratio of oxidized and reduced reactants/products, pH, and ρH_2, is required for a more complete understanding of the reaction thermodynamics from a substrate perspective (see also Section 11.2.2) [50, 66]. On a similar note, many bifurcating hydrogenases operate at rather high temperatures. It is thus important to consider contributions of the available thermal energy to catalytic steps, for example, protein motions and uphill ET steps [40, 41, 168].
- Determining ET cofactor and catalytic cofactor thermodynamics under physiological conditions is also of paramount importance. For example, temperature can significantly affect ET cofactor reduction potentials, so reduction potentials measured under typical laboratory conditions may not capture the thermodynamic reality [48, 56, 229, 230]. Additionally, studies need are needed regarding flavin protonation states (e.g. ASQ vs NSQ [173]) and the extent of potential inversion

(if any), both of which are important for understanding the finer details of the flavin cofactor reactivities [174–176].

An outcome of such work would be a handle on the energies of specific electrons at certain stages of catalysis. This can help support/disprove hypotheses and formulate new hypotheses about mechanisms of energy coupling. For example, will the results conform to the typical views of electron bifurcation with crossed potentials at the flavin cofactor [69, 170, 174]? Or do some bifurcating hydrogenases use an entirely different mechanism? This is extremely pertinent to the bifurcating Hyd enzymes where the flavin does not oxidize the mid-potential, two-electron donor like in the cytochrome bc_1, Nfn, and Etf/Fix families [9, 48, 122, 174]. Does a centralized bifurcation site hold for Hyd? Or will the notion that a set of multiple cofactors coordinate bifurcation in a more complex manner be supported [9, 122]? While the Hdr enzymes have a centralized flavin, they do not oxidize the mid-potential, two-electron donor. The role of the FAD as a centralized bifurcation core by accepting two electrons from H_2 oxidation by the Mvh module and then splitting the electron pair has reasonably been proposed [24, 26, 50, 101], but need this be the case (see also Section 11.4.2)? A more concise way to say this: electron-bifurcating hydrogenase mechanisms are often proposed based on precedent from the cytochrome bc_1/b_6f, Nfn, and Etf/Fix families, but what if Hyd and/or Hdr operate with yet undiscovered mechanistic steps of electron bifurcation?

- A structure of Hnd is needed to understand its cofactor layouts and formulate mechanistic hypotheses that relate or contrast it to the Hyd type of hydrogenases. As also mentioned in a note for Table 11.3, the Hnd cofactor composition does not meet the requirements to be a member of the Bfu family (see also Figure 11.1f, top vs. Figure 11.1e top and Figure 11.1d, bottom) [9]. A structural understanding of Hnd will thus allow it to be better understood within the context of the known classes of bifurcating enzymes.
- Regarding the hydrogenase active sites, little to nothing is known about how H_2 oxidation (or H^+ reduction) is temporally and energetically coordinated to bifurcation/confurcation. Kinetic studies targeting electron flux [231] during turnover at the hydrogenase active site and linking it to electron flux at the flavin and Fe–S cofactors (along with complementary thermodynamic information) could help determine how a hydrogenase active site is mechanistically integrated into the more conserved components of the Bfu and Hdr families.
- Electron bifurcation is often discussed in terms of the movement of electrons. Despite not being well understood (essentially at all in most cases), protons play an indispensable role in the mechanism [178, 227, 232, 233]. For example, the protonation states of a flavin and its surrounding amino acids tune the flavin's reduction potential and extent of potential inversion [178, 232, 233] and proton-based events can act as gates for reaction steps via PCET processes [31, 178, 232]. Turnover at the hydrogenase active site is also inherently described by PCET. A comprehensive mechanism for an electron-bifurcating hydrogenase cannot be proposed without the inclusion of proton and/or PCET events.
- Breakthroughs in our understanding of bifurcating hydrogenases have come from the structures (Sections 11.3 and 11.4). However, the structures have caveats:

they are predominantly low-energy conformational states of the protein samples under equilibrium conditions. Therefore, they might not represent productive conformational states sampled during a single turnover or, more generally, higher-energy states amenable to reactivity (e.g. ET, bond making or breaking, substrate product and release) [165, 168, 234]. An incredibly daunting but worthwhile long-term goal should be capturing higher-energy structures along the reaction coordinate [235, 236]. Similarly, many of the known spectroscopic signatures represent ground-state properties of the Fe–S clusters and other cofactors, whereas there may be higher-energy forms populated along the reaction coordinate. Another very daunting but worthwhile goal is determining the electronic properties of these high-energy forms by clever applications of spectroscopic methods [176, 237–240].

- Last, but not least: the mechanistic underpinnings of ET steps facilitated by the Fe–S clusters warrant detailed investigations, especially within the context of enzyme dynamics. Indeed, both the Hdr and HydBC modules appear to undergo small- and/or large-scale motions which, at a minimum, modulate the distances between ET cofactors (see also Sections 11.4.2 and 11.4.3.1). Understandably, this is often interpreted as enhancing ET rates from a distance perspective entirely because, generally speaking, when redox cofactors are closer together ET between them should be faster [39, 40, 241, 242]. But what about other aspects of ET theory, particularly a possible dependence of the electronic coupling term on the precise protein structure, e.g. the intervening medium, between two redox cofactors [37, 39, 242]? Specifically, the precise tunneling pathway and thus the rate of ET between an Fe–S cluster and another redox partner (e.g. another Fe–S cluster or a flavin) could depend on factors such as amino acid identity [39, 198, 243], conformational sampling of the protein [203–206, 244, 245], and water molecule placement [198, 243]. Overall: determining finer details of ET parameters between cofactors [37, 39, 201, 202, 241, 242, 246], the electronic coupling element used as the example here, could help better understand ET principles within the incredibly complex ET relays used by electron–bifurcation hydrogenases.

11.6 Summary and Outlook

The main purposes of this chapter were to (i) overview and discuss the biochemistry and general reactivity of electron-bifurcating hydrogenases, (ii) overview and discuss the mechanistic proposals, and (iii) outline important topics for future work. Along the way, we included some commentary and/or speculation about certain topics, such as the subunit architectures and cofactor compositions (Section 11.2.2), activity assays (Section 11.2.2), ET routes (notably in Mvh–Hdr, Section 11.4.3.1), and motivations and/or rationalizations for future areas of study (Section 11.5). A wealth of knowledge is known about these systems, including from studies that occurred prior to knowledge about their bifurcation abilities, but there is still much to be learned.

Considering electron bifurcation more broadly, information acquired from the electron-bifurcating hydrogenases should have general applicability to structure–function studies of bifurcating Hyd-type [9, 122, 139, 217, 247–249] and HdrABC-containing [24, 25, 217, 250–252] hydrogenases and their nonbifurcating counterparts [7, 10–12, 20, 21, 99, 123]. On a more general note, the various dynamics employed by these enzymes poise them as excellent systems for understanding enzyme motions and interrelated electrostatic contributions to catalysis, which are major areas of study in the broader mechanistic enzymology field [165, 167, 168, 201, 253–255].

Acknowledgments

This work was authored in part by the Alliance for Sustainable Energy, LLC, the manager, and operator of the National Renewable Energy Laboratory for the U.S. Department of Energy (DOE) under Contract No. DEAC36-08GO28308. G.E. Vansuch was supported by a Laboratory Directed Research and Development (LDRD) Director's Fellowship. J.R. Humphreys and C.E. Lubner were supported by the Laboratory Directed Research and Development (LDRD) Program at NREL. C.E. Lubner was supported by the U.S. Department of Energy, Office of Science Early Career Program. E.C. Kisgeropoulos, D.W. Mulder, and P.W. King were supported by the Office of Basic Energy Sciences, Division of Chemical Sciences, Geosciences, and Biosciences, Photosynthetic Systems and Physical Biosciences Programs. The U.S. Government and the publisher, by accepting the article for publication, acknowledges that the U.S. Government retains a nonexclusive, paid-up, irrevocable, worldwide license to publish or reproduce the published form of this work, or allow others to do so, for U.S. Government purposes.

G.E. Vansuch thanks Michael W.W. Adams and Gerrit Schut for critically reading and commenting on an early draft of Section 11.2.2 and Seth A. Wiley and Syed M.S. Imran for helpful discussions.

Author Contributions

Research – G.E. Vansuch with contributions from E.C. Kisgeropoulos, J.R. Humphreys, and P.W. King; Original draft, writing – G.E. Vansuch (Sections 11.1, 11.2, 11.4–11.6 and parts of 3), E.C. Kisgeropoulos (Section 11.3), J.R. Humphreys (Section 11.2), C.E. Lubner (Sections 11.4 and 11.5), P.W. King (Section 11.3 and a portion of Section 11.2); sequence alignments/tables and also accession code compilation for Table 11.3 – P.W. King; Tables/Data Tabulations – G.E. Vansuch (Section 11.2), E.C. Kisgeropoulos (Section 11.3); Figures – G.E. Vansuch (Figures 11.1, 11.2, 11.5–11.8) and D.W. Mulder (Figures 11.3 and 11.4); Revisions – G.E. Vansuch, E.C. Kisgeropoulos, C.E. Lubner, D.W. Mulder, and P.W. King; Final Draft – G.E. Vansuch.

Note added during the editorial and proof processes: Since the submission of this chapter, four important papers have been published in direct relation to contents of this chapter and are cited here. Regarding putative members of the Bfu hydrogenase family is Ref. [256]. Regarding dynamics in electron bifurcation is Ref. [257]. Regarding ferredoxin binding in the Bfu family are Refs. [259, 260].

References

1 Fontecilla-Camps, J.C., Volbeda, A., Cavazza, C., and Nicolet, Y. (2007). Structure/function relationships of [NiFe]- and [FeFe]-hydrogenases. *Chem. Rev.* 107 (10): 4273–4303.

2 Schwartz, E., Fritsch, J., and Friedrich, B. (2013). H_2-metabolizing prokaryotes. In: *The Prokaryotes: Prokaryotic Physiology and Biochemistry* (ed. E. Rosenberg, E.F. DeLong, S. Lory, et al.), 119–199. Berlin, Heidelberg: Springer Berlin Heidelberg.

3 Greening, C., Biswas, A., Carere, C.R. et al. (2016). Genomic and metagenomic surveys of hydrogenase distribution indicate H_2 is a widely utilised energy source for microbial growth and survival. *ISME J.* 10 (3): 761–777.

4 Vignais, P.M., Billoud, B., and Meyer, J. (2001). Classification and phylogeny of hydrogenases. *FEMS Microbiol. Rev.* 25 (4): 455–501.

5 Søndergaard, D., Pedersen, C.N.S., and Greening, C. (2016). HydDB: a web tool for hydrogenase classification and analysis. *Sci. Rep.* 6 (1): 34212.

6 Peters, J.W., Lanzilotta, W.N., Lemon, B.J., and Seefeldt, L.C. (1998). X-ray crystal structure of the Fe-only hydrogenase (*Cp*I) from *Clostridium pasteurianum* to 1.8 angstrom resolution. *Science* 282 (5395): 1853–1858.

7 Losey, N.A., Mus, F., Peters, J.W. et al. (2017). *Syntrophomonas wolfei* uses an NADH-dependent, ferredoxin-independent [FeFe]-hydrogenase to reoxidize NADH. *Appl. Environ. Microbiol.* 83 (20): e01335-17.

8 Verhagen, M.F.J.M., O'Rourke, T., and Adams, M.W.W. (1999). The hyperthermophilic bacterium, *Thermotoga maritima*, contains an unusually complex iron-hydrogenase: amino acid sequence analyses versus biochemical characterization. *Biochim. Biophys. Acta, Bioenerg.* 1412 (3): 212–229.

9 Schut, G.J., Haja, D.K., Feng, X. et al. (2022). An abundant and diverse new family of electron bifurcating enzymes with a non-canonical catalytic mechanism. *Front. Microbiol.* 13: 946711.

10 Schuchmann, K., Chowdhury, N.P., and Müller, V. (2018). Complex multimeric [FeFe] hydrogenases: biochemistry, physiology and new opportunities for the hydrogen economy. *Front. Microbiol.* 9: 2911.

11 Losey, N.A., Poudel, S., Boyd, E.S., and McInerney, M.J. (2020). The beta subunit of non-bifurcating NADH-dependent [FeFe]-hydrogenases differs from those of multimeric electron-bifurcating [FeFe]-hydrogenases. *Front. Microbiol.* 11: 1109.

12 Zuchan, K., Baymann, F., Baffert, C. et al. (2021). The dyad of the Y-junction- and a flavin module unites diverse redox enzymes. *Biochim. Biophys. Acta, Bioenerg.* 1862 (6): 148401.

13 Poudel, S., Tokmina-Lukaszewska, M., Colman, D.R. et al. (2016). Unification of [FeFe]-hydrogenases into three structural and functional groups. *Biochim. Biophys. Acta, Gen. Subj.* 1860 (9): 1910–1921.

14 Furlan, C., Chongdar, N., Gupta, P. et al. (2022). Structural insight on the mechanism of an electron-bifurcating [FeFe] hydrogenase. *elife* 11: e79361.

15 Kpebe, A., Benvenuti, M., Guendon, C. et al. (2018). A new mechanistic model for an O_2-protected electron-bifurcating hydrogenase, Hnd from *Desulfovibrio fructosovorans*. *Biochim. Biophys. Acta, Bioenerg.* 1859 (12): 1302–1312.

16 Malki, S., Saimmaime, I., De Luca, G. et al. (1995). Characterization of an operon encoding an NADP-reducing hydrogenase in *Desulfovibrio fructosovorans*. *J. Bacteriol.* 177 (10): 2628–2636.

17 Volbeda, A., Charon, M.-H., Piras, C. et al. (1995). Crystal structure of the nickel–iron hydrogenase from *Desulfovibrio gigas*. *Nature* 373 (6515): 580–587.

18 Higuchi, Y., Okamoto, T., Fujimoto, K. et al. (1994). Location of active sites of NiFe hydrogenase determined by the combination of multiple isomorphous replacement and multiwavelength anomalous-diffraction methods. *Acta Cryst. D* 50 (5): 781–785.

19 Silva, P.J., de Castro, B., and Hagen, W.R. (1999). On the prosthetic groups of the NiFe sulfhydrogenase from *Pyrococcus furiosus*: topology, structure, and temperature-dependent redox chemistry. *J. Biol. Inorg. Chem.* 4 (3): 284–291.

20 Horch, M., Lauterbach, L., Lenz, O. et al. (2012). NAD(H)-coupled hydrogen cycling – structure–function relationships of bidirectional [NiFe] hydrogenases. *FEBS Lett.* 586 (5): 545–556.

21 Shomura, Y., Taketa, M., Nakashima, H. et al. (2017). Structural basis of the redox switches in the NAD^+-reducing soluble [NiFe]-hydrogenase. *Science* 357 (6354): 928–932.

22 Feng, X., Schut, G.J., Haja, D.K. et al. (2022). Structure and electron transfer pathways of an electron-bifurcating NiFe-hydrogenase. *Sci. Adv.* 8 (8): eabm7546.

23 Kaster, A.-K., Moll, J., Parey, K., and Thauer, R.K. (2011). Coupling of ferredoxin and heterodisulfide reduction via electron bifurcation in hydrogenotrophic methanogenic archaea. *Proc. Natl. Acad. Sci. U.S.A.* 108 (7): 2981–2986.

24 Appel, L., Willistein, M., Dahl, C. et al. (2021). Functional diversity of prokaryotic HdrA(BC) modules: role in flavin-based electron bifurcation processes and beyond. *Biochim. Biophys. Acta, Bioenerg.* 1862 (4): 148379.

25 Yan, Z. and Ferry, J.G. (2018). Electron bifurcation and confurcation in methanogenesis and reverse methanogenesis. *Front. Microbiol.* 9: 1322.

26 Wagner, T., Koch, J., Ermler, U., and Shima, S. (2017). Methanogenic heterodisulfide reductase (HdrABC-MvhAGD) uses two noncubane [4Fe-4S] clusters for reduction. *Science* 357 (6352): 699–703.

27 Thauer, R.K., Kaster, A.-K., Goenrich, M. et al. (2010). Hydrogenases from methanogenic archaea, nickel, a novel cofactor, and H_2 storage. *Annu. Rev. Biochem.* 79 (1): 507–536.

28 Milton, R.D., Ruth, J.C., Deutzmann, J.S., and Spormann, A.M. (2018). *Methanococcus maripaludis* employs three functional heterodisulfide reductase complexes for flavin-based electron bifurcation using hydrogen and formate. *Biochemistry* 57 (32): 4848–4857.

29 Costa, K.C., Lie, T.J., Xia, Q., and Leigh, J.A. (2013). VhuD facilitates electron flow from H_2 or formate to heterodisulfide reductase in *Methanococcus maripaludis*. *J. Bacteriol.* 195 (22): 5160–5165.

30 Schut, G.J. and Adams, M.W.W. (2009). The iron-hydrogenase of *Thermotoga maritima* utilizes ferredoxin and NADH synergistically: a new perspective on anaerobic hydrogen production. *J. Bacteriol.* 191 (13): 4451–4457.

31 Brandt, U. (1996). Bifurcated ubihydroquinone oxidation in the cytochrome bc_1 complex by proton-gated charge transfer. *FEBS Lett.* 387 (1): 1–6.

32 Herrmann, G., Jayamani, E., Mai, G., and Buckel, W. (2008). Energy conservation via electron-transferring flavoprotein in anaerobic bacteria. *J. Bacteriol.* 190 (3): 784–791.

33 Li, F., Hinderberger, J., Seedorf, H. et al. (2008). Coupled ferredoxin and crotonyl coenzyme A (CoA) reduction with NADH catalyzed by the butyryl-CoA dehydrogenase/Etf complex from *Clostridium kluyveri*. *J. Bacteriol.* 190 (3): 843–850.

34 Thauer, R.K., Kaster, A.-K., Seedorf, H. et al. (2008). Methanogenic archaea: ecologically relevant differences in energy conservation. *Nat. Rev. Microbiol.* 6 (8): 579–591.

35 Costa, K.C., Wong, P.M., Wang, T. et al. (2010). Protein complexing in a methanogen suggests electron bifurcation and electron delivery from formate to heterodisulfide reductase. *Proc. Natl. Acad. Sci. U.S.A.* 107 (24): 11050–11055.

36 Pereira, I.A.C., Ramos, A. R., Grein, F. et al. (2011). A comparative genomic analysis of energy metabolism in sulfate reducing bacteria and archaea. *Front. Microbiol.* 2: 69.

37 Cordes, M. and Giese, B. (2009). Electron transfer in peptides and proteins. *Chem. Soc. Rev.* 38 (4): 892–901.

38 Vanoni, M.A. (2021). Iron-sulfur flavoenzymes: the added value of making the most ancient redox cofactors and the versatile flavins work together. *Open Biol.* 11 (5): 210010.

39 Gray, H.B. and Winkler, J.R. (2003). Electron tunneling through proteins. *Q. Rev. Biophys.* 36 (3): 341–372.

40 Page, C.C., Moser, C.C., Chen, X., and Dutton, P.L. (1999). Natural engineering principles of electron tunnelling in biological oxidation–reduction. *Nature* 402 (6757): 47–52.

41 Crofts, A.R. and Rose, S. (2007). Marcus treatment of endergonic reactions: a commentary. *Biochim. Biophys. Acta, Bioenerg.* 1767 (10): 1228–1232.

42 Thauer, R.K., Jungermann, K., and Decker, K. (1977). Energy conservation in chemotrophic anaerobic bacteria. *Bacteriol. Rev.* 41 (1): 100–180.

43 Schoepp-Cothenet, B., van Lis, R., Atteia, A. et al. (2013). On the universal core of bioenergetics. *Biochim. Biophys. Acta, Bioenerg.* 1827 (2): 79–93.

44 Nitschke, W., Schoepp-Cothenet, B., Duval, S. et al. (2022). Aqueous electrochemistry: the toolbox for life's emergence from redox disequilibria. *Electrochem. Sci. Adv.* 3 (2): e2100192.

45 Sousa, F.L., Thiergart, T., Landan, G. et al. (2013). Early bioenergetic evolution. *Philos. Trans. R. Soc. Lond. Ser. B Biol. Sci.* 368 (1622): 20130088.

46 Bifurcate Meaning of Bifurcate by Lexico, https://www.lexico.com/definition/bifurcate (accessed 8 July 2024).

47 Das, A., Hessin, C., Ren, Y., and Desage-El, M.M. (2020). Biological concepts for catalysis and reactivity: empowering bioinspiration. *Chem. Soc. Rev.* 49 (23): 8840–8867.

48 Wise, C.E., Ledinina, A.E., Yuly, J.L. et al. (2021). The role of thermodynamic features on the functional activity of electron bifurcating enzymes. *Biochim. Biophys. Acta, Bioenerg.* 1862 (4): 148377.

49 Müller, V., Chowdhury, N.P., and Basen, M. (2018). Electron bifurcation: a long-hidden energy-coupling mechanism. *Ann. Rev. Microbiol.* 72 (1): 331–353.

50 Buckel, W. and Thauer, R.K. (2013). Energy conservation via electron bifurcating ferredoxin reduction and proton/Na^+ translocating ferredoxin oxidation. *Biochim. Biophys. Acta, Bioenerg.* 1827 (2): 94–113.

51 Peters, J.W., Miller, A.-F., Jones, A.K. et al. (2016). Electron bifurcation. *Curr. Opin. Chem. Biol.* 31: 146–152.

52 Buckel, W. and Thauer, R.K. (2018). Flavin-based electron bifurcation, ferredoxin, flavodoxin, and anaerobic respiration with protons (Ech) or NAD^+ (Rnf) as electron acceptors: a historical review. *Front. Microbiol.* 9: 401.

53 Mitchell, P. (1975). The protonmotive Q cycle: a general formulation. *FEBS Lett.* 59 (2): 137–139.

54 Wang, S., Huang, H., Moll, J., and Thauer, R.K. (2010). $NADP^+$ reduction with reduced ferredoxin and $NADP^+$ reduction with NADH are coupled via an electron-bifurcating enzyme complex in *Clostridium kluyveri*. *J. Bacteriol.* 192 (19): 5115–5123.

55 Stephens, P.J., Jollie, D.R., and Warshel, A. (1996). Protein control of redox potentials of iron–sulfur proteins. *Chem. Rev.* 96 (7): 2491–2514.

56 Smith, E.T., Blamey, J.M., Zhou, Z.H., and Adams, M.W.W. (1995). A variable-temperature direct electrochemical study of metalloproteins from hyperthermophilic microorganisms involved in hydrogen production from pyruvate. *Biochemistry* 34 (21): 7161–7169.

57 Katsyv, A., Essig, M., Bedendi, G. et al. (2023). Characterization of ferredoxins from the thermophilic, acetogenic bacterium *Thermoanaerobacter kivui*. *FEBS J.* 290 (16): 4107–4125.

58 Thauer, R.K. (2011). Hydrogenases and the global H_2 cycle. *Eur. J. Inorg. Chem.* 2011 (7): 919–921.

59 Burton, K. and Wilson, T.H. (1953). The free-energy changes for the reduction of diphosphopyridine nucleotide and the dehydrogenation of l-malate and l-glycerol 1-phosphate. *Biochem. J.* 54 (1): 86–94.

60 Chenault, H.K. and Whitesides, G.M. (1987). Regeneration of nicotinamide cofactors for use in organic synthesis. *Appl. Biochem. Biotechnol.* 14 (2): 147–197.

61 Rodkey, F.L. (1955). Oxidation-reduction potentials of the diphosphopyridine nucleotide system. *J. Biol. Chem.* 213 (2): 777–786.

62 Rodkey, F.L. (1959). The effect of temperature on the oxidation-reduction potential of the diphosphopyridine nucleotide system. *J. Biol. Chem.* 234 (1): 188–190.

63 Rodkey, F.L. and Donovan, J.A. (1959). Oxidation-reduction potentials of the triphosphopyridine nucleotide system. *J. Biol. Chem.* 234 (3): 677–680.

64 Tietze, M., Beuchle, A., Lamla, I. et al. (2003). Redox potentials of methanophenazine and CoB-S-S-CoM, factors involved in electron transport in methanogenic archaea. *ChemBioChem* 4 (4): 333–335.

65 Laird, M.G., Adlung, N., Koivisto, J.J., and Scheller, S. (2024). Thiol-disulfide exchange kinetics and redox potential of the coenzyme M and coenzyme B heterodisulfide, an electron acceptor coupled to energy conservation in methanogenic archaea. *ChemBioChem* 25 (1): e202300595.

66 Jay, Z.J., Hunt, K.A., Chou, K.J. et al. (2020). Integrated thermodynamic analysis of electron bifurcating [FeFe]-hydrogenase to inform anaerobic metabolism and H_2 production. *Biochim. Biophys. Acta, Bioenerg.* 1861 (1): 148087.

67 Boyd, E.S., Schut, G.J., Adams, M.W.W., and Peters, J.W. (2014). Hydrogen metabolism and the evolution of biological respiration. Two separate families of enzymes that oxidize hydrogen and also produce it arose through convergent evolution. *Microbe Mag.* 9: 361–367.

68 Poudel, S., Dunham, E.C., Lindsay, M.R. et al. (2018). Origin and evolution of flavin-based electron bifurcating enzymes. *Front. Microbiol.* 9: 1762.

69 Baymann, F., Schoepp-Cothenet, B., Duval, S. et al. (2018). On the natural history of flavin-based electron bifurcation. *Front. Microbiol.* 9: 1357.

70 Schuchmann, K. and Müller, V. (2014). Autotrophy at the thermodynamic limit of life: a model for energy conservation in acetogenic bacteria. *Nat. Rev. Microbiol.* 12 (12): 809–821.

71 Selig, M., Xavier, K.B., Santos, H., and Schönheit, P. (1997). Comparative analysis of Embden-Meyerhof and Entner-Doudoroff glycolytic pathways in hyperthermophilic archaea and the bacterium *Thermotoga*. *Arch. Microbiol.* 167 (4): 217–232.

72 Wang, S., Huang, H., Kahnt, J., and Thauer, R.K. (2013). A reversible electron-bifurcating ferredoxin- and NAD-dependent [FeFe]-hydrogenase (HydABC) in *Moorella thermoacetica*. *J. Bacteriol.* 195 (6): 1267–1275.

73 Schröder, C., Selig, M., and Schönheit, P. (1994). Glucose fermentation to acetate, CO_2 and H_2 in the anaerobic hyperthermophilic eubacterium *Thermotoga maritima*: involvement of the Embden-Meyerhof pathway. *Arch. Microbiol.* 161 (6): 460–470.

74 Thauer, R. (1977). Limitation of microbial H_2-formation via fermentation. In: *Microbial Energy Conversion* (ed. H.G. Schlegel and J. Barnea), 201–204. Pergamon.

75 Singh, R., White, D., Demirel, Y. et al. (2018). Uncoupling fermentative synthesis of molecular hydrogen from biomass formation in *Thermotoga maritima*. *Appl. Environ. Microbiol.* 84 (17): e00998-18.

76 Schuchmann, K. and Müller, V. (2016). Energetics and application of heterotrophy in acetogenic bacteria. *Appl. Environ. Microbiol.* 82 (14): 4056–4069.

77 Huang, H., Wang, S., Moll, J., and Thauer, R.K. (2012). Electron bifurcation involved in the energy metabolism of the acetogenic bacterium *Moorella thermoacetica* growing on glucose or H_2 plus CO_2. *J. Bacteriol.* 194 (14): 3689–3699.

78 Poehlein, A., Schmidt, S., Kaster, A.-K. et al. (2012). An ancient pathway combining carbon dioxide fixation with the generation and utilization of a sodium ion gradient for ATP synthesis. *PLoS One* 7 (3): e33439.

79 Schuchmann, K. and Müller, V. (2012). A bacterial electron-bifurcating hydrogenase. *J. Biol. Chem.* 287 (37): 31165–31171.

80 Mayer, F. and Müller, V. (2014). Adaptations of anaerobic archaea to life under extreme energy limitation. *FEMS Microbiol. Rev.* 38 (3): 449–472.

81 Mand, T.D. and Metcalf, W.W. (2019). Energy conservation and hydrogenase function in methanogenic archaea, in particular the genus *Methanosarcina*. *Microbiol. Mol. Biol. Rev.* 83 (4): e00020-19.

82 Deppenmeier, U. and Müller, V. (2008). Life close to the thermodynamic limit: how methanogenic archaea conserve energy. In: *Bioenergetics: Energy Conservation and Conversion* (ed. G. Schäfer and H.S. Penefsky), 123–152. Berlin, Heidelberg: Springer Berlin Heidelberg.

83 Lie, T.J., Costa, K.C., Lupa, B. et al. (2012). Essential anaplerotic role for the energy-converting hydrogenase Eha in hydrogenotrophic methanogenesis. *Proc. Natl. Acad. Sci. U.S.A.* 109 (38): 15473–15478.

84 Chongdar, N., Pawlak, K., Rüdiger, O. et al. (2020). Spectroscopic and biochemical insight into an electron-bifurcating [FeFe] hydrogenase. *J. Biol. Inorg. Chem.* 25 (1): 135–149.

85 Katsyv, A., Kumar, A., Saura, P. et al. (2023). Molecular basis of the electron bifurcation mechanism in the [FeFe]-hydrogenase complex HydABC. *J. Am. Chem. Soc.* 145 (10): 5696–5709.

86 Wang, S., Huang, H., Kahnt, J. et al. (2013). NADP-specific electron-bifurcating [FeFe]-hydrogenase in a functional complex with formate dehydrogenase in *Clostridium autoethanogenum* grown on CO. *J. Bacteriol.* 195 (19): 4373–4386.

87 Mock, J., Zheng, Y., Mueller, A.P. et al. (2015). Energy conservation associated with ethanol formation from H_2 and CO_2 in *Clostridium autoethanogenum* involving electron bifurcation. *J. Bacteriol.* 197 (18): 2965–2980.

88 Zheng, Y., Kahnt, J., Kwon IH et al. (2014). Hydrogen formation and its regulation in *Ruminococcus albus*: involvement of an electron-bifurcating [FeFe]-hydrogenase, of a non-electron-bifurcating [FeFe]-hydrogenase, and of a putative hydrogen-sensing [FeFe]-hydrogenase. *J. Bacteriol.* 196 (22): 3840–3852.

89 Katsyv, A., Jain, S., Basen, M., and Müller, V. (2021). Electron carriers involved in autotrophic and heterotrophic acetogenesis in the thermophilic bacterium *Thermoanaerobacter kivui*. *Extremophiles* 25 (5): 513–526.

90 Payne, N., Kpebe, A., Guendon, C. et al. (2022). The electron-bifurcating FeFe-hydrogenase Hnd is involved in ethanol metabolism in *Desulfovibrio fructosovorans* grown on pyruvate. *Mol. Microbiol.* 117 (4): 907–920.

91 Kpebe, A., Guendon, C., Payne, N. et al. (2023). An essential role of the reversible electron-bifurcating hydrogenase Hnd for ethanol oxidation in *Solidesulfovibrio fructosivorans*. *Front. Microbiol.* 14: 1139276.

92 Jacq-Bailly, A., Benvenuti, M., Payne, N. et al. (2021). Electrochemical characterization of a complex FeFe hydrogenase, the electron-bifurcating Hnd from *Desulfovibrio fructosovorans*. *Front. Chem.* 8: 573305.

93 Nomura, S., Paczia, N., Kahnt, J., and Shima, S. (2024). Isolation of an H_2-dependent electron-bifurcating CO_2-reducing megacomplex with MvhB polyferredoxin from *Methanothermobacter marburgensis*. *FEBS J.* 291 (11): 2449–2460.

94 Setzke, E., Hedderich, R., Heiden, S., and Thauer, R.K. (1994). H_2: heterodisulfide oxidoreductase complex from *Methanobacterium thermoautotrophicum*. Composition and properties. *Eur. J. Biochem.* 220 (1): 139–148.

95 Stojanowic, A., Mander, G.J., Duin, E.C., and Hedderich, R. (2003). Physiological role of the F_{420}-non-reducing hydrogenase (Mvh) from *Methanothermobacter marburgensis*. *Arch. Microbiol.* 180 (3): 194–203.

96 Pfeiffer, M., Bestgen, H., Bürger, A., and Klein, A. (1998). The *vhuU* gene encoding a small subunit of a selenium-containing [NiFe]-hydrogenase in *Methanococcus voltae* appears to be essential for the cell. *Arch. Microbiol.* 170 (6): 418–426.

97 Sorgenfrei, O., Linder, D., Karas, M., and Klein, A. (1993). A novel very small subunit of a selenium containing [NiFe] hydrogenease of *Methanococcus voltae* is postranslationally processed by cleavage at a defined position. *Eur. J. Biochem.* 213 (3): 1355–1358.

98 Baltazar, C.S.A., Marques, M.C., Soares, C.M. et al. (2011). Nickel–iron–selenium hydrogenases – an overview. *Eur. J. Inorg. Chem.* 2011 (7): 948–962.

99 Rosenbaum, F.P. and Müller, V. (2024). A novel hexameric $NADP^+$-reducing [FeFe] hydrogenase from *Moorella thermoacetica*. *FEBS J.* 291 (3): 596–608.

100 Dermoun, Z., De Luca, G., Asso, M. et al. (2002). The NADP-reducing hydrogenase from *Desulfovibrio fructosovorans*: functional interaction between the C-terminal region of HndA and the N-terminal region of HndD subunits. *Biochim. Biophys. Acta, Bioenerg.* 1556 (2): 217–225.

101 Watanabe, T., Pfeil-Gardiner, O., Kahnt, J. et al. (2021). Three-megadalton complex of methanogenic electron-bifurcating and CO_2-fixing enzymes. *Science* 373 (6559): 1151–1156.

102 Arnér, E.S.J. (2010). Selenoproteins—what unique properties can arise with selenocysteine in place of cysteine? *Exp. Cell Res.* 316 (8): 1296–1303.

103 Pfeiffer, M., Bingemann, R., and Klein, A. (1998). Fusion of two subunits does not impair the function of a [NiFeSe]-hydrogenase in the archaeon *Methanococcus voltae*. *Eur. J. Biochem.* 256 (2): 447–452.

104 Lubitz, W., Ogata, H., Rüdiger, O., and Reijerse, E. (2014). Hydrogenases. *Chem. Rev.* 114 (8): 4081–4148.

105 Evans, R.M., Krahn, N., Weiss, J. et al. (2024). Replacing a cysteine ligand by selenocysteine in a [NiFe]-hydrogenase unlocks hydrogen production activity and addresses the role of concerted proton-coupled electron transfer in electrocatalytic reversibility. *J. Am. Chem. Soc.* 146 (25): 16971–16976.

106 Marques, M.C., Tapia, C., Gutiérrez-Sanz, O. et al. (2017). The direct role of selenocysteine in [NiFeSe] hydrogenase maturation and catalysis. *Nat. Chem. Biol.* 13 (5): 544–550.

107 Baltazar, C.S.A., Teixeira, V.H., and Soares, C.M. (2012). Structural features of [NiFeSe] and [NiFe] hydrogenases determining their different properties: a computational approach. *J. Biol. Inorg. Chem.* 17 (4): 543–555.

108 Stein, M. and Lubitz, W. (2001). The electronic structure of the catalytic intermediate Ni-C in [NiFe] and [NiFeSe] hydrogenases. *Phys. Chem. Chem. Phys.* 3 (23): 5115–5120.

109 Hedderich, R., Albracht, S.P.J., Linder, D. et al. (1992). Isolation and characterization of polyferredoxin from *Methanobacterium thermoautotrophicum.* The *mvhb* gene product of the methylviologen-reducing hydrogenase operon. *FEBS Lett.* 298 (1): 65–68.

110 Steigerwald, V.J., Pihl, T.D., and Reeve, J.N. (1992). Identification and isolation of the polyferredoxin from *Methanobacterium thermoautotrophicum* strain ΔH. *Proc. Natl. Acad. Sci. U.S.A.* 89 (15): 6929–6933.

111 Reeve, J.N., Beckler, G.S., Cram, D.S. et al. (1989). A hydrogenase-linked gene in *Methanobacterium thermoautotrophicum* strain ΔH encodes a polyferredoxin. *Proc. Natl. Acad. Sci. U.S.A.* 86 (9): 3031–3035.

112 Watanabe, T. and Shima, S. (2020). MvhB-type polyferredoxin as an electron-transfer chain in putative redox-enzyme complexes. *Chem. Lett.* 50 (2): 353–360.

113 Furlan, C. (2023). *Cryogenic Electron Microscopy Studies on Hydrogenases.* University of York.

114 Soboh, B., Linder, D., and Hedderich, R. (2004). A multisubunit membrane-bound [NiFe] hydrogenase and an NADH-dependent Fe-only hydrogenase in the fermenting bacterium *Thermoanaerobacter tengcongensis. Microbiology* 150 (7): 2451–2463.

115 Carere, C.R., Rydzak, T., Verbeke, T.J. et al. (2012). Linking genome content to biofuel production yields: a meta-analysis of major catabolic pathways among select H_2 and ethanol-producing bacteria. *BMC Microbiol.* 12 (1): 295.

116 Kelly, C.L., Pinske, C., Murphy, B.J. et al. (2015). Integration of an [FeFe]-hydrogenase into the anaerobic metabolism of *Escherichia coli. Biotechnol. Rep.* 8: 94–104.

117 Cha, M., Chung, D., and Westpheling, J. (2016). Deletion of a gene cluster for [Ni-Fe] hydrogenase maturation in the anaerobic hyperthermophilic bacterium *Caldicellulosiruptor bescii* identifies its role in hydrogen metabolism. *Appl. Microbiol. Biotechnol.* 100 (4): 1823–1831.

118 Kröninger, L., Steiniger, F., Berger, S. et al. (2019). Energy conservation in the gut microbe *Methanomassiliicoccus luminyensis* is based on membrane-bound ferredoxin oxidation coupled to heterodisulfide reduction. *FEBS J.* 286 (19): 3831–3843.

119 Cheng, Y. (2018). Single-particle cryo-EM—how did it get here and where will it go. *Science* 361 (6405): 876–880.

120 Juszczak, A., Aono, S., and Adams, M.W.W. (1991). The extremely thermophilic eubacterium, *Thermotoga maritima*, contains a novel iron-hydrogenase whose cellular activity is dependent upon tungsten. *J. Biol. Chem.* 266 (21): 13834–13841.

121 Madeira, F., Madhusoodanan, N., Lee, J. et al. (2024). The EMBL-EBI Job Dispatcher sequence analysis tools framework in 2024. *Nucleic Acids Res.* gkae241.

122 Feng, X., Schut, G.J., Adams, M.W.W., and Li, H. (2024). Structures and electron transport paths in the four families of flavin-based electron bifurcation enzymes. In: *Macromolecular Protein Complexes V: Structure and Function* (ed. J.R. Harris and J. Marles-Wright), 383–408. Cham: Springer International Publishing.

123 Radon, C., Mittelstädt, G., Duffus, B.R. et al. (2020). Cryo-EM structures reveal intricate Fe-S cluster arrangement and charging in *Rhodobacter capsulatus* formate dehydrogenase. *Nat. Commun.* 11 (1): 1912.

124 Maret, W. and Li, Y. (2009). Coordination dynamics of zinc in proteins. *Chem. Rev.* 109 (10): 4682–4707.

125 Chen, J., Calderone, L.A., Pan, L. et al. (2023). The Fe and Zn cofactor dilemma. *Biochim. Biophys. Acta, Proteins Proteom.* 1871 (5): 140931.

126 Kumar, H. and Leimkühler, S. (2023). Changing the electron acceptor specificity of *Rhodobacter capsulatus* formate dehydrogenase from NAD^+ to $NADP^+$. *Int. J. Mol. Sci.* 24 (22): 16067.

127 Hagen, W.R. (1992). EPR spectroscopy of iron-sulfur proteins. *Adv. Inorg. Chem.* 38: 165–222.

128 Guigliarelli, B. and Bertrand, P. (1999). Application of EPR spectroscopy to the structural and functional study of iron-sulfur proteins. *Adv. Inorg. Chem.* 47: 421–497.

129 Hagen, W.R. (2006). EPR spectroscopy as a probe of metal centres in biological systems. *Dalton Trans.* (37): 4415–4434.

130 Smith, E.T. and Adams, M.W.W. (1994). Identification of an unusual paramagnetic species and of three [2Fe-2S] clusters in the iron-only hydrogenase from the hyperthermophilic bacterium *Thermotoga maritima*. *Biochim. Biophys. Acta, Proteins Proteom.* 1206 (1): 105–112.

131 Cammack, R. (1992). Iron-sulfur clusters in enzymes: themes and variations. *Adv. Inorg. Chem.* 38: 281–322.

132 Kisgeropoulos, E.C., Artz, J.H., Blahut, M. et al. (2024). Properties of the iron-sulfur cluster electron transfer relay in an [FeFe]-hydrogenase that is tuned for H_2 oxidation catalysis. *J. Biol. Chem.* 300 (6): 107292.

133 Mathews, R., Charlton, S., Sands, R.H., and Palmer, G. (1974). On the nature of the spin coupling between the iron-sulfur clusters in the eight-iron ferredoxins. *J. Biol. Chem.* 249 (13): 4326–4328.

134 Fournel, A., Gambarelli, S., Guigliarelli, B. et al. (1998). Magnetic interactions between a $[4Fe–4S]^{1+}$ cluster and a flavin mononucleotide radical in the enzyme trimethylamine dehydrogenase: a high-field electron paramagnetic resonance study. *J. Chem. Phys.* 109 (24): 10905–10913.

135 Birrell, J.A., Laurich, C., Reijerse, E.J. et al. (2016). Importance of hydrogen bonding in fine tuning the [2Fe-2S] cluster redox potential of HydC from *Thermotoga maritima*. *Biochemistry* 55 (31): 4344–4355.

136 Verhagen, M.F.J.M., O'Rourke, T.W., Menon, A.L., and Adams, M.W.W. (2001). Heterologous expression and properties of the γ-subunit of the Fe-only hydrogenase from *Thermotoga maritima*. *Biochim. Biophys. Acta, Bioenerg.* 1505 (2): 209–219.

137 De Luca, G., Asso, M., Bélaïch, J.-P., and Dermoun, Z. (1998). Purification and characterization of the HndA subunit of NADP-reducing hydrogenase from

Desulfovibrio fructosovorans overproduced in *Escherichia coli*. *Biochemistry* 37 (8): 2660–2665.

138 Salerno, J.C. (1985). Electron-paramagnetic-resonance lineshapes of biological molecules: some effects of distributed parameters. *Biochem. Soc. Trans.* 13 (3): 611–615.

139 Kumar, A., Kremp, F., Roth, J. et al. (2023). Molecular architecture and electron transfer pathway of the Stn family transhydrogenase. *Nat. Commun.* 14 (1): 5484.

140 Zambrano, I.C., Kowal, A.T., Mortenson, L.E. et al. (1989). Magnetic circular dichroism and electron paramagnetic resonance studies of hydrogenases I and II from *Clostridium pasteurianum*. *J. Biol. Chem.* 264 (35): 20974–20983.

141 Lubner, C.E., Artz, J.H., Mulder, D.W. et al. (2022). A site-differentiated [4Fe–4S] cluster controls electron transfer reactivity of *Clostridium acetobutylicum* [FeFe]-hydrogenase I. *Chem. Sci.* 13 (16): 4581–4588.

142 Gamelin, D.R., Bominaar, E.L., Kirk, M.L. et al. (1996). Excited-state contributions to ground-state properties of mixed-valence dimers: spectral and electronic-structural studies of $[Fe_2(OH)_3(tmtacn)_2]^{2+}$ related to the $[Fe_2S_2]^+$ active sites of plant-type ferredoxins. *J. Am. Chem. Soc.* 118 (34): 8085–8097.

143 Noodleman, L. (1991). Exchange coupling and resonance delocalization in reduced iron-sulfur $[Fe_4S_4]^+$ and iron-selenium $[Fe_4Se_4]^+$ clusters. 2. A generalized nonlinear model for spin-state energies and EPR and hyperfine properties. *Inorg. Chem.* 30 (2): 256–264.

144 Lindahl, P.A., Day, E.P., Kent, T.A. et al. (1985). Mössbauer, EPR, and magnetization studies of the *Azotobacter vinelandii* Fe protein. Evidence for a $[4Fe\text{-}4S]^{1+}$ cluster with spin S = 3/2. *J. Biol. Chem.* 260 (20): 11160–11173.

145 Onate, Y.A., Finnegan, M.G., Hales, B.J., and Johnson, M.K. (1993). Variable temperature magnetic circular dichroism studies of reduced nitrogenase iron proteins and $[4Fe\text{-}4S]^+$ synthetic analog clusters. *Biochim. Biophys. Acta, Protein Struct. Mol. Enzymol.* 1164 (2): 113–123.

146 Greene, B.L., Schut, G.J., Adams, M.W.W., and Dyer, R.B. (2017). Pre-steady-state kinetics of catalytic intermediates of an [FeFe]-hydrogenase. *ACS Catal.* 7 (3): 2145–2150.

147 Silakov, A., Kamp, C., Reijerse, E. et al. (2009). Spectroelectrochemical characterization of the active site of the [FeFe] hydrogenase HydA1 from *Chlamydomonas reinhardtii*. *Biochemistry* 48 (33): 7780–7786.

148 Sommer, C., Adamska-Venkatesh, A., Pawlak, K. et al. (2017). Proton coupled electronic rearrangement within the H-cluster as an essential step in the catalytic cycle of [FeFe] hydrogenases. *J. Am. Chem. Soc.* 139 (4): 1440–1443.

149 Winkler, M., Senger, M., Duan, J. et al. (2017). Accumulating the hydride state in the catalytic cycle of [FeFe]-hydrogenases. *Nat. Commun.* 8 (1): 16115.

150 Bertrand, P., Gayda, J.-P., Fee, J.A. et al. (1987). Comparison of the spin-lattice relaxation properties of the two classes of [2Fe-2S] clusters in proteins. *Biochim. Biophys. Acta, Protein Struct. Mol. Enzymol.* 916 (1): 24–28.

151 Verdaguer, M. and Robert, V. (2013). 8.04 - Fundamentals, principles, and concepts of molecular magnetism and spintronics. In: *Comprehensive Inorganic*

Chemistry II, 2e (ed. J. Reedijk and K. Poeppelmeier), 131–189. Amsterdam: Elsevier.

152 Darensbourg, M.Y., Lyon, E.J., and Smee, J.J. (2000). The bio-organometallic chemistry of active site iron in hydrogenases. *Coord. Chem. Rev.* 206–207: 533–561.

153 Kamp, C., Silakov, A., Winkler, M. et al. (2008). Isolation and first EPR characterization of the [FeFe]-hydrogenases from green algae. *Biochim. Biophys. Acta, Bioenerg.* 1777 (5): 410–416.

154 Lorent, C., Katz, S., Duan, J. et al. (2020). Shedding light on proton and electron dynamics in [FeFe] hydrogenases. *J. Am. Chem. Soc.* 142 (12): 5493–5497.

155 Hendrickson, E., Kaul, R., Zhou, Y. et al. (2004). Complete genome sequence of the genetically tractable hydrogenotrophic methanogen *Methanococcus maripaludis*. *J. Bacteriol.* 186 (20): 6956–6969.

156 Pandelia, M.-E., Nitschke, W., Infossi, P. et al. (2011). Characterization of a unique [Fe–S] cluster in the electron transfer chain of the oxygen tolerant [NiFe] hydrogenase from *Aquifex aeolicus*. *Proc. Natl. Acad. Sci. U.S.A.* 108 (15): 6097–6102.

157 Roessler, M.M., Evans, R.M., Davies, R.A. et al. (2012). EPR spectroscopic studies of the Fe–S clusters in the O_2-tolerant [NiFe]-hydrogenase Hyd-1 from *Escherichia coli* and characterization of the unique [4Fe–3S] cluster by HYSCORE. *J. Am. Chem. Soc.* 134 (37): 15581–15594.

158 Yamazaki, S. (1982). A selenium-containing hydrogenase from *Methanococcus vannielii*. Identification of the selenium moiety as a selenocysteine residue. *J. Biol. Chem.* 257 (14): 7926–7929.

159 Jones, J. and Stadtman, T. (1981). Selenium-dependent and selenium-independent formate dehydrogenases of *Methanococcus vannielii*. Separation of the two forms and characterization of the purified selenium-independent form. *J. Biol. Chem.* 256 (2): 656–663.

160 Wells, M., Basu, P., and Stolz, J.F. (2021). The physiology and evolution of microbial selenium metabolism. *Metallomics* 13 (6): mfab024.

161 Hohn, M.J., Palioura, S., Su, D. et al. (2011). Genetic analysis of selenocysteine biosynthesis in the archaeon *Methanococcus maripaludis*. *Mol. Microbiol.* 81 (1): 249–258.

162 Böck, A. and Rother, M. (2005). A pseudo-SECIS element in *Methanococcus voltae* documents evolution of a selenoprotein into a sulphur-containing homologue. *Arch. Microbiol.* 183 (2): 148–150.

163 Coremans, J.M.C.C., Van der Zwaan, J.W., and Albracht, S.P.J. (1989). Redox behaviour of nickel in hydrogenase from *Methanobacterium thermoautotrophicum* (strain Marburg). Correlation between the nickel valence state and enzyme activity. *Biochim. Biophys. Acta, Protein Struct. Mol. Enzymol.* 997 (3): 256–267.

164 Albracht, S.P.J. (1994). Nickel hydrogenases: in search of the active site. *Biochim. Biophys. Acta, Bioenerg.* 1188 (3): 167–204.

165 Henzler-Wildman, K. and Kern, D. (2007). Dynamic personalities of proteins. *Nature* 450 (7172): 964–972.

166 Frauenfelder, H., Sligar, S.G., and Wolynes, P.G. (1991). The energy landscapes and motions of proteins. *Science* 254 (5038): 1598–1603.

167 Gora, A., Brezovsky, J., and Damborsky, J. (2013). Gates of enzymes. *Chem. Rev.* 113 (8): 5871–5923.

168 Klinman, J.P. (2023). Dynamical activation of function in metalloenzymes. *FEBS Lett.* 597 (1): 79–91.

169 Evans, D.H. (2008). One-electron and two-electron transfers in electrochemistry and homogeneous solution reactions. *Chem. Rev.* 108 (7): 2113–2144.

170 Nitschke, W. and Russell, M.J. (2012). Redox bifurcations: mechanisms and importance to life now, and at its origin. A widespread means of energy conversion in biology unfolds…. *BioEssays* 34 (2): 106–109.

171 Peters, J.W., Beratan, D.N., Schut, G.J., and Adams, M.W.W. (2018). On the nature of organic and inorganic centers that bifurcate electrons, coupling exergonic and endergonic oxidation–reduction reactions. *Chem. Commun.* 54 (33): 4091–4099.

172 Imran, S.M.S., Wiley, S.A., and Lubner, C.E. (2024). Electrochemistry of flavin-based electron bifurcation: 'current' past and 'potential' futures. *Curr. Opin. Electrochem.* 47: 101536.

173 Miura, R. (2001). Versatility and specificity in flavoenzymes: control mechanisms of flavin reactivity. *Chem. Rec.* 1 (3): 183–194.

174 Kayastha, K., Vitt, S., Buckel, W., and Ermler, U. (2021). Flavins in the electron bifurcation process. *Arch. Biochem. Biophys.* 701: 108796.

175 Wise, C.E., Ledinina, A.E., Mulder, D.W. et al. (2022). An uncharacteristically low-potential flavin governs the energy landscape of electron bifurcation. *Proc. Natl. Acad. Sci. U.S.A.* 119 (12): e2117882119.

176 Lubner, C.E., Jennings, D.P., Mulder, D.W. et al. (2017). Mechanistic insights into energy conservation by flavin-based electron bifurcation. *Nat. Chem. Biol.* 13 (6): 655–659.

177 Berry, L., Poudel, S., Tokmina-Lukaszewska, M. et al. (2018). H/D exchange mass spectrometry and statistical coupling analysis reveal a role for allostery in a ferredoxin-dependent bifurcating transhydrogenase catalytic cycle. *Biochim. Biophys. Acta, Gen. Subj.* 1862 (1): 9–17.

178 Yuly, J.L., Lubner, C.E., Zhang, P. et al. (2019). Electron bifurcation: progress and grand challenges. *Chem. Commun.* 55 (79): 11823–11832.

179 Demmer, J.K., Chowdhury, N.P., Selmer, T. et al. (2017). The semiquinone swing in the bifurcating electron transferring flavoprotein/butyryl-CoA dehydrogenase complex from *Clostridium difficile*. *Nat. Commun.* 8 (1): 1577.

180 Hoben, J.P., Lubner, C.E., Ratzloff, M.W. et al. (2017). Equilibrium and ultrafast kinetic studies manipulating electron transfer: a short-lived flavin semiquinone is not sufficient for electron bifurcation. *J. Biol. Chem.* 292 (34): 14039–14049.

181 Wise, C.E., Ledinina, A.E., and Lubner, C.E. (2022). Site-differentiated iron-sulfur cluster ligation affects flavin-based electron bifurcation activity. *Metabolites* 12 (9): 823.

182 Osyczka, A., Moser, C.C., Daldal, F., and Dutton, P.L. (2004). Reversible redox energy coupling in electron transfer chains. *Nature* 427 (6975): 607–612.

183 Yuly, J.L., Zhang, P., Ru, X. et al. (2021). Efficient and reversible electron bifurcation with either normal or inverted potentials at the bifurcating cofactor. *Chem* 7 (7): 1870–1886.

184 Yuly, J.L., Zhang, P., and Beratan, D.N. (2021). Energy transduction by reversible electron bifurcation. *Curr. Opin. Electrochem.* 29: 100767.

185 Fu, W., Drozdzewski, P.M., Morgan, T.V. et al. (1993). Resonance Raman studies of iron-only hydrogenases. *Biochemistry* 32 (18): 4813–4819.

186 Greene, B.L., Wu, C.-H., McTernan, P.M. et al. (2015). Proton-coupled electron transfer dynamics in the catalytic mechanism of a [NiFe]-hydrogenase. *J. Am. Chem. Soc.* 137 (13): 4558–4566.

187 Ash, P.A., Kendall-Price, S.E.T., and Vincent, K.A. (2019). Unifying activity, structure, and spectroscopy of [NiFe] hydrogenases: combining techniques to clarify mechanistic understanding. *Acc. Chem. Res.* 52 (11): 3120–3131.

188 Chongdar, N., Birrell, J.A., Pawlak, K. et al. (2018). Unique spectroscopic properties of the H-Cluster in a putative sensory [FeFe] hydrogenase. *J. Am. Chem. Soc.* 140 (3): 1057–1068.

189 Corrigan, P.S., Majer, S.H., and Silakov, A. (2023). Evidence of atypical structural flexibility of the active site surrounding of an [FeFe] hydrogenase from *Clostridium beijerinkii. J. Am. Chem. Soc.* 145 (20): 11033–11044.

190 Preissler, J., Wahlefeld, S., Lorent, C. et al. (2018). Enzymatic and spectroscopic properties of a thermostable [NiFe]-hydrogenase performing H_2-driven NAD^+-reduction in the presence of O_2. *Biochim. Biophys. Acta, Bioenerg.* 1859 (1): 8–18.

191 Tokmina-Lukaszewska, M., Huang, Q., Berry, L. et al. (2023). Fe protein docking transduces conformational changes to MoFe nitrogenase active site in a nucleotide-dependent manner. *Commun. Chem.* 6 (1): 254.

192 Rodríguez-Maciá, P., Pawlak, K., Rüdiger, O. et al. (2017). Intercluster redox coupling influences protonation at the H-cluster in [FeFe] hydrogenases. *J. Am. Chem. Soc.* 139 (42): 15122–15134.

193 Zhu, W., Iavarone, A.T., and Klinman, J.P. (2024). Hydrogen–deuterium exchange mass spectrometry identifies local and long-distance interactions within the multicomponent radical SAM enzyme, PqqE. *ACS Cent. Sci.* 10 (2): 251–263.

194 Sazanov, L.A. and Hinchliffe, P. (2006). Structure of the hydrophilic domain of respiratory complex I from *Thermus thermophilus. Science* 311 (5766): 1430–1436.

195 Pan, X., Cao, D., Xie, F. et al. (2020). Structural basis for electron transport mechanism of complex I-like photosynthetic NAD(P)H dehydrogenase. *Nat. Commun.* 11 (1): 610.

196 Schuller, J.M., Birrell, J.A., Tanaka, H. et al. (2019). Structural adaptations of photosynthetic complex I enable ferredoxin-dependent electron transfer. *Science* 363 (6424): 257–260.

197 PyMOL (2010). *The PyMOL Molecular Graphics System, Version 1.20.* New York: Schrödinger, LLC.

198 Hayashi, T. and Stuchebrukhov, A.A. (2011). Quantum electron tunneling in respiratory complex I. *J. Phys. Chem. B* 115 (18): 5354–5364.

199 Kümmerle, R., Gaillard, J., Kyritsis, P., and Moulis, J.-M. (2001). Intramolecular electron transfer in [4Fe-4S] proteins: estimates of the reorganization energy and electronic coupling in *Chromatium vinosum* ferredoxin. *J. Biol. Inorg. Chem.* 6 (4): 446–451.

200 Daizadeh, I., Medvedev, D.M., and Stuchebrukhov, A.A. (2002). Electron transfer in ferredoxin: are tunneling pathways evolutionarily conserved? *Mol. Biol. Evol.* 19 (4): 406–415.

201 Davidson, V.L. (2008). Protein control of true, gated, and coupled electron transfer reactions. *Acc. Chem. Res.* 41 (6): 730–738.

202 Davidson, V.L. (2000). What controls the rates of interprotein electron-transfer reactions. *Acc. Chem. Res.* 33 (2): 87–93.

203 Harris, T.K., Davidson, V.L., Chen, L. et al. (1994). Ionic strength dependence of the reaction between methanol dehydrogenase and cytochrome *c*-551i: evidence of conformationally coupled electron transfer. *Biochemistry* 33 (42): 12600–12608.

204 Balabin, I.A., Beratan, D.N., and Skourtis, S.S. (2008). Persistence of structure over fluctuations in biological electron-transfer reactions. *Phys. Rev. Lett.* 101 (15): 158102.

205 Kretchmer, J.S., Boekelheide, N., Warren, J.J. et al. (2018). Fluctuating hydrogen-bond networks govern anomalous electron transfer kinetics in a blue copper protein. *Proc. Natl. Acad. Sci. U.S.A.* 115 (24): 6129–6134.

206 Shen, W., Teo, R.D., Beratan, D.N., and Warren, J.J. (2022). Cofactor dynamics couples the protein surface to the heme in cytochrome *c*, facilitating electron transfer. *J. Phys. Chem. B* 126 (19): 3522–3529.

207 Leferink, N.G.H., Pudney, C.R., Brenner, S. et al. (2012). Gating mechanisms for biological electron transfer: integrating structure with biophysics reveals the nature of redox control in cytochrome P450 reductase and copper-dependent nitrite reductase. *FEBS Lett.* 586 (5): 578–584.

208 Zuchan, K., Nitschke, W., Baymann, F. et al. (2022). NAD(H)-induced flavosemiquinone stabilization in the electron-confurcating [FeFe]-hydrogenase (HydABC) of *Thermotoga maritima*. *Biochim. Biophys. Acta, Bioenerg.* 1863: 148758.

209 Zhang, Z., Huang, L., Shulmeister, V.M. et al. (1998). Electron transfer by domain movement in cytochrome bc_1. *Nature* 392 (6677): 677–684.

210 Darrouzet, E., Valkova-Valchanova, M., Moser, C.C. et al. (2000). Uncovering the [2Fe2S] domain movement in cytochrome bc_1 and its implications for energy conversion. *Proc. Natl. Acad. Sci. U.S.A.* 97 (9): 4567–4572.

211 Crofts, A.R. (2021). The modified Q-cycle: a look back at its development and forward to a functional model. *Biochim. Biophys. Acta, Bioenerg.* 1862 (8): 148417.

212 Kayastha, K., Katsyv, A., Himmrich, C. et al. (2022). Structure-based electron-confurcation mechanism of the Ldh-EtfAB complex. *elife* 11: e77095.

213 Murray, D.T., Ge, X., Schut, G.J. et al. (2024). Correlating conformational equilibria with catalysis in the electron bifurcating EtfABCX of *Thermotoga maritima*. *Biochemistry* 63 (1): 128–140.

214 Hamann, N., Mander, G.J., Shokes, J.E. et al. (2007). A cysteine-rich CCG domain contains a novel [4Fe-4S] cluster binding motif as deduced from studies

with subunit B of heterodisulfide reductase from *Methanothermobacter marburgensis*. *Biochemistry* 46 (44): 12875–12885.

215 Pelmenschikov, V., Ferreira, D., Venceslau, S.S. et al. (2023). Substrate-dependent conformational switch of the noncubane [4Fe-4S] cluster in heterodisulfide reductase HdrB. *J. Am. Chem. Soc.* 145 (1): 7–11.

216 Coremans J.M.C.C., van Garderen C.J., and Albracht S.P.J. (1992). On the redox equilibrium between H_2 and hydrogenase. *Biochim. Biophys. Acta, Protein Struct. Mol. Enzymol.* 1119 (2): 148–156.

217 Huwiler, S.G., Löffler, C., Anselmann, S.E.L. et al. (2019). One-megadalton metalloenzyme complex in *Geobacter metallireducens* involved in benzene ring reduction beyond the biological redox window. *Proc. Natl. Acad. Sci. U.S.A.* 116 (6): 2259–2264.

218 Léger, C., Lederer, F., Guigliarelli, B., and Bertrand, P. (2006). Electron flow in multicenter enzymes: theory, applications, and consequences on the natural design of redox chains. *J. Am. Chem. Soc.* 128 (1): 180–187.

219 Wang, M., Gao, J., Müller, P., and Giese, B. (2009). Electron transfer in peptides with cysteine and methionine as relay amino acids. *Angew. Chem. Int. Ed.* 48 (23): 4232–4234.

220 Banci, L., Bertini, I., Calderone, V. et al. (2012). An electron-transfer path through an extended disulfide relay system: the case of the redox protein ALR. *J. Am. Chem. Soc.* 134 (3): 1442–1445.

221 Surdhar, P.S. and Armstrong, D.A. (1987). Reduction potentials and exchange reactions of thiyl radicals and disulfide anion radicals. *J. Phys. Chem.* 91 (26): 6532–6537.

222 Madadi-Kahkesh, S., Duin, E.C., Heim, S. et al. (2001). A paramagnetic species with unique EPR characteristics in the active site of heterodisulfide reductase from methanogenic archaea. *Eur. J. Biochem.* 268 (9): 2566–2577.

223 Duin, E.C., Madadi-Kahkesh, S., Hedderich, R. et al. (2002). Heterodisulfide reductase from *Methanothermobacter marburgensis* contains an active-site [4Fe–4S] cluster that is directly involved in mediating heterodisulfide reduction. *FEBS Lett.* 512 (1): 263–268.

224 Duin, E.C., Bauer, C., Jaun, B., and Hedderich, R. (2003). Coenzyme M binds to a [4Fe–4S] cluster in the active site of heterodisulfide reductase as deduced from EPR studies with the [^{33}S]coenzyme M-treated enzyme. *FEBS Lett.* 538 (1): 81–84.

225 Bennati, M., Weiden, N., Dinse, K.-P., and Hedderich, R. (2004). ^{57}Fe ENDOR spectroscopy on the iron–sulfur cluster involved in substrate reduction of heterodisulfide reductase. *J. Am. Chem. Soc.* 126 (27): 8378–8379.

226 Shokes, J.E., Duin, E.C., Bauer, C. et al. (2005). Direct interaction of coenzyme M with the active-site Fe–S cluster of heterodisulfide reductase. *FEBS Lett.* 579 (7): 1741–1744.

227 Wu, J. and Chen, S.-L. (2022). Key piece in the Wolfe cycle of methanogenesis: the S–S bond dissociation conducted by noncubane [Fe_4S_4] cluster-dependent heterodisulfide reductase. *ACS Catal.* 12 (4): 2606–2622.

228 Hedderich, R., Hamann, N., and Bennati, M. (2005). Heterodisulfide reductase from methanogenic archaea: a new catalytic role for an iron-sulfur cluster. *Biol. Chem.* 386 (10): 961–970.

229 Mukund, S. and Adams, M.W.W. (1990). Characterization of a tungsten-iron-sulfur protein exhibiting novel spectroscopic and redox properties from the hyperthermophilic archaebacterium *Pyrococcus furiosus*. *J. Biol. Chem.* 265 (20): 11508–11516.

230 Adams, M.W.W. (1992). Novel iron-sulfur centers in metalloenzymes and redox proteins from extremely thermophilic bacteria. *Adv. Inorg. Chem.* 38: 341–396.

231 de Vries, S., Dörner, K., Strampraad, M.J.F., and Friedrich, T. (2015). Electron tunneling rates in respiratory complex I are tuned for efficient energy conversion. *Angew. Chem. Int. Ed.* 54 (9): 2844–2848.

232 Agarwal, R.G., Coste, S.C., Groff, B.D. et al. (2022). Free energies of proton-coupled electron transfer reagents and their applications. *Chem. Rev.* 122 (1): 1–49.

233 Tyburski, R., Liu, T., Glover, S.D., and Hammarström, L. (2021). Proton-coupled electron transfer guidelines, fair and square. *J. Am. Chem. Soc.* 143 (2): 560–576.

234 Reddish, M.J., Vaughn, M.B., Fu, R., and Dyer, R.B. (2016). Ligand-dependent conformational dynamics of dihydrofolate reductase. *Biochemistry* 55 (10): 1485–1493.

235 Smith, N., Dasgupta, M., Wych, D.C. et al. (2024). Changes in an enzyme ensemble during catalysis observed by high-resolution XFEL crystallography. *Sci. Adv.* 10 (13): eadk7201.

236 Stiller, J.B., Otten, R., Häussinger, D. et al. (2022). Structure determination of high-energy states in a dynamic protein ensemble. *Nature* 603 (7901): 528–535.

237 Maiuri, M., Delfino, I., Cerullo, G. et al. (2015). Low frequency dynamics of the nitrogenase MoFe protein via femtosecond pump probe spectroscopy—observation of a candidate promoting vibration. *J. Inorg. Biochem.* 153: 128–135.

238 Mao, Z., Carroll, E.C., Kim, P.W. et al. (2017). Ultrafast charge-transfer dynamics in the iron–sulfur complex of *Rhodobacter capsulatus* ferredoxin VI. *J. Phys. Chem. Lett.* 8 (18): 4498–4503.

239 Mao, Z., Liou, S.-H., Khadka, N. et al. (2018). Cluster-dependent charge-transfer dynamics in iron–sulfur proteins. *Biochemistry* 57 (6): 978–990.

240 Galinato, M.G.I., Kleingardner, J.G., Bowman, S.E.J. et al. (2012). Heme-protein vibrational couplings in cytochrome *c* provide a dynamic link that connects the heme-iron and the protein surface. *Proc. Natl. Acad. Sci. U.S.A.* 109 (23): 8896–8900.

241 Moser, C.C., Keske, J.M., Warncke, K. et al. (1992). Nature of biological electron transfer. *Nature* 355 (6363): 796–802.

242 Stubbe, J., Nocera, D.G., Yee, C.S., and Chang, M.C.Y. (2003). Radical initiation in the class I ribonucleotide reductase: long-range proton-coupled electron transfer? *Chem. Rev.* 103 (6): 2167–2202.

243 Hayashi, T. and Stuchebrukhov, A.A. (2010). Electron tunneling in respiratory complex I. *Proc. Natl. Acad. Sci. U.S.A.* 107 (45): 19157–19162.

244 Berstis, L., Beckham, G.T., and Crowley, M.F. (2015). Electronic coupling through natural amino acids. *J. Chem. Phys.* 143 (22): 225102.

245 Falzon, L. and Davidson, V.L. (1996). Intramolecular electron transfer in trimethylamine dehydrogenase: a thermodynamic analysis. *Biochemistry* 35 (37): 12111–12118.

246 Rees, D.C. (1985). Electrostatic influence on energetics of electron transfer reactions. *Proc. Natl. Acad. Sci. U.S.A.* 82 (10): 3082–3085.

247 Wang, S., Huang, H., Kahnt, J., and Thauer, R.K. (2013). *Clostridium acidurici* electron-bifurcating formate dehydrogenase. *Appl. Environ. Microbiol.* 79 (19): 6176–6179.

248 Schut, G.J., Thorgersen, M.P., Poole, F.L. et al. (2021). Tungsten enzymes play a role in detoxifying food and antimicrobial aldehydes in the human gut microbiome. *Proc. Natl. Acad. Sci. U.S.A.* 118 (43): e2109008118.

249 Kremp, F., Roth, J., and Müller, V. (2020). The *Sporomusa* type Nfn is a novel type of electron-bifurcating transhydrogenase that links the redox pools in acetogenic bacteria. *Sci. Rep.* 10 (1): 14872.

250 Yan, Z., Wang, M., and Ferry, J.G. (2017). A ferredoxin- and $F_{420}H_2$-dependent, electron-bifurcating, heterodisulfide reductase with homologs in the domains *bacteria* and *archaea*. *mBio* 8 (1): e02285-16.

251 Abdul Halim, M.F., Day, L.A., and Costa, K.C. (2021). Formate-dependent heterodisulfide reduction in a *Methanomicrobiales* archaeon. *Appl. Environ. Microbiol.* 87 (6): e02698-20.

252 Abdul Halim, M.F., Fonseca, D.R., Niehaus, T.D., and Costa, K.C. (2024). Functionally redundant formate dehydrogenases enable formate-dependent growth in *Methanococcus maripaludis*. *J. Biol. Chem.* 300 (1): 105550.

253 Fried, S.D. and Boxer, S.G. (2017). Electric fields and enzyme catalysis. *Annu. Rev. Biochem.* 86 (1): 387–415.

254 Kraut, D.A., Carroll, K.S., and Herschlag, D. (2003). Challenges in enzyme mechanism and energetics. *Annu. Rev. Biochem.* 72: 517–571.

255 Hammes, G.G., Benkovic, S.J., and Hammes-Schiffer, S. (2011). Flexibility, diversity, and cooperativity: pillars of enzyme catalysis. *Biochemistry* 50 (48): 10422–10430.

256 Zheng, X. and Huang L. (2024). Diverse non-canonical electron bifurcating [FeFe]-hydrogenases of separate evolutionary origins in *Hydrogenedentota*. *mSystems* 9 (9): e0099924.

257 Khan A.S., Hicks A., Leite W.C. et al. (2024). Extended conformations of bifurcating electron transfer flavoprotein constitute up to half the population, possibly mediating conformational change. *Chem. Sci.* 15 (45): 18796–18817.

258 de Luca, G., de Philip, P., Rousset, M. et al. (1998). The NADP-reducing hydrogenase of *Desulfovibrio fructosovorans*: evidence for a native complex with hydrogen-dependent methyl-viologen-reducing activity. *Biochem. Biophys. Res. Commun.* 248 (3): 591–596.

259 Li, H., Schut, G.J., Feng, X. et al. (2025). Cryo-EM reveals a composite flavobicluster electron bifurcation site in the Bfu family member NfnABC. *Commun. Biol.* 8: 239.

260 Xiao, X., Schut, G.J., Feng, X. et al. (2025). Cryo-EM structures define the electron bifurcating flavobicluster and ferredoxin binding site in an archaeal Nfn-Bfu transhydrogenase. *J. Biol. Chem.* 301 (4): 108410.

12

Inhibition of [FeFe]-Hydrogenases by Small Molecules

Claudia Brocks and Thomas Happe

Ruhr University Bochum, Faculty of Biology and Biotechnology, Photobiotechnology, Universitätsstraße 150, 44801 Bochum, Germany

Chapter Goals

- Understanding the stepwise progress of research: from discovery to complex enzyme understanding.
- Getting an overview of the complex catalytic mechanisms in [FeFe]-hydrogenase.
- Overview of known inhibitory molecules and gases, the so-called small molecules.
- Understanding how O_2 can attack the catalytic center of [FeFe]-hydrogenases.
- Recognition of how different the inhibitory mechanisms of small molecules can be and that some of the small molecules can be used as O_2 protectants.
- Understanding the mechanism of a naturally occurring O_2-protected [FeFe]-hydrogenase.
- Motivation to further investigate the inhibitory mechanisms and (naturally occurring) O_2-protecting effects.

12.1 Historical Perspective of [FeFe]-Hydrogenase Research

Hydrogenases are metalloenzymes that are involved in the metabolism of molecular hydrogen (H_2). They are categorized into [Fe]-hydrogenases, [NiFe]-hydrogenases, and [FeFe]-hydrogenases, depending on the metal content of their active site cofactors [1]. [NiFe]- and [FeFe]-hydrogenases catalyze the reversible reduction of protons to H_2 ($2H^+ + 2e^- \leftrightarrow H_2$), and [FeFe]-hydrogenases display very high hydrogen production rates exceeding turnover numbers of $10\,000\,s^{-1}$ [2]. How these enzymes achieve this efficiency is intensely studied, and also it makes them interesting for industrial applications.

The first scientific description of the metalloenzyme class of hydrogenases was published in 1931 [3]. Over the years, intensive research has revealed many details on the structure and catalytic mechanism of [FeFe]-hydrogenases. A selection of

Iron-Sulfur Clusters: Biogenesis and Biochemistry, Volume 2, First Edition.
Edited by Silke Leimkühler, Günter Schwarz, Oliver Lenz, and Oliver Einsle.

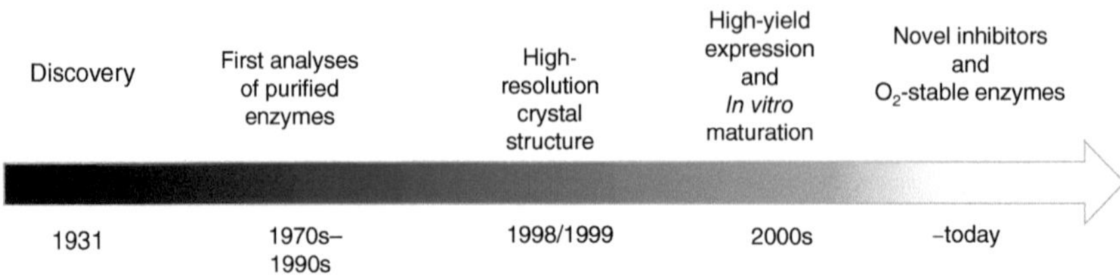

Figure 12.1 Timeline of key findings for [FeFe]-hydrogenase research. Selected findings that were key events for advancing [FeFe]-hydrogenase research are ordered chronologically from the discovery of H_2 metabolizing enzymes almost hundred years ago to rather recent and new directions of characterizing novel types of (O_2 stable) [FeFe]-hydrogenases and molecular details of the effects of inhibitors.

the key findings is summarized in Figure 12.1. One of the first [FeFe]-hydrogenases to be isolated and characterized was CpI of the anaerobic bacterium *Clostridium pasteurianum* [4], and, important for all future research, both artificial and natural electron donors were identified, such as methyl viologen, benzyl viologen, methylene blue, and ferredoxin [5]. The [FeFe]-hydrogenase from the eukaryotic unicellular green alga *Chlamydomonas reinhardtii*, CrHydA1, that, due to its small architecture, has become an important model for many spectroscopic techniques and has been analyzed in its partially purified form since the 1980s [6–8]. In the 1990s, homogenous CrHydA1 preparations were obtained and used for detailed biochemical and biophysical analyses of this enzyme. The determination of the N-terminal amino acid sequence of CrHydA1 allowed us to compare the sequences of [FeFe]-hydrogenases from eukaryotes and prokaryotes, and the low sequence homology indicated different [FeFe]-hydrogenase classes [9].

In 1998 and 1999, the first crystal structures of [FeFe]-hydrogenases were solved to quite high resolutions, namely that of CpI with a resolution of 1.8 Å (protein data bank [PDB] entry 1FEH) [10] and that of the enzyme DdH from the sulfate-reducing bacterium *Desulfovibrio desulfuricans* with a resolution of 1.6 Å (PDB entry 1HFE) [11]. With these, detailed information about the protein structure and the active center of this enzyme family was accessible for the first time. As will be elaborated on below, the active site cofactor of [FeFe]-hydrogenases, the H-cluster, consists of a standard [4Fe–4S] cluster coupled to a biologically unique subcluster of two iron ions – eponymic for these enzymes – that are coordinated to carbon monoxide (CO) and cyanide (CN^-) ligands and bridged by an azadithiolate (ADT) bridge. This unusual metal cluster cannot be synthesized in organisms that do not naturally possess [FeFe]-hydrogenases such as the well-established heterologous expression host *Escherichia coli*. This hampered research on these enzymes, especially when methods that require a lot of proteins were sought. A very important discovery in this regard was the identification of the dedicated assembly factors, or maturases, of the diiron subcluster in *C. reinhardtii* [12]. With the help of this knowledge, an expression system was established that yields high amounts of holo-[FeFe]-hydrogenases [13]: Heterologous expression is carried out in *E. coli* under anaerobic conditions, and the use of a special *E. coli* strain deficient for a negative regulator of Fe–S cluster biosynthesis, iscR [14], as well as supporting medium supplements such as iron and cysteine results in yields of up to 30 mg holo-protein per liter expression culture. The complete maturation of the [FeFe]-hydrogenase enzymes is achieved by coexpressing the genes for the three dedicated maturase enzymes HydE, HydF, and HydG [12, 15, 16]. This enzyme-based maturation process of [FeFe]-hydrogenases is referred to as *in vivo* maturation, and the molecular details of the process and the three maturating enzymes, respectively, are subject to intensive research [16–21].

A few years later, artificial maturation, known as *in vitro* maturation, was discovered. Indeed, a historical perspective on [FeFe]-hydrogenase research should also include the progress that chemists made in synthesizing chemical analogs of hydrogenase cofactors [22, 23]. For the purpose of this review, however, we only cite the study of Li and Rauchfuss [24] who synthesized a cofactor mimic that is, as is known today, indeed an almost exact copy of the H-clusters diiron subcluster. It was shown that this chemical analog could be loaded onto one of the aforementioned dedicated maturase enzymes, HydF [25], which was a highly interesting finding [18]. With regard to [FeFe]-hydrogenase research, the second highlight from this study was that such loaded HydF was capable of transferring the mimic to a so-termed apo-[FeFe]-hydrogenase [25] – an [FeFe]-hydrogenase precursor as synthesized in *E. coli* without coexpression of the maturase sequences, which only contains the [4Fe–4S]-subcluster of the H-cluster [16]. Soon thereafter it was, rather surprisingly, discovered that the synthetically produced diiron-site analog can "maturate" apo-[FeFe]-hydrogenases all by itself, that is, in the absence of HydF [26, 27]. The mimic inserts itself into the apo-enzymes and couples spontaneously with high efficiency to the [4Fe–4S] cluster. The resulting active holo-[FeFe]-hydrogenases are, so far, not distinguishable from enzymes maturated *in vivo*, neither regarding activity nor their structures [27]. Nowadays, because of its simplicity and efficiency, this *in vitro* maturation system is a widely used method that can be used to maturate high amounts of protein for diverse applications.

Based on the progress of knowledge and techniques, [FeFe]-hydrogenase research has strongly advanced in the last decade. For example, the elucidation of the amino acid-based proton transfer pathway (PTP) leading from the protein surface to the active site was aided by the availability of high protein amounts and established crystallization pipelines that allowed structural and spectroscopic analyses as well as simulations [28–32]. In the topical context of this chapter, recent studies on the interactions of inhibitor molecules with [FeFe]-hydrogenases are highlighted. For example, for most [FeFe]-hydrogenases, such as the most studied model [FeFe]-hydrogenases CpI, CrHydA1, and DdH, molecular oxygen (O_2) is an irreversible inhibitor because it results in the oxidative degradation of the H-cluster. Insights into the mechanistic details of this process have only recently been gained [33–35]. Notably, an [FeFe]-hydrogenase from the organism *Clostridium beijerinckii*, termed CbA5H, was identified in 2016 and shown to be O_2 stable [36]. This opened up new avenues of [FeFe]-hydrogenase research and much information on the mechanism of O_2-resistance could be gained recently [37–39]. In contrast to O_2, many inhibitors of [FeFe]-hydrogenases act reversibly and sometimes influence specific regions of the protein structure. The mechanisms of their actions have been subject to much attention: Interactions of sulfide molecules [40, 41], cyanide [42], formaldehyde [43], and zinc [32] with various representatives of [FeFe]-hydrogenases are being investigated, and while of relevance for applications of [FeFe]-hydrogenases, knowledge about types and action of inhibitors is equally important for understanding the physiology of H_2-metabolizing microbes that shape many biotopes. Not least, studying inhibitor–hydrogenase interactions has, in some instances, yielded unexpected insights into catalytic properties. Due to their importance, Sections 12.3 and 12.4 focus on these small molecules.

12.2 Characteristics of [FeFe]-Hydrogenases: How They Do What They Do

[FeFe]-hydrogenases are found in bacteria and are, according to current knowledge, the only hydrogenase types found in eukaryotes [44, 45]. In the latter, their occurrence is restricted to anaerobic or facultatively anaerobic eukaryotic microbes such as microalgae and protists. In anaerobic bacteria, they are mainly found in clostridia and sulfate reducers, but they also occur in Gram-negative species such as *Shewanella oneidensis* [44–46]. All [FeFe]-hydrogenases have in common the so-termed H-domain, a 48 kDa highly conserved protein domain that hosts the catalytic center, the H-cluster (Figure 12.2). Bacterial [FeFe]-hydrogenases have additional protein domains that contain accessory electron transfer centers. To date, five main structural types, termed M1 to M5, are distinguished [45, 47]. The group of M1-type hydrogenases contains the structurally most compact monomeric [FeFe]-hydrogenases, which exclusively comprise the H-domain. The model enzyme of this type is CrHydA1 from the green alga *C. reinhardtii*. Based on the simple structure of CrHydA1, it is often used for the characterization of catalytic and spectroscopic properties of the active center, which is the only cofactor present in M1-type hydrogenases [48].

The [FeFe]-hydrogenase subtypes M2 to M5 are categorized according to the increasing complexity of additional domains or subunits that harbor accessory Fe–S clusters. A characteristic of M2- to M5-type [FeFe]-hydrogenases is the N-terminal F-domain that contains ferredoxin-type Fe–S clusters (Figure 12.2a). In M2-type [FeFe]-hydrogenases, the F-domain contains two cubane [4Fe–4S] clusters and is similar to a bacterial 2[4Fe–4S] ferredoxin [45, 49]. The Fe–S clusters located proximal and distal with regard to the H-cluster are called FSA and FSB, respectively [50] (Figure 12.2). A representative of this group is DdH from *D. desulfuricans,* a heterodimeric enzyme with a size of 53 kDa [45]. Clostridial enzymes such as the monomeric 64 kDa [FeFe]-hydrogenase CpI from *Clostridium pasteurianum* contain the 2[4Fe–4S]-ferredoxin domain and in addition, N-terminal of that, a [4Fe–4S]-cluster coordinated by a histidine and three cysteines as well as a [2Fe–2S]-cluster which is similar to that in plant-type ferredoxins [10, 45]. M4- and M5-type [FeFe]-hydrogenases contain additional Fe–S clusters in the C-terminus or in extra subunits [45] and will not be discussed in detail here.

As briefly described above, the catalytic [6Fe–6S] cofactor of [FeFe]-hydrogenases, the H-cluster, consists of a cubane [4Fe–4S] cluster (often termed $4Fe_H$ in the

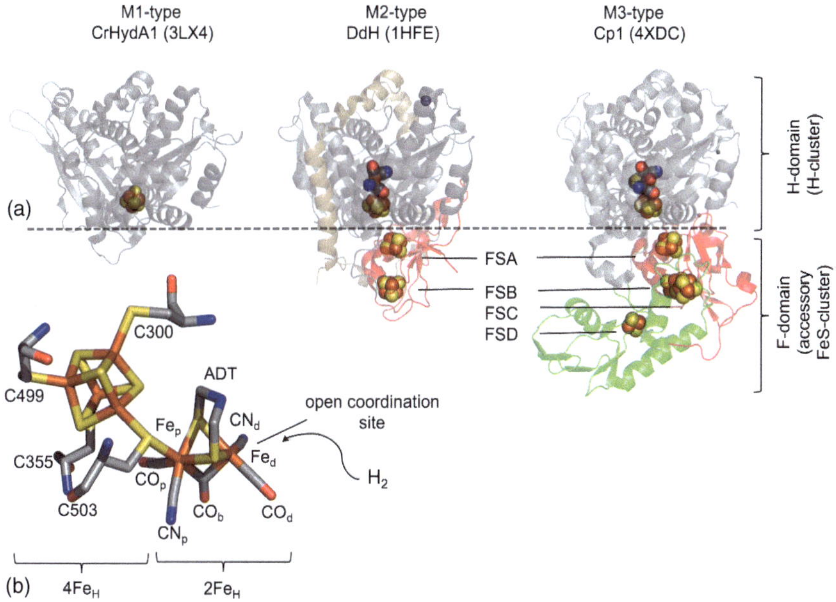

Figure 12.2 [FeFe]-hydrogenase subtypes M1, M2, and M3 and the H-cluster. (a) Crystal structures of model enzymes CrHydA1 (PDB: 3LX4) from *Chlamydomonas reinhardtii* [16], DdH from *Desulfovibrio desulfuricans* (PDB: 1HFE) [11], and CpI from *Clostridium pasteurianum* (PDB: 4XDC) [27]. The H-domain is shown in gray and contains the H-cluster (spheres). The F-domain is colored red or green and shows the respective Fe–S clusters as spheres. The red-colored domain in DdH contains two accessory [4Fe–4S] clusters. DdH is a heterodimer consisting of a catalytic subunit (gray and red) and a small subunit (sand-colored). In CpI, two additional Fe–S clusters are found in the N-terminus (green): a [4Fe–4S] cluster and a [2Fe–2S]-cluster. Electrons are transferred from the F- to the H-domain from the surface-exposed Fe–S cluster to the inner Fe–S clusters and finally to the H-cluster. (b) The [6Fe–6S] cofactor of [FeFe]-hydrogenases, the H-cluster, consists of a [4Fe–4S] cluster (4FeH) and a diiron cluster (2FeH). The 4FeH-cluster is coordinated by four cysteines. One of these, cysteine residue 503 of CpI (C503Cp1), connects the 4FeH and 2FeH subclusters. An ADT group bridges the two Fe-atoms via two sulfur atoms. In addition to the ADT headgroup, 2FeH is characterized by biologically unusual CO and CN^- ligands. At the Fe ion proximal to 4FeH, Fep, COp, and CN_p are coordinated. Fe_p is bridged to the distal iron (Fe_d) via CO_b (or COμ). The ligands CO_d and CN_d are bound to Fe_d. The open coordination site responsible for H_2 turnover as well as the binding of inhibitor molecules is located at Fe_d. The H-cluster is shown as a stick model based on the CpI structure PDB: 4XDC [27]. (a) and (b): The figures were created in PyMOL. Ions of clusters are colored according to elements: C: gray, N: blue; O: red; S: yellow: Fe: brown.

[FeFe]-hydrogenase literature) and an unusual diiron moiety (termed $2Fe_H$). Both subclusters are connected through the thiol group of a cysteine residue (Figure 12.2). The $4Fe_H$ subcluster is covalently anchored within the protein matrix by this and three additional cysteines. The two Fe ions of the $2Fe_H$ subcluster are distinguished as proximal (Fe_p) and distal (Fe_d) with regard to the $4Fe_H$ cluster. They are embedded into the protein environment of the active site niche and octahedrally coordinated [10, 11]. The two Fe ions are bridged through two sulfur atoms of an ADT ligand (ADT; [S-CH_2-NH-CH_2-S]) [25], and in addition, each Fe ion is coordinated by one CN^- and one CO ligand (Figure 12.2b). A third CO ligand (called CO_b or CO_μ) is coordinated in a bridging manner between Fe_p and Fe_d in most catalytic states of the cofactor [11]. The open coordination site is located at Fe_d and near the ADT headgroup, a secondary amine. At this site, the actual catalysis – H_2 oxidation or proton reduction – takes place [51–53]. The open coordination site can be blocked by small molecules, which then leads to a catalytically inactive or, in the case of O_2, an irreversibly destroyed protein [33, 41, 42, 54].

The maturation of the [FeFe]-hydrogenases begins with the assembly of the $4Fe_H$ subcluster, which, according to current knowledge, is accomplished by the standard Fe–S cluster biosynthetic machinery [21]. In [FeFe]-hydrogenase research, the enzyme that only carries $4Fe_H$, for example after heterologous production in *E. coli* without coexpression of the maturase sequences [13, 19], is termed "apo" despite the presence of a part of the H-cluster. *In vivo*, as mentioned above, the completion from this "apo" state to the active holo state, that is, the assembly and integration of the $2Fe_H$ subcluster, occurs through a complex maturation process that requires the highly specific H-cluster maturases HydE, HydF, and HydG [16, 19, 21]. However, it can also be carried out very efficiently *in vitro* by simply incubating the apo enzyme with a synthetic $2Fe_H$ analog (termed $2Fe_H^{mimic}$ in the following) [26]. This system is also employed to investigate the final stages of [FeFe]-hydrogenase maturation, that is, the actual transfer and integration of $2Fe_H$ into the "apo" [FeFe]-hydrogenase precursor. For example, a mutagenesis study analyzed the maturation channel and the uptake process of $2Fe_H^{mimic}$ from the surface to the active site. While certain positively charged residues enable the uptake and final fixation of the cofactor in the binding niche, channel closure occurs through displacement and remodeling of the dynamic structural elements "lid," "lock," and "plug" [16, 17].

In the maturation process, the $2Fe_H$ subcluster is connected to the $4Fe_H$ subcluster through the aforementioned cysteine and additionally forced into the entatic state necessary for optimal catalysis. The unusual CO and CN^- ligands play a crucial role in the high catalytic activity of the H-cluster as they mediate the low spin and oxidation states of the two Fe ions [48, 51, 55]. During catalysis, a number of catalytic states with individual electron distribution or redox states could be observed [56–59]. The catalytic cycle of [FeFe]-hydrogenases is complex, and some redox and/or protonation states are still under debate. A detailed review of these states is beyond the scope of this chapter, but three states will be named because they are of importance below. For H_2 production, a coupled transfer of two electrons and protons is necessary. In the beginning of the catalytic cycle, the H-cluster is in the so-termed oxidized state H_{ox} ($[4Fe–4S]^{2+}$-Fe_p(I)Fe_d(II)). Upon distinct reduction and protonation events, the H-cluster accumulates specific reduced and/or protonated states, for example H_{red} ($[4Fe–4S]^{2+}$-Fe_p(I)Fe_d(I)) or H_{sred}, in which the redox state of the $4Fe_H$ cluster is additionally affected ($[4Fe–4S]^+$-Fe_p(I)Fe_d(I)). After the release of H_2, the oxidized state H_{ox} is formed again. By using Fourier transform infrared (FTIR) spectroscopy, individual catalytic states of the cofactor can be identified based on specific vibrational signals of the state-specific CO^- and CN^- ligand configurations. The vibrational signals can be observed in the range between 2100 and 1800 cm^{-1} and quantitative ratios of these individual states in dynamic equilibrium can be determined [17, 35, 60]. The vibrational signals of the CO ligands are especially sensitive toward redox changes of the H-cluster, while the CN^- ligands have H-bonding interactions with the highly conserved protein environment and thereby tightly anchor $2Fe_H$ [27, 61, 62].

The catalytic activity of hydrogenases requires the transport of electrons, protons, or H_2 to and from the active site. For receiving and releasing electrons, [FeFe]-hydrogenases interact with external electron donors and acceptors. For *in vitro* studies, both physiological redox partners such as ferredoxins, flavodoxins, or cytochromes [63–65] and small redox molecules such as methyl viologen [63, 66] are employed frequently. In turn, these are routinely reduced by the strong reductant sodium dithionite (NaDT), although NaDT has recently been suggested to influence catalytic states of [FeFe]-hydrogenases [67]. However, [FeFe]-hydrogenases can also be adsorbed on a pyrolytic graphite electrode and analyzed by protein-film electrochemistry, during which direct electron transfer occurs between the electrode and the hydrogenase without additional electron mediators [53, 68, 69].

In the case of M1-type hydrogenases, electrons are transferred directly to the H-cluster [47, 64]. In all other [FeFe]-hydrogenases, electron transfer occurs through an electron transfer chain built from the additional Fe–S clusters in the F-domain (Figure 12.2a). These Fe–S clusters are arranged in such a fashion that electron transfer from a surface-located Fe–S cluster to the H-cluster is possible. The distances between the Fe–S clusters are 8–11 Å, allowing direct electron transfer between the clusters [10]. In the case of CpI or the closely related M3-type [FeFe]-hydrogenase CaI from *Clostridium acetobutylicum*, it is still discussed whether the His-ligated [4Fe–4S] cluster or the plant-type ferredoxin-like [2Fe–2S] cluster is the first acceptor of electrons from external donors [50, 70]. Nevertheless, a mutagenesis study by Lubner et al. demonstrated the important and special properties of the His-ligated Fe–S cluster and the specific importance of the imidazole ring of the histidine residue for the electron transfer [71].

The energetically preferred synchronous transfer of both substrates, electrons and protons, also known as proton-coupled electron transfer, is essential for the high catalytic efficiency and reversibility of [FeFe]-hydrogenases [72, 73]. An amino acid-based PTP for proton transfer from the surrounding solvent to the open coordination site of the $2Fe_H$ subsite was characterized based on the crystal structures of CpI and CrHydA1 [28] (Figure 12.3a). Viewed from protein surface to H-cluster, this pathway starts at an arginine on the protein surface (R286 in CpI; note that the residue numbers are those of CpI, unless stated otherwise) and transports protons through E282, S319, E279, two structurally conserved water molecules (water molecules 826 and 1120 in chain B of the CpI structure PDB: 4XDC), and finally to cysteine C299 that is located near the ADT bridge of the $2Fe_H$ subsite [28, 30, 74] (Figure 12.3a). It is assumed that protons are transported through this pathway to the secondary amine group of the $2Fe_H$ subcluster according to a Grotthuss mechanism [28, 76]. If individual amino acid side chains of this PTP are lost, water molecules can take over the proton transport function as long as sufficient space or accessibility for water molecules is given and a hydrogen bond distance to neighboring positions of less than 4 Å can be guaranteed [28]. Indeed,

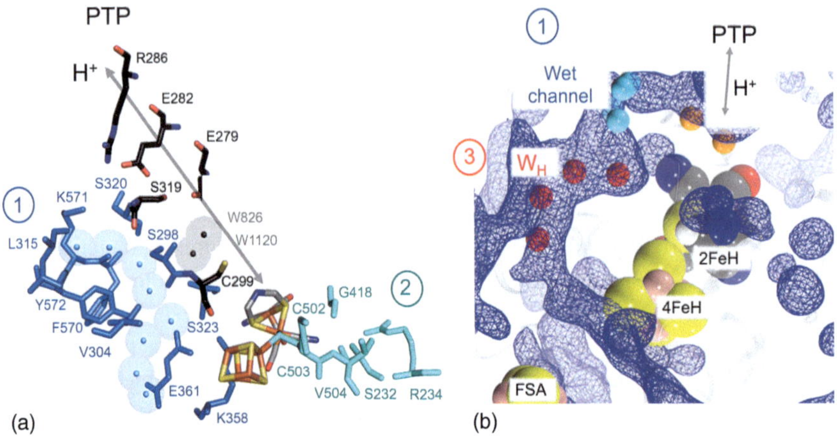

Figure 12.3 Proton transfer pathways (PTPs) and water channels in CpI. (a) Presentation of the PTP [28] and hydrophilic pathways of the H-domain. Amino acid residues are shown as sticks, and water molecules as spheres. The residues and water molecules of the PTP are colored in black. The identified hydrophilic pathways are numbered 1 and 2 and residues and water molecules are colored blue and turquoise, respectively. Pathway (1) depicts the so-termed wet channel [74]. This pathway also includes the "water pathway" that begins at position E361 and transports protons through K358 to the CN_d ligand of the active site [30]. Pathway (2) shows the "dynamic water pathway." During electron transfer from the Fe–S cluster FSA to the H-cluster, an opening of a water channel was simulated, which could mediate the transfer of protons from the surface to C299 and further to the active site [73]. (b) Molecular dynamics (MD) simulations identified dynamic water cavities around the H-cluster that are highlighted as blue mesh. Shown is the H-cluster water pathway termed "W_H" (numbered as (3)) that leads from the surface to $4Fe_H$ and further on to $2Fe_H$ [75]. Water molecules are shown as spheres and colored red when they were assigned to pathway W_H, cyan in the case of the wet channel (1); already depicted in (a), and orange in the case of the two structurally conserved water molecules of the PTP.

specific water channels close to the PTP could be identified (Figure 12.3). Three of these water channels, or water pathways, have been distinguished to date, namely, (1) the so-termed wet channel which is static [74] (Figure 12.3a,b), (2) a dynamic pathway detected by simulations that opens upon electron transfer to the active site [73] (Figure 12.3a), and (3) dynamic water cavities around the H-cluster termed water channel "W_H" [75] (Figure 12.3b). The "wet channel" (1) was detected during a crystal structure-based analysis using the program CAVER (a tool for detecting and visualizing tunnels and channels in protein structures) [77] in the three model [FeFe]-hydrogenases CrHydA1, CpI, and DdH [74]. Upon closer analysis of the crystal structure of CpI, a chain of water molecules could be assigned to this "wet channel," three of which are near the H-cluster [74] (Figure 12.3a). This channel can be divided into two sections. One section starts close to the surface-exposed lysine 571 and leads through the protein matrix to E361, located near $4Fe_H$ (≈8 Å distance). From there, water molecules can diffuse into the so-called central cavity, which forms the direct protein environment of the $2Fe_H$ subcluster and offers space for substrates to interact with the ADT headgroup or with the substrate binding site at Fe_d [30, 74]. The second section of the "wet channel" starts close to position

E361, which is part of the already described wet-channel section (Figure 12.3a). This channel section leads to residue K358 and employs protein-coordinated water molecules, which can migrate through the central cavity to CN_d. However, the function and/or role of this pathway and the mechanism of proton transfer, if it occurs at all, by which the proton is transferred from K358 to the H-cluster is still unclear [30].

In addition to the "wet channel," a dynamic pathway (2) (Figure 12.3a) was simulated to be open only during electron transfer from the Fe–S cluster FSA to the H-cluster. This "dynamic water pathway" might be involved in proton-coupled electron transfer by mediating the transfer of protons from the surface to the cysteine that lies directly adjacent to the H-cluster and is part of the PTP ($C299_{CpI}$) [73] (Figure 12.3a). The third hydrophilic pathway, "W_H," was identified by molecular dynamics simulation based on the crystal structure of CpI with the PDB accession 4XDC. This water pathway begins at the protein surface near to the entry of the "wet channel" and leads to $4Fe_H$ and further to $2Fe_H$, where it connects with the "wet channel" (Figure 12.3b). By manipulating W_H by site-directed mutagenesis of residues close to $4Fe_H$, it was found that the dynamic water cavity W_H is also a diffusion pathway for O_2 [75].

Finally, gas channels have been identified in [FeFe]-hydrogenases. These are important for the diffusion of H_2 but have mostly been analyzed in the context of the diffusion of gaseous inhibitors [41, 74, 78–80]. Therefore, the gas channel networks will be described in the context of small molecule inhibitors below.

12.3 Small Molecules Inhibit [FeFe]-Hydrogenases

Various small molecules have been identified that affect the catalytic activity of [FeFe]-hydrogenases (Figure 12.4a). Many of these are (ir)reversible inhibitors, whereas some can also protect the enzyme from the irreversible destruction caused by the binding of, and reaction with O_2. To date, mostly O_2, CO, sulfide, zinc (Zn^{2+}), CN^-, and formaldehyde (HCHO) have been analyzed regarding their effect on [FeFe]-hydrogenases (Table 12.1). All of these molecules affect the enzyme by blocking the open coordination site or by interrupting the PTP. The inhibitors tested to date are mostly extrinsic molecules, but one of these molecules, sulfide, could be identified as being part of a native O_2 protection mechanism of a rare O_2-resistant [FeFe]-hydrogenase termed Cba5H. In Sections 12.3.1–12.3.4, known inhibitor diffusion pathways will be summarized first, followed by a description of the details of the mechanisms involved in inhibition or protection. Regarding the latter, we have divided the inhibitor molecules into three groups. The attack by the O_2 molecule is described first because it is irreversible in the case of most [FeFe]-hydrogenases studied to date in that it destroys the H-cluster. The second group describes all small molecules that affect the (region of the) open coordination site reversibly, and which are therefore potential O_2-protecting molecules. The third section briefly summarizes small molecules that influence the PTP.

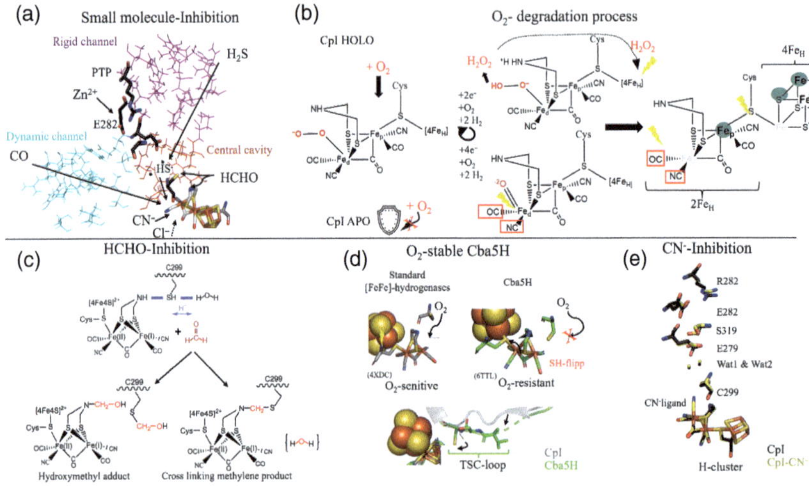

Figure 12.4 Inhibition of [FeFe]-hydrogenases. (a) The sites of attack of the small molecules CO, Zn^{2+}, H_2S, HCHO, CN^-, and Cl^-. CO can use several diffusion pathways to reach Fe_d. Highlighted is the diffusion of CO through gas channel B (cyan, line-structure) [74, 80]. CN^- and Cl^- bind to Fe_d [42, 81]. Their diffusion pathways are still unknown. H_2S uses the gas channel A, reacts to HS^-, and finally also binds to Fe_d [41]. It is hypothesized that Zn^{2+} interacts with the PTP amino acid E282 (black stick model) [32]. (b) Irreversible degradation of the H-cluster by O_2. The sequences of events (from left to right) are described in the text. The Fe ions that were not affected in O_2-treated CpI crystals [33] are highlighted in green. (c) The proposed reaction mechanism for HCHO [43]. (d) The protective mechanism of the O_2-stable [FeFe]-hydrogenase Cba5H [37, 38] in comparison to O_2-sensitive hydrogenases such as CpI. (e) Shown are the amino acid conformational changes within the PTP observed upon CN^- binding (overlay of untreated CpI (PDB: 4XDC; black) and CN^--treated CpI (PDB: 8AP2; [yellow]) [42]. (a) to (e) Details are described in the text.

12.3.1 Diffusion of Small Molecules to the H-cluster

As described above, [FeFe]-hydrogenases employ multiple pathways for transporting substrates to and from the active site [34, 74]. Unfortunately, these pathways are also open for inhibitor molecules. The mutual use of pathways of substrates or products and inhibitors has been deeply studied for various enzymes such as globins, cholesterol oxidase [82], carbamoyl phosphate synthetase (CPS) [83], acetyl-CoA synthase/CO dehydrogenase (ACS/CODH) [84], glutamine phosphoribosyl pyrophosphate (RPPP), amidotransferase (GPATase) [85], and [NiFe]-[86–90], and [FeFe]-hydrogenases. In [FeFe]-hydrogenases, gas molecules use different diffusion pathways depending on their size and/or hydrophobicity. This has been shown for two predominantly hydrophobic gas channels in the H-domain. One was discovered as a tunnel in the crystal structure of DdH from *D. desulfuricans* [11]. This gas channel was termed "rigid" because the cavities could be identified based on the crystal structure. The diffusion through this rigid pathway was initially thought to be continuous, but the behavior of gas molecules in MD simulations showed a stepwise diffusion from cavity to cavity toward the active site [78].

Table 12.1 Overview of known inhibitor molecules of [FeFe]-hydrogenases and their reaction behavior.

Molecule	Protein	Inhibition behavior	Molecule binding-site	O_2 protectant
O_2	"Standard" [FeFe]-hydrogenases	Irreversible	Fe_d	
	Cba5H	Reversible	—	
CO	[FeFe]-hydrogenases	Reversible	Fe_d	Yes
Sulfide	[FeFe]-hydrogenases	Reversible	H_2S: water site of PTP HS^-: Fe_d	Yes
HCHO	CrHydA1 Cba5H	Reversible	PTP cysteine	Yes
	CpI	Reversible (red.) Irreversible (ox.)	Amine-base and PTP cysteine	
	(CaI)			
CN^-	CpI DdH	Irreversible	Fe_d	No
Zn^{2+}	CpI	Unknown	PTP glutamate (E282)	Partly
Cl^- & Br^-	CrHydA1 CaI	Reversible	Fe_d is proposed	Partly

Details are described in the text.

In addition to this rather open rigid gas channel, MD simulations identified a second pathway that appears to be dynamic [78]. Upon closer inspection of both pathways, a putative gas storage site was discovered based on xenon-binding studies in the rigid channel [79, 91]. Recent MD simulations suggest that the inhibitor H_2S may diffuse through this one [41]. A comparative MD simulation of the diffusion behavior of O_2 and H_2 molecules along the mentioned gas channels using the crystal structure of CpI (PDB: 3C8Y) showed that the smaller H_2 molecules (1.3 Å) have a broader diffusion spectrum through the protein structure than the larger O_2 molecules (1.6 Å) [78].

More recent MD studies that analyzed the diffusion sites of O_2 and the gaseous, reversible inhibitor CO based on the CpI crystal structure filed under PDB: 1FEH created a diffusion map based on the free energy levels of the gas molecules at individual positions of the network [79, 80]. These MD simulations took all hydrophobic and hydrophilic regions of the protein into account and indicated that, in addition to the rigid and dynamic hydrophobic gas channels in the H-domain, these inhibitors can permeate along many other hydrophobic, but also hydrophilic paths within the H- as well as the F-domain [79, 80]. A comparison of the mapping of the free energy barriers for O_2 and CO showed both common and different diffusion pathways [79, 80]. The observed differences are due to differences in the molecular size and the physicochemical properties of the two molecules. CO is a polar gas and has a van der Waals volume of 16.20 cm^3/mol, while O_2 is an apolar gas with a van der Waals volume of 13.00 cm^3/mol [80].

The theoretical studies by Mohammadi and Vashisth and Liu et al. suggested hydrophilic pathways for the diffusion of O_2, which was a bit unexpected for an apolar molecule [79, 80]. However, recent experimental results corroborate this finding, in those variants of CpI, in which amino acids in the region of the water channel W_H (Figure 12.3b) were exchanged, showed a higher resistance toward O_2 [75]. High-resolution crystal structures of these variants could be obtained (PDB entries 7QHF and 8CJY) and used for MD simulations. These revealed that the amino acid exchanges resulted in a blockage of the hydrophilic channel W_H, indicating that O_2 diffuses through this hydrophilic region in the wild-type enzyme that is more O_2 sensitive [75]. Notably, similar O_2 diffusion principles appear to be present in the [NiFeSe]-hydrogenase from *Desulfovibrio vulgaris* Hildenborough. Here, the diffusion of O_2 or reactive oxygen species (ROS) through hydrophilic regions was also impaired through amino acid exchanges, and this resulted in diminished oxidation of cysteine residues near the active-site metal cluster [87].

Currently, how the other small molecules discussed herein reach the active site of hydrogenases is much less studied. One reason for the deep study of diffusion pathways of O_2 is the search for positions within [FeFe]-hydrogenases that might be altered in order to restrict O_2 accessibility to protect the active site against harmful attack [74].

12.3.2 Irreversible Attack of O_2 Molecules

O_2 inhibits [FeFe]-hydrogenases irreversibly in that it leads to a degradation of the H-cluster. The O_2 molecule, after diffusing through the protein, binds at the open coordination site at Fe_d [34, 92] and initiates an oxidative degradation of the active site metal cluster [92–94]. Different studies observed a predominant degradation of either the diiron subcluster $2Fe_H$ [95] or the [4Fe–4S] subcluster $4Fe_H$ [92, 96, 97]. By investigating the degradation process based on CpI crystal structures, it was possible to integrate these somewhat contradictory conclusions of previous studies [33] (Figure 12.4b). CpI is particularly suitable for this purpose because X-ray crystal structures are available of both the fully maturated holo-enzyme (PDB: 4XDC) and the so-termed apo-form, which contains all Fe–S clusters except the diiron subsite ($2Fe_H$) of the H-cluster (PDB: 4XDD) [27]. To study the O_2-induced degradation process, Esselborn and coauthors treated CpI crystals with air and analyzed the decrease in the electron density of the H-cluster atoms after defined time points in over 100 CpI crystals [33]. In the apo-form, i.e. in the absence of $2Fe_H$, the $4Fe_H$ cluster did not show significant damage under the influence of O_2. However, in the fully mature holo-form, both $2Fe_H$ (loss of Fe_d) and $4Fe_H$ (degradation to a [2Fe–2S] cluster) showed damage after treatment with air [33] (Figure 12.4b). This observation was corroborated by activity measurements: After adding the synthetic $2Fe_H^{mimic}$ complex to O_2-treated CpI wild-type protein, a reactivation of approximately 50% compared to the initial activity could be achieved. Because the functional attachment of the diiron site requires a preformed and intact $4Fe_H$ subcluster [19], reactivation is only possible if $4Fe_H$ is still intact. The observation that only about a half of the O_2-treated enzyme population could be reactivated suggests that the two H-cluster subunits are attacked differently [33].

Esselborn and coauthors proposed two alternative ROS-dependent redox reactions, each of which selectively attacks one of the subclusters. In both cases, the degradation is initiated by O_2 binding to Fe_d and forming a primary Fe_d–O_2 adduct in which the O_2 molecule is most likely to be converted to a superoxide ligand (O_2^-) [97, 98]. In one of the proposed pathways leading to H-cluster degradation, the superoxide can be converted to a hydroperoxyl radical (Fe_d-HO_2^-) upon protonation. This can lead to the formation of hydrogen peroxide (H_2O_2) by further protonation. After diffusion to the $4Fe_H$ subcluster, H_2O_2 can attack this cluster [33, 94, 99]. In the second possible route leading to O_2-induced H-cluster degradation suggested by Esselborn et al. and others [93, 94, 100], the same Fe_d-HO_2^- species would be disproportionate to H_2O and an iron-oxo complex ($Fe(IV) = O_2^-$). This species would oxidize the nearby CO_d and CN_d ligands to form CO_2 or OCN^-. The two diatomic ligands would subsequently be replaced by water molecules and the thus solvated Fe_d would eventually be released from the complex [33] (Figure 12.4b). An additional attack of the formed ROS on amino acids in the immediate H-cluster environment is in principle conceivable, too [100]. Evidence for such an oxidation was described for cysteine 169 (CrHydA1 numbering) from crystallized O_2-exposed CrHydA1 [95]. This, however, was not observed in O_2-treated crystals of CpI [33].

Kinetically, the described sequences of the O_2 degradation mechanism can be summarized by the following equation (Eq. (12.1)): After the diffusion of O_2 to the open coordination site of the H-cluster (Fe_d) of the active protein (A) and the binding of O_2 to Fe_d as superoxide (O_2^-) (rate constant k_1), an inhibited protein intermediate (I) is formed that can be either reactivated (k_a) or irreversibly inactivated (k_3) [68]. Reactivation is not commonly observed but can be achieved by the reduction of O_2 to H_2O in a few O_2-tolerant [NiFe]-hydrogenases (see end of this section). If there is insufficient reduction of O_2 to H_2O due to electron limitation, and in all O_2-sensitive [FeFe]-hydrogenases studied to date, the O_2-bound form I leads to the degradation of the H-cluster (D), since the damage caused by ROS leads to an irreparable loss of iron and sulfur atoms [68, 100] (Eq. (12.1)).

$$A \underset{k_a}{\overset{k_i}{\rightleftharpoons}} I \xrightarrow{k_3} D \qquad (12.1)$$

However, this seemingly simple process description can be influenced before, during, and after O_2 binding. Before O_2 can bind to Fe_d, it must diffuse through the gas channel network. As mentioned in Sections 12.3.1 and 12.3.2, limiting O_2 accessibility to the active site can increase the O_2 stability of the enzyme, for example by blocking the hydrophilic cavity in CpI [75]. The generation of the Fe_d-HO_2^- species through protonation of the primary Fe_d-O_2 adduct is assumed to initiate the degradation processes discussed to date. This protonation step requires the PTP to be confirmed experimentally using the CrHydA1 variant C169A. The exchange of this cysteine, which mediates the final step of proton transport to the diiron subcluster, to alanine prohibits proton transport [28] and also prevents the O_2 degradation process by trapping the O_2-bound primary adduct [98]. Blocking proton transfer to bound O_2 could apparently also be achieved artificially for CrHydA1 and CpI by lyophilization [97]. Lyophilized [FeFe]-hydrogenases are very O_2 stable as long as the lyophilized enzyme is stored upon the strict exclusion of water, and it was suggested that, as a result of freeze drying, surface water and therewith any proton source is lost. In addition to the availability of protons, high electron pressure may also influence the irreversible degradation of the H-cluster by O_2. One protective mechanism observed in O_2-tolerant [NiFe]-hydrogenases is the complete reduction of bound O_2 to H_2O by direct transfer of four electrons and four protons [101, 102]. However, this mechanism could only be demonstrated to a much lesser extent in [FeFe]-hydrogenases [94, 100]. In addition to the options summarized above, several small molecules can protect the H-cluster from irreversible O_2-induced degradation. These are described in the following.

12.3.3 Inhibitor Molecules that Protect [FeFe]-Hydrogenases Against O_2

12.3.3.1 CO Protects Against the O_2-Initiated H-Cluster Degradation

In [FeFe]-hydrogenases, carbon monoxide (CO) is an essential part of the $2Fe_H$ subcluster [10, 11, 27, 50], but as a free diffusing molecule, CO can inhibit the enzyme [54]. To reach the active site, CO diffuses from the surface through gas channels into the interior of the protein and finally to the H-cluster [80]. The inhibition mechanism of CO is similar to the first step of irreversible inactivation by O_2 but can be reversible depending on the redox state of the H-cluster. Both CO and O_2 bind to Fe_d of the $2Fe_H$ subcluster, as demonstrated for CO by X-ray crystallography and for CO and O_2 by X-ray absorption spectroscopy (EXAFS) [54, 92]. Both O_2 and CO attack the H-cluster in the oxidized so-termed H_{ox} state ($[4Fe–4S]^{2+}$-$Fe_p(I)Fe_d(II)$) more effectively than in the reduced states H_{red}/H_{sred} [96, 103, 104]. In the oxidized state ($H_{ox}CO$), CO is a reversible inhibitor. Notably, although the two gas molecules display such similar binding modes, CO has an 80 to 200 times higher binding affinity to Fe_d than O_2 so that the $H_{ox}CO$ state can protect the H-cluster against irreversible degradation by O_2 [103]. The CO-inhibited enzyme can be reactivated by intense purging with H_2 or by illumination. However, if CO binds to the H-cluster in its reduced state H_{red}, the $H_{red}CO$ state is accumulated, in which the connection of $2Fe_H$ to the bridging sulfur atom is destroyed [104]. Reactivation of the hydrogenase from this stably CO-bound $2Fe_H$ cluster is not possible, rendering CO an irreversible inhibitor under this condition. Through protein film electrochemistry, a potential range of −0.36 to −0.47 V was identified for this process [104]. However, if the H-cluster is in its further reduced H_{sred} state, which is formed below −0.6 V, the inhibition is reversible again [104].

12.3.3.2 Formaldehyde Attacks Catalytically Important Sites of [FeFe]-Hydrogenases

As described above, CO can protect the H-cluster against O_2, particularly in the H_{ox} state. Another small molecule, formaldehyde (HCHO), can protect the cluster reversibly, preferably under reducing conditions. HCHO is a very reactive molecule and in protein structures, thiol groups of cysteine residues or amide groups of lysine residues are the main attack sites of HCHO. Due to its high reactivity toward polypeptides, HCHO is commonly used as a crosslinking agent [105]. It was first shown in 2011 by protein film electrochemistry that HCHO inhibits [FeFe]-hydrogenases [106]. Subsequent research showed that the molecule exerts its inhibiting effect directly at the H-cluster and not, or hardly, through crosslinking within the polypeptide. Density functional theory (DFT) calculations already predicted that this molecule binds to the open coordination site (Fe_d) in the two-electron-reduced state of the H-cluster [106, 107]. Advanced electron paramagnetic resonance (EPR) spectroscopy on HCHO-treated [FeFe]-hydrogenase proteins under reducing conditions likewise suggested that the HCHO molecule binds close to the H-cluster [108]. A recent study showed that HCHO binds to an Fe ion of a chemically synthesized $2Fe_H$ derivative that is similar to the [FeFe]-hydrogenase $2Fe_H$ cofactor [109]. New insights into the effect of HCHO on [FeFe]-hydrogenases could be gained through an electrochemical study on CrHydA1, Cba5H, and CpI, combined with crystal structures of HCHO-treated CpI [43]. Cyclic voltammetry (CV) measurements under the influence of HCHO showed a clear, significantly stronger inhibition in the direction of H_2 production for all wild-type proteins examined [43, 106]. Chronoamperometry tests under reducing conditions ($E < -0.36$ V, pH 6) could confirm this strong inhibition, which was reversible [43]. Under oxidative conditions ($E > -0.36$, pH 6), the inhibitory response toward HCHO of CpI is slow and weak, but irreversible [43], similar to previous studies on CaI [106]. Compared to the CpI enzyme, HCHO inhibits CrHydA1 and Cba5H more strongly and quickly. However, for these proteins, the inhibition was observed to be partially reversible [43]. This difference has not yet been understood or clarified experimentally. Structural changes of CpI caused by HCHO were investigated in detail employing high-resolution crystal structures of CpI treated with this inhibitor molecule (PDB: 8QM3) [43].

A direct binding of HCHO to Fe_d could not be confirmed. Instead, HCHO seems to have two major attack sites, namely the proton-transporting cysteine ($C299_{CpI}$) and the adjacent amine base of $2Fe_H$ (Figure 12.4c). According to the electron densities of HCHO-treated CpI crystals, a hydroxymethyl adduct ($-CH_2-OH$) is formed at the amine base, which, in turn, appears to form a crosslinking product with the neighboring cysteine residue, which could happen after further protonation and dehydration. Alternatively, besides the reaction at the amine base, HCHO reacts with the free thiol residue of the PTP cysteine and thereby forms a second hydroxymethyl adduct ($-CH_2-OH$). In the case of CrHydA1 and Cba5H, a reaction of HCHO only with the PTP cysteine was assumed due to their largely reversible reaction observed upon protein film electrochemistry [43]. It is tempting to speculate that HCHO could be a protectant against O_2, for CrHydA1 as shown by Wait and coworkers but also for Cba5H [106]. The observed differences in the sensitivity of the three hydrogenases to HCHO are attributed to differences in the protein matrix, as these can influence the accessibility of HCHO to the H-cluster [43], similar to what was predicted for other inhibitor molecules [41, 110, 111].

12.3.3.3 The Role of Sulfide in [FeFe]-Hydrogenases

Sulfide is an additional example of an [FeFe]-hydrogenase inhibitor that can protect these enzymes from irreversible attack by O_2. Interestingly, this effect was observed rather by chance under physiological conditions, when, many years ago, sulfide inhibition was recognized for hydrogenases of sulfate-reducing bacteria such as *D. desulfuricans* or *D. vulgaris*. [FeFe]-hydrogenases of these organisms are in a specific O_2-stable but inactive state (termed H_{inact}) after anaerobic protein isolation [112, 113]. However, after initial reductive activation, this protective mechanism is permanently lost [113, 114]. This native (but transient) O_2 stability might be a consequence of the high sulfide amount added to the cultures of these sulfate reducers because this sulfide-induced inactive state can also be artificially generated by treating isolated model [FeFe]-hydrogenases (CrHydA1, CpI, DdH) with sulfide (Na_2S) or H_2S under oxidizing and anaerobic conditions [40] (Figure 12.4a). A crystallographic and spectroscopic study investigated the sulfide-mediated properties of DdH [115]. This study could demonstrate that a sulfide-ligand bound to Fe_d is responsible for the inactive, but O_2-stable H_{inact} state of this enzyme [115]. According to MD simulations, this ligand is a HS^- molecule [41]. Additional MD simulations and DFT calculations indicate that several steps precede this binding event: To reach the active site, simulations suggest that H_2S diffuses through the rigid gas channel of the H-domain up to a binding niche that is still 8 Å away from the H-cluster and close to the PTP positions $C178_{DdH}$ and $E156_{DdH}$. One of the PTP water molecules is displaced by H_2S, so that the proton transfer to the H-cluster is interrupted and the enzyme is inactivated [41]. As long as sulfide is located in this distant binding niche, it is termed a noncompetitive inhibitor [41]. However, the protonatable PTP amino acid $E156_{DdH}$ is located close to the H_2S-binding site, so that a HS^- molecule is formed by deprotonation, which diffuses to the open coordination site and binds to Fe_d. A reactivation of the sulfide-inhibited protein can be reached via two reduction steps. A one-electron transfer step leads from the H_{inact} state to the so-termed H_{trans} state, in which the cubane $4Fe_H$ subcluster is reduced by one electron ($[4Fe–4S]^+$-Fe_p(II)Fe_d(II)) [59]. The second reduction step is combined with a protonation event and converts the H-cluster back to an active state, which resumes its normal catalytic H_2 turnover. In principle, this sulfide-inhibited but O_2-stable state can be achieved for all purified [FeFe]-hydrogenases and should always be reversible if contamination with polysulfides is avoided as these molecules can damage the enzyme irreversibly [41]. Due to differences in the protein matrix and architecture of [FeFe]-hydrogenases, different inhibition kinetics have to be assumed [41]. For example, the [FeFe]-hydrogenases Me (*Megasphaera elsdenii*), CaI, and CpI show 100-fold slower inhibition kinetics compared to CrHydA1 and DdH [41]. This kinetic property is similar to what was observed regarding inhibition by CO and O_2 of these [FeFe]-hydrogenases [78–80, 111].

12.3.3.4 Cba5H – A [FeFe]-Hydrogenase that Protects Itself by Its Own Sulfide

For applications that aim to harness the high efficiency of [FeFe]-hydrogenase, O_2-stable hydrogenases would alleviate many constraints of the whole industrial process, starting with protein production on a large scale. Much research has focused on rendering known and well-analyzed [FeFe]-hydrogenases such as CrHydA1, DdH, or CpI O_2 stable, however, although some approaches resulted in somewhat more O_2-tolerant enzymes, these enzymes are not suitable for any completely aerobic process step yet [34, 74]. However, only a couple of years ago, a new [FeFe]-hydrogenase type was discovered that is naturally O_2 stable, namely an M2-type enzyme from *Clostridium beijerinckii*, named Cba5H [36]. CbA5H can be treated with air and O_2, respectively, to reach a state in which the enzyme is inactive, but from which it can be reactivated by, for example, purging with H_2 [36, 37]. FTIR analyses that, as mentioned above, are sensitive to the vibrations of the CO^- and CN^- ligands and the redox states of the H-cluster, respectively, showed that CbA5H assumes a certain state, termed H_{inact} state, upon exposure to O_2 as well as to oxidants such as thionine [36, 37, 116]. This state was first shown to be adopted in CbA5H treated with reversible inhibitors and then with O_2, and a reactivation of the H_{inact} state to the (active) H_{ox} state was reached by reduction with, for example, sodium dithionite or by intense H_2 purging [36, 37]. However, surprisingly, and new in the [FeFe]-hydrogenase research field, actually no addition of external inhibitors was necessary for this protective effect against O_2 in CbA5H. Comparisons of the structure obtained from an aerobically generated crystal of CbA5H with structures of anoxically maintained crystals of CpI and DdH showed a significant reduction of the distance between the thiol group of cysteine 367 in CbA5H, which corresponds to the proton transporting cysteine adjacent to $2Fe_H$ (C299 in CpI), and Fe_d of $2Fe_H$ in the Cba5H protein (Figure 12.4d). This short distance indicates that, in air-treated Cba5H, the thiol-group of C367 indeed coordinates to Fe_d, blocking this site for incoming O_2 molecules and thus preventing the degradation of the H-cluster. This leads to a protected but inactive hydrogenase protein. Interestingly, this protection of Fe_d observed in CbA5H is similar to the previously described inhibition with extrinsic sulfide in canonical [FeFe]-hydrogenases such as CrHydA1, DdH, or CpI (Figure 12.4a).

Further structural comparison of Cba5H, CpI, and DdH revealed, in addition to the "switched" cysteine residue, a shift of a protein loop. The conformational shift of this loop, termed "TSC" loop as it encompasses the CbA5H residues T365, S366, and the protecting C367, appears to be responsible for the changed orientation of C367 in O_2-exposed CbA5H (Figure 12.4d). Combining results from protein film electrochemistry, FTIR spectroscopy, X-ray structures, and site-directed mutagenesis of CbA5H, the high flexibility of the TSC loop seems to require a specific amino acid residue motif, L364, P386, and A561 (so-termed LPA motif) far away from the H-cluster. Therefore, this native mechanism that confers O_2 stability appears to be based on a long-range effect [38]. This motif is not present in known O_2-sensitive [FeFe]-hydrogenases such as CpI, in which the corresponding residues are F296, L318, and F493 [37]. The exact function of the LPA motif in CbA5H and perhaps additional O_2-stable [FeFe]-hydrogenases needs to be investigated further. The identification of CbA5H also raises hopes that [FeFe]-hydrogenases might be identified in the vast and still growing sequence databases that employ different mechanisms and that are, perhaps, even more O_2 stable or even active in the presence of air.

12.3.4 Influences of Small Molecules on the Proton Transfer Pathway

The small molecules described above interact reversibly with the H-cluster or its direct environment. Other small molecules inhibit the hydrogenase irreversibly or have their attack sites far away from the H-cluster as described in the following. Recent studies analyzed the reactions of exogenous cyanide (CN^-) or the metal cation Zn^{2+} with [FeFe]-hydrogenases. In the case of CN^-, its binding to the open coordination site (Fe_d) could be identified using high-resolution protein crystal structures of the CpI enzyme (PDB: 8AP2) [42], as it was already known for external CO [54, 92]. In general, the crystal structure of CN^--treated CpI is very similar to that of untreated CpI, which is in the H_{ox} state (PDB: 4XDC) [27], or the CO-inhibited enzyme (PDB: 8ALN) [42, 54]). Since the binding of CN^- to the open coordination site blocks the catalytic site of [FeFe]-hydrogenase [28], the binding of CN^- may be regarded as another means to protect the H-cluster from harmful O_2 attack. However, extrinsic CN^- binds irreversibly to the H-cluster, at least at pH 8 [110], so that it is not helpful as a protecting agent. Spectroscopic characterizations of the DdH [FeFe]-hydrogenase as well as of CN^--treated protein crystals of CpI and or CN^--treated protein films CrHydA1 were able to detect a specific third Fe–CN band (2112 cm^{-1} in the case of CpI) in addition to a general absorption shift of the already known ligand signals to higher energies [42, 110]. The described absorption shift indicates further oxidation of the $2Fe_H$ subsite [42]. Based on their spectroscopic studies on [FeFe]-hydrogenases, Martini et al. were able to identify two new redox states namely H_{trans}-like and H_{inact}-like states [110]. The extrinsic CN^- preferentially binds to Fe_d in the H_{ox} state, thereby accumulating the H_{trans}-like state, which changes to the H_{inact}-like state through oxidation. It should be noted that besides extrinsic CN^-, CN^--"cannibalization" can happen during the *in vitro* maturation process with $2Fe_H{}^{mimic}$ [110]. In addition to the identification of the CN^- ligand at Fe_d and the spectroscopic characterization of CN^--treated crystals, additional structural changes in the protein environment of CpI were detected [42]. The PTP shows unusual conformational changes compared to the structure of untreated CpI (PDB: 4XDC) [27] as presented in Figure 12.4e. The most important change is a significant decrease in the distance of the amino acid residue E279 to the serine residue 319 by about 1 Å [42]. It had been suggested before that the distance between these two residues results in a proton transport barrier, which could be overcome only by flexible conformational changes of the residues [28, 29]. However, the necessary conformational changes of these side chains had not been observed structurally in untreated CpI wild-type structures. The necessary conformation of E279 and S319 has, so far, only been observed in CN^--treated CpI crystals, although the reason for that remains to be analyzed in the future [42].

Zn^{2+} is an additional inhibitor of [FeFe]-hydrogenases that affects the PTP. The presence of 0.05 mM Zn^{2+} diminishes the H_2 production activity of CpI by 70–80%, demonstrating a significant inhibitory effect, which was less pronounced when lower Zn^{2+} concentrations were used [32]. Up to now, it remains unclear if the inhibition is reversible or not and whether it can be employed to use Zn^{2+} as an O_2 protectant. The exact mechanism of the inhibition caused by Zn^{2+} is not yet known.

The cation probably binds to the glutamic acid residue $E282_{CpI}$, which is part of the PTP. The negatively charged residue is surface exposed and thereby a good target for Zn^{2+} (Figure 12.4a). Replacement of glutamic acid with leucine ($E282L_{CpI}$) resulted in a Zn^{2+}-insensitive protein variant, and this suggests a coordination of the cation at position E282, which could affect proton transfer [32]. By testing the effects $CaCl_2$ or NaCl on H_2 production activity, which were negligible, Cl^- as the inhibiting agent could be excluded, although chloride (Cl^-) as well as bromide (Br^-) ions can inhibit [FeFe]-hydrogenases at high potential. According to the current state of knowledge, CN^- and Zn^{2+} would not be useful for protecting the H-cluster from O_2-induced degradation, but both appear to influence the PTP and thus represent possible probes to study the PTP in more detail.

12.4 Summary and Outlook

In Section 12.3, information on the various inhibition mechanisms of [FeFe]-hydrogenases by small molecules has been provided. Each molecule has its inhibition specificity, some of them are reversible some are partly reversible, some are irreversible, and some are unknown (Table 12.1). However, none of these molecules can damage the active site as much as O_2. An important finding in hydrogenase research was the identification of the new clostridial O_2-resistant [FeFe]-hydrogenase CbaSH, which has a natural O_2 resistance. A detailed understanding of this native O_2-resistance mechanism, the role of sulfide, and the associated identification of specific amino acid motifs may allow the identification of other native O_2-stable [FeFe]-hydrogenases in the future and further increase the industrial interest in this enzyme family. The fact that the influence of extrinsic sulfide can also protect standard [FeFe]-hydrogenases from harmful O_2 also allows better handling of these enzyme types, which increases their application possibilities. In addition, the cost-intensive production processes need to be further simplified to increase economic efficiency, for which the investigation of new expression systems might be useful in the future.

References

1 Vignais, P.M., Billoud, B., and Meyer, J. (2001). Classification and phylogeny of hydrogenases. *FEMS Microbiol. Rev.* 25 (4): 455–501.

2 Hexter, S.V., Grey, F., Happe, T. et al. (2012). Electrocatalytic mechanism of reversible hydrogen cycling by enzymes and distinctions between the major classes of hydrogenases. *Proc. Natl. Acad. Sci. USA* 109 (29): 11516–11521.

3 Stephenson, M. and Stickland, L.H. (1931). Hydrogenase: A bacterial enzyme activating molecular hydrogen: The properties of the enzyme. *Biochem. J.* 25 (1): 205–214.

4 Nakos, G. and Mortenson, L.E. (1971). Structural properties of hydrogenase from *Clostridium pasteurianum* W5. *Biochemistry* 10 (13): 2442–2449.

5 Chen, J.S. and Mortenson, L.E. (1974). Purification and properties of hydrogenase from *Clostridium pasteurianum* W5. *Biochim. Biophys. Acta* 371 (2): 283–298.

6 Roessler, P. and Lien, S. (1982). Anionic modulation of the catalytic activity of hydrogenase from *Chlamydomonas reinhardtii*. *Arch. Biochem. Biophys.* 213 (1): 37–44.

7 Roessler, P.G. and Lien, S. (1984). Purification of Hydrogenase from *Chlamydomonas reinhardtii*. *Plant Physiol.* 75 (3): 705–709.

8 Roessler, P. and Lien, S. (1984). Effects of electron mediator charge properties on the reaction kinetics of hydrogenase from *Chlamydomonas*. *Arch. Biochem. Biophys.* 230 (1): 103–109.

9 Happe, T. and Naber, J.D. (1993). Isolation, characterization and N-terminal amino acid sequence of hydrogenase from the green alga *Chlamydomonas reinhardtii*. *Eur. J. Biochem.* 214 (2): 475–481.

10 Peters, J.W., Lanzilotta, W.N., Lemon, B.J., and Seefeldt, L.C. (1998). X-ray crystal structure of the Fe-only hydrogenase (CpI) from *Clostridium pasteurianum* to 1.8 angstrom resolution. *Science* 282 (5395): 1853–1858.

11 Nicolet, Y., Piras, C., Legrand, P. et al. (1999). *Desulfovibrio desulfuricans* iron hydrogenase: The structure shows unusual coordination to an active site Fe binuclear center. *Structure* 7 (1): 13–23.

12 Posewitz, M.C., King, P.W., Smolinski, S.L. et al. (2004). Discovery of two novel radical S-adenosylmethionine proteins required for the assembly of an active [Fe]-Hydrogenase. *J. Biol. Chem.* 279 (24): 25711–25720.

13 Kuchenreuther, J.M., Grady-Smith, C.S., Bingham, A.S. et al. (2010). High-yield expression of heterologous [FeFe]-Hydrogenases in *Escherichia coli*. *PLoS ONE* 5 (11): e15491.

14 Schwartz, C.J., Giel, J.L., Patschkowski, T. et al. (2001). IscR, an Fe-S cluster-containing transcription factor, represses expression of *Escherichia coli* genes encoding Fe-S cluster assembly proteins. *Proc. Natl. Acad. Sci. USA* 98 (26): 14895–14900.

15 King, P.W., Posewitz, M.C., Ghirardi, M.L., and Seibert, M. (2006). Functional studies of [FeFe]-Hydrogenase maturation in an *Escherichia coli* biosynthetic system. *J. Bacteriol.* 188 (6): 2163–2172.

16 Mulder, D.W., Boyd, E.S., Sarma, R. et al. (2010). Stepwise [FeFe]-Hydrogenase H-cluster assembly revealed in the structure of HydA$^{(DeltaEFG)}$. *Nature* 465 (7295): 248–251.

17 Lampret, O., Esselborn, J., Haas, R. et al. (2019). The final steps of [FeFe]-Hydrogenase maturation. *Proc. Natl. Acad. Sci. USA* 116 (32): 15802–15810.

18 Britt, R.D., Rao, G., and Tao, L. (2020). Biosynthesis of the catalytic H-cluster of [FeFe]-Hydrogenase: The roles of the Fe-S maturase proteins HydE, HydF, and HydG. *Chem. Sci.* 11 (38): 10313–10323.

19 Mulder, D.W., Ortillo, D.O., Gardenghi, D.J. et al. (2009). Activation of HydA$^{(DeltaEFG)}$ requires a preformed [4Fe–4S] cluster. *Biochemistry* 48 (26): 6240–6248.

20 Peters, J.W., Schut, G.J., Boyd, E.S. et al. (2015). [FeFe]- and [NiFe]-Hydrogenase diversity, mechanism, and maturation. *Biochim. Biophys. Acta* 1853 (6): 1350–1369.

21 Pagnier, A., Balci, B., Shepard, E.M. et al. (2022). [FeFe]-Hydrogenase *in vitro* maturation. *Angew. Chem. Int. Ed. Engl.* 61 (49): e202212074.

22 Watanabe, M., Honda, Y., Hagiwara, H., and Ishihara, T. (2017). [FeFe]-Hydrogenase and its organic molecule mimics—Artificial and bio-engineering application for hydrogen production. *J. Photochem. Photobiol. C Photchem. Rev.* 33: 1–26.

23 Schilter, D., Camara, J.M., Huynh, M.T. et al. (2016). Hydrogenase enzymes and their synthetic models: The role of metal hydrides. *Chem. Rev.* 116 (15): 8693–8749.

24 Li, H. and Rauchfuss, T.B. (2002). Iron carbonyl sulfides, formaldehyde, and amines condense to give the proposed azadithiolate cofactor of the Fe-only hydrogenases. *J. Am. Chem. Soc.* 124 (5): 726–727.

25 Berggren, G., Adamska, A., Lambertz, C. et al. (2013). Biomimetic assembly and activation of [FeFe]-Hydrogenases. *Nature* 499 (7456): 66–69.

26 Esselborn, J., Lambertz, C., Adamska-Venkates, A. et al. (2013). Spontaneous activation of [FeFe]-Hydrogenases by an inorganic 2Fe active site mimic. *Nat. Chem. Biol.* 9 (10): 607–609.

27 Esselborn, J., Muraki, N., Klein, K. et al. (2016). A structural view of synthetic cofactor integration into [FeFe]-Hydrogenases. *Chem. Sci.* 7 (2): 959–968.

28 Duan, J., Senger, M., Esselborn, J. et al. (2018). Crystallographic and spectroscopic assignment of the proton transfer pathway in [FeFe]-Hydrogenases. *Nat. Commun.* 9 (1): 4726.

29 Puthenkalathil, R.C. and Ensing, B. (2022). Fast proton transport in [FeFe]-Hydrogenase via a flexible channel and a proton hole mechanism. *J. Phys. Chem. B* 126 (2): 403–411.

30 Long, H., King, P.W., and Chang, C.H. (2014). Proton transport in *Clostridium pasteurianum* [FeFe]-Hydrogenase I: A computational study. *J. Phys. Chem. B* 118 (4): 890–900.

31 Hong, G., Cornish, A.J., Hegg, E.L., and Pachter, R. (2011). On understanding proton transfer to the biocatalytic $[Fe\text{-}Fe]_H$ sub-cluster in $[Fe\text{-}Fe]H_2$ases: QM/MM MD simulations. *Biochim. Biophys. Acta* 1807 (5): 510–517.

32 Cornish, A.J., Gärtner, K., Yang, H. et al. (2011). Mechanism of proton transfer in [FeFe]-Hydrogenase from *Clostridium pasteurianum*. *J. Biol. Chem.* 286 (44): 38341–38347.

33 Esselborn, J., Kertess, L., Apfel, U.-P. et al. (2019). Loss of specific active-site iron atoms in oxygen-exposed [FeFe]-Hydrogenase determined by detailed X-ray structure analyses. *J. Am. Chem. Soc.* 141 (44): 17721–17728.

34 Lu, Y. and Koo, J. (2019). O_2 sensitivity and H_2 production activity of hydrogenases-A review. *Biotechnol. Bioeng.* 116 (11): 3124–3135.

35 Sidabras, J.W. and Stripp, S.T. (2023). A personal account on 25 years of scientific literature on [FeFe]-Hydrogenase. *J. Biol. Inorg. Chem.* 28 (4): 355–378.

36 Morra, S., Arizzi, M., Valetti, F., and Gilardi, G. (2016). Oxygen stability in the new [FeFe]-Hydrogenase from *Clostridium beijerinckii* SM10 (CbA5H). *Biochemistry* 55 (42): 5897–5900.

37 Winkler, M., Duan, J., Rutz, A. et al. (2021). A safety cap protects hydrogenase from oxygen attack. *Nat. Commun.* 12 (1): 756.

38 Rutz, A., Das, C.K., Fasano, A. et al. (2023). Increasing the O_2 resistance of the [FeFe]-Hydrogenase CbA5H through enhanced protein flexibility. *ACS Catal.* 13 (2): 856–865.

39 Heghmanns, M., Rutz, A., Kutin, Y. et al. (2022). The oxygen-resistant [FeFe]-Hydrogenase CbA5H harbors an unknown radical signal. *Chem. Sci.* 13 (24): 7289–7294.

40 Rodríguez-Maciá, P., Reijerse, E.J., van Gastel, M. et al. (2018). Sulfide protects [FeFe]-Hydrogenases from O_2. *J. Am. Chem. Soc.* 140 (30): 9346–9350.

41 Felbek, C., Arrigoni, F., de Sancho, D. et al. (2021). Mechanism of hydrogen sulfide-dependent inhibition of [FeFe]-Hydrogenase. *ACS Catal.* 11 (24): 15162–15176.

42 Duan, J., Hemschemeier, A., Burr, D.J. et al. (2023). Cyanide binding to [FeFe]-Hydrogenase stabilizes the alternative configuration of the proton transfer pathway. *Angew. Chem. Int. Ed. Engl.* 62 (7): e202216903.

43 Duan, J., Veliju, A., Lampret, O. et al. (2023). Insights into the molecular mechanism of formaldehyde inhibition of [FeFe]-hydrogenases. *J. Am. Chem. Soc.* 145 (48): 26068–26074.

44 Morra, S. (2022). Fantastic [FeFe]-Hydrogenases and where to find them. *Front. Microbiol.* 13: 853626.

45 Meyer, J. (2007). [FeFe]-hydrogenases and their evolution: A genomic perspective. *Cell. Mol. Life Sci.* 64 (9): 1063–1084.

46 Fontecilla-Camps, J.C., Amara, P., Cavazza, C. et al. (2009). Structure-function relationships of anaerobic gas-processing metalloenzymes. *Nature* 460 (7257): 814–822.

47 Winkler, M., Esselborn, J., and Happe, T. (2013). Molecular basis of [FeFe]-Hydrogenase function: An insight into the complex interplay between protein and catalytic cofactor. *Biochim. Biophys. Acta* 1827 (8-9): 974–985.

48 Silakov, A., Kamp, C., Reijerse, E. et al. (2009). Spectroelectrochemical characterization of the active site of the [FeFe]-Hydrogenase HydA1 from *Chlamydomonas reinhardtii. Biochemistry* 48 (33): 7780–7786.

49 Nicolet, Y., Lemon, B.J., Fontecilla-Camps, J.C., and Peters, J.W. (2000). A novel Fe–S cluster in Fe-only hydrogenases. *Trends Biochem. Sci.* 25 (3): 138–143.

50 Artz, J.H., Mulder, D.W., Ratzloff, M.W. et al. (2017). Reduction potentials of [FeFe]-Hydrogenase accessory iron-sulfur clusters provide insights into the energetics of proton reduction catalysis. *J. Am. Chem. Soc.* 139 (28): 9544–9550.

51 Bruschi, M., Greco, C., Bertini, L. et al. (2010). Functionally relevant interplay between the Fe_4–S_4 cluster and CN^- ligands in the active site of [FeFe]-Hydrogenases. *J. Am. Chem. Soc.* 132 (14): 4992–4993.

52 Bruschi, M., Greco, C., Kaukonen, M. et al. (2009). Influence of the 2FeH subcluster environment on the properties of key intermediates in the catalytic cycle of [FeFe]-Hydrogenases: Hints for the rational design of synthetic catalysts. *Angew. Chem. Int. Ed. Engl.* 48 (19): 3503–3506.

53 Armstrong, F.A., Evans, R.M., Hexter, S.V. et al. (2016). Guiding principles of hydrogenase catalysis instigated and clarified by protein film electrochemistry. *Acc. Chem. Res.* 49 (5): 884–892.

54 Lemon, B.J. and Peters, J.W. (1999). Binding of exogenously added carbon monoxide at the active site of the iron-only hydrogenase (CpI) from *Clostridium pasteurianum*. *Biochemistry* 38 (40): 12969–12973.

55 Silakov, A., Reijerse, E.J., Albracht, S.P.J. et al. (2007). The electronic structure of the H-cluster in the [FeFe]-Hydrogenase from *Desulfovibrio desulfuricans*: A Q-band ^{57}Fe-ENDOR and HYSCORE study. *J. Am. Chem. Soc.* 129 (37): 11447–11458.

56 Lorent, C., Katz, S., Duan, J. et al. (2020). Shedding light on proton and electron dynamics in [FeFe]-Hydrogenases. *J. Am. Chem. Soc.* 142 (12): 5493–5497.

57 Birrell, J.A., Pelmenschikov, V., Mishra, N. et al. (2020). Spectroscopic and computational evidence that [FeFe]-Hydrogenases operate exclusively with CO-bridged intermediates. *J. Am. Chem. Soc.* 142 (1): 222–232.

58 Mebs, S., Senger, M., Duan, J. et al. (2017). Bridging hydride at reduced H-cluster species in [FeFe]-Hydrogenases revealed by infrared spectroscopy, isotope editing, and quantum chemistry. *J. Am. Chem. Soc.* 139 (35): 12157–12160.

59 Lubitz, W., Ogata, H., Rüdiger, O., and Reijerse, E. (2014). Hydrogenases. *Chem. Rev.* 114 (8): 4081–4148.

60 Morra, S., Duan, J., Winkler, M. et al. (2021). Electrochemical control of [FeFe]-Hydrogenase single crystals reveals complex redox populations at the catalytic site. *Dalton Trans.* 50 (36): 12655–12663.

61 Lampret, O., Adamska-Venkatesh, A., Konegger, H. et al. (2017). Interplay between CN^- ligands and the secondary coordination sphere of the H-cluster in [FeFe]-Hydrogenases. *J. Am. Chem. Soc.* 139 (50): 18222–18230.

62 Knörzer, P., Silakov, A., Foster, C.E. et al. (2012). Importance of the protein framework for catalytic activity of [FeFe]-Hydrogenases. *J. Biol. Chem.* 287 (2): 1489–1499.

63 Adams, M.W., Mortenson, L.E., and Chen, J.S. (1980). Hydrogenase. *Biochim. Biophys. Acta* 594 (2-3): 105–176.

64 Winkler, M., Kuhlgert, S., Hippler, M., and Happe, T. (2009). Characterization of the key step for light-driven hydrogen evolution in green algae. *J. Biol. Chem.* 284 (52): 36620–36627.

65 Günzel, A., Engelbrecht, V., and Happe, T. (2022). Changing the tracks: Screening for electron transfer proteins to support hydrogen production. *J. Biol. Inorg. Chem.* 27 (7): 631–640.

66 Kosem, N. (2023). H_2 production from methyl viologen-dependent hydrogenase activity monitored by gas chromatography. *Bio. Protoc.* 13 (23): e4895.

67 Martini, M.A., Rüdiger, O., Breuer, N. et al. (2021). The nonphysiological reductant sodium dithionite and [FeFe]-hydrogenase: Influence on the enzyme mechanism. *J. Am. Chem. Soc.* 143 (43): 18159–18171.

68 Orain, C., Saujet, L., Gauquelin, C. et al. (2015). Electrochemical measurements of the kinetics of inhibition of two [FeFe]-Hydrogenases by O_2 demonstrate that the reaction is partly reversible. *J. Am. Chem. Soc.* 137 (39): 12580–12587.

69 Armstrong, F.A., Belsey, N.A., Cracknell, J.A. et al. (2009). Dynamic electrochemical investigations of hydrogen oxidation and production by enzymes and implications for future technology. *Chem. Soc. Rev.* 38 (1): 36–51.

70 Gauquelin, C., Baffert, C., Richaud, P. et al. (2018). Roles of the F-domain in [FeFe]-Hydrogenase. *Biochim. Biophys. Acta Bioenerg.* 1859 (2): 69–77.

71 Lubner, C.E., Artz, J.H., Mulder, D.W. et al. (2022). A site-differentiated [4Fe-4S] cluster controls electron transfer reactivity of *Clostridium acetobutylicum* [FeFe]-Hydrogenase I. *Chem. Sci.* 13 (16): 4581–4588.

72 Lampret, O., Duan, J., Hofmann, E. et al. (2020). The roles of long-range proton-coupled electron transfer in the directionality and efficiency of [FeFe]-Hydrogenases. *Proc. Natl. Acad. Sci. USA* 117 (34): 20520–20529.

73 McCullagh, M. and Voth, G.A. (2013). Unraveling the role of the protein environment for [FeFe]-Hydrogenase: A new application of coarse-graining. *J. Phys. Chem. B* 117 (15): 4062–4071.

74 Lautier, T., Ezanno, P., Baffert, C. et al. (2011). The quest for a functional substrate access tunnel in [FeFe]-Hydrogenase. *Faraday Discuss.* 148: 385–407. discussion 421–41.

75 Brocks, C., Das, C.K., Duan, J. et al. (2024). A dynamic water channel affects O_2 stability in [FeFe]-Hydrogenases. *ChemSusChem* 17 (3): e202301365.

76 Cukierman, S. (2006). Et tu, Grotthuss! and other unfinished stories. *Biochim. Biophys. Acta* 1757 (8): 876–885.

77 Petrek, M., Otyepka, M., Banás, P. et al. (2006). CAVER: A new tool to explore routes from protein clefts, pockets and cavities. *BMC Bioinform.* 7: 316.

78 Cohen, J., Kim, K., King, P. et al. (2005). Finding gas diffusion pathways in proteins: Application to O_2 and H_2 transport in CpI [FeFe]-Hydrogenase and the role of packing defects. *Structure* 13 (9): 1321–1329.

79 Mohammadi, M. and Vashisth, H. (2017). Pathways and thermodynamics of oxygen diffusion in [FeFe]-Hydrogenase. *J. Phys. Chem. B* 121 (43): 10007–10017.

80 Liu, Y., Mohammadi, M., and Vashisth, H. (2018). Diffusion network of CO in [FeFe]-Hydrogenase. *J. Chem. Phys.* 149 (20): 204108.

81 Del Barrio, M., Sensi, M., Fradale, L. et al. (2018). Interaction of the H-cluster of [FeFe]-Hydrogenase with Halides. *J. Am. Chem. Soc.* 140 (16): 5485–5492.

82 Coulombe, R., Yue, K.Q., Ghisla, S., and Vrielink, A. (2001). Oxygen access to the active site of cholesterol oxidase through a narrow channel is gated by an Arg-Glu pair. *J. Biol. Chem.* 276 (32): 30435–30441.

83 Thoden, J.B., Holden, H.M., Wesenberg, G. et al. (1997). Structure of carbamoyl phosphate synthetase: A journey of 96 A from substrate to product. *Biochemistry* 36 (21): 6305–6316.

84 Tan, X., Volbeda, A., Fontecilla-Camps, J.C., and Lindahl, P.A. (2006). Function of the tunnel in acetylcoenzyme A synthase/carbon monoxide dehydrogenase. *J. Biol. Inorg. Chem.* 11 (3): 371–378.

85 Krahn, J.M., Kim, J.H., Burns, M.R. et al. (1997). Coupled formation of an amidotransferase interdomain ammonia channel and a phosphoribosyltransferase active site. *Biochemistry* 36 (37): 11061–11068.

86 Kalms, J., Schmidt, A., Frielingsdorf, S. et al. (2018). Tracking the route of molecular oxygen in O_2-tolerant membrane-bound [NiFe]-Hydrogenase. *Proc. Natl. Acad. Sci. USA* 115 (10): E2229–E2237.

87 Zacarias, S., Temporão, A., Del Barrio, M. et al. (2019). A hydrophilic channel is involved in oxidative inactivation of a [NiFeSe]-Hydrogenase. *ACS Catal.* 9 (9): 8509–8519.

88 Zacarias, S., Temporão, A., Carpentier, P. et al. (2020). Exploring the gas access routes in a [NiFeSe]-Hydrogenase using crystals pressurized with krypton and oxygen. *J. Biol. Inorg. Chem.* 25 (6): 863–874.

89 Barbosa, T.M., Baltazar, C.S.A., Cruz, D.R. et al. (2020). Studying O_2 pathways in [NiFe]- and [NiFeSe]-Hydrogenases. *Sci. Rep.* 10 (1): 10540.

90 Baltazar, C.S.A., Marques, M.C., Soares, C.M. et al. (2011). Nickel–iron–selenium hydrogenases – An overview. *Eur. J. Inorg. Chem.* 2011 (7): 948–962.

91 Nicolet, Y. (2001). Etude cristallographique de l'hydrogénase à fer de la bactérie sulfato-réductrice *Desulfovibrio desulfuricans* ATCC 7757. PhD thesis. Universite Joseph Fourier, Grenoble.

92 Lambertz, C., Leidel, N., Havelius, K.G.V. et al. (2011). O_2 reactions at the six-iron active site (H-cluster) in [FeFe]-Hydrogenase. *J. Biol. Chem.* 286 (47): 40614–40623.

93 Stiebritz, M.T. and Reiher, M. (2009). Theoretical study of dioxygen induced inhibition of [FeFe]-Hydrogenase. *Inorg. Chem.* 48 (15): 7127–7140.

94 Finkelmann, A.R., Stiebritz, M.T., and Reiher, M. (2014). Activation barriers of oxygen transformation at the active site of [FeFe]-Hydrogenases. *Inorg. Chem.* 53 (22): 11890–11902.

95 Swanson, K.D., Ratzloff, M.W., Mulder, D.W. et al. (2015). [FeFe]-Hydrogenase oxygen inactivation is initiated at the H-cluster 2Fe subcluster. *J. Am. Chem. Soc.* 137 (5): 1809–1816.

96 Stripp, S.T., Goldet, G., Brandmayr, C. et al. (2009). How oxygen attacks [FeFe]-Hydrogenases from photosynthetic organisms. *Proc. Natl. Acad. Sci. USA* 106 (41): 17331–17336.

97 Noth, J., Kositzki, R., Klein, K. et al. (2015). Lyophilization protects [FeFe]-Hydrogenases against O_2-induced H-cluster degradation. *Sci. Rep.* 5: 13978.

98 Mebs, S., Kositzki, R., Duan, J. et al. (2018). Hydrogen and oxygen trapping at the H-cluster of [FeFe]-Hydrogenase revealed by site-selective spectroscopy and QM/MM calculations. *Biochim. Biophys. Acta Bioenerg.* 1859 (1): 28–41.

99 Bruska, M.K., Stiebritz, M.T., and Reiher, M. (2011). Regioselectivity of H-cluster oxidation. *J. Am. Chem. Soc.* 133 (50): 20588–20603.

100 Kubas, A., Orain, C., de Sancho, D. et al. (2017). Mechanism of O_2 diffusion and reduction in [FeFe]-Hydrogenases. *Nat. Chem.* 9 (1): 88–95.

101 Goris, T., Wait, A.F., Saggu, M. et al. (2011). A unique iron-sulfur cluster is crucial for oxygen tolerance of a [NiFe]-Hydrogenase. *Nat. Chem. Biol.* 7 (5): 310–318.

102 Fritsch, J., Scheerer, P., Frielingsdorf, S. et al. (2011). The crystal structure of an oxygen-tolerant hydrogenase uncovers a novel iron-sulphur centre. *Nature* 479 (7372): 249–252.

103 Goldet, G., Brandmayr, C., Stripp, S.T. et al. (2009). Electrochemical kinetic investigations of the reactions of [FeFe]-Hydrogenases with carbon monoxide and oxygen: Comparing the importance of gas tunnels and active-site electronic/redox effects. *J. Am. Chem. Soc.* 131 (41): 14979–14989.

104 Baffert, C., Bertini, L., Lautier, T. et al. (2011). CO disrupts the reduced H-cluster of [FeFe]-Hydrogenase. A combined DFT and protein film voltammetry study. *J. Am. Chem. Soc.* 133 (7): 2096–2099.

105 Metz, B., Kersten, G.F.A., Hoogerhout, P. et al. (2004). Identification of formaldehyde-induced modifications in proteins: Reactions with model peptides. *J. Biol. Chem.* 279 (8): 6235–6243.

106 Wait, A.F., Brandmayr, C., Stripp, S.T. et al. (2011). Formaldehyde—a rapid and reversible inhibitor of hydrogen production by [FeFe]-Hydrogenases. *J. Am. Chem. Soc.* 133 (5): 1282–1285.

107 Foster, C.E., Krämer, T., Wait, A.F. et al. (2012). Inhibition of [FeFe]-Hydrogenases by formaldehyde and wider mechanistic implications for biohydrogen activation. *J. Am. Chem. Soc.* 134 (17): 7553–7557.

108 Bachmeier, A., Esselborn, J., Hexter, S.V. et al. (2015). How formaldehyde inhibits hydrogen evolution by [FeFe]-Hydrogenases: Determination by ^{13}C ENDOR of direct Fe-C coordination and order of electron and proton transfers. *J. Am. Chem. Soc.* 137 (16): 5381–5389.

109 Zhang, F., Woods, T.J., Zhu, L., and Rauchfuss, T.B. (2021). Inhibition of [FeFe]-Hydrogenase by formaldehyde: Proposed mechanism and reactivity of FeFe alkyl complexes. *Chem. Sci.* 12 (47): 15673–15681.

110 Martini, M.A., Bikbaev, K., Pang, Y. et al. (2023). Binding of exogenous cyanide reveals new active-site states in [FeFe]-Hydrogenases. *Chem. Sci.* 14 (11): 2826–2838.

111 Cohen, J., Kim, K., Posewitz, M. et al. (2005). Molecular dynamics and experimental investigation of H_2 and O_2 diffusion in [Fe]-hydrogenase. *Biochem. Soc. Trans.* 33 (Pt 1): 80–82.

112 Glick, B.R., Martin, W.G., and Martin, S.M. (1980). Purification and properties of the periplasmic hydrogenase from *Desulfovibrio desulfuricans*. *Can. J. Microbiol.* 26 (10): 1214–1223.

113 Roseboom, W., de Lacey, A.L., Fernandez, V.M. et al. (2006). The active site of the [FeFe]-Hydrogenase from *Desulfovibrio desulfuricans*. II. Redox properties, light sensitivity and CO-ligand exchange as observed by infrared spectroscopy. *J. Biol. Inorg. Chem.* 11 (1): 102–118.

114 Hatchikian, E.C., Forget, N., Fernandez, V.M. et al. (1992). Further characterization of the [Fe]-Hydrogenase from *Desulfovibrio desulfuricans* ATCC 7757. *Eur. J. Biochem.* 209 (1): 357–365.

115 Rodríguez-Maciá, P., Galle, L.M., Bjornsson, R. et al. (2020). Caught in the H_{inact} : Crystal structure and spectroscopy reveal a sulfur bound to the active site of an O_2-stable state of [FeFe]-Hydrogenase. *Angew. Chem. Int. Ed. Engl.* 59 (38): 16786–16794.

116 Corrigan, P.S., Tirsch, J.L., and Silakov, A. (2020). Investigation of the unusual ability of the [FeFe]-Hydrogenase from *Clostridium beijerinckii* to access an O_2-protected state. *J. Am. Chem. Soc.* 142 (28): 12409–12419.

13

Biosynthesis of [NiFe]-Hydrogenase

Oliver Lenz and Giorgio Caserta

Technische Universität Berlin, Institute of Chemistry, Biophysical Chemistry, Straße des 17. Juni 135, 10623 Berlin, Germany

Chapter Goals

- Getting an overview of the complex biosynthesis of [NiFe]-hydrogenase
- Understanding how a field of research develops from initial observations to a detailed picture of a complex biosynthesis network
- Motivation for future studies on unresolved questions of [NiFe]-hydrogenase biosynthesis

Main Chapter Points

- Identification of the basic set of maturation proteins and their functions responsible for the incorporation of the inorganic $NiFe(CN)_2CO$ cofactor into the large subunit of hydrogenase
- Overview of Fe–S cluster insertion into the small hydrogenase subunit and its oligomerization with the large subunit to form catalytically active [NiFe]-hydrogenase
- Recognition of exceptions from the canonical biosynthesis and additional functions for the biosynthesis of certain [NiFe]-hydrogenase types

13.1 [NiFe]-Hydrogenases and Their Function

Like [FeFe]-hydrogenases, [NiFe]-hydrogenases also catalyze the reversible cleavage of molecular hydrogen (H_2) into protons and electrons ($H_2 \rightleftharpoons 2\,H^+ + 2\,e^-$) (Chapter 15). However, instead of two iron ions, their catalytic center consists of one iron and one nickel ion, which is about 600-fold less abundant in the earth's crust than iron. The increased complexity of the catalytic center is also reflected in the fact that more accessory proteins are required for the assembly of the

Iron-Sulfur Clusters: Biogenesis and Biochemistry, Volume 2, First Edition.
Edited by Silke Leimkühler, Günter Schwarz, Oliver Lenz, and Oliver Einsle.

inorganic cofactor, as will be described below. So, what is the benefit of nickel? [NiFe]-hydrogenases generally have a much higher affinity for their substrate H_2 than [FeFe]-hydrogenases [1, 2], which is consistent with H_2 uptake as their typical physiological function in nature. Consequently, [NiFe]-hydrogenases can maintain their catalytic function at much lower environmental H_2 concentrations than [FeFe]-hydrogenases. This is probably a consequence of the fact that H_2 was much more abundant on the early Earth and its concentration in the atmosphere decreased dramatically even before the great oxidation event [3], which required the evolution of enzymes with higher affinity for H_2. The great oxidation event has even exerted a second evolutionary pressure on the hydrogenases: molecular oxygen (O_2). While [FeFe]-hydrogenases are generally irreversibly inactivated in the presence of both a reductant (e.g. H_2) and O_2 (Chapter 12 and Section 15.2), some [NiFe]-hydrogenases are able to sustain H_2 cleavage in the presence of O_2 (Section 15.3). An impressive example is the widespread high-affinity [NiFe]-hydrogenases, which enable their aerobic hosts to extract energy from atmospheric H_2 at concentrations as low as 0.5 ppmv [4]. Here, the responsible [NiFe]-hydrogenase is connected on the inside of the cytoplasmic membrane to the respiratory chain, which generates a proton motive force based on the electrons obtained from H_2 and where O_2 is ultimately reduced to water [5]. Another class of [NiFe]-hydrogenases feed H_2-derived electrons from the periplasmic side of the membrane into the membranous electron transport chain to fuel the metabolism through aerobic or anaerobic (sulfate, nitrate, iron, etc.) respiration. The enrichment of the periplasmic space with protons originating from H_2 contributes to the proton motive force [6]. The anaerobic reduction of CO_2 to methane by hydrogenotrophic methanogenic archaea is also fueled by H_2, and various [NiFe]-hydrogenases and [Fe]-hydrogenase (Chapter 15) are involved in this geochemically important process. One of them is the cofactor F_{420}-reducing [NiFe]-hydrogenase, which directly couples H_2 oxidation with the reduction of the F_{420} cofactor, which in turn serves as an intracellular reducing agent for C1 intermediates of methanogenesis (Chapter 15) [7]. Similar energy-converting enzymes are the $NAD(P)^+$-reducing [NiFe]-hydrogenases, which directly supply reducing equivalents for both metabolism and the respiratory chain [8, 9]. A [NiFe]-hydrogenase that is not directly involved in H_2-driven energy conversion is the so-called H_2-sensing or regulatory hydrogenase. This enzyme controls the biosynthesis of energy-converting [NiFe]-hydrogenases via an unconventional two-component regulatory system at the level of gene expression in response to the presence of H_2 [10]. Among the numerous [NiFe]-hydrogenases that catalyze H_2 oxidation in the living cell, there are some members that mediate H_2 evolution to get rid of excess reducing equivalents or fermentation products under anaerobic conditions. A prominent example is the hydrogenase-3 of *Escherichia coli*, which is part of the membrane-bound formate-hydrogen lyase complex and couples formate oxidation with H_2 production [11]. The phylogeny and the various physiological functions of [NiFe]-hydrogenases have been excellently reviewed in [6, 12–14]. Furthermore, the catalytic cycle of [NiFe]-hydrogenases and O_2 tolerance mechanisms are discussed in Section 15.3.

13.2 The Inorganic Catalytic Center in the [NiFe]-Hydrogenase Basic Module

The minimal H_2-converting module of [NiFe]-hydrogenases consists of a large subunit harboring the heterobimetallic [NiFe] active site and a small, electron-transferring subunit bearing at least one [4Fe–4S] cluster, as present in some members the heteromultimeric, cofactor-coupled [NiFe]-hydrogenases of group 3 ([13], Figure 13.1). This minimal module is reminiscent of the H-cluster of [FeFe]-hydrogenase, where the diiron center is associated with a [4Fe–4S] cluster (Chapter 12). However, while a bridging cysteine residue is involved in coordinating both two metal moieties in [FeFe]-hydrogenases (Chapters 12 and 15), the catalytic center and the proximal Fe–S cluster of [NiFe]-hydrogenases are located in two different subunits (Figure 13.1).

The presence of only one Fe–S cluster in the small subunit is a rare exception. The vast majority of [NiFe]-hydrogenases contain three Fe–S clusters in the electron-transferring subunit (Figure 13.1a), which differ in the number of iron and sulfide ions as well as in the type and number of coordinating amino acid residues. Assembly and insertion of the Fe–S clusters into the small subunit of [NiFe]-hydrogenases occurs independently of the metal center insertion into the large subunit and is most likely mediated by the housekeeping Isc and Suf machineries (Section 13.11 and Chapters 1 and 2) [15, 16].

Four highly conserved cysteine residues are involved in the covalent coordination of the inorganic metal center in the large subunit (Figure 13.1). All four cysteine-derived thiolates are ligands of the nickel ion, and two of them also serve as bridging ligands for the iron. In addition, two cyanides (CN^-) and a carbon monoxide (CO) belong to the ligand sphere of the iron, which is thus kept in a low-spin Fe(II) state. Notably, the coordination of the active site iron ion(s) by diatomic ligands is a recurring feature in all three types of hydrogenases (Chapters 12, 14, and 15). Except for the fact that one of the four cysteines is exchanged by selenocysteine in a subgroup of [NiFe]-hydrogenases, the so-called [NiFeSe]-hydrogenases [17, 18], the first and also the second coordination sphere of the catalytic center seems to be highly conserved in all [NiFe]-hydrogenases. This also includes the amino acid residues that interact with the oxygen and nitrogen atoms of the CN^- and CO ligands [19]. The apparent identity of the catalytic center including the second coordination sphere in all [NiFe]-hydrogenases already indicates that the assembly of the $NiFe(CN)_2CO$ cofactor and its insertion into the apo-form of the hydrogenase large subunit occurs in a rather standardized manner.

13.3 Genetic Basis of [NiFe]-Hydrogenase Maturation

The two hydrogenase subunits forming the basic module are generally encoded by two adjacent structural genes as a transcriptional unit in the genome. In most cases, these genes are accompanied by additional genes encoding, for example, the primary electron acceptors and further functional moieties of multisubunit

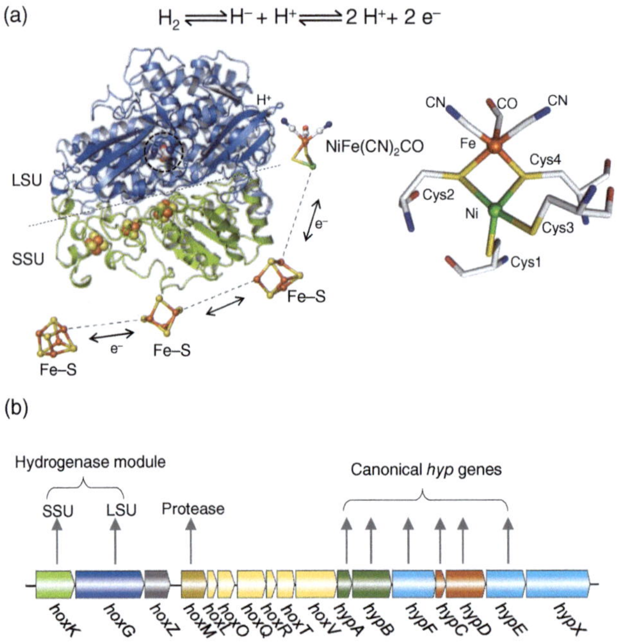

Figure 13.1 (a) Typical [NiFe]-hydrogenase module, consisting of a large subunit (LSU) containing the $NiFe(CN)_2CO$ active site (dotted black circle) and an electron-transferring small subunit (SSU) harboring three Fe–S clusters. The enlarged catalytic center is shown on the right. The $Fe(CN)_2CO$ unit and the nickel ion are coordinated to the protein by two bridging (Cys2 and Cys4) and two terminal cysteine residues (Cys1 and Cys3). The numbering of the cysteine residues varies between the different large subunits of hydrogenase; here, they are numbered according to their occurrence in the amino acid sequence. (b) Operon structure of the hydrogenase genes required for the biosynthesis of the membrane-bound [NiFe]-hydrogenase (MBH) from *Cupriavidus necator* [20]. The genes *hoxK* and *hoxG* encode the SSU and LSU of MBH, while *hoxZ* encodes the primary electron acceptor, a membrane-integral *b*-type cytochrome. The gene products of *hypA–hypE* are found in all organisms that synthesize active [NiFe]-hydrogenases and are necessary for the biosynthesis of the $NiFe(CN)_2CO$ cofactor and its incorporation into the premature LSU. Another common feature is the presence of a gene (*hoxM* in the case of *C. necator* MBH) whose product mediates the proteolytic processing of the C-terminal extension of the large hydrogenase subunit after the insertion of the [NiFe] site. The genes *hoxL*, *O*, *Q*, *R*, *T*, and *V* as well as *hypX* have special functions in [NiFe]-hydrogenase biosynthesis under aerobic conditions and are usually less frequent or even absent in gene clusters encoding [NiFe]-hydrogenases that are produced and catalytically active exclusively under anaerobic conditions (see text for details).

[NiFe]-hydrogenases (Figure 13.1b) [21]. With a few exceptions (see Section 13.9), each [NiFe]-hydrogenase large subunit comes with an associated endopeptidase, which is also encoded in the corresponding hydrogenase gene cluster. Microorganisms synthesizing functional [NiFe]-hydrogenases invariably also encode six auxiliary proteins, designated HypA, B, C, D, E, and F (Figure 13.1b). The term "Hyp" was coined in 1990 by August Böck, the pioneer in investigating the maturation of [NiFe]-hydrogenases [22]. He and his coworkers found that deletions of the *hyp* genes, which are part of the operon encoding hydrogenase 3 (Hyd-3) of *E. coli*, have a pleiotropic effect (*hyp* – *hy*drogenase *p*leiotropic) on the catalytic function of all three [NiFe]-hydrogenases of *E. coli* [23]. Later, a seventh *hyp* gene, *hypX*, has been identified, which mainly occurs in genomes of microorganisms that synthesize [NiFe]-hydrogenases under aerobic conditions (Figure 13.1b) [24].

The Hyp proteins are responsible for the sequential bioassembly of the $NiFe(CN)_2CO$ cofactor, which includes the biosynthesis of the CO and CN^- ligands, assembly of the $Fe(CN)_2CO$ synthon, and its insertion into the apo-form of the large subunit, followed by the incorporation of nickel [22, 25]. The overall process is illustrated in Figure 13.2 and the role of the individual Hyp functions is outlined in the following sections.

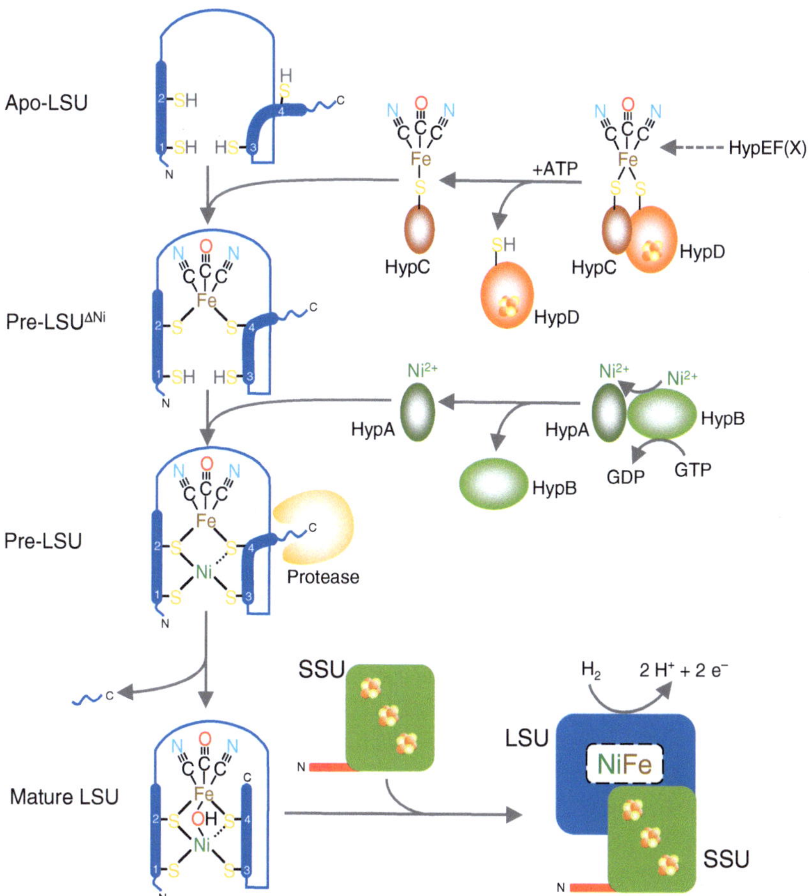

Figure 13.2 Current scheme showing the successive steps of the biosynthesis of the large subunit (LSU) of [NiFe]-hydrogenase and its subsequent oligomerization with the small subunit (SSU) to form the catalytically active enzyme. The red-colored N-terminal signal peptide, which is not present in all SSUs, directs periplasmic [NiFe]-hydrogenases through the cytoplasmic membrane via the Tat apparatus (see text for details). The white numbers 1–4 represent the four cysteine residues coordinating the $NiFe(CN)_2CO$ cofactor to the large subunit. Details of the various reactions are described in this chapter.

13.4 Cyanide Synthesis by HypE and HypF

The molecular precursor of forming the cyanide ligands has been known since 2000, when Böck and coworkers discovered, based on a previous observation by others [26], that *E. coli* cells unable to produce carbamoyl phosphate, a molecule required for arginine and pyrimidine biosynthesis, do not synthesize active [NiFe]-hydrogenases [27]. Accordingly, the purified HypF protein from *E. coli* catalyzed both the dephosphorylation of carbamoyl phosphate and the carbamoyl phosphate-dependent hydrolysis of ATP into AMP and pyrophosphate [28]. This suggested the formation of a carbamoyl adenylate [29] as an intermediate of cyanide synthesis. Indeed, HypF also showed a transcarbamylase activity and transferred the carbamoyl group to the thiolate of the C-terminal cysteine of HypE [30]. The resulting thiocarboxamide is then converted into thiocyanate in an ATP-dependent dehydration reaction catalyzed by HypE.

Within a good decade, a complex reaction scheme for the synthesis of cyanide ligands by the two maturases HypE and HypF has emerged (Figure 13.3). The prototypic HypF protein comprises four domains [31, 32]. The N-terminal acyl phosphatase (ACP) domain acts as a carbamoyl phosphate phosphatase, and the produced carbamate is thought to traverse a tunnel that extends from the ACP domain through a zinc-finger domain into the YrdC-like middle domain. There, the carbamate is converted into carbamoyl adenylate in a nucleotide-binding pocket in an ATP-dependent reaction, which releases pyrophosphate [28]. The carbamoyl adenylate, whose existence has been verified in the HypF structure of *E. coli* [31], then migrates through a cavity to the active site in the C-terminal KaeI-like domain, which acts as a carbamoyl transferase (see below). It should be mentioned here that there are exceptions to the canonical HypF structure. Some organisms contain a truncated HypF version that consists only of the KaeI-like *O*-carbamoyltransferase domain. A prominent member is the aerobic H_2 oxidizer *C. necator* (formerly *Ralstonia eutropha* and *Alcaligenes eutrophus*), which synthesizes both a canonical and a truncated version of HypF, and the latter is sufficient for the full activity of the two energy-converting [NiFe]-hydrogenases of this organism [33]. This suggests that these organisms might synthesize carbamoyl adenylate by a different reaction, which can be used directly by the truncated HypF versions.

Once bound to the conserved metal-containing (Zn or Fe, [31, 32, 34]) nucleotide-binding site of the *O*-carbamoyltransferase domain, the carbamoyl group of carbamoyl adenylate is then transferred to the C-terminal cysteine of the HypE protein. This results in a HypE-thiocarboxamide, as the Böck group has demonstrated by mass spectrometric analysis of the reaction product of carbamoyl phosphate, ATP, HypF, and HypE of *E. coli* [30]. HypE belongs to the PurM family of proteins that catalyze the ATP-dependent dehydration of their substrates. Purified HypE from *E. coli* has also been shown to have low ATPase activity, which is required to dehydrate the carboxamido moiety to CN, revealing HypE-bound thiocyanate (Figure 13.3) [30, 35].

The reactions described earlier necessitate a concerted reaction between the HypF and HypE proteins, and their interaction has been shown by copurification

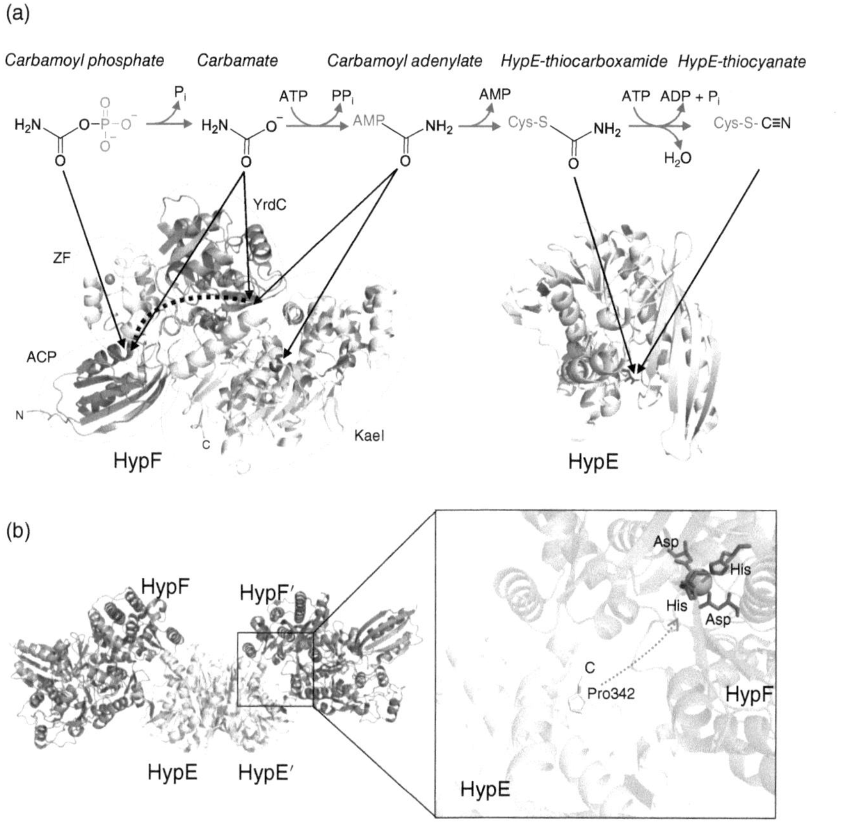

Figure 13.3 Synthesis of the cyanide ligands of the $NiFe(CN)_2CO$ cofactor catalyzed by the HypF and HypE maturases. (a) Sequential steps in the biosynthesis of HypE-bound thiocyanate from carbamoyl phosphate. The arrows indicate the locations in HypF (PDB code 3VTH) and HypE (only one monomer of the usual homodimer is shown, PDB code 3WJR) where the respective reactions take place or the intermediates are located. The acyl phosphatase (ACP), zinc finger (ZF), YrdC, and Kael domains of HypF are indicated by dotted circles, and the C-terminal cysteine of HypE is shown as a stick model. The bold dashed line in HypF indicates a tunnel that allows the labile carbamate to be transported from the ACP domain to the YrdC domain. (b) Structure of the $HypE_2$–$HypF_2$ complex from *Caldanaerobacter subterraneus* (left, PDB code 3VTI). The enlarged section suggests that the C terminus of HypE, which is carbamoylated by HypF, is located near the Fe-containing carbamoyl transferase site of HypF (the iron ion and its coordinating amino acid residues are highlighted). The last nine amino acid residues of the C terminus (dotted arrow), including the C-terminal cysteine of HypE, are not resolved in the structure, and Pro342 is, therefore, the last residue shown. Source: Adapted from [32].

experiments [35, 36]. Later, the crystal structures of HypE [37–39] and HypF–HypE complexes [32] have been described. HypE generally forms a homodimeric structure. To mimic the carbamoyltransferase activity of HypF on HypE, the Miki group has incubated HypE crystals with potassium cyanate, which resulted in the carbamoylation of the C-terminal cysteine, which is the only cysteine in HypE of *Thermococcus kodakarensis* [38]. The subsequent addition of ATP initialized the dehydration of the carboxamido moiety, which led to the formation of the cyano group on the C-terminal cysteine of HypE. The underlying proposed reaction mechanism involves a conserved deprotonated lysine residue and glutamate as the basis for the required deprotonation reactions [38].

One crystal structure is available for the HypEF complex, namely that of the thermophilic anaerobic bacterium *C. subterraneus* [32]. The typical dimeric HypE structure, in which the N-termini of the HypE monomers interact closely with each other, is retained, and each HypE monomer forms a complex with a HypF monomer, resulting in an elongated $HypE_2$–$HypF_2$ structure (Figure 13.3b). The Kael-like O-carbamoyltransferase domain of the HypF proteins forms the majority of the interaction sites with the HypE proteins. Unfortunately, the last nine amino acid residues of HypE, including the C-terminal cysteine, were not resolved in the structure. However, the most C-terminal structurally resolved proline residue was found near (ca. 23 Å) the iron-containing carbamoyl transferase site of HypF, suggesting that the C-terminal cysteine of HypE can reach the Fe site in HypF to accept the carbamoyl group (Figure 13.3b) [32]. Notably, the C terminus of HypE in the complex with HypF (and HypCD, see Section 13.5) adopts an outward-facing conformation that allows its carbamoylation by HypF, while the dehydration of the thiocarboxamide to thiocyanate is catalyzed in the inward-facing conformation [38].

This completes the current picture of the cyanide group synthesis from carbamoyl phosphate, which is catalyzed by the two maturases HypF and HypE, with a C-terminal thiocyanate on HypE as the product.

13.5 Assembly of the $Fe(CN)_2CO$ Synthon on the HypCD Complex

HypE forms a ternary complex with the maturases HypC and HypD [36, 40, 41], suggesting that the cyanide group is delivered by HypE to HypCD (Figure 13.4a). Since the thiocyanate is initially buried in the HypE–HypF complex, the cyanated HypE must leave the complex in order to transfer the cyanide group to the HypCD complex. This explains why a super-complex between all four maturases HypC, HypD, HypE, and HypF has never been identified. However, the mechanism that leads to the dissociation of the HypEF complex after HypE cyanization is unknown [37].

The transfer of the ^{14}CN label on HypE to the HypCD complex, but not to the individual HypC and HypD proteins, has been shown already in 2004 [40], which was the first experimental direct evidence for the cyanide transfer. In this study, the presence of an [4Fe–4S] cluster in HypD has been experimentally verified by Mössbauer and UV–Vis spectroscopy, and the authors speculated that the cluster might deliver the electrons required for CN^-/CO ligand attachment (two electrons per ligand are required) to a single iron Fe jointly bound by the HypC and HypD proteins. It was also speculated that the cluster might be involved in the biosynthesis of CO.

In fact, it has been shown later that the isolated HypCD complex serves as a scaffold for the assembly of the $Fe(CN)_2CO$ synthon of the active site of [NiFe]-hydrogenase [42, 43]. This was possible by infrared (IR) spectroscopic analysis of

(a)

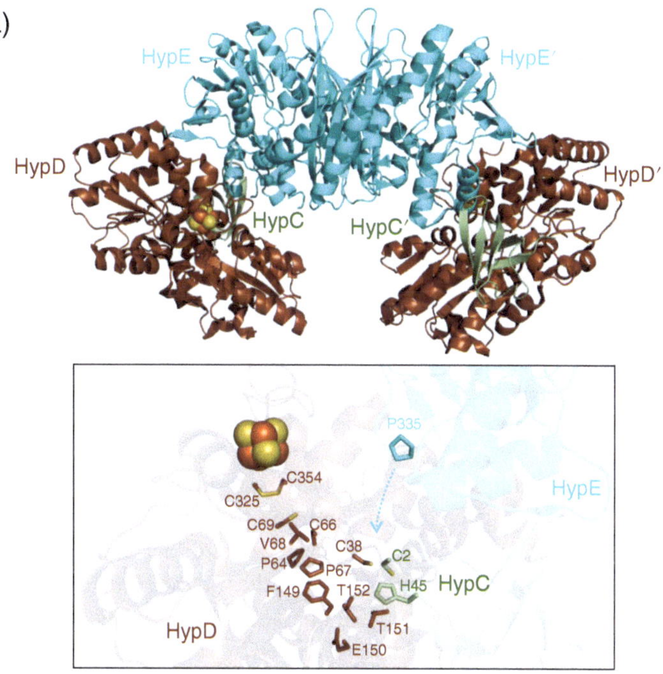

Figure 13.4 (a) Crystal structure of the ternary complex consisting of HypC, HypD, and HypE. Source: Adapted from [41]. The lower part shows that the two cysteine residues of HypC (C2) and HypD (C38), which are most likely to coordinate the iron ion equipped with two cyanides and one CO ligand, are close to each other and that the cyanated C terminus of HypE (which is not fully resolved in the structure, with Pro335 being the most C-terminal residue) can protrude (dashed arrow) into the iron-binding cavity of the HypCD heterodimer [45]. The side chains (stick representation) of conserved amino acids are labeled with single-letter code and colored according to the protein. (b) Complex between HypC from *E. coli* and the apo-large subunit HoxC of the regulatory hydrogenase from *C. necator*, as predicted by AlphaFold 3. The calculation has been recently performed [46] by including an iron (brown ball), which is now possible with AlphaFold 3 [47]. The calculated complex shows a tight interaction between the two proteins, which signifies that HypC (green) must leave the HypCD complex to deliver the Fe(CN)$_2$CO synthon to the apo-large subunit (blue). Interestingly, Cys2 of HypC, which is thought to carry the Fe(CN)$_2$CO synthon, protrudes into the [NiFe]-binding motif of HoxC, comprising the residues Cys1–3, while the C terminus of HoxC, which includes Cys4 is displaced from the active site core. The active site-coordinating cysteine residues as well as Cys2 from HypC are shown in stick representation and labeled in the color of the corresponding protein. Note the different positions of the histidine (stick representation), which becomes the C-terminal amino acid upon cleavage of the C-terminal extension.

overproduced *E. coli* and *C. necator* HypCD complexes. The position of the vibrational bands in the IR spectrum was consistent with two CN^- ligands and one CO attached to an iron ion. According to a follow-up study, HypD alone might also have the ability to carry the Fe(CN)$_2$CO moiety [44]. However, this is a rather puzzling result, since it is currently thought that the HypC protein delivers the Fe(CN)$_2$CO synthon to the apo-large subunit of the [NiFe]-hydrogenase (see Section 13.7).

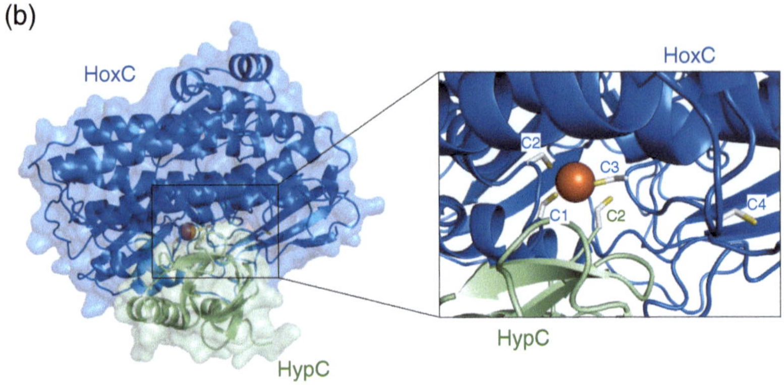

Figure 13.4 (*Continued*)

According to a long-standing model, conserved cysteine residues of both HypC (or a HypC ortholog) and HypD jointly coordinate an iron ion (whose origin is briefly discussed in Section 13.11) that receives first two cyanides and then a carbonyl ligand. The first evidence was provided by mutagenesis studies on the HypC and HypD proteins of *E. coli*. The (independent) exchange of the N-terminal cysteine residue of HypC (Cys2) and a conserved cysteine residue of HypD (Cys41) with other residues interrupted the maturation of the [NiFe]-hydrogenase and prevented the two protein variants from interacting with the protein partner [48, 49]. The importance of the corresponding Cys residue (Cys36) for the function of HypD of *C. necator* has also been shown [42]. The involvement of the two invariant cysteine residues from HypC and HypD in iron binding has been strongly supported by the crystal structure of the HypCD complex from *T. kodakarensis* [41]. The N-terminal β-barrel domain of the small HypC protein (ca. 10 kDa) is tightly bound to HypD (ca. 40 kDa), bringing the N-terminal Cys2 of HypC very close to the invariant Cys38 of HypD to bind an iron ion. However, a HypCD crystal structure carrying the bound single iron or even the entire $Fe(CN)_2CO$ synthon has not yet been described.

Remarkably, also the crystal structure of ternary complex of HypC, HypD, and HypE has been resolved by the Miki group [41]. Here too, HypE has retained its dimeric structure, and two HypCD modules connect from the side, as was observed in the HypEF complex (see Section 13.4). Although the structure is not fully resolved, it allows the interpretation that the C-terminal arm of HypE, which harbors the C-terminal cysteine residue, which is cyanized in its loaded form, protrudes into the HypD cavity carrying the iron ion. This would allow the cyanide transfer from C-terminal cysteine of HypE to the iron. Miki and coworkers developed a model in which the thiol of an additional conserved cysteine residue in the HypD assembly site attacks the thiocyanate of HypE, leading to the transfer of the CN^- group to the iron ion bound by HypCD (Figure 13.4a) [41].

The ligand stoichiometry of the $Fe(CN)_2CO$ synthon assembled in HypCD and the structure of the HypCDE complex require two successive interactions of cyanated HypE with iron-containing HypCD. How this is achieved is currently unknown. One possibility would be that a HypCD module first forms a complex with a HypE monomer of the dimeric HypE complex and receives the first CN^-. Then, the HypCD module moves over to the second HypE monomer, where it receives the second CN^-. As mentioned earlier, the conserved [4Fe–4S] cluster in HypD presumably provides the reducing power necessary for ligand attachment [40]. However, the native electron donor that supplies the necessary electrons to the Fe–S cluster remains unknown.

Despite some controversy over the order of binding of the diatomic ligands to the iron ion in HypCD, i.e. first cyanide, then CO [42] vs. first CO, then cyanide [50], the current model assumes that the $Fe(CN)_2$ intermediate fragment is completed by the attachment of CO to the $Fe(CN)_2CO$ synthon. The origin of the CO ligand is discussed in Section 13.6.

13.6 Origin and Synthesis of the Carbon Monoxide Ligand

While the biosynthesis of the cyanide ligands is well understood, the biogenesis of the CO ligand under anaerobic conditions – where most [NiFe]-hydrogenases are physiologically active – remains elusive. Carbamoyl phosphate [27] and CO_2, the latter of which was proposed to be reduced to CO at the [4Fe–4S] cluster of HypD [51, 52], have been discussed as precursors. However, these possibilities have apparently been excluded by *in vivo* labeling experiments [53, 54]. There, possible precursors labeled with the stable ^{13}C isotope were added to *E. coli* (mutant) cultures, and the incorporation of (parts of) these precursors into [NiFe]-hydrogenases was tested by IR spectroscopic analyses of the purified proteins. The addition of ^{13}C-citrulline, which is converted by ornithine transcarbamylase into ornithine and carbamoyl phosphate, the substrate of HypF, led to the specific isotope labeling of only the cyanide ligands of the catalytic center. Conversely, the addition of ^{13}CO gas to the cultures resulted in specific labeling of only the CO ligand of the active site [53, 54]. Furthermore, supplementation of the cells with $NaH^{13}CO_3$, the substrate of carbamoyl phosphate synthetase, also led to the specific and complete labeling of the cyanide residues [53]. These results clearly showed that HypF in conjunction with HypE synthesizes exclusively the cyanide ligands of the [NiFe]-hydrogenase active site and that endogenously produced gaseous CO might serve as the CO ligand, which was later corroborated by titration experiments of *C. necator* cultures with ^{13}CO gas and subsequent IR analysis of purified [NiFe]-hydrogenase [55]. As HCO_3^- and CO_2 are in equilibrium in living cells (maintained by carbonic anhydrase), the aforementioned experiments seem to exclude CO_2 as a precursor of the CO ligand of [NiFe]-hydrogenase. N^{10}-formyltetrahydrofolate, formyl-CoA, and formyl-phosphate have also been discussed as metabolic sources for the CO ligand under anaerobic conditions [54], but corresponding experimental evidence is lacking.

A seventh Hyp maturase, HypX, comes into play for the specific synthesis of the CO ligand of the $Fe(CN)_2CO$ synthon under aerobic conditions, under which the general CO synthesis pathway is not functional [55]. The *hypX* gene occurs in microbes, which produce [NiFe]-hydrogenases under aerobic conditions, and one prominent member is *C. necator* [56]. [NiFe]-hydrogenase-dependent growth of *C. necator* under aerobic conditions with H_2 as the energy source and CO_2 as a carbon source is severely delayed in a *hypX* deletion mutant, and the addition of CO gas led to wild-type growth of the mutant, which suggested that HypX is involved in CO ligand formation of the [NiFe] active site. Furthermore, supplementation of an aerobic culture of a HypX-synthesizing *C. necator* strain with $^{13}C_2$-glycine, a C1 precursor for the canonic tetrahydrofolate (THF)-based C1 metabolism, led to the appearance of the isotope labeling in the active site CO ligand of [NiFe]-hydrogenase. Based on the primary amino acid sequence of HypX, which consists of an N-terminal N^{10}-formyl-THF transferase module and a C-terminal enoyl-CoA hydratases/isomerase module [24], the results suggested that N^{10}-formyl-THF is converted into CO by HypX [56].

This was later confirmed by the crystal structure of HypX from the thermophilic aerobic H_2 oxidizer *Aquifex aeolicus* [57] and the biochemical characterization of purified HypX from *C. necator* [58]. In fact, HypX transfers the formyl group of N^{10}-formyltetrahydrofolate first to coenzyme A (CoA). In the second step, formyl-CoA is decarbonylated and CO gas is released (Figure 13.5). Although the exact delivery process is still unclear, the CO ligand is proposed to be attached to the same iron center in HypCD that previously received the cyanide ligands to furnish the $Fe(CN)_2CO$ synthon.

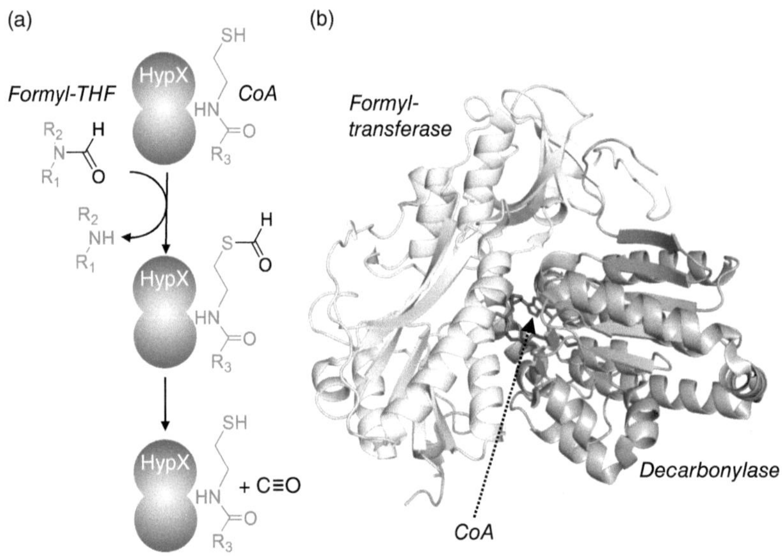

Figure 13.5 (a) Formyl-tetrahydrofolate (formyl-THF)-dependent production of CO gas by the auxiliary protein HypX, in which the formyl group of formyl-THF is first transferred to coenzyme A (CoA) bound to both modules of HypX. It is assumed that the loaded CoA then swings over to the C-terminal module, where the formyl group is decarbonylated to CO. (b) This assumption is consistent with the location of a CoA molecule (stick representation) between the formyl transferase and decarbonylase modules in the crystal structure of HypX from *A. aeolicus*. Source: Adapted from [57, 58].

13.7 Transfer of the $Fe(CN)_2CO$ Synthon from the HypCD Complex to Apo-Hydrogenase

In 2002, Blokesch and Böck published a pioneering study on the formation of HypC–HypD complexes, from which the current model of $Fe(CN)_2CO$ synthon transfer has already been derived [48]. Here, condition-dependent protein complex formation was monitored by immunological analysis of *E. coli* (mutant) extracts separated on native PAGE gels that preserve potential protein–protein interactions between Hyp maturases and also between Hyp maturase(s) and the large hydrogenase subunit. Importantly, the observed interactions were based on proteins produced in wild-type levels, whereas crystal structure and biochemical/spectroscopic analyses of protein complexes require overproduction, which can severely affect the equilibrium of complex formation. Blokesch and Böck observed the accumulation of the HypCD complex only in cells in which either the supply of cyanide to HypCD was interrupted or the recipient of the $Fe(CN)_2CO$ synthon, the hydrogenase large subunit, was missing. In cells in which cyanide supply was normal and the hydrogenase large subunit was present, the HypCD complex was not detected, but here the HypC protein was found in a complex with the immature large subunit. From these results, it was inferred that the HypCD complex serves as the assembly site of the $Fe(CN)_2CO$ synthon and that the loaded HypC protein leaves the complex to deliver the synthon to the [NiFe]-hydrogenase large subunit (Figure 13.2) [48].

The first experimental evidence for the existence of a HypC protein carrying the $Fe(CN)_2CO$ synthon was provided only 20 years later by native mass spectrometry (MS) experiments on the HybG–HypD complex of *E. coli* [59]. In this context, it is important to note that prokaryotes synthesizing more than one [NiFe]-hydrogenase often encode more than one HypC-like protein. *E. coli*, for example, has three [NiFe]-hydrogenases, Hyd-1, Hyd-2, and Hyd-3, with different physiological functions, and it has been shown that HypC interacts with the large subunit of Hyd-3 (and to some extent with Hyd-1), while the HypC homolog HybG interacts with the large subunits of Hyd-1 and Hyd-2 [60]. Hence, HypC–HypD and HybG–HypD complexes also exist in the cells. Native MS experiments on the $Fe(CN)_2CO$-loaded HybG–HypD complex separated by collision-induced dissociation led to the identification of an HybG species that carried the $Fe(CN)_2CO$ synthon. It was also shown by native MS that coincubation of the HybG–HypD complex in the presence of the large subunit of Hyd-2 resulted in the dissociation of HybG–HypD and the concomitant formation of a complex of HybG and the large subunit [59].

The transfer of the $Fe(CN)_2CO$ synthon from HypCD to the hydrogenase was also demonstrated using an *in vitro* maturation system for Hyd-1 from *E. coli* [61]. Remarkably, this *in vitro* maturation system contained only purified proteins, i.e. HypF, HypE, the HybG–HypD complex, the small and large subunits of Hyd-1, and the endopeptidase that specifically cleaves the C-terminal extension of the large subunit upon complete metal center insertion (see Section 13.9), as well as an activation mix containing the accessory components ATP, carbamoyl phosphate, $NiCl_2$, and sodium dithionite as reducing agent. When mixing all the components, the authors observed the complete maturation of catalytically active Hyd-1 by means of H_2-dependent benzyl viologen reduction, the determination of the metal content, and *in situ* IR difference spectroscopy to verify the presence of the $NiFe(CN)_2CO$ cofactor. If one of the assay components was omitted, only a very low or no hydrogenase activity could be detected. The transfer of the $Fe(CN)_2CO$ synthon from the HybG–HypD complex to the large subunit of Hyd-1 was successfully demonstrated using the ^{57}Fe-labeled HybG–HypD complex and subsequent metal analysis of the large subunit. The incorporation of the $Fe(CN)_2CO$ synthon into the large subunit also occurred when $NiCl_2$ was omitted, which provided further evidence that the incorporation of the iron fragment precedes the incorporation of nickel [61]. Surprisingly, the *in vitro* maturation assay revealed the formation of active Hyd-1 only by the addition of $NiCl_2$, while in living cells the two maturases HypA and HybB are necessary for nickel insertion (see Section 13.8). Furthermore, the addition of carbamoyl phosphate was necessary in addition to HypE and HypF, suggesting that the HybG–HypD complex employed was not fully equipped with the $Fe(CN)_2CO$ synthon. Apart from the CO already present on the HybG–HypD complex, no further source of the CO was added to the assay.

Another interesting discovery was that the HypC–HypD and HybG–HypD complexes of *E. coli* possess ATPase activity [62, 63], and the Sawers group identified amino acid residues essential for this hydrolysis reaction. Interestingly, these residues are very close to the proposed binding site of the $Fe(CN)_2CO$ synthon. However, the functional implications of the ATPase activity remained unclear, and the authors discussed functions either for synthesis of the CO ligand of the $Fe(CN)_2CO$ unit by ATP-supported reduction of CO_2 [62] or for ATP-assisted transfer of the $Fe(CN)_2CO$ synthon to the hydrogenase large subunit [63].

Recently, by using the *E. coli* HypCD complex loaded with the $Fe(CN)_2CO$ synthon and the purified apo-form of the large subunit of the regulatory [NiFe]-hydrogenase from *C. necator*, it has been demonstrated that the $Fe(CN)_2CO$ synthon transfer to the large subunit requires ATP, which was verified by IR spectroscopy [46]. While almost no transfer was observed in the absence of ATP (or in the presence of GTP), the addition of a nonhydrolyzable ATP analog also led to the shuttling of the synthon to the large hydrogenase subunit, suggesting that ATP binding, but not ATP hydrolysis, initializes the transfer. Moreover, *in situ* protein–ligand docking calculations and AlphaFold structure predictions provided information about the ATP-binding site in HypCD, which is in agreement with the previous mutagenesis experiments [63]. Furthermore, MS cross-linking experiments suggested that ATP binding causes significant conformational changes in the HypC protein within the HypCD complex. Remarkably, simultaneous binding of the $Fe(CN)_2CO$ fragment and ATP in the assembly site of HypCD appears to be impossible, implying that ATP displaces the $Fe(CN)_2CO$ synthon from the HypCD-binding site, which, in the presence of the hydrogenase large subunit, is likely to lead to the release of $Fe(CN)_2CO$-loaded HypC, which can then insert the cofactor fragment into the apo-form of the large subunit (Figures 13.4b and 13.6) [46].

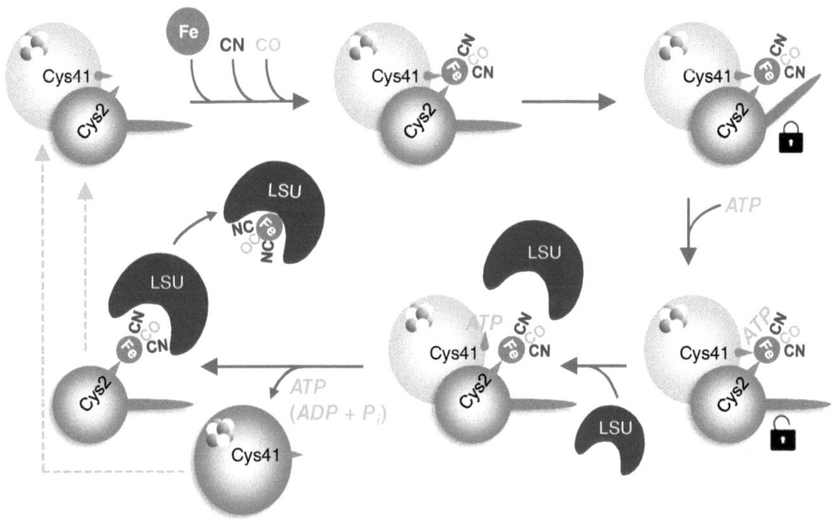

Figure 13.6 Scheme of the HypCD-mediated ATP-dependent transfer of the $Fe(CN)_2CO$ synthon to the apo-form of the hydrogenase large subunit (LSU) as revealed by a recent biochemical, spectroscopic, and computational study. The C-terminal part of HypC (represented by a dark gray rod) can adopt locked and unlocked conformations, which depend on the presence of both the $Fe(CN)_2CO$ synthon and ATP in the HypCD complex. Notably, ATP alone does not lead to the release of the synthon-loaded HypC protein from the HypCD complex, which is required for proper interaction with the large subunit. This process appears to be triggered by the simultaneous presence of ATP and the large subunit. The numbering of the Cys residues of HypC and HypD corresponds to that of the *E. coli* proteins. Source: Adapted from [46].

13.8 Mobilization and Insertion of Nickel by HypAB

Long before iron has been identified as a constituent of the catalytic center of [NiFe]-hydrogenase [64], nickel was identified as a necessary metal to confer hydrogenase activity in microorganisms [65, 66]. So, nickel is an essential component of the active site and its incorporation, like all subsequent steps of maturation, takes place at the stage of the premature large hydrogenase subunit. After the $Fe(CN)_2CO$ synthon has been attached to one or more of the four cysteines coordinating the [NiFe] active site [67], the nickel is inserted (Figure 13.2) [25].

Two genes, *hypA* and *hypB*, have been identified first in *E. coli*, whose gene products appeared to facilitate nickel incorporation [68–70]. Also here it should be mentioned that *E. coli*, as in the case of HypC (see Section 13.7), encodes two HypA-like proteins, with HypA being involved in nickel delivery to Hyd-3 and its paralog HybF fulfilling the same task for Hyd-1 and Hyd-2 [69, 70]. Mutations in the corresponding genes led to immature hydrogenases, and this detrimental effect could be partially overcome by supplementing the cell cultures with high amounts of nickel [68–70]. This already indicated that HypA (HybF) and HypB interact during nickel insertion [70] and explain why the *in vitro* maturation system described in Section 13.7 did not require the addition of purified HypA and HypB proteins [61]. Both HypA and HypB, the latter of which has NTPase activity, were identified as nickel- and zinc-binding proteins, and the potential metal-binding sites were determined by site-directed mutagenesis experiments (reviewed in [22, 25]). HypA and HypB proteins from different microbial species have been analyzed using various techniques, including molecular biological, biochemical, spectroscopic, and structural methods, revealing significant differences in their architectural, metal binding, and NTP utilization (GTP vs. ATP) properties [25, 45]. However, it has been concluded that there is a fundamental functional relationship among all HypA and HypB proteins [45], and therefore the interaction between the HypA and HypB metallochaperones primarily from *E. coli*, to which the Zamble group has made major contributions, will be briefly discussed here. Pull-down experiments with affinity-tagged hydrogenase-related proteins revealed an interaction of HypA with the hydrogenase large subunit HycE of Hyd-3 from *E. coli*, which was independent of the presence of HypB [71]. The HypA–HycE interaction was much weaker in an *E. coli hypC* mutant, indicating that HypA forms a complex primarily with the large subunit that already contains the $Fe(CN)_2CO$ unit. HycE was also found in a complex with HypA, HypB, and SlyB, and a complex consisting of the latter three metallochaperones independent of the large subunit was also identified. SlyD, a metal-binding peptidyl-prolyl cis/trans-isomerase, has previously been found to bind to HypB, and the corresponding gene deletion resulted in a marked decrease of hydrogenase activity in *E. coli*, which could be compensated for by the addition of nickel [72]. This led to the proposal that SlyD is involved in nickel homeostasis and nickel delivery to HypB of *E. coli* [73]. However, a role of SlyD in [NiFe]-hydrogenase maturation has only been observed in *E. coli* and *Helicobacter pylori* [74] and is therefore not considered further here.

The nickel-driven complex formation of HypA with the $Fe(CN)_2CO$-loaded hydrogenase large subunit was the first evidence that HypA might be the actual nickel donor for the catalytic center. The question remained how HypA manages its nickel acquisition and donation capabilities and what role HypB plays in this. HypA (and HybF) contains both a zinc site, which is proposed to have a structural function, and a conserved N-terminal nickel-binding site [70, 75]. HypB, on the other hand, also contains two metal binding sites, including a less well-conserved N-terminal binding site with a high affinity for nickel [76] and an invariant nickel/zinc-binding site in the GTPase domain of HypB, referred to as the low-affinity nickel-binding site [77]. HypB usually occurs as a dimer in solution and in crystal structures [77–80], and the nickel in the low-affinity site is coordinated by four conserved amino acid residues, two from each HypB monomer. Thus, HypB dimerization is necessary for nickel binding at the low-affinity site [81]. GTP promotes the dimer formation of HypB and also leads to nickel binding with low micromolar affinity, whereas the presence of GDP enhances the complex formation with HypA. GDP also accelerated the transfer of nickel from the low-affinity binding site of HypB to HypA [82, 83]. Nickel binding to HypA, in turn, weakened the complex formation with HybB, leading to the conclusion that GTP hydrolysis by HypB enables the unidirectional transfer of nickel from HypB to HypA (Figure 13.7a) [82]. Further details about the nickel transfer mechanism were obtained by X-ray absorption spectroscopy analysis of the HypAB complex of *E. coli* [84]. The nickel coordination at the low-affinity binding site of HypB, which is presumably formed by two conserved cysteine thiols and two histidine nitrogens, was not altered by the presence of GDP or HypA in the complex. Instead, the nucleotide-loaded state of HypB induces a conformational change in the N-terminal nickel-binding site of HypA. Nickel-loaded HypA obviously exists in two mixed conformations, one induced by HypB and one occurring in free HypA [84]. This conclusion is strongly supported by the crystal structure of the archaeal HypAB complex of *T. kodakarensis* [80]. Archaeal HypA and HypB proteins differ slightly from their bacterial counterparts [45] and the HypB protein from *T. kodakarensis* uses ATP instead of GTP. Nevertheless, *T. kodakarensis* HypB also forms a dimer that associates with two HypA monomers, thus forming a heterotetrametric complex. In the crystal structure, no metal was observed in the HypB dimer and only one HypA monomer carried a nickel ion at its N terminus, while the other monomer was equipped with zinc in the zinc-binding domain. Interestingly, complex formation with HypB induced significant conformational changes in HypA, including the formation of an (alternative) nickel-binding site in the N terminus. Moreover, isothermal titration calorimetry experiments revealed that the binding affinity for nickel by HypA increases by three orders of magnitude (from 7 μM to 7 nM) in the presence of HypB and ATP [80], which is in perfect agreement with the mixed binding mode for *E. coli* HypA (Figure 13.7a) [84].

The functions of HypA and HypB reflect a delicate interplay of metal-binding affinities that are modulated by nucleotide hydrolysis in combination with dynamic structural changes, thus enabling targeted metal ion transfer. The tight binding of metals by special metallochaperones is indispensable, since free transition metal ions can be poisonous in the cellular context. In the case of HypA, the presence of different nickel-binding modes might serve this purpose, since HypA first has to control nickel acquisition, then nickel trafficking and finally nickel release to the premature large hydrogenase subunit.

The exact mechanism by which HypA delivers nickel to the hydrogenase active site is currently unknown, but the cocrystal structure of HypA and the large subunit of *T. kodakarensis* provided a first clue (Figure 13.7b) [85]. For their crystallization studies, the authors used the premature, metal-free hydrogenase large subunit, which, according to the current maturation model, should not be the target for nickel delivery by HypA, since it still lacks the $Fe(CN)_2CO$ synthon. However, apart from the absence of the [NiFe] active site, the structure showed marked differences compared to the fully mature large hydrogenase subunit. Hydrogenase large subunits are usually synthesized with a characteristic C-terminal extension that is cleaved by a specific endopeptidase after the assembly of the catalytic center has been completed (see Section 13.9). This C-terminal extension protrudes from the premature hydrogenase subunit and thereby also changes the native position of a short amino acid stretch that carries one of the four cysteine residues (Cys4) that coordinate the active site [85]. The extension adopts a position that is normally occupied by the N terminus in the mature protein, causing the N terminus to swing out and now serve as a docking site for the HypA protein (Figure 13.7b). Through this protein–protein interaction, the nickel-binding N terminus of HypA is brought close to the active site

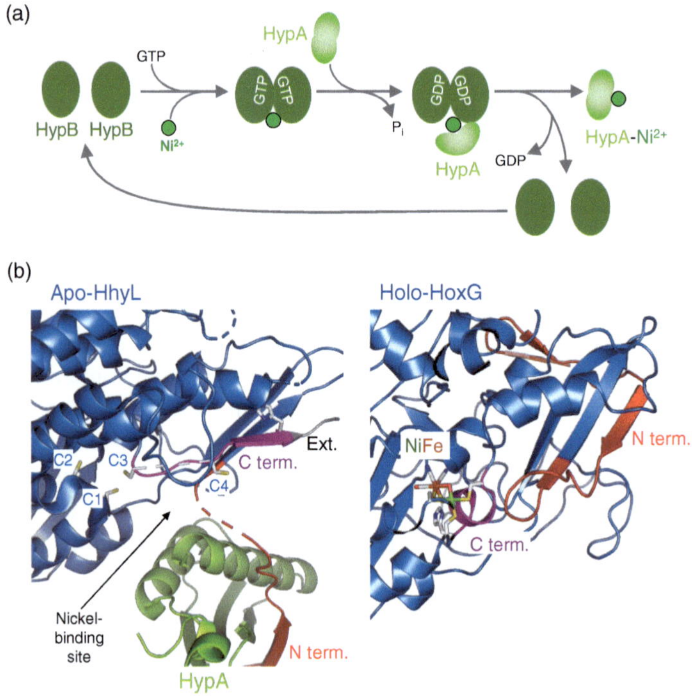

Figure 13.7 (a) Simplified mechanism of nickel transfer from HypB to HypA based on the proposals of the Zamble and Miki groups. Source: Adapted from [80, 84]. GTP binding leads to the formation of a HypB homodimer, which allows the two HypB monomers to jointly coordinate a nickel ion. Subsequent GTP hydrolysis weakens the nickel affinity and promotes the binding of HypA, which in turn drastically changes its conformation, thereby forming a high-affinity nickel-binding site that takes over the nickel ion from the HypB dimer. The absence of nickel destabilizes the HypB dimer, which then releases the HypA protein. HypA adopts the original conformation in which the nickel ion is bound with low affinity and can now transfer the metal to the preform of the large hydrogenase subunit. The release of GDP from HypB reinitializes the cycle. For the sake of clarity, the N-terminal high-affinity nickel-binding site, which occurs in many HybB proteins, and the zinc site of HypA (see text) are not shown in this scheme. (b) Cartoon representation of the cocrystal structure of the apo-form of the [NiFe]-hydrogenase large subunit (HyhL) from *T. kodakarensis* with HypA (left) (Source: Adapted from [85]) in comparison with the mature large subunit HoxG of the membrane-bound hydrogenase from *C. necator* (right). Source: Adapted from [86]. The C terminus of apo-HyhL (magenta) including the C-terminal extension (grey) interacts with a domain of the large subunit (blue) at a location that is occupied by the N terminus of the protein (red) in the mature protein. The liberated N terminus of HyhL interacts with HypA (green) in a way that brings the nickel-binding site of HypA close to the cysteine residues (labeled 1–3, stick representation) of the active site. It is assumed that the endoproteolytic cleavage of the C-terminal extension (green) after insertion of the $NiFe(CN)_2CO$ cofactor allows the C-terminal peptide (magenta) to enter the active site, where Cys4 contributes to the coordination of the metal ions (see text).

cysteine residues of the large subunit and HypA can release the nickel [85], which of course also requires an appropriate interplay of the binding affinities of the two nickel-binding sites. Nickel delivery completes the metalation of the catalytic center of the hydrogenase large subunit (Figure 13.2).

13.9 Role of the C-Terminal Extension of the Premature Large Subunit

In the early 1990s, several groups observed that large subunits of [NiFe]-hydrogenases migrated in two slightly different forms in SDS-PAGE gels, with the premature form being slightly larger than the native protein. The native protein was found to be truncated at its C terminus, and the cleavage site was identified as the position after the last residue of the highly conserved DPCxxCxxH/R motif, in which the two cysteines (Cys3 and Cys4) are coordinating ligands of the $NiFe(CN)_2CO$ site [87–90]. According to a (meta)genomic survey, most [NiFe]-hydrogenase large subunits are synthesized with such a C-terminal extension comprising 3–68 amino acids, and the hydrogenase gene clusters encoding such extended hydrogenase large subunits also contain a gene, encoding a dedicated endopeptidase responsible for the posttranslational cleavage of the C-terminal extension [21]. What role does this C-terminal extension play during the maturation process? The Böck group has recognized early on that the endoproteolytic cleavage of the C-terminal extension of the large hydrogenase subunit HycE of *E. coli* Hyd-3 by the cognate endopeptidase HycI (HycH was originally thought to be the protease, which was corrected later) ultimately requires the prior incorporation of nickel into HycE. This conclusion was based on the observation that only a nickel-loaded HycE precursor purified from cells lacking the endopeptidase was specifically cleaved by partially purified HycI, whereas a precursor isolated from cells in which *hypB*, *C*, *D*, and *E* were deleted was not cleaved by HycI [90]. Although each [NiFe]-hydrogenase large subunit carrying a C-terminal extension requires its own endopeptidase, indicating specific interactions of the hydrogenase-endopeptidase pairs, the endopeptidases share high amino acid sequence similarity [21, 91]. This suggests a similar cleavage mechanism for all these peptidases. Further clues to the possible mechanism were provided by the crystal structure of the endopeptidase HybD, which is responsible for the nickel-dependent cleavage of the C-terminal extension of Hyd-2 from *E. coli* [91]. Although the HybD structure did not provide details about the interaction with the hydrogenase large subunit and the cleavage mechanism itself, evidence for nickel recognition was found. At the bottom of a shallow cleft near the protein surface, a conserved metal-binding site was identified that was occupied by a cadmium ion from the crystallization buffer. The purified HybD protein was found to be metal-free [91] but site-directed exchanges of amino acids in the endopeptidase HycI corresponding to those coordinating the cadmium in the HybD structure abolished the proteolytic activity of HycI [92]. These results led to the speculation that the metal-binding site in the endopeptidase recognizes nickel, which initializes the cleavage reaction [93].

Although certain modifications of the C-terminal extension, including amino acid replacements and truncations, are tolerated, its complete removal by genetic engineering generally results in the formation of inactive hydrogenase [94–97]. Interestingly, in some cases, genetic removal of the entire C-terminal extension allowed the formation of hydrogenase with canonical subunit composition, as observed for the NAD^+-reducing [NiFe]-hydrogenase of *C. necator* [95] and Hyd-2 from *E. coli* [97]. Nevertheless, in both cases, the large subunit was at least deficient in nickel, resulting in hydrogenase complexes without catalytic function, providing further evidence for the essential role of the C-terminal extension of the large subunit in active site maturation. This conclusion was recently challenged by the observation that the large subunit of the MBH from *C. necator*, whose C-terminal extension had been genetically deleted, was still equipped with the $NiFe(CN)_2CO$ cofactor by the Hyp machinery. Moreover, it formed a catalytically competent hydrogenase heterodimer with its small subunit, which did not differ from the native enzyme in terms of activity and IR spectroscopic properties. However, the removal of the C-terminal extension led to significantly lowered [NiFe]-hydrogenase levels in the cells, which supported much slower H_2-dependent growth than observed for wild-type cells. It has been concluded that, although not essential for maturation, the presence of the C-terminal extension significantly optimizes the efficiency of metal center incorporation into the premature hydrogenase large subunit [98]. The same conclusion can be drawn from the results after the removal of the C-terminal extension of the large subunit of [NiFe]-hydrogenase from *Desulfomicrobium baculatum* [99].

It should be pointed out that although most [NiFe]-hydrogenase large subunits are synthesized with a C-terminal extension, there are also numerous examples without a C-terminal extension that appear to exploit the canonical maturation machinery for metal center insertion. These include the H_2-sensing regulatory hydrogenases, CO dehydrogenase-associated hydrogenases, Ech-type hydrogenases, and certain so far uncharacterized hydrogenases [21]. The presence of a C-terminal extension thus appears to be a common, but not essential, feature of the [NiFe]-hydrogenase large subunits.

Another interesting observation was the genetic transfer of the C-terminal extension of the large subunit of Hyd-1 from *E. coli* to the large subunit of Hyd-2, which did not affect the interaction of the protein with both the HypC paralog HybG and the Hyd-2-specific endopeptidase HypD [97, 100]. This suggests that the C-terminal extension may not serve as a recognition site for the maturation machinery, but merely to keep the active site of the premature large subunit in a conformation that allows the Hyp proteins to insert the metal center. Such a scenario is consistent with the observations made from the cocrystal structure of HypA in complex with the large subunit of *T. kodakarensis* [85] (see Section 13.8), in which the C terminus displaces the N terminus from its position in the native protein and the latter functions as a contact site for HypA (Figure 13.7b).

The endoproteolytic cleavage of the C-terminal extension allows the most C-terminal cysteine of the DPCxxCxxH/R motif to move into the active site cavity where it serves as the last residue to bridge the nickel and iron ions. This results in the fully mature large subunit, which is now able to oligomerize with the small subunit to form a catalytically active [NiFe]-hydrogenase (Figure 13.2).

13.10 Hydrogenase Maturation Intermediates Isolated from Living Cells

Apart from the investigation of the function of the Hyp proteins themselves and their interaction with the premature [NiFe]-hydrogenase large subunit described earlier, the examination of the structural, spectroscopic, and catalytic properties of the various maturation intermediates of the large subunit is also of crucial importance for understanding the incorporation process of the active site components. In literature, there are only few reports on the characterization of isolated large subunits of [NiFe]-hydrogenases. In 2005, the large subunit of the regulatory [NiFe]-hydrogenase, one of the few examples that do not contain the typical C-terminal extension (see Section 13.9), was purified from *C. necator* and shown to lack catalytic H_2 oxidation activity. Metal determination revealed a stoichiometric iron content, while nickel was bound in substoichiometric amounts, from which it was concluded that the $Fe(CN)_2CO$ incorporation actually precedes the incorporation of nickel [101]. That the isolated large subunit remains catalytically inactive was also observed in electrochemical experiments on the heterotetrameric soluble hydrogenase from *Pyrococcus furiosus*. While the heterodimeric hydrogenase module immobilized on carbon nanotubes mediated H_2 turnover, the isolated subunit, whose metal content was not characterized, showed no electrochemical response [102].

Fifteen years after its initial characterization [101], the catalytic and spectroscopic properties of the isolated large subunit HoxC of the regulatory [NiFe]-hydrogenase from *C. necator* were reinvestigated [103]. Using an optimized cultivation procedure, a fully mature HoxC protein could be isolated, as shown by metal determination as well as UV–Vis, Mössbauer, and IR spectroscopy. Hydrogen–deuterium exchange ($H_2 + D_2O \rightleftharpoons HD$ and $HDO/HD + D_2O \rightleftharpoons D_2 + HDO$) measurements revealed that the isolated HoxC subunit is capable of H_2 activation. However, the activity was 200 times lower than that of the native regulatory [NiFe]-hydrogenase, which shows that the oligomerization with the hydrogenase small subunit is necessary to attain native catalytic activity. The potential reason for this has also been uncovered. IR spectroscopy showed that the $NiFe(CN)_2CO$ center of HoxC primarily exists in a redox state where a hydroxy ligand is bound in the bridging position between nickel and iron, i.e. at the site where H_2 binding and cleavage normally take place. This hydroxy ligand is most likely the result of the deprotonation of a water molecule, and the shielding of the catalytic center from the solvent by the small subunit of the hydrogenase probably prevents water from accessing the active site of mature [NiFe]-hydrogenase [103, 104].

The isolation of [NiFe]-hydrogenase subunits still carrying their C-terminal extension has also been described. Sawers and coworkers characterized the isolated large subunit of Hyd-2 from *E. coli*, which carried an additional pentapeptide at its

C terminus, resulting in an extended C-terminal extension [105]. This extension prevented the insertion of the native NiFe$(CN)_2$CO cofactor, endoproteolytic cleavage by the cognate protease, and thus the oligomerization with the small subunit. However, the large subunit isolated in this way contained substoichiometric amounts of iron, as was demonstrated by iron determination and UV–Vis spectroscopy, but no acid-labile sulfide. It was also shown that the incorporation of these iron species occurs independently of the Hyp machinery and that the cysteines in the active center are necessary for iron binding [105]. These results show how sensitive the intricate maturation system is to changes that are unrelated to the structure of the mature protein, and that even small changes can lead to improper metal insertion.

Also, the premature large subunit pre-HoxG of the O_2-tolerant MBH from *C. necator* has been isolated in pure form [106]. Due to the deletion of the specific endopeptidase gene, the protein still carried the typical C-terminal extension. As expected, the pre-HoxG contained the Fe$(CN)_2$CO fragment and also nickel, albeit substoichiometrically. The cofactor-containing pre-HoxG did not catalyze H_2-dependent reduction of artificial electron acceptors known to interact with native MBH of *C. necator*. However, as in the case of the large subunit of the regulatory [NiFe]-hydrogenase from *C. necator* [103], pre-HoxG also performed the hydrogen–deuterium exchange reaction at significantly lower rates than the native MBH [106]. These findings clearly underpinned the model that the proteolytic cleavage of the C-terminal extension represents the last step before the formation of the bipartite catalytic hydrogenase module.

A comprehensive picture of the cofactor content and its coordination in the different maturation intermediates has been recently obtained by the purification of four premature versions of the MBH large subunit HoxG from *C. necator* [67]. These were obtained by deleting the appropriate maturation genes in the corresponding *C. necator* strains, from which the intermediates were purified. These included apo-HoxG (isolated from a Δ*hypD* strain), a pre-HoxG$^{\Delta Ni}$ version only containing the Fe$(CN)_2$CO fragment of the catalytic site (Δ*hypA*), pre-HoxG, carrying the entire NiFe$(CN)_2$CO cofactor but still the C-terminal extension (deletion of the endopeptidase gene), and fully mature HoxG (deletion of the small subunit gene) (Figure 13.2). As expected, apo-HoxG contained only substoichiometric amounts of iron and no nickel, while the other three HoxG versions contained roughly one iron per protein as well as the iron-bound CN^- and CO ligands, as detected by metal determination and IR spectroscopy. Only pre-HoxG and mature HoxG also contained nickel almost stoichiometrically, while pre-HoxG$^{\Delta Ni}$ was devoid of this metal. Detailed investigations using IR, nuclear resonance vibrational, and Mössbauer spectroscopy revealed different spectroscopic features of the three metal-containing variants, which demonstrate the presence of the Fe$(CN)_2$CO fragment in different coordination environments, while the iron was always kept in a low-spin Fe^{II} state, as observed in mature [NiFe]-hydrogenases. Interestingly, the spectroscopic properties of pre-HoxG and mature HoxG differed significantly, although both intermediates had the same cofactor content, which in turn shows the crucial role of the C-terminal extension in determining the metal coordination by the active center during the maturation process. This was also reflected by the redox reactivity of the maturation intermediates. While pre-HoxG$^{\Delta Ni}$ and pre-HoxG were inert to treatment with oxidizing and reducing agents, only mature HoxG showed a redox response [67], which was consistent with previous observations for the isolated large subunit of the regulatory [NiFe]-hydrogenase of *C. necator* [103]. In fact, the data obtained from nuclear resonance vibrational and X-ray absorption spectroscopy indicated the presence of a hydroxy ligand in the bridging position between nickel and iron in mature HoxG, which was absent in preHoxG [67].

Based on the currently available data, the $Fe(CN)_2CO$ synthon is delivered by HypC and connected as an entire unit to Cys2 (and Cys4) of the active site (Figure 13.2). Since the actual numbering of the active site cysteines is different for the various hydrogenase large subunits, they are numbered here according to their occurrence in the primary protein sequence. In the next step, the nickel ion is inserted and presumably coordinated by Cys1, 2, and 3, while Cys4, which bridges the Ni and Fe ion in mature [NiFe]-hydrogenases, remains at a nonbonding distance, consistent with the position of the C terminus in the crystal structure of the apo-large subunit of *T. kodakarensis*, where Cys4 is not located in the vicinity of the other active site cysteines (Figure 13.7b) [85]. This assumption is also supported by the fact that Cys4 is not required for nickel coordination in the large subunits of *E. coli* Hyd-3 as well as the NAD^+-reducing hydrogenase and the regulatory hydrogenase of *C. necator* [107–109]. The C-terminal extension is cleaved only after insertion of nickel, and its removal leads to a restructuring of the active site, which becomes redox active and includes a hydroxy ligand between Ni and Fe. This hydroxy ligand presumably protects the active site from interaction with O_2, for example, during maturation under aerobic conditions. The fully mature large subunit is competent to form a complex with the small electron-transferring subunit to form an active [NiFe]-hydrogenase.

13.11 Iron–Sulfur Cluster Insertion into the Small Subunit

The typical small subunit of a [NiFe]-hydrogenase accommodates three Fe–S clusters in a row that, depending on the physiological conditions in the cell and the function of the respective hydrogenase, conduct the electrons released at the active site to specific electron acceptors or vice versa. These Fe–S clusters are usually of the [4Fe–4S] type in the proximal and distal positions relative to the [NiFe] active site, while a [3Fe–4S] cluster occupies the medial position between the other two clusters. However, there are numerous exceptions to this “canonical” Fe–S cluster arrangement, ranging from subunits containing either three [4Fe–4S] or three [3Fe–4S] clusters to subunits carrying only one [4Fe–4S] cluster, and even subunits harboring an unusual [4Fe–3S] cluster in the proximal position. Apart from the hydrogenase small subunits, also the maturase HypD contains a [4Fe–4S] cluster.

Compared to the knowledge about NiFe$(CN)_2$CO cofactor assembly, little is known about the incorporation of Fe–S clusters in hydrogenase-related proteins. This is probably because the Fe–S clusters are believed to be incorporated by the canonical Isc (iron–sulfur-cluster) and Suf (sulfur mobilization) machineries of the hosts [110]. In fact, the only reports on the insertion of Fe–S clusters appear to come from the Sawers group and focus on the three different [NiFe]-hydrogenases from the model organism *E. coli*. They initially showed that Hyd-1 and Hyd-2 were inactive and the activity of Hyd-3 was drastically reduced when both the ferrous iron transport system FeoABC and the ferric iron uptake system dependent on the siderophore enterobactin (EntC) were knocked out, thus identifying the main iron transport systems required for hydrogenase synthesis in *E. coli* [111]. In a subsequent study, it was investigated whether the Fe–S cluster biogenesis pathways Isc and/or Suf are responsible for furnishing the hydrogenase small subunits with preformed Fe–S clusters (Chapters 1 and 2) [112]. Mutants lacking either of the A-type carrier (ATC) proteins SufA, IscA, and ErpA [113] were used for this. Wild-type-like activities of Hyd-1 and Hyd-2 were observed in a *sufA* mutant, while both hydrogenases showed less than 5% of wild-type activities in *iscA* and *erpA* mutants. These results indicated that both ErpA and IscA are required for the development of hydrogenase activity and that the Isc machinery, but not the Suf machinery, is likely responsible for the delivery of Fe–S clusters to Hyd-1 and Hyd-2 of *E. coli*, as the small subunits of both hydrogenases were undetectable in the *iscA* and *erpA* mutants [113]. Later, it was shown that the Isc system is also necessary for the assembly and function of Hyd-3. In a mutant lacking the scaffold protein IscU, no Hyd-3 activity was detected at all, but in contrast to the observations for Hyd-1 and Hyd-2, the IscA and ErpA mutants were able to compensate for each other [114]. A recent study of the Sawers group addressed the origin of the single iron ion that is coordinated by two conserved cysteine residues from the HypCD complex by analyzing complex formation in different Isc/Suf mutant backgrounds using native MS [115]. Tandem MS/MS experiments involving collision-induced dissociation led to the separation of HypCD isolated from wild-type cells into HypD and HypC, the latter of which carried the typical chemical modification previously observed for HybG from the HybG–HypD complex (see Section 13.7) [59] and which is assumed to be the Fe$(CN)_2$CO synthon attached to the N-terminal cysteine of HypC. This HypC modification was not observed in HypCD complexes purified from an *iscU* mutant, while it was still present in HypCD complexes isolated from *sufA* or *erpA-iscA* mutants. These results suggest that the iron of the Fe$(CN)_2$CO group on HypCD is delivered by the Isc machinery. Somewhat surprisingly, the [4Fe–4S] cluster in HypD was found to be present, albeit in some cases in lower amounts, in the HypCD preparations of all *E. coli* strains carrying mutations in one or more proteins involved in Fe–S cluster synthesis [115]. Although further research is needed, the Isc machinery appears to be the main player in the insertion of Fe–S clusters into the small subunit, at least in *E. coli*, and it also appears to be the source of iron in the active center of [NiFe]-hydrogenases.

13.12 Subunit Oligomerization and Transport Across the Cytoplasmic Membrane

Insertion of the NiFe$(CN)_2$CO cofactor into the [NiFe]-hydrogenase large subunit (Sections 13.4–13.8) is the prerequisite for subsequent endoproteolytic cleavage of the C-terminal extension (Section 13.9), which finally allows the fully mature large subunit to form a complex with the corresponding small subunit (Figure 13.2). It is assumed that the subunit oligomerization of the heterodimeric [NiFe]-hydrogenase module occurs without the help of further auxiliary functions. This has been shown using *in vitro* assembly systems for Hyd-1 and Hyd-2 from *E. coli*, in which, in addition to the purified Hyp maturases and the specific endopeptidase, only the large and small hydrogenase subunits were present, which formed a catalytically active hydrogenase module with the spectroscopic properties typical for native [NiFe]-hydrogenases [61, 96]. Three years later, the *in vitro* assembly of the independently purified subunits of the regulatory [NiFe]-hydrogenase from *C. necator* has been demonstrated [116]. Here, the Fe–S cluster-containing small subunit was heterologously produced and purified from *E. coli*, while the fully mature large subunit, carrying a (^{57}Fe-labeled) NiFe$(CN)_2$CO cofactor, was isolated from a *C. necator* strain lacking the small subunit. Subsequent incubation of the two purified subunits under anaerobic conditions in the presence of H_2 and reducing equivalents led to the formation of a heterodimeric enzyme that was indistinguishable from the native regulatory [NiFe]-hydrogenase in terms of its catalytic properties and redox behavior in spectroscopy. Notably, this *in vitro* assembly strategy allows detailed investigation of catalytically relevant [NiFe]-hydrogenase intermediates using ^{57}Fe-sensitive spectroscopic techniques without the interference of Fe–S cluster signals of the small subunit (Chapters 15 and 17) [116]. However, it should be mentioned that a ratio of 5 : 1 between the small and large subunits was necessary to obtain full catalytic activity, which was due to the fact that the small subunit preparations were less homogeneous with regard to the metal sites than that of the large subunit [116], as was also observed in the aforementioned *in vitro* system for Hyd-1 [96]. This suggests that, although no further proteins are required for subunit oligomerization, an internal proofreading mechanism may exist that only allows the formation of complexes between the subunits with native cofactor content. Recently, it has been shown that is even possible to create a [NiFe]-hydrogenase heterodimer composed of the subunits from two different hydrogenases. By adapting the subunit interfaces through specific amino acid exchanges, a chimeric enzyme consisting of the large subunit of Hyd-1 and the small subunit of Hyd-2 from *E. coli* was generated that showed mixed catalytic properties of both [NiFe]-hydrogenases [117]. This supports the notion that specific protein–protein interaction between the subunits is the main determinant for proper oligomerization and does not require further functions.

Subunit oligomerization represents the final step in the maturation of those [NiFe]-hydrogenases that remain in the cellular cytoplasm (Figure 13.2). In these cases, the fully mature heterodimeric hydrogenase modules form complexes with their dedicated electron acceptor/donor proteins or regulatory proteins in the case of H_2-sensing [NiFe]-hydrogenases, in order to achieve their physiological function [6, 12]. However, there are many [NiFe]-hydrogenases that are active in the periplasm, where they are either bound to the periplasmic side of the membrane or remain soluble in the periplasm, and usually catalyze H_2 oxidation to provide reducing equivalents for anaerobic and aerobic respiration [6, 12]. The equipment of the [NiFe]-hydrogenase subunits with their metal cofactors and the subsequent subunit oligomerization takes place exclusively in the cytoplasmic space. How are the assembled [NiFe]-hydrogenases transported across the cytoplasmatic membrane? The ubiquitous Sec system, which transports selectively unfolded amino acid chains through the cytoplasmic membrane [118], can be excluded.

In 1999, Wu and collaborators investigated the membrane transport of Hyd-2 from *E. coli*, which belongs to the [NiFe]-hydrogenases attached to the periplasmic side of the cytoplasmic membrane [119]. The determinant for membrane translocation was identified as a specific N-terminal peptide of the hydrogenase small subunit, which contains a consensus twin-arginine translocation (Tat) motif (S-R-R-x-F-L-K) (Figure 13.2), which had previously been associated with a novel membrane transport mechanism for folded, cofactor-containing proteins [120]. Replacing the Tat peptide with a sequence recognized by the Sec system completely prevented the transport of Hyd-2 through the membrane [119]. It was also observed that neither of the two hydrogenase subunits is transported to the periplasm if the other subunit is missing. Since the large subunit does not contain a signal peptide for membrane transport and the small subunit carries the Tat peptide but is not targeted to the membrane in the absence of the large subunit, it was concluded that the formation of the Hyd-2 heterodimer in the cytoplasm leads to a conformational change of the Tat signal peptide, which is only then recognized by the Tat translocase and induces the joint transport of the small and large hydrogenase subunits by a hitchhiker mechanism [119]. A parallel study confirmed that *E. coli* Hyd-2, Hyd-1, and other periplasmatic, cofactor-containing enzymes are indeed translocated by the Tat system [121], composed of the core components TatA, TatB, and TatC [122]. Later, the Tat dependence of the translocation of other MBHs has been shown, including those from *Wolinella succinogenes* [123], *C. necator* [124], *Rhizobium leguminosarum* [125], *Azotobacter chroococcum* [126], *Campylobacter jejuni* [127], and *H. pylori* [128].

In addition to the general function of the Tat apparatus, whose strong dynamics complicate its investigation [129], an important aspect for further studies is the structural dynamics of the N-terminal signal peptide of the hydrogenase small subunit, which ensures that transport is only initiated when the two fully mature subunits form a complex. Using a bacterial two-hybrid assay, the Sargent group demonstrated the protein–protein interaction of the Tat-signal peptide-containing precursors of the small subunits of Hyd-1 and Hyd-2 from *E. coli* with the previously uncharacterized chaperones HyaE and HybE. Based on these results, it was speculated that both proteins have a proofreading role in preventing premature transport by the Tat apparatus [130]. This was later substantiated by the observation that the small subunit of Hyd-2 was erroneously targeted to the membrane in a *hybE* mutant, while the nickel-containing large subunit remained in the cytoplasm [131]. A similar conclusion has been drawn for the HyaE ortholog HoxO from *C. necator*, which, together with another auxiliary protein HoxQ (see also Section 13.13),

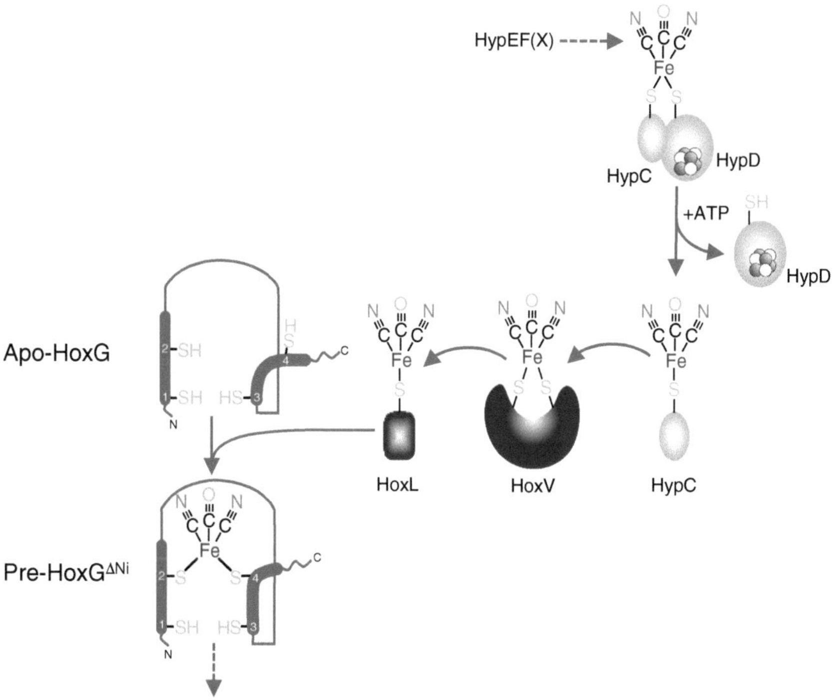

Figure 13.8 Presumed role of the auxiliary maturation proteins HoxV and HoxL in the transfer of the $Fe(CN)_2CO$ synthon to the apo-large subunit HoxG of the MBH from *C. necator*. The canonical HypCD complex first delivers the synthon to the HoxV protein, which shares similarities with [NiFe]-hydrogenase large subunits. The HypC-like HoxL protein is proposed to mediate the final transfer of the synthon from HoxV to apo-HoxG. Source: Adapted from [133].

formed a complex with the small subunit precursor HoxK of the *C. necator* MBH *in vivo* and *in vitro*, and the presence of the Tat signal peptide was required for this interaction. Furthermore, both proteins were shown to be essential for the formation of active MBH [132]. Although these results provide a first indication of how the Tat-mediated export of [NiFe]-hydrogenase is coordinated by auxiliary proteins, it must be emphasized that these proteins are only found in a few species, especially those that produce [NiFe]-hydrogenases under aerobic conditions (see Section 13.13). Thus, the heterodimeric hydrogenase module of most organisms that possess periplasmic [NiFe]-hydrogenases therefore appears to be transported by the Tat system without specific auxiliary proteins.

13.13 Special Case: Maturation of O_2-Tolerant Membrane-Bound [NiFe]-Hydrogenases

The basic genetic equipment for the synthesis of a heterodimeric [NiFe]-hydrogenase module consists of the genes that encode the two hydrogenase subunits, a gene that encodes the specific endopeptidase, and the six *hypA–F* genes. Together with suitable transcription factors, this should be sufficient to synthesize an active hydrogenase under anaerobic conditions. Under aerobic conditions, however, where the so-called O_2-tolerant MBHs are synthesized, additional functions are required to guarantee the synthesis and catalytic function of the hydrogenase in the presence of O_2. The enzyme for which these additional functions have been most intensively studied is the O_2-tolerant MBH of *C. necator*, which will therefore be considered further here. In addition to the genes *hoxK* and hoxG, which encode the MBH small and large subunits, the protease-encoding gene *hoxM* and the *hypA–F* genes, MBH biosynthesis under aerobic conditions requires the *hypX* gene as well as the genes *hoxL, O, Q, R, T*, and *V* (Figure 13.1b) [20]. As described in Section 13.6, HypX is required for the formation of the CO ligand of the $NiFe(CN)_2CO$ cofactor under aerobic conditions [56]. There are two other proteins that interfere with the Hyp machinery in *C. necator*, namely HoxL, a HypC paralog, and HoxV, which shares limited similarity with [NiFe]-hydrogenase large subunits, including two of the four active site cysteines (Figure 13.8) [133]. Both proteins are essential for the biosynthesis of active MBH. Copurification experiments revealed that HypC does not form a stable complex with the large subunit pre-HoxG but with HoxV. The HypC paralog HoxL, on the other hand, formed complexes with both pre-HoxG and HoxV. IR spectroscopic analysis of the HoxV protein purified from a Δ*hoxG* background revealed the appearance of the characteristic bands for CN^- and CO bands, indicating the presence of the $Fe(CN)_2CO$ fragment. In the presence of HoxG, the IR absorptions of HoxV were significantly lower, while no signals were observed in the absence of the Hyp apparatus. These results were interpreted as the canonical formation of the $Fe(CN)_2CO$ synthon by the HypCD complex and the subsequent transfer of the synthon to HoxV, which serves as an intermediate storage site. HoxL then takes over the $Fe(CN)_2CO$ synthon from HoxV and subsequently inserts it into apo-HoxG [133]. Thus, the $Fe(CN)_2CO$ synthon takes a detour via HoxV on its way from HypCD to the large subunit of hydrogenase (Figure 13.8). The reason for this is still unclear. *C. necator* synthesizes four different [NiFe]-hydrogenases, all of which are O_2-tolerant, but MBH is the only one that requires HoxLV. Structural reasons for the presence of HoxLV are therefore more likely than O_2-protective functions. This is also consistent with the observation that the HoxV ortholog HupK is required for the synthesis of the MBH of *R. leguminosarum* under both microaerobic and anaerobic, nitrogen-fixing conditions [134]. As mentioned in Section 13.12, the two proteins HoxO and HoxQ, which are essential for the synthesis of active MBH under aerobic conditions, were isolated in complex with the small MBH subunit HoxK, and one interpretation was that they have a proofreading function to prevent the premature transport of HoxK through the Tat system [132]. An alternative, perhaps more likely function has also been discussed. This is the protection of the Fe–S clusters in HoxK, in particular, the proximal [4Fe–3S] cluster coordinated by six cysteines [86, 135], from inactivation by O_2 during assembly [20]. This hypothesis was supported by the observation that the absence of the proteins HupG (a HoxO ortholog) and HupH (a HoxQ ortholog) in *R. leguminosarum* led to a complete loss of hydrogenase activity in the presence of O_2 and that this effect could be (partially) reversed by lowering the O_2 concentration [136]. A similar effect was also observed for Δ*hoxR* and Δ*hoxT* mutants of *C. necator*. Cells grown in the presence of O_2 showed only traces of MBH activity, while O_2-limited cultures exhibited almost wild-type activity [137], which clearly indicated a protective function of the HoxR and HoxT proteins against O_2. While HoxT shows sequence similarity with HybE, which is thought to be involved in the Tat-mediated transport of Hyd-2 from *E. coli* [130], HoxR can be classified as a rubredoxin-like protein, but without an iron cofactor [138]. When purified from *C. necator*, HoxR was found to be associated with the MBH small subunit HoxK. Moreover, MBH purified from the Δ*hoxR* mutant grown in the presence of O_2 showed only 20% of the activity of the native enzyme, which was consistent with spectroscopic properties that could be correlated with (i) a partially damaged Cys_6[4Fe–3S] cluster in the small MBH subunit and (ii) an unusual spectroscopic behavior of the [NiFe] site. These results suggest that HoxR prevents oxidative damage to the metal cofactors in MBH during maturation, in particular, the Cys_6[4Fe–3S] cluster in the small subunit [138].

In summary, the canonical Hyp machinery (and an endopeptidase) is sufficient for the maturation of [NiFe]-hydrogenase under anaerobic conditions, whereas additional auxiliary proteins are required for biosynthesis of certain O_2-tolerant [NiFe]-hydrogenases under aerobic conditions, which appear to be primarily responsible for the correct insertion of the metal cofactors into both [NiFe]-hydrogenase subunits and the protection of the (unusual) Fe–S clusters in the small subunit.

13.14 Conclusions

Hydrogenases, which are highly abundant in the microbial world, catalyze one of the simplest chemical reactions, but the selective activation of H_2 requires elaborate inorganic metal centers in these enzymes, whose assembly requires sophisticated maturation mechanisms. In this chapter, we describe the machinery responsible for the class of [NiFe]-hydrogenases and discuss that at least six Hyp proteins and, in most cases, also a specific endopeptidase are sufficient to insert the $NiFe(CN)_2CO$ center into the large subunit of the basic heterodimeric [NiFe]-hydrogenase module. Our current knowledge of this process is based on intensive research by numerous research groups and includes the synthesis of the cyanide ligands by HypF and HypE, the assembly of the $Fe(CN)_2CO$ synthon on the HypC–HypD scaffold complex, the transfer of the synthon to the apo-large subunit, and the subsequent insertion of the nickel ion, mediated by the proteins HypA and HypB. Upon insertion of the metal center, the large subunit usually undergoes a proteolytic modification that eventually leads to its oligomerization with the hydrogenase small subunit, whose Fe–S clusters are likely to be inserted by the Isc machinery of the cell. Subunit oligomerization gives rise to catalytically active [NiFe]-hydrogenase (Figure 13.2), which is then targeted to its site of action in the cell. In many cases, this requires the transport of the mature enzymes through the cytoplasmic membrane via the Tat translocase. However, there are also [NiFe]-hydrogenases that require additional biosynthesis proteins, which, for example, ensure that the enzyme is synthesized in an active form under aerobic conditions.

Despite the large amount of data already available for the [NiFe]-hydrogenase biosynthesis, many questions remain unanswered. These include, among other things, the origin and biosynthesis of the CO ligand of the $Fe(CN)_2CO$ synthon under anaerobic conditions, how the Fe ligand stoichiometry of two CN^- and one CO is realized, and the detailed molecular mechanisms of both the maturases and the large hydrogenase subunit that guide the metals to their dedicated positions in the active site.

Knowledge of the [NiFe]-hydrogenase biosynthesis process is not only of high interest from a scientific point of view but also provides the necessary background for the heterologous overproduction of [NiFe]-hydrogenases for biotechnological/industrial applications. Moreover, with the detailed insights into metal content and coordination in the different maturation intermediates, it may be possible to create (semi-)artificial [NiFe]-hydrogenases equipped with chemically synthesized inorganic cofactors, as has been successfully demonstrated in the case of [FeFe]- and [Fe]-hydrogenases [139–141].

Acknowledgments

This work was supported by the Deutsche Forschungsgemeinschaft, Priority Program Iron–Sulfur for Life (SPP1927, LE 2934/1-1/2) and the cluster of excellence "UniSysCat" under Germany's Excellence Strategy-EXC2008-390540038. The authors thank Gary Sawers for critically reviewing the manuscript.

References

1 Fourmond, V., Baffert, C., Sybirna, K. et al. (2013). The mechanism of inhibition by H_2 of H_2-evolution by hydrogenases. *Chem. Commun.* 49 (61): 6840.

2 van Haaster, D.J., Hagedoorn, P.-L., Jongejan, J.A. et al. (2005). On the relationship between affinity for molecular hydrogen and the physiological directionality of hydrogenases. *Biochem. Soc. Trans.* 33 (1): 12–14.

3 Hsia, C.C.W., Schmitz, A., Lambertz, M. et al. (2013). Evolution of air breathing: oxygen homeostasis and the transitions from water to land and sky. *Compr. Physiol.* 3 (2): 849–915.

4 Greening, C. and Grinter, R. (2022). Microbial oxidation of atmospheric trace gases. *Nat. Rev. Microbiol.* 20 (9): 513–528.

5 Grinter, R., Kropp, A., Venugopal, H. et al. (2023). Structural basis for bacterial energy extraction from atmospheric hydrogen. *Nature* 615 (7952): 541–547.

6 Schwartz, E., Fritsch, J., and Friedrich, B. (2013). H_2-metabolizing prokaryotes. In: *The Prokaryotes* (ed. E. Rosenberg, E.F. DeLong, S. Lory, et al.), 119–199. Berlin, Heidelberg: Springer Berlin Heidelberg.

7 Vitt, S., Ma, K., Warkentin, E. et al. (2014). The F_{420}-reducing [NiFe]-hydrogenase complex from *Methanothermobacter marburgensis*, the first X-ray structure of a group 3 family member. *J. Mol. Biol.* 426 (15): 2813–2826.

8 Wu, C.-H., McTernan, P.M., Walter, M.E., and Adams, M.W.W. (2015). Production and application of a soluble hydrogenase from *Pyrococcus furiosus*. *Archaea* 2015: 1–8.
9 Lauterbach, L. and Lenz, O. (2013). Catalytic production of hydrogen peroxide and water by oxygen-tolerant [NiFe]-hydrogenase during H_2 cycling in the presence of O_2. *J. Am. Chem. Soc.* 135 (47): 17897–17905.
10 Buhrke, T., Lenz, O., Porthun, A., and Friedrich, B. (2004). The H_2-sensing complex of *Ralstonia eutropha*: interaction between a regulatory [NiFe] hydrogenase and a histidine protein kinase. *Mol. Microbiol.* 51 (6): 1677–1689.
11 Peters, K. and Sargent, F. (2023). Formate hydrogenlyase, formic acid translocation and hydrogen production: dynamic membrane biology during fermentation. *Biochim. Biophys. Acta Bioenerg.* 1864 (1): 148919.
12 Vignais, P.M. and Billoud, B. (2007). Occurrence, classification, and biological function of hydrogenases: an overview. *Chem. Rev.* 107 (10): 4206–4272.
13 Søndergaard, D., Pedersen, C.N.S., and Greening, C. (2016). HydDB: a web tool for hydrogenase classification and analysis. *Sci. Rep.* 6 (1): 34212.
14 Lubitz, W., Ogata, H., Rüdiger, O., and Reijerse, E. (2014). Hydrogenases. *Chem. Rev.* 114 (8): 4081–4148.
15 Braymer, J.J., Freibert, S.A., Rakwalska-Bange, M., and Lill, R. (2021). Mechanistic concepts of iron-sulfur protein biogenesis in biology. *Biochim. Biophys. Acta Mol. Cell Res.* 1868 (1): 118863.
16 Esquilin-Lebron, K., Dubrac, S., Barras, F., and Boyd, J.M. (2021). Bacterial approaches for assembling iron-sulfur proteins. *mBio* 12 (6): e02425-21.
17 Garcin, E., Vernede, X., Hatchikian, E. et al. (1999). The crystal structure of a reduced [NiFeSe] hydrogenase provides an image of the activated catalytic center. *Structure* 7 (5): 557–566.
18 Zacarias, S., Vélez, M., Pita, M. et al. (2018). Characterization of the [NiFeSe] hydrogenase from *Desulfovibrio vulgaris* Hildenborough. *Methods Enzymol.* 613: 169–201.
19 Ogata, H., Lubitz, W., and Higuchi, Y. (2016). Structure and function of [NiFe] hydrogenases. *J. Biochem.* 160 (5): 251–258.
20 Fritsch, J., Lenz, O., and Friedrich, B. (2013). Structure, function and biosynthesis of O_2-tolerant hydrogenases. *Nat. Rev. Microbiol.* 11 (2): 106–114.
21 Greening, C., Biswas, A., Carere, C.R. et al. (2016). Genomic and metagenomic surveys of hydrogenase distribution indicate H_2 is a widely utilised energy source for microbial growth and survival. *ISME J.* 10 (3): 761–777.
22 Böck, A., King, P.W., Blokesch, M., and Posewitz, M.C. (2006). Maturation of hydrogenases. In: *Advances in Microbial Physiology*, vol. 51 (ed. R.K. Poole), 1–225. Elsevier.
23 Lutz, S., Jacobi, A., Schlensog, V. et al. (1991). Molecular characterization of an operon (*hyp*) necessary for the activity of the three hydrogenase isoenzymes in *Escherichia coli*. *Mol. Microbiol.* 5 (1): 123–135.
24 Rey, L., Fernández, D., Brito, B. et al. (1996). The hydrogenase gene cluster of *Rhizobium leguminosarum* bv.viciae contains an additional gene (*hypX*), which encodes a protein with sequence similarity to the

N_{10}-formyltetrahydrofolate-dependent enzyme family and is required for nickel-dependent hydrogenase processing and activity. *Mol. Gen. Genet.* 252 (3): 237–248.

25 Lacasse, M.J. and Zamble, D.B. (2016). [NiFe]-hydrogenase maturation. *Biochemistry* 55 (12): 1689–1701.

26 Barrett, E.L., Kwan, H.S., and Macy, J. (1984). Anaerobiosis, formate, nitrate, and *pyrA* are involved in the regulation of formate hydrogenlyase in *Salmonella typhimurium. J. Bacteriol.* 158 (3): 972–977.

27 Paschos, A., Glass, R.S., and Böck, A. (2001). Carbamoylphosphate requirement for synthesis of the active center of [NiFe]-hydrogenases. *FEBS Lett.* 488 (1–2): 9–12.

28 Paschos, A., Bauer, A., Zimmermann, A. et al. (2002). HypF, a carbamoyl phosphate-converting enzyme involved in [NiFe] hydrogenase maturation. *J. Biol. Chem.* 277 (51): 49945–49951.

29 Wang, T.-T., Bishop, S.H., and Himoe, A. (1972). Detection of carbamate as a product of the carbamate kinase-catalyzed reaction by stopped flow spectrophotometry. *J. Biol. Chem.* 247 (14): 4437–4440.

30 Reissmann, S., Hochleitner, E., Wang, H. et al. (2003). Taming of a poison: biosynthesis of the NiFe-hydrogenase cyanide ligands. *Science* 299 (5609): 1067–1070.

31 Petkun, S., Shi, R., Li, Y. et al. (2011). Structure of hydrogenase maturation protein HypF with reaction intermediates shows two active sites. *Structure* 19 (12): 1773–1783.

32 Shomura, Y. and Higuchi, Y. (2012). Structural basis for the reaction mechanism of S-carbamoylation of HypE by HypF in the maturation of [NiFe]-hydrogenases. *J. Biol. Chem.* 287 (34): 28409–28419.

33 Wolf, I., Buhrke, T., Dernedde, J. et al. (1998). Duplication of *hyp* genes involved in maturation of [NiFe] hydrogenases in *Alcaligenes eutrophus* H16. *Arch. Microbiol.* 170 (6): 451–459.

34 Tominaga, T., Watanabe, S., Matsumi, R. et al. (2012). Structure of the [NiFe]-hydrogenase maturation protein HypF from *Thermococcus kodakarensis* KOD1. *Acta Crystallogr. F* 68 (10): 1153–1157.

35 Blokesch, M., Paschos, A., Bauer, A. et al. (2004). Analysis of the transcarbamoylation-dehydration reaction catalyzed by the hydrogenase maturation proteins HypF and HypE. *Eur. J. Biochem.* 271 (16): 3428–3436.

36 Jones, A.K., Lenz, O., Strack, A. et al. (2004). NiFe hydrogenase active site biosynthesis: identification of Hyp protein complexes in *Ralstonia eutropha. Biochemistry* 43 (42): 13467–13477.

37 Rangarajan, E.S., Asinas, A., Proteau, A. et al. (2008). Structure of [NiFe] hydrogenase maturation protein HypE from *Escherichia coli* and its interaction with HypF. *J. Bacteriol.* 190 (4): 1447–1458.

38 Tominaga, T., Watanabe, S., Matsumi, R. et al. (2013). Crystal structures of the carbamoylated and cyanated forms of HypE for [NiFe] hydrogenase maturation. *Proc. Natl. Acad. Sci. U.S.A.* 110 (51): 20485–20490.

39 Shomura, Y., Komori, H., Miyabe, N. et al. (2007). Crystal structures of hydrogenase maturation protein HypE in the apo and ATP-bound forms. *J. Mol. Biol.* 372 (4): 1045–1054.

40 Blokesch, M., Albracht, S.P.J., Matzanke, B.F. et al. (2004). The complex between hydrogenase-maturation proteins HypC and HypD is an intermediate in the supply of cyanide to the active site iron of [NiFe]-hydrogenases. *J. Mol. Biol.* 344 (1): 155–167.

41 Watanabe, S., Matsumi, R., Atomi, H. et al. (2012). Crystal structures of the HypCD complex and the HypCDE ternary complex: transient intermediate complexes during [NiFe] hydrogenase maturation. *Structure* 20 (12): 2124–2137.

42 Bürstel, I., Siebert, E., Winter, G. et al. (2012). A universal scaffold for synthesis of the $Fe(CN)_2(CO)$ moiety of [NiFe] hydrogenase. *J. Biol. Chem.* 287 (46): 38845–38853.

43 Soboh, B., Stripp, S.T., Muhr, E. et al. (2012). [NiFe]-hydrogenase maturation: isolation of a HypC–HypD complex carrying diatomic CO and CN^- ligands. *FEBS Lett.* 586 (21): 3882–3887.

44 Stripp, S.T., Soboh, B., Lindenstrauss, U. et al. (2013). HypD is the scaffold protein for $Fe\text{-}(CN)_2CO$ cofactor assembly in [NiFe]-hydrogenase maturation. *Biochemistry* 52 (19): 3289–3296.

45 Miki, K., Atomi, H., and Watanabe, S. (2020). Structural insight into [NiFe] hydrogenase maturation by transient complexes between Hyp proteins. *Acc. Chem. Res.* 53 (4): 875–886.

46 Kwiatkowski, A., Caserta, G., Schulz, A.-C. et al. (2024). ATP-triggered $Fe(CN)_2CO$ synthon transfer from the maturase HypCD to the active site of apo-[NiFe]-hydrogenase. *J. Am. Chem. Soc.* 146 (45): 30976–30989.

47 Abramson, J., Adler, J., Dunger, J. et al. (2024). Accurate structure prediction of biomolecular interactions with AlphaFold 3. *Nature* 630 (8016): 493–500.

48 Blokesch, M. and Böck, A. (2002). Maturation of [NiFe]-hydrogenases in *Escherichia coli*: the HypC cycle. *J. Mol. Biol.* 324 (2): 287–296.

49 Blokesch, M. and Böck, A. (2006). Properties of the [NiFe]-hydrogenase maturation protein HypD. *FEBS Lett.* 580 (17): 4065–4068.

50 Stripp, S.T., Lindenstrauss, U., Granich, C. et al. (2014). The influence of oxygen on [NiFe]–hydrogenase cofactor biosynthesis and how ligation of carbon monoxide precedes cyanation. *PLoS ONE* 9 (9): e107488.

51 Roseboom, W., Blokesch, M., Böck, A., and Albracht, S.P.J. (2005). The biosynthetic routes for carbon monoxide and cyanide in the Ni–Fe active site of hydrogenases are different. *FEBS Lett.* 579 (2): 469–472.

52 Soboh, B., Stripp, S.T., Bielak, C. et al. (2013). The [NiFe]-hydrogenase accessory chaperones HypC and HybG of *Escherichia coli* are iron- and carbon dioxide-binding proteins. *FEBS Lett.* 587 (16): 2512–2516.

53 Lenz, O., Zebger, I., Hamann, J. et al. (2007). Carbamoylphosphate serves as the source of CN^-, but not of the intrinsic CO in the active site of the regulatory [NiFe]-hydrogenase from *Ralstonia eutropha*. *FEBS Lett.* 581 (17): 3322–3326.

54 Forzi, L., Hellwig, P., Thauer, R.K., and Sawers, R.G. (2007). The CO and CN^- ligands to the active site Fe in [NiFe]-hydrogenase of *Escherichia coli* have different metabolic origins. *FEBS Lett.* 581 (17): 3317–3321.

55 Bürstel, I., Hummel, P., Siebert, E. et al. (2011). Probing the origin of the metabolic precursor of the CO ligand in the catalytic center of [NiFe] hydrogenase. *J. Biol. Chem.* 286 (52): 44937–44944.

56 Bürstel, I., Siebert, E., Frielingsdorf, S. et al. (2016). CO synthesized from the central one-carbon pool as source for the iron carbonyl in O_2-tolerant [NiFe]-hydrogenase. *Proc. Natl. Acad. Sci. U.S.A.* 113 (51): 14722–14726.

57 Muraki, N., Ishii, K., Uchiyama, S. et al. (2019). Structural characterization of HypX responsible for CO biosynthesis in the maturation of NiFe-hydrogenase. *Commun. Biol.* 2 (1): 385.

58 Schulz, A.-C., Frielingsdorf, S., Pommerening, P. et al. (2020). Formyltetrahydrofolate decarbonylase synthesizes the active site CO ligand of O_2-tolerant [NiFe] hydrogenase. *J. Am. Chem. Soc.* 142 (3): 1457–1464.

59 Arlt, C., Nutschan, K., Haase, A. et al. (2021). Native mass spectrometry identifies the HybG chaperone as carrier of the $Fe(CN)_2CO$ group during maturation of *E. coli* [NiFe]-hydrogenase 2. *Sci. Rep.* 11 (1): 24362.

60 Blokesch, M., Magalon, A., and Böck, A. (2001). Interplay between the specific chaperone-like proteins HybG and HypC in maturation of hydrogenases 1, 2, and 3 from *Escherichia coli*. *J. Bacteriol.* 183 (9): 2817–2822.

61 Soboh, B., Adrian, L., and Stripp, S.T. (2022). An *in vitro* reconstitution system to monitor iron transfer to the active site during the maturation of [NiFe]-hydrogenase. *J. Biol. Chem.* 298 (9): 102291.

62 Nutschan, K., Golbik, R.P., and Sawers, R.G. (2019). The iron–sulfur-containing HypC-HypD scaffold complex of the [NiFe]-hydrogenase maturation machinery is an ATPase. *FEBS Open Bio.* 9 (12): 2072–2079.

63 Haase, A. and Sawers, R.G. (2023). A redox-active HybG-HypD scaffold complex is required for optimal ATPase activity during [NiFe]-hydrogenase maturation in *Escherichia coli*. *FEBS Open Bio.* 13 (2): 341–351.

64 Volbeda, A., Charon, M.-H., Piras, C. et al. (1995). Crystal structure of the nickel–iron hydrogenase from *Desulfovibrio gigas*. *Nature* 373 (6515): 580–587.

65 Friedrich, B., Heine, E., Finck, A., and Friedrich, C.G. (1981). Nickel requirement for active hydrogenase formation in *Alcaligenes eutrophus*. *J. Bacteriol.* 145 (3): 1144–1149.

66 Graf, E.-G. and Thauer, R.K. (1981). Hydrogenase from *Methanobacterium thermoautotrophicum*, a nickel-containing enzyme. *FEBS Lett.* 136 (1): 165–169.

67 Caserta, G., Hartmann, S., Van Stappen, C. et al. (2023). Stepwise assembly of the active site of [NiFe]-hydrogenase. *Nat. Chem. Biol.* 19 (4): 498–506.

68 Waugh, R. and Boxer, D.H. (1986). Pleiotropic hydrogenase mutants of *Escherichia coli* K12: growth in the presence of nickel can restore hydrogenase activity. *Biochimie* 68 (1): 157–166.

69 Jacobi, A., Rossmann, R., and Böck, A. (1992). The *hyp* operon gene products are required for the maturation of catalytically active hydrogenase isoenzymes in *Escherichia coli*. *Arch. Microbiol.* 158 (6): 444–451.

70 Blokesch, M., Rohrmoser, M., Rode, S., and Böck, A. (2004). HybF, a zinc-containing protein involved in NiFe hydrogenase maturation. *J. Bacteriol.* 186 (9): 2603–2611.

71 Chan Chung, K.C. and Zamble, D.B. (2011). Protein interactions and localization of the *Escherichia coli* accessory protein HypA during nickel insertion to [NiFe] hydrogenase. *J. Biol. Chem.* 286 (50): 43081–43090.

72 Zhang, J.W., Butland, G., Greenblatt, J.F. et al. (2005). A role for SlyD in the *Escherichia coli* hydrogenase biosynthetic pathway. *J. Biol. Chem.* 280 (6): 4360–4366.

73 Kaluarachchi, H., Sutherland, D.E.K., Young, A. et al. (2009). The Ni(II)-binding properties of the metallochaperone SlyD. *J. Am. Chem. Soc.* 131 (51): 18489–18500.

74 Cheng, T., Li, H., Yang, X. et al. (2013). Interaction of SlyD with HypB of *Helicobacter pylori* facilitates nickel trafficking. *Metallomics* 5 (7): 804.

75 Atanassova, A. and Zamble, D.B. (2005). *Escherichia coli* HypA is a zinc metalloprotein with a weak affinity for nickel. *J. Bacteriol.* 187 (14): 4689–4697.

76 Leach, M.R., Sandal, S., Sun, H., and Zamble, D.B. (2005). Metal binding activity of the *Escherichia coli* hydrogenase maturation factor HypB. *Biochemistry* 44 (36): 12229–12238.

77 Sydor, A.M., Lebrette, H., Ariyakumaran, R. et al. (2014). Relationship between Ni(II) and Zn(II) coordination and nucleotide binding by the *Helicobacter pylori* [NiFe]-hydrogenase and urease maturation factor HypB. *J. Biol. Chem.* 289 (7): 3828–3841.

78 Gasper, R., Scrima, A., and Wittinghofer, A. (2006). Structural insights into HypB, a GTP-binding protein that regulates metal binding. *J. Biol. Chem.* 281 (37): 27492–27502.

79 Chan, K.-H., Li, T., Wong, C.-O., and Wong, K.-B. (2012). Structural basis for GTP-dependent dimerization of hydrogenase maturation factor HypB. *PLoS ONE* 7 (1): e30547.

80 Watanabe, S., Kawashima, T., Nishitani, Y. et al. (2015). Structural basis of a Ni acquisition cycle for [NiFe] hydrogenase by Ni-metallochaperone HypA and its enhancer. *Proc. Natl. Acad. Sci. U.S.A.* 112 (25): 7701–7706.

81 Dias, A.V., Mulvihill, C.M., Leach, M.R. et al. (2008). Structural and biological analysis of the metal sites of *Escherichia coli* hydrogenase accessory protein HypB. *Biochemistry* 47 (46): 11981–11991.

82 Lacasse, M.J., Douglas, C.D., and Zamble, D.B. (2016). Mechanism of selective nickel transfer from HypB to HypA, *Escherichia coli* [NiFe]-hydrogenase accessory proteins. *Biochemistry* 55 (49): 6821–6831.

83 Douglas, C.D., Ngu, T.T., Kaluarachchi, H., and Zamble, D.B. (2013). Metal transfer within the *Escherichia coli* HypB–HypA complex of hydrogenase accessory proteins. *Biochemistry* 52 (35): 6030–6039.

84 Lacasse, M.J., Summers, K.L., Khorasani-Motlagh, M. et al. (2019). Bimodal nickel-binding site on *Escherichia coli* [NiFe]-hydrogenase metallochaperone HypA. *Inorg. Chem.* 58 (20): 13604–13618.

85 Kwon, S., Watanabe, S., Nishitani, Y. et al. (2018). Crystal structures of a [NiFe] hydrogenase large subunit HyhL in an immature state in complex with a Ni chaperone HypA. *Proc. Natl. Acad. Sci. U.S.A.* 115 (27): 7045–7050.

86 Fritsch, J., Scheerer, P., Frielingsdorf, S. et al. (2011). The crystal structure of an oxygen-tolerant hydrogenase uncovers a novel iron-sulphur centre. *Nature* 479 (7372): 249–252.

87 Gollin, D.J., Mortenson, L.E., and Robson, R.L. (1992). Carboxyl-terminal processing may be essential for production of active NiFe hydrogenase in *Azotobacter vinelandii*. *FEBS Letters* 309 (3): 371–375.

88 Sorgenfrei, O., Linder, D., Karas, M., and Klein, A. (1993). A novel very small subunit of a selenium containing [NiFe] hydrogenase of *Methanococcus voltae* is postranslationally processed by cleavage at a defined position. *Eur. J. Biochem.* 213 (3): 1355–1358.

89 Menon, N.K., Robbins, J., Der Vartanian, M. et al. (1993). Carboxy-terminal processing of the large subunit of [NiFe] hydrogenases. *FEBS Lett.* 331 (1–2): 91–95.

90 Rossmann, R., Sauter, M., Lottspeich, F., and Böck, A. (1994). Maturation of the large subunit (HYCE) of *Escherichia coli* hydrogenase 3 requires nickel incorporation followed by C-terminal processing at Arg537. *Eur. J. Biochem.* 220 (2): 377–384.

91 Fritsche, E., Paschos, A., Beisel, H.G. et al. (1999). Crystal structure of the hydrogenase maturating endopeptidase HybD from *Escherichia coli*. *J. Mol. Biol.* 288 (5): 989–998.

92 Theodoratou, E., Paschos, A., Magalon, A. et al. (2000). Nickel serves as a substrate recognition motif for the endopeptidase involved in hydrogenase maturation. *Eur. J. Biochem.* 267 (7): 1995–1999.

93 Theodoratou, E., Huber, R., and Böck, A. (2005). [NiFe]-hydrogenase maturation endopeptidase: structure and function. *Biochem. Soc. Trans.* 33 (1): 108–111.

94 Theodoratou, E., Paschos, A., Mintz-Weber, S., and Böck, A. (2000). Analysis of the cleavage site specificity of the endopeptidase involved in the maturation of the large subunit of hydrogenase 3 from *Escherichia coli*. *Arch. Microbiol.* 173 (2): 110–116.

95 Massanz, C., Fernandezz, V.M., and Friedrich, B. (1997). C-terminal extension of the H_2-activating subunit, HoxH, directs maturation of the NAD-reducing hydrogenase in *Alcaligenes eutrophus*. *Eur. J. Biochem.* 245 (2): 441–448.

96 Senger, M., Stripp, S.T., and Soboh, B. (2017). Proteolytic cleavage orchestrates cofactor insertion and protein assembly in [NiFe]-hydrogenase biosynthesis. *J. Biol. Chem.* 292 (28): 11670–11681.

97 Thomas, C., Muhr, E., and Sawers, R.G. (2015). Coordination of synthesis and assembly of a modular membrane-associated [NiFe]-hydrogenase is determined by cleavage of the C-terminal peptide. *J. Bacteriol.* 197 (18): 2989–2998.

98 Hartmann, S., Frielingsdorf, S., Caserta, G., and Lenz, O. (2020). A membrane-bound [NiFe]-hydrogenase large subunit precursor whose C-terminal extension is not essential for cofactor incorporation but guarantees optimal maturation. *MicrobiologyOpen* 9 (6): 1197–1206.

99 Witkowska, M., Jedrzejczak, R.P., Joachimiak, A. et al. (2023). Promising approaches for the assembly of the catalytically active, recombinant *Desulfomicrobium baculatum* hydrogenase with substitutions at the active site. *Microb. Cell. Fact.* 22 (1): 134.

100 Pinske, C., Thomas, C., Nutschan, K., and Sawers, R.G. (2019). Delimiting the function of the C-terminal extension of the *Escherichia coli* [NiFe]-hydrogenase 2 large subunit precursor. *Front. Microbiol.* 10: 2223.

101 Winter, G., Buhrke, T., Lenz, O. et al. (2005). A model system for [NiFe] hydrogenase maturation studies: purification of an active site-containing hydrogenase large subunit without small subunit. *FEBS Lett.* 579 (20): 4292–4296.

102 Wang, Y., Song, Y., Ma, C. et al. (2021). Electrochemical characterization of a truncated hydrogenase from *Pyrococcus furiosus*. *Electrochim. Acta* 387: 138502.

103 Caserta, G., Lorent, C., Ciaccafava, A. et al. (2020). The large subunit of the regulatory [NiFe]-hydrogenase from *Ralstonia eutropha* – a minimal hydrogenase? *Chem. Sci.* 11 (21): 5453–5465.

104 Caserta, G., Pelmenschikov, V., Lorent, C. et al. (2021). Hydroxy-bridged resting states of a [NiFe]-hydrogenase unraveled by cryogenic vibrational spectroscopy and DFT computations. *Chem. Sci.* 12 (6): 2189–2197.

105 Soboh, B., Kuhns, M., Braussemann, M. et al. (2012). Evidence for an oxygen-sensitive iron–sulfur cluster in an immature large subunit species of *Escherichia coli* [NiFe]-hydrogenase 2. *Biochem. Biophys. Res. Commun.* 424 (1): 158–163.

106 Hartmann, S., Frielingsdorf, S., Ciaccafava, A. et al. (2018). O_2-tolerant H_2 activation by an isolated large subunit of a [NiFe] hydrogenase. *Biochemistry* 57 (36): 5339–5349.

107 Magalon, A. and Böck, A. (2000). Analysis of the HypC-HycE complex, a key intermediate in the assembly of the metal center of the *Escherichia coli* Hydrogenase 3. *J. Biol. Chem.* 275 (28): 21114–21120.

108 Massanz, C. and Friedrich, B. (1999). Amino acid replacements at the H_2-activating site of the NAD-reducing hydrogenase from *Alcaligenes eutrophus*. *Biochemistry* 38 (43): 14330–14337.

109 Winter, G., Buhrke, T., Jones, A.K., and Friedrich, B. (2004). The role of the active site-coordinating cysteine residues in the maturation of the H_2-sensing [NiFe] hydrogenase from *Ralstonia eutropha* H16. *Arch. Microbiol.* 182 (2–3): 138–146.

110 Pinske, C. and Sawers, R.G. (2014). The importance of iron in the biosynthesis and assembly of [NiFe]-hydrogenases. *Biomol. Concepts* 5 (1): 55–70.

111 Pinske, C. and Sawers, G. (2011). Iron restriction induces preferential down-regulation of H_2-consuming over H_2-evolving reactions during fermentative growth of *Escherichia coli*. *BMC Microbiol.* 11 (1): 196.

112 Pinske, C. and Sawers, R.G. (2012). Delivery of iron-sulfur clusters to the hydrogen-oxidizing [NiFe]-hydrogenases in *Escherichia coli* requires the A-type carrier proteins ErpA and IscA. *PLoS ONE* 7 (2): e31755.

113 Vinella, D., Brochier-Armanet, C., Loiseau, L. et al. (2009). Iron-sulfur (Fe–S) protein biogenesis: phylogenomic and genetic studies of A-type carriers. *PLoS Genet.* 5 (5): e1000497.

114 Pinske, C., Jaroschinsky, M., and Sawers, R.G. (2013). Levels of control exerted by the Isc iron-sulfur cluster system on biosynthesis of the formate hydrogenlyase complex. *Microbiology* 159 (Pt_6): 1179–1189.

115 Haase, A., Arlt, C., Sinz, A., and Sawers, R.G. (2024). Evidence the Isc iron-sulfur cluster biogenesis machinery is the source of iron for [NiFe]-cofactor biosynthesis in *Escherichia coli*. *Sci. Rep.* 14 (1): 3026.

116 Caserta, G., Lorent, C., Pelmenschikov, V. et al. (2020). *In vitro* assembly as a tool to investigate catalytic intermediates of [NiFe]-hydrogenase. *ACS Catal.* 10 (23): 13890–13894.

117 Fasano, A., Guendon, C., Jacq-Bailly, A. et al. (2023). A chimeric NiFe hydrogenase heterodimer to assess the role of the electron transfer chain in tuning the enzyme's catalytic bias and oxygen tolerance. *J. Am. Chem. Soc.* 145 (36): 20021–20030.

118 Allen, W.J. and Collinson, I. (2023). A unifying mechanism for protein transport through the core bacterial Sec machinery. *Open Biol.* 13 (8): 230166.

119 Rodrigue, A., Chanal, A., Beck, K. et al. (1999). Co-translocation of a periplasmic enzyme complex by a hitchhiker mechanism through the bacterial Tat pathway. *J. Biol. Chem.* 274 (19): 13223–13228.

120 Berks, B.C. (1996). A common export pathway for proteins binding complex redox cofactors? *Mol. Microbiol.* 22 (3): 393–404.

121 Bogsch, E.G., Sargent, F., Stanley, N.R. et al. (1998). An essential component of a novel bacterial protein export system with homologues in plastids and mitochondria. *J. Biol. Chem.* 273 (29): 18003–18006.

122 Voordouw, G. (2000). A universal system for the transport of redox proteins: early roots and latest developments. *Biophys. Chem.* 86 (2–3): 131–140.

123 Gross, R., Simon, J., and Kröger, A. (1999). The role of the twin-arginine motif in the signal peptide encoded by the *hydA* gene of the hydrogenase from *Wolinella succinogenes*. *Arch. Microbiol.* 172 (4): 227–232.

124 Bernhard, M., Friedrich, B., and Siddiqui, R.A. (2000). *Ralstonia eutropha* TF93 Is blocked in Tat-mediated protein export. *J. Bacteriol.* 182 (3): 581–588.

125 Meloni, S., Rey, L., Sidler, S. et al. (2003). The twin-arginine translocation (Tat) system is essential for *Rhizobium*–legume symbiosis. *Mol. Microbiol.* 48 (5): 1195–1207.

126 Maltempi De Souza, E., De Oliveira Pedrosa, F., Wassem, R. et al. (2007). Genes involved in Sec-independent membrane targeting of hydrogenase in *Azotobacter chroococcum*. *Res. Microbiol.* 158 (3): 272–278.

127 Hitchcock, A., Hall, S.J., Myers, J.D. et al. (2010). Roles of the twin-arginine translocase and associated chaperones in the biogenesis of the electron transport chains of the human pathogen *Campylobacter jejuni*. *Microbiology* 156 (10): 2994–3010.

128 Benoit, S.L. and Maier, R.J. (2014). Twin-arginine translocation system in *Helicobacter pylori*: TatC, but not TatB, is essential for viability. *mBio* 5 (1): e01016-13.

129 Palmer, T. and Stansfeld, P.J. (2020). Targeting of proteins to the twin-arginine translocation pathway. *Mol. Microbiol.* 113 (5): 861–871.

130 Dubini, A. and Sargent, F. (2003). Assembly of Tat-dependent [NiFe] hydrogenases: identification of precursor-binding accessory proteins. *FEBS Lett.* 549 (1–3): 141–146.

131 Jack, R.L., Buchanan, G., Dubini, A. et al. (2004). Coordinating assembly and export of complex bacterial proteins. *EMBO J.* 23 (20): 3962–3972.

132 Schubert, T., Lenz, O., Krause, E. et al. (2007). Chaperones specific for the membrane-bound [NiFe]-hydrogenase interact with the Tat signal peptide of the small subunit precursor in *Ralstonia eutropha* H16. *Mol. Microbiol.* 66 (2): 453–467.

133 Ludwig, M., Schubert, T., Zebger, I. et al. (2009). Concerted action of two novel auxiliary proteins in assembly of the active site in a membrane-bound [NiFe] hydrogenase. *J. Biol. Chem.* 284 (4): 2159–2168.

134 Albareda, M., Pacios, L.F., Manyani, H. et al. (2014). Maturation of *Rhizobium leguminosarum* hydrogenase in the presence of oxygen requires the interaction of the chaperone HypC and the scaffolding protein HupK. *J. Biol. Chem.* 289 (31): 21217–21229.

135 Frielingsdorf, S., Fritsch, J., Schmidt, A. et al. (2014). Reversible [4Fe-3S] cluster morphing in an O_2-tolerant [NiFe] hydrogenase. *Nat. Chem. Biol.* 10 (5): 378–385.

136 Manyani, H., Rey, L., Palacios, J.M. et al. (2005). Gene products of the *hupGHIJ* operon are involved in maturation of the iron-sulfur subunit of the [NiFe] hydrogenase from *Rhizobium leguminosarum* bv. viciae. *J. Bacteriol.* 187 (20): 7018–7026.

137 Fritsch, J., Lenz, O., and Friedrich, B. (2011). The maturation factors HoxR and HoxT contribute to oxygen tolerance of membrane-bound [NiFe] hydrogenase in *Ralstonia eutropha* H16. *J. Bacteriol.* 193 (10): 2487–2497.

138 Fritsch, J., Siebert, E., Priebe, J. et al. (2014). Rubredoxin-related maturation factor guarantees metal cofactor integrity during aerobic biosynthesis of membrane-bound [NiFe] hydrogenase. *J. Biol. Chem.* 289 (11): 7982–7993.

139 Berggren, G., Adamska, A., Lambertz, C. et al. (2013). Biomimetic assembly and activation of [FeFe]-hydrogenases. *Nature* 499 (7456): 66–69.

140 Esselborn, J., Lambertz, C., Adamska-Venkatesh, A. et al. (2013). Spontaneous activation of [FeFe]-hydrogenases by an inorganic [2Fe] active site mimic. *Nat. Chem. Biol.* 9 (10): 607–609.

141 Shima, S., Chen, D., Xu, T. et al. (2015). Reconstitution of [Fe]-hydrogenase using model complexes. *Nat. Chem.* 7 (12): 995–1002.

14

[Fe]-Hydrogenase and the FeGP Cofactor Involved in the CO_2-Reducing Hydrogenotrophic Methanogenic Pathway

Seigo Shima, Joao Pedro Fernandes-Queiroz, and Masanori Kaneko

Max Planck Institute for Terrestrial Microbiology, Microbial Protein Structure Group, Karl-von-Frisch-Straße 10, 35043 Marburg, Germany

Chapter Goals

- This chapter focuses on the structure and function of [Fe]-hydrogenase and the biosynthesis of its prosthetic group, the iron-guanylylpyridinol (FeGP) cofactor.
- The unique coenzymes, enzymes, and cofactors involved in the hydrogenotrophic methanogenic pathway are first presented, and then the structure and function of Hmd and the FeGP cofactor are discussed in detail.

Main Chapter Points

- As the main chapter, we describe the biosynthetic pathway of the FeGP cofactor, which was investigated by structure-to-function and *in vitro* biosynthesis analyses.

14.1 Introduction

[Fe]-hydrogenase (Hmd) is involved in the methanogenesis from CO_2 and H_2 [1–3], which is called the hydrogenotrophic methanogenic pathway. Whereas most methanogenic archaea commonly metabolize CO_2 and H_2 as substrates for methanogenesis, some of them can also produce methane from methyl compounds such as methanol, methylthiols, and methylamines. Acetic acid and methoxy compounds are also used as methanogenic substrates by some methanogens [4]. In the CO_2-reducing hydrogenotrophic methanogenic pathway, four H_2 molecules provide eight electrons for the reduction of CO_2 to methane, with [NiFe]-hydrogenases and [Fe]-hydrogenase involved in the oxidation of H_2 [2] (Figure 14.1).

Iron-Sulfur Clusters: Biogenesis and Biochemistry, Volume 2, First Edition.
Edited by Silke Leimkühler, Günter Schwarz, Oliver Lenz, and Oliver Einsle.

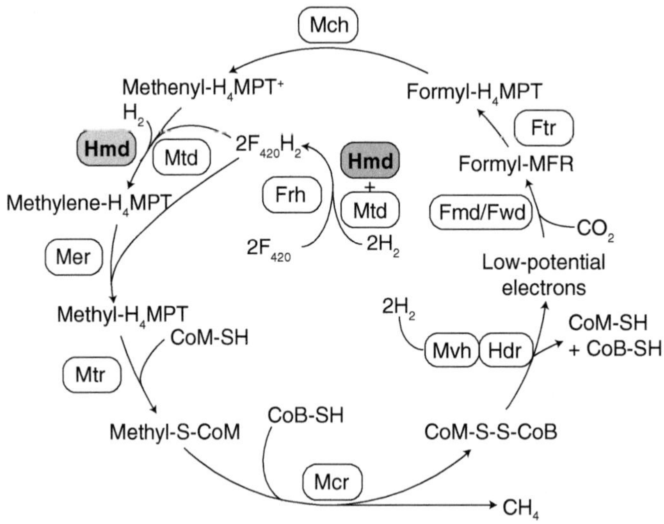

Figure 14.1 The hydrogenotrophic methanogenic pathway. Abbreviations: MFR, methanofuran; H_4MPT, tetrahydromethanopterin; $F_{420}H_2$, the reduced form of F_{420}; CoM-SH, coenzyme M; CoB-SH, coenzyme B; CoM-S-S-CoB, heterodisulfide of CoM-SH and CoB-SH; Fmd/Fwd, molybdopterin/tungstopterin dependent formylmethanofuran dehydrogenase isoenzymes; Ftr, formyltransferase; Mch, methenyl-H_4MPT^+ cyclohydrolase; Mtd, F_{420}-dependent methylene-H_4MPT dehydrogenase; Hmd, H_2-forming methylene-H_4MPT dehydrogenase ([Fe]-hydrogenase); Mer, methylene-H_4MPT reductase; Mtr, energy-conserving membrane-bound methyltransferase (MtrA-H). Mcr, methyl–coenzyme M reductase; Frh, F_{420}-reducing [NiFe]-hydrogenase; Hdr, heterodisulfide reductase; Mvh, heterodisulfide reductase-associating [NiFe]-hydrogenase. Source: Adapted from [5].

14.2 Unique Coenzymes Found in Methanogens

Methanofuran (MFR) is an aminomethylfuran derivative with a long side chain. The aminomethyl group of the furan ring is bound with a formyl group as a C_1 unit [6]. Tetrahydromethanopterin (H_4MPT), an analog of tetrahydrofolate (H_4F), also serves as a C_1 carrier and contains a pterin ring as the head group. The pterin rings of H_4MPT and H_4F are attached to different side chains containing a para-aminobenzoate (PABA) and para-aniline moiety, respectively [6]. The pK_a value of the para-amino group (N10) of these pterin derivatives varies greatly depending on the presence or absence of a carbonyl group on the PABA moiety, resulting in a difference in the redox properties of the two pterin coenzymes at the redox potential of the methenyl-/methylene and methylene-/methyl couples [7]. The small compound 2-mercaptoethanesulfonic acid, also known as coenzyme M (CoM-SH), acts as an additional C_1 carrier with a methyl group attached at the thiol group. Methyl-S-CoM is the substrate for the methanogenic reaction, in which methyl-S-CoM is reduced by 7-mercaptoheptanoyl threonine phosphate, called coenzyme B (CoB-SH). The deazaflavin coenzyme F_{420} is a hydride carrier with a ring structure similar to flavin but with redox properties close to that of NAD(P) [6].

The reduced form of F_{420} ($F_{420}H_2$) is used as a hydride donor in two reactions of hydrogenotrophic methanogenesis.

In addition to organic cofactors, the metallocofactors involved in methanogenesis enzymes are unique. The nickel porphinoid F_{430} is the cofactor of methyl-coenzyme M reductase, which catalyzes the methane-forming reaction [8]. The membrane-bound methyltransferase (MtrA-H) contains a vitamin B12 analog, 5-hydroxybenzimidazolyl cobamide [9]. [Fe]-hydrogenase contains the FeGP cofactor [10].

14.3 Enzymes Involved in the Hydrogenotrophic Methanogenic Pathway

During hydrogenotrophic methanogenesis, CO_2 is reduced and fixed as formyl-MFR. This reaction is catalyzed by formylmethanofuran dehydrogenase (Fmd and Fwd) [11]. Fmd and Fwd contain molybdopterin and tungstopterin as the prosthetic group, respectively. Fwd is constitutively synthesized while the production of Fmd is controlled by the molybdenum concentration in the medium [12–14]. These enzymes have two active sites, and CO_2 is first reduced to formate at the molybdopterin/tungstopterin-containing active site [15]. Formate is then transferred through a channel to the second active site, a dinuclear zinc center. Here, formate conjugates with the aminomethyl group of MFR to form formyl-MFR. Electrons for CO_2 reduction are thought to be provided by ferredoxin [1, 16].

The formyl group of formyl-MFR is transferred to H_4MPT to form formyl-H_4MPT, which is catalyzed by formyltransferase (Ftr), an enzyme that contains no prosthetic group. The crystal structures of Ftr from several methanogens and a sulfate-reducing archaeon have been reported [17–19]. Based on the substrate-bound crystal structure, the following catalytic mechanism has been proposed [17]. The reaction is initiated by the nucleophilic attack of N5 on the formyl carbon of formyl-H_4MPT, resulting in a tetrahedral anion intermediate. This intermediate is stabilized by a conserved glutamate residue, acting as an acid/base catalyst. The active site of Ftr is located at the interface region between the two monomers of the dimeric structure, where formyl-MFR is bound to one monomer and H_4MPT is bound to the other monomer [17]. Thus, the dimeric structure is a prerequisite for activity. Ftr from *Methanopyrus kandleri* showed an equilibrium between monomer, dimer, and tetramer [20], and this equilibrium was shifted by the concentration of lyotropic salts in the buffer solution, which is a reflection of the high salt concentration in the cell of this methanogenic archaea. In line with the bipartite active site, the properties of *M. kandleri* Ftr in different salt concentrations indicated that only the dimer and the tetramer are enzymatically active and that tetramerization increases thermostability [21].

Methenyl-H_4MPT^+ cyclohydrolase (Mch) catalyzes the conversion of formyl-H_4MPT to methenyl-H_4MPT^+, whereby a condensation reaction forms the methenyl (-N5-CH-N10-) ring of methenyl-H_4MPT^+ [22]. Crystal structures of Mch from *M. kandleri* and *Archaeoglobus fulgidus* have been reported, the latter in complex with

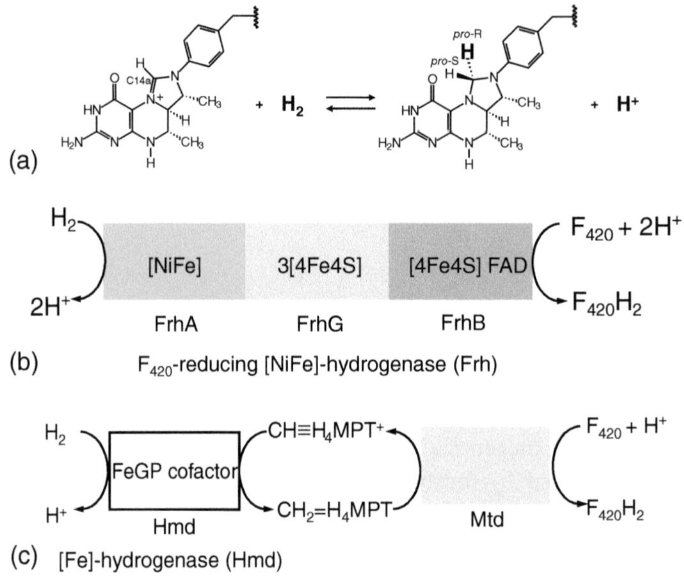

Figure 14.2 F_{420}-reducing reactions with H_2 involved in the hydrogenotrophic methanogenic pathway. (a) The reaction is catalyzed by [Fe]-hydrogenase (Hmd). H_2 is heterolytically cleaved and a hydride is transferred into the *pro-R* position of methylene-H_4MPT. Schematic presentation of the structure and the catalytic function of F_{420}-reducing [NiFe]-hydrogenase composed of the FrhA, FrhG, and FrhB subunits (b) and a coupled reaction of Hmd and F_{420}-dependent methylene-tetrahydromethanopterin (CH_2-H_4MPT) dehydrogenase (Mtd) (c).

H_4MPT and formyl-H_4MPT [23, 24]. The catalytic mechanism has been proposed on the basis of this complex structure. Mch is a homotrimeric enzyme without a prosthetic group. The active site is located in a cleft formed between two domains of a monomer. A glutamic acid residue acts as an acid/base catalyst, and the function of key amino acid residues was confirmed by site-directed mutagenesis [23].

The methenyl group of methenyl-H_4MPT^+ is reduced by a reaction catalyzed by two methylene-H_4MPT dehydrogenases: an F_{420}-dependent enzyme (Mtd) [25] and an H_2-dependent enzyme (Hmd) [26, 27] (Figure 14.2). Mtd has no prosthetic group, and hydrogenotrophic methanogens commonly use Mtd. The crystal structure of Mtd from *M. kandleri* has been determined [28, 29]. In the homotrimeric enzyme, the active site is located in the cleft between the two domains of a monomer. The ring structures of the head groups of $F_{420}H_2$ and methenyl-H_4MPT^+ are parallel to each other, in which the *Si* face of $F_{420}H_2$ and the *Re* face of methenyl-H_4MPT^+ are 2.7 Å apart. The hydride of $F_{420}H_2$ is transferred directly to the C14a carbon of methenyl-H_4MPT^+.

Most hydrogenotrophic methanogens belonging to *Methanopyrales*, *Methanococcales*, *Methanobacteriales*, and some *Methanomicrobiales* contain Hmd, while *Methanosarcinales* and other methanogenic archaea do not [30]. Hmd catalyzes the activation of H_2 and the hydride transfer of H_2 to methenyl-H_4MPT^+. In the reverse reaction, H_2 is formed from methylene-H_4MPT and a proton (Figure 14.2a)

[26, 27]. Due to the enzymatic properties and the prosthetic group in this enzyme, Hmd is also referred to as [Fe]-hydrogenase. A detailed description of this new type of hydrogenase is given later in Section 14.5.

Methylene-H_4MPT is reduced to methyl-H_4MPT by a reaction catalyzed by methylene-H_4MPT reductase (Mer) using $F_{420}H_2$ as the hydride donor [31, 32]. The kinetic analysis indicated that Mer catalyzes the hydride transfer reaction with a ternary complex mechanism [32]. The binding mode of F_{420} has been identified in the crystal structure [33]; however, the binding mode of methylene/methyl-H_4MPT on Mer is not known. Recently, a catalytic mechanism has been proposed based on structural comparison with methylene-H_4F reductases [34].

The methyl group of methyl-H_4MPT is then transferred to CoM-SH to form methyl-S-CoM. This reaction is catalyzed by the membrane-integral methyltransferase (Mtr) and coupled to the transfer of sodium ions across the plasma membrane, the only energy-conserving reaction in the hydrogenotrophic methanogenic pathway [9, 35–38]. The sodium ion gradient formed by the reaction is used for ATP generation by ATP synthase. Mtr contains a vitamin B12-like cofactor, 5-hydroxybenzimidazolyl cobamide, which has a cobalt-coordinating tetrapyrrole structure [9]. Biochemical experiments indicate that Mtr catalyzes a two-step reaction, in which the methyl group of methyl-H_4MPT is first transferred to the Co(I) ion of the cofactor in the MtrA subunit, and the methyl group is then transferred to CoM-SH [37]. The second methyl group transfer reaction is coupled to the transport of sodium ions. The crystal structure of the soluble domain of MtrA suggests that a separate polypeptide segment, linked by hydrogen bonds and located in the lower axial ligand region of the cobalt, is involved in transferring the energy released during demethylation to the sodium translocation [39]. Recently, the cryo-electron microscopy (Cryo-EM) structure of the Mtr complex has been reported [38].

Methyl-CoM reductase (Mcr) catalyzes the formation of methane from methyl-S-CoM with CoB-SH as the reducing agent. The by-product of the reaction is heterodisulfide of CoM-SH and CoB-SH [8]. Mcr contains the nickel porphinoid F_{430} as a prosthetic group. Mcr is folded in a $(\alpha\beta\gamma)_2$ conformation, in which F_{430} is embedded deeply inside the protein core of the heterotrimer, but one of the axial ligands of F_{430} is ligated with a glutamine residue of the α-subunit of another trimer [40]. Several crystal structures of Mcr have been reported [41–43]. In one crystal structure, the thiol group of CoM-SH binds to Ni in F_{430}, while CoB-SH binds to the entrance of the active site channel with the thiol group facing F_{430} at a distance of 6.2 Å. In another structure, the heterodisulfide is bound at the active site, in which the CoB-SH binding position is the same as in the other structure, and the CoM-SH moiety of the heterodisulfide is bound to the nickel of F_{430} with sulfonate oxygen. The former and latter structures are analogous to the structure of the substrates before the reaction and the product after the reaction, respectively [40]. The nickel ion of F_{430} can assume three redox states, Ni(I), Ni(II), and Ni(III), and all states are involved in the catalytic cycle of F_{430} [8]. There are two major catalytic mechanisms proposed, but in both cases, methyl-S-CoM interacts with F_{430} in the NI(I) state. The Thauer group originally proposed a mechanism involving a methyl-Ni(III) intermediate resulting from the nucleophilic attack of F_{430} in the

Ni(I) state to methyl-S-CoM [40]. The other mechanism, proposed by the Siegbahn group, based on density functional theory (DFT) calculations, suggests that Ni(I) of F_{430} attacks the thioether sulfur of methyl-S-CoM, resulting in a hemolytic cleavage with the formation of a methyl radical [44]. The transient methyl radical abstracts a hydrogen atom from CoB-SH to form methane and the CoB-S· radical. The Ragsdale group used spectroscopic techniques to demonstrate the presence of a CoM-S-Ni(II) intermediate in the Mcr reaction, which supports the methyl radical mechanism [45].

The heterodisulfide produced by the Mcr reaction is an important electron acceptor in the hydrogenotrophic methanogenic pathway. The heterodisulfide reductase (Hdr) catalyzes the reduction of the heterodisulfide and ferredoxin using a flavin-based electron-bifurcating (FBEB) mechanism [46]. Hdr is complexed with a heterodisulfide-associated electron-donating module, either a [NiFe]-hydrogenase (Mvh) or an F_{420}-dependent formate dehydrogenase (Fdh) [47–49]. The Mvh–Hdr and Fdh–Hdr complexes use the FBEB mechanism to provide low-potential electrons for the reduction of ferredoxin while using the heterodisulfide as a high-potential electron acceptor. The crystal structure of the Mvh–Hdr complex from *Methanothermococcus thermolithotrophicus* [50] and the Cryo-EM structure of the Fdh–Hdr complex from *Methanospirillum hungatei* [49] have been reported. The HdrA subunit contains FAD, which is responsible for the electron bifurcation reaction. HdrB contains two non-cubane [4Fe–4S] clusters to reduce heterodisulfide in a single-electron reduction mechanism, which is required for the FBEB mechanism. An HdrA domain with a 2 × [4Fe–4S] ferredoxin fold is proposed to be the site of interaction with external ferredoxin [50]. Fdh–Hdr from *M. hungatei* has been purified as a complex with Fmd and forms a dimeric 1-MDa structure $(Fdh–Hdr–Fmd)_2$ and a hexameric 3-MDa structure $(Fdh–Hdr–Fmd)_6$ [49]. In both megacomplexes, the low-potential-electron output domain is linked via the polyferredoxin FwdF to the CO_2-reducing active site in the FmdBD subunits. It was recently reported that Mvh–Hdr from *Methanothermobacter marburgensis* also forms a complex with Fmd or Fwd [5]. In the Mvh–Hdr–Fmd (Fwd) complex, the MvhB polyferredoxin is likely involved in connecting the redox-active catalytic centers of the enzyme complex.

14.4 [NiFe]-Hydrogenases Involved in the Hydrogenotrophic Methanogenesis

In addition to heterodisulfide reductase-associated [NiFe]-hydrogenase Mvh, three [NiFe]-hydrogenases are involved in hydrogenotrophic methanogenesis without cytochromes. Cytosolic F_{420}-reducing [NiFe]-hydrogenase (Frh) catalyzes the reversible reduction of coenzyme F_{420} with H_2. Membrane-bound [NiFe]-hydrogenases (Eha and Ehb) have been found in the genome of hydrogenotrophic methanogens [51].

The [NiFe]-hydrogenase Frh consists of three subunits, FrhA, FrhG, and FrhB. FrhA is the hydrogenase large subunit, which contains the standard dinuclear [NiFe] active site for the oxidation of H_2 to two electrons and two protons (Figure 14.2b). Crystal structures of Frh from *M. marburgensis* and *Methanosarcina barkeri* and a cryo-EM structure of Frh from *M. marburgensis* have been reported [52–54]. FrhG is the hydrogenase small subunit, in which three [4Fe–4S] clusters form an electron chain connecting the [NiFe] site with FrhB. FrhB contains one [4Fe–4S] cluster and one FAD molecule. FAD acts as a one-to-two electron switch, which enables the formation of a hydride consisting of two electrons and one proton. The hydride on the reduced FAD is transferred to F_{420}, which leads to the formation of $F_{420}H_2$. Structure analysis revealed that the Frh heterotrimer (FrhAGB) occurs in a $(FrhAGB)_2$ dimer, of which six modules oligomerize into a 1.25-MDa octadodecamer $(FrhAGB)_{12}$. Furthermore, the complex $(FrhAGB)_{12}$ structure is also found in other methanogens, as shown by low-resolution electron microscopy with negative staining [55, 56]. However, the physiological role of the 1.25-MDa hexadodecameric Frh remains unknown.

The structure of Mvh at the atomic level has been elucidated by the crystal structure analysis of the Mvh–Hdr complex from *M. thermolithotrophicus* [50]. The Mvh gene cluster in methanogens encodes the four proteins MvhAGDB [57]. MvhA and MvhG represent the large and small subunits of [NiFe]-hydrogenase. MvhD contains a [2Fe–2S] cluster and functions as an electron relay within the Mvh–Hdr complex. MvhB is a polyferredoxin containing 12 [4Fe–4S] clusters and is proposed to be a diffusible electron carrier that transfers low-potential electrons from the Hdr complex to Fmd/Fwd for CO_2 reduction [46, 58]. Recently, the function of MvhB as an additional electron relay between Hdr and Fmd (or Fwd) has been proposed [59], and the involvement of MvhB in the Mvh–Hdr–Fmd(Fwd) complex has also been reported [5], although the structure and the location of MvhB in the Mvh–Hdr–Fmd(Fwd) complex are still unknown.

Neither Eha nor Ehb has been purified from methanogens, and heterologous production of these membrane hydrogenases has not been reported. The function of Eha and Ehb has been inferred only from the gene sequences and the function of biochemically characterized homologous Ech hydrogenase from the methanogens with cytochromes (i.e. *M. barkeri*) [2]. The membrane-associated nature of some subunits of Eha and Ehb strongly suggests that the H_2-activating function of enzymes may be coupled to ion transport across the cytosolic membrane. It has been proposed that endergonic reduction of ferredoxin with H_2 is achieved by exploiting the sodium ion gradient formed during methanogenesis [2]. Mutational analysis of the *eha* gene in *Methanococcus maripaludis* has shown that this gene is essential for providing low-potential electrons for biosynthesis [60] and suggests that Ehb is involved in anabolic CO_2 assimilation [61].

14.5 Function of [Fe]-Hydrogenase (Hmd)

In 1990, the Thauer group first reported the isolation of Hmd from *M. marburgensis* [26, 27]. Enzymatic activity of Hmd was also detected in *Methanopyrales* and *Methanobacteriales* methanogens but not in some *Methanosarcinales* and *Methanomicrobiales* methanogens [62, 63]. These results suggested that Hmd is only present in a few hydrogenotrophic methanogens. Later genome sequencing of various microbes revealed a broader distribution of Hmd in hydrogenotrophic methanogens, including some members of the order *Methanomicrobiales*. Genomic analyses also revealed the presence of Hmd paralogs (Hmd II and Hmd III) in a number of methanogenic archaea and sulfur-reducing bacteria [30, 59, 64].

Hmd production is upregulated under nickel-limiting conditions. When the Ni^{2+} concentration in the medium is below 0.2 μM, Hmd production increases sixfold [65]. Under nickel-limiting conditions, the production of Mtd is also upregulated, while the [NiFe]-hydrogenase Frh is no longer produced. Thus, the F_{420} reduction by Frh is switched to the coupled reaction of Hmd and Mtd (Figure 14.2c). In the Hmd-Mtd system, Hmd reduces methenyl-H_4MPT^+ to methylene-H_4MPT, which is oxidized by Mtd to form $F_{420}H_2$. Under nickel-limiting conditions, Mvh production also drops to 20%, and the doubling time increases from about two hours in nickel-sufficient batch culture conditions to about nine hours in a nickel-limited continuous culture.

Hmd is a dimeric enzyme composed of two 38-kDa monomers, each of which harbors the FeGP cofactor. Although the presence of iron ions in the enzyme was described in the first publication on the biochemical properties of Hmd [26], Hmd was considered a metal-free protein catalyst in the early stages of research [7]. Despite this misinterpretation, the originally discovered kinetic features are still valid and fundamental to our understanding of this enzyme, so it is worth reviewing them in this chapter. Kinetic analysis of the enzyme activity indicated that Hmd catalyzes the hydride transfer reaction with a ternary complex mechanism [27]. In addition to the hydride transfer to methenyl-H_4MPT^+, Hmd also catalyzes the so-called proton exchange reaction, i.e. the production of HD and D_2 from D_2O and H_2 in the gas phase [66, 67]. These results indicate that Hmd performs not only single proton exchange reactions but also double exchange reactions. In addition, Hmd also catalyzes ortho/para H_2 exchange reactions [68]. Importantly, these exchange reactions only occur in the presence of the substrate methenyl-H_4MPT^+. This suggests that Hmd can catalyze the heterolytic cleavage of H_2 only in the presence of the native hydride acceptor [7]. This finding led to the intriguing hypothesis that Hmd itself does not catalyze the H_2 activation reaction and that methenyl-H_4MPT^+ represents the site of H_2 activation [69]. This hypothesis was based on the reversible formation of carbocations and H_2 from alkanes under superacid discovered by Olah [70]. The proposed Hmd catalytic mechanism required the presence of the C14a carbocation of methenyl-H_4MPT^+. Accordingly, it was hypothesized that methenyl-H_4MPT^+ binds to the active site of Hmd, where protein–substrate interactions would induce the carbocationic state of C14a by protonation of the substrate by suitable amino acid residues [69].

14.6 Reconstitution of the Hmd Holoenzyme with the Extracted Cofactor

If Hmd was a truly pure metal-free protein catalyst, the active enzyme could be heterologously produced in a recombinant host such as *Escherichia coli* without additional requirements. However, although the Hmd protein was produced as

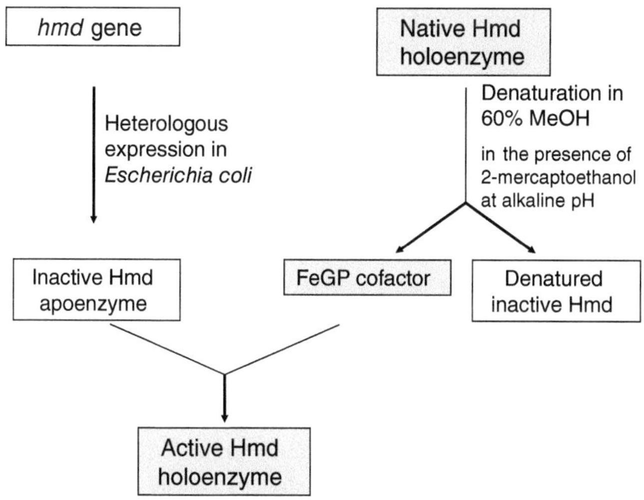

Figure 14.3 Extraction of the FeGP cofactor from the native Hmd holoenzyme of *M. marburgensis* and reconstitution of the [Fe]-hydrogenase holoenzyme from the heterologously produced apoenzyme and the isolated FeGP cofactor. Source: Adapted from [71].

a soluble protein in *E. coli*, no enzymatic activity could be detected. This led researchers to consider the presence of an organic cofactor in Hmd. Buurman et al. reported the extraction of a functional cofactor from native Hmd purified from *M. marburgensis* (Figure 14.3) [71, 72]. In this extraction method, purified Hmd was denatured in the presence of 2-mercaptoethanol and 6 M guanidinium chloride, 8 M urea, or 60% methanol [71, 72]. The resulting extract was ultrafiltrated (10 kDa cutoff) to separate small molecules from denatured Hmd. The filtrate containing the small molecules was then added to Hmd proteins from *Methanocaldococcus jannaschii* or *M. kandleri*, which had been heterologously prepared from *E. coli*. Simply mixing the filtrate and the Hmd apoenzymes yielded holoenzymes with catalytic activities similar to those of Hmd proteins purified from the native hosts (Figure 14.3). The activity was observed in the fraction containing compounds below 1000 Da in size exclusion chromatography. The metal content in the filtrate was analyzed, but no metal ions were detected. The discovery of an iron ion in the cofactor had to await the discovery of the photolysis of the Hmd cofactor, which led to the dissociation of the iron ion from the protein [72]. The prosthetic group of Hmd was later designated as the FeGP cofactor.

The discovery of the FeGP cofactor and the reconstitution of the Hmd holoenzyme (Figure 14.3) was a breakthrough in the structural and biochemical analysis of Hmd. Thanks to the reconstitution protocol, we produced the native and variant Hmd holoenzymes with the cofactor extracted from *M. marburgensis*, which enabled their spectroscopic, biochemical, and structural analysis [73]. In the initial work, the Hmd holoenzyme was denatured using high concentrations of guanidinium chloride or urea [71]. Since such salts are difficult to remove from the extracted solution containing the FeGP cofactor, the denaturation step was modified using 60% methanol as the denaturant. In the original extraction method, 2-mercaptoethanol was already added to the extraction solution. It was later found that the thiol leads to dissociation of the -S-Fe- bond between the protein and the FeGP cofactor. In the absence of the thiol compounds, the FeGP cofactor cannot be dissociated from the protein. Ammonia (1%) is added to the extraction solution to form 2-mercaptoethanol deprotonated, supporting its function as a thiolate. After the addition of methanol, ammonia, and 2-mercaptoethanol, a certain incubation time is required. Higher temperatures increase the rate of extraction, but the cofactor becomes unstable. The standard extraction condition is a 15-minute incubation at 40 °C. The denatured protein and the extracted cofactor are separated by ultrafiltration (10 kDa cutoff). The filtrate containing the FeGP cofactor is evaporated and concentrated. The extracted cofactor is used after dilution with an alkaline buffer (e.g. 10 mM ammonium carbonate buffer pH 9.0) containing 1 mM 2-mercaptoethanol.

There are two types of reconstitution methods [73]. The first method is used to measure the activity of the FeGP cofactor or heterologously produced Hmd apoenzyme. The specific activity (U/mg protein) of the apoenzyme can be measured in the assay buffer in a quartz cuvette for spectroscopic analysis containing an excess of the FeGP cofactor compared to the apoenzyme. Conversely, the specific activity of the cofactor (U/mol of the FeGP cofactor) can be determined from the enzyme activity in an assay buffer containing an excess of Hmd apoenzyme compared to the FeGP cofactor. In both cases, the assay components are added to the quartz cuvette one after the other, first the Hmd apoenzyme, then the cofactor, and finally the substrate (i.e. methenyl- or methylene-H_4MPT). The reason for this is because the assay buffer does not contain thiol compounds that would stabilize the cofactor. The concentration of the FeGP cofactor used for reconstitution can be determined by the UV–Vis absorbance of the cofactor solution or by an activity assay after reconstitution. The second method is to produce large amounts of holoenzyme and to mix 0.1–0.4 mM apo-Hmd with a 1.4- to 3-fold excess of FeGP and incubate briefly at room temperature [73, 74].

14.7 Light Sensitivity of Hmd and Finding of the Functional Iron in the Cofactor

The photosensitivity of Hmd was reported in 2004 by Lyon et al. [72]. Highly active Hmd showed an absorbance peak at 360 nm in addition to a shoulder at 300 nm and a smaller peak at 260 nm [73]. The older Hmd preparation method decreased the enzyme activity as well as the height of the 360 nm peak and increased the absorbance at 300 nm, suggesting a correlation between the activity and the chromophore of this enzyme. Exposure to white light inactivated Hmd in a time-dependent manner, decreasing the absorbance at 360 nm and increasing the height of the 300 nm shoulder. Detailed analysis using light sources of different wavelengths showed that UV-A (320–400 nm) and blue light (400–500 nm) irradiation caused inactivation [72]. On the other hand, light with a wavelength of more than 500 nm did not affect the peak absorbance and activity of Hmd. This wavelength-dependent inactivation is similar to the absorbance spectrum of Hmd. These data suggest that UV-A and blue light degrade the chromophore of the Hmd active site. Given the presence of two iron atoms in homodimeric Hmd [26], one would expect the predicted chromophore to contain iron. However, photo-inactivation of Hmd alone did not reduce the iron content of Hmd. The addition of ethylenediaminetetraacetic acid (EDTA) only caused iron ions to be released from the inactivated Hmd. In the absence of light irradiation, EDTA neither inactivated Hmd nor released the iron from the protein, suggesting that the binding affinity of iron to Hmd is weakened by light-induced inactivation. The purified FeGP cofactor showed a similar light sensitivity, indicating that it contains functional iron [72].

14.8 Structure and Properties of the FeGP Cofactor

The discovery of functional iron in the FeGP cofactor motivated the researchers to study the structure and properties of the cofactor in comparison with the prosthetic group of [NiFe]- and [FeFe]-hydrogenases. First, a common inhibitor was tested. Carbon monoxide (CO) and cyanide (CN^-) are known inhibitors of [NiFe]- and [FeFe]-hydrogenases, and, indeed, CO and CN^- inhibited Hmd activity, although the affinity of the inhibitors is much lower than that for the other hydrogenases [72]. The cofactors of both [NiFe]- and [FeFe]-hydrogenases contain CO and CN^- ligands on the iron ion of the dinuclear center [75]. These nonprotein ligands can be detected by infrared (IR) spectroscopy [76]. The IR spectrum of Hmd shows two peaks at 1944 and 2011 cm^{-1}. Incubation of Hmd with CO or CN^- produced a new absorbance peak, which was, based on the wavenumber, interpreted as the binding of an external CO or CN^- ligand. The IR spectrum of the isolated FeGP cofactor showed similar but not identical IR bands at 1972 and 2031 cm^{-1} [76]. Light inactivation of Hmd decreased the intensity of the internal CO bands, indicating that the CO ligands of the FeGP cofactor are dissociated from the iron site by light irradiation. During light decomposition of the FeGP cofactor in the Hmd holoenzyme, an additional IR band is formed, which was interpreted as the binding of dissociated CO to an intact cofactor of another Hmd molecule [76].

Electron paramagnetic resonance (EPR) spectroscopy is a method for characterizing paramagnetic metal-containing active sites in metalloenzymes. [NiFe]- and [FeFe]-hydrogenases have been characterized by this method and many redox states of the catalytic cycle have been elucidated [75]. The iron site of Hmd was revealed to be diamagnetic and does not show an EPR signal. Therefore, Mössbauer spectroscopy was employed as a method to analyze the diamagnetic iron [77]. This method can detect signals in samples enriched with ^{57}Fe, including diamagnetic iron species. To enrich the FeGP cofactor with ^{57}Fe, *M. marburgensis* was cultured in the presence of $^{57}FeCl_2$. Hmd was purified from the cells, and the purified enzyme and the isolated FeGP cofactor were subjected to Mössbauer spectroscopy. Simple twin peaks derived from a single iron species were observed in both Hmd and the cofactor. The data indicated that the iron site was in the Fe(II) or Fe(0) low-spin state. Incubation with CO and CN^- caused a shift of the twin peaks but did not change the redox and spin states of the iron. This indicates that these inhibitors fully occupied the iron site but did not alter its redox properties. Mössbauer data of the isolated FeGP cofactor showed that the redox and spin states of the iron remained largely unchanged by extraction from the protein [77].

Although the FeGP cofactor contains low-spin diamagnetic iron species, a small amount of paramagnetic iron was present in the sample, which may explain why no reasonable nuclear magnetic resonance (NMR) signal for the FeGP cofactor could be detected [78]. On the other hand, NMR spectra were obtained from the FeGP cofactor degraded by light irradiation and purified after EDTA treatment. ^{1}H, ^{13}C, ^{15}N, and ^{31}P NMR data provided the chemical structure of the organic moiety [78]. The atomic composition was further confirmed by high-resolution mass spectrometry and analysis of exchangeable protons in the structure. The chemical structure of the photolyzed structure is shown in Figure 14.4d. The main organic product of photolysis is guanylylpyridinol (GP). Later, GP was found to be a precursor of the biosynthesis of the FeGP cofactor (see Section 14.12).

The binding of the iron site of the FeGP cofactor to a cysteine residue of the Hmd protein was first demonstrated by X-ray absorption spectroscopy [79]. Extended

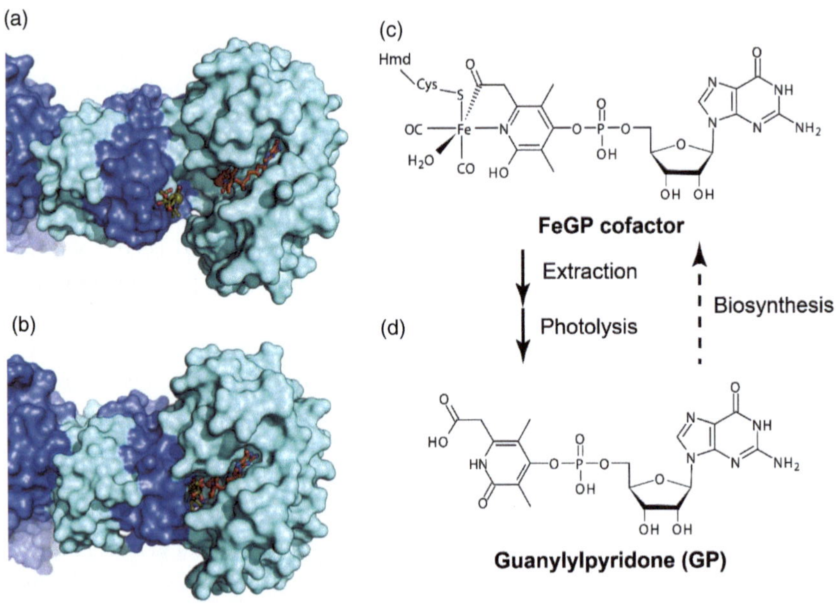

Figure 14.4 Structure of [Fe]-hydrogenase and the FeGP cofactor. (a and b) The open/closed conformations of [Fe]-hydrogenase. Two monomers in the homodimer are depicted in light-green and light-blue models. The C-terminal domains intertwine and form a central domain. The N-terminal domain and the central domain form an active site cleft, in which the FeGP cofactor (a stick model) is bound to the N-terminal domain. Binding of methenyl-H_4MPT (stick model) to the central domain in the open conformation (a) induces the closed conformation (b). (c) The FeGP cofactor is bound to [Fe]-hydrogenase. (d) The FeGP cofactor can be extracted from the holoenzyme and is photolyzed to 6-carboxymethyl-3,4-dimethyl-4-guanylyl-2-pyridone (precursor **3** or GP), which is a precursor of the biosynthesis of the FeGP cofactor.

X-ray absorption fine structure (EXAFS) showed a peak at 2.3 Å away from the iron site. Site-directed exchange of one (Cys176) of the three conserved cysteine residues in Hmd from *M. jannaschii* resulted in reconstituted Hmd with dramatically reduced enzyme activity (1% residual activity), indicating that Cys176 is the protein ligand for the FeGP cofactor. In addition, EXAFS analysis detected at least one light atom (N or C), suggesting the coordination of the organic part of the cofactor to the iron. Although the binding of external CO and CN^- inhibitors to the iron site was detectable, the total number of ligands of the iron could not be determined by EXAFS analysis.

14.9 Photolysis Mechanism

One of the unique features of Hmd and the FeGP cofactor that distinguishes them from the [NiFe]- and [FeFe]-hydrogenases and their dinuclear metal centers is the photosensitivity [72]. Comparing the metal complex structures of the hydrogenases, the FeGP cofactor structurally differs from the other hydrogenases in the presence of a pyridinol ligand whose pyridinol nitrogen and acyl carbon anchor the iron ion. This suggests that the acyl ligand and/or the pyridinol nitrogen ligand are the photosensitive sites of the FeGP cofactor. Experiments indicated that the acyl ligand is dissociated and converted to a carboxylate by light, suggesting that it is the site of photolysis [80] (Figure 14.4c,d).

A time-resolved IR analysis of the photolysis of the FeGP cofactor has been reported [81]. IR bands corresponding to the three $Fe(CO)_3$ species were detected at the beginning of photolysis, which decreased with prolonged light exposure, and the product consisted of GP and decarboxylated GP. Time-resolved IR data and photolysis of model complexes suggest that the acyl ligand is dissociated by light as CO, and the released CO binds to intact FeGP cofactor in the system, forming an acyl-$Fe(CO)_3$ intermediate [81]. Experiments on the *in vitro* biosynthesis of the FeGP cofactor (see Section 14.12) combined with mass spectrometry showed that scrambling of the CO and acyl ligands occurs during photolysis and not during biosynthesis.

14.10 Crystal Structures of Hmd Apoenzymes

The Hmd from *M. marburgensis* is well characterized biochemically, but for a long time it was only crystallized as a twin crystal, which prevented the solution of the crystal structure. The first crystal structure of Hmd was solved using the heterologously produced Hmd apoenzymes from *M. jannaschii* and *M. kandleri* in an apo-form [82]. These homodimeric proteins contain three domains: two N-terminal domains and a central domain. The central domain consists of two C-terminal domains that are intertwined, forming two clefts between the N-terminal and central domains. In the crystal structure of the Hmd apoenzyme from *M. jannaschii*, the clefts are in a closed conformation. In contrast, the cleft of the *M. kandleri* Hmd apoenzyme is open. The presence of two conformations in homologous enzymes suggests that the Hmd protein is flexible in the hinge region of the cleft and that its conformation is fixed by crystal contacts. This finding also suggests that the active site of Hmd is located in this cleft. Such an open-closed conformational change in the active site cleft is known as the induced-fit mechanism.

The structure of the N-terminal domain shows similarity to proteins containing a Rossmann fold, which binds nucleotide derivatives such as NAD(P). Modeling showed that the GP part of the FeGP cofactor fits into the nucleotide-binding site, suggesting an interaction with Hmd via its base and ribose moieties [82]. Furthermore, in this model, Cys176 is located near the pyridinol ring of GP, supporting the hypothesis that the pyridinol ring coordinates the iron ion and the iron complex is coordinated by Cys176, as predicted by mutational and EXAFS analyses [79].

14.10.1 Crystal Structure of Reconstituted Hmd Holoenzymes

The first crystal structure of Hmd holoenzyme was determined using the reconstituted enzyme of Hmd from *M. jannaschii* [74]. The domain structure was identical to that of the apoenzyme, but the structure of the apo-Hmd was in a closed cleft conformation, whereas the structure of the Hmd holoenzyme from *M. jannaschii* was in an open cleft conformation. Cys176 is bound to the iron ion and anchors the cofactor to the protein. The model for the FeGP cofactor was constructed based on spectroscopic data and information from the chemical structure of the light-decomposed product (GP) (Figure 14.4d). Two nonprotein ligands of the iron were identified, which were interpreted as two CO ligands, and only the pyridinol nitrogen was bound to the iron.

The initial model of the holoenzyme structure of Hmd was later corrected based on the crystal structure of the Cys176Ala variant of reconstituted Hmd of *M. jannaschii* [83]. Cys176 binds to the iron moiety of the FeGP cofactor. In the Cys176Ala variant of Hmd reconstituted with the FeGP cofactor, dithiothreitol, which was added to the apoenzyme sample, binds to the iron moiety via its thiolate and hydroxyl groups. The bulky dithiothreitol molecule occupied the space of the 6-carboxymethyl group [83], which was assumed to rotate in the initial model structure [74]. This result indicated that the 6-carboxymethyl group cannot be modeled at the originally proposed position [74]. To resolve this discrepancy, the pyridinol moiety of the initial model was rotated by 180°. In the new model, the 6-carboxymethyl group was replaced by a 6-acylmethyl group and the acyl carbon was ligated to the iron to match the electron density [83]. The presence of an acyl group was later supported by experiments showing that the 6-carboxymethyl group is the product of degradation of the acyl ligand [80]. The compatibility of the proposed Hmd structure with an acyl ligand at the iron has been confirmed by EXAFS analysis. Thus, the presence of an acyl ligand in the FeGP cofactor was proposed. Rearrangement of the ligand structure of the FeGP cofactor by the addition of the acyl ligand resulted in the remodeling of the binding site of the CO ligand as shown in Figure 14.4c [83].

14.10.2 Crystal Structure of Reconstituted Hmd with the Substrate

Hiromoto et al. reported the crystal structure of the open conformation of the reconstituted C176A variant of *M. jannaschii* Hmd in complex with methylene-H_4MPT

[84]. In this structure, methylene-H_4MPT binds to the central domain, and the distance between the substrate C14a and the iron of the FeGP cofactor bound to the N-terminal domain was 9 Å, too large for direct hydride transfer. Based on this finding and the presence of the open/closed structure of the Hmd protein, an open-closed conformational change during catalysis was proposed.

The presence of the acyl ligand was finally confirmed by the high-resolution crystal structure of the reconstituted Hmd holoenzyme from *Methanococcus aeolicus* [85]. The 1.06 Å crystal structure clearly indicated the presence of the acyl ligand and the geometry of the FeGP cofactor in Hmd (Figures 14.4a). The conformation of the iron complex of the FeGP cofactor was very similar to that of a model compound containing the same iron species and ligands [86]. Binding of methenyl-H_4MPT^+ induced the closed conformation of Hmd, in which the solvent ligand at the iron site near C14a of methenyl-H_4MPT^+ dissociates. This result suggests that H_2 is bound to this open iron site after solvent dissociation. This hypothesis was further supported by the fact that the possible H_2-binding site is connected to the bulk solvent by a narrow channel in the protein. Only in the presence of methenyl-H_4MPT^+, Hmd adopt the closed conformation in which the solvent molecule was expected to dissociate and thus enable H_2 binding [85]. This activation mechanism explains the above-described dependence of the exchange reaction on methenyl-H_4MPT^+.

14.10.3 The Other Crystal Structures of Hmd and its Homologs

As mentioned earlier, crystallographic analysis of native Hmd from *M. marburgensis* was hampered by crystal twinning [87]. We have found the crystallization condition for a single crystal of this enzyme. The corresponding crystal structure revealed a homodimer similar to the Hmd structures from *M. jannaschii* and *M. kandleri*. However, three homodimers of *M. marburgensis* Hmd are linked together to form a hexameric structure [88]. A particularly important difference is that the C-terminus of one dimer unfolds from the protein core and enters the neighboring dimer, where the carboxy-terminal oxygen is bound to the open iron site. This structure suggests that *M. marburgensis* Hmd forms such an intertwined hexamer in the absence of methenyl-H_4MPT^+ and that the active site iron is protected from oxidative stress and photolysis. These hypotheses were supported by biochemical experiments [88]. The hexameric form of *M. marburgensis* Hmd may also be one of the reasons for the aforementioned crystal twinning. Cocrystallization of the protein with the inhibitor isocyanide yielded a standard homodimer in which the site trans to the acyl ligand is occupied by an isocyanide carbon [87].

Many hydrogenotrophic methanogens have two homologs of Hmd (HmdII and HmdIII), and the crystal structure of a reconstituted HmdII holoenzyme from *M. jannaschii* has been reported [89]. HmdII, like Hmd, forms a homodimer with the central domain consisting of an intertwined C-terminal domain. The N-terminal domain is slightly smaller than that of Hmd but binds the FeGP cofactor in the same way. HmdII shows very low catalytic activity and a sensory rather than a catalytic function has been proposed [59, 89]. At the C-terminal end of HmdII, there is an extension of helices, which, as proposed by Oza et al., could be a site of interaction with nucleic acids [90]. HmdII from a sulfur-reducing bacterium *Desulfurobacterium thermolithotrophum* uses methenyl-H_4F^+ instead of methenyl-H_4MPT^+ as the substrate. It might be interesting to analyze the binding mode of methenyl-H_4F^+ by mutational analysis to create an archaeal Hmd that uses an H_4F derivative instead of the H_4MPT derivative as the substrate. This could be of interest for biotechnological applications to enable Hmd catalysis in bacteria or eukaryotes.

Another crystal structure of Hmd was obtained from heterologously produced enzymes from *Methanolacinia paynteri* [91]. The reconstituted holoenzyme was crystallized as an asymmetric homodimer, with one monomer in an open conformation with the FeGP cofactor and the other in a closed conformation without cofactor. Thus, binding of FeGP cofactor resulted in an open conformation in one of the homodimers. On the other hand, cocrystallization of *M. paynteri* Hmd with GMP and GP, which is part of the FeGP cofactor, did not induce an open conformation. This indicates that the binding of the iron complex of the FeGP cofactor to the cysteine thiolate is required to induce an open conformation. The active site lysine residue interacts with the 2-hydroxy group of the pyridinol ring of GP, but not with the FeGP cofactor of the holoenzyme. This finding suggested that this lysine residue may be involved in the binding of the FeGP cofactor. This hypothesis was confirmed by analyzing the binding kinetics of the FeGP cofactor to the protein, using a variant in which the lysine residue was replaced by alanine [91].

14.11 ESI-MS Analysis for Detection of the CO/Acyl Ligands

The presence of acyl ligands was first proposed based on modeling of the iron complex structure into the electron density of the FeGP cofactor bound to the Cys176Ala variant of Hmd [83] and verified by a high-resolution crystal structure of the Hmd holoenzyme [85]. The presence of the acyl ligand in the FeGP cofactor was also demonstrated by biophysical analysis, including mass spectrometry and IR spectroscopy as well as by the FeGP photolysis products [80]. Fourier transform ion cyclotron resonance mass spectrometry (ESI-FT-ICR-MS) revealed two intense peaks (695.0429 and 635.0221 m/z) using the FeGP cofactor extracted with acetic acid in the negative mode. The larger mass peak was assigned to the deprotonated FeGP cofactor in complex with acetate. The smaller mass peak corresponds to the FeGP cofactor without acetate ligand. The estimated elemental composition corresponds to that of the FeGP cofactor with the acyl ligand ($C_{21}H_{20}N_6O_{12}P_1Fe_1$). The calculated mass was 635.0232 m/z. MS/MS analysis of the 695 m/z species yielded three peaks (635.0237, 607.0288, and 551.0389 m/z) with mass differences between these peaks of 27.9949 and 55.9899, corresponding to the dissociation of one and two CO units, respectively. This MS/MS analysis indicated that the FeGP cofactor contains three CO units, which is consistent with the presence of two CO and one acyl ligand, since the mass difference caused by the dissociation of CO and acyl is the same.

Chemical experiments showed that 6-carboxylated GP is formed spontaneously by photolysis of the 6-acylmethyl substituent of the FeGP cofactor. The photolysis reaction in the presence of [^{18}O]-H_2O enriched the carboxy group of GP with ^{18}O, indicating that photolysis hydrolyzes the acyl group to form a carboxy group [80]. Furthermore, when GP was incubated in the presence of [^{18}O]-H_2O under illumination, GP was not labeled with ^{18}O. These experiments indicated that the acyl group of the FeGP cofactor was converted to a carboxyl group by photolysis. In addition, photolysis of the acyl ligand of the FeGP in the presence of methanol and ethanol yielded the corresponding methyl and ethyl esters of GP, which also provided evidence for the reactivity of the acyl ligand [80]. In the same report, the presence of the acyl ligand was demonstrated by IR spectroscopy.

14.12 Biosynthesis of the FeGP Cofactor

14.12.1 Stable Isotope Labeling Experiments

To analyze the reactions occurring during the biosynthesis of the FeGP cofactor, stable-isotope labeling experiments have been performed [92]. Hydrogenotrophic methanogens were grown in the presence of ^{13}C- or ^{2}H-labeled compounds. To verify the incorporation of the isotopes into the FeGP cofactor, ^{13}C- and ^{2}H-NMR, matrix-assisted laser desorption/ionization time-of-flight mass spectrometry (MALDI-TOF-MS), ESI-FT-ICR-MS, and IR spectroscopy have been used. Single and two-dimensional NMR spectroscopic analyses can determine the position of the incorporated ^{13}C and ^{2}H in the chemical structure. Mass spectrometric analysis reveals the number of labeled atoms in the compound. The number of atoms incorporated into the compound can be determined from the increase in the mass. In addition, fragmentation experiments using MS–MS analysis provided information about the position of the isotopes in the compounds. The incorporation of ^{13}C into the CO ligands of the FeGP cofactor can be verified using IR spectroscopy.

The intact FeGP cofactor can be analyzed by ESI-FT-ICR-MS, whereby the FeGP cofactor is detected as the form without the stabilizing reagents used for extraction (i.e. dithiothreitol, 2-mercaptoethanol, and acetate). The organic part (GP) decomposed by light and the decarboxylated form of GP were detected using MALDI-TOF-MS. The difference in the number of labeled carbon atoms between the FeGP cofactor and GP indicates the number of labeled CO ligands, and the difference between GP and the decarboxylated GP indicates the isotope labeling of carbon in the carboxy group of GP. The carboxy carbon of GP originates from the acyl carbon of the FeGP cofactor [80].

Two methanogenic archaea were used for the stable isotope labeling experiments. *M. marburgensis* grows autotrophically under an H_2 and CO_2 atmosphere in a fermenter, yielding ~80 g wet cell from a 10 L culture [73]. Under nickel-limiting conditions, the production of Hmd is upregulated sevenfold in this methanogen, with the proportion of Hmd protein in the total cellular protein being about 10% [65, 72, 73]. Thus, the yield of Hmd from cells obtained by fermenter cultures is high enough to purify micromolar amounts of the FeGP cofactor. Furthermore, *M. marburgensis* cells can take up acetate, pyruvate, and methionine, but only to a limited extent, so the degree of labeling is low [92]. These properties of *M. marburgensis* are particularly well suited for NMR analysis to detect low levels of ^{13}C and ^{2}H in the cofactor. *Methanobrevibacter smithii* is an acetate-dependent methanogen that lacks the nickel-dependent CO dehydrogenase. This methanogen uses acetate as the carbon source [93]. In addition, *M. smithii* can take up pyruvate, glycine, propionate, L-methionine, succinate, and formate, which is an advantage for isotopic analyses compared to *M. marburgensis*. However, as the growth yield for *M. smithii* is poor, the amounts of the labeled FeGP cofactor isolated from this organism are not sufficient for NMR analysis. Therefore, *M. marburgensis* FeGP samples were used for stable isotope labeling analysis using NMR, while *M. smithii* samples were analyzed by mass spectrometry [92].

^{13}C-NMR analyses of the FeGP cofactor isolated from *M. marburgensis* indicated that labeled atoms derived from [1–^{13}C]acetate, [2–^{13}C]acetate, and [1–^{13}C]pyruvate were incorporated into the FeGP cofactor, as indicated in the labeling positions of the carbon atoms shown in Figure 14.5a. Mass spectrometric analysis confirmed these results. The methyl group of the C3 position of the pyridinol ring is labeled by the methyl group of methionine, which suggests that the methyl group at C3 is derived from the *S*-adenosylmethionine (SAM)-dependent methyltransferase reaction. The C2 of the pyridinol is labeled with the C1 of acetate, which is not conjugated with a carbon atom derived from C2 of the acetate, suggesting that one acetate unit incorporated into the pyridinol ring is cleaved off during biosynthesis. The labeling pattern of the pyridinol ring is identical to the ribose moiety of the FeGP cofactor, suggesting that the precursor of the pyridinol ring can be a derivative of ribose. Schick et al. proposed 2,3-dihydroxy-4-oxo-pentanoate as a precursor for pyridinol biosynthesis, in which the part containing 6-carboxymethyl, C6, and N1 is synthesized from aspartate or β-alanine. The labeling pattern of the GMP moiety of the FeGP cofactor is in accordance with the canonical biosynthesis of methanogenic archaea. The mass spectrometric analysis of the FeGP cofactor isolated from *M. smithii* grown in the presence of [1–^{13}C]glycine, [1–^{13}C]propionate, L-[methyl-^{2}H]methionine, and [2,3-$^{13}C_2$]succinate indicated that these compounds are neither the precursors of the organic part (GP) nor of the iron complex part of the FeGP cofactor. The only exception is [1–^{13}C]formate, which is incorporated into one position in the guanine moiety as predicted by the canonical pathway.

Mass spectrometric analysis indicated that the two iron-bound CO ligands are derived from CO_2 and not via any labeled organic compounds tested, including C1 or C2 of acetate or pyruvate, which excluded the possibility that tyrosine (through dehydroglycine) is the CO precursor, as has been reported for CO ligand biosynthesis in H-cluster of [FeFe]-hydrogenase [94]. Although the CO ligand of [NiFe]-hydrogenase from aerobic H_2 oxidizing bacteria is derived from N^{10}-formyl-H_4F [95, 96], which is analogous to N^5-formyl-H_4MPT, the enzyme responsible for CO biosynthesis (HypX) is not present in methanogenic archaea. Upon incubation of *M. smithii* with ^{13}CO, the molecular mass of the isolated FeGP was increased by (+3), indicating that two CO and one acyl ligands are enriched

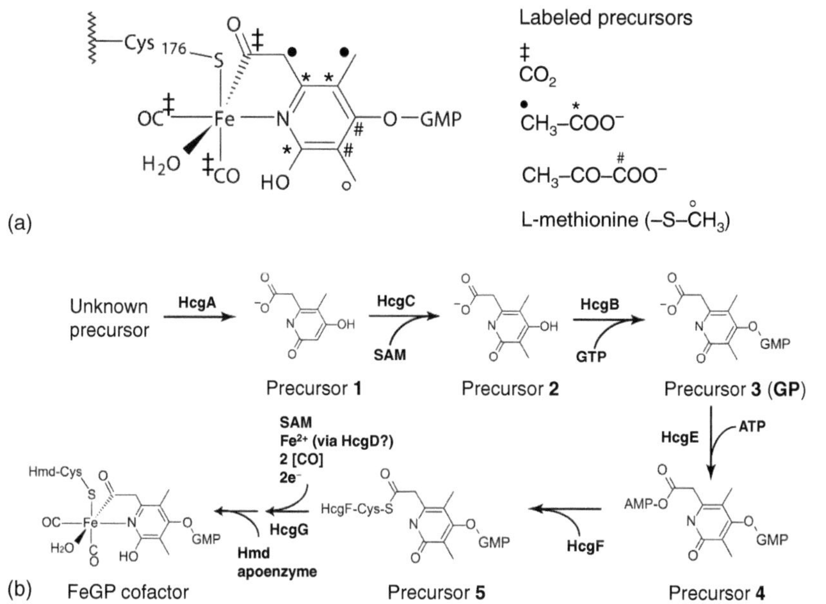

Figure 14.5 Biosynthesis of the FeGP cofactor. (a) Stable-isotope labeling pattern in the FeGP cofactor. Stable isotope-enriched carbon atoms are marked. (b) Sequence of the reactions involved in the biosynthesis of the FeGP cofactor.

with [^{13}C]-CO, suggesting that these iron ligands can be formed from external CO gas. However, the enrichment of ^{13}C in GP (76%) was much higher than the enrichment of the third (+3) peak (20%) of the FeGP cofactor. The authors of this study discussed the possibility that the CO and acyl ligands of the cofactor are scrambled during biosynthesis, as has been observed in model compounds of the FeGP cofactor [86]. However, we have recently found that the CO and acyl ligands are scrambled only during photolysis of the FeGP cofactor, not during biosynthesis [81]. This finding indicates that the higher enrichment in GP is a result of scrambling in the acyl and CO ligands in the FeGP cofactor upon photolysis, while biosynthetic incorporation of CO is minor, which can be synthesized from salvage biosynthesis, using GP obtained from decomposed ^{13}C-labeled FeGP cofactor.

14.12.2 The *hcg* Gene Cluster, Sequence Similarities, and Mutational Analysis

Genome analysis has shown that Hmd is present in a wider range of microorganisms including methanogens belonging to *Methanomicrobiales*, in which *Methanocorpusculum labreanum* is the first *Methanomicrobiales* methanogen found that contains the *hmd* gene in its genome [97]. Phylogenetic analysis indicated that Hmd from *M. labreanum* is similar to the Hmd proteins within *Methanobacteriales*, which suggested that the *hmd* gene from *M. labreanum* was horizontally transferred from a *Methanobacteriales* methanogen [98]. The neighboring seven genes in the *hmd*

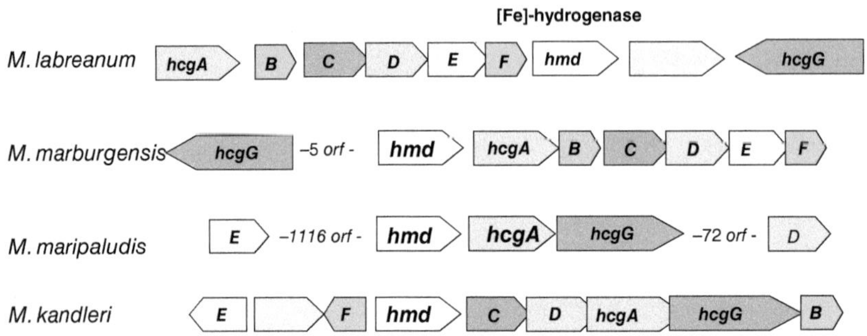

Figure 14.6 Gene clusters of the *hmd* co-occurring (*hcg*) genes in methanogenic archaea. Source: Adapted from [2].

gene cluster of *M. labreanum* are also conserved in the *hmd* region of the genomes of *Methanobacteriales*, and these seven genes were classified into the same phylogenetic position. Based on this finding, it was proposed that the seven genes (*hcgA–G*) that co-occur with *hmd* are related to the function of Hmd, i.e. they are FeGP cofactor biosynthesis genes, and may have been horizontally transferred together. The seven *hcg* genes also occur as a gene cluster in *Methanopyrales*. In *Methanococcales* methanogens, however, only *hcgA* and *hcgG* are clustered with *hmd*, while the other five *hcg* genes are scattered throughout the genome (Figure 14.6).

A sequence similarity search was carried out to predict the function of the *hcg* genes [2]. HcgA is homologous to BioB, a radical SAM enzyme catalyzing biotin biosynthesis. HcgB is similar to nucleotide triphosphatases, which cleave the pyrophosphate unit from nucleoside triphosphate. HcgC shows no similarity to any proteins. HcgD is similar to the protein sequences of the Nif3-like protein family with unknown function. HcgE is similar to MoaE, an enzyme involved in molybdopterin biosynthesis. HcgF has no similarity to known proteins. HcgG has very low similarity to fibrillarin-like proteins. These similarity data did not correlate with the function of the Hcg proteins in the biosynthesis of the FeGP cofactor.

First evidence for the function of the *hcg* genes was provided by mutagenesis experiments with *M. maripaludis* [99]. The effect of the *hcg* gene deletion was evaluated by measuring the growth characteristics of the mutants. Deletion of the *hmd* gene in *M. maripaludis* mutants (Δ*fru*Δ*frc*), lacking the F_{420}-reducing hydrogenase, resulted in a long lag time in growth under H_2 and CO_2. This lag time was used as physiological evidence for the absence of active Hmd in *M. maripaludis*. Each *hcg* gene deletion resulted in a similar lag time as for the Δ*hmd* mutant. The lag time was not observed in any of the mutants grown with formate, as the function of Hmd was replaced by F_{420}-dependent formate dehydrogenase. These experiments provided indirect phenotypic evidence that the *hcg* genes are involved in the biosynthesis of the FeGP cofactor. Later, Fujishiro et al. reported that *M. mariparudis* strains lacking *hcgB* or *hcgC* do not exhibit Hmd activity [100].

14.12.3 Structure to Function Analysis of Hcg Proteins

The function of HcgB-F proteins was investigated by structure–function analysis. In this method, the crystal structures of the Hcg proteins were solved and then the three-dimensional structures were compared with those of other enzymes to identify similarities. Based on the function of the structurally similar protein, the function of the respective Hcg protein was predicted, and the predicted function was confirmed by enzyme activity assay and cocrystal structure analysis of the Hcg protein with the predicted substrate. An overview of the sequential biosynthesis of the FeGP cofactor is shown in Figure 14.5b.

14.12.3.1 HcgA

HcgA shows sequence similarity to radical SAM enzymes such as the biotin biosynthesis enzyme BioB and the [FeFe]-hydrogenase maturation enzymes HydE and HydG [2]. The primary structure of HcgA contains a CX_5CX_2C motif that is slightly different from the canonical CX_3CX_2C motif, suggesting that HcgA carries a modified [4Fe–4S]-cluster binding structure [101]. HcgA from *M. maripaludis* was heterologously produced in *E. coli*. The purified enzyme exhibited a redox-dependent EPR signal at low temperature (12 K), and the EPR signal of reduced HcgA decreased upon the addition of SAM to the sample, indicating the presence of a [4Fe–4S] cluster in the vicinity of the SAM binding site. HcgA catalyzes the formation of 5′-deoxyadenosine, a common reaction product of a radical SAM enzyme [101]. The UV–Vis spectrum of this enzyme revealed the presence of a chromophore with an absorbance peak at 410 nm, which decreases upon reduction [102]. These properties suggest that HcgA catalyzes a radical SAM reaction. Despite extensive crystallization attempts, no crystal HcgA has been obtained. The function of HcgA has been indicated by *in vitro* biosynthesis experiments (see Section 14.12.4.4).

14.12.3.2 HcgB

The crystal structure of HcgB shows similarity to that of nucleoside triphosphatase, as predicted by sequence comparison. This enzyme cleaves the phosphodiester bond of the substrate nucleoside triphosphate (e.g. GTP) to produce nucleoside monophosphate (e.g. GMP) and pyrophosphate. If the GMP moiety of the FeGP cofactor is produced from GTP, the same reaction is required. Based on this idea, we hypothesized that HcgB catalyzes the cleavage of GTP to produce GMP and conjugation with a pyridinol precursor. To test this hypothesis, reactions were tested using commercially available pyridinol compounds and GTP [103]. In fact, HcgB catalyzed the production of guanylylpyridinol using various 2-pyridinol compounds that carry a 4-hydroxyl group on the pyridinol ring. The chemical structure of the product guanylylpyridinol was confirmed by NMR and mass spectrometry.

To study the HcgB binding site of guanylylpyridinol, an attempt was made to determine the crystal structure of the complex with the artificial guanylylpyridinol generated by the reaction of HcgB with mimetic pyridinols [103]. The cocrystal structure showed that the GMP part was bound to the corresponding position as in nucleoside triphosphatase. The pyridinol head was fixed by hydrogen bonding to the pyridinol nitrogen and its 2-hydroxy group, indicating that HcgB is specific for 2-pyridinol and that the 4-hydroxy group is required for conjugation with GMP. A water molecule occupied the 6-position of the pyridinol ring, suggesting that there is a space for a relatively large group at the 6-position of the pyridinol group and a small space at the 5-position. The photolysis product (GP) from the FeGP cofactor has a carboxymethyl group in the 6-position and a methyl group in the 5-position of the pyridinol moiety [78]. A cocrystallization study of HcgB with GP showed that the 6-carboxymethyl group was located in the empty space of HcgB (Figure 14.7a), and the carboxy oxygen formed a hydrogen bond with a serine residue (Ser132). The space at the 5-position of the pyridinol ring was filled with the methyl group of the pyridinol ring [103]. This specific interaction of the 6-carboxy group with Ser132 suggested that GP might be the physiological substrate of HcgB, which means that GP is an intermediate of FeGP cofactor biosynthesis, although GP was identified as a degradation product through illumination of the FeGP cofactor. Later, the crystal structure of HcgB from *M. mariparudis* was crystallized in complex with 6-carboxymethyl-3,5-dimethyl-4-hydroxy-2-pyridinol, which has the same structure as GP but lacks the GMP moiety [104]. The structure of the complex showed that the pyridinol compound binds to the HcgB active site at the same position as observed for the structure in complex with GP. This result strengthens the hypothesis that GP is an intermediate of FeGP cofactor biosynthesis.

14.12.3.3 HcgC

The crystal structure of HcgC from *M. jannaschii* showed similarity to SAM-dependent methyltransferases, suggesting that HcgC catalyzes a methyl transfer reaction during the biosynthesis of the FeGP cofactor. The structure of HcgC in complex with SAM or *S*-adenosylhomocysteine (SAH) was determined at a resolution of 1.7 Å [100]. This analysis revealed that HcgC is a tetramer consisting of two homodimers. As indicated by the stable isotope labeling experiments described earlier, the methyl group at the 3-position of pyridinol is synthesized from the methyl group of methionine by an SAM-dependent methyltransferase reaction [92]. The structure–function analysis of HcgB indicated that GP is an intermediate of FeGP cofactor biosynthesis [103]. On this basis, the structure of HcgC strongly suggests that this enzyme catalyzes the methyl transfer reaction of SAM to form a 3-methylpyridinol derivative.

There are two possible methyl-accepting substrates: a guanylylpyridinol and a pyridinol without the 3-methyl group. In a cocrystal structure, the binding mode of SAH was analyzed [100]. Inspection of the active site pocket revealed that the bulky guanylylpyridinol cannot be accommodated in the active site. Therefore, the substrate could be a pyridinol derivative. The binding position of the predicted pyridinol precursor was analyzed by a docking experiment, which indicated that the precursor can bind to the active site pocket near the SAH bound to the adjacent position. This docking simulation result supports the hypothesized enzyme activity. To confirm the hypothesis, the pyridinol precursor was chemically synthesized and used for an enzyme assay of the HcgC reaction [100]. Detection of the 3-methylated product of the HcgC reaction was performed by MALDI-TOF-MS and HPLC. The assay data indicated that HcgC catalyzes the methyl transfer to the synthesized pyridinol compound. NMR analysis confirmed the chemical structure of the 3-methylated pyridinol product. Thus, the enzymatic reaction of HcgC was confirmed.

Later, HcgC from *M. maripaludis* was crystallized in complex with the pyridinol precursor **1** and SAH, and the 1.7 Å crystal structure was solved (Figure 14.7b) [106].

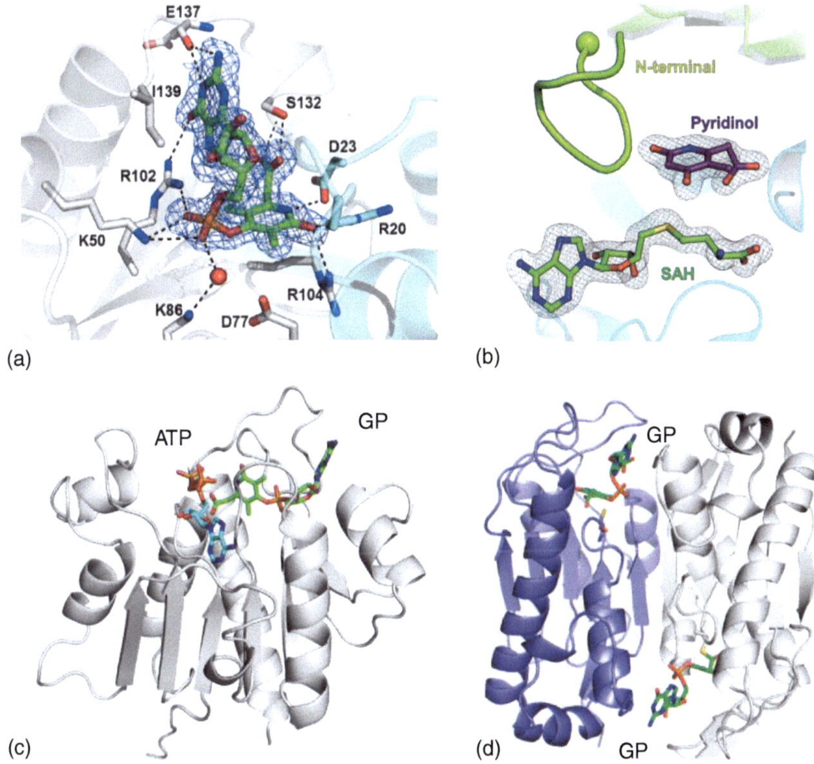

Figure 14.7 Structure to function analysis of Hcg proteins. (a) Cartoon model of the active site structure of HcgB from *M. jannaschii* in complex with GP. GP is shown by a stick model with electron density. (b) The active site structure of HcgC from *M. maripaludis* in complex with *S*-adenosylhomocysteine (SAH) and pyridinol precursor **1** (pyridinol), which is shown by a stick model with electron density. (c) Cartoon model of the structure of HcgE from *M. marburgensis* in complex with ATP and GP. (d) Cartoon model of the structure of HcgF from *M. jannaschii* in complex with GP; two monomers are shown in blue and white cartoon models. Source: Adapted from [103–105].

The protein structure was identical to the tetrameric HcgC structure from *M. jannaschii*. In the *M. maripaludis* HcgC structure, all four monomers bind SAH and two of the four monomers contain an electron density corresponding to the pyridinol substrate. For the other two monomers, linear compounds were observed in the active site, while only SAH was found in one active site. It was suspected that the linear compound was a degradation product of the pyridinol substrate by a side reaction of HcgC or a spontaneous reaction. This interpretation was supported by the fact that rapid crystallization (two days) or soaking of the HcgC crystals with SAH and the pyridinol substrate did not reveal the presence of the linear compound in the crystal structure [106].

The pyridinol binding region of HcgC was substantiated by substrate binding. The pyridinol substrate (precursor **1**) binds to the active site through Ile115 of one monomer, Ile5 and Val9 of another monomer, and SAH, where the distance

(4.2 Å) between the C3 of pyridinol and the sulfur atom of SAH is appropriate for the methyl transfer reaction [106]. The pyridinol substrate is primarily fixed by hydrogen bonding of the 6-carboxymethyl group to the main-chain nitrogen and side-chain OH groups of Thr179. In addition, the carboxy groups of SAH, Met178, and Tyr51 each bridge one water molecule to the 6-carboxymethyl group of the substrate. The pyridinol ring is further linked to the protein by three water molecules attached to hydrophilic residues. The pyridinol nitrogen is stabilized by a water molecule attached to Ser233 and Glu209. The 2-OH group of pyridinol is bridged by two water molecules attached to Ile5, Ile6, and Glu209. The residues linked to the water molecules are completely conserved in the HcgC proteins [106]. Thus, the crystal structure of HcgC with SAH and the pyridinol compound indicated that the direct binding of the side chain of Thr179 and the interactions with the water molecules bridging the pyridinol substrate to the protein are important for the binding and the catalysis of HcgC. The exchange of Thr179 for valine and Glu209 for glutamine led to the complete inactivation of the enzyme. In addition, the replacement of Thr6, Tyr51, and Ser233 drastically reduced the catalytic efficiency. These results of the mutation analysis supported the importance of the direct interactions of the 6-carboxymethyl group with Thr179 and the interactions with the water molecules [106].

Based on the crystal structure of HcgC with SAH and precursor **1**, and the site-directed amino acid exchanges, a catalytic mechanism is proposed, in which the water molecules that link the pyridinol precursor to the protein play an important role in localizing an electron pair on C3 by a resonance effect [106]. Water molecules appear to be ideally suited to maximize the probability of electron density at C3 of the pyridinol substrate by maintaining a delicate energetic balance between deprotonation/protonation and keto/enolate intermediate structures. This study clarified why the methyl-transfer reaction takes place at the 4-hydroxy-2-pyridinol stage (precursor **1**) and not at the 4-guanylyl-2-pyridinol stage. This is because the latter is energetically more difficult for the reaction due to the smaller number of resonance structures.

14.12.3.4 HcgD

HcgD is a member of the Nif3 (NGG1p-interacting factor 3)-like protein family, based on the discovery of a member of this protein family that was first isolated with the yeast dual regulator protein NGG1p [107]. Mouse Nif3-like proteins may be involved in transcriptional regulation and neuronal differentiation, although the exact physiological functions of this protein family are not known. Crystal structures of Nif3-like proteins suggest that the binuclear metal-binding sites are located in interdomain clefts [108]. In humans and mice, Nif3 family proteins are implicated in neuronal differentiation. It has been reported that neuronal differentiation in the developing hippocampus is impaired by perinatal iron deficiency [109], which may indirectly indicate the relationship between the Nif3 protein family and iron.

HcgD from *M. jannaschii* was heterologously produced in *E. coli*. The purified HcgD shows a slight yellow color under air, suggesting the presence of a chromophore attached to the protein [108]. The UV–visible spectrum of purified HcgD in the oxidized form showed broad shoulder peaks at 320 and 360 nm in addition to the protein absorption peak at 280 nm. A colorimetric assay revealed the presence of approximately two iron ions in an HcgD monomer. An anomalous iron absorption edge was observed in the HcgD crystal by X-ray fluorescence spectroscopy. Although Nif3-like proteins may also contain zinc in the dinuclear metal site, no zinc was detected. The UV–visible spectrum of HcgD changed upon reduction with dithionite, suggesting that the chromophore of HcgD is redox active [108].

Crystal structures of oxidized and dithionite-reduced HcgD were solved [108]. The structure was similar to that of the other Nif3-like proteins, and the protein forms a trimer of homodimers, resulting in a hexameric toroidal structure. The binuclear iron binding site is located at the bottom of a wide cleft formed between two domains of each HcgD monomer, and the binuclear iron site is located within the toroid. In the oxidized HcgD structure, one iron site is occupied about half as strongly as the other (Fe1/Fe2 = 0.48/1). In contrast, the occupancy of the iron sites in reduced HcgD is almost the same (Fe1/Fe2 = 0.85/1). The protein ligands bound to the iron ions are fully conserved in the HcgD proteins. Nonprotein ligands (chloride and phosphate) were modeled to Fe1 of the dinuclear iron site of oxidized and reduced HcgD. These nonprotein ligands interact with a positively charged histidine residue, consistent with the anionic nature of the ligands. A negatively charged nonprotein ligand has not been observed in other Nif3-like protein structures. The binding property of the dinuclear iron site is interesting because the possible biosynthesis intermediate GP contains a negatively charged 6-carboxymethyl group, which could be the site of insertion of an iron ion into the guanylypyridinol precursor **3** (GP). However, cocrystallization of HcgD with GP failed [108].

Addition of a chelating reagent, citrate or EDTA, decreased the electron density of an iron site (Fe1) where citrate is bound to the other iron (Fe2) [108]. UV–visible spectroscopy and iron determination confirmed the decrease in iron content of the HcgD preparation treated with chelating reagents and subsequently washed by ultrafiltration. These results indicated that the binding affinity of one of the iron ions (Fe1) of the dinuclear iron site is lower than that of the other iron ion (Fe2) and that the affinity of iron in the oxidized state is lower than that in the reduced state. A plausible iron transport function for HcgD has been proposed, based on the chelate-dependent detachability of an iron ion at the Fe1 site and the binding affinity of the negatively charged ions for the other iron (Fe2). Thus, HcgD may play a role as an iron chaperone in the biosynthetic sequence of the FeGP cofactor [108].

14.12.3.5 HcgE

HcgE from *M. marburgensis* was heterologously produced in *E. coli*, and the purified protein was subjected to crystallization. As already determined in the primary structure comparison, the crystal structure of HcgE at 1.6 Å resolution showed similarity to E1-like enzymes, which catalyze the adenylylation of the carboxy group of ubiquitin, the C-terminal carboxy group of other proteins and other compounds with ATP [105]. E1-like enzymes are used in the ubiquitin system and in the biosynthesis of molybdopterin, thiamine, and antibiotics [110]. The conserved N-terminal domain of E1-like enzymes contains the binding site for ATP, and the C-terminal domain of each E1-like enzyme is responsible for the binding of the specific substrate. The structure of the N-terminal domain of HcgE is very similar to that of the E1-like enzymes, and ATP binds at the predicted binding site. On the other hand, the structure of the C-terminal domain was different from the others, suggesting that the C-terminal domain of HcgE may bind a different substrate than other E1-like enzymes.

Based on the possible reaction sequence of the FeGP cofactor and the predicted reaction catalyzed by HcgE, the hypothesis was put forward that GP could be the substrate of the HcgE reaction, in which the 6-carboxymethyl group of GP could be adenylylated. To test this hypothesis, HcgE was crystallized with ATP and GP in the absence of divalent cations in the crystallization solution to prevent the reaction (Figure 14.7c) [105]. The crystal structure of HcgE–ATP–GP at 2.8 Å resolution revealed the binding of GP to the C-terminal domain, as predicted. GP binding was not obtained in the absence of ATP. The position of the carboxy group of GP corresponds to that of the carboxy group of the C-terminal amino acid residue of MoaD on MoeB, which are involved in molybdopterin biosynthesis. The structure of the HcgE–ATP–GP complex indicated that HcgE catalyzes the adenylylation of the 6-carboxymethyl group of GP by ATP to form adenylated GP (AMP-GP or precursor **4**) [105].

The predicted reaction of HcgE was tested by detecting the production of AMP-GP (precursor **4**) and pyrophosphate by the reaction of HcgE with ATP, GP, and Mg^{2+} [105]. The product was analyzed by mass spectrometry, and MS–MS analysis of the mass peak of AMP-GP, which confirmed that the molecule is composed of AMP and GP moieties. The production of pyrophosphate was also detected, and the reaction rate increased with the amount of HcgE added to the assay. A catalytic mechanism was proposed based on the HcgE–ATP–GP complex structure, in which the carboxyl group of GP is located near the α-phosphate of ATP as in a transition state. The carboxyl group of GP attacks the α-phosphate of ATP in an S_N2 reaction, in which AMP-GP could be formed via a pentavalent intermediate [105].

14.12.3.6 HcgF

Heterologously produced HcgF from *M. jannaschii* was purified and crystallized, and the crystal structure was solved at 2.0 Å resolution [105]. The homodimeric HcgF structure was similar to that of the nicotinamide mononucleotide (NMN) deaminase PncC [111], although HcgF did not show NMN deaminase activity. Based on the similarity, it was hypothesized that HcgF may bind a similar substrate with a ring structure at the possible active site cleft. Based on the possible reaction sequence, GP was suggested as a possible substrate. The crystal structure of HcgF in complex with GP showed that GP binds the active site cleft of both monomers of the homodimer (Figure 14.7d) [105]. GP, which was bound to one of the two HcgF monomers, formed a covalent thioester bond between its carboxyl carbon and the thiolate of Cys9 of HcgF in half of the molecules, with the second half not being covalently bound.

Based on the standard reactions involving the E1-like enzymes and their substrates, it was hypothesized that AMP-GP formed by the HcgE reaction could react with HcgF to form a thioester bond at the carboxy group [105]. This hypothesis suggested that AMP-GP is the substrate of the reaction to form HcgF-Cys9-GP (precursor **5**). The space in the active site of HcgF allows the binding of AMP-GP. To test this catalytic reaction, an HcgF assay was devised that contained HcgE, ATP, GP, and Mg^{2+}. HcgF accelerated the production of AMP in the assay [105]. The structure of the active site of the HcgF-GP complex allowed the proposal of a catalytic mechanism in which the reaction starts from the attack of the adenylylated carboxy group of AMP-GP by the thiolate of Cys9 in an S_N2 reaction. Through this reaction, a thioester bond is formed and AMP is released.

14.12.3.7 HcgG

Heterologous production of HcgG in *E. coli* in a soluble form has failed. However, it has recently been reported that His-tagged HcgG from *M. maripaludis* was homologously produced in the native host and purified by Ni-NTA affinity chromatography [102]. The purified enzyme exhibited UV–visible absorbance peaks at 350, 420, and 600 nm, and the peak height at 420 nm was slightly decreased by reduction, suggesting that HcgG contains more than one chromophore and that at least one of the chromophores is redox active. The tertiary structure of HcgG is not known. The function of HcgG was proposed on the basis of *in vitro* biosynthesis experiments and is described below.

14.12.4 *In Vitro* Biosynthesis

The crystal structures of the HcgB-F proteins were solved, and the functions of HcgB, HcgC, HcgE, and HcgF were predicted based on the crystal structures and structures with the potential precursors. The predicted functions were then verified by enzyme activity tests as described earlier. The crystal structure of HcgD suggested a possible iron-trafficking function of this protein. Thus, the structure–function strategy was successfully applied for the analysis of biosynthesis of the FeGP cofactor. However, this strategy could not be applied to HcgA and HcgG. HcgA was successfully heterologously overproduced in *E. coli*. However, the purified protein could not be crystallized. Until recently, HcgG production in *E. coli* resulted in inclusion bodies, which cannot be used for crystallization and enzyme assays. Therefore, other strategies had to be developed for HcgA and HcgG to study their functions [102, 112].

The unknown enzyme reactions involved in the biosynthesis of the FeGP cofactor are a reaction for the biosynthesis of the first pyridinol precursor, 6-carboxymethyl-4-hydroxy-5-methyl-2-pyridinol (precursor **1**), and the final reaction of the biosynthesis of the FeGP cofactor from HcgF-bound GP (precursor **5**) and an iron precursor. The latter reaction may involve the precursor **5**, an iron ion, and a CO precursor. In addition, an electron carrier may be required to provide two electrons for the formation of the acyl ligand from the thioester group of precursor **5**. Assuming that the Hcg proteins known so far are all proteins required for FeGP cofactor biosynthesis, HcgA or HcgG should catalyze the first and last reactions [102].

In addition, the structure of the precursor substrates involved in the biosynthetic sequence should be verified, since the structure–function analysis and the subsequent enzyme assay cannot rule out the possibility that the Hcg proteins can also use other substrates [112]. For example, it has to be verified whether the 6-carboxymethyl group of GP is actually the precursor of the acyl ligand since this hypothesis can be challenged by the finding that external CO can be used for the biosynthesis of the acyl ligand. If CO is the precursor of the acyl ligand, it is possible that an $Fe(CO)_3$ intermediate is involved in the biosynthesis sequence and reacts with decarboxylated GP. In addition, it is necessary to elucidate the precursor of the CO ligands of the FeGP cofactor.

To analyze the function of HcgA and HcgG and the nature of the native precursors in FeGP cofactor biosynthesis, an *in vitro* biosynthesis assay was developed. This assay contains cell extracts of *M. maripaludis hcg* mutants, heterologously produced HcgA, HcgB, and HcgC, homologously produced HcgG, and stable isotope-labeled precursors. As described below, the *in vitro* biosynthesis experiments elucidated the precursor structures, the function of HcgA and HcgG, the presence of an organic CO donor, and the electron donor requirement for biosynthesis.

14.12.4.1 Design of the *In Vitro* Biosynthesis Assay

The *in vitro* biosynthesis assay of the FeGP cofactor was based on extracts of *M. maripaludis* cells cultivated in a standard medium with formate as a growth substrate (Figure 14.8) [112]. The basic idea is that the cell extract wild-type *M. maripaludis* contains all the enzymes necessary for the biosynthesis of FeGP cofactor, even after the cells have been disrupted and the membrane proteins have been removed by ultracentrifugation. The FeGP cofactor is therefore synthesized by cytoplasmic enzymes using endogenous precursors present in the cell extract and precursor molecules added to the assay. Since a previous report indicated that the Hmd holoenzyme and the apoenzyme of *M. maripaludis* are unstable

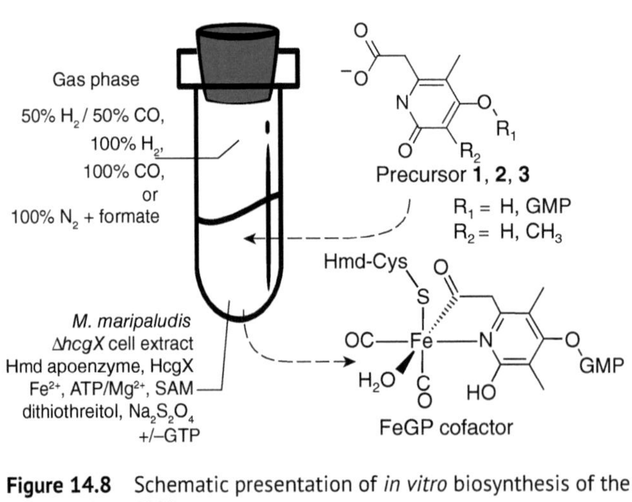

Figure 14.8 Schematic presentation of *in vitro* biosynthesis of the FeGP cofactor. Source: Adapted from [10].

and lose activity rapidly, the stable Hmd apoenzyme from *M. jannaschii* was used for the assay to reconstitute the Hmd holoenzyme. The production of the FeGP cofactor in the *in vitro* experiments was detected using the enzyme assay with methylene-H_4MPT as a substrate. The structure of the FeGP cofactor was analyzed by mass spectrometry. To detect the relatively low activity of the Hmd reconstituted with the newly biosynthesized FeGP cofactor, *M. maripaludis* mutants were used that lack one or two *hcg* genes and are unable to produce the FeGP cofactor.

The small molecules added to the reaction cocktail were selected from the structure–function studies described earlier. The precursor of the organic moiety of the FeGP cofactor was GP (precursor **3**), which was prepared by photolysis of the FeGP cofactor isolated from native Hmd purified from *M. marburgensis*. Since a possible radical SAM enzyme, HcgA or HcgG, could catalyze one of the biosynthesis reactions, SAM was included in the assay. ATP and Mg^{2+} were added for the HcgE reaction. Dithiothreitol and sodium dithionite were added to stabilize the FeGP cofactor produced in the assay and to reduce the assay solution, respectively. $Fe(SO_4)_2(NH_4)_2$ was added as a source of ferrous iron. From the stable isotope labeling experiment in the cells, CO and CO_2 were known as sources of the CO ligands. However, since CO_2 did not support *in vitro* biosynthesis, 50% H_2/50% CO gas was used as the standard gas mixture. The *in vitro* assay was carried out at 40 °C, which is the optimal growth temperature for *M. maripaludis*.

14.12.4.2 Confirmation of the Precursors by *In Vitro* Biosynthesis

The *in vitro* biosynthesis assay was first tested using the *M. maripaludis* Δ*hcgB*Δ*hcgC* strain. Therefore, GP was included as a precursor, which is the reaction product of HcgB and HcgC (see Figure 14.5b) [112]. Furthermore, the pyridinol precursors **1** and **2** were also used as precursors, in which purified HcgB and/or HcgC were added to the assay to enable the biosynthesis of GP, respectively. The Hmd activity increased in a time-dependent manner, indicating the biosynthesis of the FeGP cofactor proceeds in the *in vitro* assay. Mass spectrometric analysis confirmed the formation of the FeGP cofactor.

To confirm the necessity of the 6-carboxymethyl group of the pyridinol precursor, a derivative of precursor **2** lacking the carboxy group of the 6-carboxymethyl group (precursor **2′**) was synthesized. In addition, a GP-like derivative (precursor **3′**), which lacks the 3-methyl group of the pyridinol ring, was enzymologically synthesized by the HcgB reaction with precursor **2′**. No Hmd activity was detected *in vitro* biosynthesis under standard conditions by using **2′** and **3′**. These results confirmed the necessity of the 6-carboxymethyl group of GP for the biosynthesis of FeGP cofactor [112].

No in vitro biosynthesis activity was observed under 100% N_2, but a low and significant activity under 100% H_2. This finding suggests that H_2 acts as an electron donor in the *in vitro* biosynthesis. Since *M. maripaludis* uses formate as an electron donor for growth, the effect of formate was also tested. The addition of formate to the *in vitro* biosynthesis assay under N_2 showed a certain activity. These results indicate that electron donors (i.e. H_2 or formate) are required for the *in vitro* biosynthesis [112]. Since the *M. maripaludis* Δ*hcgB*Δ*hcgC* strain has been derived from the strain containing Δ*frc*Δ*fru* mutations [99], it lacks the F_{420}-reducing hydrogenase (Frh), and the only cytoplasmic hydrogenase of the strain is the heterodisulfide reductase (Hdr)-associated hydrogenase (Mvh). In this methanogen, formate is oxidized by F_{420}-dependent formate dehydrogenase, which forms a complex with Hdr. Based on the results of the H_2- and formate-dependent biosynthetic activity, it can be speculated that Hdr or its catalytic reduction product (reduced ferredoxin and CoB-SH) may be involved in the biosynthesis of the FeGP cofactor as electron donors for the biosynthesis of the acyl ligand.

The standard *in vitro* biosynthesis assay was performed with a gas phase that contained 50% H_2 and 50% CO. The addition of CO to the gas phase resulted in a dramatic increase in activity, and titration experiments (CO in N_2) showed that the activity increased with the concentration of CO in the gas phase [112]. This finding suggests that external CO gas acts as a precursor of the CO ligands and as an electron donor. In hydrogenotrophic methanogens, CO dehydrogenase catalyzes the oxidation of CO to CO_2 using ferredoxin as the electron acceptor. If CO is the electron donor, ferredoxin could be considered the electron carrier.

The incorporation of ^{13}CO gas into the FeGP cofactor was confirmed by mass spectrometric analysis of the FeGP cofactor, which was synthesized in the *in vitro* assay under 50% H_2/50% [^{13}C]-CO [112]. The results showed that 18% of the biosynthesized cofactor contained two ^{13}C atoms and 30% of the cofactor contained one ^{13}C atom, but there was no FeGP cofactor with three ^{13}C atoms. Thus, about half of the cofactor was not labeled with ^{13}C, although the gas phase did not contain [^{12}C]-CO. This result indicated that the majority of the CO ligands of the FeGP cofactor was biosynthesized from an organic compound in the cell extract.

14.12.4.3 Basis of the *In Vitro* Complementation Assay

HcgA and HcgG should catalyze the biosynthesis of precursor **1** or complete the biosynthesis of the FeGP cofactor by forming an iron complex with precursor **5**. To identify the function of HcgA and HcgG, an adapted *in vitro* biosynthesis assay was performed [102], which was based on the complementation of cell extracts of *M. maripaludis* mutants that lack the *hcgA* or *hcgG* gene. Experiments were performed to test whether the activity of the *in vitro* biosynthesis assay was complemented by the addition of precursors **1**–**3**.

14.12.4.4 *In Vitro* Complementation of HcgA

The cell extract of the *M. mariparudis* Δ*hcgA* mutant shows no Hmd activity and the absence of HcgA protein was confirmed by mass spectrometric proteomic analysis. The *in vitro* biosynthesis assay using the *M. mariparudis* Δ*hcgA* mutant under standard conditions without the addition of an external precursor shows no Hmd activity. This result confirmed that HcgA is required for the biosynthesis of the FeGP cofactor. When the *in vitro* assay was supplemented with GP (precursor **3**), the FeGP cofactor was formed [102]. This finding suggests that HcgA catalyzes reactions that precede GP formation. Furthermore, the addition of pyridinol precursors (**1** or **2**) also enabled FeGP cofactor biosynthesis when GTP was added to the HcgB reaction, which was not added in the standard assay. In the absence of an external precursor, the addition of HcgA and GTP also enabled the production of the FeGP cofactor, suggesting that HcgA catalyzes a reaction that forms precursor **1** from an unknown compound present in the cell extract.

The HcgA catalyzed reaction was confirmed by adding the filtrate (3 kDa cutoff) of the cell extract from the *M. maripaludis* Δ*hcgA* mutant to the assay [102]. Proteins larger than 3 kDa were removed by filtration, but the smaller compounds containing the possible HcgA substrate were retained in the filtrated fraction. Since HcgA was proposed to be a radical SAM enzyme from the previous studies [101], the HcgA reaction was tested in the presence of the cell extract filtrate and SAM. As expected, this enzyme assay produced precursor **1**, which was detected by mass spectrometric analysis of the assay solution. The structure of the product was also confirmed by MS–MS analysis, which revealed the same fragmentation pattern as the chemically synthesized authentic precursor **1**. The structure was further verified by the subsequent reaction with HcgC, which resulted in the formation of a compound whose molecular mass was identical to that of precursor **2**. In the absence of externally added SAM, the lower enzyme activity of HcgA was detected, indicating that endogenous SAM is fractionated in the filtrate of the cell extract. The HcgA activity without external SAM was inhibited by about half by the addition of 1 mM SAH, an inhibitor of radical SAM enzymes. These results support that HcgA catalyzes a radical SAM reaction, producing precursor **1** from SAM and an unknown substrate in the cell extract [102].

A kinetic analysis of the HcgA reaction showed that precursor **1** was produced in a time-dependent manner. The common by-product of the radical SAM reaction, 5′-deoxyadenosine, was formed, providing further evidence that HcgA catalyzes a radical SAM reaction. Notably, the production rate of 5′-deoxyadenosine in the HcgA reaction was about three times that of precursor **1**, and its production slowed down before the production of precursor **1** was completed. This may indicate a mechanism specific to HcgA. Although no proteins were detected in the filtrate by proteome analysis, it cannot be ruled out that small enzymes (<3 kDa) in the filtrate contribute to the production and consumption of 5′-deoxyadenosine [102].

Labeling experiments with stable isotopes suggested that precursor **1** might be biosynthesized by the reaction of β-alanine or aspartic acid with 2,3-dihydroxy-4-oxopentanoic acid [92]. To test this hypothesis, the HcgA reactivity was investigated in the presence of [^{15}N,^{13}C]-β-alanine or a mixture of the 20 essential [^{15}N]-labeled amino acids using an *in vitro* biosynthesis assay with cell extracts of the *M. maripaludis* Δ*hcgA* strain. No labeled FeGP cofactor was produced in this assay. This result does not support the proposed substrates of the FeGP cofactor [102]. However, to obtain conclusive results, further experiments should be carried out using an assay without cell extracts, in which only selected substrates and HcgA are included.

14.12.4.5 *In Vitro* Complementation of HcgG

The absence of Hmd activity and the HcgG protein in the cell extract from *M. maripaludis* Δ*hcgG* strain was confirmed in the same way as described in Section 14.12.4.4 on *in vitro* complementation analysis of the Δ*hcgA* strain [102]. The *in vitro* biosynthesis of FeGP carried out with extract from the Δ*hcgG* strain was not complemented by the addition of GP (precursor **3**), but was complemented by the addition of purified HcgG, indicating that HcgG catalyzes a reaction after the formation of GP [102]. The relatively high biosynthetic activity without the addition of external GP indicated that GP was accumulated in the cell extract of the Δ*hcgG* strain, but the addition of GP to the assay further increased the activity. Notably, the *in vitro* biosynthetic activity with external HcgG and GP under 100% H_2 was identical to that under the standard gas mixture (50% H_2/50% CO). This result differs from the results of the assay using the Δ*hcgB*Δ*hcgC* strain, in which the activity under a 100% H_2 atmosphere was reduced to 30% of the activity under the standard atmosphere [112]. The reason for this could be that the concentration of HcgG in the assay supplemented with purified HcgG is much higher than in the extract of the Δ*hcgB*Δ*hcgC* strain, which contains only endogenous HcgG. Titrations actually showed that a higher concentration of HcgG resulted in a higher *in vitro* biosynthetic activity [102]. This result can be interpreted to mean that an excess of HcgG stimulated the production of FeGP cofactor under 100% H_2 using an unknown organic compound as a CO precursor. It was deduced from the experimental data that HcgG uses precursor **5** as a substrate and either CO gas or an unknown organic compound as a precursor for CO ligand biosynthesis [102].

Radical SAM enzymes catalyze the formation of 5′-deoxyadenosine from SAM also in the absence of substrate, and their activity is stimulated by the addition of the substrate. Accordingly, HcgG catalyzes the formation of 5′-deoxyadenosine from SAM, and this reaction was stimulated by the addition of filtrate from cell extracts [102]. This suggests that HcgG catalyzes a radical SAM reaction that may be involved in the CO ligand biosynthesis, as is the case with the radical SAM enzyme HydG, which catalyzes the biosynthesis of CO and CN^- ligands of the H-cluster in [FeFe] hydrogenase [94]. Notably external CO inhibits the 5′-deoxyadenosine-forming activity of HcgG. One interpretation of this phenomenon is that CO is formed from an organic precursor by HcgG, and external CO inhibits the reaction by binding to the radical SAM active site. This possible mechanism of HcgG differs from that of HydG, in which dehydroglycine is formed from tyrosine via *p*-cresol in the radical SAM active site and dehydroglycine is transferred to the second active site, where the CO and CN^- ligands are formed [113, 114]. Since the *in vitro* biosynthesis assay of the FeGP cofactor is carried out with cell extracts, it is not possible to conclusively ascertain that HcgG catalyzes all reactions of acyl ligand formation and CO ligand biosynthesis. However, comparative genomic analysis revealed that the only common genes involved in FeGP cofactor biosynthesis are the seven *hcg* genes [104]. If this prediction is correct, HcgG should catalyze all reactions involved in the biosynthesis of FeGP cofactor from GP (precursor **3**) via precursors **4** and **5** [102].

The AlphaFold model of *M. jannaschii* HcgG suggests that it consists of N- and C-terminal regions and that the N-terminal region is similar to HydG, as it contains the binding site of a [4Fe–4S] cluster, responsible for the radical SAM reaction [102]. The C-terminal region is folded in a similar way to the structure of Hmd, which binds to the FeGP cofactor. Based on the structural similarities, it is hypothesized that HcgG may have two active sites: one radical SAM active site and the other is responsible for the formation of the iron complex using CO or an intermediate formed at the radical SAM active site. The predicted ability to bind the FeGP cofactor suggests that HcgG may also serve as a scaffold for the final reaction of iron complex formation.

14.12.5 Proposed Biosynthesis Sequence of the FeGP Cofactor

Structure–function analyses using heterologously produced enzymes and *in vitro* biosynthesis experiments with the *M. maripaludis* Δ*hcg* strains revealed most of the functions of the seven Hcg proteins (see Figure 14.5b). HcgA is a radical SAM enzyme that catalyzes the synthesis of the pyridinol precursor **1**. The SAM-dependent methyltransferase HcgC catalyzes the methylation of the 3-position of precursor **1** to form precursor **2**. Precursor **2** is reacted with GTP at the 4-hydroxy group to form guanylylpyridinol (GP, precursor **3**). The 6-carboxymethyl group of precursor **3** is activated by adenylylation to form AMP-GP (precursor **4**) and then the 6-carboxymethyl group is reacted with a cysteine residue of HcgF to form a thioester bond. The HcgF with thioester-bound GP (precursor **5**) is a possible substrate for the reaction catalyzed by HcgG, in which the thioester carbon is converted to an acyl ligand by a two-electron reduction. HcgG can catalyze two more reactions, the CO formation and the completion of the iron complex of the FeGP cofactor. The electrons for acyl ligand formation can be provided by H_2, formate, or CO via an unknown electron carrier. HcgG may also act as a scaffold protein by binding the FeGP precursor to the C-terminal region.

14.13 Conclusion

As described in this chapter, methanogenic archaea contain many unique metalloenzymes in the methanogenic pathway that contain molybdenum, tungsten, cobalt, nickel, iron, zinc, and/or iron–sulfur clusters. One of the metalloenzymes, Hmd, contains the FeGP cofactor, which contains a unique iron complex. Our structure–function and *in vitro* biosynthesis analyses have elucidated the catalytic function of most of the biosynthetic enzymes encoded by the *hcg* gene cluster. However, there are still several questions regarding the function of the Hcg enzymes.

The first and the last steps of the biosynthesis of the FeGP cofactor present a common challenge: the identification of the substrates used by the two radical SAM enzymes, HcgA and HcgG. In the case of HcgG, the hypothesis of an unknown precursor was developed by analogy with the radical SAM enzyme HydG, which produces CO ligands from *L*-tyrosine in the [FeFe]-hydrogenase cofactor biosynthesis. Monitoring the production of CO, 5′-deoxyadenosine, or FeGP cofactor in the presence of SAM and possible precursors (amino acids, or other organic compounds) is one way of testing HcgG activity. The situation is different with HcgA. The uniqueness of the pyridinol precursor **1**, which is produced by HcgA, makes it difficult to predict its precursor by analogy with another known reaction. The strategy for identifying the HcgA precursor could be a fractionation of the cell extract based on HcgA activity, monitored by the precursor **1** production. Site-directed mutagenesis, protein–ligand docking, and structural analysis, used in a concerted manner, can also provide potential clues to the important residues and the types of ligands that can potentially act as HcgA substrates. A retrosynthetic approach could shed light on the possible molecule(s) that are converted into this unique pyridinol compound through a radical SAM mechanism of HcgA. Since the crystal structures for HcgA and HcgG have not yet been obtained, cryo-EM offers an alternative approach. If successful, the cryo-EM structures could also provide insights into possible substrates, binding sites, cofactors, and catalytic mechanisms.

Acknowledgment

This work was supported by Deutsche Forschungsgemeinschaft, Priority Program, Iron–Sulfur for Life (SPP1927, SH87/1-1, and SH87/1-2) and by the Max Planck Society to Seigo Shima.

References

1 Thauer, R.K., Kaster, A.K., Seedorf, H. et al. (2008). Methanogenic archaea: Ecologically relevant differences in energy conservation. *Nat. Rev. Microbiol.* 6 (8): 579–591.

2 Thauer, R.K., Kaster, A.K., Goenrich, M. et al. (2010). Hydrogenases from methanogenic archaea, nickel, a novel cofactor, and H_2 storage. *Annu. Rev. Biochem.* 79: 507–536.

3 Shima, S., Huang, G., Wagner, T. et al. (2020). Structural basis of hydrogenotrophic methanogenesis. *Annu. Rev. Biochem.* 74: 713–733.

4 Kurth, J.M., Op den Camp, H.J.M., and Welte, C.U. (2020). Several ways one goal-methanogenesis from unconventional substrates. *Appl. Microbiol. Biotechnol.* 104 (16): 6839–6854.

5 Nomura, S., Paczia, N., Kahnt, J. et al. (2024). Isolation of an H_2-dependent electron-bifurcating CO_2-reducing megacomplex with MvhB polyferredoxin from *Methanothermobacter marburgensis*. *FEBS J.* 291: 2449–2424.

6 Dimarco, A.A., Bobik, T.A., and Wolfe, R.S. (1990). Unusual coenzymes of methanogenesis. *Annu. Rev. Biochem.* 59: 355–394.

7 Thauer, R.K., Klein, A.R., and Hartmann, G.C. (1996). Reactions with molecular hydrogen in microorganisms: Evidence for a purely organic hydrogenation catalyst. *Chem. Rev.* 96 (7): 3031–3042.

8 Thauer, R.K. (2019). Methyl (Alkyl)-coenzyme M reductases: Nickel F_{430}-containing enzymes involved in anaerobic methane formation and in anaerobic oxidation of methane or of short chain alkanes. *Biochemistry* 58 (52): 5198–5220.

9 Gottschalk, G. and Thauer, R.K. (2001). The Na^+-translocating methyltransferase complex from methanogenic archaea. *Biochim. Biophys. Acta* 1505 (1): 28–36.

10 Arriaza-Gallardo, F.J., Zheng, Y.C., Gehl, M. et al. (2023). [Fe]-Hydrogenase, cofactor biosynthesis and engineering. *ChemBioChem* 24 (20): e202300330.

11 Börner, G., Karrasch, M., and Thauer, R.K. (1989). Formylmethanofuran dehydrogenase-activity in cell-extracts of *Methanobacterium-thermoautotrophicum* and of *Methanosarcina barkeri*. *FEBS Lett.* 244 (1): 21–25.

12 Bertram, P.A., Schmitz, R.A., Linder, D. et al. (1994). Tungstate can substitute for molybdate in sustaining growth of *Methanobacterium thermoautotrophicum*. Identification and characterization of a tungsten isoenzyme of formylmethanofuran dehydrogenase. *Arch. Microbiol.* 161 (3): 220–228.

13 Schmitz, R.A., Albracht, S.P., and Thauer, R.K. (1992). A molybdenum and a tungsten isoenzyme of formylmethanofuran dehydrogenase in the thermophilic archaeon *Methanobacterium wolfei*. *Eur. J. Biochem.* 209 (3): 1013–1018.

14 Hochheimer, A., Hedderich, R., and Thauer, R.K. (1998). The formylmethanofuran dehydrogenase isoenzymes in *Methanobacterium wolfei* and *Methanobacterium thermoautotrophicum*: Induction of the molybdenum isoenzyme by molybdate and constitutive synthesis of the tungsten isoenzyme. *Arch. Microbiol.* 170 (5): 389–393.

15 Wagner, T., Ermler, U., and Shima, S. (2016). The methanogenic CO_2 reducing-and-fixing enzyme is bifunctional and contains 46 [4Fe-4S] clusters. *Science* 354 (6308): 114–117.

16 Wasserfallen, A. (1994). Formylmethanofuran synthesis by formylmethanofuran dehydrogenase from *Methanobacterium thermoautotrophicum* Marburg. *Biochem. Biophys. Res. Commun.* 199 (3): 1256–1261.

17 Acharya, P., Warkentin, E., Ermler, U. et al. (2006). The structure of formylmethanofuran: Tetrahydromethanopterin formyltransferase in complex with its coenzymes. *J. Mol. Biol.* 357 (3): 870–879.

18 Mamat, B., Roth, A., Grimm, C. et al. (2002). Crystal structures and enzymatic properties of three formyltransferases from archaea: Environmental adaptation and evolutionary relationship. *Protein Sci.* 11 (9): 2168–2178.

19 Ermler, U., Merckel, M.C., Thauer, R.K. et al. (1997). Formylmethanofuran:tetrahydromethanopterin formyltransferase from *Methanopyrus kandleri* – New insights into salt-dependence and thermostability. *Structure* 5 (5): 635–646.

20 Shima, S., Tziatzios, C., Schubert, D. et al. (1998). Lyotropic-salt-induced changes in monomer/dimer/tetramer association equilibrium of formyltransferase from the hyperthermophilic *Methanopyrus kandleri* in relation to the activity and thermostability of the enzyme. *Eur. J. Biochem.* 258 (1): 85–92.

21 Shima, S., Thauer, R.K., Ermler, U. et al. (2000). A mutation affecting the association equilibrium of formyltransferase from the hyperthermophilic *Methanopyrus kandleri* and its influence on the enzyme's activity and thermostability. *Eur. J. Biochem.* 267 (22): 6619–6623.

22 Te Brömmelstroet, B.W., Hensgens, C.M.H., Geerts, W.J. et al. (1990). Purification and properties of 5,10-methenyltetrahydromethanopterin cyclohydrolase from *Methanosarcina barkeri*. *J. Bacteriol.* 172 (2): 564–571.

23 Upadhyay, V., Demmer, U., Warkentin, E. et al. (2012). Structure and catalytic mechanism of N^5,N^{10}-methenyltetrahydromethanopterin cyclohydrolase. *Biochemistry* 51 (42): 8435–8443.

24 Grabarse, W., Vaupel, M., Vorholt, J.A. et al. (1999). The crystal structure of methenyltetrahydromethanopterin cyclohydrolase from the hyperthermophilic archaeon *Methanopyrus kandleri*. *Struct. Fold. Des.* 7 (10): 1257–1268.

25 Te Brommelstroet, B.W., Hensgens, C.M.H., Keltjens, J.T. et al. (1991). Purification and characterization of coenzyme-F_{420}-dependent 5,10-methylenetetrahydromethanopterin dehydrogenase from *Methanobacterium thermoautotrophicum* Strain DH. *Biochim. Biophys. Acta* 1073 (1): 77–84.

26 Zirngibl, C., Hedderich, R., and Thauer, R.K. (1990). N^5,N^{10}-methylenetetrahydromethanopterin dehydrogenase from Methanobacterium thermoautotrophicum has hydrogenase activity. *FEBS Lett.* 261 (1): 112–116.

27 Zirngibl, C., Van Dongen, W., Schwörer, B. et al. (1992). H_2-forming methylenetetrahydromethanopterin dehydrogenase, a novel type of hydrogenase without iron-sulfur clusters in methanogenic archaea. *Eur. J. Biochem.* 208 (2): 511–520.

28 Hagemeier, C.H., Shima, S., Thauer, R.K. et al. (2003). Coenzyme F_{420} dependent methylenetetrahydromethanopterin dehydrogenase (Mtd) from *Methanopyrus kandleri*: A methanogenic enzyme with an unusual quarternary structure. *J. Mol. Biol.* 332 (5): 1047–1057.

29 Ceh, K., Demmer, U., Warkentin, E. et al. (2009). Structural basis of the hydride transfer mechanism in F_{420}-dependent methylenetetrahydromethanopterin dehydrogenase. *Biochemistry* 48 (42): 10098–10105.

30 Huang, G.F., Wagner, T., Ermler, U. et al. (2020). Methanogenesis involves direct hydride transfer from H_2 to an organic substrate. *Nat. Rev. Chem.* 4 (4): 213–221.

31 Tebrommelstroet, B.W., Hensgens, C.M.H., Keltjens, J.T. et al. (1990). Purification and properties of 5,10-methylenetetrahydromethanopterin reductase, a coenzyme F_{420}-dependent enzyme, from *Methanobacterium thermoautotrophicum* strain DH. *J. Biol. Chem.* 265 (4): 1852–1857.

32 Ma, K. and Thauer, R.K. (1990). Purification and Properties of N^5,N^{10}-methylenetetrahydromethanopterin reductase from *Methanobacterium thermoautotrophicum* (Strain Marburg). *Eur. J. Biochem.* 191 (1): 187–193.

33 Aufhammer, S.W., Warkentin, E., Ermler, U. et al. (2005). Crystal structure of methylenetetrahydromethanopterin reductase (Mer) in complex with coenzyme F_{420}: Architecture of the F_{420}/FMN binding site of enzymes within the nonprolyl *cis*-peptide containing bacterial luciferase family. *Protein Sci.* 14 (7): 1840–1849.

34 Gehl, M., Demmer, U., Ermler, U. et al. (2024). Mutational and structural studies of $(ba)_8$-barrel fold methylene-tetrahydropterin reductases utilizing a common catalytic mechanism. *Protein Sci.* 33 (6): e5018.

35 Gärtner, P., Weiss, D.S., Harms, U. et al. (1994). N^5-methyltetrahydromethanopterin:coenzyme M methyltransferase from *Methanobacterium thermoautotrophicum*. Catalytic mechanism and sodium ion dependence. *Eur. J. Biochem.* 226 (2): 465–472.

36 Harms, U., Weiss, D.S., Gartner, P. et al. (1995). The energy conserving N^5-methyltetrahydromethanopterin:coenzyme M methyltransferase complex from *Methanobacterium thermoautotrophicum* is composed of eight different subunits. *Eur. J. Biochem.* 228 (3): 640–648.

37 Weiss, D.S., Gartner, P., and Thauer, R.K. (1994). The energetics and sodium-ion dependence of N^5-methyltetrahydromethanopterin:coenzyme M methyltransferase studied with cob(I)alamin as methyl acceptor and methylcob(III)alamin as methyl donor. *Eur. J. Biochem.* 226 (3): 799–809.

38 Aziz, I., Kayastha, K., Kaltwasser, S. et al. (2024). Structural and mechanistic basis of the central energy-converting methyltransferase complex of methanogenesis. *Proc. Natl. Acad. Sci. USA* 121 (14): e2315568121.

39 Wagner, T., Ermler, U., and Shima, S. (2016). MtrA of the sodium ion pumping methyltransferase binds cobalamin in a unique mode. *Sci. Rep.* 6: 28226.

40 Ermler, U., Grabarse, W., Shima, S. et al. (1997). Crystal structure of methyl coenzyme M reductase: The key enzyme of biological methane formation. *Science* 278 (5342): 1457–1462.

41 Grabarse, W., Mahlert, F., Shima, S. et al. (2000). Comparison of three methyl-coenzyme M reductases from phylogenetically distant organisms: Unusual amino acid modification, conservation and adaptation. *J. Mol. Biol.* 303 (2): 329–344.

42 Grabarse, W., Mahlert, F., Duin, E.C. et al. (2001). On the mechanism of biological methane formation: Structural evidence for conformational changes in methyl-coenzyme M reductase upon substrate binding. *J. Mol. Biol.* 309 (1): 315–330.

43 Wagner, T., Wegner, C.-E., Kahnt, J. et al. (2017). Phylogenetic and structural comparisons of the three types of methyl-coenzyme M reductase from Methanococcales and Methanobacteriales. *J. Bacteriol.* 199: e00197–e00117.

44 Pelmenschikov, V., Blomberg, M.R.A., Siegbahn, P.E.M. et al. (2002). A mechanism from quantum chemical studies for methane formation in methanogenesis. *J. Am. Chem. Soc.* 124 (15): 4039–4049.

45 Wongnate, T., Sliwa, D., Ginovska, B. et al. (2016). The radical mechanism of biological methane synthesis by methyl-coenzyme M reductase. *Science* 352 (6288): 953–958.

46 Kaster, A.K., Moll, J., Parey, K. et al. (2011). Coupling of ferredoxin and heterodisulfide reduction via electron bifurcation in hydrogenotrophic methanogenic archaea. *Proc. Natl. Acad. Sci. USA* 108: 2981–2986.

47 Setzke, E., Hedderich, R., Heiden, S. et al. (1994). H_2: Heterodisulfide oxidoreductase complex from *Methanobacterium thermoautotrophicum*: Composition and properties. *Eur. J. Biochem.* 220 (1): 139–148.

48 Costa, K.C., Wong, P.M., Wang, T. et al. (2010). Protein complexing in a methanogen suggests electron bifurcation and electron delivery from formate to heterodisulfide reductase. *Proc. Natl. Acad. Sci. USA* 107 (24): 11050–11055.

49 Watanabe, T., Pfeil-Gardiner, O., Kahnt, J. et al. (2021). Three-megadalton complex of methanogenic electron-bifurcating and CO_2-fixing enzymes. *Science* 373 (6559): 1151–1156.

50 Wagner, T., Koch, J., Ermler, U. et al. (2017). Methanogenic heterodisulfide reductase (HdrABC-MvhAGD) uses two noncubane [4Fe-4S] clusters for reduction. *Science* 357 (6352): 699–703.

51 Tersteegen, A. and Hedderich, R. (1999). *Methanobacterium thermoautotrophicum* encodes two multisubunit membrane-bound [NiFe] hydrogenases – transcription of the operons and sequence analysis of the deduced proteins. *Eur. J. Biochem.* 264 (3): 930–943.

52 Mills, D.J., Vitt, S., Strauss, M. et al. (2013). De novo modeling of the F_{420}-reducing [NiFe]-hydrogenase from a methanogenic archaeon by cryo-electron microscopy. *eLife* 2: e00218.

53 Vitt, S., Ma, K., Warkentin, E. et al. (2014). The F_{420}-reducing [NiFe]-hydrogenase complex from *Methanothermobacter marburgensis*, the first X-ray structure of a group 3 family member. *J. Mol. Biol.* 426 (15): 2813–2826.

54 Ilina, Y., Lorent, C., Katz, S. et al. (2019). X-ray crystallography and vibrational spectroscopy reveal the key determinants of biocatalytic dihydrogen cycling by [NiFe] hydrogenases. *Angew. Chem. Int. Ed.* 58 (51): 18710–18714.

55 Sprott, G.D., Shaw, K.M., and Beveridge, T.J. (1987). Properties of the particulate enzyme F_{420}-reducing hydrogenase isolated from *Methanospirillum hungatei*. *Can. J. Microbiol.* 33 (10): 896–904.

56 Wackett, L.P., Hartwieg, E.A., King, J.A. et al. (1987). Electron-microscopy of nickel-containing methanogenic enzymes – methyl reductase and F_{420}-reducing hydrogenase. *J. Bacteriol.* 169 (2): 718–727.

57 Reeve, J.N., Beckler, G.S., Cram, D.S. et al. (1989). A hydrogenase-linked gene in *Methanobacterium thermoautotrophicum* strain ΔH encodes a polyferredoxin. *Proc. Natl. Acad. Sci. USA* 86 (9): 3031–3035.

58 Hedderich, R., Albracht, S.P.J., Linder, D. et al. (1992). Isolation and characterization of polyferredoxin from *Methanobacterium thermoautotrophicum* – the Mvhb gene-product of the methylviologen-reducing hydrogenase operon. *FEBS Lett.* 298 (1): 65–68.

59 Watanabe, T., Wagner, T., Huang, G.F. et al. (2019). The bacterial [Fe]-hydrogenase paralog hmdII uses tetrahydrofolate derivatives as substrates. *Angew. Chem. Int. Ed.* 58 (11): 3506–3510.

60 Lie, T.J., Costa, K.C., Lupa, B. et al. (2012). Essential anaplerotic role for the energy-converting hydrogenase Eha in hydrogenotrophic methanogenesis. *Proc. Natl. Acad. Sci. USA* 109 (38): 15473–15478.

61 Porat, I., Kim, W., Hendrickson, E.L. et al. (2006). Disruption of the operon encoding Ehb hydrogenase limits anabolic CO_2 assimilation in the archaeon *Methanococcus maripaludis*. *J. Bacteriol.* 188 (4): 1373–1380.

62 Ma, K., Zirngibl, C., Linder, D. et al. (1991). *N^5*, *N^{10}*-Methylenetetrahydromethanopterin dehydrogenase (H_2-forming) from the extreme thermophile *Methanopyrus kandleri*. *A. Microbiol.* 156 (1): 43–48.

63 Schwörer, B. and Thauer, R.K. (1991). Activities of formylmethanofuran dehydrogenase, methylenetetrahydromethanopterin dehydrogenase, methylenetetrahydromethanopterin reductase, and heterodisulfide reductase in methanogenic bacteria. *Arch. Microbiol.* 155 (5): 459–465.

64 Kaster, A.K., Goenrich, M., Seedorf, H. et al. (2011). More than 200 genes required for methane formation from H_2 and CO_2 and energy conservation are present in *Methanothermobacter marburgensis* and *Methanothermobacter thermautotrophicus*. *Archaea* 2011: 973848.

65 Afting, C., Hochheimer, A., and Thauer, R.K. (1998). Function of H_2-forming methylenetetrahydromethanopterin dehydrogenase from *Methanobacterium thermoautotrophicum* in coenzyme F_{420} reduction with H_2. *Arch. Microbiol.* 169: 206–210.

66 Schwörer, B., Fernandez, V.M., Zirngibl, C. et al. (1993). H_2-forming *N^5*,*N^{10}*-methylenetetrahydromethanopterin dehydrogenase from *Methanobacterium thermoautotrophicum* – studies of the catalytic mechanism of H_2 formation using hydrogen isotopes. *Eur. J. Biochem.* 212 (1): 255–261.

67 Klein, A.R., Fernandez, V.M., and Thauer, R.K. (1995). H_2-forming *N^5*,*N^{10}*-methylenetetrahydromethanopterin dehydrogenase – mechanism of H_2 formation analyzed using hydrogen isotopes. *FEBS Lett.* 368 (2): 203–206.

68 Hartmann, G.C., Santamaria, E., Fernandez, V.M. et al. (1996). Studies on the catalytic mechanism of H_2-forming methylenetetrahydromethanopterin dehydrogenase: Para-ortho H_2 conversion rates in H_2O and D_2O. *J. Biol. Inorg. Chem.* 1 (5): 446–450.

69 Berkessel, A. and Thauer, R.K. (1995). On the mechanism of catalysis by a metal-free hydrogenase from methanogenic archaea – enzymatic transformation of H_2 without a metal and its analogy to the chemistry of alkanes in superacidic solution. *Angew. Chem. Int. Ed.* 34 (20): 2247–2250.

70 Olah, G.A., Halpern, Y., Shen, J. et al. (1973). Electrophilic reactions at single bonds.12. Hydrogen-deuterium exchange, protolysis (deuterolysis), and oligocondensation of alkanes with superacids. *J. Am. Chem. Soc.* 95 (15): 4960–4970.

71 Buurman, G., Shima, S., and Thauer, R.K. (2000). The metal-free hydrogenase from methanogenic archaea: Evidence for a bound cofactor. *FEBS Lett.* 485 (2–3): 200–204.

72 Lyon, E.J., Shima, S., Buurman, G. et al. (2004). UV-A/blue-light inactivation of the 'metal-free' hydrogenase (Hmd) from methanogenic archaea – the enzyme contains functional iron after all. *Eur. J. Biochem.* 271 (1): 195–204.

73 Shima, S., Schick, M., and Tamura, H. (2011). Preparation of [Fe]-hydrogenase from methanogenic archaea. *Methods Enzymol.* 494: 119–137.

74 Shima, S., Pilak, O., Vogt, S. et al. (2008). The crystal structure of [Fe]-hydrogenase reveals the geometry of the active site. *Science* 321 (5888): 572–575.

75 Lubitz, W., Ogata, H., Rudiger, O. et al. (2014). Hydrogenases. *Chem. Rev.* 114 (8): 4081–4148.

76 Lyon, E.J., Shima, S., Boecher, R. et al. (2004). Carbon monoxide as an intrinsic ligand to iron in the active site of the iron-sulfur-cluster-free hydrogenase H_2-forming methylenetetrahydromethanopterin dehydrogenase as revealed by infrared spectroscopy. *J. Am. Chem. Soc.* 126 (43): 14239–14248.

77 Shima, S., Lyon, E.J., Thauer, R.K. et al. (2005). Mössbauer studies of the iron-sulfur cluster-free hydrogenase: The electronic state of the mononuclear Fe active site. *J. Am. Chem. Soc.* 127 (29): 10430–10435.

78 Shima, S., Lyon, E.J., Sordel-Klippert, M.S. et al. (2004). The cofactor of the iron-sulfur cluster free hydrogenase Hmd: Structure of the light-inactivation product. *Angew. Chem. Int. Ed.* 43 (19): 2547–2551.

79 Korbas, M., Vogt, S., Meyer-Klaucke, W. et al. (2006). The iron-sulfur cluster-free hydrogenase (Hmd) is a metalloenzyme with a novel iron binding motif. *J. Biol. Chem.* 281 (41): 30804–30813.

80 Shima, S., Schick, M., Kahnt, J. et al. (2012). Evidence for acyl-iron ligation in the active site of [Fe]-hydrogenase provided by mass spectrometry and infrared spectroscopy. *Dalton Trans.* 41 (3): 767–771.

81 Schaupp, S., Arriaza-Gallardo, F.J., Paczia, N. et al. (2024). Acyl and CO ligands in the [Fe]-hydrogenase cofactor scramble upon photolysis. *Angew. Chem. Int. Ed.* 63 (6): e202316478.

82 Pilak, O., Mamat, B., Vogt, S. et al. (2006). The crystal structure of the apoenzyme of the iron-sulphur cluster-free hydrogenase. *J. Mol. Biol.* 358 (3): 798–809.

83 Hiromoto, T., Ataka, K., Pilak, O. et al. (2009). The crystal structure of C176A mutated [Fe]-hydrogenase suggests an acyl-iron ligation in the active site iron complex. *FEBS Lett.* 583 (3): 585–590.

84 Hiromoto, T., Warkentin, E., Moll, J. et al. (2009). The crystal structure of an [Fe]-hydrogenase-substrate complex reveals the framework for H_2 activation. *Angew. Chem. Int. Ed.* 48 (35): 6457–6460.

85 Huang, G.F., Wagner, T., Wodrich, M.D. et al. (2019). The atomic-resolution crystal structure of activated [Fe]-hydrogenase. *Nat. Catal.* 2 (6): 537–543.

86 Chen, D., Scopelliti, R., and Hu, X. (2011). A five-coordinate iron center in the active site of [Fe]-hydrogenase: Hints from a model study. *Angew. Chem. Int. Ed.* 50 (25): 5671–5673.

87 Tamura, H., Salomone-Stagni, M., Fujishiro, T. et al. (2013). Crystal structures of [Fe]-hydrogenase in complex with inhibitory isocyanides: Implications for the H_2 activation site. *Angew. Chem. Int. Ed.* 52 (37): 9656–9659.

88 Wagner, T., Huang, G.F., Ermler, U. et al. (2018). How [Fe]-hydrogenase from *Methanothermobacter* is protected against light and oxidative stress. *Angew. Chem. Int. Ed.* 57 (46): 15056–15059.

89 Fujishiro, T., Ataka, K., Ermler, U. et al. (2015). Towards a functional identification of catalytically inactive [Fe]-hydrogenase paralogs. *FEBS J.* 282 (17): 3412–3423.

90 Oza, J.P., Sowers, K.R., and Perona, J.J. (2012). Linking energy production and protein synthesis in hydrogenotrophic methanogens. *Biochemistry* 51 (12): 2378–2389.

91 Huang, G.F., Arriaza-Gallardo, F.J., Wagner, T. et al. (2020). Crystal structures of [Fe]-hydrogenase from *Methanolacinia paynteri* suggest a path of the FeGP-cofactor incorporation process. *Inorganics.* 8 (9): 50.

92 Schick, M., Xie, X., Ataka, K. et al. (2012). Biosynthesis of the iron-guanylylpyridinol cofactor of [Fe]-hydrogenase in methanogenic archaea as elucidated by stable-isotope labeling. *J. Am. Chem. Soc.* 134 (6): 3271–3280.

93 Samuel, B.S., Hansen, E.E., Manchester, J.K. et al. (2007). Genomic and metabolic adaptations of *Methanobrevibacter smithii* to the human gut. *Proc. Natl. Acad. Sci. USA* 104 (25): 10643–10648.

94 Swanson, K.D., Duffus, B.R., Beard, T.E. et al. (2011). Cyanide and carbon monoxide ligand formation in hydrogenase biosynthesis. *Eur. J. Inorg. Chem.* 7: 935–947.

95 Muraki, N., Ishii, K., Uchiyama, S. et al. (2019). Structural characterization of HypX responsible for CO biosynthesis in the maturation of NiFe-hydrogenase. *Commun. Biol.* 2: 385.

96 Schulz, A.C., Frielingsdorf, S., Pommerening, P. et al. (2020). Formyltetrahydrofolate decarbonylase synthesizes the active site CO ligand of O_2-tolerant [NiFe] hydrogenase. *J. Am. Chem. Soc.* 142 (3): 1457–1464.

97 Anderson, I.J., Sieprawska-Lupa, M., Goltsman, E. et al. (2009). Complete genome sequence of *Methanocorpusculum labreanum* type strain Z. *Stand. Genomic Sci.* 1 (2): 197–203.

98 Schick, M. (2008). *[Fe]-Hydrogenase (Hmd) in zwei methanogenen Archaea ohne Gene für die die vermuteten FeGP-Kofaktor Scaffold-Proteine Hmd II and Hmd III*. Marburg, Germany: Philipps-Universität Marburg.

99 Lie, T.J., Costa, K.C., Pak, D. et al. (2013). Phenotypic evidence that the function of the [Fe]-hydrogenase Hmd in *Methanococcus maripaludis* requires seven *hcg* (hmd co-occurring genes) but not *hmdII*. *FEMS Microbiol. Lett.* 343 (2): 156–160.

100 Fujishiro, T., Bai, L., Xu, T. et al. (2016). Identification of HcgC as a SAM-dependent pyridinol methyltransferase in [Fe]-hydrogenase cofactor biosynthesis. *Angew. Chem. Int. Ed.* 55 (33): 9648–9651.

101 McGlynn, S.E., Boyd, E.S., Shepard, E.M. et al. (2010). Identification and characterization of a novel member of the radical AdoMet enzyme superfamily and implications for the biosynthesis of the Hmd hydrogenase active site cofactor. *J. Bacteriol.* 192 (2): 595–598.

102 Arriaza-Gallardo, F.J., Schaupp, S., Zheng, Y.-C. et al. (2022). The function of two radical-SAM enzymes, HcgA and HcgG, in the biosynthesis of the [Fe]-hydrogenase cofactor. *Angew. Chem. Int. Ed.* 61: e202213239.

103 Fujishiro, T., Tamura, H., Schick, M. et al. (2013). Identification of the HcgB enzyme in [Fe]-hydrogenase-cofactor biosynthesis. *Angew. Chem. Int. Ed.* 52 (48): 12555–12558.

104 Bai, L., Fujishiro, T., Huang, G. et al. (2017). Towards artificial methanogenesis: Biosynthesis of the [Fe]-hydrogenase cofactor and characterization of the semi-synthetic hydrogenase. *Faraday Discuss.* 198: 37–58.

105 Fujishiro, T., Kahnt, J., Ermler, U. et al. (2015). Protein-pyridinol thioester precursor for biosynthesis of the organometallic acyl-iron ligand in [Fe]-hydrogenase cofactor. *Nat. Commun.* 6: 6895.

106 Bai, L., Wagner, T., Xu, T. et al. (2017). A water-bridged H-bonding network contributes to the catalysis of the SAM-dependent C-methyltransferase HcgC. *Angew. Chem. Int. Ed.* 56 (36): 10806–10809.

107 Martens, J.A., Genereaux, J., Saleh, A. et al. (1996). Transcriptional activation by yeast PDR1p is inhibited by its association with NGG1p/ADA3p. *J. Biol. Chem.* 271 (27): 15884–15890.

108 Fujishiro, T., Ermler, U., and Shima, S. (2014). A possible iron delivery function of the dinuclear iron center of HcgD in [Fe]-hydrogenase cofactor biosynthesis. *FEBS Lett.* 588 (17): 2789–2793.

109 Tran, P.V., Carlson, E.S., Fretham, S.J.B. et al. (2008). Early-Life Iron deficiency anemia alters neurotrophic factor expression and hippocampal neuron differentiation in male rats. *J. Nutr.* 138 (12): 2495–2501.

110 Schulman, B.A. and Harper, J.W. (2009). Ubiquitin-like protein activation by E1 enzymes: The apex for downstream signalling pathways. *Nat. Rev. Mol. Cell Biol.* 10 (5): 319–331.

111 Galeazzi, L., Bocci, P., Amici, A. et al. (2011). Identification of nicotinamide mononucleotide deamidase of the bacterial pyridine nucleotide cycle reveals a novel broadly conserved amidohydrolase family. *J. Biol. Chem.* 286 (46): 40365–40375.

112 Schaupp, S., Arriaza-Gallardo, F.J., Pan, H.-J. et al. (2022). *In vitro* biosynthesis of the [Fe]-hydrogenase cofactor verifies the proposed biosynthetic precursors. *Angew. Chem. Int. Ed.* 61: e202200994.

113 Kuchenreuther, J.M., Myers, W.K., Stich, T.A. et al. (2013). A radical intermediate in tyrosine scission to the CO and CN^- ligands of FeFe hydrogenase. *Science* 342 (6157): 472–475.

114 Kuchenreuther, J.M., Myers, W.K., Suess, D.L.M. et al. (2014). The HydG enzyme generates an $Fe(CO)_2(CN)$ synthon in assembly of the FeFe hydrogenase H-cluster. *Science* 343 (6169): 424–427.

15

Catalysis by Hydrogenase

Seigo Shima[1], James A. Birrell[2], Sven T. Stripp[3], Giorgio Caserta[4], and Oliver Lenz[4]

[1] *Max Planck Institute for Terrestrial Microbiology, Microbial Protein Structure Group, Karl-von-Frisch-Straße 10, 35043 Marburg, Germany*
[2] *University of Essex, School of Life Sciences, Wivenhoe Park, Colchester, CO4 3SQ, UK*
[3] *University of Potsdam, Institute of Chemistry, Spectroscopy and Biocatalysis, Karl-Liebknecht-Straße 24-25, 14476 Potsdam, Germany*
[4] *Technische Universität Berlin, Institute of Chemistry, Biophysical Chemistry, Straße des 17. Juni 135, 10623 Berlin, Germany*

Chapter Goals

- Provide a comparative overview of the different catalytic cycles of the three hydrogenase classes.
- Realise that, in addition to the inorganic active centre, the properties of the enzyme's outer sphere are essential for the catalytic function.

Main Chapter Points

- Overview of the reactions catalyzed by [Fe]-hydrogenase and enzyme inhibitors.
- Semisynthetic and structural approaches to characterize the active site of [Fe]-hydrogenase.
- Proposed catalytic mechanism of heterolytic cleavage of H_2 and subsequent hydride transfer.
- General introduction to the biology and biochemistry of [FeFe]-hydrogenase.
- Structure and spectroscopy of the catalytic active site cofactor (H-cluster).
- Description and comparison of competing models for the catalytic mechanism of [FeFe]-hydrogenase.
- Brief classification of [NiFe]-hydrogenases as well as O_2-tolerance/protection mechanisms occurring in certain hydrogenase species.

Iron-Sulfur Clusters: Biogenesis and Biochemistry, Volume 2, First Edition.
Edited by Silke Leimkühler, Günter Schwarz, Oliver Lenz, and Oliver Einsle.

- Summary of all active site intermediates observed to date, including unusual architectures of the $NiFe(CN)_2CO$ center, and their relevance to the catalytic cycle.
- Role of the first and outer coordination spheres of the [NiFe] center in catalysis.

15.1 Catalytic Cycle of [Fe]-Hydrogenase

The systematic name of [Fe]-hydrogenase is H_2-forming methylene-tetrahydromethanopterin (methylene-H_4MPT) dehydrogenase (Hmd), which catalyzes the reversible activation of H_2 and the associated hydride transfer to methenyl-H_4MPT (Figure 15.1a) [1–3]. This reaction is involved in the hydrogenotrophic methanogenic pathway [4, 5]. For details on the structure and function of Hmd and the prosthetic group, the iron guanylylpyridinol (FeGP) cofactor, see Chapter 14. One feature that distinguishes Hmd from [NiFe]- and [FeFe]-hydrogenases, is that Hmd does not catalyze the dissociation of electrons from the hydride obtained by heterolytic cleavage of molecular hydrogen (H_2). The hydride is transferred to the hydride acceptor methenyl-H_4MPT^+. Despite this mechanistic difference in the enzyme reaction, Hmd can provide electrons to the hydrogenotrophic methanogenic pathway not only by reducing methenyl-H_4MPT^+ but also by reducing F_{420} in a coupled reaction with F_{420}-dependent methylene-H_4MPT dehydrogenase (Mtd) (see Figure 14.2) [6]. In this chapter, the enzymological and biophysical properties of the active site iron of Hmd, such as kinetic analysis, interaction with inhibitors, and spectroscopic properties related to the catalytic mechanism are presented, and then the currently proposed catalytic cycle is described.

15.1.1 Electronic Properties of the Iron Site of the FeGP Cofactor

The catalytic cycle of [NiFe]- and [FeFe]-hydrogenases involves the redox reaction of metal ions [7]. Since the iron state of the FeGP cofactor in Hmd and its isolated form is diamagnetic and cannot be measured by electron paramagnetic resonance (EPR) spectroscopy, it was studied by Mössbauer spectroscopy using ^{57}Fe-enriched samples [8]. The data showed that the electronic state of the FeGP cofactor is low-spin Fe(0) or low-spin Fe(II); these two states are not distinguishable by Mössbauer spectroscopy. However, the complex structure of the FeGP cofactor in the high-resolution crystal structure of Hmd and the comparison of the mimic complexes indicate that the electronic state of the iron is low-spin Fe(II) [9]. The electronic state of the iron site of the FeGP cofactor in Hmd is always constant, independent of the presence of the substrate and H_2. This property may be related to the fact that Hmd catalyzes the heterolytic cleavage of H_2 but not the dissociation of electrons from the hydride. However, since the Mössbauer analysis was not analyzed kinetically (i.e. using time-resolved techniques such as stopped-flow), it

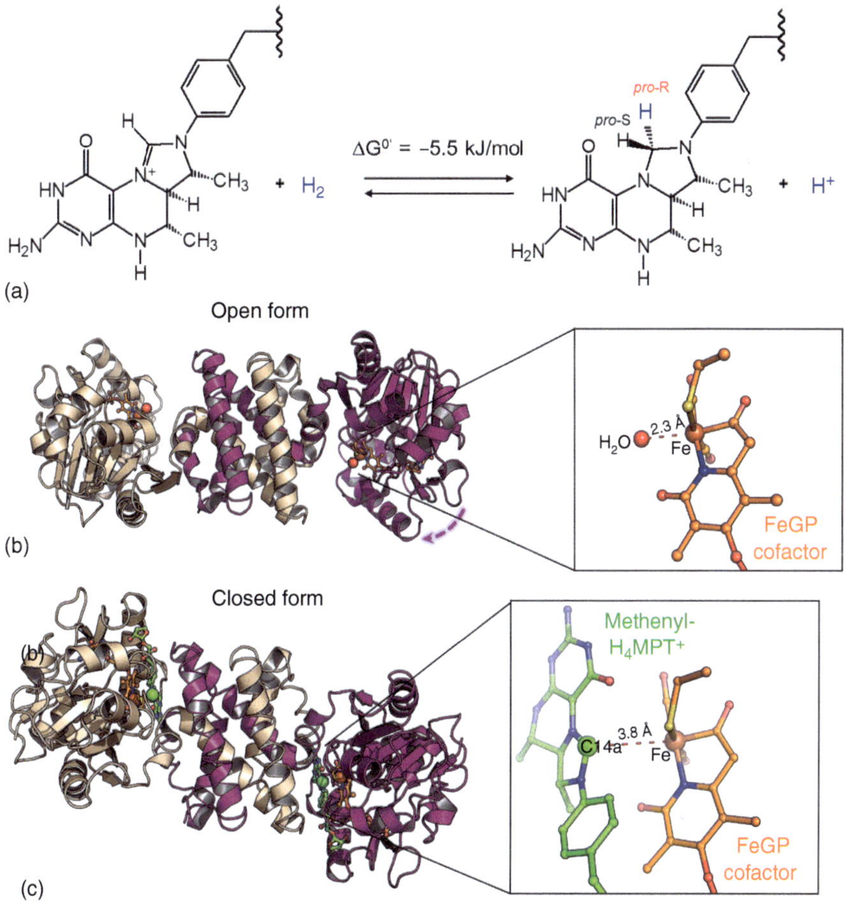

Figure 15.1 The reaction catalyzed by Hmd and the structure of the active site. (a) H_2 is heterolytically cleaved and hydride is stereospecifically transferred to C14a of methenyl-tetrahydromethanopterin (methenyl-H_4MPT^+). (b) The open from of Hmd. (c) The closed form of Hmd complexed with methenyl-H_4MPT^+. (b and c) Monomers of the Hmd homodimer are shown as wheat and purple cartoons; the FeGP cofactor and methenyl-H_4MPT^+ are shown as ball-and-stick models with orange and green carbon, respectively. The zoom-in view of the active site structures is shown in the squares. Source: Huang et al. [1]/Springer Nature.

cannot be ruled out that the iron site temporarily changes its redox state during the catalytic cycle.

15.1.2 Enzyme Reaction Kinetics

Kinetic analysis of the enzyme activity indicated that Hmd catalyzes the hydride transfer reaction with a ternary complex mechanism [2]. In addition to the hydride transfer to methenyl-H_4MPT^+, Hmd also catalyzes the proton exchange reactions to produce HD and D_2 from D_2O in the H_2 gas phase [10, 11]. These results indicate that Hmd performs not only single proton exchange reactions but also double exchange reactions. In addition, Hmd also catalyzes ortho/para-H_2 exchange reactions [12]. Importantly, these exchange reactions occur only in the presence of the substrate methenyl-H_4MPT^+. This suggests that Hmd can catalyze the heterolytic cleavage of H_2 only in the presence of this substrate [3]. At the time, before the discovery of the FeGP cofactor, this finding led to the intriguing hypothesis that the Hmd protein itself does not catalyze the H_2 activation reaction and that methenyl-H_4MPT^+ is the H_2 activation site [13]. This hypothesis was based on the reversible formation of carbocations and H_2 from alkanes under superacid conditions, as discovered by Olah et al. [14]. The proposed catalytic mechanism required the presence of the C14a carbocation of methenyl-H_4MPT^+. This substrate would bind to the active site of Hmd, and the interactions with the protein induce the carbocation of C14a by protonation at the N5 and/or N10 lone pair electrons [13]. However, recent experimental results do not support the idea that methenyl-H_4MPT^+ itself is the active site for H_2, but instead the iron site of the FeGP cofactor is activated by substrate binding (see below) [9].

15.1.3 Hmd Inhibitors and Their Contribution to the Catalytic Mechanism

Hmd is reversibly or irreversibly inhibited by organic compounds, metal ions, and reactive oxygen species. Analysis of the binding mode of the inhibitor and the mechanism of inhibition or inactivation provide clues to the structure of the active site and the catalytic mechanism of Hmd.

15.1.3.1 CO and CN^-

Most [NiFe]- and [FeFe]-hydrogenases are reversibly or irreversibly inhibited by the binding of carbon monoxide (CO) or cyanide (CN^-), respectively, to their dinuclear FeFe/NiFe active site [15]. Hmd is also inhibited by these diatomic compounds, but with very low affinity [16]. The presence of two intrinsic CO ligands at the iron site was indicated by infrared (IR) spectroscopy and, in addition, the binding of CO and CN^- to the iron site of Hmd has been confirmed by the same method [17]. IR analysis showed that CO-inhibited Hmd had three CO ligands, each binding to iron at a 90° angle. This IR spectrum is interpreted as binding of external CO to the open coordination site of the FeGP cofactor in Hmd, at which a solvent molecule is bound in the crystal structure of Hmd holoenzymes (Figure 15.1b) [18, 19]. Although the crystal structure of Hmd inhibited by CO has not been reported [16], the crystal structure of Hmd inhibited by CN^- shows an increased electron density in the solvent binding site, presumably due to CN^- [19]. These findings indicated that CO and CN^- bind to the open coordination site on the iron complex of the FeGP cofactor of Hmd [19].

15.1.3.2 Isocyanides

Isocyanides are formally isoelectronic with CO and CN^- and reversibly inhibit Hmd with very high affinity ($K_i = 1–150$ nM) [20]. These compounds are known inhibitors

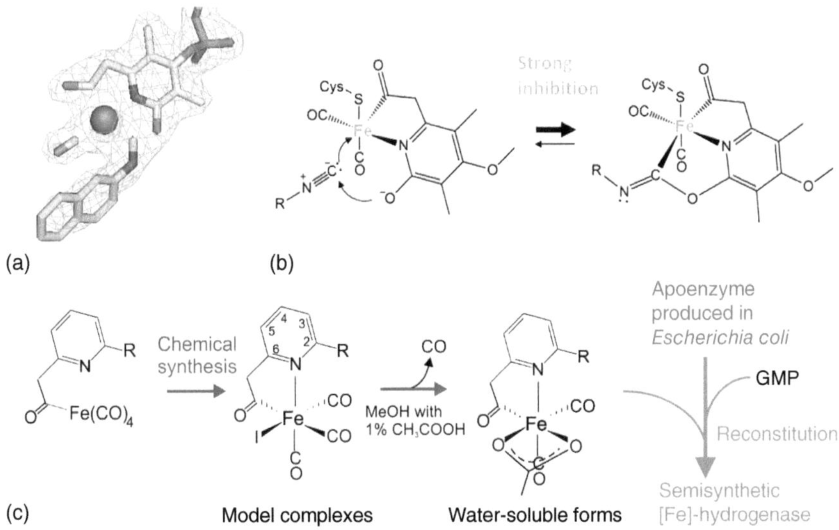

Figure 15.2 The FeGP cofactor and a mimic complex used for reconstitution of semisynthetic Hmd. (a) Crystal structure of Hmd from *M. marburgensis* inhibited by naphthyl isocyanide, a stick model with electron density. (b) Strong inhibition mechanism of Hmd with isocyanides. (c) Preparation of semisynthetic Hmd. R: OH or OCH_3 group. Source: Tamura et al. [21]/John Wiley & Sons.

of nitrogenase, CO dehydrogenase, and P450, but the affinity of isocyanides for these metalloenzymes is not very high (0.1–2 mM) and [NiFe]-hydrogenase and [FeFe]-hydrogenase tested are not inhibited by isocyanides [20]. For the inhibition of Hmd, *p*-toluenesulfonylmethyl isocyanide (TosMIC), cyclohexyl isocyanide, butyl isocyanide, methyl isocyanoacetate, and naphthyl isocyanide (NIC) have been tested. Cyclohexylisocyanide and TosMIC showed slow binding properties. UV-visible and IR spectroscopy indicated that the isocyanides bind to the iron site of the FeGP cofactor [20]. Crystal structures of Hmd from *M. marburgensis* inhibited with TosMIC (K_i = 2 nM) and NIC (K_i = 10 nM) were reported (Figure 15.2a) [21]. In the crystal structure, the isocyanide carbon binds to the FeGP cofactor by bridging between the open coordination iron site trans to the acyl ligand and the deprotonated 2-OH group of the pyridinol ring. The binding mode may be one of the reasons for the common low K_i values of isocyanides (Figure 15.2b). TosMIC is used as a standard inhibitor to test the Hmd reaction to distinguish it from the other reactions [22]. The complex structure of the enzyme inhibitor showed reactivity of the open coordination site of the iron site and the 2-OH group of the pyridinol ring with external compounds, suggesting that these two sites may also be involved in the reaction with H_2 [21].

15.1.3.3 Cu

The current standard enzyme-activity assay buffer for Hmd contains 1 mM EDTA [23], but no such chelating reagent was added in the early days [2]. However, when the enzyme activity assay for Hmd was performed without the addition of the chelating reagents, inactivation was occasionally observed. When chelating reagents (1 mM sodium citrate or 1 mM EDTA) were added, activity was constantly observed, suggesting that metal ions were involved in the inhibition. The first metal ion identified as an inhibitor was the copper ion. The addition of Cu(II) irreversibly inhibited Hmd in a time-dependent manner, with a half-maximum rate of inhibition at 1 μM with an inactivation rate constant of $1 \times 10^{-3}\ s^{-1}$ [24]. Removal of copper ions from the inhibited enzyme solution did not restore enzyme activity.

IR spectra of Hmd in the presence of Cu(II) showed a shift in the CO band, indicating that the copper ions reacted with the FeGP cofactor of Hmd [24]. EPR analysis showed that when Cu(II) and Hmd were mixed, the Cu(II) species were converted to the Cu(I) species, suggesting that the effective inhibition species is Cu(I) formed by the reduction of Cu(II). The FeGP cofactor structure contains two possible electron donors, an Fe(II) center and an acyl ligand; since mixing Cu(II) with Hmd did not change the EPR silent Fe(II) property, the EPR analysis ruled out the possibility of Cu(II) becoming Cu(I) by reduction through the Fe(II) site. Conversion of acyl ligands to carboxy groups has been observed during photolysis and decomposition by H_2O_2 (see below), so this remains a possibility.

As mentioned above, isocyanide binds to the open coordination site of the FeGP cofactor of Hmd with very high affinity [21]. Isocyanide was used as a tool to analyze inhibitor binding sites by protecting the active site from other inhibitors (e.g. inactivation by H_2O_2, see below) [24]. However, inhibition by copper was not protected by isocyanide binding to Hmd, and the IR spectra of isocyanide-bound Hmd were shifted by the addition of copper ion. These results suggest that Cu(I) probably attacks the sites of the FeGP cofactor other than the open coordination site that binds isocyanide.

15.1.3.4 Fe

The finding of inhibition of copper ion prompted an analysis of the effect of other metal ions on Hmd [24]. This analysis showed that Fe(II) strongly and Fe(III) weakly inhibited Hmd, and the inhibitory effect of Fe(II) was analyzed in detail. Inhibition by Zn(II), Ni(II), Co(II), Ca(II), Mg(II), and Mn(II) was not detected. Kinetic analysis indicated that Fe(II) reversibly inhibits with a very low inhibition constant (K_i = 40 nM) and that the inhibition is competitive with the substrate methylene-H_4MPT, suggesting that Fe(II) can bind to Hmd at the position near the methylene-H_4MPT binding site. Consistent with this finding, the IR bands of Hmd were not shifted by inhibition with Fe(II). Mössbauer spectroscopy of Hmd mixed with [^{57}Fe]-$FeSO_4$ showed no significant change in the redox state of the iron site, supporting that Fe(II) does not directly interact with the iron site of the FeGP cofactor in Hmd [24]. This result suggests that iron binding inhibits the formation of the protein structure required for substrate binding.

15.1.3.5 H_2O_2

Inhibition of Hmd by reactive oxygen species was found by experiments using xanthine oxidase (XOD) to study the effect of superoxide radicals (O_2^-) [25]. Subsequently, H_2O_2 coproduced by the XOD reaction was found to be a major irreversible inhibitor of Hmd [24]. Since the addition of Fe(II) to the assay solution containing H_2O_2 did not enhance the effect of H_2O_2, it was concluded that the hydroxyl radical produced by the Fenton reaction was not the cause of the H_2O_2 inhibition. The IR band of Hmd shifted upon addition of H_2O_2 and the peak height of CO decreased in a time-dependent manner, indicating that H_2O_2 bound to the iron site of the FeGP cofactor of Hmd and degraded the iron complex with dissociation of the CO ligand [24]. Isocyanide-inhibited Hmd was protected from inactivation by H_2O_2, suggesting that H_2O_2 may bind to the open coordination site of the iron center. H_2O_2 induced some EPR-active Fe(III) species in Hmd, indicating that oxidation of the iron site can occur. These results indicate that H_2O_2 binds to the open coordination site of the FeGP cofactor and destroys the iron complex by oxidizing the iron complex of the FeGP cofactor [24].

15.1.3.6 O_2 is Reduced to H_2O_2 by the Hmd Reaction

No significant inactivation was detected when Hmd was incubated for 72 hours on ice under 100% N_2 or 100% O_2. This result indicates that Hmd is not sensitive to O_2. However, the situation changes when the reducing substrates, methylene-H_4MPT or methenyl-H_4MPT^+ with H_2, are present in the enzyme solution [26]. In the presence of the reducing substrates, Hmd is inactivated in the presence of O_2. This phenomenon has also been confirmed in the Hmd enzyme assays under O_2; the initial rate of the Hmd reaction is the same under anoxic and oxidative conditions. However, the reaction rate slows down under the oxic conditions, and the reaction rate does not recover when the conditions are changed to anoxic. These results indicate that Hmd is inactivated under oxidative conditions, but this phenomenon is limited to the presence of reduced substrate. The UV-visible and IR spectra of the inactivated Hmd indicated that the iron moiety of the FeGP cofactor was degraded by the inactivation of Hmd [26].

Since it was known at that time that H_2O_2 inactivates Hmd [24], it was hypothesized that H_2O_2 produced by the reduction of O_2 inactivates Hmd. The formation of H_2O_2 in the tested solution was detected by an H_2O_2 assay using 10-acetyl-3, 7-dihydroxyphenoxazine [26]. The addition of catalase, which dismutates H_2O_2 to O_2 and H_2O, reduced the concentration of the dye-reacting compound in the assay, confirming that the substance formed was H_2O_2. H_2O_2 inactivates Hmd through a catalytic inactivation mechanism, by which substoichiometric very low concentrations of H_2O_2 can inactivate Hmd [26].

Given the reaction conditions, one would expect the two electrons for the formation of H_2O_2 from O_2 to be supplied directly by the reducing substrate. The common reaction intermediate for the reaction of the reducing substrates is expected to be the hydride bound to the iron moiety of the FeGP cofactor of Hmd. The authors propose a mechanism whereby O_2 binds to the iron hydride intermediate and the hydride is transferred to O_2 [26]. Based on the metal hydride-driven hydrogenation reaction by ruthenium complexes [27], the possible reaction of O_2 with hydride on the FeGP cofactor was proposed. Thus, the O_2 reduction reaction supports the presence of the proposed Fe-H intermediate in the Hmd catalytic cycle (see below). However, considering the conditions for the formation of iron hydride in Hmd, the hydride transfer reaction to O_2 is not straightforward because iron hydride is formed only when Hmd is in the closed conformation. The chance for O_2 to react with iron hydride is when the Hmd protein at least slightly opens. To explain this discrepancy, we can consider that not all iron hydride intermediates produced by the Hmd reaction react with the true hydride acceptor (methenyl-H_4MPT^+ or proton), but some remain on the iron and can react with O_2 in an open conformation.

15.1.4 Semisynthetic Hmd

Site-directed mutagenesis experiments are often used to study the catalytic site of enzymes. Variants can be easily generated by mutating the gene. However, if the catalytic site contains a cofactor, as in the case of Hmd, it is not easy to determine which part of the cofactor is responsible for the activity. In such cases, the method used is to chemically synthesize the cofactor and incorporate it into a protein to produce a semisynthetic enzyme. By comparing the characteristics of the variants, it is possible to identify the part of the cofactor responsible for the catalytic activity. An example in hydrogenase research is an experiment to identify atoms in the thiolate bridge of the active-site complex (H-cluster) of [FeFe]-hydrogenases, which connects the two iron ions. Precursors of an H-cluster can be synthesized and incorporated into a natural-type protein to produce semisynthetic [FeFe]-hydrogenases. To test the function of the thiolate bridge of the H-cluster, the bridging atom was replaced by nitrogen (NH), carbon (CH_2), or oxygen (O) [28]. By studying the activities of each of the semisynthetic enzymes, the natural bridging atom was identified as nitrogen.

In the case of Hmd, the complete FeGP coenzyme cannot yet be synthesized chemically, but analogous metal complexes containing a similar pyridinol moiety and the iron complex have been synthesized [29]. Studies of semisynthetic Hmd containing the mimic complexes have allowed the relationship between pyridinol structure and activity to be studied. This section describes the studies on semisynthetic Hmd.

A mimetic complex resembling the real FeGP cofactor was reported for the first time by the group of Xile Hu [30]. This mimic complex has a pyridinol ring providing its nitrogen and acyl ligands, and two CO ligands in the same geometry with the natural FeGP cofactor. Despite the structural similarity to the natural coenzyme, this complex and its derivatives did not exhibit H_2-activating activity. The differences from the real cofactor were the absence of the protein environment, including the cysteine protein ligand, and the fact that the pyridinol ring of the mimic compound has a modified pyridinol structure without the 3,5-methyl and 4-guanyl groups.

There are two possible reasons for the lack of activity of the mimic complex: the change in the electronic properties of the pyridinol ring due to various substitutions and the absence of a protein environment including a cysteine thiolate ligand. To investigate this further, we attempted to introduce the mimic complex into the Hmd apoenzyme from *Methanocaldococcus jannaschii* [31]. The mimic complex used in the reconstitution experiment is shown in Figure 15.2c, in which two of the three CO ligands are coordinated in the same position as the CO ligand of the FeGP cofactor, and the other CO is coordinated to the opposite side of the acyl ligand.

Iodine is ligated to the cysteine-thiolate binding site. The mimic complex was insoluble in water, but possible to be dissolved in methanol containing 1% acetic acid. IR spectra indicated that a CO ligand dissociated from the mimic complex by this treatment, and the IR spectrum of the two CO ligands was very similar to that of the FeGP cofactor extracted with acetic acid. This result suggests that the mimic complex has ligated with the acetic acid ligand and become water soluble (Figure 15.2c) [31].

When the solutions of Hmd apoenzyme and mimic complex were mixed, the mimic complex bound to the protein and the semisynthetic Hmd showed enzymatic activity [31]. The rate of reconstitution, as measured by changes in enzyme activity, showed that complete reconstitution took about one hour, which is much slower than reconstitution with the FeGP cofactor [23]. This is probably due to the absence of the GMP moiety in the mimic complex structure, which can guide the binding to the nucleotide-binding site of the Hmd protein [32]. Addition of GMP to the reconstituted solution with the mimic complex increased the final enzyme activity twofold. It is likely that the external GMP bound to the nucleotide-binding site of the Hmd protein and induced a conformational change in the protein to conform to the active site structure for higher activity. The activity of the reconstituted semisynthetic Hmd was tested by standard photometric assays of the oxidation of methylene-H_4MPT and the reduction of methenyl-H_4MPT^+ with H_2, the former reaction was also confirmed by the formation of H_2. The specific activity of the semisynthetic Hmd was about 1% of that of the native enzyme, but the catalytic activity was much higher than that of synthetic hydrogenation catalysts [31].

The CO peak observed for the acetate-bound mimic complex shifted when mixed with the Hmd apoenzyme, indicating that the mimic complex bound to the protein and the surrounding environment changed. The two IR peaks of the semisynthetic Hmd were similar to the peak of the FeGP cofactor bound to Hmd [31, 33], indicating correct binding. The additional smaller peak in the IR spectrum was interpreted as a nonspecifically bound mimic complex, and the additional IR peak was reduced by the addition of GMP, suggesting that GMP binding improved the protein environment of the cofactor. To analyze the interaction with the cysteine protein ligand (Cys176) in the semisynthetic enzyme, an Hmd apoenzyme, in which Cys176 was replaced by alanine, was prepared and mixed with the mimic complex [31]. The IR spectra of the semisynthetic Cys176Ala Hmd were different from that using native protein and showed no enzyme activity. These results support that Cys176 of the Hmd apoenzyme interacts with the mimic complex in the semisynthetic enzyme.

To investigate the reason for the low specific activity of the semisynthetic enzyme, the FeGP cofactor lacking the GMP moiety was prepared by digestion with phosphodiesterase I from snake venom [31]. This enzymatic digestion hydrolyzed the phosphate group of the GMP moiety, leaving a 4-OH group on the pyridinol ring. The occupancy of the truncated FeGP cofactor in reconstituted Hmd was only 4% of the apoenzyme molecules and the specific activity was 0.8 U/mg, indicating that the truncated FeGP cofactor reduced the specific activity of Hmd by 20-fold.

These results indicate that the presence of the GMP portion of the FeGP cofactor has a significant effect on the specific activity of Hmd.

15.1.4.1 Importance of the 2-OH Group of the Pyridinol Ring

It is hypothesized that the deprotonated 2-OH group of the pyridinol ligand of the FeGP cofactor serves as the catalytic base [34]. The function of the deprotonated 2-OH group was tested using mimic complex derivatives in which the 2-OH group of the pyridinol ring is replaced by a 2-methoxy group (CH_3-O-R) [31]. This is because the methoxy group cannot be used as a catalytic base. A mimetic complex with a 4-methoxy group was chemically synthesized and used to reconstitute the semisynthetic Hmd [31]. IR spectra of this semisynthetic enzyme confirmed that this new complex was correctly bound. Photometric enzyme activity measurements indicated that this semisynthetic Hmd was essentially inactive. This result supported the importance of the 2-OH group for the catalytic reaction of Hmd.

15.1.4.2 Semisynthetic [Mn]-Hydrogenase

The only metals used for hydrogenase in nature are nickel and iron. While other metals are used in chemically synthesized complexes that perform H_2 activation, why are only nickel and iron used in nature? This question led us to prepare a mimic complex of the FeGP cofactor containing Mn(I), which is isoelectronic with Fe(II) [35]. The structure of the synthesized [Mn] complex was identical to that of the [Fe] complex that showed activity in the semisynthetic Hmd. A notable difference is that the [Mn] complex alone already showed some H_2 activation activity, while the [Fe] complex was inactive.

Manganese-containing semisynthetic Hmd ([Mn]-hydrogenase) was prepared by the same procedure as semisynthetic Hmd with iron-containing mimetic complexes [35]. The reconstituted [Mn]-hydrogenase exhibited enzymatic activity. The occupancy of the [Mn] complex at the active site was 20%, in which the [Mn] complex was presumably specifically bound to the Hmd apoenzyme. Taking into account the occupancy, the enzyme activity was 7.5 U/mg for the reduction reaction and 0.45 U/mg for the oxidation reaction, which are much higher than in the absence of enzyme protein.

The [Mn] mimetic complex was soluble in methanol, but the complex in methanol solution did not reconstitute the active semisynthetic enzyme. Only when the [Mn] complex was solubilized in a methanol solution containing 1% acetic acid, an active enzyme was obtained. This result indicates that the acetic acid ligand is important for reconstitution. However, IR analysis showed that even when the [Mn] complex was solubilized in the presence of acetic acid, an excess amount of [Mn] complex was bound to the Hmd apoenzyme and its IR spectrum was similar to that of the original [Mn] complex. Therefore, the binding properties of the [Mn] complex could not be determined by the IR spectra.

The specific binding mode of the [Mn] complex was verified by mutational analysis [35]. Three possible binding residues of the Hmd protein that interact with the mimic complex are Thr13, D251, and C176, with C176 binding to the metal center and Thr13 and D251 binding indirectly to the complex via water molecules.

The triple mutant containing T13V, C176A, and D251A lost the activity of the semisynthetic manganese hydrogenase, but the single mutant C176A showed 8% of the semisynthetic [Mn]-hydrogenase with the native Hmd apoenzyme. This is in contrast to the loss of activity in the semisynthetic [Fe]-hydrogenase with the iron-containing mimetic complexes due to the C176A mutation [31]. This may be related to the fact that the manganese mimetic complex shows some activity in the absence of protein [35]. Thus, the semisynthetic [Mn]-hydrogenase showed significantly higher activity than that prepared with the equivalent [Fe]-mimetic complex. The reason why nature does not use manganese as the metal of Hmd is unclear; however, one possibility is that the properties of manganese and iron in biosynthesis and binding of the cofactor to the protein may be the reason for natural selection. It would be interesting to understand the properties of the MnGP cofactor, which has the same structure as the native FeGP cofactor.

15.1.5 Circular Dichroism Spectroscopy

The conformational change of Hmd upon binding to the substrate was first detected by circular dichroism (CD) spectroscopy [36]. CD spectroscopy can detect changes in chiral chromophores and achiral chromophores in a chiral environment. In protein chemistry, CD spectroscopy is primarily used to analyze the secondary structure of proteins. CD spectroscopy is also used to analyze metal complexes and metalloenzymes. The binding of CO and CN^- inhibitors to the FeGP cofactor of Hmd was clearly demonstrated by CD spectroscopy, indicating that CD spectroscopy is useful for studying Hmd. In the absence of methenyl-H_4MPT^+, simply replacing the gas phase of the Hmd sample from N_2 to H_2 had no effect on the CD spectrum. However, when methenyl-H_4MPT^+ was present and the Hmd solution was placed under the H_2 gas phase, the CD spectrum was drastically altered. This change in the CD spectrum can be interpreted as H_2 reacting with methenyl-H_4MPT in the closed conformation.

15.1.6 Crystal Structure of Hmd with the Substrate

The three-dimensional structure of the enzyme–substrate complex is essential for elucidating the catalytic cycle, but in the case of Hmd, it has been difficult to obtain such a complex crystal. Perhaps, one of the reasons is that Hmd catalyzes a reversible reaction in an anaerobic glove box containing a 95% N_2/5% H_2 gas mixture, which causes the crystallization solution to be a mixture of methenyl-H_4MPT^+ and methylene-H_4MPT. Such heterogeneous substrate mixtures can interfere with crystal growth. Our first attempt was to use enzymatically inactive Hmd; i.e. the C176A variant, which lacks the cysteine residue that binds to the FeGP cofactor. This C176A Hmd from *M. jannaschii* was successfully crystallized with methylene-H_4MPT [34]. In this Hmd complex structure, methylene-H_4MPT is bound to the central domain side of the active site cleft, and there is a distance of 9.3 Å between it and the iron site of FeGP bound to the N-terminal domain side, which is too large for direct hydride transfer, and the cleft is filled with bulk solvent.

In this situation, H_2 activation and hydride transfer are impossible. The open structure of Hmd is due to the C176 variation, which shifts the binding position of the FeGP cofactor and changes the open/closed property of this protein. We modeled the closed structure of the Hmd–substrate complex using the geometry of the previously obtained closed Hmd structure. Based on the closed structure model, we were able to model the cofactor and substrate in the appropriate positions for the reaction within the closed cleft.

Subsequently, the enzyme–substrate complexes with native and active Hmd were crystallized in an anaerobic chamber with only nitrogen gas without hydrogen. Under these conditions, *Methanococus aeolicus* Hmd was crystallized in a complex with methenyl-H_4MPT^+ and without this substrate. As predicted, the open structure in the absence of the substrate was converted to a closed structure by the binding of the substrate. In the closed structure, the distance between the methenyl carbon and the iron was 3.8 Å (Figure 15.1c), similar to that observed in the closed structure modeled from the structure of C176A Hmd. The solvent molecules were removed from the active site cleft in the closed *M. aeolicus* Hmd, and a hydrophobic environment was formed between the iron moiety and the methenyl carbon. The iron site was connected to the outside by a narrow channel, which was expected to allow molecular hydrogen to diffuse in. The closed structure was analyzed at atomic resolution (1.06 Å) and the structure of the active site was elucidated in detail [9]. In particular, the solvent molecule was bound to the opposite side of the acyl group in the open structure, but was lost in the closed structure, probably due to the proximity of the methenyl-H_4MPT^+ (Figure 15.1b,c). Thus, substrate binding induces a closed structure and the solvent molecule bound to the iron is removed. This is the first experimental demonstration of this phenomenon, although it was previously predicted using density functional theory (DFT) calculations [37]. This implies that hydrogen molecules can only bind after this substrate-dependent removal of solvent molecules. It adds an experimental explanation to the methenyl-H_4MPT^+-dependent H_2 activation mechanism shown in previous kinetic and spectroscopic studies.

The atomic-resolution crystal structure of the closed conformation of Hmd clearly shows that there are no amino acid residues that protonate N5 or N10 of methenyl-H_4MPT^+, which is a prerequisite of formation of the carbocationic character of the methenyl ring. In the closed enzyme/substrate complex structure, the methenyl ring has a planar conformation with a C14a–N5 double bond and a C14a–N10 single bond rather than a symmetrical deplanarized conformation predicted by Berkessel and Thauer [13, 38]. This methenyl-H_4MPT^+ conformation is not consistent with the carbocation hypothesis.

15.1.7 Proposed Catalytic Mechanism of Hmd

A catalytic cycle of Hmd is proposed based on enzyme kinetics, exchange reactions, electronic properties of the iron site and the structure of pyridinol, and the structure of the holoenzyme complexed with methenyl-H_4MPT^+ (Figure 15.3). Binding of methenyl-H_4MPT^+ to the enzyme induces the closed conformation, in which the

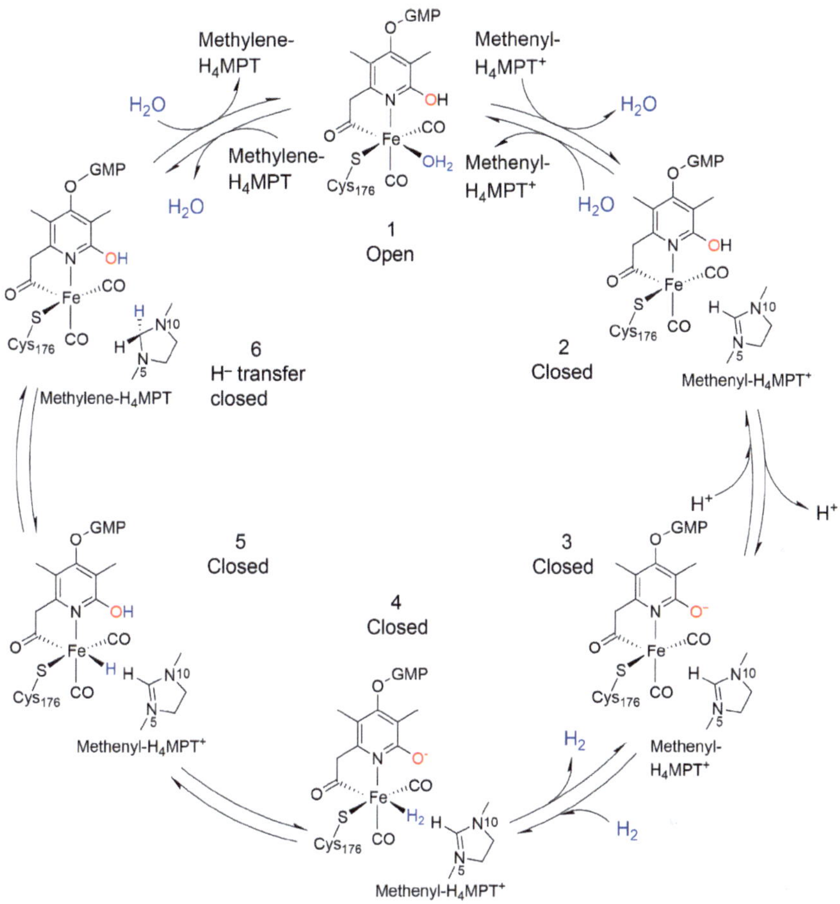

Figure 15.3 The proposed catalytic cycle of Hmd. Binding of methenyl-tetrahydromethanopterin (methenyl-H_4MPT^+) induces the closed conformation of the active site cleft, in which the solvent (H_2O) molecule is dissociated (1→2). In the closed conformation, the 2-OH group is deprotonated (2→3). H_2 is transferred through the narrow channel to bind the iron open coordination site (3→4). H_2 bound to the iron site is heterolytically cleaved by using the deprotonated 2OH group of the pyridinol of the FeGP cofactor as the catalytic base (4→5). Hydride is transferred to the methenyl-H_4MPT^+ (5→6). The product methylene-H_4MPT^+ is dissociated and a water molecule is bound to the open coordination site of the FeGP cofactor (6→1). Source: Adapted from Huang et al. [9].

solvent ligand bound to the iron site dissociates, which activates the iron site to react with H_2. H_2 diffuses to the active site through a protein channel to the iron site in the hydrophobic environment of the protein. H_2 binds to the exposed iron site. The H_2 molecule is heterolytically cleaved using the deprotonated 2-hydroxy group of the pyridinol ring of the FeGP cofactor as a catalytic base. The hydride is transiently left on the iron site and then transferred to C14a of methenyl-H_4MPT^+ and the proton is accepted by His14 next to the 2-hydroxy group of the pyridinol ring. DFT calculations based on the open/closed conformations of Hmd from *M. aeolicus* supported the catalytic mechanism.

15.1.8 Conclusion

Many types of research have contributed to the understanding of the catalytic mechanism of Hmd, in which the most important breakthrough appears to be the finding of the substrate (methenyl-H_4MPT^+)-dependent open/closed conformational change of the Hmd protein, in which the water molecule bound to the iron site is removed and the active site iron is activated by substrate binding. The proposed catalytic mechanism can explain all kinetic properties of Hmd including the substrate-dependent exchange reactions. The unique property of Hmd is the direct hydride transfer of H_2 to the organic hydride acceptor. The proposed catalytic mechanism indicates that the activation from H_2 occurs at the FeGP cofactor rather than at the substrate, methenyl-H_4MPT^+. Based on this mechanism, we can design new types of [Fe]-hydrogenases using hydride acceptors, e.g. $NAD(P)^+$, other than methenyl-H_4MPT^+, which will open a way to use [Fe]-hydrogenases as a reducing enzyme in biotechnologically important metabolism. Recent progress in understanding the biosynthetic enzymes of the FeGP cofactor (Section 14.12) will allow the use of Hmd in heterologous systems other than methanogens in the future.

15.2 Toward the Catalytic Mechanism of [FeFe]-Hydrogenase

[FeFe]-hydrogenases are enzymes containing organometallic cofactors featuring metal-carbon bonds [39–41]. Until recently, these enzymes were thought to only be found in fermentative bacteria and a few groups of eukaryotes including green algae and anaerobic fungi; however, recent work has shown that they are also widespread in archaea [42, 43]. They are distinguished from [NiFe]-hydrogenases and [Fe]-hydrogenases by the presence of a diiron cofactor [7]. The most commonly investigated [FeFe]-hydrogenases stem from *Clostridium pasteurianum* (*Cp*I), *Desulfovibrio desulfuricans* (*Dd*HydAB), and *Chlamydomonas reinhardtii* (*Cr*HydA1). These hydrogenases belong to the same phylogenetic clade ("Group A") and efficiently allow for the reversible interconversion of protons and electrons to hydrogen [7, 43, 44]. The [FeFe]-hydrogenase *Dd*HydAB, for example, shows extremely high activity with a turnover frequency of ~8000 s^{-1} for H_2 production and ~150 000 s^{-1} for H_2 oxidation under mild conditions (pH 8, 30 °C) [45, 46]. Group A also includes enzymes that are assumed to have an electron-bifurcation function [47]. Meanwhile, enzymes from Groups B–D are much less well characterized. Some examples of Group C and D enzymes have been suggested to have sensory roles [48–50]. All [FeFe]-hydrogenases operate under strictly anaerobic conditions and a complete loss of activity is observed when they are exposed to air, in most cases through a decay mechanism, which is not fully understood [51–56]. However, there are some [FeFe]-hydrogenases that are able to resist irreversible inactivation by forming inactive, O_2-protected states [57].

Understanding the structural features of [FeFe]-hydrogenase can reveal how this enzyme performs H_2 interconversion at high rates without requiring significant electric overpotential. The first crystal structures were determined from the isolated [FeFe]-hydrogenases *Cp*I as well as *Dd*HydAB [58, 59]. These structures, while obtained from different organisms, revealed essentially identical structures of the active site cofactor domain; however, it was not possible to unambiguously assign the chemical structure of the bridging ligands. Later studies converged on the same structural model with only minor discrepancies [60].

15.2.1 The Active Site Cofactor Is an Iron–Sulfur Cluster

The active site cofactor of [FeFe]-hydrogenase is referred to as hydrogen-forming or "H-cluster." This cofactor consists of a diiron site and a [4Fe–4S] cluster, designated $[2Fe]_H$ and $[4Fe–4S]_H$ (Figure 15.4a). Both clusters are covalently linked to the protein backbone *via* cysteine residues, one of which bridges $[4Fe–4S]_H$ and $[2Fe]_H$, and a range of noncovalent interactions further stabilize $[2Fe]_H$ [61]. Remarkably, the bridging cysteine residue is the only covalent attachment of the $[2Fe]_H$ moiety to the protein scaffold. The H-cluster is the defining feature of all [FeFe]-hydrogenases, with the major difference between enzymes being the number of "accessory" [4Fe–4S] clusters, which help exchange electrons with $[2Fe]_H$ [39–41]. In addition to the H-cluster, *Cp*I contains three [4Fe–4S] clusters and one [2Fe–2S] cluster, *Dd*HydAB possesses only two additional [4Fe–4S] clusters, and *Cr*HydA1 is comprised exclusively of the H-cluster [62]. The protein scaffold can be regarded as an "oversized" natural organic ligand, although it is worth mentioning that the protein environment not only serves as an anchor for the H-cluster. The protein allows for a controlled shuttling of protons (H^+) and H_2 between the active site and solvent and it defines the geometry of the H-cluster [63–65].

With respect to H_2 shuttling, the enzyme contains distinct hydrophobic gas channels that reach from the molecular surface to the active site [66]. These gas channels are not completely selective towards H_2, and the inhibitors CO and O_2 are also capable of reaching the active site through this channel [67]. Hence, the enzyme operates exclusively under anaerobic conditions. Molecular dynamics (MD) simulations helped suggest a proton channel from the structures of *Dd*HydAB and *Cp*I [68, 69]. Experimental work on *Cr*HydA1 and *Cp*I confirmed that protons travel between the active site and solvent *via* a Grotthuss-type mechanism [70–72] following a conserved proton transfer (PT) pathway established by glutamic acid, serine, arginine, and cysteine residues and "conserved" water molecules (Figure 15.4b). Although the sequence identity of Group A and the H_2-sensing [FeFe]-hydrogenases of Groups C and D is high in the region around the H-cluster, the latter two lack the conventional PT pathway [73–75].

The $[2Fe]_H$ cluster is the active catalytic center where reversible proton reduction or H_2 oxidation occurs. It is best described as a butterfly shaped [2Fe–2S] cluster (Figure 15.4b). The proximal iron ion (Fe_p, proximal to the $[4Fe–4S]_H$ cluster) is coordinated by one terminal CN^- and one terminal CO ligand and linked to the $[4Fe–4S]_H$ cluster *via* the bridging cysteine [58, 59]. The short distance between Fe_p and $[4Fe–4S]_H$ of 4 Å as well as the covalent attachment *via* a sulfur bond allows for rapid electron transfer [76] between the two halves of the H-cluster.

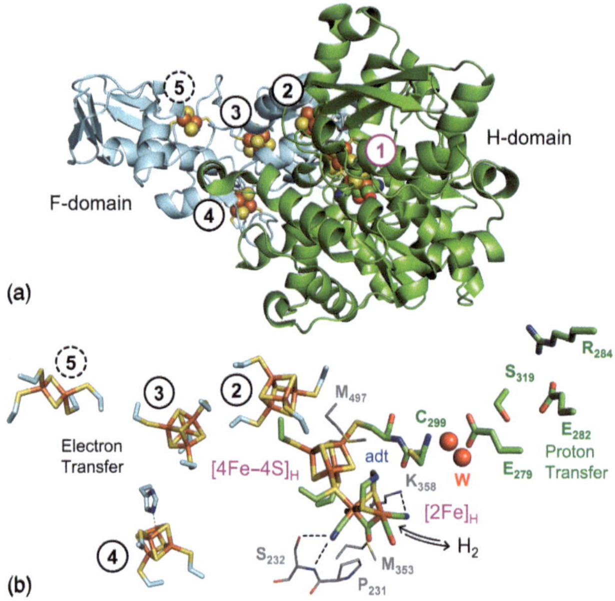

Figure 15.4 (a) Crystal structure of the [FeFe]-hydrogenase from *Clostridium pasteurianum* (PDB ID 3C8Y). (b) Close-up of the H-cluster including the electron (left) as well as proton (right) transfer pathways. The figure also shows the linking of the $[2Fe]_H$ cluster to the protein environment via the cyanides CN_p and CN_d as well as to the $[4Fe-4S]_H$ cluster via cysteine C_{503}.

The distal iron (Fe_d, distal to the $[4Fe–4S]_H$ cluster) is coordinated to one terminal CN^- and one terminal CO ligand as well but offers a square-pyramidal coordination environment with a vacant coordination site in the apical position, ideally suited for H_2 binding/formation [77–79]. In all crystallographically characterized states of the H-cluster, Fe_p and Fe_d share a bridging carbonyl ligand, μCO. The redox-dependent behavior of this ligand has been the subject of much debate in the literature and will be discussed further later in this chapter. Finally, Fe_p and Fe_d are bridged by a unique 2-azapropane-1,3-dithiolate (adt) ligand, which is thought to provide a bridging nitrogen base for heterolytic splitting of H_2 [28, 80].

15.2.2 The Electronic Structure of the H-cluster

The $[4Fe–4S]_H$ cluster can be either in the oxidized (+2) or in the reduced (+1) states and the iron ions of the $[2Fe]_H$ cluster alternate between Fe^I and Fe^{II} [39–41]. The oxidized resting state ("H_{ox}") is best described as $[4Fe–4S]^{+2}–[Fe_p^{II}-Fe_d^{I}]$ [81]. However, data from IR spectroscopy, EPR spectroscopy, and DFT calculations suggest that the H-cluster is a system of highly delocalized electron density with a high degree of covalency. Therefore, formal oxidation state assignments should only be used as approximations [82, 83].

The CO ligands of the $[2Fe]_H$ cluster are located in hydrophobic pockets (Figure 15.4b) and allow for regulation of the electronic properties of the various intermediate states of the H-cluster during catalysis – a common feature of organometallic catalysts that utilize strong field ligands [84]. The CN^- ligands have been shown to act as hydrogen-bonding acceptors within the enzyme's binding pocket, in particular at the proximal side [63], while on the distal side a lysine residue may form a salt-bridge interaction with the negatively charged CN^- ligand [85]. The protein environment is important not only because it directly controls electronic exchange but it also controls the $[2Fe]_H$ cluster's shape, i.e. into its specific active-ready geometry with an open coordination site for H_2 binding [64].

The adt ligand bridges the iron ions of the $[2Fe]_H$ cluster and is located at the end of the PT pathway (Figure 15.4b). While there has historically been much debate on the nature of this dithiolate ligand [58, 59], it is now clear that the ligand is composed of a 2-azapropane-1,3-ditholate bridge (adt, -SCH_2NHCH_2S-) [28, 80]. The combination of an open coordination site on Fe_d and a pendant nitrogen base in the adt ligand form a "frustrated Lewis pair". This special assembly can be rationalized as the origin for the unique performance of this enzyme and allows heterolytic cleavage of H_2 as well as its formation from an iron-bound hydride and an amine-centered protonation. The special structural arrangement of the $[2Fe]_H$ cofactor is called the "rotated state" (with respect to the pseudo-symmetric arrangement of the early enzyme mimetics) and forces the cofactor to display an apical, vacant coordination site on Fe_d and a characteristic μCO ligand [77]. Blocking this site with ligands such as CO, CN^-, sulfide, or even a cysteine thiolate leads to inhibition of the enzyme [86–90].

Even though the structural composition of the H-cluster is known and we have a good grasp on its biosynthesis [91, 92], the catalytic mechanism of H_2 evolution and oxidation is still highly debated. Several aspects are known to play key roles: the mixed-valence redox state H_{ox}, the conservation of a rotated state throughout catalysis, and the proton relay function of the adt ligand [93]. The formation of terminal hydrides rather than bridging hydrides [94–97] and coupled proton and electron transfer [98–102] are accepted as fundamental steps for the enzyme's unprecedented catalytic performance but the details remain the subject of controversy. Several resting and inhibited states of the $[2Fe]_H$ cofactor have been determined crystallographically [103–106]; however, it is not clear whether all important catalytic intermediates can be captured in crystal structures, in particular due to the time and conditions required to obtain suitable crystals. As such, crystal structures should be regarded as thermodynamically stable configurations of the enzyme. In contrast to single crystal X-ray diffraction of small molecules, the resolution of enzymatic crystal structures is often insufficient to reveal fine details expected between the different states [107]. Particularly problematic is that protons and hydrides, the most interesting components of the catalytic mechanism interact poorly with X-rays. Neutron diffraction and high-resolution electron microscopy (e.g. cryo-EM) may provide details about the locations of hydrogen species [108, 109] but have their own challenges. Alternatively, indirect insight into the catalytic mechanism can be obtained using spectroscopy like Mössbauer, nuclear resonance vibrational spectroscopic (NRVS), EPR, and IR spectroscopic techniques, which may allow for the investigation of specific short-lived intermediates during the catalytic process [47]. While each technique has its strengths and weaknesses, they allow for the selective real-time investigation of the chemistry occurring at the H-cluster without perturbation of the protein. For example, IR spectroscopy has been used extensively to investigate the catalytic mechanism in [FeFe]-hydrogenase based on the unique spectroscopic features of CO and CN^- ligands of the H-cluster (Figure 15.4b) that show well-resolved, intense bands between 2150 and 1750 cm^{-1} [110]. The IR bands from the CO and CN^- ligands are exquisitely sensitive to the electron density on the metal ions since the bonding strength is influenced by both ligand-to-metal σ-donation and metal-to-ligand π-backdonation. The result is that the IR bands of the CO and CN^- ligands are different for each catalytic intermediate and, fortunately, do not overlap with signals from solvent or protein. In combination with quantum mechanics/molecular mechanics (QM/MM) modeling techniques, enormous insight into the working principles of the enzymes has been generated [111].

An overview of the different catalytically relevant states, including their characteristic properties, is provided in Table 15.1. While there is no doubt about the existence of the various enzymatic states, their structural assignment and importance to the catalytic cycle is still under debate.

As both the $[4Fe–4S]_H$ and $[2Fe]_H$ clusters are redox active, it is difficult to assign oxidation states because multiple electron configurations must be considered. While

Table 15.1 Spectroscopic data of key H-cluster intermediates as observed in standard [FeFe]-hydrogenase *Cr*HydA1.

Redox species	$[4Fe–4S]_H$	Fep/Fed	Special features	νCN^-	νCO	*g*-values	References
H_{ox}	2+	II/I	—	2088, 2070	1964, 1940, 1802	2.106, 2.048, 2.001	[47, 102]
$H_{ox}H$	2+	II/I	H^+ on $[4Fe]_H$ (SO_2 on $[4Fe]_H$)	2092, 2074	1970, 1946, 1812	2.105, 2.044	[47, 102]
H_{red}' (H_{red})	1+	II/I	H^+ on $[4Fe]_H$ (no protonation)	2084, 2066	1962, 1933, 1792	Silent	[112]
H_{red} ($H_{red}H^+$)	2+	I/I	μH^- (NH_2^+)	2070, 2033	1915, 1961[a], 1891	Silent	[107, 113]
H_{sred} ($H_{sred}H^+$)	1+	I/I	μH^- (NH_2^+)	2068, 2026	1918, 1953[b], 1882	2.145, 1.860	[114]
H_{hyd}	1+	II/II	Fe_d-H^-	2088, 2076	1978, 1960, 1860	2.07, 1.94, 1.88	[114]

a) The terminal CO ligand in H_{red} appears as a μCO ligand in $H_{red}H^+$ at 1817 cm^{-1}.
b) The terminal CO ligand in Hs_{red} appears as a μCO ligand in $Hs_{red}H^+$ at 1801 cm^{-1}.

the $[4Fe–4S]_H$ cluster cycles between a 1+ and 2+ overall charge [81], the $[2Fe]_H$ cluster formally can be found in the reduced ($[Fe^IFe^I]$), mixed-valence oxidized ($[Fe^{II}Fe^I]$), and "super-oxidized" ($[Fe^{II}Fe^{II}]$) states. These two clusters may also exist in multiple protonation states (as discussed below), which further complicates the assignment and renders the formulation of a catalytic mechanism difficult. The elucidation of catalytic pathways in organometallic molecular catalysts is already a delicate undertaking. Unraveling enzymatic mechanisms is significantly more challenging due to the multitude of processes occurring along with fewer spectroscopic methods available in the (bio)chemist's toolbox. This is primarily due to the following problems:

(1) "Standard" analytical techniques like NMR spectroscopy and mass spectrometry that usually give insight into the mechanism of small molecule activation often cannot resolve the chemistry occurring at specific parts of the cofactor in a protein environment. Here, the target signals are usually small and overlapped by the signals of the protein environment (low signal-to-noise ratio).
(2) While potential intermediates can be isolated, it is always questionable if such species really take part in the catalytic pathway or represent side products.
(3) For investigating organometallic compounds, it is common to modify specific structural motifs in the second coordination sphere, e.g. to stabilize intermediates. Similar strategies are not as straightforward to perform within the enzyme. Thus, "stopping" the catalytic cycle at a specific intermediate is complicated.
(4) A direct comparison of the chemistry of small molecular catalysts with enzymatic systems is not feasible due to the absence of the protein environment in the former. [FeFe]-hydrogenases are a good example for this. Mimics of the type $[Fe_2(CO)_4(CN)_2(L)]^{2-}$ are essentially inactive under the conditions of enzymatic proton reduction [115]. Although usually adopting a reduced $[Fe^IFe^I]$ core with structural aspects of the native cofactor, due to the absence of a "rotated" geometry, the iron ions are usually not basic enough to form hydrides from protons and require further reduction (i.e. $[Fe^IFe^0]$ or $[Fe^0Fe^0]$).
(5) The limitation of working under "physiological" conditions where the enzymes are stable and active (close to neutral pH, mild applied potentials, ambient temperature/pressure, aqueous solvents) means that extreme conditions cannot be used to observe all relevant pK_a values and reduction potentials.

15.2.3 Current State of the Catalytic Mechanism of [FeFe]-hydrogenase

Following decades of intense research by numerous research groups, two related but distinct mechanisms of [FeFe]-hydrogenase catalysis have emerged and are presented in Figures 15.5 and 15.6. In both cases, H_{ox} is considered the resting state of the enzyme. The H_{ox} state possesses a μCO ligand, a vacant coordination site at Fe_d, and shows the EPR signal of a mixed-valence diiron site, $[Fe_p^{II}Fe_d^I]$ (Table 15.1). In Figure 15.5, single electron reduction of H_{ox} leads to the H_{red} state (equivalent to H_{red}' in Figure 15.6 but with a slightly different structural assignment), which is constituted by a reduced $[4Fe–4S]_H$ site and an unaltered $[2Fe]_H$ site [112]. Following

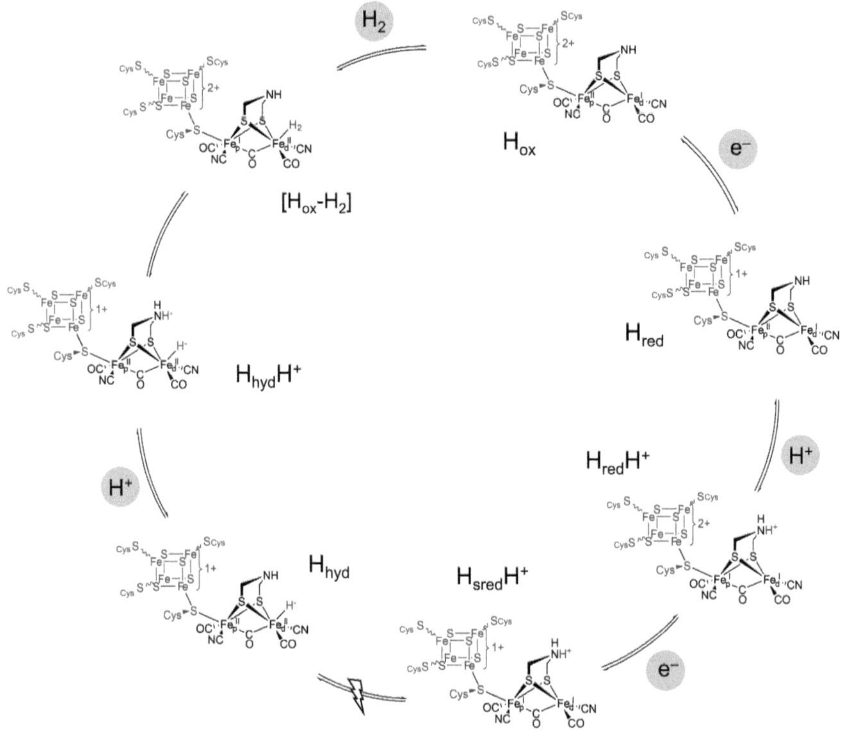

Figure 15.5 Hydrogen turnover according to a model that includes $H_{red}H^+$ and $H_{sred}H^+$ as $[2Fe]_H$-reduced and adt-protonated (NH_2^+), catalytically relevant H-cluster states.

this reduction, a proton-coupled electronic rearrangement step yields a new state, $H_{red}H^+$, where the electron density is shifted from the $[4Fe–4S]_H$ site to the $[2Fe]_H$ site, giving a reduced $[Fe_p{}^IFe_d{}^I]$ cluster. This interpretation is based on the following aspects:

(1) The observation of a pH dependence of the $H_{ox} \leftrightarrow H_{red}/H_{red}H^+$ redox potential using FTIR spectroelectrochemical redox titrations [116].
(2) A pH dependence of interconversion between H_{red} and $H_{red}H^+$ [99].
(3) No observation of $H_{red}H^+$ in [FeFe]-hydrogenases with alternative dithiolate ligands [112].
(4) A large red-shift (shift to lower energy) of the IR bands in the $H_{red}H^+$ state due to the increased electron density on $[2Fe]_H$ [98].

The protonation is proposed to occur at the amine in the adt ligand (NH_2^+). In the following step, the $[4Fe–4S]_H$ site is again reduced forming the $H_{sred}H^+$ state [117]. Internal PT from the protonated adt ligand to the $[Fe^IFe^I]$ center results in oxidative addition of the proton and the formation of a $[4Fe–4S]_H–[Fe^{II}Fe^{II}]$ intermediate (H_{hyd}) bearing a terminal hydride ligand at the distal iron ion [94, 96, 118]. This interconversion was demonstrated by illumination at cryogenic temperatures [114]. It is then assumed that a second protonation event including $H_{hyd}H^+$ [119] and the

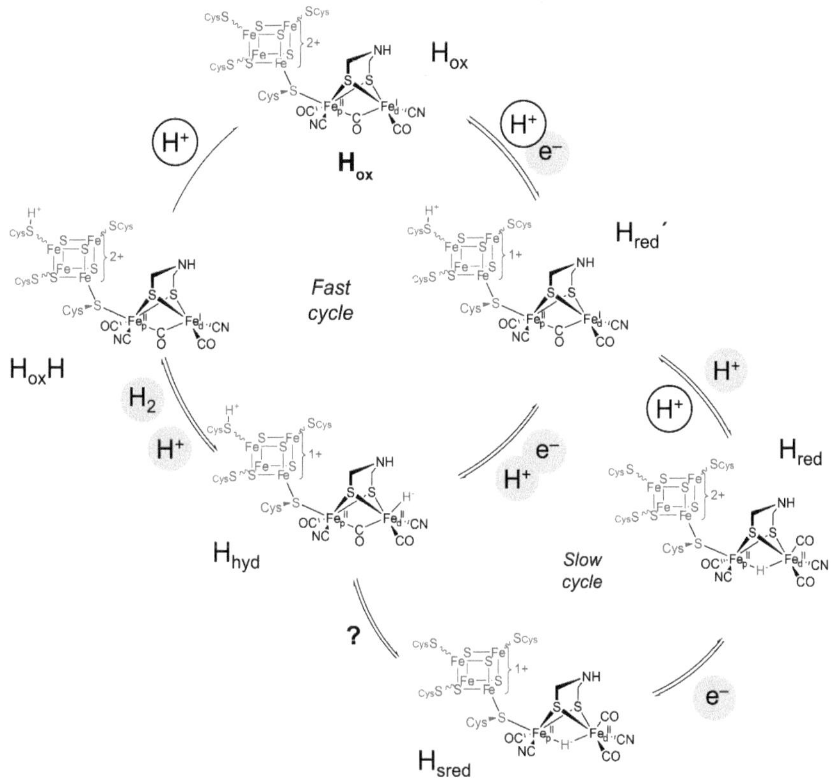

Figure 15.6 Hydrogen turnover according to a model that includes H_{red} and H_{sred} as hydrogen-inhibited H-cluster states in the "slow" cycle.

theoretical species H_{ox}-H_2 as intermediates then completes the $2e^-/2H^+$ cycle forming H_2 [120]. Based on the specific activity of the enzyme, the cycle may operate in the same way for proton reduction and H_2 oxidation.

In contrast to the catalytic cycle presented in Figure 15.5, the catalytic cycle in Figure 15.6 differentiates between a "slow" and a "fast" trajectory [93]. Starting with the transition of H_{ox} into H_{red}' (equivalent to H_{red} in Figure 15.6 but with a protonated $[4Fe–4S]_H$), single electron reduction of the $[4Fe–4S]_H$ cluster is suggested to be associated with a protonation of a cysteine at the $[4Fe–4S]_H$ cluster. This interpretation is based on the observation of a pH dependence of the H_{ox}/H_{red}' redox potential using FTIR spectroelectrochemical redox titrations [101], the simulation of the H-cluster spectra using DFT calculations [102], and the dependence of state populations on pH [121, 122]. Notably, a conserved PT pathway toward the $[4Fe–4S]_H$ site was found to further support this idea. It is assumed that this "regulatory" pathway allows exploiting proton-coupled electron transfer (PCET) and charge compensation to promote the reduction of the $[4Fe–4S]_H$ rather than the $[2Fe]_H$ cluster [101]. From the H_{red}' state, the cycle branches into the fast or slow trajectory. In the fast cycle, a second step of PCET is assumed to form the H_{hyd} state [93]. Here, the $[4Fe–4S]_H$ cluster remains reduced and protonated, and the additional electron and proton are added to $[2Fe]_H$ forming a terminal hydride ligand and a "super-oxidized" diiron site, $[Fe^{II}Fe^{II}]$. Similar to the mechanism provided in Figure 15.5, it is believed that a third protonation takes place at the amine of the adt ligand that reacts with the hydride to form H_2 (transient states $H_{hyd}H^+$ and H_{ox}-H_2, not shown in Figure 15.6). At low pH, the oxidized H-cluster remains protonated at the "regulatory" cysteine, yielding the $H_{ox}H$ state [102, 122]. $H_{ox}H$ might convert to H_{ox} upon deprotonation before entering a new round of catalysis, as shown in Figure 15.6. Alternatively, it is conceivable that $H_{ox}H$, H_{red}', and H_{hyd} interconvert directly, forming a catalytic cycle that is characterized by a protonated $[4Fe–4S]_H$ cluster [93].

Entering the slow cycle, Figure 15.6 depicts how deprotonation of the H_{red}' state can lead to electron spill-over toward the diiron site, which then binds a proton to form a bridging hydride ligand (μH^-) giving a state with three terminal CO ligands known as H_{red} [95]. Since H_{red}' and H_{red} are both proposed to be one-electron more reduced and have one additional proton relative to H_{ox}, they are tautomers. Interconversion of tautomers does not involve an overall gain or loss of protons, and it is still unclear why there is a pH dependence for the H_{red}'/H_{red} transition. The H_{red} state can be further reduced at the $[4Fe–4S]_H$ site to the H_{sred} state. As the interconversion of $H_{sred}H^+$ and H_{hyd} has been demonstrated [114], a similar transition between H_{sred} and H_{hyd} cannot be ruled out ("?" in Figure 15.6).

H_{red} and H_{sred} in Figure 15.6 have largely similar FTIR band assignments to those of $H_{red}H^+$ and $H_{sred}H^+$ in Figure 15.6, except that the missing μCO band gives rise to a terminal CO band at around 1961 cm^{-1} (H_{red}) and 1953 cm^{-1} (H_{sred}). Cryogenic FTIR spectroscopy shows the appearance of a bridging CO band at 1817 cm^{-1} in the $H_{red}H^+$ state [114], interestingly at higher energies than in the H_{ox} state (Table 15.1). This result has two possible interpretations: (i) the μCO ligand in the $H_{red}H^+/H_{sred}H^+$ states is quite flexible at room temperature so that a well-defined peak can only be observed by reducing the flexibility at cryogenic temperature [123, 124]; (ii) there are two different states, one at low temperature that has a protonated amine ($H_{red}H^+/H_{sred}H^+$) and one at room temperature that carries a μH^- ligand (H_{red}/H_{sred}) [125]. In support of the former interpretation, the 1961 and 1953 cm^{-1} bands remain observable at low temperature, where the bridging CO band is observed [123]. Due to crowded spectra, a unique assignment of these bands to the $[2Fe]_H$-reduced states is not possible. In support of the latter interpretation, the energies of the bands in FTIR and XAS spectroscopy of both H_{ox} (as a reference) and H_{red}/H_{sred} can be reproduced by DFT calculations on a μH^- containing structure [54, 95, 126]. Moreover, site-selective isotope editing of the μCO ligand [86] results in a red-shift of the 1953 and 1961 cm^{-1} bands upon reduction into H_{red}/H_{sred} corroborating a change in geometry [95]. Direct support for a μH^- ligand has not been obtained yet. In support of the former interpretation, the IR signature of the room temperature and low temperature H_{red} states is essentially identical apart from the μCO band. If there were a change from a $[Fe^{II}Fe^{II}]$ to an $[Fe^{I}Fe^{I}]$ oxidation state, one may have expected a shift in the IR bands due to changes in metal-to-ligand π-backdonation – as generally observed for other changes in the redox state of the H-cluster. However, the distribution of charge across the H-cluster is a known phenomenon [81], so the formal differences between a hydride-binding $[2Fe]_H{}^{4+}$ and a protonated $[2Fe]_H{}^{2+}$ species may result in similar spectra after all [125]. Subsequent calculations have shown that the energies of the IR bands from $H_{red}H^+/H_{sred}H^+$ can be reproduced using a μCO-binding, protonated adt ligand model [124]. The ~15 cm^{-1} upshift of the μCO band upon reduction from H_{ox} to $H_{red}H^+$ remains unexplained.

15.2.4 Toward a Consensus Catalytic Mechanism

To resolve the controversies and settle on a consensus mechanism, we urgently need new experiments. First, time-resolved spectroscopic and structural studies under catalytic turnover conditions are needed to identify which intermediates appear and disappear on the timescale of a single catalytic turnover and assign their structures. In this direction, attempts have been made to study [FeFe]-hydrogenases with time-resolved IR spectroscopy [127–129] but since these enzymes are incredibly fast, diffusion limitation of electron mediators is a difficult issue to overcome. As of yet, no attempts have been made to perform time-resolved X-ray crystallography of [FeFe]-hydrogenases, e.g. using serial synchrotron X-ray crystallography (SSX) or serial femtosecond X-ray crystallography (SFX) at an X-ray free-electron laser (XFEL) – although an XFEL structure is available [130]. This is likely due to difficulties in providing suitable photo-triggers for initiating catalysis in crystals [131]. Another advantage of SFX comprises recording diffraction data at room temperature [132], which may help in understanding how temperature affects the reduced H-cluster. At the very least, combining X-ray crystallography and IR spectroscopy to match states to structures is needed for all states in the catalytic cycle, and efforts in this direction are underway [133].

In this chapter, the focus has been quite extensively on spectroscopy and structure. However, activity experiments may also provide important clues to understanding the catalytic mechanism and whichever mechanism is chosen as the consensus based on spectroscopy and/or structures must also agree with activity data. Traditionally, chemical reductants like sodium dithionite are exploited to drive H_2 production followed by semicontinuous gas chromatography or monitoring the reduction of redox mediators like methyl viologen by electronic spectroscopy upon hydrogenase-catalyzed H_2 oxidation. An important, more quantitative technique in this regard is protein-film electrochemistry (PFE) as pioneered by Fraser Armstrong [44] and several groups have provided important insights into the catalytic mechanism of [FeFe]-hydrogenases using this approach [134–137]. Recent results have been supportive of the catalytic mechanism presented in Figure 15.5 [138], although a curious result of reversible enzyme inactivation at low applied electrical potentials [139] is more in line with the mechanistic proposal in Figure 15.6 where the H_{sred} state is thought to be inactive. However, without spectroscopic characterization of the states observed during electrochemistry, results are difficult to rationalize and depend on extensive modeling. While coupling of electrochemistry to spectroscopy has been achieved in many studies [95, 96, 99, 133], these studies have relied on diffusion of the enzyme to/from the electrode and/or use of redox mediators, which limits the catalytic rates as mentioned above. Attempts to couple PFE to spectroscopy, e.g. using surface-enhanced IR spectroscopy, are underway and may ultimately provide the most detailed level of insight possible into the catalytic cycle of [FeFe]-hydrogenases.

15.2.5 Conclusion

Overall, [FeFe]-hydrogenases are truly fascinating examples of how nature has combined an organometallic complex with a protein scaffold and perfectly tuned the properties for highly efficient catalysis. [FeFe]-hydrogenases are incredibly active enzymes, and their measured turnover frequencies may only reflect a fraction of their true rates. Understanding their mechanism and how the protein scaffold and active site H-cluster are harmonized may provide future chemists with design criteria for producing state-of-the-art molecular catalysts for industrial H_2 production and consumption technologies. Two distinct mechanistic proposals currently exist in the literature, each with its own merits and supporting evidence. Future experiments may reveal additional insight helping to build a complete picture of the catalytic mechanism. This will not only benefit the hydrogenase and molecular catalyst communities but the field of chemistry at large as it is the very rules of nature that are under investigation. If we can completely understand the simplest of reactions, the combination of protons and electrons to make H_2, then this gives us hope that more complicated chemistry such as CO_2 and N_2 reduction discussed in other chapters of this book are also tractable.

15.3 Catalytic Cycle of [NiFe]-Hydrogenase

H_2 possesses significant potential to function as an energy carrier in renewable energy systems. In nature, H_2 cycling is orchestrated by metalloenzymes known as hydrogenases. Among them, [NiFe]-hydrogenases catalyze the reversible reaction $H_2 \rightleftarrows 2H^+ + 2e^-$ under ambient conditions, using the earth-abundant transition metals nickel and iron [7, 140]. Catalysis takes place at the biologically unique [NiFe] center (Figure 15.7a,b) [141], synthesized by the synergetic interaction of at least six auxiliary proteins (HypABCDEF, Chapter 13) [142, 143]. [NiFe]-hydrogenases exhibit remarkable catalytic rates of up to $10^3\,s^{-1}$ for H_2 oxidation, which is made possible by a complex protein infrastructure, including an intricate gas channel network for H_2 diffusion between the protein surface and the active site [144]. Electrons are channeled to and from the [NiFe] site through a series of Fe–S clusters, which vary in number and type, and in some cases communicating with electron-mediating organic cofactors such as nicotinamide adenine dinucleotide (NAD), flavin adenine dinucleotide (FAD), and flavin mononucleotide (FMN). The remarkable catalytic rates of hydrogenases are also based on efficient proton transport (PT). A few studies discussed pathways for long-range PT and

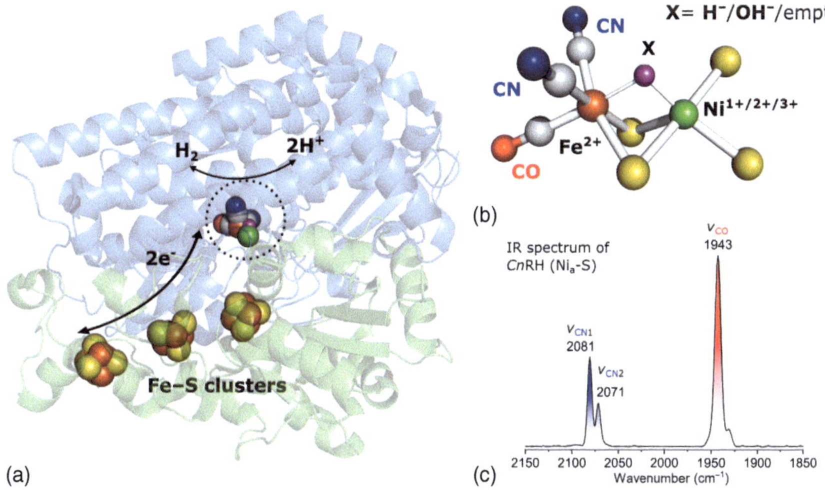

Figure 15.7 [NiFe]-hydrogenase functional unit. (a) Cartoon representation of the heterodimeric [NiFe]-hydrogenase functional unit from *Desulfovibrio vulgaris Miyazaki F* (PDB: 4U9I) comprising the catalytic large subunit (blue) and the Fe–S cluster-containing small subunit (green). (b) Ball and stick representation of the [NiFe] active site, Ni in green, Fe in brown, oxygen in red, nitrogen in blue, thiolates from coordinating cysteines in yellow, and bridging ligand in magenta. (c) IR absorbance spectrum of the Ni_a-S intermediate of the regulatory [NiFe]-hydrogenase (RH) from *Cupriavidus necator*. The ν_{CO} (area in red) and ν_{CN1} and ν_{CN2} bands (area in blue) related to the CO and CN stretching vibrations of the diatomic active site ligands are labeled with their corresponding wavenumbers.

identified conserved amino acids playing a role in H^+ gating during catalysis [65, 145–147].

The [NiFe] center is deeply buried in the heterodimeric [NiFe]-hydrogenase functional unit and anchored to the large subunit via two cysteines bound terminally to Ni and two cysteines bridging the Ni and Fe ions. The Fe is additionally ligated by one CO and two CN^- strong-field ligands. Depending on the observed redox state, a third bridging ligand (i.e. a hydride or a hydroxide anion) can be present (Figure 15.7b) [148–150]. Unusual active site arrangements/ligands are briefly introduced in Section 15.3.3.

Vibrational spectroscopy, and in particular IR spectroscopy (Chapter 17), has played a pivotal role in hydrogenase research and has greatly contributed to our understanding of the catalytic mechanism of hydrogenases [110, 150]. In fact, the various (active and inhibited) states of hydrogenases can easily be distinguished by IR spectroscopy, with the absorptions of the CO ligands between 1800 and 2030 cm^{-1} and those of the CN^- ligands between 2030 and 2150 cm^{-1}, a spectral region essentially free of protein and solvent contributions (Figure 15.7c). Historically, IR spectroscopy enabled the first identification of the diatomic CO/CN^- ligands, thereby correcting the assignment of three water molecules bound to the active site Fe in an early crystal structure of the [NiFe]-hydrogenase from *Desulfovibrio gigas* [151–153].

15.3.1 Classification of [NiFe]-Hydrogenases: The Complex Trait of O_2 Tolerance

Before introducing structural details and mechanistic aspects of the hydrogen activation in [NiFe]-hydrogenases, it is important to introduce certain facets of their biological classification which, in turn, will help the reader to appreciate the complexity of these systems. [NiFe]-hydrogenases are classified into 4 groups containing 29 subgroups based on (i) amino acid sequence similarity, (ii) metal/cofactor-binding motifs, (iii) genetic organization, and (iv) reported biochemical features [154]. Comprehensive surveys and a web database for hydrogenase classification are available and might help the reader to further dig into this field [43, 49, 155].

Some of the best-studied [NiFe]-hydrogenases belong to Group *1*, encompassing H_2-uptake membrane-bound hydrogenases (MBHs). Among them are the ancestral (subgroup *1a*) and prototypical (subgroup *1b*) enzymes, e.g. the MBHs from *Desulfovibrio vulgaris* Miyazaki F [*Dv*MF] and *Desulfovibrio gigas*. These enzymes liberate electrons for, e.g. sulfate/nitrate respiration and are usually sensitive to O_2, a common inhibitor for all hydrogenases [154]. Group *1* also features certain O_2-tolerant variants (subgroup *1d*) that maintain H_2 oxidation in the presence of O_2, such as MBH from *Cupriavidus necator* (*Cn*MBH) and Hyd-1 from *Escherichia coli* (*Ec*Hyd-1) [142, 156]. The O_2-tolerance properties of this subgroup, which is associated with the respiratory chain, correlate with the presence of a unique high-potential [4Fe–3S] cluster in the position "proximal" to the [NiFe] site in the electron-transferring hydrogenase small subunit [157–159]. This [4Fe–3S] center is coordinated by six Cys residues and has been shown to consecutively transfer two electrons ($[4Fe–3S]^{5+/4+/3+}$) within a physiological potential range by undergoing redox-dependent reversible transformations [160, 161]. This is in contrast with the standard [4Fe–4S] cubane, which is present in the proximal position of O_2-sensitive [NiFe]-hydrogenases and is coordinated by four Cys residues, and can only mediate the transfer of a single electron ($[4Fe–4S]^{2+/1+}$). Importantly, the peculiar redox chemistry of the [4Fe–3S] cluster has been correlated with the ability of these hydrogenases to detoxify O_2 into harmless water, a process that requires the immediate availability of $4H^+$ and $4e^-$ [162]. Whether O_2 reduction takes place at the [NiFe] active site or one of the Fe–S clusters (e.g. the proximal [4Fe–3S] cluster) is a matter of debate [163]. Importantly, activity of *Cn*MBH is switched off under electron-deficient conditions through the insertion of a hydroxy ligand into the vacant coordination site between Ni and Fe, thereby preventing the active site from interacting with O_2 [164].

The so far described mechanisms of O_2 tolerance in [NiFe]-hydrogenases are summarized in Table 15.2. [NiFe]-hydrogenases from subgroup *1c* are predicted to be O_2 sensitive [49]. However, the [NiFe]-hydrogenase from *Citrobacter* sp S-77 is stable to O_2 exposure and retains catalysis in the presence of 1% O_2, exhibiting partially O_2-tolerant catalysis [170, 171]. X-ray structural data on this particular hydrogenase revealed a flexible aspartate residue that aids the proximal [4Fe–4S] cluster to adopt different (redox-dependent) conformations [165], which led to

Table 15.2 Proposed mechanisms of O_2 tolerance and/or O_2 protection employed by O_2-tolerant [NiFe]-hydrogenases.

[NiFe]-hydrogenase	Proposed mechanisms of O_2 tolerance/O_2 protection
*Ec*Hyd-1, *Cn*MBH (subgroup *1d*)	Catalytic detoxification of O_2:
	$O_2 + 4\,H^+ + 4\,e^- \rightarrow 2\,H_2O$ [162]
	Additional protection under electron-limiting conditions: Occupation of the free binding site of the catalytic center by a bridging OH^- ligand derived from H_2O [164]
[NiFe]-hydrogenase from *Citrobacter* sp. S-77 (subgroup *1c*)	Catalytic detoxification of O_2[a)]:
	$O_2 + 4\,H^+ + 4\,e^- \rightarrow 2\,H_2O$ [165]
C. necator AH (subgroup *1h*) *and M. smegmatis* Huc (subgroup *2a*)	Narrow gas channels that restrict O_2 access to the catalytic center [35, 36]
*Cn*RH (subgroup *2b*)	Narrow gas channel that restricts O_2 access to the catalytic center [37]
*Cn*SH/*Ht*SH (subgroup *3d*)	Catalytic detoxification of O_2:
	$O_2 + H^+ + H_2 + NADH \rightarrow 2\,H_2O + NAD^+$
	NADH-dependent (per)oxidase activity involving the transient formation of sulfenate species [38, 39]
	O_2 protection through saturation of the binding capacity of Ni [166] either in a Ni^{IV}/Fe^{II} or in a spin-coupled Ni^{III}/Fe^{III} form [167, 168]
*Pf*SH1 (subgroup *3b*)	Unclear, a suggestion is reminiscent of the O_2-tolerance mechanism of *Cn*SH [169]

a) O_2 detoxification has not been shown.

the hypothesis that it might stabilize a super-oxidized $[4Fe–4S]^{3+}$ state, which is a common trait for high-potential iron–sulfur proteins (HiPIP) [172]. The observed dynamics of the proximal [4Fe–4S] cluster led the authors to hypothesize a *Cn*MBH-like mechanism that detoxifies O_2 into water (Table 15.2) which, however, has not been supported by experimental data [162].

Particularly interesting are also [NiFe]-hydrogenases belonging to subgroup *1h*, usually referred to as "Actinobacterial-type (AH) hydrogenases." These enzymes can scavenge electrons from tropospheric H_2 (ca. 530 parts per billion by volume [ppbv]) to sustain aerobic respiration during starvation [173]. These hydrogenases are usually characterized by higher apparent affinities for H_2 ($K_m \approx 30–200$ nM) and appear insensitive to inhibition by O_2, which make them attractive for enzyme-based fuel cells that oxidize H_2 in ambient air [174]. A recent structural and functional characterization of the related Huc hydrogenase (subgroup *2a*) from *Mycobacterium smegmatis* unraveled how this complex enzyme transfers reducing power from trace H_2 using three [3Fe–4S] clusters and a long-range quinone transport system [175]. The Huc hydrogenase appears to restrict access of O_2 molecules through narrow gas channels (Table 15.2), a mechanism of O_2 tolerance that has previously been suggested for the AH from *C. necator* [176, 177].

[NiFe]-hydrogenases belonging to groups *2* and *3* have also been investigated with respect to O_2 tolerance, comprising cytosolic H_2 uptake and the so-called bidirectional enzymes. Among the best-studied are the regulatory [NiFe]-hydrogenase (*Cn*RH) (subgroup *2b*) and the NAD^+-reducing soluble [NiFe] hydrogenase (*Cn*SH) (subgroup *3d*) from *C. necator* as well as the $NADP^+$-reducing soluble hydrogenase (*Pf*SH1) (subgroup *3b*) from the anaerobic hyperthermophilic archaeon *Pyrococcus furiosus*. *Cn*RH and *Cn*SH have been shown to retain catalysis in the presence of O_2, but the underlying mechanisms of O_2 tolerance differ. The *Cn*RH has been proposed to restrict O_2 access via narrow gas channels [178], a mechanism also suggested for [NiFe]-hydrogenases belonging to subgroup *1h* (Table 15.2). *Cn*RH is the only known hydrogenase that, after purification in air, exists in a nearly stoichiometrically catalytic state (Ni_a-S, see Section 15.3.3.1) without the need for a reductive activation procedure. The *Cn*SH, on the other hand, has been proposed to detoxify O_2 at its [NiFe] active site by means of an NADH-dependent (per)oxidase reaction involving the formation of cysteine sulfenates (Table 15.2) [179, 180]. Interestingly, the SH counterpart isolated from *Hydrogenophilus thermoluteolus* (*Ht*SH) also exhibits an unusual architecture at its [NiFe] site in the presence of O_2, in which glutamate acts as a bidentate ligand for the nickel ion (see Section 15.3.4 for further details on this unusual structural arrangement) [166]. *Pf*SH1 is also surprisingly tolerant to O_2, retaining about 30% of its maximal H_2 oxidation capacity in the presence of 1% O_2. The underlying mechanism, however, is not fully understood, and unconventional structural arrangements of the [NiFe] active site (e.g. formation of [cysteine]S–O species) as those observed for *Cn*SH and *Ht*SH might play a role (see Section 15.3.4) [169].

[NiFe]-hydrogenases of Group *4* have been less investigated. Among them, the membrane-bound formate hydrogenlyase (FHL) complex from *E. coli* is of particular interest. This two-enzyme complex links formate oxidation by a molybdopterin-containing formate dehydrogenase (FDH) to H^+ reduction by a [NiFe]-hydrogenase [181]. Significantly, FHL has been shown to account for almost all hydrogen production *in vivo* in *E. coli* [182], and its biochemical characterization revealed a significant tolerance to CO, which usually inhibits standard [NiFe]-hydrogenases (see the next section) [183].

15.3.2 [NiFe]-Hydrogenase Active Site States

The IR absorptions of the CO and CN^- ligands have been shown to be extremely sensitive to changes in the electron density distribution at the [NiFe] center. As stated above, the Fe center is coordinated by one CO and two CN^- ligands. Therefore, the IR spectrum of each hydrogenase state displays three bond-stretching vibrations (Figure 15.7c). These include a CO bond stretching (ν_{CO}) and two stretchings related to the CN^- ligands (ν_{CN1}, ν_{CN2}). It is important to mention that the ν_{CN1} and ν_{CN2} bands cannot be attributed to the individual CN^- ligands as they both contain contributions from the two ligands (see Chapter 17 for details). These three reporter bands are excellent fingerprints of the different hydrogenase redox states. The position of the ν_{CO} is particularly diagnostic because an increase in electron density at the [NiFe] site increases back-donation to π*orbitals of the ligand, which weakens the CO bond and moves the observed band to a lower energy. For instance, Figure 15.7c displays the IR spectrum of the as-isolated *Cn*RH in the Ni_a-S state (see also below). Upon incubation with the substrate H_2, reducing agents (e.g. sodium dithionite), and inhibitors (e.g. O_2 and CO), the [NiFe] active site passes through several intermediate states, which result in distinct changes in the CO and CN^- ligand stretching vibrations. Significantly, the Fe ion retains a low-spin (S = 0) ferrous form in all catalytic and inhibited states observed so far [184, 185], while the Ni ion is the active player varying its oxidation state (Ni^I, Ni^{II}, and Ni^{III}). Importantly, redox species with Ni^I and Ni^{III} formal oxidation states carry an unpaired electron in their outermost orbital and can be detected by EPR spectroscopy [81].

Each redox intermediate is generally depicted as Ni_x-Y, where the subscript "x" indicates whether the species is catalytically active, resting, or inactive. Subscript "a" stands for active species, "r" for ready, "u" for unready, and "ia" for inactive species. The label "Y" indicates the redox state, which is derived from their original designations given when they were first discovered by EPR spectroscopy (Ni-A, Ni-B, and Ni-C). An exception is the Ni-L species, where L stands for light as this intermediate was originally trapped upon light exposure of the reduced [NiFe]-hydrogenase under cryogenic conditions (see Section 15.3.3.4 for details of Ni_a-L intermediate). All states denoted by "S" are EPR-"silent" as the nickel ion resides in the diamagnetic Ni^{II} state. A certain number of redox states (e.g. Ni_a-L, Ni_a-SR, and Ni_r-S) have been shown to be enriched in two or three subforms. These are usually denoted with additional subscripts "**I, II, III**", "**1, 2, 3**" or superscripts "′, ″" (e.g. Ni_a-L1, Ni_a-SR″, and Ni_r-S_I) [148, 149]. The literature is not always consistent concerning this nomenclature and care must be taken when comparing the various subforms from different species. The following subsections present the structural elements of the [NiFe] active site arrangements that are not directly related to the catalytic cycle and are schematically shown in Figure 15.8.

15.3.2.1 CO-Bound States

CO gas is a typical inhibitor for standard [NiFe]-hydrogenases, and certain active site states react with CO, which binds terminally to the open coordination site of the nickel (Figure 15.8a). Two electronically distinct CO-inhibited states have been observed; an EPR-silent one originating from the binding of CO to the Ni_a-S intermediate, which comprises two different forms of the same redox state designated Ni-S_{CO} and Ni-$(S_{CO})_{Red}$, and a paramagnetic Ni-CO species, originating from the binding of CO to the Ni_a-L state (Figure 15.8a). X-ray structural and spectroscopic data from various enzymes are available for both CO-inhibited states [7, 186]. Remarkably, O_2-tolerant [NiFe]-hydrogenases of subgroups *1d, 1h, 2b*, and *3d* are not subject to CO inhibition. The presumably O_2-tolerant S77 [NiFe]-hydrogenase, on the other hand, has been shown to weakly bind an exogenous CO molecule [187].

15.3.2.2 Hydroxy-Bridged States

Ni_u-A and Ni_r-B are oxidized [NiFe]-hydrogenase redox states, featuring a bridging OH ligand between the Ni^{III} and Fe^{II} ions (Figure 15.8b,c). Both redox states

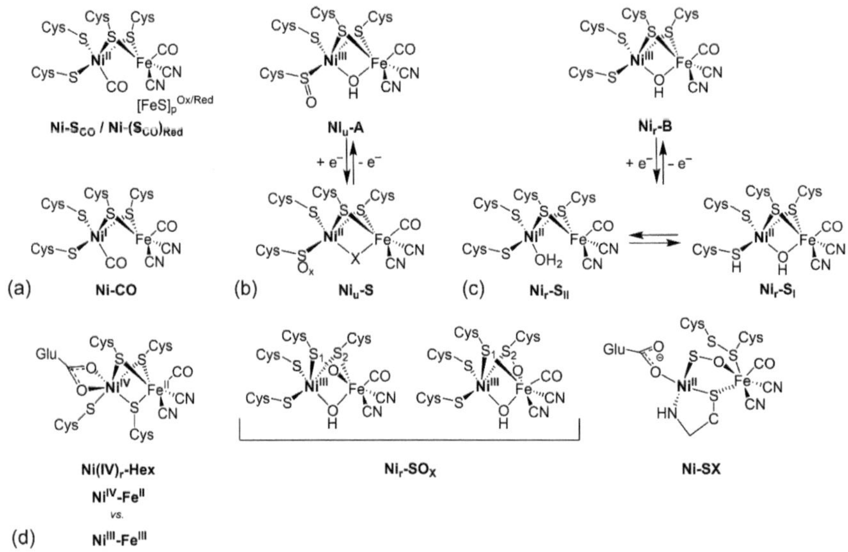

Figure 15.8 Atypical architectures of the [NiFe]-hydrogenase active site. (a) CO-bound inhibited species featuring an exogenous CO ligand terminally bound to the nickel ion. $[FeS]_p$ represents the proximal [4Fe–4S] cluster. (b) Proposed structures for the Ni_u-A and Ni_u-S redox species featuring oxidative modification of the cysteine thiolate ligands. (c) Proposed structures for the Ni_r-B and Ni_r-S_I and Ni_r-S_{II} redox species. (d) Proposed structures for the O_2-shielded Ni(IV)$_r$-Hex (left), the two Ni_r-SO_x species responsible for IR split spectrum in *Cn*SH enzyme (middle structures), and the inhibited Ni-SX species (right structure).

are accumulated under aerobic and anaerobic oxidizing conditions and differ in the rate of reactivation. The origin of the bridging OH ligand is still controversial, with a water molecule being the most likely candidate [188, 189]. Ni_r-B can be easily reactivated (within seconds/minutes) with H_2, whereas Ni_u-A requires a longer time, depending on the hydrogenase species, and specific conditions. The IR and EPR spectra of Ni_u-A and Ni_r-B are also very similar as the ligands at the iron ion and the corresponding electron density do not change significantly. Exemplarily, the EPR spectra of Ni_u-A and Ni_r-B of the O_2-sensitive membrane-bound [NiFe]-hydrogenase from *Desulfovibrio vulgaris* Miyazaki F (*Dv*MF) differ mainly in the g_y tensor component, suggesting a specific alteration of one ligand (sitting plausibly in the xz plane) near the nickel center [7]. High-resolution X-ray crystallographic data and theoretical calculations proposed oxidative modifications of the cysteine thiolate ligands (cys-sulfoxide, S=O, or cys-sulfenic acid, S–OH) for the Ni_u-A state (Figure 15.8b), which is a plausible reason for the slower reactivation kinetics [149]. O_2-tolerant [NiFe]-hydrogenases do not typically accumulate Ni_u-A species, which correlates with the absence of CO-inhibited species.

One-electron reduction of the Ni_u-A and Ni_r-B states produces the unready Ni_u-S and ready Ni_r-S states (Figure 15.8b,c). As both states are EPR silent, their identification was exclusively based on IR spectroscopic data and DFT computations. While the nature of Ni_u-S is still debated, Ni_r-S has been proposed to retain an OH bridge. The active site-containing large subunits, HoxC and HoxG proteins, of the O_2-tolerant regulatory and membrane-bound [NiFe]-hydrogenases from *C. necator* have been isolated independently of their corresponding Fe–S cluster-containing small subunits [141, 190]. These isolated hydrogenase large subunits were significantly enriched in two Ni_r-S subforms. The absence of interfering Fe–S clusters allowed a detailed spectroscopic analysis of the catalytic center using ^{57}Fe-sensitive techniques such as NRVS (see Chapter 17), which revealed the arrangement of the Ni_r-S_I substrate by detecting Fe/Ni-OH modes of the [NiFe] active site (Figure 15.8c) [191].

15.3.2.3 Unusual [NiFe] Site Arrangements

The following section summarizes atypical architectures of the [NiFe] center. Some of them have been correlated with specific O_2-tolerance mechanisms and/or protective functions used by the hydrogenase to prevent O_2 binding at the [NiFe] site. Others represent essentially inhibited redox states.

In 2017, Shomura et al. reported the X-ray structure of the O_2-tolerant NAD^+-reducing soluble [NiFe]-hydrogenase (SH) from *Hydrogenophilus thermoluteolus* TH-1^T (*Ht*SH). The oxidized active site featured a unique six-coordinated Ni center coordinated by three bridging cysteines, one terminal cysteine, and the bidentate carboxylate group of a glutamate residue ($Ni(IV)_r$-Hex, Figure 15.8d) [166]. This coordinatively saturated structural arrangement has been proposed to protect the [NiFe] site from O_2 attack. In view of the diamagnetic properties of this redox state, a Ni^{IV}/Fe^{II} and a spin-coupled Ni^{III}/Fe^{III} electronic arrangement are discussed [167, 168].

In 2023, Higuchi et al. reported high-resolution (1.04 Å) X-ray data of the *Dv*MF hydrogenase in its oxidized state, which exhibits a square-planar geometry for the Ni ion coordinated by a cysteine sulfenate, a deprotonated amide N and monodentate carboxylate of a glutamate [192]. This structural arrangement possibly represents the so far elusive Ni-SX redox species (Figure 15.8d). This arrangement is also reminiscent of the active site structure of the V74C variant of the *Solidesulfovibrio fructosivorans* hydrogenase, which is characterized by an improved O_2 tolerance and an unexpected structural difference in the first coordination sphere of the Ni ion. Instead of four coordinating standard cysteine thiolates, the Ni ligands in the V74C hydrogenase variant are proposed as two terminal cysteine thiolates in combination with a bridging SH^- ligand and carboxamido N atom of the amino acid main chain [193]. Another peculiar [NiFe] active site arrangement was recently reported by Armstrong, Carr, and coworkers, who carried out a detailed structural characterization of the Arg-to-Lys variant of [NiFe]-hydrogenase 2 from *E. coli*. The purified protein variant exhibited a very tightly bound diatomic ligand, which was bound *end-on* to the Ni ion and stabilized by hydrogen bonds to the N_ζ atom of the pendant lysine and a water molecule [194]. The nature of the bound molecule that inhibits catalysis is still unresolved.

The aerobically isolated soluble *Cn*SH exhibits an unusual IR "split-spectrum", characterized by one intense ν_{CO} at about 1957 cm^{-1} and four ν_{CN} located at 2070, 2081, 2090, and 2098 cm^{-1}. This observation led to the misinterpretation that the active site of *Cn*SH is equipped with two additional cyanide ligands. Detailed spectroscopic and computational works, however, revealed that the IR split spectrum reflects the contributions of two active site species (with overlapping ν_{CO} bands) both exhibiting oxygenation at one of the two bridging cysteines (Ni$_r$-SO$_X$, Figure 15.8d). These two species in *Cn*SH have been proposed as intermediates of the O_2 detoxification pathway (Table 15.2) [179].

15.3.3 Catalytic Intermediates

Mechanistic studies have focused on a limited subset of enzymes, with the prevailing assumption that the same mechanism applies to all [NiFe]-hydrogenases. The consensus catalytic mechanism of [NiFe]-hydrogenases is the result of the joint efforts of various research groups over the last few decades, who have mainly used steady-state and time-resolved IR spectroscopy methods, EPR spectroscopy, electrochemistry, and X-ray structure data. H_2 activation and H^+ reduction take place at the [NiFe] site, and a catalytic cycle includes at least the four intermediates Ni$_a$-S, Ni$_a$-SR, Ni$_a$-C, and Ni$_a$-L, which are shown schematically in Figure 15.9a [148–150]. In brief, H_2 is bound to the active site in the Ni$_a$-S state and subsequently heterolytically cleaved yielding the diamagnetic Ni$_a$-SR state. This state is characterized by a hydride bridging the Ni and Fe ions and a proton that is likely located on one of the Ni-bound cysteine residues. The removal of one proton and one electron from the active site results in the Ni$_a$-C state, which retains the bridging hydride and subsequently converts to the Ni$_a$-L, a tautomeric form of Ni$_a$-C. The removal of a second proton and electron leads to the recovery of the Ni$_a$-S state, ready to activate another H_2 molecule. In the following sections, we summarize the spectroscopic and structural evidence for all known catalytic intermediates and describe the importance of the outer coordination sphere in fine-tuning of the proton and electron dynamics at the [NiFe] site.

15.3.3.1 Ni$_a$-S

The Ni$_a$-S intermediate (often termed Ni$_a$-SI, where "I" is an abbreviation that stands for intermediate) is among the best-characterized [NiFe]-hydrogenase intermediates. This species has been shown to react with H_2, which is presumably cleaved heterolytically. However, recent computations suggest that a homolytic mechanism for the activation of H_2 would proceed with a lower energetic barrier [195]. The exact location of the H_2 binding site is still very controversial, as both Ni and Fe ions could be involved in the formation of a dihydrogen–metal sigma complex (also known as Michaelis complex).

Detailed computational studies and the isolation of Ni-bound CO-inhibited states favor Ni as the initial H_2 binding site [196, 197]. Crystallographic data of the Ni$_a$-S intermediate are available for the F_{420}-reducing [NiFe]-hydrogenase from *Methanosarcina barkeri* (*Mb*FRH), which belongs to subgroup *3a* [198]. Comprehensive spectroscopic data (IR, resonance Raman [RR], and/or NRVS) for the Ni$_a$-S intermediate have been collected for various hydrogenases, including the abovementioned *Mb*FRH and *Cn*RH [199, 200]. The active site adopts a Ni^{II}-[]-Fe^{II}

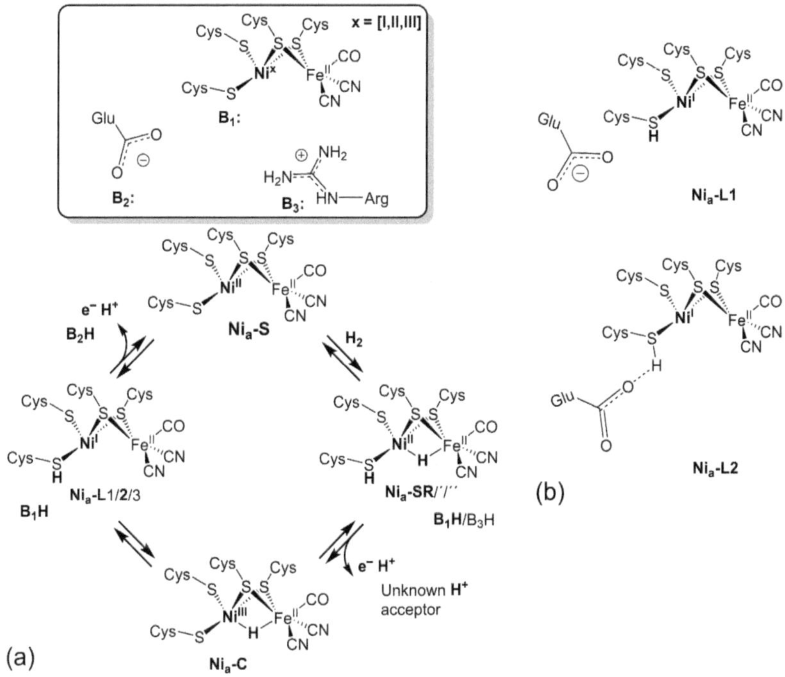

Figure 15.9 Consensus catalytic cycle of [NiFe]-hydrogenase. (a) Schematic representation of the active site of [NiFe]-hydrogenases (top) and the proposed catalytic cycle involving the intermediates Ni$_a$-S, Ni$_a$-SR, Ni$_a$-C, and Ni$_a$-L (bottom). During catalysis, only nickel is redox active, while the iron maintains an Fe^{II} low-spin configuration. The intermediates Ni$_a$-SR and Ni$_a$-L comprise three subforms. The proposed proton acceptors in the vicinity of the active site are a Ni-bound terminal cysteine (Cys, B_1), a conserved glutamate (Glu, B_2), and an arginine (Arg, B_3), which are displayed in the top panel in their protonated/deprotonated forms as under physiological conditions. (b) Proposed structural elements that distinguish the Ni$_a$-L1 and Ni$_a$-L2 subforms.

electronic configuration with a vacant bridging site. Accordingly, the *Mb*FRH active site exhibits a peculiar seesaw geometry with trans S–Ni–S bond angles of ca. 107° and 171° and a Ni–Fe distance of 2.7 Å, a geometrical/electronic arrangement that is in line with the results of spectroscopic and computational studies.

The electron density distribution of the Ni$_a$-S intermediate is modulated by the outer coordination sphere of the [NiFe] site. IR spectroscopic data on *Pf*SH1 hydrogenase revealed small shifts of the ν_{CO} band of the Ni$_a$-S state as a function of pH, which correlates with an acid–base equilibrium that may involve a glutamate residue located within hydrogen-bonding (H-B) distance of one of the Ni-coordinating cysteines [201]. The importance of this glutamate for catalysis has been confirmed for various [NiFe]-hydrogenases. Site-directed exchanges of the Glu for the isosteric glutamine (Gln) resulted in a dramatic decrease in the hydrogenase activities of *Ec*Hyd1, the *S. fructosivorans* hydrogenase, and *Pf*SH1 [202–204].

The outer coordination sphere of the [NiFe]-hydrogenase active site also contains a strictly conserved arginine, whose guanidine headgroup is 4–5 Å away from the nickel and iron atoms. The exchange of the Arg509 in *Ec*Hyd1 with lysine led to a 100-fold lower activity than the native enzyme [205]. These data prompted the Armstrong group to propose the guanidine group as the general base in H_2 activation by [NiFe]-hydrogenases, which acts together with the [NiFe] site as a frustrated Lewis pair system. Although this mechanism is reminiscent to that of [FeFe]-hydrogenases (which make use of a protonable azadithiolate bridge, Section 15.2), there is still a lack of further evidence for changes in the protonation state of the conserved arginine in [NiFe]-hydrogenases. In a more recent report from the Dyer group, it was shown that replacement of the corresponding Arg in *Pf*SH1 (Arg355) by lysine affects H_2 oxidation, H^+ reduction, and even the catalytic H/D exchange reaction, which is largely based on the Ni_a-S $\rightleftarrows$ Ni_a-SR interconversion [206]. Reduced activity correlated with a disturbed tautomeric equilibrium between the Ni_a-C and Ni_a-L intermediates, which is important for the proton/hydride handling and thereby critical for substrate processing.

15.3.3.2 Ni_a-SR

Ni_a-SR represents possibly the most controversial intermediate in [NiFe]-hydrogenases. Since Ni_a-SR states are diamagnetic (Ni^{II}), they are not accessible to conventional EPR spectroscopic methods, and in 2007 Lubitz and Tumas pointed out limitations in X-ray crystallography of this intermediate, which arise from weak X-ray scattering of H nuclei [207]. A few [NiFe] active site arrangements have been proposed based on calculations, ranging from HNi(μ-H)Fe dihydrides to NiFe(η^2-H_2) dihydrogen complexes, in which the hydrogen species are bound to either the Ni or the Fe fragment [208].

The first "native" structure of the Ni_a-SR intermediate has nevertheless been obtained by X-ray crystallography. In 2015, the Lubitz group reported an ultrahigh-resolution structure (0.89 Å) of the O_2-sensitive *Dv*MF [NiFe]-hydrogenase in the Ni_a-SR state [209]. The authors identified both products of the heterolytic splitting of dihydrogen, i.e. a hydride bridging Ni and Fe and a proton attached to the sulfur of a Ni-bound cysteine ligand. The bridging hydride was found to be slightly asymmetric with Ni–H and Fe–H bond lengths of 1.58 and 1.78 Å, respectively. This arrangement was confirmed by the Cramer group using NRVS, where the authors detected a small band at 675 cm^{-1}, which was shown to be sensitive to H/D isotope exchange and assigned to a wagging Ni–H–Fe vibrational mode [210].

Remarkably, the observed Ni–Fe distance of 2.6 Å in the Ni_a-SR state was not significantly different from that in other catalytic intermediates (e.g. Ni_a-S) [198]. This rigidity of the catalytic center implies minimal structural rearrangements during H_2 turnover, which may contribute to the remarkable catalytic efficiency of these enzymes. However, the situation is less clear than it might appear, as it has been shown that the Ni_a-SR intermediate consists of three sub-forms (Ni_a-SR_I, Ni_a-SR_{II}, and Ni_a-SR_{III}, Figure 15.9a) which can be distinguished by IR spectroscopy based on the energy of their ν_{CO} stretching vibrations [148]. DFT calculations by the Neese group using models for the active site in the Ni_a-SR state, which contain either a protonated or deprotonated cysteine thiolate [208], have accurately reproduced the ν_{CO} of Ni_a-SR_I and Ni_a-SR_{III}, respectively. This suggests that these two forms represent

[NiFe] site states with identical electronic configurations but consecutive PT steps after the heterolytic splitting of H_2.

15.3.3.3 Ni_a-C

The incubation of [NiFe]-hydrogenases with their substrate H_2 leads to the enrichment of varying amounts of a paramagnetic species known as Ni_a-C. This redox state was the first catalytic [NiFe]-hydrogenase intermediate detected by EPR spectroscopy, and the first report dates back to 1982 [211]. The associated configuration of the active site had been highly controversial until the pioneering work of the Hoffman laboratory, which detected a strong ^{1}H-hyperfine interaction of an exchangeable proton with the nickel of the [NiFe]-hydrogenase from *D. gigas* in the Ni_a-C state [212]. This study paved the way for a detailed characterization of the bridging hydride using isotope-sensitive (H/D) pulsed EPR techniques [213]. Additionally, EPR spectra of several [NiFe]-hydrogenases in the Ni_a-C state recorded at very low temperatures (<20 K) exhibit the so-called "split" signal [150], arising from a magnetic interaction between the paramagnetic Ni^{III} species and the singly reduced proximal $[4Fe–4S]^+$ cluster. In summary, (EPR) spectroscopic and computational studies propose a Ni^{III}-H-Fe^{II} electronic arrangement for the active site in the Ni_a-C state, which loses one electron (taken up by the proximal [4Fe–4S] cluster) and one proton in comparison to the Ni_a-SR analog. In a recent study by the Hirota group, it has been proposed that the H^+ bound to a cysteine thiolate in the Ni_a-SR state is transferred to a conserved glutamate during the conversion to Ni_a-C [214].

In contrast to EPR, the characterization of the bridging hydride in the Ni_a-C state using vibrational spectroscopy turned out to be difficult. While the IR CO/CN fingerprints of the Ni_a-C intermediate could be easily identified [7, 148], the vibrations of the bridging hydride could not be resolved due to their weak absorption as well as the overlap with protein/solvent modes. H/D isotope labeling of the bridging hydride also did not cause any significant changes in the CO/CN signals of [NiFe]-hydrogenases, in contrast to the well-known "trans effect" observed for [FeFe]-hydrogenases [98]. There, the terminal binding of D^- instead of H^- to the "distal" Fe ion of the diiron center results in a shift of the ν_{CO} of the bridging CO ligand in the trans position to the H/D by 5–10 cm^{-1}.

RR spectroscopy has also been employed in hydrogenase research to detect metal-ligand vibrations of both the [NiFe] center (e.g. Fe–CO/CN and Ni–S) and the Fe–S clusters [215]. H_2-reduced samples of the *Cn*RH and *Cn*MBH were also subjected to a detailed RR characterization. However, no specific Ni_a-C absorptions were detected, which is due to the photolysis of the bridging hydride by the laser beam, leading to the accumulation of the subsequent Ni_a-L intermediate (see below) [200, 216]. Ni_a-C-enriched *Cn*RH was also subjected to NRVS spectroscopy, which does not lead to the photoconversion of the Ni_a-C state. Indeed, changes in the Fe–CO/CN absorptions were observed in the spectra after H/D exchange of the bridging ligand, but not the expected Ni-H-Fe wagging mode of the corresponding bridging hydride [217]. In addition to the [NiFe] site itself, the outer coordination sphere determines the geometric/electronic environment, which affects the properties of all intermediates of the active site, including Ni_a-C. Spectroscopic data on the photoconversion of Ni_a-C to Ni_a-L in the *Dv*MF hydrogenase, for example, revealed spectral features of a protonated glutamate residue in both intermediates [214]. Recent data on *Cn*RH, however, show that the same glutamate remains deprotonated in both intermediate states, suggesting that the reaction mechanisms of [NiFe]-hydrogenases may be slightly different [218].

15.3.3.4 Ni_a-L

The photoconversion of Ni_a-C into Ni_a-L species has been shown for several [NiFe]-hydrogenases. This process has been correlated with the loss of the bridging hydride between Ni and Fe, which was substantiated by the disappearance of signals assigned to the exchangeable hydride/deuteride ligand in advanced EPR (ENDOR/HYSCORE) experiments [7, 213].

Ni_a-L represents a tautomeric form of Ni_a-C in which the hydride electrons are stored on the nickel, resulting in a formal Ni^I species, while the remaining proton is bound to a nearby residue (possibly a cysteine thiolate). The Ni_a-L intermediate was first observed for the [NiFe]-hydrogenase *Allochromatium vinosum* by EPR spectroscopy [219], but convincing evidence for its catalytic relevance was only provided three decades later by complementary investigations on the O_2-sensitive *Dv*MF hydrogenase and O_2-tolerant *Ec*Hyd1 and *Pf*SH1 [201, 220, 221]. Various Ni_a-L subforms have been identified for all three enzymes using both IR and EPR spectroscopy. Although the way these subforms have been classified is not always consistent, the three main species Ni_a-L1, Ni_a-L2, and Ni_a-L3 have been described, which are reminiscent of the three subforms observed for Ni_a-SR (see above). Insights from IR, RR, and computational works suggest that in some of these states one of the Ni-bound terminal cysteines is protonated [196, 200, 214]. Indeed, a recent IR spectroscopic investigation on *Cn*RH provided experimental evidence that in the Ni_a-L1 and Ni_a-L2 subforms (Figure 15.9b) a proton is localized at a nearby cysteine thiolate (Cys479 in *Cn*RH), as isotope-sensitive ν_{SH} stretching vibrations were detected [218]. For the Ni_a-L2 intermediate, it has been proposed that the protonated thiolate forms a hydrogen bond with the proton-accepting glutamate (Glu13 in *Cn*RH) (Figure 15.9b), which resulted in considerable spectral changes of the IR signals of the conserved glutamate. The proposed structural elements of the Ni_a-L1/L2 intermediates in *Cn*RH have also been discussed for *Pf*SH1 [201]. In contrast, IR spectroscopic experiments on *Ec*Hyd-1 revealed no change in the protonation state of the corresponding cysteine [222]. This discrepancy suggests that different proton acceptors exist in [NiFe]-hydrogenases. Consistently, *Ec*Hyd-1 accumulates two Ni_a-L species whose ν_{CO} bands are red-shifted by about 30–40 cm^{-1} compared to those observed in the *Dv*MF hydrogenase, *Pf*SH1 and *Cn*RH. The Hirota group showed that under certain conditions, the *Dv*MF hydrogenase accumulates a deprotonated form of Ni_a-L, named Ni_a-L3, whose ν_{CO} is found at lower wavenumbers (ca. 20 cm^{-1}) compared to the protonated analog Ni_a-L2 [223]. In view of these data and previous findings on a protonated/deprotonated [NiFe]-hydrogenase model complex [224], it appears plausible that *Ec*Hyd-1 accumulates deprotonated forms of the Ni_a-L intermediate, whose protons may be bound to the conserved glutamate (E28 in *Ec*Hyd-1) thereby resulting in active site intermediates with identical electronic configurations, but in an advanced PT step

[148]. On the other hand, as ν_{SH} bands are difficult to detect due to their relatively small cross section, it is therefore also possible that ν_{SH} stretching vibrations in *Ec*Hyd-1 escaped detection due to inhomogeneous enrichment of the Ni_a-L intermediate or poorer signal-to-noise ratios.

15.3.4 Conclusions

The catalytic mechanism of [NiFe]-hydrogenases has been thoroughly investigated to gain a detailed understanding of the structure–function relationship that could serve as a blueprint for a "perfect" transition metal catalyst for H_2 activation [225]. Among the numerous members of the [NiFe]-hydrogenase family, O_2-tolerant enzymes are the focus of current research, as they maintain catalysis in the presence of O_2 and some of them can even scavenge electrons from H_2 in the air [174]. Comprehensive structural, biochemical, and spectroscopic work has revealed important properties of the [NiFe] active site in both the catalytically active and inhibited states. Among these is a considerable rigidity of the $NiFe(CN)_2CO$ center, which is maintained during turnover, a factor that likely contributes to the remarkable catalytic efficiency of [NiFe]-hydrogenases. Four intermediate states of the catalytic cycle have been reported, designated Ni_a-S, Ni_a-SR, Ni_a-C, and Ni_a-L. The active site iron maintains a low-spin Fe^{II} configuration throughout the catalytic cycle, established by an unusual coordination sphere with two CN^- and a CO molecule, while the Ni cycles through oxidation states of 1+, 2+, and 3+ during catalysis [7, 150]. H_2 has been predicted to bind to the Ni ion of the hydrogenase in the Ni_a-S state to form a dihydrogen–metal sigma complex [196, 197]. The bound H_2 molecule is then heterolytically cleaved, resulting in the Ni_a-SR state characterized by a bridging hydride and a bound proton. Two proton-binding sites have been proposed, either a Ni-bound cysteine thiolate or an invariant arginine residue, which forms a frustrated Lewis pair with the active site metal(s) [205, 209]. The observation of different Ni_a-SR subforms and the availability of high-resolution structural data support protonation at a Ni-bound thiolate [209]. The importance of the second and outer coordination spheres has been studied by different groups, with important residues (e.g. glutamate and arginine) acting as H^+ acceptors/gates [203, 204, 218]. The released electrons are transferred via an array of one to three Fe–S clusters. Among them, the proximal [4Fe–3S] cluster of O_2-tolerant [NiFe]-hydrogenase (subgroup *1d*) has received particular attention, as its redox chemistry appears to correlate with O_2 detoxification properties of the corresponding hydrogenases [162]. From the analysis of various O_2-tolerant [NiFe]-hydrogenases (Table 15.2), it can be clearly concluded that the mechanisms of O_2 tolerance are diverse and, most importantly, they are even different for hydrogenases from the same organism.

Acknowledgment

This work was supported by the Deutsche Forschungsgemeinschaft, Priority Program Iron-Sulfur for Life (SPP1927, LE 2934/1-1/2, SH 87/1-1/2). S.T.S and J.A.B. thank Ulf-Peter Apfel for help with an early version of the manuscript. G.C.

and O.L. thank the cluster of excellence “UniSysCat” under Germany’s Excellence Strategy-EXC2008-390540038 for support. S.S. thanks for support the Max Planck Society.

References

1 Huang, G., Wagner, T., Ermler, U., and Shima, S. (2020). Methanogenesis involves direct hydride transfer from H_2 to an organic substrate. *Nat. Rev. Chem.* 4 (4): 213–221.

2 Zirngibl, C., Van Dongen, W., Schwörer, B. et al. (1992). H_2-forming methylenetetrahydromethanopterin dehydrogenase, a novel type of hydrogenase without iron-sulfur clusters in methanogenic archaea. *Eur. J. Biochem.* 208 (2): 511–520.

3 Thauer, R.K., Klein, A.R., and Hartmann, G.C. (1996). Reactions with molecular hydrogen in microorganisms: evidence for a purely organic hydrogenation catalyst. *Chem. Rev.* 96 (7): 3031–3042.

4 Thauer, R.K., Kaster, A.-K., Goenrich, M. et al. (2010). Hydrogenases from methanogenic archaea, nickel, a novel cofactor, and H_2 storage. *Annu. Rev. Biochem.* 79 (1): 507–536.

5 Shima, S., Huang, G., Wagner, T., and Ermler, U. (2020). Structural basis of hydrogenotrophic methanogenesis. *Ann. Rev. Microbiol.* 74 (1): 713–733.

6 Afting, C., Hochheimer, A., and Thauer, R.K. (1998). Function of H_2-forming methylenetetrahydromethanopterin dehydrogenase from *Methanobacterium thermoautotrophicum* in coenzyme F420 reduction with H_2. *Arch. Microbiol.* 169 (3): 206–210.

7 Lubitz, W., Ogata, H., Rüdiger, O., and Reijerse, E. (2014). Hydrogenases. *Chem. Rev.* 114 (8): 4081–4148.

8 Shima, S., Lyon, E.J., Thauer, R.K. et al. (2005). Mössbauer studies of the iron–sulfur cluster-free hydrogenase: the electronic state of the mononuclear Fe active site. *J. Am. Chem. Soc.* 127 (29): 10430–10435.

9 Huang, G., Wagner, T., Wodrich, M.D. et al. (2019). The atomic-resolution crystal structure of activated [Fe]-hydrogenase. *Nat. Catal.* 2 (6): 537–543.

10 Schwörer, B., Fernandez, V.M., Zirngibl, C., and Thauer, R.K. (1993). H_2-forming N^5, N^{10}-methylenetetrahydromethanopterin dehydrogenase from *Methanobacterium thermoautotrophicum*: studies of the catalytic mechanism of H_2 formation using hydrogen isotopes. *Eur. J. Biochem.* 212 (1): 255–261.

11 Klein, A.R., Fernández, V.M., and Thauer, R.K. (1995). H_2-forming N^5, N^{10}-methylenetetrahydromethanopterin dehydrogenase: mechanism of H_2 formation analyzed using hydrogen isotopes. *FEBS Lett.* 368 (2): 203–206.

12 Hartmann, G.C., Santamaria, E., Fernández, V.M., and Thauer, R.K. (1996). Studies on the catalytic mechanism of H_2-forming methylenetetrahydromethanopterin dehydrogenase: para-ortho H_2 conversion rates in H_2O and D_2O. *J. Biol. Inorg. Chem.* 1 (5): 446–450.

13 Berkessel, A. and Thauer, R.K. (1995). On the mechanism of catalysis by a metal-free hydrogenase from methanogenic archaea: enzymatic transformation of H_2 without a metal and its analogy to the chemistry of alkanes in superacidic solution. *Angew. Chem. Int. Ed. Engl.* 34 (20): 2247–2250.

14 Olah, G.A., Halpern, Y., Shen, J., and Mo, Y.K. (1973). Electrophilic reactions at single bonds. XII. Hydrogen-deuterium exchange, protolysis (deuterolysis), and oligocondensation of alkanes with superacids. *J. Am. Chem. Soc.* 95 (15): 4960–4970.

15 Adams, M.W.W., Mortenson, L.E., and Chen, J.-S. (1980). Hydrogenase. *Biochim. Biophys. Acta BBA - Rev. Bioenerg.* 594 (2–3): 105–176.

16 Lyon, E.J., Shima, S., Buurman, G. et al. (2004). UV-A/blue-light inactivation of the 'metal-free' hydrogenase (Hmd) from methanogenic archaea: the enzyme contains functional iron after all. *Eur. J. Biochem.* 271 (1): 195–204.

17 Lyon, E.J., Shima, S., Boecher, R. et al. (2004). Carbon monoxide as an intrinsic ligand to iron in the active site of the iron–sulfur-cluster-free hydrogenase H_2-forming methylenetetrahydromethanopterin dehydrogenase as revealed by infrared spectroscopy. *J. Am. Chem. Soc.* 126 (43): 14239–14248.

18 Hiromoto, T., Ataka, K., Pilak, O. et al. (2009). The crystal structure of C176A mutated [Fe]-hydrogenase suggests an acyl-iron ligation in the active site iron complex. *FEBS Lett.* 583 (3): 585–590.

19 Shima, S., Pilak, O., Vogt, S. et al. (2008). The crystal structure of [Fe]-hydrogenase reveals the geometry of the active site. *Science* 321 (5888): 572–575.

20 Shima, S. and Ataka, K. (2011). Isocyanides inhibit [Fe]-hydrogenase with very high affinity. *FEBS Lett.* 585 (2): 353–356.

21 Tamura, H., Salomone-Stagni, M., Fujishiro, T. et al. (2013). Crystal structures of [Fe]-hydrogenase in complex with inhibitory isocyanides: implications for the H_2-activation site. *Angew. Chem. Int. Ed.* 52 (37): 9656–9659.

22 Schaupp, S., Arriaza-Gallardo, F.J., Pan, H. et al. (2022). In vitro biosynthesis of the [Fe]-hydrogenase cofactor verifies the proposed biosynthetic precursors. *Angew. Chem. Int. Ed.* 61 (22): e202200994.

23 Shima, S., Schick, M., and Tamura, H. (2011). Preparation of [Fe]-hydrogenase from methanogenic archaea. In: Rosenzweig, A. C., Ragsdale S. W., editors. *Methods in Enzymology*, vol. 494, 119–137. Academic Press.

24 Hidese, R., Ataka, K., Bill, E., and Shima, S. (2015). Cu^I and H_2O_2 inactivate and Fe^{II} inhibits [Fe]-hydrogenase at very low concentrations. *ChemBioChem* 16 (13): 1861–1865.

25 Shima, S., Thauer, R.K., and Ermler, U. (2015). Carbon monoxide as intrinsic ligand to iron in the active site of [Fe]-hydrogenase. In: *Metal-Carbon Bonds in Enzymes and Cofactors* (ed. H. Sigel, A. Sigel, and R.K.O. Sigel), 219–240. De Gruyter.

26 Huang, G., Wagner, T., Ermler, U. et al. (2018). Dioxygen sensitivity of [Fe]-hydrogenase in the presence of reducing substrates. *Angew. Chem. Int. Ed.* 57 (18): 4917–4920.

27 Clapham, S.E., Hadzovic, A., and Morris, R.H. (2004). Mechanisms of the H_2-hydrogenation and transfer hydrogenation of polar bonds catalyzed by ruthenium hydride complexes. *Coord. Chem. Rev.* 248 (21–24): 2201–2237.

28 Berggren, G., Adamska, A., Lambertz, C. et al. (2013). Biomimetic assembly and activation of [FeFe]-hydrogenases. *Nature* 499 (7456): 66–69.

29 Wang, C., Lai, Z., Huang, G., and Pan, H. (2022). Current state of [Fe]-hydrogenase and its biomimetic models. *Chem. Eur. J.* 28 (57): e202201499.

30 Chen, D., Scopelliti, R., and Hu, X. (2010). [Fe]-hydrogenase models featuring acylmethylpyridinyl ligands. *Angew. Chem. Int. Ed.* 49 (41): 7512–7515.

31 Shima, S., Chen, D., Xu, T. et al. (2015). Reconstitution of [Fe]-hydrogenase using model complexes. *Nat. Chem.* 7 (12): 995–1002.

32 Huang, G., Arriaza-Gallardo, F.J., Wagner, T., and Shima, S. (2020). Crystal structures of [Fe]-hydrogenase from *Methanolacinia paynteri* suggest a path of the FeGP-cofactor incorporation process. *Inorganics* 8 (9): 50.

33 Shima, S., Schick, M., Kahnt, J. et al. (2012). Evidence for acyl–iron ligation in the active site of [Fe]-hydrogenase provided by mass spectrometry and infrared spectroscopy. *Dalton Trans.* 41 (3): 767–771.

34 Hiromoto, T., Warkentin, E., Moll, J. et al. (2009). The crystal structure of an [Fe]-hydrogenase–substrate complex reveals the framework for H_2 activation. *Angew. Chem. Int. Ed.* 48 (35): 6457–6460.

35 Pan, H.-J., Huang, G., Wodrich, M.D. et al. (2019). A catalytically active [Mn]-hydrogenase incorporating a non-native metal cofactor. *Nat. Chem.* 11 (7): 669–675.

36 Shima, S., Vogt, S., Göbels, A., and Bill, E. (2010). Iron-chromophore circular dichroism of [Fe]-hydrogenase: the conformational change required for H_2 activation. *Angew. Chem. Int. Ed.* 49 (51): 9917–9921.

37 Dey, A. (2010). Density functional theory calculations on the mononuclear non-heme iron active site of Hmd hydrogenase: role of the internal ligands in tuning external ligand binding and driving H_2 heterolysis. *J. Am. Chem. Soc.* 132 (39): 13892–13901.

38 Berkessel, A. (2001). Activation of dihydrogen without transition metals. *Curr. Opin. Chem. Biol.* 5 (5): 486–490.

39 Birrell, J.A., Rodríguez-Maciá, P., Reijerse, E.J. et al. (2021). The catalytic cycle of [FeFe] hydrogenase: a tale of two sites. *Coord. Chem. Rev.* 449: 214191.

40 Kleinhaus, J.T., Wittkamp, F., Yadav, S. et al. (2021). [FeFe]-Hydrogenases: maturation and reactivity of enzymatic systems and overview of biomimetic models. *Chem. Soc. Rev.* 50 (3): 1668–1784.

41 Sidabras, J.W. and Stripp, S.T. (2023). A personal account on 25 years of scientific literature on [FeFe]-hydrogenase. *J. Biol. Inorg. Chem.* 28 (4): 355–378.

42 Greening, C., Cabotaje, P.R., Valentin Alvarado, L.E. et al. (2024). Minimal and hybrid hydrogenases are active from archaea. *Cell* 187 (13): 3357–3372.e19.

43 Vignais, P.M. and Billoud, B. (2007). Occurrence, classification, and biological function of hydrogenases: an overview. *Chem. Rev.* 107 (10): 4206–4272.

44 Vincent, K.A., Parkin, A., and Armstrong, F.A. (2007). Investigating and exploiting the electrocatalytic properties of hydrogenases. *Chem. Rev.* 107 (10): 4366–4413.

45 Glick, B.R., Martin, W.G., and Martin, S.M. (1980). Purification and properties of the periplasmic hydrogenase from *Desulfovibrio desulfuricans*. *Can. J. Microbiol.* 26 (10): 1214–1223.

46 Hatchikian, E.C., Forget, N., Fernandez, V.M. et al. (1992). Further characterization of the [Fe]-hydrogenase from *Desulfovibrio desulfuricans* ATCC 7757. *Eur. J. Biochem.* 209 (1): 357–365.

47 Land, H., Senger, M., Berggren, G., and Stripp, S.T. (2020). Current state of [FeFe]-hydrogenase research: biodiversity and spectroscopic investigations. *ACS Catal.* 10 (13): 7069–7086.

48 Zheng, D., Wang, M., Chen, L. et al. (2014). The influence of a S-to-S bridge in diiron dithiolate models on the oxidation reaction: a mimic of the $Hair_{ox}$ state of [FeFe]-hydrogenases. *Chem. Commun.* 50 (66): 9255–9258.

49 Søndergaard, D., Pedersen, C.N.S., and Greening, C. (2016). HydDB: a web tool for hydrogenase classification and analysis. *Sci. Rep.* 6 (1): 34212.

50 Chongdar, N., Birrell, J.A., Pawlak, K. et al. (2018). Unique spectroscopic properties of the H-cluster in a putative sensory [FeFe] hydrogenase. *J. Am. Chem. Soc.* 140 (3): 1057–1068.

51 Stripp, S.T., Goldet, G., Brandmayr, C. et al. (2009). How oxygen attacks [FeFe] hydrogenases from photosynthetic organisms. *Proc. Natl. Acad. Sci.* 106 (41): 17331–17336.

52 Swanson, K.D., Ratzloff, M.W., Mulder, D.W. et al. (2015). [FeFe]-hydrogenase oxygen inactivation is initiated at the H cluster 2Fe subcluster. *J. Am. Chem. Soc.* 137 (5): 1809–1816.

53 Kubas, A., Orain, C., De Sancho, D. et al. (2017). Mechanism of O_2 diffusion and reduction in FeFe hydrogenases. *Nat. Chem.* 9 (1): 88–95.

54 Mebs, S., Kositzki, R., Duan, J. et al. (2018). Hydrogen and oxygen trapping at the H-cluster of [FeFe]-hydrogenase revealed by site-selective spectroscopy and QM/MM calculations. *Biochim. Biophys. Acta BBA - Bioenerg.* 1859 (1): 28–41.

55 Esselborn, J., Kertess, L., Apfel, U.-P. et al. (2019). Loss of specific active-site iron atoms in oxygen-exposed [FeFe]-hydrogenase determined by detailed x-ray structure analyses. *J. Am. Chem. Soc.* 141 (44): 17721–17728.

56 Yadav, S., Haas, R., Boydas, E.B. et al. (2024). Oxygen sensitivity of [FeFe]-hydrogenase: a comparative study of active site mimics inside *vs.* outside the enzyme. *Phys. Chem. Chem. Phys.* 26 (28): 19105–19116.

57 Morra, S., Arizzi, M., Valetti, F., and Gilardi, G. (2016). Oxygen stability in the new [FeFe]-hydrogenase from *Clostridium beijerinckii* SM10 (CbA5H). *Biochemistry* 55 (42): 5897–5900.

58 Peters, J.W., Lanzilotta, W.N., Lemon, B.J., and Seefeldt, L.C. (1998). X-ray crystal structure of the Fe-only hydrogenase (CpI) from *Clostridium pasteurianum* to 1.8Ångstrom resolution. *Science* 282 (5395): 1853–1858.

59 Nicolet, Y., Piras, C., Legrand, P. et al. (1999). *Desulfovibrio desulfuricans* iron hydrogenase: the structure shows unusual coordination to an active site Fe binuclear center. *Structure* 7 (1): 13–23.

60 Nicolet, Y., Lemon, B.J., Fontecilla-Camps, J.C., and Peters, J.W. (2000). A novel Fe–S cluster in Fe-only hydrogenases. *Trends Biochem. Sci.* 25 (3): 138–143.

61 Stripp, S.T., Duffus, B.R., Fourmond, V. et al. (2022). Second and outer coordination sphere effects in nitrogenase, hydrogenase, formate dehydrogenase, and CO dehydrogenase. *Chem. Rev.* 122 (14): 11900–11973.

62 Stripp, S.T. and Happe, T. (2009). How algae produce hydrogen—news from the photosynthetic hydrogenase. *Dalton Trans.* 45: 9960.

63 Lampret, O., Adamska-Venkatesh, A., Konegger, H. et al. (2017). Interplay between CN^- ligands and the secondary coordination sphere of the H-cluster in [FeFe]-hydrogenases. *J. Am. Chem. Soc.* 139 (50): 18222–18230.

64 Duan, J., Mebs, S., Laun, K. et al. (2019). Geometry of the catalytic active site in [FeFe]-hydrogenase is determined by hydrogen bonding and proton transfer. *ACS Catal.* 9 (10): 9140–9149.

65 Tai, H., Hirota, S., and Stripp, S.T. (2021). Proton transfer mechanisms in bimetallic hydrogenases. *Acc. Chem. Res.* 54 (1): 232–241.

66 Cohen, J., Kim, K., King, P. et al. (2005). Finding gas diffusion pathways in proteins: application to O_2 and H_2 transport in CpI [FeFe]-hydrogenase and the role of packing defects. *Structure* 13 (9): 1321–1329.

67 Goldet, G., Brandmayr, C., Stripp, S.T. et al. (2009). Electrochemical kinetic investigations of the reactions of [FeFe]-hydrogenases with carbon monoxide and oxygen: comparing the importance of gas tunnels and active-site electronic/redox effects. *J. Am. Chem. Soc.* 131 (41): 14979–14989.

68 Ginovska-Pangovska, B., Ho, M.-H., Linehan, J.C. et al. (2014). Molecular dynamics study of the proposed proton transport pathways in [FeFe]-hydrogenase. *Biochim. Biophys. Acta BBA - Bioenerg.* 1837 (1): 131–138.

69 Cornish, A.J., Gärtner, K., Yang, H. et al. (2011). Mechanism of proton transfer in [FeFe]-hydrogenase from *Clostridium pasteurianum*. *J. Biol. Chem.* 286 (44): 38341–38347.

70 Duan, J., Senger, M., Esselborn, J. et al. (2018). Crystallographic and spectroscopic assignment of the proton transfer pathway in [FeFe]-hydrogenases. *Nat. Commun.* 9 (1): 4726.

71 Senger, M., Eichmann, V., Laun, K. et al. (2019). How [FeFe]-hydrogenase facilitates bidirectional proton transfer. *J. Am. Chem. Soc.* 141 (43): 17394–17403.

72 Cornish, A.J., Ginovska, B., Thelen, A. et al. (2016). Single-amino acid modifications reveal additional controls on the proton pathway of [FeFe]-hydrogenase. *Biochemistry* 55 (22): 3165–3173.

73 Cabotaje, P.R., Walter, K., Zamader, A. et al. (2023). Probing substrate transport effects on enzymatic hydrogen catalysis: an alternative proton transfer pathway in putatively sensory [FeFe] hydrogenase. *ACS Catal.* 13 (15): 10435–10446.

74 Chongdar, N., Rodríguez-Maciá, P., Reijerse, E.J. et al. (2023). Redox tuning of the H-cluster by second coordination sphere amino acids in the sensory [FeFe] hydrogenase from *Thermotoga maritima*. *Chem. Sci.* 14 (13): 3682–3692.

75 Land, H., Sekretareva, A., Huang, P. et al. (2020). Characterization of a putative sensory [FeFe]-hydrogenase provides new insight into the role of the active site architecture. *Chem. Sci.* 11 (47): 12789–12801.

76 Beinert, H. (2000). Iron-sulfur proteins: ancient structures, still full of surprises. *J. Biol. Inorg. Chem.* 5 (1): 2–15.

77 Liu, T. and Darensbourg, M.Y. (2007). A mixed-valent, Fe(II)Fe(I), diiron complex reproduces the unique rotated state of the [FeFe]hydrogenase active site. *J. Am. Chem. Soc.* 129 (22): 7008–7009.

78 Singleton, M.L., Bhuvanesh, N., Reibenspies, J.H., and Darensbourg, M.Y. (2008). Synthetic support of de novo design: sterically bulky [FeFe]-hydrogenase models. *Angew. Chem. Int. Ed.* 47 (49): 9492–9495.

79 Gloaguen, F. and Rauchfuss, T.B. (2009). Small molecule mimics of hydrogenases: hydrides and redox. *Chem. Soc. Rev.* 38 (1): 100–108.

80 Esselborn, J., Lambertz, C., Adamska-Venkatesh, A. et al. (2013). Spontaneous activation of [FeFe]-hydrogenases by an inorganic [2Fe] active site mimic. *Nat. Chem. Biol.* 9 (10): 607–609.

81 Lubitz, W., Reijerse, E., and Van Gastel, M. (2007). [NiFe] and [FeFe] hydrogenases studied by advanced magnetic resonance techniques. *Chem. Rev.* 107 (10): 4331–4365.

82 Schwab, D.E., Tard, C., Brecht, E. et al. (2006). On the electronic structure of the hydrogenase H-cluster. *Chem. Commun.* (35): 3696.

83 Reijerse, E.J., Pelmenschikov, V., Birrell, J.A. et al. (2019). Asymmetry in the ligand coordination sphere of the [FeFe] hydrogenase active site is reflected in the magnetic spin interactions of the aza-propanedithiolate ligand. *J. Phys. Chem. Lett.* 10 (21): 6794–6799.

84 Gordon, J.C. and Kubas, G.J. (2010). Perspectives on how nature employs the principles of organometallic chemistry in dihydrogen activation in hydrogenases. *Organometallics* 29 (21): 4682–4701.

85 Silakov, A., Wenk, B., Reijerse, E., and Lubitz, W. (2009). ^{14}N HYSCORE investigation of the H-cluster of [FeFe] hydrogenase: evidence for a nitrogen in the dithiol bridge. *Phys. Chem. Chem. Phys.* 11 (31): 6592.

86 Senger, M., Mebs, S., Duan, J. et al. (2016). Stepwise isotope editing of [FeFe]-hydrogenases exposes cofactor dynamics. *Proc. Natl. Acad. Sci.* 113 (30): 8454–8459.

87 Rodríguez-Maciá, P., Reijerse, E.J., Van Gastel, M. et al. (2018). Sulfide protects [FeFe] hydrogenases from O_2. *J. Am. Chem. Soc.* 140 (30): 9346–9350.

88 Winkler, M., Duan, J., Rutz, A. et al. (2021). A safety cap protects hydrogenase from oxygen attack. *Nat. Commun.* 12 (1): 756.

89 Rutz, A., Das, C.K., Fasano, A. et al. (2023). Increasing the O_2 resistance of the [FeFe]-hydrogenase CbA5H through enhanced protein flexibility. *ACS Catal.* 13 (2): 856–865.

90 Martini, M.A., Bikbaev, K., Pang, Y. et al. (2023). Binding of exogenous cyanide reveals new active-site states in [FeFe] hydrogenases. *Chem. Sci.* 14 (11): 2826–2838.

91 Britt, R.D., Tao, L., Rao, G. et al. (2022). Proposed mechanism for the biosynthesis of the [FeFe] hydrogenase H-cluster: central roles for the radical SAM enzymes HydG and HydE. *ACS Bio. Med. Chem. Au.* 2 (1): 11–21.

92 Pagnier, A., Balci, B., Shepard, E.M. et al. (2022). [FeFe]-Hydrogenase in vitro maturation. *Angew. Chem. Int. Ed.* 61 (49): e202212074.

93 Haumann, M. and Stripp, S.T. (2018). The molecular proceedings of biological hydrogen turnover. *Acc. Chem. Res.* 51 (8): 1755–1763.

94 Reijerse, E.J., Pham, C.C., Pelmenschikov, V. et al. (2017). Direct observation of an iron-bound terminal hydride in [FeFe]-hydrogenase by nuclear resonance vibrational spectroscopy. *J. Am. Chem. Soc.* 139 (12): 4306–4309.

95 Mebs, S., Senger, M., Duan, J. et al. (2017). Bridging hydride at reduced H-cluster species in [FeFe]-hydrogenases revealed by infrared spectroscopy, isotope editing, and quantum chemistry. *J. Am. Chem. Soc.* 139 (35): 12157–12160.

96 Mulder, D.W., Guo, Y., Ratzloff, M.W., and King, P.W. (2017). Identification of a catalytic iron-hydride at the H-cluster of [FeFe]-hydrogenase. *J. Am. Chem. Soc.* 139 (1): 83–86.

97 Rumpel, S., Sommer, C., Reijerse, E. et al. (2018). Direct detection of the terminal hydride intermediate in [FeFe] hydrogenase by NMR spectroscopy. *J. Am. Chem. Soc.* 140 (11): 3863–3866.

98 Mulder, D.W., Ratzloff, M.W., Bruschi, M. et al. (2014). Investigations on the role of proton-coupled electron transfer in hydrogen activation by [FeFe]-hydrogenase. *J. Am. Chem. Soc.* 136 (43): 15394–15402.

99 Sommer, C., Adamska-Venkatesh, A., Pawlak, K. et al. (2017). Proton coupled electronic rearrangement within the H-cluster as an essential step in the catalytic cycle of [FeFe] hydrogenases. *J. Am. Chem. Soc.* 139 (4): 1440–1443.

100 Rodríguez-Maciá, P., Pawlak, K., Rüdiger, O. et al. (2017). Intercluster redox coupling influences protonation at the H-cluster in [FeFe] hydrogenases. *J. Am. Chem. Soc.* 139 (42): 15122–15134.

101 Senger, M., Laun, K., Wittkamp, F. et al. (2017). Proton-coupled reduction of the catalytic [4Fe-4S] cluster in [FeFe]-hydrogenases. *Angew. Chem. Int. Ed.* 56 (52): 16503–16506.

102 Senger, M., Mebs, S., Duan, J. et al. (2018). Protonation/reduction dynamics at the [4Fe–4S] cluster of the hydrogen-forming cofactor in [FeFe]-hydrogenases. *Phys. Chem. Chem. Phys.* 20 (5): 3128–3140.

103 Lemon, B.J. and Peters, J.W. (1999). Binding of exogenously added carbon monoxide at the active site of the iron-only hydrogenase (CpI) from *Clostridium pasteurianum*. *Biochemistry* 38 (40): 12969–12973.

104 Rodríguez-Maciá, P., Galle, L.M., Bjornsson, R. et al. (2020). Caught in the H_{inact}: crystal structure and spectroscopy reveal a sulfur bound to the active site of an O_2-stable state of [FeFe] hydrogenase. *Angew. Chem. Int. Ed.* 59 (38): 16786–16794.

105 Duan, J., Hemschemeier, A., Burr, D.J. et al. (2023). Cyanide binding to [FeFe]-hydrogenase stabilizes the alternative configuration of the proton transfer pathway. *Angew. Chem. Int. Ed.* 62 (7): e202216903.

106 Esselborn, J., Muraki, N., Klein, K. et al. (2016). A structural view of synthetic cofactor integration into [FeFe]-hydrogenases. *Chem. Sci.* 7 (2): 959–968.

107 Nicolet, Y., De Lacey, A.L., Vernède, X. et al. (2001). Crystallographic and FTIR spectroscopic evidence of changes in Fe coordination upon reduction of the active site of the Fe-only hydrogenase from *Desulfovibrio desulfuricans. J. Am. Chem. Soc.* 123 (8): 1596–1601.

108 Kwon, H., Basran, J., Devos, J.M. et al. (2020). Visualizing the protons in a metalloenzyme electron proton transfer pathway. *Proc. Natl. Acad. Sci.* 117 (12): 6484–6490.

109 Maki-Yonekura, S., Kawakami, K., Takaba, K. et al. (2023). Measurement of charges and chemical bonding in a cryo-EM structure. *Commun. Chem.* 6 (1): 98.

110 Stripp, S.T. (2021). In situ infrared spectroscopy for the analysis of gas-processing metalloenzymes. *ACS Catal.* 11 (13): 7845–7862.

111 Ahmadi, S., Barrios Herrera, L., Chehelamirani, M. et al. (2018). Multiscale modeling of enzymes: QM-cluster, QM/MM, and QM/MM/MD: A tutorial review. *Int. J. Quantum Chem.* 118 (9): e25558.

112 Adamska-Venkatesh, A., Krawietz, D., Siebel, J. et al. (2014). New redox states observed in [FeFe] hydrogenases reveal redox coupling within the H-cluster. *J. Am. Chem. Soc.* 136 (32): 11339–11346.

113 Patil, D.S., Moura, J.J., He, S.H. et al. (1988). EPR-detectable redox centers of the periplasmic hydrogenase from *Desulfovibrio vulgaris. J. Biol. Chem.* 263 (35): 18732–18738.

114 Lorent, C., Katz, S., Duan, J. et al. (2020). Shedding light on proton and electron dynamics in [FeFe] hydrogenases. *J. Am. Chem. Soc.* 142 (12): 5493–5497.

115 Felton, G.A.N., Mebi, C.A., Petro, B.J. et al. (2009). Review of electrochemical studies of complexes containing the Fe_2S_2 core characteristic of [FeFe]-hydrogenases including catalysis by these complexes of the reduction of acids to form dihydrogen. *J. Organomet. Chem.* 694 (17): 2681–2699.

116 Roseboom, W., De Lacey, A.L., Fernandez, V.M. et al. (2006). The active site of the [FeFe]-hydrogenase from *Desulfovibrio desulfuricans*. II. Redox properties, light sensitivity and CO-ligand exchange as observed by infrared spectroscopy. *J. Biol. Inorg. Chem.* 11 (1): 102–118.

117 Adamska, A., Silakov, A., Lambertz, C. et al. (2012). Identification and characterization of the "super-reduced" state of the H-cluster in [FeFe] hydrogenase: A new building block for the catalytic cycle? *Angew. Chem. Int. Ed.* 51 (46): 11458–11462.

118 Winkler, M., Senger, M., Duan, J. et al. (2017). Accumulating the hydride state in the catalytic cycle of [FeFe]-hydrogenases. *Nat. Commun.* 8 (1): 16115.

119 Mészáros, L.S., Ceccaldi, P., Lorenzi, M. et al. (2020). Spectroscopic investigations under whole-cell conditions provide new insight into the metal hydride chemistry of [FeFe]-hydrogenase. *Chem. Sci.* 11 (18): 4608–4617.

120 Pelmenschikov, V., Birrell, J.A., Pham, C.C. et al. (2017). Reaction coordinate leading to H_2 production in [FeFe]-hydrogenase identified by nuclear resonance vibrational spectroscopy and density functional theory. *J. Am. Chem. Soc.* 139 (46): 16894–16902.

121 Laun, K., Baranova, I., Duan, J. et al. (2021). Site-selective protonation of the one-electron reduced cofactor in [FeFe]-hydrogenase. *Dalton Trans.* 50 (10): 3641–3650.

122 Senger, M., Duan, J., Pavliuk, M.V. et al. (2022). Trapping an oxidized and protonated intermediate of the [FeFe]-hydrogenase cofactor under mildly reducing conditions. *Inorg. Chem.* 61 (26): 10036–10042.

123 Ratzloff, M.W., Artz, J.H., Mulder, D.W. et al. (2018). CO-bridged H-cluster intermediates in the catalytic mechanism of [FeFe]-hydrogenase CaI. *J. Am. Chem. Soc.* 140 (24): 7623–7628.

124 Birrell, J.A., Pelmenschikov, V., Mishra, N. et al. (2020). Spectroscopic and computational evidence that [FeFe] hydrogenases operate exclusively with CO-bridged intermediates. *J. Am. Chem. Soc.* 142 (1): 222–232.

125 Stripp, S.T., Mebs, S., and Haumann, M. (2020). Temperature dependence of structural dynamics at the catalytic cofactor of [FeFe]-hydrogenase. *Inorg. Chem.* 59 (22): 16474–16488.

126 Chernev, P., Lambertz, C., Brünje, A. et al. (2014). Hydride binding to the active site of [FeFe]-hydrogenase. *Inorg. Chem.* 53 (22): 12164–12177.

127 Mirmohades, M., Adamska-Venkatesh, A., Sommer, C. et al. (2016). Following [FeFe] hydrogenase active site intermediates by time-resolved Mid-IR spectroscopy. *J. Phys. Chem. Lett.* 7 (16): 3290–3293.

128 Greene, B.L., Schut, G.J., Adams, M.W.W., and Dyer, R.B. (2017). Pre-steady-state kinetics of catalytic intermediates of an [FeFe]-hydrogenase. *ACS Catal.* 7 (3): 2145–2150.

129 Sanchez, M.L.K., Sommer, C., Reijerse, E. et al. (2019). Investigating the kinetic competency of *Cr*HydA1 [FeFe] hydrogenase intermediate states via time-resolved infrared spectroscopy. *J. Am. Chem. Soc.* 141 (40): 16064–16070.

130 Artz, J.H., Zadvornyy, O.A., Mulder, D.W. et al. (2020). Tuning catalytic bias of hydrogen gas producing hydrogenases. *J. Am. Chem. Soc.* 142 (3): 1227–1235.

131 Bourgeois, D. and Weik, M. (2009). Kinetic protein crystallography: a tool to watch proteins in action. *Crystallogr. Rev.* 15 (2): 87–118.

132 Neutze, R., Wouts, R., Van Der Spoel, D. et al. (2000). Potential for biomolecular imaging with femtosecond X-ray pulses. *Nature* 406 (6797): 752–757.

133 Morra, S., Duan, J., Winkler, M. et al. (2021). Electrochemical control of [FeFe]-hydrogenase single crystals reveals complex redox populations at the catalytic site. *Dalton Trans.* 50 (36): 12655–12663.

134 Khushvakov, J., Nussbaum, R., Cadoux, C. et al. (2021). Following electroenzymatic hydrogen production by rotating ring–disk electrochemistry and mass spectrometry. *Angew. Chem. Int. Ed.* 60 (18): 10001–10006.

135 Oughli, A.A., Conzuelo, F., Winkler, M. et al. (2015). A redox hydrogel protects the O_2-sensitive [FeFe]-hydrogenase from *Chlamydomonas reinhardtii* from oxidative damage. *Angew. Chem. Int. Ed.* 54 (42): 12329–12333.

136 Ash, P.A., Reeve, H.A., Quinson, J. et al. (2016). Synchrotron-based infrared microanalysis of biological redox processes under electrochemical control. *Anal. Chem.* 88 (13): 6666–6671.

137 Fasano, A., Land, H., Fourmond, V. et al. (2021). Reversible or irreversible catalysis of H^+/H_2 conversion by FeFe hydrogenases. *J. Am. Chem. Soc.* 143 (48): 20320–20325.

138 Fasano, A., Baffert, C., Schumann, C. et al. (2024). Kinetic modeling of the reversible or irreversible electrochemical responses of FeFe-hydrogenases. *J. Am. Chem. Soc.* 146 (2): 1455–1466.

139 Hajj, V., Baffert, C., Sybirna, K. et al. (2014). FeFe hydrogenase reductive inactivation and implication for catalysis. *Energy Environ. Sci.* 7 (2): 715–719.

140 Shafaat, H.S., Rüdiger, O., Ogata, H., and Lubitz, W. (2013). [NiFe] hydrogenases: a common active site for hydrogen metabolism under diverse conditions. *Biochim. Biophys. Acta BBA - Bioenerg.* 1827 (8–9): 986–1002.

141 Caserta, G., Hartmann, S., Van Stappen, C. et al. (2023). Stepwise assembly of the active site of [NiFe]-hydrogenase. *Nat. Chem. Biol.* 19 (4): 498–506.

142 Fritsch, J., Lenz, O., and Friedrich, B. (2013). Structure, function and biosynthesis of O_2-tolerant hydrogenases. *Nat. Rev. Microbiol.* 11 (2): 106–114.

143 Lacasse, M.J. and Zamble, D.B. (2016). [NiFe]-hydrogenase maturation. *Biochemistry* 55 (12): 1689–1701.

144 Liebgott, P.-P., Leroux, F., Burlat, B. et al. (2010). Relating diffusion along the substrate tunnel and oxygen sensitivity in hydrogenase. *Nat. Chem. Biol.* 6 (1): 63–70.

145 Greco, C., Fourmond, V., Baffert, C. et al. (2014). Combining experimental and theoretical methods to learn about the reactivity of gas-processing metalloenzymes. *Energy Environ. Sci.* 7 (11): 3543–3573.

146 Tombolelli, D. and Mroginski, M.A. (2019). Proton transfer pathways between active sites and proximal clusters in the membrane-bound [NiFe] hydrogenase. *J. Phys. Chem. B* 123 (16): 3409–3420.

147 Siegbahn, P.E.M. (2004). Proton and electron transfers in [NiFe] hydrogenase. *Adv. Inorg. Chem.* 56, 101–125.

148 Ash, P.A., Hidalgo, R., and Vincent, K.A. (2017). Proton transfer in the catalytic cycle of [NiFe] hydrogenases: insight from vibrational spectroscopy. *ACS Catal.* 7 (4): 2471–2485.

149 Tai, H., Higuchi, Y., and Hirota, S. (2018). Comprehensive reaction mechanisms at and near the Ni–Fe active sites of [NiFe] hydrogenases. *Dalton Trans.* 47 (13): 4408–4423.

150 Pandelia, M., Ogata, H., and Lubitz, W. (2010). Intermediates in the catalytic cycle of [NiFe] hydrogenase: functional spectroscopy of the active site. *ChemPhysChem* 11 (6): 1127–1140.

151 Volbeda, A., Charon, M.-H., Piras, C. et al. (1995). Crystal structure of the nickel–iron hydrogenase from *Desulfovibrio gigas*. *Nature* 373 (6515): 580–587.

152 Volbeda, A., Garcin, E., Piras, C. et al. (1996). Structure of the [NiFe] hydrogenase active site: evidence for biologically uncommon Fe ligands. *J. Am. Chem. Soc.* 118 (51): 12989–12996.

153 Happe, R.P., Roseboom, W., Pierik, A.J. et al. (1997). Biological activition of hydrogen. *Nature* 385 (6612): 126–126.

154 Greening, C., Biswas, A., Carere, C.R. et al. (2016). Genomic and metagenomic surveys of hydrogenase distribution indicate H_2 is a widely utilised energy source for microbial growth and survival. *ISME J.* 10 (3): 761–777.

155 Vignais, P. (2001). Classification and phylogeny of hydrogenases. *FEMS Microbiol. Rev.* 25 (4): 455–501.

156 Lenz, O., Lauterbach, L., and Frielingsdorf, S. (2018). O_2-tolerant [NiFe]-hydrogenases of *Ralstonia eutropha* H16: physiology, molecular biology, purification, and biochemical analysis. In: Armstrong, F., editor. *Methods in Enzymology*, vol. 613, 117–151. Academic Press.

157 Fritsch, J., Scheerer, P., Frielingsdorf, S. et al. (2011). The crystal structure of an oxygen-tolerant hydrogenase uncovers a novel iron-sulphur centre. *Nature* 479 (7372): 249–252.

158 Volbeda, A., Amara, P., Darnault, C. et al. (2012). X-ray crystallographic and computational studies of the O_2-tolerant [NiFe]-hydrogenase 1 from *Escherichia coli*. *Proc. Natl. Acad. Sci.* 109 (14): 5305–5310.

159 Shomura, Y., Yoon, K.-S., Nishihara, H., and Higuchi, Y. (2011). Structural basis for a [4Fe-3S] cluster in the oxygen-tolerant membrane-bound [NiFe]-hydrogenase. *Nature* 479 (7372): 253–256.

160 Frielingsdorf, S., Fritsch, J., Schmidt, A. et al. (2014). Reversible [4Fe-3S] cluster morphing in an O_2-tolerant [NiFe] hydrogenase. *Nat. Chem. Biol.* 10 (5): 378–385.

161 Roessler, M.M., Evans, R.M., Davies, R.A. et al. (2012). EPR spectroscopic studies of the Fe–S clusters in the O_2-tolerant [NiFe]-hydrogenase Hyd-1 from *Escherichia coli* and characterization of the unique [4Fe–3S] cluster by HYSCORE. *J. Am. Chem. Soc.* 134 (37): 15581–15594.

162 Wulff, P., Day, C.C., Sargent, F., and Armstrong, F.A. (2014). How oxygen reacts with oxygen-tolerant respiratory [NiFe]-hydrogenases. *Proc. Natl. Acad. Sci.* 111 (18): 6606–6611.

163 Abou Hamdan, A., Burlat, B., Gutiérrez-Sanz, O. et al. (2013). O_2-independent formation of the inactive states of NiFe hydrogenase. *Nat. Chem. Biol.* 9 (1): 15–17.

164 Schmidt, A., Kalms, J., Lorent, C. et al. (2023). Stepwise conversion of the Cys_6[4Fe–3S] to a Cys_4[4Fe–4S] cluster and its impact on the oxygen tolerance of [NiFe]-hydrogenase. *Chem. Sci.* 14 (40): 11105–11120.

165 Noor, N.D.M., Matsuura, H., Nishikawa, K. et al. (2018). Redox-dependent conformational changes of a proximal [4Fe–4S] cluster in Hyb-type [NiFe]-hydrogenase to protect the active site from O_2. *Chem. Commun.* 54 (87): 12385–12388.

166 Shomura, Y., Taketa, M., Nakashima, H. et al. (2017). Structural basis of the redox switches in the NAD^+-reducing soluble [NiFe]-hydrogenase. *Science* 357 (6354): 928–932.

167 Kulka-Peschke, C.J., Schulz, A.-C., Lorent, C. et al. (2022). Reversible glutamate coordination to high-valent nickel protects the active site of a [NiFe] hydrogenase from oxygen. *J. Am. Chem. Soc.* 144 (37): 17022–17032.

168 Kumar, R. and Stein, M. (2023). The fully oxidized state of the glutamate coordinated O_2-tolerant [NiFe]-hydrogenase shows a Ni(III)/Fe(III) open-shell singlet ground state. *J. Am. Chem. Soc.* 145 (20): 10954–10959.

169 Kwan, P., McIntosh, C.L., Jennings, D.P. et al. (2015). The [NiFe]-hydrogenase of *Pyrococcus furiosus* exhibits a new type of oxygen tolerance. *J. Am. Chem. Soc.* 137 (42): 13556–13565.

170 Eguchi, S., Yoon, K.-S., and Ogo, S. (2012). O_2-stable membrane-bound [NiFe]hydrogenase from a newly isolated *Citrobacter* sp. S-77. *J. Biosci. Bioeng.* 114 (5): 479–484.

171 Muhd Noor, N.D., Nishikawa, K., Nishihara, H. et al. (2016). Improved purification, crystallization and crystallographic study of Hyd-2-type [NiFe]-hydrogenase from *Citrobacter* sp. S-77. *Acta Crystallogr. Sect. F Struct. Biol. Commun.* 72 (1): 53–58.

172 Carter, C.W. (2006). High potential iron sulfur proteins. In: Messerschmidt, A., Huber, R., Poulas, T., Wieghardt, K., Cygler, M., Bode, W., editors. *Handbook of Metalloproteins*. Wiley.

173 Greening, C. and Grinter, R. (2022). Microbial oxidation of atmospheric trace gases. *Nat. Rev. Microbiol.* 20 (9): 513–528.

174 Greening, C., Kropp, A., Vincent, K., and Grinter, R. (2023). Developing high-affinity, oxygen-insensitive [NiFe]-hydrogenases as biocatalysts for energy conversion. *Biochem. Soc. Trans.* 51 (5): 1921–1933.

175 Grinter, R., Kropp, A., Venugopal, H. et al. (2023). Structural basis for bacterial energy extraction from atmospheric hydrogen. *Nature* 615 (7952): 541–547.

176 Schäfer, C., Bommer, M., Hennig, S.E. et al. (2016). Structure of an actinobacterial-type [NiFe]-hydrogenase reveals insight into O_2-tolerant H_2 oxidation. *Structure* 24 (2): 285–292.

177 Schäfer, C., Friedrich, B., and Lenz, O. (2013). Novel, oxygen-insensitive group 5 [NiFe]-hydrogenase in *Ralstonia eutropha*. *Appl. Environ. Microbiol.* 79 (17): 5137–5145.

178 Buhrke, T., Lenz, O., Krauss, N., and Friedrich, B. (2005). Oxygen tolerance of the H_2-sensing [NiFe] hydrogenase from *Ralstonia eutropha* H16 is based on limited access of oxygen to the active site. *J. Biol. Chem.* 280 (25): 23791–23796.

179 Horch, M., Lauterbach, L., Mroginski, M.A. et al. (2015). Reversible active site sulfoxygenation can explain the oxygen tolerance of a NAD^+-reducing [NiFe] hydrogenase and its unusual infrared spectroscopic properties. *J. Am. Chem. Soc.* 137 (7): 2555–2564.

180 Lauterbach, L. and Lenz, O. (2013). Catalytic production of hydrogen peroxide and water by oxygen-tolerant [NiFe]-hydrogenase during H_2 cycling in the presence of O_2. *J. Am. Chem. Soc.* 135 (47): 17897–17905.

181 Steinhilper, R., Höff, G., Heider, J., and Murphy, B.J. (2022). Structure of the membrane-bound formate hydrogenlyase complex from *Escherichia coli*. *Nat. Commun.* 13 (1): 5395.

182 Maeda, T., Sanchez-Torres, V., and Wood, T.K. (2007). *Escherichia coli* hydrogenase 3 is a reversible enzyme possessing hydrogen uptake and synthesis activities. *Appl. Microbiol. Biotechnol.* 76 (5): 1035–1042.

183 McDowall, J.S., Murphy, B.J., Haumann, M. et al. (2014). Bacterial formate hydrogenlyase complex. *Proc. Natl. Acad. Sci.* 111 (38): E3948-56.

184 Huyett, J.E., Carepo, M., Pamplona, A. et al. (1997). ^{57}Fe Q-band pulsed ENDOR of the hetero-dinuclear site of nickel hydrogenase: comparison of the NiA, NiB, and NiC states. *J. Am. Chem. Soc.* 119 (39): 9291–9292.

185 Roncaroli, F., Bill, E., Friedrich, B. et al. (2015). Cofactor composition and function of a H_2-sensing regulatory hydrogenase as revealed by Mössbauer and EPR spectroscopy. *Chem. Sci.* 6 (8): 4495–4507.

186 Pandelia, M.-E., Ogata, H., Currell, L.J. et al. (2010). Inhibition of the [NiFe] hydrogenase from *Desulfovibrio vulgaris* Miyazaki F by carbon monoxide: an FTIR and EPR spectroscopic study. *Biochim. Biophys. Acta BBA - Bioenerg.* 1797 (2): 304–313.

187 Imanishi, T., Nishikawa, K., Taketa, M. et al. (2022). Structural and spectroscopic characterization of CO inhibition of [NiFe]-hydrogenase from *Citrobacter* sp. S-77. *Acta Crystallogr. Sect. F Struct. Biol. Commun.* 78 (2): 66–74.

188 Carepo, M., Tierney, D.L., Brondino, C.D. et al. (2002). ^{17}O ENDOR detection of a solvent-derived Ni–(OH*x*)–Fe bridge that is lost upon activation of the hydrogenase from *Desulfovibrio gigas*. *J. Am. Chem. Soc.* 124 (2): 281–286.

189 Van Der Zwaan, J.W., Coremans, J.M.C.C., Bouwens, E.C.M., and Albracht, S.P.J. (1990). Effect of $^{17}O_2$ and ^{13}CO on EPR spectra of nickel in hydrogenase from *Chromatium vinosum*. *Biochim. Biophys. Acta BBA - Protein Struct. Mol. Enzymol.* 1041 (2): 101–110.

190 Caserta, G., Lorent, C., Ciaccafava, A. et al. (2020). The large subunit of the regulatory [NiFe]-hydrogenase from *Ralstonia eutropha* – a minimal hydrogenase? *Chem. Sci.* 11 (21): 5453–5465.

191 Caserta, G., Pelmenschikov, V., Lorent, C. et al. (2021). Hydroxy-bridged resting states of a [NiFe]-hydrogenase unraveled by cryogenic vibrational spectroscopy and DFT computations. *Chem. Sci.* 12 (6): 2189–2197.

192 Hiromoto, T., Nishikawa, K., Inoue, S. et al. (2023). New insights into the oxidation process from neutron and X-ray crystal structures of an O_2-sensitive [NiFe]-hydrogenase. *Chem. Sci.* 14 (35): 9306–9315.

193 Volbeda, A., Martin, L., Liebgott, P.-P. et al. (2015). [NiFe]-hydrogenases revisited: nickel–carboxamido bond formation in a variant with accrued O_2-tolerance and a tentative re-interpretation of Ni-SI states. *Metallomics* 7 (4): 710–718.

194 Evans, R.M., Beaton, S.E., Rodriguez Macia, P. et al. (2023). Comprehensive structural, infrared spectroscopic and kinetic investigations of the roles of the active-site arginine in bidirectional hydrogen activation by the [NiFe]-hydrogenase 'Hyd-2' from *Escherichia coli*. *Chem. Sci.* 14 (32): 8531–8551.

195 Siegbahn, P.E.M. and Liao, R.-Z. (2020). The energetics of hydrogen molecule oxidation in NiFe-hydrogenase. *ACS Catal.* 10 (10): 5603–5613.

196 Dong, G., Phung, Q.M., Pierloot, K., and Ryde, U. (2018). Reaction mechanism of [NiFe] hydrogenase studied by computational methods. *Inorg. Chem.* 57 (24): 15289–15298.

197 Bruschi, M., Tiberti, M., Guerra, A., and De Gioia, L. (2014). Disclosure of key stereoelectronic factors for efficient H_2 binding and cleavage in the active site of [NiFe]-hydrogenases. *J. Am. Chem. Soc.* 136 (5): 1803–1814.

198 Ilina, Y., Lorent, C., Katz, S. et al. (2019). X-ray crystallography and vibrational spectroscopy reveal the key determinants of biocatalytic dihydrogen cycling by [NiFe] hydrogenases. *Angew. Chem. Int. Ed.* 58 (51): 18710–18714.

199 Caserta, G., Lorent, C., Pelmenschikov, V. et al. (2020). *In vitro* assembly as a tool to investigate catalytic intermediates of [NiFe]-hydrogenase. *ACS Catal.* 10 (23): 13890–13894.

200 Horch, M., Schoknecht, J., Mroginski, M.A. et al. (2014). Resonance Raman spectroscopy on [NiFe] hydrogenase provides structural insights into catalytic intermediates and reactions. *J. Am. Chem. Soc.* 136 (28): 9870–9873.

201 Greene, B.L., Wu, C.-H., Vansuch, G.E. et al. (2016). Proton inventory and dynamics in the Ni_a-S to Ni_a-C transition of a [NiFe] hydrogenase. *Biochemistry* 55 (12): 1813–1825.

202 Dementin, S., Burlat, B., De Lacey, A.L. et al. (2004). A glutamate is the essential proton transfer gate during the catalytic cycle of the [NiFe] hydrogenase. *J. Biol. Chem.* 279 (11): 10508–10513.

203 Greene, B.L., Vansuch, G.E., Wu, C.-H. et al. (2016). Glutamate gated proton-coupled electron transfer activity of a [NiFe]-hydrogenase. *J. Am. Chem. Soc.* 138 (39): 13013–13021.

204 Evans, R.M., Ash, P.A., Beaton, S.E. et al. (2018). Mechanistic exploitation of a self-repairing, blocked proton transfer pathway in an O_2-tolerant [NiFe]-hydrogenase. *J. Am. Chem. Soc.* 140 (32): 10208–10220.

205 Evans, R.M., Brooke, E.J., Wehlin, S.A.M. et al. (2016). Mechanism of hydrogen activation by [NiFe] hydrogenases. *Nat. Chem. Biol.* 12 (1): 46–50.

206 Vansuch, G.E., Wu, C.-H., Haja, D.K. et al. (2020). Metal–ligand cooperativity in the soluble hydrogenase-1 from *Pyrococcus furiosus*. *Chem. Sci.* 11 (32): 8572–8581.

207 Lubitz, W. and Tumas, W. (2007). Hydrogen: an overview. *Chem. Rev.* 107 (10): 3900–3903.

208 Krämer, T., Kampa, M., Lubitz, W. et al. (2013). Theoretical spectroscopy of the Ni^{II} intermediate states in the catalytic cycle and the activation of [NiFe] Hydrogenases. *ChemBioChem* 14 (14): 1898–1905.

209 Ogata, H., Nishikawa, K., and Lubitz, W. (2015). Hydrogens detected by subatomic resolution protein crystallography in a [NiFe] hydrogenase. *Nature* 520 (7548): 571–574.

210 Ogata, H., Krämer, T., Wang, H. et al. (2015). Hydride bridge in [NiFe]-hydrogenase observed by nuclear resonance vibrational spectroscopy. *Nat. Commun.* 6 (1): 7890.

211 Moura, J.J.G., Moura, I., Huynh, B.H. et al. (1982). Unambiguous identification of the nickel EPR signal in ^{61}Ni-enriched *Desulfovibrio gigas* hydrogenase. *Biochem. Biophys. Res. Commun.* 108 (4): 1388–1393.

212 Fan, C., Teixeira, M., Moura, J. et al. (1991). Detection and characterization of exchangeable protons bound to the hydrogen-activation nickel site of *Desulfovibrio gigas* hydrogenase: a proton and deuterium Q-band ENDOR study. *J. Am. Chem. Soc.* 113 (1): 20–24.

213 Brecht, M., Van Gastel, M., Buhrke, T. et al. (2003). Direct detection of a hydrogen ligand in the [NiFe] center of the regulatory H_2-sensing hydrogenase from *Ralstonia eutropha* in its reduced state by HYSCORE and ENDOR spectroscopy. *J. Am. Chem. Soc.* 125 (43): 13075–13083.

214 Tai, H., Nishikawa, K., Higuchi, Y. et al. (2019). Cysteine SH and glutamate COOH contributions to [NiFe] hydrogenase proton transfer revealed by highly sensitive FTIR spectroscopy. *Angew. Chem. Int. Ed.* 58 (38): 13285–13290.

215 Caserta, G., Zuccarello, L., Barbosa, C. et al. (2022). Unusual structures and unknown roles of Fe–S clusters in metalloenzymes seen from a resonance Raman spectroscopic perspective. *Coord. Chem. Rev.* 452: 214287.

216 Siebert, E., Horch, M., Rippers, Y. et al. (2013). Resonance Raman spectroscopy as a tool to monitor the active site of hydrogenases. *Angew. Chem. Int. Ed.* 52 (19): 5162–5165.

217 Lorent, C., Pelmenschikov, V., Frielingsdorf, S. et al. (2021). Exploring structure and function of redox intermediates in [NiFe]-hydrogenases by an advanced experimental approach for solvated, lyophilized and crystallized metalloenzymes. *Angew. Chem. Int. Ed.* 60 (29): 15854–15862.

218 Waffo, A.F.T., Lorent, C., Katz, S. et al. (2023). Structural determinants of the catalytic Ni_a-L intermediate of [NiFe]-hydrogenase. *J. Am. Chem. Soc.* 145 (25): 13674–13685.

219 Van Der Zwaan, J.W., Albracht, S.P.J., Fontijn, R.D., and Slater, E.C. (1985). Monovalent nickel in hydrogenase from *Chromatium vinosum*: light sensitivity and evidence for direct interaction with hydrogen. *FEBS Lett.* 179 (2): 271–277.

220 Hidalgo, R., Ash, P.A., Healy, A.J., and Vincent, K.A. (2015). Infrared spectroscopy during electrocatalytic turnover reveals the Ni-L active site state during H_2 oxidation by a NiFe hydrogenase. *Angew. Chem. Int. Ed.* 54 (24): 7110–7113.

221 Tai, H., Nishikawa, K., Suzuki, M. et al. (2014). Control of the transition between Ni-C and Ni-SI_a states by the redox state of the proximal Fe-S cluster in the catalytic cycle of [NiFe] hydrogenase. *Angew. Chem. Int. Ed.* 53 (50): 13817–13820.

222 Ash, P.A., Kendall-Price, S.E.T., Evans, R.M. et al. (2021). The crystalline state as a dynamic system: IR microspectroscopy under electrochemical control for a [NiFe] hydrogenase. *Chem. Sci.* 12 (39): 12959–12970.

223 Tai, H., Nishikawa, K., Inoue, S. et al. (2015). FT-IR characterization of the light-induced Ni-L2 and Ni-L3 states of [NiFe] hydrogenase from *Desulfovibrio vulgaris* Miyazaki F. *J. Phys. Chem. B* 119 (43): 13668–13674.

224 Weber, K., Krämer, T., Shafaat, H.S. et al. (2012). A functional [NiFe]-hydrogenase model compound that undergoes biologically relevant reversible thiolate protonation. *J. Am. Chem. Soc.* 134 (51): 20745–20755.

225 Simmons, T.R., Berggren, G., Bacchi, M., and at al. (2014). Mimicking hydrogenases: from biomimetics to artificial enzymes. *Coord. Chem. Rev.* 270–271: 127–150.

16

Macromolecular Crystallography

Konstantin Bikbaev and Ingrid Span

Friedrich-Alexander-Universität Erlangen-Nürnberg, Bioinorganic Chemistry, Egerlandstr. 1, 91058 Erlangen, Germany

Chapter Goals

- Provides an understanding of the basic concepts of macromolecular crystallography
- Takes the reader on a short tour through the history of X-ray crystallography
- Highlights exciting new developments in time-resolved crystallography
- Explains aspects that are specific for the structure determination of Fe–S proteins
- Ideal for beginners without prior knowledge of the area

Main Chapter Points

- *Introduction*: General aspects of the basic principles and important historical milestones in the development of X-ray crystallography.
- *Crystallization of macromolecules*: An overview of the nature of crystallization, methods for crystallizing proteins, and the preparation of a sample for a diffraction experiment.
- *Symmetry and the unit cell*: Basic overview of crystal symmetry, explaining the concepts of the unit cell, asymmetric unit, Bravais lattice, and space group.
- *Diffraction and interpretation of diffraction patterns*: Basics of X-ray diffraction, how diffraction patterns are interpreted, and key concepts like Bragg's law.
- *Data collection and processing*: Principles of the processes involved in collecting and processing X-ray diffraction data.
- *X-ray structure analysis*: Phase determination methods of diffracted X-rays to obtain electron density maps and solution of the three-dimensional structure of proteins from diffraction data.
- *Time-resolved crystallography*: Introduction to time-resolved crystallography and its significance in capturing dynamic structural changes in proteins.
- *Nuances of Fe–S protein crystallography*: Discussion of the specific challenges in crystallizing and handling Fe–S proteins and strategies to address these issues.

Iron-Sulfur Clusters: Biogenesis and Biochemistry, Volume 2, First Edition.
Edited by Silke Leimkühler, Günter Schwarz, Oliver Lenz, and Oliver Einsle.

16.1 Introduction

X-ray crystallography is a technique that uses X-ray diffraction patterns to determine high-resolution, three-dimensional structures of molecules. It can be used to solve the structure of small molecules, materials, and biomolecules. To be able to carry out X-ray diffraction experiments, the substance of interest must be in crystalline form, which can be particularly challenging for proteins or nucleic acids.

The structure of a macromolecule consists of an arrangement of all atoms except hydrogen and allows insights into the folding of the polypeptide chain and the architecture of the active site. In 2024, the number of macromolecular structures deposited in the Research Collaboratory for Structural Bioinformatics (RCSB) Protein Data Bank (PDB) [1] exceeds 200 000 entries with the vast majority determined using X-ray crystallography. This highlights that despite tremendous breakthroughs in several disciplines of structure biology, especially in the field of single particle cryo-electron microscopy (cryo-EM) and the prediction of protein structures using Alphafold2, X-ray crystallography remains one of the most powerful tools for determining protein structures. It enables us to study the processes that take place in living organisms and visualize the biomolecules at atomic resolution. In addition, it plays important roles in related fields such as drug design.

Structural biology comprises a wide range of techniques. In addition to X-ray crystallography, cryo-EM, and structure prediction tools, it also includes nuclear magnetic resonance spectroscopy, X-ray solution scattering, neutron diffraction, electron diffraction, and various spectroscopic techniques. Rather than viewing the different structural biology techniques as mutually exclusive, we should focus on learning how they complement each other and how we can combine different methods to obtain meaningful information about biological processes.

Each method has unique advantages and caveats. The advantages of crystallography are that it allows us to determine structures with very high resolution and to locate small molecules, such as water, ligands, metals, or cofactors in the structure. Data collection at different wavelengths also enables the identification of metals in the structure, which is an invaluable tool for investigating metalloproteins. Additionally, X-ray crystallography can be combined with many other techniques, which allows us to obtain complementary information on the same crystalline sample. Arguably, the biggest caveat is that obtaining high-quality crystals can be challenging.

The central object of crystallography is the crystal [2]. Hereby, the regularities of the crystalline state, the arrangement of the building blocks in crystals, physical and chemical properties, synthesis, and growths of crystals are essential. Crystals play an important role in many disciplines, including mineralogy, inorganic and organic chemistry, physical chemistry, physics, metallurgy, material science, geology, geophysics, biology, and medicine. The relevance of crystals may become even more clear when looking at where they occur. Almost all naturally formed minerals are crystals. Rocks, icebergs, solid chemicals, and biomolecules consist of crystals or can form crystalline states. The hard substances of teeth and bones are crystalline and calcification processes of vessels and muscles in humans and animals can be traced back to the formation of crystals. Many viruses also have a crystalline structure.

Crystals have fascinated mankind for thousands of years, reflected by their use to create expensive jewelry. It was only in the twentieth century that we began to understand their nature. The early days of crystallography began with the discovery of X-rays in 1895 by Wilhelm Conrad Röntgen in Germany [3, 4]. The next major discovery was that the wavelength of X-rays might be comparable with the interatomic distances in 1912 by Max von Laue [5]. He observed diffraction images when exposing copper sulfate pentahydrate crystals to X-ray radiation and provided a physical explanation for the diffraction images. A breakthrough was achieved by Sir William Henry Bragg and his son Sir William Lawrence Bragg, who developed an elegant mathematical explanation of the images generated by Laue. The famous Bragg's law describes the relationship between the angles of diffraction, the wavelength of the X-rays, and the interplanar spacing in the crystal lattice [6]. Bragg's law and the work of father and son on the diffraction of electromagnetic waves were crucial to the introduction of diffraction as a tool for crystal structure determination [7–10]. In 1913, W. H. Bragg built the first X-ray spectrometer and reported the first crystal structure of diamond [11].

The successful application of X-ray diffraction to protein structure is relatively recent [12]. Despite early reports of crystallites in blood preparations in 1840, which were in fact hemoglobin crystals [13], it took almost 100 years before the first X-ray photograph of a protein crystal was taken [14]. The first three-dimensional structures of a protein were reported in 1958 by Sir John Kendrew, who determined the structure of sperm whale myoglobin [15]. Only two years later, his colleague Max Ferdinand Perutz solved the crystal structure of horse hemoglobin to a resolution of 5.5 Å [16]. However, the contributions of Perutz, who developed the method of isomorphous replacement, were crucial to solving the crystal structures of both proteins [17].

Although Dorothy Crowfoot Hodgkin was not the first to determine the crystal structure of a protein, her contributions to combining biology and crystallography were monumental. Already in 1934, she was able to record the first X-ray diffraction images of pepsin crystals together with Bernal [14]. In 1935, she started working on insulin, which became her lifelong interest. After reporting the first X-ray diffraction images in the same year, it took another 34 years before the crystal structure of this essential peptide hormone was solved to a resolution of 2.8 Å in 1969 [18]. She continued her structural studies of porcine insulin and in 1988 obtained a crystal structure of 1.5 Å at room temperature from X-ray data [14]. In addition to her studies of insulin, she used X-ray crystallography to determine the structure of many biologically relevant molecules, including cholesterol, penicillin, and vitamin B_{12} [19–21].

One of the most influential chemists of all time, Linus Pauling considered himself a crystallographer. Despite never solving a macromolecular structure, his contributions to the fundamental structural motifs of proteins before observed in crystal structures have been essential for the field [22]. He was intrigued by the structure and dynamics of the chemical bond [23]. The planarity of the peptide group in combination with the role of hydrogen bonds that he anticipated and the logical assumption that regular structure should arise from the repetition of stable motifs, he deduced the structure of the α-helix [24]. In 1951, Pauling and Corey published the structure of the α-helix, even though the model was in slight disagreement with diffraction data on α-keratin, and later they also proposed the β-sheet structure, composed of extended protein chains [25, 26]. Interestingly, their prediction of a DNA structure with the backbone on the inside and the bases on the outside turned out to be entirely incorrect [27]. Nevertheless, Pauling did not only leave his mark on science, but he was also an untiring advocate of peace.

This brief summary of the most important achievements of crystallographers that led to major advances in our understanding of the structure and function of biological macromolecules leads us to the question of what we mean when speaking of a structure. In common language, a structure can mean a building or other object made up of several parts or the arrangement of and relations between the parts or elements of something complex. In proper scientific terms, however, there is only one way to precisely describe the structure of a molecule, by specifying the coordinates in three-dimensional space of each point within the molecule, each with respect to some defined and agreed-upon system of axes in space, namely, a coordinate system [28]. The intrinsic structure of the molecule is invariant and remains the same no matter what spatial coordinate system or origin is chosen. Therefore, the relative positions of points within the molecule remain the same even if its constituent points are transformed from one coordinate system to another.

Therefore, if the structure of a molecule is defined by giving the spatial coordinates x_j, y_j, and z_j of each atom j in the molecule, atom 1 has the same spatial relationship to atom 2, regardless of the coordinate system. To define the structure of a molecule in a precise way, we must therefore produce a list of atomic coordinates x_j, y_j, and z_j. This set of ordered triples x_j, y_j, and z_j, one for each atom, becomes a molecular structure. These spatial coordinates are then suitable for translation into a visual representation, such as pictures or images. When we talk about solving the structure of any kind of molecule, from proteins or nucleic acids to large assemblies such as viruses or ribosomes, we are ultimately trying to identify and specify the x_j, y_j, and z_j coordinates of each atom in the molecule. In this chapter, we want to give an overview of how to get from a pure and homogeneous protein solution to the atomic model of the structure (Figure 16.1).

16.2 Crystallization of Macromolecules

Solid matter can exist either as unordered structures, which are called amorphous solids (e.g. glass), or as highly ordered ones called crystalline solids (e.g. diamond). In the former case, the substituents (atoms, ions, or molecules) of a solid material do not exhibit any symmetry. On the contrary, in the case of crystals, they are highly ordered. A crystal might be represented as a set of small identical fractions, usually of a parallelepiped shape, that are called unit cells. Many substances can arrange themselves into crystalline structures under certain circumstances, and proteins are not an exception. Protein crystals exhibit all properties of a crystal; however, the

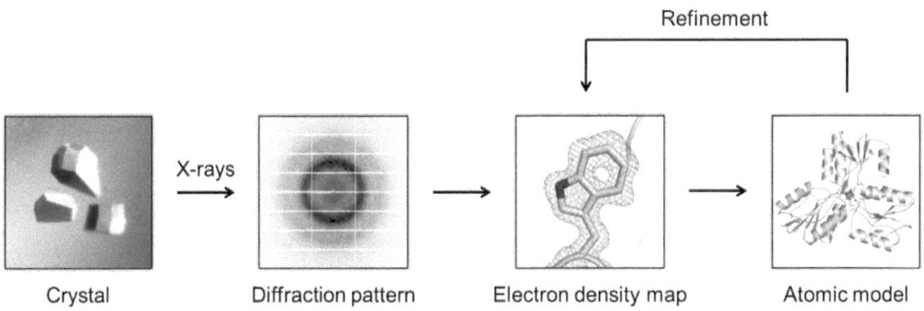

Figure 16.1 A scheme of a crystallography experiment. Crystals are irradiated with X-rays yielding a set of diffraction patterns collected at different angles. Then, the obtained diffraction images are processed resulting in an electron density map. An appropriate atomic model is created to fit the electron density, and this model is then iteratively improved by multiple steps of refinement.

Figure 16.2 Comparison of a low- and highly-mosaic crystalline structure. Each square represents a cluster of multiple unit cells that are ordered well within one square but worse relative to the unit cells from another square.

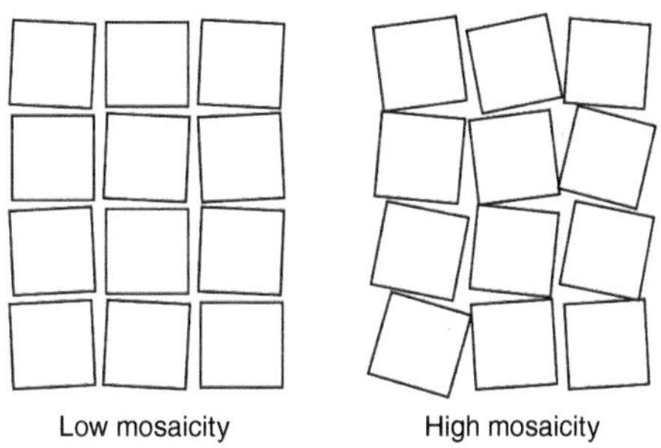

forces that hold the protein molecules together are very weak (hydrogen bonds, ionic and hydrophilic interactions, etc.), which makes protein crystals very fragile, soft, and highly mosaic (Figure 16.2).

But why should such highly ordered structures as crystals form spontaneously? The answer lies within the fields of kinetics and thermodynamics. Every system proceeds toward the minimization of the Gibbs free energy and maximization of entropy. The decrease in the Gibbs free energy is achieved, for example, by the formation of chemical bonds and physicochemical interactions between molecules, such as those holding them together in a crystal. To crystallize a protein, one must find the optimal conditions to ensure the maximum intermolecular interactions that make crystal formation energetically more favorable compared to the formation of the amorphous precipitate. Crystals are of course systems of lower entropy compared to the dissolved matter. However, this fact does not contradict the second law of thermodynamics stating that the entropy of an isolated system cannot decrease. The entropy of the crystal's surroundings increases even more upon the process of crystallization due to the hydrating water molecules being released from the protein surface.

Atoms are able to bend waves of different natures (e.g. X-rays, electrons, and neutrons). This phenomenon is called diffraction. If the atoms are positioned in a periodic fashion, the diffracted beams under certain conditions can interfere and produce waves of higher amplitude which are possible to detect. Crystal lattice provides a set of highly ordered molecules that can produce diffraction patterns upon irradiation with waves of wavelengths similar to the intermolecular distances. For electromagnetic radiation, the waves of such wavelengths correspond to X-rays. The collected diffraction images are used then to build a three-dimensional structure of a studied substance.

Crystallizing a protein is not a trivial task and is often the bottleneck of an experiment. Compared to classical nonbiological crystallography, protein crystallographers' toolkit is severely limited due to the fragility of protein molecules. They are very sensitive to temperature, pH, ionic strength, any other solvent apart from water, and many other variables. Any deviations from a protein's comfort zone disturbs its tertiary and quaternary structures, which leads to the impossibility of crystallization. Proteins are usually of highly flexible structure, and the probability of crystal formation is significantly lower than that for small molecules. Obtaining a highly pure protein sample is crucial for crystallization, whereas, in many cases of small molecules, crystallization itself might be used as a purification technique. Moreover, not every protein can be crystallized. The vast majority of the solved protein structures are those of globular water-soluble proteins (lysozyme, hemoglobin, albumin), as they have stable tertiary structures and spherical shapes favorable for crystallization. Other classes of proteins such as disordered (casein), membrane (ATP synthase), and fibrous proteins (collagen) are much more challenging to crystallize due to their structural characteristics; however, there are examples of such proteins that have been crystallized. The success of a crystallization experiment also depends on the size of a protein. Large assemblies are typically harder to crystallize due to the increased number of degrees of freedom in their structures. Nevertheless, there are multiple examples of large assemblies (100 kDa and larger) and even viral particles that have been crystallized.

To initiate the process of crystallization, the solution of a substance must be brought into a supersaturated state. This can be achieved by either increasing the protein concentration or decreasing its solubility by adjusting the temperature, pH, precipitant concentration, or other conditions. The supersaturation state can be separated into three zones: precipitation, nucleation, and metastable zones (Figure 16.3). Within the precipitation zone, only amorphous precipitate is formed. In both the nucleation and metastable zones, crystals can grow, but only in the nucleation zone solid–liquid phase separation occurs with microcrystals as the solid phase. As the crystals grow, the concentration of the protein decreases drawing the system toward the under-saturation zone till it reaches the equilibrium at the border between the metastable and under-saturation zones.

There are several techniques available for macromolecular crystallization. The main difference from the crystallization techniques used for small molecules and complexes is that the crystallization time of macromolecules must be significantly lower. Proteins are noticeably larger than small molecules; thus, they require longer times to self-organize into a crystal. For this reason, the approaches of protein crystallization strive to slow down solvent evaporation as much as possible. One of the most commonly used methods is vapor diffusion. The setup is a plate with typically 24, 48, or 96 wells. A drop of a protein solution, typically 1–5 μL, is placed either on

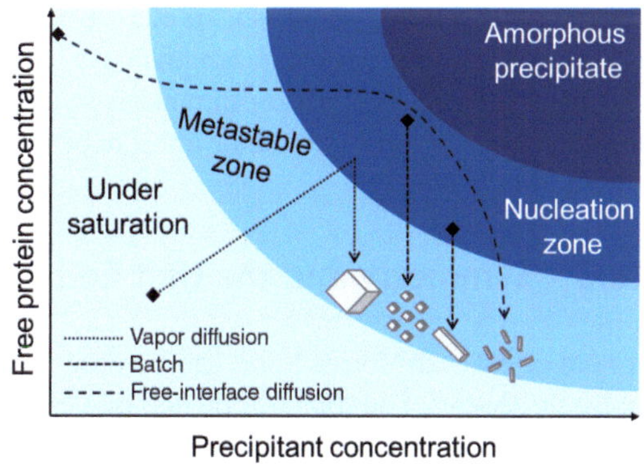

Figure 16.3 The crystallization phase diagram consists of four zones: under saturation, metastable, nucleation, and amorphous precipitate. The pathways done by the crystallization system using various crystallization techniques are depicted by dashed and dotted lines. The size and the number of crystals often depend on where in the phase diagram the system started its way.

a post inside a well (then it is called a sitting drop vapor diffusion technique) or on a glass slide in a hanging position facing the inside part of a well (hanging drop vapor diffusion). Each well has a reservoir for a precipitant solution, whose volume is typically significantly higher than the drop volume (50–1000 μL). Then, some amount of the precipitant is mixed into the protein drop and the system is hermetically closed. Common protein–precipitant ratios are 1 : 1, 2 : 3, or 3 : 2. Since the precipitant is diluted by the protein solution, its concentration in the drop is lower than in the reservoir, and the system will strive to establish an equilibrium. Water will leave the drop and condense in the reservoir solution, thereby increasing the concentration of the protein and precipitant in the drop and driving the system into the nucleation and metastable zones. The batch method has some similarities with vapor diffusion; however, in the batch plates, there is only space for protein–precipitant drops and no reservoirs. To prevent quick water evaporation, the wells are covered with mineral or silicon oil or their mixture. Mineral oil is hardly or not at all penetrable by water, whereas silicon oil lets water slowly diffuse through itself. The speed of evaporation can be regulated by changing the ratio of the mineral and silicon oils in the covering mixture [29]. Another common crystallization approach is called free-interface diffusion, which is usually performed in a capillary. A precipitant and protein solutions are layered onto each other, and the capillary is then sealed from both sides. The solutions slowly diffuse in the interface area, and crystals are formed.

Once the crystals have been obtained, they need to be properly prepared for the experiment. To preserve the protein crystals, they must be flash-frozen in liquid nitrogen. Low temperatures slow down all chemical processes significantly, preventing the protein structure from changing for a long time. This is especially important for metalloenzymes, which are often oxygen sensitive. Moreover, low temperatures minimize the radiation damage caused by the beam during the data collection. Protein crystals contain a high percentage of water and in the vast majority of cases are grown in water solutions. Upon freezing, the water solidifies inside and around a protein crystal introducing a disorder that might affect the protein crystal quality significantly. For this reason, crystals are usually soaked in a substance that vitrifies after flash-freezing and introduces fewer defects in a protein crystal. Usually, the cryoprotectant chosen is also a precipitant used for crystallization, but it may be a different substance if it better preserves a particular protein crystal. Common cryoprotectants are glycerol, sugars (25–50%), salts (50–90% solutions), polyethylene glycols of different chain lengths (w/v 25–100%), oils, polyols, and alcohols [30]. After freezing, crystals are shipped at cryogenic temperatures in a dry-shipping dewar to the beamline where they await to be irradiated.

16.3 Symmetry and the Unit Cell

Protein crystals are formed and can be described by the same principles as small molecule crystals. The smallest fraction that can build the whole crystal by applying crystallographic symmetry operations is called an asymmetric unit. This unit does not possess crystallographic symmetry within itself; therefore, it is called asymmetric. An asymmetric unit is an object of interest for crystallographers, and it is what is deposited in the PDB [1]. An asymmetric unit can be comprised of a single molecule or a multimolecular assembly; however, if an asymmetric unit is an oligomer, it does not mean that the protein exists in this form in solution as well (Figure 16.4).

A unit cell is constructed by applying symmetry operations to an asymmetric unit. In protein crystals, symmetry operations such as 2-, 3-, 4-, and 6-fold rotation and screw axes within a unit cell, as well as translation operations applied to unit cells are possible. Reflection, inversion, and glide reflection are prohibited due to the presence of asymmetric α-carbons in amino acids. Because of this limitation, only 65 of 230 possible space groups can be observed in a protein crystal. A space group is a combination of 14 Bravais lattices (Figure 16.5), describing the shape of the unit cell and the centering of the lattice points, and 32 crystallographic point groups, describing the symmetry within a unit cell.

According to Bragg's law, which we will discuss later in this chapter, a crystal can be considered multiple sets of parallel planes reflecting X-rays. The obvious planes that we can clearly see in lattice models are the faces of unit cell parallelepipeds. However, these are not the only planes that can be found in crystals. The parallel

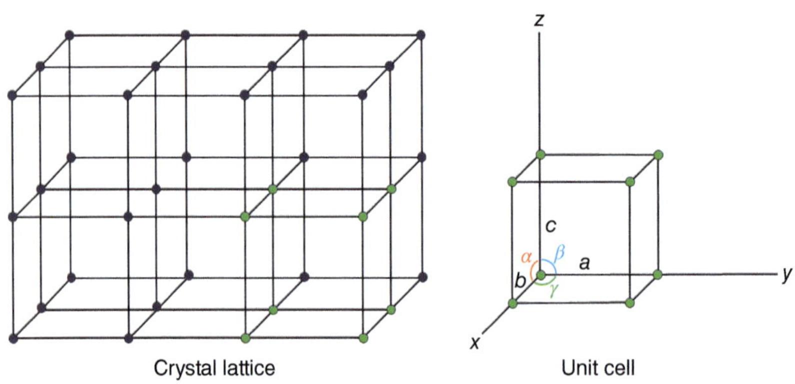

Figure 16.4 A parallelepiped-shaped primitive unit cell with edges a, b, and c and angles α, β, and γ. Multiple unit cells form a crystal lattice when translated to all directions of space.

planes can also pass through a unit cell more than once and do not have to be parallel or perpendicular to the *x, y,* and *z* axes. To describe these planes, three indices *hkl* called Miller indices are used. They show how many planes per unit cell cut through a corresponding axis. For example, index *h* corresponds to axis *x*, *k* to *y*, and *l* to *z*. They are always integers and can be negative. Miller indices are written as three digits in parentheses, for example, (210). If a negative digit is present, then the minus is written above the digit ($\bar{2}10$). To find Miller indices of a set of planes, one has to put a unit cell on a coordinate system, identify where the plane intersects each axis, and then take reciprocals of these numbers. If a plane does not intersect an axis (in other words, is parallel to it), then the value it gets is infinity, and its reciprocal is zero. For instance, if a plane crosses each axis in points (0.5, 0, 0), (0, 0.5, 0), and (0, 0, 0.5), its indices will be $(\frac{1}{0.5}\frac{1}{0.5}\frac{1}{0.5})$ or (222). When a plane is a unit cell face, two of its indices will be zeros, since a face is parallel to two axes, and the third will be one since it intersects the third axis once per unit cell. The interplanar space for a plane family (*hkl*) is designated d_{hkl}.

Bravais lattices can be divided into seven crystal systems. Those describe only the shape of the unit cell, specifically edges *a, b, c* and angles α, β, γ of the unit cell, parallelepiped. For example, the crystal system with the lowest symmetry is triclinic. All of its edges and angles are not equal to each other. On the other hand, the highest symmetry belongs to the cubic crystals system, where $a = b = c$, and $\alpha = \beta = \gamma$. In addition to describing the shape of a unit cell, Bravais lattices also tell us where the lattice points are positioned. If the lattice points are only found at the corners of the unit cell, we call it a primitive unit cell. In this case, each unit cell possesses only one lattice point. Although we see eight lattice points for primitive cells in Figure 16.5 (except hexagonal which has 14, but it can be represented as three merged parallelepipeds with a threefold rotation axis in the middle), it is important to remember that each unit cell is surrounded by six equivalent ones. They share lattice points with each other, resulting in a 1 : 1 ratio of lattice points and unit cells. When there

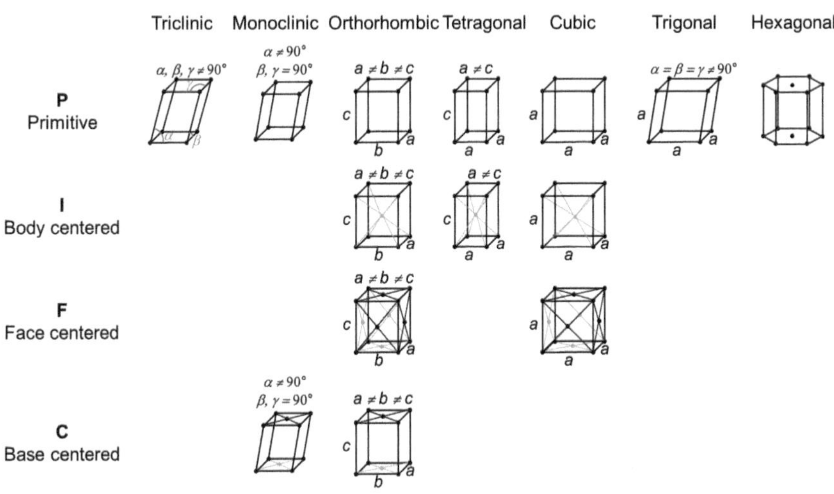

Figure 16.5 The 14 Bravais lattices.

is an extra lattice point at the center of the unit cell, it becomes a body-centered unit cell. Lattice points can also be found on the faces of the unit cell. If all faces have lattice points, it is called a face-centered unit cell. If only two opposite faces have lattice points, it is a base-centered unit cell. All nonprimitive unit cells have more than one lattice point per unit cell.

16.4 Diffraction and Interpretation of Diffraction Patterns

Diffraction is a property of a wave of any nature to bend around obstacles and apertures. It is important to distinguish between diffraction and other ways a wave interacts with the matter, such as reflection, refraction, and scattering. Reflection occurs on a surface and unlike diffraction; it throws the reflected wave back at the same angle as it fell on the surface. Refraction takes place when a wave passes through the border between two media, for example, air and water, which makes it redirect.

Diffraction and scattering are similar phenomena occurring when a wave hits an obstacle and then is propagated by it in other directions. The underlying physics behind both of them share common ground, and diffraction can be considered a special case of scattering [31]. These terms sometimes are used interchangeably, and atoms are called scatterers in the context of diffraction. The term "diffraction" is used when the scattered waves interfere and produce characteristic diffraction patterns, whereas "scattering" is usually a general definition of the physical phenomenon not necessarily implying the formation of the diffraction pattern.

A diffraction pattern (Figure 16.1) can be observed when multiple waves overlap with each other. In a diffraction pattern, one can see empty and colored areas. The latter occurs due to the phenomenon called constructive interference. The waves of the same wavelength and phase angle combine in one with the same parameters, but increased amplitude. Higher amplitude means a stronger signal, allowing the dark spots or lines to appear in the diffraction pattern. If the interfering waves are out of phase, they quench each other resulting in a wave of zero amplitude. This occurrence is referred to as destructive interference. If two waves are not perfectly in or out of phase, they can still interfere producing an intermediate amplitude, depending on the phase angles. In X-ray crystallography, the dark spots in a diffraction pattern are often called reflections due to Bragg's approximation considering one spot as a set of the constructively interfered X-rays reflected from one set of imaginary planes of the crystal.

A wave can diffract on an obstacle only if their sizes are comparable. In the case of crystals and X-rays, the wavelength range is, however, quite large and can be several orders of magnitude. X-rays are waves of choice for diffraction on crystals since they satisfy Bragg's law (Figure 16.6) and, therefore, produce diffraction spots after being scattered by the electron clouds in a crystal. The law was postulated by William Henry Bragg and his son Lawrence Bragg in 1913 [8]. They considered crystals as multiple sets of imaginary parallel planes able to reflect X-rays. The law establishes

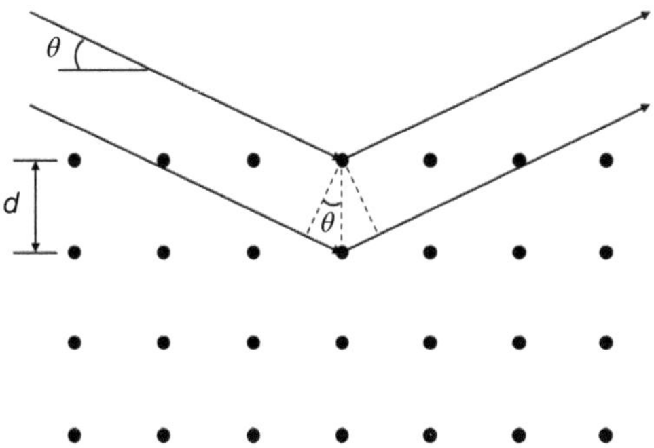

Figure 16.6 A schematic representation of Bragg's law describing the condition under which a strong diffracted beam can occur. X-rays are reflected by imaginary planes formed by lattice points at the angle θ, and the parallel planes are separated by the distance d.

the relationship between the wavelength of X-rays λ and the product of interplanar space distances d_{hkl} and the sine of the angle θ at which a beam falls and is reflected

$$n\lambda = 2d_{hkl} \sin \theta$$

where n is an integer. If the conditions are met, a plane indexed as hkl will produce a diffracted beam, and together with other parallel planes with the same indexing they will produce a diffraction spot. In the opposite case, not meeting conditions of Bragg's law means that the diffracted beams interfere destructively quenching each other, and no resulting beam emerges; therefore, no spot can be detected.

Electromagnetic radiation, such as X-rays or visible light, interacts with electrons, meaning that diffraction also occurs on the electron cloud of atoms. For this reason, the intensity of the diffraction spots correlates with the number of electrons moving around the set of planes producing the reflection. If there is minimal electron density within a set of planes, despite being permitted by Bragg's law, this ray will appear faint or may go undetected.

An important property of the sin of diffraction angle θ is that it is inversely related to the interplanar spacing d_{hkl}

$$\sin \theta \propto \frac{1}{d_{hkl}}$$

This suggests that large unit cells, featuring wider spacings, result in smaller diffraction angles, leading to numerous reflections. In contrast, smaller unit cells yield larger diffraction angles, resulting in fewer measurable reflections. Large unit cells contain more atoms, and they produce more reflections in the diffraction pattern, and small unit cells have fewer atoms and produce fewer reflections [32]. This feature is very convenient when one is not sure if the crystal they obtained is a protein crystal. This might happen when the crystallization mixture contains salts. Unit cells of proteins are significantly larger than those of salts; therefore, in most cases, it is easy to differentiate them by only looking at the diffraction pattern without processing the data.

Protein molecules in a crystal exist in real space, whereas reflections exist in another, reciprocal space. The spacings between unit cells in real space and between reflections in reciprocal space are inversely proportional, making it possible to calculate the dimensions of a crystal unit cell by measuring distances between the reflections produced by the diffracted on that crystal X-rays. In the reciprocal space, each reflection is designated with a triad of integer numbers h, k, and l. They work exactly like x, y, and z coordinates in real space. The central position of a diffraction pattern is considered the origin with $(h, k, l) = (0, 0, 0)$. However, this position is always a blind spot in a diffraction pattern, since it is obscured by the beam stop preventing the detector from being damaged by the direct nondiffracted beam. The remaining reflections are designated with coordinates h, k, and l, which are counted from the origin.

16.5 Data Collection and Processing

The main goal of the data collection step is to irradiate a crystal of the protein of interest with X-rays and record as many reflections as possible, as they contain intensities that are crucial for the structure determination. We aim to collect every reflection from every possible set of planes satisfying Bragg's law.

To collect diffraction data, one requires a source of X-rays that will be diffracted by a crystal. Nowadays, there are two main types of instruments in crystallographers' inventory. Those are diffractometers and synchrotrons. While the latter are smaller, less expensive, and easier-to-maintain X-ray sources, they lack the necessary intensity for efficient macromolecular crystallography. Using a diffractometer, the data collection can be extremely time-consuming, sometimes requiring months or even years to gather enough diffraction data, whereas using synchrotron radiation, this process usually takes seconds. Additionally, long exposure times increase the radiation dose taken by the protein molecules, which decreases the quality and reliability of the data. For that reason, synchrotrons are in most cases the instruments of choice for protein crystallography. A synchrotron is a large electron accelerator buried underground that is able to produce high-intensity X-ray radiation. Scientists use this radiation for research in many fields, including structural biology (Figure 16.7).

Protein crystals are shipped to a synchrotron at the temperature of liquid nitrogen, which is around 77K, and the data collection is performed under the stream of cold nitrogen usually at around 100K to minimize the radiation damage and slow down all chemical processes occurring in the crystal. During the experiment, a single crystal is rotated by a goniometer orientating it at different angles in the beam. The diffraction images are collected while the crystal is rotating. Usually, one collects several thousand images from 360°, but this number can be smaller if a crystal produces high-quality images with enough indexed reflections. After the data collection, the crystal is usually not usable anymore due to the radiation damage that it has taken.

The diffracted beams are registered by a detector placed behind the crystal. A classical detector is an X-ray-sensitive film; however, nowadays much more advanced solid-state charge-coupled devices are used. The more photons hit a certain position on the detector, the brighter the spot, making it possible not only to determine the reflection positions but also to measure their intensities required for calculating amplitudes [28]. The reflections that are more distant from the detector

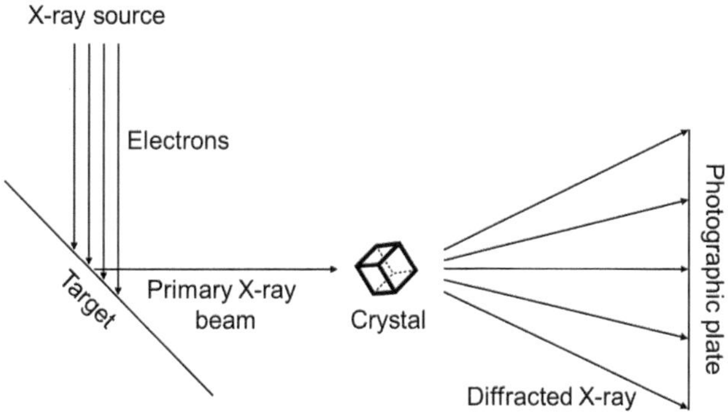

Figure 16.7 A schematic representation of a diffraction experiment. Electrons accelerated in a synchrotron ring collide with a target made of pure metal, for example copper, which produces a high-intensity X-ray beam directed into a crystal. The X-rays diffract by interacting with the electron clouds of atoms in the crystal, and the diffracted beams are registered by a photographic plate positioned behind the crystal.

center correspond to higher resolution. The reflections that are at the greatest distance from the center determine the resolution of a structure that is stated in the PDB or scientific reports. Before collecting diffraction data, the distance between the sample and the detector can be altered depending on the expected resolution. Shorter distances ensure the registration of high-resolution spots, but they might decrease the quality of the low-resolution data. That is why it is important to move the detector further from the sample if no high-resolution reflections are expected. The data are recorded as a set of diffraction patterns along with a file containing the measurement parameters important for the following data processing.

While collecting data, a researcher operating the beamline is able to customize various beam settings according to their needs. This includes parameters like focus, transmission, and exposure time to regulate the amount of radiation a crystal receives. Sometimes increasing the dosage may be necessary, although this may compromise the stability of the protein, to ensure a sufficient number of reflections collected from the crystal. Data collection conditions depend on the crystal being studied and can vary between crystals. The best radiation dosage for each specific protein crystal can often be determined only experimentally. Multiple beamlines offer the possibility to change the wavelength of the produced X-ray allowing an experimenter to collect anomalous dispersion data from heavy atoms or adjust the resolution limit.

Once the diffraction snapshots are collected, they are ready for processing. Data that can be extracted from the reciprocal space are reflection positions *hkl* and their intensities, which can then be transformed into information about the unit cell parameters and amplitudes. Data processing aims to obtain the amplitude information from the collected reflections. To yield a macromolecular structure, this step is followed by solving the phase problem and refinement, which will be discussed later in this chapter. Processing of diffraction images is performed by specialized software packages, such as Mosflm [33], HKL2000 [34], XDS [35], and DIALS [36]. The former two require additional software for scaling and merging, while the latter two do all the processing independently. The whole processing can be broken down into three steps:

1. Spot finding and indexing
2. Parameter refinement
3. Scaling and merging [37]

The processing begins with identifying the reflections and giving them names consisting of three integer numbers h, k, l (Miller indices). These indices are not necessarily related to those used for assigning the planes. The indexed reflections are then used to determine the space group and the geometry of the unit cell, which includes distances a, b, c and angles α, β, γ. Although indexing identifies the parameters of the unit cell only approximately, this step is crucial for the following accurate integration of the intensities. Crucial data needed for successful indexing is the distance from the crystal to the detector and the X-ray wavelength used in the experiment. Another important piece of information provided by indexing is mosaicity. In reality, crystals are never perfectly ordered and consist of better-ordered clusters (usually 10–100 μm) that are slightly shifted in relation to each other. If the degree of mosaicity is too high, the data cannot be processed and the researcher may need to reconsider the crystallization conditions. Typically, a mosaicity of less than 0.5° is considered good [38].

Measuring the intensities of pinpointed and indexed reflections is called integration, as the process is basically finding the area under the curve if a reflection is considered a peak. To accurately determine the intensity value of specific spots, it is necessary to measure the background surrounding those spots. By establishing this baseline background value, one can then interpolate the expected background values beneath the spots. Subtracting these interpolated background values from the spot intensity yields a more precise and reliable measurement [37].

The integrated data are not consistently scaled due to various physical factors. These factors include changes in radiation intensity, variations in crystal volume illuminated by X-rays, crystal movement during the experiment, and irregular detector response. Scaling and merging are methods used to address these variations. Typical software used for these purposes are Aimless [39] from the CCP4 suite [40, 41] and Scalepack [34]. After this, the complete and accurate statistics of the data are provided, allowing a user to evaluate the quality of the processed data, and make corrections in the processing step if the statistics are unsatisfactory. This is a consensus in the crystallographers' society that certain parameters should not be too low, otherwise, the dataset is considered not reliable enough.

16.6 X-ray Structure Analysis

In the X-ray diffraction experiment, we obtain all information required to determine the crystal lattice and the three-dimensional structure. The primary result of an X-ray diffraction experiment is a map of the electron density within the crystal. The electron density can be obtained from the structure factors (F), the volume of the cell (V), and the phases.

In a diffraction experiment, we collect patterns of the diffracted waves (reflections), which are recorded by a detector [42]. Each reflection is characterized by its amplitude and phase. However, only reflection amplitudes can be obtained from the measured intensities and no direct information about reflection phases is provided by the diffraction experiment. To obtain the structure of the individual diffracting motif, it is necessary to calculate the Fourier transformation of the structure factors. In our case, the diffracting motif is the distribution of electrons in the asymmetric part of the crystal unit cell. The structure factors F represent the reflection amplitudes and phases. Here, it becomes obvious that we need to obtain information on phases to solve the structure of a molecule. To solve this so-called phase problem in protein crystallography, several methods have been developed to determine the phases. Typically, they lead to an initial approximate electron-density distribution in the crystal, which can be improved iteratively, eventually converging at a faithful structural model of the protein. Methods to obtain phases from the diffracted waves are isomorphous replacement, anomalous diffraction, and molecular replacement (MR).

16.6.1 Isomorphous Replacement

The first historically important method of phase determination is isomorphous replacement. Hereby, heavy atoms such as bromine and iodine are introduced into the protein before crystallization. Heavy atoms have significantly more electrons than the lighter atoms of amino acids, which are mostly composed of carbon, nitrogen, oxygen, and hydrogen. Therefore, they diffract X-rays better, which translates into higher relative intensities in the diffraction pattern. These heavy atoms can form covalent or coordinative bonds with the protein or find a defined location within the protein. The enrichment must be isomorphous, which means that the structure of the protein and the packing of the molecule in the unit cell must remain the same. Thus, we obtain the same diffraction pattern for the native crystal as for the isomorphous crystal, except for the intensities of the reflections originating from the heavy atoms. To use isomorphous methods, it is crucial to obtain crystals with the same lattice parameters and diffraction pattern. Otherwise, the crystals cannot be used to resolve the phase ambiguity.

In cases, when we have isomorphous crystals, the Patterson method can be used [43]. This mathematical function named after Arthus Lindo Patterson can be used to obtain a model structure without knowing phases. It has a form similar to the equation for electron density except that all phases are effectively zero. A Fourier map is constructed using the intensities from the diffraction experiment or more precisely the squares of the moduli of the structure factors. The resulting maps contain peaks that represent the vectors between atoms rather than the atoms themselves. A skilled crystallographer can interpret these Patterson maps and use this information to suggest a structural model that fits this set of vectors.

Isomorphous replacement methods can be divided into two different approaches: single isomorphous replacement (SIR) and multiple isomorphous replacement (MIR). The SIR method uses one crystal that has been derivatized using a heavy atom. Hereby, the errors in the electron density maps are high and incorrect or incomplete structures are obtained. To improve these data, the approach has been extended to include more crystals that have been derivatized with multiple heavy atoms. Using multiple datasets from derivatized crystals is characteristic of the MIR method and allows the combination of structural information. The more independent diffraction images of derivatives that can be included in the structure solution, the smaller the errors for solving the phase problem.

16.6.2 Anomalous Diffraction

For many years, experimental phasing was dominated by anomalous diffraction methods. These methods can be divided into multi- and single-wavelength anomalous diffraction (MAD and SAD) and were established with the increasing availability of extremely intense synchrotron X-ray beams and more importantly, with the ability to tune the wavelength of the sources to virtually any desired value. Along with the availability of intense synchrotron radiation, technical improvements in X-ray detectors and collection methods have also contributed greatly to the measured accuracy of otherwise small anomalous effects.

The experimental problem was always to amplify the difference between $F_{\bar{h}\bar{k}\bar{l}}$ and $F_{\bar{h}\bar{k}\bar{l}}$ and obtain its most accurate measure. In the structure solution of small molecules, the phase problem is solved by mathematics-based direct methods. Although these methods have proven inadequate for macromolecules, they can be used to solve the selenium substructures in SAD analyses of selenomethionyl proteins. Hereby, special cell strains and expression techniques can be used to replace the sulfur in methionine with selenium, which is a more efficient anomalous scatterer than sulfur. SAD and MAD phasing of selenomethionyl proteins were the mainstays of structural genomics efforts to populate the PDB with novel proteins. In addition, native SAD, exploiting the intrinsic S and P atoms of biomolecules, has become routine.

The tuning of the wavelength, which is also a measure of X-ray energy, enables data collection near the absorption edge of anomalous scatters. Hereby, metalloproteins naturally contain metals that are suitable for anomalous data collection. Alternatively, anomalous scatterers can be introduced into the protein crystal. The energy of the X-ray radiation is in close proximity to the absorption edge of one type of atom in the crystal, and the radiation leads to the excitation or ionization of these atoms in the crystal. This leads to the emission of Kα radiation of the excited atoms, and the X-ray radiation interacts strongly with these atoms. This phenomenon is called anomalous dispersion or anomalous scattering and leads to changes in the amplitude and phase. These differences in intensities provide information about the phase. Anomalous dispersion is observed if the molecules in the crystals contain one or more heavy atoms. As a consequence, the intensities of the Friedel pairs $I(\bar{h}kl)$ and $I(\bar{h}\bar{k}\bar{l})$ are different. Friedel's Law describes the relationship between the intensities of Friedel pairs, which are Bragg reflections related by inversion through the origin. Friedel's Law states that members of a Friedel pair have equal amplitude and opposite phase. Whenever there is anomalous scattering, Friedel's Law is broken. The atom type and the wavelength both impact the intensity difference, e.g. the effect for sulfur atoms is very small, whereas it is significantly larger for heavy metal ions.

The MAD method involves data collection at multiple wavelengths, including above and below the absorption edge of heavy atoms. These measurements are typically carried out at cryogenic temperatures to avoid excessive radiation damage. For a MAD experiment, at least three datasets at different wavelengths are collected. In addition to the native dataset, datasets at the peak of the absorption edge and at the inflection point below the absorption edge are collected. From the differences in these datasets, the coordinates of the heavy atoms can be derived as well as the phase. The phase information can then be used to solve the three-dimensional structure of the macromolecule.

16.6.3 Molecular Replacement

The method that is commonly used to solve the phase problem in macromolecular crystallography is MR [44]. Hereby, a known molecular model is used to solve the unknown crystal structure of a related molecule. For this approach, it is crucial to know the structure of a related protein or even the same protein obtained from a different crystal form. The structure of the related protein contains atomic coordinates, but the number of molecules and the orientation in the unit cell may differ. We assume that this starting model approximates the unknown structure. The next step is to try all possible orientations and positions of the model in the unknown crystal and compare the calculated diffractions with the observed diffractions. The best fit is the structure that is closest to the unknown crystal. The model can then be used to calculate the phases from the model and use them as initial phases for the unknown structure. Next, the initial map is calculated with the phases obtained by the starting model and the amplitudes obtained from the diffraction experiment. This way, the model is rebuilt using the experimental information to me closer to the unknown structure. From here on, the next steps include the refinement of the starting model to fit the unknown structure.

The recent advances in protein structure prediction by deep-learning programs such as AlphaFold [45] and RoseTTAFold [46] have gained significant attention in the field of structural biology. The accuracy of protein prediction is impressive, but they are still not as accurate as experimental structures [47]. However, the predicted models bring advantages, including insights into improved construct design, the generation of compelling hypotheses about interacting proteins or domains, and the acceleration of structure solution by MR for crystallography. Structure prediction tools can be used to generate accurate search models for MR and will likely decrease the dependence on experimental phasing.

16.6.4 Structure Refinement

The primary result of an X-ray diffraction experiment is a map of electron density within the crystal. Based on this electron density map, the position of individual atoms and molecules is deduced, which leads to the structural model (Figure 16.8).

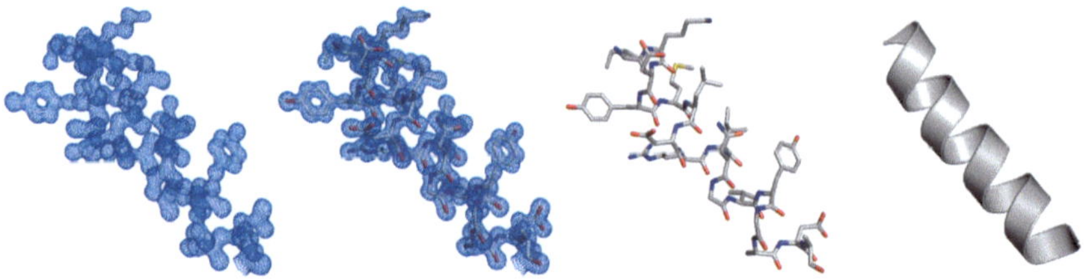

Figure 16.8 Electron density map ($2F_{obs} - F_{calc}$, φ_{calc}) at 1σ contour level and the refined model in different representations.

It is important to keep in mind that the atomic coordinates and the structural model are only an interpretation of the experimental data. The quality and accuracy of the atomic model depend on the quality of the diffraction data and the skills of the crystallographer. After obtaining phases, the atomic coordinates are refined by varying all model parameters to achieve the best agreement between the observed reflection amplitudes (F_{obs}) and those calculated from the model (F_{calc}).

The basic electron density map is obtained by Fourier transformation of the set of experimentally observed reflection amplitudes (F_{obs}) and their phases. As the phases cannot be obtained experimentally, they have to be calculated from the model. Therefore, the electron density map (F_{obs}, φ_{calc}) is an approximation of the true structure, and the quality of the map depends on the accuracy of the calculated phases. This type of map (F_{obs}, φ_{calc}) is typically not shown for crystallographic data; instead, difference maps such as ($F_{obs} - F_{calc}$, φ_{calc}) and ($2F_{obs} - F_{calc}$, φ_{calc}) are used to display the electron density map. The difference map ($F_{obs} - F_{calc}$, φ_{calc}) shows the difference between the true structure and the current model. Hereby, atoms or molecules that are present in the true structure but not included in the model appear in the positive map (typically in green color) and atoms or molecules that are absent in the true structure but present in the model appear in the negative map (typically in red color). This type of map provides useful information in the process of refining the model to be closer to the true structure. However, it is not intuitive to display the electron density of the protein or parts of the protein. The commonly used type of electron density map is the ($2F_{obs} - F_{calc}$, φ_{calc}) map. This map shows the electron density map of the structure and additionally reveals the positive and negative maps to guide the crystallographer where the model needs improvement.

Another important aspect to consider when showing electron density maps is the level of contouring. A sensible contour level for ($2F_{obs} - F_{calc}$, φ_{calc}) maps is approximately 1σ. The value σ corresponds to the root-mean-square deviation of all map points from the average value. Below this value, the noise resulting from the errors of the experimental and calculated data becomes very high. For other types of electron density maps, such as ($F_{obs} - F_{calc}$, φ_{calc}) or maps obtained at different wavelengths, higher contour levels may be better suited. Higher contour levels may also be used to highlight certain features for specific reasons.

The omit map is another type of electron density map that can be useful when the calculated phases are not accurate enough or when ligands or cofactors are cocrystallized with the protein. As the Fourier maps depend more on the phases than on the amplitude, inaccurate phases may introduce an unwanted map bias. The prong phases will reproduce the same errors in the model, which again produces the same inaccurately calculated phases. Therefore, these types of bias are challenging to overcome during the refinement process. A way out of this tricky situation is to remove questionable parts, ligands, or cofactors from the structure and calculate a map from this model, where some parts have been omitted. The refinement of the model without the questionable part is supposed to lead to more accurate phases and reveal the unbiased electron density of the omitted part.

When publishing a crystal structure, it is important to show electron density maps in the area of interest to support claims based on the structure with experimental data and not only rely on the model, as this is only an interpretation of the data. For ligands or cofactors, it is recommended to show unbiased electron density as provided by an omit map.

How good the electron density looks ultimately depends on the quality of the diffraction data and phases. The question is how to objectively assess the quality of data. The most important indicators of data quality are resolution, completeness, signal-to-noise ratio ($I/\sigma(I)$), and the R-factor [42]. Another important factor for the quality of the map is the resolution at which the map was calculated. Resolution is defined as the ability to distinguish between neighboring features in an electron density map. The resolution can be calculated using the equation:

$$d_{\min} = \frac{\lambda}{2 \sin \theta_{\max}}$$

Thus, the resolution depends on the maximum scattering angle ($\theta_{\max}$) and the wavelength (λ). Hereby, the maximum scattering angle again depends on the quality of the crystal and the type of atoms in the crystal. The crystal therefore predetermines these parameters. However, the wavelength of the radiation at synchrotron sources is variable; therefore, higher resolution can be obtained.

Lab-scale X-ray instruments frequently use copper K-α an X-ray energy. The energy is 8.04 keV, which corresponds to an X-ray wavelength of 1.5406 Å, therefore limiting the resolution of the data obtained from these sources. A commonly used energy for data collection at synchrotron beamline is 12.00 keV, which corresponds to an X-ray wavelength of 1.034 Å. Consequently, data collection at synchrotron sources can result in higher resolution for high-quality crystals. In cases where the setup at the beamline prevents data collection at 1.034 Å due to technical restrictions, for example, if the crystal diffracts below 1.034 Å, collecting data at higher energies, such as 14.00 or 16.00 keV, can enable collecting data with even higher resolution.

The resolution affects the clarity of the electron density map and therefore the information that can be obtained. Structures with a resolution of 5.0 Å show the overall shape of the protein and secondary structural elements, such as α-helices and β-sheets can probably be identified within the electron density map [48]. At a resolution of 2.5 Å, the structure of α-helices and β-sheets and the positioning of amino acid side chains can be assigned. Structures with resolutions above 2.0 Å allow us to determine the conformation of the peptide bond and longer side chains, but they do not show atoms as individual ellipsoids or spheres. This can only be achieved at very high resolutions of 1.0 Å. In electron density maps with atomic resolution, unusual features such as rare conformations and very short hydrogen bonds can be detected.

When looking at electron density maps, some of the regions in the structure are well defined, while other regions are disordered. This is due to the varying degrees of flexibility in a protein. Some amino acid residues adopt multiple stable conformations, resulting in a disordered or absent electron density. Alternatively, the protein has regions with higher mobility or vibrations, which also results in high disorder. As an example, unstructured termini of proteins or affinity tags that have been introduced for easier protein isolation can be so flexible that poor or no electron density is observed in the map.

The completeness of a dataset is the coverage of all theoretically possible unique reflections. This value should be very close to 100%, which means that all possible unique reflections have been detected. Incomplete data in the highest resolution shell means that the actual resolution for structural refinement may need to be adjusted. Incomplete data collection may be due to a poor data collection strategy, for example, the rotation during data collection is not sufficient for the space group of the crystal. Another reason could be that the crystal did not survive the exposure to high doses of radiation and the diffraction decays before a complete dataset can be collected. Hereby, it helps to calculate a data collection strategy after collecting a few test images to increase the chances of collecting a complete dataset before the diffraction of the crystal decays. The average signal-to-noise ratio ($I/\sigma(I)$) is also an important parameter to assess data quality. However, it is quite difficult to determine the uncertainties of the measurements ($\sigma(I)$).

The initial electron density map that is obtained right after solving the phase problem is not optimal yet. It is improved during an iterative refinement process that involves alternating rounds of automated optimization according to least-squares or maximum-likelihood methods and manual modifications that improve the agreement between the model and the true structure. During the refinement process, the accuracy of the phases is improved, which leads to a better electron density. Hereby, *a priori* knowledge of the stereochemistry of individual amino acids and peptide groups is used to locate these protein building blocks. There is one question that arises here. How do I know if my model is good?

How well the model agrees with the true structure can be assessed by the R-factor. The R-factor is a global indicator of the quality of diffraction data. It reflects the difference between the observed reflection amplitudes (F_{obs}) and those calculated from the model (F_{calc}). The R-factor can be calculated using the equation:

$$R = \frac{\sum \left| |F_{obs}(h,k,l)| - |F_{calc}(h,k,l)| \right|}{\sum |F_{obs}(h,k,l)|}$$

If the model and the true structure are identical, the difference between the structure factors obtained from experimental data and the calculated structure factors would be zero, which results in an R-factor of zero. In reality, this is not the case, but the R-factors for protein structures are in the range of 0.15 and 0.20. These discrepancies derive from errors in the diffraction experiment and the differences in the protein molecules. These differences are a result of different conformations of amino acids and protein regions and different orientations of protein domains within the crystal. Therefore, the crystal structure represents an average of all conformations within the protein.

When looking at the files provided by the PDB [1], different types of R-factors are listed. The most commonly reported descriptor of data quality is the conventional merging R-factor R_{merge}. However, R_{merge} depends on the multiplicity, sometimes also referred to as redundancy, of the data. The higher the multiplicity, the higher the R_{merge} becomes. Multiplicity reflects the average number of measurements per individual, symmetrically unique reflection. Because of the symmetry of most crystals, some reflections are expected to be equivalent. The more reflections are collected, the higher the multiplicity and as a result, the more accurate the data. Therefore, a dataset with higher multiplicity results in a higher R_{merge}, which indicates a lower quality of the dataset despite the higher accuracy obtained by including more experimental data. A multiplicity-independent version of the R-factor, R_{meas}, was therefore proposed to be an improved factor for assessing the quality of the data [49–51].

Protein structures contain regions with higher flexibility, which is reflected in the temperature factor or B-factor. This factor is a measure of how much a protein oscillates around its origin and is expressed in $Å^2$. Rigid parts of the protein structure commonly have B-factors of 20 $Å^2$, while domains with significantly higher values are considered flexible. This increased oscillation and displacement of atoms will also result in poor electron density for these atoms or regions.

What is the resolution of my dataset? This central question arises during the refinement process and the answer to this question has changed over the past years [52, 53]. Due to the centrosymmetric nature of diffraction, the resolution limit is determined by the edge of the detector. After initial screenshots, it is obvious how well the crystal diffracts in certain positions. At this point, it is crucial to determine the highest resolution that is possible and modify the data collection strategy if necessary. In practical terms, the detector distance or wavelength should be adjusted, if diffraction is observed in the corners of the detector. The most critical parameter for determining the resolution limit is the completeness of the data. However, there is no specific value for completeness that enables the data cutoff.

For a long time, it seems that the parameter chosen for determining the resolution limit was based on personal preference. More recently, studies on how to use data quality indicators to select a high-resolution cutoff that leads to the best model present an effort to unify this choice. Traditionally, the signal-to-noise ratio was used to determine the limit of the dataset. The cutoff for a dataset was chosen to result in a $I/\sigma(I)$ value of 2.0. Another widespread data quality indicator that was used to decide to which resolution to refine the data is the R_{merge} value. This value depends on the multiplicity of data and is therefore poorly suited to determine the high-resolution limit. These criteria result in high-resolution cutoffs that are too conservative and discard much useful data.

To overcome this problem, the Pearson correlation coefficient (CC) was introduced [53]. The CC value can potentially assess both data accuracy and the agreement of the model and data on a common scale. Modifications of this value are $CC_{1/2}$ and CC*. To determine the $CC_{1/2}$ value, the unmerged experimental data is divided into two parts, each of them containing a half dataset of randomly selected unique reflections. The CC values are then calculated for each half dataset. A high correlation results in a $CC_{1/2}$ value of 1.0, which is usually observed at low resolution and can decrease to 0.1 at high resolution. CC* can be calculated according to the equation:

$$CC^* = \sqrt{\frac{2CC_{1/2}}{1 + CC_{1/2}}}$$

CC* and $CC_{1/2}$ are considered robust, statistically informative values for defining the high-resolution cutoff in macromolecular crystallography. Since the process of building and refining a model of biomolecular structures depends on data quality and the crystallographer, it is important to carefully validate such models. Using information on chemical bond lengths and angles as well as stereochemistry, the accuracy of the model depends not only on how well the model fits the electron density map but also on how correct the chemistry is.

The most common programs for refinement are CNS [54], SHELXL [55], REFMAC5 [56], or PHENIX [57]. They utilize stereochemical-restraint libraries that contain information on the lengths and angles of the peptide bond [58]. Root-mean-square deviations (rmsd) from standard stereochemistry indicate the extent to which a refined model deviates from the geometrical targets found in dictionaries. Hereby, a value of approximately 0.02 Å is expected for the rmsd of bonds and values in the range of 0.5° and 2.0° for the rmsd of angles. One of the most useful tools in structure validation is the Ramachandran plot. It shows the pairs of torsion angles of the peptide backbone and indicates the allowed and expected areas. It is recommended that more than 98% of the values should be in the allowed areas. In addition, some specific programs can be used to validate certain aspects of the structure. A very useful tool to validate metal binding is CheckMyMetal. In cases where no experimental evidence for the binding or the location of the metal is available, this tool can be used to detect poorly modeled metal-binding sites [59].

Validation of the structural model is required when depositing new structures to the PDB. During this process, several tools for validation are implemented, which guide the crystallographer on how to optimize the model. The results of the validation are summarized in a validation report, which must be submitted to many journals before the reviewing process. The validation reports contain a lot of useful information on how well the model agrees with the true structure. Only good models can teach us something about the function of a biomolecule.

16.7 Time-Resolved Crystallography

So far, this chapter has focused on traditional rotational crystallography at cryogenic temperatures. This approach is well suited to determining the structure of the protein at equilibrium and helps to better understand its architecture or the fine details of interactions within an active site or bound ligands. By trapping different states of the protein, it is also possible to gain insight into the intermediates of a chemical reaction. However, most intermediates have a short lifetime. It is therefore essential to be able to obtain snapshots of enzymes during catalysis in order to understand the mechanism that enables them to function.

The development of time-resolved serial crystallography using X-ray free-electron lasers (XFELs) and synchrotrons has revolutionized the field of macromolecular crystallography [60–62]. Whereas traditional single crystal rotational crystallography typically involves collecting a dataset on a single macroscopic crystal, serial crystallography uses many microscopic crystals and only one image is measured per crystal. There are several advantages to this approach. First, small crystals are easier to obtain than large crystals. Second, exposing crystals to high levels of energy leads to radiation damage and uncontrolled reactions, especially in metalloenzymes. By collecting only one frame per crystal, radiation damage is dramatically minimized. Third, because of the low radiation damage, data collection can be performed at room temperature or even higher, closer to the physiologically relevant temperature. Finally, it allows data to be collected at different points in the catalytic cycle, providing time-resolved information on many important biological processes.

Time-resolved serial crystallography can be performed using XFEL or synchrotron radiation. The development of these two techniques allows us to study processes from femtosecond to minute timescales. Photoactivated processes are very fast and femtosecond time resolution is required to observe intermediates of these reactions. Many enzymes, on the other hand, operate on slower timescales that are easily accessible by serial synchrotron crystallography (SSX).

XFELs are linear accelerator-based X-ray sources that deliver femtosecond coherent X-ray pulses with a peak brilliance that is orders of magnitude higher than that of modern synchrotron sources [61]. XFELs have enabled fast data collection for time-resolved measurements. Technical and methodological breakthroughs have been achieved in serial crystallography, including new sample-delivery methods and strategies for reconstructing datasets from single frames.

The success of serial crystallography experiments depends heavily on the ability to grow large quantities of well-ordered small crystals with a homogeneous size distribution [63]. The size of the crystals should be in the range of micro- to nanometers. The main goal of macromolecular crystallography has always been to obtain large, well-diffracting crystals, which can be challenging or even impossible at times. In particular, solubilized membrane proteins have been unsuccessful in producing large crystals. When screening for crystals, the first hits from sparse matrix screens are often microcrystalline showers. Subsequently, optimization of the crystallization conditions can lead to crystals that are large enough for X-rays or even neutron crystallography.

Serial crystallography has led to an increased interest in microcrystals, and new techniques have been developed to achieve uniform microcrystal suspensions. Small crystals can be obtained using the batch method, which can be easily scaled from micro- to milliliters. Alternatively, sitting-drop setups can yield reasonable amounts of microcrystals. In both cases, the number and size distribution of crystals can be affected by the speed and degree of mixing of the protein and precipitant solutions. Free-interface diffusion or density centrifugation has also been reported to yield sufficient microcrystals for serial crystallography. Figure 16.3 illustrates the phase diagram, indicating that a significant degree of supersaturation is required to cultivate a substantial quantity of microcrystals. Microseeding techniques can be extremely useful in overcoming nucleation limitations and increasing the likelihood of growing isomorphic microcrystals.

Although obtaining small crystals is easier than obtaining large ones, detecting microcrystals is more challenging than detecting their larger counterparts because they often appear spherical [61]. Optical methods can be used to identify large crystals, while ultraviolet (UV) light can distinguish between salt and macromolecules. Second-order nonlinear imaging of chiral crystals (SONICC) is a valuable tool for sensitive and selective imaging of protein microcrystals [64]. Proteins consisting of submicron particles can also be visualized through nanoparticle tracking analysis (NanoSight) [65] or dynamic light scattering (DLS) [63] in solution. Alternatively, transmission electron microscopy (TEM) [66] or X-ray powder diffraction can be used. These methods do not provide the same resolution as single crystal diffraction at an XFEL, but they can determine whether the sample is crystalline and estimate the diffraction quality between different crystal batches.

Data collection on small crystals requires new techniques for crystal growth and detection as well as new methods for sample delivery [67]. Macroscopic crystals can be mounted on a goniometer using conventional loop or capillary mounts. However, these are not suitable for microcrystals. The methods developed for transporting the crystal suspension into the interaction zone can be classified into two groups: liquid-jet setups or delivery on solid supports. Delivery using liquid jets can be achieved through the use of a gas dynamic virtual nozzle to generate microscopic droplet streams [68] or through acoustic droplet ejection [69]. These setups provide a convenient method for high-throughput replenishment of crystals into the X-ray beam location. However, they require large sample volumes and, therefore, large sample quantities due to their low hit rates. The hit rate is a crucial statistic for determining data acquisition efficiency. It is defined as the mean number of crystals per shot. In an ideal experiment, new crystals would be delivered continuously at every X-ray exposure point. However, in liquid-jet setups, most crystals are not hit by the beam, resulting in low hit rates.

Sample delivery on solid supports, such as fixed target or chip technology, can in principle achieve 100% hit rates [70]. However, fixed-target or chip setups require protection of the crystals from drying out. Depending on the support, the macromolecules may have a preferential orientation, resulting in insufficient sampling of the reciprocal space.

Data collection strategies for small and large crystals differ significantly from each other. Large crystals can be characterized before data collection, and the data collection strategy can be optimized in terms of resolution, mosaicity, and orientation. Data is then collected while rotating the crystal, yielding fully integrated reflections. Small crystals interact with the X-ray beam briefly and are destroyed upon exposure to high flux density, resulting in only partial reflections. Therefore, a large number of fresh crystals injected into the X-ray beam in random orientation are required to obtain a complete dataset.

Datasets from serial crystallography consist of independent frames obtained from randomly oriented crystals. Unlike traditional crystallography, which collects data on a single crystal with specific properties, serial data collection involves crystals of varying sizes and qualities, resulting in significant fluctuations in recorded intensities. To address the challenges related to indexing and integrating large numbers of diffraction patterns in an automated manner, new software suites were developed. Currently, there are three different programs available for the analysis of serial diffraction data: CrystFEL [71], cctbx.xfel [72], and nXDS [73]. These programs differ in their approaches and algorithms.

In conclusion, serial crystallography at XFEL and synchrotrons has revolutionized macromolecular crystallography over the past decade. The rapid development in this field has transitioned structural biology from static pictures of frozen molecules to the collection of molecular movies on biomolecules.

16.8 Nuances of Fe–S Protein Crystallography

Fe–S proteins bind an inorganic cluster to the protein backbone. The presence of metal ions that bind to defined sites in proteins offers great advantages for structure determination using X-ray crystallography. However, there are also challenges associated with metallocofactors such as iron–sulfur clusters.

First, it is important to note that most Fe–S clusters are sensitive to oxygen. Therefore, all steps during protein production and purification must be conducted under an inert atmosphere. Oxygen exposure can result in a reduced Fe–S cluster occupancy or even the complete decomposition of the clusters. In some cases, the degradation of Fe–S clusters leads to significant structural changes, which in turn makes it impossible to carry out crystallization or any other functional analysis of the protein. The handling of oxygen-sensitive proteins typically requires the use of anaerobic chambers operated with nitrogen or argon gas.

In many cases, Fe–S proteins require specialized bacterial strains to ensure the maturation of the Fe–S protein. Examples of such strains include BL21(DE3)Δ*iscR* [74] and BL21(DE3)Suf^{++} [75], which use different approaches to enhance Fe–S biosynthesis.

In addition, the structure of inorganic clusters and the ligands bound to them are more challenging to predict. The current prediction software, including Alphafold2 and RoseTTAFold, is not yet capable of accurately predicting the location of clusters within protein structures. Furthermore, the docking software employed for drug design is considerably less effective when applied to proteins that contain metal ions in their active sites. Despite ongoing efforts to develop docking approaches for metalloproteins, this remains a highly challenging task, particularly in the case of iron–sulfur proteins [75]. This underscores the continued relevance of experimental approaches for studying Fe–S and other metalloproteins.

Nevertheless, the presence of heavy atoms, such as iron, in proteins is of significant benefit for the application of X-ray crystallography. As previously stated, electron-rich sites within a protein, such as iron or sulfur atoms, result in anomalous scattering at specific wavelengths. Anomalous scattering can be exploited for experimental phasing or to determine the location of the special atoms (substructure). The majority of proteins lack sufficient electron-rich atoms for experimental phasing, therefore require the introduction of heavy atoms, such as selenomethionine or the derivatization of crystals with heavy atoms. The determination of the substructure of the heavy atoms is also beneficial in cases where the architecture of the cofactor is unknown or when partial occupancy of metals gives rise to uncertainties regarding the metal sites.

The theoretical values of the absorption edges can be used to determine the specific wavelength at which anomalous data should be collected. Nevertheless, it is recommended that these values be measured experimentally if the beamline in use permits the measurement of X-ray absorption spectra. This is due to the fact that the absorption edges of the same elements can vary depending on their surrounding environment.

16.9 Conclusions

This chapter presents a summary of the fundamental principles of macromolecular crystallography and its application to research on Fe–S proteins. The theoretical background of crystallography is introduced, and each step of the crystallographic experiment is explained in detail, starting with protein crystallization and continuing through data processing, phasing, and model building. X-ray crystallography remains a key technique for determining protein structures, and it is particularly effective for studying metalloproteins. The presence of heavy atoms in Fe–S proteins is highly beneficial for experimental phasing and determining substructures. The advent of innovative crystallographic techniques, including serial crystallography, electron diffraction, and neutron diffraction, has paved the way for new avenues of structural investigation, thereby enhancing our understanding of Fe–S proteins and their roles in various biological processes.

References

1 Berman, H.M., Westbrook, J., Feng, Z. et al. (2000). The protein data bank. *Nucleic Acids Res.* 28: 235–242.
2 Borchardt-Ott, W. and Sowa, H. (2018). *Kristallographie*. Berlin, Heidelberg, Springer Berlin Heidelberg.
3 Röntgen, W.C. (1896). On a new kind of rays. *Science* 3, 59: 227–231.
4 Wiechmann, F.G. (1896). A new form of radiation. *Science* 3 (72): 726–730.

5 Friedrich, W., Knipping, P., and von Laue, M. (1912). *Interferenz-Erscheinungen bei Röntgenstrahlen*, 303–322. Klasse der Königlich Bayer Akad der Wissenschaften zu München.

6 Bragg, W.H. (1912). X-rays and crystals. *Nature* 90 (2243): 219.

7 Bragg, W.L. (1913). The diffraction of short electromagnetic waves by a crystal. *Proceedings of the Cambridge Philosophical Society*, 43–57.

8 Bragg, W.H. and Bragg, W.L. (1913). The reflection of X-rays by crystals. *Proc. R. Soc. Lond. Ser. A Contain Pap. Math. Phys. Character* 88 (605): 428–438.

9 Bragg, W.H. (1913). The reflection of X-rays by crystals (II.). *Proc. R. Soc. Lond. Ser. A Contain Pap. Math. Phys. Character* 89 (610): 246–248.

10 Bragg, W.L. (1913). The structure of some crystals as indicated by their diffraction of X-rays. *Proc. R. Soc. Lond. Ser. A Contain Pap. Math. Phys. Character* 89 (610): 248–277.

11 Bragg, W.L.H.L. (1913). The structure of the diamond. *Proc. R. Soc. Lond. Ser. A Contain Pap. Math. Phys. Character* 89 (610): 277–291.

12 Jaskolski, M., Dauter, Z., and Wlodawer, A. (2014). A brief history of macromolecular crystallography, illustrated by a family tree and its Nobel fruits. *FEBS J.* 281 (18): 3985–4009.

13 Hünefeld, F. (1840). Der Chemismus in der thierischen Organisation: physiologisch-chemische Untersuchungen der materiellen Veränderungen oder des Bildungslebens im. https://books.google.com/books?hl=en&lr=&id=qgc_AAAAcAAJ&oi=fnd&pg=PR7&ots=CUGDnsQHLe&sig=dovMKxc4qYn_BkbDIghIPWMgS5A (accessed 27 Mar 2024).

14 Bernal, J.D. and Crowfoot, D. (1934). X-ray photographs of crystalline pepsin. *Nature* 133 (3369): 794–795.

15 Kendrew, J.C., Bodo, G., Dintzis, H.M. et al. (1958). A three-dimensional model of the myoglobin molecule obtained by x-ray analysis. *Nature* 181 (4610): 662–666.

16 Perutz, M.F., Rossmann, M.G., Cullis, A.F. et al. (1960). Structure of hæmoglobin: a three-dimensional fourier synthesis at 5.5-Å. Resolution, obtained by X-ray analysis. *Nature* 185 (4711): 416–422.

17 Green, D.W., Ingram, V.M., and Perutz, M.F. (1954). The structure of haemoglobin – IV. Sign determination by the isomorphous replacement method. *Proc. R. Soc. Lond. Ser. A Math. Phys. Sci.* 225 (1162): 287–307.

18 Adams, M.J., Blundell, T.L., Dodson, E.J. et al. (1969). Structure of rhombohedral 2 zinc insulin crystals. *Nature* 224 (5218): 491–495.

19 Carlisle, C.H. and Crowfoot, D. (1945). The crystal structure of cholesteryl iodide. *Proc. R. Soc. Lond. Ser. A Math. Phys. Sci.* 184 (996): 64–83.

20 Crowfoot, D., Bunn, C.W., Rogers-Low, B.W., and Turner-Jones, A. (1949). The X-ray crystallographic investigation of the structure of penicillin. In: *Chemistry of Penicillin* (ed. H.T. Clarke), 310–366. Princeton: Princeton University Press.

21 Hodgkin, D.C., Kamper, J., Lindsey, J. et al. (1957). The structure of vitamin B_{12}. I. An outline of the crystallographic investigation of vitamin B_{12}. *Proc. R. Soc. Lond. Ser. A Math. Phys. Sci.* 242 (1229): 228–263.

22 Pauling, L. (1993). How my interest in proteins developed. *Protein Sci.* 2 (6): 1060–1063.

23 Pauling, L. (1992). How I became interested in the chemical bond: a reminiscence. In: *The Chemical Bond: Structure and Dynamics* (ed. A. Zewail), 99–109. Boston: Academic Press.

24 Edison, A.S. (2001). Linus Pauling and the planar peptide bond. *Nat. Struct. Biol.* 8 (3): 201–202.

25 Pauling, L., Corey, R.B., and Branson, H.R. (1951). The structure of proteins; two hydrogen-bonded helical configurations of the polypeptide chain. *Proc. Natl. Acad. Sci. USA* 37 (4): 205–211.

26 Pauling, L. and Corey, R.B. (1951). The pleated sheet, a new layer configuration of polypeptide chains. *Proc. Natl. Acad. Sci. USA* 37 (5): 251–256.

27 Pauling, L. and Corey, R.B. (1953). A Proposed Structure For The Nucleic Acids. *Proc. Natl. Acad. Sci. USA* 39 (2): 84–97.

28 McPherson, A. (2009). *Introduction to Macromolecular Crystallography*. Wiley.

29 D'Arcy, A., Elmore, C., Stihle, M., and Johnston, J.E. (1996). A novel approach to crystallising proteins under oil. *J. Cryst. Growth* 168 (1–4): 175–180.

30 Jang, K., Kim, H.G., Hlaing, S.H.S. et al. (2022). A short review on cryoprotectants for 3D protein structure analysis. *Crystals* 12 (2): 138.

31 Berg, M.J. and Sorensen, C.M. (2018). A review and reassessment of diffraction, scattering, and shadows in electrodynamics. *J. Quant. Spectrosc. Radiat. Transf.* 1 (210): 225–239.

32 Rhodes, G. (2006). Crystallography made crystal clear: a guide for users of macromolecular models, 3e, 1–306. http://www.sciencedirect.com:5070/book/9780125870733/crystallography-made-crystal-clear?via=ihub= (accessed 31 Mar 2024).

33 Battye, T.G.G., Kontogiannis, L., Johnson, O. et al. (2011). iMOSFLM: a new graphical interface for diffraction-image processing with MOSFLM. *Acta Crystallogr. D Biol. Crystallogr.* 67 (4): 271–281.

34 Otwinowski, Z. and Minor, W. (1997). [20] processing of X-ray diffraction data collected in oscillation mode. *Methods Enzymol.* 1 (276): 307–326.

35 Kabsch, W. (2010). XDS. *Acta Crystallogr. D Biol. Crystallogr.* 66 (Pt 2): 125.

36 Winter, G., Beilsten-Edmands, J., Devenish, N. et al. (2022). DIALS as a toolkit. *Protein Sci.* 31 (1): 232–250.

37 Powell, H.R. (2017). X-ray data processing. *Biosci. Rep.* 37 (5): BSR20170227.

38 Powell, H.R. (2021). A beginner's guide to X-ray data processing. *Biochem (Lond).* 43 (3): 46–50.

39 Evans, P.R. and Murshudov, G.N. (2013). How good are my data and what is the resolution? *Acta Crystallogr. D Biol. Crystallogr.* 69 (7): 1204–1214.

40 Winn, M.D., Ballard, C.C., Cowtan, K.D. et al. (2011). Overview of the *CCP* 4 suite and current developments. *Acta Crystallogr. D Biol. Crystallogr.* 67 (4): 235–242.

41 Potterton, L., Agirre, J., Ballard, C. et al. (2018). *CCP* 4 *i* 2: the new graphical user interface to the *CCP* 4 program suite. *Acta Crystallogr. D Struct. Biol.* 74 (2): 68–84.

42 Wlodawer, A., Minor, W., Dauter, Z., and Jaskolski, M. (2008). Protein crystallography for non-crystallographers, or how to get the best (but not more) from published macromolecular structures. *FEBS J.* 275 (1): 1–21.

43 Patterson, A.L. (1934). A Fourier series method for the determination of the components of interatomic distances in crystals. *Phys. Rev.* 46 (5): 372.

44 Evans, P. and McCoy, A. (2007). An introduction to molecular replacement. *Acta Crystallogr. D Biol. Crystallogr.* 64 (1): 1–10.

45 Jumper, J., Evans, R., Pritzel, A. et al. (2021). Highly accurate protein structure prediction with AlphaFold. *Nature* 596 (7873): 583–589.

46 Baek, M., DiMaio, F., Anishchenko, I. et al. (2021). Accurate prediction of protein structures and interactions using a three-track neural network. *Science* 373 (6557): 871–876.

47 Read, R.J., Baker, E.N., Bond, C.S. et al. (2023). AlphaFold and the future of structural biology. *IUCrJ* 10 (4): 377–379.

48 Herres-Pawlis, S. and Kluefers, P. (2016). *Bioanorganische Chemie: Metalloproteine, Methoden und Modelle.* Wiley-VCH.

49 Diederichs, K. and Karplus, P.A. (1997). Improved R-factors for diffraction data analysis in macromolecular crystallography. *Nat. Struct. Biol.* 4 (4): 269–275.

50 Weiss, M.S. and Hilgenfeld, R. (1997). On the use of the merging R factor as a quality indicator for X-ray data. *J. Appl. Crystallogr.* 30 (2): 203–205.

51 Weiss, M.S. (2001). Global indicators of X-ray data quality. *J. Appl. Crystallogr.* 34 (2): 130–135.

52 Karplus, P.A. and Diederichs, K. (2015). Assessing and maximizing data quality in macromolecular crystallography. *Curr. Opin. Struct. Biol.* 1 (34): 60–68.

53 Karplus, P.A. and Diederichs, K. (2012). Linking crystallographic model and data quality. *Science* 336 (6084): 1030–1033.

54 Brünger, A.T., Adams, P.D., Clore, G.M. et al. (1998). Crystallography & NMR system: a new software suite for macromolecular structure determination. *Acta Crystallogr. D Biol. Crystallogr.* 54 (Pt 5): 905–921.

55 Sheldrick, G.M. (2015). Crystal structure refinement with SHELXL. *Acta Crystallogr. C Struct. Chem.* 1 (71): 3–8.

56 Murshudov, G.N., Skubák, P., Lebedev, A.A. et al. (2011). REFMAC5 for the refinement of macromolecular crystal structures. *Acta Crystallogr. D Biol. Crystallogr.* 67 (Pt 4): 355–367.

57 Adams, P.D., Grosse-Kunstleve, R.W., Hung, L.W. et al. (2002). PHENIX: building new software for automated crystallographic structure determination. *Acta Crystallogr. D Biol. Crystallogr.* 58 (Pt 11): 1948–1954.

58 Wlodawer, A. (2017). Stereochemistry and validation of macromolecular structures. *Methods Mol. Biol.* 1607: 595.

59 Zheng, H., Cooper, D.R., Porebski, P.J. et al. (2017). CheckMyMetal: a macromolecular metal-binding validation tool. *Acta Crystallogr. D Struct. Biol.* 73 (3): 223–233.

60 Chapman, H.N., Fromme, P., Barty, A. et al. (2011). Femtosecond X-ray protein nanocrystallography. *Nature* 470 (7332): 73–78.

61 Schlichting, I. (2015). Serial femtosecond crystallography: the first five years. *IUCrJ* 2 (2): 246–255.

62 Caramello, N. and Royant, A. (2024). From femtoseconds to minutes: time-resolved macromolecular crystallography at XFELs and synchrotrons. *Acta Crystallogr. D Struct. Biol.* 80: 60–79.

63 Kupitz, C., Grotjohann, I., Conrad, C.E. et al. (2014). Microcrystallization techniques for serial femtosecond crystallography using photosystem II from Thermosynechococcus elongatus as a model system. *Philos. Trans. R. Soc. B Biol. Sci.* 369 (1647): 20130316.

64 Wampler, R.D., Kissick, D.J., Dehen, C.J. et al. (2008). Selective detection of protein crystals by second harmonic microscopy. *J. Am. Chem. Soc.* 130 (43): 14076–14077.

65 Tian, X., Nejadnik, M.R., Baunsgaard, D. et al. (2016). A comprehensive evaluation of nanoparticle tracking analysis (nanosight) for characterization of proteinaceous submicron particles. *J. Pharm. Sci.* 105 (11): 3366–3375.

66 Stevenson, H.P., Makhov, A.M., Calero, M. et al. (2014). Use of transmission electron microscopy to identify nanocrystals of challenging protein targets. *Proc. Natl. Acad. Sci. USA* 111 (23): 8470–8475.

67 Zhao, F.Z., Zhang, B., Yan, E.K. et al. (2019). A guide to sample delivery systems for serial crystallography. *FEBS J.* 286 (22): 4402–4417.

68 DePonte, D.P., Weierstall, U., Schmidt, K. et al. (2008). Gas dynamic virtual nozzle for generation of microscopic droplet streams. *J. Phys. D. Appl. Phys.* 41 (19): 195505.

69 Roessler, C.G., Kuczewski, A., Stearns, R. et al. (2013). Acoustic methods for high-throughput protein crystal mounting at next-generation macromolecular crystallographic beamlines. *J. Synchrotron Radiat.* 20 (5): 805–808.

70 Oghbaey, S., Sarracini, A., Ginn, H.M. et al. (2016). Fixed target combined with spectral mapping: approaching 100% hit rates for serial crystallography. *Acta Crystallogr. D Struct. Biol.* 72 (8): 944–955.

71 White, T.A., Kirian, R.A., Martin, A.V. et al. (2012). CrystFEL: a software suite for snapshot serial crystallography. *J. Appl. Crystallogr.* 45 (2): 335–341.

72 Sauter, N.K., Hattne, J., Grosse-Kunstleve, R.W., and Echols, N. (2013). New Python-based methods for data processing. *Acta Crystallogr. D Biol. Crystallogr.* 69 (7): 1274–1282.

73 Kabsch, W. (2014). Processing of X-ray snapshots from crystals in random orientations. *Acta Crystallogr. D Biol. Crystallogr.* 70 (8): 2204–2216.

74 Schwartz, C.J., Giel, J.L., Patschkowski, T. et al. (2001). IscR, an Fe-S cluster-containing transcription factor, represses expression of *Escherichia coli* genes encoding Fe-S cluster assembly proteins. *Proc. Natl. Acad. Sci. USA* 98 (26): 14895–14900.

75 Corless, E.I., Mettert, E.L., Kiley, P.J., and Antony, E. (2019). Elevated expression of a functional suf pathway in *Escherichia coli* BL21(DE3) enhances recombinant production of an iron-sulfur cluster-containing protein. *J. Bacteriol.* 202 (3): e00496-19.

17

Vibrational and Mössbauer Spectroscopic Techniques to Study Iron–Sulfur Clusters

Christian Lorent[1], Giorgio Caserta[1], Volker Schünemann[2], and Ingo Zebger[1]

[1]Technische Universität Berlin, Institut für Chemie, Sekr. PC14, Straße des 17. Juni 135, Berlin, 10623, Germany
[2]University of Kaiserslautern-Landau, Department of Physics, Biophysics and Medical Physics, Erwin-Schrödinger-Str. 46, Kaiserslautern, 67663, Germany

Chapter Goals

- This chapter provides a broad overview of key spectroscopic methods used to study [Fe–S] clusters, offering numerous illustrative examples of how these techniques can be applied.
- Readers will be equipped to develop analytical skills for interpreting vibrational and Mössbauer spectroscopic data, identifying e.g., redox states, spin configurations and ligand environments in [Fe–S]-containing systems.
- The main goal is to highlight the importance of vibrational and Mössbauer spectroscopy in bioinorganic research, particularly at the intersection of chemistry, biology, and physics.

Main Chapter Points

- Vibrational spectroscopy is a powerful tool to elucidate structural and electronic properties of [Fe–S] clusters (e.g., structural distortions, oxidation states and coordination changes).
- Vibrational spectroscopic methods such as IR and (resonance) Raman play an important role in exploring the binding of substrates and/or inhibitors to enzymes (e.g., hydrogenases and nitrogenases), contributing to the identification of important reaction intermediates.
- Mössbauer spectroscopy has been shown to be very valuable in exploring structural and electronic properties of [Fe–S] centers, particularly their complex spin states.

Iron-Sulfur Clusters: Biogenesis and Biochemistry, Volume 2, First Edition.
Edited by Silke Leimkühler, Günter Schwarz, Oliver Lenz, and Oliver Einsle.

- Using synchrotron radiation, nuclear resonance vibrational spectroscopy (NRVS) exploits the Mössbauer effect to uncover molecular vibrations involving Mössbauer-active nuclei like ^{57}Fe.

17.1 Vibrational Spectroscopy

This section concisely illustrates the basics of molecular vibrations, which are explained in more detail in the textbooks by Wilson, Decius, and Cross as well as Siebert and Hildebrandt [1–4]. For a diatomic molecule, a molecular vibration can be understood as the continuous motion of point masses connected via an elastic spring. If one point mass is displaced from the equilibrium position by Δx, a restoring force F_x appears, according to Hooke's law, which depends on the stiffness of the spring, illustrated by the force constant f. In a diatomic molecule, the latter constant corresponds to the chemical bond strength, Δx is related to the change in the chemical bond length and the point masses symbolize the vibrating atoms.

$$F_x = -f\Delta x \tag{17.1}$$

The frequency of such a stretching vibration can be expressed in terms of classical physics as follows:

$$\nu = \frac{1}{2\pi}\sqrt{\frac{f}{\mu}} \tag{17.2}$$

where the reduced mass μ is defined as

$$\mu = \frac{m_\alpha m_\beta}{m_\alpha + m_\beta} \tag{17.3}$$

From Eq. (17.2) it becomes apparent that for a diatomic harmonic oscillator the frequency of a vibration contains specific chemical information on molecular properties, as it depends on the bond strength f and the reduced mass μ of the atoms contributing to this vibration. This qualifies vibrational spectroscopy to specifically explore the structure of molecules and consequently also chemical transformations.

Vibrations in a molecule with N connected atoms, which may also involve non-covalent interactions like hydrogen bonds, depend on the degrees of freedom of the system. Essentially, each atom moves in three directions in a Cartesian coordinate system and therefore features three degrees of freedom ($3N$). For each molecule, three forms of motion are possible: translation, rotation, and vibration. Accordingly, three degrees of freedom correspond to the translational displacement of all atoms by the same distance in the respective x-, y-, and z-direction. Within the approximation of point masses, three (two for linear molecules) degrees of freedom are observed for the rotation of a molecule, depending on its moment of inertia around three perpendicular axes. In the case of linear molecules, the moment of inertia is zero around the molecular axis, reducing the rotational degrees of freedom to 2. Therefore, the number of molecular vibrations for a molecule is given by $3N - 6$

($3N-5$ for linear molecules) degrees of freedom, corresponding to the respective normal modes.

17.1.1 Normal Modes

The concept of normal modes can fundamentally describe the molecular vibrations from simple to very complex systems and is often used to calculate vibrational spectra. A normal mode ascribes the vibration of all atoms in phase with different amplitudes, but identical frequency, while the center of mass is conserved. This way, each normal mode is characterized by one vibrational frequency and a specific pattern of atomic motions.

To analyze normal modes, it is appropriate to introduce mass-weighted Cartesian displacement coordinates q_i,

$$q_i = \sqrt{m_i} d_i \tag{17.4}$$

The Cartesian displacement coordinates (Δx, Δy, Δz) are summarized as d_i, while m_i is the mass of the displaced atoms. Thus, the kinetic energy T can be simply expressed as

$$T = \frac{1}{2}\sum_{i=1}^{3N} \dot{q}_i^2 \tag{17.5}$$

To utilize a mathematically convenient expression for treating normal modes, especially for quantum chemical calculations, the mass-weighted Cartesian coordinates can be transformed into the so-called normal coordinates using an orthogonal transformation.

$$Q_k = \sum_{i=1}^{3N} l_{ik} q_i \tag{17.6}$$

The coefficients l_{ik} are chosen so that the potential energy includes no cross product of Q_k, while the kinetic energy remains in the form of Eq. (17.5). Thereby, each normal coordinate corresponds to one specific normal mode.

17.1.2 Vibrational Spectroscopic Techniques

Vibrational spectroscopy is used in various fields of physics, chemistry, and biology. Especially for biological systems like metalloenzymes, infrared (IR), resonance Raman (RR), and nuclear resonance vibrational (NRVS) spectroscopies provided important structural and functional details by specifically monitoring normal modes related to their active site. To identify characteristic vibrations, isotopic labeling is commonly applied. Due to the larger mass of, e.g. ^{34}S compared to the most abundant ^{32}S, Fe–^{34}S bands shift to a lower frequency. Although all the aforementioned techniques probe molecular vibrations, the underlying phenomena to induce transitions between the initial i and final f vibrational states and the energy of the radiation used for excitation are quite different (Figure 17.1). Additionally, the signal intensities observed for the probed normal modes depend on the chosen

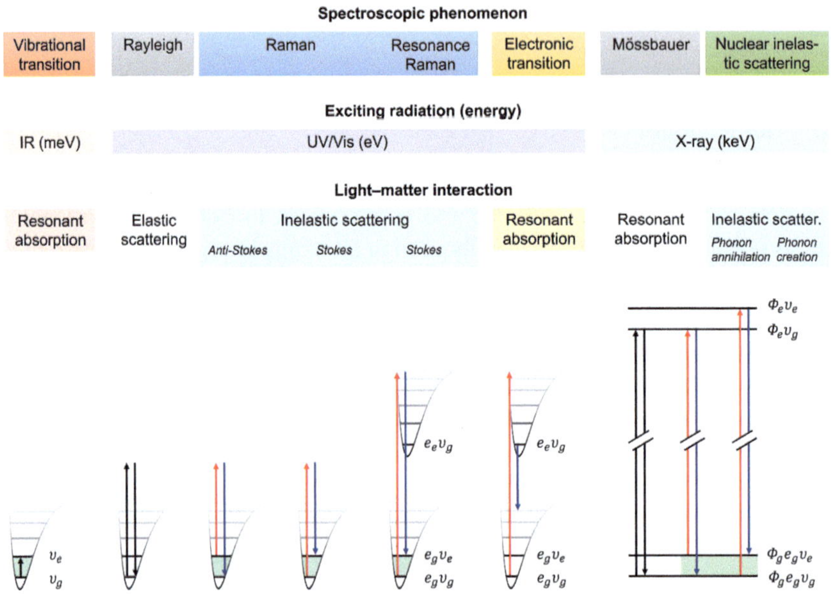

Figure 17.1 Overview of different spectroscopic phenomena. The radiation used for excitation, the corresponding fundamental light–matter interaction, and the respective quantum transitions (from top to bottom). The energy levels of the ground and excited states for vibrational, electronic, and nuclear transitions are exemplified. Upward arrows indicate excitation, downward arrows deexcitation. For inelastic transitions, these arrows are shown in red and blue. Vibrational transitions discussed in this chapter are highlighted in green. Source: Adapted from the PhD thesis of Christian Lorent [4]. Visual source: Christian Lorent (co-author).

methodology. The intensity I is, based on Fermi's golden rule, proportional to the transition rate P_{fi}, the latter being proportional to the square of the transition moment M_{fi}.

$$I \propto P_{fi} \propto |M_{fi}|^2 \tag{17.7}$$

The transition moment is given by the following equation:

$$M_{fi} = \langle \psi_f | \widehat{\Omega} | \psi_i \rangle \tag{17.8}$$

Here, $\widehat{\Omega}$ is the perturbation operator that characterizes the interaction (perturbation) of the molecule with the electromagnetic wave, while the wavefunctions ψ_i and ψ_f correspond to the initial and final states of the molecule, e.g. the vibrational levels of the transition (Figure 17.1).

17.2 Infrared Spectroscopy

The astronomer William Herschel documented the existence of IR radiation over 200 years ago when he identified the "invisible heat rays" in the sunlight [5]. When utilizing IR spectroscopy, the related part of the electromagnetic spectrum is chosen for a resonant vibrational excitation. Hereby, the corresponding photon absorption induces a transition between the initial and final vibrational levels (Figure 17.1). The transition moment, as described in Eq. (17.8), is derived from the electrical dipole moment operator $\hat{\mu}_q$ occurring in IR spectroscopy.

$$\hat{\mu}_q = \sum_{\alpha} e_\alpha q_\alpha \tag{17.9}$$

hereby e_α relates to the effective charge of the atom α, while q_α belongs to the respective Cartesian position vector. Referring to Eq. (17.7), the respective vibrational transition monitored by IR spectroscopy necessitates a transition dipole moment, which differs from zero. When utilizing a harmonic oscillator, the transition dipole moment operator $\hat{\mu}_q$ is expressed with respect to normal coordinates:

$$[\hat{\mu}_q]_{fi} = \langle \psi_f | \hat{\mu}_q | \psi_i \rangle = \sum_{k=1}^{3N-6} \hat{\mu}_q^k \langle \psi_f | Q_k | \psi_i \rangle \tag{17.10}$$

A normal mode is IR active, when the derivative of the corresponding dipole moment $\hat{\mu}_q^k$ is nonzero, which suggests that the relative atomic displacement that is associated with the normal coordinate Q_k involves a change in the molecular dipole moment. Furthermore, the corresponding integral $\langle \psi_f | Q_k | \psi_i \rangle$ differs from zero when the related vibrational quantum numbers f and i are separated by one.

17.2.1 MIR Spectroscopy

Mid-IR (MIR) spectroscopy focuses on the spectral region from 400 to 4000 cm^{-1}. This technique is particularly suited to investigate (metallo)proteins, by monitoring, e.g. intramolecular stretching vibrations of metal-bound inorganic ligands like cyanide (CN^-) and/or carbon monoxide (CO) between 1750 and 2200 cm^{-1}, secondary structural changes of amide I and amide II bands between 1700 and 1500 cm^{-1} as well as the contribution of single amino acids, mostly between 1750 and 1000 cm^{-1} [6, 7]. The strong IR absorbers CO and CN^- are often part of the metal active site or bind as a substrate or inhibitor in gas-converting metalloenzymes [8–11]. MIR has also been used to monitor nitrosylation of [Fe–S] clusters by probing the characteristic NO stretching vibrations between 1720 and 1820 cm^{-1}. However, these bands can overlap with the carbonyl stretching vibrations from protonated glutamates and aspartates, preventing an unambiguous band assignment [12].

H_2-metabolizing enzymes (i.e. hydrogenases) represent an outstanding example where MIR spectroscopy has greatly advanced the understanding of the structure and function of their catalytic sites. A detailed description of these metalloenzymes is provided in other chapters of this book. The coordination sphere of the metal ions in [NiFe]- and [FeFe]-hydrogenases is very similar, involving cysteinyl thiolates and the strong-field diatomic CO and CN^- ligands. [NiFe]- and [FeFe]-hydrogenases exhibit a huge variety of spectroscopically characterized catalytic or noncatalytic redox states [8]. The IR spectrum of the hydrogen-binding intermediate (known as Ni_a-S) of the regulatory [NiFe]-hydrogenase from *Cupriavidus necator* is displayed in Figure 17.2a. It exhibits one band from a CO stretching vibration at 1943 cm^{-1}

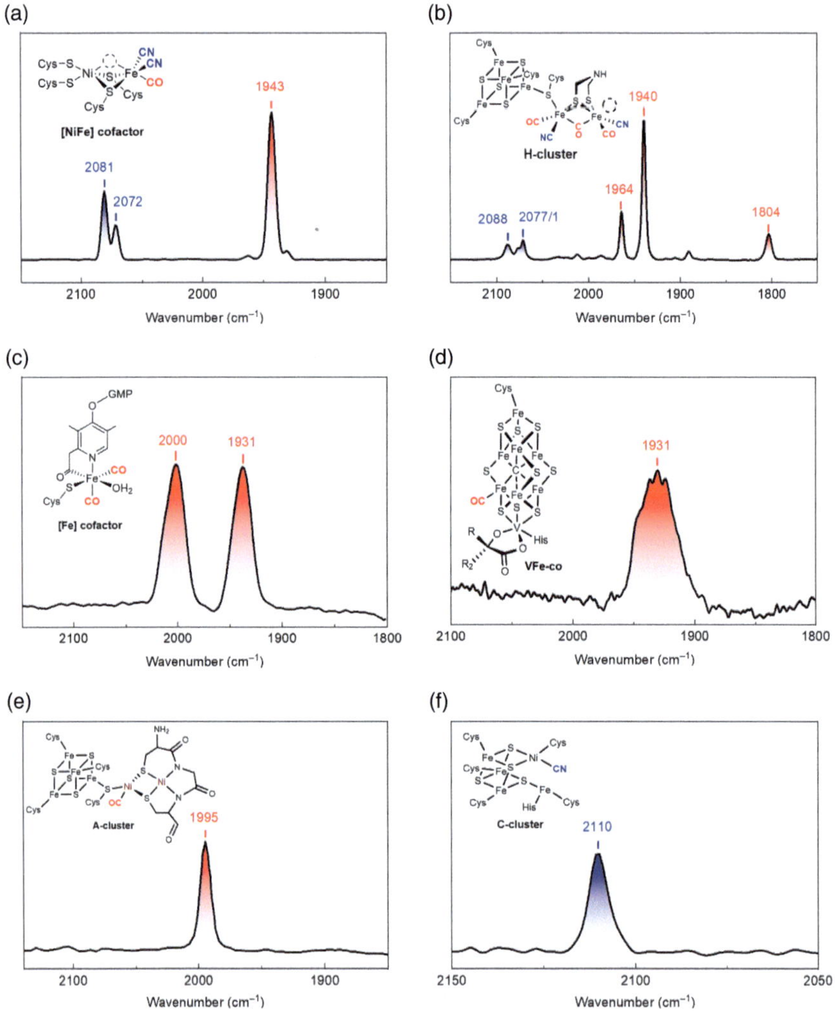

Figure 17.2 Representative MIR spectra with IR bands from CO or CN^- ligands bound to active sites in (a) regulatory [NiFe]-hydrogenase from *Cupriavidus necator* [13], (b) [FeFe]-hydrogenase HydA1 from *Chlamydomonas reinhardtii* [14], (c) [Fe]-hydrogenase from *Methanococcus aeolicus* [15], (d) V-nitrogenase from *Azotobacter vinelandii* [9], (e) monofunctional ACS from *Carboxydothermus hydrogenoformans* [10], (f) CODH II from *Carboxydothermus hydrogenoformans* [11]. Metal-bound CO and CN^- ligands are highlighted in red and blue, respectively. For panels a–c, these ligands are an integral part of the active site, while for panels d–f they represent a bound substrate or inhibitor. The software WebPlotDigitizer was utilized to extract the data used for panels c and e from the indicated references. Visual source: Christian Lorent (co-author).

and two signals at 2072 and 2081 cm^{-1}, originating from the antisymmetric and symmetric stretching modes of the CN^- ligands [13]. Due to the larger number of CO and CN^- ligands in [FeFe]-hydrogenases, the corresponding IR spectra are more complex. In Figure 17.2b, the spectrum of the redox intermediate with a vacant H_2-biding site at the distal iron of the [FeFe]-hydrogenase HydA1 (known as H_{ox}) from *Chlamydomonas reinhardtii* is shown [14]. The CO absorption from the bridging CO is detected at 1804 cm^{-1}, while the vibrational modes from the terminal CO ligands are related to the signals at 1940 and 1964 cm^{-1}. Main features from CN vibrations are observed at 2071 and 2088 cm^{-1}. The IR spectrum of the [Fe]-hydrogenase from *Methanococcus aeolicus* (Figure 17.2c) exhibits intense bands at 1931 and 2000 cm^{-1} [15], which are assigned to the two terminal CO ligands of the iron-guanylylpyridinol (FeGP) cofactor.

Nitrogenases are another class of metalloenzymes, which are responsible for biological nitrogen fixation. Due to the high stability of the inert dinitrogen triple bond, this is one of the most challenging chemical transformations. To accomplish the activation of N_2, these enzymes employ a two-component system, comprising multiple iron–sulfur clusters. Among them are an unusual [8Fe–7S] complex called the P-cluster, and a complex [Me–7Fe–9S–C] metal center (Me = Mo, V, or Fe) called the M-cluster, also known as FeMo-co for the most extensively studied Mo-nitrogenase [16]. CO can act as a noncompetitive inhibitor for dinitrogen reduction. Its binding to the FeMo-co as terminal or/and bridging ligand has been studied in protein solutions using stopped-flow IR [17] and in immobilized enzymes under electrochemical control accessed via the attenuated total reflection (ATR) IR mode [18, 19]. The vibrations of the terminal Fe-bound COs are located between 1970 and 1870 cm^{-1}, while those of the bridging CO species are found between 1720 and 1680 cm^{-1} [20]. Vanadium-containing nitrogenases can also reduce CO to ethylene. This implies the co-presence of at least two CO at the FeV-co active site, enabling the C–C coupling. The presence of two CO ligands has been verified by X-ray crystallography and IR spectroscopy, where a very broad absorption at 1931 cm^{-1} from a terminal bound CO (Figure 17.2d) as well as indications for bridging CO species have been observed [9].

CO dehydrogenase/acetyl-CoA synthase (CODH/ACS) constitutes a bifunctional enzyme, in which the reduction of CO_2 to CO is catalyzed by the CODH module, while the acetyl-CoA synthesis takes place in the ACS domain. The latter utilizes the CO produced by the CODH, coenzyme A and a methyl cation, donated from a methylated corrinoid iron–sulfur protein (CH_3–CoCFeSP) to form acetyl-CoA [21]. The catalytic A-cluster in ACS incorporates two Ni ions coordinated via thiol bridges to a [4Fe–4S] site (Figure 17.2e). Binding of CO to the A-cluster has been shown by stopped-flow IR and rapid-freeze-quench EPR spectroscopy [10]. The characteristic band at 1995 cm^{-1} was assigned to the stretching mode of a terminally Ni-coordinated CO molecule [10]. The active site of CODHs, i.e. the C-cluster, was also spectroscopically studied utilizing the inhibitor CN^- as an isoelectronic and isosteric analog of CO. CN^- has been shown to bind to the C-cluster as a characteristic CN stretching vibration at 2110 cm^{-1} was observed upon CN^- incubation of CODH (Figure 17.2f). Cyanate (^-OCN) that is isoelectronic with CO_2 could also bind to CODH (ν_{OCN} at 2168 cm^{-1}). Notably, the reduction of the bound cyanate led to the formation of the CN^--inhibited state [11, 22].

Iron–sulfur clusters are very sensitive toward nitric oxide (NO), which can cause multistep cluster degradation. While in whole cells nitrosylation is regulated and even repaired, [Fe–S] clusters from isolated proteins degrade irreversibly forming nitrosyl iron complexes such as mononitrosyl iron complexes (MNICs) and dinitrosyl iron complexes (DNICs). To monitor this process, IR spectroscopy can be a powerful tool. In particular, the stretching vibration of the NO molecule, when bound to the Fe, represents a valuable maker band with absorptions between 1720 and 1820 cm^{-1} [23, 24].

17.2.2 FIR Spectroscopy

The far-infrared (FIR) spectral regime between 400 and 50 cm^{-1} is typically used for studying interactions such as ion-pair formation, mixing of protic ionic liquids as well as the selective identification of color pigments based on their spectroscopic fingerprint at low frequency [25, 26]. However, a few insights into the structural and electronic identity of biological systems have been derived with this technique. The latter involves the investigation of structural changes and/or conformations in peptides [27–29], as FIR spectroscopy is susceptible to vibrations from the peptide backbone as well as hydrogen bonds [30]. FIR spectroscopy has also been extended to the study of metalloproteins, targeting vibrations from [Fe–S] clusters that are usually probed using RR spectroscopy (see Section 17.3). Considering the different selection rules and that resonance enhancement (see below) is not required, it is possible to record spectroscopic data near room temperature and easily monitor reduced [Fe–S] clusters. However, due to the overlap with background absorptions, e.g. water, FIR is less selective than other spectroscopic techniques such as RR, electron paramagnetic resonance (EPR), or NRVS. The sensitivity of FIR, however, can be increased by either utilizing a synchrotron source or applying difference spectroscopy in a spectroelectrochemical approach [31].

FIR absorbance spectra of three [Fe–S]-containing proteins, i.e. the rubredoxin from *Clostridium pasteurianum* (Rb), a Ferredoxin from *spinach* (Fd), and one from *C. pasteurianum* (Figure 17.3a) were recorded at an infrared synchrotron beamline using the ATR mode [31]. Rubredoxins are mono-iron-containing proteins that are classified as iron–sulfur proteins but do not contain inorganic sulfides. These proteins function as electron mediators using a single $Fe(S{-}Cys)_4$ metal site, changing its oxidation state between the +2 and +3. Spinach ferredoxin harbors a [2Fe–2S] cluster ligated with four cysteines and is utilized at low potentials as the final acceptor of electrons from Photosystem I [32], switching between +2 and +1 oxidation states. The ferredoxin from *C. pasteurianum* contains two cubane-type [4Fe–4S] clusters, each coordinated by four cysteines, changing their oxidation states from +2 to +1. This protein serves as a two-electron carrier often involved in pyruvate oxidation and hydrogen evolution in anaerobic organisms [33]. The most intense absorptions for the aforementioned three [Fe–S] proteins are found at 370, 362, and 354 cm^{-1}, respectively. These signals have been attributed (based on the corresponding RR band assignment) to the coupled bridging and terminal Fe–S stretching modes, arising from the breathing vibrations of the iron centers. Besides the main absorptions mentioned earlier, additional bands are found at higher and lower energies that are characterized by a lower intensity. These have been assigned to Fe–S stretching vibrations and deformation modes of the [Fe–S] center.

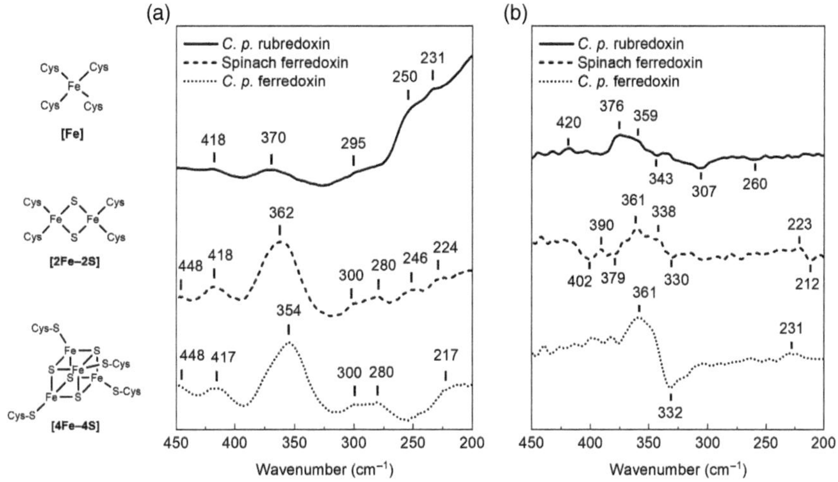

Figure 17.3 FIR spectra of exemplary [Fe–S] clusters. (a) Absorbance spectra of oxidized rubredoxin from *Clostridium pasteurianum* (solid line), spinach ferredoxin (dashed line) and the ferredoxin from *C. pasteurianum* (dotted line). (b) Corresponding oxidized-*minus*-reduced FIR difference spectra. The software WebPlotDigitizer was utilized to extract the data from the indicated references. Source: Adapted from El Koury et al. [31]. Visual source: Christian Lorent.

The corresponding potential-dependent difference spectra, recorded in transmission mode, are displayed in Figure 17.3b. All tentative band assignments for the oxidized and reduced cofactor species are summarized in Table 17.1. The various [Fe–S] cluster species exhibit distinct spectral signatures. Upon reduction, a downshift for signals related to the Fe–S modes is observed, which is consistent with an increased Fe–S bond length, due to the lowered charge at the iron(s).

17.2.3 Conclusion

MIR spectroscopy has provided valuable insights into redox and structural changes at the active sites of several metalloenzymes. Among them, the analysis of CO/CN stretching vibrations and their structural and electronic implications in, e.g. hydrogenases, CODHs, and nitrogenases has been proven particularly successful. FIR spectroscopy, on the contrary, showed limited yet important applications in the study of [Fe–S] clusters. Here, potential dependent difference spectra of typical [Fe–S] clusters give an overview on their vibrational modes in different redox states. The observed bands have been tentatively assigned with the help of Raman spectroscopic data and theoretical approaches.

17.3 Raman Spectroscopy

In 1928, C. V. Raman and K. S. Krishnan provided experimental evidence for the frequency-shifted scattering of light [34]. This phenomenon is now known as the

Table 17.1 Tentative assignments of FIR and RR bands for redox-sensitive metal–ligand modes detected between 450 and 200 cm^{-1} in samples from a $Fe(S-Cys)_4$ center in *Clostridium pasteurianum* rubredoxin, a [2Fe–2S] cluster in *spinach* ferredoxin, and two [4Fe–4S] clusters in *C. pasteurianum* ferredoxin.

C. pasteurianum rubredoxin	Spinach ferredoxin	*C. pasteurianum* ferredoxin	Assignments
420 (ox)			$\nu(Fe–S)^{as}/\nu(Fe–S^t)$
	402 (red)		$\nu(Fe–S^b)$
	390 (ox)		$\nu(Fe–S^b)$
376 (ox)/343 (red)	379 (red)		$\nu(Fe–S)$
359 (ox)/307 (red)	361 (ox) /338 (ox)		$\nu(Fe–S)/\nu(Fe–S^b) + \nu(Fe–S^t)$
		361 (ox)/332 (red)	$\nu(Fe–S^t)$/ mode along Fe–Fe direction
260 (red)			$\delta(Fe–S–C) + \delta(Fe–S–Fe)$
	223 (ox)		$\delta(Fe–S–C) + \delta(Fe–S–Fe)$
	212 (red)	231 (ox)	$\delta(Fe–S–C) + \delta(Fe–S–Fe)$

Abbreviations for the different types of vibrations: ν (stretching), δ (bending), as (antisymmetric), b (bridging), and t (terminal). Source: Adapted from El Koury et al. [31].

Raman effect, which is well described in a textbook by Derek A. Long [35]. In terms of classical physics, Raman scattering can be described by an induced electrical dipole moment μ_{ind}. The oscillating electric field $\vec{E}$ of the incident electromagnetic wave interacts with the electron cloud of a scattering molecule, thereby inducing an oscillating dipole in the molecule. The properties of the resulting dipole depend on the polarizability of the system, which can be expressed as a 3×3 tensor $\tilde{\alpha}$.

$$\vec{\mu}_{ind} = \tilde{\alpha}\vec{E} \tag{17.11}$$

Accordingly, the elements of the polarizability tensor contain the Cartesian components of the induced dipole moment and the electric field. Since the vibrating atoms in a molecule change the electron distribution with the time t, the polarizability with respect to a normal mode frequency ν_k can be written as follows:

$$\tilde{\alpha}(\nu) = \tilde{\alpha}_0(\nu_0) + 2\pi\nu_k t\left(\frac{\partial\tilde{\alpha}}{\partial Q_k}\right)_0 \tag{17.12}$$

The combination of Eq. (17.11) with (17.12) and the incorporation of the field vector $\vec{E}_0$ and frequency ν_0 of the incoming light then finally yield

$$\begin{aligned} \vec{\mu}_{ind} &= \vec{E}_0\tilde{\alpha}_0\cos(2\pi\nu_0 t) && \text{Rayleigh scattering} \\ &+\vec{E}_0\left(\frac{\partial\tilde{\alpha}}{\partial Q_k}\right)_0 Q_k\cos[2\pi(\nu_0+\nu_k)t] && \text{anti – Stokes Raman scattering} \\ &+\vec{E}_0\left(\frac{\partial\tilde{\alpha}}{\partial Q_k}\right)_0 Q_k\cos[2\pi(\nu_0-\nu_k)t] && \text{Stokes Raman scattering} \end{aligned} \tag{17.13}$$

The first term of this equation refers to the elastic Rayleigh scattering, where the incoming and scattered radiation have the same frequency (Figure 17.1). The other two terms represent the inelastic Raman scattering. Anti-Stokes Raman scattering appears if the frequency of the scattered radiation is higher than that of the incident light. Thus, the molecule remains in a vibrational state lower in energy than the initial state. In contrast, if the frequency of the scattered light is lower than that of the incident radiation, the system is transferred to a vibrationally excited state. This process is called Stokes Raman scattering. The aforementioned terms have to be different from zero to fulfill the underlying selection rule of Raman spectroscopy. Consequently, the polarizability changes with the normal coordinate.

A deeper comprehension of the Raman scattering is provided by a quantum mechanical treatment. According to an approach by G. Placzek, based on the dispersion theory of H. A. Kramers, W. K. Heisenberg, and P. M. A. Dirac, a two-photon process can be described using the second-order perturbation theory [36–38]. Accordingly, the scattering tensor is the sum of the connecting vibronic states $|r\rangle$ and is given by

$$(\alpha_{\rho\sigma})_{fi} = \frac{1}{h}\sum_{r\neq i,f}\left(\frac{\langle f|\hat{M}_\rho|r\rangle\langle r|\hat{M}_\sigma|i\rangle}{\nu_r - \nu_i - \nu_0 - i\Gamma_r} + \frac{\langle f|\hat{M}_\sigma|r\rangle\langle r|\hat{M}_\rho|i\rangle}{\nu_r - \nu_f + \nu_0 + i\Gamma_r}\right) \tag{17.14}$$

The indices ρ and σ refer to the Cartesian components of the transition polarizability $(\alpha_{\rho\sigma})_{fi}$, a second rank tensor. $\hat{M}_\rho$ and $\hat{M}_\sigma$ denote the respective electronic transition dipole moments, while Γ_r is the so-called damping rate, which characterizes the lifetime of $|r\rangle$. As stated in Eq. (17.14), the transition probabilities of all vibronic states contribute to the transition polarizability and therefore to the Raman intensity.

17.3.1 Resonance Raman Spectroscopy

If the incident radiation approaches the frequency of an electronic transition in a molecule (Figure 17.1), the intensity of certain vibrations of the chromophore is enhanced. This phenomenon is called the RR effect. In this case, Eq. (17.14) simplifies, since the second term becomes very small and can be neglected. Additionally, the application of the Born–Oppenheimer approximation enables the separation of the vibrational and electronic wave functions, which yields

$$(\alpha_{\rho\sigma})_{fi} = \frac{1}{h}\sum_{r\neq i,f}\left(\frac{\langle \upsilon_f \mid \upsilon_r\rangle\langle \upsilon_r \mid \upsilon_i\rangle M_{gr,\rho}M_{gr,\sigma}}{\nu_r - \nu_i - \nu_0 - i\Gamma_r}\right) \tag{17.15}$$

The integrals $\langle \upsilon_f \mid \upsilon_r\rangle$ and $\langle \upsilon_r \mid \upsilon_i\rangle$ are the Franck–Condon factors, while the electric dipole moments $M_{gr,\rho}$ and $M_{gr,\sigma}$ correspond to the electronic transition from the ground (g) to the resonant excited state (r). The latter components can be expanded in a Taylor series regarding the normal coordinate Q_k. According to the work of A. C. Albrecht, several terms may contribute to the scattering tensor [35, 39].

$$(\alpha_{\rho\sigma})_{fi} = A + B + C + D \tag{17.16}$$

All these components comprise different scattering mechanisms. However, usually the A- and the B-terms are dominant. The A-term can be expressed as follows:

$$A \cong \frac{1}{h}\sum_r \left(\frac{\langle v_f \mid v_r\rangle\langle v_r \mid v_i\rangle M^0_{gr,\rho} M^0_{gr,\sigma}}{\nu_r - \nu_i - \nu_0 - i\Gamma_r} \right) \tag{17.17}$$

Here, the electric dipole moments $M^0_{gr,\rho}$ and $M^0_{gr,\sigma}$ refer to a vertical electronic transition. The scattering tensor is dominated by the A-term if the electronic transition has a substantial oscillator strength, which leads to a large transition dipole moment. In this scenario, the resonance enhancement mostly depends on the product of the Franck–Condon factors $\langle v_f \mid v_r\rangle\langle v_r \mid v_i\rangle$ and thus is influenced by the geometry of the resonant electronic excited state. The B-term enhancement can be considered for weaker electronic transitions (see below).

$$B \cong \frac{1}{h}\sum_r \left(\frac{\langle v_f|Q_k|v_r\rangle\langle v_r \mid v_i\rangle \left(\frac{\partial M_{gr,\rho}}{\partial Q_k}\right)_0 M^0_{gr,\sigma}}{\nu_r - \nu_i - \nu_0 - i\Gamma_r} + \frac{\langle v_r|Q_k|v_i\rangle\langle v_f \mid v_r\rangle \left(\frac{\partial M_{gr,\sigma}}{\partial Q_k}\right)_0 M^0_{gr,\rho}}{\nu_r - \nu_f - \nu_0 - i\Gamma_r} \right) \tag{17.18}$$

The change of the electronic transition dipole moment with respect to the normal coordinate $\frac{\partial M_{gr}}{\partial Q_k}$ enables the so-called Herzberg–Teller vibronic coupling between the electronic ground and excited state(s). The B-term enhancement can be significant when the oscillator strength is "borrowed" via strong vibronic coupling from a nearby strong electronic transition.

17.3.2 Studying [Fe–S] Clusters by Resonance Raman Spectroscopy

Iron–sulfur clusters have been extensively characterized using RR spectroscopy. To observe resonance enhancement, the energy of the laser used for excitation needs to be similar in energy to an allowed electronic transition of the cluster (Figure 17.1). [Fe–S] clusters usually exhibit strong S → Fe ligand to metal charge transfer (LMCT) transitions, which appear as broad absorptions (400–600 nm) in the UV/Vis spectra (Figure 17.4). In general, the magnitude of the resonance enhancement is strongly dependent on the energy and intensity of the electronic transition as well as the coupling to the normal modes of the metal cluster. Due to the strong dipole moment of the LMCT transitions in [Fe–S] clusters, the A-term mechanism is considered the dominant one, enhancing the low-frequency stretching vibrations of the Fe–S bonds observed between 200 and 450 cm^{-1} [53, 54]. The characteristic spectral signatures of the UV/Vis and RR spectra for conventional [Fe–S] clusters are illustrated below.

17.3.3 Rubredoxin

Rubredoxins represent the simplest structure of an iron–sulfur cluster, comprising a single iron ion coordinated by four cysteines in a tetrahedral arrangement. This metal site in the oxidized form exhibits three intense absorption bands in the UV/Vis spectrum (Figure 17.4a), as exemplified by data recorded using the isolated protein from thermophilic green sulfur bacterium *Chlorobaculum tepidum*, which functions as an electron acceptor for pyruvate ferredoxin oxidoreductase [40]. If the iron center is reduced to the ferrous Fe(II) state, no absorption bands are present in the visible range of the spectrum as the energy of the electronic transitions increases and shifts in the UV region. For the same reason, a strong RR enhancement of the Fe–S bands, using lasers with wavelengths in the visible range, is only observed for the ferric Fe(III) state. The RR spectrum of oxidized rubredoxin from *Desulfovibrio gigas* (Figure 17.4b) is dominated by the intense band at 314 cm^{-1}, which has been assigned to the breathing mode of the tetrahedrally coordinated iron center. The much less intense bands at 348, 363, and 376 cm^{-1} were identified as asymmetric stretching vibrations based on ^{34}S isotopic labeling [41].

17.3.4 [2Fe–2S] Clusters

The diiron center of ferredoxins containing [2Fe–2S] clusters are usually coordinated by four cysteines, two of which can be replaced by one or two histidines, forming the NEET-type and Rieske-type [2Fe–2S] clusters, respectively. Under oxidizing conditions, both iron ions are in the ferric Fe(III) state. The UV/Vis spectrum of the oxidized all-Cys coordinated [2Fe–2S] cluster in spinach ferredoxin features two absorption bands at 423 and 465 nm in the visible range (Figure 17.4c) [55]. Reduction of the cluster by one electron yields the paramagnetic mixed valence Fe(II)Fe(III) state, which still exhibits clear UV/Vis bands, while the super-reduced Fe(II)Fe(II) state is mostly featureless in the visible range. In line with the previously introduced A-term for the resonance enhancement, the presence of strong electronic transitions in the visible region enables resonant excitation and enhancement. In contrast to rubredoxins, RR spectra of one-electron reduced [2Fe–2S] clusters can therefore be detected but the resonance enhancement can be up to one order of magnitude less. The main band in the RR spectrum of reduced spinach ferredoxin (Figure 17.4d) appears at 325 cm^{-1} and was correlated to the terminal Fe(III)-Cys stretching vibration. Obviously, the relative intensities of the different vibrational Fe–S modes are highly dependent on the wavelength of the exciting laser, so a comparison of relative band intensities should be ideally done using the same excitation line.

Contrary to the spinach ferredoxin, the UV/Vis spectrum of oxidized [2Fe–2S](His)$(Cys)_3$ cluster in the outer mitochondrial membrane protein mitoNEET exhibits bands at 460 and 540 nm, similar to the [2Fe–2S]$(His)_2(Cys)_2$ cluster in the Rieske protein from *Thermus thermophilus*, which exhibits absorptions at 458 and 560 nm (Figure 17.4e) [44, 46]. The RR spectra of all three [2Fe–2S] cluster types show quite different vibrational patterns (Figure 17.4f), underlying their different ligations. The intense bands at 283 and 394 cm^{-1} in the RR spectrum of oxidized spinach ferredoxin originate from the bridging Fe–S stretching vibrations from inorganic sulfur. Less intense features at 329, 336, and 366 cm^{-1}, on the contrary, have been assigned to terminal Fe–S stretching vibrations involving the cysteine residues [45, 56]. The (His)$(Cys)_3$ coordination in mitoNEET creates a rather labile [2Fe–2S] cluster and causes obvious alterations of the geometry

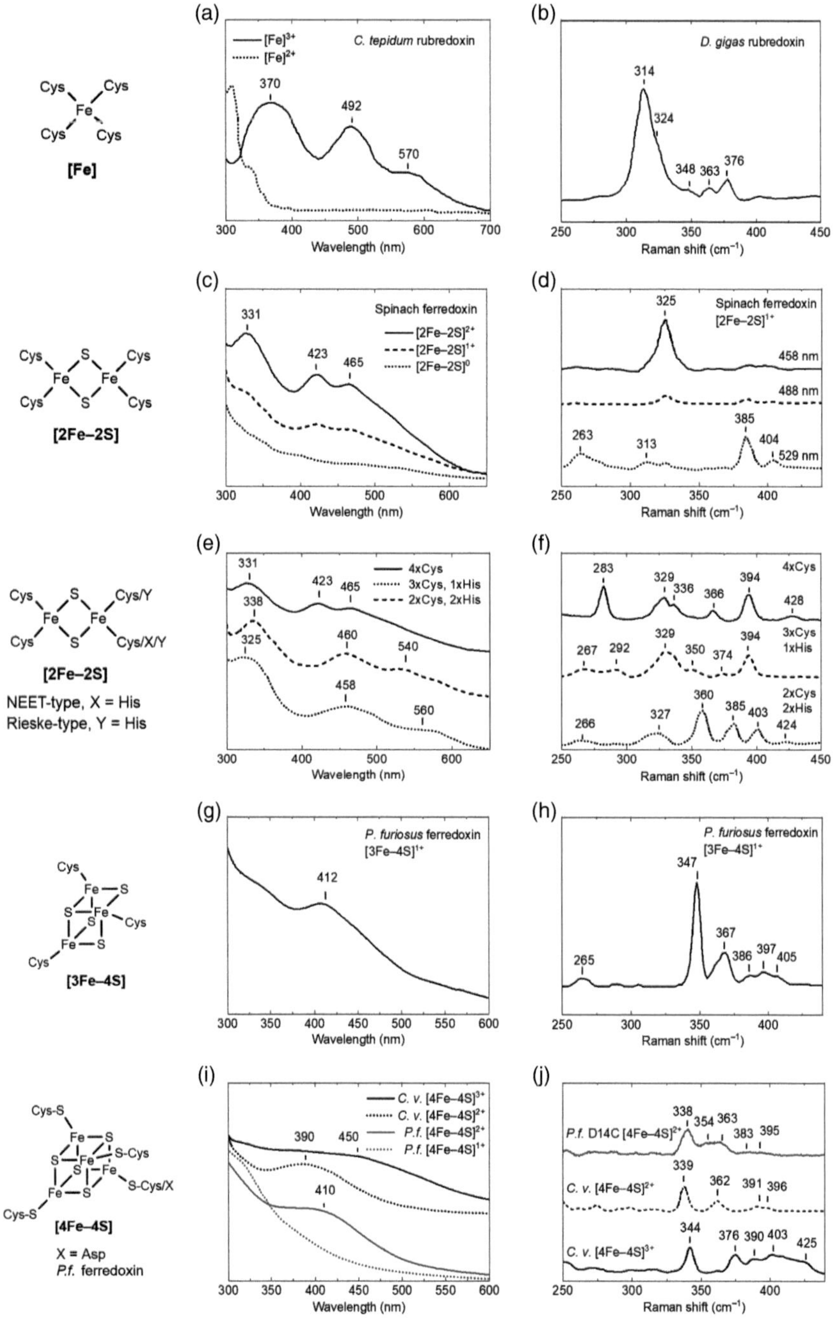
[Fe]
[2Fe–2S]
[2Fe–2S]
NEET-type, X = His
Rieske-type, Y = His
[3Fe–4S]
[4Fe–4S]
X = Asp
P.f. ferredoxin
(a)
C. tepidum rubredoxin
[Fe]3+
[Fe]2+
Wavelength (nm)
(b)
D. gigas rubredoxin
Raman shift (cm−1)
(c)
Spinach ferredoxin
[2Fe–2S]2+
[2Fe–2S]1+
[2Fe–2S]0
Wavelength (nm)
(d)
Spinach ferredoxin
[2Fe–2S]1+
458 nm
488 nm
529 nm
Raman shift (cm−1)
(e)
4xCys
3xCys, 1xHis
2xCys, 2xHis
Wavelength (nm)
(f)
4xCys
3xCys 1xHis
2xCys 2xHis
Raman shift (cm−1)
(g)
P. furiosus ferredoxin
[3Fe–4S]1+
Wavelength (nm)
(h)
P. furiosus ferredoxin
[3Fe–4S]1+
Raman shift (cm−1)
(i)
C. v. [4Fe–4S]3+
C. v. [4Fe–4S]2+
P.f. [4Fe–4S]2+
P.f. [4Fe–4S]1+
Wavelength (nm)
(j)
P.f. D14C [4Fe–4S]2+
C. v. [4Fe–4S]2+
C. v. [4Fe–4S]3+
Raman shift (cm−1)

Figure 17.4 UV/Vis and RR spectra of [Fe–S] clusters of various nuclearity. (a) UV/Vis spectra of oxidized (solid line) and reduced (dotted line) rubredoxin from *Chlorobaculum tepidum* [40]. (b) RR spectrum of the oxidized rubredoxin from *Desulfovibrio gigas* was recorded using a 497 nm excitation line [41]. (c) UV/Vis spectra of the oxidized (solid line), reduced (dashed line), and super-reduced (dotted line) [2Fe–2S] cluster from the spinach ferredoxin [42]. (d) RR spectra of reduced spinach ferredoxin were recorded using 458 (solid line), 488 (dashed line), and 529 nm (dotted line) lines [43]. (e) UV/Vis spectra of the different oxidized [2Fe–2S] clusters exhibiting 4Cys (spinach ferredoxin, solid line), 3Cys:1His (NEET-type, dashed line), and 2Cys:2His (Rieske type from *Thermus thermophilus*, dotted line) ligations [44]. (f) RR spectrum of the oxidized [2Fe–2S] from spinach ferredoxin (solid line) was recorded using a 458 nm line, mitoNEET (dashed line) recorded using a 514 nm excitation, and a Rieske protein from *Thermus thermophilus* (dotted line) recorded using a 488 nm line [45, 46]. (g) UV/Vis spectrum of the oxidized [3Fe–4S] cluster of the ferredoxin from *Pyrococcus furiosus* [47]. (h) RR spectrum of the oxidized [3Fe–4S] cluster of the ferredoxin from *Pyrococcus furiosus* recorded using a 458 nm line [47, 48]. (i) UV/Vis spectra of the superoxidized 3+ (black, solid) and oxidized 2+ (black, dotted) [4Fe–4S] cluster from the HiPIP from *Chromatium vinosum* [49] and the oxidized 2+ (gray, solid) and reduced 1+ (gray, dotted) [4Fe–4S] cluster of the ferredoxin from *Pyrococcus furiosus* [50]. (j) RR spectra of the superoxidized (black, solid) and oxidized (black, dashed) [4Fe–4S] clusters from the HiPIP from *Chromatium vinosum* (488 nm line) [51] and the oxidized (gray, solid) [4Fe–4S] cluster of the D14C variant ferredoxin from *Pyrococcus furiosus* (458 nm line) [47, 49, 50, 52]. The software WebPlotDigitizer was utilized to extract the data from the indicated references. To improve readability and avoid overlay of the spectra, the UV/Vis data in panels e and i are separated by a constant offset and the RR band intensities in panels f and j are scaled. Visual source: Christian Lorent (co-author).

and symmetry of the metal site, thereby changing the vibrational spectrum. Most prominently, no intense band from a bridging Fe–S stretching vibration is present below 300 cm^{-1}. Instead, at least three overlapping bands were detected between 250 and 300 cm^{-1}. These signals exhibit a strong pH dependency, possibly depending on the local structure of the Fe–His bond [46]. The [2Fe–2S] cluster ligated by two histidines in the Rieske protein from *Thermus thermophilus* features like mitoNEET a lower level of symmetry, which results in spectral changes in the Raman spectrum [44]. The RR spectra exhibit a significant pH dependence, with marked changes at basic conditions. Notably, the low-frequency band at 267 cm^{-1} could be assigned to the Fe(III)–N(His) stretching motion. This band disappears upon reduction of the metal site, indicating that the additional electron plausibly resides at the histidine-ligated iron [42].

17.3.5 [3Fe–4S] Clusters

[3Fe–4S] are usually observed upon removal of one iron from oxidized [4Fe–4S] clusters as observed in several ferredoxins like Fd II from *Desulfovibrio gigas*, Fd III from *Desulfovibrio africanus*, and the ferredoxin from *Pyrococcus furiosus*. However, [3Fe–4S] clusters are also an integral part of a few electron transfer chains as in the case of [NiFe]-hydrogenases [8]. These metal clusters usually show one redox transition (+1/0) within the physiological range. Notably, the "super-" (−1) and "hyper-reduced" all-ferrous state form (−2) have been enriched for several ferredoxins from *Sulfolobus acidocaldarius*, *Desulfovibrio africanus*, and *Azotobacter vinelandii*. The 1+, 0, and 1− forms exhibit a clear absorption band in the visible range between 400 and 450 nm with their relative intensities systematically decreasing upon reduction events at the [3Fe–4S] center [57]. The UV/Vis spectrum of the oxidized $[3Fe–4S]^{1+}$ ferredoxin cluster from the thermophilic archaeon *Pyrococcus furiosus* exhibits one broadband at 412 nm (Figure 17.4g). In the corresponding RR spectrum, a weak band at 265 cm^{-1} and an intense one at 347 cm^{-1} have been assigned to modes from bridging sulfur atoms, while bands above 347 cm^{-1} are associated with vibrations from the terminal cysteine ligands (Figure 17.4h) [48, 58]. In line with a weaker electronic transition, RR spectra of the (one electron reduced) $[3Fe–4S]^0$ cluster from the same species exhibit a weak enhancement of the Fe–S bands. The spectral signature is remarkably different from the oxidized form and shows stronger enhanced bridging Fe–S vibrations at 245, 271, 287, 368, and 379 cm^{-1} and weakly enhanced signals from terminal stretching modes at 319, 334, and 349 cm^{-1} [59].

17.3.6 [4Fe–4S] Clusters

Cuban-type [4Fe–4S] clusters are probably among the most abundant types of iron–sulfur clusters in biological systems. Very often they are an essential part of electron transfer chains in complex proteins or act as intermolecular electron mediators (e.g. ferredoxins) as a crucial element of the biological redox chemistry. Usually, [4Fe–4S] clusters show a 2+ to 1+ net charge redox transition at rather low potentials, while a specific class of so-called high potential iron–sulfur proteins (HiPIP) typically exhibit a physiological 3+/2+ redox transition. It was proposed that hydrogen bonds involving the protein (amide) backbone may be the origin for the stabilization of the high instead of the low potential transition [51]. The UV/Vis spectrum of the fully oxidized $[4Fe–4S]^{3+}$ cluster in the HiPIP from *Chromatium vinosum* is dominated by a very broad absorption at 450 nm (Figure 17.4i) [49]. Upon one electron reduction, a new band at 390 nm appears, which resembles the absorption at 410 nm observed for the diamagnetic $[4Fe–4S]^{2+}$ cluster in the ferredoxin from *Pyrococcus furiosus* [50]. No UV–Vis absorptions can be detected in the visible range for $[4Fe–4S]^{1+}$ species. Consequently, no strong RR enhancement is expected and to the best of our knowledge unambiguous RR spectra for a $[4Fe–4S]^{1+}$ cluster have not been detected so far. The oxidized $[4Fe–4S]^{2+}$ state on the other hand exhibits clear RR bands (Figure 17.4j). Exemplarily, the [4Fe–4S] cluster in the D14C variant of the ferredoxin from *Pyrococcus furiosus* is coordinated by four cysteines instead of three cysteines and one aspartate compared to the native protein. For this prototypical cluster, the main RR band at 338 cm^{-1} and a weaker feature at 383 cm^{-1} originate from bridging sulfur vibrations. The other bands between 350 and 400 cm^{-1} are related to terminal Fe–S(Cys) modes [52]. Like the UV/Vis, the RR spectrum of the one electron reduced HiPIP from *Chromatium vinosum* is very similar to the oxidized ferredoxin from *Pyrococcus furiosus*, indicating a similar electronic and geometric structure. When the HiPIP is oxidized to the 3+ state, a general small upshift of the characteristic modes is observed, which can be rationalized by the overall shortening of the Fe–S bonds due to an increased charge at the irons. In addition, the Fe–S modes at higher energy significantly increase in intensity.

17.3.7 Studying Hydrogenases by Resonance Raman Spectroscopy

[NiFe]- and [FeFe]-hydrogenases incorporate specialized (hetero)bimetallic centers for the hydrogen conversion as well as conventional [Fe–S] clusters. Both types of metal cofactors give rise to characteristic Raman bands. Cofactor-specific resonance enhancement is therefore crucial to selectively probe the active sites of [NiFe]- or [FeFe]-hydrogenases. By tuning the wavelength of the incident laser radiation, the maximum enhancement of the normal modes associated with metal-ligand vibrations from the active site, mostly Fe–CO and Fe–CN in the spectral region between 400 and 700 cm^{-1}, can be achieved. The underlying electronic transitions from metal to ligand charge transfer (MLCT), e.g. Fe → CO are rather weak, so a B-term mechanism likely contributes significantly to the total resonance enhancement. The monitored Fe–CO/CN modes are characteristic of structurally and electronically defined redox states (see also Chapter 15) and have been used to study the enzymatic reaction cycles of hydrogenases [13, 60].

The sharp RR bands in the spectrum of the regulatory [NiFe]-hydrogenase from *Cupriavidus necator* (*Cn*RH) are associated with complex Fe–CN/CO vibrations from the active site (Figure 17.5a). These bands are characteristic of the hydrogen binding intermediate, which has an open coordination site in the bridging position between the Ni(II) and Fe(II) ions. Under reducing conditions, these bands exhibit a specific upshift of 1–13 cm^{-1}, typical for a more reduced redox intermediate with a Ni(I)Fe(II) configuration. At higher excitation wavelengths (568 and 647 nm), even Ni–S(Cys) vibrations can be observed in the region between 250 and 400 cm^{-1}. Since *Cn*RH incorporates three [4Fe–4S] clusters as an electron transfer chain,

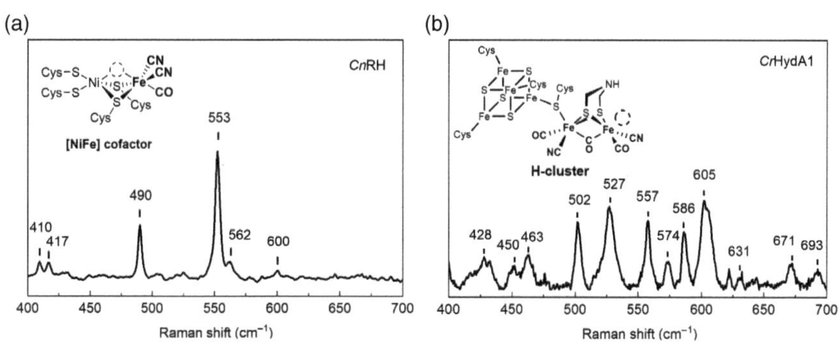

Figure 17.5 RR spectra of hydrogenases. (a) RR spectrum of the regulatory [NiFe]-hydrogenase from *Cupriavidus necator* was recorded using a 568 nm excitation line [13]. (b) RR spectrum of the [FeFe]-hydrogenase HydA1 from *Chlamydomonas reinhardtii* was recorded using 488 nm excitation [60]. For both hydrogenases, the shown RR spectra characterize active site species with a vacant coordination site, ready to bind hydrogen or inhibitors. Visual source: Christian Lorent (co-author).

this spectral window is usually dominated by intense Fe–S vibrational bands from oxidized clusters, when using lasers with lower wavelength for excitation (406–488 nm) [13].

The RR spectrum of the [FeFe]-hydrogenase from the photosynthetic alga *Chlamydomonas reinhardtii* (*Cr*HydA1) shows a much more complex pattern in the Fe–CN/CO region (Figure 17.5b), which is expected based on the larger number and more diverse coordination of the CO and CN^- ligands compared to [NiFe]-hydrogenases. The observed marker bands represent the spectroscopic signature of the H_2-binding catalytic state of the active site (H-cluster), characterized by a mixed-valence Fe(I)Fe(II) site and an oxidized $[4Fe–4S]^{2+}$ cluster. As evidenced by a high intensity of the [4Fe–4S] cluster-related modes and supported by IR spectroscopy (see Section 17.1), RR provided also insights into another redox intermediate characterized by conserved Fe(I)Fe(II) configuration at the [2Fe] center, but a reduced $[4Fe–4S]^{1+}$ cluster [60].

17.3.8 Conclusion

RR spectroscopy is undoubtedly an important technique to study iron–sulfur clusters. The resonance enhancement, however, relies on strong electronic transitions in the visible spectral range. Based on the characteristic Fe–S vibrations, the identity, geometry, and electronic structure of single- or multicofactor-containing biological systems can be derived. Recently, RR spectroscopy was also introduced as a powerful tool to investigate the active sites of [NiFe]- and [FeFe]-hydrogenases, enabling direct access to explore the structure and function of several catalytic intermediates.

17.4 Mössbauer-Based Spectroscopic Techniques

17.4.1 Mössbauer Spectroscopy

Mössbauer spectroscopy has been used to investigate iron–sulfur proteins already a few years after Rudolf L. Mößbauer observed the phenomenon of what he called “nuclear resonance absorption of γ-radiation,” later known as the Mössbauer effect [61, 62]. Among the early work was the investigation of [2Fe–2S] ferredoxin from spinach by Johnson and Hall [63] followed by the reports on [4Fe–4S] ferredoxin [64, 65] and the iron–sulfur protein rubredoxin in its oxidized and reduced forms [66, 67]. Since then, many Mössbauer studies on iron–sulfur proteins have also been performed [68, 69] on enzymes having more complex [Fe–S] sites like hydrogenases [70–73] and nitrogenases [74, 75].

Mössbauer spectroscopy is a technique that relies on energetic shifts of nuclear transitions induced by the electron shell of the Mössbauer atom. The shifts of the nuclear transition energies are measured with a very high precision of some neV. This makes it possible to detect changes in the electronic structure of the Mössbauer atom, because the nuclei and the electrons exhibit very small yet detectable interactions, the so-called hyperfine interactions [76]. In the case of iron, the relevant isotope for Mössbauer spectroscopy is ^{57}Fe with a natural abundance of 2.1%. ^{57}Fe Mössbauer spectroscopy detects all ^{57}Fe in the sample, which is a big advantage compared, e.g. with EPR spectroscopy [77], which relies on unpaired electron spins and can therefore not detect diamagnetic iron–sulfur centers and many integer spin systems. For that reason, Mössbauer spectroscopy is still a key technique to investigate [Fe–S] proteins [78]. Nevertheless, it should be mentioned that it is very often advantageous to use both Mössbauer spectroscopy and EPR in a complementary fashion, especially when iron–sulfur sites with half-integer spins are investigated [79]. This strategy has been applied to many iron–sulfur proteins and helped to identify and characterize phenomena such as [Fe–S] cluster conversion/degradation [80], and the binding site of substrates such as citrate for the [4Fe–4S] cluster of aconitase [81].

Based on experiments with ^{191}Ir foils, Rudolf L. Mößbauer discovered that there is a certain probability for the recoilless emission and absorption of γ-quanta [61]. This phenomenon was later observed for ^{57}Fe, which majorly advanced the application of the technique in many fields. Qualitatively, the effect can be explained by a simple image: A γ-quantum with the energy E_γ hits a nucleus, which has an excited nuclear state with an energy E. After absorption of a γ-quantum, the recoil energy of the absorbing nucleus, once embedded in a solid, is absorbed by the entire solid. Thus, the expression for the recoil energy E_R of the nucleus Eq. (17.19) no longer contains the mass M of the nucleus, but the mass of the entire solid M_solid:

$$E_\mathrm{R} = \frac{E_\gamma^2}{2Mc^2} \approx \frac{E^2}{2M_{solid}c^2} \cong 0 \tag{17.19}$$

Since $M_{solid} \gg M$ the recoil energy E_R approaches zero for solids. In terms of quantum mechanics, there is a probability for a zero-phonon process, in which no lattice oscillations are excited. Rudolf L. Mößbauer verified that the probability for a zero-phonon process is given by the value later referred to as the Lamb–Mössbauer factor f.

$$f = \exp\left\{ -k^2 < x^2 > \right\} \tag{17.20}$$

Hereby, $<x^2>$ denotes the mean square displacement of the Mössbauer nucleus in the direction of the incoming γ-ray and $k = 2\pi/\lambda$ is the wavenumber of the γ-ray. The decrease in the mean square displacement with decreasing temperature leads to an increase in the Lamb–Mössbauer factor and thus the Mössbauer effect.

For ^{57}Fe-Mössbauer spectroscopy in transmission mode, the γ-rays emitted by a radioactive ^{57}Co source pass through a defined sample and a detector measures the intensity of the beam transmitted. This technique is successfully applied to characterize powders, foils, and frozen solutions. To record a Mössbauer spectrum, the energy of the emitted radiation is modulated, according to the Doppler effect, by moving the source relative to the absorber at a speed v. In the resulting spectra, the γ-ray intensity is plotted as a function of v.

Three parameters are usually used to describe the nuclear interactions: the isomer shift, the quadrupole splitting, and the magnetic hyperfine interaction.

Table 17.2 Isomer shifts and quadrupole splitting of selected iron–sulfur sites in proteins.

Type	Formal valence	δ (mms^{-1})	ΔE_Q (mms^{-1})	Total spin	Reference
$[1Fe]^{3+}{}_{rubredoxin}$	Fe^{3+}	0.32	0.5	5/2	[66]
$[1Fe]^{2+}{}_{rubredoxin}$	Fe^{2+}	0.70	3.25	2	[66]
$[2Fe–2S]^{2+}{}_{Rieske}$	$Fe^{3+}–Fe^{3+}$	0.24, 0.32	0.52, 0.91	0	[44]
$[2Fe–2S]^{2+}{}_{4Cys}$	$Fe^{3+}–Fe^{3+}$	0.27	0.60	0	[82]
$[2Fe–2S]^{1+}{}_{Rieske}$	$Fe^{3+}–Fe^{2+}$	0.31, 0.74	0.63, 3.25	1/2	[44]
$[2Fe–2S]^{1+}{}_{4Cys}$	$Fe^{3+}–Fe^{2+}$	0.35, 0.60	0.65, 2.70	1/2	[82]
$[3Fe–4S]^{0}$	$Fe^{2.5+}–Fe^{2.5+}$	0.46	1.47	2	[83]
	Fe^{3+}	0.32	0.52		
$[4Fe–4S]^{1+}$	$Fe^{2.5+}–Fe^{2.5+}$	0.50	1.32	1/2	[65]
	$Fe^{2+}–Fe^{2+}$	0.58	1.89		
$[4Fe–4S]^{2+}$	$Fe^{2.5+}–Fe^{2.5+}$	0.42	1.26	0	[84]
	$Fe^{2.5+}–Fe^{2.5+}$	0.42	1.26		

The isomer shift δ is proportional to the s-electron density at the iron nuclei in the absorber $|\phi(0)_{absorber}|^2$ and the source $|\phi(0)_{source}|^2$ and can be expressed as follows:

$$\delta = \frac{4}{5}\pi Ze^2R^2\frac{\delta R}{R}\left[|\phi(0)_{absorber}|^2 - |\phi(0)_{source}|^2\right] \tag{17.21}$$

with $R = \frac{1}{2}(R_e + R_g)$ and $\delta R = R_e - R_g$, where R_g and R_e are the nuclear radii of the ground and excited state, respectively. For iron, $\delta R = R_e - R_g$ is negative and therefore a more positive isomer shift implies a lower value of $|\phi(0)_{absorber}|^2$. This may be caused either by a decrease in the total s-electron population or by shielding of the s-electrons by the 3d electrons. The isomer shift is sensitive to the oxidation state of the iron. Typical values of various iron–sulfur centers are given in Table 17.2.

The electric quadrupole moment Q of the nuclear ground state of ^{57}Fe with the nuclear spin of the ground state $I_g = 1/2$ is zero. Therefore, the quadrupole interaction does not affect the nuclear ground state but splits the first excited state with $I_e = 3/2$ into two sublevels. In the case of ^{57}Fe, it produces a two-line Mössbauer absorption pattern (Figure 17.6) with a corresponding quadrupole splitting ΔE_Q, which is given by

$$\Delta E_Q = \frac{1}{2}eQV_{zz}\sqrt{1+\frac{\eta^2}{3}} \tag{17.22}$$

Here, V_{zz} denotes the largest component of the electric field gradient (EFG) tensor V_{zz}, which is present at the ^{57}Fe nucleus. The coordinate frame of the EFG-tensor $\overleftrightarrow{V}$ is mostly chosen in such a way that $|V_{xx}| \le |V_{yy}| \le |V_{zz}|$. The asymmetry parameter is given by $\eta = (V_{xx} - V_{yy})/V_{zz}$. For a powder sample, η and the sign of ΔE_Q can only be determined, if a magnetic interaction is present in addition to the EFG at the ^{57}Fe nucleus. Both ΔE_Q and η reflect the occupation of the iron 3d orbitals as well as the ligand net charges surrounding the iron. A more detailed description can be found

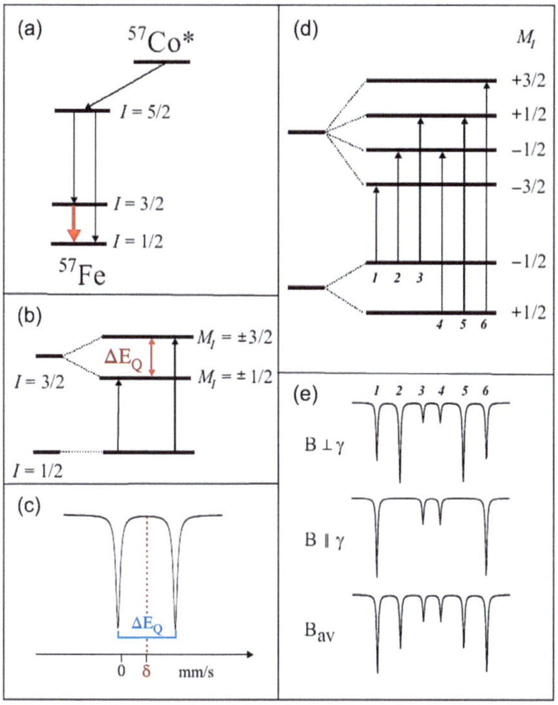

Figure 17.6 Schematic overview on Mössbauer transitions and spectra. (a) Decay scheme of a ^{57}Co source that produces the 14.4 keV radiation to excite the ^{57}Fe nuclei in the sample. (b) Mössbauer transitions in the presence of an electric field gradient leading to a two-line pattern with the quadrupole splitting in the Mössbauer spectrum shown in (c). The position of the spectrum is given by the isomer shift. In the case of a magnetic field at the ^{57}Fe nucleus, the energy levels of the ^{57}Fe nucleus split according to the Zeeman effect (d) and dipole selection rules for electromagnetic radiation allow six transitions causing a six-line absorption pattern shown in (e). Because of the dipole selection rules, the magnetic Mössbauer spectra depend on the orientations of the magnetic field relative to the propagation direction of γ-ray. The quadrupole interaction has been neglected in d and e. Source: Reprinted with permission from Krebs et al. [85]/Copyright 2005 American Chemical Society. Visual source: Volker Schünemann (co-author).

in the book of Gütlich, Bill, and Trautwein [76] and in a review on iron porphyrins by Debrunner [86].

If a magnetic field is present at the ^{57}Fe nucleus, it splits both its ground and its excited states according to the Zeeman effect. The nuclear spins couple both to the external magnetic field $\vec{B}_0$ and to the internal magnetic hyperfine field $\vec{B}_{hf}$. In the case of paramagnetic iron centers in proteins (like iron–sulfur clusters), the magnetic hyperfine interaction is often described in terms of the hyperfine coupling tensor $\overleftrightarrow{A}$. With the spin-expectation value $\langle\vec{S}\rangle$ of the corresponding electronic state, the internal magnetic hyperfine field is given, with g_N being the nuclear g-factor and μ_N the nuclear magneton as

$$\vec{B}_{\rm hf} = -\langle\vec{S}\rangle \cdot \overleftrightarrow{A}/(g_N\mu_N) \qquad (17.23)$$

In Mössbauer spectroscopy, the hyperfine coupling $\overleftrightarrow{A}$ tensor is often given in units of Tesla but also found in MHz units. For ^{57}Fe, 1 T corresponds to 1.36 MHz.

17.4.2 Mössbauer Spectroscopy from Single Fe(S–Cys)$_4$ Sites up to [4Fe–4S] Clusters

The Fe(S–Cys)$_4$ center of reduced rubredoxin displays a doublet with $\delta = 0.70$ mms^{-1} and $\Delta E_Q = 3.25$ mms^{-1} at 4.2 K (Figure 17.7a). This isomer shift is typical of high-spin ($S = 2$) ferrous iron tetrahedrally coordinated with four sulfur ligands. Although reduced rubredoxin is paramagnetic, there is no magnetic splitting observable in zero external fields. This is because the spin expectation values $\langle \vec{S} \rangle$ of the $S = 2$ systems need to be induced by external magnetic fields. This results in magnetically split Mössbauer spectra (Figure 17.7b), which can be analyzed with the spin Hamiltonian formalism to determine parameters like the zero-field splitting D and the rhombicity parameter E/D [76, 79, 85].

The Mössbauer spectrum of high-spin ($S = 5/2$) ferric iron like that in oxidized rubredoxin displays a complicated magnetic structure, because the spin expectation values of the three Kramers doublets of the $S = 5/2$ spin system are nonzero even in the zero applied field (Figure 17.7c,d). The relaxation rates of $S = 5/2$ spin systems are slow with respect to the spectroscopic Mössbauer time window given by the Lamor precession time of the nuclear spin. Therefore, in the magnetic hyperfine field, even measurements at 200 K would lead to magnetically broadened Mössbauer spectra. Only in case of significant electronic spin–spin relaxation, simple doublets can be observed. This is the case for Fe–S$_4$ model complexes. Nevertheless, even for magnetically disturbed Mössbauer spectra, the isomer shift can be determined using a simple Lorentzian line shape analysis. Oxidized rubredoxins have an isomer shift of ~0.3 mms^{-1}, which is characteristic of a tetrahedral sulfur coordination and significantly smaller than observed for octahedral N/O coordination ($\delta \sim 0.5$–0.6 mms^{-1}).

The bishistidinyl-coordinated [2Fe–2S] Rieske protein displays two doublets in a ratio of 1 : 1, caused by the difference in ligand geometry of the two iron sites (Figure 17.7e). One doublet exhibits $\delta = 0.24$ mms^{-1} and $\Delta E_Q = 0.52$ mms^{-1} and has been assigned to the Fe^{3+} site, tetrahedrally coordinated by two bridging sulfides and two cysteine thiolates. The bishistidinyl coordination of the second Fe^{3+} site leads to a doublet with an increase in both the isomer shift and quadrupole splitting ($\delta = 0.32$ mms^{-1} and $\Delta E_Q = 0.91$ mms^{-1}). Magnetic exchange coupling between the two high spin ferric sites leads to a diamagnetic ground state. Thus, no magnetic splitting in the Mössbauer spectrum is observed in low field, while using high external fields a magnetic splitting can be seen (Figure 17.7f). It should be noted that in such a case an analysis with a sextet of Lorentzian line shape is not sufficient since powder averaging has to be performed by the simulation of a Mössbauer spectrum for every orientation of the EFG tensor, with respect to the external field. Software packages allowing the full diagonalization of the nuclear Hamiltonian are available for these analyses [89, 90]. The Mössbauer spectrum of cysteine-ligated oxidized $[2Fe–2S]^{2+}$ ferredoxin exhibits a single doublet with $\delta = 0.27$ mms^{-1} and

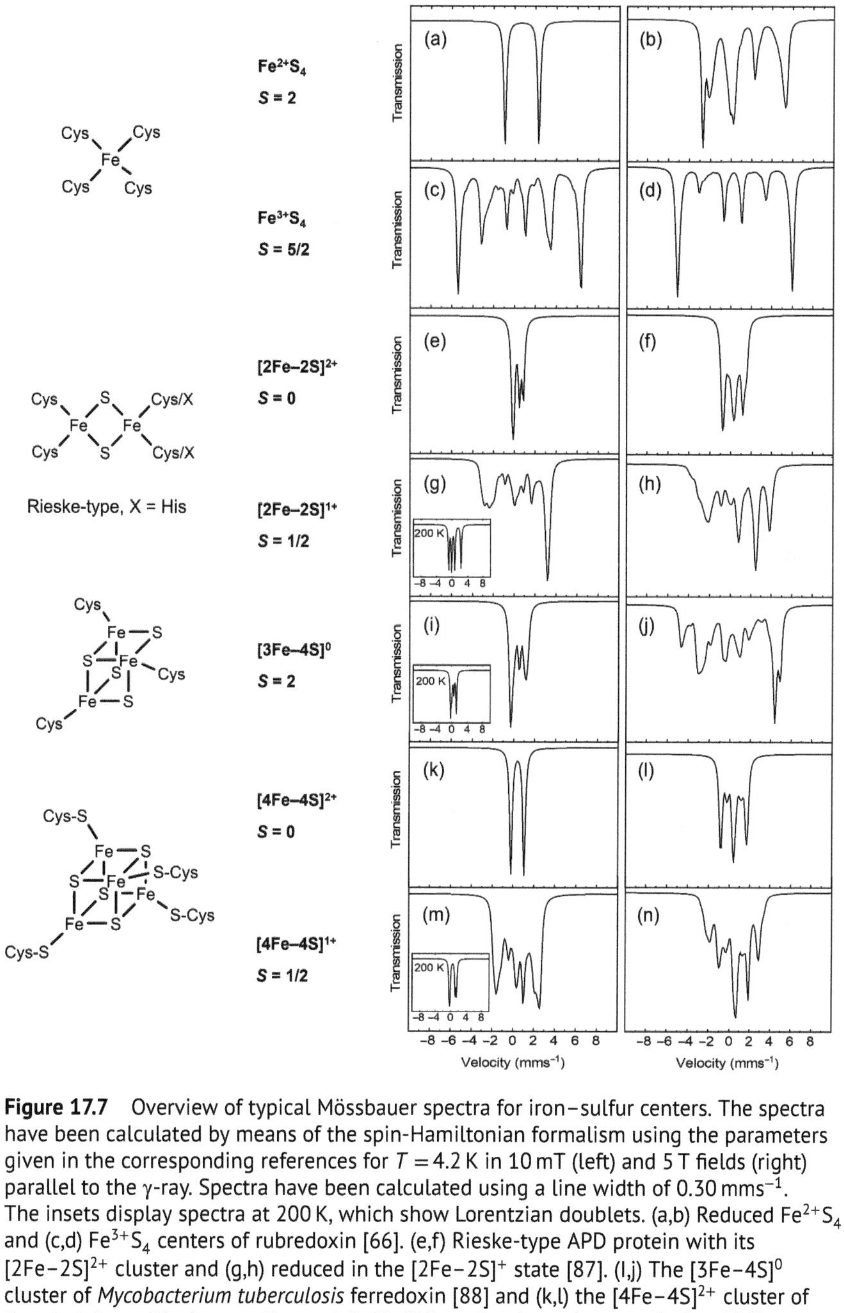

Figure 17.7 Overview of typical Mössbauer spectra for iron–sulfur centers. The spectra have been calculated by means of the spin-Hamiltonian formalism using the parameters given in the corresponding references for $T = 4.2$ K in 10 mT (left) and 5 T fields (right) parallel to the γ-ray. Spectra have been calculated using a line width of 0.30 mms^{-1}. The insets display spectra at 200 K, which show Lorentzian doublets. (a,b) Reduced $Fe^{2+}S_4$ and (c,d) $Fe^{3+}S_4$ centers of rubredoxin [66]. (e,f) Rieske-type APD protein with its $[2Fe-2S]^{2+}$ cluster and (g,h) reduced in the $[2Fe-2S]^{+}$ state [87]. (I,j) The $[3Fe-4S]^{0}$ cluster of *Mycobacterium tuberculosis* ferredoxin [88] and (k,l) the $[4Fe-4S]^{2+}$ cluster of the photolyase-like iron–sulfur protein PhrB from *Agrobacterium tumefaciens* [84] in comparison to (m,n) the $[4Fe-4S]^{1+}$ cluster of reduced ferredoxin from *B. stearothermophilus* [65]. We thank Maren H. Hoock for the support in the preparation of this figure. Visual source: Volker Schünemann (co-author).

$\Delta E_Q = 0.60\,\text{mms}^{-1}$, indicating two structurally equivalent tetrahedrally sulfur coordinated Fe^{3+} sites. One electron-reduced ferredoxins and Rieske-type proteins all show a complex magnetic hyperfine pattern, caused by the magnetic hyperfine field of the $S = ½$ total spin of the cluster (Figure 17.7g,h). In contrast to single $Fe^{3+}S_4$ sites, the spin relaxation in these bimetallic centers can be significantly accelerated by increasing the temperature. At temperatures up to 230 K, two doublets with a relative area of 1 : 1 are observed with one of them displaying δ- and ΔE_Q-values as in the oxidized state, thus representing the $Fe^{3+}S_4$ site of the cluster. For $[2Fe–2S]^{1+}$ ferredoxins, the reduced $Fe^{2+}S_4$ site shows $\delta \approx 0.55–0.59\,\text{mms}^{-1}$ (at 200 K), whereas the bishistidinyl coordination in the case of the Rieske cluster leads to $\delta \approx 0.68\,\text{mms}^{-1}$ (at 230 K).

The Mössbauer spectrum of [3Fe–4S] clusters in their oxidized 1+ state shows one doublet with $\delta = 0.28\,\text{mms}^{-1}$ and $\Delta E_Q = 0.53\,\text{mms}^{-1}$ at 100 K, indicating that all three Fe^{3+} sites are structurally equivalent. Spin coupling between the individual $S = 5/2$ spins leads to a paramagnetic ground state with $S = ½$, featuring a broad magnetic Mössbauer pattern at 4.2 K. The reduced $[3Fe–4S]^0$ clusters exhibit two doublets, which are slightly broadened due to the magnetic hyperfine field caused by the $S = 2$ system (Figure 17.7i). A doublet with $\delta = 0.32\,\text{mms}^{-1}$ and $\Delta E_Q = 0.52\,\text{mms}^{-1}$ originates from the Fe^{3+} site of the cluster, while a second doublet with twice the area of the first iron site shows $\delta = 0.46\,\text{mms}^{-1}$ and $\Delta E_Q = 1.47\,\text{mms}^{-1}$. This doublet arises from a $Fe^{2+}Fe^{3+}$ pair with a delocalized surplus electron. Such a type of pair is commonly denoted as a $Fe^{2.5+}Fe^{2.5+}$ pair and in fact the corresponding isomer shift is the mean of the δ-values of a high-spin ferric and a high-spin ferrous $Fe–S_4$ site. The delocalized electron leads to parallel spin pairing and a spin state of $S = 9/2$ results from the double exchange mechanism. Antiparallel exchange coupling with the $S = 5/2$ spin of the third iron leads to a total spin of $S = 2$ and a distinct magnetically split pattern can be produced by applying fields of 1 T (or larger) at 4.2 K (Figure 17.7j).

Ferredoxins with [4Fe–4S] clusters in their 2+ state show one doublet with $\delta = 0.45\,\text{mms}^{-1}$ and thus contain two delocalized $Fe^{2.5+}Fe^{2.5+}$ pairs (Figure 17.7k). The two $S = 9/2$ pair spins couple antiparallelly, which causes a diamagnetic ($S = 0$) ground state of the cluster. When applying external magnetic fields, the diamagnetic character of the cluster can be inferred from the fact that the observed magnetic splitting in the Mössbauer spectrum is caused by the external field only (Figure 17.7l). Reduced [4Fe–4S] clusters in their 1+ state have $S = ½$. This leads to a magnetically disturbed Mössbauer spectrum, in both low and high external fields (Figure 17.7m,n). As in the case of $[2Fe–2S]^{1+}$ clusters, raising the temperature of the measurement leads to acceleration of the electronic spin relaxation rate and two doublets with an intensity ratio of 1 : 1 can be observed (inset of Figure 17.7m). For reduced ferredoxin from *B. stearothermophilus* [65], two doublets are observed already at 77 K. A doublet with $\delta = 0.59\,\text{mms}^{-1}$ and $\Delta E_Q = 1.84\,\text{mms}^{-1}$ represents the $Fe^{2+}Fe^{2+}$ pair and a doublet with $\delta = 0.49\,\text{mms}^{-1}$ and $\Delta E_Q = 1.2\,\text{mms}^{-1}$ the $Fe^{2.5+}Fe^{2.5+}$ pair of the $[4Fe–4S]^{1+}$ cluster.

Mössbauer spectroscopy also indicated that the four iron centers in the $[4Fe–4S]^{2+}$ cluster of isoprenoid synthase IspH (also called LytB) revealed that in its substrate-free form the structure of the iron–sulfur cluster is not that of a conventional ferredoxin-type $[4Fe–4S]^{2+}$ cluster. Instead, one of the iron sites displays $\delta = 0.89\,mms^{-1}$. This value is identical within experimental error to that of an unusual fourth iron site in the citrate-bound form of aconitase [91]. The Mössbauer parameters of this site are influenced by the binding of the substrate and/or inhibitor molecules, which is a clear indication that the substrate binds directly to this special iron site of the $[4Fe–4S]^{2+}$ cluster. In the substrate-free enzyme form, the coordination sphere of this special iron site is hexa-coordinated bound by three inorganic sulfur atoms from the cluster and surprisingly three water ligands. This unusual coordination of one iron site was also verified using NRVS [92].

17.4.3 Complex Iron–Sulfur Clusters: Hydrogenases and Nitrogenases

The catalytically relevant iron site of the iron–sulfur cluster-free [Fe]-hydrogenase (Hmd) from methanogenic archaea has also been characterized using Mössbauer spectroscopy. Bill and coworkers reported $\delta = 0.06\,mms^{-1}$ and $\Delta E_Q = 0.65\,mms^{-1}$ for this iron site [70]. The low isomer shift is indicative for a ferrous low-spin Fe^{2+} site, which was also confirmed by field-dependent Mössbauer spectroscopy. Interestingly, the extracted cofactor from the protein matrix had a significantly different quadrupole splitting ($\delta = 0.03\,mms^{-1}$ and $\Delta E_Q = 0.43\,mms^{-1}$), pointing to a special role of the protein matrix for structural stabilization of this site.

The hydride-bound state (H_{hyd}) of the [FeFe]-hydrogenase from *Chlamydomonas reinhardtii* has been shown to exhibit a reduced $[4Fe–4S]^{1+}$ cluster coupled to a special diferrous two iron site with a terminally bound Fe-hydride. In fact, Mulder et al. could resolve in addition to the paramagnetic Mössbauer pattern from the $[4Fe–4S]^{1+}$ cluster two doublets in a relative ratio of 1 : 1 at 4.2 K and a field of 50 mT [72]. One doublet shows $\delta = 0.23\,mms^{-1}$ and $\Delta E_Q = 1.06\,mms^{-1}$. The other one has $\delta = 0.03\,mms^{-1}$ and $\Delta E_Q = 0.68\,mms^{-1}$. Both components show diamagnetic behavior in high external fields, consistent with two ferrous low spin sites. Based on the low value of the isomer shift, it was concluded that the latter iron site is coordinated with a hydride. Research on [FeFe]-hydrogenases using ^{57}Fe-sensitive techniques was further facilitated by the development of chemical procedures enabling selective labeling of the [2Fe] subsite using the synthetic $[^{57}Fe_2(adt)(CN)_2(CO)_4]^{2-}$ compound [93]. The assembly of the active site of [NiFe]-hydrogenases has very recently been followed also by Mössbauer spectroscopy [73]. The fully assembled $NiFe(CN)_2CO$ center of the catalytic subunit of the membrane-bound O_2-tolerant [NiFe]-hydrogenase from *Cupriavidus necator* has $\delta = 0.14\,mms^{-1}$ and $\Delta E_Q = 0.62\,mms^{-1}$. For the nickel-free hydrogenase, $\delta = 0.08\,mms^{-1}$ and $\Delta E_Q = 1.02\,mms^{-1}$ were reported. Both species are attributed to Fe(II) low-spin centers, which is in line with the proposed mechanism of maturation of these enzymes comprising the initial insertion of a low-spin $Fe^{II}(CN)_2CO$ fragment into the apo-protein.

The MoFe nitrogenase protein is one of the most complex iron enzymes and harbors the eight-iron containing P-cluster in addition to the N_2-binding FeMo cofactor with seven iron and one Mo site. The possibility of selective ^{57}Fe

enrichment of the FeMo cofactor of nitrogenase allowed a detailed Mössbauer spectroscopic study performed by the Münck group [75]. An even more elegant approach developed very recently by the Suess group is the site-selective ^{57}Fe labeling of the FeMo cofactor itself [74]. They found that iron sites of the FeMo cofactor numbered 6 and 7 both have a single quadrupole doublet with the isomer shift of 0.38 - 0.39 mms^{-1} in the M^N state of the FeMo cofactor. In contrast, iron site one (Fe_1) shows $\delta = 0.49\,mms^{-1}$ at 77 K. At liquid helium temperatures, the isomer shift increases slightly ($\delta = 0.54\,mms^{-1}$; $\Delta E_Q = 1.32\,mms^{-1}$), which points to a temperature-dependent electron delocalization at higher T. When the protein is oxidized to the M^{ox} state, Fe_1 gets oxidized because both the isomer shift and the quadrupole splitting are now typical for an Fe^{3+} site ($\delta = 0.36\,mms^{-1}$; $\Delta E_Q = 0.77\,mms^{-1}$). Under turnover conditions, the Mössbauer spectrum of Fe_1 shows $\delta = 0.53\,mms^{-1}$ and $\Delta E_Q = 1.54\,mms^{-1}$, which are very similar to the resting M^{ox} state. Based on the value of the isomer shift, the authors conclude that Fe_1 is part of a mixed-valent pair of $Fe^{2.5+}$ ions.

17.4.4 Conclusion

Mössbauer spectroscopy will continue to be an invaluable tool in iron–sulfur protein research. The capacity to selectively probe iron sites is not only employed for the identification of iron–sulfur centers in proteins, as demonstrated in this section. Consequently, the utilization of Mössbauer spectroscopy for the exploration of iron–sulfur assembly and iron-centered reaction mechanisms will persist in the future [94]. In particular, the combination of laboratory-based Mössbauer spectroscopy and synchrotron-based NRVS (see the next section) on ^{57}Fe-labeled iron–sulfur systems promises to yield highly interesting new insights in the future.

17.5 Nuclear Resonance Vibrational Spectroscopy

Nuclear resonance vibrational spectroscopy (NRVS), also known as nuclear inelastic scattering (NIS), is a synchrotron-based technique that discloses the vibrational dynamics for Mössbauer-active nuclei such as ^{57}Fe [76, 95]. ^{57}Fe-NRVS has been recognized as an excellent tool in bioinorganic chemistry, and over the past two decades significant achievements were reported in the study of heme-containing proteins [96], [Fe–S] enzymes, and intricate systems involved in the activation of small molecules such as hydrogenases and nitrogenases [97–99]. NRVS exhibits a selectivity, reminiscent of RR spectroscopy, but with the significant advantage that it is not subject to the selection rules of conventional Raman and IR spectroscopy. NRVS is exceptionally suited for studying low-frequency metal-based vibrations and can inspect bond strengths as well as the identity of the various ligands making use of isotope labeling. Primarily, the power of NRVS derives directly from the observed specificity and sensitivity as this technique does not experience contribution from the sample background and/or solvent thereby resulting in minimal spectroscopic noise. This section surveys the fundamentals of NRVS, focusing on its application in studying important [Fe–S] enzymes.

17.5.1 Theoretical Background

NRVS relies on the inelastic but still resonant scattering of X-ray radiation at atomic nuclei. As a sole requirement, nuclei need to show recoilless absorption and emission of gamma rays, i.e. the Mössbauer effect. Theoretical works and early experiments exploring vibrational dynamics associated with nuclear transitions were already reported in the 1960s [100, 101]. However, it took three more decades (1995, W. Sturhahn and M. Seto) to experimentally verify inelastic nuclear scattering, enabling the investigations of transitions with quantized recoil fraction [102, 103].

The theoretical background of NRVS is based on the coupling of nuclear resonant excitation to lattice vibrations (i.e. phonons) [102–106]. Analogously to the Stokes and Anti-Stokes phenomena described earlier for Raman spectroscopy, NRVS monitors the inelastic scattering related to phonon creation and annihilation (Figure 17.1). The energy to promote a nuclear transition is several orders of magnitude larger compared to an electronic transition. Exemplarily, for the ^{57}Fe isotope, 14.4 keV are required. The energy width of the ^{57}Fe resonance is only 4.66 neV, while vibrational transitions are typically in the order of meV. This situation hindered the direct detection of the recoil fraction until important advances at the synchrotron facilities (e.g. improvements in X-ray optics) established NRVS as a novel spectroscopic technique [107].

NRVS selectively probes normal modes related to the motion of Mössbauer-active nuclei. This single selection rule allows monitoring all vibrational information of an oscillating core (e.g. the catalytic site of a metalloenzyme), contrary to other vibrational techniques like IR and Raman spectroscopy.

Since in NRVS, the vibrational transition is associated with a nuclear transition (Figure 17.1), the transition moment [8] needs to be expanded, including the initial Φ_i and final Φ_f nuclear states.

$$M_{fi} = \langle \upsilon_f | \langle \Phi_f | \hat{\Omega} | \Phi_i \rangle | \upsilon_i \rangle \tag{17.24}$$

Accordingly, the operator $\hat{\Omega}$ comprises the interaction of the external X-ray field represented by the vector potential A and the nuclear current operator $\hat{j}$ in the Cartesian coordinate system of the nucleons (x).

$$\hat{\Omega} = \int \hat{j}(x) A(x) d^3x \tag{17.25}$$

To obtain a representation for the nuclear vibration, the interaction operator $\hat{\Omega}$ can be rewritten, separating the center of mass r and the internal nuclear coordinates y using $r = x - y$.

$$\hat{\Omega} = e^{ik\hat{r}} a \hat{J}(-k) \tag{17.26}$$

Here, $\hat{J}$ represents the nuclear current operator after spatial Fourier transformation in the center-of-momentum frame, k is the wave vector of the X-ray radiation, and a is the field vector, which can be derived from the vector potential.

$$a = \frac{A(x)}{e^{ikx}} \tag{17.27}$$

This way, the vibrational and nuclear transition can be factorized.

$$\langle \upsilon_f | \langle \Phi_f | \hat{\Omega} | \Phi_i \rangle | \upsilon_i \rangle = \underbrace{\langle \upsilon_f | e^{ik\hat{r}} | \upsilon_i \rangle}_{\text{vibrational}} \underbrace{\langle \Phi_f | a\hat{J}(-k) | \Phi_i \rangle}_{\text{nuclear}} \tag{17.28}$$

The summation over all final states and averaging over all initial states yields, after normalization to the photon flux of the incident radiation, the energy-dependent absorption cross section $\sigma(E)$

$$\sigma(E) = \left\langle \sum_f \left| \langle \upsilon_f | e^{ik\hat{r}} | \upsilon_i \rangle \right|^2 \delta(\varepsilon_{fi} - E) \right\rangle \frac{2\pi}{\hbar |a|^2} \left\langle \sum_f \left| \langle \Phi_f | a\hat{J}(-k) | \Phi_i \rangle \right|^2 \right\rangle \tag{17.29}$$

where E refers to the difference in energy between the incident X-ray radiation and the nuclear transition. ε_{fi} is the energy separation of the vibrational states.

An important parameter is the transition probability density $S(E, k)$, which is influenced by the nucleus environment and is defined as

$$S(E,k) = \int \frac{dt}{2\pi\hbar} e^{\frac{iEt}{\hbar}} \left\langle e^{ik\hat{r}(t)} e^{ik\hat{r}(0)} \right\rangle \tag{17.30}$$

By integrating Eq. (17.29) over all energies, a compact expression for $\sigma(E)$ can be obtained that features the nuclear resonant cross section σ_N and the energy width of the nuclear level Γ_N.

$$\sigma(E) = \frac{\pi}{2} \sigma_N \Gamma_N S(E) \tag{17.31}$$

Within the approximation of a harmonic oscillating lattice, $S(E)$ can be expanded for single- and multiphonon contributions. This way, the elastic nuclear resonant Mössbauer transition $\delta_\Gamma(E)$ and the inelastic nuclear transitions, fundamental for NRVS, can be separated.

$$S(E) = f_{\text{LM}} \left(\underbrace{\delta_\Gamma(E)}_{\text{Mössbauer}} + \underbrace{\underbrace{\left((\overline{n}(E) + 1) \frac{E_R}{E} D_j(|E|) \right)}_{\text{single-phonon transition } (S_1)} + \sum_{n=2}^{\infty} S_n(E)}_{\text{NRVS}} \right) \tag{17.32}$$

Here, f_{LM} is the Lamb–Mössbauer factor, $\overline{n}$ is the average of the thermal mode occupation numbers, E_R is the recoil energy, and D_j is the partial vibrational density of states (PVDOS) of nucleus j. The PVDOS is regularly used to deconvolute experimental data in NRVS and contains a direct expression for the normal mode composition factors.

The following sections single out a few [Fe–S] enzymes, where NRVS was successfully employed to characterize the metal active sites.

17.5.2 [Fe(S–Cys)$_4$] Metal Site: The Case Study of Rubredoxin

Rubredoxins are a class of low-molecular-weight (ca. 50 amino acids, 6 kDa) iron-containing proteins that are classified as iron–sulfur proteins, but do not contain inorganic sulfide. These proteins function as electron mediators using a single Fe(S–Cys)$_4$ redox center, which changes between the +2 and +3 oxidation states accessing a redox potential window typically in the range +50 to −50 mV (vs. SHE). Rubredoxins have been extensively characterized by X-ray crystallography and optical spectroscopy (e.g. UV–Vis, Raman, and FIR) revealing the structural arrangement of the tetra-thiolate functional group as well as important aspects such as hydrogen bonding network and structural distortions due to perturbations in the outer coordination sphere. In 2005, the Cramer group reported NRVS spectra for the oxidized and reduced Rd from *Pyrococcus furiosus* (*Pf*) [108]. Intense (asymmetric) Fe–S stretching modes were observed between 340 and 380 cm^{-1}, which shifted upon reduction to 300–340 cm^{-1} (Figure 17.8a). These latter absorptions are not accessible by conventional RR spectroscopy (see Section 17.3). Additional broad bands were observed between 100 and 200 cm^{-1}, deriving primarily from S–Fe–S and Fe–S–C bending motions, and near 70 cm^{-1} which were assigned as a collective motion of all cysteine residues. NRVS data have also been reported for the reduced and oxidized form of the rubredoxin from *Pyrococcus abyssii* [112].

17.5.3 [2Fe–2S] Clusters: 4xCys, 2xCys–2xHis, and 3xCys–1xHis Ligations

[2Fe–2S] clusters represent the simplest polymetallic system constituted by two iron ions bridged by two sulfide ions [113]. They are coordinated by four cysteinyl ligands in ferredoxins, two Cys and two His in Rieske-type or three Cys and one His in NEET-type proteins. These proteins are usually responsible for electron transfer in biological systems, having reduction potentials from −150 to −400 mV (vs. SHE) in Cys-ligated clusters, while the presence of His residues shifts the reduction potential to more positive values. In fact, MitoNEET proteins have potentials from −100 to +50 mV while the reduction potentials of [2Fe–2S] Rieske clusters span from −150 to +400 mV. NRVS data for ferredoxin, Rieske-type, and MitoNEET [2Fe–2S] clusters have been reported. The [2Fe–2S](Cys)$_4$ sites in oxidized and reduced ferredoxins from *Rhodobacter capsulatus* (*Rc*FdVI) and *Aquifex aeolicus* (*Aa*Fd5) underwent detailed examination [114]. In their oxidized forms, strong and broad bands from Fe–S stretching modes were observed around 335 cm^{-1}, followed by minor features at ca. 390 and 420 cm^{-1}. This latter absorption exhibited large ($\Delta\nu$ = −16 cm^{-1}) ^{36}S isotope shifts and was attributed to the asymmetric Fe–S^{b} stretching vibration involving the bridging sulfides (S^{b}). Significant absorptions were also observed at lower energies with a complex envelope of bending, torsion, and protein modes with maxima centered at 78 and 190 cm^{-1}, which were deconvoluted using full protein normal-mode analysis. In the spectra for the reduced [2Fe–2S] cluster, the bands at 420 and 393 cm^{-1} redshifted to 400 and 382 cm^{-1}, respectively.

The oxidized Rieske-type [2Fe–2S] clusters of the toluene/o-xylene monooxygenase component C (ToMOC) from *Pseudomonas sp.* OX1 exhibited similar NRVS

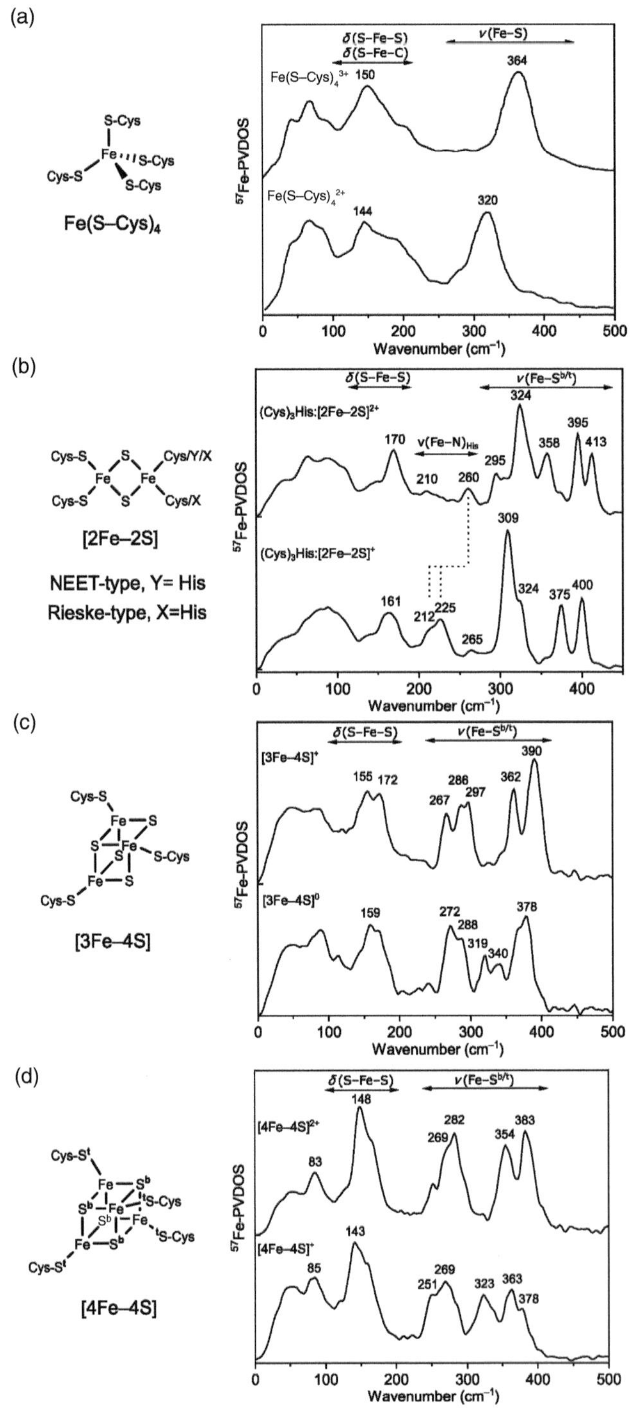

Figure 17.8 NRVS data of [Fe–S] clusters and model complexes of various nuclearity. (a) $[Fe(S–Cys)_4]$ site of the rubredoxin from *Pyrococcus furiosus* [108]. (b) [2Fe–2S] cluster of MitoNEET enzyme [109]. (c) [3Fe–4S] cluster of the D14C variant of *Pyrococcus furiosus* ferredoxin treated with ferricyanide [110]. (d) [4Fe–4S] cluster of the D14C variant of *Pyrococcus furiosus* ferredoxin [111]. Processed NRVS data were provided by Leland B. Gee and Vladimir Pelmenshikov. The software WebPlotDigitizer was utilized to extract the data used for panel a. Visual source: Giorgio Caserta (co-author).

data compared to those of *Rc*FdVI and *Aa*Fd5 proteins [115]. Well-resolved peaks at 332, 361, 410, and 425 cm^{-1} (asymmetric and symmetric Fe–S stretching modes) were observed together with a weaker signal near 280 cm^{-1} that is absent in ferredoxins and possibly containing contributions of a terminal ligand (His or Cys). NRVS spectra of MitoNEET exhibit similar band patterns (Figure 17.8b) [109]. The low-frequency bands in the range of 50–120 cm^{-1} correspond to displacements of [2Fe–2S] cluster relative to the protein framework and torsional modes, while the 120–180 cm^{-1} bands involve significant bending contributions. Bands around 400 cm^{-1} involve Fe–S stretching of S^b. Finally, the broad absorption at 260 cm^{-1} (Figure 17.8b, no analogy in Fd spectra) has been predicted to contain contributions of the His ligands. Notably, Fe(II)–N_δ(His) vibrational modes were resolved in the spectra of the reduced sample at 225 and 212 cm^{-1}.

17.5.4 [3Fe–4S] Clusters

[3Fe–4S] clusters are usually found in metalloproteins such as succinate dehydrogenases, fumarate reductases, and periplasmic [NiFe]-hydrogenases. They usually undergo their 1+/0 redox transition at higher potential (ca. +250 mV) than $[4Fe–4S]^{2+/+}$ clusters [116]. A recent study from the Greening group characterized the Huc [NiFe]-hydrogenase from *Mycobacterium smegmatis* (*Ms*), which uses surprisingly three [3Fe–4S] clusters to transfer reducing power from H_2 molecules to a long-range quinone transport system [117]. In 2016, the D14C variant of *Pyrococcus furiosus* Fd, which shows an intact and redox-active $[3Fe–4S]^{+/0}$ cluster upon ferricyanide treatment, was investigated by means of NRVS [110]. NRVS data of this cluster feature acoustic and torsional modes below 100 cm^{-1}, breathing (Fe–Fe) and S–Fe–S bending modes around 100–200 cm^{-1}, and Fe–S stretching bands in the 200–420 cm^{-1} region comprising contributions from inorganic bridging sulfides, Fe–S^b, in the 200–300 cm^{-1} region and terminal cysteine thiolates, Fe–S^t, in the 300–420 cm^{-1} regime (Figure 17.8c). An important spectral feature of the [3Fe–4S] cluster is the split-band pattern of the stretching Fe–S^b mode region (267, 286, and 297 cm^{-1} bands for $[3Fe–4S]^{1+}$) in comparison to the well-defined peaks at 269/281 cm^{-1} for the [4Fe–4S] counterpart (Figure 17.8c,d). This split originates from the loss of symmetry in the [3Fe–4S] cluster upon removal of one of the Fe ions. Notably, the spectra of both the oxidized and reduced [3Fe–4S] clusters display a blueshift within ~20 cm^{-1} when compared to the counterpart bands of the [4Fe–4S] cluster, which agrees with preceding RR investigations [118].

17.5.5 [4Fe–4S] Clusters

[4Fe–4S] clusters are possibly the most abundant [Fe–S] centers present in all domains of life. They are usually ligated by cysteines of the polypeptide chain, although coordination via other residues (His, Tyr, Ser, Glu, Lys, and Asp) has been reported. These cofactors are found within various proteins responsible for functions that include electron transfer, regulation of gene expression, substrate binding and activation, DNA repair, sulfur insertion in t-RNAs, and many others [54]. [4Fe–4S] clusters usually adopt +2/+1 oxidation states, with reduction potentials spanning from −0.2 to −0.6 V (vs. SHE).

NRVS data have been reported for the [4Fe–4S] cluster of the ferredoxin from *Pyrococcus furiosus* (D14C variant) [111], the nitrogenase Fe protein from *Azotobacter vinelandii* [119], the IspH protein from *E. coli* [92], and NO-sensing proteins WhiD and NsrR from *S. coelicolor* [120]. NRVS data of all these proteins with their [4Fe–4S] in the +2 oxidation state are very similar and resemble that of a synthetic [4Fe–4S] analog suggesting a minor influence of protein modes on the cluster vibrations. Below 200 cm^{-1}, the intensity derives from normal modes involving torsion and breathing of the [4Fe–4S] cluster as well as its motion along with the peptide backbone (Figure 17.8d). Here, the stronger contribution is around 150 cm^{-1}, which shows negligible changes in $^{32}S/^{34}S$ isotope labeling experiments. Normal mode analysis assigned this major band to bridging Fe–S^b modes. The spectral region between 200 and 300 cm^{-1} exhibits several modes involving predominantly bending and Fe–Fe stretches contributing to the major absorption near 280 cm^{-1}. Modes between 300 and 400 cm^{-1} have been predicted to contain stretching motions from both Fe–S^b and Fe–S^t. Significantly, bands in this region exhibit larger shifts upon $^{34}S/^{36}S$ substitutions, which aids the normal mode identification. Noteworthy are also data of the *Azotobacter vinelandii* nitrogenase Fe protein (Av2) [119], which were collected at three different oxidation levels including the all-ferrous form of the $[4Fe–4S]^0$ cluster. The reported NRVS data revealed a systematic shift of ca. 10–30 cm^{-1} to lower energies of Fe–S stretching frequencies upon reduction events of the [4Fe–4S] cluster, while the localization of lower energy S–Fe–S bending modes (between 120 and 155 cm^{-1}) remains invariant to the [4Fe–4S] oxidation level.

An unusual [4Fe–4S] cluster in LytB/IspH protein has been shown to be responsible for the biosynthesis of essential terpenoids in most pathogenic bacteria. The Schünemann group investigated this metal cofactor by means of NRVS [92], unraveling that the apical Fe ion of the $[4Fe–4S]^{2+}$ site is plausibly interacting with a network of three water molecules. In fact, the spectral region between 240 and 340 cm^{-1} contains a distinctive band at 256 cm^{-1} caused by several degenerate modes displaying considerable S–Fe–O stretching character. The binding of the substrate and/or inhibitors at the apical Fe ion of the [4Fe–4S] cluster has also been investigated using NRVS [92, 121].

17.5.6 Binding of NO to [Fe–S] Clusters

Reaction of NO with [Fe–S] cluster proteins and synthetic analogs usually results in the disassembly of the [Fe–S] core and formation of nitrosyl iron complexes such as MNICs and DNICs [122]. These species are easily spotted by NRVS through vibrational modes involving iron-bound nitrosyl ligands, which are observed in the high-energy region between 500 and 700 cm^{-1} [123–125]. These bands originate from the bending and symmetric/asymmetric stretching modes of the N–Fe–N unit and are accompanied by a decrease in the intensities of the Fe–S bands. Noteworthy are the pioneering studies from the Cramer and Lippard groups, which used NRVS to characterize various nitrosyl species (e.g. Roussin's black salt, RBS) formed from alterations of standard Cys-ligated [4Fe–4S] clusters as well as [2Fe–2S] Rieske centers [115, 123]. The reaction of protein-bound [Fe–S] clusters with NO has been examined also in NO-sensing proteins WhiD and NsrR [120]. By making use of $^{14}NO/^{15}NO$ and $^{32}S/^{34}S$ isotope labeling, it could be shown that nitrosylation of both [4Fe–4S] clusters yields a range of products including the multi-iron Roussin's red ester (RRE), and RBS-like species in which one or more of the inorganic sulfides are substituted by a Cys thiolate, resulting in the so-called Roussin's Black Ester (RBE).

17.5.7 The Nitrogenase FeMe Cofactor

The biological activation of dinitrogen (N_2) is accomplished by nitrogenases [16]. This is one of the most challenging cellular reactions, as it requires 16 molecules of ATP for the activation of a single N_2 molecule according to the following stoichiometry: $N_2 + 8\,e^- + 8H^+ + 16\,ATP \rightarrow 2\,NH_3 + H_2 + 16\,ATP + 16\,P_i$. Catalysis is carried out by an unusual [Fe–S] cluster, the bimetallic FeMe-cofactor (Metal = V, Mo, Fe) with the [Me–7Fe–9S–C–(homocitrate)] composition. This active site receives electrons from another atypical [Fe–S] center, the [8Fe–7S] P-cluster. Nitrogenases have been extensively characterized [20], with the first NRVS studies initiated by the Cramer group [98, 126]. NRVS spectra for the FeMe-co and P-cluster have been reported for all three classes of nitrogenases housing V, Mo, and Fe at their active site [127]. Notably, data have been collected even on single crystals using the MoFe protein from *Azotobacter vinelandii* [128].

Several Fe–S stretching modes are observed between 250 and 400 cm^{-1}, while S–Fe–S bending modes are found at lower energies (e.g. absorption band near 140 cm^{-1}). Significantly, the FeMo cofactor exhibits a strong signal near 190 cm^{-1}, whereas conventional Fe–S clusters have weak bands (Figure 17.9a). This band has been assigned to a symmetric Fe_6C stretch and referred to as a "breathing" mode of FeMo-co. This absorption has also been identified in the nitrogenase maturase NifB, a member of radical-SAM enzyme family that catalyzes the fusion of two [4Fe–4S] centers and the insertion of carbide and sulfide ions to build the NifB cofactor, a [8Fe–9S–C] precursor of the FeMo-cofactor known as the L-cluster [132]. Recently, De Beer and coworkers identified an absorption at 422 cm^{-1} for FeMo-co that was assigned via quantum mechanics/molecular mechanics calculations to an Fe–S stretching mode involving the belt "S2B" sulfide bridging Fe2 and Fe6 (Figure 17.9a) [127]. This belt sulfide has been shown to be replaced by CO (Fe–CO stretches/bends have also been detected using NRVS [133]) and a protonated nitrogen in X-ray structures [9, 134], thereby representing the plausible binding site for N_2.

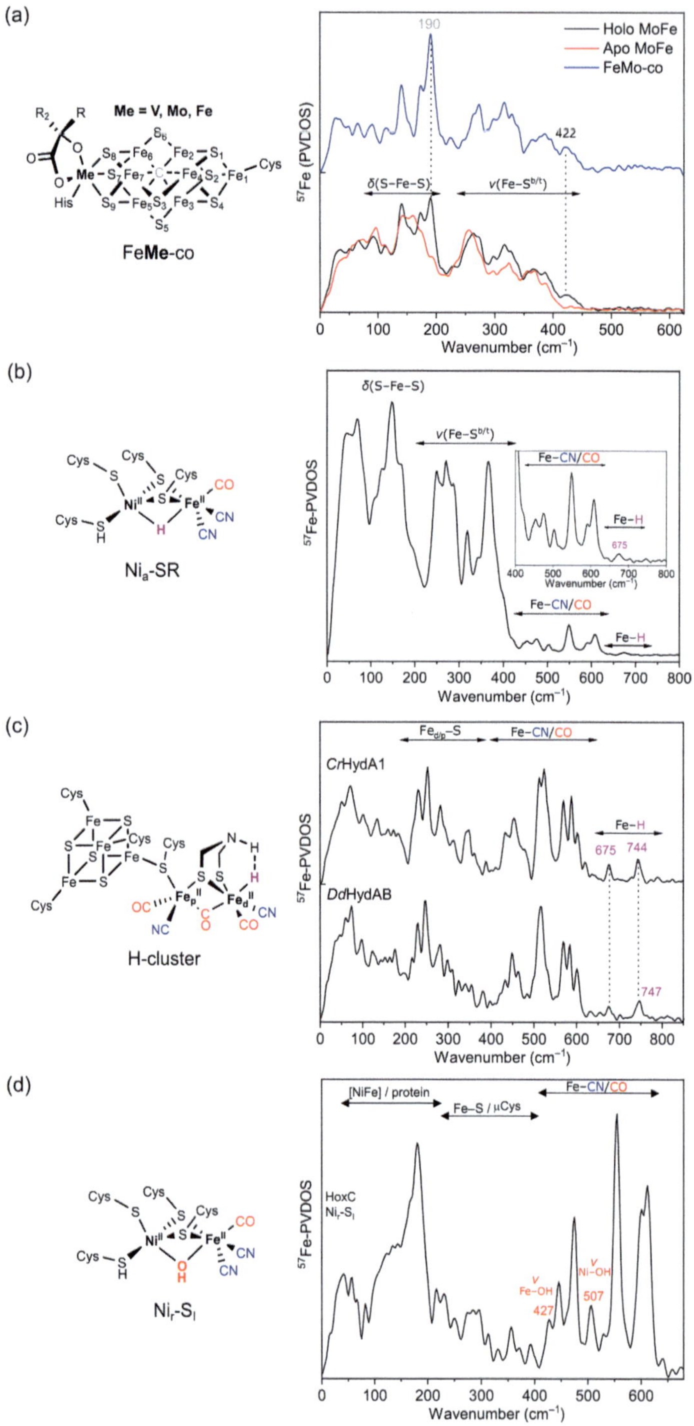
(a)
Me = V, Mo, Fe
FeMe-co
Holo MoFe
Apo MoFe
FeMo-co
190
422
δ(S–Fe–S)
ν(Fe–S^b/t)
57Fe (PVDOS)
Wavenumber (cm−1)
(b)
Nia-SR
δ(S–Fe–S)
ν(Fe–S^b/t)
Fe–CN/CO
Fe–H
675
57Fe-PVDOS
Wavenumber (cm−1)
(c)
H-cluster
Fed/p–S
Fe–CN/CO
CrHydA1
DdHydAB
Fe–H
675
744
747
57Fe-PVDOS
Wavenumber (cm−1)
(d)
Nir-SI
[NiFe] / protein
Fe–S / μCys
Fe–CN/CO
HoxC
Nir-SI
ν Fe–OH
427
ν Ni–OH
507
57Fe-PVDOS
Wavenumber (cm−1)

Figure 17.9 NRVS data of complex [Fe–S] containing enzymes involved in the activation of small molecules. (a) The nitrogenase FeMo cofactor [127]. (b) The [NiFe]-hydrogenase metal cofactor featuring a bridging hydride species (Ni_a-SR) [129]. (c) The [FeFe]-hydrogenase H-cluster featuring terminal hydride species (H_{hyd}) [130]. (d) The [NiFe]-hydrogenase active site featuring a bridging hydroxy ligand (Ni_r-S_I) [131]. Processed NRVS data in a, b, c, and d were provided by Casey Van Stappen, Hongxin Wang, James Birrell, and Caserta Giorgio. Visual source: Giorgio Caserta (co-author).

17.5.8 [NiFe]- and [FeFe]-Hydrogenases

NRVS has greatly contributed to the understanding of [NiFe]- and [FeFe]-hydrogenases, which catalyze the reversible conversion of H_2 according to the following stoichiometry: $H_2 \rightleftarrows 2H^+ + 2e^-$ [8]. Both hydrogenase classes bear [Fe–S] clusters, which channel electrons toward or away from the H_2-converting site, i.e. the [FeFe]-hydrogenase H-cluster (a [4Fe–4S] cluster bridged to a $[Fe_2(adt)(CN)_2(CO)_4]$ subsite) and the heterobimetallic $NiFe(CN)_2CO$ cofactor in [NiFe]-hydrogenases.

The first NRVS investigations for both hydrogenase classes date back more than ten years ago [135, 136]. Typical NRVS data comprise dominant signals of Fe–S(–Fe) stretching and bending modes in the region between 100 and 420 cm^{-1}, which are mostly related to the [Fe–S] clusters [137, 138] as well as Fe–CO/CN stretching and bending modes of their active site in the 400–650 cm^{-1} spectral region [136, 139]. Considering that most of the irons are located in the [Fe–S] clusters; the spectral contribution of the hydrogenase active site is usually surpassed by the [Fe–S] cluster contribution. Nonetheless, high protein concentrations and sample homogeneity allowed the first observation of metal-hydride vibration, a bridging Ni–H–Fe wag (675 cm^{-1}), in the [NiFe]-hydrogenase from *Desulfovibrio vulgaris* Miyazaki F (*Dv*MF) that is displayed in Figure 17.9b [129, 140]. The development of biochemical and/or synthetic procedures for the specific ^{57}Fe-labeling of the hydrogenase active sites made NRVS completely selective for the normal modes of these metal sites [93, 141]. This enabled important discoveries identifying, e.g. weak (isotope-sensitive) Fe–H bending modes (Figure 17.9c) for the catalytic terminal hydride species from the [FeFe]-hydrogenases from *Chlamydomonas reinhardtii* and *Desulfovibrio desulfuricans* [130]. Notably, the observed Fe–H modes appear to be vibrationally coupled to the active site environment such as residues of the H^+ transfer pathway and influenced by alterations of the azadithiolate functionality [142–144].

Besides catalytic intermediates, NRVS contributed to untangle structural elements of inhibited and/or resting species as well as [NiFe]-hydrogenase maturation intermediates [73, 131, 145, 146]. Noteworthy is the detection of isotope-sensitive Fe-^{32}S/^{34}S absorptions (340–360 cm^{-1}) for the O_2 protected [FeFe]-hydrogenase H_{inact} [145] as well as Ni–µOH (507 cm^{-1}) and Fe–µOH (427 cm^{-1}) stretching vibrations of a [NiFe]-hydrogenase intermediate with a hydroxy ligand in the bridging position between Ni and Fe (Figure 17.9d) [131].

17.5.9 Conclusion

NRVS has been shown to provide unique site selectivity for Mössbauer-active nuclei. This fostered the application of ^{57}Fe-NRVS in bioinorganic chemistry. Considerable progress in recent years has been made in the study of complex metalloenzymes and, herein, we have outlined the role played by NRVS in the characterization of [Fe–S] clusters of different nuclearity including complex cofactors activating H_2 and N_2 molecules. Finally, the growing availability of ^{57}Fe-labeled biological systems will foster the applicability of NRVS, promising significant new insights in the near future.

References

1 Wilson, E.B., Decius, J.C., and Cross, P.C. (1980). *Molecular Vibrations: The Theory of Infrared and Raman Vibrational Spectra*. New York: Dover Publications, 388 p.

2 Siebert, F. and Hildebrandt, P. (2008). *Vibrational Spectroscopy in Life Science*. Weinheim: Wiley-VCH, 310 p. (Tutorials in biophysics).

3 Horch, M. (2015) Structure-function relationships of metalloenzymes [cited 2024 Apr 26]; https://depositonce.tu-berlin.de/handle/11303/4815.

4 Lorent C. (2022) Combining electron paramagnetic resonance and vibrational spectroscopy to explore catalysis of hydrogen-converting enzymes [cited 2024 Apr 26]; https://depositonce.tu-berlin.de/handle/11303/17790.

5 Herschel, W. (1800). XIX. Experiments on the solar, and on the terrestrial rays that occasion heat; with a comparative view of the laws to which light and heat, or rather the rays which occasion them, are subject, in order to determine whether they are the same, or different. *Phil. Trans. R. Soc.* 31 (90): 437–538.

6 Barth, A. (2007). Infrared spectroscopy of proteins. *Biochim. Biophys. Acta Bioenerg.* 1767 (9): 1073–1101.

7 Lorenz-Fonfria, V.A. (2020). Infrared difference spectroscopy of proteins: from bands to bonds. *Chem. Rev.* 120 (7): 3466–3576.

8 Lubitz, W., Ogata, H., Rüdiger, O., and Reijerse, E. (2014). Hydrogenases. *Chem. Rev.* 114 (8): 4081–4148.

9 Rohde, M., Laun, K., Zebger, I. et al. (2021). Two ligand-binding sites in CO-reducing V nitrogenase reveal a general mechanistic principle. *Sci. Adv.* 7 (22): eabg4474.

10 George, S.J., Seravalli, J., and Ragsdale, S.W. (2005). EPR and infrared spectroscopic evidence that a kinetically competent paramagnetic intermediate is formed when acetyl-coenzyme A synthase reacts with CO. *J. Am. Chem. Soc.* 127 (39): 13500–13501.

11 Ciaccafava, A., Tombolelli, D., Domnik, L. et al. (2016). When the inhibitor tells more than the substrate: the cyanide-bound state of a carbon monoxide dehydrogenase. *Chem. Sci.* 7 (5): 3162–3171.

12 Barth, A. (2000). The infrared absorption of amino acid side chains. *Prog. Biophys. Mol. Biol.* 74 (3–5): 141–173.

13 Horch, M., Schoknecht, J., Mroginski, M.A. et al. (2014). Resonance Raman spectroscopy on [NiFe] hydrogenase provides structural insights into catalytic intermediates and reactions. *J. Am. Chem. Soc.* 136 (28): 9870–9873.

14 Senger, M., Mebs, S., Duan, J. et al. (2018). Protonation/reduction dynamics at the [4Fe–4S] cluster of the hydrogen-forming cofactor in [FeFe]-hydrogenases. *Phys. Chem. Chem. Phys.* 20 (5): 3128–3140.

15 Huang, G., Wagner, T., Wodrich, M.D. et al. (2019). The atomic-resolution crystal structure of activated [Fe]-hydrogenase. *Nat. Catal.* 2 (6): 537–543.

16 Einsle, O. and Rees, D.C. (2020). Structural enzymology of nitrogenase enzymes. *Chem. Rev.* 120 (12): 4969–5004.

17 George, S.J., Ashby, G.A., Wharton, C.W., and Thorneley, R.N.F. (1997). Time-resolved binding of carbon monoxide to nitrogenase monitored by stopped-flow infrared spectroscopy. *J. Am. Chem. Soc.* 119 (27): 6450–6451.

18 Paengnakorn, P., Ash, P.A., Shaw, S. et al. (2017). Infrared spectroscopy of the nitrogenase MoFe protein under electrochemical control: potential-triggered CO binding. *Chem. Sci.* 8 (2): 1500–1505.

19 Milosevic, M. (2004). Internal reflection and ATR spectroscopy. *Appl. Spectrosc. Rev.* 39 (3): 365–384.

20 Van Stappen, C., Decamps, L., Cutsail, G.E. et al. (2020). The spectroscopy of nitrogenases. *Chem. Rev.* 120 (12): 5005–5081.

21 Svetlitchnaia, T., Svetlitchnyi, V., Meyer, O., and Dobbek, H. (2006). Structural insights into methyltransfer reactions of a corrinoid iron–sulfur protein involved in acetyl-CoA synthesis. *Proc. Natl Acad. Sci. USA* 103 (39): 14331–14336.

22 Ciaccafava, A., Tombolelli, D., Domnik, L. et al. (2017). Carbon monoxide dehydrogenase reduces cyanate to cyanide. *Angew. Chem. Int. Ed.* 56 (26): 7398–7401.

23 Grabarczyk, D.B., Ash, P.A., and Vincent, K.A. (2014). Infrared spectroscopy provides insight into the role of dioxygen in the nitrosylation pathway of a [2Fe2S] cluster iron–sulfur protein. *J. Am. Chem. Soc.* 136 (32): 11236–11239.

24 Dodd, E.L. and Le Brun, N.E. (2024). Probing the mechanism of the dedicated NO sensor [4Fe-4S] NsrR: the effect of cluster ligand environment. *J. Inorg. Biochem.* 252 (1): 112457–112471.

25 Ludwig, R. (2018). Studying interaction, ion-pair formation, and mixing behavior of protic ionic liquids by means of far-infrared spectroscopy. In: *Frontiers and Advances in Molecular Spectroscopy*, 527–567. Elsevier. https://linkinghub.elsevier.com/retrieve/pii/B9780128112205000174.

26 Giménez, P., Linares, A., Sessa, C. et al. (2022). Capability of Far-Infrared for the selective identification of red and black pigments in paint layers. *Spectrochim. Acta A Mol. Biomol. Spectrosc.* 266: 120411.

27 Itoh, K., Nakahara, T., Shimanouchi, T. et al. (1968). Far-infrared spectra of polyalanines with α-helical and β-form structures. *Biopolymers* 6 (12): 1759–1766.

28 Moore, W.H. and Krimm, S. (1976). Vibrational analysis of peptides, polypeptides, and proteins. I. Polyglycine I. *Biopolymers* 15 (12): 2439–2464.

29 Moore, W.H. and Krimm, S. (1976). Vibrational analysis of peptides, polypeptides, and proteins. II. β-Poly(L-alanine) and β-poly(L-alanylglycine). *Biopolymers* 15 (12): 2465–2483.

30 Stehle, C.U., Abuillan, W., Gompf, B., and Dressel, M. (2012). Far-infrared spectroscopy on free-standing protein films under defined temperature and hydration control. *J. Chem. Phys.* 136 (7): 075102.

31 El Khoury, Y. and Hellwig, P. (2011). A combined far-infrared spectroscopic and electrochemical approach for the study of iron–sulfur proteins. *ChemPhysChem* 12 (14): 2669–2674.

32 Knaff, D.B. (1996). Ferredoxin and ferredoxin-dependent enzymes. In: *Oxygenic Photosynthesis: The Light Reactions* (ed. D.R. Ort, C.F. Yocum, and I.F. Heichel), 333–361. Dordrecht, Netherlands: Springer (Govindjee, Amesz J, Aro EM, Barber J, Blankenship RE, Murata N, et al., editors. Advances in Photosynthesis and Respiration; vol. 4). https://link.springer.com/10.1007/0-306-48127-8_17.

33 Sobel, B.E. and Lovenberg, W. (1966). Characteristics of *Clostridium pasteurianum* ferredoxin in oxidation-reduction reactions. *Biochemistry* 5 (1): 6–13.

34 Raman, C.V. and Krishnan, K.S. (1928). A new type of secondary radiation. *Nature* 121 (3048): 501–502.

35 Long, D.A. (2002). *The Raman effect: A Unified Treatment of the Theory of Raman Scattering by Molecules*. Chichester, Weinheim: Wiley, 597 p.

36 Kramers, H.A. and Heisenberg, W. (1925). Über die Streuung von Strahlung durch Atome. *Z Physik.* 31 (1): 681–708.

37 Dirac, P.A.M. (1927). The quantum theory of the emission and absorption of radiation. *Proc. R. Soc. Lond. A* 114 (767): 243–265.

38 Placzek G. Rayleigh-Streuung und Raman-effekt, in MarX Handbuch der Radiologie, Band IV, Quantenmechanik der Materie und Strahlung. Teil II Leipzig: Akademische.

39 Albrecht, A.C. (1961). On the theory of Raman intensities. *J. Chem. Phys.* 34 (5): 1476–1484.

40 Yoon, K.S., Hille, R., Hemann, C., and Tabita, F.R. (1999). Rubredoxin from the green sulfur bacterium *Chlorobium tepidum* functions as an electron acceptor for pyruvate ferredoxin oxidoreductase. *J. Biol. Chem.* 274 (42): 29772–29778.

41 Czernuszewicz, R.S., LeGall, J., Moura, I., and Spiro, T.G. (1986). Resonance Raman spectra of rubredoxin: new assignments and vibrational coupling mechanism from iron-54/iron-56 isotope shifts and variable-wavelength excitation. *Inorg. Chem.* 25 (5): 696–700.

42 Kuila, D., Schoonover, J.R., Dyer, R.B. et al. (1992). Resonance Raman studies of Rieske-type proteins. *Biochim. Biophys. Acta Bioenerg.* 1140 (2): 175–183.

43 Fu, W., Drozdzewski, P.M., Davies, M.D. et al. (1992). Resonance Raman and magnetic circular dichroism studies of reduced [2Fe-2S] proteins. *J. Biol. Chem.* 267 (22): 15502–15510.

44 Fee, J.A., Findling, K.L., Yoshida, T. et al. (1984). Purification and characterization of the Rieske iron-sulfur protein from *Thermus thermophilus*. Evidence for a [2Fe-2S] cluster having non-cysteine ligands. *J. Biol. Chem.* 259 (1): 124–133.

45 Kuila, D., Fee, J.A., Schoonover, J.R. et al. (1987). Resonance Raman spectra of the [2Fe-2S] clusters of the Rieske protein from Thermus and phthalate dioxygenase from Pseudomonas. *J. Am. Chem. Soc.* 109 (5): 1559–1561.

46 Tirrell, T.F., Paddock, M.L., Conlan, A.R. et al. (2009). Resonance Raman studies of the (His)$(Cys)_3$ 2Fe-2S cluster of MitoNEET: comparison to the $(Cys)_4$ mutant and implications of the effects of pH on the Labile Metal Center. *Biochemistry* 48 (22): 4747–4752.

47 Brereton, P.S., Verhagen, M.F.J.M., Zhou, Z.H., and Adams, M.W.W. (1998). Effect of iron-sulfur cluster environment in modulating the thermodynamic properties and biological function of ferredoxin from *Pyrococcus furiosus*. *Biochemistry* 37 (20): 7351–7362.

48 Duderstadt, R.E., Staples, C.R., Brereton, P.S. et al. (1999). Effects of mutations in Aspartate 14 on the spectroscopic properties of the $[Fe_3S_4]^{+,0}$ clusters in *Pyrococcus furiosus* ferredoxin. *Biochemistry* 38 (32): 10585–10593.

49 Dus, K., De Klerk, H., Sletten, K., and Bartsch, R.G. (1967). Chemical characterization of high potential iron proteins from *Chromatium* and *Rhodopseudomonas gelatinosa*. *Biochim. Biophys. Acta Protein Struct.* 140 (2): 291–311.

50 Aono, S., Bryant, F.O., and Adams, M.W. (1989). A novel and remarkably thermostable ferredoxin from the hyperthermophilic archaebacterium *Pyrococcus furiosus*. *J. Bacteriol.* 171 (6): 3433–3439.

51 Backes, G., Mino, Y., Loehr, T.M. et al. (1991). The environment of Fe4S4 clusters in ferredoxins and high-potential iron proteins. New information from x-ray crystallography and resonance Raman spectroscopy. *J. Am. Chem. Soc.* 113 (6): 2055–2064.

52 Brereton, P.S., Duderstadt, R.E., Staples, C.R. et al. (1999). Effect of serinate ligation at each of the iron sites of the $[Fe_4S_4]$ cluster of *Pyrococcus furiosus* ferredoxin on the redox, spectroscopic, and biological properties. *Biochemistry* 38 (32): 10594–10605.

53 Spiro, T.G. and Czernuszewicz, R.S. (1995). [18] resonance Raman spectroscopy of metalloproteins. In: *Methods in Enzymology*, 416–460. Elsevier. https://linkinghub.elsevier.com/retrieve/pii/0076687995460209.

54 Caserta, G., Zuccarello, L., Barbosa, C. et al. (2022). Unusual structures and unknown roles of Fe–S clusters in metalloenzymes seen from a resonance Raman spectroscopic perspective. *Coord. Chem. Rev.* 452: 214287.

55 Im, S.C., Kohzuma, T., McFarlane, W. et al. (1997). Formation, properties, and characterization of a fully reduced $Fe^{II}Fe^{II}$ form of spinach (and parsley) [2Fe-2S] ferredoxin with the macrocyclic complex $[Cr(15\text{-}aneN_4)(H_2O)_2]^{2+}$ as reductant. *Inorg. Chem.* 36 (7): 1388–1396.

56 Yachandra, V.K., Hare, J., Gewirth, A. et al. (1983). Resonance Raman spectra of spinach ferredoxin and adrenodoxin and of analog complexes. *J. Am. Chem. Soc.* 105 (21): 6462–6469.

57 Duff, J.L.C., Breton, J.L.J., Butt, J.N. et al. (1996). Novel redox chemistry of [3Fe–4S] clusters: electrochemical characterization of the All-Fe(II) form of the [3Fe–4S] cluster generated reversibly in various proteins and its spectroscopic investigation in *Sulfolobus acidocaldarius* ferredoxin. *J. Am. Chem. Soc.* 118 (36): 8593–8603.

58 Johnson, M.K., Czernuszewicz, R.S., Spiro, T.G. et al. (1983). Resonance Raman spectroscopic evidence for a common [3-iron-4-sulfur] structure among proteins containing three-iron centers. *J. Am. Chem. Soc.* 105 (22): 6671–6678.

59 Johnson, M.K., Duderstadt, R.E., and Duin, E.C. (1999). Biological and synthetic [Fe3S4] clusters. In: *Advances in Inorganic Chemistry*, 1–82. Elsevier. https://linkinghub.elsevier.com/retrieve/pii/S0898883808600768.

60 Katz, S., Noth, J., Horch, M. et al. (2016). Vibrational spectroscopy reveals the initial steps of biological hydrogen evolution. *Chem. Sci.* 7 (11): 6746–6752.

61 Mössbauer, R.L. (1959). Kernresonanzabsorption von γ-Strahlung in Ir^{191}. *Zeitschrift für Naturforschung A* 14 (3): 211–216.

62 Mössbauer, R.L. (1958). Kernresonanzfluoreszenz von Gammastrahlung in Ir^{191}. *Z Physik.* 151 (2): 124–143.

63 Johnson, C.E. and Hall, D.O. (1968). Mössbauer effect study of the state of iron in spinach ferredoxin. *Nature* 217 (5127): 446–448.

64 Mullinger, R.N., Cammack, R., Rao, K.K. et al. (1975). Physicochemical characterization of the four-iron-four-sulphide ferredoxin from *Bacillus stearothermophilus. Biochem. J.* 151 (1): 75–83.

65 Middleton, P., Dickson, D.P.E., Johnson, C.E., and Rush, J.D. (1978). Interpretation of the Mossbauer spectra of the four-iron ferredoxin from *Bacillus stearothermophilus. Eur. J. Biochem.* 88 (1): 135–141.

66 Schulz, C. and Debrunner, P.G. (1976). Rubredoxin, a simple iron-sulfur protein: its spin Hamiltoman and hyperfine parameters. *J. Phys. Colloques* 37 (C6): C6-153–C6-158.

67 Phillips, W.D., Poe, M., Weiher, J.F. et al. (1970). Proton magnetic resonance, magnetic susceptibility and Mössbauer studies of *Clostridium pasteurianum* rubredoxin. *Nature* 227 (5258): 574–577.

68 Beinert, H., Holm, R.H., and Münck, E. (1997). Iron-sulfur clusters: nature's modular, multipurpose structures. *Science* 277 (5326): 653–659.

69 Pandelia, M.E., Lanz, N.D., Booker, S.J., and Krebs, C. (2015). Mössbauer spectroscopy of Fe–S proteins. *Biochim. Biophys. Acta Mol. Cell Res.* 1853 (6): 1395–1405.

70 Shima, S., Lyon, E.J., Thauer, R.K. et al. (2005). Mössbauer studies of the iron–sulfur cluster-free hydrogenase: the electronic state of the mononuclear Fe active site. *J. Am. Chem. Soc.* 127 (29): 10430–10435.

71 Pereira, A.S., Tavares, P., Moura, I. et al. (2001). Mössbauer characterization of the iron–sulfur clusters in *Desulfovibrio vulgaris* hydrogenase. *J. Am. Chem. Soc.* 123 (12): 2771–2782.

72 Mulder, D.W., Guo, Y., Ratzloff, M.W., and King, P.W. (2017). Identification of a catalytic iron-hydride at the H-cluster of [FeFe]-hydrogenase. *J. Am. Chem. Soc.* 139 (1): 83–86.

73 Caserta, G., Hartmann, S., Van Stappen, C. et al. (2023). Stepwise assembly of the active site of [NiFe]-hydrogenase. *Nat. Chem. Biol.* 19 (4): 498–506.

74 Badding, E.D., Srisantitham, S., Lukoyanov, D.A. et al. (2023). Connecting the geometric and electronic structures of the nitrogenase iron–molybdenum cofactor through site-selective 57Fe labelling. *Nat. Chem.* 15 (5): 658–665.

75 Yoo, S.J., Angove, H.C., Papaefthymiou, V. et al. (2000). Mössbauer study of the MoFe protein of nitrogenase from *Azotobacter vinelandii* using selective ^{57}Fe enrichment of the M-centers. *J. Am. Chem. Soc.* 122 (20): 4926–4936.

76 Gütlich, P., Bill, E., and Trautwein, A.X. (2011). *Mössbauer Spectroscopy and Transition Metal Chemistry*. Berlin, Heidelberg: Springer. http://link.springer.com/10.1007/978-3-540-88428-6.

77 Stich, T.A. (2021). Characterization of paramagnetic iron-sulfur clusters using electron paramagnetic resonance spectroscopy. In: *Fe-S Proteins* (ed. P.C. Dos Santos), 259–280. New York, NY: Springer (Methods in Molecular Biology; vol. 2353). https://link.springer.com/10.1007/978-1-0716-1605-5_14.

78 Ueda, C., Langton, M., and Pandelia, M.E. (2021). Characterization of Fe-S clusters in proteins by Mössbauer spectroscopy. In: *Fe-S Proteins* (ed. P.C. Dos Santos), 281–305. New York, NY: Springer (Methods in Molecular Biology; vol. 2353). https://link.springer.com/10.1007/978-1-0716-1605-5_15.

79 Trautwein, A.X., Bill, E., Bominaar, E.L., and Winkler, H. (1991). Iron-containing proteins and related analogs — complementary Mössbauer, EPR and magnetic susceptibility studies. In: *Bioinorganic Chemistry*, 1–95. Berlin, Heidelberg: Springer (Clarke MJ, Goodenough JB, Ibers JA, Jørgensen CK, Mingos DMP, Neilands JB, et al., editors. Structure and Bonding; vol. 78). http://link.springer.com/10.1007/3-540-54261-2_1.

80 Moura, J.J., Moura, I., Kent, T.A. et al. (1982). Interconversions of [3Fe-3S] and [4Fe-4S] clusters. Mössbauer and electron paramagnetic resonance studies of *Desulfovibrio gigas* ferredoxin II. *J. Biol. Chem.* 257 (11): 6259–6267.

81 Emptage, M.H., Kent, T.A., Kennedy, M.C. et al. (1983). Mössbauer and EPR studies of activated aconitase: development of a localized valence state at a subsite of the [4Fe-4S] cluster on binding of citrate. *Proc. Natl. Acad. Sci. U.S.A.* 80 (15): 4674–4678.

82 Muenck, E., Debrunner, P.G., Tsibris, J.C.M., and Gunsalus, I.C. (1972). Moessbauer parameters of putidaredoxin and its selenium analog. *Biochemistry* 11 (5): 855–863.

83 Papaefthymiou, V., Girerd, J.J., Moura, I. et al. (1987). Moessbauer study of *D. gigas* ferredoxin II and spin-coupling model for Fe3S4 cluster with valence delocalization. *J. Am. Chem. Soc.* 109 (15): 4703–4710.

84 Bauer TO, Graf, D., Lamparter, T., and Schünemann, V. (2014). Characterization of the photolyase-like iron sulfur protein PhrB from *Agrobacterium tumefaciens* by Mössbauer spectroscopy. *Hyperfine Interact.* 226 (1–3): 445–449.

85 Krebs, C., Price, J.C., Baldwin, J. et al. (2005). Rapid freeze-quench ^{57}Fe Mössbauer spectroscopy: monitoring changes of an iron-containing active site during a biochemical reaction. *Inorg. Chem.* 44 (4): 742–757.

86 Debrunner, P.G. (1989). Mössbauer spectroscopy of iron porphyrins. In: *Iron Porphyrins Part III*, Physical Bioinorganic Chemistry Series (ed. A.B.P. Lever and H.B. Gray), 139–234. New York: VCH Publishers, Inc.

87 Stegmaier, K., Blinn, C.M., Bechtel, D.F. et al. (2019). Apd1 and Aim32 are prototypes of bishistidinyl-coordinated non-Rieske [2Fe–2S] proteins. *J. Am. Chem. Soc.* 141 (14): 5753–5765.

88 Müller, C.S., Bechtel, D.F., Auerbach, H. et al. (2019). Characterization of *Mycobacterium tuberculosis* ferredoxin with Mössbauer spectroscopy. *Hyperfine Interact.* 240 (1): 117.

89 Gunnlaugsson, H.P. (2016). Spreadsheet based analysis of Mössbauer spectra. *Hyperfine Interact.* 237 (1): 79.

90 Prisecaru I. WMOSS4 Mössbauer Spectral Analysis Software, www.wmoss.org, 2009–2016.

91 Seemann, M., Janthawornpong, K., Schweizer, J. et al. (2009). Isoprenoid biosynthesis via the MEP pathway: *in vivo* Mössbauer spectroscopy identifies a $[4Fe\text{-}4S]^{2+}$ center with unusual coordination sphere in the LytB protein. *J. Am. Chem. Soc.* 131 (37): 13184–13185.

92 Faus, I., Reinhard, A., Rackwitz, S. et al. (2015). Isoprenoid biosynthesis in pathogenic bacteria: nuclear resonance vibrational spectroscopy provides insight into the unusual [4Fe-4S] cluster of the *E. coli* LytB/IspH protein. *Angew. Chem. Int. Ed.* 54 (43): 12584–12587.

93 Gilbert-Wilson, R., Siebel, J.F., Adamska-Venkatesh, A. et al. (2015). Spectroscopic investigations of [FeFe] hydrogenase maturated with $[^{57}Fe_2(adt)(CN)_{2(CO)4}]^{2-}$. *J. Am. Chem. Soc.* 137 (28): 8998–9005.

94 Schulz, V., Steinhilper, R., Oltmanns, J. et al. (2024). Mechanism and structural dynamics of sulfur transfer during de novo [2Fe-2S] cluster assembly on ISCU2. *Nat. Commun.* 15 (1): 3269.

95 Paulsen, H., Schünemann, V., Trautwein, A.X., and Winkler, H. (2005). Mössbauer studies of coordination compounds using synchrotron radiation. *Coord. Chem. Rev.* 249 (1–2): 255–272.

96 Scheidt, W.R., Li, J., and Sage, J.T. (2017). What can be learned from nuclear resonance vibrational spectroscopy: vibrational dynamics and hemes. *Chem. Rev.* 117 (19): 12532–12563.

97 Gee, L.B., Wang, H., and Cramer, S.P. (2018). NRVS for Fe in biology: experiment and basic interpretation. In: *Methods in Enzymology*, 409–425. Elsevier. https://linkinghub.elsevier.com/retrieve/pii/S0076687917303403.

98 Cramer, S.P., Xiao, Y., Wang, H. et al. (2006). Nuclear resonance vibrational spectroscopy (NRVS) of Fe–S model compounds, Fe–S proteins, and nitrogenase. *Hyperfine Interact.* 170 (1–3): 47–54.

99 Wang, H., Alp, E.E., Yoda, Y., and Cramer, S.P. (2014). A practical guide for nuclear resonance vibrational spectroscopy (NRVS) of biochemical samples and model compounds. In: *Metalloproteins* (ed. J.C. Fontecilla-Camps and Y. Nicolet), 125–137. Totowa, NJ: Humana Press (Methods in Molecular Biology; vol. 1122). https://link.springer.com/10.1007/978-1-62703-794-5_9.

100 Visscher, W.M. (1960). Study of lattice vibrations by resonance absorption of nuclear gamma rays. *Ann. Phys.* 9 (2): 194–210.

101 Singwi, K.S. and Sjölander, A. (1960). Resonance absorption of nuclear gamma rays and the dynamics of atomic motions. *Phys. Rev.* 120 (4): 1093–1102.

102 Sturhahn, W., Toellner, T.S., Alp, E.E. et al. (1995). Phonon density of states measured by inelastic nuclear resonant scattering. *Phys. Rev. Lett.* 74 (19): 3832–3835.

103 Seto, M., Yoda, Y., Kikuta, S. et al. (1995). Observation of nuclear resonant scattering accompanied by phonon excitation using synchrotron radiation. *Phys. Rev. Lett.* 74 (19): 3828–3831.

104 Chumakov, A.I., Rüffer, R., Grünsteudel, H. et al. (1995). Energy dependence of nuclear recoil measured with incoherent nuclear scattering of synchrotron radiation. *Europhys. Lett.* 30 (7): 427–432.

105 Chumakov, A.I. and Sturhahn, W. (1999). Experimental aspects of inelastic nuclear resonance scattering. *Hyperfine Interact.* 123/124 (1/4): 781–808.

106 Alp, E.E., Sturhahn, W., Toellner, T.S. et al. (2002). Vibrational dynamics studies by nuclear resonant inelastic X-ray scattering. *Hyperfine Interact.* 144/145 (1/4): 3–20.

107 Sturhahn, W. (2004). Nuclear resonant spectroscopy. *J. Phys. Condens. Matter* 16 (5): S497–S530.

108 Xiao, Y., Wang, H., George, S.J. et al. (2005). Normal mode analysis of *Pyrococcus furiosus* rubredoxin via nuclear resonance vibrational spectroscopy (NRVS) and resonance Raman spectroscopy. *J. Am. Chem. Soc.* 127 (42): 14596–14606.

109 Gee, L.B., Pelmenschikov, V., Mons, C. et al. (2021). NRVS and DFT of MitoNEET: understanding the special vibrational structure of a [2Fe-2S] cluster with $(Cys)_3$ $(His)_1$ ligation. *Biochemistry* 60 (31): 2419–2424.

110 Lauterbach, L., Gee, L.B., Pelmenschikov, V. et al. (2016). Characterization of the $[3Fe–4S]^{0/1+}$ cluster from the D14C variant of *Pyrococcus furiosus* ferredoxin via combined NRVS and DFT analyses. *Dalton Trans.* 45 (17): 7215–7219.

111 Mitra, D., Pelmenschikov, V., Guo, Y. et al. (2011). Dynamics of the [4Fe-4S] cluster in *Pyrococcus furiosus* D14C ferredoxin via nuclear resonance vibrational and resonance Raman spectroscopies, force field simulations, and density functional theory calculations. *Biochemistry* 50 (23): 5220–5235.

112 Paulsen, H., Trautwein, A.X., Wegner, P. et al. (2011). Interpretation of nuclear resonant vibrational spectra of rubredoxin using a combined quantum mechanics and molecular mechanics approach. *ChemPhysChem* 12 (17): 3434–3441.

113 Johnson, D.C., Dean, D.R., Smith, A.D., and Johnson, M.K. (2005). Structure, function, and formation of biological iron-sulfur clusters. *Annu. Rev. Biochem.* 74 (1): 247–281.

114 Xiao, Y., Tan, M.L., Ichiye, T. et al. (2008). Dynamics of *Rhodobacter capsulatus* [2Fe-2S] Ferredoxin VI and *Aquifex aeolicus* Ferredoxin 5 via nuclear resonance vibrational spectroscopy (NRVS) and resonance Raman spectroscopy. *Biochemistry* 47 (25): 6612–6627.

115 Tinberg, C.E., Tonzetich, Z.J., Wang, H. et al. (2010). Characterization of iron dinitrosyl species formed in the reaction of nitric oxide with a biological Rieske center. *J. Am. Chem. Soc.* 132 (51): 18168–18176.

116 Zanello, P. (2018). Structure and electrochemistry of proteins harboring iron-sulfur clusters of different nuclearities. Part II. [4Fe-4S] and [3Fe-4S] iron-sulfur proteins. *J. Struct. Biol.* 202 (3): 250–263.

117 Grinter, R., Kropp, A., Venugopal, H. et al. (2023). Structural basis for bacterial energy extraction from atmospheric hydrogen. *Nature* 615 (7952): 541–547.

118 Conover, R.C., Kowal, A.T., Fu, W.G. et al. (1990). Spectroscopic characterization of the novel iron-sulfur cluster in *Pyrococcus furiosus* ferredoxin. *J. Biol. Chem.* 265 (15): 8533–8541.

119 Mitra, D., George, S.J., Guo, Y. et al. (2013). Characterization of [4Fe-4S] cluster vibrations and structure in nitrogenase Fe protein at three oxidation levels via combined NRVS, EXAFS, and DFT analyses. *J. Am. Chem. Soc.* 135 (7): 2530–2543.

120 Serrano, P.N., Wang, H., Crack, J.C. et al. (2016). Nitrosylation of nitric-oxide-sensing regulatory proteins containing [4Fe-4S] clusters gives rise to multiple iron–nitrosyl complexes. *Angew. Chem. Int. Ed.* 55 (47): 14575–14579.

121 O'Dowd, B., Williams, S., Wang, H. et al. (2017). Spectroscopic and computational investigations of ligand binding to IspH: discovery of non-diphosphate inhibitors. *ChemBioChem* 18 (10): 914–920.

122 Crack, J.C., Green, J., Thomson, A.J., and Brun, N.E.L. (2014). Iron–sulfur clusters as biological sensors: the chemistry of reactions with molecular oxygen and nitric oxide. *Acc. Chem. Res.* 47 (10): 3196–3205.

123 Tonzetich, Z.J., Wang, H., Mitra, D. et al. (2010). Identification of protein-bound dinitrosyl iron complexes by nuclear resonance vibrational spectroscopy. *J. Am. Chem. Soc.* 132 (20): 6914–6916.

124 Schiewer, C.E., Müller, C.S., Dechert, S. et al. (2019). Effect of oxidation and protonation states on [2Fe–2S] cluster nitrosylation giving $\{Fe(NO)_2\}^9$ dinitrosyl iron complexes (DNICs). *Inorg. Chem.* 58 (1): 769–784.

125 Tseng, Y.T., Pelmenschikov, V., Iffland-Mühlhaus, L. et al. (2023). Substrate-gated transformation of a pre-catalyst into an iron-hydride intermediate $[(NO)_2(CO)Fe(\mu\text{-}H)Fe(CO)(NO)_2]^-$ for catalytic dehydrogenation of dimethylamine borane. *Inorg. Chem.* 62 (2): 769–781.

126 Xiao, Y., Fisher, K., Smith, M.C. et al. (2006). How nitrogenase shakes – initial information about P–cluster and FeMo-cofactor normal modes from nuclear resonance vibrational spectroscopy (NRVS). *J. Am. Chem. Soc.* 128 (23): 7608–7612.

127 Van Stappen, C., Benediktsson, B., Rana, A. et al. (2023). Structural correlations of nitrogenase active sites using nuclear resonance vibrational spectroscopy and QM/MM calculations. *Faraday Discuss.* 243: 253–269.

128 Guo, Y., Brecht, E., Aznavour, K. et al. (2013). Nuclear resonance vibrational spectroscopy (NRVS) of rubredoxin and MoFe protein crystals. *Hyperfine Interact.* 222 (S2): 77–90.

129 Ogata, H., Krämer, T., Wang, H. et al. (2015). Hydride bridge in [NiFe]-hydrogenase observed by nuclear resonance vibrational spectroscopy. *Nat. Commun.* 6 (1): 7890.

130 Pelmenschikov, V., Birrell, J.A., Pham, C.C. et al. (2017). Reaction coordinate leading to H_2 production in [FeFe]-hydrogenase identified by nuclear resonance vibrational spectroscopy and density functional theory. *J. Am. Chem. Soc.* 139 (46): 16894–16902.

131 Caserta, G., Pelmenschikov, V., Lorent, C. et al. (2021). Hydroxy-bridged resting states of a [NiFe]-hydrogenase unraveled by cryogenic vibrational spectroscopy and DFT computations. *Chem. Sci.* 12 (6): 2189–2197.

132 George, S.J., Igarashi, R.Y., Xiao, Y. et al. (2008). Extended X-ray absorption fine structure and nuclear resonance vibrational spectroscopy reveal that NifB-co, a FeMo-co precursor, comprises a 6Fe core with an interstitial light atom. *J. Am. Chem. Soc.* 130 (17): 5673–5680.

133 Scott, A.D., Pelmenschikov, V., Guo, Y. et al. (2014). Structural characterization of CO-Inhibited Mo-nitrogenase by combined application of nuclear resonance vibrational spectroscopy, extended X-ray absorption fine structure, and density functional theory: new insights into the effects of CO binding and the role of the interstitial atom. *J. Am. Chem. Soc.* 136 (45): 15942–15954.

134 Sippel, D., Rohde, M., Netzer, J. et al. (2018). A bound reaction intermediate sheds light on the mechanism of nitrogenase. *Science* 359 (6383): 1484–1489.

135 Kuchenreuther, J.M., Guo, Y., Wang, H. et al. (2013). Nuclear resonance vibrational spectroscopy and electron paramagnetic resonance spectroscopy of ^{57}Fe-enriched [FeFe] hydrogenase indicate stepwise assembly of the H-cluster. *Biochemistry* 52 (5): 818–826.

136 Kamali, S., Wang, H., Mitra, D. et al. (2013). Observation of the Fe-CN and Fe-CO vibrations in the active site of [NiFe] hydrogenase by nuclear resonance vibrational spectroscopy. *Angew. Chem. Int. Ed.* 52 (2): 724–728.

137 Lauterbach, L., Wang, H., Horch, M. et al. (2015). Nuclear resonance vibrational spectroscopy reveals the Fe–S cluster composition and active site vibrational properties of an O_2-tolerant NAD $^+$-reducing [NiFe] hydrogenase. *Chem. Sci.* 6 (2): 1055–1060.

138 Preissler, J., Wahlefeld, S., Lorent, C. et al. (2018). Enzymatic and spectroscopic properties of a thermostable [NiFe]-hydrogenase performing H_2-driven NAD^+-reduction in the presence of O_2. *Biochim. Biophys. Acta Bioenerg.* 1859 (1): 8–18.

139 Lorent, C., Pelmenschikov, V., Frielingsdorf, S. et al. (2021). Exploring structure and function of redox intermediates in [NiFe]-hydrogenases by an advanced experimental approach for solvated, lyophilized and crystallized metalloenzymes. *Angew. Chem. Int. Ed.* 60 (29): 15854–15862.

140 Wang, H., Yoda, Y., Ogata, H. et al. (2015). A strenuous experimental journey searching for spectroscopic evidence of a bridging nickel–iron–hydride in [NiFe] hydrogenase. *J. Synchrotron Radiat.* 22 (6): 1334–1344.

141 Caserta, G., Lorent, C., Pelmenschikov, V. et al. (2020). *In vitro* assembly as a tool to investigate catalytic intermediates of [NiFe]-hydrogenase. *ACS Catal.* 10 (23): 13890–13894.

142 Pham, C.C., Mulder, D.W., Pelmenschikov, V. et al. (2018). Terminal hydride species in [FeFe]-hydrogenases are vibrationally coupled to the active site environment. *Angew. Chem. Int. Ed.* 57 (33): 10605–10609.

143 Pelmenschikov, V., Birrell, J.A., Gee, L.B. et al. (2021). Vibrational perturbation of the [FeFe] hydrogenase H-cluster revealed by $^{13}C^{2}H$-ADT labeling. *J. Am. Chem. Soc.* 143 (22): 8237–8243.

144 Reijerse, E.J., Pham, C.C., Pelmenschikov, V. et al. (2017). Direct observation of an iron-bound terminal hydride in [FeFe]-hydrogenase by nuclear resonance vibrational spectroscopy. *J. Am. Chem. Soc.* 139 (12): 4306–4309.

145 Rodríguez-Maciá, P., Galle, L.M., Bjornsson, R. et al. (2020). Caught in the H_{inact}: crystal structure and spectroscopy reveal a sulfur bound to the active site of an O_2-stable state of [FeFe] hydrogenase. *Angew. Chem. Int. Ed.* 59 (38): 16786–16794.

146 Kwiatkowski, A., Caserta, G., Schulz, A.C. et al. (2024). ATP-triggered $Fe(CN)_2CO$ synthon transfer from the maturase HypCD to the active site of apo-[NiFe]-hydrogenase. *J. Am. Chem. Soc.* 146 (45): 30976–30989.

18

The Multisite Microstate Model: A Theoretical Approach for Analyzing Charge Transfer in Proteins

G. Matthias Ullmann and Rajeev Ranjan Roy

University of Bayreuth, Theoretical and Computational Biochemistry, Universitätsstr. 30, NW I, 95447 Bayreuth, Germany

Main Chapter Points

- Outline of the multisite microstate model to mathematically describe binding, protonation, and redox equilibria and kinetics on the same principles
- Introduction of the Poisson–Boltzmann equation as a practical means to calculate the energetic parameter required for the multisite microstate model
- Exemplifying the principle of the model on iron–sulfur proteins

18.1 Introduction

Structural biology shapes our thinking in modern biochemistry. Much of our understanding of biochemical mechanisms relies on biomolecular structures. Structures alone, however, do not answer biochemical questions. Only in combination with data from other methods such as spectroscopy, electrochemistry, kinetics, and thermodynamics, structural data help us to draw a more complete picture of the underlying molecular mechanism. Theoretical approaches allow us to check if our ideas of the molecular mechanism are consistent with the laws of physics. Therefore, theoretical approaches are becoming increasingly popular also in biological science. However, because of the complexity of biology and biochemistry, theory in this field is still in its infancy compared to physics. Therefore, it is probably worth to take a step back and analyze what is the role of theory in enzymology and how theory promotes our understanding of nature through idealized models.

Theory in science plays several important roles:

- *Understanding and explanation*: Theories help us to understand different phenomena by providing explanations for observed facts or patterns by organizing and making sense of empirical data.
- *Guidance for research and practice*: Theories provide frameworks for conducting research and designing interventions or solutions. They offer researchers and

Iron-Sulfur Clusters: Biogenesis and Biochemistry, Volume 2, First Edition.
Edited by Silke Leimkühler, Günter Schwarz, Oliver Lenz, and Oliver Einsle.

practitioners a roadmap for investigating phenomena, developing hypotheses, designing experiments, and evaluating outcomes.

- *Prediction*: A good theory not only explains existing observations but also allows us to predict future events or behaviors. By understanding the underlying principles proposed by a theory, we can anticipate how things will behave under different circumstances.
- *Integration of knowledge*: Theories often integrate diverse observations and findings into a coherent framework. They help connect seemingly unrelated facts or ideas, fostering interdisciplinary understanding and collaboration.
- *Simplification*: Theories simplify complex phenomena by identifying key variables and relationships. By abstracting away unnecessary details, theories make it easier to comprehend and work with complex systems.
- *Critical thinking*: Theories encourage critical thinking by prompting us to question assumptions, test hypotheses, and evaluate evidence. They provide a basis for assessing the validity of arguments and claims.
- *Progress and innovation*: Theories drive progress and innovation by inspiring new ideas and discoveries. They challenge existing paradigms, stimulate debate, and motivate further inquiry, leading to advancements in knowledge and technology.
- *Communication*: Theories provide a common language and conceptual framework for communication within and across disciplines. They facilitate the exchange of ideas, enabling researchers and practitioners to communicate effectively and build upon each other's work.

Thus, theoretical approaches are central to the scientific approach and are also the basis for designing experiments and enabling their interpretation.

When a real object is analyzed, it is required to construct an abstract model of this object. A model functions as a generalized hypothetical framework used to analyze or elucidate an object. It offers a simplified portrayal of a real object, aimed at improving our comprehension, prediction, and control of its behavior. However, every model entails approximations, rendering it an idealized representation of reality. It captures only specific facets of the real object, particularly those pertinent to the properties being investigated. Other aspects of the same object might not be accurately described by the model, as it was originally tailored for different objectives. A model must elucidate experimental data and forecast the outcomes of new experiments. For predictive capability, a model must be sufficiently intricate to encompass all crucial aspects of the real object at an appropriate level of detail. Yet, the complexity of a model inversely affects the ease of interpreting its results. Thus, a model should be as complex as necessary, but not more, to offer insights into the behavior of the real system. Above all, a model should foster our comprehension of nature.

Constructing a model of a real system typically involves several stages (Figure 18.1). The initial and perhaps most crucial step is creating the conceptual model. Here, the real object is transformed into an idealized model object. By identifying which properties are of interest, essential features are selected to replicate these properties, determining how they describe the desired aspects of the object.

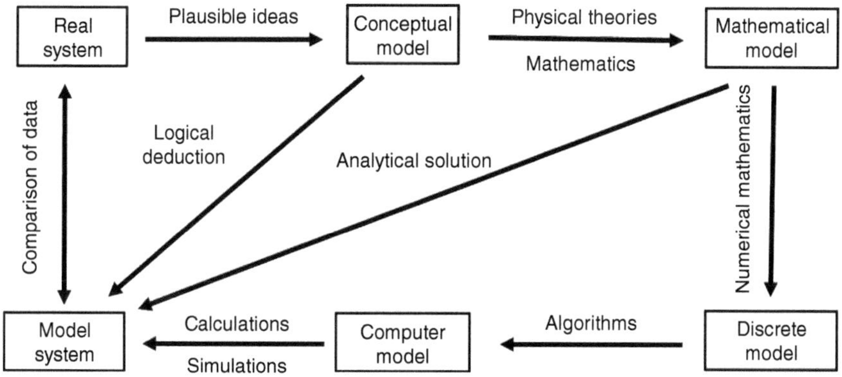

Figure 18.1 Developing theoretical frameworks for actual systems. This scheme illustrates the steps involved in constructing theoretical frameworks aimed at analyzing real-world systems. In the first step, a conceptual model needs to be constructed, which can already be used to make qualitative conclusions using just logical deduction. This first step is the most important step since it really determines the power of the subsequent mathematical model. A mathematical model can help formulate questions and answers more stringent. If the model can be solved analytically, the parameter of the model can be analyzed using just paper and pencil. If an analytical solution is not possible, numerical simulations can help understand the system behavior.

This initial stage necessitates a thorough examination of the object to be modeled, discerning the necessary details for the desired representation. When a qualitative understanding suffices, the conceptual model often serves to visualize the object, representing the ultimate goal of the inquiry. Conversely, when a quantitative or semiquantitative understanding is sought, the conceptual model must be translated into a mathematical model.

A concrete physical theory is required to formulate the mathematical model outlined in the conceptual model. If the mathematical model can be analytically solved, the objective is achieved, enabling a comparison of the model behavior with that of the real object. However, mathematical models often prove too complex to be solved analytically. In such instances, the mathematical model is translated into a discrete mathematical model using techniques from numerical mathematics. This discrete model can then be implemented in computer programs. Subsequently, the computer model facilitates simulations and calculations using sets of parameters. Comparing resulting simulation data with experimental data enables assessment of the model quality. Quantities derived from the model can enhance understanding of the real system. Additionally, analyzing the impact of well-defined changes in the initial parameters can yield deeper insights into the real behavior of the system.

For modeling charge transfer in biomolecules, we combine several theoretical approaches. At the most basic and abstract level, we rely on the system theory [1] and statistical thermodynamics [2, 3]. Based on these approaches, the multisite microstate model [4–7] was constructed, in which a biomolecular system is modeled as a collection of sites that can adopt several states. To actually model a biomolecular system, an energy function needs to be defined. We rely on many aspects of the calculations on the Poisson–Boltzmann equation [8–10] although also other ways to calculate energies can be used.

In this paper, we first describe the binding theory of a single ligand to a receptor and underline the importance of the chemical potential in this theory. The term ligand is used here in a generalized sense and protons or electrons can also be considered ligands. Next, we outline the approach to describe a biomolecule as a system of binding sites and how this approach is used to describe the thermodynamics and kinetics of biomolecular systems. Most of the practical calculations of the multisite microstate model rely on the Poisson–Boltzmann model of proteins in an ionic solution, which is introduced next and then applied to calculate the energetic parameter of the microstate model. At the end, some practical hints are given to construct concrete models of biomolecular systems.

18.2 Binding of Ligands to a Receptor

We are analyzing the binding of single ligand L to a (macromolecular) receptor M

$$M + L \rightleftharpoons ML \tag{18.1}$$

The free energy released during binding ΔG_{bind} is given by the difference in the chemical potentials.

$$\Delta G_{\text{bind}} = \mu_{ML} - (\mu_M + \mu_L) \tag{18.2}$$

where μ_{ML}, μ_M, and μ_L are the chemical potential of the respective chemical species. The chemical potential [11, 12] is defined by Eq. (18.3)

$$\mu_A = \left(\frac{\partial G}{\partial n_A}\right)_{T,p,n_{i \neq A}} \tag{18.3}$$

i.e. the chemical potential μ_A of compound A is the free energy change ∂G that is required to change the number n_A of particles of compound A while keeping temperature T, pressure p, and the number of all other compounds $n_{i \neq A}$ constant. From Eq. (18.3) together with the ideal gas law, one can derive another commonly used equation for the chemical potential

$$\mu_A = \mu_A^{\circ} + RT \ln \frac{\gamma_A [A]}{C^{\circ}} \tag{18.4}$$

$$= \mu_A^{\circ} + RT \ln(a_A) \tag{18.5}$$

where γ_A is the activity coefficient of compound A, $[A]$ is its concentration, and C° is the standard concentration. These three quantities are combined into the activity a_A of compound A: R is the general gas constant, T is the absolute temperature, and μ_A° is the standard chemical potential of compound A [13]. For convenience, we define a new quantity μ'_A that depends only on the activity of compound A.

$$\mu'_A = \mu_A - \mu_A^{\circ} = RT \ln(a_A) \tag{18.6}$$

Combining Eqs. (18.5) and (18.2), one obtains

$$\Delta G_{\text{bind}} = \mu^{\circ}_{ML} - (\mu^{\circ}_{M} + \mu^{\circ}_{L}) + RT\ln\frac{a_{ML}}{a_M a_L} \tag{18.7}$$

$$= \Delta G^{\circ}_{\text{bind}} + RT\ln\frac{a_{ML}}{a_M a_L} \tag{18.8}$$

where $\Delta G^{\circ}_{\text{bind}} = \mu^{\circ}_{ML} - (\mu^{\circ}_{M} + \mu^{\circ}_{L})$ is defined as the standard free energy of binding ΔG°. At equilibrium, ΔG_{bind} becomes zero

$$0 = \Delta G^{\circ}_{\text{bind}} + RT\ln\frac{a^{\text{eq}}_{ML}}{a^{eq}_M a^{\text{eq}}_L} \tag{18.9}$$

i.e. the activity of the various compounds adopts their respective equilibrium value a^{eq}_A and the two terms cancel each other. Using the probability of finding the ligand bound to M or not, i.e. p_{ML} and $p_M = 1 - p_{ML}$ instead of the respective activity, i.e. p_{ML} and p_M are the probabilities of two states, in one the ligand is bound to the molecule, in the other not, and using the definition in Eq. (18.6), one gets

$$0 = \Delta G^{\circ}_{\text{bind}} + RT\ln\frac{a_{ML}}{a_M} - \mu'_L \tag{18.10}$$

$$= \Delta G^{\circ}_{\text{bind}} + RT\ln\frac{p_{ML}}{p_M} - \mu'_L \tag{18.11}$$

$$= \Delta G^{\circ}_{\text{bind}} + RT\ln\frac{p_{ML}}{1 - p_{ML}} - \mu'_L \tag{18.12}$$

Rearranging this equation, one obtains

$$p_{ML} = \frac{e^{-(\Delta G^{\circ}_{\text{bind}} - \mu'_L)/RT}}{1 + e^{-(\Delta G^{\circ}_{\text{bind}} - \mu'_L)/RT}} \tag{18.13}$$

This equation describes the sigmoidal titration curve typical for ligand binding. Note that the denominator of Eq. (18.13) corresponds to a two-state partition function and consequently $\Delta G^{\circ}_{\text{bind}} - \mu'_L$ is the energy of the ligand-bound state relative to the state without bound ligand.

Two common biological ligands are protons and electrons. Their chemical potentials are important parameters for biochemical systems. However, these chemical potentials are commonly known as pH value and electrode potential E (or solution redox potential). They are defined as

$$\text{pH} = -\lg(a_{\text{H}^+}) \tag{18.14}$$

$$E = -\ln(a_{\text{e}^-}) \tag{18.15}$$

Probably for historical reasons, the sign of these "chemical potentials" is negative, i.e. a high proton concentration – or actually activity – corresponds to a low pH value and vice versa. The same is true for the electrode potential, i.e. a low electrode potential corresponds to many available electrons in the electrode. Thus, the respective chemical potentials are

$$\mu_{\text{H}^+} = -RT\ln(10)\text{pH} \tag{18.16}$$

$$\mu_{\text{e}^-} = -FE \tag{18.17}$$

where F is the Faraday constant. Note that $\mu^\circ_{H^+}$ and $\mu^\circ_{e^-}$ are usually by definition zero. The standard free energy for binding a proton or an electron to a receptor are, respectively,

$$\Delta G^\circ_{bind,H^+} = -RT\ln(10)pKa \quad (18.18)$$

$$\Delta G^\circ_{bind,e^-} = -FE^\circ \quad (18.19)$$

Using the chemical potential instead of pH and redox potential has the advantage that all binding phenomena can be described using the same theoretical framework.

Binding of ligands can be generalized, i.e. one can consider more than one binding site and more than one ligand type, which can lead to nonsigmoidal titration curves and other irregularities that seems surprising at first sight, but can be well explained on the basis of the microstate model [14–16].

18.3 Analyzing Biomolecular Systems

18.3.1 General Considerations

The physicist and philosopher Carl Friedrich von Weizsäcker [17] once wrote "*The world is not composed of objects, only the finite mind of man breaks up the whole – to which he belongs himself – into objects, in order to find his way.*"[1] Thus, the division of the world into several objects is a prerequisite for our analysis of the world. We try to reconstruct the wholeness of the world by considering the interactions between the objects. This reconstruction leads us to a theoretical description of the world, which helps us to describe it.

In the middle of the last century, Ludwig von Bertalanffy started to develop a general systems theory [1, 18]. Although this theory is not fully worked out, it can serve as a general guideline to construct a model of biomolecular systems. In general, a system can be defined as a set of objects and the relation between them (see Figure 18.2). It can be useful to combine several objects into subsystems and describe the system as a collection of subsystems and their interactions. Thus, often a system can have a hierarchical structure. For instance, a physical model can be defined as a collection of elementary particles and their interactions. If we consider chemical systems, it is more practical to combine several elementary particles into nuclei and the system can then be described as electrons, nuclei, and their interactions. It might even be useful to combine more constituents of the molecular system into functional groups, amino acid residues, nucleic acid bases, or metal centers coordinated by several functional groups. The way how we describe a system depends on the purpose of the model and the theoretical tools that are available.

A system as described above can adopt several states. A state of a system refers to the configuration of the system encompassing all relevant parameters, variables,

1 Original in German: "*Die Welt ist nicht aus Objekten zusammengesetzt, nur der endliche Verstand des Menschen zerlegt das Ganze, zu dem er selbst gehört, in Objekte, um sich zurechtzufinden.*" from Ref. [17].

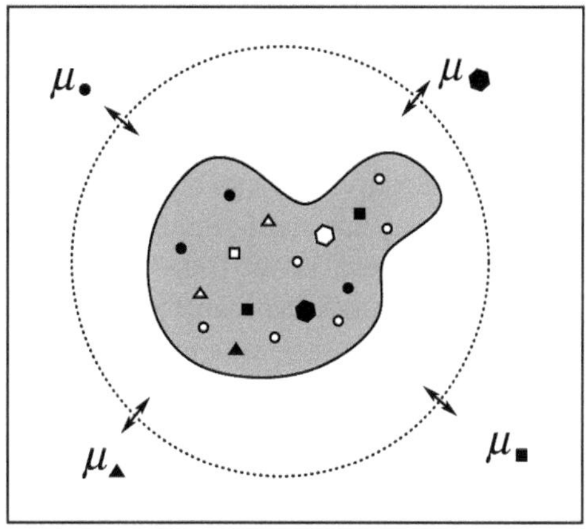

Figure 18.2 Schematic representation of a molecular system. The system consists of the macromolecule (gray area) together with some solvent layers (represented by the dotted line). The macromolecule has several binding sites (different symbols) for ligand different types of ligand. These binding sites can be either filled or empty. The ligand molecules can be taken up form the environment where the ligand molecules have a well-defined chemical potential μ.

and properties that describe its characteristics. Each state is associated with a certain energy that the system possesses in this state. At the most basic level, the state of the system can be a quantum state. For practical purposes, it is however useful to categorize several quantum states into microstates. Thus, the classification of the states of a system can also have a hierarchical order. This idea was already formulated in the conformational substate model by Frauenfelder et al. [19]. Microstates can also be grouped into macrostates. In contrast to microstates, macrostates are characterized by general system features such as the number of particles. Microstates instead have a higher degree of similarity. For instance, in a molecular system, a microstate can be in a particular protonation state in a given conformation. A macrostate instead can be characterized by the total number of protons bound to the molecule.

A system has an environment that can influence the system (for instance, by electrostatic fields). With respect to its environment, a system can be isolated (no exchange of energy or matter with its environment), closed (exchange of energy but not matter), or open (exchange of energy or matter) in a thermodynamic sense. The most general systems are open systems. Such a system can exchange molecules (or in general particles) with its environment. The exchange of molecules with the environment can quantified by the binding theory outlined above. Namely, the energy for binding a ligand molecule of type i to the system is proportional to the chemical potential μ_i of this type of ligand molecule in the environment.

In the multisite microstate model, we describe a macromolecular system (such as a protein in aqueous solution) that can bind several ligands such as protons, electrons, ions, or small molecules. At a basic level, such a molecular system is a collection of electrons and nuclei. An exact theoretical description of such a system would require a full quantum mechanical treatment, which is out of the reach for the computational power available today. Instead, we divide the system into several subsystems, namely into several sites, which are able to bind ligands, and the background, which remains unchanged. Concretely, the purpose of our model here is to describe a (bio-)molecular system, which is, for instance, able to perform charge transfer reactions such as proteins that are involved in biological energy transduction. Such proteins often bind cofactors such as iron–sulfur clusters or heme centers that allow electron transfer. In addition, there are many proton binding sites, usually the sidechain of charged amino acids, that facilitate proton transfer. Moreover, certain cofactors such as flavins or quinones can bind both protons and electrons such as flavins or quinones. Thus, a site in a protein can be, for instance, the sidechain of glutamate, that can bind a proton to its γ-carboxyl group, or a site can be an $[Fe_2–S_2]$ cluster together with the sidechains of the coordinating residues that can bind electrons. These different sites (subsystems) themself can also exist in different states, for instance, protonated or deprotonated, and oxidized or reduced.

To formalize the multisite microstate description, the following approach can be used. The (micro)state of the system that is divided into N sites and an unchangeable background is characterized by an N dimensional state vector $\vec{s}_c$

$$\vec{s}_c = (\ldots, i_u, \ldots, j_v, \ldots) \tag{18.20}$$

where i_u and j_v characterizes the state of site i, which is in state u, and site j, which is in state v; the subscript c of the state vector $\vec{s}_c$ characterizes the conformation of the system. Note that there are two different kinds of states, namely site states and system states. The energy $E(\vec{s}_c)$ of the state of such a system can be calculated as a sum of site energies $E_c^{\text{site}}(i_u)$ and the interaction $W_c(i_u,j_v)$ between the sites. In addition, there is an energy of the unchangeable background ΔG_c^{conf}. Thus, the energy of the state $\vec{s}_c$ of the system is given by

$$E(\vec{s}_c) = \sum_{i=1}^{N} \left(E_c^{\text{site}}(i_u) - \sum_{l=1}^{L} n_l(i_u)\, \mu_l \right) + \frac{1}{2} \sum_{i=1}^{N} \sum_{j=1}^{N} W_c(i_u,j_v) + \Delta G_c^{\text{conf}} \tag{18.21}$$

where $E_c^{\text{site}}(i_u)$ is the site energy of site i in the state u when the whole system is in conformation c; L is the total number of ligand types that can bind to the molecule; μ_l is the chemical potential of ligand type l; $n_l(i_u)$ is the number of ligand molecules of type l bound to site i in state u; $W_c(i_u,j_v)$ is the interaction between site i in state u and site j in state v; ΔG_c^{conf} is the energy of the background in conformation c. That means for different overall conformations of the macromolecule, we have different $E_c^{\text{site}}(i_u)$, $W_c(i_u,j_v)$, and ΔG_c^{conf}. This energy can be used in the thermodynamic and kinetic analysis of biomolecular systems.

18.3.2 Thermodynamic Analysis

One key concept in the understanding of statistical thermodynamics is the ensemble. In thermodynamics, the term ensemble refers to a collection of systems that share certain macroscopic properties, such as energy, volume, and the number of particles. An ensemble serves as a key theoretical framework for describing the statistical behavior of systems. The three most commonly used ensembles are the microcanonical ensemble (isolated systems), the canonical ensemble (closed systems), and the grand canonical ensemble (open systems). A dilute protein solution with a buffered solvent is a good approximation of a grand canonical ensemble, where each protein molecule with its surrounding solvent represents a system. Since the solution is dilute, the different protein molecules do not interact and since the solution is buffered (for instance, pH and redox buffered) the chemical potential of protons and electrons is nearly constant. Each system in the ensemble adopts a certain (micro)state. The number of states Ω of a system that consists of N sites, where each site can exist in η_i site states and can exist in C different global conformations, is given by

$$\Omega = C\prod_{i=1}^{N}\eta_i \tag{18.22}$$

A measured property $\langle A \rangle$ of an ensemble is the average of this property over the whole example. However, since the system can be classified into states, the average property can be obtained as a weighted average of this property A_ω in state ω over all states, and the weight is given by the probability of the state, i.e.

$$\langle A \rangle = \sum_{\omega=1}^{\Omega} P_\omega A_\omega \tag{18.23}$$

If the ensemble reaches equilibrium, the probabilities of the states adopt a Boltzmann distribution and thus we obtain for the probability of state i

$$P_i = \frac{\mathrm{e}^{-(E_i)/RT}}{\sum_{\omega=1}^{\Omega} \mathrm{e}^{-(E_\omega)/RT}} \tag{18.24}$$

With this approach, equilibrium probabilities such as protonation probabilities can be calculated. But it is also possible to calculate, for instance, pH-dependent spectra of protein solutions considering the spectra of the individual species.

18.3.3 Kinetic Analysis

The nonequilibrium behavior of a system as described earlier can also be analyzed using a microstate model. In the multisite microstate formalism, reactions are depicted as transitions between two well-defined microstates of the system. The temporal evolution of each microstate population can be simulated using a master equation.

$$\frac{\mathrm{d}}{\mathrm{d}t}P_\eta(t) = \sum_{\eta=1}^{\Omega} k_{\eta\zeta}P_\zeta(t) - \sum_{\zeta=1}^{\Omega} k_{\zeta\eta}P_\eta(t) \tag{18.25}$$

where $P_\eta(t)$ represents the probability that the system is in state η at time t, $k_{\eta\zeta}$ denotes the transition probability per unit time from state ζ to η. In Eq. (18.25), the first sum covers all reactions generating state η, while the second sum encompasses reactions depleting state η. Simulating reaction networks using Eq. (18.25) assumes these processes adhere to a Markovian stochastic process, where the probability of a given reaction solely depends on the current state of the system, independent of the path leading to that state. Eq. (18.25) is a system of coupled linear differential equations with constant coefficients, admitting an analytical solution [6, 20]. This equation delineates the time evolution of the probability distribution of the microstates of the system. The required rate constants can be calculated using electrostatic methods relying on the Marcus theory [6, 20, 21] and quantum chemical calculations relying on the transition state theory [20, 22]. Dynamical Monte Carlo Simulations, relying on the same theoretical principles, can also be used to simulate such systems [23–25] by solving the master equation numerically.

18.4 Modeling Protein Electrostatics Using the Poisson–Boltzmann Equation

In principle, the energy of each microstate of the system can be calculated by quantum chemical methods or by molecular mechanics. However, since a protein needs to be solvated, many water molecule have to be considered in such calculations. To obtain reliable (free) energies of the microstates, a sampling over various solvent configurations would be required, which is computationally expensive. Therefore, methods in which the solvent degrees of freedom are treated implicitly are a good alternative. Among these methods, Poisson–Boltzmann electrostatics is among the most reliable. This method is outlined below.

18.4.1 Conceptual Model

The conceptual model that lies underneath the protein electrostatics in an aqueous ionic solution described by the Poisson–Boltzmann Equation is as follows: The system consists of a protein immersed in a solution containing ions. The ionic strength of the solution is relatively low, meaning it contains a small concentration of ions. The protein has a specific conformation, i.e. a specific three-dimensional structure, which determines the spatial distribution of its charged groups. Besides the charged groups, permanent dipoles that originate, for instance, from the peptide bonds also contribute to the charge distribution within the protein. For example, α-helices contribute a strong dipole to the electrostatic potential [26]. The aqueous solvent is highly polarizable, i.e. the water molecules react to an external field by reorienting themself – positive charges will point toward a negative field and vice versa (orientational polarization). Thus, the solvent molecules create a field as a reaction to the external field, the so-called reaction field, that diminishes the original field. In addition, there is a much smaller polarization effect that originates from a redistribution of the electron density in the molecule as a reaction to the external field. This later effect is present in the solvent as well as in the protein molecule (the solute). The orientational polarization does not play a major role in the protein, since it is assumed that the protein can adopt only a single conformation with small conformational flexibility. In summary, the solvent is a region with a high polarizability, while the protein is a region with a low polarizability. Thus, there is a spatially varying polarizability, i.e. a spatially varying permittivity (sometimes also called dielectric constant). The ions are only present in the solvent, where they adopt an equilibrium distribution, i.e. cations will have a higher concentration in regions with negative electrostatic potential while anions will have a higher concentration in regions with positive electrostatic potential. This conceptual model is the basis for the mathematical model that will be derived in Section 18.4.2.

18.4.2 The Mathematical Model

The electrostatic potential $\phi(\mathbf{r})$ of the charge distribution $\rho(\mathbf{r})$ in vacuum at a position $\mathbf{r}$ is given by Eq. (18.26) [27].

$$\phi(\mathbf{r}) = \int \frac{\rho(\mathbf{r})}{4\pi\varepsilon_0\mathbf{r}}\mathrm{d}\mathbf{r} \tag{18.26}$$

where ε_0 is the vacuum dielectric constant. The electrostatic field $\mathbf{E}$ is the negative gradient of the potential ϕ (Eq. (18.27)), ∇ is the gradient operator.

$$\mathbf{E}(\mathbf{r}) = -\nabla\phi(\mathbf{r}) \tag{18.27}$$

In a dielectric medium, the electrostatic field is shielded due to the reorientation of permanent and induced dipoles. It is useful to define a dielectric displacement vector $\mathbf{D}$ (Eq. (18.28)).

$$\mathbf{D}(\mathbf{r}) = \varepsilon(\mathbf{r})\mathbf{E}(\mathbf{r}) \tag{18.28}$$

where $\varepsilon(\mathbf{r})$ is the spatially varying permittivity. Applying the Gauß law to a charge density $\rho(\mathbf{r})$ in a medium with the spatially varying permittivity $\varepsilon(\mathbf{r})$, one obtains

$$\nabla\mathbf{D}(\mathbf{r}) = \nabla\left[\varepsilon(\mathbf{r})\mathbf{E}(\mathbf{r})\right] = 4\pi\rho(\mathbf{r}) \tag{18.29}$$

With Eq. (18.27), one obtains the Poisson equation for a medium with a spatially varying permittivity $\varepsilon(\mathbf{r})$.

$$\nabla\left[\varepsilon(\mathbf{r})\nabla\phi(\mathbf{r})\right] = -4\pi\rho(\mathbf{r}) \tag{18.30}$$

As outlined earlier, the protein is assumed to have a fixed structure. The charge density $\rho(\mathbf{r})$ of the system therefore consists of two parts, the spatially fixed charges $\rho_p(\mathbf{r})$ of the protein and the mobile charges $\rho_{\mathrm{ion}}(\mathbf{r})$ of the solution (ions) (Eq. (18.31))

$$\rho(\mathbf{r}) = \rho_p(\mathbf{r}) + \rho_{\mathrm{ion}}(\mathbf{r}) \tag{18.31}$$

The mobile charges of the solution adopt an equilibrium distribution. The mean concentration $\langle c_i(\mathbf{r})\rangle$ of ions of type i at the position $\mathbf{r}$ is given by Eq. (18.32)

$$\langle c_i(\mathbf{r})\rangle = c_i^{\mathrm{bulk}}\exp(-\mathcal{E}_i(\mathbf{r})/RT) \tag{18.32}$$

where c_i^{bulk} is the bulk concentration of the ionic species i and $\mathcal{E}_i(\mathbf{r})$ is the energy of an ion of type i at position $\mathbf{r}$, which is given by Eq. (18.33) for a dilute solution.

$$\mathcal{E}_i(\mathbf{r}) = Z_i\phi(\mathbf{r}) \tag{18.33}$$

where Z_i is the charge of the ion of type i. The charge density $\rho_{\mathrm{ion}}(\mathbf{r})$ is then given by a sum over the charge density of all K ionic species in the solution (Eq. (18.34)).

$$\rho_{\mathrm{ion}}(\mathbf{r}) = \sum_{i=1}^{K} c_i^{\mathrm{bulk}} Z_i \exp\left(\frac{-Z_i\phi(\mathbf{r})}{RT}\right) \tag{18.34}$$

The Poisson–Boltzmann equation for a protein in ionic solution is then given by combining Eqs. (18.30), (18.31), and (18.34) to Eq. (18.35)

$$\nabla\left[\varepsilon(\mathbf{r})\nabla\phi(\mathbf{r})\right] = -4\pi\left(\rho_p(\mathbf{r}) + \sum_{i=1}^{K} c_i^{\text{bulk}} Z_i \exp\left(\frac{-Z_i\phi(\mathbf{r})}{RT}\right)\right) \tag{18.35}$$

The Poisson–Boltzmann equation is a nonlinear partial differential equation due to the presence of the potential $\phi(\mathbf{r})$ on both sides of the equation. Specifically, the nonlinearity arises from the exponential term on the right side of the equation, which describes the ion distribution around the protein. Nonlinear partial differential equations are difficult to solve, even numerically. For relatively small electrostatic potentials and not too high ion concentrations, the Poisson–Boltzmann equation can be linearized by expanding the exponential up to the linear term (Eq. (18.36)).

$$\sum_{i=1}^{K} c_i^{\text{bulk}} Z_i \exp\left(\frac{-Z_i\phi(\mathbf{r})}{RT}\right) \cong \sum_{i=1}^{K} c_i^{\text{bulk}} Z_i - \sum_{i=1}^{K} c_i^{\text{bulk}} Z_i^2 \frac{\phi(\mathbf{r})}{RT} \tag{18.36}$$

The first term in Eq. (18.36) vanishes because of electroneutrality of the solution. It is useful to define the ionic strength I (Eq. (18.37)) and the inverse Debye length κ (Eq. (18.38)).

$$I = \frac{1}{2}\sum_{i=1}^{K} c_i^{\text{bulk}} Z_i^2 \tag{18.37}$$

$$\kappa = \sqrt{\frac{8\pi N_A I}{\varepsilon(\mathrm{r}) k_B T}} \tag{18.38}$$

With the definitions in Eqs. (18.37) and (18.38), and with Eqs. (18.35) and (18.36), one obtains the linearized Poisson–Boltzmann equation (Eq. (18.39)).

$$\nabla\left[\varepsilon(\mathbf{r})\nabla\phi(\mathbf{r})\right] = -4\pi\rho_p(\mathbf{r}) + \varepsilon(\mathbf{r})\kappa^2(\mathbf{r})\phi(\mathbf{r}) \tag{18.39}$$

The linearity of Eq. (18.39) implies that the potentials resulting from two separate fixed charge distributions, $\rho_1(\mathbf{r})$ and $\rho_2(\mathbf{r})$, are additive, provided that the spatial distribution of the dielectric permittivity remains constant. In other words, for a combined charge distribution $\rho_1(\mathbf{r})$ and $\rho_2(\mathbf{r})$, the total potential $\phi(\mathbf{r})$ can be obtained by summing the individual potentials $\phi_1(\mathbf{r})$ and $\phi_2(\mathbf{r})$, as long as the dielectric permittivity $\varepsilon(\mathbf{r})$ across the space remains unchanged. This property holds significant implications for various applications. For example, linearity enables the determination of relative binding constants by continuum electrostatic calculations. The ability to add potentials simplifies many calculations that rely on the superposition principle.

Analytical solutions of the Poisson–Boltzmann equation exist only for simple geometries such as charged spheres, charged cylinders, or charged planes [28, 29]. For irregular geometries, solutions can be obtained by numerical methods. Most often, the Poisson–Boltzmann equation is solved by finite difference methods [30]. But also other methods to solve partial differential equations such as boundary element methods [31] or multigrid-based methods [32] can be used to solve the Poisson–Boltzmann equation.

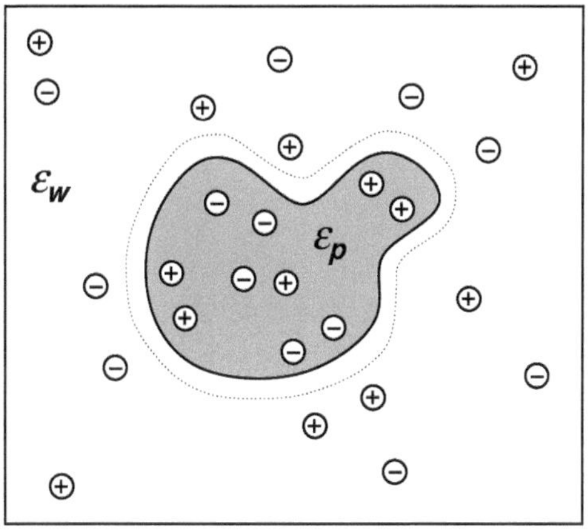

Figure 18.3 Conceptual model of the electrostatics of a protein. The protein with spatially fixed charges (gray area) is embedded in an ionic aqueous solution in which the mobile charges (ions) adopt an equilibrium distribution. The dotted line represents the Stern layer. The polarizability of the solvent ε_W is high due to the high mobility of water molecules, which moreover possess large permanent dipoles. The polarizability in the protein ε_p is low, since it adopts a structure with only limited mobility.

18.4.3 Electrostatic Potentials and Electrostatic Energies

The solution to the Poisson–Boltzmann equation yields the electrostatic potential $\phi(\mathbf{r})$, which can be decomposed into two distinct components.

$$\phi(\mathbf{r}) = \sum_{i=1}^{M} \frac{q_i}{4\pi\varepsilon_p \mid \mathbf{r} - \mathbf{r}'_i \mid} + \phi_{\text{rf}}(\mathbf{r}) \tag{18.40}$$

The first component represents the Coulomb electrostatic potential at position $\mathbf{r}$, arising from M point charges q_i located at positions $\mathbf{r}'_i$ within a medium characterized by a permittivity ε_p.

The second component $\phi_{\text{rf}}(\mathbf{r})$ accounts for the reaction field potential. This potential arises from the interplay of the charge distribution within the protein, the dielectric boundary between the protein and the solvent, and the distribution of ions in the solution (see Figure 18.3). The reaction field always acts in opposition to the field of the solute, effectively shielding it. This reaction field plays a crucial role in elucidating the structural and functional properties of proteins.

The electrostatic potential $\phi(\mathbf{r})$ can be used to give quantitative insights into biochemical mechanisms (Figure 18.4). One can distinguish between two different types of electrostatic energies: interaction energies between two charge distributions and reaction field energies, i.e. the interaction of the charge distribution with its own reaction field.

The interaction energy G_{inter} arises from charging a set of charges represented by ρ_2 in the presence of the electrostatic potential induced by the charge distribution ρ_1. If we consider ρ_2 as comprising a single point charge q_f, the interaction energy can be expressed as

$$G_{\text{inter}} = \int_0^{q_f} \phi(\rho_1, \mathbf{r}_q)\mathrm{d}q = \phi(\rho_1, \mathbf{r}_q)q_f \tag{18.41}$$

Because the potential $\phi(\rho_1, \mathbf{r}_q)$ at the position $\mathbf{r}_q$ of the charge q_f is entirely independent of the specific value of q_f itself, the integration in Eq. (18.41) simplifies to a

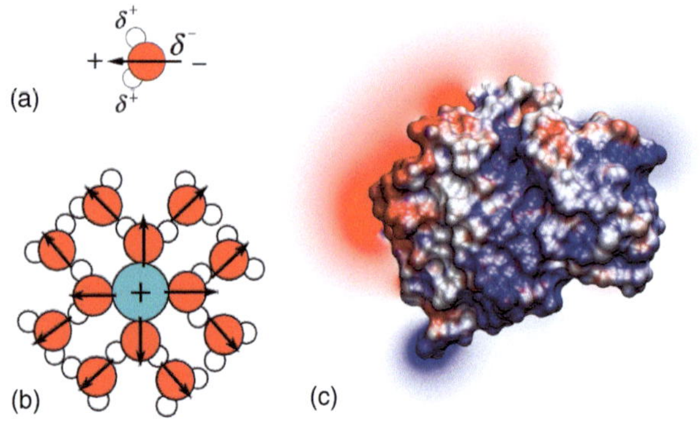

Figure 18.4 Electrostatic potential in an aqueous solution. The electrostatic potential in an aqueous solution arises from two components: the potential due to the charges of the solute and the potential due to the reaction field which arises from the reorientation of solvent molecule. The large dipole in a water molecule (a) leads to the orientation of negative end of the dipole toward a positive charge. The collective effect of this arrangement of the water molecule leads to the reaction field (b). The potential of a protein is often mapped to its surface, as shown here for the Rieske domain of the cytochrome bc_1 complex (c). The potential ranges from $-3\ e_\circ/RT$ (deep red) to $+3\ e_\circ/RT$ (deep blue). In addition, the picture shows the potential that reaches out into the space as a slice through space.

multiplication. Thus Eq. (18.41) can be used to describe the interaction between two distinct sets of point charges $\{q\}$ and $\{Q\}$ as follows:

$$G_{\text{inter}} = \sum_{i=1}^{N_q} q_i\ \phi(\{Q\}, \mathbf{r}_{q_i}) = \sum_{i=1}^{N_Q} Q_i\ \phi(\{q\}, \mathbf{r}_{Q_i}) \tag{18.42}$$

where N_q and N_Q represent the number of charges in the charge sets $\{q\}$ and $\{Q\}$, respectively. $\phi(\{Q\}, \mathbf{r}_{q_i})$ denotes the potential originating from the charge set $\{Q\}$ at the location of the charge q_i, while $\phi(\{q\}, \mathbf{r}_{p_i})$ represents the potential originating from the charge set $\{q\}$ at the position of the charge Q_i. Note that the interaction energy is symmetric as indicated in Eq. (18.42).

The reaction field energy, sometimes referred to as self-energy, is the interaction energy between the charge set $\{q\}$ with its own reaction field potential ϕ_{rf}. To calculate this energy, one envisions the process of charging a particle within a dielectric medium and seeks to determine the energy associated with this charging process. Similar to Eq. (18.41), we can write

$$G_{\text{rf}} = \int_0^{q_f} \phi_{\text{rf}}(q, \mathbf{r}_q) \mathrm{d}q \tag{18.43}$$

However here, the reaction field potential ϕ_{rf} varies according to the charge of the particle. Usually, one assumes a linear response, i.e.

$$\phi_{\text{rf}} = Cq_f \tag{18.44}$$

where C is the proportionality constant. From Eq. (18.43), one obtains

$$G_{\mathrm{rf}} = \int_0^{q_f} Cq \, dq = \frac{1}{2} C q_f^2 \tag{18.45}$$

Using Eq. (18.44), Eq. (18.45) becomes

$$G_{\mathrm{rf}} = \frac{1}{2} \phi_{\mathrm{rf}}(q_f, \mathbf{r}_q) q_f \tag{18.46}$$

Note that the factor $\frac{1}{2}$ originates from the linear response ansatz. Equation (18.46) can be generalized for a point charge set $\{q\}$

$$G_{\mathrm{rf}} = \frac{1}{2} \sum_{i=1}^{N_q} q_i \, \phi_{\mathrm{rf}}(\{q\}, \mathbf{r}_{q_i}) \tag{18.47}$$

Though initially appearing simple, the continuum electrostatic model proves remarkably effective in elucidating the characteristics and phenomena associated with the electrostatics of biomolecules. These phenomena include protein association, redox reactions, and pH titration behavior. Its efficacy likely stems from its ability to account for the solvent degrees of freedom, which are typically challenging to sample in molecular dynamics simulations, by assuming a dielectric continuum description of the medium. To fully harness the potential of continuum electrostatics, it must be integrated with other statistical thermodynamics techniques as elaborated below.

18.4.4 Defining the Low-Dielectric Cavity of the Protein

One crucial step in solving the Poisson–Boltzmann equation is defining the dielectric regions, i.e. specifying the regions of high- and low-dielectric permittivity. For small molecules, the region inside the molecule can be defined by the van der Waals radii of the atoms. However, for medium-sized molecules, gaps often appear between spatially neighboring, chemically nonbonded atoms, creating void regions within the molecule. This issue is even more pronounced in proteins (see Figure 18.5). Namely, water molecules will not be able to touch the whole van der Waals surface of a molecule.

To address this problem, the solvent-accessible surface was introduced in molecular modeling [33]. This surface can be defined by rolling a probe sphere of the size of a solvent molecule over the surface of the atoms of the molecule [34]. The solvent-accessible surface is defined by all possible centers of the probe spheres that can roll over the van der Waals surface without penetrating any atom. This surface helps in defining the molecular surface, also known as the solvent-excluded surface. It is the boundary formed by the surface of the probe sphere when it is in contact with the van der Waals surface of the molecule. In other words, it is the locus of points on the probe sphere that just touches the van der Waals spheres of the atoms in the molecule. The region inside the molecular surface is the molecular volume or the solvent-inaccessible volume. In protein electrostatics, the solvent-inaccessible volume usually defines the low-dielectric cavity. However, this definition has some difficulties and requires assumptions that are only approximately true.

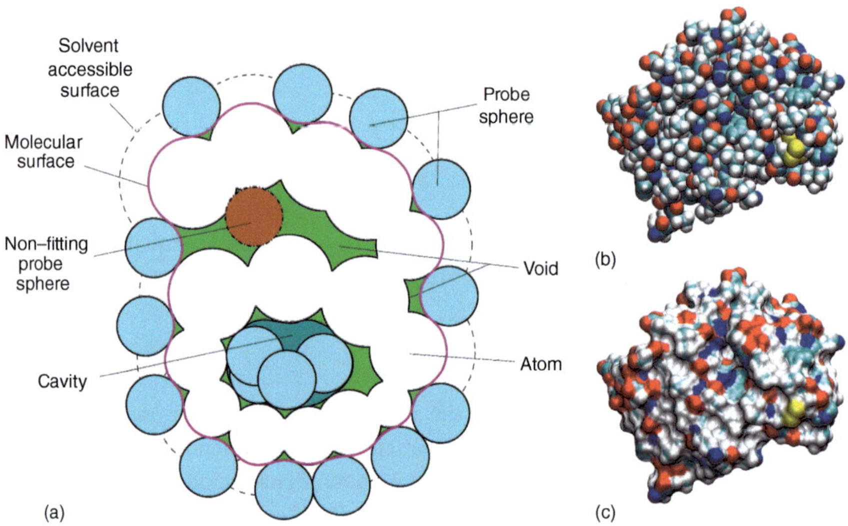

Figure 18.5 Defining the dielectric regions. The low-dielectric region inside a protein is defined by a probe sphere representing a solvent molecule (cyan spheres). (a) The protein atoms are represented as white spheres. The probe sphere is rolled over the surface of the protein and so defines the solvent accessible surface (dashed line), which is defined by the center of the probe sphere, and the molecular surface (magenta line), which is defined by the closest possible contact of probe sphere and the atoms of the molecule. The procedure leads to the definition of voids, i.e. regions inside the protein that are too small to fit a solvent molecule. The upper green area depicts such a void, since the probe sphere (shown in brown here) is to large to fit in this void. However, there can also be internal cavities, i.e. regions that are large enough to fit at lease one probe sphere (dark cyan region). (b) Rieske protein in a van-der-Waals sphere representation. (c) Molecular Surface of the Rieske protein. While in the van-der-Waals sphere representation one can see gaps between the sphere, these gaps vanish if one considers the molecular surface.

One problem is defining the radius of a water molecule. Although a water molecule is not spherical, the strong polarization of the hydrogen–oxygen covalent bond means the deviation is not very high. The commonly used radius is 1.4 Å. This value originates from pair-correlation functions of water measured by neutron scattering data, which have a peak at about 2.8 Å, corresponding to the center-to-center distance of the two nearest neighbor water molecules [35].

The van der Waals radii of atoms are also only approximations. The most commonly used set of radii was determined by Bondi [36] by analyzing crystal structures. This set was later extended by Truhlar and coworkers using quantum chemical calculations [37]. Some other sets of radii were suggested and different strategies were used to derive them [38–40]. Often, the radius is just assigned to the element and only in some cases, the chemical bonding of the atom is considered [40]. The latter approach considers that due to the polarization within the bond, the radii of the atom may deviate, which is especially important for bonds involving hydrogen atoms. Namely, hydrogen atoms in polar bonds have a considerably smaller radius than those in nonpolar bonds.

The probe sphere radius and the radii of the atoms define the solvent-excluded volume, which is the complement to the solvent accessible volume. The solvent accessible volume can be divided into the bulk solvent and internal protein cavities, i.e. parts of the volume that are separated by the bulk solvent by solvent excluded volume [41]. In continuum electrostatic calculations, these internal protein cavities are usually filled with high-dielectric media. In reality, the internal cavities may or may not be filled with water [42]. Whether the cavity is filled with water or not makes a difference, since a cavity filled with water will give rise to a reaction field, while an empty cavity will not. However, in continuum electrostatic calculations, a high permittivity is usually assigned to all cavities, whether they are filled or not. This way of treating internal cavities is a potential source of error in continuum electrostatic calculations that is often overlooked. Moreover, different sets of radii may or may not be able to detect cavities and thus potential sources of reaction fields. The inclusion of explicit water molecules in continuum electrostatic calculations is usually also not advisable, since the water molecules will adopt one particular orientation, which cannot change during the calculation. Thus, the modeling of water-filled cavities as regions with a high permittivity is usually the better choice. Taken together, the choice of the set of radii is an important step in defining the dielectric boundaries in continuum electrostatic calculations. This calculation plays probably a more important role than commonly thought.

18.4.5 Fitting Quantum Chemical Electrostatic Potentials to Point Charges

In continuum electrostatics calculations as in molecular mechanics, the charge distribution within the protein is usually represented by atom-centered point charges [43]. The charges are the so-called partial charges, i.e. noninteger charges when measured in elementary charge units. The partial charges are used to describe the asymmetric distribution of electrons in chemical bonds. Nevertheless, partial charges are not quantum-theoretical observables and can thus not be unambiguously determined. In fact, there are several ways to calculate partial charges from molecular structures. Some charges, like Mulliken or Löwdin charges, are derived by partitioning the molecular wave function using orbital based schemes [44]. Other methods distribute the (calculated) electron density according to certain schemes (for instance, used for Bader charges, Hirshfeld charges, or Voronoi Deformation Density charges) [44]. The most frequently used methods to derive molecular partial charges rely on the fitting of point charges to the quantum-chemically calculated electrostatic potential (CHELPG, Merz-Kollmann, RESP) [43]. The general philosophy of these latter methods is discussed in this section.

In quantum chemistry, the vacuum electrostatic potential $\phi_{\text{qc}}(\mathbf{r})$ of a molecule consisting of N atoms having the nuclear coordinates $\mathbf{R}_j$ depends on the charge of the nuclei of the molecule Z_j and the electron density $\rho(\mathbf{r}')$ at a point $\mathbf{r}$ is given by

$$\phi_{\text{qc}}(\mathbf{r}) = \sum_{j=1}^{N} \frac{Z_j}{|\mathbf{r} - \mathbf{R}_j|} - \int \frac{\rho(\mathbf{r}')}{|\mathbf{r} - \mathbf{r}'|} \mathrm{d}\mathbf{r}' \tag{18.48}$$

The goal of the fit is now to represent this potential as closely as possible by point charges located at the nuclear positions. This classical Coulomb potential $\phi_{\text{cl}}(\mathbf{r})$ arising from the set of point charges $\{q_j\}$ at the position of the nuclei is given by

$$\phi_{\text{cl}}(\mathbf{r}) = \sum_{j=1}^{N} \frac{q_j}{|\mathbf{r} - \mathbf{R}_j|} \tag{18.49}$$

To map the quantum chemical electrostatic potential ϕ_{qc} to the classical potential ϕ_{cl}, least square fit techniques can be used to obtain point charges q_j. For this purpose, a function $f(\{q_j\})$ can be defined that quantifies the absolute deviation between the quantum chemical and the classical electrostatic potential at M points around the molecule; γ_i is a weight given to point i, which can adopt different positive values by choice. Often it is advantageous to include a second term that restraints the charges that are supposed to fit the potential to partial charges obtained from other methods (such as Mullikan charges, Löwdin charges, or Hirshfield charges). In the example here, a harmonic restraint is used to keep the charge $\{q_j\}$ close to the target charge set $\{q_j^\circ\}$ using restraint constant K_j.

$$f(\{q_j\}) = \sum_{i=1}^{M} \gamma_i \left(\sum_{j=1}^{N} \frac{q_j}{|\mathbf{r}_i - \mathbf{R}_j|} - \phi_{\text{qm}}(\mathbf{r}_i) \right)^2 + \sum_{j=1}^{N} K_j (q_j - q_j^\circ)^2 \tag{18.50}$$

The function $f(q_j)$ needs to be minimized with respect to q_j to obtain the optimal point charges that fit the quantum chemical electrostatic potential, i.e.

$$0 = \frac{\partial f(\{q_j\})}{\partial q_k} \tag{18.51}$$

Equation (18.51) leads to N interdependent linear equations, one for the derivation of Eq. (18.50) with respect to each charge. This N-dimensional system of linear equations can be written in a matrix form of equation

$$\mathbf{0} = \mathbf{A}\,\mathbf{q} - \mathbf{b} \tag{18.52}$$

where $\mathbf{0}$ is a null vector, $\mathbf{q}$ is the vector containing the N point charges that ought to fit the potential, the matrix $\mathbf{A}$ is a matrix containing distance information and the restraints, $\mathbf{b}$ is a vector containing information on the quantum chemical electrostatic potential and the restraint, namely the matrix $\mathbf{A}$ has the following diagonal elements A_{kk} and off-diagonal elements A_{kl}, and the vector $\mathbf{b}$ has the elements b_k:

$$A_{kk} = \sum_{i=1}^{M} \frac{\gamma_i}{|\mathbf{r}_i - \mathbf{R}_k|^2} + K_k \tag{18.53}$$

$$A_{kl} = \sum_{i=1}^{M} \frac{\gamma_i}{|\mathbf{r}_i - \mathbf{R}_k||\mathbf{r}_i - \mathbf{R}_l|} \tag{18.54}$$

$$b_k = \sum_{i=1}^{M} \frac{\gamma_i \phi_{\text{qm}}(\mathbf{r}_i)}{|\mathbf{r}_i - \mathbf{R}_k|} + K_k q_k^\circ \tag{18.55}$$

Obviously, the matrix $\mathbf{A}$ is symmetric, since $A_{kl} = A_{lk}$. Note that the matrix $\mathbf{A}$ is often ill-conditioned and thus the solution of Eq. (18.51) is often prone to large numerical errors. This numerical problem can be overcome by using singular value decomposition or a generalized inverse matrix [45–47].

In addition to the restraints on the charges discussed above, also hard constraints on charges can be employed. These constraints are usually introduced by using Lagrange multipliers [48]. Most commonly, the sum of all partial charges of the molecule is restrained to the overall charge of the molecule. In addition, also the dipole or higher multipoles can be used as restraints. Some force fields, such as CHARMM, group the charges of atoms to charge groups with an integer charge such that, for instance, the partial charges of methyl groups sum up to zero, or partial charges of a carboxylate group sum to −1. Also, these requirements can be enforced by Lagrange multipliers in the charge fit.

One crucial point in the charge fitting procedure is the choice of the points at which the potential will be calculated. Usually, no points are chosen within the molecule, since the potential inside the molecule is dominated by the nuclear charges and the direct electron cloud. Since the point charges are used to describe the interaction with other charges that are outside of the molecules van der Waals radii, only the potential at points outside of the molecule is relevant. Furthermore, the points should also be not too far from the molecule, since the potential far away from the molecule is dominated by the overall charge and dipole of the molecule and not by the details of the local charge distribution.

18.4.6 Problems and Shortcomings of the Poisson–Boltzmann Model

At first glance, the big success of the Poisson–Boltzmann equation in the description of biomacromolecules can be surprising, since a macroscopic model, namely continuum electrostatics, is used at a molecular level. The reason for this success is most likely that the continuum approach describes very well the averaging of the solvent degree of freedom over a longer time scale. Moreover, the protein usually adopts only very few global conformations that represent the conformational landscape of the protein reasonably well. Nevertheless, there are several shortcomings of the model and it is important to be aware of these limitations.

The most serious disadvantage of the model is that the electrostatic potential is always calculated for one defined structure of the protein and only the solvent degrees of freedom are implicitly averaged. To model the conformational degrees of freedom, calculations for several different conformations are required. Another potential problem lies in the choice of the dielectric constants. While there is not much debate on the dielectric constant of water, which is assumed to be about 80 at room temperature, the dielectric constant of proteins or other biomacromolecules is still discussed. Although there are good arguments that a reasonable value is about 4 [49], at the same time values as low as 2 and as high as 20 are discussed as well [50]. Most likely, it is not correct to assume a single dielectric constant for the whole protein but rather to derive a dielectric permittivity function that slightly varies in different positions of the protein [51–54]. Another problem already discussed in Section 18.4.4 is that the radii, which are used to define the borders between the low-dielectric region inside the molecule and the high-dielectric solvent, do not have a strict physical meaning.

Besides the problems concerning the parameters used to describe the Poisson–Boltzmann model, there are two problems that are inherent to the physical model itself. First, there is an explicit temperature dependence in the Poisson–Boltzmann equation. This temperature dependence affects only the distribution of the ions in the solution. However, there is another implicit temperature dependence that is related to the mobility of the molecule and the solvent, which is the temperature dependence of the dielectric constant. For instance, the dielectric constant of water decreases with increasing temperature [55]. The second inherent problem lies in the linear response ansatz for the calculation of the solvation energy that is described earlier (see Section 18.4.3), which may not always be justified, especially if the electrostatic potential becomes very large [56].

In Section 18.4.5, it is outlined how atom-centered point charges are usually obtained. However, it is possible that atom-centered point charges do not represent the quantum-chemical electrostatic potential well. The direct use of electron densities obtained from a wave function approximated by basis sets have the weakness that they have no sharp border and thus the electron density would reach out into the high-dielectric region. The determination of appropriate point charges is a crucial step in calculating electrostatic potentials and should not be underestimated.

Despite these weaknesses, the Poisson–Boltzmann model provides reliable energies if properly applied. It is therefore important to understand the underlying assumptions of the model and its limitations. The Poisson–Boltzmann model is at the moment the only practical possibility to calculate the energies of complex systems in the context of the multisite microstate model as outlined below.

18.5 Electrostatic Calculations of the Energy Parameter

In Section 18.3.1, a general equation for the energy of the state of a system is introduced (Eq. (18.21)), which separates the energy of the system into site energies, interaction energies, energies connected to conformational changes, and energies connected to the release or uptake of particles. The last of these terms is related to the chemical potential, which is outlined in Section 18.2. The chemical potentials of the various ligands can be treated as independent variables. The other energy terms can be obtained from electrostatic calculations in combination with experimental binding constants as outlined below.

According to the model described above, a site can exist in several site states. Such site states can be, for instance, different redox states, protonation states, or tautomeric states. The site energy $E_c^{\text{site}}(i_u)$ is the energy of the site i adopting site state u while the whole macromolecule is in conformation c. A common approach is to define a model compound, usually a small molecule, which can represent the site without the rest of the macromolecule present. For instance, an acetic acid can represent the protonatable group of a glutamate residue in a protein. The model compound can adopt several states, for instance, protonated or deprotonated. To each state of the model compound, an energy is assigned. Since energy is only defined up to a constant, the model compound energy of one of the states can be set to zero and the energy of the other states is given relative to this state. This model compound energy is altered by the environment within the macromolecule. On the one hand, the solvation changes when the model compound is transferred from the aqueous solution into the macromolecule. On the other hand, the model compound interacts with charges within the protein that do not belong to any other sites in the macromolecule. Thus, one can write

$$E_c^{\text{site}}(i_u) = G_{\text{model}}(i_u) + \Delta G_{\text{rf,c}}(i_u) + \Delta G_{\text{back,c}}(i_u) \tag{18.56}$$

where $G_{\text{model}}(i_u)$ is the model energy of site i in state u; $\Delta G_{\text{rf,c}}(i_u)$ is the change of interaction energy of site i in state u with its reaction field when transferred from the solvent into the macromolecule; $\Delta G_{\text{back,c}}(i_u)$ is the change of the interaction of the site with the background charges when transferring the model compound into the macromolecule. The background charges are the charges of the macromolecule that do not belong to any site.

The model energy $G_{\text{model}}(i_u)$ is usually taken from experimental values or high-level theory quantum chemical calculations on small molecular models. This energy term is best explained by an example. Suppose we have a carboxylic group in the protein. A model for this carboxylic group can be an acetic acid. This molecule has two protonation states, protonated (index p) and deprotoned (index d). The energy difference between these two states of the model compound in solution is related to its $\text{p}K_\text{a}$ value

$$\Delta G_{\text{prot}} = -RT \ln 10 \ \text{p}K_\text{a} = G_{\text{model}}(i_p) - G_{\text{model}}(i_d) \tag{18.57}$$

If we set $G_{\text{model}}(i_p)$ to zero, we obtain for $G_{\text{model}}(i_d)$

$$G_{\text{model}}(i_d) = RT \ln 10 \ \text{p}K_\text{a} \tag{18.58}$$

Note that the calculations can be more complicated if the states can exist in multiple tautomeric states or if more than one proton or different types of ligands bind to the site.

The site energy shift due to the change in the solvation energy is given by

$$\Delta G_{\text{rf,c}}(i_u) = \frac{1}{2} \sum_{k=1}^{N_{i_u}} Q_k(i_u) \left(\phi_p\left(\mathbf{r}_k; \{Q(i_u)\}\right) - \phi_m\left(\mathbf{r}_k; \{Q(i_u)\}\right)\right) \tag{18.59}$$

where $\{Q(i_u)\}$ is the set of charges assigned to state u of site i consisting of N_{i_u} charges and $Q_k(i_u)$ is one of these charges; $\phi_p\left(\mathbf{r}_k; \{Q(i_u)\}\right)$ is the electrostatic potential at the position $\mathbf{r}_k$ of charge $Q_k(i_u)$. The potential $\phi_p\left(\mathbf{r}_k; \{Q(i_u)\}\right)$ is calculated using the charge set $\{Q(i_u)\}$ and the positions and radii of the atoms of the protein (hence the subscript p) to define the dielectric boundary and the ion-excluded region. The potential $\phi_m\left(\mathbf{r}_k; \{Q(i_u)\}\right)$ instead is calculated using the positions of the atoms of the model compound (hence the subscript m) to define the dielectric boundary and the ion-excluded region. Note that the positions of the charges should be the same for the calculation of two electrostatic potentials and also the grids for calculating these potentials (i.e. positioning, size, and spacing) have to be the same in order to minimize numerical errors and to cancel the grid artifact [57].

The change in the interaction of the site with the background charges when transferring the model compound into the macromolecule is given by

$$\Delta G_{\text{back,c}}(i_u) = \sum_{k=1}^{N_p} q_k \phi_p\left(\mathbf{r}_k; \{Q(i_u)\}\right) - \sum_{l=1}^{N_{m_i}} q_l \phi_m\left(\mathbf{r}_k; \{Q(i_u)\}\right) \tag{18.60}$$

the meaning of $\phi_p\left(\mathbf{r}_k; \{Q(i_u)\}\right)$ and $\phi_m\left(\mathbf{r}_k; \{Q(i_u)\}\right)$ is stated above. There are N_{m_i} charges q_l that belong to the background in the model compound and there are N_p charges q_k that belong to the background in the protein. If the model compound does not contain any background charges, the second term in Eq. (18.60) is simply zero.

The interaction between two sites i and j in the respective states n and m is given by

$$W_c(i_u, j_v) = \sum_{k=1}^{N_{j_m}} Q_k(i_u)\phi_p\left(\mathbf{r}_k; \{Q(j_v)\}\right) \tag{18.61}$$

$$= \sum_{k=1}^{N_{i_n}} Q_k(j_v)\phi_p\left(\mathbf{r}_k; \{Q(i_u)\}\right) \tag{18.62}$$

The meaning of the charges and potentials is the same as above. Note that the energy should be symmetric, i.e. the interaction of the charges of site state i_u with the potential due to the charges of site state j_v is the same as the interaction of the charges of site state j_v with the potential due to the charges of site state i_u. Differences between the two numbers are due to numerical errors and can be used as a quality measure for the numerical solution of the Poisson–Boltzmann equation.

A macromolecule can exist in several conformations. In the context of the microstate model, it is assumed that there is only a small number of different conformations. Since the different conformations cause a change in the distribution of the dielectric permittivity, there are different sets of site energies and interaction energies for each conformation. These terms cover the changes in the interaction energies and the changes in the solvation energies of the sites due to the conformational change. However, there are additional energies that need to be taken into account. One energetic contribution $\Delta G_c^{\text{rf,back}}$ is the change of the solvation energy due to the altered distribution of the charges and of the dielectric permittivity. The other energetic contribution is connected to steric energy of the different conformation.

$$\Delta G_c^{\text{conf}} = \Delta G_c^{\text{rf,back}} + \Delta G_c^{\text{MM,back}} \tag{18.63}$$

The solvation energy contribution is calculated as

$$\Delta G_c^{\text{rf,back}} = \frac{1}{2}\sum_{k=1}^{N_p} q_k\left(\phi_p\left(\mathbf{r}_k; \{q\}\right) - \phi_p^{\text{vac}}\left(\mathbf{r}_k; \{q\}\right)\right) \tag{18.64}$$

where $\phi_p\left(\mathbf{r}_k; \{q\}\right)$ is the potential calculated using the background charge set $\{q\}$ and a high-dielectric constant for the solvent and a low-dielectric constant for the protein; $\phi_p^{\text{vac}}\left(\mathbf{r}_k; \{q\}\right)$ is the potential calculated with the same dielectric constant for the protein and a dielectric constant of 1 assigned to the solvent region. The $\Delta G_c^{\mathrm{MM,back}}$ energy term can be calculated using a force field. Note that in the force field energy, it can be advantageous to not consider the "stiff terms" (bond energy, angle energy, improper energy, and Lennard–Jones energy) and only consider the "soft terms" (electrostatic energy and torsion energy). The "stiff terms" energy terms are very sensitive to small changes in the coordinates and thus these energies often lead to unwanted behavior if only a small number of conformations are used.

18.6 Practical Hints to Construct a Multisite System

The construction of a multisite model of large proteins has many facets. In this section, we discuss many considerations of the actual model construction and how to divide the system into sites. A protein that consists only of amino acids without any cofactor or covalent modifications is relatively easy to construct and is done in many studies [58]. The construction of more complex systems containing several cofactors is much less trivial and requires a detailed analysis of the system as well as the development of new software.

18.6.1 Molecular Structures as a Basis of Detailed Calculations

The calculations of charge transfer reactions using the microstate model rely on molecular structures. The quality of these structures is therefore crucial for the quality of the calculations as we have shown in a recent paper [59]. To judge the quality of a structure, several parameters need to be considered. For structures from X-ray crystallography and cryo-electron microscopy, the resolution is the most crucial parameter. From our experience, structures with a resolution of about 2 Å or better are reliable for theoretical calculations. However, there are other criteria to judge crystal structure such as R-factors and many other quality parameters [60]. Structures determined by NMR are in our opinion often not well enough defined to allow reliable calculations. Structures in the Protein Data Bank are however of mix quality and some structural models can greatly benefit from additional refinement steps [61]. Fortunately, these additional refinements are possible because for a couple of years it is required to not only submit the structural model to the Protein Data Bank but also the experimental structure factors. In the case of unusual cofactors, ligands, or metal ions are bound to the protein, an additional analysis is mandatory, since often the geometric restraints used during the original refinement were not optimal. For instance, [Fe_4–S_4] clusters are cubic in some structures due to wrong restraints that were frequently used [62]. There are also specialized web servers that allow to validate the metal coordination in proteins [63]. It should also be kept in mind that Fourier transform artifacts in the electron density reconstruction can occur especially next to electron-rich atoms [64]. Another known problem in crystal structures is that crystal contacts may locally distort the structure. Besides visual inspection of the crystal structure in programs such as Coot [65], also servers like Evolutionary Protein–Protein Interface Classifier [66] can help in the analysis of potentially disruptive crystal contacts. For all these reasons, a careful analysis of structures and electron densities is required before the actual calculations can be started.

The electrostatic calculations often require some additional structural modeling. Usually at least hydrogen atom positions need to be constructed, since most crystal structures lack hydrogen atoms. After the geometric construction of the hydrogen atoms, it is usually advantageous to energetically optimize their positions using classical force fields. During optimization, nonhydrogen atoms should be kept fixed to stay as close as possible to the determined crystal structure. Moreover, in order to apply a classical force field, a certain protonation state has to be assigned to the protein. The optimization will then stabilize this assigned protonation state, which may not be the only relevant state, leading to biased calculations [67].

18.6.2 How to Choose Sites, Model Compounds, and Model Compound Energies

To perform calculations with the multisite microstate model, the system needs to be divided into multiple sites that are well separated from each other. In case, such a model is constructed for a protein, the division has to cut through covalent bonds. When the protein consists only of amino acids without any bound cofactors, the division is relatively simple. A division can simply be done by cutting through the C_α–C_β bond and the functional group. For instance, the carboxylate group of the aspartate can be considered a site (Figure 18.6a). In this case, the states of the site are just two – protonated and unprotonated. In the case of histidine, the situation is a bit more complex, since a proton can bind to either of the nitrogen atoms of the imidazole ring. Moreover, there is a possibility that the imidazole ring deprotonates completely leading to an imidazolate or protonates completely leading to an imidazolium. Thus, there are four possible states for a histidine site (Figure 18.6b).

There are also sites that involve more than one amino acid residue. For instance, the redox center of plant-type ferredoxin involves four cysteine residues of the protein and an inorganic $[Fe_2–S_2]$ core. In a biologically relevant oxidized state, both iron ions are formally Fe(III). In the biologically relevant reduced state, one of the iron ions is formally reduced to Fe(II). In the protein, the two ions are usually not equivalent and the reducing electron is thus localized either on one or the other iron ion. Thus, in total, there are three possible states for the $[Fe_2–S_2]$ core.

The situation is even more complex for the mitoNEET-like proteins [68]. The redox-active center of these proteins also have an $[Fe_2–S_2]$ core, but this core is coordinated by three cysteines and one histidine (see Figure 18.6c). The $[Fe_2–S_2]$ core can bind an electron, and the imidazole ring can deprotonate at $N_{\epsilon 2}$. Since the $[Fe_2–S_2]$ core and the imidazole ring are not electronically isolated, i.e. the electrons are delocalized over the whole mitoNEET center, it has to be considered one site. Thus, a mitoNEET center can exist in six possible states. The imidazole ring can be protonated or not, and each of these two protonation states can exist in three redox states for the $[Fe_2–S_2]$ core.

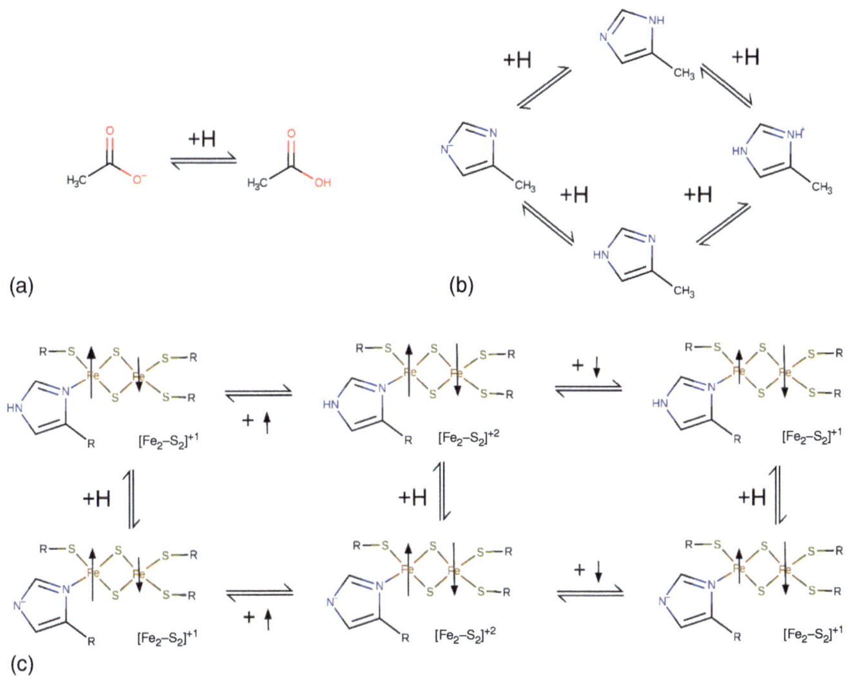

Figure 18.6 Examples for different model compounds. (a) Acetate has two possible protonation states. (b) Imidazole has four possible protonation states of which two states are tautomeric isomers. (c) The mitoNEET center has six possible states. The upward and downward pointing arrows at the iron atoms indicate the spin at the irons, which are antiferromagnetically coupled. In the oxidized state, both iron atoms are formally high-spin Fe(III) and have thus five unpaired electrons. The reducing electron can then be added to either of the iron atoms leading to two different reduced states. MarvinSketch Version 24.3.0 by ChemAxon (http://www.chemaxon.com) was used for drawing the chemical structures and reactions.

In general, sites can be separated from the protein background and from each other if they are electronically well isolated, which is the case to a good approximation when a nonpolar single bond is located between the site and the protein background. The definition of the site boundaries follows a similar philosophy as the definition of QM/MM centers [69]. In some cases, it is possible that a site can have many possible states for instance due to multiple protonation and redox states. Our software package GMCT [70] is able to calculate the energies of such complex systems.

A crucial energy parameter is the model energy $G_{\text{model}}(i_u)$. In many cases, the model energy can be taken from experimental model values such as pK_a values or redox potentials of the so-called model compounds, which resemble the site in the protein. Examples of such model compounds are imidazole for histidine, acetate for aspartate, or phenol for tyrosine. However, in some cases, model compounds do not exist or it is impossible to study the model compound in aqueous solution as, for instance, synthetic analogs of iron–sulfur clusters [71]. To determine model compound energies for such sites, we proposed a method that we call

Virtual Model Compound Approach (VMCA) [59]. If the redox potentials of several proteins having the same redox-active site but differing in amino acid sequences are available, the model redox potential can be inferred through electrostatic calculations. This idea can be expressed as

$$E_{\text{exp,i}} = E_{\text{model}} + \Delta E_{\text{shift,i}} \tag{18.65}$$

where $E_{\text{exp,i}}$ represents the measured redox potential of protein i, E_{model} is the redox potential of the redox-active site, and $\Delta E_{\text{shift,i}}$ denotes the shift due to the protein environment, calculable through methods like continuum electrostatics. Since the redox center is the same across all proteins considered, E_{model} should be identical, although its precise value is unknown. With experimental redox potentials available for N proteins, we employed the method of least squares to determine E_{model}. The calculation yields

$$E_{\text{model}} = \frac{1}{N}\sum_{i=1}^{N}\left(E_{\text{exp,i}} - \Delta E_{\text{shift,i}}\right) \tag{18.66}$$

Here, E_{model} represents the model redox potential, calculated as the average difference between measured and calculated redox potentials. Graphically, these data can be represented in a scatter plot of $\Delta E_{\text{shift,i}}$ against $E_{\text{exp,i}}$. The intercept of the regression line with a slope of one on the y-axis provides the value of E_{model}. It is important to note that a general linear least squares fit of $E_{\text{exp,i}}$ vs. $\Delta E_{\text{shift,i}}$ with a free slope is not justified due to Eq. (18.65).

The structures and redox potentials used in the VMCA need to be carefully selected [59]. The structures should have a high resolution and have to be well refined. As far as possible, the proteins should be small. It is advantageous, though not required, if the protein has only one redox-active site. The redox potential should have been determined in aqueous solution. Redox potentials determined by protein film voltammetry and similar methods [72] have the disadvantage that the proteins are embedded in a matrix and thus the dielectric surrounding of the protein is not in a high-dielectric aqueous solution.

The determination of the reference energy of model compound is an important step, since it directly shift the site energy. Investing time and effort to search the relevant literature and analyze the data helps to save time in later steps and to obtain reliable simulations of the charge transfer equilibria and reactions.

18.7 Conclusions

In this paper, we have summarized the multisite microstate model, a theoretical framework for investigating charge transfer in proteins. By combining system theory and statistical thermodynamics, biomolecular systems can be effectively modeled as collections of interacting sites, each capable of adopting multiple states. This approach, rooted in the Poisson–Boltzmann equation, provides a comprehensive method for calculating the electrostatic parameters that govern site interactions and energy distributions within proteins.

Our description extends beyond the binding of single ligands to encompass more complex scenarios involving multiple binding sites and various ligand types. This generalization allows for a deeper exploration of the intricate thermodynamics and kinetics inherent in biomolecular systems. The detailed methodology we have provided for constructing molecular models is intended to guide researchers in balancing the complexity of their models with the need for interpretability and practical application.

The multisite microstate model not only enhances our understanding of molecular mechanisms but also offers predictive capabilities that can inform experimental design and interpretation. By elucidating the conditions and parameters critical to charge transfer processes, this model paves the way for advancements in fields such as drug design, enzyme catalysis, and the development of biomolecular sensors.

Overall, our work underscores the importance of theoretical models in biochemistry, emphasizing their role in simplifying complex systems and providing insights that are crucial for both fundamental research and practical applications. The multisite microstate model has been applied to study the thermodynamics of charge transfer complexes [73, 74]; however, this method has also been applied already to study the kinetics of these complexes [5, 6]. Although in the past, the model has mainly been used to study charge transfer reactions, it is also possible to extend the method to investigate more complex enzymatic reactions.

Acknowledgement

We gratefully acknowledge Deutsche Forschungsgemeinschaft (DFG) Priority Programme "Iron–Sulfur for Life: Cooperative Function of Iron-Sulfur Centers in Assembly, Biosynthesis, Catalysis and Disease" (SPP 1927; UL-174-9; Project number 311144407) for the financial support.

References

1 von Bertalanffy, L., Beier, W., and Laue, R. (1977). *Biophysik des Fliessgleichgewichts*, 2e. Berlin: Akademie-Verlag.

2 Hill, T.L. (1960). *An Introduction to Statistical Thermodynamics*. Addison-Wesley Publishing Company, Inc.

3 Dill, K.A., Bromberg, S., and Stigter, D. (2010). *Molecular Driving Forces: Statistical Thermodynamics in Biology, Chemistry, Physics, and Nanoscience*. Garland Science. https://doi.org/10.4324/9780203809075.

4 Bashford, D. and Karplus, M. (1990). p*K*a's of ionizable groups in proteins: atomic detail from a continuum electrostatic model. *Biochemistry* 29 (44): 10219–10225. https://doi.org/10.1021/bi00496a010.

5 Becker, T., Ullmann, R.T., and Ullmann, G.M. (2007). Simulation of the electron transfer between the tetraheme subunit and the special pair of the

photosynthetic reaction center using a microstate description. *J. Phys. Chem. B* 111 (11): 2957–2968. https://doi.org/10.1021/jp066264a.

6 Bombarda, E. and Ullmann, G.M. (2011). Continuum electrostatic investigations of charge transfer processes in biological molecules using a microstate description. *Faraday Discuss.* 148: 173–193. https://doi.org/10.1039/C003905E.

7 Ullmann, G.M., Mueller, L., and Bombarda, E. (2016). *Theoretical Analysis of Electron Transfer in Proteins: From Simple Proteins to Complex Machineries*, Advances in Photosynthesis and Respiration. Cytochrome Complexes: Evolution, Structures, Energy Transduction, and Signaling, 99–127. Springer Netherlands. https://doi.org/10.1007/978-94-017-7481-9_6.

8 Bashford, D. and Karplus, M. (1991). Multiple-site titration curves of proteins: an analysis of exact and approximate methods for their calculation. *J. Phys. Chem.* 30: 74–98. https://doi.org/10.1021/j100176a093.

9 Bashford, D. (2004). Macroscopic electrostatic models for protonation states in proteins. *Front. Biosci.* 9: 1082–1099. https://doi.org/10.2741/1187.

10 Ullmann, G.M. and Bombarda, E. (2014). Continuum electrostatic analysis of proteins. In: *Protein Modelling*, 135–163. Springer International Publishing. https://doi.org/10.1007/978-3-319-09976-7_6.

11 Job, G. and Herrmann, F. (2006). Chemical potential—a quantity in search of recognition. *Eur. J. Phys.* 27 (2): 353–371. https://doi.org/10.1088/0143-0807/27/2/018.

12 Tanford, C. (1981). Chemical potential of bound ligand, an important parameter for free energy transduction. *Proc. Natl. Acad. Sci. U.S.A.* 78 (1): 270–273. https://doi.org/10.1073/pnas.78.1.270.

13 Gilson, M.K., Given, J.A., Bush, B.L., and McCammon, J.A. (1997). The statistical thermodynamic basis for computation of binding affinities: a critical review. *Biophys. J.* 72: 1047–1069. https://doi.org/10.1016/S0006-3495(97)78756-3.

14 Onufriev, A., Case, D.A., and Ullmann, G.M. (2001). A novel view of pH titration in biomolecules. *Biochemistry* 40 (12): 3413–3419. https://doi.org/10.1021/bi002740q.

15 Bombarda, E. and Ullmann, G.M. (2010). pH-dependent pK_a values in proteins—a theoretical analysis of protonation energies with practical consequences for enzymatic reactions. *J. Phys. Chem. B* 114 (5): 1994–2003. https://doi.org/10.1021/jp908926w.

16 Klingen, A.R., Bombarda, E., and Ullmann, G.M. (2006). Theoretical investigation of the behavior of titratable groups in proteins. *Photochem. Photobiol. Sci.* 5 (6): 588–596. https://doi.org/10.1039/b515479k.

17 von Weizsäcker, C.F. (1989). Welt, Wissenschaft, Wirklichkeit. In: *Geist und Natur*, vol. 4 of 5, Chapter 1 (ed. H.P. Dürr and W.C. Zimmerli), 17–27. Bern: Scherz Verlag.

18 von Bertalanffy, L. (1969). *General System Theory: Foundations, Development, Applications*, 1e. New York: George Braziller, Inc.

19 Frauenfelder, H., Sligar, S.G., and Wolynes, P.G. (1991). The energy landscapes and motions of proteins. *Science* 254 (5038): 1598–1603. https://doi.org/10.1126/science.1749933.

20 Ferreira, A.M. and Bashford, D. (2006). Model for proton transport coupled to protein conformational change: application to proton pumping in the bacteriorhodopsin photocycle. *J. Am. Chem. Soc.* 128 (51): 16778–16790. https://doi.org/10.1021/ja060742d.

21 Sham, Y.Y., Muegge, I., and Warshel, A. (1999). Simulating proton translocations in proteins: probing proton transfer pathways in the *Rhodobacter sphaeroides* reaction center. *Proteins* 36 (4): 484–500. https://doi.org/10.1002/(SICI)1097-0134(19990901)36:4<484::AID-PROT13>3.0.CO;2-R.

22 Truhlar, D.G. (2015). Transition state theory for enzyme kinetics. *Arch. Biochem. Biophys.* 582: 10–17. https://doi.org/10.1016/j.abb.2015.05.004.

23 Till, M.S., Essigke, T., Becker, T., and Ullmann, G.M. (2008). Simulating the proton transfer in gramicidin a by a sequential dynamical Monte Carlo method. *J. Phys. Chem. B* 112 (42): 13401–13410. https://doi.org/10.1021/jp801477b.

24 Pflock, T.J., Oellerich, S., Krapf, L. et al. (2011). The electronically excited states of LH2 complexes from *Rhodopseudomonas acidophila* strain 10050 studied by time-resolved spectroscopy and dynamic Monte Carlo simulations. II. Homo-arrays of LH2 complexes reconstituted into phospholipid model membranes. *J. Phys. Chem. B* 115 (28): 8821–8831. https://doi.org/10.1021/jp2023583.

25 Beyer, S.R., Müller, L., Southall, J. et al. (2015). The open, the closed, and the empty: time-resolved fluorescence spectroscopy and computational analysis of RC-LH1 complexes from *Rhodopseudomonas palustris*. *J. Phys. Chem. B* 119 (4): 1362–1373. https://doi.org/10.1021/jp510822k.

26 Sengupta, D., Behera, R.N., Smith, J.C., and Ullmann, G.M. (2005). The α-helix dipole: screened out? *Structure* 13 (6): 849–855. https://doi.org/10.1016/j.str.2005.03.010.

27 Jackson, J.D. (1999). *Classical Electrodynamics*, 3e. New York: Wiley.

28 Kirkwood, J.G. (1934). Theory of solutions of molecules containing widely separated charges with special application to zwitterions. *J. Chem. Phys.* 2 (7): 351–361. https://doi.org/10.1063/1.1749489.

29 Daune, M. (1999). *Molecular Biophysics: Structures in Motion*, 1e. Oxford University Press.

30 Honig, B. and Nicholls, A. (1995). Classical electrostatics in biology and chemistry. *Science* 268 (5214): 1144–1149. https://doi.org/10.1126/SCIENCE.7761829.

31 Zauhar, R.Y. and Varnek, A. (1996). A fast and space-efficient boundary element method for computing electrostatic and hydration effects in large molecules. *J. Comput. Chem.* 17: 864–877. https://doi.org/10.1002/(SICI)1096-987X(199605)17:7<864::AID-JCC10>3.0.CO;2-B.

32 Holst, M.J., Kozack, R.E., Saied, F., and Subramiam, S. (1994). Treatment of electrostatic effect in proteins: multigrid-based newton itertative method for solution of the full nonlinear Poisson–Boltzmann equation. *Proteins* 18: 231–245. https://doi.org/10.1002/prot.340180304.

33 Lee, B. and Richards, F. (1971). The interpretation of protein structures: estimation of static accessibility. *J. Mol. Biol.* 55 (3): 379–400. https://doi.org/10.1016/0022-2836(71)90324-X.

34 Shrake, A. and Rupley, J. (1973). Environment and exposure to solvent of protein atoms. Lysozyme and insulin. *J. Mol. Biol.* 79 (2): 351–371. https://doi.org/10.1016/0022-2836(73)90011-9.

35 Amann-Winkel, K., Bellissent-Funel, M.C., Bove, L.E. et al. (2016). X-ray and neutron scattering of water. *Chem. Rev.* 116 (13): 7570–7589. https://doi.org/10.1021/acs.chemrev.5b00663.

36 Bondi, A. (1964). van der Waals Volumes and Radii. *J. Phys. Chem.* 68 (3): 441–451. https://doi.org/10.1021/j100785a001.

37 Mantina, M., Chamberlin, A.C., Valero, R. et al. (2009). Consistent van der Waals radii for the whole main group. *J. Phys. Chem. A* 113 (19): 5806–5812. https://doi.org/10.1021/jp8111556.

38 Nina, M., Beglov, D., and Roux, B. (1997). Atomic radii for continuum electrostatics calculations based on molecular dynamics free energy simulations. *J. Phys. Chem. B* 101 (26): 5239–5248. https://doi.org/10.1021/jp970736r.

39 Banavali, N.K. and Roux, B. (2002). Atomic radii for continuum electrostatics calculations on nucleic acids. *J. Phys. Chem. B* 106 (42): 11026–11035. https://doi.org/10.1021/jp025852v.

40 Seeliger, D. and de Groot, B.L. (2007). Atomic contacts in protein structures. A detailed analysis of atomic radii, packing, and overlaps. *Proteins* 68 (3): 595–601. https://doi.org/10.1002/prot.21447.

41 Maurer, M. and Oostenbrink, C. (2019). Water in protein hydration and ligand recognition. *J. Mol. Recognit.* 32 (12): e2810. https://doi.org/10.1002/jmr.2810.

42 Matthews, B.W. and Liu, L. (2009). A review about nothing: are apolar cavities in proteins really empty? *Protein Sci.* 18 (3): 494–502. https://doi.org/10.1002/pro.61.

43 Francl, M.M. and Chirlian, L.E. (2000). The pluses and minuses of mapping atomic charges to electrostatic potentials. *Rev. Comput. Chem.* 14: 1–31. https://doi.org/10.1002/9780470125915.ch1.

44 Cornell, W.D. and Chipot, C. (1998). Charge distribution calculations: alternative approaches. In: *Encyclopedia of Computational Chemistry* (ed. P. von Rague Schleyer), 284–301. New York: Wiley.

45 Mouesca, J.M., Chen, J.L., Noodleman, L. et al. (1994). Density functional/Poisson–Boltzmann calculations of redox potentials for iron–sulfur clusters. *J. Am. Chem. Soc.* 116 (26): 11898–11914.

46 Hinsen, K. and Roux, B. (1997). A potential function for computer simulation studies of proton transfer in acetylacetone. *J. Comput. Chem.* 18 (3): 368–380. https://doi.org/10.1002/(SICI)1096-987X(199702)18:3<368::AID-JCC7>3.0.CO;2-S.

47 Sigfridsson, E. and Ryde, U. (1998). Comparison of methods for deriving atomic charges from the electrostatic potential and moments. *J. Comput. Chem.* 19 (4): 377–395. https://doi.org/10.1002/(SICI)1096-987X(199803)19:4<377::AID-JCC1>3.0.CO;2-P.

48 Leach, A.R. (1996). *Molecular Modelling: Principles and Applications*. Essex, UK: Addison Wesley Longman.

49 Gilson, M.K. and Honig, B. (1988). Calculation of the total electrostatic energy of a macromolecular system: solvation energies, binding energies, and conformational analysis. *Proteins* 4 (1): 7–18. https://doi.org/10.1002/prot.340040104.

50 Antosiewicz, J., McCammon, J.A., and Gilson, M.K. (1994). Prediction of pH-dependent properties of proteins. *J. Mol. Biol.* 238: 415–436. https://doi.org/10.1006/jmbi.1994.1301.

51 Simonson, T., Perahia, D., and Bricogne, G. (1992). Intramolecular dielectric screening in proteins. *J. Mol. Biol.* 218: 859–886. https://doi.org/10.1016/0022-2836(91)90273-9.

52 Simonson, T. and Perahia, D. (1995). Internal and interfacial dielectric properties of cytochrome c from molecular dynamics in aqueous solution. *Proc. Natl. Acad. Sci. U.S.A.* 92: 1082–1086. https://doi.org/10.1073/pnas.92.4.1082.

53 Simonson, T. and Perahia, D. (1995). Microscopic dielectric properties of cytochrome c from molecular dynamics simulation in aqueous solution. *J. Am. Chem. Soc.* 117: 7987–8000. https://doi.org/10.1021/ja00135a018.

54 Li, L., Li, C., Zhang, Z., and Alexov, E. (2013). On the dielectric "constant" of proteins: smooth dielectric function for macromolecular modeling and its implementation in DelPhi. *J. Chem. Theory Comput.* 9 (4): 2126–2136. https://doi.org/10.1021/ct400065j.

55 Catenaccio, A., Daruich, Y., and Magallanes, C. (2003). Temperature dependence of the permittivity of water. *Chem. Phys. Lett.* 367 (5–6): 669–671. https://doi.org/10.1016/S0009-2614(02)01735-9.

56 Bragg, A.E., Cavanagh, M.C., and Schwartz, B.J. (2008). Linear response breakdown in solvation dynamics induced by atomic electron-transfer reactions. *Science* 321 (5897): 1817–1822. https://doi.org/10.1126/science.1161511.

57 Klapper, I., Hagstrom, R., Fine, R. et al. (1986). Focusing of electric fields in the active site of Cu-Zn superoxide dismutase: effects of ionic strength and amino-acid modification. *Proteins* 1 (1): 47–59. https://doi.org/10.1002/prot.340010109.

58 Alexov, E., Mehler, E.L., Baker, N. et al. (2011). Progress in the prediction of p*K*a values in proteins. *Proteins* 79 (12): 3260–3275. https://doi.org/10.1002/prot.23189.

59 Roy, R.R. and Ullmann, G.M. (2023). Virtual model compound approach for calculating redox potentials of $[Fe_2S_2]$-Cys_4 centers in proteins –structure quality matters. *J. Chem. Theory Comput.* 19 (23): 8930–8941. https://doi.org/10.1021/acs.jctc.3c00779.

60 Wlodawer, A., Minor, W., Dauter, Z., and Jaskolski, M. (2007). Protein crystallography for non-crystallographers, or how to get the best (but not more) from published macromolecular structures. *FEBS J.* 275 (1): 1–21. https://doi.org/10.1111/j.1742-4658.2007.06178.x.

61 Gao, Y., Thorn, V., and Thorn, A. (2023). Errors in structural biology are not the exception. *Acta Crystallogr. D* 79 (3): 206–211. https://doi.org/10.1107/s2059798322011901.

62 Moriarty, N.W. and Adams, P.D. (2019). Iron–sulfur clusters have no right angles. *Acta Crystallogr. D* 75 (1): 16–20. https://doi.org/10.1107/S205979831801519X.

63 Zheng, H., Cooper, D.R., Porebski, P.J. et al. (2017). CheckMyMetal: a macromolecular metal-binding validation tool. *Acta Crystallogr. D* 73: 223–233. https://doi.org/10.1107/S2059798317001061.

64 Burger, E.M., Andrade, S.L., and Einsle, O. (2015). Active sites without restraints: high-resolution analysis of metal cofactors. *Curr. Opin. Struct. Biol.* 35: 32–40. https://doi.org/10.1016/j.sbi.2015.07.016.

65 Emsley, P., Lohkamp, B., Scott, W.G., and Cowtan, K. (2010). Features and development of Coot. *Acta Crystallogr. D* 66 (4): 486–501. https://doi.org/10.1107/S0907444910007493.

66 Duarte, J.M., Srebniak, A., Schärer, M.A., and Capitani, G. (2012). Protein interface classification by evolutionary analysis. *BMC Bioinformatics* 13 (1): 334. https://doi.org/10.1186/1471-2105-13-334.

67 Wlodek, S., Antosiewicz, J., and Mccammon, J. (1997). Prediction of titration properties of structures of a protein derived from molecular dynamics trajectories. *Protein Sci.* 6 (2): 373–382. https://doi.org/10.1002/pro.5560060213.

68 Paddock, M.L., Wiley, S.E., Axelrod, H.L. et al. (2007). MitoNEET is a uniquely folded 2Fe–2S outer mitochondrial membrane protein stabilized by pioglitazone. *Proc. Natl. Acad. Sci. U.S.A.* 104 (36): 14342–14347. https://doi.org/10.1073/pnas.0707189104.

69 Culka, M., Gisdon, F.J., and Ullmann, G.M. (2017). Computational biochemistry—enzyme mechanisms explored. *Adv. Prot. Chem. Struct. Biol.* 109: 77–112. https://doi.org/10.1016/bs.apcsb.2017.04.004.

70 Ullmann, R.T. and Ullmann, G.M. (2012). GMCT: a Monte Carlo simulation package for macromolecular receptors. *J. Comput. Chem.* 33 (8): 887–900. https://doi.org/10.1002/jcc.22919.

71 Venkateswara Rao, P. and Holm, R.H. (2003). Synthetic analogues of the active sites of iron–sulfur proteins. *Chem. Rev.* 104 (2): 527–560. https://doi.org/10.1021/cr020615.

72 Leger, C. and Bertrand, P. (2008). Direct electrochemistry of redox enzymes as a tool for mechanistic studies. *Chem. Rev.* 108 (7): 2379–2438. https://doi.org/10.1021/cr0680742.

73 Klingen, A.R., Palsdottir, H., Hunte, C., and Ullmann, G.M. (2007). Redox-linked protonation state changes in cytochrome *bc*1 identified by Poisson–Boltzmann electrostatics calculations. *Biochim. Biophys. Acta, Bioenerg.* 1767 (3): 204–221. https://doi.org/10.1016/j.bbabio.2007.01.016.

74 Ullmann, G.M. (2000). The coupling of protonation and reduction in proteins with multiple redox centers: theory, computational method, and application to cytochrome *c*3. *J. Phys. Chem. B* 104: 6293–6301. https://doi.org/10.1021/jp000711j.

19

Structural and Functional Bioinorganic Model Chemistry of Fe–S Clusters – Synthesis and Analysis

Benedict Josua Elvers and Carola Schulzke

Universität Greifswald, Institut für Biochemie, Felix-Hausdorff-Straße 4, 17489 Greifswald, Germany

Chapter Goals

- To provide insight into the active sites of four groups of iron-sulfur proteins ranging from comparably simple (rubredoxin) to excessively complex (e.g. nitrogenase) with a focus on their molecular structures and electrochemical properties.
- To give an overview of the bioinorganic model chemistry in the context of these active sites with regard to synthetic strategies, structural evaluations and electrochemical behavior as compared to the artificial complexes' natural templates.

Main Chapter Points

- General features of iron–sulfur motifs, such as their natural occurrence, their reactivity, and the challenges and opportunities that synthetic chemists face when attempting to model and investigate their chemical and electronic structures.
- The biological relevance and the synthetic and analytical chemistry of [FeS_4] centers and their model complexes including electrochemical considerations.
- The biological relevance and the synthetic and analytical chemistry of [Fe_2S_2] centers and their model complexes including electrochemical considerations.
- The biological relevance and the synthetic and analytical chemistry of [Fe_4S_4] and [Fe_3S_4] centers and their model complexes including electrochemical considerations.
- The biological relevance and the synthetic and analytical chemistry of iron–sulfur clusters of higher nuclearity and their model complexes including electrochemical considerations.

19.1 Introduction

Nature serves as an abundant source of inspiration for chemists. Structural motifs and reactivities demonstrated by natural cofactors challenge synthetic chemists to

Iron-Sulfur Clusters: Biogenesis and Biochemistry, Volume 2, First Edition.
Edited by Silke Leimkühler, Günter Schwarz, Oliver Lenz, and Oliver Einsle.

create synthetic model complexes with increasingly better similarity or efficiency. However, the complexity of biological systems often impedes achieving these goals. Using nature as further inspiration with regard to biosynthesis and/or biocatalysis might necessitate the combined expertise of biochemists, synthetic chemists, spectroscopists, theoretical chemists, and physicists to push the boundaries of current knowledge and achievement. However, artificial/synthetic systems may also come with novel possibilities, e.g. enabling the design of structures with unprecedented attributes. For instance, artificial Fe–S clusters can be synthesized, for which no actual counterpart in nature exists, and the observed respective differences in properties may be quite informative as well.

Iron–sulfur (Fe–S) clusters play a fundamental role as essential cofactors across all domains of life. Recent research has revealed their multifaceted functions, from acting as catalytic active sites and generating radicals to sulfur donation for cofactor biosynthesis. Beyond these vital roles, Fe–S clusters may also hold significance in regulating gene expression and facilitating DNA repair processes [1]. Moreover, these clusters are prominent components of one of nature's particularly intriguing enzyme families: the nitrogenases which are capable of transforming unreactive dinitrogen into ammonia [2]. The clusters' most fundamental significance lies in their pivotal role in electron transfer, and even whole respective relay chains, rendering them indispensable for numerous metalloenzymes. Working in concert with cytochromes, they enable the efficient transport of electrons across membranes and proteins, facilitating cellular respiration and sustaining life [1]. An efficient transfer of electrons depends on a delicate interplay of redox potentials. A suitably matching redox potential is crucial for the electron to reach its intended cofactor. Fe–S clusters can adopt various forms (Fe_xS_x) and coordination modes and thereby finely tune the redox potential to suit the particular active site target.

To fully understand the role of specific Fe–S clusters in electron transfer, their individual redox potentials must be determined. The best option for an electrochemical characterization of molecular entities such as Fe–S clusters is to apply cyclic voltammetry. The recorded potentials must be quantified/referenced and the method for that depends on the employed solvent system. In the case of an aqueous system, the redox potential can be determined vs. a normal hydrogen electrode (NHE), a silver/silver chloride (Ag/AgCl), or a standard calomel electrode (SCE). Complexes not soluble in water are referenced vs. the ferrocene/ferrocenium (Fc/Fc^+) redox couple as an internal standard. Since molecular complexes and their redox potential also depend on their interactions with the solvent(s) used, a rigorous recalculation is not a suitable approach to compare all systems [3]. Instead, the overall potential range comparing similar systems should be considered. Redox potentials provided in Sections 19.3–19.5 are commonly derived from such electrochemical investigations.

In nature, Fe–S clusters exhibit a remarkable diversity based on three primary shapes: rhombic $[Fe_2S_2]$ clusters, cuboidal $[Fe_3S_4]$ clusters, and cuboidal $[Fe_4S_4]$ clusters (Figure 19.1). Mononuclear rubredoxins, $[Fe(Cys)_4]$, are, strictly speaking, no clusters and necessarily come with a different shape; as Fe–S-dependent electron transporters they are often still considered to belong to this cofactor class, though [4]. Structural motifs of the relatively more simple Fe–S clusters serve as building blocks

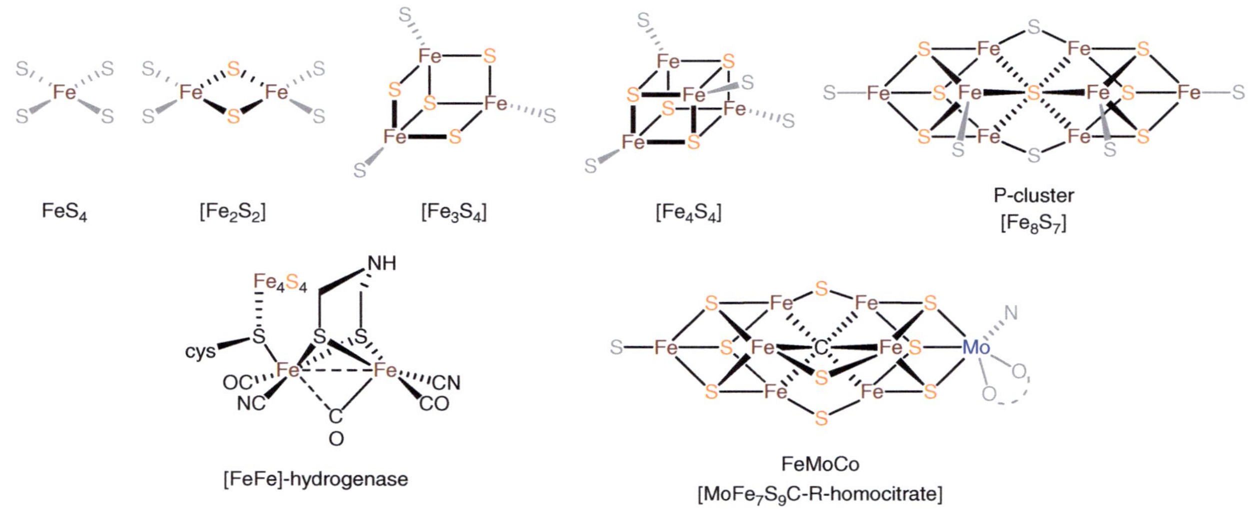

Figure 19.1 Chemical structures of known Fe–S clusters and FeS_4 centers, which are fastened to the protein environment by amino acid coordination. The inorganic sulfides are shown in yellow and protein-based ligands in gray. The most common cysteine ligation may in some cases be replaced by histidine, aspartic acid, glutamic acid, arginine, or serine. [FeFe]-hydrogenase is illustrated as the preeminent example among the structurally characterized hydrogenase enzymes. The same is true for the molybdenum-bearing form of nitrogenases. Source: Adapted from [1, 4].

for more complex structures, such as the H-cluster in [FeFe]-hydrogenase and the P-cluster of nitrogenase. Typically, the Fe centers adopt a tetrahedral coordination geometry. The Fe atoms are bridged by inorganic sulfide and tethered to the protein environment by functional groups of specific amino acids – most often cysteine. In general, Fe–S clusters are predominantly coordinated by sulfur-donor ligands. Other Lewis basic amino acids such as histidine, aspartic acid, glutamic acid, arginine, and serine may also participate in Fe–S cluster coordination but do so much more rarely. Amino acid coordination further stabilizes the Fe–S clusters within their natural environment [1].

In recent years, biochemists, synthetic chemists, spectroscopists, and theoretical chemists have worked in tandem to understand the complex interplay and relationships between the structure and function of these cofactors. Synthetic model complexes for mimicking structural or functional aspects were synthesized to fully understand the chemical and physical principles and concepts utilized by nature.

This chapter is dedicated to (bio)inorganic aspects of synthetic model complexes for Fe–S cofactors, focusing on exemplary synthetic procedures and notable milestones. General methodologies to analyze their electrochemical, spectroscopic, and structural properties are being deliberated. Wherever possible, a direct comparison of the available model redox data and what is known in this respect from the biological template is provided.

19.2 Fe–S Cluster Motifs: Natural Occurrence, Reactivity, and Artificial Synthesis

Synthetic Fe–S cluster model chemistry can be both, very very simple and efficient as well as incredibly challenging, depending on the cofactor to be modeled.

General methods for the synthesis of Fe–S clusters involve a sequence of methodologies recognized over the course of many years of dedicated research, namely:

(i) spontaneous self-assembly,
(ii) fragment condensation,
(iii) core rearrangement, and
(iv) fragmentation/dissociation.

The spontaneous self-assembly (i) process involves combining mononuclear precursors (such as FeS_4 centers) to construct an Fe–S cluster. In the case of cluster formation, covalent bonds form between metals and between metals and ligands, potentially accompanied by electron transfers. Fragment condensation (ii) refers to the fusion of an existing di- or polynuclear cluster with itself or another mono- or polynuclear species, creating more complex multinuclear species. This recombination may yield one or multiple products of significance. Core rearrangement (iii) signifies the conversion of a cluster into a distinct core geometry (geometrical isomerization) while preserving the overall nuclearity. This transformation can be achieved through processes such as thermal treatment, redox reactions, or the introduction/substitution of ligands, with the latter causing a substantial alteration in the core composition. In a fragmentation (iv) reaction, the initial cluster gives rise to new clusters with reduced nuclearity. This reaction type can be conceptualized as the reverse process of fragment condensation, leading to the formation of smaller clusters from a larger precursor [4, 5].

The syntheses of most clusters utilize $Fe^{2+/3+}$ ions in conjunction with other metal compounds, a core-bridging ligand or its precursor (i.e. sulfur, sulfide, or a nitrogen anion salt), and a carefully selected terminal ligand, such as thiolate, alcoholate, imide, or a (pseudo)halide. These self-assembly systems are predominantly formed in methanol or nonprotic polar solvents, including acetonitrile, dichloromethane, dimethylformamide (DMF), and dimethylsulfoxide (DMSO), while the use of nonpolar solvents, e.g. toluene, benzene, or tetrahydrofuran (THF), was also reported. The resulting typical cationic or anionic clusters are isolated as salts, with BF_4^-, BPh_4^-, R_4N^+, or Ph_4P^+ counterions, while neutral clusters have been reported too. A general observation is that nonpolar solvents tend to allow the formation of neutral clusters, while polar solvents facilitate the formation of charged compounds [5].

Sections 19.3–19.6 will briefly introduce the structure and reactivity of natural Fe–S clusters and their biological relevance. Subsequently, selected examples of synthetic model complexes are presented. A particular emphasis is placed on the synthetic strategy and the clusters' electrochemical properties for their pivotal role in electron transport chains of enzymatic reactions and, where appropriate, correlated with available structural data.

19.3 $[FeS_4]$

Rubredoxins belong to the first structurally characterized proteins with active sites utilizing cofactors based on iron in a sulfur environment [6]. Their $[FeS_4]$ center consists of a mononuclear $Fe^{2+/3+}$ coordinated by four cysteines (Figure 19.2).

Since no inorganic sulfide is bound and only a single iron ion present, these centers are not really "Fe–S clusters." They are, hence, referred to as "$[FeS_4]$ center" rather than a cluster in this chapter. Still, they exhibit the same tetrahedral coordination geometry around iron and thereby resemble the ubiquitous $[Fe_2S_2]$ and $[Fe_4S_4]$ clusters to quite some extent. These centers have very similar redox activity compared to the clusters and the respective synthetic models are pivotal for fragment condensation reactions to obtain clusters of higher nuclearity.

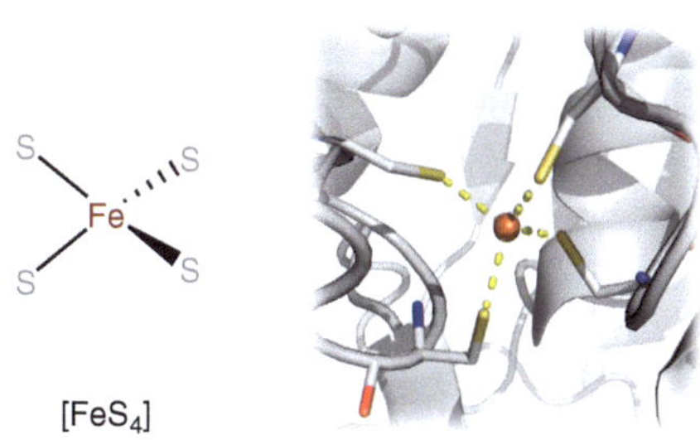

$[FeS_4]$

Figure 19.2 (a) Chemical structure of a tetrahedral $[FeS_4]$ center with truncated thiolate ligands (in the enzyme this would be cysteine). (b) Molecular structure of the electron transfer complex rubredoxin reductase from *Pseudomonas aeruginosa* at 2.45 Å resolution (PDB: 2V3B). Source: Adapted from [7].

19.3.1 Biological Relevance

$[FeS_4]$ centers are crucial components in various proteins, including rubredoxins, ruberythrins, desulforedoxin, and superoxide reductase [8]. Among these proteins, rubredoxins are the most common and extensively studied $[FeS_4]$ center-containing proteins, occurring predominantly in bacteria and archaea [8, 9]. $[FeS_4]$ centers play a vital role in electron transport and are involved in hydrocarbon and alkane oxidation and the reduction of molecular oxygen and reactive oxygen species (ROS) during oxidative stress [1, 10].

Rubredoxins derive their name from their deep red color, indicative of an Fe^{3+} center, and their reduction potential ranges from −100 to +100 mV vs. NHE [9]. This relatively narrow potential window is attributed to the low pH dependency of naturally occurring $[FeS_4]$ centers [1]. Notably, the overall coordination geometry remains intact during the transition between $Fe^{2+/3+}$ oxidation states.

19.3.2 Synthetic Model Complexes

Reduced state models are easily obtained according to the following equation (with RS representing any deprotonated organic thiolate; in general R = organic residue):

$$FeCl_2 + 4\,RS^- \longrightarrow [Fe(SR)_4]^{2-} + 2\,Cl^-$$

Typical solvents for these reactions are methanol and ethanol, which provide enough solubility while inhibiting iron oxide formation. A typical isolation procedure includes using nonpolar counter ions such as tetraphenylphosphonium (PPh_4^+, Ph = phenyl) or tetraalkylammonium ($N(alkyl)_4^+$) for precipitation and recrystallization. This method was also used for the synthesis and isolation of $[Fe(S_2\text{-}o\text{-}xyl)_2]^{2-}$ using a chelating dithiolate ligand (S_2-*o*-xyl = *o*-xylyl-α,α′-dithiolate; in the reaction directly below given as S^SR), which represents one of the first synthetic model complexes for rubredoxins [11]. This complex can be used for further substitution reactions introducing monodentate ligands [11].

$$[Fe(S^\wedge SR)_2]^{2-} + 4\,RS^- \longrightarrow [Fe(SR)_4]^{2-} + 2\,S^\wedge SR^{2-}$$

Regrettably, the stability of these complexes in solution relies on sufficient steric bulk; if the size of the substituent (R) is too small, multiple $[FeS_4]$ centers may spontaneously condense to create a cluster via a fragment condensation reaction [12]. This was shown, e.g. for the reaction of $FeCl_3$ with 3.5 eq. of NaSPh in ethanol, forming $[Fe_4(SPh)_{10}]^{2-}$, unless a precipitating salt is present to remove the $[FeS_4]$ complex from solution [13].

This instability is presumably associated with a redox reaction of the Fe^{3+} centers with the coordinated sulfur(−II) donor atoms leading to Fe^{2+} and disulfide (RSSR). To avoid this "bleaching" of Fe^{3+}, an alternative route involves preligated iron complexes for ligand substitution reactions [4]. Here, a phenolate-ligated complex with a high(er) reduction potential is used and reacted with a thiol, releasing the substituted phenol [14]. Using these reaction conditions, even sterically less hindered complexes can be obtained [14]. Notably, realising the Fe^{2+} complexes is not necessarily as straightforward. Changing the iron salt precursor to $FeSO_4$, $FeCl_2$, or

$(Et_4N)_2[FeCl_4]$ is a suitable approach, but the overall stability again depends on the used ligand(s) [15]. One example of a successful synthesis of such a complex is the synthesis of $(NEt_4)_2[Fe(S\text{-}^tBu)_4]$ (^{t}Bu = *tert*-butyl). The complex can be utilized in a ligand exchange reaction to obtain peptide-coordinated iron complexes from acetonitrile solution of peptide and metal precursor [16]. Another successful example comprises the synthesis of water-soluble $Ba[Fe(SCH_2CH_2OH)_4]$, which could be prepared by reacting $FeSO_4$, $Ba(OH)_2$, and mercaptoethanol in water. To obtain the desired complex, $BaSO_4$ was removed by filtration, and the [FeS_4] complex was isolated by extraction and subsequent crystallization from acetone [15a].

Using cysteine-bearing peptides as a source of thiolate ligands resulted in unstable complexes in either oxidation state (Fe^{3+} or Fe^{2+}). Despite their lability, the *in situ* generated complexes were characterized spectroscopically [17]. The challenge in synthesizing the cysteine-coordinated complex underscores nature's remarkable ability to stabilize these complexes within protein scaffolds by design [18].

Coordinating sterically undemanding bidentate dithiolate ligands often results in stacked dimers in the solid phase unless an additional ligand is used. For example, the utilization of two redox noninnocent dithiolene ligands results in the formation of complexes of type $[(Fe(dt)_2)_2]$ (dt = dithiolene ligand, with an ene-dithiolate functional group) unless the second coordination site is occupied by a different ligand (not a dithiolene, that is) [19]. These heteroleptic complexes often exhibit at least one additional ligand occupying a fifth coordination site, thereby impeding tetrahedral coordination. Still, these complexes are of substantial interest as *functional* model complexes for small molecule activation, e.g. hydrogenase model complexes [19c, 20].

Native rubredoxins exhibit redox potentials within a range of +100 to −100 mV vs. NHE, which are tuned by the availability of hydrogen bonding and the overall protein environment [9, 21]. The redox potentials of rubredoxin analogs vary comparatively widely depending on the substituent. Employing a strongly donating alkyl substituent, the shift is significantly negative, e.g. $[Fe(SEt)_4]^{-/2-}$ exhibits a redox potential of $E_{1/2}$ = −1.08 V vs. SCE (SEt^- = ethanethiolate), which is far away from observed enzyme potentials [14b]. Employing aromatic substituents, the complexes' potential for the redox transition $[Fe(SR)_4]^{-/2-}$ (here: R = aromatic substituents) spans from $E_{1/2}{}^{Fe(II)/Fe(III)}$ = −0.53 V (for RS = $S\text{-}C_6H_5$) to +0.03 V (for S = S-2,6-$[CH_3CONH]_2C_6H_3$) vs. SCE [15c]. The latter finding underlines that redox properties can be tuned by hydrogen bonding. Introducing an intramolecular NH···S hydrogen bond results in a more positive shift of the redox potential [15c]. This example demonstrates that by careful ligand design, it is possible to deliberately decrease the electron density of the sulfur, mimicking the effect of the protein environment. N.B.: almost generally, the redox potentials of Fe–S model complexes are much lower than those of their templates. This goes back to the absence or presence of hydrogen bonds, charges, charge distribution, solvent/environment, and similar.

The water-soluble complex $Ba[Fe(SCH_2CH_2OH)_4]$ exhibits a redox transition for $[Fe(SCH_2CH_2OH)_4]^{-/2-}$ of $E_{1/2}$ = −0.11 V vs. SHE (−0.35 V vs. SCE), which is close to the +100 to −100 mV range of native rubredoxins [15a]. In DMSO, the potential of this complex is much more negative, with $E_{1/2}$ = −0.81 V vs. SCE, emphasizing the influence of solvent on the analyte. In a more biomimetic approach, polypeptide-coordinated iron complexes exhibited potentials ranging from–0.54 to −0.30 V vs. SCE in acetonitrile with longer peptides facilitating the higher (less negative) potentials [16b].

A comprehensive analysis of the available electrochemical data indicates that the redox potentials observed in biomimetic complexes are governed by similar principles as those of proteinic rubredoxins. This involves a complex interplay of solvent effects plus influences of first and secondary spheres. Utilizing structural data can significantly support understanding the influence of the coordination environment. Structural analysis by single crystal X-ray diffraction (SCXRD) revealed a tetrahedral geometry around the central iron ion in an early tetrathiolate model complex [13b]. More detailed metrical structure parameters of closely related complexes and data from a biological sample are provided in Figure 19.3 [4]. The chosen examples cover both oxidation states, Fe^{2+} and Fe^{3+}. A direct comparison reveals only minor differences in the bonding distances between the different models and the active sites. More notable are the differences in the impact of electron transfer on the S–Fe–S

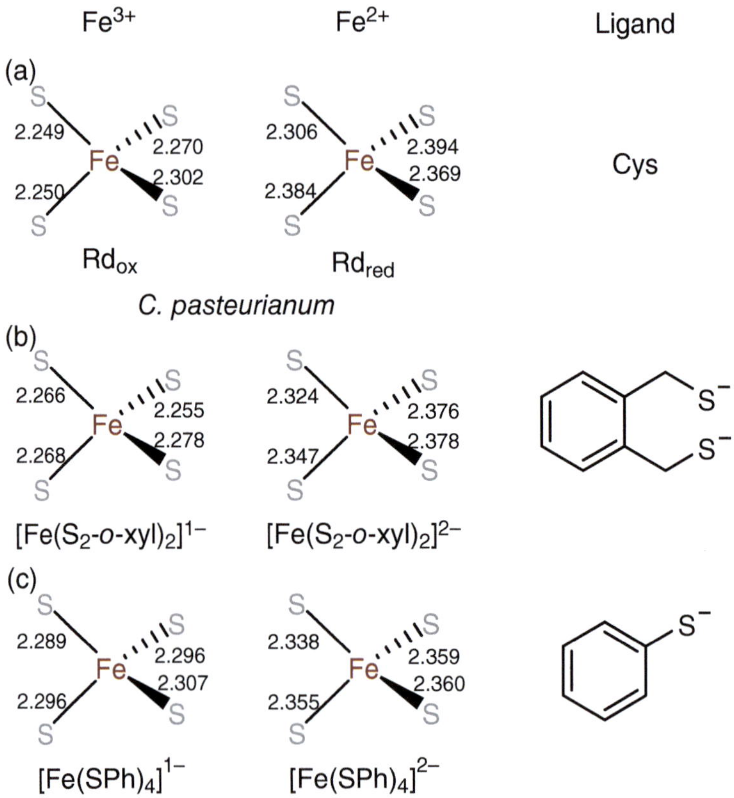

Figure 19.3 Structural data of selected biological [22] and biomimetic [14b, 23] [FeS_4] centers. Ranges for the S–Fe–S angle: (a) Fe^{2+}: 101.6–114.7°, Fe^{3+}: 103.9–112.8°, (b) Fe^{2+}: 106.2–112.3°, Fe^{3+}: 103.5–114.9°, (c) Fe^{2+}: 103.7–115.2°, Fe^{3+}: 97.9–119.0°. These and further data are more comprehensively summarized in a review article. Source: Adapted from Rao and Holm [4].

angles, which are much more rigid in the proteins for being fixed by the protein backbone. The structural parameter accuracy of these examples varies, in particular because the experiments were carried out at different temperatures [4]. Therefore, the compiled bond lengths must be considered with caution and only general trends should be deliberated.

In the case of the more aliphatic ligand (Figure 19.3b), the general Fe–S bonding distance is shorter than in the more aromatic one (Figure 19.3c). Simultaneously, $[Fe(S_2\text{-}o\text{-}xyl)_2]^{-/2-}$ exhibits a redox potential of −1.03 V vs. SCE in DMF while $[Fe(SPh)_4]^{-/2-}$ exhibits a potential of −0.53 V vs. SCE in MeCN [11b, 15c]. Despite the different solvent systems, this can be considered a significant difference, which is also in accordance with the stronger electron-donating character of the more aliphatic backbone which supports oxidation of the iron center.

19.4 $[Fe_2S_2]$ Clusters

$[Fe_2S_2]$ clusters are essential cofactors found in blue-green bacteria, green algae, higher plants (e.g. in chloroplasts), and higher organisms. The protein motif is highly conserved, comprising around 96–98 amino acids, with four amino acid residues coordinating an $[Fe_2S_2]$ core unit in tetrahedral iron geometries. Cysteines are by far the most commonly employed amino acids for cluster coordination. Histidine as an alternative ligand is not unusual with various examples available while coordination environments involving aspartic acid or arginine residues are also known [1].

This gives rise to distinct coordination motifs among the various $[Fe_2S_2]$ clusters requiring further classification. A classical cluster coordinated by four cysteines represents a ferredoxin (Fd). With histidine substituting one cysteine, the cluster is classified as a mitoNEET center. With histidine replacing two cysteines on the same iron, the resulting species is the so-called Rieske center. These respective coordination motifs are shown in Figure 19.4 [24].

The different coordination environments have a notable influence on the overall (redox) properties of these active sites, as outlined in the following.

19.4.1 Biological Relevance of $[Fe_2S_2]$ Clusters

$[Fe_2S_2]$ clusters are engaged in crucial electron transport, sulfur donation, and sensing activities. When involved in electron transport, they can access two different physiologically relevant oxidation states (Fe^{3+}/Fe^{2+} with a total spin of ½ and Fe^{3+}/Fe^{3+} with a spin of 0) and therefore transfer single electrons [4]. Upon coordination by four cysteines, they are referred to as dinuclear ferredoxins, which exhibit a potential range of −50 to −450 mV vs. NHE.

Related Rieske proteins are coordinated by two cysteines (Cys) and two histidines (His) [25]. Rieske-type clusters serve in the electron transport chain of, e.g. respiration, bacterial oxygenase enzymes that catalyze oxidative hydroxylation or in the bifurcated Q-cycle of the quinol-oxidizing cytochrome complexes [26]. One advantage of utilizing histidine is the possibility of coupling the electron transport to proton transport (PCET), as exemplified for the reaction of a diferric (i.e. Fe^{3+}/Fe^{3+})

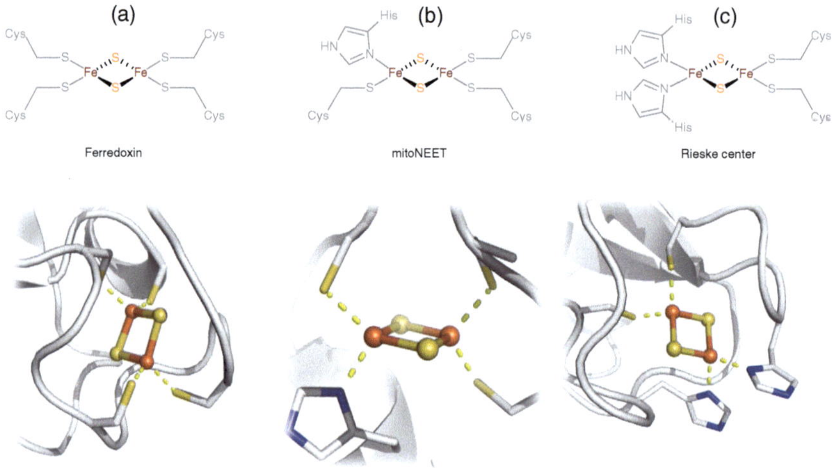

Figure 19.4 Coordination motifs in $[Fe_2S_2]$ clusters shown as chemical structures and as molecular structures from X-ray structural analysis. Source: Adapted from [1]. (a) Structure of ferredoxin at a resolution of 1.25 Å, sourced from the thermophilic cyanobacterium *M. laminosus* (PDB: 1RFK). Source: Adapted from [24a]. (b) Structure of a mitoNEET center at 1.5 Å resolution, originating from a uniquely folded outer mitochondrial membrane protein from *Homo sapiens* stabilized by pioglitazone (PDB: 2QH7). Source: Adapted from [24b]. (c) Structure of the Rieske center, at 1.5 Å resolution, obtained from the bovine heart mitochondrial cytochrome BC1 complex (PDB: 1RIE). Source: Adapted from [24c].

Rieske cluster with hydroquinone substrates [26c, 27]. The redox potentials of Rieske-type clusters significantly shift to a more positive range of −100 to +490 mV vs. NHE [28] compared to ferredoxins. Rieske-type proteins also show a substantial dependency on the pH for their redox potential due to the possibility of protonating the imine function of the His. The formation of an additional N—H bond removes electron density from the N–Fe contact and, hence, from the metal center. This in turn raises the potential as reducing iron becomes more feasible. In a quite notable example, the PCET reactivity enables Rieske clusters via a redox and protonation state-dependent structural change to act as an "electron gateway" in oxygenase enzymes [26a].

The mitoNEET protein is another intriguing example in nature, in which the cluster is coordinated by one histidine and three cysteine residues [29]. In their biological environment, the clusters' redox potentials range from +40 to −160 mV vs. SHE, depending on the pH value (ranging from 6 to 11) [30], while they strongly respond to a change in the hydrogen-bonding pattern [31]. The actual function and relevance of this arrangement is still being debated. It is hypothesized that it serves a vital role in sensing redox events [1].

Sulfur donation originates from the decomposition of an $[Fe_2S_2]$ cluster and plays a role in S-adenosylmethionine (SAM)-mediated biotin synthesis [32]. Additionally, due to their inherent sensitivity toward O_2/ROS, which can be exploited for gene regulation/expression, $[Fe_2S_2]$ clusters are also implicated for a role as respective sensors [33].

19.4.2 Synthetic Model Complexes of $[Fe_2S_2]$ Clusters

The synthesis of a first example for a synthetic $[Fe_2S_2]$ cluster precedes the synthesis of its $[FeS_4]$ counterpart and even the discovery of its biological "templates" [34]. In 1973, the Holm group prepared $[Fe_2S_2(S_2\text{-}o\text{-xyl})_2]^{2-}$ (S_2-*o*-xyl = *o*-xylyldithiolat; represented by the more general SR in the reaction shown below) in a spontaneous self-assembly reaction. Reacting $FeCl_3$, NaHS, methanolate, and the corresponding thiol in methanol results in the formation of the desired product as shown below:

$$2\,FeCl_3 + 4\,RS^- + 2\,HS^- + 2\,MeO^- \longrightarrow [Fe_2S_2(SR)_4]^{2-} + 6\,Cl^- + 2\,MeOH$$

Methanolate is added to deprotonate the thiol, which works well since thiols are generally more Brønsted acidic than alcohols. The thiol used by Holm and coworkers in this first example was a bidentate one with two thiol/thiolate functional groups on each ligand. The product was isolated by recrystallization from DMF/methanol using NEt_4^+ or $AsPh_4^+$ as counterions. The obtained complex is fully oxidized and exhibits a planar, rhombic core [34]. Unfortunately, this approach does not work when employing monodentate ligands. Syntheses with monodentate thiols result in the cubic $[Fe_4S_4(SR)_4]^{2-}$ complexes, which will be discussed below.

Instead of hydrogen sulfide, elemental sulfur can be used as a source for the bridging sulfides but it needs to be reduced in the process. The electrons are provided by the oxidation of the thiol sulfur atoms from –II to –I resulting in a disulfide byproduct. In what is known as a fragment condensation reaction, the two iron centers can then be linked to each other by the two sulfide ions [35]:

$$2[Fe(SR)_4]^{2-} + \frac{1}{4}S_8 \longrightarrow [Fe_2S_2(SR)_4]^{2-} + RSSR + 2\,RS^-$$

Employing Fe^{3+} instead of Fe^{2+} is possible, but then further reduction equivalents are needed and the amount of disulfide byproduct doubles [4, 36]:

$$2[Fe(SR)_4]^{-} + \frac{1}{4}S_8 \longrightarrow [Fe_2S_2(SR)_4]^{2-} + 2\,RSSR$$

After the initial synthesis of $[Fe_2S_2(S_2\text{-}o\text{-xyl})_2]^{2-}$ (S_2-*o*-xyl = *o*-xylyldithiolat), it was subsequently shown that this complex undergoes a ligand exchange reaction in the presence of an alternative thiol in solution [11a].

$$[Fe_2S_2(SR)_4]^{2-} + 4\,R'SH \rightleftharpoons \left[Fe_2S_2(SR')_4\right]^{2-} + RSH$$

For this reaction to occur smoothly with a homogenous product, the R′SH thiol must be either given in significant excess or the RSH thiol from the precursor needs to be removed from the reaction mixture. This approach is more efficient with more acidic thiols which lead to a complete replacement [4].

Treatment of $[Fe_2S_2(SR)_4]^{2-}$ complexes with four equivalents of an acid halide affords $[Fe_2S_2X_4]^{2-}$ [37]. Such products are useful for the synthesis of $[Fe_2S_2]$ clusters with more complex thiolates, e.g. cysteine peptides [38]. Essentially, this is also an exchange of thiolate ligands but now involving an isolated intermediate species.

$$[Fe_2S_2X_4]^{2-} + 4\,RS^- \longrightarrow [Fe_2S_2(SR)_4]^{2-} + 4\,X^-$$

Notably, this method facilitates heteroligated complexes with, for instance, alcoholates (e.g. phenolate, PhO^-) or bidentate ligands [39].

Generally, heteroligated complexes are of interest due to the natural occurrence of Rieske-type clusters in proteins, namely $[Fe_2S_2(Cys)_2(His)_2]^z$ (z = charge, depending on the oxidation states of the iron centers and protonation degree of the histidines), and mitoNEET centers [39b, 40]. Mimicking these centers requires the introduction of nitrogen-donor ligands.

The first model complex for a Rieske center $(NEt_4)_2[(pmmi)Fe_2S_2(S_2\text{-}o\text{-}xyl)]$ (pmmi = 2,2′-(phenyl methylene)bis(3-methyl-1H-indole), S_2-*o*-xyl = *o*-xylyl-α,α′-dithiolate) was published in 2008. The late appearance of a respective model underlines the difficulty that is associated with their preparation. The complex was obtained in a one-pot reaction between $[FeS_2Cl_4](Et_4N)_2$, *in-situ* deprotonated 2,2′-(phenyl methylene)bis(3-methyl-1H-indole) and subsequent addition of the deprotonated S_2-*o*-xyl [41]. The twist, that facilitated the successful preparation, was to introduce in stoichiometric amount the very bulky N-donor ligand first. Its presence hinders the approach of a second bulky ligand. After this step, the S-donor ligand with much lesser steric bulk was added and was able to bind to the available coordination sites without a steric conflict with the N-donor ligand. For the resulting complex, no equilibrium between its homoleptic and heteroleptic form was observed, which is a major advantage compared to earlier systems [41]. The reaction is shown in Scheme 19.1.

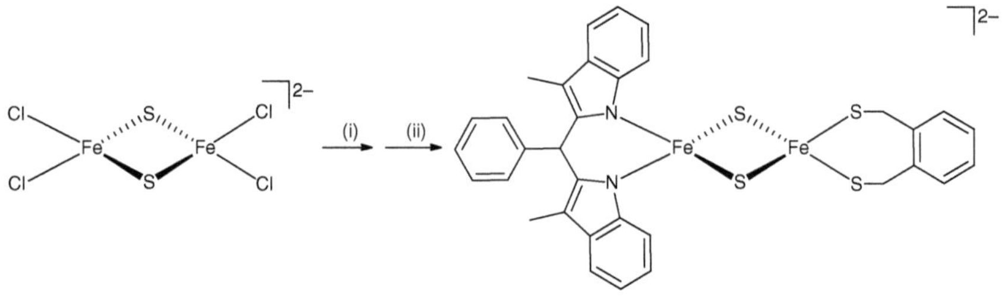

Scheme 19.1 Synthesis of $(NEt_4)_2[(pmmi)Fe_2S_2(S_2\text{-}o\text{-}xyl)]$ (pmmi = 2,2′-(phenylmethylene)bis(3-methyl-1H-indole), S_2-*o*-xyl = *o*-xylyl-α,α′-dithiolate). (i): pmmiH_2, *n*-BuLi, 0 °C to RT; dropwise addition to $[Fe_2S_2Cl_4](NEt_4)_2$ at −40 °C. (ii): S_2-*o*-xylH_2, *n*-BuLi, 0 °C; dropwise addition at −40 °C. Source: Adapted from [41].

For a subsequent complex with a similar architecture of the N-donor ligand, instead of indole N-donors, benzimidazole N-donors were used (2,2′-phenylmethylen-bis(benzimidazolate); short: pmbi) (Figure 19.5a) [42]. The ligand in $[Fe_2S_2(pmbi)_2]^{2-}$ can be protonated on the non-coordinating nitrogen atoms. Based on this ability, the benzimidazole backbone stabilizes both oxidation states ($Fe^{3+}Fe^{3+}/Fe^{2+}Fe^{3+}$), i.e. even the reduced one [42]. For the absence of the S-donors, this complex is considered a functional rather than a structural model. A closely related functional model complex was synthesized with a slightly altered method. The homoleptic complex ($[Fe_2S_2(^{Pr}bbim)_2]^{2-}$) (^{Pr}bbim = 2,2-bis(heptane)-bis(benzimidazolate), Figure 19.5c) can be obtained by a reaction between $(Et_4N)_2[FeS_2Cl_4]$ and the thallium salt of the ligand ($Tl_2{}^{Pr}bbim$) [44]. The resulting complex is the first example of a carbonyl-free [Fe–S] cluster demonstrating

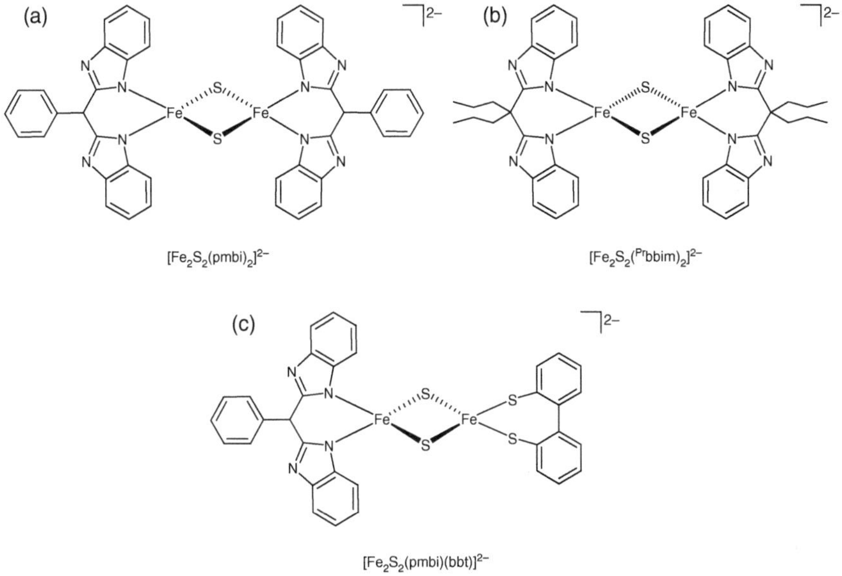

Figure 19.5 Functional and/or structural Rieske-type model complexes. (a) [$Fe_2S_2(pmbi)_2]^{2-}$ (pmbi = 2,2′-phenylmethylen-bis(benzimidazolate). Source: Adapted from [42, 43b, c]. (b) [$Fe_2S_2(^{Pr}bbim)_2]^{2-}$ (^{Pr}bbim = 2,2-bis(heptan)-bis(benzimidazolate)). (c) [Fe_2S_2(pmbi)(bbt)$]^{2-}$ (bbt = 1,1′-biphenyl-2,2′-bis(thiolate). Source: [26d]/American Chemical Society. All complexes are shown in their diferric state.

concerted proton-electron transfer (CPET) reactivity, which was later also shown for the older pmbi complex [43, 44]. The study comprised the preparation and characterization of all four congeners of the bis(benzimidazolate)-ligated [Fe_2S_2] clusters, facilitating insights into their electrochemistry and CPET reactivity (see discussion below Figure 19.6) [44].

To advance from the functional models with four N-donors to structural and functional models, the pmbi ligand was utilized in the same synthetic approach as for the first model from 2008 and combined with the 1,1′-biphenyl-2,2′-bis(thiolate) ligand (Figure 19.5c) [26d, 43b]. A very detailed investigation of the CPET reactivity of the complex and its electrochemistry highlighted the interplay of the protonation states

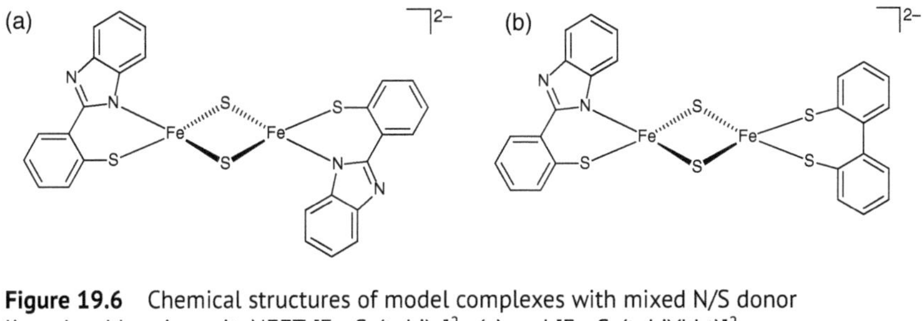

Figure 19.6 Chemical structures of model complexes with mixed N/S donor ligands addressing mitoNEET [$Fe_2S_2(tpbi)_2]^{2-}$ (a) and [Fe_2S_2(tpbi)(bbt)$]^{2-}$ (b, ptbi = 2-(2-phenylthiolate)benzimidazolate, bbt = 1,1′-biphenyl-2,2′-bis(thiolate). Both complexes are shown in their diferric state. Source: [45a]/American Chemical Society.

and redox properties. This is the first system that features a biomimetic His_2Cys_2-like coordination motif while concomitantly exhibiting CPET activity [26d, 43c].

To mimic the $[Fe_2S_2]$ cluster coordinated by three cysteines and one histidine in mitoNEET, a chelating mixed donor benzimidazole-thiolate ligand was used in combination with a dithiolate ligand [45]. During the investigation, homoleptic $[Fe_2S_2(ptbi)_2]^{2-}$ (ptbi = 2-[2-phenylthiolate]benzimidazolate, Figure 19.6a) and heteroleptic $[Fe_2S_2(tpbi)(bbt)]^{2-}$ (bbt = 1,1′-biphenyl-2,2′-bis(thiolate), Figure 19.6b) were obtained by deprotonating the ligand precursor(s) with potassium hydride and subsequent mixing with $(Et_4N)_2[FeS_2Cl_4]$ in the cold [45a].

The span of redox potential is notably depending on the used donor atoms and ligand protonation status (of histidine, or N-donor ligand) in both proteins and model complexes [28]. The presence or absence of a proton modulates the electron density of histidine including the coordinating nitrogen atom, which subsequently affects the electron density of the coordinated iron ion. Protonation of the ligand therefore facilitates reduction of the complex (since the iron receives less electron density from the ligand) and thereby raises the redox potential. The same effect is observable across the various model complexes with suitable N-donor ligands.

Synthetic $[Fe_2S_2(SR)_4]^{2-}$ clusters and their relatives with N-donor ligands may undergo two reductions, that transform the diferric state to a diferrous state, which is unlikely to occur in a biological species operating within the naturally accessible potential window, because the second reduction is at very low potential.

Depending on the ligand, even relatively simple ferredoxin-type model complexes cover a wide range of redox potentials. $[Fe_2S_2(S_2\text{-}o\text{-}xyl)_2]^{2-/3-}$ exhibits a redox potential of −1.49 V, and the second reduction $[Fe_2S_2(S_2\text{-}o\text{-}xyl)_2]^{3-/4-}$ has a potential of −1.73 V vs. NHE in DMF. Using an aromatic ligand backbone, less negative potentials can be realized. $[Fe_2S_2(SPh_4)_4]^{2-/3-/4-}$ exhibits redox potentials at −1.09 V (−1.04 V in MeCN vs. SCE) and −1.37 V vs. NHE in DMF, respectively [11a]. The difference in these three examples is in line with the available electron density for the central iron ion, following the same principles as $[FeS_4]$ centers. A more notable shift toward positive potential is observed upon incorporating nitrogen into the backbones of S-donor dithiolate ligands. A closely related series of $[Fe_2S_2]$ cluster model complexes with slight modifications in the amine-bearing scaffold $[Fe_2S_2(SC_6H_3\text{-}2,6\text{-}[NHCOR]_2)_4]^{2-}$ shows quasi-reversible reductions in MeCN with potentials ranging from −0.79 V (R = -COC$(CH_3)_3$), over −0.64 V (R = -$COCH_3$) to −0.59 V (R = -$COCF_3$) vs. SCE. This effect was attributed to intramolecular N–H⋯S hydrogen bonding, which was found to be additive depending on the number of hydrogen bonds [46]. The first reduction of $[Fe_2S_2(SR)_4]^{2-}$ results in $[Fe_2S_2(SR)_4]^{3-}$. The mixed iron oxidation state in $[Fe_2S_2(SR)_4]^{3-}$ upon reduction is mirrored in the larger cluster $[Fe_4S_4(SR)_4]^{2-}$. The $[Fe_4S_4]$ cluster appears more stable, though. A potential rearrangement with concomitant loss of thiolate ligands is assumed to be the reason for the lability. This rearrangement was experimentally observed for $[Fe_2S_2(SPh)_4]^{2-/3-}$ and can also take place spontaneously under protic conditions by intramolecular redox reactions/scrambling with concomitant disulfide (RS-SR) formation [13a, 47]. Such intramolecular reorganization also explains the problems in isolating $[Fe_2S_2]^{+/0}$ clusters and is in accordance with the requirement of a protein environment for stabilization.

The indole-based Rieske-type model complex $(NEt_4)_2[(pmmi)Fe_2S_2(S_2\text{-}o\text{-}xyl)]$ exhibits a reduction $[(pmmi)Fe_2S_2(S_2\text{-}o\text{-}xyl)]^{2-/3-}$ at −1.35 V and a second reduction $[(pmmi)Fe_2S_2(S_2\text{-}o\text{-}xyl)]^{3-/4-}$ at approximately −2.0 V vs. DFc/DFc^+ (DFc = Decamethylferrocene) that is irreversible [41]. In MeCN, the complex exhibits only one reduction event at approximately $E_{1/2}$ = −1.75 V vs. Fc/Fc^+ at −25 °C [41]. These potentials are very distinct from the biological potentials of Rieske-type centers. This goes back to the dianionic nature of the indole-based ligand, which renders the addition of electrons (reduction) to the cluster less feasible.

In the bis(benzimidazole)-based model complexes, the shift is slightly more positive, most likely based on the delocalization of the negative charge. $[Fe_2S_2(pmbi)_2]^{2-}$ exhibits a first reduction at $E_{1/2}$ = −1.14 V and a second reduction at −2.10 V vs. Fc/Fc^+ (corresponding to the $[Fe_2S_2]^{1+/2+}$ and $[Fe_2S_2]^{0/1+}$ couples) [43b]. $[Fe_2S_2(^{Pr}bbim)_2]^{2-}$ has one reduction potential at −1.225 V and a second reduction at −2.198 V vs. Fc/Fc^+. For the latter, the signals become less pronounced in the presence of one equivalent of a strong acid, but the shift is then significantly more positive [44]. The more electron-donating character of the propyl groups (in ^{Pr}bbim) compared to the phenyl substituent (in pmbi) is responsible for the slightly lower potentials of the $[Fe_2S_2(^{Pr}bbim)_2]^{2-}$ complex.

Introducing the dithiolate ligand for a better structural modeling with $[Fe_2S_2(pmbi)(bbt)]^{2-}$ (Figure 19.5c) results in a more negative potential. Two reduction events are observed at −1.43 and −2.19 V vs. Fc/Fc^+ in MeCN, reflecting the better electron donation of a thiolate compared to the benzimidazole nitrogen [26d]. Adding one equivalent of acid results in a significant shift to positive potential for the first reduction event of more than 100 mV (~−1.28 V vs. Fc/Fc^+) and another 300 mV shift (~−1.0 V vs. Fc/Fc^+) occurs after addition of a second equivalent of protons [26d]. Hence, the electrochemical response of the Rieske-type model complex to protonation closely reflects the properties of the actual proteins [26d].

The redox behavior of mitoNEET-type model complexes, comprising homoleptic $[Fe_2S_2(tpbi)_2]^{2-}$ and heteroleptic $[Fe_2S_2(tpbi)(bbt)]^{2-}$ (shown in Figure 19.6), was investigated in MeCN at −15 °C. The homoleptic complex has a first reduction potential at −1.407 V and a subsequent reduction potential at −2.227 V vs. Fc/Fc^+. Introducing a dithiolate ligand induces a cathodic shift. In $[Fe_2S_2(tpbi)(bbt)]^{2-}$, the first reversible reduction occurs at $E_{1/2}(Fe^{3+,3+}/Fe^{3+,2+})$ = −1.647 V, followed by a second reduction event at $E_{1/2}(Fe^{3+,2+}/Fe^{2+,2+})$ = −2.375 V vs. Fc/Fc^+. A direct comparison between $[Fe_2S_2(tpbi)_2]^{2-}$, $[Fe_2S_2(tpbi)(bbt)]^{2-}$, and $[Fe_2S_2(pmbi)(bbt)]^{2-}$ (Figure 19.5c, −1.43 V vs. Fc/Fc^+) highlights the relative electron density donation toward the iron centers of N- and S-donor atoms. The mitoNEET-type model complex $[Fe_2S_2(tpbi)(bbt)]^{2-}$ partially decomposes upon the addition of an equivalent of acid. A small anodic shift can be observed, but the initial signal cannot be restored entirely upon adding one equivalent of the base. Therefore, it was not possible to draw any further conclusions regarding CPET processes for this species.

Biological $[Fe_2S_2]$ clusters exhibit a slightly distorted tetrahedral geometry around the Fe^{3+} ion atoms organized as rhombs, as shown in Figure 19.4. In model complexes, the angles and distances depend on the overall donor ability of the terminal thiolates, or nitrogen donor ligands in Rieske- and mitoNEET-type complexes,

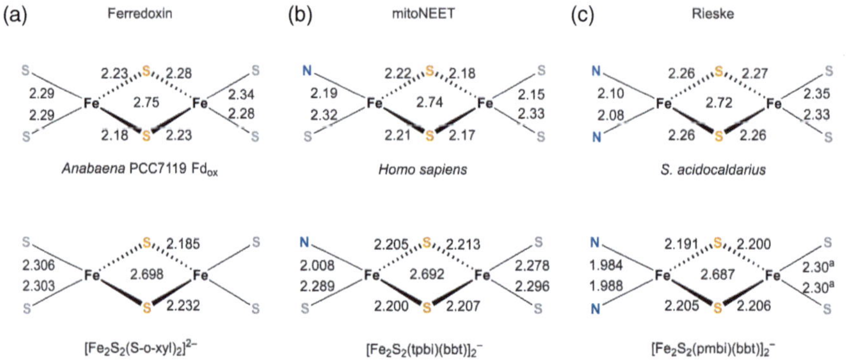

Figure 19.7 Bond lengths (in Å) of selected $[Fe_2S_2]$ clusters. (a) Ferredoxin from *Anabaena* PCC7119 and a ferredoxin-type $[Fe_2S_2]$ cluster, $[Fe_2S_2(S\text{-}o\text{-}xyl)_2]^{2-}$ (Fe–μS–Fe: 75.3°, μS–Fe–μS: 104.7°). Source: Adapted from [4]. (b) Human mitoNEET protein (PDB: 2QH7, Fe–μS–Fe: 76.9°/77.3°, μS–Fe_S-μS: 104.1°, μS–$Fe_{S/N}$-μS: 101.7°). Source: Adapted from [24b] and a mitoNEET-type $[Fe_2S_2]$ cluster, $[Fe_2S_2(tpbi)(bbt)]^{2-}$ (Fe–μS–Fe: 75.1°/75.3°, μS–$Fe_{S/N}$-μS: 104.9°, μS–$Fe_{S/S}$-μS: 104.4°). Source: Adapted from [45a]. (c) Rieske protein active site from *S. acidocaldarius* (PDB: 1JM1, Fe_N-μS–Fe_S: 73.9°, μS–Fe_N-μS: 106.2°, μS–Fe_S-μS: 105.8°). Source: Adapted from [40] and a Rieske-type $[Fe_2S_2]$ model cluster, $[Fe_2S_2(pmbi)(bbt)]^{2-}$ (Fe_N-μS–Fe_S: 75.0°/75.4°, μS–Fe_S-μS: 104.6°, μS–Fe_N-μS: 104.9°). Source: Adapted from [26d]. Color coding: thiolate (gray), inorganic sulfide (yellow), nitrogen (blue), and iron (black). Fe...Fe distances are provided in the centers of the rhombs. [a] A disorder over two positions only allows for an approximation of the actual bond distance [26d].

respectively. The metrical parameters of the rhomb are mainly determined by the core sulfides, though [48]. Important parameters of selected biological $[Fe_2S_2]$ clusters and selected models thereof are provided in Figure 19.7 (and its caption). Bond lengths and angles in the protein structures can be significantly influenced by hydrogen bonding, and the ligands greatly influence the metrical parameters in the models. Crystallographic data were not measured at the same temperature, and the resolutions vary, which renders a direct comparison problematic. Therefore, only more general trends are highlighted in the following.

The influence of the protein is particularly evident in the molecular structure of ferredoxin from *Anabaena* PCC7119 (Fd_{ox}) in the asymmetric Fe–$S_{thiolate}$ bond lengths (Figure 19.7, top left), which are symmetric in the respective model complex (Figure 19.7, bottom left). The influence of the protein backbone vs. model dithiolate ligand can also be observed for the mitoNEET-type $[Fe_2S_2]$ cluster (Figure 19.7b) but is less pronounced in the Rieske-type cluster (Figure 19.7c).

The Fe—N bond lengths in model complexes tend to be shorter than in the proteins. This goes back to the negative charge of the ligands in the models, which concomitantly stabilizes the Fe–N bond. Histidine coordinates as a neutral ligand in proteins, resulting in a longer Fe—N bond (and higher redox potentials). While the bonding angles within the $[Fe_2S_2]$ rhombs are not unaffected by this, the overall differences are relatively small, especially when considering the impact of the ligand-type and the inherent rigidity of the protein backbone.

19.5 [Fe_4S_4] and [Fe_3S_4] Clusters

Model [Fe_4S_4] clusters were the first synthesized analogs of Fe–S clusters [49] after these centers were identified from protein crystallographic data [6, 50]. Inside proteins, [Fe_4S_4] clusters adopt cubane-like structures, as shown in Figure 19.8a, in which the iron centers and sulfide occupy alternate edges of a cube. More than 100 synthetic complexes are available in the literature with all kinds of coordinated ligands, while the vast majority is ligated by four thiolates similar to the protein active sites [1, 4].

Protein-bound trinuclear sites were first recognized in the 7-Fe protein of *A. vinelandii* Fd I in 1980 by Mössbauer spectroscopy and subsequent protein crystallography [53]. Another protein-bound 3-Fe site was found in the crystal structure of the inactive form of the enzyme aconitase, which was deemed a degradation product [54]. In both cases, the sulfur content and the exact cuboidal structure could only be determined years later [55]. Formally, the cuboidal [Fe_3S_4] cluster consists of three edge-shared Fe_2S_2 rhombs. Essentially, one iron edge is missing from a complete [Fe_4S_4] cubane. A schematic representation of a cuboidal [Fe_3S_4] cluster and a molecular structure of a protein-bound [Fe_3S_4] cluster are shown in Figure 19.8b.

19.5.1 Biological Relevance of [Fe_4S_4] and [Fe_3S_4] Clusters

The most abundant of all Fe–S clusters are the [Fe_4S_4] centers with a cuboidal structure and four amino acid residues (often cysteines) coordinated to the iron ions that hold the cluster in place. The other four edges of the cubane are occupied by sulfides, which are, hence, bound to three iron centers each. The iron centers, consequently, are also coordinated by three sulfides each with the amino acid coordination completing the tetrahedral coordination geometry. Besides their crucial role in electron transport, [Fe_4S_4] clusters also act as essential cofactors in catalysis, sulfur donation, and sensing activity, fulfilling complex roles in different

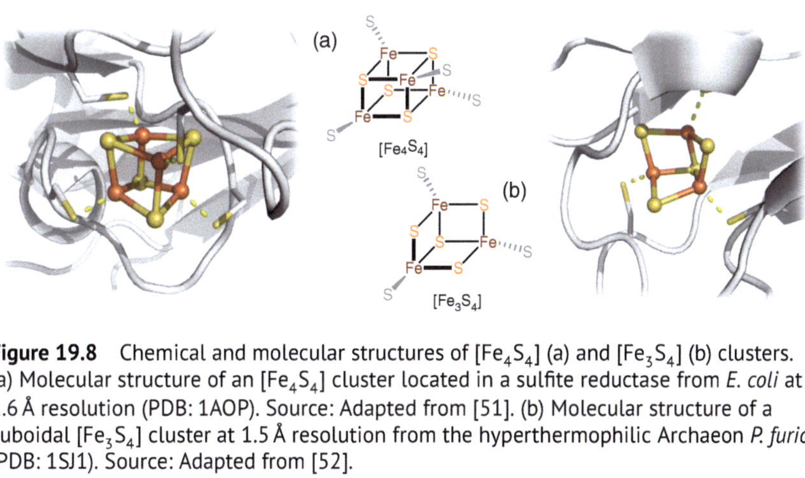

Figure 19.8 Chemical and molecular structures of [Fe_4S_4] (a) and [Fe_3S_4] (b) clusters. (a) Molecular structure of an [Fe_4S_4] cluster located in a sulfite reductase from *E. coli* at 1.6 Å resolution (PDB: 1AOP). Source: Adapted from [51]. (b) Molecular structure of a cuboidal [Fe_3S_4] cluster at 1.5 Å resolution from the hyperthermophilic Archaeon *P. furiosus* (PDB: 1SJ1). Source: Adapted from [52].

processes. Under physiological conditions, these clusters undergo redox cycling between $[Fe_4S_4]^{1+}$ and $[Fe_4S_4]^{2+}$ states. However, in the case of their high-potential analogs (high potential iron protein: HiPIP), they cycle between the $[Fe_4S_4]^{2+}$ and $[Fe_4S_4]^{3+}$ oxidation states. The all-ferrous state has been observed in the Fe protein of the nitrogenase under highly reducing conditions; however, for the unnatural conditions, this fully reduced form (with four Fe^{2+}) is not necessarily physiologically relevant [5].

As ferredoxins, $[Fe_4S_4]$ clusters are predominantly found in anaerobic bacteria, transferring electrons from substrates to various electron acceptors, including $NADP^+$, FMN, FAD, riboflavin, and sulfite [10]. The typical (or low-potential) ferredoxins have redox potentials in a range of −250 to −650 mV while the HiPIPs exhibit a range of +100 to +400 mV. The substantial difference goes back to second sphere effects, since the $[Fe_4S_4]$ clusters in both cases are essentially identical and coordinated by cysteines only. In HiPIPs, the cluster resides in a hydrophobic pocket within a relatively highly charged protein [1]. The absence or presence of hydrogen bonding and the charge have an impact on the electron density of the iron centers via their donor atoms. In the HiPIPs, the cluster environment facilitates access to a higher oxidation state so that they operate with $[Fe_4S_4]^{2+}$ as a reduced form, which equals the oxidized form of the ferredoxins. Relatively, the reduced HiPIP cluster is more stabilized than the oxidized one. Therefore, reduction is preferred over oxidation and the potential is raised. In the ferredoxins, in contrast, the oxidized form is relatively more stabilized and their potential is, hence, substantially lower than that of the HiPIPs.

Numerous enzymes with catalytically active $[Fe_4S_4]$ clusters have been identified. The radical SAM enzyme superfamily is best characterized, exhibiting a diverse array of radical catalysis (SAM = S-adenosyl methionine) [56]. These $[Fe_4S_4]$-dependent enzymes use redox cycling to facilitate the reductive cleavage of SAM [57]. Sulfur donation may occur similarly to the case of $[Fe_2S_2]$ clusters (see Section 19.4), but depends on the enzyme family. For instance, the lipoyl synthase likely utilizes an auxiliary $[Fe_4S_4]$ for its sulfur insertion chemistry [58]. Proteins containing $[Fe_4S_4]$ clusters are further implicated in sensing O_2 and NO levels to generate an appropriate response via gene expression [59]. For instance, the fumarate nitrate reduction regulator functions in sensing O_2 and differentially regulates genes under aerobic and anaerobic conditions [59a, 60].

$[Fe_3S_4]$ clusters are much scarcer compared to $[Fe_2S_2]$ and $[Fe_4S_4]$ clusters. In $[Fe_3S_4]$ clusters, each iron center is coordinated by three sulfides and one protein residue. Despite the missing iron edge, $[Fe_3S_4]$ clusters usually adopt a similar cuboidal structure as the $[Fe_4S_4]$ clusters. The absence of one iron can result from partial decomposition or oxidation of $[Fe_4S_4]$ clusters, which might be essential for activating catalytic abilities [61]. One important example of a catalytically inactive $[Fe_3S_4]$ cluster that originates from a (catalytically active) $[Fe_4S_4]$ cluster is found in aconitase, which catalyzes the conversion of citrate to *iso*-citrate [54c]. Other $[Fe_3S_4]$ clusters are known for their role in electron transport as ferredoxines with a potential in the range of −50 and −450 mV vs. NHE or gene expression regulation, as observed in *R. leguminosarum* and *D. shibae* [62]. An even more rarely occuring $[Fe_4S_3]$ cluster is involved in oxygen protection of the [NiFe]-hydrogenase, while hydrogenases are normally extremely sensitive toward dioxygen [63]. In this enzyme, three oxidation states for the $[Fe_4S_3]$ cluster have been identified: $[Fe_4S_3]^{3+/4+/5+}$, ranging from $[Fe^{2+}{}_3Fe^{3+}]$ to $[Fe^{2+}Fe^{3+}{}_3]$ [63b]. It is assumed that the protective function results from the reduction of oxygen to water at the hydrogenase active site, which is supported by this $[Fe_4S_3]$ cluster [63, 64].

19.5.2 Synthetic Model Complexes of $[Fe_4S_4]$

The multitude of known artificial $[Fe_4S_4]$ clusters highlights their typical inherent stability and the availability of different synthetic methods. Notably, it is generally easier to synthesize the cubanes than the dinuclear $[Fe_2S_2]$ species. The commonly applied synthetic route for nearly all homo- and heterometallic cubane clusters is based on a self-assembly reaction. The self-assembly requires (i) an Fe^{2+} or Fe^{3+} to react with (ii) a core bridging ligand or its corresponding precursor, e.g. S_8, S^{2-}, or a nitrogen anion salt, and (iii) a suitable terminal ligand or its precursor, e.g. a thiolate (RS^-) or a (pseudo)halide (e.g. Cl^-, CN^-) which later occupies the edge of the cubane.

Conventionally, the reaction medium is methanol or another, non-protic, polar solvent, e.g. CH_3CN, CH_2Cl_2, DMF, or DMSO. To obtain the targeted product clusters, weakly coordinating anions (e.g. BF_4^-, BPh_4^-, and triflate (OTf^-)) or cations (e.g. R_4N^+ or Ph_4P^+) are used to precipitate or crystallize the corresponding salts. Other synthetic pathways are described in the literature but they are much less commonly applied [65].

One of the first synthesized, isolated, and fully characterized $[Fe_4S_4]$ clusters, $[Fe_4S_4(SCH_2Ph)_4]^{2-}$, was obtained from self-assembly reaction with an $[Fe^{3+}{}_2Fe^{2+}{}_2S_4]$ core [49a]. The reaction sequence can be generalized as follows:

$$4\,FeCl_3 + 6\,RS^- + 4\,HS^- + 4\,MeO^- \longrightarrow [Fe_4S_4(SR)_4]^{2-} + 12\,Cl^- + 4\,MeOH + RSSR$$

Also, lithium salts and elemental sulfur can be utilized, but a greater excess of thiolates is necessary to obtain inorganic sulfide from sulfur [66]:

$$4\,FeCl_3 + 14\,LiRS + {}^1\!/_2\,S_8 \longrightarrow Li_2[Fe_4S_4(SR)_4] + 12\,LiCl + 5\,RSSR$$

Markedly, due to the relatively convenient access to this cluster type, it was deduced that the "success of the self-assembly approach proved that proteins are unnecessary for the synthesis of this cluster type" [5].

Considering these clusters' pivotal biological role, numerous additional synthetic approaches have emerged to yield oxidation state varied, site-differentiated, and heteroleptic model complexes. While $[Fe_4S_4(SCH_2Ph)_4]^{2-}$ with a mixed oxidation state is easily obtained in the self-assembly reaction as outlined above, the fully reduced all-ferrous ($[Fe^{2+}{}_4S_4]^0$) and the fully oxidized all-ferric state ($[Fe^{3+}{}_4S_4]^{+4}$) are more difficult to realize and require specific characteristics of their ligands.

All-ferrous cubane-type $[Fe_4S_4]$ clusters are evidenced by spectroscopic and electrochemical investigations of physiologically relevant Fe–S clusters [67].

One example of an artificial complex mimicking the all-ferrous state utilizes cyanide ligands, $[Fe_4S_4CN_4]^{4-}$, and its molecular structure exhibits comparable metrical parameters as in the protein [68]. The synthesis starts with a cubane cluster obtained in a typical self-assembly reaction as described above ($FeCl_3$, sulfur, and NaSEt). The obtained $[Fe_4S_4(SEt)_4]^{2-}$ is then *in situ* transformed via replacing the terminal thiolate ligands by chloride. The resulting intermediate product can then be used for further modification since the terminal chloride ligands are easily replaceable, e.g. by phosphines (Scheme 19.2). The phosphine itself serves as a ligand and as a reducing agent [70] to yield $[Fe_4S_4(PR_3)_4]^+$ with an $[Fe^{2+}{}_3Fe^{3+}S_4]^+$ core that is susceptible to ligand exchange with cyanide. Alternatively, in the absence of phosphine, a strong reducing agent such as a benzophenoketyl salt $(K(Ph_2CO))$ can be added prior to a cyanide source to obtain $[Fe_4S_4(CN)_4]^{3-}$ [69]. Further addition of one equivalent of $K(Ph_2CO)$ resulted in the isolation of the all-ferrous $[Fe_4S_4(CN)_4]^{4-}$, which was, unsurprisingly, shown to be extremely sensitive toward oxidation. The key for isolating the all-ferrous state, in the first place, was using the strongly electron-withdrawing cyanide ligands to result in a suitable redox potential for the subsequent reduction [68].

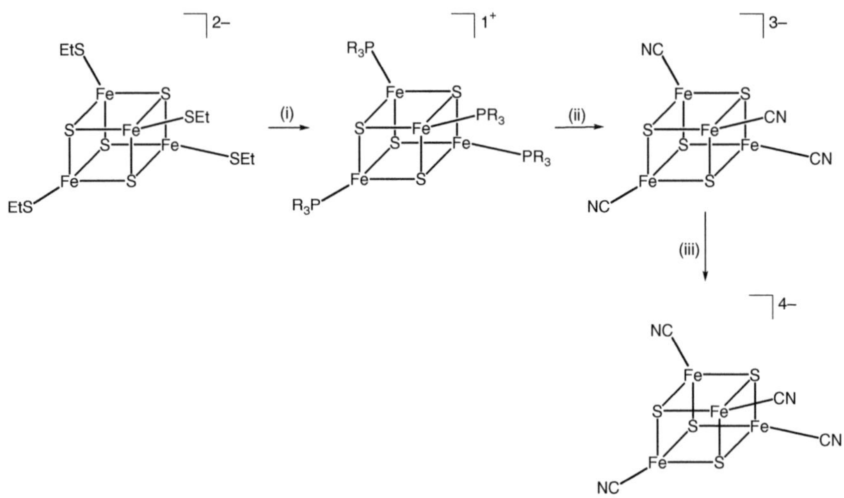

Scheme 19.2 Synthesis of a cyanide coordinated complex with an all-ferrous $[Fe_4S_4]^0$ cluster core. (i) RC(O)Cl, then excess of a suitable PR_3. (ii) $(Bu_4N)(CN)$, THF, $Na(BPh_4)$. (iii) $K(Ph_2CO)$, MeCN. Source: Adapted from [68, 69].

The proof of principle with the strong cyanide ligands inspired further syntheses of all-ferrous $[Fe_4S_4]$ clusters employing *N*-heterocyclic carbenes (NHCs). Starting with a solution of $[FeCl_2(PR)_2]$ ($R = {}^iPr$) followed by the addition of two equivalents of $S(Me_3Si)_2$ forms a cubane-type complex *in situ*; then at least two equivalents of carbene are added. Upon recrystallization, this combination yields the complex shown in Scheme 19.3 with strong NHC ligands that can be fine-tuned with regard to steric bulk close to the coordinated iron center [71]. Without the excess of carbene and $S(Me_3Si)_2$, clusters of higher nuclearity might result, which may be transformed into $[Fe_4S_4]$ clusters by further addition of carbene ligands ("carbene cleavage," i.e. carbene-driven cluster cleavage) [71].

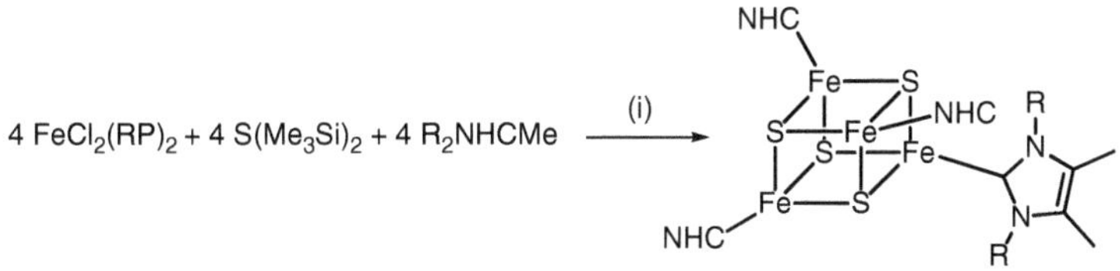

Scheme 19.3 Synthesis of an all-ferrous $[Fe_4S_4]$ cluster featuring *N*-heterocyclic carbenes as terminal ligands. (i): 2 eq. $S(Me_3Si)_2$, 2.2 eq. R_2NHCMe, benzene. R = ^{i}Pr. Source: Adapted from [71].

The obtained all-ferrous NHC-coordinated complex is more stable than its cyanide predecessor and, notably, was the first example of an [FeS] cluster with Fe–C σ-bonds [71].

Concerning the all-ferric state, a natural $[Fe_4S_4]^{4+}$ site in proteins has not been observed as of yet. Despite its apparent absence in nature, a respective synthetic example, $[Fe_4S_4(N(SiMe_3)_2)_4]$, was obtained via a self-assembly reaction. Here exactly opposite type terminal ligands had to be chosen compared to the all-ferrous cluster. There are two options known for facilitating a weak field ligand coordinated all-ferric model complex: (i) by reacting $[Fe(N(SiMe_3)_2)_2]$ with sulfur in toluene [72] or (ii) by using $[FeCl(N(SiMe_3)_2)_2(THF)]$ with NaSH in THF (Scheme 19.4) [73].

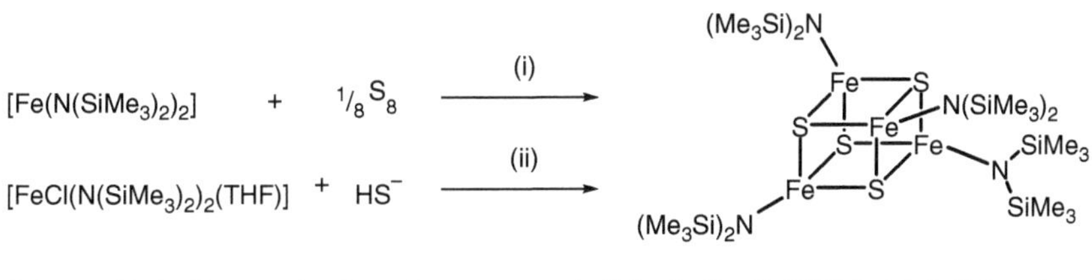

Scheme 19.4 Two synthetic pathways toward all-ferric $[Fe_4S_4(N(SiMe_3)_2)_4]$. (i) In toluene. Source: Adapted from [72]. (ii) In THF plus multiple recrystallizations from *n*-pentane at −33 °C. Source: Adapted from [73].

Since sulfur in oxidation state –II is susceptible to being oxidized, excess or further addition of a sulfide salt results in the formation of the reduced complex ($[Fe_4S_4(N(SiMe_3)_2)_4]^{1-/2-}$). Therefore, a very careful control of the reaction conditions is required.

$[Fe_4S_4(N(SiMe_3)_2)_4]$ is also a suitable precursor to obtain a thiolate-coordinated all-ferric $[Fe_4S_4]$ cluster. The addition of four equivalents of very bulky 2,4,6-tris(bis(trimethylsilyl)methyl)benzenethiol (HSTbt) to a solution of $[Fe_4S_4(N(SiMe_3)_2)_4]$ results in the formation of $[Fe_4S_4(STbt)_4]$ (Scheme 19.5). An alternative route involves the direct use of $[Fe(N(SiMe_3)_2)]$ as a starting material. The protonated HSTbt will be deprotonated by the amide $N(SiMe_3)_2$, which also functions as an internal base in both synthetic approaches. The bulky hydrocarbon ligands protect the clusters sterically and provide a hydrophobic environment around the Fe^{3+} centers, which further shields the central iron ion from nucleophiles [74].

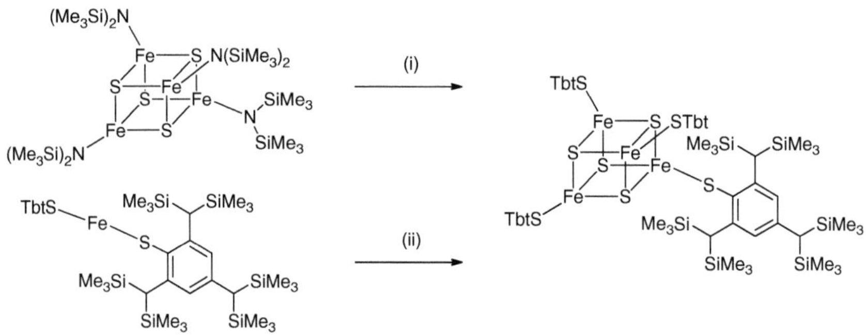

Scheme 19.5 Synthesis of the all-ferric $[Fe_4S_4(STbt)_4]$. (i) In toluene, RT. (ii) Fe complex in *n*-hexane, addition of 1 eq. sulfur in toluene. Source: Adapted from [74].

While there is significant interest in $[Fe_4S_4]$ clusters with the same sort of terminal ligands (homoleptic clusters), it is important to note that in nature the overall protein environment of these clusters may be heteroleptic. A small fraction of native $[Fe_4S_4]$ clusters occur in a 3 : 1 site-differentiated form, generally written as $[Fe_4S_4(Cys)_3L]$, where L is an endogenous or exogenous ligand. These site-differentiated clusters are often associated with catalytic or electron transfer functions. Prominent examples comprise the aconitase, which binds H_2O/OH^-, substrate intermediates (citrate), and its product (isocitrate) during the catalytic cycle [5, 54c] and an $[Fe_4S_4(Cys)_3(His)]$ cluster, which is responsible for the electron transfer to a [NiFe]-hydrogenase [75]. Consequently, there is considerable interest in site-differentiated synthetic complexes as well. Site differentiation results in an uneven electron density distribution within the cluster. To understand this and the impact it has on the overall redox properties in more detail, artificial model clusters are invaluable tools.

One way to obtain site-differentiated clusters relies on using bulky, polydentate ligands to enforce a 3 : 1 site differentiation. Two examples of three-dentate ligands and their respective complexes are shown in Scheme 19.6.

Adding a tridentate thiol, e.g. LS_3A or LS_3B (Scheme 19.6, right), to an $[Fe_4S_4(SEt)_4]^{2-}$ precursor results in the substitution of three terminal thiolate ligands [77]. The remaining terminal ethanthiolate (EtS^-) can be replaced by chloride using Me_3COCl or by another more acidic thiol [76, 77]. This method was also successfully employed to obtain a manifold of different terminal site-specific ligated clusters to observe the effect of different monodentate, bridging, or polydentate ligands on the molecules' overall properties [4, 78]. For instance, it could be quantified that the intrinsic potential difference between an all cysteine bound cluster (lower potential) and the one in which one cysteine is replaced by a histidine (higher potential) amounts to 320 mV, which is quite substantial.

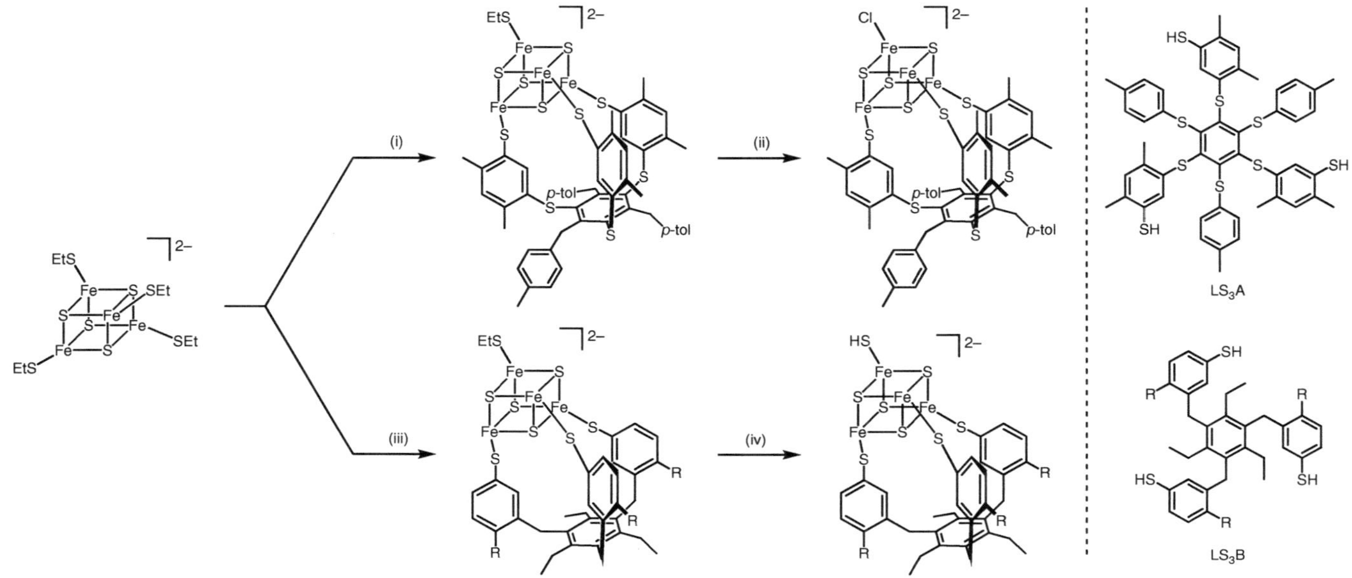

Scheme 19.6 Synthesis of site-differentiated $[Fe_4S_4]$ clusters. (i): *In situ* deprotection of LS_3A. Source: Adapted from [76]. (ii) Me_3COCl. Source: Adapted from [76a]. (iii) Dropwise addition of LS_3B in THF to Fe-complex in MeCN. R = -F, -OMe. (iv) H_2S exposure. Source: Adapted from [77].

Another synthetic strategy employs sterically demanding monodentate ligands to obtain a 3 : 1 site-differentiated cluster with only monodentate terminal ligands (i.e. nontemplated; Scheme 19.7).

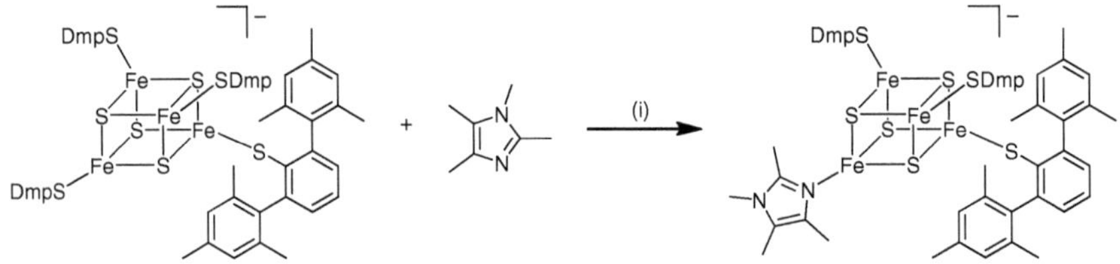

Scheme 19.7 Synthesis of a nontemplated site-differentiated $[Fe_4S_4]$ cluster mimicking a HiPIP. (i) In toluene. Source: Adapted from [79].

Treatment of $[Fe_4S_4(N(SiMe_3)_2)]$ with a highly sterically hindered thiol (HSDmp = 2,6-bis(mesityl)benzene-1-thiol) results in a substitution reaction and release of $(Me_3Si)_2NH$, followed by a core reduction, and subsequent thiolate binding. Addition of THF results in the corresponding THF-coordinated cluster. The THF molecules coordinate only transiently to eventually facilitate binding of tetramethylimidzaole. Starting from the one-electron reduced complex, however, allows a direct substitution of the thiolate ligand, exactly as shown in Scheme 19.7 [79]. More electrons on the cluster core support the binding of the neutral imidazole moiety better compared to the anionic thiolate functional group, and this is also in agreement with the HSAB theory (HSAB = hard and soft Lewis acids and bases).

Further methods to obtain site-differentiated 3 : 1 clusters comprise substitution reactions of phosphine-coordinated $[Fe_4S_4]^+$ clusters [80] or the redox-initiated fragmentation reaction of an all-ferrous tetracubane (Scheme 19.8a and b) [81]. One particularly remarkable example uses a fragment condensation approach to obtain a biomimetic complex of the H-cluster framework of [FeFe]-hydrogenases (see Section 19.6). The $[Fe_4S_4(SEt)_4]^{2-}$ precursor is simply docked onto a bulky thiolate ligand site which is *in situ* deprotected to facilitate a bridging coordination [83]. Another interesting possibility is the use of a tridentate nitrogen ligand of substantial steric bulk as demonstrated, e.g. for $[Fe_4S_4(L(NIm^{p\text{-tol}})_3)Cl]^-$ (Scheme 19.8c) [82].

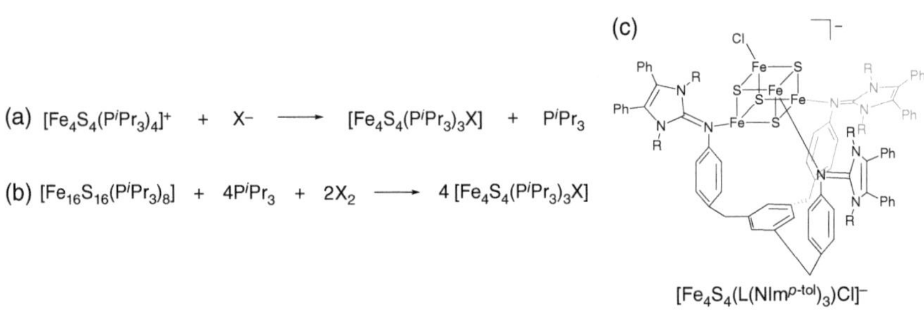

Scheme 19.8 Syntheses of site-differentiated phosphine-coordinated $[Fe_4S_4]$ clusters (a and b) and the chemical structure of $[Fe_4S_4(L(NIm^{p\text{-tol}})_3)Cl]^-$ (c). Source: Adapted from [80–82].

The available large family of biomimetic $[Fe_4S_4]$ clusters provides a wealth of electrochemical data. For the sake of clarity, the following discussion will focus solely on some specific exemplary complexes (Figure 19.9) and their respective electrochemistry.

The manifold of ligands that can be coordinated to $[Fe_4S_4]$ clusters allows realizing various cluster oxidation states. This diversity is particularly evident in the synthesis of all-ferrous and all-ferric cluster complexes. However, choosing ligands that stabilize the more extreme oxidation states restricts the available redox transitions of the products. This aligns with the HSAB theory, wherein strong and/or anionic ligands better stabilize high-valent metal centers, while softer/weaker/neutral ligands better stabilize lower metal oxidation states. As a result, different ligands and combinations thereof enable access to varying distinct ranges of redox transitions, as evident from the data in Table 19.1.

The chemically and structurally simplest example of an $[Fe_4S_4]$ cluster for which electrochemical data is available is $[Fe_4S_4(SPh)_4]^{2-}$ [47]. The total charge of the core cluster can vary between $[Fe_4S_4]^{2+/+/0}$ and the two half-wave potentials were found at −1.0 and −1.73 V vs. SCE, respectively [47, 84]. The introduction of one or two intramolecular hydrogen bonds by implementing NH moieties in the terminal S-donor ligand results in a positive shift of the redox potential as evidenced with $[Fe_4S_4(Spda)_4]^{2-}$ (−0.72 V vs. SCE in MeCN). Removing one intramolecular hydrogen bond from $[Fe_4S_4(Spda)_4]^{2-}$ results in a negative shift of 90 mV (to −0.83 V vs. SCE in MeCN), which mirrors the redox modulation effect as already described for the mononuclear $[FeS_4]$ centers (see respective Section 19.3) [46a].

Half-wave potentials for the species that include the all-ferrous CN-coordinated complex ($[Fe_4S_4(CN)_4]^{4-}$) are reported as −0.44 and −1.42 V vs. SCE in DMF [69]. The low half-step potential of −1.42 V for the $[Fe_4S_4]^{1+/0}$ redox transition (i.e. reduction to the all-ferrous form) is in accordance with the notable sensitivity toward aerobic conditions and oxidation in general. However, this potential is substantially positive shifted compared to $[Fe_4S_4(SPh)_4]^{2-}$ with which the all-ferrous form can be electrochemically accessed but not stabilized efficiently enough for an isolation/purification. The most significant chemical difference between the two systems is the strong electron-withdrawing character of the cyanide ligands that effects the redox potentials of both observed redox transitions with an anodic (positive) shift for the cyanide complexes and facilitates the isolation of the all-ferrous form in this case [5, 69].

In contrast, the all-ferric STbt-coordinated complex can be reduced easily. $[Fe_4S_4(STbt)_4]^0$ (chemically accessed from $[Fe_4S_4(STbt)_4]^{2-}$; listed as such in Table 19.1) exhibits a first reduction at $E_{1/2}$ = −0.048 V and a second at approximately −1 V vs. $Ag/AgNO_3$ in CH_2Cl_2. The complex core can exhibit three different oxidation states: $[Fe_4S_4]^{4+/3+2+}$ [74]. Previously reported complexes with bulky thiolate ligands ($[Fe_4S_4(SDmp)_4]^{2-}$) exhibit comparable redox potentials, which are merely shifted due to the applied solvent system [79].

These two examples, $[Fe_4S_4(CN)_4]^{3-}$ and $[Fe_4S_4(STbt)_4]^{2-}$, show how substantially terminal ligands can tune the redox reactivity in biomimetic complexes while stabilizing different oxidation states of the $[Fe_4S_4]$ clusters. Cysteinate just

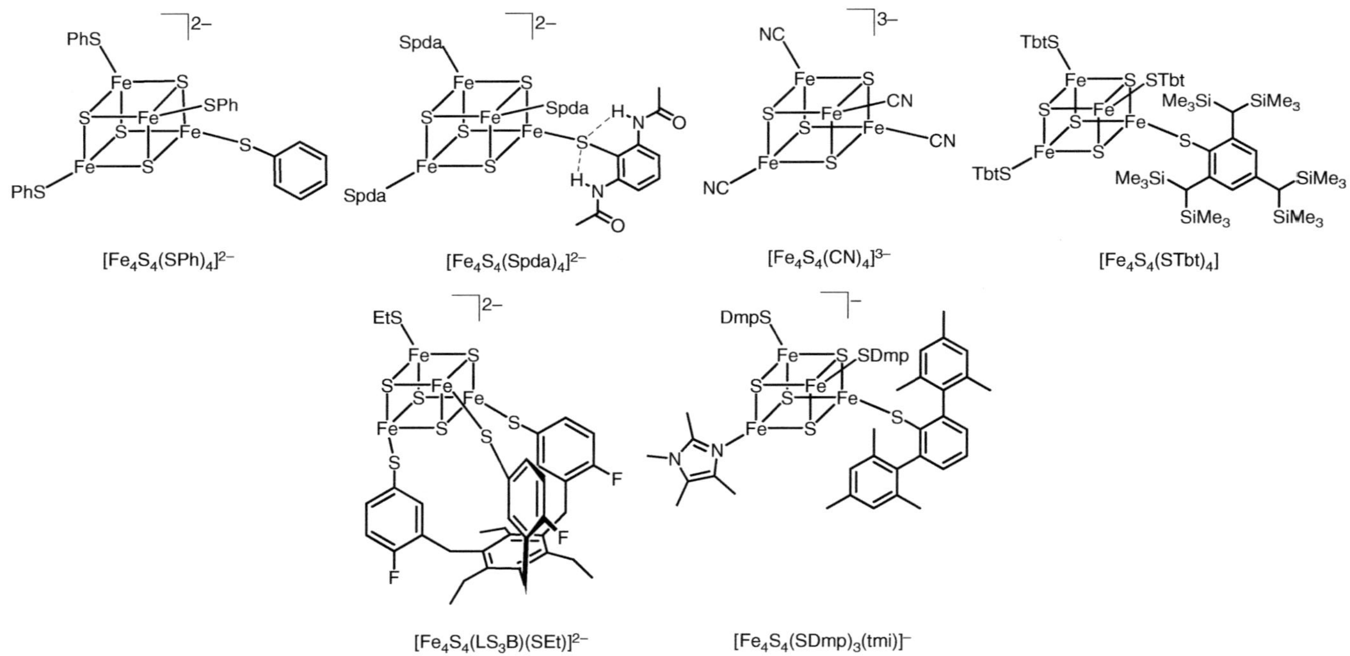

Figure 19.9 Selected examples of $[Fe_4S_4]$ clusters the following discussion of electrochemical properties is based on.

Table 19.1 Half-wave potentials (in V) of some selected homoleptic and 3 : 1 site-differentiated [Fe_4S_4] clusters and their reported redox states.

	Redox states	$E_{1/2}$ (1)	$E_{1/2}$ (2)[a)]	Ref.	Solvent
$[Fe_4S_4(SPh)_4]^{2-}$	$[Fe_4S_4]^{2+/+/0}$	−1.0	−1.73	[47, 84]	MeCN[b)]
$[Fe_4S_4(Spda)_4]^n$	$[Fe_4S_4]^{2+/+}$	−0.72	/	[46a]	MeCN[b)]
$[Fe_4S_4(CN)_4]^{3-}$	$[Fe_4S_4]^{2+/+/0}$	−0.44	−1.42	[68]	DMF[b)]
$[Fe_4S_4(STbt)_4]^{2-}$	$[Fe_4S_4]^{4+/3+/2+}$	−0.048	−0.997	[74]	CH_2Cl_2[c)]
$[Fe_4S_4(LS_3B)(SEt)]^{2-}$	$[Fe_4S_4]^{2+/+}$	−1.47	/	[77]	MeCN[c)]
$[Fe_4S_4(SDmp)_4]^{2-}$	$[Fe_4S_4]^{4+/3+/2+}$	+0.06	−0.82	[79]	THF[c)]
$[Fe_4S_4(SDmp)_3(tmi)]^-$	$[Fe_4S_4]^{3+/2+}$	−0.38	/	[79]	THF[c)]

a) In cases where three different oxidation states are accessible, the redox potentials for both transitions are provided (1) and (2) if available in the literature; / = redox transition not accessible or not reported.
b) vs. SCE.
c) vs. Ag/$AgNO_3$.

as its surrogate thiolate ligands in models exhibit a strong σ-Lewis basicity (which equals strong electron donation), which in artifitial ligands may be accompanied by π-donation, while cyanide ligands are also σ- donors, but, more importantly, in addition very good π-acceptor ligands with the opposite effect of re-taking electron density from the coordinated metal center. Also, steric features such as protection by bulk can be easily implemented via backbone modification in the thiolate ligands. Alone or in combination these effects can efficiently or at least to some extent stabilize up to three different redox states including those extreme ones with all-ferric or all-ferrous iron centers.

For the 3 : 1 site-differentiated complexes $[Fe_4S_4(LS_3B)(SEt)]^{2-}$ and $[Fe_4S_4(SDmp)_3(tmi)]^-$, only one reduction potential is reported each [77, 79]. The thiolate coordinated complex $[Fe_4S_4(LS_3B)(SEt)]^{2-}$ exhibits a half-wave potential of −1.47 V vs. Ag/$AgNO_3$ in MeCN, which can be assigned to the $[Fe_4S_4]^{2+/+}$ redox transition. It is similar to those of simpler $[Fe_4S_4]^{2+}$ clusters with four thiolate ligands but is relatively strongly negative. A direct comparison of the 3 : 1 site-differentiated tetramethylimidazol (Me_4Im) coordinated $[Fe_4S_4(SDmp)_3(tmi)]^-$ complex with its homoleptic precursor $[Fe_4S_4(SDmp)_4]^{2-}$ reveals a major shift in the redox potential. The homoleptic complex exhibits half-wave potentials of +0.06 and −0.82 V vs. Ag/$AgNO_3$ in THF, which can be ascribed to $[Fe_4S_4]^{4+/3+/2+}$ transitions. $[Fe_4S_4(SDmp)_3(tmi)]^-$ can only be reduced once and exhibits a chemically reversible half-wave potential of −0.38 V vs. Ag/$AgNO_3$ in THF, which was assigned to an $[Fe_4S_4]^{3+/2+}$ redox transition [79]. The homoleptic complex serves as a HiPIP model, while the site-differentiated complex mimics an $[Fe_4S_4(Cys)_3(His)]$ site. Their respective $[Fe_4S_4]^{3+/2+}$ transition is approximately 440 mV apart. The positive shift, when one thiolate ($^-$SDmp) is replaced by the neutral Me_4Im, aligns well with changes in electron density donor strengths of the respective exchanged ligands.

The anionic thiolate provides significantly more electron density than the neutral Me_4Im ligand. A notably higher redox potential in the heteroleptic compared to the homoleptic complex suggests that exchanging one cysteine ligand for a histidine in the protein alone is sufficient to raise the redox potential to a level which is, for instance, necessary for the Ni–Fe active site in [NiFe]-hydrogenases [79].

With regard to the molecular structures of these compounds, the initially used term *cubane* or *cuboidal* refers to all $[Fe_4S_4]$ clusters comprising six edge-shared nonplanar $\{Fe_2S_2\}$ arrangements, regardless of their respective distortions from idealized T_d symmetry. The influence of packing effects in the solid state and symmetry-induced distortions on the metrical parameters renders a comprehensive evaluation of $[Fe_4S_4]$ cluster structures rather difficult. For example, none of the synthetic clusters shown in Figure 19.9 exhibits perfect T_d symmetry. The more frequently encountered compressed tetragonal distortion exhibits four shorter Fe–S bonds and eight longer Fe–S bonds, remaining approximately parallel and perpendicular, respectively, relative to each other (D_{2d} distortion mode). What makes comparing parameters from proteins and parameters from artificial model complexes even more difficult is the typical difference in resolution. While small molecular complexes often have a resolution around 0.72–0.84 Å, protein data can have a far larger range, from 0.80 Å to more than 2 Å resolution. However, one general recognizable trend is the correlation of a reduction of the clusters with an increase in the bond lengths between iron centers and terminal thiolate S-donor atoms evidencing weaker bonds [4]. Again, this agrees with the HSAB theory of Fe^{3+} being harder than Fe^{2+} and therefore more accepting of the anionic thiolate coordination.

Another vital parameter often associated with analyzing $[Fe_4S_4]$ clusters is the volume spanned by their cuboidal structure. The reduction generally leads to an increased volume since the Fe–S distances elongate, but these effects are modulated easily by intramolecular and intermolecular interactions in the crystalline structure (packing effects) [4, 85]. Therefore, these parameters always have to be considered with the utmost caution.

19.5.3 Synthetic Model Complexes of $[Fe_3S_4]$

Compared to $[Fe_4S_4]$ clusters, $[Fe_3S_4]$ clusters are notably rarer in natural occurrences as in their synthetic mimics. An obvious route to acquire artificial $[Fe_3S_4]$ clusters involves the removal of one iron edge of an $[Fe_4S_4]$ cluster. Unless a chelating ligand stabilizes and supports the resulting $[Fe_3S_4]$ cluster, this approach typically affords mixtures of thermally unstable [Fe–S] clusters [86]. Such approach was employed in the first reported synthesis of an $[Fe_3S_4]$ cluster. The synthesis depends on the semirigid trifunctional cavitand ligand LS_3A (see Scheme 19.6) [86a, b]. This ligand templates the 3 : 1 site-differentiated cluster $[Fe_4S_4(LS_3A)(SEt)]^{2-}$ (EtS^- = ethanethiolate), which allows for a site-specific modification of the unique iron site. Upon addition of one equivalent of trifluoromethanesulfonic acid (HOTf), the ethanethiolate ligand is being protonated and subsequently replaced by a labile triflate ligand (OTf^-), which can be easily displaced by N-methylimidodiacetate (Meida). The addition of a second equivalent of Meida results in the removal of iron to afford the $[Fe_3S_4]$ cluster as shown in Scheme 19.9 [87].

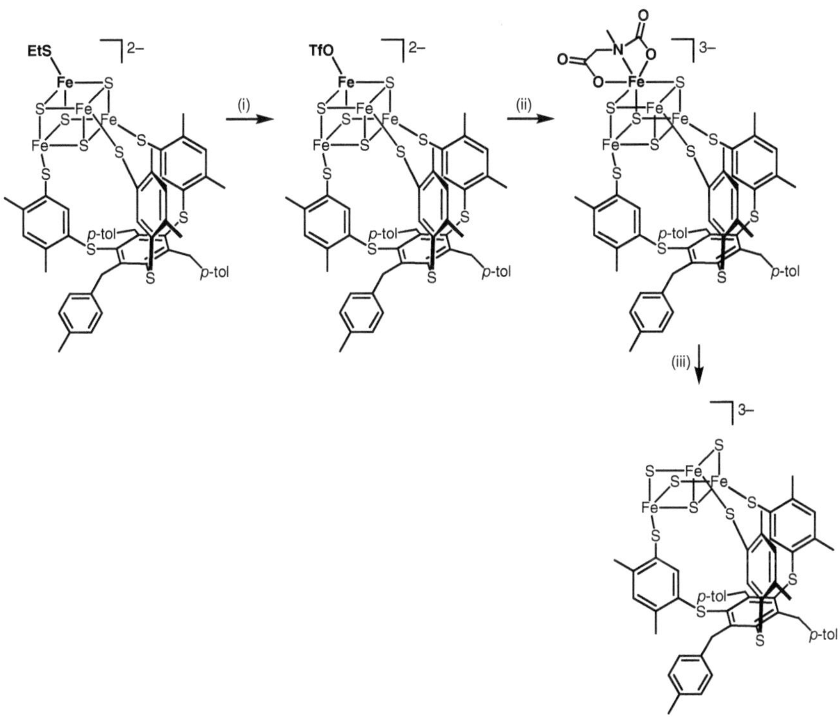

Scheme 19.9 Synthesis of $[Fe_3S_4(LS_3A)]^{3-}$ via a templated approach. (i) $(Et_3NH)(OTf)$; (ii) and (iii) 1 eq. $(Et_4N)_2(Meida)$. Source: Adapted from [87].

Interestingly, the resulting $[Fe_3S_4]$ cluster can be used to obtain heterometallic, cuboidal clusters with Ti, Co, Ni, Cu, Mo, W, and Ag or be reconstituted to $[Fe_4S_4]$ using an appropriate iron ion donor (e.g. $[FeCl_4]^{2-}$) [88].

Eventually, it was shown that even with monodentate ligands an open-cuboidal $[Fe_3S_4]$ cluster, $[Fe_3S_4(IMes)_3]^{+/0}$, can be prepared and both biological oxidation states accessed $[Fe_3S_4]^{+/0}$ [89]. This is facilitated by the use of weakly coordinating anions on the labile iron edge and an increased steric bulk around the remaining iron centers, which are coordinated by strongly binding NHCs. First, the unique iron site is removed by a powerful chelator (tris(pyrazolyl)borate) generating an inner-sphere KPF_6 complex (plus KCl and Tp_2Fe) as evidenced by SCXRD (Scheme 19.10) [89]. Subsequent oxidation with $[Cp_2Fe][PF_6]$ decreases the affinity between cluster and potassium ion and induces precipitation of KPF_6. In a salt metathesis reaction with $NaBAr^F{}_4$, an analytically pure $[Fe_3S_4]^+$ cluster can be obtained. Alternatively, a reduction with $[Cp^*{}_2Co]$ results in the $[Fe_3S_4]^0$ cluster.

The removal of one iron center has specific implications for the overall electrochemical redox behavior. $[Fe_3S_4(LS_3A)]^{3-}$ exhibits three oxidation states of the core cluster $[Fe_3S_4]^{+/0/-}$. The chemically reversible redox events occur at −0.793 and −1.722 V vs. SCE in MeCN [86a]. The low potential for the second reduction event already indicates that it might not be bio-relevant. In proteinic $[Fe_3S_4]$ clusters, the redox transitions are commonly limited down to the $[Fe_3S_4(Cys)_3]^{3-}$ state ($[Fe_4S_4]^0$),

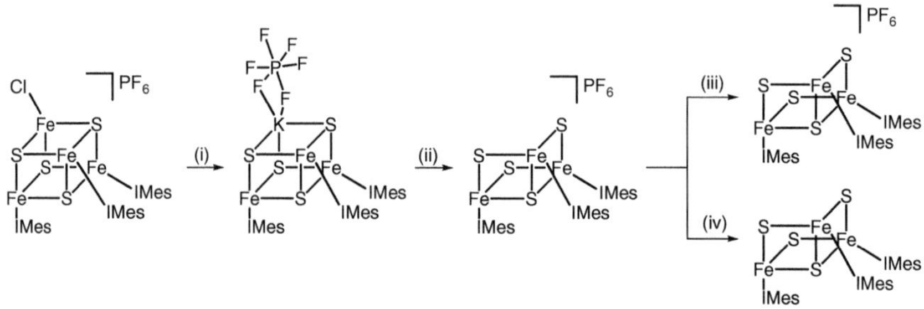

Scheme 19.10 Synthesis of open-cuboidal $[Fe_3S_4]^+$ and $[Fe_3S_4]^0$ clusters with monodentate NHC ligands. (i) 2 KTp in DFB; (ii) $[Cp_2Fe][PF_6]$ in DFB; (iii) $NaBAr^F{}_4$; (iv) $[Cp^*{}_2Co]$. Tp = tris(pyrazolyl)borate, DFB = 1,2-difluorobenzene. Source: [89]/American Chemical Society.

while the accessibility of states is also influenced by the pH [90]. Open-cuboidal model $[Fe_3S_4(IMes)_3]$ similarly exhibits two half-wave potentials $E_{1/2} = -1.17$ and -2.32 V vs. Fc/Fc^+ in *ortho*-difluorobenzene (DFB) ascribed to $[Fe_3S_4]^{+/0/-}$ transitions. The potentials of the NHC coordinated cluster complex are shifted to more negative potential, which renders oxidation comparatively easier. This is in agreement with a comparably harder donor atom in the NHC as in the thiolate. However, comparability of the respective data is to some extent compromised for the used distinct solvents and references. Despite the electrochemical reversibility observed in the $[Fe_3S_4(IMes)_3]^{0/-}$ couple, isolating $[Fe_3S_4(IMes)_3]^-$ has remained elusive [89].

Direct comparison with an $[Fe_4S_4]$ cluster, $[Fe_4S_4(SPh)_4]^{2-}$ ($[Fe_4S_4]^{2+/+/0}$, $E_{1/2} = -1.0$ and -1.73 V vs. SCE) [47, 84], reveals that the loss of one iron ion drastically affects the availability of redox transitions. The $[Fe_4S_4]$ cluster can be more readily oxidized compared to the $[Fe_3S_4]$ clusters, since, simply speaking, the reduction in electron density of the cluster can be buffered by all four iron centers. Reduction, on the other hand, is comparatively easier for the $[Fe_3S_4]$ clusters because one electron-rich ligand is missing as well, the presence of which in $[Fe_4S_4]$ would repel to some extent the incoming electron. However, in general the various terminal ligands have of course a quite notable effect on the redox properties themselves and such a comparison should be considered with caution unless identical ligands can be evaluated for the different cluster types.

The available structural data reveal some interesting features of the synthetic models, which depend on oxidation states and ligand environments in some contrast with protein data. The molecular structures of the oxidized $[Fe_3S_4]^+$ (Fd_{ox}) and reduced $[Fe_3S_4]^0$ (Fd_{red}) cluster of *A. vinelandii* [91] (Figure 19.10, left) do not differ significantly in metric parameters within the experimental error, highlighting nature's ability to stabilize a specific coordination geometry and to, thereby, influence reactivity via the protein backbone (*entatic state*) [91].

The thiolate-coordinated model complex $[Fe_3S_4(LS_3A)]^{3-}$ has overall comparable bond lengths for the terminal thiolate ligand and inner Fe–S distances. Interestingly, the averaged Fe–Fe distances are slightly longer compared to *A. vinelandii* Fd_{ox} and

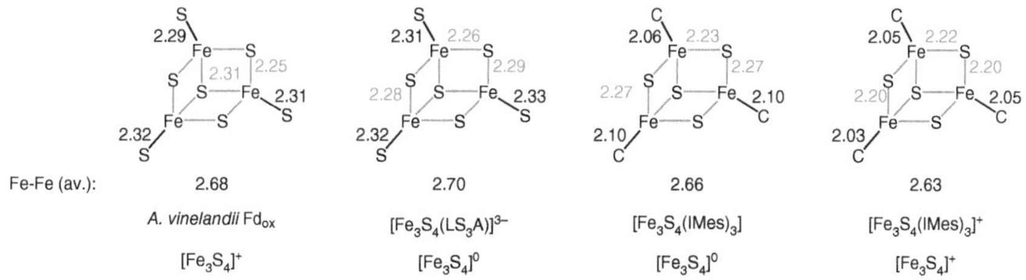

Figure 19.10 Bonding distances (in Å) in $[Fe_3S_4]$ clusters from *A. vinelandii*. Source: Adapted from [91] and exemplary biomimetic complexes. Source: Adapted from [86a, 89], with an indication of the core cluster oxidation state. Distances in gray are averaged Fe–S bond distances within the core cluster.

to $[Fe_3S_4(IMes)_3]^{+/0}$. This may go back to some strain exerted by the chelating tridentate ligand scaffold.

In $[Fe_3S_4(IMes)_3]^0$ and its oxidized derivative (Figure 19.10), the average Fe–S distances correlate with the Fe–C distances. Upon oxidation, when at least formally one Fe^{2+} is turned into an Fe^{3+} center and thereby reduced in size, the Fe–C distances shorten as do the Fe–S bonds. In $[Fe_3S_4(IMes)_3]^0$, the iron centers have oxidation states of Fe^{3+}, Fe^{3+}, and Fe^{2+}. The reduced and one of the oxidized iron centers form a mixed valence pair (charge delocalization), while the remaining iron center is strictly Fe^{3+} (localized charge). In the respective crystal structure, one iron ion exhibits significantly shorter Fe—C and Fe—S bond distances (Fe–C : 2.06 Å vs. 2.10 Å, Fe–S : 2.23 Å vs.2.27 Å) compared to the other two and can thereby clearly be identified as the iron center with the localized charge. In $[Fe_3S_4(IMes)_3]^+$, all three iron centers are oxidized, which is reflected in the very similar metrical parameters for all [89]. The oxidized cluster is more contracted, and the bonding distances exhibit a more narrow range. Interestingly, this is not the case in the molecular structure of *A. vinelandii*, where Fd_{red} and Fd_{ox} exhibit very similar bond lengths [91]. The absence of protein-based structural restraints apparently allows the artificial cluster, $[Fe_3S_4(IMes)_3]^{0/+}$, to exhibit more flexibility upon redox changes.

19.6 Clusters of Higher Nuclearity

Larger Fe–S clusters play a crucial role as active sites in proteins engaged in more intricate forms of catalysis, such as the reactions catalyzed by hydrogenases and nitrogenases. These proteins house multiple Fe–S clusters to facilitate electron transport relays and catalytic activities within the enzymes.

Two examples of complex higher nuclear [Fe–S] clusters are shown in Figure 19.11. The P-cluster (a) is located in the FeMo protein of nitrogenases and is responsible for the electron transport from the Fe protein to the FeMo cofactor (FeMoCo). The FeMoCo (b) is the active site in the nitrogenase catalyzing the important yet difficult transformation of dinitrogen to ammonia.

The complex nature of these extended clusters makes it challenging to comprehensively compare structural and functional models and their natural templates.

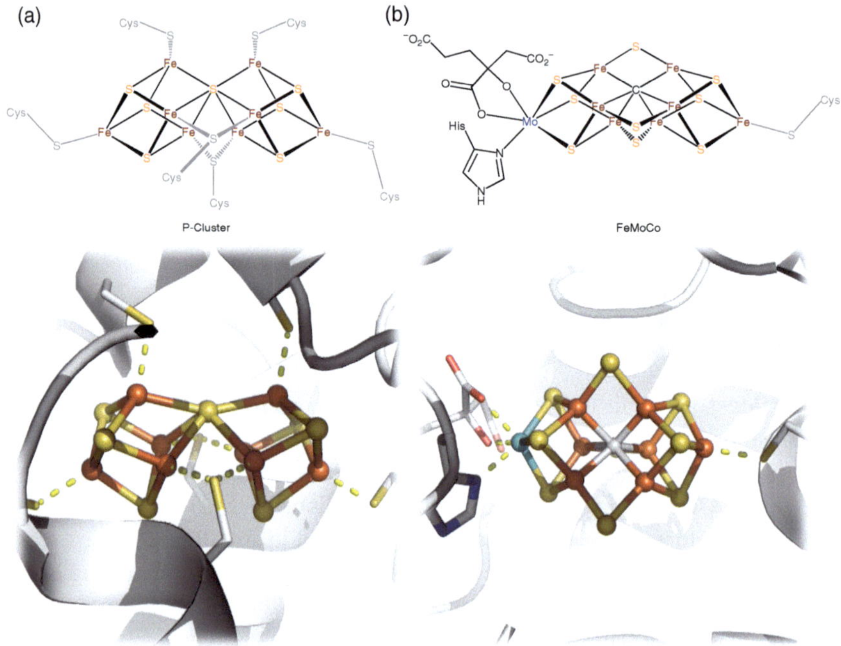

Figure 19.11 Chemical and molecular structures of (a) the reduced P-cluster ($[Fe_8S_7(Cys)_6]$, MoFe protein) and (b) FeMoCo ($[MoFe_7S_9C(Cys)(His)(HC)]$, HC = homocitrate) in a nitrogenase from *A. vinelandii* (PDB: 1M1N) at 1.16 Å resolution. The interstitial atom in the FeMoCo was later identified as a carbide. Color coding on top: yellow sulfur: sulfide; gray sulfur: cysteine/thiol. Source: Adapted from [92].

The following discussion will, hence, focus on selected available electrochemical and structural data that facilitate valuable insights into properties, dependencies, and (re-)activities.

19.6.1 Hydrogenase and Nitrogenase

Hydrogenases possess the remarkable ability to reversibly catalyze the evolution of H_2 from protons. They have been, and still are investigated intensely with a focus on molecular hydrogen generation as a fuel and respective "storage" strategies for (future) industrial energy supply concepts. Three general types of hydrogenases are known as of yet. These are hydrogenases with [NiFe], [FeFe], or [Fe] active sites.

The hydrogen-activating so-called H-cluster found universally in all [FeFe]-hydrogenases has a quite notable composition. It is a highly unusual organometallic structure comprising an $[Fe_4S_4]$ cluster linked to a diiron site via a bridging cysteine. Within the diiron site, low-valent iron ions are interconnected by an azadithiolate ligand and coordinated by potent π-accepting ligands, namely cyanide (CN^-) and carbon monoxide (CO) (Figure 19.12a) [93]. These ligands form very strong bonds with electron-bearing metal centers (Fe^{3+} is a d^5 metal, Fe^{2+} a d^6 metal) and are normally considered very toxic. It is quite extraordinary that they are found to be

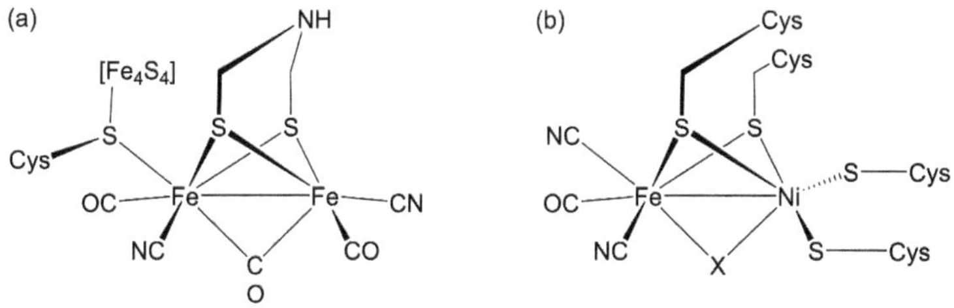

Figure 19.12 Chemical structures of the [FeFe]- (a) and [NiFe]-hydrogenase active sites (b). X = oxide, sulfur, hydroperoxide, or hydroxide.

indispensable in a biological context. Recently, it was suggested that the strong bonds result in a very rigid structure that is geometrically not much affected by redox events, which again would facilitate the transformation of the very small proton and H_2 species. This is still a matter of investigation and, while reasonable, has not been proven finally as of yet.

The catalytic centers of [NiFe]-hydrogenases comprise one nickel and one iron ion, coordinated by four conserved cysteine ligands. The iron is additionally ligated by a carbonyl group (CO) and two cyanide groups (CN^-) (Figure 19.12b) [94]. The source of electrons is an [Fe_4S_4] cluster located in the smaller subunit of the protein. The [Fe] hydrogenase is not a cluster but an iron-coordinated complex with no [Fe–S] cluster close to the active site [95].

[FeFe]-hydrogenase exhibits the highest H_2 formation rate, boasting an impressive 8000 μmol H_2 min^{-1} mg^{-1} of protein. It consists of multiple protein subunits, while the H-cluster is nestled within a complex protein environment. To enable H_2 gas transfer, a hydrophobic and relatively tight channel connects the active site with the protein surface. Depending on the organism from which the [FeFe]-hydrogenase originates, additional non-essential Fe–S clusters, termed F-clusters, may also be observed, that can contribute to electron transport.

The H-cluster has a quite fascinating structure. The Fe atoms of the dinuclear center are bridged by two μ_2-S atoms and a bridging CO ligand [1, 96]. The proximal (as in: close to the electron relay system) Fe is coordinated by a Cys residue that also links to an [Fe_4S_4] cluster. The distal (as in: further away) Fe atom retains an open coordination site for substrate binding. Remarkably, the [FeFe] site and the [Fe_4S_4] site have a distance to each other of only approximately 4 Å, which renders electron transfer very efficient. In principle, six physiological redox states can be accessed by the H-cluster via a combination of oxidation states of its two components. For the [Fe_4S_4] cluster, two oxidation states can be assigned: $[Fe_4S_4]^{2+}$ and $[Fe_4S_4]^{+}$. The [FeFe] cluster can adopt the [$Fe^{2+}Fe^{2+}$], [$Fe^{2+}Fe^{1+}$], and the fully reduced [$Fe^{1+}Fe^{1+}$] state; the latter two are exceptionally strongly reduced. Despite this multitude of available redox states, only two resting states are currently known: the "active oxidized" state H_{ox} ($[Fe_4S_4]^{2+}$ with [$Fe^{1+}Fe^{2+}$]) and the "active reduced" state H_{red} ($[Fe_4S_4]^{2+}$ with [$Fe^{1+}Fe^{1+}$]). Despite extensive research, some aspects of the catalytic mechanism of the [FeFe] hydrogenase remain unclear, while model chemistry may facilitate better insights into these fascinating enzymes.

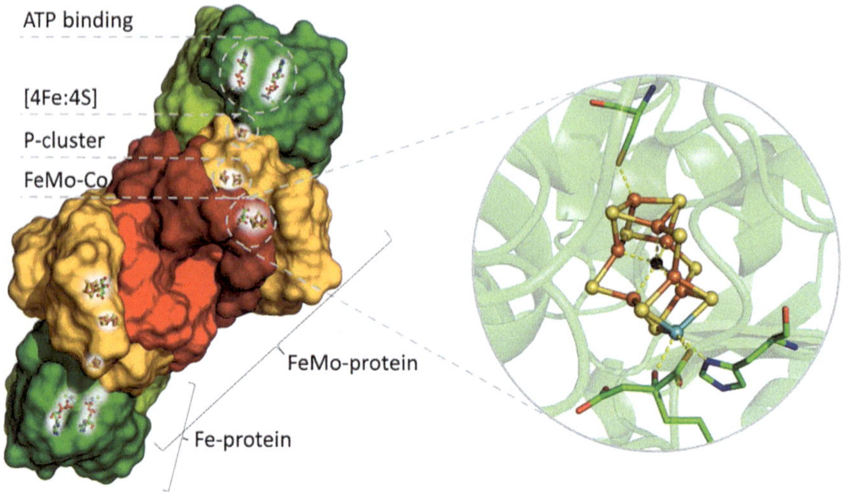

Figure 19.13 (a) To the left a PyMol representation of dimeric nitrogenase protein (PDB: 1N2C) from *A. vinelandii* at 3 Å resolution, consisting of the Fe and FeMo proteins. Source: Adapted from [98, 99]. (b) To the right the active site of the nitrogenase from *A. vinelandii* at 1.16 Å resolution (PDB: 1M1N). Source: Adapted from [92a]. The interstitial atom was later unambiguously identified as a carbide. Source: Adapted from [92b, c].

Nitrogenases are similarly important, unusual, and intriguing enzymes as hydrogenases. They play a crucial role in catalyzing the formation of ammonia from molecular nitrogen, utilizing protons, electrons, and energy equivalents (ATP). The yields of ammonia may vary depending on the specific active nitrogenase type. Among the nitrogenase cofactors, the most efficient form is the one that exhibits an [FeMo] cofactor [92a], which is superior to the closely related [FeFe] and [FeV] cofactors [97]. These cofactors are deeply embedded within their respective proteins to protect them from exposure to aerobic conditions and hydrolysis of intermediates when exposed to water so that (uncontrolled) decomposition can be avoided (Figure 19.13).

Two proteins work together to split the inherently very stable triple bond of dinitrogen: the Fe protein and the MoFe protein. The Fe protein contains an $[Fe_4S_4]$ cluster, coordinated by two Cys residues from the Fe protein and the MoFe protein each. When bound to the MoFe protein, it transfers electrons from the $[Fe_4S_4]$ cluster to the P-cluster in the MoFe protein, which in turn transfers electrons to the active site. The FeMo protein houses two Fe–S clusters. The P-cluster is a double-cubane $[Fe_8S_7]$ site, where two $[Fe_4S_3]$ subclusters are bridged by a sixfold bound sulfide. The second Fe–S cluster in the most efficient form of nitrogenase has a heterocubane structural motif: $[MoFe_7S_9C$-*R*-homocitrate]. In this structure, molybdenum is coordinated by homocitrate and an additional histidine residue, while a cysteine residue coordinates the terminal Fe site on the opposite end. The remaining six Fe atoms are coordinated by sulfides, forming a trigonal-prismatic structure that also bears an interstitial carbide (C^{4-}) equidistantly to the six iron in its center. This interstitial carbide renders artificial, *structural* model complexes a very challenging target, but many *functional* model complexes are known for a plethora of coordinated metals [100].

19.6.2 Model Chemistry for Clusters of Higher Nuclearity

The synthetic model chemistry for clusters of higher nuclearity can roughly be divided into three main approaches: (i) synthesis of homonuclear Fe–S containing clusters with different sizes, (ii) synthesis of heteronuclear Fe–S containing clusters (with different sizes), and (iii) heteroleptic ligand insertion. The latter may also contain non-iron metals in order to mimic the nitrogenase cofactor.

Homonuclear cluster complexes of higher nuclearity (i) can be obtained by different means depending on the desired product. The most common approach is a fragment condensation or a spontaneous self-assembly reaction, which leads to di- and tetracubane species. For example, using hydrogen sulfide or thiolate results in the formation of dicubane (Figure 19.14a,b, respectively), the subclusters of which are connected by an unsupported sulfido [77] or thiolato [101] bridge.

The bridging sulfide ligand can be introduced to a site-differentiated $[Fe_4S_4]$ cluster (e.g. Scheme 19.6, p. 18) by replacing ethanethiolate (EtS^-) by HS^- with a hydrogen sulfide salt followed by the reaction of a second cluster with the exposed Fe–SH that replaces the EtS^- ligand of the other cluster to result in a μ_2-S bridged dicubane.

Spontaneous self-assembly to a cluster of higher nuclearity (Figure 19.14c) can be also realized via careful control of the reaction stoichiometry in a single step. For example, the synthesis of the $[Fe_4S_4]$ cluster shown in Scheme 19.3 can be performed with one equivalent of carbene ligand instead of two to obtain an $[Fe_8S_8]$ cluster

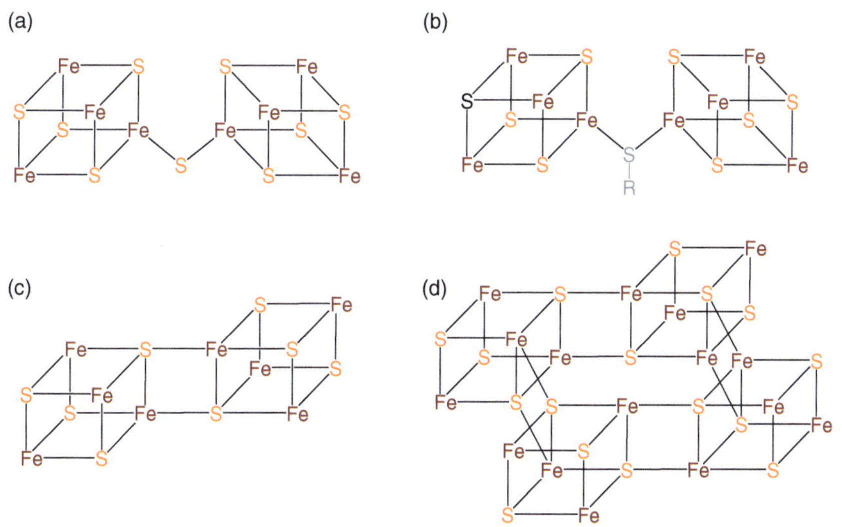

Figure 19.14 Chemical structures of di- (a–c) and tetracubane clusters (d) obtained from spontaneous self-assembly or fragment condensation reactions. Source: Adapted from [71, 77, 101]. Supporting ligands on the Fe sites are not shown for clarity reasons.

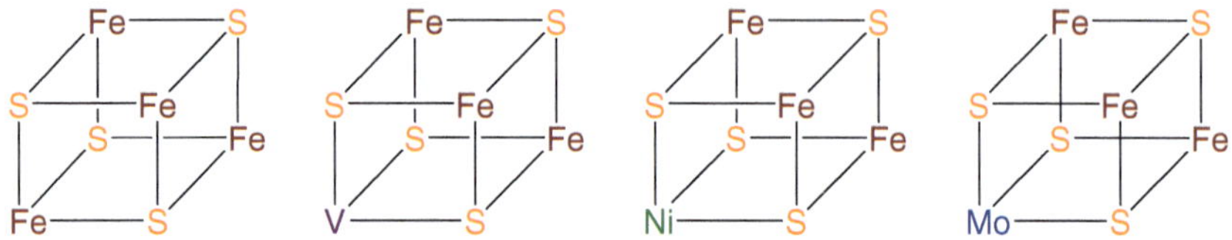

Figure 19.15 Schematic depiction of examples for homo- and heteronuclear cubane $[MFe_3S_4]$ clusters with potential biological relevance. V and Mo are present within nitrogenases, while Ni is part of the [NiFe] hydrogenase and carbon monoxide dehydrogenase. Source: Adapted from [5]. Terminal ligands are not shown for clarity reasons.

that is fused over the Fe–S edge lengths [71]. The synthesis of an $[Fe_{16}(\mu_3\text{-}S)_8(\mu_4\text{-}S)_8]$ cluster (Figure 19.14d) can be achieved after the reduction-initiated loss of phosphine ligands from the phosphine coordinated precursor complexes shown in Scheme 19.2 [80].

The active site chemical structures of hydrogenases and nitrogenases suggested and, at some point, necessitated the inclusion of heterometals into biomimetic Fe–S species and led to investigations into how an alternate metal might impact the overall (redox) properties of respective [Fe–S] clusters. The synthetic strategy toward heteronuclear clusters of higher nuclearity (ii) is most often based on the utilization of $[Fe_3S_4]$ clusters upon removing one iron edge from $[Fe_4S_4]$ clusters and subsequent replacement by a different metal ion, which yields 3 : 1-differentiated $[MFe_3S_4]$ clusters [86a]. The electrophile (metal fragment) is typically inserted bearing labile leaving groups, which is then followed by coordination of a stabilizing chelate ligand [88, 102]. Well-investigated examples of this type of reaction comprise employed heterometals vanadium, nickel, and molybdenum as found in nitrogenases, hydrogenases [5], and carbon monoxide dehydrogenases [101, 102] ($NiFe_3S_4$ cluster active site) (Figure 19.15). Tungsten is employed based on its chemical similarity to its lighter congener molybdenum, while also cobalt-, copper-, silver-, palladium-, platinum-, and thallium-containing heterocubanes are known [36b, 88, 102, 103].

Notable examples for the syntheses of stable 3 : 1 differentiated heteronuclear cubane species comprise a self-assembly reaction (for molybdenum) or a ligand substitution (for vanadium) (Scheme 19.11) [104].

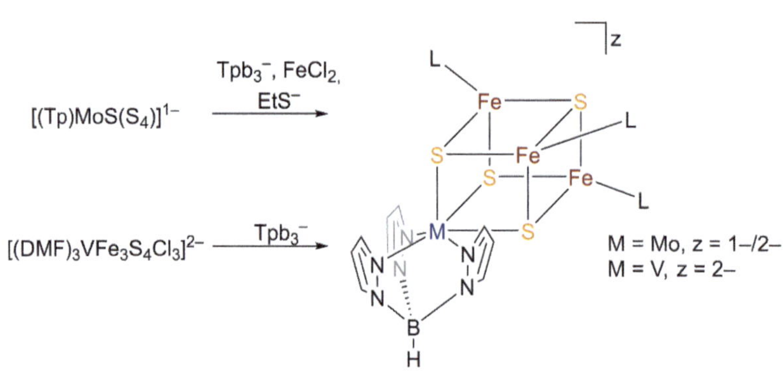

Scheme 19.11 Syntheses of heteronuclear cubanes in a self-assembly reaction (M = Mo) or via a substitution reaction (M = V). Tpb_3^- = tris(pyrazolyl)hydroborate(1−); L = halide, N_3^-, CN^-, RS^-. Source: Adapted from [104, 105].

Introduction of phosphine ligands to replace the monoanionic ligands (L^- in Scheme 19.11) followed by reduction with tetraborohydrate results in the case of molybdenum in the formation of a symmetric Fe–S edge-fused dicubane [105]. Subsequent substitution of the phosphine ligands by (hydrogen-)sulfide yields a heterometallic complex with three different coordination modes of the bridging sulfides ($[Mo_2Fe_6(\mu_2\text{-}S)_2(\mu_3\text{-}S)_6(\mu_6\text{-}S)]$, Scheme 19.12) [106]. The core exhibits a C_{2v} symmetry and can be conceptualized as two distorted $[MFe_3S_4]$ cubanes fused via a common hexavalent sulfur supported by two sulfide bridges. The overall structural motif resembles the P-cluster [92a, 107] shown in Figure 19.11, while not exhibiting an all-iron motif. This cluster has a notable similarity to the P-cluster of nitrogenase with regard to the overall structure and also mimics the FeMoCo due to the inclusion of a molybdenum edge in the cubane moieties. However, it is not an exceptionally good model for either because important differences to the natural templates remain. The realization of such species was a substantial achievement, though, and the respective syntheses and handling are anything but easy. Other bridging ligands, such as methanolate, can also facilitate the formation of closely related clusters. The dicubane subsequently served as a precursor for numerous other cluster topologies [108]. The formation of a respective vanadium-bearing complex was also reported [104c, 109].

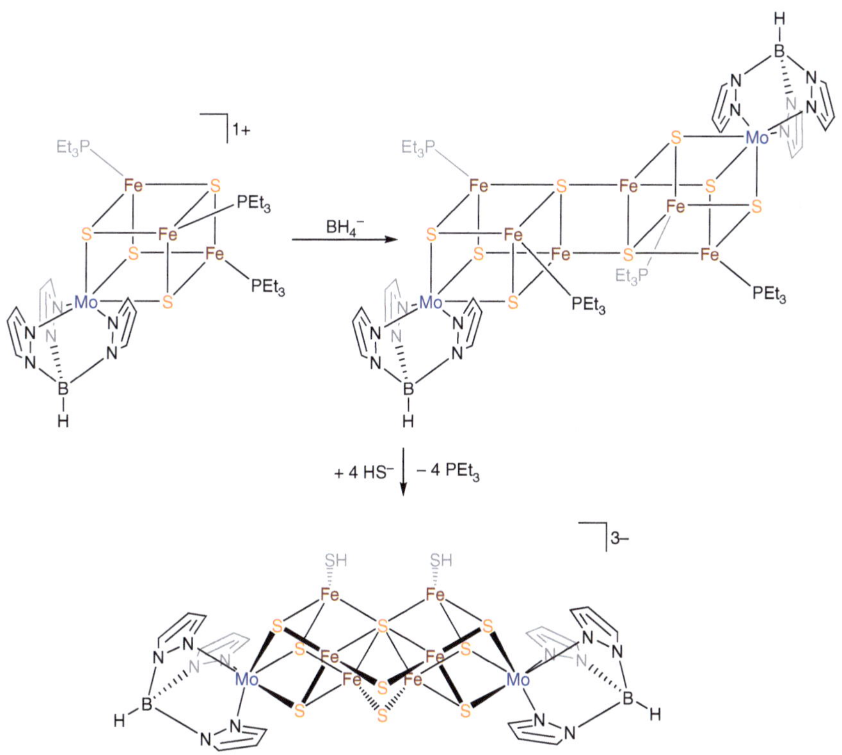

Scheme 19.12 Core rearrangement upon reduction of $[MoFe_3S_4]$ to form an edge-length-bridged double cubane followed by a ligand substitution reaction (and another core rearrangement) to obtain a P-cluster-type heterometallic complex. Source: Adapted from [106].

An all-iron structural model of the P-cluster can be obtained following a self-assembly approach using different thiolate ligands in a one-pot reaction. Employing $[Fe(N(SiMe_3)_2)_2]$ as an iron source and two different, sterically demanding thiolate ligands in the presence of sulfur, two different $[Fe_7S_8]$ clusters can be obtained (Scheme 19.13) [110]. Again, an excess of thiolate is used as a source of reduction equivalents for the elemental sulfur.

A particularly interesting P-cluster model, $[Fe_8S_7(\mu_2\text{-}S_{thiolate})_2]$, was prepared very recently via two alternative routes: (i) in a one-pot procedure from an all-ferric $[Fe_4S_4(SR)_4]$ cubane synthon as a modular building block (Scheme 19.14, top) or (ii) using the isolated dinuclear intermediate/byproduct from the reaction of $Fe\{N(SiMe_3)_2\}_2$ with HSTbt as a reactant with the all-ferric cluster (Scheme 19.14, bottom) [111]. The latter approach gave better yields.

Notably, instead of hydrogen sulfide (HS^-), hydrogen selenide (HSe^-) can be used as a nucleophile in the formation of certain clusters. This allows for more detailed insights into the fragment condensation reactions with regard to the identification of sulfides that might later function as μ_2-S, μ_3-S, or μ_6-S bridging atoms [112]. Since it was established that at least one of the sulfide ligands of nitrogenase are naturally labile and are actually removed from the cluster during turnover, such insights are particularly valuable, as it is not entirely proven as of yet which sulfide(s) exactly engage in such behavior and how.

Site-differentiated clusters can further be used to synthesize artificial H-cluster frameworks reminiscent of the [FeFe]-hydrogenase active site. An intriguing example for an [FeFe]-hydrogenase model comprises the realization of a protected coordination site for an [FeFe] precursor, which is then activated to couple a site-differentiated $[Fe_4S_4]$ cluster in a fragment condensation approach as shown in Scheme 19.15 [83].

During the synthesis, one S-donor atom of the tridentate thiolate ligand is protected by acylation. Heating in toluene with a relatively high boiling temperature allows the chelating ligands unprotected S-donor atoms to substitute CO ligands of the $[Fe_3(CO)_{12}]$ precursor, which loses one iron center in the process. When adding another cluster with a more basic ligand, a saponification reaction occurs at the protected thiolate, which deprotects it and allows it to replace another CO from a terminal position and to coordinate to the vacant Fe site. Simultaneously, this S-donor forms a bridge with the cubane complex that initiates the deprotection. Such reactivity can be achieved with simple $[Fe_4S_4(EtS)_4]^{2-}$, $[Fe_4S_4(SR)_{4-n}(Fe_2(CH_3C(CH_2S)_3(CO)_5)_n]^{2-}$ (n = 1, 2, 3, 4) species, or with a site-differentiated $[Fe_4S_4]$ cluster featuring a polydentate ligand [83]. In case the cubane precursor has not much steric bulk in its terminal ligands, several substitution patterns become feasible, i.e. several dinuclear iron moieties may be attached to the cubane moiety. This renders controlling the reaction difficult and makes purification of the target species very challenging. However, a monosubstituted variant was unambiguously proven via gas-phase photoelectron spectroscopy [83].

Mimicking the active site structure of nitrogenase is very demanding due to the inherent complexity of a heterocubane cluster (Figure 19.11b) and the presence of an interstitial carbide, as now established beyond doubt, adds another layer of

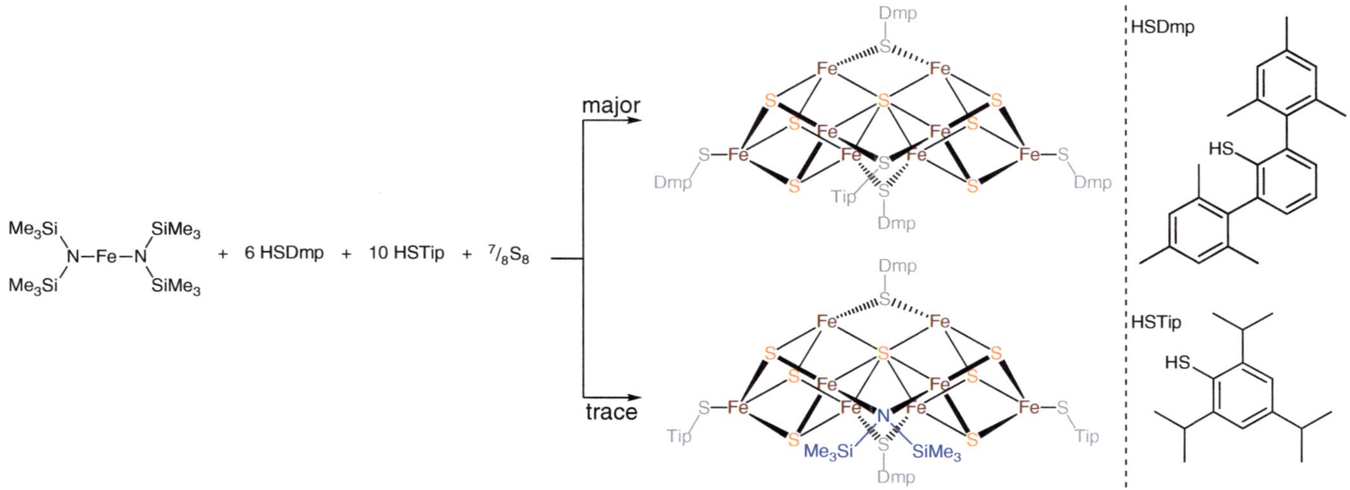

Scheme 19.13 Synthesis of two different [Fe_7S_8] clusters as P-cluster analogs in one-pot self-assembly reactions. Source: Adapted from [110].

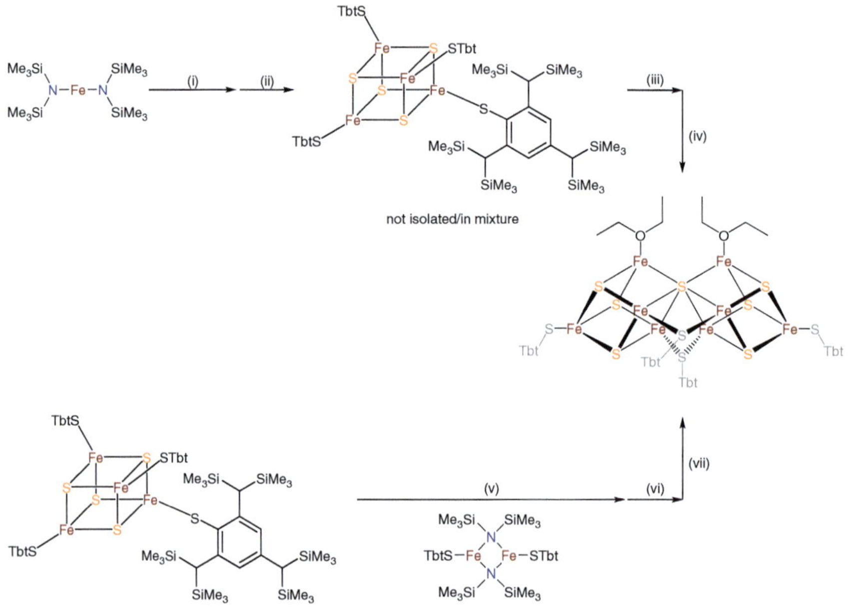

Scheme 19.14 Synthetic approaches for a sophisticated P-cluster model complex. (i) HSTbt in toluene; (ii) 7/8 S_8 in toluene, (iii) stirring in toluene for 24 h, (iv) crystallization in Et_2O, (v) in toluene, (vi) 24 h stirring in toluene, (vii) crystallization in Et_2O. Source: Adapted from [111].

difficulties [92b, c]. There is a plethora of functional models known, with which it was more or less successfully tried to copy the overall reactivity of nitrogenases. These models are generally not based on cluster species but rather on mono- or dinuclear molecular complexes. Even though it is most likely iron, which is responsible for activating the molecular nitrogen enzymatically, other metals are substantially more frequently used for the functional models and all these species fall beyond the scope of this chapter. For readers interested in this context, two articles shall be recommended here [100, 113].

When structurally modeling the FeMoCo, one faces three major challenges: (i) Weak-field, four-coordinate iron environments tend to be inherently unstable, often undergoing rearrangements to achieve a more saturated ligand environment. This is even more the case under mechanistically ill-defined cluster assembly conditions. (ii) The FeMoCo exhibits a polyanionic structure (featuring sulfide, oxide, nitride, carbide, and their respective substitutions), precluding the use of chelate ligand designs to control core hetero ligand compositions. (iii) The variations in physical and chemical properties within the potential ligand environment pose challenges in determining reactivity and stability. For instance, the increasing basicity of 2p anions complicates formulating a viable strategy to incorporate heterometallic fragments and a heteroligand environment into FeMoCo model complexes.

After the discovery of the interstitial atom of FeMoCo at the beginning of the century, it was for some time unclear whether this light atom was carbon,

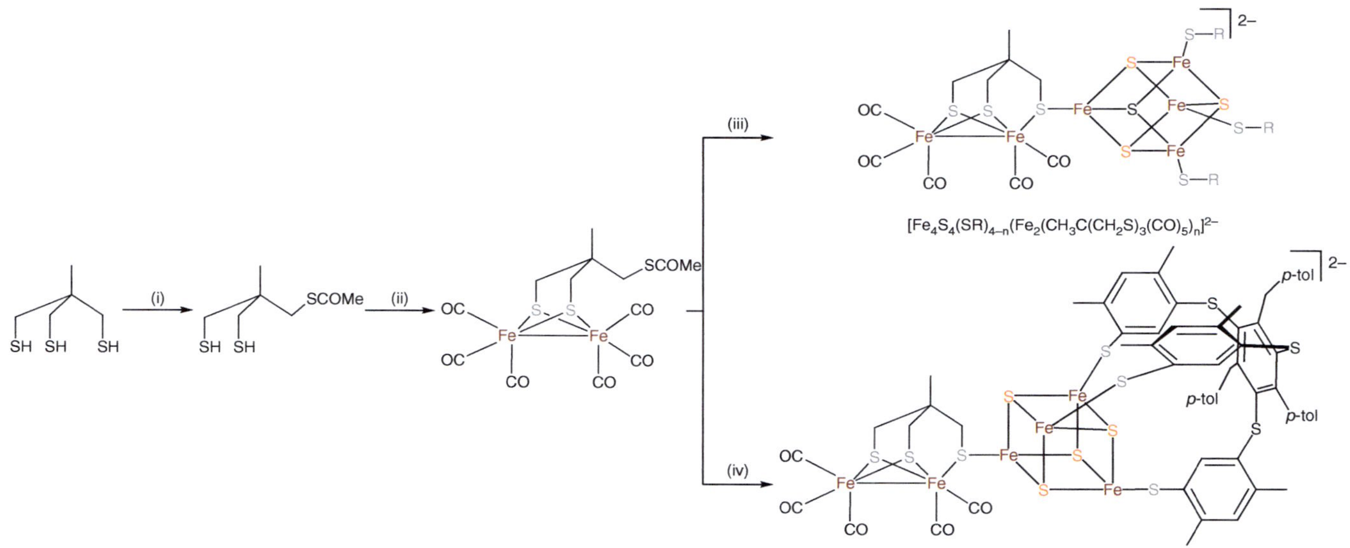

Scheme 19.15 Synthetic pathways toward an artificial H-cluster model complex via a fragment condensation. (i) Acetic anhydride, $NaHCO_3$, Et_2O; (ii) $[Fe_3(CO)_{12}]$, toluene; (iii) $(NBu_4)_2[Fe_4S_4(EtS)_4]$, MeCN; (iv): LS_3A, MeCN. Source: [83]/Springer Nature.

nitrogen, or oxygen. Modeling attempts around that time were also made to help in identifying the identity of this central light atom. Therefore, relatively earlier models incorporated edge-substituted [Fe–S] clusters with ligands of nitrogen and oxygen [4, 114], while heterometallic complexes with substituted edges were also reported [112b, 115]. Efforts were further made to obtain an interstitial oxygen high-nuclearity cluster [116]. Generally, the syntheses of such species followed a fragment condensation approach to assemble more complex clusters. After carefully choosing the metal fragment precursors and optimizing the reaction conditions, a manifold of different structural model complexes was obtained. Considering that it is now established that the interstitial light atom is carbon, the following discussion will not cover models aimed at O or N coordination, though.

One example of an iron cluster, $[Fe_6C(CO)_{16})]^{2-}$, which was not specifically designed for this purpose, with six iron ions caging a central carbide, has been known since the 1970s [117]. While Fe–Fe separations (2.67 A) and Fe–$C_{carbide}$ (1.89 A) bond lengths are comparable to the FeMoCo (~2.60 A; 2.00 A, respectively), the host polyhedron is octahedral rather than trigonal prismatic and the iron oxidation state is, interestingly, more reduced compared to the natural cofactor [118]. Also, one has to note the absence of any sulfur in this model.

Another example mimicking the overall composition of the nitrogenase, albeit without the carbide, comprises the [MoFeS] cluster shown in Scheme 19.16. The synthetic approach follows a self-assembly reaction paired with excess hydrogen sulfide as reduction equivalents.

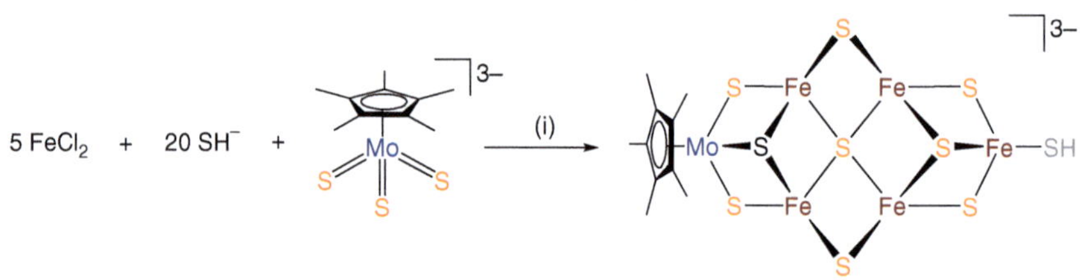

Scheme 19.16 Synthetic route toward $[Cp^*MoFe_5S_9(SH)]^{3-}$; (Cp* = Pentamethylcyclopentadienyl). (i) In MeCN. Source: [119]/John Wiley & Sons.

Despite its non-cubane structure, this biomimetic complex can be viewed as an artificial nitrogenase active site mimic. In fact, it was even tested for protein incorporation (successfully) and reactivity (to some extent successfully; no N_2 reduction but small C_2H_2 or cyanide substrates could be transformed) [119, 120]. These studies highlight the complexity of structure–activity-relationships. Such a semi-artificial enzyme approach was previously very successful for hydrogenase models and enzymes, while the interplay between model complexes and protein biochemistry is for nitrogenases decidedly more challenging. At present, it appears unlikely that a structural FeMoCo model will possess the ability to activate dinitrogen inside or outside of the enzyme.

The combined influence on the properties and (re-)activities of Fe–S clusters of various factors, such as ligand spheres, media, pH, and hydrogen bonding, poses a rather complex challenge for understanding them in detail [121]. This complexity amplifies when considering clusters of higher nuclearity like those in nitrogenase and hydrogenase models. For instance, the redox potential of nitrogenase is contingent not only on the coordinated metal (FeMo, FeV, FeFe) but also significantly affected by the overall protein environment, including pH values [122]. Hence, conducting an in-depth comparison of electrochemical properties or a direct structural analysis proves challenging, or rather nearly impossible. Implications stemming from backbone rigidity and specific influences within the secondary ligand sphere have a substantial influence on the active site properties. Recently, it was shown on a mononuclear molybdenum complex how much a second sphere effect that is nothing but a noncovalent ligand–ligand cooperativity can impact the redox properties of a species [123]. The induced π–π-stacking of two aromatic rings in the backbones of two individual ligands shifted a reduction potential by approximately 300 mV, which is a stunning value for an isolated event. In the active sites of enzymes, many more such interactions may take place simultaneously. The number of possibilities of course also increases with the nuclearity of the cofactors. Considering the size of the Fe–S cluster components in nitrogenases and the number of redox-active sites in models and cofactors, electrochemical comparisons are not much informative. Structural comparisons are available in the literature where they merely serve to highlight the similarity in metrical parameters without allowing us to draw any further conclusions (in contrast to the case of the smaller compounds discussed in Sections 19.3–19.5). For these reasons, a respective discussion is not included in this chapter. In the future, with even more advanced model chemistry it may become possible to delve into such details regarding the relationship between structures and electrochemical properties, and, hence, activity even for the very complex model compounds for nitrogenases. In any case, the respective experimental and computational chemistry of these fascinating species will continue to inspire and challenge the respective bioinorganic community for quite some time to come.

19.7 Conclusion

In summary, it can be stated that the essential biological systems based on iron and sulfur range from relatively simple (rubredoxin) to extremely complex (e.g. nitrogenase). This is also accompanied by an increasing complexity of chemical and molecular structures, redox behavior, kinetics, and, where applicable, catalysis. Due to the enormous variability of iron–sulfur based structures, broad potential ranges can be covered for electron transfer by means of the redox pair Fe^{2+}/Fe^{3+}, which even extends beyond the "biological range" in the sense of the actual natural limits set by water reduction (H^+ to H^0 at −0.43 V at natural pH) and water oxidation (O^{2-} to O^0 at 0.84 V). Extending beyond these limits is only possible for the evolution of peptides that can, for instance, exclude water from the active sites or utilize it in distinct transformations. A protein environment may enforce geometries of the coordination spheres that deviate from those of a relaxed complex dissolved in water or an organic solvent. Further modulations can be achieved by the specific amino acids, which covalently bind the prosthetic group (e.g. Rieske centers), reside at or close to the active site, or line substrate channels. The peptide of a protein or enzyme most often has a decisive role. This renders bioinorganic model chemistry challenging, to say the least, as it is typically impossible to include a good peptide mimic in the experimental model complexes. Still, model chemistry has helped in deciphering many questions posed by the discovery of the various distinct biological Fe–S species with regard to the chemical and electronic structures, the redox properties, and reactivities. In addition, the natural enzyme catalysts provide inspiration for biomimetic approaches toward novel catalytic systems, which is most relevant in the field of renewable energies (nitrogenases, hydrogenases). While the ferredoxins, the most ancient Fe–S species, in former times were perceived to engage almost exclusively in one-electron transfer, unexpectedly, more and more actual (re-)activities in atom/moiety transfer, catalysis, and regulation have emerged lately, and they continue to emerge. This guarantees that, in all likelihood, Nature will continue indefinitely to challenge the bioinorganic chemists for the synthesis of model compounds, the instrumental analyses of models, proteins, and enzymes, and the investigation and understanding of all their functions. We, as bioinorganic chemists, are very much looking forward to all the things to be discovered, to be understood in detail, and possibly to later be utilized for the benefit of our existence in this constantly changing world.

References

1 Boncella, A.E., Sabo, E.T., Santore, R.M. et al. (2022). The expanding utility of iron-sulfur clusters: Their functional roles in biology, synthetic small molecules, maquettes and artificial proteins, biomimetic materials, and therapeutic strategies. *Coord. Chem. Rev.* 453: 214229.

2 Einsle, O. and Rees, D.C. (2020). Structural enzymology of nitrogenase enzymes. *Chem. Rev.* 120: 4969–5004.

3 (a) Heinze, J. (1984). Cyclic voltammetry—"Electrochemical Spectroscopy". *Angew. Chem. Int. Ed.* 23: 831–847. (b) Connelly, N.G. and Geiger, W.E. (1996). Chemical redox agents for organometallic chemistry. *Chem. Rev.* 96: 877–910. (c) Elgrishi, N., Rountree, K.J., McCarthy, B.D. et al. (2018). A practical beginner's guide to cyclic voltammetry. *J. Chem. Educ.* 95: 197–206.

4 Venkateswara Rao, P. and Holm, R.H. (2004). Synthetic analogues of the active sites of iron-sulfur proteins. *Chem. Rev.* 104: 527–559.

5 Lee, S.C., Lo, W., and Holm, R.H. (2014). Developments in the biomimetic chemistry of cubane-type and higher nuclearity iron-sulfur clusters. *Chem. Rev.* 114: 3579–3600.

6 Herriott, J.R., Sieker, L.C., Jensen, L.H., and Lovenberg, W. (1970). Structure of rubredoxin: An x-ray study to 2.5 A resolution. *J. Mol. Biol.* 50: 391–406.

7 Hagelueken, G., Wiehlmann, L., Adams, T.M. et al. (2007). Crystal structure of the electron transfer complex rubredoxin rubredoxin reductase of *Pseudomonas aeruginosa*. *Proc. Natl. Acad. Sci.* 104: 12276–12281.

8 Liu, J., Chakraborty, S., Hosseinzadeh, P. et al. (2014). Metalloproteins containing cytochrome, iron-sulfur, or copper redox centers. *Chem. Rev.* 114: 4366–4469.

9 Zanello, P. (2013). The competition between chemistry and biology in assembling iron–sulfur derivatives. Molecular structures and electrochemistry. Part I. {Fe(S^{γ}Cys)$_4$} proteins. *Coord. Chem. Rev.* 257: 1777–1805.

10 Meyer, J. (2008). Iron-sulfur protein folds, iron-sulfur chemistry, and evolution. *J. Biol. Inorg. Chem.* 13: 157–170.

11 (a) Mayerle, J.J., Denmark, S.E., DePamphilis, B.V. et al. (1975). Synthetic analogs of the active sites of iron-sulfur proteins. XI. Synthesis and properties of complexes containing the iron sulfide Fe_2S_2 core and the structures of bis[*o*-xylyl-α,α′-dithiolato-μ-sulfido-ferrate(III)] and bis[*p*-tolylthiolato-*μ*-sulfido-ferrate(III)] dianions. *J. Am. Chem. Soc.* 97: 1032–1045. (b) Lane, R.W., Ibers, J.A., Frankel, R.B. et al. (1977). Synthetic analogues of the active sites of iron-sulfur proteins. 14. Synthesis, properties, and structures of bis[*o*-xylyl-α,α′-dithiolato]ferrate(II, III) anions, analogues of oxidized and reduced rubredoxin sites. *J. Am. Chem. Soc.* 99: 84–98.

12 Millar, M., Lee, J.F., O'Sullivan, T. et al. (1996). Models for the iron-sulfur protein rubredoxin: The use of sterically hindered thiolate ligands to stabilize $[Fe(SR)_4]^{1-}$ complexes; some considerations of the structure of the $[Fe(S\text{-}Cys)_4]$ centers in oxidized rubredoxins. *Inorg. Chim. Acta* 243: 333–343.

13 (a) Hagen, K.S., Reynolds, J.G., and Holm, R.H. (1981). Definition of reaction sequences resulting in self-assembly of $[Fe_4S_4(SR)_4]^{2-}$ clusters from simple reactants. *J. Am. Chem. Soc.* 103: 4054–4063. (b) Millar, M., Lee, J.F., Koch, S.A., and Fikar, R. (1982). Synthetic models for the iron-sulfur protein rubredoxin: Synthesis, structure, and properties of a highly symmetric iron(III) tetrathiolate anion. *Inorg. Chem.* 21: 4105–4106.

14 (a) Koch, S.A., Maelia, L.E., and Millar, M. (1983). Syntheses and structures of the $[Fe^{III}(SPh)_4]^-$ and $[Fe^{III}(SEt)_4]^-$ anions. A general route to iron(III) tetrathiolate complexes. *J. Am. Chem. Soc.* 105: 5944–5945. (b) Maelia, L.E., Millar, M., and Koch, S.A. (1992). General synthesis of iron(III) tetrathiolate complexes. Structural and spectroscopic models for the $[Fe(Cys\text{-}S)_4]$ center in oxidized rubredoxin. *Inorg. Chem.* 31: 4594–4600.

15 (a) Werth, M.T., Kurtz, D.M., Howes, B.D., and Boi, H.H. (1989). Observation of S = 2 EPR signals from ferrous iron-thiolate complexes. Relevance to rubredoxin-type sites in proteins. *Inorg. Chem.* 28: 1357–1361. (b) Chung, W.P., Dewan, J.C., Tuckerman, M., and Walters, M.A. (1999). Effect of N–H···S amide–thiolate hydrogen bonding on bond lengths in rubredoxin models $[(CH_3)_3NCH_2CONH_2]_2[M(S_2\text{-}o\text{-}xyl)_2]$. *Inorg. Chim. Acta* 291: 388–394. (c) Okamura Ta, T.A., Takamizawa, S., Ueyama, N., and Nakamura, A. (1998). Novel rubredoxin model tetrathiolato iron(II) and cobalt(II) complexes containing intramolecular single and double NH.S hydrogen bonds. *Inorg. Chem.* 37: 18–28. (d) Mascharak, P.K., Smith, M.C., Armstrong, W.H. et al. (1982). Fluorine-19 chemical shifts as structural probes of metal-sulfur clusters and the cofactor of nitrogenase. *Proc. Natl. Acad. Sci.* 79: 7056–7060.

16 (a) Ueyama, N., Sun, W.Y., and Nakamura, A. (1992). Evidence for intramolecular NH---S hydrogen bonds from deuterium-NMR spectroscopy of reduced-rubredoxin model iron(II) complexes with bidentate peptide ligands. *Inorg. Chem.* 31: 4053–4059. (b) Sun, W.Y., Ueyama, N., and Nakamura, A. (1991). Reduced rubredoxin models containing *Z*-Cys-Pro-Leu-Cys-Gly-NH-C_6H_4-*p*-X (X = MeO, H, F, CN): Electronic influence by a distant para substituent through NH---S hydrogen bonds. *Inorg. Chem.* 30: 4026–4031.

17 (a) Christou, G., Ridge, B., and Rydon, H.N. (1977). Direct formation of peptide analogues of rubredoxins and four-iron ferredoxins from their components. *J. Chem. Soc. Chem. Commun.* 24: 908–909. (b) Ueyama, N., Nakata, M., and Nakamura, A. (1981). Oxidized rubredoxin models. Iron(III) complexes of Z–Cys–Ala–Ala–Cys–OMe and Z–Ala–Cys–OMe. *Bull. Chem. Soc. Jpn.* 54: 1727–1730. (c) Anglin, J.R. and Davison, A. (1975). Iron(II) and cobalt(II) complexes of Boc-(Gly-L-Cys-Gly)$_4$-NH_2 as analogs for the active site of the iron-sulfur protein rubredoxin. *Inorg. Chem.* 14: 234–237. (d) Nivorozhkin, A.L., Segal, B.M., Musgrave, K.B. et al. (2000). Metallocyclopeptide complexes with MII(S Cys)$_4$ chromophores. *Inorg. Chem.* 39: 2306–2313.

18 Bonfio, C. (2021). The curious case of peptide-coordinated iron-sulfur clusters: Prebiotic and biomimetic insights. *Dalton Trans.* 50: 801–807.

19 (a) Balch, A.L. and Miller, J. (2002). 1,2-Dithiolene complexes of ruthenium and iron. *Inorg. Chem.* 10: 1410–1415. (b) Balch, A.L., Dance, I.G., and Holm, R.H. (2002). Characterization of dimeric dithiolene complexes. *J. Am. Chem. Soc.* 90: 1139–1145. (c) Roy, S., Mazinani, S.K., Groy, T.L. et al. (2014). Catalytic hydrogen evolution by Fe(II) carbonyls featuring a dithiolate and a chelating phosphine. *Inorg. Chem.* 53: 8919–8929. (d) Lv, H., Ruberu, T.P., Fleischauer, V.E. et al. (2016). Catalytic light-driven generation of hydrogen from water by iron dithiolene complexes. *J. Am. Chem. Soc.* 138: 11654–11663.

20 Begum, A. and Sarkar, S. (2011). An Iron(III) dithiolene complex as a functional model of iron hydrogenase. *Eur. J. Inorg. Chem.* 2012: 40–43.

21 (a) Maiti, B.K., Almeida, R.M., Moura, I., and Moura, J.J.G. (2017). Rubredoxins derivatives: Simple sulphur-rich coordination metal sites and its relevance for biology and chemistry. *Coord. Chem. Rev.* 352: 379–397. (b) Lin, I.J., Gebel, E.B., Machonkin, T.E. et al. (2005). Changes in hydrogen-bond strengths explain reduction potentials in 10 rubredoxin variants. *Proc. Natl. Acad. Sci.* 102: 14581–14586. (c) Gamiz-Hernandez, A.P., Kieseritzky, G., Ishikita, H., and Knapp, E.W. (2011). Rubredoxin function: Redox behavior from electrostatics. *J. Chem. Theory Comput.* 7: 742–752.

22 Min, T., Ergenekan, C.E., Eidsness, M.K. et al. (2001). Leucine 41 is a gate for water entry in the reduction of *Clostridium pasteurianum* rubredoxin. *Protein Sci.* 10: 613–621.

23 (a) Carrell, H.L., Glusker, J.P., Job, R., and Bruice, T.C. (1977). A synthetic tetranuclear iron-sulfur complex with ionized side chains: The crystal structure of $(Fe_4S_4(S(CH_2)_2COO)_4)^{6-}$ $(Na_5\ N(C_4H_9)_4)^{6+}$ 5 C_5H_9NO. *J. Am. Chem. Soc.* 99: 3683–3690. (b) Coucouvanis, D., Swenson, D., Baenziger, N.C. et al. (1981). Tetrahedral complexes containing the Fe(II)S_4 core. The syntheses, ground-state electronic structures and crystal and molecular structures of the bis(tetraphenylphosphonium) tetrakis (thiophenolato)ferrate (II) and bis (tetrapheny lphosphonium) bis(dithiosquarato)ferrate (II) complexes. An analog for the active site in reduced rubredoxins (Rdred). *J. Am. Chem. Soc.* 103: 3350–3362.

24 (a) Fish, A., Danieli, T., Ohad, I. et al. (2005). Structural basis for the thermostability of ferredoxin from the cyanobacterium *Mastigocladus laminosus*. *J. Mol. Biol.* 350: 599–608. (b) Paddock, M.L., Wiley, S.E., Axelrod, H.L. et al. (2007). MitoNEET is a uniquely folded 2Fe 2S outer mitochondrial membrane protein stabilized by pioglitazone. *Proc. Natl. Acad. Sci.* 104: 14342–14347. (c) Iwata, S., Saynovits, M., Link, T.A., and Michel, H. (1996). Structure of a water soluble fragment of the 'Rieske' iron-sulfur protein of the bovine heart mitochondrial cytochrome bc1 complex determined by MAD phasing at 1.5 A resolution. *Structure* 4: 567–579. (d) Dicus, M.M., Conlan, A., Nechushtai, R. et al. (2010). Binding of histidine in the $(Cys)_3(His)_1$-coordinated [2Fe-2S] cluster of human mitoNEET. *J. Am. Chem. Soc.* 132: 2037–2049.

25 Zhou, X., Gao, Y., Wang, W. et al. (2021). Architecture of the mycobacterial succinate dehydrogenase with a membrane-embedded Rieske Fe–S cluster. *Proc. Natl. Acad. Sci.* 118: e2022308118.

26 (a) Ferraro, D.J., Gakhar, L., and Ramaswamy, S. (2005). Rieske business: Structure-function of Rieske non-heme oxygenases. *Biochem. Biophys. Res. Commun.* 338: 175–190. (b) Link, T.A. (1999). The structures of rieske and rieske-type proteins. *Adv. Inorg. Chem.* 47: 83–157. (c) Crofts, A.R. (2004). Proton-coupled electron transfer at the Qo-site of the bc1 complex controls the rate of ubihydroquinone oxidation. *Biochim. Biophys. Acta* 1655: 77–92. (d) Albers, A., Demeshko, S., Dechert, S. et al. (2014). Fast proton-coupled electron transfer observed for a high-fidelity structural and functional [2Fe-2S] Rieske model. *J. Am. Chem. Soc.* 136: 3946–3954.

27 (a) Bak, D.W. and Elliott, S.J. (2014). Alternative Fe–S cluster ligands: Tuning redox potentials and chemistry. *Curr. Opin. Chem. Biol.* 19: 50–58. (b) Link, T.A. (1997). The role of the 'Rieske' iron sulfur protein in the hydroquinone oxidation (Q(P)) site of the cytochrome bc1 complex. The 'proton-gated affinity change' mechanism. *FEBS Lett.* 412: 257–264.

28 Zanello, P. (2014). The competition between chemistry and biology in assembling iron–sulfur derivatives. Molecular structures and electrochemistry. Part II. $\{[Fe_2S_2](S^{\gamma}Cys)_4\}$ proteins. *Coord. Chem. Rev.* 280: 54–83.

29 Lin, J., Zhou, T., Ye, K., and Wang, J. (2007). Crystal structure of human mitoNEET reveals distinct groups of iron sulfur proteins. *Proc. Natl. Acad. Sci.* 104: 14640–14645.

30 Bak, D.W., Zuris, J.A., Paddock, M.L. et al. (2009). Redox characterization of the Fe–S protein MitoNEET and impact of thiazolidinedione drug binding. *Biochemistry* 48: 10193–10195.

31 Zuris, J.A., Halim, D.A., Conlan, A.R. et al. (2010). Engineering the redox potential over a wide range within a new class of Fe–S proteins. *J. Am. Chem. Soc.* 132: 13120–13122.

32 Jarrett, J.T. (2017). *Characterization, Properties and Applications*, vol. 1, 223–248. De Gruyter.

33 (a) Schwartz, C.J., Giel, J.L., Patschkowski, T. et al. (2001). IscR, an Fe-S cluster-containing transcription factor, represses expression of *Escherichia coli* genes encoding Fe-S cluster assembly proteins. *Proc. Natl. Acad. Sci.* 98: 14895–14900. (b) Johnson, D.C., Dean, D.R., Smith, A.D., and Johnson, M.K. (2005). Structure, function, and formation of biological iron-sulfur clusters. *Annu. Rev. Biochem.* 74: 247–281. (c) Mettert, E.L., Perna, N.T., and Kiley, P.J. (2017). *Biochemistry, Biosynthesis and Human Diseases*, vol. 2, 75–96. de Gruyter.

34 (a) Mayerle, J.J., Frankel, R.B., Holm, R.H. et al. (1973). Synthetic analogs of the active sites of iron-sulfur proteins. Structure and properties of bis[*o*-xylyldithiolato-μ_2-sulfidoferrate(III)], an analog of the 2Fe-2S proteins. *Proc. Natl. Acad. Sci.* 70: 2429–2433. (b) Tsukihara, T., Fukuyama, K., Tahara, H. et al. (1978). X-ray analysis of ferredoxin from Spirulina platensis. II. Chelate structure of active center. *J. Biochem.* 84: 1645–1647.

35 Hagen, K.S., Watson, A.D., and Holm, R.H. (1983). Synthetic routes to iron sulfide (Fe_2S_2, Fe_3S_4, Fe_4S_4, and Fe_6S_9), clusters from the common precursor tetrakis(ethanethiolate)ferrate(2-) ion ($[Fe(SC_2H_5)_4]^{2-}$): Structures and properties of $[Fe_3S_4(SR)_4]^{3-}$ and bis(ethanethiolate)nonathioxohexaferrate(4-) ion ($[Fe_6S_9(SC_2H_5)_2]^{4-}$), examples of the newest types of Fe-S-SR clusters. *J. Am. Chem. Soc.* 105: 3905–3913.

36 (a) Gillum, W.O., Frankel, R.B., Foner, S., and Holm, R.H. (1976). Synthetic analogues of the active sites of iron-sulfur proteins. XIII. Further electronic structural relationships between the analogues $[Fe_2S_2(SR)_4]^{2-}$ and the active sites of oxidized 2Fe-2S* proteins. *Inorg. Chem.* 15: 1095–1100. (b) Zhou, J., Scott, M.J., Hu, Z. et al. (1992). Synthesis and comparative reactivity and electronic structural features of $[MFe_3S_4]^{z+}$ cubane-type clusters (M = iron, cobalt, nickel). *J. Am. Chem. Soc.* 114: 10843–10854.

37 Wong, G.B., Bobrik, M.A., and Holm, R.H. (1978). Inorganic derivatives of iron sulfide thiolate dimers and tetramers: Synthesis and properties of the halide series $[Fe_2S_2X_4]^{2-}$ and $[Fe_4S_4X_4]^{2-}$ (X = chlorine, bromine, iodine). *Inorg. Chem.* 17: 578–584.

38 (a) Nakamura, A. and Ueyama, N. (1989). Cysteine-containing oligopeptide model complexes of iron-sulfur proteins. *Adv. Inorganic Chem.* 33: 39–67. (b) Ueyama, N., Ueno, S., Nakamura, A. et al. (1992). A synthetic analogue for the active site of plant-type ferredoxin: Two different coordination isomers by a four-cys-containing [20]-peptide. *Biopolymers* 32: 1535–1544.

39 (a) Cleland, W.E. and Averill, B.A. (1984). Effects of phenoxide ligation on iron-sulfur clusters. 2. Preparation and properties of $[Fe_2S_2(OAr)_4]^{2-}$ ions. *Inorg. Chem.* 23: 4192–4197. (b) Salifoglou, A., Simopoulos, A., Kostikas, A. et al. (1988). Dimeric complexes containing the disulfidodiron(2+) cores coordinated by non-sulfur terminal ligands. Synthesis, structural characterization, and spectroscopic properties of $[Et_4N]_2[Fe_2S_2(o,o'\text{-}C_{12}H_8O_2)_2]$, $[Et_4N]_2[Fe_2S_2(C_4H_4N)_4]$, and $[Et_4N]_2[Fe_2S_2(O\text{-}o\text{-}C_6H_4CH(n\text{-}C_4H_9)NHC_6H_4\text{-}o\text{-}S)_2]$

and the structure of $[Ph_4P]_2[Fe_2S_2(OC_6H_4\text{-}p\text{-}CH_3)_4]$. *Inorganic Chem.* 27: 3394–3406.

40 Bonisch, H., Schmidt, C.L., Schafer, G., and Ladenstein, R. (2002). The structure of the soluble domain of an archaeal Rieske iron-sulfur protein at 1.1 A resolution. *J. Mol. Biol.* 319: 791–805.

41 Ballmann, J., Albers, A., Demeshko, S. et al. (2008). A synthetic analogue of Rieske-type [2Fe-2S] clusters. *Angew. Chem. Int. Ed.* 47: 9537–9541.

42 Albers, A., Demeshko, S., Dechert, S. et al. (2011). The complete characterization of a reduced biomimetic [2Fe-2S] cluster. *Angew. Chem. Int. Ed.* 50: 9191–9194.

43 (a) Olsen, M.T., Rauchfuss, T.B., and Wilson, S.R. (2010). Role of the azadithiolate cofactor in models for [FeFe]-hydrogenase: Novel structures and catalytic implications. *J. Am. Chem. Soc.* 132: 17733–17740. (b) Albers, A., Demeshko, S., Propper, K. et al. (2013). A super-reduced diferrous [2Fe-2S] cluster. *J. Am. Chem. Soc.* 135: 1704–1707. (c) Albers, A., Bayer, T., Demeshko, S. et al. (2013). A functional model for the Rieske center: Full characterization of a biomimetic N-ligated [2Fe-2S] cluster in different protonation states. *Chem. Eur. J.* 19: 10101–10106.

44 Saouma, C.T., Kaminsky, W., and Mayer, J.M. (2012). Protonation and concerted proton-electron transfer reactivity of a bis-benzimidazolate ligated [2Fe-2S] model for Rieske clusters. *J. Am. Chem. Soc.* 134: 7293–7296.

45 (a) Bergner, M., Dechert, S., Demeshko, S. et al. (2017). Model of the MitoNEET [2Fe-2S] cluster shows proton coupled electron transfer. *J. Am. Chem. Soc.* 139: 701–707. (b) Beardwood, P. and Gibson, J.F. (1992). Iron–sulfur dimers with benzimidazolate–thiolate, –phenolate or bis(benzimidazolate) terminal chelating ligands. Models for Rieske-type proteins. *Dalton Trans.* 16: 2457–2466.

46 (a) Ueyama, N., Yamada, Y., Okamura Ta, T.A. et al. (1996). Structure and properties of $[Fe_4S_4\{2,6\text{-bis(acylamino)benzenethiolato-S}\}_4]^{2-}$ and $[Fe_2S_2\{2,6\text{-bis(acylamino)benzenethiolato-S}\}_4]^{2-}$: Protection of the Fe–S bond by double NHS hydrogen bonds. *Inorg. Chem.* 35: 6473–6484. (b) Ohno, R., Ueyama, N., and Nakamura, A. (2002). Influence of the distal para substituent through NH—S hydrogen bonds on the positive shift of the reduction potentials of $[Fe_4S_4(\text{Z-cys-Gly-NHC}_6H_4\text{-p-X})_4]^{2-}$ (X = H, OMe, F, Cl, CN) complexes. *Inorg. Chem.* 30: 4887–4891.

47 Cambray, J., Lane, R.W., Wedd, A.G. et al. (1977). Chemical and electrochemical interrelationships of the 1-Fe, 2-Fe, and 4-Fe analogs of the active sites of iron-sulfur proteins. *Inorg. Chem.* 16: 2565–2571.

48 Maelia, L.E. and Koch, S.A. (1986). Gallium analogs of iron-sulfide-thiolate compounds. Analysis of the structural parameters in gallium(III) and iron(III) chalcogenide compounds. *Inorg. Chem.* 25: 1896–1904.

49 (a) Herskovitz, T., Averill, B.A., Holm, R.H. et al. (1972). Structure and properties of a synthetic analogue of bacterial iron--sulfur proteins. *Proc. Natl. Acad. Sci.* 69: 2437–2441. (b) Averill, B.A., Herskovitz, T., Holm, R.H., and Ibers, J.A. (1973). Synthetic analogs of the active sites of iron-sulfur proteins. II. Synthesis

and structure of the tetra(mercapto-μ_3-sulfido-iron) clusters, $[Fe_4S_4(SR)_4]^{2-}$. *J. Am. Chem. Soc.* 95: 3523–3534.
50 (a) Sieker, L.C., Adman, E., and Jensen, L.H. (1972). Structure of the Fe-S complex in a bacterial ferredoxin. *Nature* 235: 40–42. (b) Carter, C.W. Jr., Kraut, J., Freer, S.T. et al. (1972). A comparison of Fe_4S_4 clusters in high-potential iron protein and in ferredoxin. *Proc. Natl. Acad. Sci.* 69: 3526–3529.
51 Crane, B.R., Siegel, L.M., and Getzoff, E.D. (1995). Sulfite reductase structure at 1.6 A: Evolution and catalysis for reduction of inorganic anions. *Science* 270: 59–67.
52 Nielsen, M.S., Harris, P., Ooi, B.L., and Christensen, H.E. (2004). The 1.5 A resolution crystal structure of $[Fe_3S_4]$-ferredoxin from the hyperthermophilic archaeon *Pyrococcus furiosus*. *Biochemistry* 43: 5188–5194.
53 Emptage, M.H., Kent, T.A., Huynh, B.H. et al. (1980). On the nature of the iron-sulfur centers in a ferredoxin from *Azotobacter vinelandii*. Mössbauer studies and cluster displacement experiments. *J. Biol. Chem.* 255: 1793–1796.
54 (a) Robbins, A.H. and Stout, C.D. (1985). Iron-sulfur cluster in aconitase. Crystallographic evidence for a three-iron center. *J. Biol. Chem.* 260: 2328–2333. (b) Robbins, A.H. and Stout, C.D. (1989). Structure of activated aconitase: Formation of the [4Fe-4S] cluster in the crystal. *Proc. Natl. Acad. Sci.* 86: 3639–3643. (c) Beinert, H., Kennedy, M.C., and Stout, C.D. (1996). Aconitase as ironminus sign sulfur protein, enzyme, and iron-regulatory protein. *Chem. Rev.* 96: 2335–2374.
55 (a) Stout, C.D. (1988). 7-Iron ferredoxin revisited. *J. Biol. Chem.* 263: 9256–9260. (b) Stout, G.H., Turley, S., Sieker, L.C., and Jensen, L.H. (1988). Structure of ferredoxin I from *Azotobacter vinelandii*. *Proc. Natl. Acad. Sci.* 85: 1020–1022. (c) Stout, C.D. (1989). Refinement of the 7 Fe ferredoxin from *Azotobacter vinelandii* at 1.9 A resolution. *J. Mol. Biol.* 205: 545–555. (d) Robbins, A.H. and Stout, C.D. (1989). The structure of aconitase. *Proteins* 5: 289–312.
56 Broderick, J.B., Duffus, B.R., Duschene, K.S., and Shepard, E.M. (2014). Radical S-adenosylmethionine enzymes. *Chem. Rev.* 114: 4229–4317.
57 Goldman, P.J., Grove, T.L., Sites, L.A. et al. (2013). X-ray structure of an AdoMet radical activase reveals an anaerobic solution for formylglycine post-translational modification. *Proc. Natl. Acad. Sci.* 110: 8519–8524.
58 Lanz, N.D., Pandelia, M.E., Kakar, E.S. et al. (2014). Evidence for a catalytically and kinetically competent enzyme-substrate cross-linked intermediate in catalysis by lipoyl synthase. *Biochemistry* 53: 4557–4572.
59 (a) Crack, J.C., Green, J., Thomson, A.J., and Le Brun, N.E. (2014). Iron-sulfur clusters as biological sensors: The chemistry of reactions with molecular oxygen and nitric oxide. *Acc. Chem. Res.* 47: 3196–3205. (b) Nisbett, L.M. and Boon, E.M. (2016). Nitric oxide regulation of H-NOX signaling pathways in bacteria. *Biochemistry* 55: 4873–4884. (c) Lehnert, N., Kim, E., Dong, H.T. et al. (2021). The biologically relevant coordination chemistry of iron and nitric oxide: Electronic structure and reactivity. *Chem. Rev.* 121: 14682–14905.
60 Nicolet, Y. and Fontecilla-Camps, J.C. (2017). *Characterization, Properties and Applications*, vol. 1, 359–386. De Gruyter.

61 Fabrizi de Biani, F. and Zanello, P. (2017). The competition between chemistry and biology in assembling iron-sulfur derivatives. Molecular structures and electrochemistry. Part IV. {$[Fe_3S_4](S^{\gamma}Cys)_3$} proteins. *Inorg. Chim. Acta* 455: 319–328.

62 (a) Pellicer Martinez, M.T., Martinez, A.B., Crack, J.C. et al. (2017). Sensing iron availability via the fragile [4Fe-4S] cluster of the bacterial transcriptional repressor RirA. *Chem. Sci.* 8: 8451–8463. (b) Pellicer Martinez, M.T., Crack, J.C., Stewart, M.Y. et al. (2019). Mechanisms of iron- and O_2-sensing by the [4Fe-4S] cluster of the global iron regulator RirA. *eLife* 8: e47804. (c) Behringer, M., Plotzky, L., Baabe, D. et al. (2020). RirA of *Dinoroseobacter shibae* senses iron via a [3Fe-4S]1+ cluster co-ordinated by three cysteine residues. *Biochem. J.* 477: 191–212.

63 (a) Fritsch, J., Scheerer, P., Frielingsdorf, S. et al. (2011). The crystal structure of an oxygen-tolerant hydrogenase uncovers a novel iron-sulphur centre. *Nature* 479: 249–252. (b) Mouesca, J.-M., Fontecilla-Camps, J.C., and Amara, P. (2013). The structural plasticity of the proximal $[4Fe_3S]$ cluster is responsible for the O_2 tolerance of membrane-bound [NiFe] Hydrogenases. *Angew. Chem.* 125: 2056–2060.

64 Goris, T., Wait, A.F., Saggu, M. et al. (2011). A unique iron-sulfur cluster is crucial for oxygen tolerance of a [NiFe]-hydrogenase. *Nat. Chem. Biol.* 7: 310–318.

65 Bierbach, U., Saak, W., Haase, D., and Pohl, S. (1991). Neutrale und kationische Eisen-Schwefel-Komplexe und -cluster synthese und Kristallstrukturen von $[Fe(SR)_2L_2]\cdot PhMe$ und $[Fe_4S_4(SR)_2L_2]$ sowie zur Bildung von $[Fe(SR)L_3]^+$, $[FeL_4]^{2+}$, $[Fe_4S_4(SR)L_3]^+$ und $[Fe_4S_4L_4]^{2+}$ (R = 2,4,6-i-$Pr_3C_6H_2$; L = $SC(NMe_2)_2$)/ neutral and cationic iron-sulfur complexes and clusters syntheses and crystal structures of $[Fe(SR),L_2]\cdot PhMe$ and $[Fe_4S_4(SR)_2L_2]$, and on the formation of $[Fe(SR)L_3]^+$, $[FeL_4]^{2+}$, $[Fe_4S_4(SR)L_3]^+$ and $[Fe_4S_4L_4]^{2+}$ (R = 2,4,6-i-$Pr_3C_6H_2$; L = $SC(NMe_{22})$. *Z. Naturforsch. B* 46: 1629–1634.

66 Christou, G. and Garner, C.D. (1979). A convenient synthesis of tetrakis[thiolato-μ_3-sulphido-iron]$^{2-}$clusters. *J. Chem. Soc. Dalton Trans.* 1093–1094.

67 (a) Angove, H.C., Yoo, S.J., Burgess, B.K., and Münck, E. (1997). Mössbauer and EPR evidence for an All-ferrous Fe_4S_4 cluster with S = 4 in the Fe protein of nitrogenase. *J. Am. Chem. Soc.* 119: 8730–8731. (b) Hans, M., Buckel, W., and Bill, E. (2008). Spectroscopic evidence for an all-ferrous $[4Fe\text{-}4S]^0$ cluster in the superreduced activator of 2-hydroxyglutaryl-CoA dehydratase from *Acidaminococcus fermentans. J. Biol. Inorg. Chem.* 13: 563–574.

68 Scott, T.A., Berlinguette, C.P., Holm, R.H., and Zhou, H.C. (2005). Initial synthesis and structure of an all-ferrous analogue of the fully reduced $[Fe_4S_4]^0$ cluster of the nitrogenase iron protein. *Proc. Natl. Acad. Sci.* 102: 9741–9744.

69 Scott, T.A. and Zhou, H.C. (2004). The first all-cyanide Fe_4S_4 cluster: $[Fe_4S_4(CN)_4]^{3-}$. *Angew. Chem. Int. Ed.* 43: 5628–5631.

70 Goh, C., Segal, B.M., Huang, J. et al. (1996). Polycubane clusters: Synthesis of $[Fe_4S_4(PR_3)_4]^{1+,0}$ (R = But, Cy, Pri) and $[Fe_4S_4]^0$ core aggregation upon loss of phosphine. *J. Am. Chem. Soc.* 118: 11844–11853.

71 Deng, L. and Holm, R.H. (2008). Stabilization of fully reduced iron-sulfur clusters by carbene ligation: The $[Fe_nS_n]^0$ oxidation levels (n = 4, 8). *J. Am. Chem. Soc.* 130: 9878–9886.

72 Ohki, Y., Sunada, Y., and Tatsumi, K. (2005). Synthesis of [2Fe–2S] and [4Fe–4S] clusters having terminal amide ligands from an Iron(II) amide complex. *Chem. Lett.* 34: 172–173.

73 Sharp, C.R., Duncan, J.S., and Lee, S.C. (2010). $[Fe_4S_4]^q$ cubane clusters (q = 4+, 3+, 2+) with terminal amide ligands. *Inorg. Chem.* 49: 6697–6705.

74 Moula, G., Matsumoto, T., Miehlich, M.E. et al. (2018). Synthesis of an all-ferric cuboidal iron-sulfur cluster $[Fe^{III4}S_4(SAr)_4]$. *Angew. Chem. Int. Ed.* 57: 11594–11597.

75 Volbeda, A., Charon, M.H., Piras, C. et al. (1995). Crystal structure of the nickel-iron hydrogenase from *Desulfovibrio gigas*. *Nature* 373: 580–587.

76 (a) Stack, T.D.P. and Holm, R.H. (1987). Subsite-specific functionalization of the $[4Fe\text{-}4S]^{2+}$ analog of iron-sulfur protein clusters. *J. Am. Chem. Soc.* 109: 2546–2547. (b) Stack, T.D.P. and Holm, R.H. (1988). Subsite-differentiated analogs of biological $[4Fe\text{-}4S]^{2+}$ clusters: Synthesis, solution and solid-state structures, and subsite-specific reactions. *J. Am. Chem. Soc.* 110: 2484–2494. (c) Daley, C.J. and Holm, R.H. (2003). Reactions of site-differentiated $[Fe_4S_4]^{2+,\ 1+}$ clusters with sulfonium cations: Reactivity analogues of biotin synthase and other members of the S-adenosylmethionine enzyme family. *J. Inorg. Biochem.* 97: 287–298.

77 Terada, T., Wakimoto, T., Nakamura, T. et al. (2012). Tridentate thiolate ligands: Application to the synthesis of the site-differentiated [4Fe-4S] cluster having a hydrosulfide ligand at the unique iron center. *Chem. Asian J.* 7: 920–929.

78 Zhou, C. and Holm, R.H. (1997). Comparative isotropic shifts, redox potentials, and ligand binding propensities of [1:3] site-differentiated cubane-type $[Fe_4Q_4]^{2+}$ clusters (Q = S, Se). *Inorg. Chem.* 36: 4066–4077.

79 Ohki, Y., Tanifuji, K., Yamada, N. et al. (2011). Synthetic analogues of $[Fe_4S_4(Cys)_3(His)]$ in hydrogenases and $[Fe_4S_4(Cys)_4]$ in HiPIP derived from all-ferric $[Fe_4S_4(N(SiMe_3)_2)_4]$. *Proc. Natl. Acad. Sci.* 108: 12635–12640.

80 Zhou, H.C. and Holm, R.H. (2003). Synthesis and reactions of cubane-type iron-sulfur-phosphine clusters, including soluble clusters of nuclearities 8 and 16. *Inorg. Chem.* 42: 11–21.

81 Deng, L., Majumdar, A., Lo, W., and Holm, R.H. (2010). Stabilization of 3:1 site-differentiated cubane-type clusters in the $[Fe_4S_4]^{1+}$ core oxidation state by tertiary phosphine ligation: Synthesis, core structural diversity, and S = 1/2 ground states. *Inorg. Chem.* 49: 11118–11126.

82 McSkimming, A. and Suess, D.L.M. (2018). Selective synthesis of site-differentiated Fe_4S_4 and Fe_6S_6 clusters. *Inorg. Chem.* 57: 14904–14912.

83 Tard, C., Liu, X., Ibrahim, S.K. et al. (2005). Synthesis of the H-cluster framework of iron-only hydrogenase. *Nature* 433: 610–613.

84 Zhou, C., Raebiger, J.W., Segal, B.M., and Holm, R.H. (2000). The influence of net charge on the redox potentials of Fe_4S_4 cubane-type clusters in aprotic solvents. *Inorg. Chim. Acta* 300-302: 892–902.

85 (a) Dauter, Z., Wilson, K.S., Sieker, L.C. et al. (1997). Atomic resolution (0.94 A) structure of *Clostridium acidurici* ferredoxin. Detailed geometry of [4Fe-4S] clusters in a protein. *Biochemistry* 36: 16065–16073. (b) Berg, J.M., Hodgson, K.O., and Holm, R.H. (1979). Crystal structure of $[(C_2H_5)_4N]_3[Fe_4S_4(SCH_2Ph)_4]$, a reduced ferredoxin site analog with a nontetragonal Fe_4S_4 core structure in the solid state. *J. Am. Chem. Soc.* 101: 4586–4593.

86 (a) Zhou, J., Hu, Z., Münck, E., and Holm, R.H. (1996). The cuboidal Fe_3S_4 cluster: Synthesis, stability, and geometric and electronic structures in a non-protein environment. *J. Am. Chem. Soc.* 118: 1966–1980. (b) Zhou, J. and Holm, R.H. (1995). Synthesis and metal ion incorporation reactions of the cuboidal Fe_3S_4 cluster. *J. Am. Chem. Soc.* 117: 11353–11354. (c) Roth, E.K.H. and Jordanov, J. (1992). Oxdiation reactions of $[Fe_4S_4(S\text{-}2,4,6\text{-}[iso\text{-}Pr]_3C_6H_2)_4]^{2-}$ and an oxidative conversion of the Fe_4S_4 core into an Fe_3S_4 center. *Inorg. Chem.* 31: 240–243. (d) Weterings, J.P., Kent, T.A., and Prins, R. (1987). Low-temperature oxidation of [4Fe-4S] analogs. Generation of a iron/sulfur cluster spectroscopically similar to the 3-Fe centers in the 3-Fe ferredoxins. *Inorg. Chem.* 26: 324–329.

87 Raebiger, J.W., Crawford, C.A., Zhou, J., and Holm, R.H. (1997). Reactivity of cubane-type $[(OC)_3MFe_3S_4(SR)_3]^{3-}$ clusters (M = Mo, W): Interconversion with cuboidal $[Fe_3S_4]^0$ clusters and electron transfer. *Inorg. Chem.* 36: 994–1003.

88 Zhou, J., Raebiger, J.W., Crawford, C.A., and Holm, R.H. (1997). Metal ion incorporation reactions of the cluster $[Fe_3S_4(LS_3)]^{3-}$, containing the cuboidal $[Fe_3S_4]^0$ core. *J. Am. Chem. Soc.* 119: 6242–6250.

89 Brown, A.C. and Suess, D.L.M. (2023). An open-cuboidal $[Fe_3S_4]$ cluster characterized in both biologically relevant redox states. *J. Am. Chem. Soc.* 145: 2075–2080.

90 (a) Fawcett, S.E., Davis, D., Breton, J.L. et al. (1998). Voltammetric studies of the reactions of iron-sulphur clusters ([3Fe-4S] or [M3Fe-4S]) formed in Pyrococcus furiosus ferredoxin. *Biochem. J.* 335 (Pt 2): 357–368. (b) Chen, K., Hirst, J., Camba, R. et al. (2000). Atomically defined mechanism for proton transfer to a buried redox centre in a protein. *Nature* 405: 814–817.

91 Schipke, C.G., Goodin, D.B., McRee, D.E., and Stout, C.D. (1999). Oxidized and reduced *Azotobacter vinelandii* ferredoxin I at 1.4 A resolution: Conformational change of surface residues without significant change in the $[3Fe\text{-}4S]^{+/0}$ cluster. *Biochemistry* 38: 8228–8239.

92 (a) Einsle, O., Tezcan, F.A., Andrade, S.L.A. et al. (2002). Nitrogenase MoFe-protein at 1.16 angstrom resolution: A central ligand in the FeMo-cofactor. *Science* 297: 1696–1700. (b) Spatzal, T., Aksoyoglu, M., Zhang, L. et al. (2011). Evidence for interstitial carbon in nitrogenase FeMo cofactor. *Science* 334: 940. (c) Lancaster, K.M., Roemelt, M., Ettenhuber, P. et al. (2011). X-ray emission spectroscopy evidences a central carbon in the nitrogenase iron-molybdenum cofactor. *Science* 334: 974–977.

93 (a) Esselborn, J., Muraki, N., Klein, K. et al. (2016). A structural view of synthetic cofactor integration into [FeFe]-hydrogenases. *Chem. Sci.* 7: 959–968. (b) Wittkamp, F., Senger, M., Stripp, S.T., and Apfel, U.P. (2018).

[FeFe]-Hydrogenases: Recent developments and future perspectives. *Chem. Commun.* 54: 5934–5942. (c) Land, H., Senger, M., Berggren, G., and Stripp, S.T. (2020). Current state of [FeFe]-hydrogenase research: Biodiversity and spectroscopic investigations. *ACS Catal.* 10: 7069–7086.

94 Lauterbach, L. and Lenz, O. (2013). Catalytic production of hydrogen peroxide and water by oxygen-tolerant [NiFe]-hydrogenase during H_2 cycling in the presence of O_2. *J. Am. Chem. Soc.* 135: 17897–17905.

95 Wang, C., Lai, Z., Huang, G., and Pan, H.J. (2022). Current state of [Fe]-hydrogenase and its biomimetic models. *Chem. Eur. J.* 28: e202201499.

96 Peters, J.W., Lanzilotta, W.N., Lemon, B.J., and Seefeldt, L.C. (1998). X-ray crystal structure of the Fe-only hydrogenase (CpI) from *Clostridium pasteurianum* to 1.8 angstrom resolution. *Science* 282: 1853–1858.

97 Sippel, D. and Einsle, O. (2017). The structure of vanadium nitrogenase reveals an unusual bridging ligand. *Nat. Chem. Biol.* 13: 956–960.

98 Elvers, B.J. (2022). *Über die redoxbasierte Reaktivität von phosphanstabilisierten Monodithiolenkomplexen des Wolframs und des Molybdäns für die Kleinmolekülaktivierung (small molecule activation).* University of Greifswald.

99 Schindelin, H., Kisker, C., Schlessman, J.L. et al. (1997). Structure of ADP x AIF4(−)-stabilized nitrogenase complex and its implications for signal transduction. *Nature* 387: 370–376.

100 Nishibayashi, Y. (2019). *Transition Metal-Dinitrogen Complexes: Preparation and Reactivity.* Wiley.

101 Rao, P.V., Bhaduri, S., Jiang, J. et al. (2005). On $[Fe_4S_4]^{2+}$ $-(\mu_2\text{-SR})\text{-}M^{II}$ bridge formation in the synthesis of an A-cluster analogue of carbon monoxide dehydrogenase/acetylcoenzyme A synthase. *J. Am. Chem. Soc.* 127: 1933–1945.

102 Panda, R., Berlinguette, C.P., Zhang, Y., and Holm, R.H. (2005). Synthesis of MFe_3S_4 clusters containing a planar M^{II} site (M = Ni, Pd, Pt), a structural element in the C-cluster of carbon monoxide dehydrogenase. *J. Am. Chem. Soc.* 127: 11092–11101.

103 Cen, W., Lee, S.C., Li, J. et al. (1993). Heterometal dependence of electron distributions and redox potentials in clusters with cubane-type $[MFe_3S_4]_2$ cores. *J. Am. Chem. Soc.* 115: 9515–9523.

104 (a) Demadis, K.D., Campana, C.F., and Coucouvanis, D. (1995). Synthesis and structural characterization of the new $Mo_2Fe_6S_8(PR_3)_6(Cl_4\text{-cat})_2$ clusters. Double cubanes containing two edge-linked $[MoFe_3S_4]^{2+}$ reduced cores. *J. Am. Chem. Soc.* 117: 7832–7833. (b) Fomitchev, D.V., McLauchlan, C.C., and Holm, R.H. (2002). Heterometal cubane-type MFe_3S_4 clusters (M = Mo, V) trigonally symmetrized with hydrotris(pyrazolyl)borate(1-) and tris(pyrazolyl)methanesulfonate(1–) capping ligands. *Inorg. Chem.* 41: 958–966. (c) Malinak, S.M., Demadis, K.D., and Coucouvanis, D. (1995). Catalytic reduction of hydrazine to ammonia by the VFe_3S_4 cubanes. Further evidence for the direct involvement of the heterometal in the reduction of nitrogenase substrates and possible relevance to the vanadium nitrogenases. *J. Am. Chem. Soc.* 117: 3126–3133.

105 (a) Berlinguette, C.P., Miyaji, T., Zhang, Y., and Holm, R.H. (2006). Precursors to clusters with the topology of the P^N cluster of nitrogenase: Edge-bridged double cubane clusters $[(Tp)_2Mo_2Fe_6S_8L_4]^z$: Synthesis, structures, and electron transfer series. *Inorg. Chem.* 45: 1997–2007. (b) Pesavento, R.P., Berlinguette, C.P., and Holm, R.H. (2007). Stabilization of reduced molybdenum-iron-sulfur single- and double-cubane clusters by cyanide ligation. *Inorg. Chem.* 46: 510–516.

106 Zhang, Y. and Holm, R.H. (2003). Synthesis of a molecular $Mo_2Fe_6S_9$ cluster with the topology of the PN cluster of nitrogenase by rearrangement of an edge-bridged $Mo_2Fe_6S_8$ double cubane. *J. Am. Chem. Soc.* 125: 3910–3920.

107 (a) Peters, J.W., Stowell, M.H., Soltis, S.M. et al. (1997). Redox-dependent structural changes in the nitrogenase P-cluster. *Biochemistry* 36: 1181–1187. (b) Mayer, S.M., Lawson, D.M., Gormal, C.A. et al. (1999). New insights into structure-function relationships in nitrogenase: A 1.6 A resolution X-ray crystallographic study of Klebsiella pneumoniae MoFe-protein. *J. Mol. Biol.* 292: 871–891.

108 Zhang, Y. and Holm, R.H. (2004). Structural conversions of molybdenum-iron-sulfur edge-bridged double cubanes and P^n-type clusters topologically related to the nitrogenase P-cluster. *Inorg. Chem.* 43: 674–682.

109 (a) Zuo, J.L., Zhou, H.C., and Holm, R.H. (2003). Vanadium-iron-sulfur clusters containing the cubane-type $[VFe_3S_4]$ core unit: Synthesis of a cluster with the topology of the PN cluster of nitrogenase. *Inorg. Chem.* 42: 4624–4631. (b) Scott, T.A. and Holm, R.H. (2008). VFe_3S_4 single and double cubane clusters: Synthesis, structures, and dependence of redox potentials and electron distribution on ligation and heterometal. *Inorg. Chem.* 47: 3426–3432.

110 Ohki, Y., Ikagawa, Y., and Tatsumi, K. (2007). Synthesis of new [8Fe-7S] clusters: A topological link between the core structures of P-cluster, FeMo-co, and FeFe-co of nitrogenases. *J. Am. Chem. Soc.* 129: 10457–10465.

111 Moula, G., Nagasaki, A., Matsumoto, T. et al. (2021). Synthesis of a nitrogenase P(N)-cluster model with $[Fe_8S_7(\mu\text{-}St_{hiolate})_2]$ core from the all-ferric $[Fe_4S_4(S_{thiolate})_4]$ cubane synthon. *Angew. Chem. Int. Ed.* 60: 15792–15797.

112 (a) Berlinguette, C.P. and Holm, R.H. (2006). Edge-bridged $Mo_2Fe_6S_8$ to pN-type $Mo_2Fe_6S_9$ cluster conversion: Structural fate of the attacking sulfide/selenide nucleophile. *J. Am. Chem. Soc.* 128: 11993–12000. (b) Zheng, B., Chen, X.D., Zheng, S.L., and Holm, R.H. (2012). Selenium as a structural surrogate of sulfur: Template-assisted assembly of five types of tungsten-iron-sulfur/selenium clusters and the structural fate of chalcogenide reactants. *J. Am. Chem. Soc.* 134: 6479–6490.

113 (a) Mézailles, N. (2019). *Transition Metal-Dinitrogen Complexes* (ed. Y. Nishibayashi), 221–269. Wiley. (b) Stucke, N., Flöser, B.M., Weyrich, T., and Tuczek, F. (2018). Nitrogen fixation catalyzed by transition metal complexes: Recent developments. *Eur. J. Inorg. Chem.* 2018: 1337–1355. (c) van Alten, R.S., Watjen, F., Demeshko, S. et al. (2020). (Electro-)chemical splitting of dinitrogen with a rhenium pincer complex. *Eur. J. Inorg. Chem.* 2020: 1402–1410. (d) Macleod, K.C. and Holland, P.L. (2013). Recent developments in the homogeneous reduction of dinitrogen by molybdenum and iron. *Nat. Chem.* 5: 559–565. (e) Mus, F., Alleman, A.B., Pence, N. et al. (2018). Exploring the alternatives of biological nitrogen fixation. *Metallomics* 10: 523–538.

114 (a) Duncan, J.S., Zdilla, M.J., and Lee, S.C. (2007). Synthesis and elaboration of the dinuclear iron-imide cluster core $[Fe_2(\mu\text{-}NR)_2]^{2+}$. *Inorg. Chem.* 46: 1071–1080. (b) Chen, X.D., Duncan, J.S., Verma, A.K., and Lee, S.C. (2010). Selective syntheses of iron-imide-sulfide cubanes, including a partial representation of the Fe-S-X environment in the FeMo cofactor. *J. Am. Chem. Soc.* 132: 15884–15886. (c) Chen, X.D., Zhang, W., Duncan, J.S., and Lee, S.C. (2012). Iron-amide-sulfide and iron-imide-sulfide clusters: Heteroligated core environments relevant to the nitrogenase FeMo cofactor. *Inorg. Chem.* 51: 12891–12904. (d) Xu, G., Zhou, J., Wang, Z. et al. (2019). Controlled incorporation of nitrides into W-Fe-S clusters. *Angew. Chem. Int. Ed.* 58: 16469–16473. (e) Xue, L., Xu, G., Wang, Z., and Chen, X.D. (2021). Nitride-incorporated W-Fe-S double cubane clusters: Terminal ligand substitutions and redox behaviors. *Dalton Trans.* 50: 6840–6847.

115 Han, J., Koutmos, M., Ahmad, S.A., and Coucouvanis, D. (2001). Rational synthesis of high nuclearity Mo/Fe/S clusters: The reductive coupling approach in the convenient synthesis of $(Cl_4\text{-cat})_2Mo_2Fe_6S_8(PR_3)_6$ [R = Et, nPr, nBu] and the new $[(Cl_4\text{-cat})_2Mo_2Fe_2S_3O(PEt_3)_3Cl]$-1/2$(Fe(PEt_3)_2(MeCN)_4)$ and $(Cl_4\text{-cat})_2Mo_2Fe_3S_5(PEt_3)_5$ clusters. *Inorg. Chem.* 40: 5985–5999.

116 Ohta, S., Ohki, Y., Hashimoto, T. et al. (2012). A nitrogenase cluster model $[Fe_8S_6O]$ with an oxygen unsymmetrically bridging two protoFe_4S_3 cubes: Relevancy to the substrate binding mode of the FeMo cofactor. *Inorg. Chem.* 51: 11217–11219.

117 Churchill, M.R. and Wormald, J. (1974). Crystal and molecular structure of tetramethylammonium carbidohexadecacarbonylhexaferrate(2−), $[Me_4N]_2[Fe_6(CO)_{16}C]$, a hexanuclear iron cluster complex with an encapsulated six-co-ordinate carbon atom. *J. Chem. Soc. Dalton Trans.* 22: 2410–2415.

118 Bjornsson, R., Lima, F.A., Spatzal, T. et al. (2014). Identification of a spin-coupled Mo(III) in the nitrogenase iron-molybdenum cofactor. *Chem. Sci.* 5: 3096–3103.

119 Tanifuji, K., Sickerman, N., Lee, C.C. et al. (2016). Structure and reactivity of an asymmetric synthetic mimic of nitrogenase cofactor. *Angew. Chem. Int. Ed.* 55: 15633–15636.

120 Tanifuji, K., Jasniewski, A.J., Lee, C.C. et al. (2022). Incorporation of an asymmetric Mo-Fe-S cluster as an artificial cofactor into nitrogenase. *ChemBioChem* 23: e202200384.

121 Ohta, S. and Ohki, Y. (2017). Impact of ligands and media on the structure and properties of biological and biomimetic iron-sulfur clusters. *Coord. Chem. Rev.* 338: 207–225.

122 Zanello, P. (2019). Structure and electrochemistry of proteins harboring iron-sulfur clusters of different nuclearities. Part V. nitrogenases. *Coord. Chem. Rev.* 398: 113004.

123 Elvers, B.J., Pätsch, S., Bandaru, S.S.M. et al. (2023). Fine-tuning redox properties of heteroleptic molybdenum complexes through ligand-ligand-cooperativity. *Angew. Chem. Int. Ed.* 62: e202303151.

20

Metal-Dependent Formate Dehydrogenases and Their Interplay and Relationship to Iron–Sulfur Clusters

Benjamin R. Duffus

University of Potsdam, Institute of Biochemistry and Biology, Department of Molecular Enzymology, Karl-Liebknecht-Str. 24-25, 14476 Potsdam, Germany

Chapter Goals

- To provide a topical overview regarding metal-dependent formate dehydrogenases and the iron–sulfur clusters associated with their function.
- To describe the interconnectedness between metal-dependent formate dehydrogenases and iron–sulfur clusters, with respect to catalysis, biosynthesis, and electron transfer.
- To differentiate biochemical properties of metal-dependent formate dehydrogenases to the iron–sulfur clusters they interact with, in the secondary and outer coordination sphere relative to the bis-MGD active site cofactor.

Main Chapter Points

- Metal-dependent FDHs are bacterial and archaeal iron–sulfur molybdo- or tungstoflavoenzymes that require iron–sulfur clusters for biosynthesis, structure, and function.
- Metal-dependent FDHs interface with iron–sulfur clusters as a central redox carrier.
- Formate-based metabolism provides an alternative to H_2 that sometimes provides contextual advantages.
- FDHs are associated with enzymes playing key roles in anaerobic respiration.

Formate dehydrogenases (FDHs) represent a diverse group of enzymes in bacteria, archaea, and eukaryotes that catalyze the reversible two-electron and one-proton abstraction of formate to produce the greenhouse gas carbon dioxide (CO_2). Formate oxidation to CO_2 has a similar oxidation potential to that of molecular dihydrogen (H_2) oxidation [1]. In this respect, formate can be used as an energy source or a way to deposit reducing equivalents by a wide range of microorganisms in a fashion similar

Iron-Sulfur Clusters: Biogenesis and Biochemistry, Volume 2, First Edition.
Edited by Silke Leimkühler, Günter Schwarz, Oliver Lenz, and Oliver Einsle.

to H_2. Under certain biological contexts, generation of formate via CO_2 reduction represents a first step in the fixation of carbon, such as via the Wood–Ljungdahl pathway, or en route to generate methane [2]. Biology has exploited the property that CO_2 and proton reduction occur at very similar reduction potentials. Given the current societal need to mitigate CO_2 levels through carbon sequestration, an appealing strategy has been to apply the functions of CO_2 processing enzymes as bioinspired models of catalysis, as a carbon-neutral or carbon-negative strategy for fuel production [3]. Metal-independent FDHs catalyze CO_2 reduction and formate oxidation in conjunction with nicotinamide adenine dinucleotide (NAD^+) as a cosubstrate. However, metal-dependent FDHs catalyze this reaction at a pterin-based metalloenzyme coordinating either a molybdenum or tungsten metal ion at the active site. While both enzyme classes have been investigated and applied toward improving CO_2 reduction, catalysis of metal-dependent FDHs is dependent upon the biosynthesis and assembly of iron–sulfur clusters, not only to serve to build electron transfer chains but also to incorporate into ligands of the active site to render them catalytically competent. Given the significant interdependence between metal-dependent FDHs and iron–sulfur cluster biogenesis, a survey of the different FDHs isolated and characterized to date is presented here, with an emphasis toward metal-dependent FDHs characterized recently. The involvement of iron–sulfur proteins in formate metabolism has been previously reviewed elsewhere [4], in addition to several literature surveys [5] and reviews highlighting more niche aspects of FDH catalysis [6–12], and the reader therefore is referred to these and other previous reviews for historical details. However, in the years since the first review focused on the relationship between Fe–S clusters and FDH [4], substantial insights regarding their interconnectedness have been achieved, by which new insights can be made. In turn, the aim of this review is to correlate and expound on the substantial quantity of research directed toward metal-dependent FDHs in the context of iron–sulfur cluster biogenesis and biochemistry.

20.1 Formate: General Considerations

Formic acid, also known as methanoic acid, is a colorless, corrosive, and pungent liquid that exists in antiquity as an early known organic chemical characterized [13]. The name formate derives itself ants (latin formica), the source by which it was originally discovered, and the first scientific study on it was published in approximately the year 1670 [14]. Even though formic acid is relatively stable at room temperature, it can undergo thermal decomposition, producing the highly toxic gas carbon monoxide [13]. Formic acid itself is primarily toxic due to its corrosive effect on the skin and mucous membranes, as it can be readily absorbed [15]. It also exhibits inhibitory binding of mitochondrial cytochrome c oxidase. Though formate is a relatively weak inhibitor of mitochondrial cytochrome c oxidase with a K_i in the mM range [16], it can still impart detrimental effects, as it has been identified to affect the central nervous system in rabbits [17, 18], and is known to cause hypoxic labilization of the lysosomal complex in rats due to generation of histotoxic hypoxia [19]. It has been shown in rats that prolonged inhalation of formic acid vapors can lead to death [18]. Toxic amounts of formic acid are also generated accompanying methanol poisoning, resulting in metabolic acidosis that results in calcium influx and cellular damage [20, 21]. In the environment, formic acid can be degraded rapidly and completely; however, it typically accompanies a change in pH upon biodegradation. As a reduced C_1 compound relative to CO_2, formic acid and its related alkali salts are highly useful in industrial and agricultural applications, including but not limited to biomass preservation, textile production, feed preservation, pharmaceutical and food additives, rubber coagulation, and other oilfield applications [13].

By comparison, CO_2 is a known greenhouse gas that has been accumulating on the Earth and is contributing toward climate change due to anthropomorphic energy and material needs. It is a byproduct from currently applied chemical processes affecting societal daily life and contributes to the global average temperature increases observed. In biology, at least six different autotrophic carbon fixation pathways are known to exist, and the most common is via the Calvin–Benson–Bassham or the 3-hydroxypropionate 4-hydroxybutyrate cycle, reflecting a carboxylation–reduction reaction [22]. By comparison, the reductive acetyl-CoA pathway also fixes carbon, but via reduction prior to carboxylation, representing a primordial metabolic pathway and a simpler way to generate energy. In this pathway, CO_2 is assimilated via two branches, the methyl branch and the carbonyl branch, producing the components that are incorporated as acetyl-CoA [23, 24]. Interestingly, hydrogenotrophic methanogens and autotrophic acetogens employ an acetyl-CoA pathway in their metabolism to derive energy and carbon from H_2 and CO_2 [2]. To achieve such reactions, a concerted set of metalloenzymes, including hydrogenases, carbon monoxide dehydrogenases (CODH), acetyl-CoA synthase, heterodisulfide reductase, as well as FDH (by way of formylmethanofuran dehydrogenase) are employed. While FDH plays a central role in the reductive acetyl-CoA pathway in that it catalyzes the first step of the methyl branch of the pathway, reducing CO_2 to formate into acetyl-CoA, it also can serve as an alternative electron source to electrons generated from hydrogenases [25].

20.2 Metal-Independent FDHs

FDHs are primarily differentiated by their involvement of an active site metal at the active site and the additional cofactors or other prosthetic groups bound. For the sake of completeness, metal-independent FDHs such as from bacteria, yeast, fungi, and plants depend upon $NAD(P)^+$ as a cosubstrate and as a hydride acceptor for converting formate to CO_2 [26]. Several reviews dedicated more toward metal-independent FDHs have been published, and the reader is directed there for more detailed information regarding them [27–32]. In methylotrophic yeast and bacteria, metal-independent FDHs are thought to catalyze the terminal step in the catabolism of C_1 compounds, to provide organisms with energy and reducing equivalents. A representative metal-independent FDH from the yeast *Candida boidinii* [33] is commercially available, either from the host organism or heterologously expressed in *Escherichia coli*. A large focus toward metal-independent, NAD^+-dependent FDHs has been dedicated to adapt its selectivity between accepting NAD^+/NADH vs $NADP^+$/NADPH for cofactor recycling in bifunctional enzymatic reaction cascades [30, 31, 34], although work directed toward the reverse CO_2 reduction reaction has been targeted more recently [35–39]. As a commercially available product, metal-independent FDHs have seen substantial application also in electrode and device materials, either in conjunction with the coupling of enzymes that consume $NAD(P)^+$ or NAD(P)H or more recently with devices targeting CO_2 capture. Noted examples of metal-independent FDHs have been genetically engineered for heterologous expression in *E. coli* for isolation of the enzyme or to genetically modify prokaryotic metabolism by engineering overproduction of the metal-independent FDH as part of the reductive glycine pathway [40, 41]. These aspects are largely outside the scope of this chapter with reviews that can be found elsewhere [42, 43].

Metal-independent FDHs interconvert formate and CO_2 via a hydride transfer mechanism by which the reduced formate anion or oxidized CO_2 molecule binds in close proximity to the C4-(pyridine ring) position of the oxidized NAD^+ or reduced NADH molecule, respectively, forming a ternary complex comprising the enzyme, formate, and NAD^+ [44]. The formate C—H bond cleavage/forming step has been shown to be the rate-determining step of the reaction, as shown by an apparent kinetic isotope effect with deuterioformate. It should be emphasized that electron transfer is between the reduced or oxidized reactant/product species and that for a new catalytic cycle to occur, a new NAD^+/NADH molecule needs to occupy the active site. Perhaps for this reason, metal-independent FDHs show relatively sluggish CO_2 reduction activities that are on order of 1–10 min^{-1}, despite substantial work toward engineering the active site to improve CO_2 reduction efficiency. Predominantly, work toward improving activity has been dedicated toward improving the recycling of $NAD(P)^+$ to NAD(P)H as a coupled chemical reaction.

20.3 Metal-Dependent FDHs

In contrast with metal-independent FDHs, metal-dependent FDHs typically represent protein complexes of relatively high molecular weight that bind several prosthetic groups, including Fe–S clusters, heme groups, flavins, and quinones, in addition to the active site bis-molybdopterin guanine dinucleotide (bis-MGD) cofactor (for example, see Figure 20.1). In comparison to metal-independent FDHs, the presence of multicenter redox cofactors and extant electron transfer chains supplant the need for an active site cosubstrate to serve as an electron acceptor or donor, with activity generally dependent upon the efficient ability of the bis-MGD cofactor to shuttle between the Mo/W(IV) and Mo/W(VI) oxidation states. *In vitro* enzymatic activity of metal-dependent FDHs is detected biochemically via generation or consumption of electron acceptors or donors

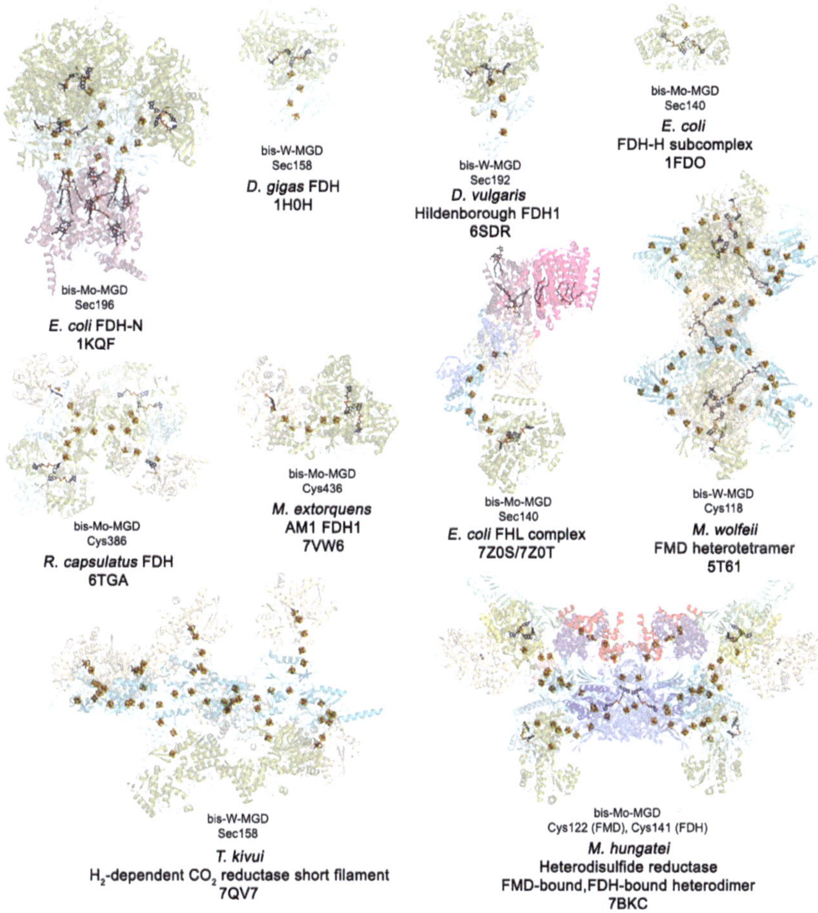

Figure 20.1 Diversity in structure, subunit organization, and iron–sulfur cluster arrangement in metal-dependent FDHs and formylmethanofuran dehydrogenases that have been structurally characterized. For each structure represented, the associated PDB code and the active site residue binding either Mo or W at the active site are indicated. The FDH and formylmethanofuran dehydrogenase subunits are depicted in deep olive color in structures by which one bis-MGD active site is depicted, while supercomplex structures that feature both subunits are depicted in deep olive and olive color, respectively. The Zn_2-associated subunit of formylmethanofuran dehydrogenase is depicted in wheat color, and diaphorase, non-F_{420}-associated [NiFe]-hydrogenase subunits, and [FeFe]-hydrogenase H-cluster subunits are depicted in light orange color. F_{420}-associated [NiFe]-hydrogenase subunits and subunits largely featuring electron transfer chains are depicted in pale cyan, deep teal, and slate colors. Membrane anchor-associated subunits are depicted in raspberry and hot pink colors. Heterodisulfide reductase subunits are depicted in violet and red colors. All other associated cap-associated and miscellaneous subunits are depicted in pale green or smudge colors. The depicted *E. coli* FHL complex structure depicts an overlay of two structures, with the bis-MGD cofactor overlaid in the anaerobically isolated structure that lacked the bis-MGD cofactor. The depicted heterodisulfide reductase and HDCR complexes depict the bound FDH subunits in an apo (bis-MGD) state, although are expected to bind a bis-MGD cofactor similar to the other structures shown in the figure. Images were prepared in Pymol 2.5.7.

that can be detected spectrophotometrically, most commonly via redox dyes such as methyl viologen, benzyl viologen, and 2,6-dichlorophenolindophenol (DCPIP), or via absorbing byproducts such as NAD(P)H or ferredoxin. Other redox partners such as cofactors F_{420}, quinones, and cytochromes have been identified to interact with FDHs, and for FDHs that are able to transfer electrons to or from NAD(P)H and/or ferredoxin, additional modular domains to receive such electron donors/acceptors have been identified. Moreover, established methods for formate generation are known that can be detected spectrophotometrically by Lang and Lang, or via GC–MS methods [229] as well as via infrared spectroscopy. Alternatively, enzymatic activity for several metal-dependent FDHs has been achieved via electrochemical methods, via adsorption of FDH onto an electrode.

Molybdoenzymes such as metal-dependent FDHs that coordinate a bis-MGD cofactor belong to the DMSO reductase subfamily of molybdenum and tungsten enzymes, named for by the enzyme by which the cofactor was first identified and characterized [45, 46]. Biosynthesis of the bis-MGD cofactor and the catalytically relevant sulfido ligand in metal-dependent FDHs constitutes a process that requires Fe–S clusters by way of the biosynthetic proteins involved. Details regarding the involvement of Fe–S clusters in the biosynthesis of the bis-MGD cofactor can be found in Chapter 7.

An association between Mo and Fe and the expression of *E. coli* FDHs was preliminarily identified in 1944 [47] that was later confirmed in 1954 along with the requirement of selenite [48], although Mo and Se were reported to not have an influence on the growth of the *E. coli* strains used. As reviewed more recently [49], evidence for different FDHs expressed by *E. coli* was warranted in apparent FDH activity in different fraction types isolated [50, 51], which was later shown to reflect distinct respiratory expression conditions [52, 53] that represented different gene operons expressed [53–55].

Most of the metal-dependent FDHs that were isolated biochemically in the 1960s–1980s were known to be iron–sulfur enzymes, with variable detection of flavin. While the apparent detection of molybdenum and tungsten coincided with improved methods for isolating the enzymes, their putative involvement at the active site was not fully appreciated until characterization of molybdoenzymes by way of the pterin ligand coordinated was performed, such as the Mo-MPT in sulfite oxidase [56, 57], the sulfurated Mo-MPT in xanthine dehydrogenase/xanthine oxidase [56, 57] or with an additional cytosine dinucleotide bound in Mo/Cu carbon monoxide dehydrogenase [58], and bis-MGD cofactor detected in DMSO reductase [45]. An association of FDH as a pterin-binding enzyme was subsequently identified in *Methanobacterium formicicum* [59–61], *Moorella thermoacetica* [62], and *Clostridium pasteurianum* [63] and formylmethanofuran dehydrogenase from *Methanosarcina barkeri* [64] that was later confirmed to be a bis-MGD cofactor type [46, 65, 66]. Interestingly, some metal-dependent FDHs that were characterized in this time are presumed to bind either a Mo or W metal ion at the active site due to their similarity to other metal-dependent FDHs (e.g. FDH from *Pseudomonas oxalaticus*), but have not been studied in detail since [67–69].

20.4 Metal-Dependent Formate Dehydrogenases: Common Modular Catalytic Unit

The first crystallographic characterization of FDH was published in 1997 on the FDH-associated subcomplex of the formate hydrogenlyase (FHL) complex, of the FdhF subunit from *E. coli*. It served as one of the first molybdoenzymes structurally characterized, and along with the crystal structure of DMSO reductase from *Rhodobacter sphaeroides* [70], was among the first molybdoenzymes structurally characterized to coordinate a bis-MGD cofactor (Figure 20.1) [71, 72]. Structures of the subunit in an oxidized (2.8 Å resolution) and a formate-reduced (2.3 Å resolution) state were elucidated. The active site Mo was originally thought to have a hydroxyl ligand [72] that was supported by X-ray absorption spectroscopy [73, 74], but the Mo-OH bond was later interpreted to be of a Mo-SH type. [75]. The original formate-reduced structure had a mistraced loop region that included the active site residues Sec140 and His141, where upon reinterpretation an apparent binding/unbinding of Sec140 to the active site Mo was observed.

Despite apparent differences in the active site metal ion and the active site ligand Cys/Sec residues, metal-dependent FDHs share a minimum catalytic subunit supporting the formate oxidation and CO_2 reduction activity, that is based upon the *E. coli* FdhF crystal structure described above, is shown in Figure 20.2, representing the cryogenic electron microscopy (cryo-EM) structure of *Rhodobacter capsulatus* FDH. Catalysis occurs at bis-metal-binding pterin (Mo or W) guanine dinucleotide (bis-MGD) cofactor, in conjunction with a sulfido ligand bound to the active site metal ion. Each respective pterin ligand provides a bidendate, noninnocent dithiolene/dithiolate ligand with both components of the ligand that are redox active. All of these components of the bis-MGD cofactor contribute toward its rich redox states that characterize mononuclear Mo and W enzymes [76, 77]. The bidentate coordination of two discrete pterin ligands was reminiscent to the structure of the W-coordinating aldehyde ferredoxin oxidoreductase from *Pyrococcus furiosus* [78], with the exception that the guanine dinucleotides are present, relative to a tungsten bis-molybdopterin (bis-W-MPT) cofactor characterized. In addition, either a cysteine or selenocysteine residue coordinates directly to the active site metal. Also in the primary coordination sphere are active site residues that partake in the substrate or product protonation/deprotonation and control substrate affinity to the active site [72]. A characteristic trait observed in all metal-dependent FDHs is the coordination of a [4Fe–4S] cluster in proximity to the tetrahydropterin-like Q pterin from the bis-MGD cofactor. Among FDHs structurally characterized, the [4Fe–4S] cluster is within electron transfer distance to the Q pterin ring, reflecting a putative coupling between the redox centers. In the FDH cryo-EM structure shown in Figure 20.2, the Q pterin is approximately 6 Å (edge to edge) from the FS0 [4Fe–4S] cluster. Given their proximity to one another, respective reduction potentials are affected due to their proximity, a property that has also been observed for other molybdoenzymes coordinating a bis-MGD cofactor [79].

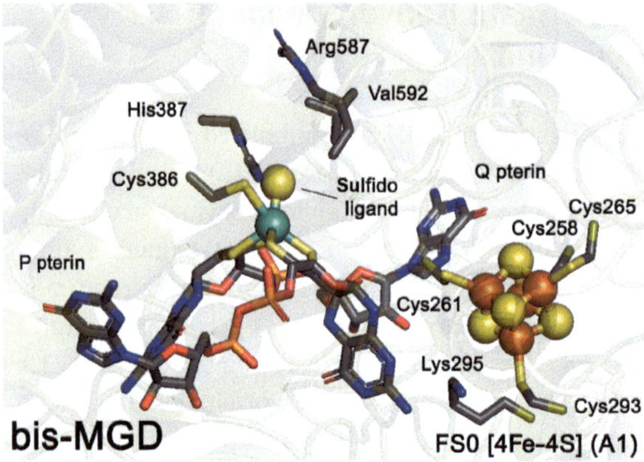

Figure 20.2 Representative structural depiction of the bis-MGD cofactor, constituting the active site of metal-dependent FDHs. The structure represents the cryo-EM map of FDH from *R. capsulatus* (PDB code 6TGA, 3.3 Å resolution) with a coordinated bis-Mo-MGD cofactor with a bound sulfido ligand and the FS0 [4Fe–4S] cluster (A1) that is in proximity to the cofactor. Primarily variation in the active site metal (Mo vs. W) and the amino acid coordinated to the metal (Cys vs. Sec) exists among different FDHs characterized to date. Active site residues that partake in substrate protonation and deprotonation and/or coordination are depicted. In addition, residues that have been studied to investigate the coupling between the bis-MGD cofactor and the FS0 [4Fe–4S] cluster have been noted. The FdsA subunit coordinating the bis-MGD cofactor and the FS0 [4Fe–4S] cluster is depicted in deep olive color, while the neighboring FdsB subunit is depicted in light orange color. Image was prepared in Pymol 2.5.7.

20.5 FDH and Link of bis-MGD to a [4Fe–4S] Cluster

Metal-dependent FDHs are representative members of mononuclear molybdo- and tungstoenzymes in prokaryotes and archaea that coordinate a bis-MGD cofactor at the active site. An important feature exhibited in most of all metal-dependent FDHs characterized to date is a [4Fe–4S] cluster within electron transfer distance of the Q pterin, that is believed to be important to the bis-MGD as an electron transfer conduit to or from the active site, and is bound in the first αβ domain of the polypeptide (see Figures 20.1 and 20.2) [79, 80]. The linked [4Fe–4S] cluster is observed in virtually all of the metal-dependent FDHs characterized to date and, aside from a few notable exceptions, is observed as conserved across several of the DMSO reductase enzyme family members [79]. Two different class types of binding motifs ligating this Fe–S cluster are known. Most of the currently characterized metal-dependent FDHs possess a CX_2CX_3C motif, while *E. coli* nitrate reductase NarGHI and DMSO reductase DmsABC coordinate this cluster with a C/HX_3CX_3C motif, and in all cases the fourth Cys is positionally dependent on the individual molybdoenzyme sequence [81]. For the latter class motif, this results in a reduced $[4Fe\text{–}4S]^+$ cluster with a high-spin ground state $S = 3/2$ that can be detected in the low-field region of an electron paramagnetic resonance (EPR) spectrum, and consequentially not at the $g = 2$ region that is typically observed for the former class motif [80, 82, 83]. For this reason, this [4Fe–4S] cluster is typically noted as "FS0," as a vestige of NarGHI and DmsABC. It is the first Fe–S cluster that typically constitutes the beginning or end of the Fe–S cluster electron transfer chain, depending upon the direction of catalysis. This Fe–S cluster is also analogous to the N7 [4Fe–4S] cluster in the Nqo3 subunit of the respiratory complex I from *Thermus thermophilus* (*Tt*RC I) [84]. As was discovered in the crystal structure of *Tt*RC I, the N7 Fe–S cluster was found to be too far away to partake in electron transfer with the main redox chain, whereby it was proposed that it confers structural stability on the fold in domain I of Nqo3. By comparison, molybdoenzymes such as FDH have evolved the electron transfer chain, resulting in additional coordination of the bis-MGD cofactor [79].

Because the proximal [4Fe–4S] cluster is in close proximity to the bis-MGD cofactor in FDH, and because it is redox active, the two redox centers typically undergo coupling or together influence the redox properties of the individual components of the [4Fe–4S] cluster–bis-MGD unit. In context to molybdoenzymes, this has primarily been characterized in bacterial nitrate reductases, and more recently in FDH from *R. capsulatus*. In all cases, altering the electronic coupling between the two redox centers results in shifted potentials, creating a larger mismatch that (typically) results in lowered apparent activity and altered spectroscopic properties for the individual redox centers, largely due to the bis-MGD and the FS0 Fe–S cluster consequentially behaving as separate, distinct redox species. In almost all cases, altering the residues between the tetrahydropterin-like Q pterin of the bis-MGD and the FS0 Fe–S cluster results in a shift in the latter's reduction potential substantially more negative. With respect to the effect of the coupling between FS0 Fe–S cluster and the respective bis-MGD Mo(VI) → Mo(V) and Mo(V) → Mo(IV) one-electron reduction potentials, different shifts have been observed, although it is important to note that limited studies have been performed, and the presence vs absence of the sulfido ligand at the bis-MGD active site can also additionally affect these one-electron reduction potentials as well [85]. For example, the sulfurated bis-MGD cofactor of FDH from *R. capsulatus* with the native Lys295 residue between respective cofactors results in Mo(VI) → Mo(V) and Mo(V) → Mo(IV) one-electron reduction potentials of -408 ± 12 mV and -477 ± 10 mV relative to the A1 (FS0) [4Fe–4S] cluster reduction potential of -337 ± 5 mV vs. SHE. When Lys295 is substituted for Ala, the Mo(VI) → Mo(V) reduction potential shifts more positive to -349 ± 4 mV and the A1 [4Fe–4S] shifts substantially more negative to -466 ± 6 mV; the Mo(V) → Mo(IV) reduction potential was more negative than could be measured under the conditions selected [86]. By comparison, substitution of the analogous Lys residue in periplasmic nitrate reductase from *R. sphaeroides* to Met resulted in a shift of the redox potential of the [4Fe–4S] cluster by 400 mV more negative that coincided with an altered activity, pH profile, and altered reduction kinetics of the [4Fe–4S] cluster relative to nonsubstituted enzyme [87]; a similar effect was observed on the activity of *Cupriavidus necator* nitrate reductase, upon substitution [88]. A similar, but less drastic effect on the [4Fe–4S] cluster has been observed for the dissimilatory nitrate reductase NarGHI, substituting Arg94 to Ser resulted in a shift of the redox potential of the [4Fe–4S] cluster by

115 mV, causing an apparent decrease in enzymatic activity without causing a substantial structural or redox change on the bis-MGD cofactor [89]. An important frame of reference here is that these trends have mostly been observed with the FS0 [4Fe–4S] cluster coordinated by four Cys residues, and studies investigating the altered coordination of FS0 (e.g. the His-bound [4Fe–4S] cluster in dissimilatory nitrate reductase NarGHI from *E. coli*) have shown additionally that the hydrogen bonding network between the bis-MGD cofactor and the FS0 cluster is important [90].

20.6 Link of the FDH bis-MGD Cofactor with Extended Fe-S Cluster Electron Chains

Concomitant with their characterization, metal-dependent FDHs typically possess extended Fe–S cluster electron transfer chains that serve to link either modular components or distinct active sites together beyond the FS0 [4Fe–4S] cluster (see Figure 20.1). Historically, this typically was described as simply being non-heme iron that was bound to complexes that could be isolated *in vitro* that was in substantial excess relative to molybdenum or tungsten, for example the FDH from *Wolinella succinogenes* [91]. The iron in FDH was unable to react with 4-chloromercuriphenyl sulfonate [91], suggesting that it was constitutively linked to the major peptide, a characteristic that appeared to be common to metal-dependent FDHs [91–93]. Later, with the elucidation of crystal structures of FDH, in conjunction with *E. coli* nitrate reductase and respiratory complex I, the excess nonheme iron bound on FDH was shown to be Fe–S cluster electron transfer chains. For example, *E. coli* FDH-N (FdnGHI) [94] and the *E. coli* FHL complex [95] have been structurally characterized to have extended Fe–S cluster chains that lead to a second modular active site unit as membrane-bound supercomplexes. Similar to *W. succinogenes* FDH, these structurally characterized FDHs represent membrane-bound complexes, although these Fe–S cluster electron transfer chains are also observed in soluble multicenter FDH-associated complexes. For example, the crystal structure of FDH from *Desulfovibrio gigas* showed coordination of the bis-MGD cofactor along with 4 × [4Fe–4S] clusters in a dimeric structure, but the associated FdhC subunit is predicted to interact with FdhAB as a multiheme cytochrome [96]. In all cases, because these Fe–S clusters are involved in electron transfer processes to or from the bound catalytic subunits, they typically possess redox potentials that are accessible relative to the redox reaction catalyzed by the enzyme. EPR spectroscopy and spectroelectrochemical methods have been typically employed to identify the redox potentials of these Fe–S clusters, but for many FDHs redox characterization of the electron transfer chain Fe–S clusters remains relatively undercharacterized due to their similar redox potentials, similar g-tensor parameters experimentally detected and complexity associated with spectral deconvolution.

20.7 FDH that Employ Ferredoxin as Electron Acceptor

Given the diversity in subunit and electron transfer chain structure, and in different organic and metallocofactors bound to metal-dependent FDHs that can serve as electron acceptors or donors, it is not unexpected that discrete Fe–S cluster complexes as electron transfer carriers can also interface with metal-dependent FDHs, in relation to constitutive Fe–S clusters serving in a structural capacity. One of the first examples by which CO_2 could be directly reduced to formate was the CO_2 reductase from *C. pasteurianum*, which did not require NAD(P)H, but rather required either reduced ferredoxin or one-electron reduced methyl viologen [97]. Several studies showed it was a molybdenum-containing Fe–S enzyme, it performed catalysis reversibly, it did not bind selenium, and it lacked bound flavin following protein purification [98–102]. The enzyme was found isolated as a heterodimer, with the FdhA subunit coordinating the bis-MGD cofactor and the FS0 [4Fe–4S] cluster, and the FdhB subunit predicted to coordinate 4×[4Fe–4S] clusters [102]. In retrospect, spectroscopic studies on the enzyme however showed that the bis-MGD cofactor was exceptionally labile, with the enzyme being both O_2- and light-sensitive, that has to date limited characterization of this FDH [63, 103, 104].

More recently, the Fdh3 from *Thermococcus onnurineus* NA1 was biochemically characterized and was shown to employ reduced ferredoxin as an electron donor, exhibiting a reported CO_2 reduction-specific activity of 0.728 ± 0.094 U/mg [105]. This FDH was found to coordinate a bis-W-MGD cofactor and was also shown to use $NADP^+$ as an electron acceptor. As one of three FDH isoenzymes encoded by the organism, the *fdh3* operon encodes for a Fdh3A subunit that binds the bis-MGD cofactor and encodes for a Fdh3B that likely is a diaphorase module with a bound FAD cofactor. In addition, the operon encodes for two Fe–S cluster subunits (Fdh3G1 and Fdh3G2) that have been proposed to serve as the Fe–S electron transfer chain between the two modules.

20.8 *E. coli* Metal-Dependent Formate Dehydrogenases

The facultative anaerobic bacterium *E. coli* is known to adapt to different environmental conditions and can use a diverse array of small molecule electron acceptors for respiration and to produce energy [106–108]. *E. coli* is able to metabolize formate via three different membrane-associated FDHs that is produced from pyruvate during anaerobic fermentation via the pyruvate–formate lyase complex (PFL) [109]. As reported elsewhere, during fermentation, approximately one third of the carbon derived from glucose is converted to formate, and when left to accumulate in the cytoplasm, it would result in acidification and uncouple the proton gradient of *E. coli* [110, 111]. Thus, *E. coli* encodes for several FDHs that are expressed as dependent on the environmental growth conditions. FDH-H (FdhF) is associated with hydrogen production by way of the FHL enzyme complex, while FDH-N

(FdnGHI) is linked to nitrate reductase, and FDH-O (FdoGHI) is detected under aerobic growth. Historically, the activities of FDH-H, FDH-N, and FDH-O were differentiated by their associated electron acceptor, such as benzyl viologen, phenazine methosulfate with DCPIP, and oxygen, respectively. Recently, based upon revisiting an old experimental result [112], it has been discovered that in the absence of cognate [NiFe]-hydrogenases, FDH-N and FDH-O feature a degree of nonspecificity with respect to the electron donor used, whereby both exhibit H_2:benzyl viologen oxidoreductase activity [113].

As noted above, formic acid and its conjugate base formate are known as corrosive agents [13] and under elevated concentrations are known to be toxic to bacteria. Organisms such as *E. coli* employ a dedicated secretion mechanism to efflux and influx generated formate, involving the gating of a membrane transporter and the symport of a formate anion with a proton [114]. The principal formate transporter in *E. coli* has been identified as FocA; its integral membrane structure with six transmembrane domains has suggested that it functions as a channel as opposed to a transporter [115–117]. The individual FocA monomers are thought to function separately from the larger complex, since it has been shown that the monomeric unit can directly interact with PFL [118, 119].

20.9 Formate Hydrogenlyase Complex

The efficient production of H_2 from biomass by *E. coli* under anaerobic growth conditions via mixed-acid fermentation is facilitated through the production and conversion of formate (a byproduct of fermentative conversion of pyruvate to acetyl-CoA) via the FHL complex [120–122]. FDH-H in *E. coli* is encoded by the *fdhF* gene [54] as part of the FHL complex with the operon *hycABCDEFGHI*, and is expressed primarily under fermentative and anaerobic conditions [49]. The overall complex is membrane-bound in the direction of the cytoplasm and features several linked Fe–S clusters as membrane-integral electron transfer components in the HycB, HycC, HycD, HycF, and HycG subunits, along with a coordinated bis-MGD cofactor in the FdhF subunit, and a [NiFe]-hydrogenase active site in the HycE subunit. Interestingly, the FdhF subunit can be isolated individually as a soluble complex [65, 123, 124], which has served as an anchoring model protein for understanding FDH-based catalysis [73, 125–130].

More recently, isolation of the *E. coli* FHL complex in total [131] has allowed for clearer assessment of the involvement and coupling of the FDH and [NiFe]-hydrogenase components and has provided additional molecular insights toward a process first identified over 100 years ago [132]. The intact, membrane-bound complex has been reported to be stable [131], and in addition to the formate disproportionation reaction classically observed, the FHL complex has been demonstrated to exhibit an H_2-driven CO_2 reduction reaction [133–135]. Recently, a cryo-EM structure on the anaerobically isolated complex has recently been solved (Figure 20.1) [95]. A noteworthy structural characteristic of the FDH subunit of the FHL complex is its apparent similarity to the respiratory complex I subunit Nqo3 from *T. thermophilus* (or human mitochondrial NDUFS1), which was predicted by sequence homology [84] and was recently confirmed by structure [95]. The electron transfer pathway engaged by the FdhF subunit to the hydrogenase module (HycG/HycE) is facilitated by 4 × [4Fe–4S] clusters in the HycB subunit and 2 × [4Fe–4S] clusters in the HycF subunit. Interestingly, an additional metal ion assigned as Fe is bound in a CxxC-x_{34}-CPxC motif in HycF that is stabilized by the HycF Arg122 residue, which has been proposed to help improve the affinity of the FdhF module to the FHL complex [95].

FDH-N in *E. coli* is encoded by the *fdnGHI* operon and is coexpressed with the associated *narGHI* genes that encode for a membrane-bound nitrate reductase. A crystal structure of FDH-N has been determined at 1.6 Å resolution, by which the subunits form a mushroom-shaped $(\alpha\beta\gamma)_3$ structure with the bis-MGD subunit oriented toward the periplasm (Figure 20.1) [94]. The FdnH subunit possesses 4 × [4Fe–4S] clusters that serve to link the bis-MGD cofactor with a diheme site in the FdnI subunit. The heme site serves as the menaquinone binding site for electron transfer to the nitrate reductase, with its own set of Fe–S clusters, electron transfer to the NarG subunit houses a bis-MGD cofactor that is oriented toward the cytoplasm. The orientation of the 4 × [4Fe–4S] clusters of FDH-N is similar to that observed in NarGHI [136], [NiFe]-hydrogenase, and respiratory complex I. Similar to FDH-H, the bis-MGD cofactor coordinates a Mo ion at the active site, and it is bound by Sec196. The sixth ligand was originally modeled to be a hydroxyl ligand, but considering the reinterpretation of the FDH-H crystal structure [75], it likely represents a bound sulfido ligand.

In addition to FDH-N, a second FDH isoenzyme is encoded by the *fdoGHI* operon, that is associated with an expressed FDH under either aerobic or nitrate-respiring conditions [53]. It is constitutively expressed and can be found at low levels under these, in addition to fermentative conditions [53, 137]. Similar to FDH-N, FDH-O is predicted to have a similar structural arrangement, and the *fdoGHI* genes are coexpressed with the analogous nitrate reductase *narZYV* genes, forming a second formate:nitrate oxidoreductase complex in the membrane [138, 139]. FDH-O is the least extensively characterized FDH isoenzyme in *E. coli*, but interaction partners to stabilize FDH-O along with FDH-N in the periplasm, following maturation and translocation from the cytoplasm to the inner membrane have been identified, that include the Skp chaperone protein, the L,D-transpeptidase ErfK, the oligopeptide-binding protein OppA, and TolB [140].

20.10 NAD$^+$-Reducing, Metal-Dependent FDHs

A subclass of metal-dependent FDHs are representative as complexes that have a built-in diaphorase subunit linked to the bis-MGD cofactor by an electron transfer chain of Fe–S clusters, which allows for the use of NAD(P)$^+$/NAD(P)H as an electron acceptor or donor, respectively [6]. While this subgroup of FDHs is similar in principle to metal-independent FDHs with respect to stoichiometric NAD(P)$^+$/NAD(P)H consumption relative to formate oxidation/CO_2 reduction (respectively), it interfaces with the bound FAD or FMN present, as formate or CO_2 does not interact catalytically with the bound nicotinamide adenine dinucleotide. These FDHs typically harbor the diaphorase module in a separate subunit relative to the subunit binding the bis-MGD cofactor [141], although notable exceptions exist. As noted elsewhere in this chapter, metal-dependent FDHs featuring the binding of FAD/FMN in a nearby, related subunit as part of a larger protein complex with distinct Fe–S cluster active sites are also known.

With respect to metal-dependent FDHs featuring only the bis-MGD, FAD/FMN, and respective Fe–S cluster chains, several have been biochemically and structurally characterized to understand their mode of function, including the FDH from *C. necator* [142–146], the FDH from *Rhodopseudomonas palustris* [147], the FDH from *Methylosinus trichosporium* Ob3b [148–150], the FDH1 from *Methylobacterium extorquens* AM1 [151–153], and the FDH from *R. capsulatus* [86, 141, 154, 155]. For example, the first FDH structurally characterized with a bound flavin was via the cryo-EM structure of *R. capsulatus* FDH (Figure 20.1). The structure revealed a 360 kDa dimer of FdsABGD heterotetramers, in which the FMN-containing FdsB subunit along with FdsG is bound to the FdsA subunit [156]. Separate structural characterization of the FdsBG subcomplex from *C. necator* FDH in the absence of FdsA has confirmed the modularity between the active site units [157]. This subcomplex offers similar protein folds and similar cofactor arrangements to those observed in the subunits Nqo1 and Nqo2 of respiratory complex I from *T. thermophilus* (*Tt*RC I) [84]. Despite the similarity of the FdsA subunit to the *E. coli* FDH-H (FdhF) structure [72], to the HoxU subunit of NAD^+-reducing [NiFe] hydrogenases [158], and to the Nqo3 subunit of *Tt*RC I, the overall structure is distinct, in that a dimer of FdsABGD heterotetramers is observed, with two [4Fe–4S] clusters (one from each monomeric unit) creating an interface between the dimers, featuring Cys_3His coordination within electron transfer distance [156]. Recently, it has been shown that the reduction potential of this [4Fe–4S] cluster is exceptionally low, whereby electron transfer likely does not occur, since the isolation of the monomeric unit exhibits comparable formate oxidation activity relative to the dimer [86].

Within the subgroup of metal-dependent FDHs featuring a bis-MGD cofactor and a terminal diaphorase subunit, some differences in the protein fold structure near the bis-MGD persist. The holoenzyme of *R. capsulatus* FDH features the FdsD subunit that serves as a cap or plug, which holds the bis-MGD cofactor in place. As noted previously, almost all FDH complexes that contain FdsD have the residue Trp917 (*R. capsulatus* FDH amino acid numbering) in FdsA, that is at the tip of the cap loop that interfaces with FdsD, including the FDHs from *C. necator*, *M. trichosporium*, *R. palustris*, and *Rhizobium radiobacter* [156]. A notable exception is the FDH1 from *M. extorquens* AM1. Recent cryo-EM structures of this FDH have shown that such an additional subunit domain is not necessary, but differences in the FDH1α subunit (equivalent to FdsA) between the homologous [159] and recombinant [160] expression have suggested that bis-MGD cofactor binding is associated with conformational changes in a similar interfacial domain that FdsD binds (Figure 20.1).

20.11 W-Containing FDHs from Sulfate-Reducing Bacteria

Sulfate-reducing bacteria (SRB) are anaerobic prokaryotes that respire with sulfate as a terminal electron acceptor, producing sulfide as the end product [161–163]. These bacteria can use a wide variety of substrates to reduce sulfate, including lactate, formate, H_2, malate, ethanol, pyruvate, and fumarate. The first FDH characterized from SRB was the formate:ferricytochrome c_{553} oxidoreductase from *Desulfovibrio vulgaris* Miyazaki [164]. Several FDHs of this type have been biochemically characterized, including *Desulfovibrio desulfuricans* ATCC 27774 [165–167], *Desulfovibrio alaskensis* NCIMB 13491 [168], *D. gigas* [169–171], and *Desulfovibrio vulgaris* Hildenborough [172–175]. It is interesting to note that all of these organisms encode for multiple FDHs and have been shown to exhibit different preferences with respect to Mo or W bound, across those that have been characterized, and have shown to be very robust with respect to their isolation in the presence of O_2.

The sulfate-reducing bacterium *D. gigas* has been extensively studied and is known to use sulfate as a main terminal electron acceptor to degrade organic compounds [161]. Historically, this bacterium has been of interest to study several enzymes involved in energy transfer, and FDH was an early target. *D. gigas* FDH was found to be a periplasmic enzyme [176], although the genome of *D. gigas* encodes for two FDHs [177]; the second one (DGI_3334 and DGI_3335) remains uncharacterized. *D. gigas* FDH was targeted for biochemical characterization due to its apparent O_2 insensitivity [171]. Partial purification of the enzyme was previously shown to be stable in air, as long as thiols such as DTT or 2-mercaptoethanol were present in the purification buffer during purification [169, 170]. Similar to aldehyde oxidoreductase from the same organism, the FDH was shown to specifically incorporate W over Mo, even when the media was not depleted of Mo [171, 178].

D. gigas FDH was the first W-FDH structurally characterized, and preliminarily a surprise was that it coordinated a Sec residue, since a primary amino acid sequence was not available at the time (Figure 20.1) [96, 179]. The W ion is bound to two MPT moieties, to a selenocysteine, and to an inorganic S or O ligand [96, 179]. The crystallographic data favor a S atom for the sixth ligand, although the resolution of the data (1.8 Å) nevertheless did not permit unambiguous discrimination between O and S. Similar to *E. coli* FdhF, the bis-MGD cofactor is buried in the protein, stabilized by an extensive network of hydrogen bonding interactions. Interestingly, a surface-exposed disulfide bridge (Cys817-Cys844) was observed in proximity to one of the guanidine ligands of bis-MGD, but was removed from the electron transfer pathway of the enzyme, as not actively involved in catalysis [96].

20.12 FDH: *D. vulgaris* Hildenborough

FDH from *D. vulgaris* Hildenborough was first purified in 1995, and with developed tools to overexpress recombinant *D. vulgaris* Hildenborough, two of three isoforms have been extensively biochemically characterized. In addition, several crystal structures of the FDH isoenzyme #1 (DVU0587, DVU0588) have recently been obtained under various redox states (Figure 20.1) [174, 180–182]. In its oxidized form, the W ion is hexacoordinated to four S atoms from the two dithiolenes, a Se atom from Sec192, and an indeterminate sulfur ligand (W-SH or W=S) [174]. Cocrystallized glycerol molecules were found present in the structure and have been proposed to likely occupy formate binding sites at the active site, which were later confirmed in time-dependent formate soaking of as-isolated FDH crystals [174, 181]. By comparison, the formate-reduced form that exhibited a conformational rearrangement was observed, involving I191, Sec192, and His193 that was similarly observed in the presence of $Na_2S_2O_4$ [174, 180]. The catalytically relevant His193 moves away from the active site, forming a distinct *cis* conformation with Arg441, a water molecule that occupied the former position of the His193 imidazole ring, while the Se atom remained coordinated to the W ion [174, 181]. A Sec192 loop shift is propagated to neighboring amino acids from different domains that promote the tilt of the side chains of Trp533, Phe537, and Phe160. Interestingly, the disulfide bridge observed previously for *D. gigas* FDH is also present here in FDH isoenzyme #1, and substitution to remove the bridge followed by anaerobic purification resulted in a highly O_2-sensitive enzyme that could not be activated, with conformation changes near His193 and Arg441 that resulted in an altered hydrogen bonding network [182].

20.13 Tungsten vs. Molybdenum FDH

Thus far in the chapter, examples of metal-dependent FDHs with respect to the metal at the active site (as Mo or W) have been described without a lot of concern with respect to the selectivity or specificity of the metal at the active site. Despite the similar atomic and ionic radii of either metal, substantial differences in the assimilation and incorporation of Mo or W metal ions into metalloenzymes have been noted for some time [48, 183–186]. Despite tungsten primarily serving as an antagonist against molybdenum in the maturation of the bis-MGD cofactor, regulation of both metal ions is important for the biosynthesis and assembly of the cofactor, in organisms that encode for a tungsten-containing FDH (see Chapter 7 for additional contextual details). This has been ultimately reflected not only in the relative inability to incorporate tungsten into nitrogenase [187] but also in the relative inability of some bis-MGD-containing enzymes to incorporate tungsten is noted, such as nitrate reductase [188, 189], in addition to contextual examples of FDH. For example, the FDH from *M. formicicum* is characteristically a Mo-containing enzyme, and in the presence of tungstate, isolated FDH possessed molybdopterin cofactor without productive coordination of the W ion [190].

20.14 Metal-Dependent Formate Dehydrogenase from *M. thermoacetica*: Variance and Codependence of Metal Ions

Some anaerobic bacteria are able to assimilate CO_2 by way of FDH. One of the first tungstoenzymes characterized was the FDH from *M. thermoacetica* (*Clostridium thermoaceticum*). Concomitant with the early FDH characterization studies in *E. coli*, work toward characterizing the metabolism of *M. thermoacetica* showed an unique metabolic trait of the acetogen, namely in that formate was an intermediate in the reduction of CO_2 to acetate, in that whole cells catalyzed a rapid exchange between CO_2 and formate in radiotracer studies, and in that addition of unlabeled formate to cell-free extracts resulted in decreased radiolabel incorporated into acetate, reflecting formate's intermediacy in the metabolism [191–193]. Isolation of the enzyme was attempted and early on was shown to be an $NADP^+$-consuming and O_2-sensitive enzyme, requiring purification in the presence of sulfhydryl reagents. Catalysis was demonstrated to be reversible with $NADP^+$/NADPH, although the CO_2 reduction reaction required an additional NADPH regeneration system [194, 195]. Substantial effort was directed toward improving the purification and handling of the enzyme, spanning almost 20 years [62, 194, 196, 197]. Interestingly, selenite, molybdate, and tungstate were found to be critical components associated with the enzyme, whereby the addition of tungstate stimulated activity more than molybdate, and whereby coaddition was additive [198]. More recently, the genome sequence of *M. thermoacetica* has been obtained [199] that confirmed the differential and required function of both Mo and W in the synthesis of the bis-MGD cofactor at a single active site [198].

20.15 Formylmethanofuran Dehydrogenases

Methanogens serve an important role in the global carbon cycle, to convert CO_2 to methane. Methane is mostly produced from acetate, CO_2, and H_2 [2]. Methanogenesis starts from the reduction of CO_2 to form formylmethanofuran, with electrons that are supplied by a reduced ferredoxin [200]. Methanogens employ a formylmethanofuran dehydrogenase to catalyze the initial CO_2 reduction reaction en route to methane, producing formylmethanofuran. This reaction was first demonstrated in cell extracts of *Methanobacterium thermoautotrophicum* (first described as the carbon dioxide reduction (CDR) factor) [201] and in other methanogenic bacteria. In subsequent work, cell extracts from *M. thermoautotrophicum* and *M. barkeri* that mediated the reduction of methyl viologen with formylmethanofuran instead of formate constituted an enzyme of a very high molecular mass and showed cyanide inactivation despite an apparent lack of cyanide inhibition in cell suspensions of methanogenic bacteria [202]. This, in addition to the association of *M. thermoautotrophicum* with the metal molybdenum for growth [203], suggested that this reaction could be catalyzed by a molybdoenzyme that was subsequently confirmed [204] and later shown to be pterin-containing [64] and of a bis-MGD cofactor form [66]. In retrospect, because the first confirmation of an FDH possessing the bis-MGD cofactor was identified the following year from *M. formicicum* [46], and because different cofactor forms were detected in formylmethanofuran dehydrogenases that were characterized in parallel (e.g. *M. thermoautotrophicum* Marburg) with distinct pterin dinucleotides [205] that coincided with the discovery of the molybdopterin cytosine dinucleotide cofactor in Mo/Cu-CODH from *Pseudomonas carboxydoflava*, resistance to associate formylmethanofuran dehydrogenase as a CO_2 reductase akin to metal-dependent FDHs persisted. For example, while it was discovered that formylmethanofuran dehydrogenases are associated with a bis-MGD cofactor as similar to FDH due to the detection of FormA-GMP, and with characterized examples having differences with respect to the metal incorporated, the respective reaction for formylmethanofuran dehydrogenase was thought to be distinct relative to FDH [202]. However, upon structural characterization of formylmethanofuran dehydrogenase, the enzymatic catalytic function could be better assigned (Figure 20.1) [206]. Under the context described above, recently a number of metal-dependent FDHs and formylmethanofuran dehydrogenases from methanogenic archaea have been biochemically characterized to provide insights regarding regulation and control of the bis-MGD biosynthesis with respect to the metal ion incorporated [207]. While the incorporation of either metal is largely dependent upon the environmental niche that the originating microorganism thrives in, physiological metal uptake and incorporation of respective metals are typically achieved at the level of the principal metal uptake transporters [208–210]. Nevertheless, some noteworthy organisms have the ability to incorporate either metal and encode for redundant FDH-based structural genes and molybdopterin biosynthetic components with multiple metal-differentiated uptake genes.

20.16 FDHs in Methanogens

Despite the relative wealth of information regarding the regulation of FDH in well-studied bacteria such as *E. coli*, how the enzyme is regulated in different bacterial or archaeal niches, as similar or different remains understudied and is of current interest. In recent, efforts in several laboratories have been dedicated toward the development of genetic tools in methanogenic and acetogenic bacteria to achieve such comparisons [211]. For example, while it has been known for some time that methanogenic bacteria encode for multiple isoenzymes of FDH that have a distinct (lack of) selectivity between Mo or W at the active site (either as FDH or FDMH), clarity regarding how these modular complexes are regulated by partner proteins that are functioning either as soluble, individual complexes or together as larger protein supercomplexes is largely unclear, due to a lack of complete information regarding the key proteins involved. The genome sequence of an organism can in principle provide insights regarding the putative participation of FDHs of interest, but the linear nature of genome structure often can misconstrue the putative involvement of enzymes such as FDH, by which biochemical characterization is warranted.

In addition to FDH serving as a modular component in formylmethanofuran dehydrogenase catalysis, it also can serve as an alternative H_2 substrate in the heterodisulfide reductase reaction. For example, methanogens from the order *Methanomicrobiales* are capable of growth with H_2, but some lack genes encoding for hydrogenases that are typically observed in hydrogenotrophic methanogens. Preliminary results from *Methanococcus maripaludis* showed that either H_2 or formate could reduce the heterodisulfide-H_2 via the F_{420}-nonreducing hydrogenase (Vhu) or formate dehydrogenase (Fdh), respectively [212]. Isolation of discrete complexes could be obtained, either as respective $(Fdh)_2Hdr$ or $(Vhu)_2Hdr$ homocomplexes or as a (Fdh)(Vhu)Hdr heterocomplex [213]. In *Methanoculleus thermophilus*, biochemical characterization of its heterodisulfide reductase was determined to copurify with FDH and exhibited selective activity with formate over H_2 as an electron donor [214]. It is important to note that whole cells of *M. thermophilus* exhibited an FHL reaction, reflecting the presence of distinct copies of energy-conserving and F_{420}-reducing hydrogenases and FDHs. In addition to the presence of the formylmethanofuran dehydrogenase operon, *M. thermophilus* additionally harbors three copies of the *fdhAB* operon in its genome, reflecting formate's central incorporation into the methanogenic pathway.

Recently, isolation and structural characterization of stabilized complexes of the heterodisulfide reductase in the presence of hydrogenase and FDH has afforded insight regarding the interaction of such modular catalytic units, in addition to understanding how flavin-based electron bifurcation (FBEB) operates in such a large protein complex (Figure 20.1) [206, 215–217]. The "relatively" simpler Hdr-Mvh [215] and Fmd (structure depicts the tungsten-based bis-MGD cofactor in the analogous Fwd subunits) complexes [206] exhibit modular units as dimeric structures; the former harbors a MvhD subunit that coordinates a single Fe–S cluster that is independent of, but interfaces with the hydrogenase module in the Hdr-Mvh structure [218]; for some organisms, the MvhD subunit is fused into the Hdr-based subcomplex [219]. Recently, characterization of the Fdh–Hdr–Fmd complex (both as a dimer and as a hexameric form) showed that the Fdh module interfaces with the Hdr module via the MvhD subunit. Interestingly, the FdhB subunit harbors an exposed FAD molecule that is bound in a fold reminiscent to F_{420}-reducing hydrogenase (Frh) [220]; the related FDH from *M. formicicum* similarly binds FAD in the related FdhB subunit, at the location by which F_{420} is expected to bind [59, 221].

Recently, differentiation regarding the FDH isoforms served in the FDH-based activities in *M. thermophilus* has been achieved [222]. Previously, it was shown that Fdh1 was critical for formate-dependent growth and that Fdh2 was dispensable [223–225]; however, after a prolonged incubation of a *Δfdh1* mutant strain, growth equating to methanogenesis could be observed, suggesting a suppressor mutation. The origin of the suppressor mutation was identified to be primarily due to mutations in the *moeA* gene, and as a result a second *moeA2* gene was identified, that also affected growth with either FDH isoform present. Isolation of either Vhu–Fdh1–Hdr or Vhu–Fdh2–Hdr complex resulted in comparable F_{420} and metronidazole (alternate electron acceptor to ferredoxin) reduction, but they primarily differed in the metal content (W and Mo, respectively) bound.

20.17 FDH O_2 Sensitivity: Inhibition and Enzymatic Activation

Complex, iron–sulfur cluster-coordinating metalloenzymes, by way of the cofactors bound to the enzyme, in addition to their respective subunit structure and evolution from their respective source organism, are highly susceptible to cofactor degradation and/or loss due to their respective handling in the presence of O_2 [226]. Metal-dependent FDHs are no exception, as they are known to be exquisitely O_2-sensitive metalloenzymes that typically operate only in the absence of O_2, and are known to be exceptionally cofactor labile. As several publications relating to purifying FDHs were described as requiring strictly anaerobic purification schemes [62, 227, 228], the need for purification in the absence of O_2 and in the presence of reducing agents cannot be understated. Metal-dependent FDHs are primarily O_2 sensitive due to the sensitivity of the bound sulfido ligand at the active site to O_2 (Figure 20.2). As stated earlier, the first FDH crystal structure (*E. coli* FDH-H) lacked a sulfido ligand, by which a hydroxyl group was found to be 2.2 Å away from the Mo metal ion [72]. A lack of a sulfido ligand renders the active site catalytically incompetent, due to the inability of a hydroxyl ligand to participate in a hydride transfer reaction, to receive or hand off a hydride from formate or to CO_2 bound as bicarbonate, respectively [229, 230].

An early strategy that was employed to isolate active FDH enzyme, for example in cell-free lysates containing FDH, was to inhibit it with a reversible inhibitor. Of these, anions of the pseudohalide series (e.g. cyanide, azide, cyanate, and nitrite, in addition to nitrate) [100] were employed for purification of FDH performed under anaerobic conditions to isolate a competent, active enzyme for biochemical studies. It was preliminarily interpreted that either likely could directly interact with the bis-MGD cofactor, although it should be emphasized that in retrospect many of the tractable biochemical systems and anaerobic methods for isolation at the time were primitive, reflecting in the extreme cofactor lability observed [69, 194]. However, as more examples of FDHs were characterized from different organisms, it was shown that not all FDH enzymes exhibit the same behavior with respect to inhibiting anions, and it was shown that conventional aerobic purification methods (notably in the absence of pseudohalides) could also result in isolation of a highly active FDH enzyme [151]. With either strategy, insights regarding the intactness of the enzyme have been recently evaluated. With respect to the binding of pseudohalides, their binding and interaction have been primarily characterized by X-ray absorption and infrared spectroscopies. As previously shown for *R. capsulatus* FDH, while aerobic copurification with the pseudohalide azide can result in quantitative sulfido ligand retention at the bis-MGD cofactor upon aerobic handling [155], that itself binds in the second coordination sphere in proximity to the sulfido ligand [231], retention of the sulfido ligand in the absence of azide requires strict anaerobic handling, due to its increased lability and susceptibility to O_2 damage [155], shown previously in aerobic samples handled with the inhibitor present [154]. An interesting comparison between FDHs characterized to date, given a similar to identical active site makeup, is a difference in the inhibition of small inhibitor anions. Anions such as cyanide, azide, cyanate, nitrate, and nitrite have been primarily investigated, and many have been used in conjunction to isolate an optimally active FDH enzyme, primarily to protect the active site sulfido ligand. However, cyanide is known to be an irreversible inhibitor due to the reaction with the sufido ligand associated with the bis-MGD cofactor, generating an oxo ligand at the active site [155] and rendering the active site as a catalytic incompetent. Interestingly, differences in behavior in the FDH enzyme with respect to azide and cyanide are noted. For example, the Se-independent FDH from *Methanococcus vannielii* exhibits no apparent inhibition toward azide [232]. In addition, FDH from *D. desulfuricans* ATCC 27774 is an outlier with respect to cyanide, in that it serves as a reversible inhibitor [166].

Despite the discovery and isolation of metal-dependent FDHs that appear to be O_2-tolerant enzymes showing stability and retain activity upon aerobic purification, the FDH-based reaction at the bis-MGD cofactor itself is not remarkably O_2-tolerant. With respect to isolating competently active FDH, a varying requirement for the inclusion or copurification of FDH in conjunction with sulfhydryl compounds such as 2-mercaptoethanol or DTT [62, 194, 196, 197] to activate an FDH aerobically purified without the requirement of pseudohalide inhibitors has been a curious item of observation, since this would appear to reflect an O_2-tolerant enzyme. Preliminarily, sulfhydryl compounds were thought to aid in enzyme stability, as well as to react with O_2 present [194, 196], although some FDHs that were aerobically purified in the absence of pseudohalide inhibitors (e.g. FDH1 from *M. extorquens* AM1) also do not require activation with sulfhydryl compounds [151]. Interestingly, FDH enzymatic activity from *D. gigas* or *D. desulfuricans* ATCC 27774 requires pretreatment with sulfhydryl compounds following aerobic purification without copurification with an inhibiting anion [165, 166, 170], an aspect that has been identified to be due to a surface-exposed disulfide bridge, as is described in more detail in Sections 20.11, 20.12, and 20.18 [96, 182].

20.18 FDH and Electron Bifurcation vs. O_2 Tolerance

Concomitant with the discovery that FBEB is an applicable component of anaerobic microbial metabolism [2], involvement of FDHs in coupled exergonic and endergonic oxidation–reduction reactions as a thermodynamically spontaneous process is consistent within the wide diversity of ways by which FDHs can be linked to other redox centers, including flavins operating within the FBEB scheme [233, 234]. Consistent with their identity, FDHs that engage in an identified FBEB reaction are present in the cytoplasm of archaea and bacteria, are characteristically flavoproteins, and simultaneously use two-electron donors or acceptors, with one being ferredoxin. For example, partial purification of cell-free extracts of *Clostridium acidurici* grown on purines containing FDH was shown to catalyze CO_2 reduction to formate in the presence of reduced ferredoxin and NAD^+, an activity that could not be observed with either $NADP^+$ or oxidized ferredoxin [235, 236]. More recently, this FDH was purified 130-fold, it was found to bind the flavin FMN in addition to the bis-MGD cofactor, and MALDI-TOF/MS was employed on the digested complex to identify the encoding genes [237]. Interestingly, in addition to the FDH-associated *fdh2* gene, three other genes *hylA*, *hylB*, and *hylC* were found as reminiscent to organisms with FMN-based [FeFe]-hydrogenases (e.g. *Thermatoga maritima*) [238], although *C. acidurici* lacks the [FeFe]-hydrogenase maturation genes *hydE, hydF,* and *hydG*. [237]. Despite the presence of FMN in *C. acidurici* FDH, it has been suggested that electron bifurcation can occur from the bis-MGD cofactor in a fashion similar to electron-bifurcating [FeFe]-hydrogenases and [NiFe]-hydrogenases at their respective active sites in a metal-based electron bifurcation (MBEB) reaction, via a common or a similarly conserved HydABC (recently renamed as BfuABC) protein complex that provides distinct electron transfer pathways by way of different multiple Fe–S clusters [237, 239, 240].

Recently, the periplasmic FDH isoenzyme #2 from *D. vulgaris* Hildenborough has been proposed to catalyze a MBEB reaction, in a mechanism by which the reduced bis-MGD cofactor (e.g. W(IV)) can concurrently reduce the spatially separated O_2 and cytochrome c, due to the lack of flavins and quinones bound [175]. This hypothesis is controversial, in that it has invoked challenges toward the definition of electron bifurcation [241], the predicted structure lacks an associated BfuABC protein complex similar to other electron-bifurcating enzymes and implicates the presence of a spatial junction of electron transfer [242] at the bis-MGD cofactor [175, 243] that awaits structural confirmation. However, because the FDH subunit is associated with a complex that has been proposed to be involved in aerobic respiration of *D. vulgaris* Hildenborough, the apparent formate oxidase and FDH activities observed might simply reflect a side reaction as a consequence of its adapted catalysis to function for O_2 respiration. It need not be forgotten that most FDHs that have been characterized are also exceptionally sensitive to O_2, likely due to a lack of a defense mechanism to protect the sulfido ligand at the active site in the presence of O_2, and that continued function of the enzyme likely requires repair of the active site sulfido ligand. By comparison, the characterized *D. gigas* FDH in addition to the FDH isoenzyme #1 from *D. vulgaris* Hildenborough but absent in FDH isoenzyme #2 is the presence of a surface-based cysteine-based disulfide bridge, which has been proposed to serve as a redox sensor, with reduction or substitution of the bond resulting in an optimally active enzyme [96, 182]. A similar O_2-tolerant behavior has been observed for *D. desulfuricans* ATC 27774 FDH [165, 244], as well as for *M. extorquens* AM1 FDH1 [151], although these characterized FDHs appear to lack such a disulfide bridge. Variants of FDH isoenzyme #1 from *D. vulgaris* Hildenborough lacking the disulfide bridge have been shown to have increased O_2 sensitivity in their isolation *in vitro* [182, 245], making them more similar in O_2 sensitivity to the rest of the metal-dependent FDHs characterized to date.

20.19 FDH in H_2-Dependent CO_2 Reduction (HDCR)

Concomitant with the discovery of enzymes that were able to utilize formate and H_2 as reductant sources as part of the FHL complex such as *E. coli*, it has been known that FDH and hydrogenases can exhibit coupling with one another, namely the disproportion of formate to produce CO_2 and H_2 [246, 247]. Considering the reversibility of enzymes capable of using one reductant to produce the other, FHL catalyzing an H_2-driven CO_2 reduction reaction, until recently was not investigated in detail, likely because this reaction is not physiologically favored [134, 135]. However, some organisms, such as the acetogenic bacterium *Acetobacterium woodii* are able to grow on H_2 and CO_2, which implies that an efficient carbon fixation and reduction reaction was employed by the organism [248]. Subsequently, a gene cluster as part of the Wood–Ljungdahl pathway of CO_2 fixation in *A. woodii* was found to include two genes encoding for a selenocysteine-independent and dependent FDH (*fdhF1* and *fdhF2*, respectively) in addition to genes associated with [FeFe] hydrogenase (*hydCEDBA*), with the H-cluster catalytic active site associated with the encoded *hydA1* gene [249], and with no other FDHs associated to *A. woodii*. This complex (named HDCR) was isolated and was shown to produce formate from H_2 and CO_2 and separately formate and H_2 oxidation, in a complex that was found to have Mo and Se present, reflecting the FdhF2 as present [250]. Similar to *C. pasteurianum* FDH, reduced ferredoxin could also reduce CO_2 [97] and could also be reduced by *A. woodii* CODH present [250].

HDCR complexes such as from *A. woodii* are remarkable in their production of formate with H_2 and CO_2 under whole-cell catalytic conditions [250]. While it was known that HDCR from *A. woodii* was found to form filamentous structures, the reason behind this was not well understood [251]. More recently, the HDCR from *Thermoanaerobacter kivui* was isolated and characterized by cryo-EM, which has provided insights regarding the filamental structure (Figure 20.1) [252]. It has been reported to have an exceptionally high CO_2 reductase and formate oxidation activity, to date surpassing the activity of known chemical and other biological catalysts [253, 254]. HDCR from *T. kivui* lacks participatory membrane subunits, but the HycB3 and HycB4 subunits with their associated polyferredoxin domains link the FDH-based FdhF subunit and the [FeFe]-hydrogenase H-cluster HydA2 subunit that create unit to form an extended chain of Fe–S clusters, that together forms an electron nanowire, and ultimately a filamental extended structure [251]. This arrangement in HDCR is distinct from the *E. coli* FHL-1 complex cryo-EM structure, in that multiple individual [FeFe]-hydrogenase and FDH module subunits are in close proximity to one another [252], as opposed to singular units of [NiFe]-hydrogenase and FDH that comprise the soluble arm of the membrane-bound complex [95]. The filamental unit observed in HDCR is important for enhancing the H_2-driven

CO_2 reduction reaction, as it was shown that the individual modules of HDCR expressed in the absence of HycB3, HycB4, and HydA2 were unable to generate formic acid from H_2 and CO_2, or H_2 from formic acid, but individual H_2 oxidation and formate oxidation activities could be observed [252, 253].

20.20 Conclusion

The primary goal of this chapter has been to provide a generalized overview regarding metal-dependent FDHs that have been discovered, and their interconnectedness toward iron–sulfur clusters in biological systems. While research toward metal-dependent FDHs has already spanned well over one century, new questions relating to their identification and characterization still persist in conjunction with our societal increase in anthropomorphic production of greenhouse gases, which provide inspiration toward designing efficient and sustainable catalytic systems to decrease the overall human carbon footprint on Earth. As has been demonstrated in this chapter, metal-dependent FDHs are iron–sulfur cluster molybdo(tungsto) enzymes that primarily help enable bacterial respiratory metabolism under both anaerobic and aerobic growth conditions, and often operate hand in hand with, or in place of [NiFe]- and [FeFe]-hydrogenases and/or other metalloenzymes to provide or store excess electrons under certain metabolic contexts. Moreover, the core active site of metal-dependent FDHs is a modular unit, and several examples in biology have shown that the active site module is typically found in conjunction with additional modular units, including but not exclusive to diaphorases, heterodisulfide reductases, hydrogenases, and cytochrome c oxidases. Similar to other molybdo- and tungstoenzymes characterized, the biosynthesis of the bis-MGD cofactor in FDH constitutes a largely conserved process that involves the presence of Fe–S clusters, with primary apparent differences in the apparent metal ion coordinated at the active site (as described in more detail in Chapter 7). As described herein, some organisms encode for multiple FDHs with differences in the active site ligand (Sec or Cys) and coordinating metal, each that putatively allows for overexpression under various experimental conditions. Outstanding questions relating to the regulation and specificity of isoforms toward Mo or W nevertheless persist; since trace Mo is important to induce the expression of the *moaABCDE* operon in *E. coli* that W alone is unable to perform.

A second conclusion that has been demonstrated in this chapter is that the past 30 years and its advent and development of structural (crystallographic and EM-based), spectroscopic, and bioinformatic methods have significantly impacted research toward FDH characterization, which has helped identify associated genes partaking in complex, fundamental metabolic processes in prokaryotes and archaea, and the dependence of FDH activity on Fe–S clusters. The interconnected relationship between metal-dependent FDHs and iron–sulfur clusters has made considerable headway since demonstration of ferredoxin-dependent formate oxidation to CO_2 in cell-free extracts of *Methanobacillus omelianskii* [255] and since the FDH from *P. oxalaticus* that was shown to be an Fe–S cluster flavoprotein without bound Mo or W [256]. Regardless of whether bound at the active site (as the FS0 [4Fe–4S] cluster), as part of the electron transfer relay, or serving in an electron donor/acceptor capacity, Fe–S clusters drive the high catalytic activity of the bis-MGD cofactor by absolving the need of a cosubstrate to serve as an electron donor or acceptor, as is observed for metal-independent FDHs. The direct connection of an electron transfer chain to the catalytic active site simplifies catalysis, in that the electrons generated or required for the catalytic reaction do not require coordination of a second substrate. This strategy has been exploited in biology by the multiple examples of modular enzymatic supercomplexes in biology coordinating FDH-based modules linked together by Fe–S clusters, and speaks toward the essential and tailored pairing between the bis-MGD cofactor and Fe–S clusters described in this chapter. Consistent with their diversity in bacteria and archaea, the fundamental principles of how metal-dependent FDHs function in conjunction with Fe–S clusters can serve as a foundation for identifying new roles that await discovery and can serve as inspirational models to improve CO_2 reduction catalysts of interest.

Acknowledgments

This work was funded by the DFG (German Research Foundation) under Germany's Excellence Strategy - EXC 2008-390540038 - UniSysCat.

References

1 Loach, P.A. (2018). Oxidation-reduction potentials, absorbance bands, and molar absorbance of compounds used in biochemical studies. In: *Handbook of Biochemistry and Molecular Biology*, 5e (ed. R.L. Lundblad and F.M. Macdonald), 576–582. Boca Raton, FL: CRC Press.

2 Thauer, R.K., Kaster, A.-K., Seedorf, H. et al. (2008). Methanogenic archaea: ecologically relevant differences in energy conservation. *Nat. Rev. Microbiol.* 6 (8): 579–591.

3 Appel, A.M., Bercaw, J.E., Bocarsly, A.B. et al. (2013). Frontiers, opportunities, and challenges in biochemical and chemical catalysis of CO_2 fixation. *Chem. Rev.* 113 (8): 6621–6658.

4 Thauer, R.K., Fuchs, G., and Jungermann, K. (1977). Role of iron-sulfur proteins in formate metabolism. In: *Iron-Sulfur Proteins: Structure and Metabolic Mechanisms*, vol. 3 (ed. W. Lovenberg), 121–156. New York: Academic Press.

5 Adams, M.W.W. and Mortenson, L.E. (1985). Mo reductases: nitrate reductase and formate dehydrogenase. In: *Molybdenum Enzymes. Metal Ions in Biology*, vol. 7 (ed. T.G. Spiro), 519–593. John Wiley & Sons.

6 Hartmann, T., Schwanhold, N., and Leimkühler, S. (2015). Assembly and catalysis of molybdenum or tungsten-containing formate dehydrogenases from bacteria. *Biochim. Biophys. Acta* 1854 (9): 1090–1100.

7 Maia, L.B., Moura, J.J.G., and Moura, I. (2015). Molybdenum and tungsten-dependent formate dehydrogenases. *J. Biol. Inorg. Chem.* 20 (2): 287–309.

8 Maia, L.B., Moura, I., and Moura, J.J.G. (2017). Molybdenum and tungsten-containing formate dehydrogenases: Aiming to inspire a catalyst for carbon dioxide utilization. *Inorg. Chim. Acta* 455: 350–363.

9 Niks, D. and Hille, R. (2019). Molybdenum- and tungsten-containing formate dehydrogenases and formylmethanofuran dehydrogenases: structure, mechanism, and cofactor insertion. *Protein Sci.* 28 (1): 111–122.

10 Maia, L.B., Moura, I., and Moura, J.J.G. (2021). Carbon dioxide utilisation—the formate route. In: *Enzymes for Solving Humankind's Problems: Natural and Artificial Systems in Health, Agriculture, Environment and Energy* (ed. J.J.G. Moura, I. Moura, and L.B. Maia), 29–81. Cham, Switzerland: Springer International Publishing.

11 Stripp, S.T., Duffus, B.R., Fourmond, V. et al. (2022). Second and outer coordination sphere effects in nitrogenase, hydrogenase, formate dehydrogenase, and CO dehydrogenase. *Chem. Rev.* 122 (14): 11900–11973.

12 Calzadiaz-Ramirez, L. and Meyer, A.S. (2022). Formate dehydrogenases for CO_2 utilization. *Curr. Opin. Biotechnol.* 73: 95–100.

13 Hietala, J., Vuori, A., Johnsson, P. et al. (2016). Formic acid. In: *Ullmann's Encyclopedia of Industrial Chemistry*, 1–22. Weinheim: Wiley-VCH Verlag.

14 Wray, J. (1670). Concerning some un-common observations and experiments made with an acid Juyce to be found in ants. *Philos. Trans. R. Soc. Lond.* 5: 2063–2069.

15 Malorny, G. (1969). Stoffwecheselversuche mit Natrium-formiat und Ameisensäure beim Menschen. *Z. Ernahrungswiss.* 9 (4): 340–348.

16 Nicholls, P. (1975). Formate as an inhibitor of cytochrome c oxidase. *Biochem. Biophys. Res. Commun.* 67 (2): 610–616.

17 Mayer, H. (1886). Untersuchungen über eine toxische Wirkung der niederen Fettsäuren. *Arch. Exp. Pathol. Pharmakol.* 21: 119–137.

18 von Oettingen, W.F. (1959). The aliphatic acids and their esters–toxicity and potential dangers: the saturated monobasic aliphatic acids and their esters. *A.M.A. Arch. Ind. Health* 20: 517–531.

19 Savolainen, H. and Zitting, A. (1980). Glial cell effects of subacute formic acid vapour exposure. *Acta Pharmacol. Toxicol.* 47 (3): 239–240.

20 Liesivuori, J., Kosma, V.-M., Naukkarinen, A., and Savolainen, H. (1987). Kinetics and toxic effects of repeated intravenous dosage of formic acid in rabbits. *Br. J. Exp. Pathol.* 68: 853–861.

21 Smith, S.R., Smith, S.J.M., Buckley, B.M., and Savolainen, H. (1982). Lactate and formate in methanol poisoning. *Lancet* 319 (8271): 561–562.

22 Fuchs, G. (2011). Alternative pathways of carbon dioxide fixation: insights into the early evolution of life? *Ann. Rev. Microbiol.* 65: 631–658.

23 Ragsdale, S. (1997). The Eastern and Western branches of the Wood/Ljungdahl pathway: how the East and West were won. *Biofactors* 6 (1): 3–11.

24 Drake, H.L., Küsel, K., and Matthies, C. (2002). Ecological consequences of the phylogenetic and physiological diversities of acetogens. *Antonie Van Leeuwenhoek* 81 (1): 203–213.

25 Bender, G., Pierce, E., Hill, J.A. et al. (2011). Metal centers in the anaerobic microbial metabolism of CO and CO_2. *Metallomics* 3 (8): 797–815.

26 Blanchard, J.S. and Cleland, W.W. (1980). Kinetic and chemical mechanisms of yeast formate dehydrogenase. *Biochemistry* 19 (15): 3543–3550.

27 Popov, V.N. and Lamzin, V.S. (1994). NAD^+-dependent formate dehydrogenase. *Biochem. J.* 301: 625–643.

28 Tishkov, V.I. and Popov, V.O. (2004). Catalytic mechanism and application of formate dehydrogenase. *Biochemistry (Mosc.)* 69 (11): 1252–1267.

29 Alekseeva, A.A., Savin, S.S., and Tishkov, V.I. (2011). NAD^+-dependent formate dehydrogenase from plants. *Acta Nat.* 3 (4): 38–54.

30 Alpdagtas, S. and Binay, B. (2021). $NADP^+$-dependent formate dehydrogenase: a review. *Biocatal. Biotransformation* 39 (4): 260–268.

31 Alpdagtas, S., Turunen, O., Valjakka, J., and Binay, B. (2022). The challenges of using NAD^+-dependent formate dehydrogenases for CO_2 conversion. *Crit. Rev. Biotechnol.* 42 (6): 953–972.

32 Tishkov, V.I., Pometun, A.A., and Savin, S.S. (2023). Formate dehydrogenase: from NAD(P)H regeneration to targeting pathogen biofilms, composing highly efficient hybrid biocatalysts and atmospheric CO_2 fixation. *Mosc. Univ. Chem. Bull.* 78 (4): 151–169.

33 Schütte, H., Flossdorf, J., Sahm, H., and Kula, M.R. (1976). Purification and properties of formaldehyde dehydrogenase and formate dehydrogenase from *Candida boidinii*. *Eur. J. Biochem.* 62 (1): 151–160.

34 Lu, J.X., Zhang, Y.H., Sun, D.F. et al. (2016). The development of leucine dehydrogenase and formate dehydrogenase bifunctional enzyme cascade improves the biosynthesis of L-*tert*-leucine. *Appl. Biochem. Biotechnol.* 180 (6): 1180–1195.

35 Miyaji, A. and Amao, Y. (2020). Artificial co-enzyme based on carbamoyl-modified viologen derivative cation radical for formate dehydrogenase in the catalytic CO_2 reduction to formate. *New J. Chem.* 44 (43): 18803–18812.

36 Miyaji, A. and Amao, Y. (2020). How does methylviologen cation radical supply two electrons to the formate dehydrogenase in the catalytic reduction process of CO_2 to formate? *Phys. Chem. Chem. Phys.* 22 (33): 18595–18605.

37 Sato, R. and Amao, Y. (2020). Can formate dehydrogenase from *Candida boidinii* catalytically reduce carbon dioxide, bicarbonate, or carbonate to formate? *New J. Chem.* 44 (28): 11922–11926.

38 Sato, R. and Amao, Y. (2022). No competitive inhibition of bicarbonate or carbonate for formate dehydrogenase from *Candida boidinii*-catalyzed CO_2 reduction. *New J. Chem.* 46 (33): 15820–15830.

39 Sato, R. and Amao, Y. (2022). Curious effect of isotope-labelled substrate/co-enzyme on catalytic activity of CO_2 reduction by formate dehydrogenase from *Candida boidinii*. *Bull. Chem. Soc. Jpn.* 95 (4): 556–558.

40 Bang, J. and Lee, S.Y. (2018). Assimilation of formic acid and CO_2 by engineered *Escherichia coli* equipped with reconstructed one-carbon assimilation pathways. *Proc. Natl. Acad. Sci. USA* 115 (40): E9271–E9279.

41 Kim, S., Lindner, S.N., Aslan, S. et al. (2020). Growth of *E. coli* on formate and methanol via the reductive glycine pathway. *Nat. Chem. Biol.* 16 (5): 538–545.

42 Jiang, W., Villamor, D.H., Peng, H.D. et al. (2021). Metabolic engineering strategies to enable microbial utilization of C1 feedstocks. *Nat. Chem. Biol.* 17 (8): 845–855.

43 Bierbaumer, S., Nattermann, M., Schulz, L. et al. (2023). Enzymatic conversion of CO_2: from natural to artificial utilization. *Chem. Rev.* 123 (9): 5702–5754.

44 Bandaria, J.N., Cheatum, C.M., and Kohen, A. (2009). Examination of enzymatic H-tunneling through kinetics and dynamics. *J. Am. Chem. Soc.* 131 (29): 10151–10155.

45 Johnson, J.L., Bastian, N.R., and Rajagopalan, K.V. (1990). Molybdopterin guanine dinucleotide: a modified form of molybdopterin identified in the molybdenum cofactor of dimethyl sulfoxide reductase from *Rhodobacter sphaeroides* forma specialis *denitrificans*. *Proc. Natl. Acad. Sci. USA* 87 (8): 3190–3194.

46 Johnson, J.L., Bastian, N.R., Schauer, N.L. et al. (1991). Identification of molybdopterin guanine dinucleotide in formate dehydrogenase from *Methanobacterium formicicum*. *FEMS Microbiol. Lett.* 77 (2–3): 213–216.

47 Waring, W.S. and Werkman, C.H. (1944). Iron deficiency in bacterial metabolism. *Arch. Biochem.* 4: 75–87.

48 Pinsent, J. (1954). The need for selenite and molybdate in the formation of formic dehydrogenase by members of the *Coli-aerogenes* group of bacteria. *Biochem. J.* 57 (1): 10–16.

49 Sawers, G. (1994). The hydrogenases and formate dehydrogenases of *Escherichia coli*. *Antonie Van Leeuwenhoek* 66 (1–3): 57–88.

50 Gest, H. and Peck, H.D. (1955). A study of the hydrogenlyase reaction with systems derived from normal and anaerogenic *Coli-aerogenes* bacteria. *J. Bacteriol.* 70 (3): 326–334.

51 Peck, H.D. and Gest, H. (1957). Formic dehydrogenase and the hydrogenlyase enzyme complex in *Coli-aerogenes* bacteria. *J. Bacteriol.* 73 (6): 706–721.

52 Giordano, G., Medani, C.L., Mandrand-Berthelot, M.A., and Boxer, D.H. (1983). Formate dehydrogenases from *Escherichia coli*. *FEMS Microbiol. Lett.* 17 (1–3): 171–177.

53 Sawers, G., Heider, J., Zehelein, E., and Böck, A. (1991). Expression and operon structure of the *sel* genes of *Escherichia coli* and identification of a third selenium-containing formate dehydrogenase isoenzyme. *J. Bacteriol.* 173 (16): 4983–4993.

54 Zinoni, F., Birkmann, A., Stadtman, T.C., and Böck, A. (1986). Nucleotide sequence and expression of the selenocysteine-containing polypeptide of formate dehydrogenase (formate-hydrogen-lyase-linked) from *Escherichia coli*. *Proc. Natl. Acad. Sci. USA* 83 (13): 4650–4654.

55 Berg, B.L. and Stewart, V. (1990). Structural genes for nitrate-inducible formate dehydrogenase in *Escherichia coli* K-12. *Genetics* 125 (4): 691–702.

56 Johnson, J.L., Hainline, B.E., and Rajagopalan, K.V. (1980). Characterization of the molybdenum cofactor of sulfite oxidase, xanthine oxidase, and nitrate reductase: identification of a pteridine as a structural component. *J. Biol. Chem.* 255 (5): 1783–1786.

57 Johnson, J.L., Hainline, B.E., Rajagopalan, K.V., and Arison, B.H. (1984). The pterin component of the molybdenum cofactor - structural characterization of two fluorescent derivatives. *J. Biol. Chem.* 259 (9): 5414–5422.

58 Johnson, J.L., Rajagopalan, K.V., and Meyer, O. (1990). Isolation and characterization of a second molybdopterin dinucleotide: molybdopterin cytosine dinucleotide. *Arch. Biochem. Biophys.* 283 (2): 542–545.

59 Schauer, N.L. and Ferry, J.G. (1982). Properties of formate dehydrogenase in *Methanobacterium formicicum*. *J. Bacteriol.* 150 (1): 1–7.

60 Schauer, N.L. and Ferry, J.G. (1986). Composition of the coenzyme F_{420}-dependent formate dehydrogenase from *Methanobacterium formicicum*. *J. Bacteriol.* 165 (2): 405–411.

61 May, H.D., Schauer, N.L., and Ferry, J.G. (1986). Molybdopterin cofactor from *Methanobacterium formicicum* formate dehydrogenase. *J. Bacteriol.* 166 (2): 500–504.

62 Yamamoto, I., Saiki, T., Liu, S.M., and Ljungdahl, L.G. (1983). Purification and properties of NADP-dependent formate dehydrogenase from *Clostridium thermoaceticum*, a tungsten-selenium-iron protein. *J. Biol. Chem.* 258 (3): 1826–1832.

63 Liu, C.L. and Mortenson, L.E. (1984). Formate dehydrogenase of *Clostridium pasteurianum*. *J. Bacteriol.* 159 (1): 375–380.

64 Karrasch, M., Börner, G., Enssle, M., and Thauer, R.K. (1990). The molybdoenzyme formylmethanofuran dehydrogenase from *Methanosarcina barkeri* contains a pterin cofactor. *Eur. J. Biochem.* 194 (2): 367–372.

65 Axley, M.J., Grahame, D.A., and Stadtman, T.C. (1990). *Escherichia coli* formate-hydrogen lyase - purification and properties of the selenium-dependent formate dehydrogenase component. *J. Biol. Chem.* 265 (30): 18213–18218.

66 Karrasch, M., Börner, G., and Thauer, R.K. (1990). The molybdenum cofactor of formylmethanofuran dehydrogenase from *Methanosarcina barkeri* is a molybdopterin guanine dinucleotide. *FEBS Lett.* 274 (1–2): 48–52.

67 Ruschig, U., Müller, U., Willnow, P., and Höpner, T. (1976). CO_2 reduction to formate by NADH catalyzed by formate dehydrogenase from *Pseudomonas oxalaticus*. *Eur. J. Biochem.* 70 (2): 325–330.

68 Müller, U., Willnow, P., Ruschig, U., and Höpner, T. (1978). Formate dehydrogenase from *Pseudomonas oxalaticus*. *Eur. J. Biochem.* 83 (2): 485–498.

69 Höpner, T., Ruschig, U., Müller, U., and Willnow, P. (1982). Formate dehydrogenase from *Pseudomonas oxalaticus*. *Methods Enzymol.* 89: 531–537.

70 Schindelin, H., Kisker, C., Hilton, J. et al. (1996). Crystal structure of DMSO reductase: redox-linked changes in molybdopterin coordination. *Science* 272 (5268): 1615–1621.

71 Gladyshev, V.N., Boyington, J.C., Khangulov, S.V. et al. (1996). Characterization of crystalline formate dehydrogenase H from *Escherichia coli* - stabilization, EPR spectroscopy, and preliminary crystallographic analysis. *J. Biol. Chem.* 271 (14): 8095–8100.

72 Boyington, J.C., Gladyshev, V.N., Khangulov, S.V. et al. (1997). Crystal structure of formate dehydrogenase H: catalysis involving Mo, molybdopterin, selenocysteine, and an Fe_4S_4 cluster. *Science* 275 (5304): 1305–1308.

73 George, G.N., Colangelo, C.M., Dong, J. et al. (1998). X-Ray absorption spectroscopy of the molybdenum site of *Escherichia coli* formate dehydrogenase. *J. Am. Chem. Soc.* 120 (6): 1267–1273.

74 George, G.N., Costa, C., Moura, J.J.G., and Moura, I. (1999). Observation of ligand-based redox chemistry at the active site of a molybdenum enzyme. *J. Am. Chem. Soc.* 121 (11): 2625–2626.

75 Raaijmakers, H.C.A. and Romao, M.J. (2006). Formate-reduced *E. coli* formate dehydrogenase H: the reinterpretation of the crystal structure suggests a new reaction mechanism. *J. Biol. Inorg. Chem.* 11 (7): 849–854.

76 Williams, B.R., Fu, Y., Yap, G.P.A., and Burgmayer, S.J.N. (2012). Structure and reversible pyran formation in molybdenum pyranopterin dithiolene models of the molybdenum cofactor. *J. Am. Chem. Soc.* 134 (48): 19584–19587.

77 Enemark, J.H. (2022). $\{Moco\}^n$, (n=0-8): a general formalism for describing the highly covalent molybdenum cofactor of sulfite oxidase and related Mo enzymes. *J. Inorg. Biochem.* 231: 111801.

78 Chan, M.K., Mukund, S., Kletzin, A. et al. (1995). Structure of a hyperthermophilic tungstopterin enzyme, aldehyde ferredoxin oxidoreductase. *Science* 267 (5203): 1463–1469.

79 Rothery, R.A., Workun, G.J., and Weiner, J.H. (2008). The prokaryotic complex iron-sulfur molybdoenzyme family. *Biochim. Biophys. Acta* 1778 (9): 1897–1929.

80 Blasco, F., Guigliarelli, B., Magalon, A. et al. (2001). The coordination and function of the redox centres of the membrane-bound nitrate reductases. *Cell. Mol. Life Sci.* 58 (2): 179–193.

81 Trieber, C.A., Rothery, R.A., and Weiner, J.H. (1996). Engineering a novel iron-sulfur cluster into the catalytic subunit of *Escherichia coli* dimethyl-sulfoxide reductase. *J. Biol. Chem.* 271 (9): 4620–4626.

82 Rothery, R.A., Bertero, M.G., Cammack, R. et al. (2004). The catalytic subunit of *Escherichia coli* nitrate reductase a contains a novel [4Fe-4S] cluster with a high-spin ground state. *Biochemistry* 43 (18): 5324–5333.

83 Lanciano, P., Savoyant, A., Grimaldi, S. et al. (2007). New method for the spin quantitation of $[4Fe\text{-}4S]^+$ clusters with $S = 3/2$. Application to the FS0 center of the NarGHI nitrate reductase from *Escherichia coli*. *J. Phys. Chem. B* 111 (48): 13632–13637.

84 Sazanov, L.A. and Hinchliffe, P. (2006). Structure of the hydrophilic domain of respiratory complex I from *Thermus thermophilus*. *Science* 311 (5766): 1430–1436.

85 Barber, M.J., May, H.D., and Ferry, J.G. (1986). Inactivation of formate dehydrogenase from *Methanobacterium formicicum* by cyanide. *Biochemistry* 25 (25): 8150–8155.

86 Duffus, B.R., Gauglitz, M., Teutloff, C., and Leimkühler, S. (2024). Redox potentials elucidate the electron transfer pathway of NAD^+-dependent formate dehydrogenases. *J. Inorg. Biochem.* 253: 112487.

87 Zeamari, K., Gerbaud, G., Grosse, S. et al. (2019). Tuning the redox properties of a [4Fe-4S] center to modulate the activity of Mo-bisPGD periplasmic nitrate reductase. *Biochim. Biophys. Acta* 1860 (5): 402–413.

88 Hettmann, T., Siddiqui, R.A., van Langen, J. et al. (2003). Mutagenesis study on the role of a lysine residue highly conserved in formate dehydrogenases and periplasmic nitrate reductases. *Biochem. Biophys. Res. Commun.* 310 (1): 40–47.

89 Rothery, R.A., Bertero, M.G., Spreter, T. et al. (2010). Protein crystallography reveals a role for the FS0 cluster of *Escherichia coli* nitrate reductase A (NarGHI) in enzyme maturation. *J. Biol. Chem.* 285 (12): 8801–8807.

90 Magalon, A., Asso, M., Guigliarelli, B. et al. (1998). Molybdenum cofactor properties and [Fe-S] cluster coordination in *Escherichia coli* nitrate reductase A: investigation by site-directed mutagenesis of the conserved His-50 residue in the NarG subunit. *Biochemistry* 37 (20): 7363–7370.

91 Kröger, A., Winkler, E., Innerhofer, A. et al. (1979). The formate dehydrogenase involved in electron transport from formate to fumarate in *Vibrio succinogenes*. *Eur. J. Biochem.* 94 (2): 465–475.

92 Fukuyama, T. and Ordal, E.J. (1965). Induced biosynthesis of formic hydrogenlyase in iron-deficient cells of *Escherichia coli*. *J. Bacteriol.* 90 (3): 673–680.

93 Andreesen, J.R., Schaupp, A., Neurauter, C. et al. (1973). Fermentation of glucose, fructose, and xylose by *Clostridium thermoaceticum*: effect of metals on growth yield, enzymes, and the synthesis of acetate from CO_2. *J. Bacteriol.* 114 (2): 743–751.

94 Jormakka, M., Törnroth, S., Byrne, B., and Iwata, S. (2002). Molecular basis of proton motive force generation: structure of formate dehydrogenase-N. *Science* 295 (5561): 1863–1868.

95 Steinhilper, R., Höff, G., Heider, J., and Murphy, B.J. (2022). Structure of the membrane-bound formate hydrogenlyase complex from *Escherichia coli*. *Nat. Commun.* 13 (1): 5395.

96 Raaijmakers, H., Macieira, S., Dias, J.M. et al. (2002). Gene sequence and the 1.8 Å crystal structure of the tungsten-containing formate dehydrogenase from *Desulfolvibrio gigas*. *Structure* 10 (9): 1261–1272.

97 Jungermann, K., Kirchniawy, H., and Thauer, R.K. (1970). Ferredoxin dependent CO_2 reduction to formate in *Clostridium pasteurianum*. *Biochem. Biophys. Res. Commun.* 41 (3): 682–689.

98 Thauer, R.K., Fuchs, G., Schnitker, U., and Jungermann, K. (1973). CO_2 reductase from *Clostridium pasteurianum*: molybdenum dependence of synthesis and inactivation by cyanide. *FEBS Lett.* 38 (1): 45–48.

99 Thauer, R.K., Fuchs, G., and Jungermann, K. (1974). Reduced ferredoxin: CO_2 oxidoreductase from *Clostridium pasteurianum*: its role in formate metabolism. *J. Bacteriol.* 118 (2): 758–760.

100 Thauer, R.K., Fuchs, G., and Käufer, B. (1975). Reduced ferredoxin-CO_2 oxidoreductase from *Clostridium pasteurianum*: effect of ligands to transition-metals on activity and stability of enzyme. *Hoppe Seylers Z. Physiol. Chem.* 356 (6): 653–662.

101 Thauer, R.K., Käufer, B., and Fuchs, G. (1975). Active species of 'CO_2' utilized by reduced ferredoxin:CO_2 oxidoreductase from *Clostridium pasteurianum*. *Eur. J. Biochem.* 55 (1): 111–117.

102 Scherer, P.A. and Thauer, R.K. (1978). Purification and properties of reduced ferredoxin-CO_2 oxidoreductase from *Clostridium pasteurianum*, a molybdenum iron-sulfur-protein. *Eur. J. Biochem.* 85 (1): 125–135.

103 Cramer, S.P., Liu, C.L., Mortenson, L.E. et al. (1985). Formate dehydrogenase molybdenum and tungsten sites– observation by EXAFS of structural differences. *J. Inorg. Biochem.* 23 (2): 119–124.

104 Prince, R.C., Liu, C.L., Morgan, T.V., and Mortenson, L.E. (1985). Formate dehydrogenase from *Clostridium pasteurianum*: electron paramagnetic resonance spectroscopy of the redox active centers. *FEBS Lett.* 189 (2): 263–266.

105 Yang, J.I., Lee, S.H., Ryu, J.Y. et al. (2022). A novel NADP-dependent formate dehydrogenase from the hyperthermophilic archaeon *Thermococcus onnurineus* NA1. *Front. Microbiol.* 13: 844735.

106 Stewart, V. (1988). Nitrate respiration in relation to facultative metabolism in enterobacteria. *Microbiol. Rev.* 52 (2): 190–232.

107 Unden, G., Becker, S., Bongaerts, J. et al. (1995). O_2-sensing and O_2-dependent gene regulation in facultatively anaerobic bacteria. *Arch. Microbiol.* 164 (2): 81–90.

108 McCrindle, S.L., Kappler, U., and McEwan, A.G. (2005). Microbial dimethylsulfoxide and trimethylamine-N-oxide respiration. *Adv. Microb. Physiol.* 50: 147–201.

109 Heider, J. and Bock, A. (1993). Selenium metabolism in microorganisms. *Adv. Microb. Physiol.* 35: 71–109.

110 Leonhartsberger, S., Korsa, I., and Böck, A. (2002). The molecular biology of formate metabolism in enterobacteria. *J. Mol. Microbiol. Biotechnol.* 4 (3): 269–276.

111 Sawers, R.G. (2005). Formate and its role in hydrogen production in *Escherichia coli*. *Biochem. Soc. Trans.* 33: 42–46.

112 Sawers, R.G. (1985). Membrane-bound hydrogenase isoenzymes from *Escherichia coli*. PhD thesis. University of Dundee.

113 Soboh, B., Pinske, C., Kuhns, M. et al. (2011). The respiratory molybdo-selenoprotein formate dehydrogenases of *Escherichia coli* have hydrogen: benzyl viologen oxidoreductase activity. *BMC Microbiol.* 11: 173.

114 Peters, K. and Sargent, F. (2023). Formate hydrogenlyase, formic acid translocation and hydrogen production: dynamic membrane biology during fermentation. *Biochim. Biophys. Acta* 1864 (1): 148919.

115 Suppmann, B. and Sawers, G. (1994). Isolation and characterization of hypophosphite-resistant mutants of *Escherichia coli*: identification of the FocA protein, encoded by the pfl operon, as a putative formate transporter. *Mol. Microbiol.* 11 (5): 965–982.

116 Wang, Y., Huang, Y., Wang, J. et al. (2009). Structure of the formate transporter FocA reveals a pentameric aquaporin-like channel. *Nature* 462 (7272): 467–472.

117 Waight, A.B., Love, J., and Wang, D.-N. (2010). Structure and mechanism of a pentameric formate channel. *Nat. Struct. Mol. Biol.* 17 (1): 31–37.

118 Beyer, L., Doberenz, C., Falke, D. et al. (2013). Coordination of FocA and pyruvate formate-lyase synthesis in *Escherichia coli* demonstrates preferential translocation of formate over other mixed-acid fermentation products. *J. Bacteriol.* 195 (7): 1428–1435.

119 Doberenz, C., Zorn, M., Falke, D. et al. (2014). Pyruvate formate-lyase interacts directly with the formate channel FocA to regulate formate translocation. *J. Mol. Biol.* 426 (15): 2827–2839.

120 Sawers, R.G. and Clark, D.P. (2004). Fermentative pyruvate and acetyl-coenzyme A metabolism. *EcoSal Plus* 1 (1): https://doi.org/10.1128/ecosalplus.3.5.3.

121 Rossmann, R., Sawers, G., and Böck, A. (1991). Mechanism of regulation of the formate-hydrogenlyase pathway by oxygen, nitrate, and pH: definition of the formate regulon. *Mol. Microbiol.* 5 (11): 2807–2814.

122 Redwood, M.D., Mikheenko, I.P., Sargent, F., and Macaskie, L.E. (2008). Dissecting the roles of *Escherichia coli* hydrogenases in biohydrogen production. *FEMS Microbiol. Lett.* 278 (1): 48–55.

123 Bassegoda, A., Madden, C., Wakerley, D.W. et al. (2014). Reversible interconversion of CO_2 and formate by a molybdenum-containing formate dehydrogenase. *J. Am. Chem. Soc.* 136 (44): 15473–15476.

124 Li, F.L., Scheller, S., and Lienemann, M. (2023). Comparative analysis of CO_2 reduction by soluble *Escherichia coli* formate dehydrogenase H and its selenocysteine-to-cysteine substitution variant. *J. CO2 Util.* 77: 102608.

125 Axley, M.J., Böck, A., and Stadtman, T.C. (1991). Catalytic properties of an *Escherichia coli* formate dehydrogenase mutant in which sulfur replaces selenium. *Proc. Natl. Acad. Sci. USA* 88 (19): 8450–8454.

126 Axley, M.J. and Grahame, D.A. (1991). Kinetics for formate dehydrogenase of *Escherichia coli* formate-hydrogenlyase. *J. Biol. Chem.* 266 (21): 13731–13736.

127 Gladyshev, V.N., Khangulov, S.V., Axley, M.J., and Stadtman, T.C. (1994). Coordination of selenium to molybdenum in formate dehydrogenase H from *Escherichia coli. Proc. Natl. Acad. Sci. USA* 91 (16): 7708–7711.

128 Khangulov, S.V., Gladyshev, V.N., Dismukes, G.C., and Stadtman, T.C. (1998). Selenium-containing formate dehydrogenase H from *Escherichia coli*: a molybdopterin enzyme that catalyzes formate oxidation without oxygen transfer. *Biochemistry* 37 (10): 3518–3528.

129 Robinson, W.E., Bassegoda, A., Reisner, E., and Hirst, J. (2017). Oxidation-state-dependent binding properties of the active site in a Mo-containing formate dehydrogenase. *J. Am. Chem. Soc.* 139 (29): 9927–9936.

130 Robinson, W.E., Bassegoda, A., Blaza, J.N. et al. (2020). Understanding how the rate of C-H bond cleavage affects formate oxidation catalysis by a Mo-dependent formate dehydrogenase. *J. Am. Chem. Soc.* 142 (28): 12226–12236.

131 McDowall, J.S., Murphy, B.J., Haumann, M. et al. (2014). Bacterial formate hydrogenlyase complex. *Proc. Natl. Acad. Sci. USA* 111 (38): E3948-E3956.

132 Pakes, W.C.C. and Jollyman, W.H. (1901). The bacterial decomposition of formic acid into carbon dioxide and hydrogen. *J. Chem. Soc.* 79: 386–391.

133 Pinske, C. and Sargent, F. (2016). Exploring the directionality of *Escherichia coli* formate hydrogenlyase: a membrane-bound enzyme capable of fixing carbon dioxide to organic acid. *MicrobiologyOpen* 5 (5): 721–737.

134 Roger, M., Brown, F., Gabrielli, W., and Sargent, F. (2018). Efficient hydrogen-dependent carbon dioxide reduction by *Escherichia coli*. *Curr. Biol.* 28 (1): 140–145.

135 Roger, M., Reed, T.C.P., and Sargent, F. (2021). Harnessing *Escherichia coli* for bio-based production of formate under pressurized H_2 and CO_2 gases. *Appl. Environ. Microbiol.* 87 (21): e00299-21.

136 Bertero, M.G., Rothery, R.A., Palak, M. et al. (2003). Insights into the respiratory electron transfer pathway from the structure of nitrate reductase A. *Nat. Struct. Biol.* 10 (9): 681–687.

137 Abaibou, H., Pommier, J., Benoit, S. et al. (1995). Expression and characterization of the *Escherichia coli fdo* locus and a possible physiological role for aerobic formate dehydrogenase. *J. Bacteriol.* 177 (24): 7141–7149.

138 Pommier, J., Mandrand, M.A., Holt, S.E. et al. (1992). A second phenazine methosulfate-linked formate dehydrogenase isoenzyme in *Escherichia coli*. *Biochim. Biophys. Acta* 1107 (2): 305–313.

139 Iobbi-Nivol, C., Santini, C.L., Blasco, F., and Giordano, G. (1990). Purification and further characterization of the second nitrate reductase of *Escherichia coli* K12. *Eur. J. Biochem.* 188 (3): 679–687.

140 Zorn, M., Ihling, C.H., Golbik, R. et al. (2014). Mapping cell envelope and periplasm protein interactions of *Escherichia coli* respiratory formate dehydrogenases by chemical cross-linking and mass spectrometry. *J. Proteome Res.* 13 (12): 5524–5535.

141 Hartmann, T. and Leimkühler, S. (2013). The oxygen-tolerant and NAD^+-dependent formate dehydrogenase from *Rhodobacter capsulatus* is able to catalyze the reduction of CO_2 to formate. *FEBS J.* 280 (23): 6083–6096.

142 Friedebold, J. and Bowien, B. (1993). Physiological and biochemical-characterization of the soluble formate dehydrogenase, a molybdoenzyme from *Alcaligenes eutrophus*. *J. Bacteriol.* 175 (15): 4719–4728.

143 Friedebold, J., Mayer, F., Bill, E. et al. (1995). Structural and immunological studies on the soluble formate dehydrogenase from *Alcaligenes eutrophus*. *Biol. Chem. Hoppe Seyler* 376 (9): 561–568.

144 Oh, J.I. and Bowien, B. (1998). Structural analysis of the *fds* operon encoding the NAD^+-linked formate dehydrogenase of *Ralstonia eutropha*. *J. Biol. Chem.* 273 (41): 26349–26360.

145 Niks, D., Duvvuru, J., Escalona, M., and Hille, R. (2016). Spectroscopic and kinetic properties of the molybdenum-containing, NAD^+-dependent formate dehydrogenase from *Ralstonia eutropha*. *J. Biol. Chem.* 291 (3): 1162–1174.

146 Harmer, J.R., Hakopian, S., Niks, D. et al. (2023). Redox characterization of the complex molybdenum enzyme formate dehydrogenase from *Cupriavidus necator*. *J. Am. Chem. Soc.* 145 (47): 25850–25863.

147 Yoch, D.C. and Lindstrom, E.S. (1969). Nicotinamide adenine dinucleotide-dependent formate dehydrogenase from *Rhodopseudomonas palustris*. *Arch. Mikrobiol.* 67 (2): 182–188.

148 Jollie, D.R. and Lipscomb, J.D. (1990). Formate dehydrogenase from *Methylosinus trichosporium* Ob3b. *Methods Enzymol.* 188: 331–334.

149 Yoch, D.C., Chen, Y.P., and Hardin, M.G. (1990). Formate dehydrogenase from the methane oxidizer *Methylosinus trichosporium* Ob3b. *J. Bacteriol.* 172 (8): 4456–4463.

150 Jollie, D.R. and Lipscomb, J.D. (1991). Formate dehydrogenase from *Methylosinus trichosporium* Ob3b: purification and spectroscopic characterization of the cofactors. *J. Biol. Chem.* 266 (32): 21853–21863.

151 Laukel, M., Chistoserdova, L., Lidstrom, M.E., and Vorholt, J.A. (2003). The tungsten-containing formate dehydrogenase from *Methylobacterium extorquens* AM1: purification and properties. *Eur. J. Biochem.* 270 (2): 325–333.

152 Chistoserdova, L., Laukel, M., Portais, J.C. et al. (2004). Multiple formate dehydrogenase enzymes in the facultative methylotroph *Methylobacterium extorquens* AM1 are dispensable for growth on methanol. *J. Bacteriol.* 186 (1): 22–28.

153 Chistoserdova, L., Crowther, G.J., Vorholt, J.A. et al. (2007). Identification of a fourth formate dehydrogenase in *Methylobacterium extorquens* AM1 and confirmation of the essential role of formate oxidation in methylotrophy. *J. Bacteriol.* 189 (24): 9076–9081.

154 Schrapers, P., Hartmann, T., Kositzki, R. et al. (2015). Sulfido and cysteine ligation changes at the molybdenum cofactor during substrate conversion by formate dehydrogenase (FDH) from *Rhodobacter capsulatus*. *Inorg. Chem.* 54 (7): 3260–3271.

155 Duffus, B.R., Schrapers, P., Schuth, N. et al. (2020). Anion binding and oxidative modification at the molybdenum cofactor of formate dehydrogenase from *Rhodobacter capsulatus* studied by X-ray absorption spectroscopy. *Inorg. Chem.* 59 (1): 214–225.

156 Radon, C., Mittelstädt, G., Duffus, B.R. et al. (2020). Cryo-EM structures reveal intricate Fe-S cluster arrangement and charging in *Rhodobacter capsulatus* formate dehydrogenase. *Nat. Commun.* 11 (1): 1912.

157 Young, T., Niks, D., Hakopian, S. et al. (2020). Crystallographic and kinetic analyses of the FdsBG subcomplex of the cytosolic formate dehydrogenase FdsABG from *Cupriavidus necator*. *J. Biol. Chem.* 295 (19): 6570–6585.

158 Shomura, Y., Taketa, M., Nakashima, H. et al. (2017). Structural basis of the redox switches in the NAD^+-reducing soluble [NiFe]-hydrogenase. *Science* 357 (6354): 928–931.

159 Yoshikawa, T., Makino, F., Miyata, T. et al. (2022). Multiple electron transfer pathways of tungsten-containing formate dehydrogenase in direct electron transfer-type bioelectrocatalysis. *Chem. Commun.* 58 (45): 6478–6481.

160 Park, J., Heo, Y., Jeon, B.W. et al. (2024). Structure of recombinant formate dehydrogenase from *Methylobacterium extorquens* (MeFDH1). *Sci. Rep.* 14 (1): 3819.

161 Muyzer, G. and Stams, A.J.M. (2008). The ecology and biotechnology of sulphate-reducing bacteria. *Nat. Rev. Microbiol.* 6 (6): 441–454.

162 Pereira, I.A.C., Ramos, A.R., Grein, F. et al. (2011). A comparative genomic analysis of energy metabolism in sulfate reducing bacteria and archaea. *Front. Microbiol.* 2: 69.

163 Rabus, R., Hansen, T.A., and Widdel, F. (2013). Dissimilatory sulfate- and sulfur-reducing prokaryotes. In: *The Prokaryotes: Prokaryotic Physiology and Biochemistry* (ed. E. Rosenberg, E.F. DeLong, S. Lory, et al.), 309–404. Berlin, Heidelberg: Springer Berlin Heidelberg.

164 Yagi, T. (1969). Formate:Cytochrome oxidoreductase of *Desulfovibrio vulgaris*. *J. Biochem.* 66 (4): 473–478.

165 Costa, C., Teixeira, M., LeGall, J. et al. (1997). Formate dehydrogenase from *Desulfovibrio desulfuricans* ATCC 27774: isolation and spectroscopic characterization of the active sites (heme, iron-sulfur centers and molybdenum). *J. Biol. Inorg. Chem.* 2 (2): 198–208.

166 Rivas, M.G., Gonzalez, P.J., Brondino, C.D. et al. (2007). EPR characterization of the molybdenum(V) forms of formate dehydrogenase from *Desulfovibrio desulfuricans* ATCC 27774 upon formate reduction. *J. Inorg. Biochem.* 101 (11–12): 1617–1622.

167 Mourato, C., Martins, M., da Silva, S.M., and Pereira, I.A.C. (2017). A continuous system for biocatalytic hydrogenation of CO_2 to formate. *Bioresour. Technol.* 235: 149–156.

168 Mota, C.S., Valette, O., Gonzalez, P.J. et al. (2011). Effects of molybdate and tungstate on expression levels and biochemical characteristics of formate dehydrogenases produced by *Desulfovibrio alaskensis* NCIMB 13491. *J. Bacteriol.* 193 (12): 2917–2923.

169 Riederer-Henderson, M.A. and Peck, H.D. (1986). *In vitro* requirements for formate dehydrogenase activity from *Desulfovibrio*. *Can. J. Microbiol.* 32 (5): 425–429.

170 Riederer-Henderson, M.A. and Peck, H.D. (1986). Properties of formate dehydrogenase from *Desulfovibrio gigas*. *Can. J. Microbiol.* 32 (5): 430–435.

171 Almendra, M.J., Brondino, C.D., Gavel, O. et al. (1999). Purification and characterization of a tungsten-containing formate dehydrogenase from *Desulfovibrio gigas*. *Biochemistry* 38 (49): 16366–16372.

172 Sebban, C., Blanchard, L., Bruschi, M., and Guerlesquin, F. (1995). Purification and characterization of the formate dehydrogenase from *Desulfovibrio vulgaris* Hildenborough. *FEMS Microbiol. Lett.* 133 (1–2): 143–149.

173 da Silva, S.M., Pimentel, C., Valente, F.M.A. et al. (2011). Tungsten and molybdenum regulation of formate dehydrogenase expression in *Desulfovibrio vulgaris* Hildenborough. *J. Bacteriol.* 193 (12): 2909–2916.

174 Oliveira, A.R., Mota, C., Mourato, C. et al. (2020). Toward the mechanistic understanding of enzymatic CO_2 reduction. *ACS Catal.* 10 (6): 3844–3856.

175 Graham, J.E., Niks, D., Zane, G.M. et al. (2022). How a formate dehydrogenase responds to oxygen: unexpected O_2 insensitivity of an enzyme harboring tungstopterin, selenocysteine, and [4Fe-4S] clusters. *ACS Catal.* 12 (16): 10449–10471.

176 Odom, J.M. and Peck, H.D. (1981). Localization of dehydrogenases, reductases, and electron transfer components in the sulfate-reducing bacterium *Desulfovibrio gigas*. *J. Bacteriol.* 147 (1): 161–169.

177 Morais-Silva, F.O., Rezende, A.M., Pimentel, C. et al. (2014). Genome sequence of the model sulfate reducer *Desulfovibrio gigas*: a comparative analysis within the *Desulfovibrio* genus. *MicrobiologyOpen* 3 (4): 513–530.

178 Hensgens, C.M.H., Hagen, W.R., and Hansen, T.A. (1995). Purification and characterization of a benzylviologen-linked, tungsten-containing aldehyde oxidoreductase from *Desulfovibrio gigas*. *J. Bacteriol.* 177 (21): 6195–6200.

179 Raaijmakers, H., Teixeira, S., Dias, J.M. et al. (2001). Tungsten-containing formate dehydrogenase from *Desulfovibrio gigas*: metal identification and preliminary structural data by multi-wavelength crystallography. *J. Biol. Inorg. Chem.* 6 (4): 398–404.

180 Oliveira, A.R., Mota, C., Klymanska, K. et al. (2022). Spectroscopic and structural characterization of reduced *Desulfovibrio vulgaris* Hildenborough W-FdhAB reveals stable metal coordination during catalysis. *ACS Chem. Biol.* 17 (7): 1901–1909.

181 Vilela-Alves, G., Manuel, R.R., Oliveira, A.R. et al. (2023). Tracking W-formate dehydrogenase structural changes during catalysis and enzyme reoxidation. *Int. J. Mol. Sci.* 24: 476.

182 Oliveira, A.R., Mota, C., Vilela-Alves, G. et al. (2024). An allosteric redox switch involved in oxygen protection in a CO_2 reductase. *Nat. Chem. Biol.* 20 (1): 111–119.

183 Enoch, H.G. and Lester, R.L. (1972). Effects of molybdate, tungstate, and selenium compounds on formate dehydrogenase and other enzyme systems in *Escherichia coli*. *J. Bacteriol.* 110 (3): 1032–1040.

184 Ljungdahl, L.G. (1976). Tungsten, a biologically active metal. *Trends Biochem. Sci.* 1 (3): 63–65.

185 Kletzin, A. and Adams, M.W.W. (1996). Tungsten in biological systems. *FEMS Microbiol. Rev.* 18 (1): 5–63.

186 Johnson, M.K., Rees, D.C., and Adams, M.W.W. (1996). Tungstoenzymes. *Chem. Rev.* 96 (7): 2817–2839.

187 Siemann, S., Schneider, K., Oley, M., and Müller, A. (2003). Characterization of a tungsten-substituted nitrogenase isolated from *Rhodobacter capsulatus*. *Biochemistry* 42 (13): 3846–3857.

188 Gates, A.J., Hughes, R.O., Sharp, S.R. et al. (2003). Properties of the periplasmic nitrate reductases from *Paracoccus pantotrophus* and *Escherichia coli* after growth in tungsten-supplemented media. *FEMS Microbiol. Lett.* 220 (2): 261–269.

189 de Vries, S., Momcilovic, M., Strampraad, M.J.F. et al. (2010). Adaptation to a high-tungsten environment: *Pyrobaculum aerophilum* contains an active tungsten nitrate reductase. *Biochemistry* 49 (45): 9911–9921.

190 May, H.D., Patel, P.S., and Ferry, J.G. (1988). Effect of molybdenum and tungsten on synthesis and composition of formate dehydrogenase in *Methanobacterium formicicum*. *J. Bacteriol.* 170 (8): 3384–3389.

191 Lentz, K. and Wood, H.G. (1955). Synthesis of acetate from formate and carbon dioxide by *Clostridium thermoaceticum*. *J. Biol. Chem.* 215 (2): 645–654.

192 Ljungdahl, L., Irion, E., and Wood, H.G. (1965). Total synthesis of acetate from CO_2. I. Co-methylcobyric acid and Co-(methyl)-5-methoxybenzimidazolylcobamide as intermediates with *Clostridium thermoaceticum*. *Biochemistry* 4 (12): 2771–2780.

193 Ljungdahl, L. and Wood, H.G. (1965). Incorporation of C^{14} from carbon dioxide into sugar phosphates carboxylic acids and amino acids by *Clostridium thermoaceticum*. *J. Bacteriol.* 89 (4): 1055–1064.

194 Li, L.F., Ljungdahl, L., and Wood, H.G. (1966). Properties of nicotinamide adenine dinucleotide phosphate-dependent formate dehydrogenase from *Clostridium thermoaceticum*. *J. Bacteriol.* 92 (2): 405–412.

195 Thauer, R.K. (1972). CO_2-reduction to formate by NADPH: initial step in total synthesis of acetate from CO_2 in *Clostridium thermoaceticum*. *FEBS Lett.* 27 (1): 111–115.

196 Andreesen, J.R. and Ljungdahl, L.G. (1974). Nicotinamide adenine dinucleotide phosphate-dependent formate dehydrogenase from *Clostridium thermoaceticum*: purification and properties. *J. Bacteriol.* 120 (1): 6–14.

197 Ljungdahl, L.G. and Andreesen, J.R. (1975). Tungsten, a component of active formate dehydrogenase from *Clostridium thermoaceticum*. *FEBS Lett.* 54 (2): 279–282.

198 Andreesen, J.R. and Ljungdahl, L.G. (1973). Formate dehydrogenase of *Clostridium thermoaceticum*: incorporation of selenium-75, and the effects of selenite, molybdate, and tungstate on the enzyme. *J. Bacteriol.* 116 (2): 867–873.

199 Pierce, E., Xie, G., Barabote, R.D. et al. (2008). The complete genome sequence of *Moorella thermoacetica* (f. *Clostridium thermoaceticum*). *Environ. Microbiol.* 10 (10): 2550–2573.

200 Shima, S., Huang, G.F., Wagner, T., and Ermler, U. (2020). Structural basis of hydrogenotrophic methanogenesis. *Ann. Rev. Microbiol.* 74: 713–733.

201 Romesser, J.A. and Wolfe, R.S. (1982). CDR factor, a new coenzyme required for carbon dioxide reduction to methane by extracts of *Methanobacterium thermoautotrophicum*. *Zbl. Bakt. Mik. Hyg. I C* 3 (2): 271–276.

202 Börner, G., Karrasch, M., and Thauer, R.K. (1989). Formylmethanofuran dehydrogenase activity in cell extracts of *Methanobacterium thermoautotrophicum* and of *Methanosarcina barkeri*. *FEBS Lett.* 244 (1): 21–25.

203 Schönheit, P., Moll, J., and Thauer, R.K. (1979). Nickel, cobalt, and molybdenum requirement for growth of *Methanobacterium thermoautotrophicum*. *Arch. Microbiol.* 123 (1): 105–107.

204 Karrasch, M., Börner, G., Enssle, M., and Thauer, R.K. (1989). Formylmethanofuran dehydrogenase from methanogenic bacteria, a molybdoenzyme. *FEBS Lett.* 253 (1–2): 226–230.

205 Börner, G., Karrasch, M., and Thauer, R.K. (1991). Molybdopterin adenine-dinucleotide and molybdopterin hypoxanthine dinucleotide in formyl-methanofuran dehydrogenase from *Methanobacterium thermoautotrophicum* (Marburg). *FEBS Lett.* 290 (1–2): 31–34.

206 Wagner, T., Ermler, U., and Shima, S. (2016). The methanogenic CO_2 reducing-and-fixing enzyme is bifunctional and contains 46 [4Fe-4S] clusters. *Science* 354 (6308): 114–117.

207 Vorholt, J.A. and Thauer, R.K. (2002). Molybdenum and tungsten enzymes in C1 metabolism. Molybdenum and tungsten: their roles in biological processes. *Met. Ions Biol. Syst.* 39: 571–619.

208 Rech, S., Deppenmeier, U., and Gunsalus, R.P. (1995). Regulation of the molybdate transport operon, *modABCD*, of *Escherichia coli* in response to molybdate availability. *J. Bacteriol.* 177 (4): 1023–1029.

209 Bevers, L.E., Hagedoorn, P.-L., and Hagen, W.R. (2009). The bioinorganic chemistry of tungsten. *Coord. Chem. Rev.* 253 (3–4): 269–290.

210 Seelmann, C.S., Willistein, M., Heider, J., and Boll, M. (2020). Tungstoenzymes: occurrence, catalytic diversity and cofactor synthesis. *Inorganics* 8 (8): 44.

211 Costa, K.C. and Whitman, W.B. (2023). Model organisms to study methanogenesis, a uniquely archaeal metabolism. *J. Bacteriol.* 205 (8): e00115-23.

212 Costa, K.C., Wong, P.M., Wang, T.S. et al. (2010). Protein complexing in a methanogen suggests electron bifurcation and electron delivery from formate to heterodisulfide reductase. *Proc. Natl. Acad. Sci. USA* 107 (24): 11050–11055.

213 Milton, R.D., Ruth, J.C., Deutzmann, J.S., and Spormann, A.M. (2018). *Methanococcus maripaludis* employs three functional heterodisulfide reductase complexes for flavin-based electron bifurcation using hydrogen and formate. *Biochemistry* 57 (32): 4848–4857.

214 Abdul Halim, M.F., Day, L.A., and Costa, K.C. (2021). Formate dependent heterodisulfide reduction in a *Methanomicrobiales* archaeon. *Appl. Environ. Microbiol.* 87 (6): e02698-20.

215 Wagner, T., Koch, J., Ermler, U., and Shima, S. (2017). Methanogenic heterodisulfide reductase (HdrABC-MvhAGD) uses two noncubane [4Fe-4S] clusters for reduction. *Science* 357 (6352): 699–702.

216 Watanabe, T., Pfeil-Gardiner, O., Kahnt, J. et al. (2021). Three-megadalton complex of methanogenic electron-bifurcating and CO_2-fixing enzymes. *Science* 373 (6559): 1151–1155.

217 Nomura, S., Paczia, N., Kahnt, J., and Shima, S. (2024). Isolation of an H_2-dependent electron-bifurcating CO_2-reducing megacomplex with MvhB polyferredoxin from *Methanothermobacter marburgensis*. *FEBS J.* 291 (11): 2449–2460.

218 Stojanowic, A., Mander, G.J., Duin, E.C., and Hedderich, R. (2003). Physiological role of the F_{420}-non-reducing hydrogenase (Mvh) from *Methanothermobacter marburgensis*. *Arch. Microbiol.* 180 (3): 194–203.

219 Yan, Z., Wang, M.Y., and Ferry, J.G. (2017). A ferredoxin-and F_{420} H_2-dependent, electron-bifurcating, heterodisulfide reductase with homologs in the domains *Bacteria* and *Archaea*. *MBio* 8 (1): e02285-16.

220 Vitt, S., Ma, K., Warkentin, E. et al. (2014). The F_{420}-reducing [NiFe]-hydrogenase complex from *Methanothermobacter marburgensis*, the first X-ray structure of a group 3 family member. *J. Mol. Biol.* 426 (15): 2813–2826.

221 Schauer, N.L. and Ferry, J.G. (1983). FAD requirement for the reduction of coenzyme F_{420} by formate dehydrogenase from *Methanobacterium formicicum*. *J. Bacteriol.* 155 (2): 467–472.

222 Abdul Halim, M.F., Fonseca, D.R., Niehaus, T.D., and Costa, K.C. (2024). Functionally redundant formate dehydrogenases enable formate-dependent growth in *Methanococcus maripaludis*. *J. Biol. Chem.* 300 (1): 105550.

223 Wood, G.E., Haydock, A.K., and Leigh, J.A. (2003). Function and regulation of the formate dehydrogenase genes of the methanogenic archaeon *Methanococcus maripaludis*. *J. Bacteriol.* 185 (8): 2548–2554.

224 Lupa, B., Hendrickson, E.L., Leigh, J.A., and Whitman, W.B. (2008). Formate-dependent H_2 production by the mesophilic methanogen *Methanococcus maripaludis*. *Appl. Environ. Microbiol.* 74 (21): 6584–6590.

225 Costa, K.C., Lie, T.J., Xia, Q., and Leigh, J.A. (2013). VhuD facilitates electron flow from H_2 or formate to heterodisulfide reductase in *Methanococcus maripaludis*. *J. Bacteriol.* 195 (22): 5160–5165.

226 Imlay, J. (2006). Iron-sulphur clusters and the problem with oxygen. *Mol. Microbiol.* 59 (4): 1073–1082.

227 Poston, J.M., Stadtman, T.C., and Stadtman, E.R. (1971). An anaerobic laboratory. *Methods Enzymol.* 22: 49–54.

228 Ljungdahl, L.G. and Andreesen, J.R. (1978). Formate dehydrogenase, a selenium-tungsten enzyme from *Clostridium thermoaceticum*. *Methods Enzymol.* 53: 360–372.

229 Kumar, H., Khosraneh, M., Bandaru, S.S.M. et al. (2023). The mechanism of metal-containing formate dehydrogenases revisited: the formation of bicarbonate as product intermediate provides evidence for an oxygen atom transfer mechanism. *Molecules* 28 (4): 1537.

230 Leimkühler, S. (2023). Metal-containing formate dehydrogenases, a personal view. *Molecules* 28 (14): 5338.

231 Laun, K., Duffus, B.R., Wahlefeld, S. et al. (2022). Infrared spectroscopy elucidates the inhibitor binding sites in a metal-dependent formate dehydrogenase. *Chem. Eur. J.* 14: e202201091.

232 Jones, J.B. and Stadtman, T.C. (1981). Selenium-dependent and selenium-independent formate dehydrogenases of *Methanococcus vannielii*: separation of the two forms and characterization of the purified selenium-independent form. *J. Biol. Chem.* 256 (2): 656–663.

233 Buckel, W. and Thauer, R.K. (2018). Flavin-based electron bifurcation, a new mechanism of biological energy coupling. *Chem. Rev.* 118 (7): 3862–3886.

234 Buckel, W. and Thauer, R.K. (2018). Flavin-based electron bifurcation, ferredoxin, flavodoxin, and anaerobic respiration with protons (Ech) or NAD^+ (Rnf) as electron acceptors: a historical review. *Front. Microbiol.* 9: 401.

235 Kearny, J.J. and Sagers, R.D. (1972). Formate dehydrogenase from *Clostridium acidiurici*. *J. Bacteriol.* 109 (1): 152–161.

236 Thauer, R.K. (1973). CO_2 reduction to formate in *Clostridium acidi-urici*. *J. Bacteriol.* 114 (1): 443–444.

237 Wang, S., Huang, H., Kahnt, J., and Thauer, R.K. (2013). *Clostridium acidurici* electron-bifurcating formate dehydrogenase. *Appl. Environ. Microbiol.* 79 (19): 6176–6179.

238 Furlan, C., Chongdar, N., Gupta, P. et al. (2022). Structural insight on the mechanism of an electron-bifurcating [FeFe] hydrogenase. *elife* 11: e79361.

239 Peters, J.W., Beratan, D.N., Schut, G.J., and Adams, M.W.W. (2018). On the nature of organic and inorganic centers that bifurcate electrons, coupling exergonic and endergonic oxidation-reduction reactions. *Chem. Commun.* 54 (33): 4091–4099.

240 Schut, G.J., Haja, D.K., Feng, X. et al. (2022). An abundant and diverse new family of electron bifurcating enzymes with a non-canonical catalytic mechanism. *Front. Microbiol.* 13: 946711.

241 Buckel, W. (2023). Comment on "How a formate dehydrogenase responds to oxygen: unexpected O_2 insensitivity of an enzyme harboring tungstopterin, selenocysteine, and [4Fe-4S] clusters". *ACS Catal.* 13 (12): 8237.

242 Zuchan, K., Baymann, F., Baffert, C. et al. (2021). The dyad of the Y-junction- and a flavin module unites diverse redox enzymes. *Biochim. Biophys. Acta* 1862 (6): 148401.

243 Raman, C.S. (2023). Reply to "Comment on 'How a formate dehydrogenase responds to oxygen: unexpected O_2 insensitivity of an enzyme harboring tungstopterin, selenocysteine, and [4Fe-4S] clusters'". *ACS Catal.* 13 (14): 9629–9632.

244 Maia, L.B., Fonseca, L., Moura, I., and Moura, J.J. (2016). Reduction of carbon dioxide by a molybdenum-containing formate dehydrogenase: a kinetic and mechanistic study. *J. Am. Chem. Soc.* 138 (28): 8834–8846.

245 Vilela-Alves, G., Rebelo Manuel, R., Pedrosa, N. et al. (2024). Structural and biochemical characterization of the M405S variant of *Desulfovibrio vulgaris* formate dehydrogenase. *Acta Crystallogr. Sect. F* 80 (5): 98–106.

246 Sargent, F. (2016). The model [NiFe]-hydrogenases of *Escherichia coli*. *Adv. Microb. Physiol.* 68: 433–507.

247 Finney, A.J. and Sargent, F. (2019). Formate hydrogenlyase: a group 4 [NiFe]-hydrogenase in tandem with a formate dehydrogenase. *Adv. Microb. Physiol.* 74: 465–486.

248 Balch, W.E., Schoberth, S., Tanner, R.S., and Wolfe, R.S. (1977). *Acetobacterium*, a new genus of hydrogen-oxidizing, carbon dioxide-reducing, anaerobic bacteria. *Int. J. Syst. Evol. Microbiol.* 27 (4): 355–361.

249 Poehlein, A., Schmidt, S., Kaster, A.K. et al. (2012). An ancient pathway combining carbon dioxide fixation with the generation and utilization of a sodium ion gradient for ATP synthesis. *PLoS One* 7 (3): e33439.

250 Schuchmann, K. and Müller, V. (2013). Direct and reversible hydrogenation of CO_2 to formate by a bacterial carbon dioxide reductase. *Science* 342 (6164): 1382–1385.

251 Schuchmann, K., Vonck, J., and Müller, V. (2016). A bacterial hydrogen-dependent CO_2 reductase forms filamentous structures. *FEBS J.* 283 (7): 1311–1322.

252 Dietrich, H.M., Righetto, R.D., Kumar, A. et al. (2022). Membrane-anchored HDCR nanowires drive hydrogen-powered CO_2 fixation. *Nature* 607 (7920): 823–830.

253 Schwarz, F.M., Schuchmann, K., and Müller, V. (2018). Hydrogenation of CO_2 at ambient pressure catalyzed by a highly active thermostable biocatalyst. *Biotechnol. Biofuels* 11: 237.

254 Müller, V. (2019). New horizons in acetogenic conversion of one-carbon substrates and biological hydrogen storage. *Trends Biotechnol.* 37 (12): 1344–1354.

255 Brill, W.J., Wolfe, R.S., and Wolin, E.A. (1964). Anaerobic formate oxidation: a ferredoxin-dependent reaction. *Science* 144 (361): 297–298.

256 Höpner, T. and Trautwein, A. (1972). Some properties of formate dehydrogenase. *Z. Naturforsch. Pt. B* 27 (9): 1075–1076.

Index

a

ABC transporter 58, 59, 64, 88, 89, 99, 236
acetogenic bacteria 321, 323, 722
acetylene, hydrocarbon substrates 305–306
acrylonitrile 307
aerobic H_2 oxidizer 412, 418
air-oxidization of Moco 196
aldehyde oxidase (AO) 230, 231
  drug metabolism 147–153
aldehyde oxidase crystal structure
  FAD binding site 146–147
  [2Fe-2S] clusters 145–146
  mammalian 139–141
  Mo active site 141–144
alkenes, hydrocarbon substrates 306
alkynes, hydrocarbon substrates 306
AlphaFold-based predictions 238, 239
amide hydrolysis reactions 149
amino acid analysis 278
amorphous solids 544
Anamorsin 61, 103
angle energy 639
anomalous diffraction methods 556–557
anomalous dispersion 553, 556
anomalous scattering 556, 566
anti-HA antibody 97
apo-enzymes, crystal structures
  Hmd and its homologs 461–462
  reconstituted Hmd holo-enzymes 460
  reconstituted Hmd with the substrate 460–461
apo-[FeFe]-hydrogenase 383
apo-HoxG 428, 433, 434
APS-kinase 160, 165, 167
Aquifex aeolicus IscU (AaIscU) 261
assimilatory sulfate reduction pathway 160, 164, 166, 167, 174, 179, 181
assimilatory sulfite reductase (aSir) 160, 168
asymmetric unit 541, 548
ATP dependent thiolation 244
ATP-sulfurylase 160, 165–167
A-type carrier (ATC) protein 12
A-type carrier SufA 39–41
AUX cluster 199
auxiliary [4Fe-4S] cluster 73, 233
auxiliary maturation proteins (HoxV and HoxL) role, $Fe(CN)_2CO$ synthon transfer 433
A. vinelandii NifU (AvNifU) 261
azide 207, 293, 306–307
Azotobacter vinelandii 2, 26, 257, 258, 290, 576, 586, 602, 603

b

Bacillus subtilis 27, 28, 30–32, 34
bidirectional enzymes 516
binding theory, ligands to receptor 620–622
biochemical activity assay 98
bioinformatic analysis 98, 110, 277
bioinformatic prediction 96
biological nitrogen fixation 287, 577

Iron-Sulfur Clusters: Biogenesis and Biochemistry, Volume 2, First Edition.
Edited by Silke Leimkühler, Günter Schwarz, Oliver Lenz, and Oliver Einsle.

biomolecular systems, analyzing 622–626
bipartite recognition of targets 109–112
bis-forms, definition 228
Blastocystis (Bh) 35
bond energy 320, 639
Born–Oppenheimer approximation 581
Bragg's Law 541, 543, 548, 550–552

c

Calvin–Benson–Bassham (CBB) cycle 293, 707
carbamoyl adenylate 412
  formation 412
carbon dioxide reduction (CDR) factor 721
carbon monoxide ligand origin and synthesis 417–418
catabolic APS-reductase 170–171
catalytic cycle of [Fe]-hydrogenase
  electronic property of the iron site 490–491
  enzyme reaction kinetics 491–492
  Hmd inhibitors
    CO and CN^- 492
    Cu 493–494
    Fe 494
    H_2O_2 494–495
    isocyanides 492
    O_2 is reduced to H_2O_2 495–496
  semisynthetic Hmd 496–498
    circular dichroism spectroscopy 499
    crystal structure of Hmd with substrate 499–500
    2-OH group of the pyridinol ring 498
    proposed catalytic mechanism of Hmd 500–501
    semisynthetic [Mn]-hydrogenase 498–499
catalytic cycle of [NiFe]-hydrogenase
  active site states 516–517
    catalytic intermediates 520–525
    CO-bound states 517
    hydroxy-bridged states 517–519
    unusual [NiFe] site arrangements 519–520
  classification of 514–516
  molecular hydrogen (H_2) 512
catalytic intermediates
  Ni_a-C 523–524
  Ni_a-L 524–525
  Ni_a-S intermediate 520–522
  Ni_a-SR 522–523
catalytic mechanism of [FeFe]-hydrogenase
  consensus mechanism 511–512
  current state 507–511
  electronic structure of H-cluster 504–507
  iron-sulfur cluster 503–504
Cba5H 383, 389, 390, 395–397
cellular pyridine nucleotide 318
cGrxs 89, 97, 100, 101, 104
chaperone 3, 11, 14, 31, 64, 67, 69–71, 74, 210, 211, 471, 717
chaperone-dependent transfer of ISCU2-bound cluster to [2Fe-2S] apoproteins 69–71
CIA components 88, 89, 94, 95, 97–99, 101–107, 113, 114, 117
CIA machinery 61, 72
  early CIA and late 105–106
  [2Fe-2S] protein assembly in 99–105
CIA system, components of 88, 90
client proteins 2, 6, 11, 15, 71, 258
Clostridium autoethanogenum 335
CN^--treated CpI 390, 398
CO dehydrogenase/acetyl-CoA synthase (CODH/ACS) 577
CO reduction
  by Fe nitrogenase 299
  by Mo nitrogenase 295–297
  by V nitrogenase 297–298
CO_2 reduction
  by Fe nitrogenase 302–303
  in vivo and *in vitro* 303–304
  by Mo nitrogenase 301–302
  by V nitrogenase 302

coenzyme B (CoB-SH), definition 448
core rearrangement 652, 685
CpI 387
  proton transfer pathways and water channels 388
CrHydA1 345, 382–385, 387, 388, 393–398, 502, 503, 506, 588
CryoEM structural studies 112
cryoprotectant 547, 548
crystalline solids 544
CTC 41, 89, 94–96, 105–114, 116–118
  [2Fe-2S] cluster insertion 113–114
C-terminal extension and [NiFe]-hydrogenases 425–427
C-terminal tryptophan 89, 95, 107, 110, 111, 113, 114
Cupriavidus necator 410, 412, 513, 514, 575, 576, 587, 595, 713
cyanide
  group synthesis, by HypE and HypF 412–414
  nitrogen containing substrates 307, 308
CyaY 4, 5, 9, 38
cyclic pyranopterin monophosphate (cPMP) 34, 197, 199, 231, 233–236
3′,8-cyclization of purine nucleoside/nucleotide 199
cysteine desulfurase (CD)-sulfurtransferase (ST) system 29
cysteine desulfurases 8, 32, 258, 263, 275
  complex 65
cysteine dioxygenase (CDO) 243
cysteine persulfide intermediate (Cys-S-SH) 8
cytosolic 2Fe-2S protein 97
cytosolic Fe/S components 90
cytosolic Fe/S proteins 58, 61, 62, 68, 69, 74, 87, 88, 91, 95–102, 108–111, 115, 117
cytosolic iron-sulfur protein assembly (CIA) system 58

d

damping rate 581
data collection, X-rays 552–554
data processing, X-rays 552–554
deficit spending model 291, 292
de novo assembly 8–11
de novo [2Fe-2S] cluster formation 64–69
density functional theory (DFT) 295, 395, 396, 452, 500, 501, 504, 509, 510, 518, 522
deuterated buffer 297, 302, 305
dielectric constant 626, 627, 635, 636, 638, 639
diffraction pattern 541, 542, 546, 550–553, 555, 565
dimethyl sulfoxide (DMSO) 203–210, 293, 653, 655, 667, 710, 711, 712
dissimilatory sulfate reduction pathway 160, 170
DMSO-reductases 293
DPYD 61, 115–117
drop vapor diffusion 547
drug and xenobiotic metabolism 135, 141
drug discovery process 139
DsrC binds 171
D. vulgaris Hildenborough 719–720, 726

e

elastic Rayleigh scattering 581
electric quadrupole moment 590
electromagnetic radiation 546, 551, 591
electron bifurcating hydrogenases
  [NiFe]-MvhAGD-HdrABC
    CoM-S-S-CoB binding and reduction 359–360
    electron transfer pathways 356–359
  electron transfer (ET) 320
  Hyd-type electron bifurcating hydrogenase
    HydA and HydSL subunits 339–340
    HydB subunit 340–341
    HydC subunit 341–342
  mechanistic proposals
    [FeFe]-and [NiFe]-HydABC

electron bifurcating hydrogenases (*contd.*)
electron transfer pathways 352–355
general mechanistic considerations 350–352
MvhAGD [NiFe]-hydrogenase module 349
physiological roles 321–324
prototypical 318
spectroscopy of HydABC and HndABCD type bifurcating hydrogenases
EPR and IR of the H-cluster in [FeFe]-HydABC 344–346
S. fructosivorans HndABCD 344
Tm HydABC 342–344
structural and biophysical characterizations
Hyd-type electron bifurcating hydrogenase 342
structures and cofactors of MvhAGD-HdrABC 346–349
subunit compositions and reactivities 324–337
electron bifurcation *vs.* O_2 tolerance 725–727
electron confurcation 171, 321
electron delivery process 290
electron density map 144, 541, 545, 556–560, 562
electron nuclear double resonance (ENDOR) spectroscopy 295, 359, 524
electron paramagnetic resonance (EPR) spectroscopy 295, 342, 395, 457, 490, 589, 714
electron transfer 209, 210, 212, 352–359, 384, 387–389, 396, 503, 505, 509, 585–587, 599, 602, 624, 650, 652, 656, 660, 670, 681, 691–692, 708, 712–719, 726, 729
electrostatic energies 629–631, 639
electrostatic potentials 626–631, 633–637
Enterobacteriaceae 15
Erwinia chrysanthemi 26
Escherichia coli (*E. coli*) 2, 26, 28, 29, 31, 35, 37, 39, 40, 42–44, 195, 198, 200–202, 205, 207, 208
metal-dependent FDHs 715–716
molybdoenzymes 205, 207–209
eukaryotic cytosol harbors 61
eukaryotic Fe/S protein biogenesis 74, 87, 89, 99
eukaryotic molybdenum cofactor 227, 229, 231
europium(III/II) polyaminocarboxylate complexes 296
extended X-ray absorption fine structure (EXAFS) spectroscopy 37, 295

f

face-centered unit cell 550
FAD binding site 146–147
F_{420}-dependent sulfite reductase catalyzes 177–178
FDH
electron bifurcation *vs.* O_2 tolerance 725–727
from *D. vulgaris* Hildenborough 719–720
in H_2-dependent CO_2 reduction (HDCR) 727–728
in methanogens 722–724
O_2 sensitivity 724–725
tungsten vs molybdenum 720
$Fe(CN)_2CO$ synthon assembly 414–417
$Fe(CN)_2CO$ synthon transfer, to apo-hydrogenase 419–421
$Fe(CN)_2CO$-loaded HybG-HypD complex 419
[FeFe]-hydrogenases
catalytic activity 387, 389
catalytic cycle 386
catalytic [6Fe–6S] cofactor 384–385
characteristics 384–389
CO-inhibited enzyme 394
formaldehyde attack sites 394–395
gas channels 389

inhibitor molecules 391
maturation of 383, 386
O_2 molecules, irreversible attack of 392–394
research, historical perspective 381–384
small molecule diffusion to H-cluster 390, 392
sulfide role 396
Zn^{2+} inhibitor 398–399
FeMo-cofactor 271, 577
cluster [7FeMo9SC] 5
and FeVco composition 298
Fe nitrogenase
CO reduction by 299
CO_2 reduction by 302, 303
Fe protein 288
cycle 291
ferredoxin (Fdx) 60, 73, 274
metal-dependent FDHs 715
mitochondrial 64
Fe/S biogenesis 88, 97
GSH requirement in 99–100
Fe-S cluster biogenesis pathways 34, 228, 430
Fe-S clusters 194, 209–212, 679
higher nuclearity 683–691
hydrogenase and nitrogenase 680–683
[Fe_2S_2] clusters
biological relevance of 658
synthetic model complexes of 658–664
[2Fe–2S] clusters 2, 100, 102, 103, 105, 145, 146, 583, 585, 599, 601
insertion 113–114
[3Fe–4S] clusters 37, 95, 210, 212, 429, 515, 585–586, 594, 601
[4Fe–4S] clusters 2, 103, 105, 586, 587, 602
Mössbauer spectroscopy 592–595
Raman spectroscopy 586–587
Fe/S domain and structural/spectroscopic assignment 138–139
Fe-S proteins 2, 5, 6, 7, 10, 14, 26, 34, 41–42, 257, 277, 541, 565, 566, 578, 589
Fe/S target protein 92, 96, 103, 107–108, 110, 112, 114
cytosolic and nuclear 103
^{57}Fe-labeled HybG-HypD complex 420
[2Fe–2S] chaperones 100, 102
[2Fe–2S] target proteins 89, 102
[7Fe-Mo-9S-C]-homocitrate active site 272
[Fe-S(Cys)4] metal site 599
[Fe_4S_4] and [Fe_3S_4] clusters
biological relevance of 665–667
synthetic model complexes of 667–676
[FeS_4] centers
biological relevance 653–654
synthetic model complexes 654–657, 676–679
[FeFe]-and [NiFe]-HydABC
electron transfer pathways 352–335
FhuF enzyme 26
FIR spectroscopy 578, 579
Fischer–Tropsch reaction 299
fitting quantum chemical electrostatic 633–635
flavin-based electron bifurcation (FBEB) 452, 723, 725–727
formaldehyde attack sites, of [FeFe]-hydrogenases 394–395
formate dehydrogenases (FDHs) 208, 705–728
formate hydrogenlyase complex 716–717
formic acid 706–707, 716, 728
formyl-tetrahydrofolate (formyl-THF)-dependent CO gas production 418
formylmethanofuran dehydrogenases 707, 721–722
fragment condensation 652, 653, 654, 659, 672, 683, 686, 690
fragmentation/dissociation 652

F_{420}-reducing [NiFe]-hydrogenase 408, 448, 450, 452, 520
free-interface diffusion 547, 564
Friedreich's ataxia 5

g

G-domain reaction product (MPT-AMP) 202, 240, 241
gephyrin-derived Moco 241
Gibbs free energy 545
glutamate oxaloacetate transaminase (GOT) 243
glutaredoxin GLRX5 63, 64, 70, 74
greenhouse gas 705, 707, 728

h

Haber–Bosch process 117, 287, 301
β-hairpin 32
hAOX1
  amide hydrolysis reactions 149
  inhibition of hAOX1 activity 149–150
  oxidation of aldehydes, N-aromatic heterocycles and iminium ions 148
  reductions of nitro, N-oxide, S-oxide and isothiazol compounds 149
  SNPs 150–153
harmonic oscillator 572, 575
Hcg proteins
  HcgA 467
  HcgB 467–468
  HcgC 468–471
  HcgD 470–471
  HcgE 471–472
  HcgF 472–473
  HcgG 473
H-cluster, in [FeFe]-hydrogenases 680
H_2-dependent CO_2 reduction (HDCR) 709, 727–728
HdrABC-containing complexes 335, 336
heat shock protein (HSP) 110
Herzberg–Teller vibronic coupling 582
heterodisulfide reductase reaction 723
high potential iron–sulfur proteins (HiPIP) 515, 585, 586, 666, 672, 675
His-ligated Fe-S cluster 387
H_2-metabolizing enzymes 575
Homo sapiens 88, 92, 103, 657
Hooke's law 572
horizontal gene transfer 176, 179
Hoyeraal–Hreidarsson syndrome 115
Hsp70 chaperones 11, 63, 64, 67
Hyd-type electron bifurcating hydrogenase
  HydA and HydSL subunits 339–340
  HydB subunits 340–341
  HydC subunit 341–342
HydABCD complex 324
hydride-bound state 346, 595
hydrocarbon substrates
  acetylene 305–306
  alkynes and alkenes 306
hydrogen sulfide 160, 161, 169, 181, 243, 659, 683, 685, 686, 690
hydrogenases 381
  [NiFe]-and [FeFe] 605
  nitrogenases and 595–596
hydrogenotrophic methanogenic pathway (Hmd)
  apo-enzymes, crystal structures
    Hmd and its homologs 461–462
    reconstituted Hmd holo-enzymes 460
    reconstituted Hmd with the substrate 460–461
  definition 447
  enzymes involved in 449–452
  ESI-MS analysis for detection, CO/Acyl ligands 462–463
  [Fe]-Hydrogenase function 453–454
  FeGP cofactor 457–459
    in vitro biosynthesis 473–474
    hcg gene cluster, sequence similarities and mutational analysis 465–466
    Hcg proteins, structure to function analysis 466–467, 473

stable isotope labelling experiments 463–465
structure-function analyses 479
holo-enzyme with the extracted cofactor 454–456
light sensitivity 456–457
[NiFe]-hydrogenases 452–453
photolysis mechanism 459
unique coenzymes found in methanogens 448–449
Hyp proteins 411, 426, 427, 435
HypCD complex mediated transfer, of $Fe(CN)_2CO$ synthon 419–421
HypCD complex, $Fe(CN)_2CO$ synthon assembly on 414–417
HypCD mediated ATP-dependent transfer, of $Fe(CN)_2CO$ synthon 420, 421
HypX 410, 411, 417, 418, 434, 464

i

improper energy 639
infrared spectroscopy
FIR spectroscopy 578–579
mid-IR (MIR) spectroscopy 575–578
inhibition of hAOX1 activity 149–150
inorganic catalytic center, in [NiFe]-hydrogenase basic module 409
inorganic sulfide (S^{2-}) 28
internal aldimine 8, 29, 266
in vitro biosynthesis
complementation assay 476
confirmation of precursors 475–476
design 474–475
of HcgA 476–477
of HcgG 477–479
in vitro Fe-S cluster formation 9
in vitro maturation 383, 398, 419, 420, 421
in vivo maturation 383
iron (Fe) nitrogenase 288
iron acquisition and Fe-S cluster binding 37–38
iron guanylylpyridinol (FeGP) cofactor 447, 490, 577
iron-molybdenum cofactor (FeMoco) 194, 228, 289, 290, 292, 295–299, 305, 306, 308
iron-sulfur (Fe-S) clusters 26, 577, 650
iron-sulfur (Fe/S) proteins
cellular roles of eukaryotic 59–62
iron-sulfur cluster assembly (ISC) machinery 58
iron-sulfur enzymes 710
ISC-based Fe-S protein biogenesis in bacteria
components 5–8
de novo assembly 8–11
discovery of 4–5
distribution and evolution of 14–15
Fe-S cluster delivery step 11–12
regulation 12–14
isc operon of *Myxococcus xanthus* 13
isomer shift 589, 590, 592, 594–596
isomorphous replacement method 555–556
isonitriles 306–308
isoxanthopterin-6-carboxylate 197, 198

j

Janus-intermediate 291, 292

k

kinetic analysis 153, 451, 454, 477, 490, 491, 494, 624–626
Klebsiella pneumoniae 4, 5, 258, 275

l

Lamb-Mössbauer factor 589, 598
last universal common ancestor (LUCA) 3, 15, 28, 210
β-latch motif 32
Lennard–Jones energy 639
ligand to metal charge transfer (LMCT) transitions 582
low dielectric cavity 631–633

Lowe–Thorneley scheme 291–293, 295–296, 299
lyophilized [FeFe]-hydrogenases 394

m

macromolecule crystallization 544–548
mammalian AOX crystal structures 139–141
marine thermophile *M. thermolithotrophicus* 173
mass spectrometry 8, 98, 112, 419, 458, 459, 463, 464, 467, 472, 475, 507
mature HoxG 428–429
M-cluster 577
MD simulations 390–392, 396, 503
metal-based cofactors 228
metal-binding pterin (MPT) 194, 198, 228
metal-dependent FDHs 707, 708, 710
  bis-MGD to a [4Fe-4S] cluster 712–714
  E. coli 715–716
  extended Fe–S cluster electron chains 714
  ferredoxin 715
  modular catalytic unit 711
  from *Moorella thermoacetica* 721
  NAD^+-reducing 717–718
metalloenzymes 263, 264, 287, 381, 457, 479, 499, 512, 563, 573, 575, 577, 579, 597, 606, 650, 706, 707, 720, 724, 728
methanofuran (MFR) 336, 448
methanogens, FDHs in 722–724
methanoic acid 706
Methanothrix thermoacetophila (MtIscU) 261
MFe protein 288
Mid-IR (MIR) spectroscopy 555, 556, 575–579
Miller indices 549, 554
minimal Fe-S cluster assembly machineries 3
mitochondrial amidoxime-reducing component (mARC) 214, 230, 231, 233, 244
mitochondrial Fe/S proteins 57–60, 63, 71, 88
mitochondrial ISC system
  de novo [2Fe-2S] cluster formation 64–69
  bacterial endosymbiotic ancestor 62
  chaperone-dependent transfer of ISCU2-bound cluster to [2Fe-2S] apoproteins 69–71
  reductive fusion of [2Fe-2S] clusters 71–74
mitochondrial translation 61
Mo active site 137, 141–145, 147, 150
Moco deficiency (MoCD) 236, 239, 242–245
MOCS1A 60, 231, 233–236, 244
MOCS1B-homologous bacterial protein MoaC 235
MOCS2A 227, 231, 234, 236–239
MOCS3 227, 231, 233, 238, 239, 242
Mo-dependent nitrogenase 271
Mo enzymes 136–139, 194, 195, 198, 230
Mo–iron–flavin enzymes 230
Mo nitrogenase
  CO reduction by 295–297
  CO_2 reduction by 301–302
molecular replacement (MR) 555, 557
molybdate 166, 170, 174, 193–195, 202, 203, 213, 233, 239–241, 721
molybdenum (Mo)
  in biology 193–194
  nitrogenase 288–290
molybdenum cofactor (Moco) 194
  biosynthesis pathways 194–196, 199–202, 231–233
  chemical investigation 196–199
  cyclic pyranopterin monophosphate synthesis 233–236
  deficiency of 242–244
  eukaryotic 229–231
  intimate relation 244–245

molybdenum insertion into molybdopterin 239–242
molybdoenzymes 209–211
Neurospora crassa 194
oxidation of 196
prokaryotes and eukaryotes 202–205
sulfido ligand, discovery of 205–207
thiolation of cPMP by molybdopterin synthase 236–239
molybdoenzymes 194, 196, 710
alkylated pterin derivatives 203
discovery of 194
E. coli 205, 207–209
FeS clusters 209–212
reaction catalyzed 211–212
molybdolysis 166, 170
molybdopterin (MPT) 194, 197, 198, 201
modification of 240
molybdopterin guanine dinucleotide (MGD) 203, 205
molybdopterin or metal-binding pterin (MPT) 139, 140, 142, 194, 197–209, 213, 228, 231, 233, 236–241, 244
Moorella thermoacetica 323, 710, 721
Mössbauer spectroscopy 112, 428, 457, 490, 494, 571, 588–596, 665
Fe–S(Cys)4 to [4Fe–4S] clusters 592–595
MPT-synthase complex 231, 237
Mtb SufS-SufU structure 32
M1-type hydrogenases 385, 387
multiple isomorphous replacement (MIR) 555, 557, 575–579
Mvh-Hdr class 335

n

NAD(P)$^+$-reducing [NiFe]-hydrogenases 408
NAD$^+$-reducing, metal dependent FDHs 717–718
Neurospora crassa 113, 194, 233
N-hydroxy-L-arginine 231
Ni_a-S intermediate 513, 517, 520, 521
NifB cofactor 270, 272, 276, 603
NifDK 270–273, 288
[NiFe]-and [FeFe]-hydrogenases 381, 457, 459, 490, 492, 575, 587, 588, 605, 728
$NiFe(CN)_2CO$ cofactor, cyanide ligands of 413
[NiFe]-hydrogenases 335
carbon monoxide ligand origin and synthesis 417–418
C-terminal extension, role of 425–427
cyanide group synthesis by HypE and HypF 412–414
and function 407–408
inorganic catalytic center 409
iron-sulfur cluster insertion 429–430
isolation of subunits 427–429
maturation, genetic basis of 409–411
nickel mobilization and insertion by HypAB 421–425
subunit oligomerization and transport across cytoplasmic membrane 431–433
[NiFe]-MvhAGD-HdrABC
CoM-S-S-CoB binding and reduction 359–360
electron transfer pathways 356–359
[NiFeSe]-hydrogenases 409
NifS
catalytic cycle 266
structural and catalytic features 263–267
NifU
structural and catalytic features
central domain 261–262
C′-terminal domain 262–263
N′-terminal domain 259–261
NifU and NifU-like proteins 273–274
NifUS
in aerobic nitrogen fixation 269–278
in biotechnological efforts 275–276
IscUS system 274–275
NifUS system, synthesis of Fe-S clusters 267–269
nit-1 extract assays 196

nitric oxide (NO)
with [Fe–S] cluster proteins 602–603
nitriles 293, 306–308
nitrogenases 577
active site cofactor 289, 300
alternative substrates of 292–294
FeMe-cofactor 603–605
hydrogenases and 595–596
isoforms 289
isozymes 288–290
mechanism 290–292
nitrogenase-mediated N_2-reduction 293
nitrogen containing substrates
azide 306–307
cyanide, nitriles and isonitriles 307–308
nitrogen-fixing bacteria 2, 4, 228, 277
non-polar counter ions 654
nonsynonymous SNPs (nsSNPs) 150
nuclear Fe/S proteins
diseases related to cytosolic 115–116
diseases related to maturation 116–117
nuclear inelastic scattering (NIS) 596
nuclear resonance absorption 588
nuclear resonance vibrational spectroscopy (NRVS) 295, 596–606
nucleotide excision repair (NER) 107, 115

o

O_2-initiated H-cluster degradation 394
O_2-resistant [FeFe]-hydrogenase 389, 399
O_2-sensitive [FeFe]-hydrogenases 393, 397
O_2-tolerant membrane-bound [NiFe]-hydrogenases, maturation of 433–435
O_2-tolerant [NiFe]-hydrogenases 393, 394, 515, 518, 525
oxygen atom transfer (OAT) mechanism 211–213
oxygenic photosynthetic organisms 27

p

partial charges 633–635
P-cluster 577, 595, 603, 650, 679, 680, 682, 685–688
P cluster 8Fe-7S 5
Pearson correlation coefficient 562
plant Cnx5 234, 235, 238, 241
Plasmodium falciparum 27, 34
plastid-based organelles 27
Poisson–Boltzmann equation
conceptual model 626–627
mathematical model 627–629
Poisson–Boltzmann model 620, 635–636
polarizability tensor 580
polyferredoxin (polyFd) 336, 452, 453
pre-HoxG 428, 429, 433, 434
protein crystals 398, 544, 545, 547, 548, 552
protein kinase R (PKR) 238
protein molecules 545, 546, 551, 552, 561, 625
protein-protein interactions 65, 96, 108, 117, 150, 153, 268, 419
proton-coupled electron transfer 360, 387, 389, 509
proton exchange reaction 454, 491, 492
proton transfer pathway (PTP)
small molecules influence on 398–399
pterin-based metalloenzyme 706
pyranopterin (PPT) 198
pyridoxal phosphate (PLP) cofactor 8
pyridoxal-5-phosphate (PLP) 3, 8, 28–30, 67, 258, 263, 265

r

Raman effect 580
Raman scattering 580, 581
Raman spectroscopy
[2Fe–2S] clusters 583–585
[3Fe–4S] clusters 585–586
[4Fe–4S] clusters 586–587
hydrogenases by 587–588
resonance 581–583, 585
rubredoxin 582–583
reaction field 626, 629, 630, 633, 637

recognition motif 110–112
redox flavin cofactor 38–39
resonance Raman effect 581–582
resonance Raman (RR) spectroscopy 578, 581–582
  [Fe–S] clusters 582, 585
Rhodobacter sphaeroides 203, 711
Rieske center 603, 657, 660, 692
ROS-dependent redox reactions 393
rotated state 505
Roussin's Black Ester (RBE) 603
rubredoxins 578, 582, 583, 592, 599, 650, 653–655

S

Saccharomyces cerevisiae
  Fe/S target protein 92–93
  use of 276
S-adenosyl-L-methionine (SAM) 60, 61, 63, 73, 95, 103, 199, 200, 233, 234, 244, 272
S-adenosylmethionine (SAM) 658
  dependent fusion 272
  enzymes 233
selenocysteine 209–211, 348
serial synchrotron crystallography (SSX) 511, 563
S. fructosivorans HndABCD 344
single isomorphous replacement (SIR) 555
single nucleotide polymorphisms (SNPs) 115, 136, 150–153
siroheme 113, 166, 168, 171, 178, 179
sirohydrochlorin 171
sitting drop vapor diffusion technique 547
SlyD 422
split signal 523
spontaneous self-assembly 652, 659, 683
stiff terms 639
Stokes Raman scattering 581
stress-responsive Suf regulation in bacteria 42–44
structure refinement 557–562
suf and sms systems 27
$SufBC_2D$ complex 36, 38, 39
$SufBC_2D$ scaffold complex 35
SufB intein splicing 45
SufC and SufD scaffold partners 35–37
$SufC_2D_2$ sub-complex 36
SUF function
  Fe-S cluster assembly
    iron acquisition and Fe-S cluster binding 37–38
    redox flavin cofactor 38–39
    SUF-like minimal system (SMS) 37
    SufB scaffold 35
    SufC and SufD scaffold partners 35–37
  FE-S cluster trafficking
    A-type carrier SufA 39–41
    SufT and DUF59 proteins 41–42
  sulfur delivery step
    cysteine desulfurase 34–35
    SufS cysteine desulfurase 28–30
    SufU and SufE sulfur transfer partners 30–34
SUF regulation
  stress-responsive Suf regulation in bacteria 42–44
  SufB intein splicing 45
  SufR and RisR regulation of Suf 44
SUF-like minimal system (SMS) 3, 37
SufR and RisR regulation of Suf 44
SufS cysteine desulfurase (CD) 28–30
SufS-SufE (cysteine desulfurase-sulfurtransferase) tandem 28
SufT and DUF59 proteins 41–42
SufU and SufE sulfur transfer partners 30–34
sulfate capture and activation 170
sulfate reduction metabolism
  APS-kinase 167
  ATP-sulfurylase 165–167
  catabolic APS-reductase 170–171
  dissimilatory microbial sulfate reduction 169–170
  energy conservation 171–172
  poisonous by-product PAP 167–168

sulfate reduction metabolism (*contd.*)
  sulfate capture and activation 170
  sulfite reduction 168
  sulfur assimilation and dissimilatory 172
sulfate-reducing bacteria (SRB) 396, 719
sulfate-reducing methanogen
  Fsr 179–180
  sulfate utilization 180–181
sulfido ligand, discovery of 205–207
sulfite oxidase (SOX)
  in humans and vertebrates 229
  in plants 229
sulfite reduction 168, 171–172, 177
S-sulfocysteine (SSC) 243
S-sulfonated transthyretin 242
sulfotransferase pathway 167
sulfur
  bioavailability 164
  chemical properties of 161–162
  essential cellular component 162–164
sulfur assimilation and dissimilatory 172
sulfur quinone oxidoreductase (SQR) 243
Sulfurated Moco 205

t

TAOS1 116
Tat-mediated export, of [NiFe]-hydrogenase 433–435
Tat-signal peptide-containing precursors 432
thermodynamic analysis 624–625
Thermotoga maritima (Tm HydABC) 318
thiamine 6, 9, 163, 263
thiolation of cPMP by molybdopterin synthase 236–239
thiosulfate sulfur transferase (TST) 243
3D structures of hAOX1 140
time-resolved crystallography 541, 562–565
Tm HydABC 318, 319, 342–345, 352
torsion energy 639
transition dipole moment 575, 581, 582
trithionate pathway 171
tungsten *vs.* molybdenum FDH 720

u

unconventional sulfate reduction pathway, *M. thermolithotrophicus*
  dissimilatory-like PAPS-reductase and a new PAP-phosphatase 175–177
  F_{420}-dependent sulfite reductase catalyzes 177–178
  growth on sulfate 174, 175
  methanogens and sulfur 173–174
  PAPS 174–176
unit cells 544
  symmetry and 548–550
urothione 197, 198

v

vanadium (V) nitrogenase 288
  CO reduction by 297–298
  CO_2 reduction by 302
vibrational spectroscopy
  normal modes 573
  techniques 573–574
Virtual Model Compound Approach (VMCA) 641, 642

w

W-containing FDHs 719
Wood–Ljungdahl pathway 321, 323, 706, 727

x

X-ray crystallography 32, 41, 98, 230, 298, 394, 511, 522, 541–543, 550, 565, 566, 577, 599, 639
X-ray free-electron lasers (XFELs) 563
X-rays
  structure analysis 554–562
  wavelength 543
xanthine dehydrogenase (XDH) 60, 115, 135, 136, 209, 710

xanthine oxidase family of mononuclear Mo enzymes 136–139
xanthine oxidoreductase (XO) 228–231
Xeroderma pigmentosum group D (XPD) 115

y

yeast iron homeostasis 100–102

z

Zeeman effect 591